Lecture Notes in Computer Science 11922

Steven D. Galbraith · Shiho Moriai (Eds.)

Advances in Cryptology – ASIACRYPT 2019

25th International Conference on the Theory
and Application of Cryptology and Information Security
Kobe, Japan, December 8–12, 2019
Proceedings, Part II

 Springer

Editors
Steven D. Galbraith 🆔
University of Auckland
Auckland, New Zealand

Shiho Moriai 🆔
Security Fundamentals Lab
NICT
Tokyo, Japan

ISSN 0302-9743   ISSN 1611-3349   (electronic)
Lecture Notes in Computer Science
ISBN 978-3-030-34620-1   ISBN 978-3-030-34621-8   (eBook)
https://doi.org/10.1007/978-3-030-34621-8

LNCS Sublibrary: SL4 – Security and Cryptology

This Springer imprint is published by the registered company Springer Nature Switzerland AG
The registered company address is: Gewerbestrasse 11, 6330 Cham, Switzerland

Preface

ASIACRYPT 2019, the 25th Annual International Conference on Theory and Application of Cryptology and Information Security, was held in Kobe, Japan, during December 8–12, 2019.

The conference focused on all technical aspects of cryptology, and was sponsored by the International Association for Cryptologic Research (IACR).

We received a total of 307 submissions from all over the world. This was a significantly higher number of submissions than recent Asiacrypt conferences, which necessitated a larger Program Committee (PC) than we had originally planned. We thank the seven additional PC members who accepted our invitation at extremely short notice. They are Gorjan Alagic, Giorgia Azzurra Marson, Zhenzhen Bao, Olivier Blazy, Romain Gay, Takanori Isobe, and Daniel Masny.

The PC selected 71 papers for publication in the proceedings of the conference. The two program chairs were supported by a PC consisting of 55 leading experts in aspects of cryptology. Each submission was reviewed by at least three Program Committee members (or their sub-reviewers) and five PC members were assigned to submissions co-authored by PC members. The strong conflict of interest rules imposed by the IACR ensure that papers are not handled by PC members with a close working relationship with authors. There were approximately 380 external reviewers, whose input was critical to the selection of papers.

The review process was conducted using double-blind peer review. The conference operated a two-round review system with a rebuttal phase. After the reviews and first-round discussions the PC selected 193 submissions to proceed to the second round. The authors of those 193 papers were then invited to provide a short rebuttal in response to the referee reports. The second round involved extensive discussions by the PC members. Indeed, the total number of text items in the online discussion (including reviews, rebuttals, questions to authors, and PC member comments) exceeded 3,000.

The three volumes of the conference proceedings contain the revised versions of the 71 papers that were selected, together with 1 invited paper. The final revised versions of papers were not reviewed again and the authors are responsible for their contents.

The program of Asiacrypt 2019 featured excellent invited talks by Krzysztof Pietrzak and Elaine Shi. The conference also featured a rump session which contained short presentations on the latest research results of the field.

The PC selected the work "Wave: A New Family of Trapdoor One-Way Preimage Sampleable Functions Based on Codes" by Thomas Debris-Alazard, Nicolas Sendrier, and Jean-Pierre Tillich for the Best Paper Award of Asiacrypt 2019. Two more papers were solicited to submit a full version to the *Journal of Cryptology*. They are "An LLL Algorithm for Module Lattices" by Changmin Lee, Alice Pellet-Mary, Damien Stehlé, and Alexandre Wallet, and "Numerical Method for Comparison on Homomorphically Encrypted Numbers" by Jung Hee Cheon, Dongwoo Kim, Duhyeong Kim, Hun Hee Lee, and Keewoo Lee.

The Program Chairs are delighted to recognize the outstanding work by Mark Zhandry and Shweta Agrawal, by awarding them jointly the Best PC Member Award.

Many people have contributed to the success of Asiacrypt 2019. We would like to thank the authors for submitting their research results to the conference. We are very grateful to the PC members and external reviewers for contributing their knowledge and expertise, and for the tremendous amount of work that was done with reading papers and contributing to the discussions.

We are greatly indebted to Mitsuru Matsui, the general chair, for his efforts and overall organization.

We thank Mehdi Tibouchi for expertly organizing and chairing the rump session.

We are extremely grateful to Lukas Zobernig for checking all the latex files and for assembling the files for submission to Springer.

Finally we thank Shai Halevi and the IACR for setting up and maintaining the Web Submission and Review software, used by IACR conferences for the paper submission and review process. We also thank Alfred Hofmann, Anna Kramer, Ingrid Haas, Anja Sebold, Xavier Mathew, and their colleagues at Springer for handling the publication of these conference proceedings.

December 2019   Steven Galbraith
                                                 Shiho Moriai

ASIACRYPT 2019

The 25th Annual International Conference on Theory and Application of Cryptology and Information Security

Sponsored by the International Association for Cryptologic Research (IACR)

Kobe, Japan, December 8–12, 2019

General Chair

Mitsuru Matsui   Mitsubishi Electric Corporation, Japan

Program Co-chairs

Steven Galbraith   University of Auckland, New Zealand
Shiho Moriai   NICT, Japan

Program Committee

Shweta Agrawal	IIT Madras, India
Gorjan Alagic	University of Maryland, USA
Shi Bai	Florida Atlantic University, USA
Zhenzhen Bao	Nanyang Technological University, Singapore
Paulo S. L. M. Barreto	UW Tacoma, USA
Lejla Batina	Radboud University, The Netherlands
Sonia Belaïd	CryptoExperts, France
Olivier Blazy	University of Limoges, France
Colin Boyd	NTNU, Norway
Xavier Boyen	Queensland University of Technology, Australia
Nishanth Chandran	Microsoft Research, India
Melissa Chase	Microsoft Research, USA
Yilei Chen	Visa Research, USA
Chen-Mou Cheng	Osaka University, Japan
Nils Fleischhacker	Ruhr-University Bochum, Germany
Jun Furukawa	NEC Israel Research Center, Israel
David Galindo	University of Birmingham and Fetch AI, UK
Romain Gay	UC Berkeley, USA
Jian Guo	Nanyang Technological University, Singapore
Seokhie Hong	Korea University, South Korea
Andreas Hülsing	Eindhoven University of Technology, The Netherlands
Takanori Isobe	University of Hyogo, Japan
David Jao	University of Waterloo and evolutionQ, Inc., Canada

Jérémy Jean	ANSSI, France
Elena Kirshanova	ENS Lyon, France
Virginie Lallemand	CNRS, France
Jooyoung Lee	KAIST, South Korea
Helger Lipmaa	Simula UiB, Norway
Feng-Hao Liu	Florida Atlantic University, USA
Atul Luykx	Swirlds Inc., USA
Hemanta K. Maji	Purdue, USA
Giorgia Azzurra Marson	NEC Laboratories Europe, Germany
Daniel Masny	Visa Research, USA
Takahiro Matsuda	AIST, Japan
Brice Minaud	Inria and ENS, France
David Naccache	ENS, France
Kartik Nayak	Duke University and VMware Research, USA
Khoa Nguyen	Nanyang Technological University, Singapore
Svetla Nikova	KU Leuven, Belgium
Carles Padró	UPC, Spain
Jiaxin Pan	NTNU, Norway, and KIT, Germany
Arpita Patra	Indian Institute of Science, India
Thomas Peters	UCL, Belgium
Duong Hieu Phan	University of Limoges, France
Raphael C.-W. Phan	Monash University, Malaysia
Carla Ràfols	Universitat Pompeu Fabra, Spain
Ling Ren	VMware Research and University of Illinois, Urbana-Champaign, USA
Yu Sasaki	NTT laboratories, Japan
Junji Shikata	Yokohama National University, Japan
Ron Steinfeld	Monash University, Australia
Qiang Tang	New Jersey Institute of Technology, USA
Mehdi Tibouchi	NTT Laboratories, Japan
Hoeteck Wee	CNRS and ENS, France
Mark Zhandry	Princeton University, USA
Fangguo Zhang	Sun Yat-sen University, China

Local Organizing Committee

General Chair

Mitsuru Matsui	Mitsubishi Electric Corporation, Japan

Honorary Advisor

Tsutomu Matsumoto	Yokohama National University, Japan

External Reviewers

Masayuki Abe
Parhat Abla
Victor Arribas Abril
Divesh Aggarwal
Martin Albrecht
Bar Alon
Prabhanjan Ananth
Elena Andreeva
Yoshinori Aono
Daniel Apon
Toshinori Araki
Seiko Arita
Tomer Ashur
Nuttapong Attrapadung
Man Ho Allen Au
Benedikt Auerbach
Saikrishna
   Badrinarayanan
Vivek Bagaria
Josep Balasch
Gustavo Banegas
Laasya Bangalore
Subhadeep Banik
Achiya Bar-On
Manuel Barbosa
James Bartusek
Carsten Baum
Arthur Beckers
Rouzbeh Behnia
Francesco Berti
Alexandre Berzati
Ward Beullens
Shivam Bhasin
Nina Bindel
Nicolas Bordes
Jannis Bossert
Katharina Boudgoust
Christina Boura
Florian Bourse
Zvika Brakerski
Anne Broadbent
Olivier Bronchain
Leon Groot Bruinderink

Megha Byali
Eleonora Cagli
Ignacio Cascudo
Pyrros Chaidos
Avik Chakraborti
Donghoon Chang
Hao Chen
Jie Chen
Long Chen
Ming-Shing Chen
Qian Chen
Jung Hee Cheon
Céline Chevalier
Ilaria Chillotti
Wonhee Cho
Wonseok Choi
Wutichai Chongchitmate
Jérémy Chotard
Arka Rai Choudhuri
Sherman Chow
Michele Ciampi
Michael Clear
Thomas De Cnudde
Benoît Cogliati
Sandro Coretti-Drayton
Edouard Cuvelier
Jan Czajkowski
Dana Dachman-Soled
Joan Daemen
Nilanjan Datta
Gareth T. Davies
Patrick Derbez
Apporva Deshpande
Siemen Dhooghe
Christoph Dobraunig
Rafael Dowsley
Yfke Dulek
Avijit Dutta
Sébastien Duval
Keita Emura
Thomas Espitau
Xiong Fan
Antonio Faonio

Oriol Farràs
Sebastian Faust
Prastudy Fauzi
Hanwen Feng
Samuele Ferracin
Dario Fiore
Georg Fuchsbauer
Thomas Fuhr
Eiichiro Fujisaki
Philippe Gaborit
Tatiana Galibus
Chaya Ganesh
Daniel Gardham
Luke Garratt
Pierrick Gaudry
Nicholas Genise
Esha Ghosh
Satrajit Ghosh
Kristian Gjøsteen
Aarushi Goel
Huijing Gong
Junqing Gong
Alonso González
Dahmun Goudarzi
Rishabh Goyal
Jiaxin Guan
Aurore Guillevic
Chun Guo
Kaiwen Guo
Qian Guo
Mohammad Hajiabadi
Carmit Hazay
Jingnan He
Brett Hemenway
Nadia Heninger
Javier Herranz
Shoichi Hirose
Harunaga Hiwatari
Viet Tung Hoang
Justin Holmgren
Akinori Hosoyamada
Kexin Hu
Senyang Huang

Yan Huang
Phi Hun
Aaron Hutchinson
Chloé Hébant
Kathrin Hövelmanns
Ilia Iliashenko
Mitsugu Iwamoto
Tetsu Iwata
Zahra Jafargholi
Christian Janson
Ashwin Jha
Dingding Jia
Sunghyun Jin
Charanjit S. Jutla
Mustafa Kairallah
Saqib A. Kakvi
Marc Kaplan
Emrah Karagoz
Ghassan Karame
Shuichi Katsumata
Craig Kenney
Mojtaba Khalili
Dakshita Khurana
Duhyeong Kim
Hyoseung Kim
Sam Kim
Seongkwang Kim
Taechan Kim
Agnes Kiss
Fuyuki Kitagawa
Michael Klooß
François Koeune
Lisa Kohl
Stefan Kölbl
Yashvanth Kondi
Toomas Krips
Veronika Kuchta
Nishant Kumar
Noboru Kunihiro
Po-Chun Kuo
Kaoru Kurosawa
Ben Kuykendall
Albert Kwon
Qiqi Lai
Baptiste Lambin
Roman Langrehr

Jason LeGrow
ByeongHak Lee
Changmin Lee
Keewoo Lee
Kwangsu Lee
Youngkyung Lee
Dominik Leichtle
Christopher Leonardi
Tancrède Lepoint
Gaëtan Leurent
Itamar Levi
Baiyu Li
Yanan Li
Zhe Li
Xiao Liang
Benoît Libert
Fuchun Lin
Rachel Lin
Wei-Kai Lin
Eik List
Fukang Liu
Guozhen Liu
Meicheng Liu
Qipeng Liu
Shengli Liu
Zhen Liu
Alex Lombardi
Julian Loss
Jiqiang Lu
Xianhui Lu
Yuan Lu
Lin Lyu
Fermi Ma
Gilles Macario-Rat
Urmila Mahadev
Monosij Maitra
Christian Majenz
Nikolaos Makriyannis
Giulio Malavolta
Sogol Mazaheri
Bart Mennink
Peihan Miao
Shaun Miller
Kazuhiko Minematsu
Takaaki Mizuki
Amir Moradi

Kirill Morozov
Fabrice Mouhartem
Pratyay Mukherjee
Pierrick Méaux
Yusuke Naito
Mridul Nandi
Peter Naty
María Naya-Plasencia
Anca Niculescu
Ventzi Nikov
Takashi Nishide
Ryo Nishimaki
Anca Nitulescu
Ariel Nof
Sai Lakshmi Bhavana
    Obbattu
Kazuma Ohara
Emmanuela Orsini
Elena Pagnin
Wenlun Pan
Omer Paneth
Bo Pang
Lorenz Panny
Jacques Patarin
Sikhar Patranabis
Alice Pellet-Mary
Chun-Yo Peng
Geovandro Pereira
Olivier Pereira
Léo Perrin
Naty Peter
Cécile Pierrot
Jeroen Pijnenburg
Federico Pintore
Bertram Poettering
David Pointcheval
Yuriy Polyakov
Eamonn Postlethwaite
Emmanuel Prouff
Pille Pullonen
Daniel Puzzuoli
Chen Qian
Tian Qiu
Willy Quach
Håvard Raddum
Ananth Raghunathan

Somindu Ramanna
Kim Ramchen
Shahram Rasoolzadeh
Mayank Rathee
Divya Ravi
Joost Renes
Angela Robinson
Thomas Roche
Miruna Rosca
Mélissa Rossi
Mike Rosulek
Yann Rotella
Arnab Roy
Luis Ruiz-Lopez
Ajith Suresh
Markku-Juhani
  O. Saarinen
Yusuke Sakai
Kazuo Sakiyama
Amin Sakzad
Louis Salvail
Simona Samardjiska
Pratik Sarkar
Christian Schaffner
John Schanck
Berry Schoenmakers
Peter Scholl
André Schrottenloher
Jacob Schuldt
Sven Schäge
Sruthi Sekar
Srinath Setty
Yannick Seurin
Barak Shani
Yaobin Shen
Sina Shiehian
Kazumasa Shinagawa
Janno Siim
Javier Silva
Mark Simkin

Boris Skoric
Maciej Skórski
Yongsoo Song
Pratik Soni
Claudio Soriente
Florian Speelman
Akshayaram Srinivasan
François-Xavier Standaert
Douglas Stebila
Damien Stehlé
Patrick Struck
Valentin Suder
Bing Sun
Shifeng Sun
Siwei Sun
Jaechul Sung
Daisuke Suzuki
Katsuyuki Takashima
Benjamin Hong Meng
  Tan
Stefano Tessaro
Adrian Thillard
Yan Bo Ti
Jean-Pierre Tillich
Radu Țițiu
Yosuke Todo
Junichi Tomida
Viet Cuong Trinh
Rotem Tsabary
Hikaru Tsuchida
Yi Tu
Nirvan Tyagi
Bogdan Ursu
Damien Vergnaud
Jorge Luis Villar
Srinivas Vivek
Christine van Vredendaal
Satyanarayana Vusirikala
Sameer Wagh
Hendrik Waldner

Alexandre Wallet
Michael Walter
Han Wang
Haoyang Wang
Junwei Wang
Mingyuan Wang
Ping Wang
Yuyu Wang
Zhedong Wang
Yohei Watanabe
Gaven Watson
Weiqiang Wen
Yunhua Wen
Benjamin Wesolowski
Keita Xagawa
Zejun Xiang
Hanshen Xiao
Shota Yamada
Takashi Yamakawa
Kyosuke Yamashita
Avishay Yanai
Guomin Yang
Kan Yasuda
Masaya Yasuda
Aaram Yun
Alexandros Zacharakis
Michal Zajac
Bin Zhang
Cong Zhang
En Zhang
Huang Zhang
Xiao Zhang
Zheng Zhang
Chang-An Zhao
Raymond K. Zhao
Yongjun Zhao
Yuanyuan Zhou
Jiamin Zhu
Yihong Zhu
Lukas Zobernig

Contents – Part II

Multilinear Maps

Homomorphic Encryption

Combinatorial Cryptography

Multiparty Computation (2)

Codes

Collision Resistant Hashing from Sub-exponential Learning Parity with Noise

Yu Yu[1,2,7](✉), Jiang Zhang[2](✉), Jian Weng[3], Chun Guo[4,5](✉), and Xiangxue Li[6,7](✉)

[1] Department of Computer Science and Engineering, Shanghai Jiao Tong University, Shanghai 200240, China
yuyuathk@gmail.com
[2] State Key Laboratory of Cryptology, P.O. Box 5159, Beijing 100878, China
jiangzhang09@gmail.com
[3] Jinan University, Guangzhou 510632, China
cryptjweng@gmail.com
[4] Key Laboratory of Cryptologic Technology and Information Security, Ministry of Education, Shandong University, Qingdao 266237, China
chun.guo.sc@gmail.com
[5] School of Cyber Science and Technology, Shandong University, Qingdao 266237, China
[6] School of Software Engineering, East China Normal University, Shanghai 200062, China
xiangxueli@gmail.com
[7] Westone Cryptologic Research Center, Beijing 100070, China

Abstract. The Learning Parity with Noise (LPN) problem has recently found many cryptographic applications such as authentication protocols, pseudorandom generators/functions and even asymmetric tasks including public-key encryption (PKE) schemes and oblivious transfer (OT) protocols. It however remains a long-standing open problem whether LPN implies collision resistant hash (CRH) functions. Inspired by the recent work of Applebaum et al. (ITCS 2017), we introduce a general construction of CRH from LPN for various parameter choices. We show that, just to mention a few notable ones, under any of the following hardness assumptions (for the two most common variants of LPN)

1. constant-noise LPN is $2^{n^{0.5+\varepsilon}}$-hard for any constant $\varepsilon > 0$;
2. constant-noise LPN is $2^{\Omega(n/\log n)}$-hard given $q = \mathsf{poly}(n)$ samples;
3. low-noise LPN (of noise rate $1/\sqrt{n}$) is $2^{\Omega(\sqrt{n}/\log n)}$-hard given $q = \mathsf{poly}(n)$ samples.

there exists CRH functions with constant (or even poly-logarithmic) shrinkage, which can be implemented using polynomial-size depth-3 circuits with NOT, (unbounded fan-in) AND and XOR gates. Our technical route LPN → bSVP → CRH is reminiscent of the known reductions for the large-modulus analogue, i.e., LWE → SIS → CRH, where the binary Shortest Vector Problem (bSVP) was recently introduced by Applebaum et al. (ITCS 2017) that enables CRH in a similar manner to Ajtai's CRH functions based on the Short Integer Solution (SIS) problem.

© International Association for Cryptologic Research 2019
S. D. Galbraith and S. Moriai (Eds.): ASIACRYPT 2019, LNCS 11922, pp. 3–24, 2019.
https://doi.org/10.1007/978-3-030-34621-8_1

Furthermore, under additional (arguably minimal) idealized assumptions such as small-domain random functions or random permutations (that trivially imply collision resistance), we still salvage a simple and elegant collision-resistance-preserving domain extender combining the best of the two worlds, namely, maximized (depth one) parallelizability and polynomial shrinkage. In particular, assume $2^{n^{0.5+\varepsilon}}$-hard constant-noise LPN or $2^{n^{0.25+\varepsilon}}$-hard low-noise LPN, we obtain a collision resistant hash function that evaluates in parallel only a single layer of small-domain random functions (or random permutations) and shrinks polynomially.

1   Introduction

1.1   Learning Parity with Noise

LEARNING PARITY WITH NOISE. The computational version of the Learning Parity with Noise (LPN) assumption with secret size $n \in \mathbb{N}$ and noise rate $0 < \mu < 1/2$ postulates that given any number of samples $q = \mathsf{poly}(n)$ it is computationally infeasible for any probabilistic polynomial-time (PPT) algorithm to recover the random secret $\mathbf{x} \xleftarrow{\$} \{0,1\}^n$ given $(\mathbf{A}, \ \mathbf{A} \cdot \mathbf{x} + \mathbf{e})$, where $\mathbf{A}$ is a random $q \times n$ Boolean matrix, $\mathbf{e}$ follows $\mathcal{B}_\mu^q = (\mathcal{B}_\mu)^q$, $\mathcal{B}_\mu$ denotes the Bernoulli distribution with parameter μ (taking the value 1 with probability μ and the value 0 with probability $1 - \mu$), '$\cdot$' and '$+$' denote (matrix-vector) multiplication and addition over GF(2) respectively. The decisional version of LPN simply assumes that $(\mathbf{A}, \ \mathbf{A} \cdot \mathbf{x} + \mathbf{e})$ is pseudorandom. The two versions are polynomially equivalent [6,14,39].

HARDNESS OF LPN. The computational LPN problem can be seen as the average-case analogue of the NP-complete problem "decoding random linear codes" [10]. LPN has been also extensively studied in learning theory, and it was shown in [33] that an efficient algorithm for LPN would allow to learn several important function classes such as 2-DNF formulas, juntas, and any function with a sparse Fourier spectrum. When the noise rate μ is constant (i.e., independent of secret size n), Blum, Kalai and Wasserman [15] showed how to solve LPN with time/sample complexity $2^{O(n/\log n)}$. Lyubashevsky [45] observed that one can produce almost as many LPN samples as needed using only $q = n^{1+\epsilon}$ LPN samples (of a lower noise rate), which implies a variant of the BKW attack [15] with time complexity $2^{O(n/\log\log n)}$ and sample complexity $n^{1+\epsilon}$. If one is restricted to $q = O(n)$ samples, then the best attack has exponential complexity $2^{O(n)}$ [50]. Under low noise rate $\mu = 1/\sqrt{n}$, the best attacks [9,11,18,42] solve LPN with time complexity $2^{O(\sqrt{n})}$. The low-noise LPN is mostly believed a stronger assumption than constant-noise LPN. In noise regime $\mu = 1/\sqrt{n}$, LPN can be used to build public-key encryption (PKE) schemes [2] and oblivious transfer (OT) protocols. Quantum algorithms [32] that build upon Grover search may achieve a certain level (up to quadratic) of speedup over classic ones in solving LPN, which does not change the asymptotic order (up to the constant in the exponent) of the complexity of the problem. This makes LPN a promising candidate for "post-quantum cryptography". Furthermore, LPN enjoys simplicity and is more suited for weak-power devices (e.g., RFID tags) than other quantum-secure candidates such as Learning with Errors (LWE) [58] as the many modular additions and multiplications in LWE would be simplified to AND and XOR gates in LPN.

SYMMETRIC-KEY CRYPTOGRAPHY FROM CONSTANT-NOISE LPN. LPN was used to build lightweight authentication schemes (e.g. [35,38,39], just to name a few). Kiltz et al. [41] and Dodis et al. [26] constructed randomized MACs from LPN, which implies a two-round authentication scheme with security against active adversaries. Lyubashevsky and Masny [47] gave a more efficient three-round authentication scheme from LPN and recently Cash, Kiltz, and Tessaro [19] reduced the round complexity to 2 rounds. Applebaum et al. [4] used LPN to construct efficient symmetric encryption schemes with certain key-dependent message (KDM) security. Jain et al. [37] constructed an efficient perfectly binding string commitment scheme from LPN. We refer to the survey [56] about cryptography from LPN.

PUBLIC-KEY CRYPTOGRAPHY AND MORE FROM LOW-NOISE LPN. Alekhnovich [2] established the feasibility result that public-key encryption (PKE) can be based on LPN in the low-noise regime of $\mu = 1/\sqrt{n}$. Döttling et al. [30] and Kiltz et al. [40] further showed that low-noise LPN alone already suffices for PKE schemes with CCA (and KDM [29]) security. Once we obtain a PKE, it is perhaps not so surprising to build an oblivious transfer (OT) protocol. That is, LPN-based PKE uses pseudorandom public keys (so that one can efficiently fake random public keys that are computationally indistinguishable from real ones) and in this scenario Gertner et al. [34] showed how to construct an OT protocol in a black-box manner. This observation was made explicit in [23], where universally composable OT protocols were constructed from low-noise LPN. All the above schemes are based on LPN of noise rate $1/\sqrt{n}$. The only exception seems to be the recent result by Yu and Zhang [64] that PKE and OT can also be based on constant-noise LPN with hardness $2^{n^{1/2+\epsilon}}$.

OPEN PROBLEMS AND RECENT PROGRESS. It remains open [46,56] whether LPN implies other advanced cryptographic objects, such as fully homomorphic encryption (FHE) and collision resistant hash (CRH) functions. Brakerski [16] reported some negative result that straightforward LPN-based encryptions are unlikely to achieve full homomorphism. As for LPN-based CRH, a notable progress was recently made by Applebaum et al. [5], who showed that $2^{\Omega(n/\log n)}$-hard constant-noise LPN implies CRH[1]. Based on some ideas (in particular, the bSVP assumption) from [5], we introduce a general construction of CRH from LPN with various tunable trade-offs between the parameters (e.g., noise rate, hardness, shrinkage), and then present the main feasibility results in commonly assumed noise regimes.

---

[1] More precisely, [5] obtains a win-win result that either constant-noise LPN implies CRH or one achieves arbitrary polynomial speedup over the BKW algorithm [15].

ON THE CONCURRENT WORK OF [17]. Concurrently and independent of this work, Brakerski et al. [17] used essentially the same technique as [5] and ours and constructed CRH from LPN at the (extremely low) noise rate of $\mu = \log^2 n/n$, which can be derived as a special case under our framework.

1.2 Cryptographic Hash Functions

A CRYPTOGRAPHIC HASH FUNCTION $\{0,1\}^* \rightarrow \{0,1\}^n$ is a deterministic function that maps arbitrarily (or at least sufficiently) long bit strings into digests of a fixed length. The function was originally introduced in the seminal work of Diffie and Hellman [25] to produce more efficient and compact digital signatures. As exemplified by MD5 and SHA-1/2/3, it is now one of the most widely used cryptographic primitives in security applications and protocols, such as SSL/TLS, PGP, SSH, S/MIME, IPsec and Bitcoin. Merkle [53] formulated three main security properties (that still remain in use to date) of a cryptographic hash function: preimage resistance, second preimage resistance and collision resistance, of which collision resistance seems the most essential and suffices for many aforementioned applications[2]. Similar to the mode of operations for data encryption, the design of cryptographic hash functions proceeds in two steps: one first designs a compression function that operates on fixed-length inputs and outputs, and then applies a domain extender to accept messages of arbitrary length. This dates back to the independent work of Merkle [55] and Damgård [22], who proposed a domain extender, and showed that if the underlying compression function is collision resistant then so is the hash function based on the Merkle-Damgård construction. We refer to [3] for a survey about various domain extenders for cryptographic hash functions. For the rest of this paper we will focus on such length-regular collision resistant compression functions, namely, CRH functions.

COLLISION RESISTANT HASHING. Theoretical constructions of CRH functions can be based on the hardness of factoring and discrete logarithm (via the construction of claw-free permutations [21]), which are however far from practical. Ajtai [1] introduced an elegant and (conceptually) simple construction: $f_{\mathbf{A}} : \{0,1\}^m \rightarrow \mathbb{Z}_p^n$ that for a random $\mathbf{A} \in \mathbb{Z}_p^{n \times m}$ and some (at least polynomially) large p and on input $\mathbf{z} \in \{0,1\}^m$ it computes

$$f_{\mathbf{A}}(\mathbf{z}) = \mathbf{A} \cdot \mathbf{z} \mod p, \tag{1}$$

which is collision resistant via a security reduction from the Short Integer Solution (SIS) problem, and is thus at least as hard as lattice problems such as GapSVP and SIVP. Lyubashevsky et al. [48] gave a ring-based variant of Ajtai's construction, called SWIFFT, which admits FFT and precomputation techniques for improved efficiency while preserves an asymptotic security proof from

---

[2] Unlikely collision resistance whose definition is unique and unambiguous, there are several variants of (second) preimage resistance for which people strive to find a compromise that facilitates security proofs yet captures the needs of most applications. Some variants of (second) preimage resistance are implied by collision resistance in the conventional or provisional sense [60].

ideal lattices at the same time. Despite a substantial gap between the claimed security level and the actual security bounds proved, SWIFFT [48] and its modified version (as a SHA-3 candidate) SWIFFTX [7] are among the very few hash function designs combining the best of two worlds (i.e., practical efficiency and rigorous security proof).

THE EXPAND-THEN-COMPRESS APPROACH. Recently, Applebaum et al. [5] constructed a function $h_\mathbf{M} : \{0,1\}^k \to \{0,1\}^n$ keyed by a random $n \times q$ binary matrix $\mathbf{M}$ as:

$$h_\mathbf{M}(\mathbf{y}) = \mathbf{M} \cdot \mathsf{Expand}(\mathbf{y}), \tag{2}$$

where Expand is an injective function that expands $\mathbf{y}$ into a much longer yet sparse string, i.e., for every $\mathbf{y} \in \{0,1\}^k$: $t = |\mathsf{Expand}(\mathbf{y})| < n < k < q$. Note that $h_\mathbf{M}$ can be viewed as a binary version of Ajtai's CRH (see $f_\mathbf{A}$ in (1)), where matrix $\mathbf{A}$ over $\mathbb{Z}_p$ is simplified to a binary matrix $\mathbf{M}$, and binary vector $\mathbf{z}$ is further flattened to a sparse binary vector $\mathsf{Expand}(\mathbf{y})$. Thanks to the simplification to the binary field, $h_\mathbf{M}$ can be implemented rather efficiently both in the asymptotic sense and in practice. Under certain realizations of Expand (see Lemma 3.1), $h_\mathbf{M}$ (for any specified $\mathbf{M}$) can be directly translated to a polynomial-size circuit of NOT, (unbounded fan-in) XOR and AND gates in depth 3 (or even depth 2 if the input includes not only the individual bits of $\mathbf{y}$ but also their respective complements). Interestingly, the FSB hash proposal [8] and its variant the RFSB hash [12] fall into concrete (but over optimistic) instantiations of $h_\mathbf{M}$.[3]

BINARY SVP. In order to justify the asymptotic security of the EtC hash, Applebaum et al. [5] introduced the binary Shortest Vector Problem (binary SVP or bSVP in short). Informally, the bSVP assumption asserts that given a random matrix distribution[4] $\mathbf{M} \xleftarrow{\$} \{0,1\}^{n \times q}$, it is computationally infeasible to find a non-zero $\mathbf{x} \in \{0,1\}^q$ of Hamming weight $t \ll q$ such that $\mathbf{Mx} = \mathbf{0}$. From a code-theoretic perspective, $\mathbf{M}$ specifies the $n \times q$ parity check matrix of a random binary linear code of rate $1 - n/q$, where the rows of $\mathbf{M}$ are linearly independent (except with negligible probability), and therefore the bSVP postulates that finding a short codeword is hard in the average case. We refer to [5] for discussions about meaningful regimes of (t/q) that give rise to one-way functions and collision resistant hash functions. Similar to SIS, bSVP immediately implies CRH as any efficient algorithm that comes up with a collision $h_\mathbf{M}(\mathbf{y}) = h_\mathbf{M}(\mathbf{y}')$ for $\mathbf{y} \neq \mathbf{y}'$ immediately implies a solution to bSVP, i.e., $\mathbf{M} \cdot \mathbf{x} = \mathbf{0}$, where $\mathbf{x} = \mathsf{Expand}(\mathbf{y}) - \mathsf{Expand}(\mathbf{y}')$ has weight no greater than $2t$. We mention that in the worst case, it is NP-hard to compute (or even to approximate by a constant factor) the distance of a linear code [31,63]. However, as an average-case hardness assumption, bSVP is relatively new and deserves further investigation. A shortcut and promising direction is to see whether bSVP is reducible from the learning parity with noise (LPN) problem since they are both related to random

---

[3] However, our results do not immediately constitute security proofs for the FSB-style hash functions as there remains a substantial gap between the security proved and security level claimed by the FSB instantiation.

[4] $\mathbf{M}$ in our consideration has dimension $n \times q$ instead of $\alpha n \times n$ considered by [5].

binary linear codes, and the average-case hardness of the latter is well understood. However, the work of [5] only established a weak connection between bSVP and LPN. That is, they show that at least one of the following is true:

1. One can achieve an arbitrary polynomial speedup over the BKW algorithm [15], i.e., for every constant $\varepsilon > 0$ there exists an algorithm that solves constant-noise LPN with time and sample complexity $2^{\frac{\varepsilon n}{\log n}}$ for infinitely many n's.
2. There exist CRH functions of constant shrinkage and logarithmic degree.

Otherwise stated, assume that the BKW algorithm cannot be further improved asymptotically, then bSVP (for certain parameters) and CRH are implied.

1.3   The Construction of CRH from LPN

DUALITY BETWEEN LPN AND bSVP. We explain the high-level intuition of how LPN relates (and reduces) to bSVP (deferring the choices of non-trivial parameters to next paragraph), which in turn implies CRH. Under the theme of "decoding random linear codes" where row vector $\mathbf{s}^\mathsf{T}$ is the message, $\mathbf{M}$ is an $n \times q$ generator matrix and $\mathbf{s}^\mathsf{T}\mathbf{M}$ is the codeword, the idea is to use a (sparse) column vector $\mathbf{x}$ from the corresponding parity matrix such that any (noisy) codeword multiplied by $\mathbf{x}$ is (biased to) 0, regardless of the value of $\mathbf{s}$. Informally, assume for contradiction that a useful bSVP solver succeeds in finding a sparse vector $\mathbf{x}$ for an $n \times q$ matrix $\mathbf{M}$ such that $\mathbf{Mx} = \mathbf{0}$, then this leads to a distinguishing attack against the LPN instance $(\mathbf{M}, \mathbf{s}^\mathsf{T}\mathbf{M} + \mathbf{e}^\mathsf{T})$ by computing

$$(\mathbf{s}^\mathsf{T}\mathbf{M} + \mathbf{e}^\mathsf{T}) \cdot \mathbf{x} = \mathbf{e}^\mathsf{T}\mathbf{x}$$

which is a biased bit (and thus distinguishable from uniform) due to the sparseness of $\mathbf{x}$ and $\mathbf{e}$. This already constitutes a contradiction to the decisional LPN, and one can repeat the above on sufficiently many independent samples (with a majority voting) to gain a constant advantage, and further transform it into a key-recovery attack using the same number of samples [6].

MAIN FEASIBILITY RESULTS. By exploiting the duality between LPN and bSVP, we present a general framework stated in Theorem 1.1 below (and more formally in Theorem 3.1) that enables to construct CRH from LPN for various tunable parameter choices, as stated in Corollary 1.1 (and more formally in Corollary 3.1). The constructions follow the Expand-then-Compress approach and can be implemented by a polynomial-size depth-3 circuit with NOT, (unbounded fan-in) AND and XOR gates[5]. The framework, in particular, when tuned to params #2 and #4 of Corollary 1.1, encompasses the known results obtained in [6] and the concurrent work of [17]. In addition, it establishes feasibility results for constant-noise LPN assuming much less hardness (see param #1 of Corollary 1.1) and for low-noise LPN (see param #3 of Corollary 1.1), which was not previously known. We remark that the $2^{\Omega(\sqrt{n}/\log n)}$-hardness assumption for

---

[5] The circuit falls into the class $\mathsf{AC}^0(\mathsf{MOD2})$. See Sect. 2 for a formal definition.

low-noise LPN is quite reasonable as the current best attacks need complexity $\mathsf{poly}(n) \cdot e^{\sqrt{n}}$ [43] for which even improving upon the constant in the exponent seems nontrivial. Further, the $2^{n^{0.501}}$-hardness assumed for constant-noise LPN offers even more generous security margins as the best attack goes even beyond $2^{n^{0.999}}$ [15].

Theorem 1.1 (main framework, informal). *Let n be the security parameter, and let $\mu = \mu(n)$, $k = k(n)$, $q = q(n)$, $t = t(n)$ and $T = T(n)$ such that $t^2 \leq q \leq T \approx 2^{\frac{8\mu t}{\ln 2(1-2\mu)}}$. Assume that the (decisional) LPN problem of size n and noise rate μ is T-hard given q samples, and let*

$$h_{\mathbf{M}} : \{0,1\}^k \to \{0,1\}^n, \; h_{\mathbf{M}}(\mathbf{y}) = \mathbf{M} \cdot \mathsf{Expand}(\mathbf{y}), \; \mathsf{Expand} : \{0,1\}^k \to \{0,1\}^q,$$

be functions satisfying the following conditions:

1. *($h_{\mathbf{M}}$ is compressing). $k > n$;*
2. *(Expand has sparse outputs). for all $\mathbf{y} \in \{0,1\}^k$: $|\mathsf{Expand}(\mathbf{y})| = t$;*
3. *(Expand is injective). Expand is an injection with $k \approx \log\binom{q}{t} = (1 + o(1)) \log(q/t)t > t \log q/2$ (see Fact 2), where the inequality is due to $t \leq \sqrt{q}$.*

Then, $h_{\mathbf{M}}{}^6$ is a CRH function with shrinkage factor $\frac{n}{k}$.

RATIONALE. Upon any collision $\mathbf{y} \neq \mathbf{y}'$ we get that $\mathbf{x} = \mathsf{Expand}(\mathbf{y}) - \mathsf{Expand}(\mathbf{y}')$ such that $\mathbf{e}^{\mathsf{T}}\mathbf{x}$, i.e., the XOR sum of up to $2t$ bits drawn from $\mathcal{B}_\mu$ is

$$\frac{1}{2} + \frac{2^{-(\log \frac{1}{1-2\mu})2t}}{2} \geq \frac{1}{2} + \frac{2^{-\frac{4\mu t}{\ln 2(1-2\mu)}}}{2}$$

biased to 0 by the Piling up lemma (Lemma 2.1) and inequality $\ln(1+x) \leq x$. Otherwise said, the underlying decisional LPN must be $2^{\frac{\Omega(\mu t)}{(1-2\mu)}}$-hard to counteract the aforementioned attack. We refer to Theorem 3.1 for a more formal statement and a rigorous proof. The framework allows for various trade-offs between μ, q and T (via the intermediate parameter t) and we state a few notable ones in Corollary 1.1 below. Moreover, the CRH can be contained in $\mathsf{AC}^0(\mathsf{MOD2})$ based on a parallel implementation of the underlying function Expand in AC^0.

Corollary 1.1 (LPN $\to$ CRH). *Type LPN with Hardness implies CRH with Shrinkage in $\mathsf{AC}^0(\mathsf{MOD2})$, where (Type, Hardness, Shrinkage) can be (but are not limited to) any of the following:*

1. *(Constant-noise, less hardness, poly-logarithmic shrinkage).*
   $\mu = O(1)$, $T = 2^{n^{0.5+\varepsilon}}$, $q = 2^{n^{0.5}}$ and $\frac{n}{k} < \frac{16\mu}{\ln 2(1-2\mu)n^\varepsilon} = \frac{16\mu}{\ln 2(1-2\mu)\log^{2\varepsilon}\lambda}$ *for any constant $\varepsilon > 0$.*

---

[6] More strictly speaking, the resulting CRH is either $h_{\mathbf{M}}$ itself or its domain-extended version (by a parallel repetition).

2. (Constant-noise, more hardness, constant shrinkage).
$\mu = O(1)$, $T = 2^{\frac{\varepsilon n}{\log n}}$, $q = n^{C_{\varepsilon,\mu}}$, $\frac{n}{k} < \frac{1}{2}$ for any constant $\varepsilon > 0$ and $C_{\varepsilon,\mu} = \max(\frac{32\mu}{\varepsilon \ln 2(1-\mu)}, 2)$.

3. (Low-noise, more hardness, constant shrinkage).
$\mu = 1/\sqrt{n}$, $T = 2^{\frac{\varepsilon\sqrt{n}}{\log n}}$, $q = n^{C_{\varepsilon,\mu}}$, $\frac{n}{k} < \frac{1}{2}$ for any constant $\varepsilon > 0$ and $C'_{\varepsilon,\mu} = \max(\frac{32}{\varepsilon \ln 2}, 2)$.

4. (Extremely low-noise, standard hardness, constant shrinkage).
$\mu = \frac{(\log n)^2}{n}$, $T = q > \mathsf{poly}(n)$ for every poly, and $\frac{n}{k} < \frac{1}{2}$.

INTUITIONS ABOUT PARAMETERS CHOICES. The aforementioned parameter choices are not exhaustive but they follow quite naturally from the respective noise rates. We explain the underlying intuitions for making such choices (and refer to Corollary 3.1 and its proof for formal details). For immediate efficiency we set $q = \mathsf{poly}(n)$ (s.t. the dimensions of $\mathbf{M}$ are polynomially related) and constant shrinkage factor $\frac{n}{k} < \frac{2n}{t\log q} = \frac{1}{2}$, and therefore $t = \Omega(n/\log n)$ and it requires hardness $T = 2^{\Omega(\mu n/\log n)}$. This yields the parameter settings #2, #3 and #4 for constant, low and extremely low noise rates respectively. Alternatively, in favor of minimized hardness assumed for constant-noise LPN, we let the sample complexity be nearly the same as time complexity up to a factor n^ε,[7] i.e., $\log(q) = \Omega(\frac{\log T}{n^\varepsilon}) = \Omega(\frac{t}{n^\varepsilon})$ and thus the injective condition becomes $k = \Omega(t^2/n^\varepsilon)$ and $\frac{n}{k} < \frac{n^{1+\varepsilon}}{\Omega(t^2)}$, which results in param #1 by setting $t = \Omega(n^{0.5+\varepsilon})$. However, now the issue is that the dimensions q and n of $\mathbf{M}$ are not polynomially related and thus it does not immediately give rise to an efficient CRH. This motivates us to switch to another parameter $\lambda = q = 2^{\sqrt{n}}$ such that $h_{\mathbf{M}} : \{0,1\}^{\Omega(\log^{2+2\varepsilon}\lambda)} \to \{0,1\}^{\log^2\lambda}$ for $\mathbf{M} \in \{0,1\}^{\log^2\lambda \times \lambda}$ is a $\lambda^{\Omega(\log^{2\varepsilon}\lambda)}$-hard CRH function computable in time $\mathsf{poly}(\lambda)$, which further implies a domain-extended CRH $h'_{\mathbf{M}} : \{0,1\}^{\Omega(\lambda\log^{2\varepsilon}\lambda)} \to \{0,1\}^\lambda$ by a parallel repetition.

FEASIBILITIES VS. LIMITS. Admittedly, the limits of the framework are obvious: unless under extremely low noise rate [17] the hardness assumed is much beyond polynomial (although still reasonable given the current state-of-the-art). Moreover, the parameter-switching technique (that helps to reduce hardness assumed) dramatically downgrades the security and deteriorates the shrinkage factor from polynomial to poly-logarithmic. Further, the technique only applies to constant noise: if the noise rate μ depends on n, e.g., $\mu = 1/\sqrt{n}$, then switching to a new parameter, say $\lambda = 2^{n^{0.25}}$, yields lifted noise rate $\mu = 1/\log^2\lambda$. We offer an alternative to avoid the efficiency/security loss by assuming a minimal amount of heuristics, e.g., a small domain random function. This helps to obtain a polynomially shrinking domain extender that makes only a single layer of evaluations on the underlying random function. In terms of parallelizability, this beats generic (collision-resistance-preserving) domain extenders such

---

[7] By switching to a new security parameter, we eventually obtain a CRH function with polynomial running time and super-polynomial security for which the n^ε gap factor plays a vital role.

as Merkle-Damgård [22,55] and the Merkle-tree [44,54], where to achieve polynomial shrinkage even the latter needs to evaluate a tree of depth $O(\log n)$ on length-halving CRHs. A price to pay is that we make additional (but reasonable) hardness assumptions, e.g., that the low-noise LPN problem is $2^{n^{0.25+\varepsilon}}$-hard.

Corollary 1.2 (A polynomially shrinking domain extender, informal).
Assume that (n,μ,q)-DLPN is T-hard and $R : \{0,1\}^{\log(q)} \to \{0,1\}^n$ with $\log(q) \ll n$ behaves like a random function, then for $\mathbf{y} = \mathbf{y}_1\|\cdots\|\mathbf{y}_t$ parsed as $t = k/L$ blocks, each of size $L = \log(q/t)$, we have

$$h^R : \{0,1\}^k \to \{0,1\}^n, \ h^R(\mathbf{y}) = \bigoplus_{i=1}^{t} R(i\|\mathbf{y}_i),$$

is a CRH function with shrinkage $\frac{n}{k}$, where $(\mu,T,q,\frac{n}{k})$ can be either below:

1. *(Constant-noise, less hardness, polynomial shrinkage).*
   $\mu = O(1)$, $T = 2^{n^{0.5+\varepsilon}}$, $q = 2^{n^{0.5}}$ and $\frac{n}{k} < \frac{16\mu}{\ln 2(1-2\mu)n^\varepsilon}$ *for any constant $\varepsilon > 0$.*
2. *(Low-noise, less hardness, polynomial shrinkage).*
   $\mu = 1/\sqrt{n}$, $T = 2^{n^{0.25+\varepsilon}}$, $q = 2^{n^{0.25}}$ and $\frac{n}{k} < \frac{16}{\ln 2\cdot n^\varepsilon}$ *for any constant $\varepsilon > 0$.*

ON RELATED HEURISTIC-BASED APPROACHES. It may seem trivial to obtain CRHs from idealized heuristics such as random oracles and ideal ciphers, but we stress that we only make a quite light use of idealism by assuming a small-domain random function with inputs much shorter than outputs (for which domain extension is non-trivial), which can be efficiently instantiated from practical objects such as blockciphers (assuming that a blockcipher on a public random key behaves like a random permutation). In contrast, most previous blockcipher-based compression functions (e.g. [13,55,57]) reside in the (much stronger) Ideal Cipher Model that a block cipher on every key behaves independently like a random permutation. Moreover, existing permutation-based solutions either only offer a constant shrinkage factor (typically $1/2$) [51,62], or require permutations with a large domain (e.g., [28] needs a large permutation over $\{0,1\}^{n^2}$ to obtain a CRH function with shrinkage factor $1/n$).

2   Preliminaries

NOTATIONS AND DEFINITIONS. Column vectors are represented by bold lower-case letters (e.g., $\mathbf{s}$), row vectors are denoted as their transpose (e.g., $\mathbf{s}^\mathsf{T}$), and matrices are denoted by bold capital letters (e.g., $\mathbf{A}$). $|s|$ refers to the Hamming weight of binary string s. We use $\mathcal{B}_\mu$ to denote the Bernoulli distribution with parameter μ, while $\mathcal{B}_\mu^q$ denotes the concatenation of q independent copies of $\mathcal{B}_\mu$. We use $\log(\cdot)$ to denote the binary logarithm. $\mathbf{x} \xleftarrow{\$} \mathcal{X}$ refers to drawing $\mathbf{x}$ from set $\mathcal{X}$ uniformly at random, and $\mathbf{x} \leftarrow X$ means drawing $\mathbf{x}$ according to distribution X. $\mathbf{a}\|\mathbf{b}$ denotes the concatenation of $\mathbf{a}$ and $\mathbf{b}$. A function $\mathsf{negl}(\cdot)$

is negligible if for any constant N_c we have that $\mathsf{negl}(n) < 1/\mathsf{poly}(n)$ for every polynomial poly and all sufficiently large n. AC^0 refers to the class of polynomial-size, constant-depth circuit families with unbounded fan-in AND and OR gates, where NOT gates are allowed only at input level. $\mathsf{AC}^0(\mathsf{MOD2})$ refers to the class of polynomial-size, constant-depth circuit families with unbounded fan-in AND, OR and XOR gates.

We define decisional and computational LPN problems, and we just use the decisional one due to their polynomial equivalence. In particular, there are computational-to-decisional reductions even for the same sample complexity [6].

Definition 2.1 (Learning Parity with Noise). *Let n be the security parameter, and let $\mu = \mu(n)$, $q = q(n)$ and $T = T(n)$. The decisional LPN problem with secret length n, noise rate $0 < \mu < 1/2$ and sample complexity q, denoted by (n, μ, q)-DLPN, is T-hard if every probabilistic algorithm $\mathcal{D}$ of running time T we have that the following holds for all sufficiently large n's*

$$\left| \Pr[\mathcal{D}(\mathbf{A}, \ \mathbf{A}\cdot\mathbf{x}+\mathbf{e}) = 1] - \Pr[\mathcal{D}(\mathbf{A}, \mathbf{y}) = 1] \right| \leq \frac{1}{T}, \tag{3}$$

and the computational LPN problem with the same n, μ and q, denoted by (n, μ, q)-LPN, is T-hard if for every probabilistic algorithm $\mathcal{D}$ of running time T we have that the following holds for all sufficiently large n's

$$\Pr[\ \mathcal{D}(\mathbf{A}, \ \mathbf{A}\cdot\mathbf{x}+\mathbf{e}) = \mathbf{x}\ ] \ \leq \ \frac{1}{T}, \tag{4}$$

where $q \times n$ matrix $\mathbf{A} \xleftarrow{\$} \{0,1\}^{q \times n}$ and $\mathbf{x} \xleftarrow{\$} \{0,1\}^n$, $\mathbf{y} \xleftarrow{\$} \{0,1\}^q$ and $\mathbf{e} \leftarrow \mathcal{B}_\mu^q$.

STANDARD HARDNESS. *We recall that standard polynomial hardness requires that $T > \mathsf{poly}(n), q > \mathsf{poly}(n)$ and for every poly and all sufficiently large n's.*

Unlike other primitives (such as one-way functions, pseudorandom generators and functions) whose security parameter is typically the input/key length, the security strength of collision resistant hash functions are more often represented as a function of the output length n and it is upper bounded by $2^{n/2}$ due to birthday attacks. In practice, a fixed output size (e.g. 128, 160) typically corresponds to a single function (e.g., MD5, SHA1) instead of a collection of ones[8]. One can just stick to a $h_\mathbf{M}$ for some pre-fixed random $\mathbf{M}$.

Definition 2.2 (Collision Resistant Hash Functions). *A collection of functions*

$$\mathcal{H} = \left\{ h_z : \{0,1\}^{k(n)} \rightarrow \{0,1\}^n, z \in \{0,1\}^{s(n)} \right\}$$

---

[8] Recall that a non-uniform attacker can obtain polynomial-size non-uniform advice. Thus, if every security parameter corresponds to only a single function h then the attacker can include a pair of x and x' with $h(x) = h(x')$ as part of the advice.

is a collision-resistant hash (CRH) function if the following hold:

- *(Shrinking). The shrinkage factor of $\mathcal{H}$, defined as ratio $\frac{n}{k}$, is less than 1 for every n.*
- *(Efficient). There are efficient algorithms H and $\mathcal{G}$: (1) on input $z \in \{0,1\}^s$ and $y \in \{0,1\}^k$, H outputs $h_z(y)$; and (2) given 1^n as input $\mathcal{G}$ returns an index $z \in \{0,1\}^s$.*
- *(Collision-resistant). For every probabilistic polynomial-time (PPT) adversary $\mathcal{A}$*

$$\Pr_{z \leftarrow \mathcal{G}(1^n)}[\ (y,y') \leftarrow \mathcal{A}(z) : y \neq y' \wedge h_z(y) = h_z(y')\ ] = \mathsf{negl}(n).$$

The shrinkage is linear if $n/k \leq 1 - \epsilon$, and it is poly-logarithmic (resp., polynomial) if $n/k \leq 1/\log^\epsilon n$ (resp., $n/k \leq 1/n^\epsilon$) for some positive constant $\epsilon > 0$.

T-HARDNESS. *For $T = T(n)$ we call $\mathcal{H}$ a T-hard CRH if no probabilistic adversary $\mathcal{A}$ of running time T finds any collision with probability more than $1/T$.*

The indifferentiability framework [20,49] is widely adopted to analyze and prove the security of the construction of one idealized primitive from another, typically in settings where the underlying building blocks have no secrets.

Definition 2.3 (Indifferentiability [20]). *A Turing machine C with oracle access to an ideal primitive P is (q, σ, t, ϵ)-indifferentiable from an ideal primitive R, if there exists a simulator S with oracle access to R such that for any distinguisher D that makes at most q queries, it holds that*

$$\left| \Pr[D^{C^P, P} = 1] - \Pr[D^{R, S^R} = 1] \right| \leq \epsilon,$$

where S makes σ queries and runs in time t when interacting with D and R.

The implication is that C^P can securely replace R in many scenarios. We refer to [24,59] for discussions on the (in)applicability of indifferentiability results.

Lemma 2.1 (Piling-up lemma). *For $0 < \mu < 1/2$ and random variables E_1, $E_2, \cdots, E_\ell$ that are i.i.d. to $\mathcal{B}_\mu$ we have*

$$\Pr\left[\bigoplus_{i=1}^{\ell} E_i = 0 \right] = \frac{1}{2}(1 + (1 - 2\mu)^\ell) = \frac{1}{2} + 2^{-c_\mu \ell - 1},$$

where $c_\mu = \log \frac{1}{1-2\mu}$.

Fact 1. *For any $0 \leq x \leq 1$ it holds that $\log(1 + x) \geq x$; and for any $x > -1$ we have $\log(1 + x) \leq x/\ln 2$.*

Fact 2. *For $k = o(n)$ we have $\log \binom{n}{k} = (1 + o(1))k \log \frac{n}{k}$.*

3   Collision Resistant Hash Functions

3.1   The Expand-then-Compress Construction

We give a high-level overview about the EtC construction from [5]. Fix a random $n \times q$ matrix $\mathbf{M}$ which specifies the function. On input $\mathbf{y}$, $h_{\mathbf{M}}$ first stretches it into a long-but-sparse vector, i.e., $\mathsf{Expand}(\mathbf{y})$, and then multiply it with $\mathbf{M}$, which compresses into n bits. There are many ways to instantiate $h_{\mathbf{M}}$ and we use the following one which fulfills all properties needed by our framework (cf. Theorem 1.1). In addition, Expand is highly parallel and can be efficiently implemented by a single layer of (unbounded fan-in) AND gates (assuming input includes both the individual bits of $\mathbf{y}$ and also their respective complements), and therefore $h_{\mathbf{M}}$ simply builds upon Expand by adding a layer of XOR gates. Furthermore, the Expand function can be efficiently instantiated with idealized heuristics (see Lemma 3.3).

Lemma 3.1 (A realization of the expanding function [5]). *Let n be the security parameter and let $k \leq \mathsf{poly}(n)$, $L = O(\log n)$, $t = t(n)$, $q = q(n)$ be integer-valued functions such that $k = L \cdot t$, $q = t \cdot 2^L$. Let $\mathsf{Expand} : \{0,1\}^k \to \{0,1\}^q$ be a function that parses the k-bit input into L-bit blocks as*

$$\mathbf{y} = y_1 \cdots y_L \| y_{L+1} \cdots y_{2L} \| \cdots \| y_{L(t-1)+1} \cdots y_{Lt}$$

and produces as output

$$\mathsf{Expand}(\mathbf{y}) = \mathsf{DeMul}(y_1 \cdots y_L) \| \cdots \| \mathsf{DeMul}(y_{L(t-1)+1} \cdots y_{Lt})$$

where $\mathsf{DeMul} : \{0,1\}^L \to \{0,1\}^{2^L}$ is a demultiplexer function that on input $z \in \{0,1\}^L$ outputs a 2^L-bit string which is 1 in exactly the z-th location (and 0 everywhere else). Then, we have that

1. *(Expand **has sparse outputs**). for all $\mathbf{y} \in \{0,1\}^k$: $|\mathsf{Expand}(\mathbf{y})| = t$;*
2. *(Expand **is injective**). Expand is injection with $k = L \cdot t = \log(q/t)t$.*
3. *(Expand **is parallelizable**). Expand is contained in AC^0.*

Our framework is based on the following expand-then-compress construction.

Construction 3.1. *Let $k = k(n)$ and $q = q(n)$ be integer valued functions, and let $\mathsf{Expand} : \{0,1\}^k \to \{0,1\}^q$ be an expanding function as in Lemma 3.1. A collection of functions $\mathcal{H}_{k,n} = \{h_{\mathbf{M}} : \{0,1\}^k \to \{0,1\}^n, \mathbf{M} \in \{0,1\}^{n \times q}\}$ is defined as*

$$h_{\mathbf{M}}(x) = \mathbf{M} \cdot \mathsf{Expand}(x)$$

where the key-sampler $\mathcal{G}(1^n)$ samples an $n \times q$ matrix $\mathbf{M} \xleftarrow{\$} \{0,1\}^{n \times q}$.

3.2   The Main Framework of LPN-based CRH

We state our main framework in Theorem 3.1 and then derive the main feasibility results in Corollary 3.1.

Theorem 3.1 (The main framework). *Let n be the security parameter, and let $\mu = \mu(n)$, $k = k(n)$, $q = q(n)$, $t = t(n)$ and $T = T(n)$ such that $t^2 \leq q \leq T = 2^{\frac{8\mu t}{\ln 2(1-2\mu)}}$. Assume that the (n, μ, q)-DLPN problem is T-hard, and let $h_{\mathbf{M}}$ and Expand be defined as in Lemma 3.1 and Construction 3.1 respectively. Then, for every probabilistic adversary $\mathcal{A}$ of running time $T' = 2^{\frac{4\mu t}{\ln 2(1-2\mu)}-1}$*

$$\Pr_{\mathbf{M} \xleftarrow{\$} \{0,1\}^{n \times q}} [\ (\mathbf{y}, \mathbf{y}') \leftarrow \mathcal{A}(\mathbf{M}) : \mathbf{y} \neq \mathbf{y}' \wedge h_{\mathbf{M}}(\mathbf{y}) = h_{\mathbf{M}}(\mathbf{y}')\ ] \leq \frac{1}{T'}.$$

We do not say "$h_{\mathbf{M}}$ is a T'-hard CRH" as it may not be poly(n)-time computable.

Proof. Suppose for contradiction that $\mathcal{A}$ finds out a collision with probability more than $1/T'$ s.t. $\mathbf{y} \neq \mathbf{y}'$ and $h_{\mathbf{M}}(\mathbf{y}) = h_{\mathbf{M}}(\mathbf{y}')$, then we have $\mathbf{M} \cdot \mathbf{x} = \mathbf{0}$, where $\mathbf{x} = \mathsf{Expand}(\mathbf{y}) - \mathsf{Expand}(\mathbf{y}') \neq \mathbf{0}$ due to the distinctiveness of Expand, and

$$|\mathbf{x}| \leq |\mathsf{Expand}(\mathbf{y})| + |\mathsf{Expand}(\mathbf{y}')| \leq 2t.$$

We define in Algorithm 1 below an LPN distinguisher $\mathcal{D}$ that on input $(\mathbf{M}^{\mathsf{T}}, \mathbf{z})$, where $\mathbf{M}^{\mathsf{T}} \xleftarrow{\$} \{0,1\}^{q \times n}$, and either $\mathbf{z} = \mathbf{M}^{\mathsf{T}}\mathbf{s} + \mathbf{e}$ (for $\mathbf{e} \leftarrow \mathcal{B}_{\mu}^q$) or $\mathbf{z} \xleftarrow{\$} \{0,1\}^q$, invokes $\mathcal{A}$ on $\mathbf{M}$, and if a collision $(\mathbf{y}, \mathbf{y}')$ is found, it outputs $\mathbf{x}^{\mathsf{T}}\mathbf{z}$ for $\mathbf{x} = \mathsf{Expand}(\mathbf{y}) - \mathsf{Expand}(\mathbf{y}')$, and otherwise it outputs a uniform random bit. On a successful collision, we have by Lemma 2.1 and Fact 1

$$\Pr[\mathbf{x}^{\mathsf{T}}\mathbf{z} = \mathbf{x}^{\mathsf{T}}\mathbf{e} = 0] \geq \frac{1}{2} + \frac{2^{-(\log \frac{1}{1-2\mu})2t}}{2} \geq \frac{1}{2} + \frac{2^{-\frac{4\mu t}{\ln 2(1-2\mu)}}}{2}.$$

Therefore, $\mathcal{D}$ achieves an overall advantage of

$$\Pr[\mathcal{D}(\mathbf{M}^{\mathsf{T}}, \mathbf{M}^{\mathsf{T}}\mathbf{s} + \mathbf{e}) = 0] - \Pr_{\mathbf{z} \xleftarrow{\$} \{0,1\}^q} [\mathcal{D}(\mathbf{M}^{\mathsf{T}}, \mathbf{z}) = 0]$$

$$> \frac{1}{T'} \cdot \frac{2^{-\frac{4\mu t}{\ln 2(1-2\mu)}}}{2} \geq 2^{-\frac{8\mu t}{\ln 2(1-2\mu)}},$$

which is a contradiction to the assumption.

Algorithm 1. A distinguisher $\mathcal{D}$ for (n, μ, q)-DLPN

Input: $(\mathbf{M}^{\mathsf{T}}, \mathbf{z})$, where $\mathbf{M}^{\mathsf{T}} \in \{0,1\}^{q \times n}$ and $\mathbf{z} \in \{0,1\}^q$
$(\mathbf{y}, \mathbf{y}') \leftarrow \mathcal{A}(\mathbf{M})$;
$\mathbf{x} = \mathbf{y} - \mathbf{y}'$;
if $0 < |\mathbf{x}| \leq 2t \wedge \mathbf{Mx} = \mathbf{0}$ then
  $v = \mathbf{x}^{\mathsf{T}} \mathbf{z}$
else
  $v \xleftarrow{\$} \{0,1\}$
end if
Output: v

Corollary 3.1 (Main feasibility results). *Assume that (n, μ, q)-DLPN is T-hard, then T'-hard CRH functions with shrinkage $\frac{n}{k}$ exist in $\mathsf{AC}^0(\mathsf{MOD2})$, where $(\mu, T, q, T', \frac{n}{k})$ can be any of the following:*

1. *(Constant-noise, less hardness, poly-logarithmic shrinkage).*
   $\mu = O(1)$, $T = 2^{n^{0.5+\varepsilon}}$, $q = 2^{n^{0.5}}$, $T' \approx 2^{n^{0.5+\varepsilon}}/2 = \lambda^{\log^{2\varepsilon}\lambda}/2$ *and* $\frac{n}{k} <$
   $\frac{16\mu}{\ln 2(1-2\mu)n^\varepsilon} = \frac{16\mu}{\ln 2(1-2\mu)\log^{2\varepsilon}\lambda}$ *for any constant $\varepsilon > 0$.*
2. *(Constant-noise, maximal efficiency, constant shrinkage).*
   $\mu = O(1)$, $T = 2^{\frac{\varepsilon n}{\log n}}$, $q = n^{C_{\varepsilon,\mu}}$, $T' \approx 2^{\frac{\varepsilon n}{2\log n}}$, $\frac{n}{k} < \frac{1}{2}$ *for any constant $\varepsilon > 0$*
   and $C_{\varepsilon,\mu} = \max(\frac{32\mu}{\varepsilon \ln 2(1-\mu)}, 2)$.
3. *(Low-noise, maximal efficiency, constant shrinkage).*
   $\mu = 1/\sqrt{n}$, $T = 2^{\frac{\varepsilon\sqrt{n}}{\log n}}$, $q = n^{C_{\varepsilon,\mu}}$, $T' \approx 2^{\frac{\varepsilon\sqrt{n}}{2\log n}}$, $\frac{n}{k} < \frac{1}{2}$ *for any constant $\varepsilon > 0$*
   and $C'_{\varepsilon,\mu} = \max(\frac{32}{\varepsilon \ln 2}, 2)$.
4. *(Extremely-low-noise, standard hardness, constant shrinkage).*
   $\mu = \frac{(\log n)^2}{n}$, $T = q > \mathsf{poly}(n)$ *and* $T' > \mathsf{poly}(n)$ *for every poly, and* $\frac{n}{k} < \frac{1}{2}$.

Proof. Recall that $T = 2^{\frac{8\mu t}{\ln 2(1-2\mu)}}$ and $\frac{n}{k} = \frac{n}{\log(q/t)t} < \frac{2n}{t\log q}$. To prove param
#1, we let $\frac{8\mu t}{\ln 2(1-2\mu)} = n^{0.5+\varepsilon}$ and thus $t = \ln 2(1-2\mu)n^{0.5+\varepsilon}/8\mu$, and then with
$q = 2^{\sqrt{n}}$ we get

$$\frac{n}{k} < \frac{2\sqrt{n}}{t} < \frac{16\mu}{\ln 2(1-2\mu)n^\varepsilon}.$$

However, $h_{\mathbf{M}}$ that corresponds to param #1 is not computable in $\mathsf{poly}(n)$,
and we need to switch to security parameter $\lambda = 2^{\sqrt{n}}$ s.t. $n = \log^2\lambda$, $k = \Omega(n^{1+\varepsilon}) = \Omega(\log^{2+2\varepsilon}\lambda)$, $T' = \lambda^{\log^{2\varepsilon}\lambda}/2$. The resulting $h_{\mathbf{M}} : \{0,1\}^{\Omega(\log^{2+2\varepsilon}\lambda)} \to \{0,1\}^{\log^2\lambda}$ is a T'-hard CRH function on security parameter λ but only operates on small inputs and outputs, and we use parallel repetition (Lemma 3.2)
to get a domain/range-extended CRH $h'_{\mathbf{M}} : \{0,1\}^{\Omega(\lambda\log^{2\varepsilon}\lambda)} \to \{0,1\}^\lambda$ for
$\mathbf{M} \in \{0,1\}^{\log^2\lambda \times \lambda}$, which is T'-hard and is computable in time $\mathsf{poly}(\lambda)$.

Now proceed to params #2 and #3: set $\frac{8\mu t}{\ln 2(1-2\mu)}$ to $\frac{\varepsilon n}{\log n}$ for $\mu = O(1)$ or to $\frac{\varepsilon \sqrt{n}}{\log n}$ for $\mu = 1/\sqrt{n}$, and let $\frac{2n}{t \log q} = \frac{1}{2}$ so that

$$t = \frac{\ln 2 \cdot \varepsilon(1-2\mu)n}{8\mu \log n}, \quad \log q = \frac{32\mu \log n}{\varepsilon \ln 2(1 - 2\mu)} \quad \text{for } \mu = O(1);$$

$$t = \frac{\ln 2 \cdot \varepsilon \cdot n}{8 \log n}, \quad \log q = \frac{32 \log n}{\varepsilon \ln 2} \quad \text{for } \mu = 1/\sqrt{n}.$$

Note that we also need $q \geq n^2$ in respect of the $t \leq \sqrt{q}$ condition. Finally, param #4 is seen by the following: for $\mu = \frac{(\log n)^2}{n}$, any $q = \mathsf{poly}(n)$ and t satisfying $\frac{2n}{t \log q} = 1/2$ we have that $T = 2^{\frac{8\mu t}{\ln 2(1-2\mu)}}$ is another polynomial in n.

Lemma 3.2. (Parallel repetitions of CRH). *Let $k = k(\lambda)$, $d = d(\lambda)$ and $T = T(\lambda)$ be integer valued functions. If $\mathcal{H}_{k,\lambda} = \{h_{\mathbf{s}} : \{0,1\}^k \to \{0,1\}^\lambda, \mathbf{s} \in \{0,1\}^{\mathsf{poly}(\lambda)}\}$ is a T-hard CRH function, then $\mathcal{H}'_{dk,d\lambda} = \{h'_{\mathbf{s}} : \{0,1\}^{dk} \to \{0,1\}^{d\lambda}, \mathbf{s} \in \{0,1\}^{\mathsf{poly}(\lambda)}\}$, where*

$$h'_{\mathbf{s}}(\mathbf{y}_1, \cdots, \mathbf{y}_d) = \big(h_{\mathbf{s}}(\mathbf{y}_1), \cdots, h_{\mathbf{s}}(\mathbf{y}_d)\big), \quad \mathbf{y}_1, \cdots, \mathbf{y}_d \in \{0,1\}^k,$$

is a (T/d)-hard CRH function.

3.3 Assume Less, Shrink More and in Parallel at the Same Time

Although already assuming much less hardness than previously known, the CRH immediately implied by constant-noise LPN (as specified by param #1 of Corollary 3.1) is inefficient as $\mathbf{M}$ is of dimension $n \times 2^{\sqrt{n}}$ and thus the resulting hash function has computation time far beyond polynomial. The solution by switching to another parameter $\lambda = 2^{\sqrt{n}}$ makes the hash function computable in time polynomial in λ but at the same time it dramatically downgrades the security from $2^{\Omega(n^{1/2+\epsilon})}$ to $\lambda^{\Omega(\log^{2\epsilon} \lambda)}$, and deteriorates the shrinkage factor from polynomial to poly-logarithmic. Otherwise said, we mainly establish feasibility results about basing CRH on constant-noise LPN with minimal hardness possible.

LPN+RF → MORE EFFICIENT DOMAIN EXTENDERS. In this subsection, we discuss an alternative to void the security/efficiency loss, i.e., to preserve security, polynomial shrinkage and efficiency at the same time. In addition to LPN, the construction relies on (arguably minimal) idealized assumptions such as a small-domain random function (whose domain is much smaller than the range) or a random permutation (which can be instantiated with a block cipher keyed by a random public string). Unlike the parameter-switching technique, this approach applies also to low-noise LPN with even reduced hardness. Note that idealized heuristics such as a RF trivially implies collision resistance, e.g., a RF $R : \{0,1\}^\ell \to \{0,1\}^n$ with $\ell > n$ (or otherwise truncating the output to make it compressing) is collision resistant. Therefore, based on a small-domain RF (with $\ell \ll n$) our main contribution is a simple and elegant collision-resistance-preserving domain extender combining the best of the two worlds: maximized (depth-1) parallelizability and polynomial shrinkage. More specifically, simply parse the input $\mathbf{y}$ into polynomially many blocks $y_1, \ldots, y_t$, evaluate R on them independently and in parallel, and output the XOR sum as below:

$$R : \{0,1\}^\ell \to \{0,1\}^n \quad (\ell \ll n)$$
$$h^R : \{0,1\}^k \to \{0,1\}^n \quad (k = n^{1+\varepsilon})$$
$$h^R(\mathbf{y}) = \bigoplus_{i=1}^t R(i\|\mathbf{y}_i),$$

which yields a domain extender with polynomially shrinkage, i.e., $n/k < 1/n^{\Omega(1)}$.

AN IDEALIZED REALIZATION OF $h_{\mathbf{M}}$. We recall that $h_{\mathbf{M}}(\mathbf{y}) = \mathbf{M} \cdot \mathsf{Expand}(\mathbf{y})$ for an $n \times (q = t \cdot 2^L)$ matrix $\mathbf{M}$ and that Expand parses $\mathbf{y}$ into $t = k/L$ blocks and produces same number of output blocks accordingly. We also parse $\mathbf{M}$ into t equal-size submatrices $\mathbf{M}_1, \cdots, \mathbf{M}_t$, each of dimension $n \times 2^L$. Let $R : \{0,1\}^{\log(q)} \to \{0,1\}^n$ be a random function that describes $\mathbf{M}$, i.e., for every $j \in \{0,1\}^{\log(q)}$ the output $R(j)$ corresponds to the j-th column of $\mathbf{M}$. Thus,

$$h_{\mathbf{M}}(\mathbf{y}) = \underbrace{\left[\mathbf{M}_1 \cdots \mathbf{M}_t \right]}_{\mathbf{M}} \cdot \underbrace{\begin{bmatrix} \mathsf{DeMul}(\mathbf{y}_1) \\ \vdots \\ \mathsf{DeMul}(\mathbf{y}_t) \end{bmatrix}}_{\mathsf{Expand}(\mathbf{y})} = \bigoplus_{i=1}^t R(i\|\mathbf{y}_i) \qquad (5)$$

where $R(i\|\mathbf{y}_i) = \mathbf{M}_i \cdot \mathsf{DeMul}(\mathbf{y}_i)$ simply follows the definition of R and DeMul. Therefore, computing $h_{\mathbf{M}}$ is now reduced to instantiating a small-domain random function $R : \{0,1\}^{\log(q)} \to \{0,1\}^n$ for $\log(q) \ll n$, where the access to the huge amount of randomness in $\mathbf{M}$ is efficiently implemented by R, as stated below.

Lemma 3.3. (An idealized realization of $h_{\mathbf{M}}$). *Let $k = k(n)$, $t = t(n)$, $q = q(n)$ and $L = L(n)$ be integer valued functions such that $q/t = 2^L$. Assume that $R : \{0,1\}^{\log(q)} \to \{0,1\}^n$ behaves like a random function, then $h^R(\mathbf{y}) = \bigoplus_{i=1}^t R(i\|\mathbf{y}_i)$ as defined in (5) perfectly realizes $h_{\mathbf{M}}$ specified in Construction 3.1.*

With the idealized realization of $h_{\mathbf{M}}$, we immediately obtain a simple and efficient way to extend the domain of random functions polynomially while preserving collision resistance, stated below as a corollary of Theorem 3.1.

Corollary 3.2. (A polynomially-shrinking domain extender). *Let n be the security parameter, and let $\mu = \mu(n)$, $k = k(n)$, $q = q(n)$, $t = t(n)$ and $T = T(n)$ such that $t^2 \leq q \leq T = 2^{\frac{8\mu t}{\ln 2(1-2\mu)}}$. Assume that (n, μ, q)-DLPN is T-hard and $R : \{0,1\}^{\log(q)} \to \{0,1\}^n$ behaves like a random function, then for $\mathbf{y} = \mathbf{y}_1\| \cdots \|\mathbf{y}_t$ parsed as $t = k/L$ blocks, each of size $L = \log(q/t)$, we have*

$$h^R : \{0,1\}^k \to \{0,1\}^n, \quad h^R(\mathbf{y}) = \bigoplus_{i=1}^t R(i\|\mathbf{y}_i),$$

is a T'-hard CRH function with shrinkage n/k, where $(\mu, T, q, T', \frac{n}{k})$ can be either of the following:

1. (*Constant-noise, less hardness, polynomial shrinkage*).
   $\mu = O(1)$, $T = 2^{n^{0.5+\varepsilon}}$, $q = 2^{n^{0.5}}$, $T' \approx 2^{n^{0.5+\varepsilon}}/2$ and $\frac{n}{k} < \frac{16\mu}{\ln 2(1-2\mu)n^{\varepsilon}}$ for any constant $\varepsilon > 0$.

2. (*Low-noise, less hardness, polynomial shrinkage*).
   $\mu = 1/\sqrt{n}$, $T = 2^{n^{0.25+\varepsilon}}$, $q = 2^{n^{0.25}}$, $T' \approx 2^{n^{0.25+\varepsilon}}/2$ and $\frac{n}{k} < \frac{16}{\ln 2 \cdot n^{\varepsilon}}$ for any constant $\varepsilon > 0$.

Proof. First, assume that h^R is functionally equivalent to $h_{\mathbf{M}}$. Then, param #1 is the same as the counterpart in Corollary 3.1 but we refrain from switching to a new security parameter. To prove param #2, we recall that $T = 2^{\frac{8\mu t}{\ln 2(1-2\mu)}}$ and $\frac{n}{k} = \frac{n}{\log(q/t)t} < \frac{2n}{t \log q}$ (see Theorem 3.1). Let $\frac{8\mu t}{\ln 2(1-2\mu)} = n^{0.25+\varepsilon}$ and thus $t \approx \ln 2 \cdot n^{0.75+\varepsilon}/8$, and then with $q = 2^{n^{0.25}}$ we get

$$\frac{n}{k} < \frac{2n^{0.75}}{t} < \frac{16}{\ln 2 \cdot n^{\varepsilon}}.$$

The conclusion then follows from Lemma 3.3 that h^R perfectly instantiates $h_{\mathbf{M}}$.

One may want to instantiate R with a pseudorandom function (with key made public), but the security cannot be achieved with a standard reducibility argument due to the distinction between public-coin and secret-coin CRH functions [36]. We thus resort to random permutations or idealized blockciphers.

RANDOM FUNCTIONS VS. PERMUTATIONS. The small-domain random function (to be instantiated) is not commonly found in practice, but it is implied by a large-domain random function for free, i.e., $R(x) = F(0^l\|x)$ is a random function if F is a random one. Thus, we simply consider a length-preserving random function, which can be in turn based on a random permutation (and instantiated with block ciphers). For example, for random permutations π, π_1, π_2, we have that $\pi \oplus \pi^{-1}$ [27] (or $\pi_1 \oplus \pi_2$ [52]) is indifferentiable from a length-preserving random function. This means that R on input x can be instantiated as

$$\mathsf{AES}_k(0^l\|x) \oplus \mathsf{AES}_k^{-1}(0^l\|x) \text{ or } \mathsf{AES}_{k_1}(0^l\|x) \oplus \mathsf{AES}_{k_2}(0^l\|x)$$

where $l = n - \lceil\log(q)\rceil$ bits are padded to fit into a permutation, k, k_1, and k_2 are public random keys. Intuitively, the XOR of a permutation and its inverse (or two independent permutations) is to destroy the permutation structure as its inversibility could give the adversary additional advantages in collision finding. The former instantiation relies on the assumption that a practical block cipher like AES on a random key behaves like a random permutation. We reproduce below the results by Dodis et al. [27] that $\pi \oplus \pi^{-1}$ is indifferentiable from a (length-preserving) random function. Therefore, instantiation of a random function with a blockcipher only incurs a factor of 2 in the number of calls to the underlying primitive.

Lemma 3.4. (Lemma 4 from [27]). *Let n be the security parameter, let $q = q(n)$ and let π be a random permutation over $\{0,1\}^n$. We have that $\pi \oplus \pi^{-1}$ is $(q, q, O(nq), O(\frac{q^2}{2^n}))$-indifferentiable from an n-to-n-bit random function.*

ON RELATED WORKS. We offer a new and simple construction of polynomially shrinking domain extenders from *random functions/permutations/fixed-key block ciphers*. Compared with the traditional blockcipher-based compression functions, e.g. [13,55,57], our solution avoids the key-setup costs and eliminates the need for related-key security on a large space of keys. That is, (using AES-128 as an example) we only assume that "AES on a single random key behaves like a random permutation", instead of that "AES on 2^{128} keys yields 2^{128} independent random permutations", as imposed by the Ideal Cipher Model. On the other hand, existing permutation-based solutions either only offer a constant shrinkage factor (typically $1/2$) [51,62], or require permutations with a large domain (e.g., [28] needs a large permutation on n^2-bit strings to obtain a CRH function with shrinkage factor $1/n$), and in contrast our construction runs in parallel and compresses polynomially.

4   Concluding Remarks

We construct CRH from LPN for a broad spectrum of parameter choices, and thus resolve the problem whether CRH functions can be based on the (reasonable) hardness of LPN. We also discuss how to improve the efficiency using idealized heuristics. We leave it as future work to investigate more efficient instantiation (based on Ring-LPN), and to compare it with SWIFFT/SWIFFTX.

Acknowledgments. Yu Yu was supported by the National Natural Science Foundation of China (Grant Nos. 61872236 and 61572192) and the National Cryptography Development Fund (Grant No. MMJJ20170209). Jiang Zhang is supported by the National Key Research and Development Program of China (Grant No. 2017YFB0802005, 2018YFB0804105), the National Natural Science Foundation of China (Grant Nos. 6160204661932019), and the Young Elite Scientists Sponsorship Program by CAST (2016QNRC001). Jian Weng was partially supported by National Natural Science Foundation of China (Grant Nos. 61825203, U1736203, 61732021). Chun Guo was supported by the Program of Qilu Young Scholars of Shandong University, and also partly funded by Francois-Xavier Standaert via the ERC project SWORD (724725). Xiangxue Li was supported by the National Cryptography Development Fund (Grant No. MMJJ20180106) and the National Natural Science Foundation of China (Grant Nos. 61572192, 61971192). This research is funded in part by the Anhui Initiative in Quantum Information Technologies (Grant No. AHY150100) and Sichuan Science and Technology Program (Grant No. 2017GZDZX0002).

References

1. Ajtai, M.: Generating hard instances of lattice problems (extended abstract). In: Proceedings of the Twenty-Eighth Annual ACM Symposium on the Theory of Computing (STOC 1996), pp. 99–108 (1996)
2. Alekhnovich, M.: More on average case vs approximation complexity. In: 44th Annual Symposium on Foundations of Computer Science, Cambridge, Massachusetts, pp. 298–307. IEEE, October 2003

3. Andreeva, E., Mennink, B., Preneel, B.: Security properties of domain extenders for cryptographic hash functions. J. Inf. Process. Syst. **6**(4), 453–480 (2010). https://doi.org/10.3745/JIPS.2010.6.4.453
4. Applebaum, B., Cash, D., Peikert, C., Sahai, A.: Fast cryptographic primitives and circular-secure encryption based on hard learning problems. In: Halevi, S. (ed.) CRYPTO 2009. LNCS, vol. 5677, pp. 595–618. Springer, Heidelberg (2009). https://doi.org/10.1007/978-3-642-03356-8_35
5. Applebaum, B., Haramaty, N., Ishai, Y., Kushilevitz, E., Vaikuntanathan, V.: Low-complexity cryptographic hash functions. In: Proceedings of the 2017 Conference on Innovations in Theoretical Computer Science (ITCS 2017), pp. 7:1–7:31 (2017)
6. Applebaum, B., Ishai, Y., Kushilevitz, E.: Cryptography with constant input locality. In: Menezes, A. (ed.) CRYPTO 2007. LNCS, vol. 4622, pp. 92–110. Springer, Heidelberg (2007). https://doi.org/10.1007/978-3-540-74143-5_6
7. Arbitman, Y., Dogon, G., Lyubashevsky, V., Micciancio, D., Peikert, C., Rosen, A.: SWIFFTX: a proposal for the SHA-3 standard (2009). http://www.eecs.harvard.edu/~alon/PAPERS/lattices/swifftx.pdf
8. Augot, D., Finiasz, M., Gaborit, P., Manuel, S., Sendrier, N.: SHA-3 proposal: FSB (2008). https://www.rocq.inria.fr/secret/CBCrypto/fsbdoc.pdf
9. Becker, A., Joux, A., May, A., Meurer, A.: Decoding random binary linear codes in $2^{n/20}$: how $1 + 1 = 0$ improves information set decoding. In: Pointcheval, D., Johansson, T. (eds.) EUROCRYPT 2012. LNCS, vol. 7237, pp. 520–536. Springer, Heidelberg (2012). https://doi.org/10.1007/978-3-642-29011-4_31
10. Berlekamp, E., McEliece, R.J., van Tilborg, H.: On the inherent intractability of certain coding problems. IEEE Trans. Inf. Theory **24**(3), 384–386 (1978)
11. Bernstein, D.J., Lange, T., Peters, C.: Smaller decoding exponents: ball-collision decoding. In: Rogaway, P. (ed.) CRYPTO 2011. LNCS, vol. 6841, pp. 743–760. Springer, Heidelberg (2011). https://doi.org/10.1007/978-3-642-22792-9_42
12. Bernstein, D.J., Lange, T., Peters, C., Schwabe, P.: Really fast syndrome-based hashing. In: Nitaj, A., Pointcheval, D. (eds.) AFRICACRYPT 2011. LNCS, vol. 6737, pp. 134–152. Springer, Heidelberg (2011). https://doi.org/10.1007/978-3-642-21969-6_9
13. Black, J., Rogaway, P., Shrimpton, T., Stam, M.: An analysis of the blockcipher-based hash functions from PGV. J. Cryptol. **23**(4), 519–545 (2010)
14. Blum, A., Furst, M., Kearns, M., Lipton, R.J.: Cryptographic primitives based on hard learning problems. In: Stinson, D.R. (ed.) CRYPTO 1993. LNCS, vol. 773, pp. 278–291. Springer, Heidelberg (1994). https://doi.org/10.1007/3-540-48329-2_24
15. Blum, A., Kalai, A., Wasserman, H.: Noise-tolerant learning, the parity problem, and the statistical query model. J. ACM **50**(4), 506–519 (2003)
16. Brakerski, Z.: When homomorphism becomes a liability. In: Sahai, A. (ed.) TCC 2013. LNCS, vol. 7785, pp. 143–161. Springer, Heidelberg (2013). https://doi.org/10.1007/978-3-642-36594-2_9
17. Brakerski, Z., Lyubashevsky, V., Vaikuntanathan, V., Wichs, D.: Worst-case hardness for LPN and cryptographic hashing via code smoothing. In: Ishai, Y., Rijmen, V. (eds.) EUROCRYPT 2019. LNCS, vol. 11478, pp. 619–635. Springer, Cham (2019). https://doi.org/10.1007/978-3-030-17659-4_21
18. Canteaut, A., Chabaud, F.: A new algorithm for finding minimum-weight words in a linear code: application to McEliece's cryptosystem and to narrow-sense BCH codes of length 511. IEEE Trans. Inf. Theory **44**(1), 367–378 (1998)
19. Cash, D., Kiltz, E., Tessaro, S.: Two-round man-in-the-middle security from LPN. In: Kushilevitz, E., Malkin, T. (eds.) TCC 2016. LNCS, vol. 9562, pp. 225–248. Springer, Heidelberg (2016). https://doi.org/10.1007/978-3-662-49096-9_10

20. Coron, J.S., Dodis, Y., Malinaud, C., Puniya, P.: Merkle-Damgård revisited: how to construct a hash function. In: Shoup [61], pp. 430–448
21. Damgård, I.B.: Collision free hash functions and public key signature schemes. In: Chaum, D., Price, W.L. (eds.) EUROCRYPT 1987. LNCS, vol. 304, pp. 203–216. Springer, Heidelberg (1988). https://doi.org/10.1007/3-540-39118-5_19
22. Damgård, I.B.: A design principle for hash functions. In: Brassard, G. (ed.) CRYPTO 1989. LNCS, vol. 435, pp. 416–427. Springer, New York (1990). https://doi.org/10.1007/0-387-34805-0_39
23. David, B., Dowsley, R., Nascimento, A.C.A.: Universally composable oblivious transfer based on a variant of LPN. In: Gritzalis, D., Kiayias, A., Askoxylakis, I. (eds.) CANS 2014. LNCS, vol. 8813, pp. 143–158. Springer, Cham (2014). https://doi.org/10.1007/978-3-319-12280-9_10
24. Demay, G., Gaži, P., Hirt, M., Maurer, U.: Resource-restricted indifferentiability. In: Johansson, T., Nguyen, P.Q. (eds.) EUROCRYPT 2013. LNCS, vol. 7881, pp. 664–683. Springer, Heidelberg (2013). https://doi.org/10.1007/978-3-642-38348-9_39
25. Diffie, W., Hellman, M.E.: New directions in cryptography. IEEE Trans. Inf. Theory IT−22(6), 644–654 (1976)
26. Dodis, Y., Kiltz, E., Pietrzak, K., Wichs, D.: Message authentication, revisited. In: Pointcheval, D., Johansson, T. (eds.) EUROCRYPT 2012. LNCS, vol. 7237, pp. 355–374. Springer, Heidelberg (2012). https://doi.org/10.1007/978-3•642-29011-4_22
27. Dodis, Y., Pietrzak, K., Puniya, P.: A new mode of operation for block ciphers and length-preserving MACs. In: Smart, N. (ed.) EUROCRYPT 2008. LNCS, vol. 4965, pp. 198–219. Springer, Heidelberg (2008). https://doi.org/10.1007/978-3-540-78967-3_12
28. Dodis, Y., Reyzin, L., Rivest, R.L., Shen, E.: Indifferentiability of permutation-based compression functions and tree-based modes of operation, with applications to MD6. In: Dunkelman, O. (ed.) FSE 2009. LNCS, vol. 5665, pp. 104–121. Springer, Heidelberg (2009). https://doi.org/10.1007/978-3-642-03317-9_7
29. Döttling, N.: Low noise LPN: KDM secure public key encryption and sample amplification. In: Katz, J. (ed.) PKC 2015. LNCS, vol. 9020, pp. 604–626. Springer, Heidelberg (2015). https://doi.org/10.1007/978-3-662-46447-2_27
30. Döttling, N., Müller-Quade, J., Nascimento, A.C.A.: IND-CCA secure cryptography based on a variant of the LPN problem. In: Wang, X., Sako, K. (eds.) ASIACRYPT 2012. LNCS, vol. 7658, pp. 485–503. Springer, Heidelberg (2012). https://doi.org/10.1007/978-3-642-34961-4_30
31. Dumer, I., Micciancio, D., Sudan, M.: Hardness of approximating the minimum distance of a linear code. IEEE Trans. Inf. Theory 49(1), 22–37 (2003)
32. Esser, A., Kübler, R., May, A.: LPN decoded. In: Katz, J., Shacham, H. (eds.) CRYPTO 2017, Part II. LNCS, vol. 10402, pp. 486–514. Springer, Cham (2017). https://doi.org/10.1007/978-3-319-63715-0_17
33. Feldman, V., Gopalan, P., Khot, S., Ponnuswami, A.K.: New results for learning noisy parities and halfspaces. In: 47th Symposium on Foundations of Computer Science, Berkeley, CA, USA, pp. 563–574. IEEE, 21–24 October 2006
34. Gertner, Y., Kannan, S., Malkin, T., Reingold, O., Viswanathan, M.: The relationship between public key encryption and oblivious transfer. In: Proceedings of the 41st Annual Symposium on Foundations of Computer Science, pp. 325–335 (2000). https://doi.org/10.1109/SFCS.2000.892121
35. Hopper, N.J., Blum, M.: Secure human identification protocols. In: Boyd, C. (ed.) ASIACRYPT 2001. LNCS, vol. 2248, pp. 52–66. Springer, Heidelberg (2001). https://doi.org/10.1007/3-540-45682-1_4

36. Hsiao, C.-Y., Reyzin, L.: Finding collisions on a public road, or do secure hash functions need secret coins? In: Franklin, M. (ed.) CRYPTO 2004. LNCS, vol. 3152, pp. 92–105. Springer, Heidelberg (2004). https://doi.org/10.1007/978-3-540-28628-8_6
37. Jain, A., Krenn, S., Pietrzak, K., Tentes, A.: Commitments and efficient zero-knowledge proofs from learning parity with noise. In: Wang, X., Sako, K. (eds.) ASIACRYPT 2012. LNCS, vol. 7658, pp. 663–680. Springer, Heidelberg (2012). https://doi.org/10.1007/978-3-642-34961-4_40
38. Juels, A., Weis, S.A.: Authenticating pervasive devices with human protocols. In: Shoup [61], pp. 293–308
39. Katz, J., Shin, J.S.: Parallel and concurrent security of the HB and HB$^+$ protocols. In: Vaudenay, S. (ed.) EUROCRYPT 2006. LNCS, vol. 4004, pp. 73–87. Springer, Heidelberg (2006). https://doi.org/10.1007/11761679_6
40. Kiltz, E., Masny, D., Pietrzak, K.: Simple chosen-ciphertext security from low-noise LPN. In: Krawczyk, H. (ed.) PKC 2014. LNCS, vol. 8383, pp. 1–18. Springer, Heidelberg (2014). https://doi.org/10.1007/978-3-642-54631-0_1
41. Kiltz, E., Pietrzak, K., Cash, D., Jain, A., Venturi, D.: Efficient authentication from hard learning problems. In: Paterson, K.G. (ed.) EUROCRYPT 2011. LNCS, vol. 6632, pp. 7–26. Springer, Heidelberg (2011). https://doi.org/10.1007/978-3-642-20465-4_3
42. Kirchner, P.: Improved generalized birthday attack. Cryptology ePrint Archive, Report 2011/377 (2011)
43. Kirchner, P., Fouque, P.-A.: An improved BKW algorithm for LWE with applications to cryptography and lattices. In: Gennaro, R., Robshaw, M. (eds.) CRYPTO 2015, Part I. LNCS, vol. 9215, pp. 43–62. Springer, Heidelberg (2015). https://doi.org/10.1007/978-3-662-47989-6_3
44. Lamport, L.: Constructing digital signatures from a one way function. Technical report CSL-98, SRI International, October 1979
45. Lyubashevsky, V.: The parity problem in the presence of noise, decoding random linear codes, and the subset sum problem. In: Chekuri, C., Jansen, K., Rolim, J.D.P., Trevisan, L. (eds.) APPROX/RANDOM 2005. LNCS, vol. 3624, pp. 378–389. Springer, Heidelberg (2005). https://doi.org/10.1007/11538462_32
46. Lyubashevsky, V.: The LPN problem in cryptography. In: Invited Talk at the 14th IMA International Conference on Cryptography and Coding (2013). Slides goo.gl/zpHFp7
47. Lyubashevsky, V., Masny, D.: Man-in-the-middle secure authentication schemes from LPN and weak PRFs. In: Canetti, R., Garay, J.A. (eds.) CRYPTO 2013. LNCS, vol. 8043, pp. 308–325. Springer, Heidelberg (2013). https://doi.org/10.1007/978-3-642-40084-1_18
48. Lyubashevsky, V., Micciancio, D., Peikert, C., Rosen, A.: SWIFFT: a modest proposal for FFT hashing. In: Nyberg, K. (ed.) FSE 2008. LNCS, vol. 5086, pp. 54–72. Springer, Heidelberg (2008). https://doi.org/10.1007/978-3-540-71039-4_4
49. Maurer, U., Renner, R., Holenstein, C.: Indifferentiability, impossibility results on reductions, and applications to the random oracle methodology. In: Naor, M. (ed.) TCC 2004. LNCS, vol. 2951, pp. 21–39. Springer, Heidelberg (2004). https://doi.org/10.1007/978-3-540-24638-1_2
50. May, A., Meurer, A., Thomae, E.: Decoding random linear codes in $\tilde{\mathcal{O}}(2^{0.054n})$. In: Lee, D.H., Wang, X. (eds.) ASIACRYPT 2011. LNCS, vol. 7073, pp. 107–124. Springer, Heidelberg (2011). https://doi.org/10.1007/978-3-642-25385-0_6

51. Mennink, B., Preneel, B.: Hash functions based on three permutations: a generic security analysis. In: Safavi-Naini, R., Canetti, R. (eds.) CRYPTO 2012. LNCS, vol. 7417, pp. 330–347. Springer, Heidelberg (2012). https://doi.org/10.1007/978-3-642-32009-5_20

52. Mennink, B., Preneel, B.: On the XOR of multiple random permutations. In: Malkin, T., Kolesnikov, V., Lewko, A.B., Polychronakis, M. (eds.) ACNS 2015. LNCS, vol. 9092, pp. 619–634. Springer, Cham (2015). https://doi.org/10.1007/978-3-319-28166-7_30

53. Merkle, R.: Secrecy, authentication, and public key systems. Ph.D. thesis (1979)

54. Merkle, R.C.: A digital signature based on a conventional encryption function. In: Pomerance, C. (ed.) CRYPTO 1987. LNCS, vol. 293, pp. 369–378. Springer, Heidelberg (1988). https://doi.org/10.1007/3-540-48184-2_32

55. Merkle, R.C.: One way hash functions and DES. In: Brassard, G. (ed.) CRYPTO 1989. LNCS, vol. 435, pp. 428–446. Springer, New York (1990). https://doi.org/10.1007/0-387-34805-0_40

56. Pietrzak, K.: Cryptography from learning parity with noise. In: Bieliková, M., Friedrich, G., Gottlob, G., Katzenbeisser, S., Turán, G. (eds.) SOFSEM 2012. LNCS, vol. 7147, pp. 99–114. Springer, Heidelberg (2012). https://doi.org/10.1007/978-3-642-27660-6_9

57. Preneel, B., Govaerts, R., Vandewalle, J.: Hash functions based on block ciphers: a synthetic approach. In: Stinson, D.R. (ed.) CRYPTO 1993. LNCS, vol. 773, pp. 368–378. Springer, Heidelberg (1994). https://doi.org/10.1007/3-540-48329-2_31

58. Regev, O.: On lattices, learning with errors, random linear codes, and cryptography. In: Gabow, H.N., Fagin, R. (eds.) STOC, pp. 84–93. ACM, New York (2005)

59. Ristenpart, T., Shacham, H., Shrimpton, T.: Careful with composition: limitations of the indifferentiability framework. In: Paterson, K.G. (ed.) EUROCRYPT 2011. LNCS, vol. 6632, pp. 487–506. Springer, Heidelberg (2011). https://doi.org/10.1007/978-3-642-20465-4_27

60. Rogaway, P., Shrimpton, T.: Cryptographic hash-function basics: definitions, implications, and separations for preimage resistance, second-preimage resistance, and collision resistance. In: Roy, B., Meier, W. (eds.) FSE 2004. LNCS, vol. 3017, pp. 371–388. Springer, Heidelberg (2004). https://doi.org/10.1007/978-3-540-25937-4_24

61. Shoup, V. (ed.): CRYPTO 2005. LNCS, vol. 3621. Springer, Heidelberg (2005). https://doi.org/10.1007/11535218

62. Shrimpton, T., Stam, M.: Building a collision-resistant compression function from non-compressing primitives (extended abstract). In: Aceto, L., Damgård, I., Goldberg, L.A., Halldórsson, M.M., Ingólfsdóttir, A., Walukiewicz, I. (eds.) ICALP 2008, Part II. LNCS, vol. 5126, pp. 643–654. Springer, Heidelberg (2008). https://doi.org/10.1007/978-3-540-70583-3_52

63. Vardy, A.: The intractability of computing the minimum distance of a code. IEEE Trans. Inf. Theory **43**(6), 1757–1766 (1997)

64. Yu, Y., Zhang, J.: Cryptography with auxiliary input and trapdoor from constant-noise LPN. In: Robshaw, M., Katz, J. (eds.) CRYPTO 2016. LNCS, vol. 9814, pp. 214–243. Springer, Heidelberg (2016). https://doi.org/10.1007/978-3-662-53018-4_9

New Code-Based Privacy-Preserving Cryptographic Constructions

Khoa Nguyen[(✉)], Hanh Tang, Huaxiong Wang, and Neng Zeng

School of Physical and Mathematical Sciences,
Nanyang Technological University, Singapore, Singapore
khoantt@ntu.edu.sg

Abstract. Code-based cryptography has a long history but did suffer from periods of slow development. The field has recently attracted a lot of attention as one of the major branches of post-quantum cryptography. However, its subfield of privacy-preserving cryptographic constructions is still rather underdeveloped, e.g., important building blocks such as zero-knowledge range proofs and set membership proofs, and even proofs of knowledge of a hash preimage, have not been known under code-based assumptions. Moreover, almost no substantial technical development has been introduced in the last several years.

This work introduces several new code-based privacy-preserving cryptographic constructions that considerably advance the state-of-the-art in code-based cryptography. Specifically, we present 3 major contributions, each of which potentially yields various other applications. Our first contribution is a code-based statistically hiding and computationally binding commitment scheme with companion zero-knowledge (ZK) argument of knowledge of a valid opening that can be easily extended to prove that the committed bits satisfy other relations. Our second contribution is the first code-based zero-knowledge range argument for committed values, with communication cost logarithmic in the size of the range. A special feature of our range argument is that, while previous works on range proofs/arguments (in all branches of cryptography) only address ranges of *non-negative integers*, our protocol can handle *signed fractional numbers*, and hence, can potentially find a larger scope of applications. Our third contribution is the first code-based Merkle-tree accumulator supported by ZK argument of membership, which has been known to enable various interesting applications. In particular, it allows us to obtain the first code-based ring signatures and group signatures with logarithmic signature sizes.

1 Introduction

Code-based cryptography, pioneered by McEliece [50] in 1978, is the study of cryptosystems based on conjectured hard problems from coding theory and is one of the oldest branches of public-key cryptography. The field did suffer from periods of relatively slow development, but recent years have witnessed its resurgence. On the one hand, solutions to important theoretical problems such as constructing identity-based encryption [14,33] and obtaining worst-case hardness for

© International Association for Cryptologic Research 2019
S. D. Galbraith and S. Moriai (Eds.): ASIACRYPT 2019, LNCS 11922, pp. 25–55, 2019.
https://doi.org/10.1007/978-3-030-34621-8_2

Learning Parity with Noise (LPN) [15] have been introduced. On the other hand, plausible algorithms for practical applications are being recognized by the community: with 7 PKE/KEM from codes accepted into the second round of the NIST Post-Quantum Cryptography Standardization process, the field stands together with lattice-based cryptography [2] as the two most promising candidates for post-quantum cryptography. Nevertheless, many interesting questions are still left open in the scope of code-based cryptography.

A prominent subfield of cryptography research is the designs of advanced schemes aiming to protect both privacy and security of users, namely, privacy-preserving cryptographic constructions. The major tools for building those constructions are zero-knowledge (ZK) proof [36] and argument [17,34] systems that allow to prove the truth of a statement while revealing no additional information. Almost all known zero-knowledge proof/argument systems used in code-based cryptography follow Stern's framework [63], in which the main technical idea is to employ random permutations to prove some specific properties of binary secret vectors, e.g., the secret vectors have a fixed Hamming weight. Variants of Stern's protocol have been employed to design a few privacy-preserving constructions, e.g., proofs of plaintext knowledge for code-based encryption [54], linear-size ring signatures [16,27,51,52], linear-size and sublinear-size group signatures [3,31], proofs of valid openings for LPN-based computationally hiding commitments [41] and proofs for general relations [41]. However, this line of research is still rather underdeveloped, since many important building blocks for privacy-preserving code-based cryptography are still missing, ranging from very basic ones like proof of knowledge of a hash preimage to advanced ones such as range proofs and set membership proofs. Even more worrisome is the slow progress in the field, with almost no substantial technical development being introduced in the last 6 years. This unsatisfactory state-of-affairs motivates our work.

OUR RESULTS. In this work, we introduce several new privacy-preserving constructions that we believe will considerably advance the state-of-the-art in code-based cryptography. Specifically, we provide 3 main contributions, each of which potentially yields various other applications.

First, we put forward a code-based statistically hiding and computationally binding commitment scheme with companion zero-knowledge argument of knowledge (ZKAoK) of a valid opening. The commitment scheme is based on a family of collision-resistant hash functions introduced by Augot, Finiasz and Sendrier (AFS) [6,7], similar variants of which was recently studied in [4,15,64]. The design of the scheme is quite standard, in which we plug in a randomness with sufficient min-entropy and make use of the left-over hash lemma [35]. Our non-trivial contribution here is a companion ZK argument system that makes the commitment scheme much more useful for privacy-preserving protocols. In many advanced protocols, one typically works with different sub-protocols that share a common secret, and a commitment supported by ZK proofs/arguments can greatly help in bridging these layers. In the code-based setting, such a commitment was proposed in [41], but it relies on the hardness of the LPN problem and operates in the computationally-hiding setting. In our setting, to base security on a variant of the Syndrome Decoding problem, the committed message has to be non-linearly encoded into a low-weight vector of larger dimension before being hashed. This makes proving knowledge of a valid opening quite challenging, since one has to prove that the non-linear encoding process is done correctly. We overcome this problem by employing a specific permuting technique that works in the framework of Stern's protocol [63] and that enables us to keep fine-grained control on how each bit of the message behaves in the encoding process. The fact that we can "keep track" of the secret committed bits indeed makes our protocol composable with other privacy-preserving protocols, where we can additionally prove that these bits satisfy other relations. In particular, it paves the way for our next 2 contributions.

Second, we provide zero-knowledge range arguments for *signed fractional numbers* committed via our code-based commitment. For $\ell > 0$ and $f \geq 0$, we consider fractional numbers X represented as $x_\ell x_{\ell-1} \ldots x_0 \bullet x_{-1} x_{-2} \ldots x_{-f}$, where x_ℓ is the sign bit, $x_{\ell-1}, \ldots, x_0$ are the integer bits, and $x_{-1}, \ldots, x_{-f}$ are the fractional bits. Our techniques allow to prove in zero-knowledge that a committed number X satisfies inequalities $X \leq Y$ or $X < Y$, where Y is another signed fractional number (that could be publicly given or be committed). These techniques directly yield range arguments addressing both public and hidden ranges with communication cost logarithmic in the sizes of the ranges. This not only solves an open problem in code-based cryptography but also brings a novel feature to the topic of range proofs in general. Range proofs, introduced by Brickell *et al.* [18], serve as building blocks in various applications, including anonymous credentials [22], auctions [48], e-voting [39] and anonymous e-cash [20]. Efficient constructions [19,24,26,30,37,40,46,47] have been proposed in almost all major branches of cryptography, but up to our knowledge, they only address non-negative integers. Negative numbers do often appear in our daily life in the forms of financial loss, bad reputation, medical data, etc., and it would be desirable to be able to handle them in a privacy-preserving manner. Moreover, these data values could be stored as fractional numbers, e.g., bank account balances, GPAs and tax records, and hence, a protocol addressing them directly in such forms would potentially be interesting. This inspires our investigation of range arguments for signed fractional numbers.

Our third contribution is the first code-based accumulator [9] supported by ZK arguments of valid accumulated values, which directly imply ZK arguments of set membership. Accumulators are essential building blocks in numerous authentication mechanisms, including ring and group signatures [28,29,45], anonymous credentials [1,21,23], e-cash [5], and authenticated data structures [59,60]. Accumulators with companion ZK proofs have been proposed from number-theoretic assumptions [9,57], lattice assumptions [45] and from symmetric-key primitives [13,28], but have not been known in the scope of code-based cryptography. Our construction fills in this gap and opens up a wide range of applications that have not been achieved from code-based assumptions. Our design resembles Libert et al.'s approach for lattices [45], which relies on Merkle hash trees [53] and ZKAoK of a tree path from the root to a secret leaf. However, unlike the lattice setting where smallness (and computational hardness) can be defined with respect to various metrics and the output of each hashing can be easily decomposed into binary to serve as the input of the next step, the binary linear code setting with Hamming metric makes the problem more challenging. At each step, we have to encode the hash output to a small-weight vector (with respect to its dimension) before going to the next step, and prove that the whole recursive process is done correctly. Fortunately, this difficulty can be overcome with our ZK techniques. As applications, we put forward 2 prominent anonymity-oriented constructions: ring signature [62] and group signature [25].

Our ring signature scheme is the first one from code-based assumptions that achieves signature size logarithmic in the cardinality of the ring. Previous constructions [16,27,51,52] all suffer from linear-size signatures. Designing logarithmic-size ring signatures is generally a hard problem, which usually requires a powerful supporting technique, which is - in this case - an accumulator that enables logarithmic-size arguments of set membership.

Our group signature scheme is also the first one that produces logarithmic-size signatures in the scope of code-based cryptography. Compared with previous works [3,31], our scheme not only has shorter signatures (for large groups), but also achieve the stronger notion of CCA-anonymity. The scheme is also appealing in the sense that it is the first time in all branches of cryptography a CCA-anonymous group signature scheme is achieved before a standard-model signature compatible with ZK proofs is known. (The latter is traditionally considered to be a necessary ingredient for building the former in a generic manner.)

OUR TECHNIQUES. Let us first discuss our basic techniques for proving in zero-knowledge the knowledge of a preimage of a hash, computed via the AFS hash function $\mathcal{H}_{\mathrm{afs}} : \{0,1\}^k \to \{0,1\}^n$. Let $\mathbf{B} \xleftarrow{\$} \mathbb{Z}_2^{n \times 2^c \cdot k/c}$, for some constant c dividing k. Let RE: $\{0,1\}^k \to \{0,1\}^{2^c \cdot k/c}$ be an encoding function that maps $\mathbf{x}$ to $\mathsf{RE}(\mathbf{x})$, defined as follows. First, write $\mathbf{x}$ block-wise as $\mathbf{x} = (\mathbf{x}_1 \| \ldots \| \mathbf{x}_{k/c})$, where $\mathbf{x}_j = (x_{j,1}, \ldots, x_{j,c})^\top$ for $j \in [k/c]$. Denote by $\Delta_{2^c}(\mathbf{x}_j)$ the binary vector of dimension 2^c and Hamming weight 1 whose sole 1 entry is at the t_j-th position, for $t_j = \Sigma_{i=1}^c 2^{c-i} \cdot x_{j,i} \in [0, 2^c - 1]$. Then $\mathsf{RE}(\mathbf{x})$ is defined to be $(\Delta_{2^c}(\mathbf{x}_1) \| \ldots \| \Delta_{2^c}(\mathbf{x}_{k/c}))$, and the hash output is set as $\mathbf{u} = \mathbf{B} \cdot \mathsf{RE}(\mathbf{x})$. Given $(\mathbf{B}, \mathbf{u})$, to prove that we know $\mathbf{x}$ such that $\mathcal{H}_{\mathrm{afs}}(\mathbf{x}) = \mathbf{u}$, we have to demonstrate that the encoding $\mathsf{RE}(\cdot)$ is done correctly for $\mathbf{x}$. To this end, we introduce the following permuting technique.

For every vector $\mathbf{s} = (s_1, \ldots, s_c) \in \{0,1\}^c$, define the permutation $E_\mathbf{s}$ that transforms vector $\mathbf{x} = (x_{0,0,\ldots,0}, \ldots, x_{i_1,\ldots,i_c}, \ldots, x_{1,1,\ldots,1}) \in \{0,1\}^{2^c}$ into vector $E_\mathbf{s}(\mathbf{x}) = (x'_{0,0,\ldots,0}, \ldots, x'_{1,1,\ldots,1})$, where for each $(i_1, \ldots, i_c) \in \{0,1\}^c$, we have $x_{i_1,\ldots,i_c} = x'_{i_1 \oplus s_1, \ldots, i_c \oplus s_c}$.

Note that, for any $\mathbf{s}, \mathbf{v} \in \{0,1\}^c$, we have:

$$\mathbf{x} = \Delta_{2^c}(\mathbf{v}) \iff E_\mathbf{s}(\mathbf{x}) = \Delta_{2^c}(\mathbf{v} \oplus \mathbf{s}). \qquad (1)$$

For $\mathbf{t} = (\mathbf{t}_1 \| \ldots \| \mathbf{t}_{k/c}) \in \{0,1\}^k$ consisting of k/c blocks of length c, define the permutation $E'_\mathbf{t}$ that transforms vector $\mathbf{y} = (\mathbf{y}_1 \| \ldots \| \mathbf{y}_{k/c}) \in \{0,1\}^{2^c \cdot k/c}$

consisting of k/c blocks of length 2^c into vector of the following form $E'_t(\mathbf{y}) = \left(E_{t_1}(\mathbf{y}_1)\| \dots \|E_{t_{n/c}}(\mathbf{y}_{n/c})\right)$. Note that, for any $\mathbf{t}, \mathbf{x} \in \{0,1\}^k$, we have:

$$\mathbf{y} = \mathsf{RE}(\mathbf{x}) \iff E'_t(\mathbf{y}) = \mathsf{RE}(\mathbf{x} \oplus \mathbf{t}). \tag{2}$$

In the framework of Stern's protocol [63], the equivalence in (2) enables us to prove that $\mathbf{y}$ is the correct encoding of $\mathbf{x}$, as follows. The prover samples uniformly random $\mathbf{t}$ and demonstrates to the verifier that the right-hand side of (2) holds. The verifier is thus convinced that its left-hand side also holds, while learning no additional information about $\mathbf{x}$, thanks to the "one-time pad" $\mathbf{t}$. Moreover, this permuting technique allows us to keep control over the bits of $\mathbf{x}$, in order to prove that they satisfy other relations. To this end, it suffices to use the same "one-time pad" at other appearances of $\mathbf{x}$.

Our discussed above technique then readily extends to handle the case when there are two inputs to the hash function, i.e., $\mathcal{H}_{\mathrm{afs}} : \{0,1\}^\ell \times \{0,1\}^k \to \{0,1\}^n$. If we set k sufficiently large so that the leftover hash lemma [35] applies, then we obtain a statistically hiding commitment scheme that is supported by our ZK technique. On the other hand, if we set $\ell = k = n$, then we get a function that compresses two child-inputs of n bits to one parent-output, which is then can be used to build a Merkle hash tree. Then, by combining the ZK techniques from [45] and our techniques for proving correctness of re-encoding at each step in a tree path, we get a Merkle-tree-accumulator supported by logarithmic-size zero-knowledge arguments.

To build a ring signature, we add one more level of secret under every leaf in the tree, so that each leaf corresponds to a user's public key, and define the signing process as the process of proving knowledge of an extended path from beneath a leaf up to the root of the tree. Furthermore, as in [45], by adding a CCA2-secure encryption layer supported by zero-knowledge arguments of plaintext knowledge, we can build a secure group signature. To this end, we employ the randomized McEliece scheme [58] and make it CCA2-secure in the random oracle model via the Naor-Yung transformation [55]. Both our ring and group signatures feature logarithmic-size signatures, thanks to the tree structure.

Let us next discuss our techniques for handling inequalities among signed fractional numbers, which lead to our range arguments. Comparing signed fractional numbers in zero-knowledge is highly non-trivial, due to 2 main reasons. First, unlike for non-negative numbers, the order of (binary) signed numbers is not lexicographical, e.g., for $(\ell, f) = (5, 2)$, the number $110110 \bullet 11$ is lexicographically larger than $000011 \bullet 00$, yet its decimal value is -9.25, which is smaller than the decimal value 3 of the latter. Thus, it is counterintuitive when we compare them in zero-knowledge. Second, the approach of proving $X \leq Y$ via demonstrating the existence of $Z \geq 0$ such that $X + Z = Y$ (as used in [46]) is not easily applicable here, due to the problem of overflows. For instance, the binary addition (with carries) of $011110 \bullet 11$ and $000011 \bullet 00$ would yield $100001 \bullet 11$, which is translated into an incorrect expression $30.75 + 3 = -30.25$. Therefore, we have to carefully address the complications caused by the signed bits and to ensure that overflows do not occur.

Our idea is to derive necessary and sufficient conditions for $X \leq Y$, in a way such that these conditions can be correctly and efficiently proved in ZK. Let $(x_\ell, \ldots, x_0, x_{-1}, \ldots, x_{-f})$, $(y_\ell, \ldots, y_0, y_{-1}, \ldots, y_{-f})$ be the bits representing X and Y, respectively. We observe and then formally prove that $X \leq Y$ if and only if there exist bits $z_\ell, z_{\ell-1}, \ldots, z_0, z_{-1}, \ldots, z_{-f}, c_{\ell+1}, c_\ell, c_{\ell-1}, \ldots, c_0, c_{-1}, \ldots, c_{-f+1}$ satisfying

$$\begin{cases} c_{-f+1} = x_{-f} \cdot z_{-f} \\ c_i = x_{i-1} \cdot z_{i-1} \oplus y_{i-1} \cdot c_{i-1} \oplus c_{i-1}, \quad \forall i \in [-f+2, \ell+1] \\ y_{-f} = x_{-f} \oplus z_{-f} \\ y_i = x_i \oplus z_i \oplus c_i, \quad \forall i \in [-f+1, \ell] \\ y_\ell = x_\ell \oplus c_{\ell+1}. \end{cases}$$

This simple-yet-vital result allows us to reduce the inequality relations among signed fractional numbers to simple relations among bits, which can be effectively addressed using existing techniques [44,45] for Stern's protocol.

ORGANIZATION. The rest of the paper is organized as follows. In Sect. 2, we recall the background on code-based hash functions, ZK arguments and previous Stern-like techniques. Our commitment scheme together with our techniques for proving knowledge of code-based hash preimages and committed values are described in Sect. 3. In Sect. 4, we present our treatment of signed fractional numbers and construct ZK range arguments for committed signed fractional numbers. Our accumulator and its supporting ZK argument of membership are given in Sect. 5. Applications of to ring signatures and group signatures are then discussed in Sect. 6. Due to space restriction, the descriptions and analyses of our ring and group signature schemes are deferred to the full version [56].

2   Background

2.1   Code-Based Collision-Resistant Hash Functions

This section recalls the family of code-based hash functions proposed by Augot, Finiasz and Sendrier (AFS) [6,7], which is based on the hardness of the 2-Regular Null Syndrome Decoding (2-RNSD) problem. We note that the more recent proposals of code-based hash functions [4,15,64], although relying on different assumptions, are syntactically similar to the AFS family at a high level. Working with the AFS family allows us to derive practical parameters, based on the analyses of [11,12]. Let us begin by introducing some supporting notations.

NOTATIONS. We identify $\mathbb{Z}_2$ as the set $\{0, 1\}$. The set $\{a, \ldots, b\}$ is denoted by $[a, b]$. We often write $[b]$ when $a = 1$. Let $\oplus$ denote the bit-wise addition operation modulo 2. If S is a finite set, then $x \xleftarrow{\$} S$ means that x is chosen uniformly at random from S. Throughout this paper, all vectors are column vectors. When concatenating vectors $\mathbf{x} \in \mathbb{Z}_2^m$ and $\mathbf{y} \in \mathbb{Z}_2^k$, for simplicity, we use the notation $(\mathbf{x} \| \mathbf{y}) \in \mathbb{Z}_2^{m+k}$ instead of $(\mathbf{x}^\top \| \mathbf{y}^\top)^\top$. Denote $\mathsf{B}(n, \omega)$ to be the set of all binary vectors of length n with weight ω and the symmetric group of all permutations of n elements to be S_n.

For $c \in \mathbb{Z}^+$ and for k divisible by c, define the following.

- Regular(k, c): the set of all vectors $\mathbf{w} = (\mathbf{w}_1 \| \dots \| \mathbf{w}_{k/c}) \in \{0,1\}^{2^c \cdot k/c}$ consisting of k/c blocks, each of which is an element of $B_{2^c}^1$. Here $B_{2^c}^1$ is the set that contains all the elements in $\{0,1\}^{2^c}$ with Hamming weight 1. If $\mathbf{w} \in$ Regular(k, c) for some k, c, then we call $\mathbf{w}$ a *regular word*.
- RE: $\{0,1\}^k \to \{0,1\}^{2^c \cdot k/c}$, a regular encoding function that maps $\mathbf{x}$ to RE$(\mathbf{x})$, defined as follows. Denote $\mathbf{x} = (\mathbf{x}_1 \| \dots \| \mathbf{x}_{k/c})$, where $\mathbf{x}_j = (x_{j,1}, \dots, x_{j,c})^\top$ for $j \in [k/c]$. Then compute $t_j = \Sigma_{i=1}^c 2^{c-i} \cdot x_{j,i}$. Denote by $\Delta_{2^c}(\mathbf{x}_j)$ the element in $B_{2^c}^1$ whose sole 1 entry is at the t_j-th position for $t_j \in [0, 2^c - 1]$. RE$(\mathbf{x})$ is then defined to be $(\Delta_{2^c}(\mathbf{x}_1) \| \dots \| \Delta_{2^c}(\mathbf{x}_{k/c}))$. One can check that RE$(\mathbf{x}) \in$ Regular(k, c).
- 2-Regular(k, c): the set of all vectors $\mathbf{x} \in \{0,1\}^{2^c \cdot k/c}$, such that there exist regular words $\mathbf{v}, \mathbf{w} \in$ Regular(k, c) satisfying $\mathbf{x} = \mathbf{v} \oplus \mathbf{w}$. Note that, $\mathbf{x} \in$ 2-Regular(k, c) if and only if $\mathbf{x}$ can be written as the concatenation of k/c blocks of length 2^c, each of which has Hamming weight 0 or 2. If $\mathbf{x} \in$ 2-Regular(k, c) for some k, c, then we call $\mathbf{x}$ a *2-regular word*.

The 2-RNSD Problem. Introduced by Augot, Finiasz and Sendrier [6,7], the 2-RNSD problem asks to find low-weight 2-regular codewords in random binary linear codes. This problem is closely related to the Small Codeword Problem [49] and binary Shortest Vector Problem [4], with an additional and strong constraint that the solution codeword must be 2-regular.

Definition 1. *The* 2-RNSD$_{n,k,c}$ *problem, parameterized by integers* n, k, c, *is as follows. Given a uniformly random matrix* $\mathbf{B} \in \mathbb{Z}_2^{n \times m}$, *where* $m = 2^c \cdot k/c$, *find a non-zero vector* $\mathbf{z} \in$ 2-Regular$(k, c) \subseteq \{0,1\}^m$ *such that* $\mathbf{B} \cdot \mathbf{z} = \mathbf{0}$.

The problem is shown to be NP-complete in the worst case [6]. In practice, for appropriate choices of n, k, c, the best known algorithms require exponential times in the security parameter. See [11] for a comprehensive discussion of known attacks and parameter settings.

The AFS Hash Functions. Let λ be the security parameter. The AFS family of hash functions $\mathcal{H}_{\text{afs}}$ maps $\{0,1\}^k$ to $\{0,1\}^n$, where $n, k = \Omega(\lambda)$ and $k > n$. Each function in the family is associated with a matrix $\mathbf{B} \xleftarrow{\$} \mathbb{Z}_2^{n \times 2^c \cdot k/c}$, for some properly chosen constant c dividing k. To compute the hash value of $\mathbf{x} \in \{0,1\}^k$, one encodes it to the corresponding regular word RE$(\mathbf{x}) \in \{0,1\}^{2^c \cdot k/c}$ and outputs $\mathbf{B} \cdot$ RE$(\mathbf{x})$.

The above hash functions are collision-resistant assuming the hardness of the 2-RNSD$_{n,k,c}$ problem. Suppose that the adversary can produce distinct $\mathbf{x}_0, \mathbf{x}_1$ such that $\mathbf{B} \cdot$ RE$(\mathbf{x}_0) = \mathbf{B} \cdot$ RE$(\mathbf{x}_1)$. Let $\mathbf{z} =$ RE$(\mathbf{x}_0) \oplus$ RE$(\mathbf{x}_1) \neq \mathbf{0}$ then we have $\mathbf{z} \in$ 2-Regular(k, c) and $\mathbf{B} \cdot \mathbf{z} = \mathbf{0}$. In other words, $\mathbf{z}$ is a solution to the 2-RNSD$_{n,k,c}$ problem associated with matrix $\mathbf{B}$.

In this work, we rely on the above hash function family to develop two tools for privacy-preserving code-based cryptography: (i) computationally binding and statistically hiding commitments supporting by ZK arguments of knowledge of valid openings; (ii) Cryptographic accumulators supporting by ZK arguments of accumulated values.

2.2 Zero-Knowledge Argument Systems and Stern-Like Protocols

We will work with statistical zero-knowledge argument systems, namely, interactive protocols where the zero-knowledge property holds against *any* cheating verifier, while the soundness property only holds against *computationally bounded* cheating provers. More formally, let the set of statements-witnesses $R = \{(y, w)\} \in \{0, 1\}^* \times \{0, 1\}^*$ be an NP relation. A two-party game $\langle \mathcal{P}, \mathcal{V} \rangle$ is called an interactive argument system for the relation R with soundness error e if the following conditions hold:

- **Completeness.** If $(y, w) \in R$ then $\Pr\big[\langle \mathcal{P}(y, w), \mathcal{V}(y) \rangle = 1\big] = 1$.
- **Soundness.** If $(y, w) \notin R$, then $\forall$ PPT $\widehat{\mathcal{P}}$: $\Pr[\langle \widehat{\mathcal{P}}(y, w), \mathcal{V}(y) \rangle = 1] \leq e$.

An argument system is called statistical zero-knowledge if there exists a PPT simulator $\mathcal{S}(y)$ having oracle access to any $\widehat{\mathcal{V}}(y)$ and producing a simulated transcript that is statistically close to the one of the real interaction between $\mathcal{P}(y, w)$ and $\widehat{\mathcal{V}}(y)$. A related notion is argument of knowledge, which requires the witness-extended emulation property. For protocols consisting of 3 moves (*i.e.*, commitment-challenge-response), witness-extended emulation is implied by *special soundness* [38], where the latter assumes that there exists a PPT extractor which takes as input a set of valid transcripts w.r.t. all possible values of the "challenge" to the same "commitment", and outputs w' such that $(y, w') \in R$.

Stern-Like Protocols. The statistical zero-knowledge arguments of knowledge presented in this work are Stern-like [63] protocols. In particular, they are Σ-protocols in the generalized sense defined in [10,41] (where 3 valid transcripts are needed for extraction, instead of just 2). The basic protocol consists of 3 moves: commitment, challenge, response. If we employ our first explicit construction of statistically hiding string commitment from a code-based assumption in the first move, then one obtains a statistical zero-knowledge argument of knowledge (ZKAoK) with perfect completeness, constant soundness error 2/3. In many applications, the protocol is repeated a sufficient number of times to make the soundness error negligibly small. For instance, to achieve soundness error 2^{-80}, it suffices to repeat the basic protocol 137 times.

An Abstraction of Stern's Protocols. We recall an abstraction, adapted from [43], which captures the sufficient conditions to run a Stern-like protocol. Looking ahead, this abstraction will be helpful for us in presenting our ZK argument systems: we will reduce the relations we need to prove to instances of the abstract protocol, using our specific techniques. We recall an abstraction proposed in [43]. Let K, L be positive integers, where $L \geq K$, and let VALID be a subset of $\{0,1\}^L$. Suppose that $\mathcal{S}$ is a finite set such that one can associate every $\phi \in \mathcal{S}$ with a permutation Γ_ϕ of L elements, satisfying the following conditions:

$$\begin{cases} \mathbf{w} \in \mathsf{VALID} \iff \Gamma_\phi(\mathbf{w}) \in \mathsf{VALID}, \\ \text{If } \mathbf{w} \in \mathsf{VALID} \text{ and } \phi \text{ is uniform in } \mathcal{S}, \text{ then } \Gamma_\phi(\mathbf{w}) \text{ is uniform in } \mathsf{VALID}. \end{cases} \quad (3)$$

We aim to construct a statistical ZKAoK for the following abstract relation:

$$R_{\text{abstract}} = \{(\mathbf{M}, \mathbf{v}), \mathbf{w} \in \mathbb{Z}_2^{K \times L} \times \mathbb{Z}_2^K \times \mathsf{VALID} : \mathbf{M} \cdot \mathbf{w} = \mathbf{v}\}.$$

The conditions in (3) play a crucial role in proving in ZK that $\mathbf{w} \in \mathsf{VALID}$: To do so, the prover samples $\phi \xleftarrow{\$} \mathcal{S}$ and lets the verifier check that $\Gamma_\phi(\mathbf{w}) \in \mathsf{VALID}$, while the latter cannot learn any additional information about $\mathbf{w}$ thanks to the randomness of ϕ. Furthermore, to prove in ZK that the linear equation holds, the prover samples a masking vector $\mathbf{r}_w \xleftarrow{\$} \mathbb{Z}_2^L$, and convinces the verifier instead that $\mathbf{M} \cdot (\mathbf{w} \oplus \mathbf{r}_w) = \mathbf{M} \cdot \mathbf{r}_w \oplus \mathbf{v}$.

The interaction between prover $\mathcal{P}$ and verifier $\mathcal{V}$ is described in Fig. 1. The protocol employs a statistically hiding and computationally binding string commitment scheme COM.

---

1. **Commitment:** Prover samples $\mathbf{r}_w \xleftarrow{\$} \mathbb{Z}_2^L$, $\phi \xleftarrow{\$} \mathcal{S}$ and randomness ρ_1, ρ_2, ρ_3 for COM. Then he sends $\mathsf{CMT} = (C_1, C_2, C_3)$ to the verifier, where

$$C_1 = \mathsf{COM}(\phi, \mathbf{M} \cdot \mathbf{r}_w; \rho_1), \quad C_2 = \mathsf{COM}(\Gamma_\phi(\mathbf{r}_w); \rho_2),$$
$$C_3 = \mathsf{COM}(\Gamma_\phi(\mathbf{w} \oplus \mathbf{r}_w); \rho_3).$$

2. **Challenge:** The verifier sends a challenge $Ch \xleftarrow{\$} \{1, 2, 3\}$ to the prover.
3. **Response:** Depending on Ch, the prover sends RSP computed as follows:
   - $Ch = 1$: Let $\mathbf{t}_w = \Gamma_\phi(\mathbf{w})$, $\mathbf{t}_r = \Gamma_\phi(\mathbf{r}_w)$, and $\mathsf{RSP} = (\mathbf{t}_w, \mathbf{t}_r, \rho_2, \rho_3)$.
   - $Ch = 2$: Let $\phi_2 = \phi$, $\mathbf{w}_2 = \mathbf{w} \oplus \mathbf{r}_w$, and $\mathsf{RSP} = (\phi_2, \mathbf{w}_2, \rho_1, \rho_3)$.
   - $Ch = 3$: Let $\phi_3 = \phi$, $\mathbf{w}_3 = \mathbf{r}_w$, and $\mathsf{RSP} = (\phi_3, \mathbf{w}_3, \rho_1, \rho_2)$.

Verification: Receiving RSP, the verifier proceeds as follows:

- $Ch = 1$: Check that $\mathbf{t}_w \in \mathsf{VALID}$, $C_2 = \mathsf{COM}(\mathbf{t}_r; \rho_2)$, $C_3 = \mathsf{COM}(\mathbf{t}_w \oplus \mathbf{t}_r; \rho_3)$.
- $Ch = 2$: Check that $C_1 = \mathsf{COM}(\phi_2, \mathbf{M} \cdot \mathbf{w}_2 \oplus \mathbf{v}; \rho_1)$, $C_3 = \mathsf{COM}(\Gamma_{\phi_2}(\mathbf{w}_2); \rho_3)$.
- $Ch = 3$: Check that $C_1 = \mathsf{COM}(\phi_3, \mathbf{M} \cdot \mathbf{w}_3; \rho_1)$, $C_2 = \mathsf{COM}(\Gamma_{\phi_3}(\mathbf{w}_3); \rho_2)$.

In each case, the verifier outputs 1 if and only if all the conditions hold.

Fig. 1. Stern-like ZKAoK for the relation R_{abstract}.

The properties of the protocols are summarized in Theorem 1, whose proof can be found in [43] or the full version of the present paper [56].

Theorem 1 ([43]). *Assume that* COM *is a statistically hiding and computationally binding string commitment scheme. Then, the protocol in Fig. 1 is a statistical* ZKAoK *with perfect completeness, soundness error* 2/3, *and communication cost* $\mathcal{O}(L)$. *In particular:*

- *There exists a polynomial-time simulator that, on input* $(\mathbf{M}, \mathbf{v})$, *outputs an accepted transcript statistically close to that produced by the real prover.*
- *There exists a polynomial-time knowledge extractor that, on input a commitment* CMT *and* 3 *valid responses* $(\mathrm{RSP}_1, \mathrm{RSP}_2, \mathrm{RSP}_3)$ *to all* 3 *possible values of the challenge* Ch, *outputs* $\mathbf{w}' \in$ VALID *such that* $\mathbf{M} \cdot \mathbf{w}' = \mathbf{v}$.

2.3   Previous Extending-then-Permuting Techniques

We next recall the extending-then-permuting techniques for proving in Stern's framework the knowledge of a single secret bit x and a product of 2 secret bits $x_1 \cdot x_2$, presented in [45] and [44], respectively.

Let $\oplus$ denote the bit-wise addition operation modulo 2. For any bit $b \in \{0, 1\}$, denote by $\overline{b}$ the bit $b = b \oplus 1$. Note that, for any $b, c \in \{0, 1\}$, we have $\overline{b \oplus c} = b \oplus c \oplus 1 = \overline{b} \oplus c$. For any bit b, let $\mathsf{enc}(b) = (\overline{b}, b) \in \{0, 1\}^2$.

For any bit $c \in \{0, 1\}$, define F_c as the permutation that transforms integer vector $\mathbf{v} = (v_0, v_1) \in \mathbb{Z}^2$ into vector $F_c(\mathbf{v}) = (v_c, v_{\overline{c}})$. Namely, if $c = 0$ then F_c keeps the arrangement the coordinates of $\mathbf{v}$; or swaps them if $c = 1$. Note that:

$$\mathbf{v} = \mathsf{enc}(b) \quad \Longleftrightarrow \quad F_c(\mathbf{v}) = \mathsf{enc}(b \oplus c). \tag{4}$$

The authors of [45] showed that the equivalence (4) is helpful for proving knowledge of a secret bit x that may appear in several correlated linear equations. To this end, one extends x to $\mathsf{enc}(x) \in \{0, 1\}^2$, and permutes the latter using F_c, where c is a uniformly random bit. Then one demonstrates to the verifier that the permuted vector is $\mathsf{enc}(x \oplus c)$, which implies that the original vector $\mathsf{enc}(x)$ is well-formed - which in turn implies knowledge of some bit x. Meanwhile, the bit c acts as a "one-time pad" that completely hides x.

In [44], Libert et al. proposed a method for proving the well-formedness of the product of two secret bits x_1, x_2, based on the following technique.

- For any two bits b_1, b_2, define the vector

$$\mathsf{ext}(b_1, b_2) = (\, \overline{b}_1 \cdot \overline{b}_2, \ \overline{b}_1 \cdot b_2, \ b_1 \cdot \overline{b}_2, \ b_1 \cdot b_2 \,) \in \{0, 1\}^4,$$

that is an extension of the bit product $b_1 \cdot b_2$.
- For any two bits $c_1, c_2 \in \{0, 1\}$, define T_{c_1, c_2} as the permutation that transforms vector $\mathbf{v} = (v_{0,0}, v_{0,1}, v_{1,0}, v_{1,1}) \in \mathbb{Z}^4$ into vector

$$T_{c_1, c_2}(\mathbf{v}) = \left(v_{c_1, c_2}, v_{c_1, \overline{c}_2}, v_{\overline{c}_1, c_2}, v_{\overline{c}_1, \overline{c}_2} \right) \in \mathbb{Z}^4.$$

Then, the following equivalence holds. For any bits b_1, b_2, c_1, c_2 and any vector $\mathbf{v} = (v_{0,0}, v_{0,1}, v_{1,0}, v_{1,1}) \in \mathbb{Z}^4$,

$$\mathbf{v} = \mathsf{ext}(b_1, b_2) \quad \Longleftrightarrow \quad T_{c_1, c_2}(\mathbf{v}) = \mathsf{ext}(b_1 \oplus c_1, b_2 \oplus c_2). \tag{5}$$

As a result, to prove that a bit has the form $x_1 \cdot x_2$, one can extend it to vector $\mathsf{ext}(x_1, x_2)$, then permute the latter using T_{c_1,c_2}, where c_1, c_2 are uniformly random bits. One then demonstrates to the verifier that the permuted vector is $\mathsf{ext}(x_1 \oplus c_1, x_2 \oplus c_2)$. This convinces the verifier that the original vector, i.e., $\mathsf{ext}(x_1, x_2)$, is well-formed, while learning no additional information about x_1 and x_2, thanks to the randomness of c_1 and c_2.

3   Code-Based Statistically Hiding Commitments with Companion Zero-Knowledge Protocols

In Sect. 3.1, we develop the AFS hash function to obtain a code-based computationally binding and statistically hiding commitment scheme. Then, in Sect. 3.2, we build up our techniques for proving in zero-knowledge the correct encoding of binary strings into regular words. Relying on these techniques, we put forward in Sect. 3.3 a ZKAoK of a valid opening for the given commitment scheme. This building block will further be used in other advances constructions, which will be presented later in the paper.

3.1   Our Construction

Given security parameter λ, choose $n = \mathcal{O}(\lambda)$, $k \geq n + 2\lambda + \mathcal{O}(1)$. Let the message space be $\mathcal{M} = \{0,1\}^L$, and let c be a constant dividing L and k. Let $m_0 = 2^c \cdot L/c$, $m_1 = 2^c \cdot k/c$ and $m = m_0 + m_1$. Our scheme works as follows.

- KGen: Sample $\mathbf{B}_0 \xleftarrow{\$} \mathbb{Z}_2^{n \times m_0}$, $\mathbf{B}_1 \xleftarrow{\$} \mathbb{Z}_2^{n \times m_1}$, and output commitment key $pk = \mathbf{B} = [\mathbf{B}_0 \mid \mathbf{B}_1] \in \mathbb{Z}_2^{n \times m}$.
- Com: On input a message $\mathbf{x} \in \{0,1\}^L$ and commitment key pk, sample randomness $\mathbf{s} \xleftarrow{\$} \{0,1\}^k$, compute $\mathbf{c} = \mathbf{B}_0 \cdot \mathsf{RE}(\mathbf{x}) \oplus \mathbf{B}_1 \cdot \mathsf{RE}(\mathbf{s}) \in \mathbb{Z}_2^n$, and output commitment $\mathbf{c}$ together with opening $\mathbf{s}$. Here, $\mathsf{RE}(\cdot)$ is the regular encoding function from Sect. 2.1.
- Open: On input commitment key pk, a commitment $\mathbf{c} \in \mathbb{Z}_2^n$, a message $\mathbf{x} \in \{0,1\}^L$ and an opening $\mathbf{s} \in \{0,1\}^k$, output 1 if $\mathbf{c} = \mathbf{B}_0 \cdot \mathsf{RE}(\mathbf{x}) \oplus \mathbf{B}_1 \cdot \mathsf{RE}(\mathbf{s})$, or 0 otherwise.

One can check that the proposed scheme is correct. Let us now prove the computationally binding and statistically hiding properties.

Lemma 1. *The scheme is computationally binding, assuming the hardness of the 2-RNSD$_{n,L+k,c}$ problem.*

Proof. Suppose that the adversary outputs $\mathbf{c}, \mathbf{x}, \mathbf{x}', \mathbf{s}, \mathbf{s}'$ such that $\mathbf{x} \neq \mathbf{x}'$ and

$$\mathbf{c} = \mathbf{B}_0 \cdot \mathsf{RE}(\mathbf{x}) \oplus \mathbf{B}_1 \cdot \mathsf{RE}(\mathbf{s}) = \mathbf{B}_0 \cdot \mathsf{RE}(\mathbf{x}') \oplus \mathbf{B}_1 \cdot \mathsf{RE}(\mathbf{s}').$$

Let $\mathbf{z} = \mathsf{RE}(\mathbf{x}) \oplus \mathsf{RE}(\mathbf{x}') \neq \mathbf{0}$ and $\mathbf{y} = \mathsf{RE}(\mathbf{s}) \oplus \mathsf{RE}(\mathbf{s}')$. Then $\mathbf{B}_0 \cdot \mathbf{z} + \mathbf{B}_1 \cdot \mathbf{y} = \mathbf{0}$. Next, let $\mathbf{t} = (\mathbf{z} \| \mathbf{y})$ then we have $\mathbf{t} \neq \mathbf{0}$, $\mathbf{t} \in 2\text{-Regular}(L + k, c)$ and $\mathbf{B} \cdot \mathbf{t} = \mathbf{0}$. In other words, $\mathbf{t}$ is a solution to the 2-RNSD$_{n,L+k,c}$ problem associated with uniformly random matrix $\mathbf{B}$. $\square$

The statistically hiding property of the scheme is based on the following leftover hash lemma.

Lemma 2 (Leftover hash lemma, adapted from [35]). *Let D be a distribution over $\{0,1\}^t$ with min-entropy k. For any $\epsilon > 0$ and $n \leq k - 2\log(1/\epsilon) - \mathcal{O}(1)$, the statistical distance between the joint distribution of $(\mathbf{B}, \mathbf{B} \cdot \mathbf{t})$, where $\mathbf{B} \xleftarrow{\$} \mathbb{Z}_2^{n \times t}$ and $\mathbf{t} \in \{0,1\}^t$ is drawn from distribution D, and the uniform distribution over $\mathbb{Z}_2^{n \times t} \times \mathbb{Z}_2^n$ is at most ϵ.*

If $\mathbf{t}$ is uniformly random over $\{0,1\}^k$, then the distribution of $\mathsf{RE}(\mathbf{t})$ over $\{0,1\}^{m_1}$ has min-entropy exactly k. Since $k \geq n + 2\lambda + \mathcal{O}(1)$, the distribution of $\mathbf{B}_1 \cdot \mathsf{RE}(\mathbf{s})$ is at most $2^{-\lambda}$-far from the uniform distribution over $\mathbb{Z}_2^n$. Then, for any $\mathbf{x} \in \{0,1\}^L$, the distribution of $\mathbf{c} = \mathbf{B}_0 \cdot \mathsf{RE}(\mathbf{x}) \oplus \mathbf{B}_1 \cdot \mathsf{RE}(\mathbf{s})$ is statistically close to uniform over $\mathbb{Z}_2^n$. As a result, the scheme is statistically hiding.

Remark 1. As for the lattice-based commitment scheme from [42], we can extend the message space of our scheme to $\{0,1\}^L$ for arbitrary $L = \mathsf{poly}(\lambda)$ using the Merkle-Damgard technique together with the AFS hash function.

3.2 Techniques for Handling Well-Formed Regular Words

In our ZKAoK of a valid opening for the given commitment scheme, which will be presented in Sect. 3.3, as well as in all subsequent argument systems of this paper, we need a special mechanism allowing to prove the correctness of the (non-linear) encoding process from $\mathbf{v} \in \{0,1\}^m$, for some $m \in \mathbb{Z}^+$, to regular word $\mathbf{y} = \mathsf{RE}(\mathbf{v}) \in \mathsf{Regular}(m, c)$, where c divides m. To this end, we introduce the following notations and techniques.

- Let $c \in \mathbb{Z}^+$. For every $\mathbf{s} = (s_1, \ldots, s_c) \in \{0,1\}^c$, define the permutation $E_{\mathbf{s}}$ that transforms vector $\mathbf{x} = (x_{0,0,\ldots,0}, \ldots, x_{i_1,\ldots,i_c}, \ldots, x_{1,1,\ldots,1}) \in \{0,1\}^{2^c}$ into vector $E_{\mathbf{s}}(\mathbf{x}) = (x'_{0,0,\ldots,0}, \ldots, x'_{1,1,\ldots,1})$, where for each $(i_1, \ldots, i_c) \in \{0,1\}^c$, we have $x_{i_1,\ldots,i_c} = x'_{i_1 \oplus s_1, \ldots, i_c \oplus s_c}$.
  Note that, for any $\mathbf{s}, \mathbf{v} \in \{0,1\}^c$, we have:

$$\mathbf{x} = \Delta_{2^c}(\mathbf{v}) \iff E_{\mathbf{s}}(\mathbf{x}) = \Delta_{2^c}(\mathbf{v} \oplus \mathbf{s}). \tag{6}$$

- For $\mathbf{t} = (\mathbf{t}_1 \| \ldots \| \mathbf{t}_{n/c}) \in \{0,1\}^n$ consisting of n/c blocks of length c, define the permutation $E'_{\mathbf{t}}$ that transforms vector $\mathbf{y} = (\mathbf{y}_1 \| \ldots \| \mathbf{y}_{n/c}) \in \{0,1\}^{2^c \cdot n/c}$ consisting of n/c blocks of length 2^c into vector of the following form $E'_{\mathbf{t}}(\mathbf{y}) = (E_{\mathbf{t}_1}(\mathbf{y}_1) \| \ldots \| E_{\mathbf{t}_{n/c}}(\mathbf{y}_{n/c}))$. Note that, for any $\mathbf{t}, \mathbf{v} \in \{0,1\}^n$, we have:

$$\mathbf{y} = \mathsf{RE}(\mathbf{v}) \iff E'_{\mathbf{t}}(\mathbf{y}) = \mathsf{RE}(\mathbf{v} \oplus \mathbf{t}). \tag{7}$$

The equivalence in (7) enables us to prove that $\mathbf{y}$ is the correct encoding of $\mathbf{v}$, as follows. The prover samples uniformly random $\mathbf{t}$ and demonstrates to the verifier that the right-hand side of (7) holds. The verifier is thus convinced that its left-hand side also holds, while learning no additional information about $\mathbf{v}$, thanks to the "one-time pad" $\mathbf{t}$.

3.3 ZKAoK of a Valid Opening

We now describe a ZKAoK of a valid opening for the commitment scheme from Sect. 3.1. Specifically, we consider the relation R_{com}, defined as:

$$R_{\mathsf{com}} = \big\{\big((\mathbf{B} = [\mathbf{B}_0 \mid \mathbf{B}_1], \mathbf{c}), \mathbf{x}, \mathbf{s}\big) : \mathbf{B}_0 \cdot \mathsf{RE}(\mathbf{x}) + \mathbf{B}_1 \cdot \mathsf{RE}(\mathbf{s}) = \mathbf{c}\big\}.$$

The protocol is realized based on the permuting technique of Sect. 3.2. Let $\mathbf{z} = (\mathbf{x} \parallel \mathbf{s}) \in \{0,1\}^{L+k}$ and $\mathbf{w}_{\mathsf{com}} = \mathsf{RE}(\mathbf{z}) \in \mathsf{Regular}(L+k, c) \subset \{0,1\}^{2^c \cdot (L+k)/c}$. Then the equation $\mathbf{B}_0 \cdot \mathsf{RE}(\mathbf{x}) + \mathbf{B}_1 \cdot \mathsf{RE}(\mathbf{s}) = \mathbf{c}$ can be written as $\mathbf{B} \cdot \mathbf{w}_{\mathsf{com}} = \mathbf{c}$.

Next, define the sets $\mathsf{VALID}_{\mathsf{com}} = \mathsf{Regular}(L+k, c)$ and $\mathcal{S} = \{0,1\}^{L+k}$. For $\mathbf{t} = (\mathbf{t}_1 \| \dots \| \mathbf{t}_{(L+k)/c}) \in \mathcal{S}$ consisting of $(L+k)/c$ blocks of length c, it follows from (7) that we have

$$\mathbf{w}_{\mathsf{com}} = \mathsf{RE}(\mathbf{z}) \iff E'_{\mathbf{t}}(\mathbf{w}_{\mathsf{com}}) = \mathsf{RE}(\mathbf{z} \oplus \mathbf{t}). \tag{8}$$

Moreover, if $\mathbf{t}$ is chosen uniformly at random in $\mathcal{S}$, then $E'_{\mathbf{t}}(\mathbf{w}_{\mathsf{com}})$ is uniformly random in $\mathsf{VALID}_{\mathsf{com}}$. In other words, the conditions of (3) hold, and relation R_{com} can be reduced to an instance of R_{abstract} in Sect. 2.2. As a result, we can run the interactive protocol in Fig. 1 with public input $(\mathbf{B}, \mathbf{c})$ and prover's witness $\mathbf{w}_{\mathsf{com}}$, and obtain a ZKAoK for R_{com}.

4 Range Arguments for Signed Fractional Numbers

In this section, we present our techniques for obtaining zero-knowledge arguments that *signed fractional numbers*, committed via the code-based commitment scheme of Sect. 3, belong to a (hidden or given) range. We first describe our method for handling signed fractional numbers and establish the crucial theoretical foundations of our range arguments in Sect. 4.1. Next, in Sect. 4.2, we present our protocol for proving in zero-knowledge that two committed signed fractional numbers X, Y satisfies the inequality $X \leq Y$. In Sect. 4.3, we then discuss how to handily derive various variants of range arguments, based on the results of Sects. 4.1 and 4.2.

NOTATIONS. For $a, b \in \mathbb{Z}$ and $c \in \mathbb{Q}$, we let $[a, b]$ denote the set of all integers between a and b (inclusive), and let $c \cdot [a, b]$ denote the set $\{c \cdot x \mid x \in [a, b]\}$.

4.1 A Treatment of Signed Fractional Numbers

We will work with signed fractional numbers represented in fixed-point binary format. For integers $\ell > 0$, $f \geq 0$, define the set

$$\mathbb{Q}\langle \ell \bullet f \rangle = 2^{-f} \cdot [-2^{\ell+f}, 2^{\ell+f} - 1]$$

$$= \big\{ -2^{\ell} \cdot x_{\ell} + \sum_{i=-f}^{\ell-1} 2^i \cdot x_i \mid (x_{\ell}, x_{\ell-1}, \dots, x_0, x_{-1}, \dots, x_{-f}) \in \{0,1\}^{1+\ell+f} \big\}.$$

Each element $X \in \mathbf{Q}\langle \ell \bullet f \rangle$ can be represented as $x_\ell x_{\ell-1} \ldots x_0 \bullet x_{-1} x_{-2} \ldots x_{-f}$, where x_ℓ is the sign bit, $x_{\ell-1}, \ldots, x_0$ are the integer bits, and $x_{-1}, \ldots, x_{-f}$ are the fractional bits.

The binary vector $(x_\ell, x_{\ell-1}, \ldots, x_0, x_{-1}, \ldots, x_{-f}) \in \{0,1\}^{1+\ell+f}$ representing X is denoted as $\mathsf{sbin}_{\ell,f}(A)$. For notational convenience, we write $A = \mathsf{sbin}_{\ell,f}^{-1}(\mathbf{a})$ if $\mathbf{a} = \mathsf{sbin}_{\ell,f}(A)$. In this way, we have $\mathbf{Q}\langle \ell \bullet f \rangle = \{\mathsf{sbin}_{\ell,f}^{-1}(\mathbf{a}) \mid \mathbf{a} \in \{0,1\}^{1+\ell+f}\}$.

We aim to prove in zero-knowledge inequality relations among elements of $\mathbf{Q}\langle \ell \bullet f \rangle$, e.g., to prove that $X \leq Y$ for secret/committed X, Y. As we have discussed in Sect. 1, handling inequalities over $\mathbf{Q}\langle \ell \bullet f \rangle$ is highly non-trivial, due to 2 main reasons: the existence of the signed bit and the problem of overflows.

Our idea is to derive necessary and sufficient conditions for $X \leq Y$, with $X, Y \in \mathbf{Q}\langle \ell \bullet f \rangle$, in a way such that these conditions can be correctly and efficiently proved in zero-knowledge. Theorem 2 captures this idea via the existence of $2(\ell + f + 1)$ extra bits that are related to the bits representing X and Y via $2(\ell + f) + 3$ simple equations modulo 2. This result lays the vital foundation for the argument system we will construct in Sect. 4.2.

Theorem 2. *Let* $X, Y \in \mathbf{Q}\langle \ell \bullet f \rangle$ *and let* $\mathsf{sbin}_{\ell,f}(X) = (x_\ell, \ldots, x_0, x_{-1}, \ldots, x_{-f})$, $\mathsf{sbin}_{\ell,f}(Y) = (y_\ell, \ldots, y_0, y_{-1}, \ldots, y_{-f})$. *Then,* $X \leq Y$ *if and only if there exist bits* $z_\ell, z_{\ell-1}, \ldots, z_0, z_{-1}, \ldots, z_{-f}$, $c_{\ell+1}, c_\ell, c_{\ell-1}, \ldots, c_0, c_{-1}, \ldots, c_{-f+1}$ *satisfying*

$$
\begin{cases}
c_{-f+1} = x_{-f} \cdot z_{-f} \\
c_i = x_{i-1} \cdot z_{i-1} \ \oplus \ y_{i-1} \cdot c_{i-1} \ \oplus \ c_{i-1}, \quad \forall i \in [-f+2, \ell+1] \\
y_{-f} = x_{-f} \ \oplus \ z_{-f} \\
y_i = x_i \ \oplus \ z_i \oplus c_i, \quad \forall i \in [-f+1, \ell] \\
y_\ell = x_\ell \ \oplus \ c_{\ell+1}.
\end{cases}
\tag{9}
$$

Before proving Theorem 2, we first introduce a few notations, definitions and a technical lemma.

Additions. To avoid the problem of overflows, we will treat elements of $\mathbf{Q}\langle \ell \bullet f \rangle$ as elements of $\mathbf{Q}\langle (\ell+2) \bullet f \rangle$. If $X \in \mathbf{Q}\langle \ell \bullet f \rangle$ with $\mathsf{sbin}_{\ell,f}(X) = (x_\ell, x_{\ell-1}, \ldots, x_{-f})$ then we have

$$
A = -2^\ell \cdot x_\ell + \sum_{i=-f}^{\ell-1} 2^i \cdot x_i = (-2^{\ell+2} + 2^{\ell+1} + 2^\ell) \cdot x_\ell + \sum_{i=-f}^{\ell-1} 2^i \cdot x_i,
$$

and thus $\mathsf{sbin}_{\ell+2,f}(X) = (x_\ell, x_\ell, x_\ell, x_{\ell-1}, \ldots, x_{-f}) \in \{0,1\}^{3+\ell+f}$.

Now, let $X, Z \in \mathbf{Q}\langle (\ell+2) \bullet f \rangle$. The addition of X and Z when executed in a conventional computer is indeed the addition of two fractional binary numbers whose decimal encodings are equal to X and Z. Such addition is formally defined as follows.

Definition 2 (Signed Fractional Additions in Binary). *Let* X, Z *be elements of* $\mathbf{Q}\langle (\ell+2) \bullet f \rangle$. *The sum* $\mathsf{sbin}_{\ell+2,f}(X) \boxplus_{\ell+2,f} \mathsf{sbin}_{\ell+2,f}(Z)$ *is a vector*

$\mathbf{y} = (y_{\ell+2}, y_{\ell+1}, \ldots, y_{-f})$ associated with a vector $\mathbf{c} = (c_{\ell+2}, c_{\ell+1}, \ldots, c_{-f+1})$ such that

$$\begin{cases} c_{-f+1} = x_{-f} \cdot z_{-f} \\ c_i = x_{i-1} \cdot z_{i-1} \oplus y_{i-1} \cdot c_{i-1} \oplus c_{i-1}, \quad \forall i \in [-f+2, \ell+2] \\ y_{-f} = x_{-f} \oplus z_{-f} \\ y_i = x_i \oplus z_i \oplus c_i, \quad \forall i \in [-f+1, \ell+2]. \end{cases}$$

The above definition is similar to computing sum of two's complement integers up to a scaled factor [61, Sect. 3.2]. Intuitively, Definition 2 can be viewed as a binary addition of two binary numbers $\mathbf{x}, \mathbf{y}$ while sequence $\mathbf{c}$ plays a role as a sequence of carries computed in each step (the last carry-out bit is discarded).

	$c_{\ell+2}$	$c_{\ell+1}$	$\cdots$	c_1	c_0	$\cdots$	c_{-f+1}	0
	$x_{\ell+2}$	$x_{\ell+1}$	$\cdots$	x_1	x_0	$\cdots$	x_{-f+1}	x_{-f}
$+$	$z_{\ell+2}$	$z_{\ell+1}$	$\cdots$	z_1	z_0	$\cdots$	z_{-f+1}	z_{-f}
	$y_{\ell+2}$	$y_{\ell+1}$	$\cdots$	y_1	y_0	$\cdots$	y_{-f+1}	y_{-f}

It is clear that $y_{-f} = x_{-f} \oplus z_{-f}$ and, $\forall i \in [-f+1, \ell+2]$, $y_i = x_i \oplus z_i \oplus c_i$. Regarding computing carries, $c_i = (x_{i-1} \cdot z_{i-1}) \oplus (c_{i-1} \cdot (x_{i-1} \oplus z_{i-1}))$, $\forall i \in [-f+1, \ell+2]$ and $c_{-f} = 0$. It is easy to verify that, $\forall i \in [-f+1, \ell+2]$, $c_i = x_{i-1} \cdot z_{i-1} \oplus y_{i-1} \cdot c_{i-1} \oplus c_{i-1}$.

Overflows. Note that $\mathbf{Q}\langle(\ell+2) \bullet f\rangle = 2^{-f} \cdot [-2^{2+\ell+f}, 2^{2+\ell+f} - 1]$. For $X, Y \in \mathbf{Q}\langle(\ell+2) \bullet f\rangle$, where $X+Y$ falls out of the range $2^{-f} \cdot [-2^{2+\ell+f}, 2^{2+\ell+f} - 1]$, i.e., $X + Y < -2^{2+\ell}$ or $X + Y > 2^{2+\ell} - 2^{-f}$, the addition would yield an overflow. This phenomenon is formally defined as follows.

Definition 3 (Overflows). *Let $X, Y \in \mathbf{Q}\langle(\ell+2) \bullet f\rangle$ and let $\mathbf{x} = \mathsf{sbin}_{\ell+2,f}(X)$, $\mathbf{y} = \mathsf{sbin}_{\ell+2,f}(Y)$. The signed fractional addition $\mathbf{x} \boxplus_{\ell+2,f} \mathbf{y}$ is called to cause an overflow (with respect to $\ell + 2$ and f) if and only if $X + Y < -2^{2+\ell}$ or $X + Y > 2^{2+\ell} - 2^{-f}$.*

The following lemma implies that, if we are given $X, Y \in \mathbf{Q}\langle\ell \bullet f\rangle$ but we compute their sum over $\mathbf{Q}\langle(\ell+2) \bullet f\rangle$, then we can avoid overflows, and hence, can reliably capture the inequality $X \leq Y$ via addition.

Lemma 3. *Let $X, Y \in \mathbf{Q}\langle\ell \bullet f\rangle \subset \mathbf{Q}\langle(\ell+2) \bullet f\rangle$. Then $X \leq Y$ if and only if $Z = Y - X \in 2^{-f} \cdot [0, 2^{1+\ell+f} - 1] \subset \mathbf{Q}\langle(\ell+2) \bullet f\rangle$ and $\mathsf{sbin}_{\ell+2,f}(X) \boxplus_{\ell+2,f} \mathsf{sbin}_{\ell+2,f}(Z)$ does not cause an overflow. As a corollary, $\mathsf{sbin}_{\ell+2,f}(X) \boxplus_{\ell+2,f} \mathsf{sbin}_{\ell+2,f}(Z) = \mathsf{sbin}_{\ell+2,f}(Y)$ and $\mathsf{sbin}_{\ell+2,f}(Z) = (0, 0, z_\ell, z_{\ell-1}, \ldots, z_{-f})$.*

Proof. Assume that $X \leq Y$ and let $Z = Y - X$ (over $\mathbb{Q}$). Since we have $X, Y \in 2^{-f} \cdot [-2^{\ell+f}, 2^{\ell+f} - 1]$, it follows that

$$Z \in 2^{-f} \cdot [0, 2^{1+\ell+f} - 1] \subset 2^{-f} \cdot [-2^{\ell+2+f}, 2^{\ell+2+f} - 1] = \mathbf{Q}\langle(\ell+2) \bullet f\rangle.$$

Furthermore, $X + Z \in 2^{-f} \cdot [-2^{\ell+f}, 2^{1+\ell+f} + 2^{\ell+f} - 2]$. Hence, we have $-2^{\ell+2} < X + Z < 2^{\ell+2} - 2^{-f}$. As a result, the addition $\mathsf{sbin}_{\ell+2,f}(X) \boxplus_{\ell+2,f} \mathsf{sbin}_{\ell+2,f}(Z)$ does not cause an overflow and produces the correct result $\mathsf{sbin}_{\ell+2,f}(Y)$.

For the reverse direction, if $\mathsf{sbin}_{\ell+2,f}(X) \boxplus_{\ell+2,f} \mathsf{sbin}_{\ell+2,f}(Z)$ does not cause an overflow, then $\mathsf{sbin}_{\ell+2,f}(X) \boxplus_{\ell+2,f} \mathsf{sbin}_{\ell+2,f}(Z) = \mathsf{sbin}_{\ell+2,f}(Y)$ and hence $X + Z = Y$. Moreover, since $Z \in 2^{-f} \cdot [0, 2^{1+\ell+f} - 1]$, we have $Z \geq 0$. It then follows that $X \leq Y$.

To see that the first two bits of $\mathsf{sbin}_{\ell+2,f}(Z)$ are 0, let $Z' := 2^f \cdot Z$. Then Z' is a non-negative integer in the range $[0, 2^{1+\ell+f} - 1] \subset [-2^{2+\ell+f}, 2^{2+\ell+f} - 1]$ which needs exactly $1 + \ell + f$ bits to store in place of $3 + \ell + f$ bits. Therefore, $\mathsf{sbin}_{\ell+2,f}(Z') = (0, 0, z'_{\ell+f}, z'_{\ell+f-1}, \ldots, z'_0)$ and thus $\mathsf{sbin}_{\ell+2,f}(Z) = (0, 0, z_\ell, z_{\ell-1}, \ldots, z_{-f})$. □

We are now ready to prove Theorem 2.

Proof of Theorem 2. We first assume that $X, Y \in \mathbf{Q}\langle \ell \bullet f \rangle$ and $X \leq Y$. Let $\mathbf{x} = \mathsf{sbin}_{\ell+2,f}(X) = (x_\ell, x_\ell, x_\ell, x_{\ell-1}, \ldots, x_{-f})$ and $\mathbf{y} = \mathsf{sbin}_{\ell+2,f}(Y) = (y_\ell, y_\ell, y_\ell, y_{\ell-1}, \ldots, y_{-f})$. By Lemma 3, $Z = Y - X \in \mathbf{Q}\langle (\ell + 2) \bullet f \rangle$ such that $\mathsf{sbin}_{\ell+2,f}(Z) = (0, 0, z_\ell, \ldots, z_{-f})$ and $\mathsf{sbin}_{\ell+2,f}(X) \boxplus_{\ell+2,f} \mathsf{sbin}_{\ell+2,f}(Z) = \mathsf{sbin}_{\ell+2,f}(Y)$ where the signed fractional addition does not cause an overflow. Let $\mathbf{c}$ be the sequence of carries as in Definition 2. It follows that

$$\begin{cases} c_{-f+1} = x_{-f} \cdot z_{-f} \\ c_i = x_{i-1} \cdot z_{i-1} \oplus y_{i-1} \cdot c_{i-1} \oplus c_{i-1}, \quad \forall i \in [-f+2, \ell+1] \\ c_{\ell+2} = y_\ell \cdot c_{\ell+1} \oplus c_{\ell+1} \\ y_{-f} = x_{-f} \oplus z_{-f} \\ y_i = x_i \oplus z_i \oplus c_i, \quad \forall i \in [-f+1, \ell] \\ y_\ell = x_\ell \oplus c_{\ell+i}, \quad \forall i \in [1,2]. \end{cases}$$

We can deduce that $c_{\ell+1} = y_\ell \oplus x_\ell = c_{\ell+2}$. Hence, we obtain the system of equations in (9).

We now prove the reverse direction. If there exists bits $z_\ell, z_{\ell-1}, \ldots, z_{-f}, c_{\ell+1}, c_\ell, \ldots, c_{-f+1}$ satisfying (9), then we can construct the following vectors: $\mathbf{x} = (x'_{\ell+2}, x'_{\ell+1}, \ldots, x'_{-f}) = (x_\ell, x_\ell, x_\ell, x_{\ell-1}, \ldots, x_{-f}), \mathbf{y} = (y'_{\ell+2}, y'_{\ell+1}, \ldots, y'_{-f}) = (y_\ell, y_\ell, y_\ell, y_{\ell-1}, \ldots, y_{-f}), \mathbf{z} = (z'_{\ell+2}, z'_{\ell+1}, \ldots, z'_{-f}) = (0, 0, z_\ell, z_{\ell-1}, \ldots, z_{-f})$, and $\mathbf{c} = (c'_{\ell+2}, c'_{\ell+1}, \ldots, c'_{-f+1}) = (c_{\ell+1}, c_{\ell+1}, c_\ell, c_{\ell-1}, \ldots, c_{-f+1})$. From the assumption, we deduce that

$$\begin{cases} c'_{-f+1} = x'_{-f} \cdot z'_{-f} \\ c'_i = x'_{i-1} \cdot z'_{i-1} \oplus y'_{i-1} \cdot c'_{i-1} \oplus c'_{i-1}, \quad \forall i \in [-f+2, \ell+1] \\ y'_{-f} = x'_{-f} \oplus z'_{-f} \\ y'_i = x'_i \oplus z'_i \oplus c'_i, \quad \forall i \in [-f+1, \ell+2]. \end{cases}$$

It remains to show that $c'_{\ell+2} = x'_{\ell+1} \cdot z'_{\ell+1} \oplus y'_{\ell+1} \cdot c'_{\ell+1} \oplus c'_{\ell+1}$. We have

$$y'_{\ell+1} \cdot c'_{\ell+1} \oplus c'_{\ell+1} = y_\ell \cdot c_{\ell+1} \oplus c_{\ell+1}$$
$$= y_\ell \cdot (x_\ell \cdot z_\ell \oplus y_\ell \cdot c_\ell \oplus c_\ell) \oplus c_{\ell+1} = y_\ell \cdot x_\ell \cdot z_\ell \oplus c_{\ell+1}.$$

We claim that $y_\ell \cdot x_\ell \cdot z_\ell = 0$. To prove this claim, we assume by contradiction that $y_\ell \cdot x_\ell \cdot z_\ell = 1$. This is equivalent to $x_\ell = z_\ell = y_\ell = 1$. Hence $c'_\ell = 1$

because $y'_\ell = x'_\ell \oplus z'_\ell \oplus c'_\ell$. This also implies that $x'_{\ell+1} = 1$ and $c'_{\ell+1} = 1$ because $c'_{\ell+1} = x'_\ell \cdot z'_\ell \oplus y'_\ell \cdot c'_\ell \oplus c'_\ell$. Since $y'_{\ell+1} = x'_{\ell+1} \oplus z'_{\ell+1} \oplus c'_{\ell+1}$ and $z'_{\ell+1} = 0$, we have $y'_{\ell+1} = 0 \neq y'_\ell$, which is a contradiction. Therefore, the claim follows. Thus $y'_{\ell+1} \cdot c'_{\ell+1} \oplus c'_{\ell+1} = c_{\ell+1}$ and, since $z'_{\ell+1} = 0$, we deduce that $x'_{\ell+1} \cdot z'_{\ell+1} \oplus y'_{\ell+1} \cdot c'_{\ell+1} \oplus c'_{\ell+1} = c_{\ell+1} = c'_{\ell+2}$.

By Definition 2, the above facts imply that $\mathbf{x} \boxplus_{\ell+2,f} \mathbf{z} = \mathbf{y}$. It is clear that $X = \mathsf{sbin}_{\ell+2,f}^{-1}(\mathbf{x})$ and $Y = \mathsf{sbin}_{\ell+2,f}^{-1}(\mathbf{y}) \in \mathbf{Q}\langle \ell \bullet f \rangle$. Let $Z := \mathsf{sbin}_{\ell+2,f}^{-1}(\mathbf{z}) \in 2^{-f} \cdot [0, 2^{1+\ell+f} - 1]$. By Definition 3, the addition $\mathbf{x} \boxplus_{\ell+2,f} \mathbf{z}$ does not cause an overflow. Therefore, $X + Z = Y$, and since $Z \geq 0$, we obtain that $X \leq Y$. This completes the proof. $\qquad\square$

We also obtain necessary and sufficient conditions for strict inequalities of elements in $\mathbf{Q}\langle \ell \bullet f \rangle$, which allow us to handle those of the form $X < Y$ in zero-knowledge. This result is stated in Theorem 3, whose proof follows the same lines as that of Theorem 2, and is thus omitted.

Theorem 3. *Let $X, Y \in \mathbf{Q}\langle \ell \bullet f \rangle$ and let $\mathsf{sbin}_{\ell,f}(X) = (x_\ell, \ldots, x_0, x_{-1}, \ldots, x_{-f})$, $\mathsf{sbin}_{\ell,f}(Y) = (y_\ell, \ldots, y_0, y_{-1}, \ldots, y_{-f})$. Then, $X < Y$ if and only if there exist bits $z_\ell, z_{\ell-1}, \ldots, z_0, z_{-1}, \ldots, z_{-f}, c_{\ell+1}, c_\ell, c_{\ell-1}, \ldots, c_0, c_{-1}, \ldots, c_{-f+1}$ satisfying*

$$
\begin{cases}
c_{-f+1} = x_{-f} \cdot z_{-f} \oplus y_{-f} \oplus 1 \\
c_i = x_{i-1} \cdot z_{i-1} \oplus y_{i-1} \cdot c_{i-1} \oplus c_{i-1}, \quad \forall i \in [-f+2, \ell+1] \\
y_{-f} = x_{-f} \oplus z_{-f} \oplus 1 \\
y_i = x_i \oplus z_i \oplus c_i, \quad \forall i \in [-f+1, \ell] \\
y_\ell = x_\ell \oplus c_{\ell+1}.
\end{cases}
$$

4.2 Proving Inequalities Between Committed Elements of $\mathbf{Q}\langle \ell \bullet f \rangle$

Let $\ell > 0, f \geq 0$ be integers, and let $L = 1 + \ell + f$. Consider the code-based commitment scheme of Sect. 3 with parameters n, k, c, m_0, m_1, m and L and commitment key $\mathbf{B} = [\mathbf{B}_0 \mid \mathbf{B}_1] \in \mathbb{Z}_2^{n \times m}$.

Let $X, Y \in \mathbf{Q}\langle \ell \bullet f \rangle$, whose binary representations

$$\mathbf{x} = \mathsf{sbin}_{\ell,f}(X) = (x_\ell, x_{\ell-1}, \ldots, x_0, x_{-1}, \ldots, x_{-f}) \in \{0, 1\}^L,$$
$$\mathbf{y} = \mathsf{sbin}_{\ell,f}(Y) = (y_\ell, y_{\ell-1}, \ldots, y_0, y_{-1}, \ldots, y_{-f}) \in \{0, 1\}^L.$$

are committed as

$$\mathbf{e}_x = \mathbf{B}_0 \cdot \mathsf{RE}(\mathbf{x}) \oplus \mathbf{B}_1 \cdot \mathsf{RE}(\mathbf{s}_x) \in \mathbb{Z}_2^n, \quad \mathbf{e}_y = \mathbf{B}_0 \cdot \mathsf{RE}(\mathbf{y}) \oplus \mathbf{B}_1 \cdot \mathsf{RE}(\mathbf{s}_y) \in \mathbb{Z}_2^n,$$

respectively. Our goal is to design an argument system allowing the prover to convince the verifier in zero-knowledge that the vectors $\mathbf{x}, \mathbf{y}$ committed in $\mathbf{e}_x, \mathbf{e}_y$ satisfy $\mathsf{sbin}_{\ell,f}^{-1}(\mathbf{x}) \leq \mathsf{sbin}_{\ell,f}^{-1}(\mathbf{y})$, i.e., they represent numbers $X, Y \in \mathbf{Q}\langle \ell \bullet f \rangle$ such that $X \leq Y$. Formally, we will build a ZKAoK for the relation R_{ineq} defined as follows.

$$
R_{\mathsf{ineq}} = \{((\mathbf{B} = [\mathbf{B}_0 \mid \mathbf{B}_1], \mathbf{e}_x, \mathbf{e}_y), \mathbf{x}, \mathbf{y}, \mathbf{s}_x, \mathbf{s}_y) : \mathsf{sbin}_{\ell,f}^{-1}(\mathbf{x}) \leq \mathsf{sbin}_{\ell,f}^{-1}(\mathbf{y}) \wedge \\
\mathbf{e}_x = \mathbf{B}_0 \cdot \mathsf{RE}(\mathbf{x}) \oplus \mathbf{B}_1 \cdot \mathsf{RE}(\mathbf{s}_x) \wedge \mathbf{e}_y = \mathbf{B}_0 \cdot \mathsf{RE}(\mathbf{y}) \oplus \mathbf{B}_1 \cdot \mathsf{RE}(\mathbf{s}_y)\}.
$$

To prove in zero-knowledge that the inequality $\mathsf{sbin}_{\ell,f}^{-1}(\mathbf{x}) \leq \mathsf{sbin}_{\ell,f}^{-1}(\mathbf{y})$ holds, we rely on the results established in Sect. 4.1. Specifically, based on Theorem 2, we can equivalently prove the existence of bits $z_\ell, z_{\ell-1}, \ldots, z_0, z_{-1}, \ldots, z_{-f}, c_{\ell+1}, c_\ell, c_{\ell-1}, \ldots, c_0, c_{-1}, \ldots, c_{-f+1}$ satisfying the following $2(\ell + f) + 3 = 2L + 1$ equations modulo 2:

$$\begin{cases} c_{-f+1} \ \oplus \ x_{-f} \cdot z_{-f} = 0, \\ c_i \ \oplus \ x_{i-1} \cdot z_{i-1} \ \oplus \ y_{i-1} \cdot c_{i-1} \ \oplus \ c_{i-1} = 0, \ \ \forall i \in [-f+2, \ell+1] \\ y_{-f} \ \oplus \ x_{-f} \ \oplus \ z_{-f} = 0, \\ y_i \ \oplus \ x_i \ \oplus \ z_i \ \oplus \ c_i = 0, \ \ \forall i \in [-f+1, \ell] \\ y_\ell \ \oplus \ x_\ell \ \oplus \ c_{\ell+1} = 0. \end{cases} \tag{10}$$

Now, to handle (10) in zero-knowledge, we can use the extending-then-permuting techniques of Sect. 2.3. To this end, we first perform the following extensions for each $i \in [-f, \ell]$

$$x_i \mapsto \mathbf{x}_i = \mathsf{enc}(x_i), \ y_i \mapsto \mathbf{y}_i = \mathsf{enc}(y_i), \ z_i \mapsto \mathbf{z}_i = \mathsf{enc}(z_i), \ c_{i+1} \mapsto \mathsf{enc}(c_{i+1}),$$

as well as the following extensions

$$\forall i \in [-f+1, \ell+1] : \ x_{i-1} \cdot z_{i-1} \mapsto \mathbf{t}_{i-1} = \mathsf{ext}(x_{i-1}, z_{i-1});$$
$$\forall i \in [-f+2, \ell+1] : \ y_{i-1} \cdot c_{i-1} \mapsto \mathbf{g}_{i-1} = \mathsf{ext}(y_{i-1}, c_{i-1}).$$

Let $\mathbf{M}_2 = (0, 1) \in \mathbb{Z}_2^{1 \times 2}$ and $\mathbf{M}_4 = (0, 0, 0, 1) \in \mathbb{Z}_2^{1 \times 4}$. Then (10) can be rewritten as

$$\begin{cases} \mathbf{M}_2 \cdot \mathbf{c}_{-f+1} \ \oplus \ \mathbf{M}_4 \cdot \mathbf{t}_{-f} = 0, \\ \mathbf{M}_2 \cdot \mathbf{c}_i \ \oplus \ \mathbf{M}_4 \cdot \mathbf{t}_{i-1} \ \oplus \ \mathbf{M}_4 \cdot \mathbf{g}_{i-1} \ \oplus \ \mathbf{M}_2 \cdot \mathbf{c}_{i-1} = 0, \ \ \forall i \in [-f+2, \ell+1] \\ \mathbf{M}_2 \cdot \mathbf{y}_{-f} \ \oplus \ \mathbf{M}_2 \ \mathbf{x}_{-f} \ \oplus \ \mathbf{M}_2 \cdot \mathbf{z}_{-f} = 0, \\ \mathbf{M}_2 \cdot \mathbf{y}_i \ \oplus \ \mathbf{M}_2 \cdot \mathbf{x}_i \ \oplus \ \mathbf{M}_2 \cdot \mathbf{z}_i \ \oplus \ \mathbf{M}_2 \cdot \mathbf{c}_i = 0, \ \ \forall i \in [-f+1, \ell] \\ \mathbf{M}_2 \cdot \mathbf{y}_\ell \ \oplus \ \mathbf{M}_2 \cdot \mathbf{x}_\ell \ \oplus \ \mathbf{M}_2 \cdot \mathbf{c}_{\ell+1} = 0, \end{cases}$$

which then can be combined via linear algebra into one equation of the form $\mathbf{M}_0 \cdot \mathbf{w}_0 = \mathbf{0}$, where $\mathbf{M}_0 \in \mathbb{Z}_2^{(2L+1) \times 16L}$ is a public matrix, and $\mathbf{w}_0 \in \{0, 1\}^{16L}$ has the form:

$$\begin{aligned} \mathbf{w}_0 = \ & \big(\mathbf{x}_\ell \parallel \ \cdots \ \parallel \mathbf{x}_{-f} \parallel \mathbf{y}_\ell \parallel \ \cdots \ \parallel \mathbf{y}_{-f} \parallel \mathbf{z}_\ell \parallel \ \cdots \ \parallel \mathbf{z}_{-f} \parallel \mathbf{c}_{\ell+1} \parallel \ \cdots \ \parallel \mathbf{c}_{-f+1} \parallel \\ & \mathbf{t}_\ell \parallel \ \cdots \ \parallel \mathbf{t}_{-f} \parallel \mathbf{g}_{\ell+1} \parallel \ \cdots \ \parallel \mathbf{g}_{-f+1} \big). \end{aligned} \tag{11}$$

Next, by combining equation $\mathbf{M}_0 \cdot \mathbf{w}_0 = \mathbf{0}$ with the two equations underlying the commitments $\mathbf{e}_x, \mathbf{e}_y$, we can obtain a unified equation of the form $\mathbf{M} \cdot \mathbf{w} = \mathbf{v}$, where $\mathbf{M} \in \mathbb{Z}_2^{(2L+2n+1) \times (16L+2m)}$ and $\mathbf{v} \in \mathbb{Z}_2^{2L+2n+1}$ are public, and $\mathbf{w} \in \{0, 1\}^{16L+2m}$ has the form

$$\mathbf{w} = \big(\ \mathbf{w}_0 \parallel \mathsf{RE}(\mathbf{x}) \parallel \mathsf{RE}(\mathbf{y}) \parallel \mathsf{RE}(\mathbf{s}_x) \parallel \mathsf{RE}(\mathbf{s}_y) \ \big). \tag{12}$$

At this point, we have translated our task into proving knowledge of a well-formed vector $\mathbf{w} \in \{0,1\}^{16L+2m}$ satisfying equation $\mathbf{M} \cdot \mathbf{w} = \mathbf{v}$. We next will reduce the latter task to an instance of the abstraction of Stern's protocol in Sect. 2.2. To this end, we will specify the sets $\mathsf{VALID}_{\mathsf{ineq}}, \mathcal{S}$ and permutations $\{\Gamma_\phi : \phi \in \mathcal{S}\}$ that meet the requirements in (3).

Define $\mathsf{VALID}_{\mathsf{ineq}}$ as the set of all vectors $\mathbf{w} \in \{0,1\}^{16L+2m}$ that has the form (12), where $\mathbf{s}_x, \mathbf{s}_y \in \{0,1\}^k$ and

$\diamond$ $\mathbf{x} = (x_\ell, \ldots, x_0, \ldots, x_{-f}) \in \{0,1\}^L$, $\mathbf{y} = (y_\ell, \ldots, y_0, \ldots, y_{-f}) \in \{0,1\}^L$;
$\diamond$ $\mathbf{w}_0$ has the form (11), where: for each $i \in [-f, \ell]$, there exist bits z_i, c_{i+1} satisfying
  (i) For each $i \in [-f, \ell]$, one has $\mathbf{x}_i = \mathsf{enc}(x_i)$, $\mathbf{y}_i = \mathsf{enc}(y_i)$, $\mathbf{z}_i = \mathsf{enc}(z_i)$, and $\mathbf{c}_{i+1} = \mathsf{enc}(c_{i+1})$;
  (ii) For each $i \in [-f+1, \ell+1]$, one has $\mathbf{t}_{i-1} = \mathsf{ext}(x_{i-1}, z_{i-1})$;
  (iii) For each $i \in [-f+2, \ell+1]$, one has $\mathbf{g}_{i-1} = \mathsf{ext}(y_{i-1}, c_{i-1})$.

It can be observed that the vector $\mathbf{w}$ we obtained from the above transformations is an element of this tailored set $\mathsf{VALID}_{\mathsf{ineq}}$. Next, we will employ the permuting techniques from Sects. 2.3 and 3.2 to handle the special constraint of $\mathbf{w}$.

Define $\mathcal{S}$ as the set $\{0,1\}^{4L+2k}$ and for each element $\phi \in \mathcal{S}$ of the form

$$\phi = \big(b_{x,\ell}, \ldots, b_{x,-f}, b_{y,\ell}, \ldots, b_{y,-f}, b_{z,\ell}, \ldots, b_{z,-f}, b_{c,\ell+1}, \ldots, b_{c,-f+1}, \mathbf{b}_{s,x}, \mathbf{b}_{s,y}\big),$$

where $\mathbf{b}_{s,x}, \mathbf{b}_{s,y} \in \{0,1\}^k$, define the permutation Γ_ϕ that, when applying to vector $\mathbf{t} \in \mathbb{Z}^{16L+2m}$, whose blocks are denoted as

$$\big(\; \mathbf{x}_\ell \parallel \; \cdots \; \parallel \mathbf{x}_{-f} \parallel \mathbf{y}_\ell \parallel \; \cdots \; \parallel \mathbf{y}_{-f} \parallel \mathbf{z}_\ell \parallel \; \cdots \; \parallel \mathbf{z}_{-f} \parallel \mathbf{c}_{\ell+1} \parallel \; \cdots \; \parallel \mathbf{c}_{-f+1} \parallel$$
$$\mathbf{t}_\ell \parallel \; \cdots \; \parallel \mathbf{t}_{-f} \parallel \mathbf{g}_{\ell+1} \parallel \; \cdots \; \parallel \mathbf{g}_{-f+1} \parallel \mathsf{RE}(\mathbf{x}) \parallel \mathsf{RE}(\mathbf{y}) \parallel \mathsf{RE}(\mathbf{s}_x) \parallel \mathsf{RE}(\mathbf{s}_y) \; \big),$$

it transforms $\mathbf{t}$ as follows:

$\diamond$ $\forall i \in [-f, \ell]$: $\mathbf{x}_i \mapsto F_{b_{x,i}}(\mathbf{x}_i)$; $\mathbf{y}_i \mapsto F_{b_{y,i}}(\mathbf{y}_i)$; $\mathbf{z}_i \mapsto F_{b_{z,i}}(\mathbf{z}_i)$; $\mathbf{c}_{i+1} \mapsto F_{b_{c,i+1}}(\mathbf{c}_{i+1})$
$\diamond$ $\forall i \in [-f+1, \ell+1]$: $\mathbf{t}_{i-1} \mapsto T_{b_{x,i-1}, b_{z,i-1}}(\mathbf{t}_{i-1})$
$\diamond$ $\forall i \in [-f+2, \ell+1]$: $\mathbf{g}_{i-1} \mapsto T_{b_{y,i-1}, b_{c,i-1}}(\mathbf{g}_{i-1})$
$\diamond$ $\mathsf{RE}(\mathbf{x}) \mapsto E'_{\mathbf{b}_x}(\mathsf{RE}(\mathbf{x}))$, where $\mathbf{b}_x = (b_{x,\ell}, \ldots, b_{x,-f})$.
$\diamond$ $\mathsf{RE}(\mathbf{y}) \mapsto E'_{\mathbf{b}_y}(\mathsf{RE}(\mathbf{y}))$, where $\mathbf{b}_y = (b_{y,\ell}, \ldots, b_{y,-f})$.
$\diamond$ $\mathsf{RE}(\mathbf{s}_x) \mapsto E'_{\mathbf{b}_{s,x}}(\mathsf{RE}(\mathbf{s}_x))$, $\mathsf{RE}(\mathbf{s}_y) \mapsto E'_{\mathbf{b}_{s,y}}(\mathsf{RE}(\mathbf{s}_y))$.

Based on the equivalences observed in (4), (5) and (7), one can verify that the requirements in (3) are satisfied. In other words, we have reduced the considered relation R_{ineq} to an instance of the relation R_{abstract} in Sect. 2.2.

The Interactive Protocol. Given the above preparations, our protocol now goes as follows.

- The public input consists of matrix $\mathbf{M}$ and vector $\mathbf{v}$, which are constructed from the original public input, as discussed above.
- The prover's witness consists of vector $\mathbf{w} \in \mathsf{VALID}_{\mathsf{ineq}}$, which is built from the initial secret input, as described above.

The prover and the verifier then interact as in Fig. 1. The protocol employs the statistically hiding and computationally binding string commitment scheme from Sect. 3 to obtain the desired statistical ZKAoK. As a corollary of Theorem 1, we obtain the following theorem, which summarized the properties our protocol for inequality relation between committed signed fractional numbers.

Theorem 4. *There exists a statistical zero-knowledge argument of knowledge for the relation R_{ineq} with perfect completeness, soundness error $2/3$ and communication cost $\mathcal{O}(L + m) = \mathcal{O}(\ell + f)$.*

For simulation, we simply invoke the simulator of Theorem 1. For extraction, we first run the knowledge extractor of Theorem 1 to obtain $\mathbf{w}' \in \mathsf{VALID}_{\mathsf{ineq}}$ such that $\mathbf{M} \cdot \mathbf{w}' = \mathbf{v}$. Then, by backtracking the transformations presented above, we can obtain $\mathbf{x}', \mathbf{y}', \mathbf{s}'_x, \mathbf{s}'_y$ such that $\mathsf{sbin}_{\ell,f}^{-1}(\mathbf{x}') \leq \mathsf{sbin}_{\ell,f}^{-1}(\mathbf{y}')$ and

$$\mathbf{e}_x = \mathbf{B}_0 \cdot \mathsf{RE}(\mathbf{x}') \oplus \mathbf{B}_1 \cdot \mathsf{RE}(\mathbf{s}'_x) \ \wedge \ \mathbf{e}_y = \mathbf{B}_0 \cdot \mathsf{RE}(\mathbf{y}') \oplus \mathbf{B}_1 \cdot \mathsf{RE}(\mathbf{s}'_y).$$

In particular, let $X' = \mathsf{sbin}_{\ell,f}^{-1}(\mathbf{x}') \in \mathbf{Q}\langle \ell \bullet f \rangle$ and $Y' = \mathsf{sbin}_{\ell,f}^{-1}(\mathbf{Y}') \in \mathbf{Q}\langle \ell \bullet f \rangle$, then we have $X' \leq Y'$.

4.3   Range Arguments

We now discuss how to use our results in Sects. 4.1 and 4.2 to derive various variants of range arguments for signed fractional numbers. Depending on the application contexts, different range types could be considered. Let us name a few of them.

1. **Hidden ranges with non-strict inequalities.** This requires to prove that $T \leq X \leq Y$, where T, X, Y are all committed. Such a range argument can be easily obtained by running two instances of our protocol of Sect. 4.2.
2. **Hidden ranges with strict inequalities.** In this setting, T, X, Y are hidden and the range relations are defined as $T < X < Y$, or $T \leq X < Y$. Here, a zero-knowledge argument of strict inequality is required. Such a protocol can be obtained by results of Theorem 3 and the techniques used in Sect. 4.2.
3. **Public ranges:** This type of range arguments is the easiest one, where A, B are publicly given and one proves that $A \leq X \leq B$ or $A < X < B$, for a committed number X. In fact, inequality $X \leq B$ can be handled using a simplified version of the protocol in Sect. 4.2, where the bits representing B are not required to be kept secret. Meanwhile, strict inequality $X < B$ can simply be interpreted as $X \leq B'$ for public $B' = B - 2^{-f}$.

In all cases considered above, the size of range arguments remains $\mathcal{O}(\ell + f)$, i.e., it is logarithmic in the size of the range.

5 Code-Based Accumulators and Logarithmic-Size Zero-Knowledge Arguments of Set Membership

In this section, we provide a code-based accumulator scheme supported by zero-knowledge argument of knowledge of an accumulated value. Such an argument system is essentially an argument of set membership, in which the prover convinces the verifier in zero-knowledge that his data item (e.g., his public key or his pseudonym) belongs to a given set, and is a highly desirable building block in various privacy-preserving applications. Our accumulator relies on a Merkle hash tree that is built from a suitable variant of the AFS hash function. To design a supporting zero-knowledge protocol for the hash tree, we first use the techniques for handling regular words from Sect. 3.2 to prove knowledge of hash preimages and images in the path from a leaf to the tree root, and then adapt Libert et al.'s method [45] to hide the bits determining steps in the path. As the tree depth is logarithmic in its size, we obtain an argument system for set membership with size logarithmic in the cardinality of the set.

In Sect. 5.1, we first recall the definitions and security requirement for cryptographic accumulators. Then, in Sect. 5.2, we modify the AFS hash function to support hashing with two inputs, upon which we build our Merkle-tree accumulator in Sect. 5.3. In Sect. 5.4, we describe our zero-knowledge argument system.

5.1 Cryptographic Accumulators

We recall the definitions and security requirement for accumulators.

Definition 4. *An accumulator scheme is a tuple of polynomial-time algorithms* (Setup, Accu, WitGen, Verify)*:*

- *Setup(1^λ) Given a security parameter 1^λ, outputs the public parameter pp.*
- *Accu$_{pp}(R)$ Take as input a set R with n data values as $R = \{\mathbf{d}_0, \ldots, \mathbf{d}_{N-1}\}$, outputs an accumulator value $\mathbf{u}$.*
- *WitGen$_{pp}(R, \mathbf{d})$ Take as input the set R and a value $\mathbf{d}$, outputs a witness w such that $\mathbf{d}$ is accumulated in TAcc(R), otherwise returns $\perp$ if $\mathbf{d} \notin R$.*
- *Verify$_{pp}(\mathbf{u}, (\mathbf{d}, w))$ This deterministic algorithm takes as inputs the accumulator value $\mathbf{u}$ and $(\mathbf{d}, w)$, outputs 1 if $(\mathbf{d}, w)$ is valid for the accumulator $\mathbf{u}$, otherwise returns 0 if invalid.*

Correctness. For all $pp \leftarrow$ Setup(n), the following holds:

$$\text{Verify}_{pp}\big(\text{Accu}_{pp}(R), \mathbf{d}, \text{WitGen}_{pp}(R, \mathbf{d})\big) = 1 \text{ for } \mathbf{d} \in R.$$

An accumulator is secure, as defined in [8,23], if it is infeasible to output a valid witness for a value $\mathbf{d}^*$ that is not chosen from the data value set.

Definition 5. *An accumulator scheme is secure if for all PPT adversaries $\mathcal{A}$:*

$$\Pr\big[pp \leftarrow \text{Setup}(\lambda); (R, \mathbf{d}^*, w^*) \leftarrow \mathcal{A}(pp) :$$
$$\mathbf{d}^* \notin R \wedge \text{Verify}_{pp}(\text{Accu}_{pp}(R), \mathbf{d}^*, w^*) = 1\big] = \text{negl}(\lambda).$$

5.2 Hashing with Two Inputs

We aim to build a Merkle-tree accumulator based on the AFS family of hash functions. Since in Merkle trees, every internal node is the hash of its two children nodes, we slightly modify the AFS hash functions so that the function takes two inputs instead of just one.

Definition 6. *Let* $m = 2 \cdot 2^c \cdot n/c$. *The function family* $\mathcal{H}$ *mapping* $\{0,1\}^n \times \{0,1\}^n$ *to* $\{0,1\}^n$ *is defined as* $\mathcal{H} = \{h_{\mathbf{B}} \mid \mathbf{B} \in \mathbb{Z}_2^{n \times m}\}$, *where for* $\mathbf{B} = [\mathbf{B}_0|\mathbf{B}_1]$ *with* $\mathbf{B}_0, \mathbf{B}_1 \in \mathbb{Z}_2^{n \times m/2}$, *and for any* $(\mathbf{u}_0, \mathbf{u}_1) \in \{0,1\}^n \times \{0,1\}^n$, *we have:*

$$h_{\mathbf{B}}(\mathbf{u}_0, \mathbf{u}_1) = \mathbf{B}_0 \cdot \mathsf{RE}(\mathbf{u}_0) \oplus \mathbf{B}_1 \cdot \mathsf{RE}(\mathbf{u}_1) \in \{0,1\}^n.$$

Lemma 4. *The function family* $\mathcal{H}$, *defined in Definition 6 is collision-resistant, assuming the hardness of the* 2-RNSD$_{n,2n,c}$ *problem.*

Proof. Given $\mathbf{B} = [\mathbf{B}_0|\mathbf{B}_1] \xleftarrow{\$} \mathbb{Z}_2^{n \times m}$, if one can find two distinct pairs $(\mathbf{u}_0, \mathbf{u}_1) \in \left(\{0,1\}^n\right)^2$ and $(\mathbf{v}_0, \mathbf{v}_1) \in \left(\{0,1\}^n\right)^2$ such that $h_{\mathbf{B}}(\mathbf{u}_0, \mathbf{u}_1) = h_{\mathbf{B}}(\mathbf{v}_0, \mathbf{v}_1)$, then one can obtain a non-zero vector $\mathbf{z} = \begin{pmatrix} \mathsf{RE}(\mathbf{u}_0) \oplus \mathsf{RE}(\mathbf{v}_0) \\ \mathsf{RE}(\mathbf{u}_1) \oplus \mathsf{RE}(\mathbf{v}_1) \end{pmatrix} \in$ 2-Regular$(2n, c)$ such that

$$\mathbf{B} \cdot \mathbf{z} = h_{\mathbf{B}}(\mathbf{u}_0, \mathbf{u}_1) \oplus h_{\mathbf{B}}(\mathbf{v}_0, \mathbf{v}_1) = \mathbf{0}.$$

In other words, $\mathbf{z}$ is a valid solution to the 2-RNSD$_{n,2n,c}$ problem associated with matrix $\mathbf{B}$. □

5.3 Code-Based Merkle-Tree Accumulator

We now describe our Merkle-tree accumulator based on the code-based hash functions $\mathcal{H}$ in Definition 6. The construction is adapted from the blueprint by Libert et al. [45].

Setup(λ). Given $n = \mathcal{O}(\lambda)$, $c = \mathcal{O}(1)$ and $m = 2 \cdot 2^c \cdot n/c$. Sample $\mathbf{B} \xleftarrow{\$} \mathbb{Z}_2^{n \times m}$, and output the public parameter $pp = \mathbf{B}$.

Accu$_{\mathbf{B}}(R = \{\mathbf{d}_0, \ldots, \mathbf{d}_{N-1}\} \subseteq (\{0,1\}^n)^N)$. Let the binary representation of j be $(j_1, \ldots, j_\ell) \in \{0,1\}^\ell$, re-write $\mathbf{d}_j$ as $\mathbf{u}_{j_1,\ldots,j_\ell}$. Build a binary tree with $N = 2^\ell$ leaves $\mathbf{u}_{0,0,\ldots,0}, \ldots, \mathbf{u}_{1,1,\ldots,1}$ in the following way:

1. At depth $i \in [1, \ell - 1]$, for the nodes $\mathbf{u}_{a_1,\ldots,a_i,0} \in \{0,1\}^n$ and $\mathbf{u}_{a_1,\ldots,a_i,1} \in \{0,1\}^n$, compute $h_{\mathbf{B}}(\mathbf{u}_{a_1,\ldots,a_i,0}, \mathbf{u}_{a_1,\ldots,a_i,1})$ and define it to be $\mathbf{u}_{a_1,\ldots,a_i}$ for all $(a_1, \ldots, a_i) \in \{0,1\}^i$.
2. At depth 0, for the nodes $\mathbf{u}_0 \in \{0,1\}^n$ and $\mathbf{u}_1 \in \{0,1\}^n$, compute $h_{\mathbf{B}}(\mathbf{u}_0, \mathbf{u}_1)$ and define it to be the root value $\mathbf{u}$.

Output the accumulated value $\mathbf{u}$.

WitGen$_{\mathbf{B}}(R, \mathbf{d})$. If $\mathbf{d} \notin R$, the algorithm outputs $\perp$. Otherwise, it outputs the witness w for $\mathbf{d}$ as follows.

1. Set $\mathbf{d} = \mathbf{d}_j$ for some $j \in [0, N-1]$. Re-write $\mathbf{d}_j$ as $\mathbf{u}_{j_1,\ldots,j_\ell}$ using the binary representation of the index j.

2. Consider the path from $\mathbf{u}_{j_1,\ldots,j_\ell}$ to the root $\mathbf{u}$, the witness w then consists of the binary representation $(j_1,\ldots,j_\ell)$ for j as well as all the sibling nodes of the path. Specifically,

$$w = \left((j_1,\ldots,j_\ell),(\mathbf{u}_{j_1,\ldots,j_{\ell-1},\bar{j}_\ell},\ldots,\mathbf{u}_{j_1,\bar{j}_2},\mathbf{u}_{\bar{j}_1})\right) \in \{0,1\}^\ell \times \left(\{0,1\}^n\right)^\ell,$$

$\mathsf{Verify}_\mathbf{B}(\mathbf{u},\mathbf{d},w)$. Let w be of the following form:

$$w = \left((j_1,\ldots,j_\ell),(\mathbf{w}_\ell,\ldots,\mathbf{w}_1)\right) \in \{0,1\}^\ell \times \left(\{0,1\}^n\right)^\ell.$$

This algorithm then computes $\mathbf{v}_\ell,\ldots,\mathbf{v}_0$. Let $\mathbf{v}_\ell = \mathbf{d}$ and

$$\forall i \in \{\ell-1,\ldots,1,0\}: \mathbf{v}_i = \begin{cases} h_\mathbf{B}(\mathbf{v}_{i+1},\mathbf{w}_{i+1}), & \text{if } j_{i+1}=0; \\ h_\mathbf{B}(\mathbf{w}_{i+1},\mathbf{v}_{i+1}), & \text{if } j_{i+1}=1. \end{cases}$$

Output 1 if $\mathbf{v}_0 = \mathbf{u}$ or 0 otherwise.

The correctness of the above Merkle-tree accumulator scheme follows immediately from the construction. Its security is based on the collision resistance of the hash function family $\mathcal{H}$: if an adversary can break the security of the accumulator, then one can find a solution to the 2-RNSD$_{n,2n,c}$ problem.

Theorem 5. *Assume that the* 2-RNSD$_{n,2n,c}$ *problem is hard, then the given accumulator scheme is secure.*

Proof. Assume that there exists a PPT adversary $\mathcal{B}$ who breaks the security of the given accumulator scheme with non-negligible probability. By Definition 5, $\mathcal{B}$ first receives a uniformly random matrix $\mathbf{B} \in \mathbb{Z}_2^{n \times m}$ generated by $\mathsf{Setup}(1^\lambda)$, and then outputs $(R = (\mathbf{d}_0,\ldots,\mathbf{d}_{N-1}),\mathbf{d}^*,w^*)$ such that $\mathbf{d}^* \notin R$ and $\mathsf{Verify}_\mathbf{B}(\mathbf{u}^*,\mathbf{d}^*,w^*) = 1$, where $\mathbf{u}^* = \mathsf{Accu}_\mathbf{B}(R)$.

Let $w^* = ((j_1^*,\ldots,j_\ell^*),(\mathbf{w}_\ell^*,\ldots,\mathbf{w}_1^*))$ and $(j_1^*,\ldots,j_\ell^*)$ be the binary representation of some index $j^* \in [0,N-1]$, then we can find a path that starts from $\mathbf{d}_{j^*}$ to the accumulated value $\mathbf{u}$: $\mathbf{u}_{j_1^*,\ldots,j_\ell^*} = \mathbf{d}_{j^*},\mathbf{u}_{j_1^*,\ldots,j_{\ell-1}^*},\ldots,\mathbf{u}_{j_1^*},\mathbf{u}^*$.

On the other hand, the algorithm $\mathsf{Verify}_\mathbf{B}(\mathbf{u}^*,\mathbf{d}^*,w^*)$ can compute another path: $\mathbf{k}_\ell^* = \mathbf{d}^*,\mathbf{k}_{\ell-1}^*,\ldots,\mathbf{k}_1^*,\mathbf{k}_0^* = \mathbf{u}^*$.

Since $\mathbf{d}^* \notin R$, then $\mathbf{d}^* \neq \mathbf{d}_{j^*}$. Comparing two paths, $\mathbf{d}_{j^*},\mathbf{u}_{j_1^*,\ldots,j_{\ell-1}^*},\ldots,$ $\mathbf{u}_{j_1^*},\mathbf{u}^*$ and $\mathbf{d}^*,\mathbf{k}_{\ell-1}^*,\ldots,\mathbf{k}_1^*,\mathbf{u}^*$, we can find the smallest integer $i \in [\ell]$, such that $\mathbf{k}_i^* \neq \mathbf{u}_{j_1^*,\ldots,j_i^*}$. So we obtain two distinct solutions to form a collision solution to the hash function $h_\mathbf{B}$ at the parent node of $\mathbf{u}_{j_1^*,\ldots,j_i^*}$. $\square$

5.4 Logarithmic-Size Arguments of Set Membership

In this section, we describe a statistical zero-knowledge argument that allows prover $\mathcal{P}$ to convince verifier $\mathcal{V}$ in zero-knowledge that $\mathcal{P}$ knows a value that was correctly accumulated into the root of the above code-based Merkle tree. Our protocol directly implies a logarithmic-size argument of set membership, with respect to the set of all data items accumulated into the tree root.

Given a uniformly random matrix $\mathbf{B} \in \mathbb{Z}_2^{n \times m}$ and the accumulated value $\mathbf{u} \in \{0,1\}^n$ as input, the goal of $\mathcal{P}$ is to convince $\mathcal{V}$ that it possesses a value $\mathbf{d}$ and a valid witness w. We define the associated relation $\mathrm{R}_{\mathrm{acc}}$ as follows:

$$\mathrm{R}_{\mathrm{acc}} = \Big\{ ((\mathbf{B}, \mathbf{u}) \in \mathbb{Z}_2^{n \times m} \times \{0,1\}^n; \mathbf{d} \in \{0,1\}^n, w \in \{0,1\}^\ell \times (\{0,1\}^n)^\ell) :$$

$$\mathsf{Verify}_\mathbf{B} (\mathbf{u}, \mathbf{d}, w) = 1 \Big\}.$$

Before constructing a ZKAoK for the above relation, we first present several additional permuting techniques, which are developed from [45] and from our permuting technique presented Sect. 3.2.

- For vector $\mathbf{b} = (b_1, \ldots, b_n) \in \{0,1\}^n$, where $n \in \mathbb{Z}^+$, denote by $\mathsf{Encode}(\mathbf{b})$ the vector $(\bar{b}_1, b_1, \ldots, \bar{b}_n, b_n) \in \{0,1\}^{2n}$.
- Let $\mathbf{I}_n^* \in \mathbb{Z}_2^{n \times 2n}$ be an extension of the identity matrix $\mathbf{I}_n$, obtained by inserting a zero-column $\mathbf{0}^n$ right before each of the columns of $\mathbf{I}_n$. Note that if $\mathbf{b} \in \{0,1\}^n$, then $\mathbf{b} = \mathbf{I}_n^* \cdot \mathsf{Encode}(\mathbf{b})$.
- For $\mathbf{t} = (t_1, \ldots, t_n)^\top \in \{0,1\}^n$, define the permutation $F_\mathbf{t}'$ that transforms vector $\mathbf{w} = (w_{1,0}, w_{1,1}, \ldots, w_{n,0}, w_{n,1})^\top \in \{0,1\}^{2n}$ into:

$$F_\mathbf{t}'(\mathbf{w}) = (w_{1,t_1}, w_{1,\bar{t}_1}, \ldots, w_{n,t_n}, w_{n,\bar{t}_n})^\top.$$

Note that, for any $\mathbf{t}, \mathbf{v} \in \{0,1\}^n$, we have:

$$\mathbf{w} = \mathsf{Encode}(\mathbf{b}) \iff F_\mathbf{t}'(\mathbf{w}) = \mathsf{Encode}(\mathbf{b} \oplus \mathbf{t}). \tag{13}$$

- For $b \in \{0,1\}$ and $\mathbf{v} \in \{0,1\}^{m/2}$, we denote $\mathsf{Ext}(b, \mathbf{v})$ as $\begin{pmatrix} \bar{b} \cdot \mathbf{v} \\ b \cdot \mathbf{v} \end{pmatrix}$.
- For $e \in \{0,1\}$, for $\mathbf{t} \in \{0,1\}^n$, define the permutation $\Psi_{e,\mathbf{t}}$ that acts on $\mathbf{z} = \begin{pmatrix} \mathbf{z}_0 \\ \mathbf{z}_1 \end{pmatrix} \in \{0,1\}^m$, where $\mathbf{z}_0, \mathbf{z}_1 \in \{0,1\}^{m/2}$, as follows. It transforms $\mathbf{z}$ to $\Psi_{e,\mathbf{t}}(\mathbf{z}) = \begin{pmatrix} E_\mathbf{t}'(\mathbf{z}_e) \\ E_\mathbf{t}'(\mathbf{z}_{\bar{e}}) \end{pmatrix}$. Namely, it rearranges the blocks of $\mathbf{z}$ according to e and permutes each block using $E_\mathbf{t}'$.
- For any $b, e \in \{0,1\}$ and $\mathbf{v}, \mathbf{w}, \mathbf{t} \in \{0,1\}^n$, it follows from (7) that the following equivalences hold:

$$\begin{cases} \mathbf{z} = \mathsf{Ext}(b, \mathsf{RE}(\mathbf{v})) \iff \Psi_{e,\mathbf{t}}(\mathbf{z}) = \mathsf{Ext}(b \oplus e, \mathsf{RE}(\mathbf{v} \oplus \mathbf{t})) \\ \mathbf{y} = \mathsf{Ext}(\bar{b}, \mathsf{RE}(\mathbf{w})) \iff \Psi_{\bar{e},\mathbf{t}}(\mathbf{y}) = \mathsf{Ext}(\bar{b} \oplus \bar{e}, \mathsf{RE}(\mathbf{w} \oplus \mathbf{t})). \end{cases} \tag{14}$$

Now let us examine the equations associated with the relation R_{acc}. Note that algorithm Verify computes the path $\mathbf{v}_\ell = \mathbf{d}, \mathbf{v}_{\ell-1}, \ldots, \mathbf{v}_1, \mathbf{v}_0 = \mathbf{u}$, where $\mathbf{v}_i$ for $i \in \{\ell-1, \ldots, 1, 0\}$ is computed as follows:

$$\mathbf{v}_i = \begin{cases} h_\mathbf{B}(\mathbf{v}_{i+1}, \mathbf{w}_{i+1}), & \text{if } j_{i+1} = 0; \\ h_\mathbf{B}(\mathbf{w}_{i+1}, \mathbf{v}_{i+1}), & \text{if } j_{i+1} = 1. \end{cases} \tag{15}$$

We translate Eq. (15) into the following equivalent form.

$$\begin{aligned}
\mathbf{v}_i &= \bar{j}_{i+1} \cdot h_{\mathbf{B}}(\mathbf{v}_{i+1}, \mathbf{w}_{i+1}) \oplus j_{i+1} \cdot h_{\mathbf{B}}(\mathbf{w}_{i+1}, \mathbf{v}_{i+1}) \\
&= \bar{j}_{i+1} \cdot (\mathbf{B}_0 \cdot \mathsf{RE}(\mathbf{v}_{i+1}) \oplus \mathbf{B}_1 \cdot \mathsf{RE}(\mathbf{w}_{i+1})) \oplus j_{i+1} \cdot (\mathbf{B}_0 \cdot \mathsf{RE}(\mathbf{w}_{i+1}) \oplus \mathbf{B}_1 \cdot \mathsf{RE}(\mathbf{v}_{i+1})) \\
&= \mathbf{B} \cdot \begin{pmatrix} \bar{j}_{i+1} \cdot \mathsf{RE}(\mathbf{v}_{i+1}) \\ j_{i+1} \cdot \mathsf{RE}(\mathbf{v}_{i+1}) \end{pmatrix} \oplus \mathbf{B} \cdot \begin{pmatrix} j_{i+1} \cdot \mathsf{RE}(\mathbf{w}_{i+1}) \\ \bar{j}_{i+1} \cdot \mathsf{RE}(\mathbf{w}_{i+1}) \end{pmatrix} \\
&= \mathbf{B} \cdot \mathsf{Ext}(j_{i+1}, \mathsf{RE}(\mathbf{v}_{i+1})) \oplus \mathbf{B} \cdot \mathsf{Ext}(\bar{j}_{i+1}, \mathsf{RE}(\mathbf{w}_{i+1}))
\end{aligned}$$

Therefore, to obtain a ZKAoK for R_{acc}, it is necessary and sufficient to construct an argument system in which $\mathcal{P}$ convinces $\mathcal{V}$ in zero-knowledge that $\mathcal{P}$ knows $j_1, \ldots, j_\ell \in \{0,1\}^\ell$ and $\mathbf{v}_1, \ldots, \mathbf{v}_\ell, \mathbf{w}_1, \ldots, \mathbf{w}_\ell \in \{0,1\}^n$ satisfying

$$\begin{cases}
\mathbf{B} \cdot \mathsf{Ext}(j_1, \mathsf{RE}(\mathbf{v}_1)) \oplus \mathbf{B} \cdot \mathsf{Ext}(\bar{j}_1, \mathsf{RE}(\mathbf{w}_1)) = \mathbf{u}; \\
\mathbf{B} \cdot \mathsf{Ext}(j_2, \mathsf{RE}(\mathbf{v}_2)) \oplus \mathbf{B} \cdot \mathsf{Ext}(\bar{j}_2, \mathsf{RE}(\mathbf{w}_2)) \oplus \mathbf{v}_1 = \mathbf{0}. \\
\qquad\qquad \cdots\cdots\cdots \\
\mathbf{B} \cdot \mathsf{Ext}(j_\ell, \mathsf{RE}(\mathbf{v}_\ell)) \oplus \mathbf{B} \cdot \mathsf{Ext}(\bar{j}_\ell, \mathsf{RE}(\mathbf{w}_\ell)) \oplus \mathbf{v}_{\ell-1} = \mathbf{0}.
\end{cases} \tag{16}$$

Next, we apply the function Encode defined above to vectors $\mathbf{v}_{\ell-1}, \ldots, \mathbf{v}_1$. Let $\mathbf{x}_i = \mathsf{Encode}(\mathbf{v}_i) \in \{0,1\}^{2n}$ for $i \in [\ell-1]$. Then we have $\mathbf{v}_i = \mathbf{I}_n^* \cdot \mathbf{x}_i$ for $i \in [\ell-1]$. For ease of notation, for $i \in [\ell]$, denote

$$\begin{cases}
\mathbf{y}_i = \mathsf{Ext}(j_i, \mathsf{RE}(\mathbf{v}_i)) \in \{0,1\}^m \\
\mathbf{z}_i = \mathsf{Ext}(\bar{j}_i, \mathsf{RE}(\mathbf{w}_i)) \in \{0,1\}^m
\end{cases} \tag{17}$$

Therefore, the equations in (16) is equivalent to the following.

$$\begin{cases}
\mathbf{B} \cdot \mathbf{y}_1 \oplus \mathbf{B} \cdot \mathbf{z}_1 = \mathbf{u}; \\
\mathbf{B} \cdot \mathbf{y}_2 \oplus \mathbf{B} \cdot \mathbf{z}_2 \oplus \mathbf{I}_n^* \cdot \mathbf{x}_1 = \mathbf{0}. \\
\qquad\qquad \cdots\cdots\cdots \\
\mathbf{B} \cdot \mathbf{y}_\ell \oplus \mathbf{B} \cdot \mathbf{z}_\ell \oplus \mathbf{I}_n^* \cdot \mathbf{x}_{\ell-1} = \mathbf{0}.
\end{cases} \tag{18}$$

Now, using linear algebra, we can transform the equations in (18) into a unifying equation of the form $\mathbf{M}_A \cdot \mathbf{w}_A = \mathbf{v}_A$, where $\mathbf{M}_A \in \mathbb{Z}_2^{\ell n \times L}$, $\mathbf{v}_A \in \mathbb{Z}_2^{\ell n}$ are public and $\mathbf{w}_A \in \{0,1\}^L$ is secret with $L = 2\ell m + 2(\ell-1)n$ and

$$\mathbf{w}_A = \begin{pmatrix} \mathbf{y}_1 \| \cdots \| \mathbf{y}_\ell \| \mathbf{z}_1 \| \cdots \| \mathbf{z}_\ell \| \mathbf{x}_1 \| \cdots \| \mathbf{x}_{\ell-1} \end{pmatrix} \tag{19}$$

At this point, let us specify the set VALID_A containing our secret vector $\mathbf{w}_A$, the set $\mathcal{S}_A$ and permutations $\{\Gamma_\phi : \phi \in \mathcal{S}_A\}$ such that the conditions in (3) hold. Let VALID_A be the set of all $\mathbf{w}_A' = (\mathbf{y}_1' \| \cdots \| \mathbf{y}_\ell' \| \mathbf{z}_1' \| \cdots \| \mathbf{z}_\ell' \| \mathbf{x}_1' \| \cdots \| \mathbf{x}_{\ell-1}') \in \{0,1\}^L$ satisfying the following conditions:

- For $i \in [\ell]$, there exists $\mathbf{v}_i', \mathbf{w}_i' \in \{0,1\}^n$, $j_i' \in \{0,1\}$ such that

$$\mathbf{y}_i' = \mathsf{Ext}(j_i', \mathsf{RE}(\mathbf{v}_i')) \in \{0,1\}^m, \quad \text{and} \quad \mathbf{z}_i' = \mathsf{Ext}(\bar{j}_i', \mathsf{RE}(\mathbf{w}_i')) \in \{0,1\}^m.$$

– For $i \in [\ell - 1]$, $\mathbf{x}'_i = \mathsf{Encode}(\mathbf{v}'_i) \in \{0,1\}^{2n}$.

Let $\mathcal{S}_A = \left(\{0,1\}^n\right)^\ell \times \left(\{0,1\}^n\right)^\ell \times \{0,1\}^\ell$. Then, for each element

$$\phi = \left(\, \mathbf{b}_1 \,\|\, \cdots \,\|\, \mathbf{b}_\ell \,\|\, \mathbf{e}_1 \,\|\, \cdots \,\|\, \mathbf{e}_\ell \,\|\, g_1 \,\|\, \cdots \,\|\, g_\ell \,\right) \in \mathcal{S}_A,$$

define the permutation Γ_ϕ that transforms

$$\mathbf{w}^*_A = \left(\mathbf{y}^*_1\|\cdots\|\mathbf{y}^*_\ell\|\mathbf{z}^*_1\|\cdots\|\mathbf{z}^*_\ell\|\mathbf{x}^*_1\|\cdots\|\mathbf{x}^*_{\ell-1}\right) \in \{0,1\}^L$$

with $\mathbf{y}^*_i, \mathbf{z}^*_i \in \{0,1\}^m$ for $i \in [\ell]$ and $\mathbf{x}^*_i \in \{0,1\}^{2n}$ for $i \in [\ell - 1]$ into

$$\Gamma_\phi(\mathbf{w}^*_A) = \left(\, \Psi_{g_1,\mathbf{b}_1}(\mathbf{y}^*_1)\|\cdots\|\Psi_{g_\ell,\mathbf{b}_\ell}(\mathbf{y}^*_\ell)\|\Psi_{\bar{g}_1,\mathbf{e}_1}(\mathbf{z}^*_1)\|\cdots\|\Psi_{\bar{g}_\ell,\mathbf{e}_\ell}(\mathbf{z}^*_\ell)\|\right.$$
$$\left. F'_{\mathbf{b}_1}(\mathbf{x}^*_1)\|\cdots\|F'_{\mathbf{b}_{\ell-1}}(\mathbf{x}^*_{\ell-1})\right).$$

Based on the equivalences observed in (13) and (14), it can be checked that the conditions in (3) are satisfied. We thus have reduced the considered relation into an instance of $\mathrm{R}_{\mathrm{abstract}}$.

The Interactive Protocol. Given the above preparations, our protocol goes as follows.

– The public input consists of matrix $\mathbf{M}_A$ and vector $\mathbf{v}_A$, which are constructed from the original public input, as discussed above.
– The prover's witness consists of vector $\mathbf{w}_A \in \mathsf{VALID}_A$, which is built from the initial secret input, as described above.

The prover and the verifier then interact as in Fig. 1. The protocol utilizes the statistically hiding and computationally binding string commitment scheme from Sect. 3 to obtain the desired statistical ZKAoK. The protocol has communication cost $\mathcal{O}(L) = \ell \cdot \mathcal{O}(m + n) = \mathcal{O}(\log N)$ bits and soundness error $2/3$.

6 Applications to Ring and Group Signatures

Our Merkle-tree accumulator together with its supporting zero-knowledge argument of set membership do enable a wide range of applications in code-based anonymity-oriented cryptographic protocols. In particular, these building blocks pave the way for the designs of logarithmic-size ring signatures and group signatures from code-based assumptions. In the following, we provide the high-level ideas of our constructions. The detailed descriptions and analyses can be found in the full version [56].

Ring signatures are arguably the most natural applications of accumulators, due to their decentralized setting and the observation that the ring signing procedure does capture a proof of ownership of a secret key corresponding to one of the public keys in the given ring. In our instantiation, the secret $\mathbf{x}$ of each user is an AFS hash preimage, while its image $\mathbf{d} = \mathbf{B} \cdot \mathsf{RE}(\mathbf{x})$ serves as the user's public key. To issue a signature with respect to a ring $R = \{\mathbf{d}_0, \ldots, \mathbf{d}_{N-1}\}$ containing his public key, the user builds a Merkle tree on top of R, and proves knowledge of an extended path of hash preimages from his own secret key to the leaf corresponding to his public key, and then, from there to the tree root. This can be done by extending the ZKAoK from Sect. 5.4 to handle one more layer of hashing. The obtained interactive zero-knowledge protocol is then repeated a sufficient number of times to achieve negligibly small soundness error, and then converted to a ring signature in the random oracle model via the Fiat-Shamir transformation [32]. The scheme is statistically anonymous and is unforgeable thanks to the security of the AFS hash function.

Building group signatures from accumulators is somewhat less intuitive. In fact, accumulators have been mainly used in group signatures for handling revocations. Libert et al. [45], however, showed that one in fact can design fully-anonymous group signatures from a Merkle-tree-based ring signature and a CCA2-secure encryption scheme where the latter admits a zero-knowledge argument of plaintext knowledge that is compatible with the supporting ZKAoK of the former. Since we have already obtained the ring signature block, to adapt the blueprint of [45], it remains to seek a suitable CCA2-secure encryption scheme and make them work together. To this end, we employ the Naor-Yung double encryption technique [55] to a randomized variant of the McEliece encryption scheme [50], suggested in [58]. The resulting CCA2-secure encryption mechanism is used to encrypt the identity of the signer - which is defined to be the $\log N$ bits determining the path from the tree leaf corresponding to the signer to the tree root. To complete the picture, we develop a Stern-like zero-knowledge layer for proving that such CCA2 ciphertexts are well-formed, which works smoothly with the zero-knowledge underlying the ring signature.

Acknowledgements. We thank Duong Hieu Phan, Benoît Libert, Nicolas Sendrier and Ayoub Otmani and the anonymous reviewers of ASIACRYPT 2019 for their comments and suggestions. The research is supported by the Singapore Ministry of Education under Research Grant MOE2016-T2-2-014(S). Khoa Nguyen is also supported by the Gopalakrishnan – NTU Presidential Postdoctoral Fellowship 2018.

References

1. Acar, T., Nguyen, L.: Revocation for delegatable anonymous credentials. In: Catalano, D., Fazio, N., Gennaro, R., Nicolosi, A. (eds.) PKC 2011. LNCS, vol. 6571, pp. 423–440. Springer, Heidelberg (2011). https://doi.org/10.1007/978-3-642-19379-8_26

2. Ajtai, M.: Generating hard instances of lattice problems (extended abstract). In: STOC 1996, pp. 99–108. ACM (1996)

3. Alamélou, Q., Blazy, O., Cauchie, S., Gaborit, P.: A code-based group signature scheme. Des. Codes Crypt. **82**(1–2), 469–493 (2017)

4. Applebaum, B., Haramaty, N., Ishai, Y., Kushilevitz, E., Vaikuntanathan, V.: Low-complexity cryptographic hash functions. In: ITCS 2017. LIPIcs, vol. 67, pp. 7:1–7:31. Schloss Dagstuhl - Leibniz-Zentrum fuer Informatik (2017)

5. Au, M.H., Wu, Q., Susilo, W., Mu, Y.: Compact e-cash from bounded accumulator. In: Abe, M. (ed.) CT-RSA 2007. LNCS, vol. 4377, pp. 178–195. Springer, Heidelberg (2006). https://doi.org/10.1007/11967668_12

6. Augot, D., Finiasz, M., Sendrier, N.: A fast provably secure cryptographic hash function. IACR Cryptology ePrint Archive, 2003:230 (2003)

7. Augot, D., Finiasz, M., Sendrier, N.: A family of fast syndrome based cryptographic hash functions. In: Dawson, E., Vaudenay, S. (eds.) Mycrypt 2005. LNCS, vol. 3715, pp. 64–83. Springer, Heidelberg (2005). https://doi.org/10.1007/11554868_6

8. Barić, N., Pfitzmann, B.: Collision-free accumulators and fail-stop signature schemes without trees. In: Fumy, W. (ed.) EUROCRYPT 1997. LNCS, vol. 1233, pp. 480–494. Springer, Heidelberg (1997). https://doi.org/10.1007/3-540-69053-0_33

9. Benaloh, J., de Mare, M.: One-way accumulators: a decentralized alternative to digital signatures. In: Helleseth, T. (ed.) EUROCRYPT 1993. LNCS, vol. 765, pp. 274–285. Springer, Heidelberg (1994). https://doi.org/10.1007/3-540-48285-7_24

10. Benhamouda, F., Camenisch, J., Krenn, S., Lyubashevsky, V., Neven, G.: Better zero-knowledge proofs for lattice encryption and their application to group signatures. In: Sarkar, P., Iwata, T. (eds.) ASIACRYPT 2014. LNCS, vol. 8873, pp. 551–572. Springer, Heidelberg (2014). https://doi.org/10.1007/978-3-662-45611-8_29

11. Bernstein, D.J., Lange, T., Peters, C., Schwabe, P.: Faster 2-regular information-set decoding. In: Chee, Y.M., et al. (eds.) IWCC 2011. LNCS, vol. 6639, pp. 81–98. Springer, Heidelberg (2011). https://doi.org/10.1007/978-3-642-20901-7_5

12. Bernstein, D.J., Lange, T., Peters, C., Schwabe, P.: Really fast syndrome-based hashing. In: Nitaj, A., Pointcheval, D. (eds.) AFRICACRYPT 2011. LNCS, vol. 6737, pp. 134–152. Springer, Heidelberg (2011). https://doi.org/10.1007/978-3-642-21969-6_9

13. Boneh, D., Eskandarian, S., Fisch, B.: Post-quantum group signatures from symmetric primitives. IACR Cryptology ePrint Archive, 2018:261 (2018)

14. Brakerski, Z., Lombardi, A., Segev, G., Vaikuntanathan, V.: Anonymous IBE, leakage resilience and circular security from new assumptions. In: Nielsen, J.B., Rijmen, V. (eds.) EUROCRYPT 2018. LNCS, vol. 10820, pp. 535–564. Springer, Cham (2018). https://doi.org/10.1007/978-3-319-78381-9_20

15. Brakerski, Z., Lyubashevsky, V., Vaikuntanathan, V., Wichs, D.: Worst-case hardness for LPN and cryptographic hashing via code smoothing. Electronic Collo quium on Computational Complexity (ECCC), 25:56 (2018)

16. Branco, P., Mateus, P.: A code-based linkable ring signature scheme. In: Baek, J., Susilo, W., Kim, J. (eds.) ProvSec 2018. LNCS, vol. 11192, pp. 203–219. Springer, Cham (2018). https://doi.org/10.1007/978-3-030-01446-9_12

17. Brassard, G., Chaum, D., Crépeau, C.: Minimum disclosure proofs of knowledge. J. Comput. Syst. Sci. **37**(2), 156–189 (1988)

18. Brickell, E.F., Chaum, D., Damgård, I.B., van de Graaf, J.: Gradual and verifiable release of a secret (extended abstract). In: Pomerance, C. (ed.) CRYPTO 1987. LNCS, vol. 293, pp. 156–166. Springer, Heidelberg (1988). https://doi.org/10.1007/3-540-48184-2_11

19. Camenisch, J., Chaabouni, R., shelat, a.: Efficient protocols for set membership and range proofs. In: Pieprzyk, J. (ed.) ASIACRYPT 2008. LNCS, vol. 5350, pp. 234–252. Springer, Heidelberg (2008). https://doi.org/10.1007/978-3-540-89255-7_15

20. Camenisch, J., Hohenberger, S., Lysyanskaya, A.: Compact e-cash. In: Cramer, R. (ed.) EUROCRYPT 2005. LNCS, vol. 3494, pp. 302–321. Springer, Heidelberg (2005). https://doi.org/10.1007/11426639_18

21. Camenisch, J., Kohlweiss, M., Soriente, C.: An accumulator based on bilinear maps and efficient revocation for anonymous credentials. In: Jarecki, S., Tsudik, G. (eds.) PKC 2009. LNCS, vol. 5443, pp. 481–500. Springer, Heidelberg (2009). https://doi.org/10.1007/978-3-642-00468-1_27

22. Camenisch, J., Lysyanskaya, A.: An efficient system for non-transferable anonymous credentials with optional anonymity revocation. In: Pfitzmann, B. (ed.) EUROCRYPT 2001. LNCS, vol. 2045, pp. 93–118. Springer, Heidelberg (2001). https://doi.org/10.1007/3-540-44987-6_7

23. Camenisch, J., Lysyanskaya, A.: Dynamic accumulators and application to efficient revocation of anonymous credentials. In: Yung, M. (ed.) CRYPTO 2002. LNCS, vol. 2442, pp. 61–76. Springer, Heidelberg (2002). https://doi.org/10.1007/3-540-45708-9_5

24. Chaabouni, R., Lipmaa, H., Zhang, B.: A non-interactive range proof with constant communication. In: Keromytis, A.D. (ed.) FC 2012. LNCS, vol. 7397, pp. 179–199. Springer, Heidelberg (2012). https://doi.org/10.1007/978-3-642-32946-3_14

25. Chaum, D., van Heyst, E.: Group signatures. In: Davies, D.W. (ed.) EUROCRYPT 1991. LNCS, vol. 547, pp. 257–265. Springer, Heidelberg (1991). https://doi.org/10.1007/3-540-46416-6_22

26. Couteau, G., Peters, T., Pointcheval, D.: Removing the strong RSA assumption from arguments over the integers. In: Coron, J.-S., Nielsen, J.B. (eds.) EUROCRYPT 2017. LNCS, vol. 10211, pp. 321–350. Springer, Cham (2017). https://doi.org/10.1007/978-3-319-56614-6_11

27. Dallot, L., Vergnaud, D.: Provably secure code-based threshold ring signatures. In: Parker, M.G. (ed.) IMACC 2009. LNCS, vol. 5921, pp. 222–235. Springer, Heidelberg (2009). https://doi.org/10.1007/978-3-642-10868-6_13

28. Derler, D., Ramacher, S., Slamanig, D.: Post-quantum zero-knowledge proofs for accumulators with applications to ring signatures from symmetric-key primitives. In: Lange, T., Steinwandt, R. (eds.) PQCrypto 2018. LNCS, vol. 10786, pp. 419–440. Springer, Cham (2018). https://doi.org/10.1007/978-3-319-79063-3_20

29. Dodis, Y., Kiayias, A., Nicolosi, A., Shoup, V.: Anonymous identification in *Ad Hoc* groups. In: Cachin, C., Camenisch, J.L. (eds.) EUROCRYPT 2004. LNCS, vol. 3027, pp. 609–626. Springer, Heidelberg (2004). https://doi.org/10.1007/978-3-540-24676-3_36

30. Esgin, M.F., Steinfeld, R., Liu, J.K., Liu, D.: Lattice-based zero-knowledge proofs: new techniques for shorter and faster constructions and applications. In: Boldyreva, A., Micciancio, D. (eds.) CRYPTO 2019. LNCS, vol. 11692, pp. 115–146. Springer, Cham (2019). https://doi.org/10.1007/978-3-030-26948-7_5

31. Ezerman, M.F., Lee, H.T., Ling, S., Nguyen, K., Wang, H.: A provably secure group signature scheme from code-based assumptions. In: Iwata, T., Cheon, J.H. (eds.) ASIACRYPT 2015. LNCS, vol. 9452, pp. 260–285. Springer, Heidelberg (2015). https://doi.org/10.1007/978-3-662-48797-6_12

32. Fiat, A., Shamir, A.: How to prove yourself: practical solutions to identification and signature problems. In: Odlyzko, A.M. (ed.) CRYPTO 1986. LNCS, vol. 263, pp. 186–194. Springer, Heidelberg (1987). https://doi.org/10.1007/3-540-47721-7_12

33. Gaborit, P., Hauteville, A., Phan, D.H., Tillich, J.-P.: Identity-based encryption from codes with rank metric. In: Katz, J., Shacham, H. (eds.) CRYPTO 2017. LNCS, vol. 10403, pp. 194–224. Springer, Cham (2017). https://doi.org/10.1007/978-3-319-63697-9_7

34. Goldreich, O., Micali, S., Wigderson, A.: How to prove all NP statements in zero-knowledge and a methodology of cryptographic protocol design (extended abstract). In: Odlyzko, A.M. (ed.) CRYPTO 1986. LNCS, vol. 263, pp. 171–185. Springer, Heidelberg (1987). https://doi.org/10.1007/3-540-47721-7_11

35. Goldwasser, S., Kalai, Y.T., Peikert, C., Vaikuntanathan, V.: Robustness of the learning with errors assumption. In: ICS 2010, pp. 230–240. Tsinghua University Press (2010)

36. Goldwasser, S., Micali, S., Rackoff, C.: The knowledge complexity of interactive proof systems. SIAM J. Comput. 18(1), 186–208 (1989)

37. González, A., Ráfols, C.: New techniques for non-interactive shuffle and range arguments. In: Manulis, M., Sadeghi, A.-R., Schneider, S. (eds.) ACNS 2016. LNCS, vol. 9696, pp. 427–444. Springer, Cham (2016). https://doi.org/10.1007/978-3-319-39555-5_23

38. Groth, J.: Evaluating security of voting schemes in the universal composability framework. In: Jakobsson, M., Yung, M., Zhou, J. (eds.) ACNS 2004. LNCS, vol. 3089, pp. 46–60. Springer, Heidelberg (2004). https://doi.org/10.1007/978-3-540-24852-1_4

39. Groth, J.: Non-interactive zero-knowledge arguments for voting. In: Ioannidis, J., Keromytis, A., Yung, M. (eds.) ACNS 2005. LNCS, vol. 3531, pp. 467–482. Springer, Heidelberg (2005). https://doi.org/10.1007/11496137_32

40. Groth, J.: Efficient zero-knowledge arguments from two-tiered homomorphic commitments. In: Lee, D.H., Wang, X. (eds.) ASIACRYPT 2011. LNCS, vol. 7073, pp. 431–448. Springer, Heidelberg (2011). https://doi.org/10.1007/978-3-642-25385-0_23

41. Jain, A., Krenn, S., Pietrzak, K., Tentes, A.: Commitments and efficient zero-knowledge proofs from learning parity with noise. In: Wang, X., Sako, K. (eds.) ASIACRYPT 2012. LNCS, vol. 7658, pp. 663–680. Springer, Heidelberg (2012). https://doi.org/10.1007/978-3-642-34961-4_40

42. Kawachi, A., Tanaka, K., Xagawa, K.: Concurrently secure identification schemes based on the worst-case hardness of lattice problems. In: Pieprzyk, J. (ed.) ASIACRYPT 2008. LNCS, vol. 5350, pp. 372–389. Springer, Heidelberg (2008). https://doi.org/10.1007/978-3-540-89255-7_23

43. Libert, B., Ling, S., Mouhartem, F., Nguyen, K., Wang, H.: Signature schemes with efficient protocols and dynamic group signatures from lattice assumptions. In: Cheon, J.H., Takagi, T. (eds.) ASIACRYPT 2016. LNCS, vol. 10032, pp. 373–403. Springer, Heidelberg (2016). https://doi.org/10.1007/978-3-662-53890-6_13

44. Libert, B., Ling, S., Mouhartem, F., Nguyen, K., Wang, H.: Zero-knowledge arguments for matrix-vector relations and lattice-based group encryption. In: Cheon, J.H., Takagi, T. (eds.) ASIACRYPT 2016. LNCS, vol. 10032, pp. 101–131. Springer, Heidelberg (2016). https://doi.org/10.1007/978-3-662-53890-6_4

45. Libert, B., Ling, S., Nguyen, K., Wang, H.: Zero-knowledge arguments for lattice-based accumulators: logarithmic-size ring signatures and group signatures without trapdoors. In: Fischlin, M., Coron, J.-S. (eds.) EUROCRYPT 2016. LNCS, vol. 9666, pp. 1–31. Springer, Heidelberg (2016). https://doi.org/10.1007/978-3-662-49896-5_1

46. Libert, B., Ling, S., Nguyen, K., Wang, H.: Lattice-based zero-knowledge arguments for integer relations. In: Shacham, H., Boldyreva, A. (eds.) CRYPTO 2018. LNCS, vol. 10992, pp. 700–732. Springer, Cham (2018). https://doi.org/10.1007/978-3-319-96881-0_24

47. Lipmaa, H.: On diophantine complexity and statistical zero-knowledge arguments. In: Laih, C.-S. (ed.) ASIACRYPT 2003. LNCS, vol. 2894, pp. 398–415. Springer, Heidelberg (2003). https://doi.org/10.1007/978-3-540-40061-5_26
48. Lipmaa, H., Asokan, N., Niemi, V.: Secure vickrey auctions without threshold trust. In: Blaze, M. (ed.) FC 2002. LNCS, vol. 2357, pp. 87–101. Springer, Heidelberg (2003). https://doi.org/10.1007/3-540-36504-4_7
49. Lyubashevsky, V., Micciancio, D.: Asymptotically efficient lattice-based digital signatures. J. Cryptol. **31**(3), 774–797 (2018)
50. McEliece, R.J.: A public-key cryptosystem based on algebraic coding theory. Deep Space Netw. Prog. Rep. **44**, 114–116 (1978)
51. Aguilar Melchor, C., Cayrel, P.-L., Gaborit, P.: A new efficient threshold ring signature scheme based on coding theory. In: Buchmann, J., Ding, J. (eds.) PQCrypto 2008. LNCS, vol. 5299, pp. 1–16. Springer, Heidelberg (2008). https://doi.org/10.1007/978-3-540-88403-3_1
52. Melchor, C.A., Cayrel, P.-L., Gaborit, P., Laguillaumie, F.: A new efficient threshold ring signature scheme based on coding theory. IEEE Trans. Inf. Theory **57**(7), 4833–4842 (2011)
53. Merkle, R.C.: A certified digital signature. In: Brassard, G. (ed.) CRYPTO 1989. LNCS, vol. 435, pp. 218–238. Springer, New York (1990). https://doi.org/10.1007/0-387-34805-0_21
54. Morozov, K., Takagi, T.: Zero-knowledge protocols for the McEliece encryption. In: Susilo, W., Mu, Y., Seberry, J. (eds.) ACISP 2012. LNCS, vol. 7372, pp. 180–193. Springer, Heidelberg (2012). https://doi.org/10.1007/978-3-642-31448-3_14
55. Naor, M., Yung, M.: Public-key cryptosystems provably secure against chosen ciphertext attacks. In: STOC 1990, pp. 427–437. ACM (1990)
56. Nguyen, K., Tang, H., Wang, H., Zeng, N.: New code-based privacy-preserving cryptographic constructions. IACR Cryptology ePrint Archive, 2019:513 (2019)
57. Nguyen, L.: Accumulators from bilinear pairings and applications. In: Menezes, A. (ed.) CT-RSA 2005. LNCS, vol. 3376, pp. 275–292. Springer, Heidelberg (2005). https://doi.org/10.1007/978-3-540-30574-3_19
58. Nojima, R., Imai, H., Kobara, K., Morozov, K.: Semantic security for the McEliece cryptosystem without random oracles. Des. Codes Crypt. **49**(1–3), 289–305 (2008)
59. Papamanthou, C., Shi, E., Tamassia, R., Yi, K.: Streaming authenticated data structures. In: Johansson, T., Nguyen, P.Q. (eds.) EUROCRYPT 2013. LNCS, vol. 7881, pp. 353–370. Springer, Heidelberg (2013). https://doi.org/10.1007/978-3-642-38348-9_22
60. Papamanthou, C., Tamassia, R., Triandopoulos, N.: Authenticated hash tables. In: ACM-CCS 2008, pp. 437–448. ACM (2008)
61. Patterson, D.A., Hennessy, J.L.: Computer Organization and Design, Fifth Edition: The Hardware/Software Interface, 5th edn. Morgan Kaufmann Publishers Inc., Burlington (2013)
62. Rivest, R.L., Shamir, A., Tauman, Y.: How to leak a secret. In: Boyd, C. (ed.) ASIACRYPT 2001. LNCS, vol. 2248, pp. 552–565. Springer, Heidelberg (2001). https://doi.org/10.1007/3-540-45682-1_32
63. Stern, J.: A new paradigm for public key identification. IEEE Trans. Inf. Theory **42**(6), 1757–1768 (1996)
64. Yu, Y., Zhang, J., Weng, J., Guo, C., Li, X.: Collision resistant hashing from learning parity with noise. IACR Cryptology ePrint Archive, 2017:1260 (2017)

Lattices (2)

An LLL Algorithm for Module Lattices

Changmin Lee[1], Alice Pellet-Mary[1], Damien Stehlé[1(✉)], and Alexandre Wallet[2]

[1] Univ. Lyon, EnsL, UCBL, CNRS, Inria, LIP, 69342 Lyon Cedex 07, France
changmin.lee@ens-lyon.fr
[2] NTT Secure Platform Laboratories, Tokyo, Japan

Abstract. The LLL algorithm takes as input a basis of a Euclidean lattice, and, within a polynomial number of operations, it outputs another basis of the same lattice but consisting of rather short vectors. We provide a generalization to R-modules contained in K^n for arbitrary number fields K and dimension n, with R denoting the ring of integers of K. Concretely, we introduce an algorithm that efficiently finds short vectors in rank-n modules when given access to an oracle that finds short vectors in rank-2 modules, and an algorithm that efficiently finds short vectors in rank-2 modules given access to a Closest Vector Problem oracle for a lattice that depends only on K. The second algorithm relies on quantum computations and its analysis is heuristic.

1 Introduction

The NTRU [HPS98], RingSIS [LM06, PR06], RingLWE [SSTX09, LPR10], ModuleSIS and ModuleLWE [BGV14, LS15] problems and their variants serve as security foundations of numerous cryptographic protocols. Their main advantages are their presumed quantum hardness, their flexibility for realizing advanced cryptographic functionalities, and their efficiency compared to their SIS and LWE counterparts [Ajt96, Reg09]. As an illustration of their popularity for cryptographic design, we note that 11 out of the 26 candidates at Round 2 of the NIST standardization process for post-quantum cryptography rely on these problems or variants thereof.[1] From a hardness perspective, these problems are best viewed as standard problems on Euclidean lattices, restricted to random lattices corresponding to modules over the rings of integers of number fields. Further, for some parametrizations, there exist reductions from and to standard worst-case problems for such module lattices [LS15, AD17, RSW18].

Let K be a number field and R its ring of integers. In this introduction, we will use the power-of-2 cyclotomic fields $K = \mathbb{Q}[x]/(x^d + 1)$ and their rings of integers $R = \mathbb{Z}[x]/(x^d + 1)$ as a running example (with d a power of 2). An R-module $M \subset K^n$ is a finitely generated subset of vectors in K^n that is stable under addition and multiplication by elements of R. As an example, if we consider $h \in R/qR$ for some integer q, the set $\{(f, g)^T \in R^2 : fh = g \bmod q\}$ is a module. If h is an NTRU public key, the corresponding secret key is a vector

---

[1] See https://csrc.nist.gov/projects/post-quantum-cryptography.

S. D. Galbraith and S. Moriai (Eds.): ASIACRYPT 2019, LNCS 11922, pp. 59–90, 2019.
https://doi.org/10.1007/978-3-030-34621-8_3

in that module, and its coefficients are small. Note that for $K = \mathbb{Q}$ and $R = \mathbb{Z}$, we recover Euclidean lattices in $\mathbb{Q}^n$. A first difficulty for handling modules compared to lattices is that R may not be a Euclidean domain, and, as a result, a module M may not be of the form $M = \sum_i R\mathbf{b}_i$ for some linearly independent $\mathbf{b}_i$'s in M. However, as R is a Dedekind domain, for every module M, there exist K-linearly independent $\mathbf{b}_i$'s and fractional ideals I_i such that $M = \sum I_i\mathbf{b}_i$ (see, e.g., [O'M63, Th. 81:3]). The set $((I_i, \mathbf{b}_i))_i$ is called a pseudo-basis of M. A module in K^n can always be viewed as a lattice in $\mathbb{C}^{nd}$ by mapping elements of K to $\mathbb{C}^d$ via the canonical embedding map (for our running example, it is equivalent to mapping a polynomial of degree $<d$ to the vector of its coefficients).

Standard lattice problems, such as finding a full-rank set of linearly independent short vectors in a given lattice, are presumed difficult, even in the context of quantum computations. In order to assess the security of cryptographic schemes based on NTRU/RingSIS/etc, an essential question is whether the restriction to module lattices brings vulnerabilities. Putting aside small polynomial speed-ups relying on the field automorphisms (multiplication by x in our running example), the cryptanalytic state of the art is to view the modules as arbitrary lattices, i.e., forgetting the module structure.

LLL [LLL82] is the central algorithm to manipulate lattice bases. It takes as input a basis of a given lattice, progressively updates it, and eventually outputs another basis of the same lattice that is made of relatively short vectors. Its run-time is polynomial in the input bit-length. For cryptanalysis, one typically relies on BKZ [SE94] which extends this principle to find shorter vectors at a higher cost. Finding an analogue of LLL for module lattices has been an elusive goal for at least two decades, a difficulty being to even define what that would be. Informally, it should:

- work at the field level (in particular, it should not forget the module structure and view the module just as a lattice);
- it should find relatively short module pseudo-bases by progressively updating the input pseudo-basis;
- it should run in polynomial-time with respect to the module rank n and the bit-lengths of the norms of the input vectors and ideals.

The state of the art is far from these goals. Napias [Nap96] proposed such an algorithm for fields whose rings of integers are norm-Euclidean, i.e., Euclidean for the algebraic norm. In our running example, this restricts the applicability to $d \leq 4$ (see [Cer05, Lez14] for other families of fields). Fieker and Pohst [FP96] proposed a general-purpose algorithm. However, it was not proved to provide pseudo-bases consisting of short module vectors, and a cost analysis was provided only for free modules over totally real fields. Fieker [Fie97, p. 47] suggested to use rank-2 module reduction to achieve rank-n module reduction, but there was no follow-up on this approach. Gan, Ling and Mow [GLM09] described and analyzed an LLL algorithm for Gauss integers (i.e., our running example instantiated to $d = 2$). Fieker and Stehlé [FS10] proposed to apply the LLL algorithm on the lattice corresponding to the module to find short vectors in polynomial time and reconstruct a short pseudo-basis afterwards. More recently,

Kim and Lee [KL17] described such an LLL algorithm for biquadratic fields whose rings of integers are norm-Euclidean, and provided analyses for the shortness of the output and the run-time. They also proposed an extension to arbitrary norm-Euclidean rings, still with a run-time analysis but only conjecturing and experimentally supporting the output quality.

The rank-2 restriction already captures a fundamental obstacle. The LLL algorithm for 2-dimensional lattices (which is essentially Gauss' algorithm) is a succession of divide-and-swap steps. Given two vectors $\mathbf{b}_1, \mathbf{b}_2 \in \mathbb{Q}^2$, the 'division' consists in shortening $\mathbf{b}_2$ by an integer multiple of $\mathbf{b}_1$. This integer k is the quotient of the Euclidean division of $\langle \mathbf{b}_1, \mathbf{b}_2 \rangle$ by $\|\mathbf{b}_1\|^2$. This leads to a vector $\mathbf{b}_2'$. If the latter is shorter than $\mathbf{b}_1$, then $\mathbf{b}_1$ and $\mathbf{b}_2$ are swapped and a new iteration starts. Crucial to this procedure is the fact that if the projection of $\mathbf{b}_2$ orthogonally to $\mathbf{b}_1$ is very small compared to $\|\mathbf{b}_1\|$, then the division will provide a vector $\mathbf{b}_2'$ that is shorter than $\mathbf{b}_1$. When a swap cannot be made, it means that the projection of $\mathbf{b}_2$ orthogonally to $\mathbf{b}_1$ is not too small, and hence the basis is of good quality, i.e., somewhat orthogonal and hence made of somewhat short vectors. What provides the convergence to a short basis is the Euclideanity of $\mathbb{Z}$. This is why prior works focused on this setup. Put differently, the crucial property is the fact that the covering radius of the $\mathbb{Z}$ lattice is smaller than 1: this makes it possible to shorten a vector $\mathbf{b}_2$ whose projection is sufficiently small by an appropriate integer multiple such that $\mathbf{b}_2'$ becomes smaller than $\mathbf{b}_1$. When we extend to modules, the corresponding lattice becomes R, and its covering radius has no a priori reason to be smaller than 1 (for our running example, it is $\sqrt{d}/2$). Even if we allow an infinite amount of time to find an optimal $k \in R$, the resulting $\mathbf{b}_2 - k\mathbf{b}_1$ may still be longer than $\mathbf{b}_1$, even if $\mathbf{b}_2$ is in the K-span of $\mathbf{b}_1$. This leads us to the following question: does there exist a lattice L depending only on K such that being able to solve the Closest Vector Problem (CVP) with respect to L allows to find short bases of modules in K^2?

CONTRIBUTIONS. The LLL algorithm for Euclidean lattices can be viewed as a way to leverage the ability of Gauss' algorithm to reduce 2-dimensional lattice bases, to reduce n-dimensional lattice bases for any $n \geq 2$. We propose extensions to modules of both Gauss' algorithm and of its LLL leveraging from 2 to n dimensions, hence providing a full-fledged framework for LLL-like reduction of module pseudo-bases.

Our first contribution is an oracle-based algorithm which takes as input a pseudo-basis of a module $M \subset K^n$ over the ring of integers R of an arbitrary number field K, updates it progressively in a fashion similar to the LLL algorithm, and outputs a pseudo-basis of M. The first output vector is short, and the algorithm runs in time polynomial in n and the bit-lengths of the norms of the input vectors and ideals. It makes a polynomial number of calls to an oracle that finds short vectors in rank-2 modules. This oracle-based LLL-like algorithm for modules allows us to obtain the following result for our running example (see Theorem 3.9 for a general statement).

Theorem 1.1. *Let $K = \mathbb{Q}[x]/(x^d+1)$ and $R = \mathbb{Z}[x]/(x^d+1)$, for d a power of 2. There is a polynomial-time reduction from finding a $(2\gamma\sqrt{d})^{2n-1}$-approximation* *to a shortest non-zero vector in modules in K^n (with respect to the Euclidean norm inherited from mapping an element of K^n to the concatenation of its n coefficient vectors) to finding a γ-approximation to a shortest non-zero vector in modules in K^2.*

For example, if n is constant, then the reduction allows to obtain polynomial approximation factors in modules in K^n from polynomial approximation factors in modules in K^2.

Our second contribution is a heuristic algorithm to find a very short non-zero vector in an arbitrary module in K^2, given access to a CVP oracle with respect to a lattice depending only on K. We obtain the following result for our running example (combine Corollary 4.10 with Lemma 2.3 for a general statement).

Theorem 1.2 (Heuristic). *There exists a sequence of lattices L_d and an algorithm $\mathcal{A}$ such that the following holds. Algorithm $\mathcal{A}$ takes as input a pseudo-basis of a rank-2 module $M \subset (\mathbb{Q}/(x^d + 1))^2$, and outputs a vector $\mathbf{v} \in M \setminus \{0\}$ that is no more than $2^{(\log d)^{O(1)}}$ longer than a shortest non-zero vector of M. If given access to an oracle solving CVP in L_d in polynomial time, then it runs in quantum polynomial time. Finally, for any $\eta > 0$, the lattice L_d can be chosen of dimension $O(d^{2+\eta})$.*

The quantum component of the algorithm is the decomposition of an ideal as the product of a subset of fixed ideals and a principal ideal with a generator [BS16]. By relying on [BEF+17] instead, one can obtain a dequantized variant of Theorem 1.2 relying on more heuristics and in which the algorithm runs in $2^{\widetilde{O}(\sqrt{d})}$ classical time.

We insist that the result relies on heuristics. Some are inherited from prior works (such as [PHS19]) and one is new (Heuristic 1 in Sect. 4). The new heuristic quantifies the distance to L_d of vectors in the real span of L_d that satisfy some properties. This heuristic is difficult to prove as the lattice L_d involves other lattices that are not very well understood (the log-unit lattice and the lattice of class group relations between ideals of small algebraic norms). We justify this heuristic by informal counting arguments and by some experiments in small dimensions.

Finally, we note that the dimension of L_d is near-quadratic in the degree d of the field. This is much more than the lattice dimension d of R, but we do not know how to use a CVP oracle for R to obtain such an algorithm to find short vectors in rank-2 modules. An alternative approach to obtain a similar reduction from finding short non-zero vectors in rank-2 modules to CVP with preprocessing would be as follows: to reach the goal, it suffices to find a short non-zero vector in a $(2d)$-dimensional lattice; by using the LLL algorithm and numerical approximations (see, e.g., [SMSV14]), it is possible to further assume that the bit-size of the inputs is polynomial in d; by Kannan's search-to-decision reduction for the shortest vector problem [Kan87], it suffices to obtain an algorithm that decides whether or not a lattice contains a non-zero vector of norm below 1; the latter task can be expressed as an instance of 3SAT, as the corresponding language belongs to NP; finally, 3SAT reduces to CVP with preprocessing [Mic01]. Overall, this gives an alternative to Theorem 1.2 without heuristics, but lattices L_d of much higher dimensions (which still grow polynomially in d).

TECHNICAL OVERVIEW. One of the technical difficulties of extending LLL to modules is the fact that the absolute value $|\cdot|$ over $\mathbb{Q}$ has two canonical generalizations over K: the trace norm and the algebraic norm. Let $(\sigma_i)_{i \leq d}$ denote the embedding of K into $\mathbb{C}$. The trace norm and algebraic norm of $x \in K$ are respectively defined as $(\sum_i |\sigma_i(x)|^2)^{1/2}$ and $\prod_i \sigma_i(x)$. When $K = \mathbb{Q}$, the only embedding is the identity map, and both the trace norm and the absolute value of the algebraic norm collapse to the absolute value. When the field degree is greater than 1, they do not collapse, and are convenient for diverse properties. For instance, the trace norm is convenient to measure smallness of a vector over K^n. A nice property is that the bit-size of an element of R is polynomially bounded in the bit-size of the trace norm (for a fixed field K). Oppositely, an element in R may have algebraic norm 1 (in this case, it is called a unit), but can have arbitrarily large bit-size. On the other hand, the algebraic norm is multiplicative, which interacts well with determinants. For example, the determinant of the lattice corresponding to a diagonal matrix over K is simply the product of the algebraic norms of the diagonal entries (up to a scalar depending only on the field K). LLL relies on all these properties, that are conveniently satisfied by the absolute value.

In our first contribution, i.e., the LLL-like algorithm to reduce module pseudo-bases, we crucially rely on the algebraic norm. Indeed, the progress made by the LLL algorithm is measured by the so-called potential function, which is a product of determinants. As observed in prior works [FP96,KL17], using the algebraic norm allows for a direct generalization of this potential function to module lattices. What allowed us to go beyond norm-Euclidean number fields is the black-box handling of rank-2 modules. By not considering this difficult component, we can make do with the algebraic norm for the most important parts of the algorithm. The trace norm is still used to control the bit-sizes of the module pseudo-bases occurring during the algorithm, allowing to extend the so-called size-reduction process within LLL, but is not used to "make progress". The black-boxing of the rank-2 modules requires the introduction of a modified condition for deciding which 2-dimensional pseudo-basis to consider to "make progress" on the n-dimensional pseudo-basis being reduced. This condition is expressed as the ratio between 2-determinants, which is compatible with the exclusive use of the algebraic norm to measure progress. It involves the coefficient ideals, which was unnecessary in prior works handling norm-Euclidean fields, as for such fields, all modules can be generated by a basis instead of a pseudo-basis.

Our algorithm for finding short non-zero vectors in rank-2 modules iterates divide-and-swap steps like 2-dimensional LLL (or Gauss' algorithm). The crucial component is the generalization of the Euclidean division, from $\mathbb{Z}$ to R. We are given $a \in K \setminus \{0\}$ and $b \in K$, and we would like to shorten b using R-multiples of a. In the context of $a \in \mathbb{Q} \setminus \{0\}$ and $b \in \mathbb{Q}$, a Euclidean division provides us with

$u \in \mathbb{Z}$ such that $|b+ua| \leq |a|/2$. We would like to have an analogous division in R. However, the ring R may not be Euclidean. Moreover, the covering radius of the ring R (viewed as a lattice) can be larger than 1, and hence, in most cases, there will not even exist an element $u \in R$ such that $\|b+au\| \leq \|a\|$ (here $\|\cdot\|$ refers to the trace norm). In order to shorten b using a, we also allow b to be multiplied by some element $v \in R$. For this extension to be non-trivial (and useful), we require that v is not too large (otherwise, one can always take $u = b$ and $v = -a$ for instance, if $a, b \in R$, and extend this approach for general $a, b \in K$). Hence, we are interested in finding u, v such that $\|ua + vb\| \leq \varepsilon\|a\|$ and $\|v\| \leq C$ for some $\varepsilon < 1$ and C to be determined later. Intuitively, if we allow for a large number of such multiples v (proportional to $1/\varepsilon$ and to the determinant of the lattice corresponding to R, i.e., the square root of the field discriminant), there should be one such v such that there exists $u \in R$ with $\|vb + au\| \leq \varepsilon\|a\|$. We do not know how to achieve the result with this heuristically optimal number of v's and use potentially larger v's. The astute reader will note that if we use such a v inside a divide-and-swap algorithm, we may end up computing short vectors in sub-modules of the input modules. We prevent this from happening by using the module Hermite Normal Form [BP91, Coh96, BFH17].

To find u, v such that $\|vb + au\|$ is small, we use the logarithm map Log over K. For this discussion, we do not need to explain how it is defined, but only that it "works" like the logarithm map log over $\mathbb{R}_{>0}$. In particular if $x \approx y$, then Log $x \approx$ Log y. We would actually prefer to have the converse property, but it does not hold for the standard Log over K. In Subsect. 4.1, we propose an extension $\overline{\mathrm{Log}}$ such that $\overline{\mathrm{Log}}x \approx \overline{\mathrm{Log}}y$ implies that $x \approx y$. In our context, this means that we want to find u, v such that $\overline{\mathrm{Log}}v - \overline{\mathrm{Log}}u \approx \overline{\mathrm{Log}}(b) - \overline{\mathrm{Log}}(a)$. To achieve this, we will essentially look for such u and v that are product combinations of fixed small elements in R. When applying the $\overline{\mathrm{Log}}$ function, the product combinations become integer combinations of the $\overline{\mathrm{Log}}$'s of the fixed elements. This gives us our CVP instance: the lattice is defined using the $\overline{\mathrm{Log}}$'s of the fixed elements and the target is defined using $\overline{\mathrm{Log}}(b) - \overline{\mathrm{Log}}(a)$. This description is only to provide intuition, as reality is more technical: we use the log-unit lattice and small-norm ideals rather than small-norm elements.

One advantage of using the $\overline{\mathrm{Log}}$ map is that the multiplicative structure of K is mapped to an additive structure, hence leading to a CVP instance. On the downside, one needs extreme closeness in the $\overline{\mathrm{Log}}$ space to obtain useful closeness in K (in this direction, we apply an exponential function). Put differently, we need the lattice to be very dense so that there is a lattice vector that is very close to the target vector. This is the fundamental reason why we end up with a large lattice dimension: we add a large number of $\overline{\mathrm{Log}}$'s of small-norm ideals to densify the lattice. This makes the analysis of the distance to the lattice quite cumbersome, as the Gaussian heuristic gives too crude estimates. For our running example, we have a lattice of dimension $\approx d^2$ and determinant ≈ 1, hence we would expect a 'random' target vector to be at distance $\approx d$ from the lattice. We argue for a distance of at most $\approx \sqrt{d}$ for 'specific' target vectors. Finally, we note that the lattice and its analysis share similarities with the Schnorr-Adleman lattice that Ajtai used to prove NP-hardness of SVP under randomized reductions [Ajt98, MG02] (but we do not know if there is a connection).

IMPACT. Recent works have showed that lattice problems restricted to ideals of some cyclotomic number fields can be quantumly solved faster than for arbitrary lattices, for some ranges of parameters [CDW17], and for all number fields with not too large discriminant, if allowing preprocessing that depends only on the field [PHS19]. Recall that ideal lattices are rank-1 module lattices. Our work can be viewed as a step towards assessing the existence of such weaknesses for modules of larger rank, which are those that appear when trying to cryptanalyze cryptosystems based on the NTRU, RingSIS, RingLWE, ModuleSIS and ModuleLWE problems and their variants.

Similarly to [CDW17, PHS19], our results use CVP oracles for lattices defined in terms of the number field only (i.e., defined independently of the input module). In [CDW17, PHS19], the weaknesses of rank-1 modules stemmed from two properties of these CVP instances: the lattices had dimension quasi-linear in the log-discriminant (quasi-linear in the field degree, for our running example), and either the CVP instances were easy to solve [CDW17], or approximate solutions sufficed [PHS19] and one could rely on Laarhoven's CVP with preprocessing algorithm [Laa16]. In our case, we need (almost) exact solutions to CVP instances for which we could not find any efficient algorithm, and the invariant lattice has a dimension that is more than quadratic in the log-discriminant (in the field degree, for our running example). It is not ruled out that there could be efficient CVP algorithms for such lattices, maybe for some fields, but we do not have any lead to obtain them.

As explained earlier, CVP with preprocessing is known to be NP-complete, so there always exists a fixed lattice allowing to solve the shortest vector problem in lattices of a target dimension. However, the dimension of that fixed lattice grows as a high degree polynomial in the target dimension. The fact that we only need near-quadratic dimensions (when the log-discriminant is quasi-linear in the field degree) may be viewed as a hint that finding short non-zero vectors in rank-2 modules might be easier than finding short non-zero vectors in arbitrary lattices of the same dimension.

Finally, our first result shows the generality of rank-2 modules towards finding short vectors in rank-n modules for any $n \geq 2$. The reduction allows to stay in the realm of polynomial approximation factors (with respect to the field degree) for any constant n. This tends to back the conjecture that there might be a hardness gap between rank-1 and rank-2 modules, and then a smoother transition for higher rank modules.

NOTATIONS. For two real valued functions f and g, we write $f(x) = \tilde{O}(g(x))$ if and only if there exists some constant $c > 0$ such that $f(x) = O(g(x) \cdot |\log g(x)|^c)$. By abuse of notations, we write $O(x^\alpha \mathrm{poly}(\log x))$ as $\tilde{O}(x^\alpha)$ even if $\alpha = 0$. We let $\mathbb{Z}, \mathbb{Q}, \mathbb{R}$, and $\mathbb{C}$ denote the sets of integers, rational, real, and complex numbers, respectively. For $x \in \mathbb{C}$, we let $\bar{x}$ denote its complex conjugate. We use lower-case (resp. upper-case) bold letters for vectors (resp. matrices). For vectors $\mathbf{x}_i = (x_{ij})_j$ for $i \leq k$, we write $(\mathbf{x}_1 \| \dots \| \mathbf{x}_k)$ to denote the vector obtained by concatenation. By default, the matrices are written with column vectors.

For a vector $\mathbf{x} = (x_i)_i \in \mathbb{C}^n$, we write $\|\mathbf{x}\|_i$ for $i \in \{1, 2, \infty\}$ to denote ℓ_i-norm, and we typically omit the subscript when $i = 2$. For a lattice $\Lambda \subset \mathbb{R}^n$, we let $\rho(\Lambda)$ denote the covering radius with respect to Euclidean norm.

SUPPLEMENTARY MATERIAL. Due to lack of space, some material is provided only in the full version [LPSW19]. This includes: background on computational aspects on number fields, several proofs, and reports on experiments backing the heuristic claims.

2 Preliminaries

In this section, we first recall some necessary algebraic number theory background and discuss some computational aspects. We then extend Gram-Schmidt orthogonalization to matrices over number fields. In this section, we assume that the reader is somehow familiar with the algebraic notions used in this article and in previous works. For more details on these mathematical objects, we refer the reader to [Neu99, Chapter 1] for algebraic number theory questions, to [Hop98] for anything related to modules and to [PHS19] where the same techniques were used in a simpler setting.

2.1 Algebraic Background

NUMBER FIELDS. We let K be a number field of degree d and $K_\mathbb{R} = K \otimes_\mathbb{Q} \mathbb{R}$. A number field comes with r_1 real embeddings and $2r_2$ complex embeddings σ_i's, where $r_1 + 2r_2 = d$. The field norm is defined as $\mathcal{N}(x) = \prod_{i \le d} \sigma_i(x)$ and the field trace is $\mathrm{Tr}(x) = \sum_{i \le d} \sigma_i(x)$. The canonical embedding of K is then defined as $\sigma(x) \in \mathbb{R}^{r_1} \times \mathbb{C}^{2r_2}$, where $\sigma_{r_1+i}(x) = \overline{\sigma_{r_1+r_2+i}(x)}$ for $1 \le i \le r_2$. The field trace then induces a Hermitian inner product over $K_\mathbb{R}$ whose associated Euclidean norm is $\|x\| = (\sum_{1 \le i \le d} |\sigma_i(x)|^2)^{1/2}$ for $x \in K_\mathbb{R}$. We also define $\|x\|_\infty = \max_{i \in [d]} |\sigma_i(x)|$.

In this work, elements of K are identified to their canonical embeddings. From this perspective, the set $K_\mathbb{R}$ is also identified to $\{\mathbf{y} \in \mathbb{R}^{r_1} \times \mathbb{C}^{2r_2} : \forall i \le r_2, \overline{y_{r_1+r_2+i}} = y_{r_1+i}\}$ (the embedding map σ provides a ring isomorphism between $K_\mathbb{R}$ and the latter subspace of $\mathbb{R}^{r_1} \times \mathbb{C}^{2r_2}$). We write $K_\mathbb{R}^\times$ for the subset of vectors in $K_\mathbb{R}$ with non-zero entries (it forms a group, for component-wise multiplication). We also write $K_\mathbb{R}^+$ for the subset of vectors in $K_\mathbb{R}$ with non-negative (real) coefficients. For $x \in K_\mathbb{R}$, we let $\bar{x}$ refer to the element of $K_\mathbb{R}$ obtained by complex conjugation of every coordinate.[2] We can also define a square-root $\sqrt{\cdot} : K_\mathbb{R}^+ \to K_\mathbb{R}^+$ by taking coordinate-wise square roots.

We let R be the ring of integers of K. It is a free $\mathbb{Z}$-module of rank d, and can be seen as a lattice via the canonical embedding. The discriminant Δ_K of K

---

[2] Observe that even if complex conjugation might not be well defined over K (i.e., the element $\bar{x}$ might not be in K even if x is), it is however always defined over $K_\mathbb{R}$. In this article, complex conjugation will only be used on elements of $K_\mathbb{R}$, and we make no assumption that K should be stable by conjugation.

is then the squared volume of R, i.e., $\Delta_K = \det((\sigma_i(x_j))_{ij})^2$ for any $\mathbb{Z}$-basis $(x_i)_{i \leq d}$ of R. We will often use the inequality $\log \Delta_K \geq \Omega(d)$ to simplify cost estimates.

We let $R^\times = \{u \in R \mid \exists v \in R : uv = 1\}$ denote the group of units of R. Dirichlet's unit theorem states that $R^\times$ is isomorphic to the Cartesian product of a finite cyclic group (formed by the roots of unity contained in K) with the additive group $\mathbb{Z}^{r_1+r_2-1}$. We define $\mathrm{Log} : K_\mathbb{R}^\times \to \mathbb{R}^d$ by $\mathrm{Log}(x) = (\log(|\sigma_1(x)|), \ldots, \log(|\sigma_d(x)|))^T$. Let $E = \{x \in \mathbb{R}^d \mid \forall r_1 \leq i \leq r_2 : x_i = x_{i+r_2}\}$. We have $\mathrm{Log}(K_\mathbb{R}^\times) \subseteq E$. We also define $H = \{x \in \mathbb{R}^d : \sum_{i \in [d]} x_i = 0\}$ and $\mathbf{1} = (1, \ldots, 1)^T$, which is orthogonal to H in $\mathbb{R}^d$. The set $\Lambda = \{\mathrm{Log}(u) : u \in R^\times\}$ is a lattice, called "log-unit" lattice. It has rank $r_1 + r_2 - 1$, by Dirichlet's units theorem and is full rank in $E \cap H$. Further, its minimum satisfies $\lambda_1(\Lambda) \geq (\ln d)/(6d^2)$ (see [FP06, Cor. 2]).

IDEALS. A fractional ideal I of K is an additive subgroup of K which is also stable by multiplication by any element of R, and such that $xI \subseteq R$ for some $x \in \mathbb{Z} \setminus \{0\}$. Any non-zero fractional ideal is also a free $\mathbb{Z}$-module of rank d, and can therefore be seen as a lattice in $K_\mathbb{R}$ using the canonical embedding: such lattices are called ideal lattices. The product IJ of two fractional ideals I and J is the fractional ideal generated by all elements xy with $x \in I$ and $y \in J$. Any non-zero fractional ideal I is invertible, i.e., there exists a unique ideal $I^{-1} = \{x \in K : xI \subseteq R\}$ such that $II^{-1} = R$. When $I \subseteq R$, it is said to be an integral ideal. An integral ideal $\mathfrak{p}$ is said to be prime if whenever $\mathfrak{p} = IJ$ with I and J integral, then either $I = \mathfrak{p}$ or $J = \mathfrak{p}$. For any $g \in K$, we write $\langle g \rangle = gR$ the smallest fractional ideal containing g, and we say that it is a principal ideal. The quotient of the group of non-zero fractional ideals (for ideal multiplication) by the subgroup consisting in principal ideals is the class group $\mathcal{Cl}_K$. Its cardinality h_K is called the class number. Under the GRH, there is a set of cardinality $\leq \log h_K = \widetilde{O}(\log \Delta_K)$ of prime ideals of norms $\leq 12 \log^2 \Delta_K$ that generates $\mathcal{Cl}_K$ (see, e.g., [PHS19, Se. 2.3]). We also will use the bound $h_K \cdot (\det \Lambda) \leq 2^{\widetilde{O}(\log \Delta_K)}$ (see, e.g., [PHS19, Se. 2.4]).

The algebraic norm $\mathcal{N}(I)$ of an integral ideal I is its index as a subgroup of R, and is equal to $\det(\sigma(I))/\Delta_K^{1/2}$. The algebraic norm of a prime ideal is a power of a prime number. For a principal ideal, we also have $\mathcal{N}(\langle g \rangle) = |\mathcal{N}(g)|$. The norm extends to fractional ideals using $\mathcal{N}(I) = \mathcal{N}(xI)/|\mathcal{N}(x)|$, for any $x \in R \setminus \{0\}$ such that $xI \subseteq R$. We have $\mathcal{N}(IJ) = \mathcal{N}(I)\mathcal{N}(J)$ for all fractional ideals I, J.

Lemma 2.1 ([BS96, Th. 8.7.4]). *Assume the GRH. Let $\pi_K(x)$ be the number of prime integral ideals of K of norm $\leq x$. Then there exists an absolute constant C (independent of K and x) such that $|\pi_K(x) - \mathrm{li}(x)| \leq C \cdot \sqrt{x}(d \log x + \log \Delta_K)$, where $\mathrm{li}(x) = \int_2^x \frac{dt}{\ln t} \sim \frac{x}{\ln x}$.*

MODULE LATTICES AND THEIR GEOMETRY. In this work, we call $(R\text{-})$module any set of the form $M = I_1 \mathbf{b}_1 + \ldots + I_n \mathbf{b}_n$, where the I_j's are non-zero fractional ideals of R and the $\mathbf{b}_j$'s are $K_\mathbb{R}$-linearly independent[3] vectors in $K_\mathbb{R}^m$, for some $m > 0$.

---

[3] The vectors $\mathbf{b}_j$'s are said to be $K_\mathbb{R}$-linearly independent if and only if there is no non-trivial ways to write the zero vector as a $K_\mathbb{R}$-linear combination of the $\mathbf{b}_j$'s.

The tuple of pairs $((I_1, \mathbf{b}_1), \ldots, (I_n, \mathbf{b}_n))$ is called a pseudo-basis of M, and n is its rank. Note that the notion of rank of a module is usually only defined when the module has a basis (i.e., is of the form $M = R\mathbf{b}_1 + \ldots + R\mathbf{b}_n$, with all the ideals equal to R). In this article, we consider an extension of the definition of rank, defined even if the module does not have a basis, as long as it has a pseudo-basis. In particular, fractional ideals are rank-1 modules contained in K, and sets of the form $\alpha \cdot I$ for $\alpha \in K_\mathbb{R}^\times$ and a non-zero fractional ideal I are rank-1 modules in $K_\mathbb{R}$. We refer to [Hop98] for a thorough study of R-modules, and concentrate here on the background necessary to the present work.

Two pseudo-bases $((I_1, \mathbf{b}_1), \ldots, (I_n, \mathbf{b}_n))$ and $((J_1, \mathbf{c}_1), \ldots, (J_n, \mathbf{c}_n))$ represent the same module if and only if there exists $\mathbf{U} = (u_{ij})_{i,j} \in K^{n \times n}$ invertible such that $\mathbf{C} = \mathbf{B} \cdot \mathbf{U}$; we have $u_{ij} \in I_i J_j^{-1}$ and $u'_{ij} \in J_i I_j^{-1}$ for all i, j and for $\mathbf{U}' = (u'_{ij})_{i,j} := \mathbf{U}^{-1}$. Here, the matrix $\mathbf{B}$ is the concatenation of the column vectors $\mathbf{b}_i$ (and similarly for $\mathbf{C}$). If $n > 0$, we define $\det_{K_\mathbb{R}} M = \det(\overline{\mathbf{B}}^\top \mathbf{B})^{1/2} \cdot \prod_i I_i$. It is an R-module in $K_\mathbb{R}$. Note that it is a module invariant, i.e., it is identical for all pseudo-bases of M.

We extend the canonical embedding to vectors $\mathbf{v} = (v_1, \ldots, v_m)^T \in K_\mathbb{R}^m$ by defining $\sigma(\mathbf{v})$ as the vector of $\mathbb{R}^{dm}$ obtained by concatenating the canonical embeddings of the v_i's. This extension of the canonical embedding maps any module M of rank n to a (dn)-dimensional lattice in $\mathbb{R}^{dm}$. We abuse notation and use M to refer to both the module and the lattice obtained by applying the canonical embedding.

The determinant of a module M seen as a lattice is $\det M = \Delta_K^{n/2} \cdot \mathcal{N}(\det_{K_\mathbb{R}} M)$. This matches with the module determinant definition from [FS10, Se. 2.3]. Since $\det(M) \neq 0$, this shows in particular that the diagonal coefficients r_{ii} of the R-factor are invertible in $K_\mathbb{R}$ (otherwise, one of their embedding would be 0 and so would be their norm).

We consider the following inner products for $\mathbf{a}, \mathbf{b} \in K_\mathbb{R}^m$:

$$\langle \mathbf{a}, \mathbf{b} \rangle_{K_\mathbb{R}} = \sum_{i \in [m]} a_i \overline{b}_i \in K_\mathbb{R} \quad \text{and} \quad \langle \mathbf{a}, \mathbf{b} \rangle = \text{Tr}\left(\sum_{i \in [m]} a_i \overline{b}_i \right) \in \mathbb{C}.$$

Note that we have $\langle \mathbf{v}, \mathbf{v} \rangle_{K_\mathbb{R}} \in K_\mathbb{R}^+$, as all $\sigma_i(\langle \mathbf{v}, \mathbf{v} \rangle_{K_\mathbb{R}})$'s are non-negative. For $\mathbf{v} \in K_\mathbb{R}^m$, we define $\|\mathbf{v}\|_{K_\mathbb{R}} = \sqrt{\langle \mathbf{v}, \mathbf{v} \rangle_{K_\mathbb{R}}}$ and $\|\mathbf{v}\| = \sqrt{\text{Tr}(\langle \mathbf{v}, \mathbf{v} \rangle_{K_\mathbb{R}})} = \sqrt{\langle \mathbf{v}, \mathbf{v} \rangle}$. Observe that $\|\mathbf{v}\|$ correspond to the Euclidean norm of $\mathbf{v}$ when seen as a vector of dimension dm via the canonical embedding. We extend the infinity norm to vectors $\mathbf{v} \in K_\mathbb{R}^m$ by $\|\mathbf{v}\|_\infty = \max_{i \in [m]} \|v_i\|_\infty$, where $\mathbf{v} = (v_1, \ldots, v_m)$. We also extend the algebraic norm to vectors $\mathbf{v} \in K_\mathbb{R}^m$ by setting $\mathcal{N}(\mathbf{v}) := \mathcal{N}(\|\mathbf{v}\|_{K_\mathbb{R}})$. For $m = 1$, we see that $\mathcal{N}(\mathbf{v}) = |\mathcal{N}(\mathbf{v})|$. By the arithmetic-geometric inequality, we have $\sqrt{d} \cdot \mathcal{N}(\mathbf{a})^{1/d} \leq \|\mathbf{a}\|$ for $\mathbf{a} \in K_\mathbb{R}^m$. Observe also that for any vector $\mathbf{v} = (v_1, \ldots, v_m)^T \in K_\mathbb{R}$, we have $\mathcal{N}(\mathbf{v}) \geq \max_i(\mathcal{N}(v_i))$, because for any embedding σ_j, it holds that $|\sigma_j(v_1 \overline{v_1} + \cdots + v_m \overline{v_m})| = |\sigma_j(v_1)|^2 + \cdots + |\sigma_j(v_m)|^2 \geq \max_i |\sigma_j(v_i)|^2$.

Because $K_\mathbb{R}$ is a ring and not a field, this definition is stronger than requiring that none of the $\mathbf{b}_j$'s is in the span of the others.

We define the module minimum $\lambda_1(M)$ as the norm of a shortest non-zero element of M with respect to $\|\cdot\|$. Our module-LLL algorithm will rely on the algebraic norm rather than the Euclidean norm. For this reason, we will also be interested to the minimum $\lambda_1^{\mathcal{N}}(M) = \inf(\mathcal{N}(\mathbf{v}) : \mathbf{v} \in M \setminus \{\mathbf{0}\})$. We do not know if this minimum is always reached for some vector $\mathbf{v} \in M$, but we can find an element of M whose algebraic norm is arbitrarily close to $\lambda_1^{\mathcal{N}}(M)$. The following lemma provides relationships between $\lambda_1(M)$ and $\lambda_1^{\mathcal{N}}(M)$.

Lemma 2.2. *For any rank-n module M, we have:*

$$d^{-d/2}\lambda_1(M)^d \Delta_K^{-1/2} \leq \lambda_1^{\mathcal{N}}(M) \leq d^{-d/2}\lambda_1(M)^d \leq n^{d/2}\Delta_K^{1/2}\mathcal{N}(\det{}_{K_{\mathbb{R}}} M)^{1/n}.$$

2.2 Computing over Rings

Background on field and ideal arithmetic is provided in the full version [LPSW19].

COMPUTATIONS WITH AN ORACLE. In Sect. 4, we will assume that we have access to an oracle for the Closest Vector Problem, for lattices related to K. For example, we will assume that we can solve CVP for the lattice corresponding to R, with respect to $\|\cdot\|$. This lattice has dimension d.

In a similar vein, we will use the following adaptation from [PHS19, Th. 3.4], to find short elements in rank-1 modules.

Lemma 2.3 (Heuristic). *There exists a lattice L_K (that only depends on K and has dimension $\widetilde{O}(\log \Delta_K)$) such that, given an oracle access to an algorithm that solves CVP for L_K, the following holds. There exists a heuristic quantum polynomial-time algorithm that takes as input an ideal I of K and any $\alpha \in K_{\mathbb{R}}^{\times}$, and outputs $x \in \alpha I \setminus \{0\}$ such that*

$$\|x\|_{\infty} \leq \mathrm{c} \cdot |\mathcal{N}(\alpha)|^{1/d} \cdot \mathcal{N}(I)^{1/d},$$

where $\mathrm{c} = 2^{\widetilde{O}(\log |\Delta|)/d}$. In particular, we have $\|x\|_{\infty} \leq \mathrm{c} \cdot |\mathcal{N}(x)|^{1/d}$.

The result assumes GRH and Heuristic 4 from [PHS19]. The quantum computation performed by the algorithm derives from [BS16] and consists in computing the log-unit lattice, finding a small generating set $([\mathfrak{p}_i])_i$ of the class group $\mathcal{C}l_K$ of K, and decomposing the class $[I]$ of I in $\mathcal{C}l_K$ in terms of that generating set. These quantum computations can be replaced by classical ones (e.g., [BF14, BEF+17]), at the expense of increased run-times and additional heuristic assumptions.

The lemma can be derived from [PHS19, Th. 3.4] by replacing Laarhoven's CVPP algorithm [Laa16] by an exact CVPP oracle. In [PHS19], the CVPP algorithm is used with a target vector $\mathbf{t}$ derived from the decomposition of $[I]$ on the $[\mathfrak{p}_i]$'s and the logarithm $\mathrm{Log}(g)$ of an element $g \in K$. To obtain the statement above, we replace $\mathrm{Log}(g)$ by $\mathrm{Log}(g \cdot \alpha) = \mathrm{Log}(g) + \mathrm{Log}(\alpha)$. The last lemma statement $\|x\|_{\infty} \leq \mathrm{c}\mathcal{N}(x)^{1/d}$ comes from the observation that $|\mathcal{N}(x)| \geq \mathcal{N}(\alpha) \cdot \mathcal{N}(I)$ (which holds because x belongs to $\alpha I \setminus \{0\}$).

2.3   Gram-Schmidt Orthogonalization

We extend Gram-Schmidt Orthogonalization from matrices over the real numbers to matrices over $K_{\mathbb{R}}^m$. For $(\mathbf{b}_1, \ldots, \mathbf{b}_n) \in K_{\mathbb{R}}^{m \times n}$ such that $\mathbf{b}_1, \ldots, \mathbf{b}_n$ are $K_{\mathbb{R}}$-linearly independent, we define $\mathbf{b}_1^* = \mathbf{b}_1$ and, for $1 < i \leq n$:

$$\mathbf{b}_i^* = \mathbf{b}_i - \sum_{j<i} \mu_{ij} \mathbf{b}_j^* \quad \text{with} \quad \forall j < i: \ \mu_{ij} = \frac{\langle \mathbf{b}_i, \mathbf{b}_j^* \rangle_{K_{\mathbb{R}}}}{\langle \mathbf{b}_j^*, \mathbf{b}_j^* \rangle_{K_{\mathbb{R}}}}.$$

It may be checked that $\langle \mathbf{b}_i^*, \mathbf{b}_j^* \rangle = 0$ for $i \neq j$, and that $\mathbf{b}_i^* = \operatorname{argmin}(\|\mathbf{b}_i - \sum_{j<i} y_j \mathbf{b}_j\| \mid \forall j : y_j \in K_{\mathbb{R}})$.

We also extend the QR-factorization to matrices over $K_{\mathbb{R}}$. We define $r_{ii} = \|\mathbf{b}_i^*\|_{K_{\mathbb{R}}}$ for $i \leq n$, $r_{ij} = \mu_{ji} r_{ii}$ when $i < j$, and $r_{ij} = 0$ when $i > j$. We then have $\mathbf{B} = \mathbf{Q} \cdot \mathbf{R}$, where $\mathbf{Q} \in K_{\mathbb{R}}^{m \times n}$ is the matrix whose columns are the $\mathbf{b}_i^*/\|\mathbf{b}_i^*\|_{K_{\mathbb{R}}}$'s and $\mathbf{R} = (r_{ij})_{ij}$. Note that $\overline{\mathbf{Q}}^T \mathbf{Q} = \mathbf{Id}$ and that $\mathbf{R}$ is upper-triangular with diagonal coefficients in $K_{\mathbb{R}}^+$.

The following lemma provides relationships between some module invariants and the QR-factorization.

Lemma 2.4. *Let $M \subset K_{\mathbb{R}}^m$ be a module with pseudo-basis $((I_i, \mathbf{b}_i))_{i \leq n}$. Let $\mathbf{R}$ be the R-factor of $\mathbf{B}$. Then, we have $\det_{K_{\mathbb{R}}} M = \prod_i r_{ii} I_i$ and $\det M = \Delta_K^{n/2} \prod_i \mathcal{N}(r_{ii} I_i)$. Further, for any vector $\mathbf{v} \in K_{\mathbb{R}}^m$ and fractional ideal $I \subset K$ such that $0 \subsetneq \mathbf{v} I \subseteq M$, it holds that $\mathcal{N}(\mathbf{v}) \cdot \mathcal{N}(I) \geq \min_i \mathcal{N}(r_{ii} I_i)$. This implies in particular that $\lambda_1^{\mathcal{N}}(M) = \inf_{\mathbf{s} \in M \setminus \{\mathbf{0}\}} \mathcal{N}(\mathbf{s}) \geq \min_i \mathcal{N}(r_{ii} I_i)$.*

In this work, we will mostly rely on QR-factorization. It carries the same information as Gram-Schmidt orthogonalization, but allows for simpler explanations. However, from a computational perspective, the R-factor may be difficult to represent exactly even for modules contained in K^m, because of the square roots appearing in its definition. This difficulty is circumvented by computing the Gram-Schmidt orthogonalization instead, and using it as a means to represent the R-factor. In the full version, we explain how to efficiently compute Gram-Schmidt orthogonalizations.

For lattices, if we have a basis and a full-rank family of short vectors, then we can efficiently obtain a basis of the lattice whose Gram-Schmidt vectors are no longer than those of the full-rank family of short vectors. This was generalized to modules in [FS10], relying on the extension to modules of the Hermite Normal Form [BP91,Coh96,BFH17].

Lemma 2.5 ([FS10, Th. 4]). *There exists an algorithm that takes as inputs a pseudo-basis $((I_i, \mathbf{b}_i))_{i \leq n}$ of a module $M \subset K_{\mathbb{R}}^m$ and a full-rank set of vectors $(\mathbf{s}_i)_{i \leq n}$ of M and outputs a pseudo-basis $((J_i, \mathbf{c}_i))_{i \leq n}$ such that $\mathbf{c}_i \in M$ and $\mathbf{c}_i^* = \mathbf{s}_i^*$ for all i. If $M \subset K^m$, then it terminates in polynomial-time.*

Note that the condition that $\mathbf{c}_i \in M$ implies that $R \subseteq J_i$, for all i.

3   LLL-Reduction of Module Pseudo-bases

LLL-reduction of lattice bases is defined in terms of Gram-Schmidt orthogonal-ization (or, equivalently, QR-factorization). A basis is said LLL-reduced if two conditions are satisfied. The first one, often referred to as size-reduction condi-tion, states that any off-diagonal coefficients r_{ij} of the R-factor should have a small magnitude compared to the diagonal coefficient r_{ii} on the same row. The second one, often referred to as Lovász' condition, states that the 2-dimensional vector $(r_{i,i}, 0)^T$ is no more than $1/\delta$ times longer than $(r_{i,i+1}, r_{i+1,i+1})^T$, for some parameter $\delta < 1$. The size-reduction condition allows to ensure that the norms of the vectors during the LLL execution and at its completion stay bounded. More importantly, in combination with Lovász' condition, it makes it impossible for $r_{i+1,i+1}/r_{i,i}$ to be arbitrarily small (for an LLL-reduced basis). The latter is the crux of both the LLL output quality and its fast termination.

3.1   An LLL Algorithm for Module Lattices

When extending to rings, the purpose of the size-reduction condition is better expressed in terms of the Euclidean norm $\|\cdot\|$, whereas the bounded decrease of the r_{ii}'s is better quantified in terms of the algebraic norm $\mathcal{N}(\cdot)$. This discrep-ancy makes the definition of a LLL-reduction algorithm for modules difficult. In this section, we circumvent this difficulty by directly focusing on the decrease of the r_{ii}'s, deferring to later sections the handling of the rank-2 modules of pseudo-bases $((I_i, (r_{i,i}, 0)^T), (I_{i+1}, (r_{i,i+1}, r_{i+1,i+1})^T))$. We also defer to later the bounding of bit-sizes.

Definition 3.1 (LLL-reducedness of a pseudo-basis). *A module pseudo-basis* $((I_i, \mathbf{b}_i))_{i \leq n}$ *is called LLL-reduced with respect to a parameter* $\alpha_K \geq 1$ *if, for all* $i < n$*, we have:*

$$\mathcal{N}(r_{i+1,i+1} I_{i+1}) \geq \frac{1}{\alpha_K} \cdot \mathcal{N}(r_{i,i} I_i), \tag{3.1}$$

where $\mathbf{R} = (r_{i,j})_{i,j}$ *refers to the R-factor of the matrix basis* $\mathbf{B}$*.*

We first explain that LLL-reduced pseudo-bases are of interest, and we will later discuss their computation (for some value of α_K).

Lemma 3.2. *Assume that* $((I_i, \mathbf{b}_i))_{i \leq n}$ *is an LLL-reduced pseudo-basis of a module* M*. Then:*

$$\mathcal{N}(I_1)\mathcal{N}(\mathbf{b}_1) \leq \alpha_K^{(n-1)/2} \cdot (\mathcal{N}(\det_{K_\mathbb{R}} M))^{1/n},$$
$$\mathcal{N}(I_1)\mathcal{N}(\mathbf{b}_1) \leq \alpha_K^{n-1} \cdot \lambda_1^{\mathcal{N}}(M).$$

Our LLL algorithm for modules is very similar to the one over the integers.

The algorithm proceeds by finding an approximation to a shortest non-zero element in a rank-2 module, with respect to the algebraic norm. Using

---

Algorithm 3.1. LLL-reduction over K

Input: A pseudo-basis $((I_i, \mathbf{b}_i))_{i \leq n}$ of a module $M \subset K^m$.
Output: An LLL-reduced pseudo-basis of M.
1: **while** there exists $i < n$ such that $\alpha_K \cdot \mathcal{N}(r_{i+1,i+1}I_{i+1}) < \mathcal{N}(r_{i,i}I_i)$ **do**
2:    Define M_i as the rank-2 module spanned by $((I_i, \mathbf{a}_i), (I_{i+1}, \mathbf{a}_{i+1}))$, with $\mathbf{a}_i = (r_{ii}, 0)^T$ and $\mathbf{a}_{i+1} = (r_{i,i+1}, r_{i+1,i+1})^T$;
3:    Find $\mathbf{s}_i \in M_i \setminus \{\mathbf{0}\}$ such that $\mathcal{N}(\mathbf{s}_i) \leq \gamma^d \cdot \lambda_1^{\mathcal{N}}(M_i)$;
4:    Set $\mathbf{s}_{i+1} = \mathbf{a}_i$ if it is linearly independent with $\mathbf{s}_i$, and $\mathbf{s}_{i+1} = \mathbf{a}_{i+1}$ otherwise;
5:    Call the algorithm of Lemma 2.5 with $((I_i, \mathbf{a}_i), (I_{i+1}, \mathbf{a}_{i+1}))$ and $(\mathbf{s}_i, \mathbf{s}_{i+1})$ as inputs, and let $((I'_i, \mathbf{a}'_i), (I'_{i+1}, \mathbf{a}'_{i+1}))$ denote the output;
6:    Update $I_i := I'_i$, $I_{i+1} := I'_{i+1}$ and $[\mathbf{b}_i | \mathbf{b}_{i+1}] := [\mathbf{b}_i | \mathbf{b}_{i+1}] \cdot \mathbf{A}^{-1} \cdot \mathbf{A}'$
      (where $\mathbf{A} = [\mathbf{a}_i | \mathbf{a}_{i+1}]$ and $\mathbf{A}' = [\mathbf{a}'_i | \mathbf{a}'_{i+1}]$).
7: **end while**
8: **return** $((I_i, \mathbf{b}_i))_{i \leq n}$.

Lemma 2.2, we obtain a sufficient condition on α_K such that Algorithm 3.1 terminates. In particular, if α_K is sufficiently large, then $\mathcal{N}(r_{i+1,i+1}I_{i+1}) < \frac{1}{\alpha_K}\mathcal{N}(r_{i,i}I_i)$ implies that there is a vector $\mathbf{s}$ in the local projected rank-2 module of norm significantly less than $\mathcal{N}(r_{i,i}I_i)$.

Lemma 3.3. *Take the notations of Algorithm 3.1, and consider an index $i < n$ such that $\alpha_K \cdot \mathcal{N}(r_{i+1,i+1}I_{i+1}) < \mathcal{N}(r_{i,i}I_i)$. We have $\mathcal{N}(\mathbf{s}_i) \leq \gamma^d \sqrt{\frac{2^d \Delta_K}{\alpha_K}} \mathcal{N}(r_{i,i}I_i)$.*

We are now ready to prove the main result of this section.

Theorem 3.4. *Assume that Step 3 of Algorithm 3.1 is implemented with some algorithm $\mathcal{O}$ for some parameter γ. Assume that $\alpha_K > \gamma^{2d} 2^d \Delta_K$. Then Algorithm 3.1 terminates and outputs an LLL-reduced pseudo-basis of M. Further, the number of loop iterations is bounded by*

$$\frac{n(n+1)}{\log(\alpha_K/(\gamma^{2d}2^d\Delta_K))} \cdot \log \frac{\max \mathcal{N}(r_{ii}I_i)}{\min \mathcal{N}(r_{ii}I_i)},$$

where the I_i's and r_{ii}'s are those of the input pseudo-basis.

Proof. We first show that at every stage of the algorithm, the current pseudo-basis $((I_i, \mathbf{b}_i))_{i \leq n}$ is a pseudo-basis of M. For this, it suffices to show that the operations performed on it at Step 6 preserve this property. This is provided by the fact that $\mathbf{A}^{-1} \cdot \mathbf{A}'$ maps the pseudo-basis $((I_i, \mathbf{a}_i), (I_{i+1}, \mathbf{a}_{i+1}))$ into the pseudo-basis $((I'_i, \mathbf{a}'_i), (I'_{i+1}, \mathbf{a}'_{i+1}))$ of the same rank-2 module (by Lemma 2.5). Applying the same transformation to $((I_i, \mathbf{b}_i), (I_{i+1}, \mathbf{b}_{i+1}))$ preserves the spanned rank-2 module. The correctness of Algorithm 3.1 is implied by termination and the above.

We now prove a bound on the number of loop iterations, which will in particular imply termination. Consider the quantity

$$\Pi := \prod_{i \leq n} \mathcal{N}(r_{ii}I_i)^{n-i+1}.$$

This quantity if bounded from above by $\max \mathcal{N}(r_{ii}I_i)^{n(n+1)/2}$ and from below by $\min \mathcal{N}(r_{ii}I_i)^{n(n+1)/2}$. Below, we show that Π never increases during the execution of the algorithm, and that at every iteration of the while loop, it decreases by a factor $\geq \sqrt{\alpha_K/(\gamma^{2d}2^d\Delta_K)}$. We also show that the quantity $\min \mathcal{N}(r_{ii}I_i)^{n(n+1)/2}$ can only increase during the execution of the algorithm, hence the lower bound above holds with respect to the input r_{ii} and I_i at every step of the algorithm. Combining the decrease rate with the above upper and lower bounds, this implies that the number of loop iterations is bounded by

$$\frac{n(n+1)}{\log(\alpha_K/(\gamma^{2d}2^d\Delta_K))} \cdot \log \frac{\max \mathcal{N}(r_{ii}I_i)}{\min \mathcal{N}(r_{ii}I_i)},$$

where the I_i's and r_{ii}'s are those of the input pseudo-basis.

Consider an iteration of the while loop, working at index i. We have $\alpha_K \cdot \mathcal{N}(r_{i+1,i+1}I_{i+1}) < \mathcal{N}(r_{i,i}I_i)$. Step 6 is the only one that may change Π. Observe that we have

$$\Pi = \prod_{j \leq n} \mathcal{N}\left(\det_{K_{\mathbb{R}}}\left(((I_i, \mathbf{b}_i))_{i \leq j}\right)\right).$$

During the loop iteration, none of the n modules in the expression above changes, except possibly the i-th one. Now, note that

$$\mathcal{N}\left(\det_{K_{\mathbb{R}}}\left(((I_k, \mathbf{b}_k))_{k \leq i}\right)\right) = \prod_{k \leq i} \mathcal{N}(r_{kk}I_k).$$

During the loop iteration under scope, only the i-th term in this product may change. At Step 6, it is updated from $\mathcal{N}(r_{ii}I_i)$ to $\mathcal{N}(I_i')\mathcal{N}(\mathbf{a}_i')$. By Lemma 2.5, we have $\mathcal{N}(I_i') \leq 1$ and $\mathbf{a}_i' = \mathbf{s}_i$. Now, by Lemma 3.3, we have that $\mathcal{N}(\mathbf{s}_i) \leq \gamma^d \sqrt{\frac{2^d\Delta_K}{\alpha_K}}\mathcal{N}(r_{ii}I_i)$. Overall, this gives that $\mathcal{N}(r_{ii}I_i)$ and hence Π decrease by a factor $\geq \sqrt{\alpha_K/(\gamma^{2d}2^d\Delta_K)}$.

To show that $\min \mathcal{N}(r_{ii}I_i)$ can only increase during the execution of the algorithm, observe that, during a loop iteration, only $\mathcal{N}(r_{ii}I_i)$ and $\mathcal{N}(r_{i+1,i+1}I_{i+1})$ may be modified. Let us call $\mathcal{N}(r_{ii}'I_i')$ and $\mathcal{N}(r_{i+1,i+1}'I_{i+1}')$ the corresponding values at the end of the iteration. We have seen above that $\mathcal{N}(r_{ii}'I_i') \leq \mathcal{N}(r_{ii}I_i)$, which implies that $\mathcal{N}(r_{ii}'I_i') \leq \max(\mathcal{N}(r_{ii}I_i), \mathcal{N}(r_{i+1,i+1}I_{i+1}))$. We also know from Lemma 2.4 that $\mathcal{N}(r_{ii}'I_i') \geq \min(\mathcal{N}(r_{ii}I_i), \mathcal{N}(r_{i+1,i+1}I_{i+1}))$. As the determinant of M_i is constant, we have

$$\mathcal{N}(r_{ii}'I_i') \cdot \mathcal{N}(r_{i+1,i+1}'I_{i+1}') = \mathcal{N}(r_{ii}I_i) \cdot \mathcal{N}(r_{i+1,i+1}I_{i+1}).$$

This implies that $\mathcal{N}(r_{i+1,i+1}'I_{i+1}') \geq \min(\mathcal{N}(r_{ii}I_i), \mathcal{N}(r_{i+1,i+1}I_{i+1}))$. Overall, we have that $\mathcal{N}(r_{ii}'I_i'), \mathcal{N}(r_{i+1,i+1}'I_{i+1}') \geq \min(\mathcal{N}(r_{ii}I_i), \mathcal{N}(r_{i+1,i+1}I_{i+1}))$. □

3.2 Handling Bit-Sizes

In terms of bit-sizes of the diverse quantities manipulated during the execution of the algorithm, there can be several sources of bit-size growth. Like in the classical LLL-algorithm, the Euclidean norms of off-diagonal coefficients r_{ij} for $i < j$ could grow during the execution. We handle this using a generalized size-reduction algorithm. Other annoyances are specific to the number field setup. There is too much freedom in representing a rank-1 module $I\mathbf{v}$: scaling the ideal I by some $x \in K$ and dividing $\mathbf{v}$ by the same x preserves the module. In the extreme case, it could cost an arbitrarily large amount of space, even to store a trivial rank-1 module such as $R \cdot (1, 0, \ldots, 0)^T$, if such a bad scaling is used (e.g., using such an x with large algebraic norm). Finally, even if the ideal I is "scaled", we can still multiply $\mathbf{v}$ by a unit: this preserves the rank-1 module, but makes its representation longer.[4]

Definition 3.5. *A pseudo-basis* $((I_i, \mathbf{b}_i))_{i \leq n}$*, with* $I_i \subset K$ *and* $\mathbf{b}_i \in K_{\mathbb{R}}^m$ *for all* $i \leq n$*, is said scaled if, for all* $i \leq n$*,*

$$R \subseteq I_i, \quad \mathcal{N}(I_i) \geq 2^{-d^2} \Delta_K^{-1/2} \quad and \quad \|r_{ii}\| \leq 2^d \Delta_K^{1/(2d)} \mathcal{N}(r_{ii}I_i)^{1/d}.$$

It is said size-reduced if $\|r_{ij}/r_{ii}\| \leq (4d)^d \Delta_K^{1/2}$ *for all* $i < j \leq n$*.*

Note that if $((I_i, \mathbf{b}_i))_{i \leq n}$ is scaled, then $\mathcal{N}(I_i) \leq 1$ for all $i \leq n$. Further, if the spanned module is contained in R^m, then $\mathbf{b}_i \in R^m$ for all $i \leq n$. Algorithm 3.2 transforms any pseudo-basis into a scaled pseudo-basis of the same module.

Algorithm 3.2. Scaling the ideals.

Input: A pseudo-basis $((I_i, \mathbf{b}_i))_{i \leq n}$ of a module M.
Output: A scaled pseudo-basis $((I_i', \mathbf{b}_i'))_{i \leq n}$ of M.
1: **for** $i = 1$ **to** n **do**
2:    Use LLL to find $s_i \in r_{ii} \cdot I_i \setminus \{0\}$ such that $\|s_i\| \leq 2^d \Delta_K^{1/(2d)} \mathcal{N}(r_{ii}I_i)^{1/d}$;
3:    Write $s_i = r_{ii} \cdot x_i$, with $x_i \in I_i$;
4:    Define $I_i' = I_i \cdot \langle x_i \rangle^{-1}$ and $\mathbf{b}_i' = x_i \mathbf{b}_i$.
5: **end for**
6: **return** $((I_i', \mathbf{b}_i'))_{i \leq n}$.

Lemma 3.6. *Algorithm 3.2 outputs a scaled pseudo-basis of the module M generated by the input pseudo-basis and preserves the $\mathcal{N}(r_{ii}I_i)$'s. If $M \subseteq R^m$, then it runs in time polynomial in the input bit-length and in $\log \Delta_K$.*

Algorithm 3.3 aims at size-reducing a scaled pseudo-basis. It relies on a $\lfloor \cdot \rceil_R$ operator which takes as input a $y \in K_{\mathbb{R}}$ and rounds it to some $k \in R$ by writing $y = \sum y_i r_i$ for some y_i's in $\mathbb{R}$, and rounding each y_i to the nearest integer: $k = \sum k_i r_i = \sum \lfloor y_i \rceil r_i$ (remember that the r_i's form an LLL-reduced basis of R). For computations, we will apply this operator numerically, so that we may not have $\max_i |k_i - y_i| \leq 1/2$ but, with a bounded precision computation, we can ensure that $\max_i |k_i - y_i| \leq 1$.

---

[4] Note that ideal scaling and size-reduction have been suggested in [FS10, Se. 4.1], but without a complexity analysis (polynomial complexity was claimed but not proved).

Algorithm 3.3. Size-reduction.

Input: A scaled pseudo-basis $((I_i, \mathbf{b}_i))_{i \leq n}$ of a module M.
Output: A size-reduced pseudo-basis of M.
  1: **for** $j = 1$ **to** n **do**
  2:    **for** $i = j - 1$ **to** 1 **do**
  3:       Compute $x_i = \lfloor r_{ij}/r_{ii} \rceil_R$;
  4:       $\mathbf{b}_j := \mathbf{b}_j - x_i \mathbf{b}_i$.
  5:    **end for**
  6: **end for**
  7: **return** $((I_i, \mathbf{b}_i))_{i \leq n}$.

Lemma 3.7. *Algorithm 3.3 outputs a scaled size-reduced pseudo-basis of the module M generated by the input pseudo-basis and preserves the $\mathcal{N}(r_{ii}I_i)$'s. If $M \subseteq R^m$, then it runs in time polynomial in the input bit-length and in $\log \Delta_K$.*

We now consider Algorithm 3.4, which is a variant of Algorithm 3.1 that allows us to prove a bound on the bit cost. The only difference (Step 7) is that we call Algorithms 3.2 and 3.3 at every loop iteration of Algorithm 3.1, so that we are able to master the bit-lengths during the execution. Without loss of generality, we can assume that the pseudo-basis given as input is scaled and size-reduced: if it is not the case, we can call Algorithms 3.2 and 3.3, which will produce a pseudo-basis of the same module, whose bit-length is polynomial in the input bit-length and in $\log \Delta_K$.

Algorithm 3.4. LLL-reduction over K with controlled bit-lengths

---

Input: A scaled size-reduced pseudo-basis $((I_i, \mathbf{b}_i))_{i \leq n}$ of a module $M \subseteq R^m$.
Output: An LLL-reduced pseudo-basis of M.
  1: **while** there exists $i < n$ such that $\alpha_K \cdot \mathcal{N}(r_{i+1,i+1}I_{i+1}) < \mathcal{N}(r_{i,i}I_i)$ **do**
  2:    Let M_i be the rank-2 module spanned by the pseudo-basis $((I_i, \mathbf{a}_i), (I_{i+1}, \mathbf{a}_{i+1}))$,
       with $\mathbf{a}_i = (r_{ii}, 0)^T$ and $\mathbf{a}_{i+1} = (r_{i,i+1}, r_{i+1,i+1})^T$;
  3:    Find $\mathbf{s}_i \in M_i \setminus \{\mathbf{0}\}$ such that $\mathcal{N}(\mathbf{s}_i) \leq \gamma^d \cdot \lambda_1^{\mathcal{N}}(M_i)$;
  4:    Set $\mathbf{s}_{i+1} = \mathbf{a}_i$ if it is linearly independent with $\mathbf{s}_i$, and $\mathbf{s}_{i+1} = \mathbf{a}_{i+1}$ otherwise;
  5:    Call the algorithm of Lemma 2.5 with $((I_i, \mathbf{a}_i), (I_{i+1}, \mathbf{a}_{i+1}))$ and $(\mathbf{s}_i, \mathbf{s}_{i+1})$ as
       inputs, and let $((I_i', \mathbf{a}_i'), (I_{i+1}', \mathbf{a}_{i+1}'))$ denote the output;
  6:    Update $I_i := I_i'$, $I_{i+1} := I_{i+1}'$ and $[\mathbf{b}_i|\mathbf{b}_{i+1}] := [\mathbf{b}_i|\mathbf{b}_{i+1}] \cdot \mathbf{A}^{-1} \cdot \mathbf{A}'$
       (where $\mathbf{A} = [\mathbf{a}_i|\mathbf{a}_{i+1}]$ and $\mathbf{A}' = [\mathbf{a}_i'|\mathbf{a}_{i+1}']$);
  7:    Update the current pseudo-basis by applying Algorithm 3.2 and then Algorithm 3.3 to it.
  8: **end while**
  9: **return** $((I_i, \mathbf{b}_i))_{i \leq n}$.

Theorem 3.8. *Assume that Step 3 of Algorithm 3.4 is implemented with some algorithm $\mathcal{O}$ for some parameter γ. Assume that $\alpha_K > \gamma^{2d} 2^d \Delta_K$. Given as input a scaled and size-reduced pseudo-basis of a module $M \subseteq R^m$, Algorithm 3.4*

outputs an LLL-*reduced pseudo-basis of* M *in time polynomial in the bit-length of the input pseudo-basis,* $\log \Delta_K$ *and* $1/\log(\alpha_K/(\gamma^{2d}2^d\Delta_K))$.

3.3   Finding Short Vectors for the Euclidean Norm

By Lemma 3.2 and Theorem 3.8 with $\alpha_k = (1 + c/n) \cdot \gamma^{2d}2^d\Delta_K$ for a well-chosen constant c, Algorithm 3.4 may be interpreted as a reduction from finding a $2 \cdot (\gamma^{2d}2^d\Delta_K)^n$ approximation to a vector reaching $\lambda_1^{\mathcal{N}}$ in rank-n modules, to finding a γ^d approximation to a vector reaching $\lambda_1^{\mathcal{N}}$ in rank-2 modules.

By using Lemma 2.2, we can extend the above to the Euclidean norm instead of the algebraic norm.

Theorem 3.9. *Let* $\gamma \geq 1$, *assume that* $\log \Delta_K$ *is polynomially bounded, and assume that a* $\mathbb{Z}$-*basis of* R *is known. Then there exists a polynomial-time reduction from solving* SVP$_{\gamma'}$ *in rank-n modules (with respect to* $\|\cdot\|$) *to solving* SVP$_\gamma$ *in rank-2 modules, where* $\gamma' = (2\gamma\Delta_K^{1/d})^{2n-1}$.

Proof. The reduction consists in first using Algorithm 3.4 with Step 3 implemented using the oracle solving SVP$_\gamma$ in rank-2 modules. Using the arithmetic-geometric inequality and Lemma 2.2, one can see that a vector $\mathbf{s}$ satisfying $\|\mathbf{s}\| \leq \gamma \cdot \lambda_1(M)$ also satisfies $\mathcal{N}(\mathbf{s}) \leq \gamma^d \cdot \Delta_K^{1/2} \cdot \lambda_1^{\mathcal{N}}(M)$. Hence, we have an oracle computing a $\gamma_{\mathcal{N}} = \gamma \cdot \Delta_K^{1/(2d)}$ approximation of $\lambda_1^{\mathcal{N}}(M)$. We then run Algorithm 3.4 with this oracle by setting the parameter α_K to $(1+c/n)\cdot\gamma^{2d}2^d\Delta_K^2$, where c is a constant such that $(1 + c/n)^{n-1} \leq 2$.

By Theorem 3.8, the reduction runs in in polynomial time. Further, by Lemma 3.2, the output pseudo-basis satisfies $\mathcal{N}(I_1)\mathcal{N}(\mathbf{b}_1) \leq \alpha_K^{n-1} \cdot \lambda_1^{\mathcal{N}}(M)$. By Lemma 2.2 and by definition of α_K, this gives:

$$\mathcal{N}(I_1)\mathcal{N}(\mathbf{b}_1) \leq 2(\gamma^{2d}2^d\Delta_K^2)^{n-1} \cdot d^{-d/2}\lambda_1(M)^d.$$

Now, an SVP$_\gamma$ solver for rank-2 modules directly provides an SVP$_\gamma$ solver for rank-1 module. We hence use our oracle again, on $I_1\mathbf{b}_1$. This provides a non-zero vector $\mathbf{s} \in I_1\mathbf{b}_1 \subseteq M$ such that $\|\mathbf{s}\| \leq \gamma\sqrt{d}\Delta_K^{1/(2d)} \cdot (\mathcal{N}(I_1)\mathcal{N}(\mathbf{b}_1))^{1/d}$, by Minkowski's theorem. Combining the latter with the above upper bound on $\mathcal{N}(I_1)\mathcal{N}(\mathbf{b}_1)$ provides the result.   □

4   The Divide-and-Swap Algorithm

We now focus on how to implement Step 3 of Algorithm 3.1, using a CVP oracle for a lattice depending on K only. To handle projected 2-dimensional lattices, the LLL algorithm for integer lattices proceeds like the Gauss/Lagrange reduction algorithm for 2-dimensional lattices. It relies on a divide-and-swap elementary procedure: first shorten the second vector using a $\mathbb{Z}$-multiple of the first one (using a Euclidean division, or, more pedantically, a CVP solver for the trivial lattice $\mathbb{Z}$); then swap these two vectors if the second has become (significantly) shorter than the first one. It has the effect that if this 2-dimensional basis is not reduced, then a swap occurs, and some progress is made towards reducedness of the 2-dimensional basis. This elementary step is repeated as many times as needed to achieve reduction of the lattice under scope. In this section, we generalize this process to rank-2 modules.

We first describe a lattice L that depends only on K and for which we will assume that we possess a CVP oracle. Then, we give an algorithm whose objective is to act as a Euclidean algorithm, i.e., enabling us to shorten an element of $K_\mathbb{R}$ using R-multiples of another. Once we have this generalization of the Euclidean algorithm, we finally describe a divide-and-swap algorithm for rank-2 modules.

4.1 Extending the Logarithm

The lattice L is defined using (among others) the log-unit lattice Λ. However, the Log function does not suffice for our needs. In particular, for $a, b \in K_\mathbb{R}^\times$, the closeness between a and b is not necessarily implied by the closeness of Log a and Log b, because Log does not take into account the complex arguments of the entries of the canonical embeddings of a and b. However, we will need such a property to hold. For this purpose, we hence extend the Log function. For $x \in K_\mathbb{R}^\times$, we define $\overline{\text{Log}}\, x := (\theta_1, \ldots, \theta_{r_1+r_2}, \log|\sigma_1(x)|, \ldots, \log|\sigma_d(x)|)^T$, where $\sigma_i(x) = |\sigma_i(x)| \cdot e^{I\theta_i}$ for all $i \leq r_1 + r_2$ and I is a complex root of $x^2 + 1$. The $\overline{\text{Log}}$ function takes values in $(\pi\mathbb{Z}/2\pi\mathbb{Z})^{r_1} \times (\mathbb{R}/(2\pi\mathbb{Z}))^{r_2} \times \mathbb{R}^d$.

Lemma 4.1. *For $x, y \in K_\mathbb{R}^\times$, we have:*

$$\|x - y\|_\infty \leq \left(e^{\sqrt{2}\|\overline{\text{Log}}x - \overline{\text{Log}}y\|_\infty} - 1\right) \cdot \min(\|x\|_\infty, \|y\|_\infty).$$

Observe that for $t \leq (\ln 2)/\sqrt{2}$, we have $e^{\sqrt{2}t} - 1 \leq 2\sqrt{2}t$.

4.2 The Lattice L

Let $r = \text{poly}(d)$ and $\beta > 0$ be some parameter to be chosen later. Let Λ denote the log-unit lattice. Let $\mathfrak{B}_0 = \{\mathfrak{p}_1, \ldots, \mathfrak{p}_{r_0}\}$ be a set of cardinality $r_0 \leq \log h_K$ of prime ideals generating $\mathcal{C}l_K$, with algebraic norms $\leq 2^{\delta_0}$, with $\delta_0 = O(\log\log|\Delta|)$. We will also consider another set $\mathfrak{B} = \{\mathfrak{q}_1, \ldots, \mathfrak{q}_r\}$ of cardinality r, containing prime ideals (not in $\mathfrak{B}_0$) of norms $\leq 2^\delta$, for some parameters r and $\delta \leq \delta_0$ to be chosen later. We also ask that among these ideals $\mathfrak{q}_j$, at least half of them have an algebraic norm $\geq \sqrt{2^\delta}$. Because we want r such ideals, we should make sure that the number of prime ideals of norm bounded by 2^δ in R is larger than r. This will asymptotically be satisfied if $r \leq O(2^\delta/\delta)$ (by Lemma 2.1). The constraint that at least $r/2$ ideals should have norm larger than $\sqrt{2^\delta}$ is not very limiting, as we expect that roughly $2^\delta - \sqrt{2^\delta} \geq r - \sqrt{r}$ ideals should have algebraic norm between $\sqrt{2^\delta}$ and 2^δ (forgetting about the poly(δ) terms).

We now define L as the lattice of dimension $\nu = 2(r_1 + r_2) + r_0 + r - 1$ (included in $\mathbb{R}^{\nu+1}$) spanned by the columns of the following basis matrix:

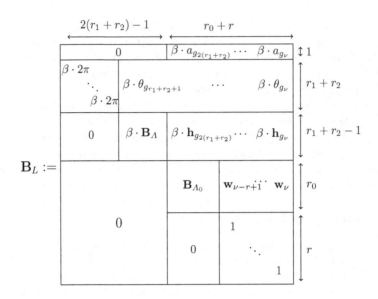

where

- $\mathbf{B}_\Lambda$ is a basis of Λ, and we let $(\mathbf{h}_i)_{r_1+r_2 < i < 2(r_1+r_2)}$ denote its columns;
- $\mathbf{B}_{\Lambda_0}$ is a basis of the lattice $\Lambda_0 := \{(x_i)_i \in \mathbb{Z}^{r_0} : \prod_i \mathfrak{p}_i^{x_i} \text{ is principal}\}$, and we let $(\mathbf{w}_i)_{2(r_1+r_2) \le i \le \nu - r}$ denote its columns;
- for any $g \in K$, we have $a_g = (\log|\mathcal{N}(g)|)/\sqrt{d}$;
- for any $g \in K$, the vector θ_g consists of the first $r_1 + r_2$ entries of $\overline{\mathrm{Log}}(g)$;
- for any $g \in K$, we have $\mathbf{h}_g = i_{H \cap E}(\Pi_H(\mathrm{Log}(g)))$, where Π_H is the orthogonal projection on H and $i_{H \cap E}$ is an isometry mapping $H \cap E$ to $\mathbb{R}^{r_1+r_2-1}$;
- for any $i > r_1 + r_2$, if we parse the bottom $r_0 + r$ coordinates of the i-th column vector as $(w_{i,1}, \ldots, w_{i,r_0}, w'_{i,1}, \ldots, w'_{i,r})$, then we have that $\langle g_i \rangle = \prod_j \mathfrak{p}_j^{w_{ij}} \cdot \prod_j \mathfrak{q}_j^{w'_{ij}}$;
- the g_i's for $i > r_1 + r_2$ are in K and, among them, $g_{r_1+r_2+1}, \ldots g_{2(r_1+r_2)-1}$ are the units of R corresponding to the columns of $\mathbf{B}_\Lambda$.

We now list a few properties satisfied by vectors in this lattice.

Lemma 4.2. *For every vector* $(\beta a \| \beta \theta \| \beta \mathbf{h} \| \mathbf{w} \| \mathbf{w}') \in L \setminus \{0\}$ *(with blocks of dimensions* $1, r_1 + r_2, r_1 + r_2 - 1, r_0$ *and* r*), there exists* $g \in K \setminus \{0\}$ *with*

- $a = (\log|\mathcal{N}(g)|)/\sqrt{d}$
- $\overline{\mathrm{Log}}(g) = (\theta' \| \mathrm{Log}(g))$ *with* $\theta' = \theta \bmod 2\pi$.
- $\mathbf{h} = i_{H \cap E}(\Pi_H(\mathrm{Log}(g)))$
- $\langle g \rangle = \prod_j \mathfrak{p}_j^{w_j} \prod_j \mathfrak{q}_j^{w'_j}$, *where* $\mathbf{w} = (w_1, \cdots, w_{r_0})$ *and* $\mathbf{w}' = (w'_1, \cdots, w'_r)$.

Further, we have that $\|\mathrm{Log}(g)\|_2 = \|(a, \mathbf{h})\|_2$.

4.3   On the Distance of Relevant Vectors to the Lattice

In this section, we make a heuristic assumption on the distance between target vectors of a specific form and the lattice L defined in the previous section. This heuristic is backed with a counting argument and numerical experiments (see the full version). As L is not full rank, we only consider target vectors $\mathbf{t}$ lying in the span of L. Also, as $\mathbf{B}_L$ contains the identity matrix in its bottom right corner, we cannot hope to have a covering radius that is much smaller than $\sqrt{r}$. In our case, the lattice dimension ν will be of the order of r, but in our application we will need a vector of L much closer to $\mathbf{t}$ than $\sqrt{r} \approx \sqrt{\nu}$. In order to go below this value, we only consider target vectors $\mathbf{t}$ whose last r coordinates are almost 0.

Heuristic 1. Assume that there exist some integer $B \leq r$ such that $B \geq 100 \cdot (\log h_K) \cdot \delta_0/\delta$ and that

$$\alpha_0 := \sqrt{2\pi} \left(\left(\frac{2B}{r^{0.96}}\right)^B \cdot \delta B (\det \Lambda) h_K \right)^{1/d} \leq \frac{\ln d}{12 d^{2.5}}.$$

Assume that the scaling parameter β in $\mathbf{B}_L$ is set to $\frac{1}{\alpha_0} \sqrt{\frac{0.01 \cdot B}{2d}}$. Then for any $\mathbf{t} \in \mathrm{Span}(L)$ whose last r coordinates $\mathbf{w}'_t$ satisfy $\|\mathbf{w}'_t\|_2 \leq 0.01 \cdot B/\sqrt{r}$, we have $\mathrm{dist}(\mathbf{t}, L) \leq \sqrt{1.05 \cdot B}$.

Discussion about Heuristic 1. We provide below a counting argument to justify Heuristic 1. We consider the following set of vectors of L, parametrized by $B \leq r$, which we view as candidates for very close vectors to such target vectors:

$$S_B := \{\mathbf{s} = (\beta a_s \| \beta \theta_s \| \beta \mathbf{h}_s \| \mathbf{w}_s \| \mathbf{w}'_s) \in L : \mathbf{w}'_s \in \{-1, 0, 1\}^r \wedge \|\mathbf{w}'_s\|_1 = B\}.$$

We argue that there is a vector in S_B that is very close to $\mathbf{t}$. Our analysis is heuristic, but we justify it with both mathematical arguments and experiments. We are going to examine the vectors $\mathbf{s} \in S_B$ such that $\mathbf{s} - \mathbf{t}$ is reduced modulo L. Let us write $\mathbf{t} = (\beta a_t \| \beta \theta_t \| \beta \mathbf{h}_t \| \mathbf{w}_t \| \mathbf{w}'_t)^T$. We define:

$$\begin{aligned}
S_{B,\mathbf{t}}^{(1)} := \{ (\beta a_s \| \beta \theta_s \| \beta \mathbf{h}_s \| \mathbf{w}_s \| \mathbf{w}'_s) \in L : \quad & \mathbf{w}'_s \in \{-1, 0, 1\}^r \wedge \|\mathbf{w}'_s\|_1 = B, \\
& \mathbf{w}_s - \lfloor \mathbf{w}_t \rceil \in \mathcal{V}(\Lambda_0), \\
& \mathbf{h}_t - \mathbf{h}_s \in \mathcal{V}(\Lambda), \\
& \theta_t - \theta_s \in (-\pi, \pi]^{r_1 + r_2} \},
\end{aligned}$$

where the notation $\mathcal{V}$ refers to the Voronoi cell (i.e., the set of points which are closer to 0 than to any other point of the lattice). The choice of $\mathbf{w}'_s$ fully determines $\mathbf{s} \in S_{B,\mathbf{t}}^{(1)}$, which gives the bound $|S_{B,\mathbf{t}}^{(1)}| = 2^B \cdot \binom{r}{B} \geq (2r/B)^B$.

We consider the following subset of $S_{B,\mathbf{t}}^{(1)}$:

$$S_{B,\mathbf{t}}^{(2)} = S_{B,\mathbf{t}}^{(1)} \cap \{ (\beta a_s \| \beta \theta_s \| \beta \mathbf{h}_s \| \mathbf{w}_s \| \mathbf{w}'_s) \in L : \mathbf{w}_s = \lfloor \mathbf{w}_t \rceil \}.$$

We heuristically assume that when we sample a uniform vector in $S_{B,\mathbf{t}}^{(1)}$, the components $\mathbf{w}_s$ of the vectors $\mathbf{s} \in S_{B,\mathbf{t}}^{(1)}$ are uniformly distributed modulo Λ_0.

Then the proportion of those for which $\mathbf{w}_s = \lfloor \mathbf{w}_t \rceil \bmod \Lambda_0$ is $1/\det(\Lambda_0) = 1/h_K$. Hence, we expect that $|S_{B,\mathbf{t}}^{(2)}| \approx |S_{B,\mathbf{t}}^{(1)}|/h_K$.

We consider the following subset of $S_{B,\mathbf{t}}^{(2)}$, parametrized by $\alpha \leq (\ln d)/(12d^{2.5})$:

$$S_{B,\alpha,\mathbf{t}}^{(3)} = S_{B,\mathbf{t}}^{(2)} \cap \{(\beta a_s \| \beta \theta_s \| \beta \mathbf{h}_s \| \mathbf{w}_s \| \mathbf{w}_s') \in L : \|(\theta_s \| \mathbf{h}_s) - (\theta_t \| \mathbf{h}_t)\|_\infty \leq \alpha\}.$$

We heuristically assume that when we sample a uniform vector in $S_{B,\mathbf{t}}^{(2)}$, the components $(\theta_s, \mathbf{h}_s)$ are uniformly distributed modulo $2\pi\mathbb{Z}^{r_1+r_2} \times \Lambda$. Observe that the first r_1 coordinates of θ_s (corresponding to real embeddings) are either 0 or π. Hence, the probability that $\theta_s = \theta_t$ on these coordinates is 2^{-r_1}. Once these first r_1 coordinates are fixed, the remaining coordinates of $(\theta_s, \mathbf{h}_s)$ have no a priori reason to be bound to a sublattice of $2\pi\mathbb{Z}^{r_2} \times \Lambda$ and we heuristically assume them to be uniformly distributed in $\mathbb{R}^{r_1+2r_2-1}/(2\pi\mathbb{Z}^{r_2} \times \Lambda)$. Overall, the probability that a vector $\mathbf{s} \in S_{B,\mathbf{t}}^{(2)}$ satisfies $\|(\theta_s, \mathbf{h}_s) - (\theta_t, \mathbf{h}_t)\|_\infty \leq \alpha$ is $\approx \frac{\alpha^{r_1+2r_2-1}}{2^{r_1}\cdot(2\pi)^{r_2}\cdot\det(\Lambda)}$. Here, we used the fact that $\sqrt{r_1 + 2r_2 - 1} \cdot \alpha$ is smaller than $\lambda_1(2\pi\mathbb{Z}^{r_2} \times \Lambda)/2$ (recall from preliminaries that $\lambda_1^{(\infty)}(\Lambda) \geq (\ln d)/(6d^2)$). We conclude that

$$|S_{B,\alpha,\mathbf{t}}^{(3)}| \approx |S_{B,\mathbf{t}}^{(2)}| \frac{\alpha^{r_1+2r_2-1}}{2^{r_1}\cdot(2\pi)^{r_2}\cdot\det(\Lambda)} \geq |S_{B,\mathbf{t}}^{(2)}| \frac{\alpha^{d-1}}{(2\pi)^{d/2}\cdot\det(\Lambda)}.$$

Finally, we consider the following subset of $S_{B,\alpha,\mathbf{t}}^{(3)}$:

$$S_{B,\alpha,\mathbf{t}}^{(4)} = S_{B,\alpha,\mathbf{t}}^{(3)} \cap \{(\beta a_s \| \beta \theta_s \| \beta \mathbf{h}_s \| \mathbf{w}_s \| \mathbf{w}_s') \in L : |a_s - a_t| \leq \alpha\}.$$

We will assume that

$$|S_{B,\alpha,\mathbf{t}}^{(4)}| \geq \frac{0.344^B \cdot \alpha\sqrt{d}}{\delta B \cdot r^{0.04 \cdot B}} \cdot |S_{B,\alpha,\mathbf{t}}^{(3)}|.$$

This assumption is backed with mathematical arguments as well as numerical experiments in the full version. Overall, we obtain that

$$
\begin{aligned}
|S_{B,\alpha,\mathbf{t}}^{(4)}| &\geq \frac{0.344^B \cdot \alpha\sqrt{d}}{\delta B \cdot r^{0.04 \cdot B}} \cdot \frac{\alpha^{d-1}}{(2\pi)^{d/2}\cdot\det(\Lambda)} \cdot \frac{1}{h_K} \cdot \left(\frac{2r}{B}\right)^B \\
&\geq \left(\frac{\alpha}{\sqrt{2\pi}}\right)^d \frac{1}{\delta B \cdot \det(\Lambda) \cdot h_K} \left(\frac{0.344 \cdot 2r}{B \cdot r^{0.04}}\right)^B \\
&\geq \left(\frac{\alpha}{\sqrt{2\pi}}\right)^d \frac{1}{\delta B \cdot \det(\Lambda) \cdot h_K} \left(\frac{r^{0.96}}{2B}\right)^B.
\end{aligned}
$$

When the above is ≥ 1, we expect that there exists $\mathbf{s} \in S_{B,\alpha,\mathbf{t}}^{(4)}$. If that is the case, then we have

$$\|\mathbf{s} - \mathbf{t}\|^2 \leq (\beta \cdot \sqrt{2d} \cdot \alpha)^2 + r_0 + \|\mathbf{w}_t' - \mathbf{w}_s'\|^2.$$

By condition on B, we know that $r_0 \leq 0.01 \cdot B$. Also, by choice of $\mathbf{w}_t'$ (and using the fact that $r \geq B$), we have that $\|\mathbf{w}_t' - \mathbf{w}_s'\|^2 \leq (\sqrt{B} + 0.01 \cdot \sqrt{B})^2 \leq 1.03 \cdot B$. Finally, choosing α minimal provides the result.

Numerical Experiments. Heuristic 1 is also backed with numerical experiments. We performed the experiments with r of the order of d^2 (looking forward, this is the value of r that will be used by our algorithm). This means that our lattice L has dimension roughly d^2, and so solving CVP in it quickly becomes impractical. We were still able to check that our heuristic seems correct in cyclotomic number fields of very small degree (up to $d = 8$). More details on these numerical experiments can be found in the full version.

4.4   A "Euclidean Division" over R

We will need the following technical observation that, given $a, b \in K_{\mathbb{R}}$, it is possible to add a small multiple ka of a to b to ensure that $\mathcal{N}(b + ka) \geq \mathcal{N}(a)$.

Lemma 4.3. *For any $a \in K_{\mathbb{R}}^{\times}$ and $b \in K_{\mathbb{R}}$, there exists $k \in [-d, d] \cap \mathbb{Z}$ such that $|\mathcal{N}(b + ka)| \geq |\mathcal{N}(a)|$.*

Note that an integer k such as in Lemma 4.3 can be found efficiently by exhaustive search.

We can now describe our "Euclidean division" algorithm over R. Our algorithm takes as input a fractional ideal $\mathfrak{a}$ and two elements $a, b \in K_{\mathbb{R}}$, and outputs a pair $(u, v) \in R \times \mathfrak{a}$. The first five steps of this algorithm aim at obtaining, for any input (a, b), a replacement (a_1, b_1) that satisfies some conditions. Namely, we would like a_1 to be balanced, i.e., $\|a_1\|$ should not be significantly more than $\mathcal{N}(a_1)^{1/d}$. We also would like b_1 to be not much larger that a_1 and $\mathcal{N}(a_1/b_1)$ to be close to 1. These conditions are obtained by multiplying the element a by an appropriate element of R, and removing a multiple of a from b. Note that we require that the output element v should not be too large. As b is not multiplied by anything, these normalization steps will not impact this output property. After these first five steps, the core of the algorithm begins. It mainly consists in the creation of a good target vector $\mathbf{t}$ in $\mathbb{R}^{\nu+1}$, followed by a CVP computation in the lattice L.

Theorem 4.4 (Heuristic). *Assume that $\mathfrak{a}$ satisfies $\mathrm{c}^{-d} \leq \mathcal{N}(\mathfrak{a}) \leq \mathrm{c}^{d}$, with c as in Lemma 2.3. Assume also that B and r are chosen so that*

$$B \geq \max \left(100 \cdot d \cdot \log[(\rho(R) + d)\mathrm{c}^4], \log h_K \cdot (103 \cdot \frac{\delta_0}{\delta})^2 \right),$$

$$\alpha_0 := \sqrt{2\pi} \left((\frac{2B}{r^{0.96}})^B \cdot \delta B (\det \Lambda) h_K \right)^{1/d} \leq \frac{\varepsilon}{43 \cdot \sqrt{d} \cdot (\rho(R) + d)\mathrm{c}^4 \cdot 2^{0.55 \cdot \delta \cdot B/d}},$$

for some $\varepsilon > 0$. Assume also that $\alpha_0 \leq (\ln d)/(12 d^{2.5})$, and set the scaling parameter β of $\mathbf{B}_L$ as in Heuristic 1. Then, under Heuristic 1 and the heuristics of Lemma 2.3, Algorithm 4.1 outputs a pair $(u, v) \in R \times \mathfrak{a}$ with

$$\|ua + vb\|_\infty \leq \varepsilon \cdot \|a\|_\infty,$$
$$\|v\|_\infty \leq \mathrm{c} \cdot 2^{0.55 \cdot \delta \cdot B/d}.$$

Algorithm 4.1. A Euclidean division over R

Input: A fractional ideal $\mathfrak{a}$, and two elements $a \in K_{\mathbb{R}}^{\times}$ and $b \in K_{\mathbb{R}}$.
Output: A pair $(u, v) \in R \times \mathfrak{a}$.

Computing a better pair (a_1, b_1)
1: Find $s \in \mathfrak{a}^{-1} \setminus \{0\}$ such that $\|s\|_{\infty} \leq c \cdot \mathcal{N}(\mathfrak{a}^{-1})^{1/d}$ as in Lemma 2.3.
2: Find $y \in R \setminus \{0\}$ such that $\|ya\|_{\infty} \leq c \cdot |\mathcal{N}(a)|^{1/d}$ as in Lemma 2.3 (with ideal $\langle a \rangle$).
    Define $a_1 = ya$.
3: Solve CVP in R to find $x \in R$ such that $\|b/(s \cdot a_1) - x\| \leq \rho(R)$.
4: Find $k \in \mathbb{Z} \cap [-d, d]$ such that $|\mathcal{N}(b - xsa_1 + ksa_1)| \geq |\mathcal{N}(sa_1)|$ (see Lemma 4.3).
5: Define $b_1 = b + (k - x)s \cdot a_1$.

Defining the target vector and solving CVP
6: Compute $(w_{t,j})_{j \leq r_0}$ and g_t such that $\mathfrak{a}^{-1} = \prod_j \mathfrak{p}_j^{w_{t,j}} \langle g_t \rangle$. Let $\mathbf{w}_t = (w_{t,j})_{j \leq r_0}$.
7: Let $a_t = (\log \mathcal{N}|b_1/(a_1 g_t)|)/\sqrt{d}$, θ_t be the first $r_1 + r_2$ coordinates of $\overline{\mathrm{Log}}(b_1/(a_1 g_t))$
    and $\mathbf{h}_t = i_{E \cap H}(\Pi_H(\mathrm{Log}(b_1/(a_1 g_t))))$.
8: Define $\mathbf{t} = (\beta a_t \| \beta \theta_t \| \beta \mathbf{h}_t \| \mathbf{w}_t \| \mathbf{0})$.
9: Solve CVP in L with target vector $\mathbf{t}$, to obtain a vector $\mathbf{s}$.

Using $\mathbf{s}$ to create a good ring element
10: Write $\mathbf{s} = (\beta a_s \| \beta \theta_s \| \beta \mathbf{h}_s \| \mathbf{w}_s \| \mathbf{w}'_s)$ and let $g_s \in K^*$ be the associated element as in
    Lemma 4.2.
11: Define the ideal $I = \mathfrak{a} \prod_{j:w_{s,j} - w_{t,j} < 0} \mathfrak{p}_j^{w_{t,j} - w_{s,j}} \prod_{j:w'_{s,j} < 0} \mathfrak{q}_j^{-w'_{s,j}}$.
12: Find $v \in I \setminus 0$ such that $\|v\|_{\infty} \leq c \cdot \mathcal{N}(I)^{1/d}$ as in Lemma 2.3.
13: Define $u' = g_s \cdot g_t \cdot v$.
14: **return** $(u'y + (k - x)sy \cdot v, v)$.

Apart from the CVP calls in R, L_K and L, Algorithm 4.1 runs in quantum polynomial time.

Proof. Throughout the proof, we keep the notations of Algorithm 4.1.

We first prove that $(u, v) \in R \times \mathfrak{a}$. As $s \in \mathfrak{a}^{-1}$ and $x, k, y \in R$, it suffices to prove that $(u', v) \in R \times \mathfrak{a}$. By definition of g_t and g_s, we have

$$\langle g_s g_t \rangle = \mathfrak{a}^{-1} \prod_j \mathfrak{p}_j^{w_{s,j} - w_{t,j}} \prod_j \mathfrak{q}_j^{w'_{s,j}} = J \cdot I^{-1},$$

with $J = \prod_{j:w_{s,j} - w_{t,j} > 0} \mathfrak{p}_j^{w_{s,j} - w_{t,j}} \prod_{j:w'_{s,j} > 0} \mathfrak{q}_j^{w'_{s,j}}$. As the $\mathfrak{p}_j$'s and $\mathfrak{q}_j$'s are integral ideals, we see that $J \subseteq R$ and $I \subseteq \mathfrak{a}$. As $v \in I$, we obtain that $v \in \mathfrak{a}$. Since $g_s \cdot g_t \in JI^{-1}$ and $v \in I$, we also have $u' = g_s g_t v \in JI^{-1}I = J \subseteq R$. This gives our first claim.

As a preliminary step towards bounding $\|ua + bv\|_{\infty} = \|u'a_1 + vb_1\|_{\infty}$, we study the sizes of a_1 and b_1. Using the equality $b_1 = b - xsa_1 + ksa_1$, we have

$$\|b_1\|_{\infty} \leq (\|b/(sa_1) - x\|_{\infty} + |k|) \cdot \|sa_1\|_{\infty} \leq (\rho(R) + d) \cdot \|s\|_{\infty} \cdot \|a_1\|_{\infty}.$$

By definition of a_1, we have $\|a_1\|_\infty \leq c\|a\|_\infty$. By assumption on $\mathfrak{a}$, we also have $\|s\|_\infty \leq c \cdot \mathcal{N}(\mathfrak{a}^{-1})^{1/d} \leq c^2$. Hence, we obtain

$$\|b_1\|_\infty \leq (\rho(R) + d)c^3\|a\|_\infty.$$

Now, by definition of a_1, we know that $\|a_1\|_\infty \leq c \cdot |\mathcal{N}(a_1)|^{1/d}$. Hence, we obtain

$$c^{-1} \leq |\mathcal{N}(b_1/a_1)|^{1/d} \leq c \cdot \frac{\|b_1\|_\infty}{\|a_1\|_\infty} \leq (\rho(R) + d) \cdot c^3.$$

The left inequality is provided by the choice of k at Step 4 (and the fact that $\mathcal{N}(s) \geq \mathcal{N}(\mathfrak{a}^{-1})$).

To bound $\|u'a_1 + vb_1\|_\infty$, we estimate the closeness of $\mathbf{t}$ and $\mathbf{s}$. If $\mathbf{t}$ was in $\mathrm{Span}(L)$, then we could apply Heuristic 1. As this is not necessarily the case, we first need to compute the distance between $\mathbf{t}$ and $\mathrm{Span}(L)$. This is done in the proof of the following lemma, which is provided in the full version.

Lemma 4.5 (Heuristic). *Under the assumptions of Theorem 4.4, we have* $\|\mathbf{s} - \mathbf{t}\|_2 \leq \sqrt{1.06 \cdot B}$.

This lemma implies that

$$\|(a_s\|\theta_s\|\mathbf{h}_s) - (a_t\|\theta_t\|\mathbf{h}_t)\|_2 \leq \sqrt{1.06 \cdot B}/\beta \leq 15 \cdot \sqrt{d} \cdot \alpha_0.$$

By definition of $\mathbf{t}$ and construction of L, this means that

$$\|\overline{\mathrm{Log}}(g_t g_s \cdot a_1/b_1)\|_2 = \|(a_s\|\theta_s\|\mathbf{h}_s) - (a_t\|\theta_t\|\mathbf{h}_t)\|_2 \leq 15 \cdot \sqrt{d} \cdot \alpha_0.$$

Recall that $u'/v = g_t g_s$. Hence we have $\|\overline{\mathrm{Log}}(u'a_1) - \overline{\mathrm{Log}}(vb_1)\|_\infty \leq 15 \cdot \sqrt{d} \cdot \alpha_0$. Using Lemma 4.1, we deduce that

$$\|u'a_1 - vb_1\|_\infty \leq (e^{15 \cdot \sqrt{2d} \cdot \alpha_0} - 1) \cdot \|b_1\|_\infty \cdot \|v\|_\infty$$
$$\leq 43 \cdot \sqrt{d} \cdot \alpha_0 \cdot \|b_1\|_\infty \cdot \|v\|_\infty,$$

where we used the fact that $\alpha_0 \leq (\ln d)/(12d^{2.5})$ and so the exponent should be smaller than $(\ln 2)/\sqrt{2}$ for d large enough. We have already bounded $\|b_1\|_\infty$. We now bound $\|v\|_\infty$. By definition of v, we have $\|v\|_\infty \leq c \cdot \mathcal{N}(I)^{1/d}$. The task is then to provide an upper bound on $\mathcal{N}(I)$. As $IJ = \mathfrak{a} \cdot \prod_j \mathfrak{p}_j^{|w_{s,j} - w_{t,j}|} \cdot \prod_j \mathfrak{q}_j^{|w'_{s,j}|}$, we have:

$$\log\mathcal{N}(IJ) = \log\mathcal{N}(\mathfrak{a}) + \sum_j |w_{s,j} - w_{t,j}|\log\mathcal{N}(\mathfrak{p}_j) + \sum_j |w'_{s,j}| \cdot \log\mathcal{N}(\mathfrak{q}_j)$$
$$\leq \log\mathcal{N}(\mathfrak{a}) + \|\mathbf{w}_s - \mathbf{w}_t\|_1 \cdot \delta_0 + \|\mathbf{w}'_s\|_1 \cdot \delta$$

Recall from Lemma 4.5 that we have $\|\mathbf{s} - \mathbf{t}\|_2 \leq \sqrt{1.06 \cdot B}$. This implies that $\|\mathbf{w}_s - \mathbf{w}_t\|_2, \|\mathbf{w}'_s\|_2 \leq \sqrt{1.06 \cdot B}$. Note that

$$\|\mathbf{w}_s - \mathbf{w}_t\|_1 \leq \sqrt{r_0} \cdot \|\mathbf{w}_s - \mathbf{w}_t\|_2 \leq 1.03 \cdot \sqrt{B \cdot r_0} \leq 0.01 \cdot \frac{\delta}{\delta_0} \cdot B,$$

by assumption on B and the fact that $r_0 \leq \log h_K$. For $\mathbf{w}'_s$, we use the fact that it has integer coordinates, to obtain $\|\mathbf{w}'_s\|_1 \leq \|\mathbf{w}'_s\|_2^2 \leq 1.06 \cdot B$. We thus obtain

$$\log\mathcal{N}(IJ) \leq \log\mathcal{N}(\mathfrak{a}) + 1.07 \cdot \delta \cdot B.$$

As J is integral, this gives an upper bound on $\mathcal{N}(I)$. However this upper bound is not sufficient for our purposes. We improve it by giving an upper bound on $\log\mathcal{N}(IJ^{-1})$, using the fact that the ideal IJ^{-1} is designed to have an algebraic norm close to the one of a_1/b_1. Recall that a_1 and b_1 are constructed so that $\mathcal{N}(a_1/b_1)$ is close to 1, which means that I and J should have roughly the same norm. More precisely, it is worth recalling that $I^{-1}J = \langle g_s g_t \rangle$, and that $\|\overline{\mathrm{Log}}(g_t g_s \cdot a_1/b_1)\|_2 \leq 15 \cdot \sqrt{d} \cdot \alpha_0$. Looking at the first coordinate of the $\overline{\mathrm{Log}}$ vector and multiplying it by $\sqrt{d}$ shows that $|\log|\mathcal{N}(g_s g_t)| + \log|\mathcal{N}(a_1/b_1)|| \leq 15 \cdot d \cdot \alpha_0$. This gives us

$$\log\mathcal{N}(IJ^{-1}) \leq |\log|\mathcal{N}(a_1/b_1)|| + 15 \cdot d \cdot \alpha_0$$

Combining the bounds on $\log\mathcal{N}(IJ)$ and $\log\mathcal{N}(IJ^{-1})$, we finally obtain that

$$\log\mathcal{N}(I) \leq \frac{1}{2} \cdot |\log\mathcal{N}(\mathfrak{a})| + 0.535 \cdot \delta \cdot B + \frac{1}{2} \cdot |\log|\mathcal{N}(a_1/b_1)|| + 7.5 \cdot d \cdot \alpha_0.$$

We have seen that $c^{-1} \leq |\mathcal{N}(b_1/a_1)|^{1/d} \leq (\rho(R) + d) \cdot c^3$. Finally, recall that $c^{-d} \leq \mathcal{N}(\mathfrak{a}) \leq c^d$. Hence, we conclude that $|\log|\mathcal{N}(a_1/b_1)|| + |\log\mathcal{N}(\mathfrak{a})| \leq d \cdot \log((\rho(R) + d) \cdot c^4) \leq 0.01 \cdot B$ by assumption on B. Recall that we assumed that $\alpha_0 \leq (\ln d)/(12d^{2.5}) \leq 1/d$. Hence, we have $d \cdot \alpha_0 \leq 1$. Using the fact that $B \geq 750$ (which is implied by the second term in the max), we obtain $7.5 \cdot d \cdot \alpha_0 \leq 0.01 \cdot B$. We conclude that

$$\log\mathcal{N}(I) \leq 0.55 \cdot \delta \cdot B.$$

Collecting terms and using the assumptions, this allows us to write

$$\|u'a_1 - vb_1\|_\infty \leq 43 \cdot \sqrt{d} \cdot \alpha_0 \cdot \|b_1\|_\infty \cdot \|v\|_\infty$$
$$\leq \alpha_0 \cdot 43 \cdot \sqrt{d} \cdot 2^{0.55 \cdot \delta \cdot B/d} \cdot (\rho(R) + d)c^4 \|a\|_\infty$$
$$\leq \varepsilon \cdot \|a\|_\infty.$$

Finally, the run-time bound follows by inspection. □

We observe that the parameters r and B of Theorem 4.4 can be instantiated as $B = \widetilde{O}(\log|\Delta| + d\log\rho(R))$ and $r^{0.96} = \Theta((1/\varepsilon)^{d/B} \cdot B \cdot 2^{0.55\delta})$. Thanks to the 0.55 in the exponent, this choice of r is compatible with the condition $r \leq O(2^\delta/\delta)$ which was required for the construction of the lattice L (recall that we want r prime ideals of norm smaller than 2^δ). We note also that the constants 0.96 and 0.55 appearing in the exponent can be chosen as close as we want to 1 and 0.5 respectively, by adapting the argument above. Hence, assuming $(1/\varepsilon)^{d/B} = O(1)$, we expect to be able to choose 2^δ as small as $B^{2+\eta}$ for any $\eta > 0$. Overall, the following corollary gives an instantiation of Theorem 4.4 with parameters that are relevant to our upcoming divide-and-swap algorithm.

Corollary 4.6 (Heuristic). *Let* $\varepsilon = 1/2^{\widetilde{O}(\log \Delta_K)/d}$. *For any* $\eta > 0$, *there exists a lattice* L' *of dimension* $\widetilde{O}((\log|\Delta_K| + d\log\rho(R))^{2+\eta})$, *an upper bound* $C = 2^{\widetilde{O}(\log|\Delta_K| + d\log\rho(R))/d}$ *and an algorithm* $\mathcal{A}$ *that achieve the following. Under Heuristic 1 and the heuristics of Lemma 2.3, algorithm* $\mathcal{A}$ *takes as inputs* $a \in K_{\mathbb{R}}^{\times}$, $b \in K_{\mathbb{R}}$ *and an ideal* $\mathfrak{a}$ *satisfying* $c^{-d} \leq \mathcal{N}(\mathfrak{a}) \leq c^d$, *and outputs* $u, v \in R \times \mathfrak{a}$ *such that*

$$\|ua + bv\|_{\infty} \leq \varepsilon \cdot \|a\|_{\infty}$$
$$\|v\|_{\infty} \leq C.$$

If given access to an oracle solving the closest vector problem in L' *in polynomial time, and when restricted to inputs* a, b *belonging to* K, *Algorithm 4.1 runs in quantum polynomial time.*

4.5 The Divide-and-Swap Algorithm

In this subsection, we describe a divide-and-swap algorithm, which takes as input a pseudo-basis of a rank-2 module and outputs a short non-zero vector of this module (for the algebraic norm). In order to do so, we will need to link the Euclidean and algebraic norms of vectors appearing during the execution, and limit the degree of freedom of the ideal coefficients. For this purpose we use a strengthening of the notion of scaled pseudo-bases from Sect. 3.2.

Definition 4.7. *A pseudo-basis* $((I_i, \mathbf{b}_i))_{i \leq n}$, *with* $I_i \subset K$ *and* $\mathbf{b}_i \in K_{\mathbb{R}}^m$ *for all* $i \leq n$, *is said strongly scaled if, for all* $i \leq n$,

$$R \subseteq I_i, \quad \mathcal{N}(I_i) \geq c^{-d} \quad and \quad \|r_{ii}\|_{\infty} \leq c \cdot \mathcal{N}(r_{ii}I_i)^{1/d},$$

where c *is as in Lemma 2.3.*

Algorithm 4.2 below strongly scales a given module pseudo-basis. It is a direct adaptation of Algorithm 3.2 in which the LLL algorithm is replaced by the algorithm from Lemma 2.3 (relying on a CVP oracle for L_K).

Algorithm 4.2. Strongly scaling the ideals.

Input: A pseudo-basis $((I_i, \mathbf{b}_i))_{i \leq n}$ of a module M.
Output: A strongly scaled pseudo-basis $((I_i', \mathbf{b}_i'))_{i \leq n}$ of M.
1: **for** $i = 1$ **to** n **do**
2:     Use Lemma 2.3 to find $s_i \in r_{ii} \cdot I_i \setminus \{0\}$ such that $\|s_i\|_{\infty} \leq c \cdot \mathcal{N}(r_{ii}I_i)^{1/d}$;
3:     Write $s_i = r_{ii} \cdot x_i$, with $x_i \in I_i$;
4:     Define $I_i' = I_i \cdot \langle x_i \rangle^{-1}$ and $\mathbf{b}_i' = x_i \mathbf{b}_i$.
5: **end for**
6: **return** $((I_i', \mathbf{b}_i'))_{i \leq n}$.

Lemma 4.8. *Algorithm 4.2 outputs a strongly scaled pseudo-basis of the module M generated by the input pseudo-basis and preserves the $\mathcal{N}(r_{ii}I_i)$'s. If given access to an oracle that solves CVP in the lattice L_K of Lemma 2.3, and if $M \subseteq R^m$, then it runs quantumly in time polynomial in the input bit-length and in $\log \Delta_K$.*

We can now describe Algorithm 4.3, our divide-and-swap algorithm. During the execution of the algorithm, the R-factor of the current matrix $(\mathbf{b}_1 | \mathbf{b}_2)$ is always computed. The algorithm is very similar to the LLL algorithm in dimension 2, except for Step 4, which is specific to this algorithm. This step ensures that when we swap the vectors, we still obtain a pseudo-basis of the input module. This seems necessary, as our Euclidean division over R involves a multiplication of the second vector by a ring element, and hence the new vector and the second pseudo-basis vector may not span the whole module anymore. At Step 4, note that the gcd is well-defined, as $\langle u \rangle$ and $\langle v \rangle \mathfrak{a}^{-1}$ are integral ideals. As an alternative to Step 4, we could use Lemma 2.5 as in Algorithm 3.1.

Algorithm 4.3. Divide-and-swap.

Input: A pseudo-basis $((\mathfrak{a}_1, \mathbf{b}_1), (\mathfrak{a}_2, \mathbf{b}_2))$ of a module $M \subset K_{\mathbb{R}}^2$.
Output: A vector $\mathbf{v} \in M$.
1: **while** $(\gamma/c)^d \mathcal{N}(r_{22}\mathfrak{a}_2) < \mathcal{N}(r_{11}\mathfrak{a}_1)$ **do**
2:     Strongly scale the pseudo-basis $((\mathfrak{a}_1, \mathbf{b}_1), (\mathfrak{a}_2, \mathbf{b}_2))$ using Algorithm 4.2.
3:     Apply Algorithm 4.1 to $(a, b, \mathfrak{a}) = (r_{11}, r_{12}, \mathfrak{a}_2 \cdot \mathfrak{a}_1^{-1})$ and $\varepsilon = 1/(4c)$. Let (u, v)
         be the output.
4:     Let $\mathfrak{b} = \gcd(\langle u \rangle, \langle v \rangle \mathfrak{a}^{-1})$, find $x \in \mathfrak{a}^{-1}\mathfrak{b}^{-1}$ and $y \in \mathfrak{b}^{-1}$ such that $uy - vx = 1$.
5:     Update $(\mathbf{b}_1, \mathbf{b}_2) \leftarrow (u\mathbf{b}_1 + v\mathbf{b}_2, x\mathbf{b}_1 + y\mathbf{b}_2)$ and $(\mathfrak{a}_1, \mathfrak{a}_2) \leftarrow (\mathfrak{a}_1\mathfrak{b}^{-1}, \mathfrak{a}_2\mathfrak{b})$.
6: **end while**
7: Strongly scale the pseudo-basis $((\mathfrak{a}_1, \mathbf{b}_1), (\mathfrak{a}_2, \mathbf{b}_2))$ using Algorithm 4.2.
8: **return** $\mathbf{b}_1$

Lemma 4.9. *Let $\gamma \geq 4 \cdot C \cdot c^2$, where C is as in Corollary 4.6. Then, given as input a pseudo-basis of a rank-2 module $M \subset K_{\mathbb{R}}^2$, Algorithm 4.3 outputs a vector $\mathbf{v} \in M \setminus \{\mathbf{0}\}$ such that $\mathcal{N}(\mathbf{v}) \leq \gamma^d \lambda_1^{\mathcal{N}}(M)$. Further, if $M \subseteq R^m$ and Algorithms 4.1 and 4.2 run in polynomial time, then Algorithm 4.3 runs in time polynomial in the input bit-length and in $\log \Delta_K$.*

Proof. We only prove here that at each loop iteration, the value $\mathcal{N}(r_{11}\mathfrak{a}_1)$ decreases by a factor at least 2^d. As in the LLL algorithm, this is the main technical part of the proof. The rest of the proof can be found in the full version.

Recall that at the end of Step 2, we have $\|r_{ii}\|_\infty \leq c \cdot \mathcal{N}(r_{ii}\mathfrak{a}_i)^{1/d}$ for $i = 1, 2$. Recall also that Algorithm 4.1 outputs u, v such that $\|ur_{11} + vr_{12}\|_\infty \leq \varepsilon \|r_{11}\|_\infty$ and $\|v\|_\infty \leq C$. The new vector $\mathbf{b}_1$ at the end of the loop iteration is $u\mathbf{b}_1 + v\mathbf{b}_2$.

We compute an upper bound on its algebraic norm:

$$
\begin{aligned}
\mathcal{N}(u\mathbf{b}_1 + v\mathbf{b}_2) &\leq (\sqrt{d})^{-d}\|u\mathbf{b}_1 + v\mathbf{b}_2\|^d = (\sqrt{d})^{-d}\left\|\begin{pmatrix} ur_{11} + vr_{12} \\ vr_{22} \end{pmatrix}\right\|^d \\
&\leq (\sqrt{d})^{-d}\left(\|ur_{11} + vr_{12}\| + \|vr_{22}\|\right)^d \\
&\leq \left(\|ur_{11} + vr_{12}\|_\infty + \|vr_{22}\|_\infty\right)^d \\
&\leq \left(\varepsilon\|r_{11}\|_\infty + \|v\|_\infty \cdot \|r_{22}\|_\infty\right)^d.
\end{aligned}
$$

Using the facts that the basis is c-strongly scaled and that the condition of Step 1 is satisfied, we have:

$$
\begin{aligned}
\mathcal{N}(u\mathbf{b}_1 + v\mathbf{b}_2) &\leq c^d \cdot \left(\varepsilon\mathcal{N}(r_{11}\mathfrak{a}_1)^{1/d} + C \cdot \mathcal{N}(r_{22}\mathfrak{a}_2)^{1/d}\right)^d \\
&\leq c^d \cdot (\varepsilon + C \cdot (c/\gamma))^d \cdot \mathcal{N}(r_{11}\mathfrak{a}_1).
\end{aligned}
$$

Now, by choice of ε and γ:

$$
\mathcal{N}(u\mathbf{b}_1 + v\mathbf{b}_2) \leq c^d \cdot \left(\frac{1}{4c} + \frac{1}{4c}\right)^d \cdot \mathcal{N}(r_{11}\mathfrak{a}_1) = 2^{-d} \cdot \mathcal{N}(r_{11}\mathfrak{a}_1).
$$

Recall that $\mathfrak{a}_1$ is also updated as $\mathfrak{a}_1\mathfrak{b}^{-1}$. Hence, to conclude, we argue that $\mathcal{N}(\mathfrak{a}_1\mathfrak{b}^{-1}) \leq 1$. Note that $\mathcal{N}(\mathfrak{a}_1) \leq 1$ holds due to scaling, and that $\mathcal{N}(\mathfrak{b}) \geq 1$ holds because $\mathfrak{b}$ is integral. Overall, we obtain that $\mathcal{N}(r_{11}\mathfrak{a}_1)$ decreases by a factor $\geq 2^d$ during a loop iteration.   □

Instantiating this lemma with the value of C obtained in Corollary 4.6, we obtain the following corollary.

Corollary 4.10 (Heuristic). *For any number field K and any $\eta > 0$, there exists a lattice L' of dimension $\widetilde{O}((\log|\Delta_K| + d\log\rho(R))^{2+\eta})$, a choice of the approximation factor $\gamma = 2^{\widetilde{O}(\log|\Delta_K| + d\log\rho(R))/d}$ and an algorithm $\mathcal{A}$ such that the following holds. Under Heuristic 1 and the heuristics of Lemma 2.3, algorithm $\mathcal{A}$ takes as input a pseudo-basis of a rank-2 module $M \subset K_{\mathbb{R}}^2$, and outputs a vector $\mathbf{v} \in M$ such that $\mathcal{N}(\mathbf{v}) \leq \gamma^d \lambda_1^{\mathcal{N}}(M)$. If given access to an oracle solving the closest vector problem in L' in polynomial time, and when restricted to modules contained in K^2, Algorithm $\mathcal{A}$ runs in quantum polynomial time.*

Acknowledgments. We thank Léo Ducas for helpful discussions. This work was supported in part by BPI-France in the context of the national project RISQ (P141580), by the European Union PROMETHEUS project (Horizon 2020 Research and Innovation Program, grant 780701) and by the LABEX MILYON (ANR-10-LABX-0070) of Université de Lyon, within the program "Investissements d'Avenir" (ANR-11-IDEX-0007) operated by the French National Research Agency (ANR).

References

[AD17] Albrecht, M.R., Deo, A.: Large modulus Ring-LWE $\geq$ Module-LWE. In: Takagi, T., Peyrin, T. (eds.) ASIACRYPT 2017. LNCS, vol. 10624, pp. 267–296. Springer, Cham (2017). https://doi.org/10.1007/978-3-319-70694-8_10

[Ajt96] Ajtai, M.: Generating hard instances of lattice problems. In: STOC (1996)

[Ajt98] Ajtai, M.: The shortest vector problem in l_2 is NP-hard for randomized reductions. In: STOC (1998)

[BEF+17] Biasse, J.-F., Espitau, T., Fouque, P.-A., Gélin, A., Kirchner, P.: Computing generator in cyclotomic integer rings. In: Coron, J.-S., Nielsen, J.B. (eds.) EUROCRYPT 2017. LNCS, vol. 10210, pp. 60–88. Springer, Cham (2017). https://doi.org/10.1007/978-3-319-56620-7_3

[BF14] Biasse, J.-F., Fieker, C.: Subexponential class group and unit group computation in large degree number fields. LMS J. Comput. Math. **17**, 385–403 (2014)

[BFH17] Biasse, J.-F., Fieker, C., Hofmann, T.: On the computation of the HNF of a module over the ring of integers of a number field. J. Symb. Comput. **80**, 581–615 (2017)

[BGV14] Brakerski, Z., Gentry, C., Vaikuntanathan, V.: (Leveled) fully homomorphic encryption without bootstrapping. ToCT **6**, 13 (2014)

[BP91] Bosma, W., Pohst, M.: Computations with finitely generated modules over Dedekind domains. In: ISSAC (1991)

[BS96] Bach, E., Shallit, J.O.: Algorithmic Number Theory: Efficient Algorithms. MIT Press, Cambridge (1996)

[BS16] Biasse, J.-F., Song, F.: Efficient quantum algorithms for computing class groups and solving the principal ideal problem in arbitrary degree number fields. In: SODA (2016)

[CDW17] Cramer, R., Ducas, L., Wesolowski, B.: Short stickelberger class relations and application to ideal-SVP. In: Coron, J.-S., Nielsen, J.B. (eds.) EUROCRYPT 2017. LNCS, vol. 10210, pp. 324–348. Springer, Cham (2017). https://doi.org/10.1007/978-3-319-56620-7_12

[Cer05] Cerri, J.-P.: Spectres euclidiens et inhomogènes des corps de nombres. Ph.D. thesis, Université Henri Poincaré, Nancy (2005)

[Coh96] Cohen, H.: Hermite and Smith normal form algorithms over Dedekind domains. Math. Comput. **65**, 1681–1699 (1996)

[Fie97] Fieker, C.: Über relative Normgleichungen in älgebraischen Zahlkörpern. Ph.D. thesis, TU Berlin (1997)

[FP96] Fieker, C., Pohst, M.E.: On lattices over number fields. In: Cohen, H. (ed.) ANTS 1996. LNCS, vol. 1122, pp. 133–139. Springer, Heidelberg (1996). https://doi.org/10.1007/3-540-61581-4_48

[FP06] Fieker, C., Pohst, M.E.: Dependency of units in number fields. Math. Comput. **75**, 1507–1518 (2006)

[FS10] Fieker, C., Stehlé, D.: Short bases of lattices over number fields. In: Hanrot, G., Morain, F., Thomé, E. (eds.) ANTS 2010. LNCS, vol. 6197, pp. 157–173. Springer, Heidelberg (2010). https://doi.org/10.1007/978-3-642-14518-6_15

[GLM09] Gan, Y.H., Ling, C., Mow, W.H.: Complex lattice reduction algorithm for low-complexity full-diversity MIMO detection. IEEE Trans. Signal Process. **57**, 2701–2710 (2009)

[Hop98] Hoppe, A.: Normal forms over Dedekind domains, efficient implementation in the computer algebra system KANT. Ph.D. thesis, TU Berlin (1998)

[HPS98] Hoffstein, J., Pipher, J., Silverman, J.H.: NTRU: a ring-based public key cryptosystem. In: Buhler, J.P. (ed.) ANTS 1998. LNCS, vol. 1423, pp. 267–288. Springer, Heidelberg (1998). https://doi.org/10.1007/BFb0054868

[Kan87] Kannan, R.: Minkowski's convex body theorem and integer programming. Math. Oper. Res. **12**, 415–440 (1987)

[KL17] Kim, Taechan, Lee, Changmin: Lattice reductions over Euclidean rings with applications to cryptanalysis. In: O'Neill, Máire (ed.) IMACC 2017. LNCS, vol. 10655, pp. 371–391. Springer, Cham (2017). https://doi.org/10.1007/978-3-319-71045-7_19

[Laa16] Laarhoven, T.: Sieving for closest lattice vectors (with preprocessing). In: Avanzi, R., Heys, H. (eds.) SAC 2016. LNCS, vol. 10532, pp. 523–542. Springer, Cham (2017). https://doi.org/10.1007/978-3-319-69453-5_28

[Lez14] Lezowski, P.: Computation of the euclidean minimum of algebraic number fields. Math. Comput. **83**(287), 1397–1426 (2014)

[LLL82] Lenstra, A.K., Lenstra Jr., H.W., Lovász, L.: Factoring polynomials with rational coefficients. Math. Ann. **261**, 515–534 (1982)

[LM06] Lyubashevsky, V., Micciancio, D.: Generalized compact knapsacks are collision resistant. In: Bugliesi, M., Preneel, B., Sassone, V., Wegener, I. (eds.) ICALP 2006. LNCS, vol. 4052, pp. 144–155. Springer, Heidelberg (2006). https://doi.org/10.1007/11787006_13

[LPR10] Lyubashevsky, V., Peikert, C., Regev, O.: On ideal lattices and learning with errors over rings. In: Gilbert, H. (ed.) EUROCRYPT 2010. LNCS, vol. 6110, pp. 1–23. Springer, Heidelberg (2010). https://doi.org/10.1007/978-3-642-13190-5_1

[LPSW19] Lee, C., Pellet-Mary, A., Stehlé, D., Wallet, A.: An LLL algorithm for module lattices (full version). Cryptology ePrint Archive (2019)

[LS15] Langlois, A., Stehlé, D.: Worst-case to average-case reductions for module lattices. Des. Codes Cryptogr. **75**, 565–599 (2015)

[MG02] Micciancio, D., Goldwasser, S.: Complexity of Lattice Problems: A Cryptographic Perspective. Kluwer Academic Press, Dordrecht (2002)

[Mic01] Micciancio, D.: The hardness of the closest vector problem with preprocessing. Trans. Inf. Theory **47**, 1212–1215 (2001)

[Nap96] Napias, H.: A generalization of the LLL-algorithm over Euclidean rings or orders. J. théorie des nombres de Bordeaux **8**, 387–396 (1996)

[Neu99] Neukirch, J.: Algebraic number theory. In: Grundlehren der Mathematischen Wissenschaften, vol. 322. Springer, Heidelberg (1999). https://doi.org/10.1007/978-3-662-03983-0

[O'M63] O'Meara, O.T.: Introduction to Quadratic Forms. Springer, Heidelberg (1963). https://doi.org/10.1007/978-3-642-62031-7

[PHS19] Pellet-Mary, A., Hanrot, G., Stehlé, D.: Approx-SVP in ideal lattices with pre-processing. In: Ishai, Y., Rijmen, V. (eds.) EUROCRYPT 2019. LNCS, vol. 11477, pp. 685–716. Springer, Cham (2019). https://doi.org/10.1007/978-3-030-17656-3_24

[PR06] Peikert, C., Rosen, A.: Efficient collision-resistant hashing from worst-case assumptions on cyclic lattices. In: Halevi, S., Rabin, T. (eds.) TCC 2006. LNCS, vol. 3876, pp. 145–166. Springer, Heidelberg (2006). https://doi.org/10.1007/11681878_8

[Reg09] Regev, O.: On lattices, learning with errors, random linear codes, and cryptography. J. ACM **56**, 34 (2009)

[RSW18] Rosca, M., Stehlé, D., Wallet, A.: On the Ring-LWE and Polynomial-LWE problems. In: Nielsen, J.B., Rijmen, V. (eds.) EUROCRYPT 2018. LNCS, vol. 10820, pp. 146–173. Springer, Cham (2018). https://doi.org/10.1007/978-3-319-78381-9_6

[SE94] Schnorr, C.-P., Euchner, M.: Lattice basis reduction: improved practical algorithms and solving subset sum problems. Math. Program. **66**, 181–199 (1994)

[SMSV14] Morel, I., Stehlé, D., Villard, G.: LLL Reducing with the most significant bits. In: ISSAC (2014)

[SSTX09] Stehlé, D., Steinfeld, R., Tanaka, K., Xagawa, K.: Efficient public key encryption based on ideal lattices. In: Matsui, M. (ed.) ASIACRYPT 2009. LNCS, vol. 5912, pp. 617–635. Springer, Heidelberg (2009). https://doi.org/10.1007/978-3-642-10366-7_36

Order-LWE and the Hardness
of Ring-LWE with Entropic Secrets

Madalina Bolboceanu[1(✉)], Zvika Brakerski[2], Renen Perlman[2],
and Devika Sharma[2]

[1] Bitdefender, Bucharest, Romania
mbolboceanu@bitdefender.com
[2] Weizmann Institute of Science, Rehovot, Israel

Abstract. We propose a generalization of the celebrated Ring Learning with Errors (RLWE) problem (Lyubashevsky, Peikert and Regev, Eurocrypt 2010, Eurocrypt 2013), wherein the ambient ring is not the ring of integers of a number field, but rather an *order* (a full rank subring). We show that our Order-LWE problem enjoys worst-case hardness with respect to short-vector problems in invertible-ideal lattices *of the order*.

The definition allows us to provide a new analysis for the hardness of the abundantly used Polynomial-LWE (PLWE) problem (Stehlé et al., Asiacrypt 2009), different from the one recently proposed by Rosca, Stehlé and Wallet (Eurocrypt 2018). This suggests that Order-LWE may be used to analyze and possibly *design* useful relaxations of RLWE.

We show that Order-LWE can naturally be harnessed to prove security for RLWE instances where the "RLWE secret" (which often corresponds to the secret-key of a cryptosystem) is not sampled uniformly as required for RLWE hardness. We start by showing worst-case hardness even if the secret is sampled from a subring of the sample space. Then, we study the case where the secret is sampled from an *ideal* of the sample space or a coset thereof (equivalently, some of its CRT coordinates are fixed or leaked). In the latter, we show an interesting threshold phenomenon where the amount of RLWE *noise* determines whether the problem is tractable.

Lastly, we address the long standing question of whether high-entropy secret is sufficient for RLWE to be intractable. Our result on sampling from ideals shows that simply requiring high entropy is insufficient. We therefore propose a broad class of distributions where we conjecture that hardness should hold, and provide evidence via reduction to a concrete lattice problem.

The full version of this work which contains details and full proofs is available at https://eprint.iacr.org/2018/494.

Bitdefender—A large portion of this study was conducted while visiting the Weizmann Institute of Science, Israel, supported by the European Union Horizon 2020 Research and Innovation Program via Project PROMETHEUS (Grant 780701).

Weizmann Institute of Science—Supported by the Israel Science Foundation (Grant No. 468/14), Binational Science Foundation (Grants No. 2016726, 2014276), and by the European Union Horizon 2020 Research and Innovation Program via ERC Project REACT (Grant 756482) and via Project PROMETHEUS (Grant 780701).

S. D. Galbraith and S. Moriai (Eds.): ASIACRYPT 2019, LNCS 11922, pp. 91–120, 2019.
https://doi.org/10.1007/978-3-030-34621-8_4

Keywords: Ring-LWE · Lattice problems · Entropic secrets

1   Introduction

The Learning with Errors (LWE) problem, as introduced by Regev [38], provides a convenient way to construct cryptographic primitives whose security is based on the hardness of *lattice problems*. The assumption that LWE is intractable was used as a basis for various cryptographic designs, including some cutting edge primitives such as fully homomorphic encryption (FHE) [14], and attribute based encryption (ABE) for general policies [9,25]. Two of the most appealing properties of the LWE problem are the existence of a reduction from *worst-case* lattice problems [12,34,36,38] (which is most relevant to this work), and its conjectured post-quantum security.

On the other hand, one of the shortcomings of the LWE assumption is the relatively high computational complexity and large instance size (as a function of the security parameter) that it induces. This results, for example, in LWE-based encryption schemes having long keys and ciphertexts, and also high encryption complexity. It was known since the introduction of the NTRU cryptosystem [29] and more rigorously in [30,37] that these aspects can be significantly improved by relying on lattices that stem from algebraic number theory.[1] In [31,32], Lyubashevsky, Peikert and Regev defined an algebraic number theoretic analog of the LWE problem, called Ring-LWE (RLWE). Similar to Regev's original result, they showed that RLWE is as hard as solving worst-case *ideal lattice* problems.

Ring-LWE and its extensions quickly became a useful resource for the construction of various cryptographic primitives [2,3,7,8,11,13,15,21,23,26] (an extremely non-exhaustive list of examples). RLWE is appealing due to its improved efficiency, and its provable security guarantee based on the hardness of worst case (ideal) lattice problems. However, in concrete instantiations, parameters are not set based on provable hardness guarantees, but rather on the minimal parameters that prevent known and conceivable attacks, in order to achieve the best possible efficiency. In the case of RLWE, parameters are set way beyond the regime where we have provable guarantee in terms of choice of security parameter and, most relevant to this work, in terms of sampling secrets from different distributions than those for which provable security applies.[2] While a gap between the provable and concrete security properties of a cryptosystem is expected, one would at least like to make sure that changing the distribution does not make the problem qualitatively easy. In other words, we would like to show that the problem remains at least asymptotically hard with the new distributions.

Over the years, it has been shown that the LWE problem is quite robust to changes in the prescribed distribution, thus providing desired evidence for the safety of using the assumption in various settings. More precisely, it was shown that LWE hardness holds even if the secret (a vector that, very roughly,

---

[1] This inefficiency is common to cryptographic constructions based on "generic" lattices. Indeed, NTRU was introduced before the LWE assumption was formulated.

[2] Sometimes this is done not for efficiency but for functionality purposes, e.g. [15].

represents the coordinates of a hidden lattice point) is not sampled uniformly, as tradition, but is rather leaked [1, 20] or is chosen from a binary distribution of sufficient entropy [12, 24] (with obvious loss coming from the secret having lower entropy). It is almost trivial to verify that if the LWE secret is chosen uniformly from a linear subspace of its prescribed space, then security degrades gracefully with the dimension of the space of secrets. (We note that there has also been much work on modifying the *noise distribution* of LWE, e.g. [6, 33]. However, the focus of this work is the distribution of secrets.)

Much less is known for RLWE. This is because its algebraic structure (which is the very reason for efficiency gains) prevents the use of techniques like randomness extraction that are instrumental to the aforementioned LWE robustness results.

This Work. Motivated by the task to investigate the behavior of the RLWE problem on non-uniform secret distributions, we present new tools to prove security in some cases and insecurity in others. The main tool that we introduce is a generalization of the RLWE problem that we call Order-LWE, and prove a worst-case hardness result for this problem.[3] We show that the Order-LWE abstraction naturally implies a new proof for a previous result in the literature [40] with new and comparable parameters.

We justify that the formulation of Order-LWE is quite useful in exploring variants of RLWE where the distribution of secrets has some algebraic structure (a special case of secrets from a subfield was studied in [22]). We prove Order-LWE hardness (and thus worst-case hardness) when the secret is sampled from any subring of the prescribed space. We use this approach to address the fundamental question of whether any distribution of secrets with sufficiently high entropy implies RLWE hardness. We show that in some settings, RLWE with uniform secrets is intractable (under conservative worst-case ideal lattice assumptions), but a slight decrease in entropy leads to a complete break. This is the case when the distribution of secrets is supported over an *ideal* (or a coset thereof). On the other hand, we show that increasing the *noise* in the RLWE instance can compensate for the deficiency in secret entropy in this setting.

Finally, we address the more ambitious goal of proving security for secrets with no algebraic structure. Since, as we mentioned above, high entropy is insufficient as condition by itself for security, we identify a family of high-entropy distributions that capture (at least approximately) many of the relaxed variants of RLWE. We show that a particular (average case) hardness assumption implies hardness for this class of distributions.

Paper Organization. We provide an overview of our results and techniques below. Section 2 contains preliminaries and definitions. The Order-LWE problem is formally defined in Sect. 3, where the worst case hardness reduction is proved as well. The new hardness result for PLWE appears in Sect. 4. We then present

---

[3] As a reader with background in algebraic number theory would speculate, this is a setting where RLWE is instantiated respective to *orders* in a number field, rather than its ring of integers.

our results on sampling secrets from subrings in Sect. 5, on sampling secrets from ideals in Sect. 6, and finally on sampling secrets from k-wise independent distributions in Sect. 7.

Due to space constraints, some material was deferred to the full version.

1.1 Background

Recall that in the LWE problem, a secret vector s is sampled from $\mathbb{Z}_q^n$ for some modulus q; an adversary gets oracle access to samples of the form $(a, b = \langle a, s \rangle + e \pmod{q})$ where each $a \in \mathbb{Z}_q^n$ is uniform and e is a small integer, say sampled from a discrete Gaussian with parameter $\ll q$. The adversary's goal is to distinguish this oracle from the one where $b \in \mathbb{Z}_q$ is random.

In the RLWE problem, the sample spaces are also vector spaces over $\mathbb{Z}_q$ but with a ring structure. In this high level overview, for the sake of simplicity of notation and algebraic structure, we restrict to the case where the ring is the ring of integers in the power-of-two cyclotomic field $\mathbb{Q}[x]/(x^n + 1)$. An interested reader may see Sect. 2 for precise definitions in the general case. The cyclotomics is a particularly simple case: the so called *ring of integers* in this case is the ring of polynomials $R = \mathbb{Z}[x]/(x^n + 1)$. In this setting, the RLWE problem with modulus q is as follows: sample a random secret $s \in R_q = R/qR$, and provide the adversary with oracle access to samples of the form $(a, as + e) \in R_q^2$, where a is uniform and e is sampled from some "small" noise distribution (for our purposes, think of e as polynomial with Gaussian coefficients $\ll q$). The arithmetics is over R_q, and the goal is to distinguish these samples from uniform R_q^2 samples.[4] For this overview, we will assume for simplicity that q is a prime, and focus on the setting (which is most commonly used in cryptography) where q splits completely as an ideal in R into a product of n distinct prime ideals. (In the case of the cyclotomics, this condition amounts to $q \equiv 1 \pmod{2n}$.) By the Chinese Remainder Theorem, the quotient $R_q := R/qR \simeq \mathbb{Z}[x]/(x^n + 1, q)$ is isomorphic to $\mathbb{Z}_q^n$ and hence an element in R_q can be represented as a vector of n elements in $\mathbb{Z}_q$ with pointwise addition and multiplication. This is called the CRT representation of elements in R_q and for $c \in R_q$, we denote its CRT coordinates by $c[1], \ldots, c[n]$.

1.2 Our Results

The Order-LWE Problem. We formulate a version of R-LWE where R is replaced by an order $\mathcal{O}$ in K. An order in a number field is a subring of R which has full rank (i.e. can be described as a $\mathbb{Z}$-span of exactly n elements).[5]

---

[4] An informed reader may notice that in the formal RLWE definition, s needs to be sampled from the dual of R_q, and e needs to be small in the so called "canonical embedding". However, in the cyclotomic setting these distinction makes little difference and our choice makes the presentation simpler. Another simplifying choice for the exposition is to only consider discrete noise distributions.

[5] The full-rank condition arises naturally in applications as we discuss below.

We furthermore generalize the modulus of the RLWE equations, and instead of taking the equations modulo (the ideal generated by) the integer q, we allow to mod out by any ideal in the order. We call this problem Order-LWE and denote it as $\mathcal{O}$-LWE. More precisely, we define two variants of the problem $\mathcal{O}$-LWE and $\mathcal{O}^\vee$-LWE which have a duality relation between them (and which are equivalent to each other when $\mathcal{O} = R$). As explained above, R-LWE is a special case of $\mathcal{O}$-LWE (and, in a different notation, of $\mathcal{O}^\vee$-LWE).

Recalling that Ring-LWE was shown to be as hard to solve as worst-case lattice problems over ideal lattices from the ring R, we propose an analogous claim for Order-LWE. Using similar techniques as those used for proving Ring-LWE hardness, but with some necessary adaptations, we show that solving $\mathcal{O}$-LWE is at least as hard as solving short-vector problems on a class of lattices that is defined by the set of invertible ideals in the order $\mathcal{O}$. This result generalizes the known result on R-LWE (note that in R all ideals are invertible). For $\mathcal{O}^\vee$-LWE, worst-case hardness follows for lattices whose *dual* is an invertible $\mathcal{O}$-ideal (again, in R this holds for all ideals). We mention that these sets of lattices coincide in the case when the dual of the order $\mathcal{O}$ is an invertible $\mathcal{O}$-ideal.

We show that using a larger order makes the $\mathcal{O}^\vee$-LWE problem harder, and in that sense R-LWE is harder than any other $\mathcal{O}^\vee$-LWE problem. This is the case even though formally the set of duals of (invertible) $\mathcal{O}$-ideals is disjoint from the set of R-ideals. We believe that this is due to the fact that any $\mathcal{O}$-ideal lattice can be (efficiently) mapped to an R ideal that contains it as a sublattice.

See Sect. 3 for a formal and general definition of $\mathcal{O}$-LWE, its dual $\mathcal{O}^\vee$-LWE and the respective worst-case hardness results.

A Corollary: New Hardness for Polynomial-LWE. Our definition of Order-LWE gives insight on the hardness of other computational problems underlying cryptographic constructions; specifically, the Polynomial-LWE problem (PLWE) [13,41]. In PLWE, s and a are simply random polynomials with integer coefficients modulo a polynomial f and an integer q, and the noise e is a polynomial with small coefficients. It is evident that the PLWE problem provides the simplest interface for LWE over polynomial rings. In many useful cases, for example the power-of-two cyclotomic case, it is straightforward to relate PLWE and RLWE. However, for general polynomials f the connection is far from immediate, since the ring of integers of an arbitrary number field does not look like $\mathbb{Z}[x]/(f)$. Recently, Rosca, Stehlé and Wallet [40] showed a reduction relating the hardness of PLWE in the general case from RLWE and thus from worst-case lattice problems.

We observe that we can straightforwardly address this problem using our Order-LWE machinery. The ambient space for the PLWE problem is the ring $\mathbb{Z}[x]/(f)$, for a polynomial $f \in \mathbb{Z}[x]$. This ring is a subring of full rank of the ring of integers of $K := \mathbb{Q}[x]/(f)$, and hence indeed an order. Therefore translation between PLWE and $\mathcal{O}$-LWE has two aspects: "reshaping" the noise distribution (identically to [40]), and syntactic mapping of the secret to the dual domain. The [40] reduction requires a few additional steps and in this sense our reduction is more direct.

Pinpointing the exact relation of our reduction to the one of [40] is not straightforward. The class of lattices for which we show worst-case hardness is different (and in fact formally disjoint) from the class of lattices in [40]. This is because the hardness result of Order-LWE deals with the worst case lattice problems on invertible $\mathcal{O}$-ideals. However, any $\mathcal{O}$-ideal can be translated into R-ideal which contains it as a sublattice. It therefore appears that R-lattices as in [40] may provide stronger evidence of intractability. On the other hand, the approximation factor achieved by our reduction is never larger and in many cases should be much smaller than that achieved by [40], depending on the specific number field.

This result suggests that perhaps it is instructive to think about orders where objects can be represented and operated on efficiently (more efficiently than over R), and in those orders $\mathcal{O}$-LWE could be a simple way to argue about the security of a cryptosystem with simpler interface than RLWE. We did not explore this avenue further. See Sect. 4 for the full details of our PLWE proof and comparison with [40].

Ring-LWE with Secrets From a Subring/Order. We consider the hardness of the Ring-LWE problem, in the setting where the secret s is sampled from some subset with algebraic structure. As we described above, the proper distribution of secrets is uniform over the ring R_q (R modulo q). In this paper we consider a subring of this ring, but we note that this subring must still contain qR, since Ring-LWE equations are taken modulo q. This naturally imposes full-rank condition on the subring and thus orders naturally arise again. Indeed, we consider distributions that are uniform over $\mathcal{O}_q = \mathcal{O}/q\mathcal{O}$ for an order $\mathcal{O}$.

To motivate the setting of sampling the secret from a subring and illustrate its importance, we start with an analogy with (standard) LWE. In the LWE context, if the secret is sampled from a k-dimensional linear subspace of $\mathbb{Z}_q^n$, the problem easily translates to an LWE instance where n is replaced by k. In the ring setting, the rich algebraic structure makes the task of defining and analyzing such straightforward transformations much more involved. Previous works [3,11,22] considered the notion of *ring-switching* which implies the hardness of RLWE when the secret is sampled from the ring of integers of a *subfield* of the field K. However, such transformations do not apply when K has no subfields of dimension k or no proper subfields at all. Our proposed setting allows to sample s from a subring of R_q that is isomorphic to $\mathbb{Z}_q^k$, for $1 \le k \le n$ and thus provides an algebraic analog of the linear subspace property.

One can view the subring property also in terms of the CRT coordinates (when q splits over R). A subring of R_q that is isomorphic to $\mathbb{Z}_q^k$ is in one-to-one correspondence with an onto mapping $\alpha : [n] \to [k]$ as follows; sample k elements $r_1, \ldots, r_k$ from $\mathbb{Z}_q$ uniformly, and set the CRT coordinates of an element $s \in R_q$ as $s[j] = r_{\alpha(j)}$, for $j \in [n]$. One can verify that this set forms a subring.

In Sect. 5, we show that the RLWE problem with secret sampled from an order $\mathcal{O}$ is harder than the $\mathcal{O}^\vee$-LWE problem, which in turn is harder than the worst case problems on the duals of (invertible) $\mathcal{O}$-ideal lattices. In fact, given two orders $\mathcal{O}' \subseteq \mathcal{O}$, we show that $\mathcal{O}^\vee$-LWE is at least as hard as $\mathcal{O}'^\vee$-LWE (albeit with increase in noise which is comparable to the norm of a minimal generating set of $\mathcal{O}^\vee$ over $\mathcal{O}'^\vee$). Since $\mathcal{O} \subseteq R$, this shows that R-LWE is harder than $\mathcal{O}^\vee$-LWE with appropriate noise increase. This result is in a similar flavor to the one of [22] which shows that R-LWE in a field K is harder than R'-LWE in a subfield $K' \subseteq K$. Since R', the ring of integers of K', is contained in R as a subring, our result implies hardness in this setting as well.

Ring-LWE with Secrets From Ideals, and High-Entropy Secrets. As we already mentioned, understanding the behavior of Ring-LWE in the setting where the secret is sampled from an arbitrary high-entropy distribution is a central subject of inquiry in the area. We show that if we sample s from a "dense" ideal (or equivalently zero-out a few of its CRT coordinates), then we may end up with a distribution that is high-entropy on one hand but makes Ring-LWE insecure on the other. More concretely, consider RLWE samples where the secret s is sampled from R_q such that its j-th CRT coordinate, $s[j]$, is uniformly chosen from $\mathbb{Z}_q$, for all $j \in T$, a randomly chosen subset of $[n]$, and $s[j] = 0$, for $j \notin T$. This is equivalent to choosing s uniformly from $\mathcal{P}_q := \mathcal{P}/qR$, where $\mathcal{P}$ is the ideal of R that contains qR.[6] If $|T| = k$ and we denote $\epsilon = \frac{k}{n}$, then the distribution of secret will have min-entropy $(1 - \epsilon)n \log q$. A good running example is ϵ which is a small constant (e.g. $\epsilon = 0.1$).

One can view this as a more structured analog of an LWE instance with a composite modulus $q = p_1 p_2$, where the secret $\mathbf{s}$ is a multiple of p_1. It is straightforward to see that, in such a LWE instance, if the magnitude of the error e is sufficiently smaller than p_1, this instance can be easily solved by dividing b by p_1 and rounding to the nearest integer, thereby yielding a noiseless set of equations modulo p_2. However, if the noise magnitude is sufficiently larger than p_1, then the instance is secure (intuitively, since one can essentially view it as scaling up of a mod p_2 LWE instance).

Things are more involved in the ring setting as we cannot just round to the nearest integer. Instead, we can interpret the factors of q as lattices. If the lattice corresponding to $\mathcal{P}$ has a good decoding basis, it means that we can recover e when small enough. We also provide a ring analog of the complementary result by showing that if the noise is sufficiently large, then RLWE hardness holds. The latter is done by viewing the instance, again, as a scaled up RLWE instance modulo $\mathcal{Q}$, only now $\mathcal{Q}$ is not an integer but an ideal. Our $\mathcal{O}$-LWE generalization of RLWE allows us to derive RLWE hardness in this setting.

Thus, we show that high entropy of secrets alone is insufficient to argue RLWE security and demonstrate an interplay between the entropy of the secret and the amplitude of noise. Interestingly, we exhibit a threshold phenomenon where the RLWE instance with secret sampled from an ideal is insecure if the error is modestly below the threshold of roughly $q^{-(1-\epsilon)}$, and secure if it is modestly above this value. The "modest" factors depend on the number field,

---

[6] As we hinted above, s is actually an element of the dual of R which is not a ring and doesn't have ideals, however there is a natural translation between the dual and primal domain that captures the CRT/ideal structure. See Sect. 6 for the formal treatment.

but correspond to a fixed polynomial in the degree n in the cyclotomic case. The formal and general analysis of these results appears in Sect. 6.

Ring-LWE with Secrets From a k-Wise Independent Distribution. Given that a general result for high-entropy distributions cannot be achieved, we consider in the final section of the paper a subclass of high-entropy distributions. These distributions do not adhere to uniform sampling from an algebraic structure but instead have the following property; the marginal distribution over any subset of k CRT coordinates is jointly (statistically close to) uniform.[7] In terms of entropy, such distributions must have min-entropy at least $k \log q$, and this entropy is also spread evenly across all k-tuples of CRT coordinates.

We speculate that the k-wise independence condition is sufficient for obtaining RLWE hardness. However, we are unable to show this via worst-case hardness. Instead, we define an average case problem, which we call Decisional Bounded Distance Decoding on a Hidden Lattice (HLBDD) and show that the RLWE problem with secret sampled from a k-wise distribution is at least as hard as this problem. In HLBDD, the adversary needs to distinguish between a random oracle on R_q and an oracle of the following form. Upon initialization of the oracle, a set $T \subseteq [n]$ of cardinality k is sampled. For every oracle call, the oracle generates elements v, e as described next, and returns $v + e \pmod{q}$. For the element v, the CRT coordinate $v[j]$ is random if $j \in T$, and 0 otherwise. The element e is a small noise element, say Gaussian. This can be viewed as the decisional version of the bounded distance decoding (BDD) problem on the ideal lattice $\mathcal{I} := \prod_{j \in T} \mathfrak{p}_j$ (where $\{\mathfrak{p}_i\}_i$ are the prime factors of the ideal qR), since the element v is sampled from $\mathcal{I}$. We stress that this is the hidden version as T is sampled randomly at the invocation of the oracle, causing $\mathcal{I}$ to be hidden. As in the standard BDD problem, we can consider HLBDD with worst-case noise and also with arbitrary noise distributions. Given the current understanding of the hardness of lattice problems, discrete Gaussian noise seems natural.

The HLBDD assumption is similar to one made in [28]. However, they only require $k = n/2$, whereas we attempt to take k to be very small, e.g. $k = n^{0.1}$. We assert that the hardness of the problem relies crucially on the set T being chosen at random in the beginning of the experiment rather than being fixed throughout. In other words, we cannot allow preprocessing that depends on T. This is because computing a good basis for the ideal lattice $\mathcal{I}$, defined by T, makes the HLBDD problem easy. It is also important to mention that T itself does not need to be known to the adversary; in this sense HLBDD resembles the approximate GCD problem [19]. Lastly, we note that it is sufficient for our purposes to limit the adversary to only make 2 oracle calls. Namely, the problem is to distinguish two samples $(v_1 + e_1, v_2 + e_2)$ from two uniform elements in R_q. Despite our efforts, we were unable to find additional corroboration to the hardness of this problem and we leave it as an interesting open problem to characterize its hardness.

Let us try to motivate and justify our assertion that the class of k-wise independent distributions is meaningful. Indeed, this class captures the spirit

---

[7] A computational variant is also possible, but needs to carefully define the indistinguishability experiment.

of some of the heuristic entropic distributions that were considered for RLWE. For example, consider the representation of the secret s as a formal polynomial modulo q (recall that $R_q \simeq \mathbb{Z}[x]/(f, q)$ is a ring of polynomials). If each coefficient of s is sampled from a Gaussian so that the total distribution has sufficient entropy (slightly above the necessary $k \log q$), then this distribution will be k-wise independent (as follows from a standard "smoothing" argument). This shows that sampling secrets with very low norm does not violate security under our new assumption. While it was previously known that sampling the secret from the noise distribution keeps security intact (also known as RLWE in Hermite Normal Form [4]), we are not aware of a proof of security when s is chosen from a narrower distribution than the error. This can be seen as a step in the direction of matching the robustness of LWE results [12, 24], that show that LWE remains hard even with high entropy binary secrets. We note that low norm secrets are of importance in the FHE literature (e.g. [11, 26, 27]). In fact, in the HElib implementation [26, 27] the secret is chosen to be a random extremely sparse polynomial. Heuristically, it seems plausible that random sparse polynomials should translate into k-wise independent distributions but we do not have a proof for this speculation as yet. (Intuitively, this follows from the fact that the translation between the coefficient and CRT representation is a linear transformation defined by a Vandemonde matrix. In order to prove k-wise independence we need to show that any subset of k rows of the Vandermonde matrix constitutes a deterministic extractor from a uniform distribution over a Hamming ball. Analogous theorems exist in other contexts, but we do not have a proof as of yet.)

Another example of an interesting k-wise independent distribution is the "entropic RLWE" formulation that came up in the obfuscation literature [15]. That setting consists of a large number of public elements $s_1, \ldots, s_m$, sampled from the noise distribution (which is Gaussian in the polynomial coefficient representation and thus can be shown to be k-wise independent in the CRT representation). The secret is generated by sampling a binary vector $\vec{z} = (z_1, \ldots, z_m)$ and outputting $s = \prod s_i^{z_i}$. Using the leftover hash lemma, one can show that so long as $\vec{z}$ has entropy sufficiently larger than $k \log q$, the resulting distribution will be k-wise independent as well. It is worth noting that in order to achieve the strongest notion of security for their obfuscator, [15] use $\vec{z}$ with entropy $\ll \log q$ to which our technique does not directly apply.

See Sect. 7 for more details on this result.

2   Preliminaries

For a vector $\mathbf{x}$ in $\mathbb{C}^n$ and $p \in [1, +\infty)$, we mean by ℓ_p norm $\|\mathbf{x}\|_p = (\sum_i |x_i|^p)^{1/p}$ and by ℓ_∞ norm $\|\mathbf{x}\|_\infty = \max_i |x_i|$. We refer to ℓ_2 norm if p is omitted. Let $\mathcal{D}$ be a distribution. When writing $x \leftarrow \mathcal{D}$ we mean sampling an element x according to the distribution $\mathcal{D}$. Similarly, for a finite set Ω, we denote by $x \xleftarrow{\$} \Omega$ sampling an element x from Ω uniformly at random. For two distributions $\mathcal{D}_1$ and $\mathcal{D}_2$ over the same measurable set Ω, we consider their *statistical distance* as $\Delta(\mathcal{D}_1, \mathcal{D}_2) =$

$\frac{1}{2}\int_\Omega |\mathcal{D}_1(x) - \mathcal{D}_2(x)| dx$. When the support of $\mathcal{D}$ is a finite set Ω, we define the *entropy* of $\mathcal{D}$ to be $H(\mathcal{D}) := \sum_{\omega \in \Omega} \mathcal{D}(\Omega) \cdot \log_2 (1/\mathcal{D}(\omega))$. In a similar way, its *min-entropy* is defined by $H_\infty(\mathcal{D}) := \min_{\omega \in \Omega} \log_2 (1/\mathcal{D}(\omega))$. It is easy to verify that $H(\mathcal{D}), H_\infty(\mathcal{D}) \le \log_2 |\Omega|$ with equality if and only if $\mathcal{D}$ is the uniform distribution over Ω. When discussing computational problems, we consider by default the standard (nonuniform) polynomial time adversarial model.

We use standard notations and definitions of lattices, Gaussians, and textbook material in algebraic number theory. See full version for detailed definitions.

Lattice Problems. Let $\mathcal{L}$ be a lattice in H represented by a basis $\mathbf{B}$ and let $\mathbf{e}+\mathcal{L}$ be a lattice coset represented by its unique representative $\overline{\mathbf{e}} = (\mathbf{e} + \mathcal{L}) \cap \mathcal{P}(\mathbf{B})$ in the fundamental parallelepiped $\mathcal{P}(\mathbf{B}) := \mathbf{B} \cdot [-1/2, 1/2)^n$ of $\mathbf{B}$. We state the standard lattice problems.

Definition 2.1 (Shortest Independent Vectors Problem). *For an approximation factor $\gamma = \gamma(n) \ge 1$ and a family of lattices $\mathfrak{L}$, the $\mathfrak{L}$-SIVP$_\gamma$ problem is: given a lattice $\mathcal{L} \in \mathfrak{L}$, output n linearly independent lattice vectors of norm at most $\gamma \cdot \lambda_n(\mathcal{L})$.*

Definition 2.2 (Discrete Gaussian Sampling). *For a family of lattices $\mathfrak{L}$ and a function γ that maps lattices from $\mathfrak{L}$ to $G := \{\mathbf{r} \in (\mathbb{R}^+)^n : r_{s_1+s_2+i} = r_{s_1+i}, \text{ for } 1 \le i \le s_2\}$, the $\mathfrak{L}$-DGS$_\gamma$ problem is: given a lattice $\mathcal{L} \in \mathfrak{L}$ and a parameter $\mathbf{r} \ge \gamma(\mathcal{L})$, output an independent sample from a distribution that is within negligible statistical distance of $D_{\mathcal{L},\mathbf{r}}$.*

Definition 2.3 (Bounded Distance Decoding). *For a family of lattices $\mathfrak{L}$ and a function δ that maps lattices from $\mathfrak{L}$ to positive reals, the $\mathfrak{L}$-BDD$_\delta$ problem is: given a lattice $\mathcal{L} \in \mathfrak{L}$, a distance bound $d \le \delta(\mathcal{L})$, and a coset $\mathbf{e} + \mathcal{L}$ where $\|\mathbf{e}\| \le d$, output $\mathbf{e}$.*

Lemma 2.4 (Babai's round-off algorithm [5], [32, Claim 2.10]). *For every family of lattices $\mathfrak{L}$, there is an efficient algorithm that given as input a lattice $\mathcal{L} \in \mathfrak{L}$, a set of linearly independent vectors $\{v_1, v_2, \ldots, v_n\}$ in $\mathcal{L}^*$ and a coset $e + \mathcal{L}$ such that $|\langle e, \overline{v_i} \rangle| \le \frac{1}{2}$, solves $\mathfrak{L}$-BDD$_\delta$ for $\delta(\mathcal{L}) = \frac{1}{2\lambda_n(\mathcal{L}^*)}$.*

Definition 2.5 (Gaussian Decoding Problem [36]). *For a lattice $\mathcal{L} \subset H$ and a Gaussian parameter $g > 0$, the $\mathsf{GDP}_{\mathcal{L},g}$ problem is: given a coset $\mathbf{e} + \mathcal{L}$ where $\mathbf{e} \in H$ was drawn from D_g, find $\mathbf{e}$.*

2.1   Algebraic Number Theory

Cancellation of Ideals. The next lemma is a generalization of [31, Lemma 2.15]. It is crucially used to make a BDD instance and a DGS sample into an Order-LWE instance in the hardness result in Sect. 3. Generally speaking, the lemma allows us to cancel invertible factors in the quotient $\mathcal{IL}/\mathcal{IJL}$ to yield an isomorphism onto $\mathcal{L}/\mathcal{JL}$ by multiplying by an appropriate "tweak" factor. The proof of the lemma uses a generalization of the Chinese Remainder Theorem adapted for ideals over orders. A proof is provided in the full version.

Lemma 2.6. *Let $\mathcal{I}, \mathcal{J}$ be integral ideals in an order $\mathcal{O}$ and let $\mathcal{L}$ be a fractional $\mathcal{O}$-ideal. Assume that $\mathcal{I}$ is invertible. Given the associated primes of $\mathcal{J}$, $\mathfrak{p}_1, \mathfrak{p}_2, \cdots, \mathfrak{p}_k$, and an element $t \in \mathcal{I} \backslash \bigcup_{i=1}^{k} \mathfrak{p}_i \mathcal{I}$ the map*

$$\theta_t : \mathcal{L}/\mathcal{J}\mathcal{L} \rightarrow \mathcal{I}\mathcal{L}/\mathcal{I}\mathcal{J}\mathcal{L}$$
$$x \mapsto t \cdot x$$

is well-defined, and induces an isomorphism of $\mathcal{O}$-modules. Moreover, θ_t is efficiently inverted given $\mathcal{I}, \mathcal{J}, \mathcal{L}$ and t. Finally, such t can be computed given $\mathcal{I}$ and $\mathfrak{p}_1, \mathfrak{p}_2, \cdots, \mathfrak{p}_k$.

We present a "counting lemma" whose proof is provided in the full version.

Lemma 2.7. *Let K be a degree n number field and $\mathcal{O}$ an order, let q be a rational prime. Let $\mathcal{Q}$ be an invertible $\mathcal{O}$-ideal and $\mathfrak{q}_1, \ldots, \mathfrak{q}_k$ be the associated primes of $q\mathcal{O}$. Then there exists $u \in \mathcal{Q}^{-1} \backslash \bigcup_i \mathcal{Q}^{-1}\mathfrak{q}_i$ of norm $\|u\|_\infty \leq O\left(\frac{n\sqrt{\log n}\Delta_{\mathcal{O}}^{1/n}}{N(\mathcal{Q})^{1/n}} \right)$.*

2.2 The Ring-LWE Problem

Let $q \geq 2$ be a (rational) integer. Let $\mathbb{T} = K_{\mathbb{R}}/R^\vee$ denote a torus in the Minkowski space. For any fractional ideal $\mathcal{I}$ of R, let $\mathcal{I}_q := \mathcal{I}/q\mathcal{I}$.

Definition 2.8 (Ring-LWE Distribution). *For $s \in R_q^\vee$, referred to as "the secret", and an error distribution ψ over $K_{\mathbb{R}}$, a sample from the R-LWE distribution $A_{s,\psi}$ over $R_q \times \mathbb{T}$ is generated by sampling $a \xleftarrow{\$} R_q$, $e \leftarrow \psi$, and outputting $(a, b = a \cdot s/q + e \mod R^\vee)$.*

Definition 2.9 (Ring-LWE, Average-Case Decision Problem). *Let φ be a distribution over $R_q^\vee$, and let Υ be a distribution over a family of error distributions, each over $K_{\mathbb{R}}$. The* average-case *Ring-LWE decision problem, denoted $R\text{-}LWE_{q,\varphi,\Upsilon}$, is to distinguish between independent samples from $A_{s,\psi}$ for a random choice of a "secret" $s \leftarrow \varphi$, and an error distribution $\psi \leftarrow \Upsilon$, and the same number of uniformly random and independent samples from $R_q \times \mathbb{T}$.*

Definition 2.10 (Following [36, Definition 6.1]). *Fix an arbitrary $f(n) = \omega(\sqrt{\log n})$. For a real $\alpha > 0$, a distribution sampled from Υ_α is an elliptical Gaussian $D_{\mathbf{r}}$, where $\mathbf{r} \in G$ is sampled as follows: for each $1 \leq i \leq s_1$, sample $x_i \leftarrow D_1$ and set $r_i^2 = \alpha^2(x_i^2 + f^2(n))/2$. For each $s_1 + 1 \leq i \leq s_1 + s_2$, sample $x_i, y_i \leftarrow D_{1/\sqrt{2}}$ and set $r_i^2 = r_{i+s_2}^2 = \alpha^2(x_i^2 + y_i^2 + f^2(n))/2$.*

Theorem 2.11 ([36, Theorem 6.2]). *Let K be an arbitrary field of degree n and $R = \mathcal{O}_K$ its ring of integers. Let $\alpha = \alpha(n) \in (0,1)$, and let $q = q(n) \geq 2$ be a (rational) integer such that $\alpha q \geq 2\omega(1)$. There is a polynomial-time quantum reduction from $\mathfrak{J}(R)\text{-}DGS_\gamma$ to $R\text{-}LWE_{q,U(R_q^\vee),\Upsilon_\alpha}$, where*

$$\gamma = \max\left\{ \eta(\mathcal{L}) \cdot \sqrt{2}/\alpha \cdot \omega(1), \sqrt{2n}/\lambda_1(\mathcal{L}^\vee) \right\} .$$

3   Order-LWE: Definition, Variants and Worst-Case Hardness

The ring of integers R of a number field K plays a central role in the definition and use of the Ring-LWE problem. However, the ring of integers is a special member of a family of rings in a number field, known as *orders*. We present a generalization of Ring-LWE which we call Order-LWE, and show that similar to Ring-LWE it also enjoys worst-case hardness, but with respect to a different set of lattices. Generalizing the problem to the setting of orders also exposes a difference between two variants of Ring-LWE that are indeed identical when considering the ring of integers, but are distinct for general orders. Some background on algebraic number theory and particularly on orders can be found in Sect. 2.1.

In the original R-LWE definition [31], the secret s was sampled from the dual of the ring of integers $R^\vee$ (modulo q), and the coefficients a were sampled from R (modulo q). We similarly define $\mathcal{O}$-LWE as a sequence of noisy linear univariate equations where the secret is sampled from $\mathcal{O}^\vee$ and the coefficients are sampled from $\mathcal{O}$. As pointed out in [31], a dual version where s is sampled from R and a from $R^\vee$ can also be defined, and is equivalent to the original one. Indeed some followup works used the alternative definition (e.g. [22]). In the context of orders, we show that this distinction can make a difference. We denote the dual version by $\mathcal{O}^\vee$-LWE. While we are able to show worst-case hardness reductions for both $\mathcal{O}$-LWE and $\mathcal{O}^\vee$-LWE, the classes of lattices for which worst-case hardness holds is different for the two variants; one is the dual of the other. Our definition also generalizes R-LWE in another dimension, by allowing to take equations modulo arbitrary ideals, and not necessarily modulo (an ideal generated by) a rational integer q. In this section we define the variants of Order-LWE and present the worst-case hardness results.

To set up the problems, let K be a number field, and let $\mathcal{O}$ be an order in it. Let $\mathcal{Q}$ be an integral $\mathcal{O}$-ideal, and let $u \in (\mathcal{O} : \mathcal{Q}) := \{x \in K : x\mathcal{Q} \subseteq \mathcal{O}\}$. For fractional $\mathcal{O}$-ideals $\mathcal{J}$ and $\mathcal{L}$, define $\mathcal{J}_\mathcal{L} := \mathcal{J}/\mathcal{J}\mathcal{L}$, and let $\mathbb{T}_{\mathcal{O}^\vee} := K_\mathbb{R}/\mathcal{O}^\vee$.

Definition 3.1 ($\mathcal{O}$-LWE Distribution). *For $s \in \mathcal{O}_\mathcal{Q}^\vee$ and an error distribution ψ over $K_\mathbb{R}$, a sample from the $\mathcal{O}$-LWE distribution $\mathcal{O}_{s,\psi,u}$ over $\mathcal{O}_\mathcal{Q} \times \mathbb{T}_{\mathcal{O}^\vee}$ is generated by sampling $a \xleftarrow{\$} \mathcal{O}_\mathcal{Q}$, $e \leftarrow \psi$ and outputting $(a, b = u \cdot (a \cdot s) + e \mod \mathcal{O}^\vee)$.*

Definition 3.2 ($\mathcal{O}$-LWE, Average-Case Decision Problem). *Let φ be a distribution over $\mathcal{O}_\mathcal{Q}^\vee$ and let Υ be a distribution over a family of error distributions, each over $K_\mathbb{R}$. The average-case $\mathcal{O}$-LWE decision problem, denoted $\mathcal{O}$-LWE$_{(\mathcal{Q},u),\varphi,\Upsilon}$, is to distinguish between independent samples from $\mathcal{O}_{s,\psi,u}$, for a random choice of a "secret" $s \leftarrow \varphi$, and an error distribution $\psi \leftarrow \Upsilon$, and the same number of uniformly random and independent samples from $\mathcal{O}_\mathcal{Q} \times \mathbb{T}_{\mathcal{O}^\vee}$.*

When the secret is sampled from the uniform distribution over $\mathcal{O}_\mathcal{Q}^\vee$, we sometimes omit it from the subscript. Observe that when $\mathcal{O} = \mathcal{O}_K$, $\mathcal{Q} = q\mathcal{O}_K$ and $u = 1/q$, the $\mathcal{O}$-LWE problem coincides with the Ring-LWE problem.

In our definition of an $\mathcal{O}$-LWE distribution, the secret $s \in \mathcal{O}_{\mathcal{Q}}^{\vee}$ and $a \in \mathcal{O}_{\mathcal{Q}}$. One can also consider a dual variant of $\mathcal{O}$-LWE where $a \in \mathcal{O}_{\mathcal{Q}}^{\vee}$ and $s \in \mathcal{O}_{\mathcal{Q}}$. In general, these two variants are not equivalent, unlike in the case of Ring-LWE (see Remark 3.5), but for special orders $\mathcal{O}$ they are, namely for orders $\mathcal{O}$ such that their duals $\mathcal{O}^{\vee}$ are invertible as $\mathcal{O}$-ideals. For example, if f is the minimal polynomial of the number field K, then the ring $\mathcal{O} = \mathbb{Z}[x]/(f)$ is an order in K, whose dual is invertible.

Definition 3.3 ($\mathcal{O}$-LWE Distribution). *For $s \in \mathcal{O}_{\mathcal{Q}}$ and an error distribution ψ over $K_{\mathbb{R}}$, a sample from the $\mathcal{O}^{\vee}$-LWE distribution $\mathcal{O}_{s,\psi,u}^{\vee}$ over $\mathcal{O}_{\mathcal{Q}}^{\vee} \times \mathbb{T}_{\mathcal{O}^{\vee}}$ is generated by sampling $a \xleftarrow{\$} \mathcal{O}_{\mathcal{Q}}^{\vee}$, $e \leftarrow \psi$, and outputting $(a, b = u \cdot a \cdot s + e \mod \mathcal{O}^{\vee})$.*

Definition 3.4 ($\mathcal{O}$-LWE, Average-Case Decision Problem). *Let φ be a distribution over $\mathcal{O}_{\mathcal{Q}}$, and let Υ be a distribution over a family of error distributions, each over $K_{\mathbb{R}}$. The* average-case $\mathcal{O}^{\vee}$-LWE decision problem, *denoted $\mathcal{O}^{\vee}$-LWE$_{(\mathcal{Q},u),\varphi,\Upsilon}$, is to distinguish between independent samples from $\mathcal{O}_{s,\psi,u}^{\vee}$, for a random choice of a "secret" $s \leftarrow \varphi$, and an error distribution $\psi \leftarrow \Upsilon$, and the same number of uniformly random and independent samples from $\mathcal{O}_{\mathcal{Q}}^{\vee} \times \mathbb{T}_{\mathcal{O}^{\vee}}$.*

As before, when the secret is sampled from the uniform distribution over $\mathcal{O}_{\mathcal{Q}}$, we sometimes omit it from the subscript. Similar to the case of $\mathcal{O}$-LWE, when $\mathcal{O} = \mathcal{O}_K$, $\mathcal{Q} = q\mathcal{O}_K$ and $u = 1/q$, the $\mathcal{O}^{\vee}$-LWE problem coincides with the variant of the Ring-LWE problem where a is sampled from $R^{\vee}/qR^{\vee}$ and s is sampled from R/qR.

Remark 3.5. The $\mathcal{O}$-LWE problem and the $\mathcal{O}^{\vee}$-LWE problem are equivalent as long as $\mathcal{O}^{\vee}$ is an invertible $\mathcal{O}$-ideal. By Lemma 2.6, the invertibility of $\mathcal{O}^{\vee}$ yields an isomorphism from $\mathcal{O}_{\mathcal{Q}}^{\vee}$ to $\mathcal{O}_{\mathcal{Q}}$ induced by multiplication by $t \in (\mathcal{O}^{\vee})^{-1}$. Therefore, the samples of the form $(a, b = u \cdot a \cdot s + e \mod \mathcal{O}^{\vee})$ are transformed to $(a' = a \cdot t, b' = b = u \cdot a' \cdot s' + e \mod \mathcal{O}^{\vee})$, where $a' = a \cdot t \in \mathcal{O}_{\mathcal{Q}}$ and $s' = s \cdot t^{-1} \in \mathcal{O}_{\mathcal{Q}}^{\vee}$. In the particular case of $\mathcal{O}$ being the ring of integers, we obtain the equivalence between Ring-LWE and the variant of Ring-LWE previously described.

The $\mathcal{O}^{\vee}$-LWE definition is inspired by [22], where the authors show that for the variant of Ring-LWE with a from the dual and s from the ring, problem becomes harder as the number field grows. In Sect. 5, we prove an analogue of this result for the set of orders under inclusion, i.e., the bigger the order is, the harder the $\mathcal{O}^{\vee}$-LWE problem is. Since the ring of integers is the maximal order in the field, the Ring-LWE problem is harder than any $\mathcal{O}^{\vee}$-LWE problem.

3.1 Worst-Case Hardness for $\mathcal{O}$-LWE and $\mathcal{O}^{\vee}$-LWE

We now state the hardness results of the $\mathcal{O}$-LWE and $\mathcal{O}^{\vee}$-LWE problems and derive the hardness of the Ring-LWE problem (see Theorem 2.11) as a special case. We begin by generalizing Definition 2.10 of the error distribution Υ_{α} to be elliptical according to u.

Definition 3.6. *Fix an arbitrary* $f(n) = \omega(\sqrt{\log n})$. *For* $\alpha > 0$ *and* $u \in K$, *a distribution sampled from* $\Upsilon_{u,\alpha}$ *is an elliptical Gaussian* $D_{\mathbf{r}}$, *where* $\mathbf{r} \in G$ *is sampled as follows: for* $i = 1, \ldots, s_1$, *sample* $x_i \leftarrow D_1$ *and set* $r_i^2 = \alpha^2(x_i^2 + (f(n) \cdot |\sigma_i(u)| / \|u\|_\infty)^2)/2$. *For* $i = s_1 + 1, \ldots, s_1 + s_2$, *sample* $x_i, y_i \leftarrow D_{1/\sqrt{2}}$ *and set* $r_i^2 = r_{i+s_2}^2 = \alpha^2(x_i^2 + y_i^2 + (f(n) \cdot |\sigma_i(u)| / \|u\|_\infty)^2)/2$.

Note that when $u \in K$ satisfies $\sigma_1(u) = \ldots = \sigma_n(u)$ (and therefore is rational), the distribution $\Upsilon_{u,\alpha}$ degenerates to Υ_α. Otherwise, $\Upsilon_{u,\alpha}$ is strictly narrower than Υ_α.

Let $\Im(\mathcal{O})$ be the set of invertible fractional ideals over the order $\mathcal{O}$. Our hardness results for $\mathcal{O}$-LWE and $\mathcal{O}^\vee$-LWE are as follows.

Theorem 3.7. *Let* K *be an arbitrary number field of degree* n *and* $\mathcal{O} \subset K$ *an order. Let* $\mathcal{Q}$ *be an integral* $\mathcal{O}$-ideal, $u \in (\mathcal{O} : \mathcal{Q})$ *and let* $\alpha \in (0,1)$ *be such that* $\alpha/\|u\|_\infty \geq 2 \cdot \omega(1)$. *There is a polynomial-time quantum reduction from* $\Im(\mathcal{O})$-DGS$_\gamma$ *to* $\mathcal{O}$-LWE$_{(\mathcal{Q},u),\Upsilon_{u,\alpha}}$, *where*

$$\gamma = \max\left\{\eta(\mathcal{QL}) \cdot \sqrt{2}\,\|u\|_\infty /\alpha \cdot \omega(1), \sqrt{2n}/\lambda_1\left(\mathcal{L}^\vee\right)\right\}. \tag{1}$$

Theorem 3.8. *Let* K *be an arbitrary number field of degree* n *and* $\mathcal{O} \subset K$ *an order. Let* $\mathcal{Q}$ *be an integral* $\mathcal{O}$-ideal, $u \in (\mathcal{O} : \mathcal{Q})$ *and let* $\alpha \in (0,1)$ *be such that* $\alpha/\|u\|_\infty \geq 2 \cdot \omega(1)$. *There is a polynomial-time quantum reduction from* $\Im(\mathcal{O}) \cdot \mathcal{O}^\vee$-DGS$_\gamma$ *to* $\mathcal{O}^\vee$-LWE$_{(\mathcal{Q},u),\Upsilon_{u,\alpha}}$, *where*

$$\gamma = \max\left\{\eta(\mathcal{QL}) \cdot \sqrt{2}\,\|u\|_\infty /\alpha \cdot \omega(1), \sqrt{2n}/\lambda_1\left(\mathcal{L}^\vee\right)\right\}. \tag{2}$$

We note that the class $\Im(\mathcal{O}) \cdot \mathcal{O}^\vee$ is exactly the class of all lattices whose dual is in $\Im(\mathcal{O})$. Thus we see that the effect of changing the domains of a and s to the dual of their previous domains is that the class of lattices for which the hardness result applies is the dual of the previous class. The classes are the same if $\mathcal{O}^\vee$ itself is an invertible ideal in $\mathcal{O}$. An equivalence between the problems can be shown in this case directly, similar to the setting in Ring-LWE.

Remark 3.9. Consider the special case where $\mathcal{O} = R$, the ideal $\mathcal{Q} = qR$ and $u = 1/q$. Then the $\mathcal{O}^\vee$-LWE$_{(\mathcal{Q},u),\Upsilon_{u,\alpha}}$ problem is equivalent to the R-LWE$_{q,\Upsilon_\alpha}$ problem as mentioned in Remark 3.5. Moreover, the sets $\Im(R) \cdot R^\vee$ and $\Im(R)$ are equal as all fractional R-ideals are invertible, and finally $\eta(\mathcal{QL}) \|u\|_\infty = \eta(\mathcal{L})$ shows that the parameters γ from Theorem 2.11, Theorems 3.7 and 3.8 coincide. In fact, the expression for γ in Theorem 3.7 is achieved in the latter two results when $\mathcal{Q} = q\mathcal{O}$ and $u = 1/q$, for any order $\mathcal{O}$.

Remark 3.10. Another important special case is the R-LWE distribution with an ideal modulus $\mathcal{Q}$ in place of the integer modulus qR. Formally, we let $\mathcal{O}$ be R, choose $u \in (R : \mathcal{Q}) = \mathcal{Q}^{-1}$ such that $\|u\|_\infty = \lambda_1^\infty(\mathcal{Q}^{-1})$ and $\alpha < \sqrt{\log n/n}$. Then the theorem above implies a reduction from $\Im(R)$-DGS$_\gamma$ to R-LWE$_{(\mathcal{Q},u),\Upsilon_{u,\alpha}}$ with γ greater than at most $\Delta_K^{1/n}$ times the γ obtained when $\mathcal{Q} = qR$ and

$u = 1/q$, as in Theorem 2.11. Making a similar comparison with modulus $\mathcal{Q}$, an invertible $\mathcal{O}$-ideal, we get reductions from $\mathfrak{I}(\mathcal{O})$-DGS$_\gamma$ to $\mathcal{O}$-LWE$_{(\mathcal{Q},u),\varUpsilon_{u,\alpha}}$ and from $\mathfrak{I}(\mathcal{O}) \cdot \mathcal{O}^\vee$-DGS$_\gamma$ to $\mathcal{O}^\vee$-LWE$_{(\mathcal{Q},u),\varUpsilon_{u,\alpha}}$ with γ greater than at most $\Delta_{\mathcal{O}}^{1/n}$ times the γ obtained when $\mathcal{Q} = q\mathcal{O}$ and $u = 1/q$ in Theorems 3.7 and 3.8.

We present an overview of the proof of Theorem 3.7 here. For a detailed proof, see the full version of this paper. The proof for Theorem 3.8 is completely analogous and follows by replacing the point of reference from lattices in $\mathfrak{I}(\mathcal{O})$ to their dual. Our proofs are related to ones in previous works and in particular to the one in [36].

Proof Overview. At the heart of the reduction, we use an iterative step that transforms discrete Gaussian samples into slightly narrower ones. Initially, we generate samples from a Gaussian distribution with large enough parameter such that these samples can be generated efficiently. Then we repeatedly apply the iterative step to generate narrower and narrower samples until we obtain the desired parameter.

The proof of the promised iteration has two components. First, we show how to transform an $\mathcal{O}$-LWE solver into a GDP solver given polynomially many discrete Gaussian sample. This lemma assumes that we are given an efficient algorithm that transforms BDD like instances into $\mathcal{O}$-LWE samples. The second step uses a quantum algorithm to generate narrower discrete Gaussian samples via a GDP solver.

4   New Worst-Case Hardness for Polynomial-LWE

The *Polynomial Learning with Errors* problem, or PLWE in short, introduced by Stehlé et al. [41][8] is closely related to both the Ring-LWE and Order-LWE problems. PLWE has an advantage of having very simple interface which is useful for manipulations and thus also for applications and implementations. In a recent work, Rosca, Stehlé and Wallet [40] showed a reduction from worst-case ideal-lattice problems to PLWE. In this section, we show that the hardness of $\mathcal{O}$-LWE that we proved in Sect. 3 implies a different worst-case hardness result for PLWE, essentially by relating it to a different class of lattices than those considered in [31,36]. In what follows we start with an informal description of the PLWE problem, the current hardness result of PLWE, our result and a comparison. This is followed by a more detailed and formal treatment.

4.1   Overview

Consider a number field K defined by an irreducible polynomial f, so that $K = \mathbb{Q}[x]/(f)$. Recall that the Ring-LWE distribution involves elements a and s of the ring of integers $R := \mathcal{O}_K$ and its dual $R^\vee$, respectively. The $\mathcal{O}$-LWE distribution is defined similarly, but with a and s coming from an arbitrary order in K and

---

[8] As "ideal-LWE". The name PLWE was used in [13].

its dual, respectively. In the PLWE setting, both a and s are elements of the ring $\mathcal{O} := \mathbb{Z}[x]/(f)$, i.e. polynomials with integer coefficients in the number field. There are number fields for which $R \neq \mathcal{O}$, however it is always true that $\mathcal{O}$ is an order of K. We highlight that in PLWE, unlike in Ring-LWE and Order-LWE, both a and s are elements of the order itself.[9]

The aforementioned [40] presented a reduction from Ring-LWE to PLWE (see Theorem 4.2 for the formal statement). Their reduction is based on the so called "Cancellation Lemma" (Lemma 2.6) which, informally, allows to "reshape" orders and ideals at the cost of increasing the size of the error. As mentioned above, in PLWE both a and s are elements of $\mathcal{O}$, whereas in Ring-LWE a and s are elements of R and $R^\vee$ respectively. The reduction of [40] applies the Cancellation Lemma to reshape both R and $R^\vee$ into $\mathcal{O}$. We mention that using the Cancellation Lemma to reshape ideals of the ring of integers R is a known technique (see [35, Section 2.3.2]). The novel contribution of [40] is both in analyzing the increase of the error and in reshaping ideals of one order into another.

We suggest an alternative reduction from $\mathcal{O}$-LWE to PLWE in Theorem 4.4. Our reduction is also based on the reshaping procedure, but with a single application of the Cancellation Lemma. More specifically, we only need to reshape $\mathcal{O}^\vee$ into $\mathcal{O}$. We show below that our reduction increases the error by a smaller factor than in the reduction of [40] from Ring-LWE. See Proposition 4.8 for the formal statement.

4.2   Hardness of PLWE

The formal definitions and hardness results follow, along with a more detailed and formal comparison of the results. We let K be a number field of degree n defined by a polynomial f. We denote $\mathcal{O} := \mathbb{Z}[x]/(f)$, and $R := \mathcal{O}_K$. The PLWE distribution and problem are defined as follows.

Definition 4.1 (PLWE Distribution and Problem [41]). *For a rational integer $q \geq 2$, a ring element $s \in \mathcal{O}_q$, and an error distribution ψ over $K_\mathbb{R}/\mathcal{O}$, the* PLWE distribution *over $\mathcal{O}_q \times K_\mathbb{R}/\mathcal{O}$, denoted by $P_{s,\psi}$, is sampled by independently choosing a uniformly random $a \xleftarrow{\$} \mathcal{O}_q$ and an error term $e \leftarrow \psi$, and outputting $(a, b = (a \cdot s)/q + e \mod \mathcal{O})$.*

For a distribution Υ over a family of error distributions, each over $K_\mathbb{R}/\mathcal{O}$, the PLWE decision problem, *denoted* $\mathrm{PLWE}_{q,\Upsilon}$, *is to distinguish between independent samples from $P_{s,\psi}$ for a random choice of $s \xleftarrow{\$} \mathcal{O}_q$, and an error distribution $\psi \leftarrow \Upsilon$, and the same number of uniformly random and independent samples from $\mathcal{O}_q \times K_\mathbb{R}/\mathcal{O}$.*

---

[9] Another difference between Ring/Order-LWE and PLWE is that in the latter, the error distribution is specified using the so called *coefficients embedding*, and not the *canonical embedding*. For the sake of simplicity, we focus on a variant of PLWE which uses the canonical embedding, (called PLWE^σ in [40]) but we call it likewise, and we avoid the distinction between the embeddings. Both hardness results in this section can be further extended to the hardness of PLWE.

For a distribution φ and an element $t \in K$ we denote by $t \cdot \varphi$ the distribution obtained by sampling an element $x \leftarrow \varphi$ and outputting $t \cdot x$. Similarly, for a family distribution Υ, we denote by $t \cdot \Upsilon$ the family obtained by multiplying each distribution by t.

We now turn to present and compare the two worst-case to average-case reductions. Let $\mathcal{C}_\mathcal{O}$ denote the conductor ideal of $\mathcal{O}$. [40] showed the following reduction from Ring-LWE to PLWE.

Theorem 4.2 ([35, Section 2.3.2][40, Theorem 4.2]). *Let $q \geq 2$ be some rational integer such that $qR + \mathcal{C}_\mathcal{O} = R$, and let Υ be a distribution over a family of error distributions, each over $K_\mathbb{R}/\mathcal{O}$. There exists a probabilistic polynomial time reduction from R-LWE$_{q,\Upsilon}$ to PLWE$_{q,t_1 t_2^2 \cdot \Upsilon}$, where $t_1 \in (R : R^\vee) \setminus \bigcup_i \mathfrak{p}_i(R : R^\vee)$ and $t_2 \in \mathcal{C}_\mathcal{O} \setminus \bigcup_i \mathfrak{p}_i \mathcal{C}_\mathcal{O}$, where $\mathfrak{p}_i$'s are the prime ideals of qR.*

Combining the reduction above with the hardness of Ring-LWE stated in Theorem 2.11 we get the following:

Corollary 4.3 (Worst-Case Hardness of PLWE from Ring-LWE). *With the same notations as above, let $\alpha \in (0,1)$ such that $\alpha q \geq 2 \left\| t_1 t_2^2 \right\|_\infty \omega(1)$. There is a reduction from $\mathfrak{J}(R)$-DGS$_\gamma$ to PLWE$_{q,\Upsilon_\alpha}$ for any*

$$\gamma = \max \left\{ \eta(\mathcal{L}) \cdot \sqrt{2}/\alpha \cdot \left\| t_1 t_2^2 \right\|_\infty \cdot \omega(1), \sqrt{2n}/\lambda_1(\mathcal{L}^\vee) \right\} .$$

We now state the hardness result based on the hardness of $\mathcal{O}$-LWE. First, using a reduction similar to the one from [35, Section 2.3.2], we obtain an analogous reduction from $\mathcal{O}$-LWE to PLWE. The proof is provided in the full version.

Theorem 4.4 *Let $q \geq 2$ be some rational integer, and let Υ be a distribution over a family of error distributions, each over $K_\mathbb{R}/\mathcal{O}$. There exists a probabilistic polynomial time reduction from $\mathcal{O}$-LWE$_{q,\Upsilon}$ to PLWE$_{q,t \cdot \Upsilon}$, where $t \in (\mathcal{O} : \mathcal{O}^\vee) \setminus \bigcup_i \tilde{\mathfrak{p}}_i(\mathcal{O} : \mathcal{O}^\vee)$, where $\tilde{\mathfrak{p}}_i$'s are the associated primes of $q\mathcal{O}$.*

Now, using the hardness of $\mathcal{O}$-LWE from Theorem 3.7 we obtain:

Corollary 4.5 (Worst-Case Hardness of PLWE from Order-LWE). *Let $q \geq 2$ be some rational integer, and let $\alpha \in (0,1)$ be such that $\alpha q \geq 2\|t\|_\infty \omega(1)$. Then there is a reduction from $\mathfrak{J}(\mathcal{O})$-DGS$_\gamma$ to PLWE$_{q,\Upsilon_\alpha}$ for any*

$$\gamma = \max \left\{ \eta(\mathcal{L}) \cdot \sqrt{2}/\alpha \cdot \|t\|_\infty \cdot \omega(1), \sqrt{2n}/\lambda_1(\mathcal{L}^\vee) \right\}.$$

4.3   On the Existence of Small Multipliers

In the following, we let α be a root of the defining polynomial f of the field K, so $\mathbb{Z}[x]/(f) = \mathbb{Z}[\alpha]$. We also denote by $\tilde{\mathfrak{p}}_1, \ldots, \tilde{\mathfrak{p}}_k$ the associated primes of $q\mathcal{O}$, where q is a rational prime. We assert that a short element t as in Theorem 4.4 exists and can be found using the combinatorial argument from Lemma 2.7. Detailed proof of the statements in this section are given in the full version.

Corollary 4.6. *There exists an element t in $(\mathcal{O} : \mathcal{O}^\vee)\backslash \bigcup_i (\mathcal{O} : \mathcal{O}^\vee)\tilde{\mathfrak{p}}_i$ whose norm is bounded by*

$$\|t\| \leq O(n \cdot \sqrt{n \log n} \cdot N(f'(\alpha))^{1/n} \cdot \Delta_{\mathcal{O}}^{1/n}).$$

A short element t as in Theorem 4.4 can also be obtained by sampling via a Gaussian distribution over $(\mathcal{O} : \mathcal{O}^\vee)$ with an appropriately small parameter, exactly as in [40, Theorem 3.1]. We refer to the full version for the formal statement and proof.

Corollary 4.7. *Assuming q is coprime to the conductor, there exists an element t in $(\mathcal{O} : \mathcal{O}^\vee)\backslash \bigcup_i (\mathcal{O} : \mathcal{O}^\vee)\tilde{\mathfrak{p}}_i$ whose norm is bounded with high probability by*

$$\|t\| \leq \sqrt{q} \cdot \sqrt{n} \cdot q^{2\delta} \cdot N(f'(\alpha))^{1/n} \cdot \Delta_{\mathcal{O}}^{1/n} \text{ , where } \delta \in \left[\frac{4n + \log \Delta_{\mathcal{O}}}{n \log q}, 1 \right].$$

The bounds from Corollaries 4.6 and 4.7 do have some common factors, namely $\Delta_{\mathcal{O}}^{1/n}$, $N(f'(\alpha))^{1/n}$ and $\sqrt{n}$. Therefore, it is enough to compare $\sqrt{q} \cdot q^{2\delta}$ and $M \cdot n \cdot \sqrt{\log n}$, respectively, where M is the hidden constant from Corollary 4.6 and $q = poly(n)$. An asymptotic comparison shows that Corollary 4.6 yields a better bound than Corollary 4.7. Also, recall that the latter result assumes q to be coprime to the conductor, whereas the earlier one is true for all rational primes q.

4.4   Comparison

Both Corollaries 4.3 and 4.5 relate PLWE to worst-case ideal lattice problems. The former result involves invertible R-ideals, whereas the family of lattices in the latter is the set of invertible $\mathcal{O}$-ideals. These two families are disjoint, as any ideal can be invertible in at most a single order. In this regard, the two results are incomparable. We note that despite being disjoint, they are known to be related by the conductor ideal, see [17] for reference. We leave exploring this connection to future work.

Another parameter for comparison is the increase of the error in both hardness results. In the proposition below we show that the element t from Theorem 4.4 can be chosen to be smaller than the product $t_1 t_2^2$ from Theorem 4.2. Before doing so we give a short description of the elements t, t_1 and t_2 in the case where qR is coprime to the conductor. We provide more details in the full version.

Let $q \geq 2$ be some rational integer such that $qR + \mathcal{C}_{\mathcal{O}} = R$, and let $qR = \prod_{i=1}^k \mathfrak{p}_i^{e_i}$ be its factorization into prime ideals in R. In this setting, the elements t, t_1, t_2 are any elements satisfying the following conditions:

1. $t \in (\mathcal{O} : \mathcal{O}^\vee)$ and $t \notin (\mathfrak{p}_i \cap \mathcal{O})(\mathcal{O} : \mathcal{O}^\vee)$ for all $i \in [k]$.
2. $t_1 \in (R : R^\vee)$ and $t_1 \notin \mathfrak{p}_i(R : R^\vee)$ for all $i \in [k]$.
3. $t_2 \in \mathcal{C}_{\mathcal{O}}$ and $t_2 \notin \mathfrak{p}_i \mathcal{C}_{\mathcal{O}}$ for all $i \in [k]$. Notice that t_2^2 satisfies same properties.

Proposition 4.8. *With the same notations as above, for any t_1 and t_2 satisfying conditions 2 and 3, the product $t^* = t_1 t_2^2$ satisfies condition 1. In particular, letting t to be the shortest satisfying condition 1, and t_1 and t_2 satisfying conditions 2 and 3, respectively, such that $t_1 t_2^2$ is the shortest, we have that $\|t\|_\infty \leq \|t_1 t_2^2\|_\infty$.*

The proof is provided in the full version of the paper.

5   Sampling Secrets from Orders

In this section, we consider a setting where the RLWE secret s is sampled from a subring of its designated space. For this purpose, it is more convenient to work with the dual version of R-LWE, which is used interchangeably in the literature but according to our notation should be denoted $R^\vee$-LWE. In this variant a is sampled from $R^\vee$ and the secret s comes from R.

More formally, we assume the following setting. Let $q \geq 2$ be a rational prime that splits completely over R.[10] Then $R_q \simeq \mathbb{Z}_q^n$, as rings, and a subring $\overline{S} \subseteq R_q$ isomorphic to $\mathbb{Z}_q^k$ corresponds to an order $\mathcal{O}$ satisfying $qR \subseteq \mathcal{O} \subseteq R$. We show that this version of the Ring-LWE problem is at least as hard as the $\mathcal{O}^\vee$-LWE problem, defined in Sect. 3. In fact, this reduction follows as a corollary of a stronger result that shows that the $\mathcal{O}^\vee$-LWE problem becomes harder as the order becomes bigger. This result can be viewed as an analogue of [22, Lemma 3.1], for the $\mathcal{O}^\vee$-LWE problem, instead of the ring variant.

Given two orders $\mathcal{O}' \subseteq \mathcal{O}$, their duals satisfy $\mathcal{O}^\vee \subseteq \mathcal{O}'^\vee$, as fractional $\mathcal{O}'$-ideals, and there exist $\{v_1, v_2, \ldots, v_m\} \subseteq \mathcal{O}'$ s.t. $\mathcal{O}^\vee = \sum_i \mathcal{O}'^\vee v_i$.[11] We will be interested in finding such set with the smallest possible norm (that is, the ℓ_2 of the concatenation of the canonical embeddings of all v_i).

Theorem 5.1. *Let $\mathcal{O}' \subseteq \mathcal{O} \subset K$ be orders, $\mathcal{Q}'$ an integral $\mathcal{O}'$-ideal and $\mathcal{Q}$ an integral $\mathcal{O}$-ideal such that $\mathcal{Q} = \mathcal{Q}'\mathcal{O}$. Let $\{v_1, v_2, \ldots, v_m\} \subseteq \mathcal{O}'$ be s.t. $\mathcal{O}^\vee = \sum_i \mathcal{O}'^\vee v_i$. Let φ be a distribution over $\mathcal{O}'_{\mathcal{Q}'}$, let Υ be a family of distributions, each over $K_\mathbb{R}/\mathcal{O}'^\vee$, and let $u \in (\mathcal{O}' : \mathcal{Q}')$. Then there is a probabilistic polynomial time reduction from $\mathcal{O}'^\vee$-LWE$_{(\mathcal{Q}',u),\varphi,\Upsilon}$ to $\mathcal{O}^\vee$-LWE$_{(\mathcal{Q},u),\varphi,\langle \Upsilon, \vec{v} \rangle}$, where $\langle \Upsilon, \vec{v} \rangle$ is the distribution (over distributions) that samples $\varphi \leftarrow \Upsilon$, and then outputs the distribution that $e_1, \ldots, e_m$ from φ and outputs $\sum_i e_i v_i$.*

Proof. We describe an efficient transformation that takes m elements from $\mathcal{O}'^\vee_{\mathcal{Q}'} \times K_\mathbb{R}/\mathcal{O}'^\vee$ and outputs an element in $\mathcal{O}^\vee_{\mathcal{Q}} \times K_\mathbb{R}/\mathcal{O}^\vee$. We show that this transformation maps uniform samples to uniform ones, and $\mathcal{O}'^\vee_{s,\psi,u}$ samples to $\mathcal{O}^\vee_{s,\langle \psi, \vec{v} \rangle, u}$ samples for any $s \leftarrow \varphi$ and $\psi \leftarrow \Upsilon$.

Given m samples $\{(a'_i, b'_i)\}_{i \in [m]}$, the transformation outputs $(a = \sum_i a'_i v_i, b = \sum_i b'_i v_i)$. Since $\mathcal{O}^\vee = \sum_i \mathcal{O}'^\vee v_i$ and $\mathcal{Q}\mathcal{O}^\vee = \sum_i \mathcal{Q}'\mathcal{O}'^\vee v_i$, this map is well-defined

---

[10] A similar argument can be stated for the general case. However, this leads to a very cumbersome statement, and we prefer to avoid it.

[11] It is even possible to do so with $m = 2$, but we will be interested in v_i with small norm, in which case it is sometimes beneficial to use larger m.

over the cosets that arise in the distributions and maps uniform distribution over $\mathcal{O}_{\mathcal{Q}'}^{\vee} \times \mathbb{T}_{\mathcal{O}'^{\vee}}$ to uniform distribution over $\mathcal{O}_{\mathcal{Q}}^{\vee} \times \mathbb{T}_{\mathcal{O}^{\vee}}$, respectively.

Now, assume that $\{(a_i', b_i')\}_{i \in [m]}$ are sampled from $\mathcal{O}_{s,\psi,u}'^{\vee}$. Then, for $i \in [m]$, the element $b_i' = u \cdot a_i' \cdot s + e_i'$, where $e_i' \leftarrow \psi$. As

$$b = \sum_i b_i' v_i = u \cdot \sum_i a_i' v_i \cdot s + \sum_i e_i v_i = u \cdot a \cdot s + e,$$

where $e = \sum_i e_i v_i$ is sampled from $\langle \psi, \vec{v} \rangle$, so the tuple (a, b) lies in $\mathcal{O}_{s,\langle \psi, \vec{v} \rangle, u}^{\vee}$. This concludes the proof.   □

Corollary 5.2. Let $\mathcal{O} \subset R$ be an order such that $qR \subseteq \mathcal{O}$, and let $\vec{v} = \{v_1, v_2, \ldots v_m\} \subset \mathcal{O}$ be short elements such that they generate $R^{\vee}$ over $\mathcal{O}^{\vee}$, i.e., $R^{\vee} = \sum_i \mathcal{O}^{\vee} v_i$. Let Υ be a family of error distributions, each over $K_{\mathbb{R}}/\mathcal{O}^{\vee}$. Then, there exists a polynomial time reduction from $\mathcal{O}^{\vee}\text{-LWE}_{(qR,1/q),\Upsilon}$ to $R\text{-LWE}_{q,U(\mathcal{O}/qR),\langle \Upsilon, \vec{v} \rangle}$.

We note that in this case, the elements $v_1, v_2, \ldots, v_m$ are generators of the conductor ideal $\mathcal{C}_{\mathcal{O}}$ as an R-ideal.

Proof. The proof follows easily as a special case of Theorem 5.1; take $\mathcal{O}' = \mathcal{O}$ and $\mathcal{O} = R$, $\mathcal{Q}' = \mathcal{Q} = qR$, and $u = 1/q$.

Important Special Cases. We now discuss a family of orders $\mathcal{O}$ that give rise to interesting secret distributions. Assume that q splits completely in R. Let $qR = \prod \mathfrak{p}_i$ denote the prime factorization of q in R. Then the Chinese remainder theorem yields the following isomorphism:

$$R_q \xrightarrow{\sim} \prod_i \left(\frac{R}{\mathfrak{p}_i} \right) \simeq \mathbb{Z}_q^n$$

$$x \mapsto (x \mod \mathfrak{p}_i)_{i \in [n]}.$$

Let $\Omega = (\Omega_1, \ldots, \Omega_k)$ be a partition of $[n]$ into k disjoint subsets. Define

$$\overline{S} := \{ \mathbf{x} \in \mathbb{Z}_q^n \mid \mathbf{x}_j = \mathbf{x}_{j'}, \text{ for } j, j' \in \Omega_i, \text{ and } i \in [k] \}.$$

Then, the set $\overline{S}$ is isomorphic to $\mathbb{Z}_q^k$ and can be written as $\mathcal{O}/qR$, for an order $\mathcal{O}$ such that $qR \subseteq \mathcal{O} \subseteq R$. Due to Corollary 5.2, we can get hardness of the Ring-LWE with the secret sampled from $\mathcal{O}/qR$ from the hardness of $\mathcal{O}^{\vee}\text{-LWE}$ and therefore, from the hardness of $\mathfrak{I}(\mathcal{O}) \cdot \mathcal{O}^{\vee}\text{-DGS}$.

In particular, if we consider K' a subfield of K, then its ring of integers $R' = \mathcal{O}_{K'}$ is a subring of R. Hence we can consider the following order $\mathcal{O} = R' + qR$ in R and see that $\mathcal{O}/qR$ corresponds to some partition of $[n]$. Using the hardness result of Ring-LWE (in K') and the comparison result in [22, Lemma 3.1], one gets that the Ring-LWE problem (in K) with the secret sampled from $\mathcal{O} = R' + qR$ is at least as hard as $\mathfrak{I}(R')\text{-DGS}$ (in K').

On the other hand, using the hardness result of $\mathcal{O}^{\vee}\text{-LWE}$ (Theorem 3.8) and Corollary 5.2, we get that the Ring-LWE problem (in K) with the secret sampled from $\mathcal{O}$ is at least as hard as $\mathfrak{I}(\mathcal{O}) \cdot \mathcal{O}^\vee$-DGS (in K). One may wonder about a relation between the sets $\mathfrak{I}(R')$ and $\mathfrak{I}(\mathcal{O}) \cdot \mathcal{O}^\vee$. It is not too hard to check that the set of invertible ideals $\mathfrak{I}(R')$ embeds into $\mathfrak{I}(\mathcal{O}) \cdot \mathcal{O}^\vee$ as follows:

$$\mathfrak{I}(R') \hookrightarrow \mathfrak{I}(\mathcal{O}) \cdot \mathcal{O}^\vee$$
$$\mathcal{L}' \mapsto \mathcal{L}'\mathcal{O} \cdot \mathcal{O}^\vee = (\mathcal{L}' + q\mathcal{L}'R) \cdot \mathcal{O}^\vee \ .$$

6 RLWE Secrets from Ideals: High Entropy is Not Enough

In this section we show, perhaps surprisingly, that sampling Ring-LWE secrets from a high-entropy distribution is not necessarily sufficient to guarantee security. Specifically, we investigate the security of Ring-LWE in the case where the distribution of secrets is uniform over an *ideal*. We note that by definition this ideal must be a factor of the ideal qR (i.e. the ideal generated by the modulus q in the number field). In many applications of RLWE it is common to choose a value of q as a prime integer which nevertheless factors (splits completely) as an ideal over R.[12] This means that elements in $R_q = R/qR$ can be represented using the Chinese Remainder Theorem as tuples of elements in $\mathbb{Z}_q = \mathbb{Z}/q\mathbb{Z}$, and the factors of qR represent elements where some of the CRT coordinates are fixed to zero. Indeed, this CRT representation allows for more efficient operations over R_q and is the reason why such values of q are chosen in the first place. It is therefore natural to investigate whether setting a subset of the CRT coordinates to 0 has an effect on security.

We show that, as mentioned above, fixing a very small ϵ fraction of the CRT coordinates (thus only eliminating ϵ fraction of entropy) could result in complete loss of security. That is, we consider a RLWE instance with uniform secret, where worst-case to average-case reductions guarantee plausible security under the current state of the art in algorithms. We then show that by fixing any ϵ fraction of the CRT coordinates, the instance becomes insecure. The value of ϵ depends on the noise level of the RLWE instance. We complement this with a positive result, showing that taking ϵ that is slightly smaller than the aforementioned prescribed value is insufficient and worst-case hardness can still be established.

Notation. We use the standard RLWE setting where K is a number field of degree n with R as its ring of integers. We usually omit the asymptotic terminology to reduce clutter of notation.

Letting $\mathcal{P} \supset qR$ be an integral ideal in R, we let $\mathcal{Q} = q\mathcal{P}^{-1}$ denote its complement with respect to qR. We note that $\mathcal{Q}$ is also an integral ideal in R. We further note that as per the above exposition, $\mathcal{P}$ (or more accurately $\mathcal{P}/qR$) represents a subset of the CRT coordinates defined by the decomposition of qR in R. Since, formally RLWE is defined with secrets distributed over $R^\vee/qR^\vee$, therefore formally we will sample our secret from $\mathcal{P}R^\vee/qR^\vee$ rather than $\mathcal{P}/qR$

---

[12] Sometimes a product of such primes is used.

itself. Note that the two spaces are isomorphic, due to the cancellation lemma (Lemma 2.6), and this distinction is mere formalism. We would also like to point out that the dual of $\frac{1}{q}\mathcal{P}R^\vee$ is $\mathcal{Q}$. To see this observe that $\mathcal{Q}^\vee = (q\mathcal{P}^{-1})^\vee = \frac{1}{q}(\mathcal{P}^{-1})^\vee = \frac{1}{q}\mathcal{P}R^\vee$.

Remark 6.1. As stated above, the results in this section capture secret distributions, or leakage scenarios, where a fixed subset of the CRT coordinates is known to be 0. We remark that all the results below generalize easily to the case where an ϵ fraction of the CRT coordinates is any fixed-value. As one would expect, fixing to some non-zero value corresponds to sampling the secret from a *coset* of an ideal.

6.1   Insecure Instances

Theorem 6.2. *Let K, R be a degree n number field and its ring of integers, respectively. Let $\mathcal{P} \supset qR$ be an integral R-ideal and $\mathcal{Q} = q\mathcal{P}^{-1}$ its complement as described above. There is a non-uniform algorithm such that for any distribution ψ satisfying $\mathrm{Pr}_{e \leftarrow \psi}[\|e\| < 1/(2\lambda_n(\mathcal{Q}))]$ is non-negligible and any distribution φ over $\mathcal{P}R^\vee/qR^\vee$, the algorithm solves search $R\text{-LWE}_{q,\varphi,\{\psi\}}$ with non-negligible probability given a single sample.*

We note that the theorem immediately implies that the same holds for $R\text{-LWE}_{q,\varphi,\Upsilon}$ where Υ is a distribution over distributions ψ so long as the probability to sample ψ as required in the theorem is non-negligible.

Proof. The algorithm will use a non-uniform advice string containing short vectors in $\mathcal{Q}$ that will be used for decoding in the lattice $\mathcal{P}$. Specifically let $V = \{v_1, \ldots, v_n\} \subset \mathcal{Q}$ be a set of $\mathbb{Z}$-linearly independent vectors satisfying $\|v_i\| \le \lambda_n(\mathcal{Q})$.

The algorithm executes as follows. Given the input (a, b), we let $\bar{a}$ denote the inverse of a over R_q. This inverse exists with high probability, and is efficiently computable. It then considers b as an element in $K_\mathbb{R}$ by taking an arbitrary representative. It further applies Babai's BDD algorithm (Lemma 2.4) on input b with respect to the lattice $\frac{1}{q}\mathcal{P}R^\vee$, and with V as the decoding basis. The BDD subroutine returns an element b' in $\frac{1}{q}\mathcal{P}R^\vee$. Finally it returns $s' = q\bar{a}b'$ $(\bmod\ qR^\vee) \in \mathcal{P}R^\vee/qR^\vee$.

We show that the algorithm succeeds whenever a is invertible and e satisfies $\|e\| < 1/(2\lambda_n(\mathcal{Q}))$. These conditions occur concurrently with non-negligible probability. We recall that $b = as/q + e \bmod R^\vee$, and note that $as/q \in \frac{1}{q}\mathcal{P}R^\vee/R^\vee$. Therefore when casting b as an element in $K_\mathbb{R}$ this element is of the form $y + e$ where $y \in \frac{1}{q}\mathcal{P}R^\vee$ and $y = as/q \pmod{R^\vee}$. We furthermore have that, for all i,

$$|Tr(e \cdot v_i)| = \left|\langle \sigma(e), \overline{\sigma(v_i)} \rangle\right| \le \|e\| \cdot \|v_i\| < 1/2 \ .$$

Therefore, recalling that $\mathcal{Q}$ is the dual of $\frac{1}{q}\mathcal{P}R^\vee$, we can apply Lemma 2.4 and deduce that the rounding algorithm recovers the value y.

Finally, the output value will be $s' = q\bar{a}y \pmod{qR^\vee} = q\bar{a}as/q$ $\pmod{qR^\vee} = s \pmod{qR^\vee}$ and the result follows. $\qquad\square$

6.2   Secure Instances

We now show that the vulnerability exposed in Theorem 6.2 can be mitigated by increasing the noise rate of the instance. Indeed, we show that sampling the secret from a distribution with lower entropy preserves worst-case hardness so long as the noise level is sufficiently high. To this end, we use our definition of Order-LWE (Definition 3.2), but with the order $\mathcal{O}$ being the ring of integers R. This definition still generalizes the classical R-LWE since it allows us to consider a "modulus" which is not necessarily an integer q but an ideal $\mathcal{Q}$. In terms of terminology, $\mathcal{O}$-LWE with $\mathcal{O} = R$ will still be denoted R-LWE, so we overload the notation of the standard RLWE problem.

Theorem 6.3. *Let K, R be a degree n number field and its ring of integers respectively. Let $\mathcal{P} \supset qR$ be an integral R-ideal and $\mathcal{Q} = q\mathcal{P}^{-1}$ its complement as described above. Let $u \in \mathcal{Q}^{-1}$ and $\varUpsilon$ be arbitrary. Then there is a polynomial time reduction from R-LWE$_{(\mathcal{Q},u),\varUpsilon}$ to R-LWE$_{q,U(\mathcal{P}R^\vee/qR^\vee),\varUpsilon}$.*

Proof. We prove the theorem by showing a (randomized) transformation T that takes as input $a \in R_\mathcal{Q}$ and outputs $\tilde{a} = T(a) \in R_q$ such that

1. If a is uniform over its domain, then so is $T(a)$ over its domain.
2. For all $s \in R^\vee/\mathcal{Q}R^\vee$, there exists $\tilde{s} \in \mathcal{P}R^\vee/qR^\vee$ s.t. $uas = \tilde{a}\tilde{s}/q \pmod{R^\vee}$, for all $a \in R_\mathcal{Q}$.

If indeed such a transformation exists, then the reduction works as follows. Start by sampling s_0 uniformly from $\mathcal{P}R^\vee/qR^\vee$. Then, given a sequence of samples (a,b) for R-LWE$_{(\mathcal{Q},u),\varUpsilon}$, apply the transformation $(a,b) \rightarrow (\tilde{a},\tilde{b}) = (T(a), b + \tilde{a}s_0/q)$ on each sample and output the resulting samples as R-LWE$_{q,U(\mathcal{P}R^\vee/qR^\vee),\varUpsilon}$ samples. By the properties of the transformation indeed $\tilde{a}$ is uniform, and $\tilde{b} = uas + e + \tilde{a}s_0/q = \tilde{a}(\tilde{s} + s_0)/q + e \pmod{R^\vee}$. Since s_0 is uniform over $\mathcal{P}R^\vee/qR^\vee$ then so is $(\tilde{s} + s_0)$ and indeed the output samples are distributed as required.

The transformation T is as follows. Given a as input, sample a random a' from $\mathcal{Q}/qR$ and output $\tilde{a} = a + a' \pmod{qR}$. The first property holds since a is uniformly distributed over all cosets of a'. As for the second property, define $\tilde{s} = qus \pmod{qR^\vee}$. Since $u \in \mathcal{Q}^{-1}$, it holds that $qu \in \mathcal{P}$ and therefore indeed $\tilde{s} \in \mathcal{P}R^\vee/qR^\vee$. We have

$$\tilde{a}\tilde{s}/q = (a + a')qus/q = uas + ua's \pmod{R^\vee}.$$

Since $a' \in \mathcal{Q}/qR$, $u \in \mathcal{Q}^{-1}$ and $s \in R^\vee/\mathcal{Q}R^\vee$ we have that $ua's = 0 \pmod{R^\vee}$ and the result follows. $\qquad\square$

6.3   A Threshold Phenomenon

Combining the results from Theorems 6.2 and 6.3, we show a (commonly used) setting where reducing the entropy of the secret results in tractability of the RLWE instance on one hand, but either using fully uniform secret (with the same noise level) or a modest increase in the noise level (with the same imperfect secret) results in the problem's intractability being resumed.

We consider the setting where K is a cyclotomic number field (so that we have a good bound on the discriminant Δ_K) and where q is a prime for which the ideal qR splits completely as an ideal over R. The latter condition is the formal description of the fact that elements in R_q (and also in $R_q^\vee$, due to Lemma 2.6) can be written in CRT form as tuples in $\mathbb{Z}_q^n$. We can thus consider secret distributions where k out of the n CRT coordinates are set to be 0, and the remaining $(n - k)$ coordinates are uniform. Naturally, this distribution has entropy $(1 - \frac{k}{n})n \log q$, i.e. $(1 - \frac{k}{n})$-fraction of the full entropy. Formally, this corresponds to sampling the RLWE secret from an ideal $\mathcal{P} \supset qR$ with algebraic norm $N(\mathcal{P}) = q^k$. The formal statement follows and we provide its proof in the full version of the paper.

Corollary 6.4. *Let K, R be a degree n number field and its ring of integers respectively, and assume furthermore that K is cyclotomic. Then for every integer $k \in [0, n]$, letting $\epsilon = k/n$, there exist $q = q_\epsilon = n^{O(1/\epsilon)}$, $\alpha = \alpha_\epsilon = \mathrm{poly}(n)/q$ and a distribution φ over $R_q^\vee$ with entropy $(1 - \epsilon)n \log q$ s.t. R-LWE$_{q,\varphi,\Upsilon_\alpha}$ is solvable in polynomial time.*

On the other hand, solving the problems R-LWE$_{q,\Upsilon_\alpha}$ and R-LWE$_{q,\varphi,\Upsilon_\beta}$ for any $\beta = \alpha \cdot \omega(n^{5/2})$ is as hard as solving $\mathfrak{I}(R)$-DGS$_\gamma$ for $\gamma = \eta \cdot \mathrm{poly}(n^{1/\epsilon})$.

Solving DGS with γ as above corresponds to approximating the Shortest Independent Vector Problem (SIVP) to within $\mathrm{poly}(n^{1/\epsilon})$ factor. At least for constant c, achieving such DGS/SIVP approximation is intractable using current state of the art algorithmic techniques. Therefore, we show a threshold effect in two different aspects. First, in terms of entropy, we show a RLWE problem which is plausibly intractable if the secret is uniformly random, becomes tractable when the entropy is slightly reduced. Second, even if the entropy is reduced, a relatively modest increase in the noise level restores intractability.

7 k-Wise Independent Secrets and Hidden Lattice BDD

In this section, we propose a class of high-entropy distributions for RLWE secrets for which we believe worst-case hardness should hold. As evidence, we show how to prove hardness based on a new average-case lattice problem. This new problem is a decision variant of the bounded distance decoding (BDD) problem. In our variant, BDD is to be solved on an ideal lattice which is sampled from a large family of ideals. It allows us to prove the hardness for distributions of RLWE secrets with norm bounded away from q and whose marginal distribution over this family of ideals (i.e. sampling an element from this distribution and taking its product with the ideal) is indistinguishable from uniform.

The Setting and Notation. In this section, we choose to use a simpler notation at the cost of some restriction on the generality of our discussion. This will allow us to present our results in a more digestible manner. In particular, we limit the discussion to cyclotomic number fields, our modulus to completely splitting, and the regime of the RLWE samples to be over $R_q \times R_q$, i.e. discrete and integral, instead of $R_q \times K_{\mathbb{R}}/R^\vee$.

Formally, we let K be a cyclotomic number field of degree n, and denote its ring of integers by R. In this case, the ideal $R^\vee$ is just a scalar multiple of R, $R^\vee = tR$, for $t \in K$ [18, Theorem 3.7]. Therefore, we can assume that the RLWE distribution is obtained by sampling s from R_q instead of from $R_q^\vee$. Moreover, we consider that the error e from a discrete Gaussian over R. This setting is quite commonly used (and is perhaps the most popular use of RLWE) and its worst case hardness is presented in [32, Lemma 2.23].

Defining k-Wise Independent Distributions. As explained above, we consider the case where q is an integer prime which splits completely, $qR = \prod_{i=1}^{n} \mathfrak{p}_i$, where each $\mathfrak{p}_i \subseteq R$ is a prime ideal. For $k \in [n]$, we define the following family of ideals

$$\mathfrak{P}_k := \left\{ \prod_{i \in T} \mathfrak{p}_i \mid T \subseteq [n], \; |T| = k \right\}.$$

Our class of (perfect/statistical/computational) k-wise independent distributions are those whose marginals are (perfectly/statistically/computationally) indistinguishable from uniform modulo any product of k prime ideals from q, so modulo any $\mathcal{P} \in \mathfrak{P}_k$. Recalling the CRT representation of R_q, this is equivalent to any k-tuple of CRT coordinates being indistinguishable from uniform.

Definition 7.1. *A distribution φ over R_q is (perfectly/statistically/computationally) k-wise independent if the random variables $(s \mod \mathcal{P})$ and $(z \mod \mathcal{P})$ are (perfectly/statistically/computationally) indistinguishable, where $s \leftarrow \varphi$, $z \leftarrow R_q$ and $\mathcal{P} \leftarrow \mathfrak{P}_k$. The asymptotics are over the dimension n and $k = k(n)$ is some integer function.*

Lemma 7.2. *Let φ be k-wise independent, and consider the following probability space. Sample $\mathcal{P} \leftarrow \mathfrak{P}_k$ and let $\mathcal{Q} = \mathcal{P}^{-1}q \in \mathfrak{P}_{n-k}$. Sample $x_1, x_2 \leftarrow \mathcal{Q}/qR$ conditioned on x_1 being invertible modulo $\mathcal{P}$, and $s \xleftarrow{\$} \varphi$. Then the distributions $(x_1, x_1 \cdot s)$ and (x_1, x_2) are indistinguishable.*

Proof. Let s' be any representative of $s \pmod{\mathcal{P}}$. Then $(x_1, x_1 \cdot s) = (x_1, x_1 \cdot s')$ since $x_1 \in \mathcal{Q}$ and $\mathcal{P} + \mathcal{Q} = qR$. Thus, Definition 7.1 implies that $(x_1, x_1 \cdot s)$ is indistinguishable from $(x_1, x_1 \cdot z)$ where z is uniform in R_q.

Now fix any $\mathcal{P}, x_1$, we will show that $x_1 z$ and x_2 are identically distributed. Since x_1 is invertible modulo $\mathcal{P}$, then $x_1 z$ is uniform modulo $\mathcal{P}$. Since $x_1 \in \mathcal{Q}/qR$ it follows that $x_1 z = 0 \pmod{\mathcal{Q}}$. Therefore $x_1 z$ is uniform in $\mathcal{Q}/qR$.

7.1   Hidden-Lattice Decision Bounded Distance Decoding

We first define the hidden lattice BDD (HLBDD) distribution, and then the decisional problem associated with it. We use Gaussian noise but other noise distributions can be considered as well, the property that we use in our proof is that the distribution is *bounded*.

Definition 7.3 (Hidden Lattice BDD Distribution). *Let $\mathcal{L}_1$ and $\mathcal{L}_2$ be two given lattices, and let $\mathfrak{L}$ be a finite family of lattices, where each member $\mathcal{L}' \in \mathfrak{L}$ satisfies $\mathcal{L}_1 \subseteq \mathcal{L}' \subseteq \mathcal{L}_2$. Let $\mathbf{r} \in G$ be a Gaussian parameter. The* Hidden Lattice BDD Distribution *over $\mathcal{L}_2/\mathcal{L}_1$, denoted by $C_{\mathcal{L}_1,\mathfrak{L},D_{\mathcal{L}_2,\mathbf{r}}}$, is sampled by choosing uniformly at random a lattice $\mathcal{L}' \xleftarrow{\$} \mathfrak{L}$, an element $x \xleftarrow{\$} \mathcal{L}'/\mathcal{L}_1$, and an error term $e \leftarrow D_{\mathcal{L}_2,\mathbf{r}}$ and outputting $y = x + e \mod \mathcal{L}_1$.*

One should think of the lattice $\mathcal{L}_2$ as the "ambient space", i.e. $\mathbb{Z}^n$ for general Euclidean setting or the ring of integer R in the algebraic setting. Note that it is possible to define the distribution with a continuous noise term e. The usual connection between discrete and continuous distribution from LWE/RLWE apply here as well (see, e.g., [32, Lemma 2.23]).

Definition 7.4 (HLBDD Problem). *Let $\mathcal{L}_1, \mathcal{L}_2, \mathfrak{L}, \mathbf{r}$ be as in Definition 7.3. The* HLBDD Problem, *denote by $\mathsf{HLBDD}_{\mathcal{L}_1,\mathfrak{L},D_{\mathcal{L}_2,\mathbf{r}}}$ is to distinguish between two samples from the distribution $C_{\mathcal{L}_1,\mathfrak{L},D_{\mathcal{L}_2,\mathbf{r}}}$, and two samples from the uniform distribution over $\mathcal{L}_2/\mathcal{L}$.*

For the purpose of this section we will set $\mathcal{L}_2 = R$ (the ring of integers of our number field) and $\mathfrak{L} = \mathfrak{P}_k$, for $k = n^{\Omega(1)}$.

Hardness and Variants. We defined HLBDD as the problem where the distinguisher only gets 2 samples from the HLBDD distribution. This is the minimal definition that is needed for our application. However, we note that we do not know of polynomial time algorithms even for weaker variants. For example, one where polynomially many samples are given to the distinguisher instead of only 2, or one where the distinguisher is provided with a (canonical) $\mathbb{Z}$-basis of the lattice $\mathcal{P}$ in addition to the samples. We note that the latter variant is at least as easy as the former since it is possible to use a hybrid argument to show that if $\mathcal{P}$ is known then indistinguishability for one sample implies indistinguishability for polynomially many samples. The connections to other problems in the literature, e.g. [28], is described in the introduction.

7.2   Stating and Proving Hardness

The reduction from HLBDD to RLWE with s from a k-wise distribution consists of two main steps. The formal theorem statement and proof are provided in the full version. We present an outline below.

Step 1. A reduction from RLWE where the adversary gets only one RLWE sample, to the version with polynomially many samples. This reduction applies to any distribution of secrets which is bounded (and is the same on both the initial and final instances). The reduction assumes in addition the hardness of the standard RLWE problem (with the usual noise distribution).

The reduction follows using a rerandomization technique from [32, Section 8.2], [13, Lemma 4]. This transformation unfortunately also requires "noise swallowing", a technique that uses the fact that adding a Gaussian with super-polynomial Gaussian parameter will mask any random variable with polynomial amplitude.

Step 2. A reduction from HLBDD to RLWE with a single sample. For this we assume that there is an adversary that can distinguish between a single RLWE sample $(a, b = as + e)$ and a uniform one.

We begin by replacing a with a decisional hidden-lattice BDD sample $(v_1 + e_1)$, where v_1 only has k nonzero CRT coordinates (randomly chosen) and e_1 is small. The decisional hidden-lattice BDD assumption asserts that this distribution will be indistinguishable from the original one. Namely, we now have $(v_1 + e_1, b = (v_1 + e_1)s + e)$. Opening the parenthesis, we have $b = v_1 s + e_1 s + e$.

We again use noise swallowing to argue that b is statistically close to $b = v_1 s + e$, i.e. we use e to swallow $e_1 s$, which can be done so long as s is small enough and e is large enough. Now we observe that since v_1 is zero on all but k CRT coordinates, and s is close to uniform in any subset of k coordinates, it follows that $v_1 s$ is statistically close to a fresh v_2 that is sampled from the same distribution as v_1 (i.e. has the same set of nonzero coordinates, but the value in each coordinate is randomly chosen). We get $b = v_2 + e$. We now apply decisional hidden-lattice BDD again to claim that $(a, b) = (v_1 + e_1, v_2 + e)$ is indistinguishable from uniform, which completes the proof.

Acknowledgments. We thank the anonymous referees for their insightful comments and useful suggestions.

References

1. Akavia, A., Goldwasser, S., Vaikuntanathan, V.: Simultaneous hardcore bits and cryptography against memory attacks. In: Reingold, O. (ed.) TCC 2009. LNCS, vol. 5444, pp. 474–495. Springer, Heidelberg (2009). https://doi.org/10.1007/978-3-642-00457-5_28
2. Alkim, E., Ducas, L., Pöppelmann, T., Schwabe, P.: Post-quantum key exchange - a new hope. In: Holz, T., Savage, S. (eds.) 25th USENIX Security Symposium, USENIX Security 16, Austin, TX, USA, 10–12 August 2016, pp. 327–343. USENIX Association (2016)
3. Alperin-Sheriff, J., Peikert, C.: Practical bootstrapping in quasilinear time. In: Canetti and Garay [16], pp. 1–20
4. Applebaum, B., Cash, D., Peikert, C., Sahai, A.: Fast cryptographic primitives and circular-secure encryption based on hard learning problems. In: Halevi, S. (ed.) CRYPTO 2009. LNCS, vol. 5677, pp. 595–618. Springer, Heidelberg (2009). https://doi.org/10.1007/978-3-642-03356-8_35
5. Babai, L.: On lovász'lattice reduction and the nearest lattice point problem. Combinatorica **6**(1), 1–13 (1986)

6. Banerjee, A., Peikert, C., Rosen, A.: Pseudorandom functions and lattices. In: Pointcheval, D., Johansson, T. (eds.) EUROCRYPT 2012. LNCS, vol. 7237, pp. 719–737. Springer, Heidelberg (2012). https://doi.org/10.1007/978-3-642-29011-4_42

7. Benhamouda, F., Krenn, S., Lyubashevsky, V., Pietrzak, K.: Efficient zero-knowledge proofs for commitments from learning with errors over rings. In: Pernul, G., Ryan, P.Y.A., Weippl, E. (eds.) ESORICS 2015. LNCS, vol. 9326, pp. 305–325. Springer, Cham (2015). https://doi.org/10.1007/978-3-319-24174-6_16

8. Boneh, D., Freeman, D.M.: Homomorphic signatures for polynomial functions. In: Paterson, K.G. (ed.) EUROCRYPT 2011. LNCS, vol. 6632, pp. 149–168. Springer, Heidelberg (2011). https://doi.org/10.1007/978-3-642-20465-4_10

9. Boneh, D., et al.: Fully key-homomorphic encryption, arithmetic circuit ABE and compact garbled circuits. In: Nguyen, P.Q., Oswald, E. (eds.) EUROCRYPT 2014. LNCS, vol. 8441, pp. 533–556. Springer, Heidelberg (2014). https://doi.org/10.1007/978-3-642-55220-5_30

10. Boneh, D., Roughgarden, T., Feigenbaum, J. (eds.): Symposium on Theory of Computing Conference, STOC 2013, Palo Alto, CA, USA 1–4 June 2013. ACM (2013)

11. Brakerski, Z., Gentry, C., Vaikuntanathan, V.: (Leveled) fully homomorphic encryption without bootstrapping. In: Goldwasser, S. (ed.) ITCS, pp. 309–325. Invited to ACM Transactions on Computation Theory. ACM (2012)

12. Brakerski, Z., Langlois, A., Peikert, C., Regev, O., Stehlé, D.: Classical hardness of learning with errors. In: Boneh et al. [10], pp. 575–584

13. Brakerski, Z., Vaikuntanathan, V.: Fully homomorphic encryption from Ring-LWE and security for key dependent messages. In: Rogaway, P. (ed.) CRYPTO 2011. LNCS, vol. 6841, pp. 505–524. Springer, Heidelberg (2011). https://doi.org/10.1007/978-3-642-22792-9_29

14. Brakerski, Z., Vaikuntanathan, V.: Efficient fully homomorphic encryption from (standard) LWE. In: Ostrovsky, R. (ed.) FOCS, pp. 97–106. IEEE (2011)

15. Brakerski, Z., Vaikuntanathan, V., Wee, H., Wichs, D.: Obfuscating conjunctions under entropic ring LWE. In: Sudan, M. (ed.) Proceedings of the 2016 ACM Conference on Innovations in Theoretical Computer Science, Cambridge, MA, USA, 14–16 January 2016, pp. 147–156. ACM (2016)

16. Canetti, Ran, Garay, Juan A. (eds.): CRYPTO 2013. LNCS, vol. 8042. Springer, Heidelberg (2013). https://doi.org/10.1007/978-3-642-40041-4

17. Conrad, K.: The conductor ideal. Expository papers/Lecture notes. http://www.math.uconn.edu/~kconrad/blurbs/gradnumthy/conductor.pdf

18. Conrad, K.: The different ideal. Expository papers/Lecture notes. http://www.math.uconn.edu/~kconrad/blurbs/gradnumthy/different.pdf

19. van Dijk, M., Gentry, C., Halevi, S., Vaikuntanathan, V.: Fully homomorphic encryption over the integers. In: Gilbert, H. (ed.) EUROCRYPT 2010. LNCS, vol. 6110, pp. 24–43. Springer, Heidelberg (2010). https://doi.org/10.1007/978-3-642-13190-5_2

20. Dodis, Y., Goldwasser, S., Tauman Kalai, Y., Peikert, C., Vaikuntanathan, V.: Public-key encryption schemes with auxiliary inputs. In: Micciancio, D. (ed.) TCC 2010. LNCS, vol. 5978, pp. 361–381. Springer, Heidelberg (2010). https://doi.org/10.1007/978-3-642-11799-2_22

21. Ducas, L., Durmus, A., Lepoint, T., Lyubashevsky, V.: Lattice signatures and bimodal Gaussians. In: Canetti and Garay [16], pp. 40–56

22. Gentry, C., Halevi, S., Peikert, C., Smart, N.P.: Field switching in BGV-style homomorphic encryption. J. Comput. Secur. 21(5), 663–684 (2013)

23. Gentry, C., Halevi, S., Smart, N.P.: Homomorphic evaluation of the AES circuit. In: Safavi-Naini, R., Canetti, R. (eds.) CRYPTO 2012. LNCS, vol. 7417, pp. 850–867. Springer, Heidelberg (2012). https://doi.org/10.1007/978-3-642-32009-5_49

24. Goldwasser, S., Kalai, Y.T., Peikert, C., Vaikuntanathan, V.: Robustness of the learning with errors assumption. In: Yao, A.C. (ed.) Innovations in Computer Science - ICS 2010, Tsinghua University, Beijing, China, 5–7 January 2010. Proceedings, pp. 230–240. Tsinghua University Press (2010)

25. Gorbunov, S., Vaikuntanathan, V., Wee, H.: Attribute-based encryption for circuits. In: Boneh et al. [10], pp. 545–554

26. Halevi, S., Shoup, V.: Algorithms in HElib. In: Garay, J.A., Gennaro, R. (eds.) CRYPTO 2014. LNCS, vol. 8616, pp. 554–571. Springer, Heidelberg (2014). https://doi.org/10.1007/978-3-662-44371-2_31

27. Halevi, S., Shoup, V.: Bootstrapping for HElib. In: Oswald, E., Fischlin, M. (eds.) EUROCRYPT 2015. LNCS, vol. 9056, pp. 641–670. Springer, Heidelberg (2015). https://doi.org/10.1007/978-3-662-46800-5_25

28. Hoffstein, J., Pipher, J., Schanck, J.M., Silverman, J.H., Whyte, W.: Practical signatures from the partial fourier recovery problem. In: Boureanu, I., Owesarski, P., Vaudenay, S. (eds.) ACNS 2014. LNCS, vol. 8479, pp. 476–493. Springer, Cham (2014). https://doi.org/10.1007/978-3-319-07536-5_28

29. Hoffstein, J., Pipher, J., Silverman, J.H.: NTRU: a ring-based public key cryptosystem. In: Buhler, J.P. (ed.) ANTS 1998. LNCS, vol. 1423, pp. 267–288. Springer, Heidelberg (1998). https://doi.org/10.1007/BFb0054868

30. Lyubashevsky, V., Micciancio, D.: Generalized compact knapsacks are collision resistant. In: Bugliesi, M., Preneel, B., Sassone, V., Wegener, I. (eds.) ICALP 2006. LNCS, vol. 4052, pp. 144–155. Springer, Heidelberg (2006). https://doi.org/10.1007/11787006_13

31. Lyubashevsky, V., Peikert, C., Regev, O.: On ideal lattices and learning with errors over rings. In: Gilbert, H. (ed.) EUROCRYPT 2010. LNCS, vol. 6110, pp. 1–23. Springer, Heidelberg (2010). https://doi.org/10.1007/978-3-642-13190-5_1

32. Lyubashevsky, V., Peikert, C., Regev, O.: A toolkit for Ring-LWE cryptography. In: Johansson, T., Nguyen, P.Q. (eds.) EUROCRYPT 2013. LNCS, vol. 7881, pp. 35–54. Springer, Heidelberg (2013). https://doi.org/10.1007/978-3-642-38348-9_3

33. Micciancio, D., Peikert, C.: Hardness of SIS and LWE with small parameters. In: Canetti and Garay [16], pp. 21–39

34. Peikert, C.: Public-key cryptosystems from the worst-case shortest vector problem: extended abstract. In: Mitzenmacher, M. (ed.) STOC, pp. 333–342. ACM (2009)

35. Peikert, C.: How (Not) to instantiate Ring-LWE. In: Zikas, V., De Prisco, R. (eds.) SCN 2016. LNCS, vol. 9841, pp. 411–430. Springer, Cham (2016). https://doi.org/10.1007/978-3-319-44618-9_22

36. Peikert, C., Regev, O., Stephens-Davidowitz, N.: Pseudorandomness of Ring-LWE for any ring and modulus. IACR Cryptology ePrint Archive 2017, vol. 258 (2017)

37. Peikert, C., Rosen, A.: Efficient collision-resistant hashing from worst-case assumptions on cyclic lattices. In: Halevi, S., Rabin, T. (eds.) TCC 2006. LNCS, vol. 3876, pp. 145–166. Springer, Heidelberg (2006). https://doi.org/10.1007/11681878_8

38. Regev, O.: On lattices, learning with errors, random linear codes, and cryptography. In: Gabow, H.N., Fagin, R. (eds.) STOC, pp. 84–93. ACM (2005). Full version in [39]

39. Regev, O.: On lattices, learning with errors, random linear codes, and cryptography. J. ACM 56(6), 34 (2009)

40. Rosca, M., Stehlé, D., Wallet, A.: On the Ring-LWE and Polynomial-LWE problems. In: Nielsen, J.B., Rijmen, V. (eds.) EUROCRYPT 2018. LNCS, vol. 10820, pp. 146–173. Springer, Cham (2018). https://doi.org/10.1007/978-3-319-78381-9_6
41. Stehlé, D., Steinfeld, R., Tanaka, K., Xagawa, K.: Efficient public key encryption based on ideal lattices. In: Matsui, M. (ed.) ASIACRYPT 2009. LNCS, vol. 5912, pp. 617–635. Springer, Heidelberg (2009). https://doi.org/10.1007/978-3-642-10366-7_36

On the Non-existence of Short Vectors in Random Module Lattices

Ngoc Khanh Nguyen[1,2](✉)

[1] IBM Research – Zurich, Rüschlikon, Switzerland
nkn@zurich.ibm.com
[2] Ruhr Universität Bochum, Bochum, Germany

Abstract. Recently, Lyubashevsky & Seiler (Eurocrypt 2018) showed that small polynomials in the cyclotomic ring $\mathbb{Z}_q[X]/(X^n + 1)$, where n is a power of two, are invertible under special congruence conditions on prime modulus q. This result has been used to prove certain security properties of lattice-based constructions against unbounded adversaries. Unfortunately, due to the special conditions, working over the corresponding cyclotomic ring does not allow for efficient use of the Number Theoretic Transform (NTT) algorithm for fast multiplication of polynomials and hence, the schemes become less practical.

In this paper, we present how to overcome this limitation by analysing zeroes in the Chinese Remainder (or NTT) representation of small polynomials. As a result, we provide upper bounds on the probabilities related to the (non)-existence of a short vector in a random module lattice with no assumptions on the prime modulus. We apply our results, along with the generic framework by Kiltz et al. (Eurocrypt 2018), to a number of lattice-based Fiat-Shamir signatures so they can both enjoy tight security in the quantum random oracle model and support fast multiplication algorithms (at the cost of slightly larger public keys and signatures), such as the Bai-Galbraith signature scheme (CT-RSA 2014), Dilithium-QROM (Kiltz et al., Eurocrypt 2018) and qTESLA (Alkim et al., PQCrypto 2017). Our techniques can also be applied to prove that recent commitment schemes by Baum et al. (SCN 2018) are statistically binding with no additional assumptions on q.

Keywords: Lattice-based cryptography · Fiat-Shamir signatures · Module lattices · Lossy identification schemes · Provable security

1 Introduction

Cryptography based on the hardness of lattice problems, such as Module-SIS or Module-LWE [16,18,21], seems to be a very likely replacement for traditional cryptography after the eventual arrival of quantum computers. With the ongoing NIST PQC Standardization Process, we are closer to using quantum-resistant encryption schemes and digital signatures in real life. For additional efficiency, many practical lattice-based constructions work over *fully-splitting* polynomial

© International Association for Cryptologic Research 2019
S. D. Galbraith and S. Moriai (Eds.): ASIACRYPT 2019, LNCS 11922, pp. 121–150, 2019.
https://doi.org/10.1007/978-3-030-34621-8_5

rings $R_q := \mathbb{Z}_q[X]/(f(X))$ where $f(X) = X^n + 1$ is a cyclotomic polynomial, n is a power of two and the prime q is selected so that $f(X)$ splits completely into n linear factors modulo q. With such a choice of parameters, multiplication in the polynomial ring can be performed very quickly using the Number Theoretic Transform (NTT), e.g. [3,10,15,24]. Indeed, one obtains a speed-up of about a factor of 5 by working over rings where $X^n + 1$ splits completely versus just 2 factors (for primes of size between 2^{20} and 2^{29} [19]). Moreover, the structure of fully-splitting rings allows us to perform various operations in parallel as well as conveniently cache and sample polynomials which also significantly improves efficiency of the schemes.

Unfortunately, it is sometimes difficult to prove security of lattice-based constructions when working over fully-splitting polynomial rings [5,8,11]. Usually, the reason is that these security proofs rely on the assumption that polynomials of small norm are invertible. Recently, Lyubashevsky and Seiler [19] (generalising [17]) showed that when n is a power of two and under certain conditions on prime modulus q, small elements of R_q are indeed invertible. The result, however, is meaningful only when $X^n + 1$ does not split into many factors modulo q (e.g. at most 32 for $n = 512$). Consequently, we cannot apply the standard NTT algorithm in such polynomial rings unless we drop the invertibility assumption[1].

In this paper, we present techniques to avoid the invertibility assumption in security proofs. This allows us to construct lattice-based primitives without any conditions on prime modulus q and consequently, we can work over fully-splitting rings and at the same time, use the NTT algorithm for fast multiplication of polynomials. We apply our results to the second-round candidates of the NIST PQC Standardization Process. Namely, we improve the efficiency of Dilithium-QROM [11] (which is the modified version of Dilithium [15] secure in the quantum random oracle model) as well as qTESLA [8]. We also briefly explain how our techniques can be applied to recent lattice-based commitment schemes [5].

1.1   Our Contribution

MAIN RESULTS. Our main technical result is an upper bound on the probability of existence of a short vector in a random module lattice (see Theorem 1.1, formally Corollary 3.9) and other related probabilities (Theorems 3.8 and 3.10). Informally, it states that the probability, over the uniformly random matrix $\mathbf{A}$, that there exists a pair of vectors $(\mathbf{z}_1, \mathbf{z}_2)$, which consists of small polynomials in R_q and $\mathbf{z}_1 \neq \mathbf{0}$, such that $\mathbf{A}\mathbf{z}_1 + \mathbf{z}_2 = \mathbf{0}$ is small (for a suitable choice of parameters). In the context of Fiat-Shamir identification and signature schemes, $\mathbf{A}$ represents a public key matrix and $\mathbf{z}_1$ (and sometimes $\mathbf{z}_2$ as well) represents a difference of two signatures/responses. Our upper bound depends on the tail function $\mathcal{T}$. For readability, we hide the concrete formula for $\mathcal{T}$ here and we refer to the formal statement in Corollary 3.9.

---

[1] Lyubashevsky and Seiler [19] showed, however, how to combine the FFT algorithm and Karatsuba multiplication in order to multiply in partially-splitting rings faster.

We recall that a similar result was presented by Kiltz et al. (e.g. Lemma 4.6 in [11]) but they only consider the case when $q \equiv 5 \pmod 8$ so that invertibility properties can be applied [17,19]. Here, we generalise their result on how to bound that probability without any assumptions on the prime modulus q.

Theorem 1.1 (Informal). *Denote* $S_\alpha := \{y \in R_q : \|y\|_\infty \leq \alpha\}$ *and let* $\ell, k, \alpha_1, \alpha_2 \in \mathbb{N}$. *Then*

$$\Pr_{\mathbf{A} \leftarrow R_q^{k \times \ell}}[\exists (\mathbf{z}_1, \mathbf{z}_2) \in S_{\alpha_1}^\ell \backslash \{\mathbf{0}\} \times S_{\alpha_2}^k : \mathbf{A}\mathbf{z}_1 + \mathbf{z}_2 = \mathbf{0}]$$

$$\leq \frac{|S_{\alpha_1}|^\ell \cdot |S_{\alpha_2}|^k}{q^{nk}} + \mathcal{T}(q, \ell, k, \alpha_1, \alpha_2), \tag{1}$$

where $\mathcal{T}(q, \ell, k, \alpha_1, \alpha_2)$ *is a function defined in Corollary 3.9.*

Figure 1 shows values of the tail function $\mathcal{T}$ for different prime moduli q. We observe that the more $f(x) = X^n + 1$ splits modulo q then the larger the value of $\mathcal{T}$. When $f(x)$ only splits into two factors, our upper bound is essentially equal to

$$\frac{|S_{\alpha_1}|^\ell \cdot |S_{\alpha_2}|^k}{q^{nk}}.$$

Indeed, in this case the value of $\mathcal{T}$ is negligible and hence, we obtain an upper bound identical to Kiltz et al. On the other hand, if we want to work over fully-splitting polynomial rings in order to apply the Number Theoretic Transform algorithm, we would have to increase q as well as the dimensions (k, ℓ) of the matrix $\mathbf{A}$ so that $\mathcal{T}(q, \ell, k, \alpha_1, \alpha_2)$ stays small. Unfortunately, this implies larger public key and signature size.

KEY TECHNIQUES. We provide an overview of the proof of Theorem 1.1. Let d be the divisor of n such that

$$X^n + 1 \equiv \prod_{i=1}^{d} f_i(X) \pmod q$$

for distinct polynomials $f_i(X)$ of degree n/d that are irreducible in $\mathbb{Z}_q[X]$. In other words, $X^n + 1$ splits into d irreducible polynomials modulo q. The proof sketch goes as follows.

Step 1: We apply the union bound:

$$\Pr_{\mathbf{A} \leftarrow R_q^{k \times \ell}}[\exists (\mathbf{z}_1, \mathbf{z}_2) \in S_{\alpha_1}^\ell \backslash \{\mathbf{0}\} \times S_{\alpha_2}^k : \mathbf{A}\mathbf{z}_1 + \mathbf{z}_2 = \mathbf{0}]$$

$$\leq \sum_{(\mathbf{z}_1, \mathbf{z}_2) \in S_{\alpha_1}^\ell \backslash \{\mathbf{0}\} \times S_{\alpha_2}^k} \Pr_{\mathbf{A} \leftarrow R_q^{k \times \ell}}[\mathbf{A}\mathbf{z}_1 + \mathbf{z}_2 = \mathbf{0}]. \tag{2}$$

Step 2: We identify the subset Z of $S_{\alpha_1}^\ell \backslash \{\mathbf{0}\} \times S_{\alpha_2}^k$ which satisfies:

$$(\mathbf{z}_1, \mathbf{z}_2) \in Z \iff \Pr_{\mathbf{A} \leftarrow R_q^{k \times \ell}}[\mathbf{A}\mathbf{z}_1 + \mathbf{z}_2 = \mathbf{0}] > 0.$$

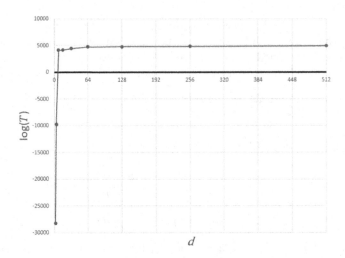

Fig. 1. Let $(n, q, \ell, k, \alpha_1, \alpha_2) = (512, \approx 2^{45}, 4, 4, 1.8 \cdot 10^6, 3.6 \cdot 10^6)$. The graph presents values of $\log(\mathcal{T}(q, \ell, k, \alpha_1, \alpha_2))$ depending on the number of irreducible polynomials d that $X^n + 1$ splits into modulo q. One notes that for prime moduli $q \approx 2^{45}$ such that $d \in \{2, 4\}$, the value of $\mathcal{T}$ is sufficiently small, hence so is the right-hand side of Eq. (1). On the other hand, values of $\mathcal{T}$ rocket for $d \geq 8$ and therefore q or dimensions (k, ℓ) of the matrix $\mathbf{A}$ must be increased in order to keep the upper bound in (1) small enough.

Hence, the probability in Eq. (1) can be bounded by

$$\sum_{(\mathbf{z}_1, \mathbf{z}_2) \in Z} \Pr_{\mathbf{A} \leftarrow R_q^{k \times \ell}}[\mathbf{A}\mathbf{z}_1 + \mathbf{z}_2 = \mathbf{0}].$$

Step 3: Next, we propose a partitioning of the set Z into subsets $Z_0, Z_1, ..., Z_d$, i.e. $Z = \bigcup_{i=0}^{d} Z_i$. Then, we show that for each $(\mathbf{z}_1, \mathbf{z}_2) \in Z_i$, the probability

$$p_i := \Pr_{\mathbf{A} \leftarrow R_q^{k \times \ell}}[\mathbf{A}\mathbf{z}_1 + \mathbf{z}_2 = \mathbf{0}]$$

is the same and we compute it. Thus, the probability in Eq. (1) can now be bounded by:

$$\sum_{i=0}^{d} \sum_{(\mathbf{z}_1, \mathbf{z}_2) \in Z_i} p_i = \sum_{i=1}^{d} |Z_i| \cdot p_i$$

Step 4: We find an upper bound on $|Z_i|$.

ZERO FUNCTION. In this paper, we will consider zeroes in the "Chinese Remainder representation"[2] of polynomials in R_q[3]. Formally, we define the following Zero function:

---

[2] Alternatively, we call it "FFT/NTT representation" in the fully-splitting case.

[3] This technique has already been investigated in the literature for e.g. constructing provably secure variants of NTRUEncrypt [25].

$$\mathsf{Zero}(y) := \{i : y \equiv 0 \ (\mathrm{mod} \ (f_i(X), q))\} \text{ and } \mathsf{Zero}(\mathbf{y}) := \bigcap_{j=1}^{k} \mathsf{Zero}(y_j),$$

where $y \in R_q$ and $\mathbf{y} = (y_1, ..., y_k) \in R_q^k$. Note that if y is invertible then $|\mathsf{Zero}(y)| = 0$. Lyubashevsky and Seiler [19] proved that whenever a non-zero y has small Euclidean norm then $|\mathsf{Zero}(y)| = 0$. Here, we extend it and provide a relationship between the Euclidean norm of y and the size of set $\mathsf{Zero}(y)$ (see Lemma 3.2). In particular, the result implies that relatively small elements of R_q have only a few zeroes in the Chinese Remainder representation. This observation will be crucial for Steps 3 and 4.

ZERO ROWS. Consider the equation $\mathbf{A}\mathbf{z}_1 + \mathbf{z}_2 = \mathbf{0}$ and let $j \in \mathsf{Zero}(\mathbf{z}_1)$. If we look at this equation modulo $(f_j(X), q)$ then we just end up with $\mathbf{z}_2 = \mathbf{0}$, i.e. $j \in \mathsf{Zero}(\mathbf{z}_2)$ and thus $j \in \mathsf{Zero}(\mathbf{z}_1 || \mathbf{z}_2)$ where $||$ denotes usual concatenation of vectors. Consequently, $\mathsf{Zero}(\mathbf{z}_1) \subseteq \mathsf{Zero}(\mathbf{z}_1 || \mathbf{z}_2)$. Clearly, we have $\mathsf{Zero}(\mathbf{z}_1 || \mathbf{z}_2) \subseteq \mathsf{Zero}(\mathbf{z}_1)$ and therefore these two sets are equal. This implies that the subset Z introduced in Step 2 can be identified as:

$$Z = \{(\mathbf{z}_1, \mathbf{z}_2) : \mathsf{Zero}(\mathbf{z}_1) = \mathsf{Zero}(\mathbf{z}_1 || \mathbf{z}_2)\}.$$

Define $Z_i = \{(\mathbf{z}_1, \mathbf{z}_2) : \mathsf{Zero}(\mathbf{z}_1) = \mathsf{Zero}(\mathbf{z}_1 || \mathbf{z}_2) \wedge |\mathsf{Zero}(\mathbf{z}_1)| = i\} \subseteq Z$ (Step 3). Informally, we say that $(\mathbf{z}_1, \mathbf{z}_2) \in Z_i$ has i zero rows, since if we write down the components of $\mathbf{z}_1$ and $\mathbf{z}_2$ in the Chinese Remainder representation, in columns, then we get exactly i rows filled with zeroes.

For fixed $(\mathbf{z}_1, \mathbf{z}_2) \in Z_i$, we compute the probability p_i defined in Step 3 by counting the number of possible $\mathbf{A}$ which satisfy $\mathbf{A}\mathbf{z}_1 + \mathbf{z}_2 = \mathbf{0}$. This could be done by considering the equation modulo $(f_j(X), q)$ for all $j \notin \mathsf{Zero}(\mathbf{z}_1)$. Indeed, for such j there is a simple way to count all $\mathbf{A} \in (\mathbb{Z}_q[X]/(f_j(x)))^{k \times \ell}$ which satisfy $\mathbf{A}\mathbf{z}_1 + \mathbf{z}_2 = \mathbf{0}$ modulo $f_j(X)$. Concretely, one of the components of $\mathbf{z}_1$, say z_u, is going to be invertible modulo $f_j(X)$ and therefore all entries of $\mathbf{A}$ not related to z_u can be chosen arbitrarily. The rest, however, will be adjusted so that the equation holds. On the other hand, if $j \in \mathsf{Zero}(\mathbf{z}_1) = \mathsf{Zero}(\mathbf{z}_1 || \mathbf{z}_2)$ then $\mathbf{A}\mathbf{z}_1 + \mathbf{z}_2$ is simply equal to $\mathbf{0}$ modulo $(f_j(X), q)$ for any $\mathbf{A}$. By applying Chinese Remainder Theorem, we obtain the total number of possible $\mathbf{A} \in R_q^{k \times \ell}$ which satisfy the equation above.

The only thing left is to provide an upper bound on $|Z_i|$ (Step 4). Firstly, we observe that if $(\mathbf{z}_1, \mathbf{z}_2) \in Z_i$ then clearly $|\mathsf{Zero}(z_j)| \geq i$ for $j = 1, ..., \ell$ where $\mathbf{z}_1 = (z_1, ..., z_\ell)$. Since each component of $\mathbf{z}_1 \in S_{\alpha_1}^\ell \setminus \{\mathbf{0}\}$ has infinity norm at most α_1, and assuming this value is relatively small, we get that each component of $\mathbf{z}_1$ has only a few zeroes in the Chinese Remainder representation (Lemma 3.2). Hence, for some larger values of i, we simply get $Z_i = \varnothing$. The second observation is that if $(\mathbf{z}_1, \mathbf{z}_2) \in Z_i$ and $\mathbf{y}_1, \mathbf{y}_2$ are vectors of some "small" polynomials then $(\mathbf{z}_1 + \mathbf{y}_1, \mathbf{z}_2 + \mathbf{y}_2)$ is likely not to have exactly i zero rows. For example, suppose that

$$\mathsf{Zero}(\mathbf{z}_1 + \mathbf{y}_1, \mathbf{z}_2 + \mathbf{y}_2) = \mathsf{Zero}(\mathbf{z}_1 + \mathbf{y}_1', \mathbf{z}_2 + \mathbf{y}_2')$$

for some other small $\mathbf{y}_1', \mathbf{y}_2'$. This implies that $(\mathbf{y}_1 - \mathbf{y}_1', \mathbf{y}_2 - \mathbf{y}_2')$ has at least i zero rows. In particular, each component of $\mathbf{y}_1 - \mathbf{y}_1'$, say $\hat{y}_j$, has at least i zeroes in the Chinese Remainder representation. However, we know that $\hat{y}_j$ is a polynomial of small norm by the choice of $\mathbf{y}_1$ and $\mathbf{y}_1'$. Therefore, $\hat{y}_j$ has only a few zeroes (by the observation above or Lemma 3.2). By picking sufficiently small $\mathbf{y}_1$ and $\mathbf{y}_1'$ we can make sure that each component $\hat{y}_j$ of $\mathbf{y}_1 - \mathbf{y}_1'$ has less than i zeroes. This would lead to a contradiction. In conclusion, our approach for bounding $|Z_i|$ is to, for each $(\mathbf{z}_1, \mathbf{z}_2) \in Z_i$, generate all pairs of form $(\mathbf{z}_1 + \mathbf{y}_1, \mathbf{z}_2 + \mathbf{y}_2) \notin Z_i$, for vectors of sufficiently small polynomials $\mathbf{y}_1, \mathbf{y}_2$, and applying the pigeonhole principle along with other simple counting arguments.

1.2 Applications

DIGITAL SIGNATURES. Kiltz et al. [11] presented a generic framework for constructing secure Fiat-Shamir signatures in the quantum random oracle model (QROM). As a concrete instantiation, they introduced a new signature scheme Dilithium-QROM, which is a modification of the original Dilithium scheme [15], and is tightly based on the hardness of Module-LWE problem in the QROM. However, in order to obtain security of Dilithium-QROM, Kiltz et al. choose the prime modulus q to be congruent to 5 modulo 8. This assumption assures that the underlying polynomial ring $\mathbb{Z}_q[X]/(X^n + 1)$ splits into two subrings modulo q and invertibility results can be applied [17,19]. Unfortunately, polynomial multiplication algorithms in such rings are not efficient. We show how to apply our probability results to the security of Dilithium-QROM[4] so that one can avoid such special assumptions on q (in particular, one could choose q so that R_q splits completely and NTT along with other optimisations can be applied). The only disadvantage is that, in order to keep the probabilities small, one should slightly increase the size of q and dimensions (k, ℓ). Unfortunately, this results in having both considerably larger public keys and signatures.

General results by Kiltz et al. can also be applied to obtain a security proof in the QROM for a number of existing Fiat-Shamir signature schemes similar to Dilithium such as the Bai-Galbraith scheme [4] (see Sect. 4) or qTESLA [8]. So far, security of the latter scheme in the quantum random oracle model is proven assuming a certain non-standard conjecture. However, one can also obtain it by applying the framework by Kiltz et al. and using our probability upper bounds. Consequently, one gets a tightly secure version of qTESLA in the QROM without any non-standard conjecture. We recall that our results allow this signature scheme to work over fully-splitting rings so that the use of NTT for polynomial multiplication is possible. However, as in the case of Dilithium-QROM, we would end up with larger public key and signature size compared to the original qTESLA (see Table 2).

COMMITMENT SCHEMES. Recently, Baum et al. [5] presented efficient commitment schemes from Module-SIS and Module-LWE. However, both their new

---

[4] We present it in the full version of this paper [20].

statistically binding commitment scheme and their improved construction from [7] rely on the general invertibility result from [19], i.e. special congruence conditions on the prime modulus q. Our probability upper bounds can be applied to prove the statistically binding property of these constructions, and consequently, one could now consider working in fully-splitting rings. As before, we observe that choosing primes q such that $X^n + 1$ splits into many factors modulo q results in having both larger commitment and proof size.

1.3 Related Works

The first asymptotically-efficient lattice-based signature scheme using the "Fiat-Shamir with Aborts" paradigm was presented in [13] which is based on the Ring-SIS problem. Later on, Lyubashevsky [14] improved the scheme by basing it on the combination of Ring-SIS and Ring-LWE. Since then, many substantial improvements have been proposed [4,8,10,15]. In the meantime, lossy identification schemes were introduced and used to construct secure digital signatures in the quantum random oracle model [1,2,11,27].

Invertibility of "small" polynomials[5] is an important property in the context of (approximate) zero-knowledge proofs based on lattices. For example, one usually needs the difference set $C - C$ to contain only invertible polynomials for extraction purposes [7,26] where C is a challenge set. Lyubashevsky and Neven [17] proved that if q is congruent to 5 modulo 8 then the polynomial ring $R_q = \mathbb{Z}_q[X]/(X^n + 1)$ splits into two subrings and elements of small infinity norm are indeed invertible. This result was generalised by Lyubashevsky and Seiler [19]. Concretely, they showed that if $q \equiv 2k + 1 \pmod{4k}$ for some k then $X^n + 1$ splits into k irreducible polynomials modulo q and also small elements in R_q are invertible. These results have been recently applied in the context of computing probabilities related to the security of lattice-based signatures and commitment schemes, e.g. [5,11].

2 Preliminaries

For $n \in \mathbb{N}$, let $[n] := \{1, \ldots, n\}$. For a set S, $|S|$ is the cardinality of S, $\mathcal{P}(S)$ is the power set of S and $\mathcal{P}_i(S)$ is the set of all subsets of S of size i. If S is finite, we denote the sampling of a uniform random element x by $x \leftarrow S$, while we denote the sampling according to some distribution $\mathfrak{D}$ by $x \leftarrow \mathfrak{D}$. By $[\![B]\!]$ we denote the bit that is 1 if the Boolean statement B is true, and 0 otherwise.

ALGORITHMS. Unless stated otherwise, we assume all our algorithms to be probabilistic. We denote by $y \leftarrow \mathsf{A}(x)$ the probabilistic computation of algorithm A on input x. If A is deterministic, we write $y := \mathsf{A}(x)$. The notation $y \in \mathsf{A}(x)$ is used to indicate all possible outcomes y of the probabilistic algorithm A on input x. We can make any probabilistic A deterministic by running it with fixed

---

[5] What we mean by "small" is that the polynomial has small infinity or Euclidean norm.

randomness. We write $y := \mathsf{A}(x; r)$ to indicate that A is run on input x with randomness r. The notation $\mathsf{A}(x) \Rightarrow y$ denotes the event that A on input x returns y. Eventually, we write $\mathrm{Time}(\mathsf{A})$ to denote the running time of A.

2.1 Cyclotomic Rings

Let n be a power of two. Denote R and R_q respectively to be the rings $\mathbb{Z}[X]/(X^n + 1)$ and $\mathbb{Z}_q[X]/(X^n + 1)$, for a prime q. We also set d to be the divisor of n such that

$$X^n + 1 \equiv \prod_{i=1}^{d} f_i(X) \pmod{q}$$

for distinct polynomials $f_i(X)$ of degree n/d that are irreducible in $\mathbb{Z}_q[X]$. Alternatively, we say that $X^n + 1$ splits into d polynomials modulo q. If $d = n$ then $X^n + 1$ *fully splits*. By default, all the equalities and congruences between ring elements in this paper are modulo q.

Regular font letters denote elements in R or R_q and bold lower-case letters represent column vectors with coefficients in R or R_q. Bold upper-case letters denote matrices. By default, all vectors are column vectors.

MODULAR REDUCTIONS. For an even (resp. odd) positive integer α, we define $r' = r \bmod^{\pm} \alpha$ to be the unique element r' in the range $-\frac{\alpha}{2} < r' \leq \frac{\alpha}{2}$ (resp. $-\frac{\alpha-1}{2} \leq r' \leq \frac{\alpha-1}{2}$) such that $r' = r \bmod \alpha$. For any positive integer α, we define $r' = r \bmod^{+} \alpha$ to be the unique element r' in the range $0 \leq r' < \alpha$ such that $r' = r \bmod \alpha$. When the exact representation is not important, we simply write $r \bmod \alpha$.

SIZES OF ELEMENTS. For an element $w \in \mathbb{Z}_q$, we write $\|w\|_{\infty}$ to mean $|w \bmod^{\pm} q|$. Define the ℓ_{∞} and ℓ_2 norms for $w = w_0 + w_1 X + \ldots + w_{n-1} X^{n-1} \in R$ as follows:

$$\|w\|_{\infty} = \max_{i} \|w_i\|_{\infty}, \quad \|w\| = \sqrt{\|w_0\|_{\infty}^2 + \ldots + \|w_{n-1}\|_{\infty}^2}.$$

Similarly, for $\mathbf{w} = (w_1, \ldots, w_k) \in R^k$, we define

$$\|\mathbf{w}\|_{\infty} = \max_{i} \|w_i\|_{\infty}, \quad \|\mathbf{w}\| = \sqrt{\|w_1\|^2 + \ldots + \|w_k\|^2}.$$

For a finite set $S \subseteq R^k$, however, we set

$$\|S\|_{\infty} = \max_{\mathbf{w} \in S} \|\mathbf{w}\|_{\infty}, \quad \|S\| = \max_{\mathbf{w} \in S} \|\mathbf{w}\|.$$

We write S_{η} to denote all elements $w \in R$ such that $\|w\|_{\infty} \leq \eta$.

EXTRACTING HIGH-ORDER AND LOW-ORDER BITS. To reduce the size of the public key, we need some algorithms that extract "higher-order" and "lower-order" bits of elements in $\mathbb{Z}_q$. The goal is that when given an arbitrary element $r \in \mathbb{Z}_q$ and another small element $z \in \mathbb{Z}_q$, we would like to be able to recover the higher order bits of $r + z$ without needing to store z. The algorithms are exactly as in [9,11], and we repeat them for completeness in Fig. 2. They are described as working on integers modulo q, but one can extend it to polynomials in R_q by simply being applied individually to each coefficient.

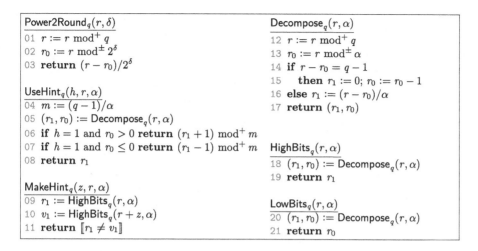

Fig. 2. Supporting algorithms for Dilithium and Dilithium-QROM [11].

Lemma 2.1. *Suppose that q and α are positive integers satisfying $q > 2\alpha$, $q \equiv 1$ (mod α) and α even. Let $\mathbf{r}$ and $\mathbf{z}$ be vectors of elements in R_q where $\|\mathbf{z}\|_\infty \le \alpha/2$, and let $\mathbf{h}, \mathbf{h}'$ be vectors of bits. Then the $\mathsf{HighBits}_q$, $\mathsf{MakeHint}_q$, and $\mathsf{UseHint}_q$ algorithms satisfy the following properties:*

1. $\mathsf{UseHint}_q(\mathsf{MakeHint}_q(\mathbf{z}, \mathbf{r}, \alpha), \mathbf{r}, \alpha) = \mathsf{HighBits}_q(\mathbf{r} + \mathbf{z}, \alpha)$.
2. *Let* $\mathbf{v}_1 = \mathsf{UseHint}_q(\mathbf{h}, \mathbf{r}, \alpha)$. *Then* $\|\mathbf{r} - \mathbf{v}_1 \cdot \alpha\|_\infty \le \alpha + 1$.
3. *For any* $\mathbf{h}, \mathbf{h}'$, *if* $\mathsf{UseHint}_q(\mathbf{h}, \mathbf{r}, \alpha) = \mathsf{UseHint}_q(\mathbf{h}', \mathbf{r}, \alpha)$, *then* $\mathbf{h} = \mathbf{h}'$.

Lemma 2.2. *If $\|\mathbf{s}\|_\infty \le \beta$ and $\|\mathsf{LowBits}_q(\mathbf{r}, \alpha)\|_\infty < \alpha/2 - \beta$, then*

$$\mathsf{HighBits}_q(\mathbf{r}, \alpha) = \mathsf{HighBits}_q(\mathbf{r} + \mathbf{s}, \alpha).$$

IDEAL LATTICES. An integer lattice of dimension n is an additive subgroup of $\mathbb{Z}^n$. For simplicity, we only consider full-rank lattices. The determinant of a full-rank lattice Λ of dimension n is equal to the size of the quotient group $\mathbb{Z}^n/\Lambda$. We denote $\lambda_1(\Lambda) = \min_{\|\mathbf{w}\| \in \Lambda} \|\mathbf{w}\|$. We say that Λ is an *ideal lattice* in R if Λ is an ideal of R. There exists a lower bound on $\lambda_1(\Lambda)$ if Λ is an ideal lattice [19,22]. Assuming that n is a power of two, we get a simplified bound.

Lemma 2.3 ([19], **Lemma 2.7**). *If Λ is an ideal lattice in R, then $\lambda_1(\Lambda) \ge \det(\Lambda)^{1/n}$.*

THE MLWE ASSUMPTION. For integers m, k, and a probability distribution $D : R_q \to [0, 1]$, we say that the advantage of algorithm A in solving the decisional $\mathrm{MLWE}_{m,k,D}$ problem over the ring R_q is

$$\mathrm{Adv}_{m,k,D}^{\mathrm{MLWE}} := \big|\Pr[\mathsf{A}(\mathbf{A}, \mathbf{t}) \Rightarrow 1 \mid \mathbf{A} \leftarrow R_q^{m \times k}; \mathbf{t} \leftarrow R_q^m]$$
$$- \Pr[\mathsf{A}(\mathbf{A}, \mathbf{As}_1 + \mathbf{s}_2) \Rightarrow 1 \mid \mathbf{A} \leftarrow R_q^{m \times k}; \mathbf{s}_1 \leftarrow D^k; \mathbf{s}_2 \leftarrow D^m]\big|.$$

The MLWE assumption states that the above advantage is negligible for all polynomial-time algorithms A. It was introduced in [12], and is a generalization of the LWE assumption from [23]. The Ring-LWE assumption [18] is a special case of MLWE where $k = 1$. Analogously to LWE and Ring-LWE, it was shown in [12] that solving the MLWE problem for certain parameters is as hard as solving certain worst-case problems in certain algebraic lattices.

3   Zeroes in the Chinese Remainder Representation

In this section, we present general results about existence of solutions $(\mathbf{A}, \mathbf{t}) \in R_q^{k \times \ell} \times R^k$ to the equation $\mathbf{Az}_1 + \mathbf{z}_2 = c\mathbf{t}$ (and other similar ones), for some $\mathbf{z}_1 \in R_q^\ell, \mathbf{z}_2 \in R_q^k, c \in R_q \backslash \{0\}$, and compute the probability of satisfying such equations for uniformly random $\mathbf{A}$ and $\mathbf{t}$. The results are crucial for security analysis of Fiat-Shamir signature schemes. For instance, security of Dilithium-QROM [11] relies heavily on the assumption that c is invertible in R_q or $\mathbf{z}_1$ contains an invertible component. In such a case, the probability can be calculated straightforwardly. Hence, q is chosen so that $q \equiv 5 \pmod 8$ because then, polynomials in R_q of small (infinity) norm are proved to be invertible [17,19]. We avoid such assumptions and analyse "zeroes in the Chinese Remainder Representation" of $\mathbf{z}_1, \mathbf{z}_2$ and c in order to provide general upper bounds on the probabilities.

3.1   Zero Rows

We start by introducing the Zero function.

Definition 3.1. *Let $y \in R_q$. We define a set*

$$\mathsf{Zero}(y) := \{i \in [d] : y \equiv 0 \pmod{f_i(X)}\}.$$

For a vector $\boldsymbol{y} = (y_1, ..., y_k) \in R_q^k$, we set $\mathsf{Zero}(\boldsymbol{y}) := \bigcap_{j=1}^k \mathsf{Zero}(y_j)$ and similarly for multiple vectors $\boldsymbol{y}_1, ..., \boldsymbol{y}_\ell$ over R_q, $\mathsf{Zero}(\boldsymbol{y}_1, ..., \boldsymbol{y}_\ell) := \bigcap_{j=1}^\ell \mathsf{Zero}(\boldsymbol{y}_j)$.

Informally, we say that y has i zeroes in the Chinese Remainder Representation if $|\mathsf{Zero}(y)| = i$. One observes that $\mathsf{Zero}(y) = \varnothing$ if and only if y is invertible, by the Chinese Remainder Theorem. Also, $\mathsf{Zero}(y) = [d] \iff y = 0$.

Lyubashevsky and Seiler [19] showed that if $\|y\| < q^{1/d}$ then y is invertible. Obviously, it is not very interesting if d is large (e.g. $d = n$). Here, we extend the result to consider the number of zeroes in the Chinese Remainder Representation.

Lemma 3.2. *Let $y \in R_q$ such that $0 < \|y\| < q^{m/d}$ for some $m \in [d]$. Then, $|\mathsf{Zero}(y)| < m$.*

Proof. Suppose that $|\mathsf{Zero}(y)| \geq m$ and pick any $i_1, ..., i_m \in \mathsf{Zero}(y)$. Define the following set:

$$\Lambda = \{z \in R : \forall j \in [m], z \equiv 0 \pmod{f_{i_j}(X)}\}.$$

Firstly, note that Λ is an additive group and $y \in \Lambda$. Moreover, for any $z \in \Lambda$, we have $z \cdot X \in \Lambda$ since each $f_{i_j}(X)$ is a factor of $X^n + 1$ modulo q. Therefore, Λ is an ideal of R, and hence an ideal lattice in the ring R. Consider the Chinese Remainder representation modulo q of all the elements in Λ. Note that they have 0 in the coefficients corresponding to $f_{i_j}(X)$ for $j \in \{1, ..., m\}$ and arbitrary values everywhere else. This implies that $\det(\Lambda) = |\mathbb{Z}^n / \Lambda| = q^{nm/d}$. Hence, by Lemma 2.3 we have $\lambda_1(\Lambda) \geq q^{m/d}$. However, we know that $||y|| > 0$, thus y is non-zero. Eventually, we obtain $||y|| < q^{m/d} \leq \lambda_1(\Lambda) \leq ||y||$ which leads to contradiction. □

The lemma above implies that if a polynomial $y \in R_q$ is short enough, then it has only a few zeroes in the Chinese Remainder Representation (but is not necessarily invertible).

We now introduce the notion of ZeroRows which will be crucial in proving the main theorem.

Definition 3.3. *Let $k \in \mathbb{N}$ and $A \subseteq R_q^k$ be a non-empty set. Then, we write* ZeroRows$_i(A)$ *to denote*

$$\text{ZeroRows}_i(A) := \{\boldsymbol{a} \in A : |\text{Zero}(\boldsymbol{a})| = i\}.$$

We say that $\boldsymbol{a} \in$ ZeroRows$_i(A)$ has i zero rows.

Name ZeroRows comes from the fact that if $\mathbf{a} = (a_1, ..., a_k) \in$ ZeroRows$_i(A)$ and if we write down the Chinese Remainder Representation of $a_1, ..., a_k$ as column vectors[6] then we get exactly i rows filled only with zeroes.

The next result gives an upper bound on ZeroRows$_i(S_\alpha^k)$ for fixed $i > 0, k$ and α. The key idea of the proof is as follows. For simplicity, consider $z' := z + X^j, z'' := z + X^\ell$ for some distinct $j, \ell \in [2n]$ and $z \in R_q$. To begin with, note that $\text{Zero}(z') \cap \text{Zero}(z'') = \varnothing$. Indeed, if there exists some $u \in \text{Zero}(z') \cap \text{Zero}(z'')$ then

$$z + X^\ell \equiv z'' \equiv 0 \equiv z' \equiv z + X^j \pmod{f_u(X)}.$$

Hence, we get a contradiction, since $X^j - X^\ell$ is invertible [6]. Therefore,

$$|\{z + X^j \in \text{ZeroRows}_i(S_\alpha) : j \in [2n]\}| \leq \lfloor d/i \rfloor.$$

This is because if size of the set is strictly larger than d/i then, by definition of ZeroRows$_i(S_\alpha)$ and the pigeonhole principle, we would have $\text{Zero}(z + X^j) \cap \text{Zero}(z + X^\ell) \neq \varnothing$ for some distinct j, ℓ. Thus, we end up with:

$$|\{z + X^j \notin \text{ZeroRows}_i(S_\alpha) : j \in [2n]\}| \geq 2n - \lfloor d/i \rfloor.$$

Our main strategy is that for each $\mathbf{z} \in$ ZeroRows$_i(S_\alpha^k)$, we count all $\mathbf{z}'$ of form $\mathbf{z} + \mathbf{y}$ (where $\mathbf{y}$ is a somewhat small polynomial) such that $\mathbf{z}' \notin$ ZeroRows$_i(S_\alpha^k)$ similarly as above, and eventually, obtain an upper bound on $|$ZeroRows$_i(S_\alpha^k)|$. The bound depends on the size of a set $W_i \subseteq R_q$, which satisfies the following

---

[6] Namely, for each a_i we define a corresponding column vector $(a'_{i,1}, ..., a'_{i,d})$, where $a'_{i,j}$ is the element of $\mathbb{Z}_q[X]/(f_j(X))$, such that $a_i \equiv a'_{i,j} \pmod{f_j(X)}$, for $j \in [d]$.

property: for any two distinct $u, v \in W_i$, $|\mathsf{Zero}(u - v)| < i^7$. Later on, we show how to use our previous result, i.e. Lemma 3.2, to construct such sets.

Lemma 3.4. *Let $k, \alpha \in \mathbb{N}, i \in [d]$ and $W_i \subseteq R_q$ be a set of polynomials in R_q such that for any two distinct $u, v \in W_i$, $|\mathsf{Zero}(u - v)| < i$. Then,*

$$|\mathsf{ZeroRows}_i(S_\alpha^k)| \le \frac{\binom{d}{i} \cdot |S_{\alpha + \|W_i\|_\infty}|^k}{|W_i|^k}.$$

Proof. Firstly, take any $\mathbf{z} = (z_1, ..., z_k) \in S_\alpha^k$ and define

$$\mathsf{Bad}(z_1, ..., z_k) := \{(z_1 + y_1, ..., z_k + y_k) \in \mathsf{ZeroRows}_i(S_\alpha^k) : y_1, ..., y_k \in W_i\}.$$

We claim that $|\mathsf{Bad}(z_1, ..., z_k)| \le \binom{d}{i}$. Indeed, suppose $|\mathsf{Bad}(z_1, ..., z_k)| > \binom{d}{i}$ and define the function

$$F : \mathsf{Bad}(z_1, ..., z_k) \to \mathcal{P}_i([d]), (z_1', ..., z_k') \longmapsto \mathsf{Zero}(z_1', ..., z_k').$$

Note that F is well-defined by definition of Bad. Also, $|\mathsf{Bad}(z_1, ..., z_k)| > \binom{d}{i} = |\mathcal{P}_i([d])|$ implies that F is not injective. Hence,

$$F(z_1 + y_1, ..., z_k + y_k) = I = F(z_1 + y_1', ..., z_k + y_k')$$

for some set $I \in \mathcal{P}_i([d])$, $y_1, ..., y_k, y_1', ..., y_k' \in W_i$ and $y_j \ne y_j'$ for some index j. Take any $u \in I$. Then, $z_j + y_j \equiv 0 \equiv z_j + y_j' \pmod{f_u(X)}$ and consequently, $y_j - y_j' \equiv 0 \pmod{f_u(X)}$. Since we picked arbitrary $u \in I$, we proved that $|\mathsf{Zero}(y_j - y_j')| \ge i$. However, this leads to a contradiction by the definition of the set W_i.

Now, define a set

$$\mathsf{Good}(z_1, ..., z_k) := \{(z_1 + y_1, ..., z_k + y_k) \notin \mathsf{ZeroRows}_i(S_\alpha^k) : y_1, ..., y_k \in W_i\}.$$

Clearly, $|\mathsf{Good}(z_1, ..., z_k)| = |W_i|^k - |\mathsf{Bad}(z_1, ..., z_k)| \ge |W_i|^k - \binom{d}{i}$. Consider the following set

$$S = \bigcup_{(z_1, ..., z_k) \in \mathsf{ZeroRows}_i(S_\alpha^k)} \mathsf{Good}(z_1, ..., z_k).$$

One observes that $S \subseteq S_{\alpha + \|W_i\|_\infty}^k \setminus \mathsf{ZeroRows}_i(S_\alpha^k)$ by definition of Good, which gives us an upper bound on $|S|$. We are now interested in finding a lower bound for $|S|$. Let $(\hat{z}_1, ..., \hat{z}_k)$ be an element of S and denote

$$\mathsf{COUNT}(\hat{z}_1, ..., \hat{z}_k) := \{(z_1, ..., z_k) \in \mathsf{ZeroRows}_i(S_\alpha^k) : (\hat{z}_1, ..., \hat{z}_k) \in \mathsf{Good}(z_1, ..., z_k)\}.$$

We claim that $|\mathsf{COUNT}(\hat{z}_1, ..., \hat{z}_k)| \le \binom{d}{i}$. Informally, this means that $(\hat{z}_1, ..., \hat{z}_k)$ belongs to at most $\binom{d}{i}$ "good" sets (out of $|\mathsf{ZeroRows}_i(S_\alpha^k)|$). Just like before, assume that $|\mathsf{COUNT}(\hat{z}_1, ..., \hat{z}_k)| > \binom{d}{i}$ and define a function

$$F : \mathsf{COUNT}(\hat{z}_1, ..., \hat{z}_k) \to \mathcal{P}_i([d]), (z_1, ..., z_k) \longmapsto \mathsf{Zero}(z_1, ..., z_k).$$

---

[7] In the example above, W_1 is represented by the set $\{X^j : j \in [2n]\}$. Indeed, $|\mathsf{Zero}(X^j - X^k)| < 1$ for all distinct j, k.

Then,
$$F(z_1, ..., z_k) = I = F(z'_1, ..., z'_k)$$
for some set $I \in \mathcal{P}_i([d])$ and $z_1, ..., z_k, z'_1, ..., z'_k \in S_\alpha$ such that there exists an index j which satisfies $z_j \neq z'_j$. Since $(\hat{z}_1, ..., \hat{z}_k) \in \mathsf{Good}(z_1, ..., z_k)$ and $(\hat{z}_1, ..., \hat{z}_k) \in \mathsf{Good}(z'_1, ..., z'_k)$, we have that $z_j + y_j = \hat{z}_j = z'_j + y'_j$ for some distinct $y_j, y'_j \in W_i$. Take any $u \in I$ and note that $z_j \equiv 0 \equiv z'_j \pmod{f_u(X)}$. Therefore,
$$y_j \equiv \hat{z}_j - z_j \equiv \hat{z}_j \equiv \hat{z}_j - z'_j \equiv y'_j \pmod{f_u(X)}.$$
Hence, $|\mathsf{Zero}(y_j - y'_j)| \geq i$. Similarly as before, we observe that this leads to a contradiction by the definition of W_i. Thus, $|\mathsf{COUNT}(\hat{z}_1, ..., \hat{z}_k)| \leq \binom{d}{i}$. This implies:
$$|S| \geq \frac{\sum_{\mathbf{z} \in \mathsf{ZeroRows}_i(S_\alpha^k)} |\mathsf{Good}(\mathbf{z})|}{\binom{d}{i}} \geq \frac{\sum_{\mathbf{z} \in \mathsf{ZeroRows}_i(S_\alpha^k)} |W_i|^k - \binom{d}{i}}{\binom{d}{i}}$$

Combining the lower bound as well as the upper bound for $|S|$ we get:
$$|S_{\alpha + \|W_i\|_\infty}|^k - |\mathsf{ZeroRows}_i(S_\alpha^k)| \geq |S|$$
$$> \frac{1}{\binom{d}{i}} |\mathsf{ZeroRows}_i(S_\alpha^k)| \cdot |W_i|^k - |\mathsf{ZeroRows}_i(S_\alpha^k)|.$$
$$(3)$$

Therefore, $|\mathsf{ZeroRows}_i(S_\alpha^k)| \leq \frac{\binom{d}{i} \cdot |S_{\alpha + \|W_i\|_\infty}|^k}{|W_i|^k}$. $\qquad \square$

We point out that the proof does not work for $i = 0$. In this case, we can use the obvious upper bound: $\mathsf{ZeroRows}_0(S_\alpha^k) \leq |S_\alpha^k|$.

Consider again the equation $\mathbf{A}\mathbf{z}_1 + \mathbf{z}_2 = c\mathbf{t}$, where $\mathbf{A}$ and $\mathbf{t}$ are variables, and denote $\mathbf{A} = (a_{i,j})$ and $\mathbf{t} = (t_1, ..., t_k)$. Clearly, we have $\mathsf{Zero}(\mathbf{z}_1, c, \mathbf{z}_2) \subseteq \mathsf{Zero}(\mathbf{z}_1, c)$. Suppose that $\mathsf{Zero}(\mathbf{z}_1, c, \mathbf{z}_2) \neq \mathsf{Zero}(\mathbf{z}_1, c)$. If we write $\mathbf{z}_1 = (z_1, ..., z_\ell)$ and $\mathbf{z}_2 = (z'_1, ..., z'_k)$, then there exist some i, j such that $i \in \mathsf{Zero}(\mathbf{z}_1, c)$ and $i \notin \mathsf{Zero}(\mathbf{z}_1, c, z'_j)$. Note that
$$\mathbf{A}\mathbf{z}_1 + \mathbf{z}_2 = c\mathbf{t} \implies a_{j,1}z_1 + ... + a_{j,\ell}z_\ell + z'_j = ct_j.$$
However,
$$0 \not\equiv z'_j \equiv a_{j,1}z_1 + ... + a_{j,\ell}z_\ell + z'_j \equiv ct_j \equiv 0 \pmod{f_i(X)},$$
which leads to a contradiction. Therefore, if $\mathsf{Zero}(\mathbf{z}_1, c, \mathbf{z}_2) \neq \mathsf{Zero}(\mathbf{z}_1, c)$ then we end up with no solutions. This motivates us to extend the $\mathsf{ZeroRows}$ function as follows.

Definition 3.5. *Let $k \in \mathbb{N}$ and $A \subseteq R_q^k, B \subseteq R_q^\ell$ be non-empty sets. Then, we define $\mathsf{ZeroRows}_i(A; B)$ to be*
$$\mathsf{ZeroRows}_i(A; B) := \{(\mathbf{a}, \mathbf{b}) \in A \times B : \mathsf{Zero}(\mathbf{a}, \mathbf{b}) = \mathsf{Zero}(\mathbf{a}) \land |\mathsf{Zero}(\mathbf{a})| = i\}.$$

Sometimes, we write $\mathsf{ZeroRows}_i(A_1, A_2; B)$ to denote $\mathsf{ZeroRows}_i(\bar{A}; B)$, where $\bar{A} = A_1 \times A_2$.

Using the same techniques as before, one can prove a similar result to Lemma 3.4 which is related to the modified ZeroRows function.

Lemma 3.6. *Let $k, \ell, \alpha_1, \alpha_2 \in \mathbb{N}, i \in [d]$ and $W_i \subseteq R_q$ be a set of polynomials in R_q such that for any two distinct $u, v \in W_i$, $|\mathsf{Zero}(u - v)| < i$. Take any set $\mathcal{D} \subseteq R_q \backslash \{0\}$ and define e to be the largest integer which satisfies $\|\mathcal{D}\| \geq q^{e/d}$. Then,*

$$|\mathsf{ZeroRows}_i(S_{\alpha_1}^\ell, \mathcal{D}; S_{\alpha_2}^k)| \leq \frac{\binom{e}{i} \cdot |S_{\alpha_1 + \|W_i\|_\infty}|^\ell \cdot |S_{\alpha_2 + \|W_i\|_\infty}|^k \cdot |\mathcal{D}|}{|W_i|^{\ell+k}}.$$

Proof. Since we follow the same strategy as in the proof of Lemma 3.4, we only provide a proof sketch. To begin with, take any $\mathbf{z}_1 = (z_1, ..., z_\ell) \in S_{\alpha_1}^\ell, c \in \mathcal{D}, \mathbf{z}_2 = (z_1', ..., z_k') \in S_{\alpha_2}^k$ and define

$$\mathsf{Bad}(\mathbf{z}_1, c, \mathbf{z}_2) := \{(\mathbf{z}_1 + \mathbf{y}, c, \mathbf{z}_2 + \mathbf{y}') \in \mathsf{ZeroRows}_i(S_{\alpha_1}^\ell, \mathcal{D}; S_{\alpha_2}^k) : \mathbf{y} \in W_i^\ell, \mathbf{y}' \in W_i^k\}.$$

We point out that c stays still. Using the same technique as before, one can prove that $|\mathsf{Bad}(\mathbf{z}_1, c, \mathbf{z}_2)| \leq \binom{e}{i}$. Informally, this is because we only consider all subsets of $\mathsf{Zero}(c)$ (instead of $[d]$ like last time) of size i and c has at most e zeroes in the Chinese Remainder Representation (Lemma 3.2).

Now, we define a set

$$\mathsf{Good}(\mathbf{z}_1, c, \mathbf{z}_2) := \{(\mathbf{z}_1 + \mathbf{y}, c, \mathbf{z}_2 + \mathbf{y}') \notin \mathsf{ZeroRows}_i(S_{\alpha_1}^\ell, \mathcal{D}; S_{\alpha_2}^k) : \mathbf{y} \in W_i^\ell, \mathbf{y}' \in W_i^k\}.$$

As before, we have $|\mathsf{Good}(\mathbf{z}_1, c, \mathbf{z}_2)| = |W_i|^{\ell+k} - |\mathsf{Bad}(\mathbf{z}_1, c, \mathbf{z}_2)| \geq |W_i|^{\ell+k} - \binom{e}{i}$. Consider the following set

$$S = \bigcup_{(\mathbf{z}_1, c, \mathbf{z}_2) \in \mathsf{ZeroRows}_i(S_{\alpha_1}^\ell, \mathcal{D}; S_{\alpha_2}^k)} \mathsf{Good}(\mathbf{z}_1, c, \mathbf{z}_2).$$

We have that

$$S \subseteq S_{\alpha_1 + \|W_i\|_\infty}^\ell \times \mathcal{D} \times S_{\alpha_2 + \|W_i\|_\infty}^k \backslash \mathsf{ZeroRows}_i(S_\alpha^k)$$

by definition of Good. Let $(\mathbf{z}_1', c, \mathbf{z}_2')$ be an element of S and denote

$$\mathsf{COUNT}(\mathbf{z}_1', c, \mathbf{z}_2') := \{(\mathbf{z}_1, c, \mathbf{z}_2) \in \mathsf{ZeroRows}_i(S_\alpha^k) : (\mathbf{z}_1', c, \mathbf{z}_2') \in \mathsf{Good}(\mathbf{z}_1, c, \mathbf{z}_2)\}.$$

Similarly as before, we can show that $|\mathsf{COUNT}(\hat{z}_1, ..., \hat{z}_k)| \leq \binom{e}{i}$. Hence, we get:

$$
\begin{aligned}
|S| &\geq \frac{\sum_{(\mathbf{z}_1, c, \mathbf{z}_2) \in \mathsf{ZeroRows}_i(S_{\alpha_1}^\ell, \mathcal{D}; S_{\alpha_2}^k)} |\mathsf{Good}(\mathbf{z}_1, c, \mathbf{z}_2)|}{\binom{e}{i}} \\
&\geq \frac{\sum_{(\mathbf{z}_1, c, \mathbf{z}_2) \in \mathsf{ZeroRows}_i(S_{\alpha_1}^\ell, \mathcal{D}; S_{\alpha_2}^k)} |W_i|^{\ell+k} - \binom{e}{i}}{\binom{e}{i}}
\end{aligned}
\tag{4}
$$

Combining the lower bound as well as upper bound for $|S|$ we get:

$$|S_{\alpha_1 + \|W_i\|_\infty}|^\ell \cdot |\mathcal{D}| \cdot |S_{\alpha_2 + \|W_i\|_\infty}|^k - |\mathsf{ZeroRows}_i(S_{\alpha_1}^\ell, \mathcal{D}; S_{\alpha_2}^k)| \geq |S|,$$

and

$$|S| \geq \frac{1}{\binom{e}{i}} |\mathsf{ZeroRows}_i(S_{\alpha_1}^\ell, \mathcal{D}; S_{\alpha_2}^k)| \cdot |W_i|^{\ell+k} - |\mathsf{ZeroRows}_i(S_{\alpha_1}^\ell, \mathcal{D}; S_{\alpha_2}^k)|.$$

Therefore, $|\mathsf{ZeroRows}_i(S_{\alpha_1}^\ell, \mathcal{D}; S_{\alpha_2}^k)| \leq \frac{\binom{e}{i} \cdot |S_{\alpha_1 + \|W_i\|_\infty}|^\ell \cdot |S_{\alpha_2 + \|W_i\|_\infty}|^k \cdot |\mathcal{D}|}{|W_i|^{\ell+k}}.$ □

Again, we note that the lemma does not hold for $i = 0$. In this case, we use a simple bound: $|\mathsf{ZeroRows}_0(S_{\alpha_1}^\ell, \mathcal{D}; S_{\alpha_2}^k)| \leq |S_{\alpha_1}|^\ell \cdot |S_{\alpha_2}|^k \cdot |\mathcal{D}|$.

In Lemma 3.6 we have an additional condition $0 \notin \mathcal{D}$. This is because otherwise we cannot define the integer e. Recall that e represents the maximal number of zeroes in the Chinese Remainder Representation that an element in $\mathcal{D}$ can have. Hence, in case $\mathcal{D} = \{0\}$, we can simply set $e = d$ and follow the strategy as in Lemma 3.6. Thus, we end up with the following corollary.

Corollary 3.7. *Let $k, \ell, \alpha_1, \alpha_2 \in \mathbb{N}, i \in [d]$ and $W_i \subseteq R_q$ be a set of polynomials in R_q such that for any two distinct $u, v \in W_i$, $|\mathsf{Zero}(u - v)| < i$. Then,*

$$|\mathsf{ZeroRows}_i(S_{\alpha_1}^\ell; S_{\alpha_2}^k)| \leq \frac{\binom{d}{i} \cdot |S_{\alpha_1 + \|W_i\|_\infty}|^\ell \cdot |S_{\alpha_2 + \|W_i\|_\infty}|^k}{|W_i|^{\ell+k}}.$$

3.2 Computing Probabilities

We state and prove the main results of our paper. The first one provides an upper bound on the probability (over $\mathbf{A}$ and $\mathbf{t}$) of existence of $(\mathbf{z}_1, \mathbf{z}_2, c)$ which satisfy $\mathbf{A}\mathbf{z}_1 + \mathbf{z}_2 = c\mathbf{t}$. This can be applied to the security analysis of the Bai-Galbraith scheme [4] or qTESLA [2,8]. The second one, however, considers a slightly different equation: $\mathbf{A}\mathbf{z}_1 + \mathbf{z}_2 = c\mathbf{t}_1 \cdot 2^\delta$ where $\mathbf{t}_1 = \mathsf{Power2Round}_q(\mathbf{t}, \delta)$ for some δ, and can be applied to the security analysis of Dilithium-QROM [11].

Theorem 3.8. *Let $\alpha_1, \alpha_2 \in \mathbb{N}$ and $\mathcal{D} \subseteq R_q \setminus \{0\}$. Also, for $i = 1, ..., d$, define $W_i \subseteq R_q$ to be a set of polynomials such that for any two distinct $u, v \in W_i$, $|\mathsf{Zero}(u - v)| < i$. Then*

$$\Pr_{\mathbf{A} \leftarrow R_q^{k \times \ell}, \mathbf{t} \leftarrow R_q^k} [\exists (\mathbf{z}_1, \mathbf{z}_2, c) \in S_{\alpha_1}^\ell \times S_{\alpha_2}^k \times \mathcal{D} : \mathbf{A}\mathbf{z}_1 + \mathbf{z}_2 = c\mathbf{t}]$$

$$\leq \frac{|S_{\alpha_1}|^\ell \cdot |S_{\alpha_2}|^k \cdot |\mathcal{D}|}{q^{nk}} + \sum_{i=1}^e \frac{\binom{e}{i} \cdot |S_{\alpha_1 + \|W_i\|_\infty}|^\ell \cdot |S_{\alpha_2 + \|W_i\|_\infty}|^k \cdot |\mathcal{D}|}{|W_i|^{\ell+k} \cdot q^{nk(1-i/d)}}$$

$$\tag{5}$$

where e is the largest integer such that $\|\mathcal{D}\| \geq q^{e/d}$.

Proof. Fix $\mathbf{z}_1 = (z_1, ..., z_\ell), \mathbf{z}_2 = (z'_1, ..., z'_k)$ and c. We first prove that

$$\mathsf{Zero}(\mathbf{z}_1, c) \neq \mathsf{Zero}(\mathbf{z}_1, c, \mathbf{z}_2) \implies \Pr_{\mathbf{A} \leftarrow R^{k \times \ell}, \mathbf{t} \leftarrow R_q^k} [\mathbf{A}\mathbf{z}_1 + \mathbf{z}_2 = c\mathbf{t}] = 0.$$

Suppose that $\mathsf{Zero}(\mathbf{z}_1, c) \neq \mathsf{Zero}(\mathbf{z}_1, c, \mathbf{z}_2)$. Then, there exists some $i \in [d]$ such that $i \in \mathsf{Zero}(\mathbf{z}_1, c)$ and $i \notin \mathsf{Zero}(\mathbf{z}_1, c, \mathbf{z}_2)$. This implies that there is some $j \in [k]$

so that $i \notin \mathsf{Zero}(\mathbf{z}_1, c, z'_j)$ (otherwise $i \in \mathsf{Zero}(\mathbf{z}_1, c, \mathbf{z}_2)$). In particular, we have $z'_j \not\equiv 0 \pmod{f_i(X)}$. Denote $\mathbf{A} = (a_{i,j})$ and $\mathbf{t} = (t_1, ..., t_k)$ and note that

$$\mathbf{A}\mathbf{z}_1 + \mathbf{z}_2 = c\mathbf{t} \implies a_{j,1}z_1 + ... + a_{j,\ell}z_\ell + z'_j = ct_j.$$

However,

$$0 \not\equiv z'_j \equiv a_{j,1}z_1 + ... + a_{j,\ell}z_\ell + z'_j \equiv ct_j \equiv 0 \pmod{f_i(X)},$$

contradiction.

Hence, there are no $\mathbf{A}, \mathbf{t}$ which satisfy $\mathbf{A}\mathbf{z}_1 + \mathbf{z}_2 = c\mathbf{t}$. Thus, we only consider $(\mathbf{z}_1, c, \mathbf{z}_2)$ such that $\mathsf{Zero}(\mathbf{z}_1, c) = \mathsf{Zero}(\mathbf{z}_1, c, \mathbf{z}_2)$, alternatively $(\mathbf{z}_1, c, \mathbf{z}_2) \in \mathsf{ZeroRows}_i(S^\ell_{\alpha_1}, \mathcal{D}; S^k_{\alpha_2})$ for some $i \leq e$. We claim that

$$\Pr_{\mathbf{A} \leftarrow R^{k \times \ell}, \mathbf{t} \leftarrow R^k_q} [\mathbf{A}\mathbf{z}_1 + \mathbf{z}_2 = c\mathbf{t}] = 1/q^{nk(1-i/d)}.$$

Note that we can write:

$$\Pr_{\mathbf{A} \leftarrow R^{k \times \ell}, \mathbf{t} \leftarrow R^k_q} [\mathbf{A}\mathbf{z}_1 + \mathbf{z}_2 = c\mathbf{t}] = \prod_{i=1}^{k} \Pr_{a_{i,1}, ..., a_{i,\ell}, t_i \leftarrow R_q} [a_{i,1}z_1 + ... + a_{i,\ell}z_\ell + z'_i = c \cdot t_i].$$

Let us fix an index i and define

$$A = \{(a_1, ..., a_\ell, t) \in R_q^{\ell+1} : \sum_{j=1}^{\ell} a_j z_j + z'_i = c \cdot t\}.$$

We want to show that $|A| = q^{n(\ell+i/d)}$. Take any $u \in [d]$ and consider the set

$$A_u = \{(a_1, ..., a_\ell, t) \in (\mathbb{Z}_q[X]/(f_u(X)))^{\ell+1} : a_1 z_1 + ... + a_\ell z_\ell + z'_i \equiv c \cdot t \pmod{f_u(X)}\}.$$

If $u \in \mathsf{Zero}(\mathbf{z}_1, c, \mathbf{z}_2)$ then any $a_1, ..., a_\ell, t$ satisfy the equation, because

$$z_1 \equiv ... \equiv z_\ell \equiv z'_i \equiv c \equiv 0 \pmod{f_u(X)}.$$

Hence, $|A_u| = q^{(l+1)\cdot n/d}$. If $u \notin \mathsf{Zero}(\mathbf{z}_1, c, \mathbf{z}_2)$ then one of $z_1, ..., z_\ell, c$ is invertible modulo $(f_u(X), q)$, without loss of generality say z_j. Then, $a_1, ..., a_{j-1}, a_{j+1}, ..., a_\ell, c$ can be chosen arbitrarily and a_j is picked such that the equation is satisfied. Therefore, $|A_u| = q^{\ell \cdot n/d}$. Now, by the Chinese Remainder Theorem we have that

$$|A| = \prod_{u=1}^{d} |A_u| = q^{i \cdot (\ell+1) \cdot n/d + (d-i) \cdot \ell \cdot n/d} = q^{n(\ell+i/d)}.$$

Hence,

$$\Pr_{a_{i,1}, ..., a_{i,\ell}, t_i \leftarrow R_q} [a_{i,1}z_1 + ... + a_{i,\ell}z_\ell + z'_i = c \cdot t_i] = \frac{|A|}{q^{(\ell+1)\cdot n}} = 1/q^{n(1-i/d)}.$$

Eventually, we obtain $\Pr_{\mathbf{A} \leftarrow R^{k \times \ell}, \mathbf{t} \leftarrow R^k_q} [\mathbf{A}\mathbf{z}_1 + \mathbf{z}_2 = c\mathbf{t}] = 1/q^{nk(1-i/d)}.$

Now, we combine the observations above and Lemma 3.6. For clarity, set $Z_i = \mathsf{ZeroRows}_i(S_{\alpha_1}^\ell, \mathcal{D}; S_{\alpha_2}^k)$. Then,

$$\Pr_{\mathbf{A} \leftarrow R^{k \times \ell}, \mathbf{t} \leftarrow R_q^k}[\exists (\mathbf{z}_1, \mathbf{z}_2, c) \in S_{\alpha_1}^\ell \times S_{\alpha_2}^k \times \mathcal{D} : \mathbf{A}\mathbf{z}_1 + \mathbf{z}_2 = c\mathbf{t}]$$

$$\leq \sum_{\mathbf{z}_1 \in S_{\alpha_1}^\ell, c \in \mathcal{D}, \mathbf{z}_2 \in S_{\alpha_2}^k} \Pr_{\mathbf{A} \leftarrow R^{k \times \ell}, \mathbf{t} \leftarrow R_q^k}[\mathbf{A}\mathbf{z}_1 + \mathbf{z}_2 = c\mathbf{t}]$$

$$\leq \sum_{i=0}^{e} \sum_{(\mathbf{z}_1, c, \mathbf{z}_2) \in Z_i} \Pr_{\mathbf{A} \leftarrow R^{k \times \ell}, \mathbf{t} \leftarrow R_q^k}[\mathbf{A}\mathbf{z}_1 + \mathbf{z}_2 = c\mathbf{t}]$$

$$\leq \sum_{i=0}^{e} \sum_{(\mathbf{z}_1, c, \mathbf{z}_2) \in Z_i} 1/q^{nk(1-i/d)} \tag{6}$$

$$\leq \sum_{i=0}^{e} |Z_i|/q^{nk(1-i/d)}$$

$$\leq \frac{|S_{\alpha_1}|^\ell \cdot |S_{\alpha_2}|^k \cdot |\mathcal{D}|}{q^{nk}} + \sum_{i=1}^{e} \frac{\binom{e}{i} \cdot |S_{\alpha_1 + \|W_i\|_\infty}|^\ell \cdot |S_{\alpha_2 + \|W_i\|_\infty}|^k \cdot |\mathcal{D}|}{|W_i|^{\ell+k} \cdot q^{nk(1-i/d)}}. \qquad \square$$

We can obtain a very similar result for $\mathcal{D} = \{0\}$ using Corollary 3.7. We just need to pick e to be the integer, such that any non-zero $(\mathbf{z}_1, \mathbf{z}_2) \in S_{\alpha_1}^\ell \times S_{\alpha_2}^k$ has at most c zero rows. Since each component of $\mathbf{z}_1$ has norm at most $\alpha_1 \sqrt{n}$, we could choose the maximal e so that $\alpha_1 \sqrt{n} \geq q^{e/d}$. We omit the proof since it is very similar to the one for Theorem 3.8.

Corollary 3.9. *Let $\alpha_1, \alpha_2 \in \mathbb{N}$. Also, for $i = 1, ..., d$, define $W_i \subseteq R_q$ to be a set of polynomials such that for any two distinct $u, v \in W_i$, $|\mathsf{Zero}(u - v)| < i$. Then*

$$\Pr_{\mathbf{A} \leftarrow R_q^{k \times \ell}}[\exists (\mathbf{z}_1, \mathbf{z}_2) \in S_{\alpha_1}^\ell \setminus \{0\} \times S_{\alpha_2}^k : \mathbf{A}\mathbf{z}_1 + \mathbf{z}_2 = 0]$$

$$\leq \frac{|S_{\alpha_1}|^\ell \cdot |S_{\alpha_2}|^k}{q^{nk}} + \sum_{i=1}^{e} \frac{\binom{d}{i} \cdot |S_{\alpha_1 + \|W_i\|_\infty}|^\ell \cdot |S_{\alpha_2 + \|W_i\|_\infty}|^k}{|W_i|^{\ell+k} \cdot q^{nk(1-i/d)}}, \tag{7}$$

where e is the largest integer such that $\alpha_1 \sqrt{n} \geq q^{e/d}$.

The next theorem considers a modified equation $\mathbf{A}\mathbf{z}_1 + \mathbf{z}_2 = c\mathbf{t}_1 \cdot 2^\delta$ where $\mathbf{t}_1 = \mathsf{Power2Round}_q(\mathbf{t}, \delta)$ for some $\delta \in \mathbb{N}$. However, we need to take a slightly different approach in order to provide a reasonable upper bound for the probability due to the appearance of $\mathsf{Power2Round}_q$ function.

Theorem 3.10. *Let $\alpha_1, \alpha_2, \delta \in \mathbb{N}$ and $\mathcal{D} \subseteq R_q \setminus \{0\}$. Also, for $i = 1, ..., d$, define $W_i \subseteq R_q$ to be a set of polynomials such that for any two distinct $u, v \in W_i$, $|\mathsf{Zero}(u - v)| < i$. Then*

$$\Pr_{\mathbf{A} \leftarrow R_q^{k \times \ell}, \mathbf{t} \leftarrow R_q^k}[\exists (\mathbf{z}_1, \mathbf{z}_2, c) \in S_{\alpha_1}^\ell \times S_{\alpha_2}^k \times \mathcal{D} : \mathbf{A}\mathbf{z}_1 + \mathbf{z}_2 = c\mathbf{t}_1 \cdot 2^\delta]$$

$$\leq |\mathcal{D}| \cdot |S_{\alpha_2}|^k \cdot \left(\left(\frac{2^\delta}{q^{(1-e_1/d)}} \right)^{nk} + \frac{|S_{\alpha_1}|^\ell}{q^{nk}} + \sum_{i=1}^{e_2} \frac{\binom{d}{i} \cdot |S_{\alpha_1 + \|W_i\|_\infty}|^\ell}{|W_i|^\ell \cdot q^{nk(1-i/d)}} \right) \tag{8}$$

where $\mathbf{t}_1 = \mathsf{Power2Round}_q(\mathbf{t}, \delta)$ and e_1 (resp. e_2) is the largest integer such that $||\mathcal{D}|| \geq q^{e_1/d}$ (resp. $\alpha_1\sqrt{N} \geq q^{e_2/d}$).

Proof. **Case 1.** suppose that $\mathbf{z}_1 = 0$. Then, the probability becomes:

$$\Pr_{\mathbf{t} \leftarrow R_q^k}[\exists(\mathbf{z}_2, c) \in S_{\alpha_2}^k \times \mathcal{D} : \mathbf{z}_2 = c\mathbf{t}_1 \cdot 2^\delta].$$

Fix $\mathbf{z}_2 = (z_1, ..., z_k), c$ and denote $\mathbf{t} = (t_1, ..., t_k)$. Consider the following probability:

$$\Pr_{\mathbf{t} \leftarrow R_q^k}[\mathbf{z}_2 = c\mathbf{t}] = \prod_{j=1}^{k} \Pr[z_j = ct_j].$$

By definition of e_1, we have $|\mathsf{Zero}(c)| \leq e_1$ by Lemma 3.2. Take arbitrary $j \in [k]$. We compute the maximal number of polynomials t_j satisfying $z_j = ct_j$. Define a set

$$T_u = \{t \in \mathbb{Z}_q[X]/(f_u(X)) : z_j \equiv ct \pmod{f_u(X)}\}.$$

Clearly, $|T_u| \leq q^{n/d}$. Let $u \notin \mathsf{Zero}(c)$. Then, c is invertible modulo $(f_u(X), q)$. Therefore, $|T_u| = 1$. By the Chinese Remainder Theorem, the number of polynomials t_j satisfying $z_j = ct_j$ is at most

$$\prod_{u=1}^{k} |T_u| \leq q^{|\mathsf{Zero}(c)| \cdot n/d} \leq q^{e_1 \cdot n/d}.$$

Hence, we end up with

$$\Pr[z_j = ct_j] \leq \frac{q^{e_1 \cdot n/d}}{q^n} = \frac{1}{q^{n(1-e_1/d)}}.$$

Thus:

$$\Pr_{\mathbf{t} \leftarrow R_q^k}[\mathbf{z}_2 = c\mathbf{t}] = \prod_{j=1}^{k} \Pr[z_j = ct_j] \leq \frac{1}{q^{nk(1-e_1/d)}}.$$

For $\mathbf{t} \in R_q^k$, the most frequent value of each coefficient of $\mathbf{t}_1$ occurs at most 2^δ times. Hence,

$$\Pr_{\mathbf{t} \leftarrow R_q^k}[\mathbf{z}_2 = c\mathbf{t}_1 \cdot 2^\delta] \leq \left(\frac{2^\delta}{q^{(1-e_1/d)}}\right)^{nk}.$$

Eventually, by the union bound we obtain:

$$\Pr_{\mathbf{t} \leftarrow R_q^k}[\exists(\mathbf{z}_2, c) \in S_{\alpha_2}^k \times \mathcal{D} : \mathbf{z}_2 = c\mathbf{t}_1 \cdot 2^\delta] \leq \sum_{\mathbf{z}_2 \in S_{\alpha_2}^k, c \in \mathcal{D}} \left(\frac{2^\delta}{q^{(1-e_1/d)}}\right)^{nk},$$

and the sum is equal to $|\mathcal{D}| \cdot |S_{\alpha_2}|^k \cdot \left(\frac{2^\delta}{q^{(1-e_1/d)}}\right)^{nk}$.

Case 2. Suppose that $\mathbf{z} = (z_1, ..., z_\ell) \neq \mathbf{0}$ and fix $\mathbf{z}_2 = (z'_1, ..., z'_k)$ and c. Also, denote $\mathbf{A} = (a_{i,j})$, $\mathbf{t} = (t_1, ..., t_k)$ and $t'_i = \mathsf{Power2Round}_q(t_i, \delta)$ for $i \in [k]$. Then,

$$\Pr_{\mathbf{A} \leftarrow R^{k \times \ell}, \mathbf{t} \leftarrow R_q^k}[\mathbf{A}\mathbf{z}_1 + \mathbf{z}_2 = c\mathbf{t}_1 \cdot 2^\delta] = \prod_{i=1}^{k} \Pr_{a_{i,1}, ..., a_{i,\ell}, t_i \leftarrow R_q}[\sum_{j=1}^{\ell} a_{i,j} z_j + z'_i = c \cdot t'_i \cdot 2^\delta].$$

Let us fix an index i and consider the set

$$A_t = \{(a_1, ..., a_\ell) \in R_q^\ell : \sum_{j=1}^{\ell} a_j z_j + z'_i = c \cdot t' \cdot 2^\delta\},$$

where $t' = \mathsf{Power2Round}_q(t)$. We want to prove that $|A_t| \leq q^{n(\ell - 1 + m/d)}$, where $m = |\mathsf{Zero}(z_1, ..., z_\ell)|$. Define

$$A_t^u = \{(a_1, ..., a_\ell) \in (\mathbb{Z}_q[X]/(f_u(X)))^\ell : \sum_{j=1}^{\ell} a_j z_j \equiv c \cdot t' \cdot 2^\delta - z'_i \pmod{f_u(X)}\}.$$

Clearly, we have $|A_t^u| \leq q^{\ell \cdot n/d}$. Consider $u \notin \mathsf{Zero}(z_1, ..., z_\ell)$. This means that z_w is invertible modulo $(f_u(X), q)$ for some $w \in [\ell]$. Hence, we can pick any possible values for $a_1, ..., a_{w-1}, a_{w+1}, ..., a_\ell$ and then adjust a_w so that it satisfies the equation. Note that for fixed $a_1, ..., a_{w-1}, a_{w+1}, ..., a_\ell$, there is exactly one such a_w. Thus, $|A_t^u| = q^{(\ell-1) \cdot n/d}$. By the Chinese Remainder Theorem, we get

$$|A_t| = \prod_{u=1}^{d} |A_t^u| \leq q^{m \cdot n\ell/d} \cdot q^{(d-m) \cdot (\ell-1)n/d} = q^{n(\ell-1+m/d)}.$$

Since we consider uniform distribution for $a_{i,1}, ..., a_{i,\ell}, t_i$, we can conclude that:

$$\Pr_{a_{i,1}, ..., a_{i,\ell}, t_i \leftarrow R_q}[\sum_{j=1}^{\ell} a_{i,j} z_j + z'_i = c \cdot t'_i \cdot 2^\delta] = \frac{\sum_{t_i \in R_q} A_{t_i}}{q^{\ell \cdot n} \cdot q^n} \leq \frac{q^{n(\ell-1+m/d)}}{q^{\ell \cdot n}} = 1/q^{n(1-m/d)}.$$

Therefore, $\Pr_{\mathbf{A} \leftarrow R^{k \times \ell}, \mathbf{t} \leftarrow R_q^k}[\mathbf{A}\mathbf{z}_1 + \mathbf{z}_2 = c\mathbf{t}_1 \cdot 2^\delta] \leq 1/q^{nk(1-m/d)}$.

Now we can apply the union bound. First of all, note that if $i > e_2$ then $\mathsf{ZeroRows}_i(S_{\alpha_1}^\ell \setminus \{\mathbf{0}\}) = \varnothing$ by Lemma 3.2. Hence,

$$S_{\alpha_1}^\ell \setminus \{\mathbf{0}\} = \bigcup_{i=0}^{d} \mathsf{ZeroRows}_i(S_{\alpha_1}^\ell \setminus \{\mathbf{0}\}) = \bigcup_{i=0}^{e_2} \mathsf{ZeroRows}_i(S_{\alpha_1}^\ell \setminus \{\mathbf{0}\}).$$

For simplicity, denote $Z_i = \mathsf{ZeroRows}_i(S_{\alpha_1}^\ell \setminus \{\mathbf{0}\})$. Then,

$$\Pr[\exists (\mathbf{z}_1, \mathbf{z}_2, c) \in S_{\alpha_1}^\ell \times S_{\alpha_2}^k \times \mathcal{D} : \mathbf{A}\mathbf{z}_1 + \mathbf{z}_2 = c\mathbf{t}_1 \cdot 2^\delta]$$

$$\leq \sum_{\mathbf{z}_1 \in S_{\alpha_1}^\ell \setminus \{\mathbf{0}\}, \mathbf{z}_2 \in S_{\alpha_2}^k, c \in \mathcal{D}} \Pr[\mathbf{A}\mathbf{z}_1 + \mathbf{z}_2 = c\mathbf{t}_1 \cdot 2^\delta]$$

$$\leq \sum_{\mathbf{z}_2 \in S_{\alpha_2}^k, c \in \mathcal{D}} \sum_{i=0}^{e_2} \sum_{\mathbf{z}_1 \in Z_i} \Pr[\mathbf{A}\mathbf{z}_1 + \mathbf{z}_2 = c\mathbf{t}_1 \cdot 2^\delta]$$

$$\leq \sum_{\mathbf{z}_2 \in S_{\alpha_2}^k, c \in \mathcal{D}} \sum_{i=0}^{e_2} \sum_{\mathbf{z}_1 \in Z_i} 1/q^{nk(1-i/d)}$$

$$\leq \sum_{\mathbf{z}_2 \in S_{\alpha_2}^k, c \in \mathcal{D}} \sum_{i=0}^{e_2} |Z_i|/q^{nk(1-i/d)}.$$

$$\tag{9}$$

By Lemma 3.4, $|Z_i| \leq \frac{\binom{d}{i} \cdot |S_{\alpha_1 + \|W_i\|_\infty}|^\ell}{|W_i|^\ell}$. Also, we have $|Z_0| \leq |S_{\alpha_1}|^\ell$. Therefore, we can bound the probability above by:

$$|\mathcal{D}| \cdot |S_{\alpha_2}|^k \cdot (|S_{\alpha_1}|^\ell / q^{nk} + \sum_{i=1}^{e_2} \frac{\binom{d}{i} \cdot |S_{\alpha_1 + \|W_i\|_\infty}|^\ell}{|W_i|^\ell \cdot q^{nk(1-i/d)}}). \tag{10}$$

The theorem now follows from combining the two cases. □

3.3   Constructing W_i

All the probability results presented in the previous subsection depend on the sizes of sets W_i. Recall that a set W_i satisfies a condition that for any two distinct $u, v \in W_i$, we have $|\mathsf{Zero}(u - v)| < i$. Based on the upper bounds obtained above, we would like to construct large sets W_i but with small infinity norm $\|W_i\|_\infty$.

Let us start by constructing W_1. We choose

$$W_1 := \{X^i : i \in [2n]\}.$$

Clearly, $X^i - X^j \in R_q$ is invertible, for $i \neq j$, so $|\mathsf{Zero}(X^i - X^j)| = 0 < 1$. Also, $|W_1| = 2n$ and $\|W_1\|_\infty = 1$.

Now, let us fix $i \geq 2$. The main idea is to set W_i to be a subset of $S = \{u \in R_q : \|u\| < \frac{1}{2}q^{i/d}\}$, i.e $\|W_i\| < \frac{1}{2}q^{i/d}$. Note that if we pick two distinct $u, v \in S$, then $0 < \|u - v\| < q^{i/d}$ by the triangle inequality. Hence, by Lemma 3.2 we get that $|\mathsf{Zero}(u - v)| < i$. Therefore, any subset of S will satisfy the condition for W_i[8].

If $t := \lfloor \frac{q^{i/d}}{2} \rfloor$ is smaller than $\sqrt{n}$ then we set

$$W_i := \{\sum_{j=1}^{t^2} \epsilon_j \cdot X^{\alpha_j} \in R_q : \epsilon_1, ..., \epsilon_{t^2} \in \{-1, 0, 1\}, \{\alpha_1, ..., \alpha_{t^2}\} \in \mathcal{P}_{t^2}([n])\}.$$

---

[8] Note that this technique can also be used for W_1 as long as $q^{1/d}$ is large enough.

Then, $||W_i||_\infty = 1, ||W_i|| = t < \frac{1}{2}q^{i/d}$ and

$$|W_i| = \sum_{j=0}^{t^2} \binom{n}{j} \cdot 2^j.$$

Suppose that $t \geq \sqrt{n}$. In this case, we provide two constructions of W_i and in the experiments we choose the one that minimises the overall probability.

1. Set $W_i := S$. Then, $||W_i||_\infty = \lfloor \frac{1}{2}q^{i/d} \rfloor$ and $|W_i| \geq V_n(\frac{1}{2}q^{i/d} - \sqrt{n})^9$ where $V_N(r)$ is the volume of an n-dimensional ball of radius r.
2. Set $W_i := S_{\lfloor t/\sqrt{n} \rfloor}$. Clearly, we have the following properties: $W_i \subseteq S$, $||W_i||_\infty = \lfloor t/\sqrt{n} \rfloor$ and $|W_i| = (2 \lfloor t/\sqrt{n} \rfloor + 1)^n$.

4 Applications to the Bai-Galbraith Scheme

We present a slightly modified version of Bai-Galbraith scheme [4] whose security is based on MLWE in the quantum random oracle model. First, we construct the corresponding lossy identification protocol[10]. Results from the previous section will be used to prove security properties of this ID scheme. Then, using the main result of [11], we obtain the secure signature scheme in the QROM. Note that identical techniques can be applied to other closely related signature schemes, such as qTESLA [2,8] or the original scheme [4]. We focus on the modified scheme because it is actually a simpler version of Dilithium-QROM. Since the highly-optimised version of Dilithium-QROM can be somewhat overwhelming to readers who are not already comfortable with such constructions, we consider its simplified version here.

4.1 The Identification Protocol

The algorithms for identification protocol $ID = (IGen, P_1, P_2, V)$ are described in Fig. 3 with the concrete parameters $par = (q, d, n, k, \ell, \gamma, \gamma', \eta, \beta)$ given later in Tables 1 and 2.

We want the challenge space in these ID and signature schemes to be a subset of the ring R, have size a little larger than 2^{256}, and consist of polynomials with small norms. In this paper, we set the dimension n of the ring R to be equal to 512. Hence, let us define the following challenge set:

$$\mathsf{ChSet} := \{c \in R \mid ||c||_\infty = 1 \text{ and } ||c|| = \sqrt{46}\}. \tag{11}$$

Hence, ChSet consists of elements in R with $-1/0/1$ coefficients that have exactly 46 non-zero coefficients. The size of this set is $\binom{n}{46} \cdot 2^{46}$, which for $n = 512$ is greater than 2^{265}.

---

[9] This can be proven similarly as in [5] by putting a box of side-length 1 centered on every integer point and checking that the ball is completely covered by these boxes.

[10] For readers not familiar with definitions of lossy and canonical identification schemes, we provide all necessary background in [20].

KEY GENERATION. The key generation starts with choosing a random 256-bit seed ρ and expanding into a matrix $\mathbf{A} \in R_q^{k \times \ell}$ by an extendable output function Sam, i.e. a function on bit strings in which the output can be extended to any desired length, modeled as a random oracle. The secret keys $(\mathbf{s}_1, \mathbf{s}_2) \in S_\eta^\ell \times S_\eta^k$ have uniformly random coefficients between $-\eta$ and η (inclusively). The value $\mathbf{t} = \mathbf{A}\mathbf{s}_1 + \mathbf{s}_2$ is then computed. The public key needed for verification is $(\rho, \mathbf{t})$ and the secret key is $(\rho, \mathbf{s}_1, \mathbf{s}_2)$.

PROTOCOL EXECUTION. The prover starts the identification protocol by reconstructing $\mathbf{A}$ from the random seed ρ. The next step has the prover sample $\mathbf{y} \leftarrow S_{\gamma'-1}^\ell$ and then compute $\mathbf{w} = \mathbf{A}\mathbf{y}$. He then writes $\mathbf{w} = 2\gamma \cdot \mathbf{w}_1 + \mathbf{w}_0$, with $\mathbf{w}_0$ between $-\gamma$ and γ (inclusively), and then sends $\mathbf{w}_1$ to the verifier.

The set ChSet is defined as in Eq. (11), and $\mathsf{ZSet} = S_{\gamma'-\beta-1}^\ell \times \{0,1\}^k$. The set of commitments WSet is defined as $\mathsf{WSet} = \{\mathbf{w}_1 \,:\, \exists \mathbf{y} \in S_{\gamma'-1}^\ell \text{ s.t. } \mathbf{w}_1 = \mathsf{HighBits}_q(\mathbf{A}\mathbf{y}, 2\gamma)\}$.

The verifier generates a random challenge $c \leftarrow \mathsf{ChSet}$ and sends it to the prover. The prover computes $\mathbf{z} = \mathbf{y} + c\mathbf{s}$. If $\mathbf{z} \notin S_{\gamma'-\beta-1}^\ell$, then the prover sets his response to $\perp$. He also replies with $\perp$ if $\mathsf{LowBits}_q(\mathbf{w} - c\mathbf{s}_2, 2\gamma) \notin S_{\gamma-\beta-1}^k$. Eventually, the verifier checks whether $\|\mathbf{z}\|_\infty < \gamma' - \beta$ and that $\mathbf{A}\mathbf{z} - c\mathbf{t}$.

4.2 Security Analysis

We omit proofs of correctness and non-abort honest verifier zero-knowledge properties since they have already been analysed in the previous works [2,4,9,11]. Instead, we focus on lossyness, min entropy and computational unique response. We recall that sets W_i are introduced in Sect. 3.3.

Lemma 4.1. *If* $\beta \geq \max_{s \in S_\eta, c \in \mathsf{ChSet}} \|cs\|_\infty$, *then* ID *is perfectly* naHVZK *and has correctness error* $\nu \approx 1 - \exp(-\beta n \cdot (k/\gamma + \ell/\gamma'))$.

IGen(par)	$P_1(sk)$
01 $\rho \leftarrow \{0,1\}^{256}$	08 $\mathbf{A} \leftarrow R_q^{k \times \ell} := \mathsf{Sam}(\rho)$
02 $\mathbf{A} \leftarrow R_q^{k \times \ell} := \mathsf{Sam}(\rho)$	09 $\mathbf{y} \leftarrow S_{\gamma'-1}^\ell$
03 $(\mathbf{s}_1, \mathbf{s}_2) \leftarrow S_\eta^\ell \times S_\eta^k$	10 $\mathbf{w} := \mathbf{A}\mathbf{y}$
04 $\mathbf{t} := \mathbf{A}\mathbf{s}_1 + \mathbf{s}_2$	11 $\mathbf{w}_1 := \mathsf{HighBits}_q(\mathbf{w}, 2\gamma)$
05 $pk = (\rho, \mathbf{t})$	12 **return** $(W = \mathbf{w}_1, St = (\mathbf{w}, \mathbf{y}))$
06 $sk = (\rho, \mathbf{s}_1, \mathbf{s}_2)$	
07 **return** (pk, sk)	$P_2(sk, W = \mathbf{w}_1, c, St = (\mathbf{w}, \mathbf{y}))$
	13 $\mathbf{z} := \mathbf{y} + c\mathbf{s}_1$
	14 **if** $\|\mathbf{z}\|_\infty \geq \gamma' - \beta$ **or** $\|\mathsf{LowBits}_q(\mathbf{w} - c\mathbf{s}_2, 2\gamma)\|_\infty \geq \gamma - \beta$
	15 **then** $\mathbf{z} := \perp$
$V(pk, W = \mathbf{w}_1, c, Z = \mathbf{z})$	16 **else return** $Z = \mathbf{z}$
17 **return** $\llbracket \|\mathbf{z}\|_\infty < \gamma' - \beta \rrbracket$ **and** $\llbracket \mathbf{w}_1 = \mathsf{HighBits}_q(\mathbf{A}\mathbf{z} - c\mathbf{t}, 2\gamma) \rrbracket$	

Fig. 3. Modified Bai-Galbraith identification protocol.

LOSSYNESS. Let us consider the scheme in which the public key is generated uniformly at random (Fig. 4), rather than as in IGen of Fig. 3. It is enough to show that even if the prover is computationally unbounded, he only has approximately a $1/|\mathsf{ChSet}|$ probability of making the verifier accept during each run of the identification scheme.

```
LossyIGen(par)
01  ρ ← {0,1}²⁵⁶; A ← R_q^{k×ℓ} := Sam(ρ)
02  t ← R_q^k
03  return pk = (ρ, t)
```

Fig. 4. The lossy instance generator LossyIGen.

Since the output of LossyIGen is uniformly random over $R_q^{k\times\ell} \times R_q^k$ and the output of IGen in Fig. 3 is $(\mathbf{A}, \mathbf{A}\mathbf{s}_1 + \mathbf{s}_2)$ where $\mathbf{A} \leftarrow R_q^{k\times\ell}$ and $(\mathbf{s}_1, \mathbf{s}_2) \leftarrow S_\eta^\ell \times S_\eta^k$, we get that

$$\mathrm{Adv}_{\mathsf{ID}}^{\mathsf{LOSS}}(\mathsf{A}) = \mathrm{Adv}_{k,\ell,D}^{\mathsf{MLWE}}(\mathsf{A}),$$

where D is the uniform distribution over S_η.

Lemma 4.2. *Let e_ℓ be the largest integer which satisfies $q^{e_\ell/d} \leq 2\sqrt{46}$. Then, ID has $\varepsilon_{\mathsf{ls}}$-lossy soundness, where*

$$\varepsilon_{\mathsf{ls}} \leq \frac{1}{|\mathsf{ChSet}|} + \frac{|S_{2(\gamma'-\beta-1)}|^\ell \cdot |S_{4\gamma+2}|^k \cdot |\mathsf{ChSet}|^2}{q^{nk}}$$
$$+ \sum_{i=1}^{e_\ell} \frac{\binom{e_\ell}{i} \cdot |S_{2(\gamma'-\beta-1)+\|W_i\|_\infty}|^\ell \cdot |S_{4\gamma+2+\|W_i\|_\infty}|^k \cdot |\mathsf{ChSet}|^2}{|W_i|^{\ell+k} \cdot q^{nk(1-i/d)}}.$$

$$(12)$$

Proof. Consider an unbounded adversary C that is executed in game LOSSY-IMP of Fig. 5.

```
GAME LOSSY-IMP:
01  pk_ls := (ρ, t) ← LossyIGen(par)
02  (w₁, St) ← C(pk_ls)
03  c ← ChSet
04  z ← C(St, c)
05  return ⟦w₁ = HighBits_q(Az − ct, 2γ)⟧ and ⟦‖z‖_∞ < γ' − β⟧
```

Fig. 5. The lossy impersonation game LOSSY-IMP.

Assume that for some $\mathbf{w}_1$, there exist two $c \neq c' \in \mathsf{ChSet}$ and two $\mathbf{z}, \mathbf{z}'$ that lead to C winning, i.e. $\|\mathbf{z}\|_\infty, \|\mathbf{z}'\|_\infty < \gamma' - \beta$ and

$$\mathbf{w}_1 = \mathsf{HighBits}_q(\mathbf{Az} - t c, 2\gamma),$$
$$\mathbf{w}_1 = \mathsf{HighBits}_q(\mathbf{Az}' - t c', 2\gamma).$$

By Lemma 2.1, we know that this implies

$$\|\mathbf{Az} - t c - \mathbf{w}_1 \cdot 2\gamma\|_\infty \leq 2\gamma + 1,$$
$$\|\mathbf{Az}' - t c' - \mathbf{w}_1 \cdot 2\gamma\|_\infty \leq 2\gamma + 1.$$

By the triangle inequality, we have that

$$\|\mathbf{A}(\mathbf{z} - \mathbf{z}') - t \cdot (c - c')\|_\infty \leq 4\gamma + 2,$$

which can be rewritten as

$$\mathbf{A}(\mathbf{z} - \mathbf{z}') + \mathbf{u} = t \cdot (c - c') \tag{13}$$

for some $\mathbf{u}$ such that $\|\mathbf{u}\|_\infty \leq 4\gamma + 2$ (and $\|\mathbf{z} - \mathbf{z}'\|_\infty \leq 2(\gamma' - \beta - 1)$).

If $\mathbf{A} \leftarrow R_q^{k \times \ell}$ and $\mathbf{t} \leftarrow R_q^k$, then, by Theorem 3.8, we have that Eq. (13) is satisfied with probability less than

$$\frac{|S_{2(\gamma'-\beta-1)}|^\ell \cdot |S_{4\gamma+2}|^k \cdot |\mathcal{D}|}{q^{nk}} + \sum_{i=1}^{e_\ell} \frac{\binom{e_\ell}{i} \cdot |S_{2(\gamma'-\beta-1)+\|W_i\|_\infty}|^\ell \cdot |S_{4\gamma+2+\|W_i\|_\infty}|^k \cdot |\mathcal{D}|}{|W_i|^{\ell+k} \cdot q^{nk(1-i/d)}},$$

where $\mathcal{D} := \{c - c' : c, c' \in \mathsf{ChSet}\} \setminus \{0\}$ and sets W_i's are defined in Sect. 3.3.

Thus, except with the above probability, for every $\mathbf{w}_1$, there is at most one possible c that allows C to win. In other words, except with the above probability, C has at most a $1/|\mathsf{ChSet}|$ chance of winning. $\square$

Note that we do not make any assumptions on the prime q. However, small d (e.g. $d = 2$ for $q \equiv 3$ or $5 \pmod 8$) implies small e_ℓ. As a consequence, the smaller d we choose, then the probability above also decreases.

MIN-ENTROPY. Now, we prove that the $\mathbf{w}_1$ sent by the honest prover in the first step is extremely likely to be distinct for every run of the protocol.

Lemma 4.3. *Let e_m be the largest integer which satisfies $q^{e_m/d} \leq 2\gamma' \sqrt{n}$. Then the identification scheme ID in Fig. 3 has*

$$\alpha > \log \left(\min \left\{ \frac{1}{M}, (2\gamma' - 1)^{n\ell} \right\} \right)$$

bits of min-entropy, where

$$M := \frac{|S_{2\gamma'}|^\ell \cdot |S_{2\gamma}|^k}{q^{nk}} + \sum_{i=1}^{e_m} \frac{\binom{d}{i} \cdot |S_{2\gamma'+\|W_i\|_\infty}|^\ell \cdot |S_{2\gamma+\|W_i\|_\infty}|^k}{|W_i|^{\ell+k} \cdot q^{nk(1-i/d)}}.$$

Proof. We claim that

$$
\Pr_{\mathbf{A} \leftarrow R_q^{k \times \ell}}[\exists \mathbf{y} \neq \mathbf{y}' \in S_{\gamma'-1}^\ell \text{ s.t. } \mathsf{HighBits}_q(\mathbf{Ay}, 2\gamma) = \mathsf{HighBits}_q(\mathbf{Ay}', 2\gamma)]
$$

$$
\leq \frac{|S_{2\gamma'}|^\ell \cdot |S_{2\gamma}|^k}{q^{nk}} + \sum_{i=1}^{e_m} \frac{\binom{d}{i} \cdot |S_{2\gamma'+\|W_i\|_\infty}|^\ell \cdot |S_{2\gamma+\|W_i\|_\infty}|^k}{|W_i|^{\ell+k} \cdot q^{nk(1-i/d)}}. \tag{14}
$$

Indeed, if we write

$$
\mathsf{Decompose}_q(\mathbf{Ay}, 2\gamma) = (\mathbf{w}_1, \mathbf{w}_0) \text{ and } \mathsf{Decompose}_q(\mathbf{Ay}', 2\gamma) = (\mathbf{w}_1', \mathbf{w}_0'),
$$

then $\mathsf{HighBits}_q(\mathbf{Ay}, 2\gamma) = \mathsf{HighBits}_q(\mathbf{Ay}', 2\gamma)$ implies that $\mathbf{Ay} = \mathbf{w}_1 \cdot 2\gamma + \mathbf{w}_0$ and $\mathbf{Ay}' = \mathbf{w}_1' \cdot 2\gamma + \mathbf{w}_0'$ with $\mathbf{w}_1 = \mathbf{w}_1'$ and $\|\mathbf{w}_0\|_\infty, \|\mathbf{w}_0'\|_\infty \leq \gamma$. Hence,

$$
\mathbf{A}(\mathbf{y} - \mathbf{y}') - (\mathbf{w}_0 - \mathbf{w}_0') = \mathbf{0} \tag{15}
$$

where

$$
\|\mathbf{y} - \mathbf{y}'\|_\infty < 2\gamma', \|\mathbf{w}_0 - \mathbf{w}_0'\|_\infty \leq 2\gamma.
$$

Corollary 3.9 shows that the probability over the choice of $\mathbf{A} \leftarrow R_q^{k \times \ell}$, that there exist two non-zero elements of norm less than 2γ and $2\gamma'$, respectively, which satisfy Eq. (15) is at most

$$
\frac{|S_{2\gamma'}|^\ell \cdot |S_{2\gamma}|^k}{q^{nk}} + \sum_{i=1}^{e_m} \frac{\binom{d}{i} \cdot |S_{2\gamma'+\|W_i\|_\infty}|^\ell \cdot |S_{2\gamma+\|W_i\|_\infty}|^k}{|W_i|^{\ell+k} \cdot q^{nk(1-i/d)}} = M.
$$

This proves Eq. (14).

Now, we know that with probability at least $1 - M$ over the choice of $\mathbf{A} \leftarrow R_q^{k \times \ell}$, each $W = \mathsf{HighBits}_q(\mathbf{Ay}, 2\gamma)$ has exactly a $\frac{1}{|S_{\gamma'-1}^\ell|} = (2\gamma' - 1)^{-n\ell}$ probability of being output. Thus, the claim in the lemma follows directly from the definition. □

COMPUTATIONAL UNIQUE RESPONSE. Here, we show the Computational Unique Response (CUR) property required for strong-unforgeability of the signature scheme.

Lemma 4.4. *Let e_c be the largest integer such that $q^{e_c/d} \leq 2(\gamma' - \beta)\sqrt{n}$. Then*

$$
\mathsf{Adv}_{\mathsf{ID}}^{\mathsf{CUR}}(\mathsf{A}) \leq \frac{|S_{2(\gamma'-\beta)}|^\ell \cdot |S_{4\gamma+2}|^k}{q^{nk}} + \sum_{i=1}^{e_c} \frac{\binom{d}{i} \cdot |S_{2(\gamma'-\beta)+\|W_i\|_\infty}|^\ell \cdot |S_{4\gamma+2+\|W_i\|_\infty}|^k}{|W_i|^{\ell+k} \cdot q^{nk(1-i/d)}}
$$

for all (even unbounded) adversaries A.

Proof. Let $(W, c, Z) = (\mathbf{w}_1, c, \mathbf{z})$ be any valid transcript and suppose A is able to generate a valid $Z' = \mathbf{z}' \neq Z$ such that $\mathsf{V}(pk = (\mathbf{A}, \mathbf{t}), \mathbf{w}_1, c, \mathbf{z}') = 1$. Thus, we have

$$
\mathbf{w}_1 = \mathsf{UseHint}_q(\mathbf{h}, \mathbf{Az} - c\mathbf{t}, 2\gamma) \text{ and } \mathbf{w}_1 = \mathsf{UseHint}_q(\mathbf{h}', \mathbf{Az}' - c\mathbf{t}, 2\gamma).
$$

Table 1. Prime moduli q for each possible value of d. We used the main result of [19] for finding q. For each case, we also provide values γ such that $2\gamma | q - 1$. Just like in [11], we set $\gamma' = \gamma$.

q	d	γ
$2^{44} - 17043$	2	592493
$2^{44} - 8583$	4	593431
$2^{44} - 13743$	8	305156
$2^{44} - 7583$	16	282832
$2^{44} - 1599$	32	285978
$2^{45} - 36991$	64	364254
$2^{45} - 58111$	128	353952
$2^{45} - 511$	256	360620
$2^{45} - 23551$	512	359769

The above two equations imply (by Lemma 2.1) that

$$\|\mathbf{A}\mathbf{z} - c\mathbf{t} - \mathbf{w}_1 \cdot 2\gamma\|_\infty \leq 2\gamma + 1 \text{ and } \|\mathbf{A}\mathbf{z}' - c\mathbf{t} - \mathbf{w}_1 \cdot 2\gamma\|_\infty \leq 2\gamma + 1.$$

By the triangle inequality, we have

$$\mathbf{A}(\mathbf{z} - \mathbf{z}') + \mathbf{u} = \mathbf{0}$$

for some $\mathbf{u}$ such that $\|\mathbf{u}\| \leq 4\gamma + 2$ and $\|\mathbf{z} - \mathbf{z}'\| < 2(\gamma' - \beta)$. Hence, by Corollary 3.9, the probability over the choice of $\mathbf{A} \leftarrow R_q^{k \times \ell}$, that there exist such $\mathbf{v}, \mathbf{u}$ is at most

$$\frac{|S_{2(\gamma'-\beta)}|^\ell \cdot |S_{4\gamma+2}|^k}{q^{nk}} + \sum_{i=1}^{e_c} \frac{\binom{d}{i} \cdot |S_{2(\gamma'-\beta)+\|W_i\|_\infty}|^\ell \cdot |S_{4\gamma+2+\|W_i\|_\infty}|^k}{|W_i|^{\ell+k} \cdot q^{nk(1-i/d)}}. \qquad \square$$

4.3   Concrete Parameteres

In this subsection, we instantiate the modified Bai-Galbraith mBG signature scheme obtained by the Fiat-Shamir transformation from ID with concrete parameters (Tables 1 and 2). We consider nine different instantiations of mBG for all possible $d \in \{2^i : i \in [9]\}$.

For each value of d, we have selected parameters (e.g. prime modulus q and γ) such that the ID scheme satisfies the following security properties: (i) $\varepsilon_{zk} = 0$, (ii) the scheme has more than 2845 bits of min-entropy, i.e. $\alpha > 2845$, (iii) $\varepsilon_{ls} \leq 2^{-264}$, (iv) $\mathrm{Adv}_{ID}^{CUR}(C) \leq 2^{-288}$. Following the steps in [11], one can prove security of the modified Bai-Galbraith scheme in the quantum random oracle model (see [20]).

We compare the nine different instantiations of the modified Bai-Galbraith scheme (Table 2) with respect to recommended parameters in Table 2.

Table 2. Parameters for the modified Bai-Galbraith scheme. Recall that ν is the maximum coefficient of secret keys s_1, s_2 and $\beta = \nu \cdot (\# \text{ of } \pm \text{ 1's in } c \in \mathsf{ChSet})$. On the other hand, variables $e_\ell, e_c, e_m, \alpha, \varepsilon_{\mathsf{ls}}, \mathrm{Adv}_{\mathsf{ID}}^{\mathsf{CUR}}(A), \nu$ are defined in Sect. 4.2.

d	2	4	8	16	32	64	128	256	512
n	512	512	512	512	512	512	512	512	512
(k, ℓ) (dimensions of $\mathbf{A}$)	$(4,4)$	$(4,4)$	$(5,5)$	$(5,5)$	$(5,5)$	$(5,5)$	$(5,5)$	$(5,5)$	$(5,5)$
# of $\pm 1's$ in $c \in \mathsf{ChSet}$	46	46	46	46	46	46	46	46	46
η (max. coeff. of s_1, s_2)	5	5	2	2	2	2	2	2	2
$\beta(= \eta \cdot (\#\text{of 1's in } c))$	230	230	92	92	92	92	92	92	92
e_ℓ (lossyness)	0	0	0	1	2	5	10	21	42
e_c (CUR)	1	2	4	8	17	34	68	136	272
e_m (min-entropy)	1	2	4	8	17	34	68	136	272
$\log(\varepsilon_{\mathsf{ls}})$	-264	-264	-264	-264	-264	-264	-264	-264	-264
$\log(\mathrm{Adv}_{\mathsf{ID}}^{\mathsf{CUR}}(A))$	-1326	-1317	-592	-924	-288	-799	-986	-766	-677
α	3373	3363	3149	3481	2845	3356	3543	3324	3235
pk size (kilobytes)	11.29	11.29	14.11	14.11	14.11	14.43	14.43	14.43	14.43
sig size (kilobytes)	5.69	5.69	6.76	6.76	6.76	6.76	6.76	6.76	6.76
Exp. Repeats $\frac{1}{1-\nu}$	4.94	4.93	4.68	5.29	5.19	3.64	3.78	3.69	3.70
BKZ block-size to break LWE	480	480	600	600	600	585	585	585	585
Best known classical bit-cost	140	140	175	175	175	171	171	171	171
Best known quantum bit-cost	127	127	159	159	159	155	155	155	155

Firstly, observe that for $d \leq 4$, we pick $q \approx 2^{44}$. In this case, we end up with public key and signature size 11.29 kB and 5.69 kB respectively.

The situation changes for $d = 8$. Interestingly, if one keeps the same parameters as for $d = 4$ then one still gets $\varepsilon_{\mathsf{ls}} \leq 2^{-264}$, hence the lossyness property is still preserved. The problem is, however, that the advantage $\mathrm{Adv}_{\mathsf{ID}}^{\mathsf{CUR}}(A)$ gets extremely big. Concretely, for parameters above we have $\log(\mathrm{Adv}_{\mathsf{ID}}^{\mathsf{CUR}}(A)) \approx 3483$. We found out that one of the compounds in the sum is actually dominating (see Lemma 4.4). Namely, we get:

$$\log\left(\frac{\binom{8}{1} \cdot |S_{2(\gamma'-\beta)+\|W_1\|_\infty}|^\ell \cdot |S_{4\gamma+2+\|W_1\|_\infty}|^k}{|W_1|^{\ell+k} \cdot q^{nk(1-1/8)}}\right) \approx 3483.$$

We believe the reason for it being so large is because for $d = 8$, $i = 1$ and $q \approx 2^{44}$ we have $t := \left\lfloor \frac{q^{i/d}}{2\sqrt{n}} \right\rfloor = 1$ (introduced in Sect. 3.3). Hence, W_1 has only 3^{512} elements. As a consequence, the value above is still big. Thus, a natural way to solve this issue would be to increase q. Unfortunately, in order to keep the MLWE problem hard, this would imply increasing the size of secret keys, i.e. η. Hence, β would also get bigger, so in order to keep the repetition rate $1/(1 - \nu)$ small, we would have to increase the value of γ (and γ'). In this case, probabilities related to the security of ID, e.g. $\varepsilon_{\mathsf{ls}}, \log(\mathrm{Adv}_{\mathsf{ID}}^{\mathsf{CUR}}(A))$, would get considerably bigger, so one would need to consider larger q again and eventually, we would end up in a vicious circle. We avoid that by increasing dimensions

$(k, \ell) = (5, 5)$ of the matrix $\mathbf{A}$. Unfortunately, this comes at a price of larger public key (14.11 kB) and signature (6.76 kB) sizes. In order to minimise such costs, we decrease the size of secret keys $\eta = 2$ and thus, we select smaller values for γ. As before, we choose $q \approx 2^{44}$. We pick almost identical parameters for $d = 16$ and $d = 32$.

Next, we consider $d \geq 64$. If we choose the parameters as for $d = 32$ then the lossyness probability ε_{ls} is no longer small and therefore, we need to increase the $q \approx 2^{45}$. We observe that the new parameters still provide much more than 128 bits of security for MLWE. The public key gets slightly larger (14.43 kB) and the signature size stays the same as before.

In order to maintain security of the Bai-Galbraith scheme in the quantum random oracle model for bigger d (i.e. $d \geq 256$), we need to increase both dimensions (k, ℓ) of the matrix $\mathbf{A}$ as well as the prime modulus q. This results in having 3.13 kB larger public key and 1.07 kB signature sizes than for $d = 2$. We remark that security parameters were chosen such that the expected number of repetitions of the protocol $1/(1 - \nu)$ is at most six. Indeed, admitting small repetition rate as well as supporting the use of the Number Theoretic Transform, efficient caching and polynomial sampling assures us that the protocol can be performed very efficiently.

Acknowledgments. The author would like to thank Vadim Lyubashevsky for fruitful discussions and anonymous reviewers for their useful comments. This work was supported by the SNSF ERC Transfer Grant CRETP2-166734 FELICITY.

References

1. Abdalla, M., Fouque, P.-A., Lyubashevsky, V., Tibouchi, M.: Tightly-secure signatures from lossy identification schemes. In: Pointcheval, D., Johansson, T. (eds.) EUROCRYPT 2012. LNCS, vol. 7237, pp. 572–590. Springer, Heidelberg (2012). https://doi.org/10.1007/978-3-642-29011-4_34
2. Alkim, E., et al.: Revisiting TESLA in the quantum random oracle model. In: Lange, T., Takagi, T. (eds.) PQCrypto 2017. LNCS, vol. 10346, pp. 143–162. Springer, Cham (2017). https://doi.org/10.1007/978-3-319-59879-6_9
3. Alkim, E., Ducas, L., Pöppelmann, T., Schwabe, P.: Post-quantum key exchange - a new hope. In: Holz, T., Savage, S. (eds.) USENIX Security 2016, pp. 327–343. USENIX Association, August 2016
4. Bai, S., Galbraith, S.D.: An improved compression technique for signatures based on learning with errors. In: Benaloh, J. (ed.) CT-RSA 2014. LNCS, vol. 8366, pp. 28–47. Springer, Cham (2014). https://doi.org/10.1007/978-3-319-04852-9_2
5. Baum, C., Damgård, I., Lyubashevsky, V., Oechsner, S., Peikert, C.: More efficient commitments from structured lattice assumptions. In: Catalano, D., De Prisco, R. (eds.) SCN 2018. LNCS, vol. 11035, pp. 368–385. Springer, Cham (2018). https://doi.org/10.1007/978-3-319-98113-0_20
6. Benhamouda, F., Camenisch, J., Krenn, S., Lyubashevsky, V., Neven, G.: Better zero-knowledge proofs for lattice encryption and their application to group signatures. In: Sarkar, P., Iwata, T. (eds.) ASIACRYPT 2014, Part I. LNCS, vol. 8873, pp. 551–572. Springer, Heidelberg (2014). https://doi.org/10.1007/978-3-662-45611-8_29

7. Benhamouda, F., Krenn, S., Lyubashevsky, V., Pietrzak, K.: Efficient zero-knowledge proofs for commitments from learning with errors over rings. In: Pernul, G., Ryan, P.Y.A., Weippl, E. (eds.) ESORICS 2015, Part I. LNCS, vol. 9326, pp. 305–325. Springer, Cham (2015). https://doi.org/10.1007/978-3-319-24174-6_16

8. Bindel, N., et al.: qTESLA. Technical report, National Institute of Standards and Technology (2017). https://csrc.nist.gov/projects/post-quantum-cryptography/round-1-submissions

9. Ducas, L., Lepoint, T., Lyubashevsky, V., Schwabe, P., Seiler, G., Stehlé, D.: CRYSTALS - Dilithium: digital signatures from module lattices. IACR Cryptology ePrint Archive 2017, 633 (2017). To appear in TCHES 2018

10. Güneysu, T., Lyubashevsky, V., Pöppelmann, T.: Practical lattice-based cryptography: a signature scheme for embedded systems. In: Prouff, E., Schaumont, P. (eds.) CHES 2012. LNCS, vol. 7428, pp. 530–547. Springer, Heidelberg (2012). https://doi.org/10.1007/978-3-642-33027-8_31

11. Kiltz, E., Lyubashevsky, V., Schaffner, C.: A concrete treatment of fiat-shamir signatures in the quantum random-oracle model. In: Nielsen, J.B., Rijmen, V. (eds.) EUROCRYPT 2018, Part III. LNCS, vol. 10822, pp. 552–586. Springer, Cham (2018). https://doi.org/10.1007/978-3-319-78372-7_18

12. Langlois, A., Stehlé, D.: Worst-case to average-case reductions for module lattices. Des. Codes Cryptogr. **75**(3), 565–599 (2015)

13. Lyubashevsky, V.: Fiat-Shamir with aborts: applications to lattice and factoring-based signatures. In: Matsui, M. (ed.) ASIACRYPT 2009. LNCS, vol. 5912, pp. 598–616. Springer, Heidelberg (2009). https://doi.org/10.1007/978-3-642-10366-7_35

14. Lyubashevsky, V.: Lattice signatures without trapdoors. In: Pointcheval, D., Johansson, T. (eds.) EUROCRYPT 2012. LNCS, vol. 7237, pp. 738–755. Springer, Heidelberg (2012). https://doi.org/10.1007/978-3-642-29011-4_43

15. Lyubashevsky, V., et al.: Crystals-dilithium. Technical report, National Institute of Standards and Technology (2017). https://csrc.nist.gov/projects/post-quantum-cryptography/round-1-submissions

16. Lyubashevsky, V., Micciancio, D.: Generalized compact knapsacks are collision resistant. In: Bugliesi, M., Preneel, B., Sassone, V., Wegener, I. (eds.) ICALP 2006, Part II. LNCS, vol. 4052, pp. 144–155. Springer, Heidelberg (2006). https://doi.org/10.1007/11787006_13

17. Lyubashevsky, V., Neven, G.: One-shot verifiable encryption from lattices. In: Coron, J.-S., Nielsen, J.B. (eds.) EUROCRYPT 2017, Part I. LNCS, vol. 10210, pp. 293–323. Springer, Cham (2017). https://doi.org/10.1007/978-3-319-56620-7_11

18. Lyubashevsky, V., Peikert, C., Regev, O.: On ideal lattices and learning with errors over rings. In: Gilbert, H. (ed.) EUROCRYPT 2010. LNCS, vol. 6110, pp. 1–23. Springer, Heidelberg (2010). https://doi.org/10.1007/978-3-642-13190-5_1

19. Lyubashevsky, V., Seiler, G.: Short, invertible elements in partially splitting cyclotomic rings and applications to lattice-based zero-knowledge proofs. In: Nielsen, J.B., Rijmen, V. (eds.) EUROCRYPT 2018, Part I. LNCS, vol. 10820, pp. 204–224. Springer, Cham (2018). https://doi.org/10.1007/978-3-319-78381-9_8

20. Nguyen, N.K.: On the non-existence of short vectors in random module lattices. Cryptology ePrint Archive, Report 2019/973 (2019). https://eprint.iacr.org/2019/973

21. Peikert, C., Rosen, A.: Efficient collision-resistant hashing from worst-case assumptions on cyclic lattices. In: Halevi, S., Rabin, T. (eds.) TCC 2006. LNCS, vol. 3876, pp. 145–166. Springer, Heidelberg (2006). https://doi.org/10.1007/11681878_8

22. Peikert, C., Rosen, A.: Lattices that admit logarithmic worst-case to average-case connection factors. In: Johnson, D.S., Feige, U. (eds.) 39th ACM STOC, pp. 478–487. ACM Press, June 2007

23. Regev, O.: On lattices, learning with errors, random linear codes, and cryptography. In: Gabow, H.N., Fagin, R. (eds.) 37th ACM STOC, pp. 84–93. ACM Press, May 2005

24. Schwabe, P., et al.: Crystals-kyber. Technical report, National Institute of Standards and Technology (2017). https://csrc.nist.gov/projects/post-quantum-cryptography/round-1-submissions

25. Stehlé, D., Steinfeld, R.: Making NTRUEncrypt and NTRUSign as secure as standard worst-case problems over ideal lattices. Cryptology ePrint Archive, Report 2013/004 (2013). http://eprint.iacr.org/2013/004

26. Stehlé, D., Steinfeld, R., Tanaka, K., Xagawa, K.: Efficient public key encryption based on ideal lattices. In: Matsui, M. (ed.) ASIACRYPT 2009. LNCS, vol. 5912, pp. 617–635. Springer, Heidelberg (2009). https://doi.org/10.1007/978-3-642-10366-7_36

27. Unruh, D.: Post-quantum security of Fiat-Shamir. In: Takagi, T., Peyrin, T. (eds.) ASIACRYPT 2017, Part I. LNCS, vol. 10624, pp. 65–95. Springer, Cham (2017). https://doi.org/10.1007/978-3-319-70694-8_3

Authenticated Encryption

Forkcipher: A New Primitive for Authenticated Encryption of Very Short Messages

Elena Andreeva[1]([⊠]), Virginie Lallemand[2], Antoon Purnal[1],
Reza Reyhanitabar[3P], Arnab Roy[4], and Damian Vizár[5]

[1] imec-COSIC, KU Leuven, Leuven, Belgium
{elena.andreeva,antoon.purnal}@esat.kuleuven.be
[2] Université de Lorraine, CNRS, Inria, LORIA, Nancy, France
virginie.lallemand@loria.fr
[3] TE Connectivity, Niederwinkling, Germany
reza.reyhanitabar@te.com
[4] University of Bristol, Bristol, UK
arnab.roy@bristol.ac.uk
[5] CSEM, Neuchâtel, Switzerland
damian.vizar@csem.ch

Abstract. Highly efficient encryption and authentication of *short* messages is an essential requirement for enabling security in constrained scenarios such as the CAN FD in automotive systems (max. message size 64 bytes), massive IoT, critical communication domains of 5G, and Narrowband IoT, to mention a few. In addition, one of the NIST lightweight cryptography project requirements is that AEAD schemes shall be "optimized to be efficient for short messages (e.g., as short as 8 bytes)".

In this work we introduce and formalize a novel primitive in symmetric cryptography called *forkcipher*. A forkcipher is a keyed primitive expanding a fixed-lenght input to a fixed-length output. We define its security as indistinguishability under a chosen ciphertext attack (for n-bit inputs to $2n$-bit outputs). We give a generic construction validation via the new *iterate-fork-iterate* design paradigm.

We then propose ForkSkinny as a concrete forkcipher instance with a public tweak and based on SKINNY: a tweakable lightweight cipher following the TWEAKEY framework. We conduct extensive cryptanalysis of ForkSkinny against classical and structure-specific attacks.

We demonstrate the applicability of forkciphers by designing three new provably-secure nonce-based AEAD modes which offer performance and security tradeoffs and are optimized for efficiency of very short messages. Considering a reference block size of 16 bytes, and ignoring possible hardware optimizations, our new AEAD schemes beat the best SKINNY-based AEAD modes. More generally, we show forkciphers are suited for lightweight applications dealing with predominantly short messages, while at the same time allowing handling arbitrary messages sizes.

Furthermore, our hardware implementation results show that when we exploit the inherent parallelism of ForkSkinny we achieve the best performance when directly compared with the most efficient mode instantiated with SKINNY.

© International Association for Cryptologic Research 2019
S. D. Galbraith and S. Moriai (Eds.): ASIACRYPT 2019, LNCS 11922, pp. 153–182, 2019.
https://doi.org/10.1007/978-3-030-34621-8_6

Keywords: Authenticated encryption · New primitive · Forkcipher · ForkSkinny · Lightweight cryptography · Short messages

1 Introduction

Authenticated encryption (AE) aims at achieving the two fundamental security goals of symmetric-key cryptography: confidentiality (privacy) and integrity (together with authentication). Historically, these two goals were achieved by the generic composition of an encryption scheme (for confidentiality) and a message authentication code (MAC) [23]. For instance, *old* versions of major security protocols such as TLS, SSH and IPsec included variants of generic composition, namely MAC-then-Encrypt, Encrypt-and-MAC and Encrypt-then-MAC schemes, respectively. But it turned out that this approach is neither the most efficient (as it needs processing the whole message twice) nor the most robust to security and implementation issues [22,43,44]; rather it is easy for practitioners to get it wrong even when using the best known method among the three, i.e. Encrypt-then-MAC, following standards [41].

The notion of AE as a primitive in its own right—integrating encryption and authentication by exposing a single abstract interface—was put forth by Bellare and Rogaway [25] and independently by Katz and Yung [34] in 2000. It was further enhanced by Rogaway [46] to authenticated encryption with associated data (AEAD). Being able to process associated data (AD) is now a default requirement for any authenticated encryption scheme; hence we use AE and AEAD interchangeably. After nearly two decades of research and standardization activities, recently fostered by the CAESAR competition (2014–2018)[26], we now have a rich set of *general-purpose* AEAD schemes, some already standardized (e.g. GCM and CCM) and some expected to be adopted by new applications and standards (e.g. the CAESAR finalists Ascon [30], ACORN [53], AEGIS-128 [55], OCB [36], COLM [9], Deoxys II [32], and MORUS [54]).

This progress may lead to the belief that the AEAD problem is "solved". However, as evidenced by the ECRYPT-CSA report in 2017 [14], several critical ongoing "Challenges in Authenticated Encryption" still need research efforts stretching years into the future. Thus, it is interesting to investigate to what extent CAESAR has resulted in solutions to these problems.

Our Target Challenge. Among the four categories of challenges—security, interface, performance, mistakes and malice—reported by the ECRYPT-CSA [14], we aim at delving into the performance regarding authenticated encryption of *very short messages*. General-purpose AEAD schemes are usually optimized for handling (moderately) long messages, and often incur some initialization and/or finalization cost that is amortized when the message is long. To quote the ECRYPT-CSA report: "The performance target is wrong ··· Another increasingly common scenario is that an authenticated cipher is applied to many small messages ··· The challenge here is to minimize overhead."

Therefore, designing efficient AEAD for short messages is an important objective as also evidenced by NIST's first call for submissions (May 14, 2018) for lightweight cryptography [42], where it is stressed as a *design requirement* that lightweight AEAD submissions shall be "optimized to be efficient for short messages (e.g., as short as 8 bytes)".

Plenty of Use Cases. The need for high-performance and low-latency processing of short messages is identified as an essential requirement in a multitude of security and safety critical use cases in various domains. Examples are Secure On board Communication (SecOC) in automotive systems [6], handling of short data bursts in critical communication and massive IoT domains of 5G [1], and Narrowband IoT (NB-IoT) [2,5] systems. For example, the new CAN FD standard (ISO 11898-1) for vehicle bus technology [3,4], which is expected to be implemented in most cars by 2020, allows for a payload up to 64 bytes. In NB-IoT standards [2,5] the maximum transport block size (TBS) is 680 bits in downlink and 1000 bits in uplink (the minimum TBS size is 16 bits in both cases). Low energy protocols also come with stringent requirements on the maximum packet size: the Bluetooth, SigFox, LoraWan and ZigBee protocols allow for maximum sizes of 47, 12, 51–255 (51 bytes for slowest data rate, 255 for the fastest), and 84 bytes packet sizes, respectively. In use cases with tight requirements on delay and latency, the typical packet sizes should be small as large packets occupy a link for more time, causing more delays to subsequent packets and increasing latency. Furthermore, in applications such as smart parking lots the data to be sent is just one bit ("free" or "occupied"), so a minimum allowed TBS size of 2 bytes (16 bits) would suit the application. Even more, most medical implant devices, such as pacemakers, permit the exchange of messages of length at most 16 bytes between the device programmer and the device.

Our Goal. Our main objective is to construct secure, modular (provably secure) AEAD schemes that excel in efficiency over previous modular AEAD constructions at processing very short inputs, while also being able to process longer inputs, albeit somewhat less efficiently. We insist that our AEAD schemes ought to be able to securely process inputs of arbitrary lengths to be fairly comparable to other general-purpose (long message centric) schemes, and to be qualified as a full-fledged variable-input-length AEAD scheme according to the requirements in NIST's call for lightweight cryptography primitives.

Towards this goal, we take an approach that can be seen as a parallel to the shift from generic composition to dedicated AEAD designs, but on the level of the primitive. We rethink the way a low level fixed-input-length (FIL) primitive is designed, and how variable-input-length (VIL) AEAD schemes are constructed from such a new primitive.

The Gap between the Primitives and AEAD. Our first observation is that there is a large gap between the high level security goal to be achieved by the VIL AEAD schemes and the security properties that the underlying FIL primitives can provide. Modular AEAD designs typically confine the AE security to the mode of operation only; the lower-level primitives, such as (tweakable) block ciphers, cryptographic permutations and compression functions, are never meant to possess any AE-like features, and in particular they are never expanding as needed to ensure ciphertext integrity in AEAD. Hence, a VIL AEAD scheme Π designed as a mode of operation for an FIL primitive F plays two roles: not only does it extend the domain of the FIL primitive but it also transforms and boosts the security property of the primitive to match the AEAD security notion. A natural question then arises, whether by explicitly decoupling these two AEAD roles we can have more efficient designs and more transparent security proofs.

The first, most obvious approach to resolving the latter question is to remove the security gap between the mode and its primitive altogether, i.e., to start from a FIL primitive F which itself is a secure FIL AEAD. This way a VIL AEAD mode will only have one role: a property-preserving domain extender for the primitive F. Property-preserving domain extension is a well-studied and popular design paradigm for other primitives such as hash functions [11,24,45].

Informally speaking, the best possible security that a FIL AEAD scheme with a *fixed* ciphertext expansion (stretch) can achieve is to be indistinguishable from a *tweakable random injective* function, i.e., to be a tweakable pseudorandom injection (PRI) [31,48]. But starting directly with a FIL tweakable PRI, we did not achieve a desirable solution in our quest for the most *efficient AEAD design for short messages.*[1] It seems that, interestingly, narrowing the security gap between the mode and its primitive, but not removing the gap entirely, is what helps us achieve our ultimate goal.

Contribution 1: Forkcipher – A New Symmetric Primitive. We introduce a novel primitive—a tweakable **forkcipher**—that yields efficient AEAD designs for short messages. A tweakable forkcipher is *nearly*, but not exactly, a FIL AE primitive; "nearly" because it produces *expanded* ciphertexts with a non-trivial redundancy, and not exactly because it has no integrity-checking mechanisms.[2] When keyed and tweaked, we show how a forkcipher maps an n-bit input block M to an output C of $2n$ bits. Intuitively, this is equivalent to evaluating *two independent* tweakable permutations on M but with an *amortized computational cost* (see Fig. 1 for an illustration of the forkcipher's high-level structure). We give a strict formalization of the security of such a forkcipher. Our new notion of *pseudorandom tweakable forked permutation* captures the game of indistinguishability of a n-bit to $2n$-bits forkcipher from a pair of random permutations in the context of chosen ciphertext attacks.

Contribution 2: Instantiating a Forkcipher. We give an efficient instance of the tweakable forkcipher and name it ForkSkinny. It is based on the lightweight tweakable block cipher SKINNY[18]. Building ForkSkinny on an existing block cipher enables us to rely on the cryptanalysis results behind SKINNY[12,13,49, 51,56,57], and in addition, helps us provide systematic analysis for the necessary forkcipher alterations. We also inherit the cipher's efficiency features and obtain a natural and consistent metric for comparison of the forkcipher performance with that of its underlying block cipher.

SKINNY comes with multiple optimization tradeoffs in area, throughput, power, efficiency and software performance in lightweight applications. Additionally, SKINNY also provides a number of choices for its block size and tweak

---

[1] See the discussion section in full version [10].

[2] We demonstrate that when used in a minimalistic mode of operation, a secure tweakable forkcipher yields a miniature FIL AEAD scheme which achieves tweakable PRI security.

size which we incorporate naturally into ForkSkinny. We have performed crypt-analyses of ForkSkinny against differential, linear, algebraic, impossible differential, MITM, integral attacks and boomerang attacks. We have taken the security analysis of ForkAES [17] into account to ensure that the same type of attacks is not possible against ForkSkinny.

To obtain ForkSkinny, we apply our newly proposed *iterate-fork-iterate* (IFI) paradigm: when encrypting a block M of n bits with a secret key and a tweak (public), we first transform M into M' using r_{init} SKINNY rounds together with the tweakey schedule. Then, we fork the encryption process by applying two parallel paths (left and right) each comprising r SKINNY rounds. Along left path the state of the cipher is processed using tweakey schedule of SKINNY, thus producing the same ciphertext as SKINNY. Along the right path the state is processed with a tweakey schedule which differs from that of the left path at each round. The IFI design strategy also provides a scope of parallelizing the implementation of the design. The IFI paradigm is conceptually easy, and supports the transference of security and performance results based on the underlying tweakable cipher. We also provide arguments for the generic security of the IFI construction paradigm assuming that the building blocks are behaving as secure pseudorandom permutations. Our generic result is indicative of the forkcipher structural soundness (but does not directly imply security, because a real forkcipher is never built from a secure pseudorandom permutation). While a forkcipher inherits some of the side-channel security features of its underlying structure, the fully-fledged side-channel security of forkciphers is out of the scope of this paper.

Contribution 3: New AEAD Modes. In our work we follow the well-established modular AE design approach for arbitrary long data in the provable security framework. There is no general consensus in the cryptographic community if AEAD schemes can claim higher merits for being modular and provably secure or not. For instance, 3 out of 7 CAESAR [26] finalists, namely ACORN, AEGIS and MORUS are monolithic designs and do not follow the provable security paradigms. Nonetheless, we trust and follow in the modular and provable security methodology for its well-known security benefits [20, 47]. Moreover, the class of provably secure AEAD designs includes all currently standardized AEAD schemes, as well as the majority of CAESAR finalists. We also emphasize that, by defining the forkcipher as a new fully-fledged primitive and building modes on top in a provable way, we clearly differentiate ourselves from the "prove-then-prune" design approaches.

Regarding the state of the art in AE designs, it appears that aiming for a *provably secure* AEAD mode that achieves the best performance for *both* long and short message scenarios is an ambitious goal. Instead, we design high-performance AEAD modes for very short inputs *whilst* maintaining the functionality and security for long ones. All our three modes, PAEF, SAEF and RPAEF can be further implemented very efficiently when instantiated with ForkSkinny.

Our first scheme **PAEF** (Parallel AEAD from a forkcipher) makes ℓ calls to a forkcipher to process a message of ℓ blocks. PAEF is fully parallelizable and thus can leverage parallel computation. We prove its *optimal* security: n bit confidentiality and n-bit authenticity (for an n-bit block input).

Our second scheme **RPAEF** (Reduced Parallel AEAD from a forkcipher) is also fully parallelizable, but in contrast to PAEF only uses the left forkcipher path for the first $(\ell-1)$ blocks, and the full (left and right) forkcipher evaluation for the final block (first block for the single block-message). When instantiated with ForkSkinny, RPAEF computes the equivalent of $(\ell-1)$ calls to SKINNY and 1 call to ForkSkinny. This general mode optimization, as compared to PAEF, comes at the cost of restrictive use of large tweaks (as large as 256 bits) and increased HW area footprint. Similarly to PAEF, we prove that RPAEF achieves optimal quantitative security.

Our third scheme **SAEF** (Sequential AEAD from a forkcipher) encrypts each block "on-the-fly" in a sequential manner (and hence is not parallelizable). SAEF lends itself well to low-overhead implementations (as it does not store the nonce and the block counter) but its security is birthday-bounded in the block size ($n/2$-bit confidentiality and authenticity for n-bit block).

Contribution 4: Hardware Performance and Comparisons. PAEF and SAEF need an equivalent of about 1 and 1.6 SKINNYevaluation per block of AD and message, respectively (both encryption and decryption). RPAEF reduces further the computational cost for all but the last message blocks to an equivalent of 1 SKINNYevaluation. When compared directly with block cipher modes instantiated with SKINNYwith a fixed tweak (to facilitate the comparison), such as the standardized GCM [40], CCM [52], and OCB [37], we outperform those significantly for predominantly short data sizes of up to four blocks. We achieve a performance gain in the range of $(10-50)\%$ for data ranging from 4 blocks down to 1 block, respectively. The additional overhead for all block-cipher-based modes is incurred by at least two additional cipher calls: one for subkey/mask generation and one for tag computation.

We provide a hardware comparison (in Sect. 7, Fig 10) of our three modes (with different ForkSkinny variants) with Sk-AEAD. The Sk-AEAD is the tweakable cipher mode TAE [38], which is same as ΘCB [37], instantiated with SKINNY-AEAD M1/M2, M5/M6 [19]. We compare on the bases of block size, nonce, and tag sizes variants. Based on the round-based implementations all of our three modes perform faster (in terms of cycles) for short data (up to 3 blocks) with about the same area. RPAEF beats its competitor for *all* message sizes at the cost of a area increase of about 20% (for only one of its variants). We further *optimize* the performances by exploiting the in-built parallelism (//) in the ForkSkinny primitive and obtain superior performance results. Namely, for messages up to three 128-bit blocks, the speed-up of PAEF and SAEF (both parallel (//)) ranges from 25% to 50%, where the advantage is largest for the single-block messages. Most importantly, the RPEAF, PAEF, and SAEF (//) instances result in fewer cycles than the ΘCB variants *for all* message sizes at a small cost in area increase. However, the relative advantage of the latter instances is more explicit for short messages; as it diminishes asymptotically with the message blocks. For message sizes up to 8 bytes, which is emphasized by NIST [42], the PAEF-FORKSKINNY-64-192 instances are more than 58% faster with also a considerably smaller implementation size.

Related Work. An AE design which bears similarities with our forkcipher idea is Manticore [8] (the CS scheme). They use the middle state of a block cipher to evaluate a polynomial hash function for authentication purposes. Yet, for a single block, Manticore needs 2 calls to the block cipher (compared to ≈ 1.6 SKINNY calls in ForkSkinny), thus failing to realize optimal efficiency for very short messages. The CS design, which has been shown insecure [50] (and fixed with an extra block cipher call), necessitates a direct cryptanalysis on the level of an AE scheme, which is a much more daunting task than dedicated cryptanalysis of a compact primitive. In [15], Avanzi proposes a somewhat similar design approach which splits an intermediate state to process them separately. More concretely, it uses a nonce addition either prior to the encryption or in the middle of the encryption rounds, specifically at the splitting phase. Yet, the fundamental difference with our design is that we use a different framework (TWEAKEY [33]) which considers the nonce and key together and injects a transformation of those *throughout* the forkcipher rounds. Moreover, it seems impossible to describe the latter designs [8,15] as neither primitives nor modes with clearly defined security goals, whereas our approach aims the opposite.

It is worth mentioning that the recent permutation based construction Farfalle [27] also has superficially similar design structure. For example, in Farfalle with a fixed input length message it is possible to produce two or more fixed length outputs. However, the design strategy of ForkSkinny and Farfalle are different in two aspects: 1. ForkSkinny follows an iterative design strategy (with round keys, round constants etc.), while Farfalle is a permutation based design, and 2. ForkSkinny has an explicit tweak input which is processed using the tweakey framework.

2 Preliminaries

All strings are binary strings. The set of all strings of length n (for a positive integer n) is denoted $\{0,1\}^n$. We let $\{0,1\}^{\leq n} = \bigcup_{i=0}^n \{0,1\}^n$. We denote by $\mathrm{Perm}(n)$ the set of all permutations of $\{0,1\}^n$. We denote by $\mathrm{Func}(m,n)$ the set of all functions with domain $\{0,1\}^m$ and range $\{0,1\}^n$, and we let $\mathrm{Inj}(m)n \subset \mathrm{Func}(m)n$ denote the set of all injective functions with the same signature.

For a string X of ℓ bits, we let $X[i]$ denote the ith bit of X for $i = 0, \ldots, \ell - 1$ (starting from the left) and $X[i \ldots j] = X[i]\|X[i+1]\| \ldots \|X[j]$ for $0 \leq i < j < \ell$. We let $\mathrm{left}_\ell(X) = X[0 \ldots (\ell-1)]$ denote the ℓ leftmost bits of X and $\mathrm{right}_r(X) = X[(|X| - r) \ldots (|X| - 1)]$ the r rightmost bits of X, such that $X = \mathrm{left}_\chi(X)\|\mathrm{right}_{|X|-\chi}(X)$ for any $0 \leq \chi \leq |X|$. Given a (possibly implicit) positive integer n and an $X \in \{0,1\}^*$, we let denote $X\|10^{n-(|X| \bmod n)-1}$ for simplicity. Given an implicit block length n, we let $\mathrm{pad10}(X) = X\|10^*$ return X if $|X| \equiv 0 \pmod{n}$ and $X\|10^*$ otherwise.

Given a string X and an integer n, we let $X_1, \ldots, X_x, X_* \xleftarrow{n} X$ denote partitioning X into n-bit blocks, such that $|X_i| = n$ for $i = 1, \ldots, x, 0 \leq |X_*| \leq n$

and $X = X_1 \| \ldots \| X_x \| X_*$, so $x = \max(0, \lfloor X/n \rfloor - 1)$. We let $|X|_n = \lceil X/n \rceil$. We let $(M', M_*) = \mathsf{msplit}_n(M)$ (as in message split) denote a splitting of a string $M \in \{0,1\}^*$ into two parts $M' \| M_* = M$, such that $|M_*| \equiv |M| \pmod{n}$ and $0 \leq |M_*| \leq n$, where $|M_*| = 0$ if and only if $|M| = 0$. We let $(C', C_*, T) = \mathsf{csplit}_n(C)$ (as in ciphertext split) denote splitting a string C of at least n bits into three parts $C' \| C_* \| T = C$, such that $|C_*| = n$, $|T| \equiv |C| \pmod{n}$, and $0 \leq |T| \leq n$, where $|T| = 0$ if and only if $|C| = n$. Finally, we let $C'_1, \ldots, C'_m, C_*, T \leftarrow \mathsf{csplit\text{-}b}_n(C)$ (as in csplit to blocks) denote the result of $\mathsf{csplit}_n(C)$ followed by partitioning of C' into $|C'|_n$ blocks of n bits, such that $C' = C'_1 \| \ldots \| C'_m$.

The symbol $\perp$ denotes an error signal, or an undefined value. We denote by $X \leftarrow_{\$} \mathcal{X}$ sampling an element X from a finite set $\mathcal{X}$ following the uniform distribution.

3   Forkcipher

We formalize the syntax and security goals of a *forkcipher*. Informally, a forkcipher is a symmetric primitive that takes as input a fixed-length block M of n bits with a secret key K and possibly a public tweak T, and expands it to an output block of fixed length greater than n bits.

In this article we formalize and instantiate the forkcipher as a tweakable keyed function which maps an n-bit input M to a $2n$-bit output block $C_0 \| C_1$. We additionally require that the input M is computable from either of the two output blocks C_0 or C_1. Also, given one half of the output C_0, the other half C_1 should be *reconstructible* from it, and vice versa. These are the basic properties imposed in the syntax of our n-bit to $2n$-bit forkcipher.

When used with a random key, the *ideal* forkcipher implements a *pair* of independent random permutations π_0 and π_1 for every tweak T, namely $C_0 = \pi_0(M)$ and $C_1 = \pi_1(M)$. We define a secure forkcipher to be computationally indistiguishable from such an idealized object - a tweak-indexed collection of *pairs* of random permutations.

A Trivial Forkcipher. It may be clear at this point that the security notion towards which we are headed can be achieved with two instances of a secure tweakable block cipher that are used in parallel. One could thus instantiate a forkcipher by a secure tweakable block cipher used with two independent keys (or a tweak-space separation mechanism).

The main novelty in a forkcipher is that it provides the same security as a pair of tweakable block ciphers at a reduced cost. Yet this reduction of complexity has nothing to do with the security goals and syntax; these only model the kind of object a forkcipher inevitably is, and which security properties it aspires to achieve.

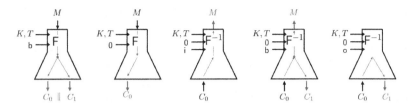

Fig. 1. Forkcipher encryption (two leftmost): the output selector s outputs both output blocks C_0, C_1 if $s = $ b, the "left" ciphertext block C_0 if $s = 0$ (if $s = $ b then C_1). Forkcipher decryption (three rightmost): the first indicator $b = 0$ denotes the left ciphertext block is input ($b = 1$ when right). The second output selector $s = $ i when the ciphertext is inverted to block M (middle); $s = $ b when both blocks M, C' are output; and $s = $ o when the other ciphertext block C' is output.

3.1   Syntax

A forkcipher is a pair of deterministic algorithms, the encryption[3] algorithm:

$$F : \{0,1\}^k \times \mathcal{T} \times \{0,1\}^n \times \{0,1,\mathsf{b}\} \to \{0,1\}^n \cup \{0,1\}^n \times \{0,1\}^n$$

and the inversion algorithm:

$$F^{-1}\{0,1\}^k \times \mathcal{T} \times \{0,1\}^n \times \{0,1\} \times \{\mathsf{i,o,b}\} \to \{0,1\}^n \cup \{0,1\}^n \times \{0,1\}^n.$$

The encryption algorithm takes a key K, a tweak $\mathsf{T} \in \mathcal{T}$, a plaintext block M and an output selector s, and outputs the "left" n-bit ciphertext block C_0 if $s = 0$, the "right" n-bit ciphertext block C_1 if $s = 1$, and a both blocks C_0, C_1 if $s = $ b. We write $F(K, \mathsf{T}, M, s) = F_K(\mathsf{T}, M, s) = F_K^{\mathsf{T}}(M, s) = F_K^{\mathsf{T},s}(M)$ interchangeably. The decryption algorithm takes a key K, a tweak T, a ciphertext block C (left/right half of output block), an indicator b of whether this is the left or the right ciphertext block and an output selector s, and outputs the plaintext (or inverse) block M if $s = $ i, the other ciphertext block C' if $s = $ o, and both blocks M, C' if $s = $ b. We write $F^{-1}(K, \mathsf{T}, M, b, s) = F^{-1}_K(\mathsf{T}, M, b, s) = F^{-1}{}^{\mathsf{T}}_K(M, b, s) = F_K^{\mathsf{T},b,s}(M)$ interchangeably. We call k, n and $\mathcal{T}$ the keysize, blocksize and tweak space of F, respectively.

A tweakable forkcipher F meets the *correctness condition*, if for every $K \in \{0,1\}^k, \mathsf{T} \in \mathcal{T}, M \in \{0,1\}^n$ and $\beta \in \{0,1\}$ all of the following conditions are met:

1. $F^{-1}(K, \mathsf{T}, F(K, \mathsf{T}, M, \beta), \beta, \mathsf{i}) = M$
2. $F^{-1}(K, \mathsf{T}, F(K, \mathsf{T}, M, \beta), \beta, \mathsf{o}) = F(K, \mathsf{T}, M, \beta \oplus 1)$
3. $(F(K, \mathsf{T}, M, 0), F(K, \mathsf{T}, M, 1)) = F(K, \mathsf{T}, M, \mathsf{b})$
4. $\big(F^{-1}(K, \mathsf{T}, C, \beta, \mathsf{i}), F^{-1}(K, \mathsf{T}, C, \beta, \mathsf{o})\big) = F^{-1}(K, \mathsf{T}, C, \beta, \mathsf{b})$

In other words, for each pair of key and tweak, the forkcipher applies two independent permutations to the input to produce the two output blocks. We focus on a specific form of $\mathcal{T}$ only: when $\mathcal{T} = \{0,1\}^t$ for some positive t.

---

[3] We again conflate the label for the primitive with the label of the encryption algorithm.

The formalization we just gave faithfully models how a forkcipher is used to realize its full potential. As explained in the discussion section of the full version [10], the most suitable FIL expanding cipher to construct modes of operation is a forkcipher, which implements two parallel tweakable permutations. Such a primitive can be formalized with a simpler syntax and equivalent functionality, such as by fixing the selector to b in both the algorithms (one could discard an unneeded output block). Yet, such a syntax would not align well with the way a forkcipher is used (for example in Sect. 6): our syntax of choice allows the user of a forkcipher to precisely select what gets computed, to do so more efficiently when both output blocks are needed, and without wasting computations if only one output block is required. This will become clear upon inspection of ForkSkinny in Sect. 4.

3.2   Security Definition

We define the security of forkciphers by indistiguishability from the closest, most natural idealized version of the primitive, a pseudorandom tweakable forked permutation, with the help of security games in Fig. 2. A forked permutation is a pair of oracles, that make use of two permutations, s.t. the two permutations are always used with the same preimage, no matter if the query is made in the forward or the backward direction.

An adversary $\mathcal{A}$ that aims at breaking a tweakable forkcipher F plays the games **prtfp-real** and **prtfp-ideal**. We define the advantage of $\mathcal{A}$ at distinguishing F from a pair of random tweakable permutations in a *chosen ciphertext attack* as

$$\mathbf{Adv}_F^{\mathrm{prtfp}}(\mathcal{A}) = \Pr[\mathcal{A}^{\mathbf{prtfp\text{-}real}_F} \Rightarrow 1] - \Pr[\mathcal{A}^{\mathbf{prtfp\text{-}ideal}_F} \Rightarrow 1].$$

3.3   Iterate-Fork-Iterate

One approach to build a forkcipher from an existing iterated tweakable cipher is by applying our novel *iterate-fork-iterate* (IFI) paradigm. Following the IFI, in encryption a fixed length message block M is transformed via a fixed number of rounds or *iterations* of a tweakable cipher to M'. Then, M' is *forked* and two copies of the internal state are created, which are *iterated* to produce C_0 and C_1. Two of the main objectives of designing forkcipher in the IFI paradigm are (partial) transference of security results and maintaining forkcipher security without increasing the original cipher key size. In order to rule out that the IFI design succumbs to *generic* attacks (i.e., attacks that treat the primitive as a blackbox), we carry out a provable generic analysis. This result indicates structural soundness in the sense that no additional exploitable weakness are introduced, but does not directly imply security of IFI forkciphers, because a real forkcipher never uses a number of rounds in the partial iteration that is a secure pseudorandom permutation.

Game **prtfp-real$_F$**	Game **prtfp-ideal$_F$**
$K \leftarrow\$ \{0,1\}^k$	**for** $T \in \mathcal{T}$ **do** $\pi_{T,0}, \pi_{T,1} \leftarrow\$ \mathrm{Perm}(n)$
$b \leftarrow \mathcal{A}^{\mathrm{ENC},\mathrm{DEC}}$	$b \leftarrow \mathcal{A}^{\mathrm{ENC},\mathrm{DEC}}$
return b	**return** b
Oracle $\mathrm{ENC}(T, M, s)$	Oracle $\mathrm{ENC}(T, M, s)$
return $F(K, T, M, s)$	**if** $s = 0$ **then return** $\pi_{T,0}(M)$
	if $s = 1$ **then return** $\pi_{T,1}(M)$
	if $s = b$ **then return** $\pi_{T,0}(M),$
	$\pi_{T,1}(M)$
Oracle $\mathrm{DEC}(T, C, \beta, s)$	
return $F^{-1}(K, T, C, \beta, s)$	Oracle $\mathrm{DEC}(T, C, \beta, s)$
	if $s = i$ **then return** $\pi_{T,\beta}^{-1}(C)$
	if $s = o$ **then return** $\pi_{T,(\beta \oplus 1)}(\pi_{T,\beta}^{-1}(C))$
	if $s = b$ **then return** $\pi_{T,\beta}^{-1}(C),$
	$\pi_{T,(\beta \oplus 1)}(\pi_{T,\beta}^{-1}(C))$

Fig. 2. Games **prtfp-real** and **prtfp-ideal** defining the security of a (strong) forkcipher.

IFI Generic Validation. We show that a IFI forkcipher is a structurally sound construction as long as the three components: three tweak-indexed collections of permutations are ideal tweak permutations in the full version of the paper. Fix the block length n and the tweak length t. Formally, for three tweakable random permutations p, p_0, p_1 (i.e. $p = (p_T \leftarrow\$ \mathrm{Perm}(n))_{T \in \{0,1\}^t}$ is a collection of independent uniform elements of $\mathrm{Perm}(n)$ indexed by the elements of $T \in \{0,1\}^t$, and similar applies for p_0 and p_1), the forkcipher $F = \mathrm{IFI}[p, p_0, p_1]$ is defined by $F^{T,b}(M) = p_{T,0}(p_T(M)), p_{T,1}(p_T(M))$, and by $F^{-1T,b,b}(C) = p_T^{-1}(p_{T,b}^{-1}(C)), p_{T,b \oplus 1}(p_{T,b}^{-1}(C))$ (the rest follows from the correctness). We note that the three tweakable random permutations act as a key for $\mathrm{IFI}[p, p_0, p_1]$ and we omit them for the sake of simplicity. In the full version [10], we prove the indistinguishability of the IFI construction from a single *forked* random permutation in the information-theoretic setting.

Our IFI Instantiation. IFI is motivated by the most popular design strategy for block cipher design - *iterative* or round-based structure where the round functions are typically identical, up to round keys and constants. In forkcipher, after an initial number of rounds r_{init} two copies of the internal state are processed with different tweakeys. The number of rounds after the forking step, r_0 (left) and r_1 (right), are determined from the cryptanalytic assurances of the IFI block cipher instantiation. The block cipher round functions instantiate the forkcipher round functions (both before and after forking), again up to constants and round key addition. The single (secret) key SK security of both (left and right) forward $F^{T,0}$, $F^{T,1}$ and inverse $F^{-1T,0,i}$ (resp. $F^{-1T,1,i}$) forkcipher transformations, and the related-key (RK) security of $F^{T,1}$ follow easily from the underlying security of the block cipher. We further perform the SK and RK analysis for $\mathsf{F}^{T,0}$ and the reconstruction $\mathsf{F}^{-1^{T,0,\circ}}$ (resp. $\mathsf{F}^{-1^{T,1,\circ}}$) transformations.

In our instantiation, $r_0 = r_1$ as a direct consequence of the IFI design approach. Suppose, in the SK model $\mathsf{F}^{T,0}$ is secure using $r_{\text{init}} + r_0$ number of rounds. Such $\mathsf{F}^{T,0}$ can be instantiated using any existing (secure) off-the-shelf tweakable block cipher, which is the approach taken here. Then, having $r_{\text{init}} + r_1$ rounds, where $r_1 < r_0$, for $\mathsf{F}^{T,1}$ will obviously weaken the security of the forkcipher. This is true, assuming that we apply the same round function in both forking branches. In this article we choose a tweakable SPN-based block cipher to construct a forkcipher.

4   ForkSkinny

We design the forkcipher ForkSkinny using the recently published lightweight tweakable block cipher SKINNY [18]. As detailed in Table 1, we propose several instances, with various block and tweakey sizes, in order to fit the different use cases. For simplifying the notation, in the rest of this section we will denote the transformations $C_b \leftarrow \mathsf{ForkSkinny}_K^{T,b}(M)$ as $\mathsf{ForkSkinny}_b$, where $b = 0$ or 1 and the corresponding inverse transformations $\mathsf{ForkSkinny}^{-1^{T,b,i}_K}$ as $\mathsf{ForkSkinny}_b^{-1}$.

4.1   Specification

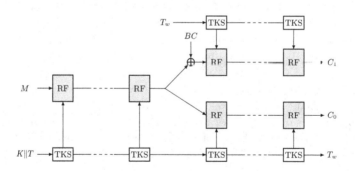

Fig. 3. ForkSkinny encryption with selector $s = \mathsf{b}$. A plaintext M, a key K and a tweak T (blue) are used to compute a ciphertext $C = C_0 \| C_1$ (red) of twice the size of the plaintext. RF is a single round function of SKINNY (with modified round constant), TKS is round tweakey update function [18], and BC is a branch constant that we introduce. (Color figure online)

Overall Structure. We illustrate our design in Fig. 3 for ForkSkinny-128-192. This version takes a 128-bit plaintext M, a 64-bit tweak T and a 128-bit secret key K as input, and outputs two 128-bit ciphertext blocks C_0 and C_1 (i.e., $\mathsf{ForkSkinny}(K, T, M, \mathsf{b}) = C_0, C_1$). The first $r_{\text{init}} = 21$ rounds of ForkSkinny are almost identical to the one of SKINNY and only differ in the value of the constant added to the internal state. After that, the encryption is *forked*, which means that two copies of the internal state are further modified with different sets of tweakeys. For reasons that we detail below, a constant denoted by BC (Branch Constant) is added to the internal state used to compute C_1, right after forking. Then, ForkSkinny$_0$ iterates $r_0 = 27$ rounds and ForkSkinny$_1$ iterates $r_1 = 27$ rounds. As illustrated in Fig. 3, after forking, the tweakeys for the round functions of ForkSkinny$_0$ are computed from the tweakey state obtained after r_{init} rounds, while the tweakeys for the round functions of ForkSkinny$_1$ are derived from the tweakey state at the end of $r_{\text{init}} + r_0$ rounds (denoted by T_w). Figure 4 details the ForkSkinny construction, where Enc-Skinny$_r(\cdot, \cdot)$ denotes the SKINNYencryption using r round functions taking as input a plaintext or state together with a tweakey. Similarly, Dec-Skinny$_r(\cdot, \cdot)$ denotes the corresponding decryption algorithm using r rounds.

1: **function** ForkSkinnyEnc(M, K, T, s)
2:     $tk \leftarrow K \| T$
3:     $L \leftarrow$ Enc-Skinny$_{r_{\text{init}}}(M, tk)$
4:     **if** $s = 0$ **or** $s = $ b **then**
5:         $C_0 \leftarrow$ Enc-Skinny$_{r_0}(L, \text{TKS}_{r_{\text{init}}}(tk))$
6:     **end if**
7:     **if** $s = 1$ **or** $s = $ b **then**
8:         $tk' \leftarrow \text{TKS}_{r_{\text{init}}+r_0}(tk)$
9:         $C_1 \leftarrow$ Enc-Skinny$_{r_1}(L \oplus BC, tk')$
10:     **end if**
11:     **if** $s = 0$ **return** C_0
12:     **if** $s = 1$ **return** C_1
13:     **if** $s = $ b **return** C_0, C_1
14: **end function**

1: **function** ForkSkinnyDec(C, K, T, b, s)
2:     $tk \leftarrow K \| T$
3:     $tk' \leftarrow \text{TKS}_{r_{\text{init}}}(tk)$
4:     **if** $b = 0$ **then**
5:         $L \leftarrow$ Dec-Skinny$_{r_0}(C, tk')$
6:     **else if** $b = 1$ **then**
7:         $tk'' \leftarrow \text{TKS}_{r_0}(tk')$
8:         $L \leftarrow$ Dec-Skinny$_{r_1}(C_b, tk'') \oplus BC$
9:     **end if**
10:     **if** $s = $ i **or** $s = $ b **then**
11:         $M \leftarrow$ Dec-Skinny$_{r_{\text{init}}}(L, tk)$
12:     **end if**
13:     **if** $s = $ o **or** $s = $ b **then**
14:         **if** $b = 0$ **then** $tk' \leftarrow \text{TKS}_{r_0}(tk')$
15:         $C' \leftarrow$ Enc-Skinny$_{r_{b \oplus 1}}(L, tk')$
16:     **end if**
17:     **if** $s = $ i **return** M
18:     **if** $s = $ o **return** C'
19:     **if** $s = $ b **return** M, C'
20: **end function**

Fig. 4. ForkSkinny encryption and decryption algorithms. Here TKS denotes the round tweakey scheduling function of SKINNY. TKS$_r$ depicts r rounds of TKS.

Round Function. As stated previously, the round function used in ForkSkinny is derived from the one of SKINNY and can be described as:

$$\mathcal{R}_i = \texttt{Mixcolumn} \circ \texttt{Addconstant} \circ \texttt{Addroundtweakey} \circ \texttt{Shiftrow} \circ \texttt{Subcell}$$

where Subcell, Shiftrow and Mixcolumn are identical to the ones of SKINNY. The Addroundtweakey function and the tweakey schedule are also left unchanged. Note that in ForkSkinny more tweakeys than in SKINNY are produced since we have $r_{\text{init}} + r_0 + r_1$ rounds. To keep the content short, we leave the details of these operations to full version [10] of this article.

The only change we made in the round function of ForkSkinny stands in the AddConstants step. Instead of using 6 bit round constants (generated with an LFSR), we use 7 bit ones. This change was required to avoid adding the same round constant to different rounds, as 6 bit round constants only provides 64 different values while some of our instances require a number of iterations higher than that. These 7 bit round constants may be chosen randomly and fixed. In our implementation we use an affine 7 bit LFSR to generate the round constant. The update function is defined as:

$$(rc_6||rc_5||\ldots||rc_0) \rightarrow (rc_5||rc_4||\ldots||rc_0||rc_6 \oplus rc_5 \oplus 1)$$

The 7 bits are initialized to 0 and updated before using in the round function. The bits from the LFSR are used exactly the same way as in Skinny. The 4×4 array

$$\begin{pmatrix} c_0 & 0 & 0 & 0 \\ c_1 & 0 & 0 & 0 \\ c_2 & 0 & 0 & 0 \\ 0 & 0 & 0 & 0 \end{pmatrix}$$

is constructed depending on the size of the internal state, where $c_2 = \texttt{0x2}$ and

$(c_0, c_1) = (rc_3||rc_2||rc_1||rc_0, 0||rc_6||rc_5||rc_4)$ when each cell is 4 bits

$(c_0, c_1) = (0||0||0||0||rc_3||rc_2||rc_1||rc_0, 0||0||0||0||0||rc_6||rc_5||rc_4)$ when each cell is 8 bits.

Branch Constant. We introduce constants to be added right after the forking point. When each cell is made of 4 bits we add BC_4, and when each cell is a byte we add BC_8, where:

$$BC_4 = \begin{pmatrix} 1 & 2 & 4 & 9 \\ 3 & 6 & d & a \\ 5 & b & 7 & f \\ e & c & 8 & 1 \end{pmatrix} \qquad BC_8 = \begin{pmatrix} 01 & 02 & 04 & 08 \\ 10 & 20 & 41 & 82 \\ 05 & 0a & 14 & 28 \\ 51 & a2 & 44 & 88 \end{pmatrix}.$$

This addition is made right after forking, to the right branch leading to C_1. Note that these constants are generated by clocking LFSRs, given by: $(x_3||x_2||x_1||x_0) \rightarrow (x_2||x_1||x_0||x_3 \oplus x_2)$, and initialised with $x_0 = 1$, $x_1 = x_2 = x_3 = 0$ for BC_4, and with the LFSR $(x_7||x_6||x_5||x_4||x_3||x_2||x_1||x_0) \rightarrow (x_6||x_5||x_4||x_3||x_2||x_1||x_0||x_7 \oplus x_5)$, again initialised with $x_0 = 1$ and all the other bits equal to 0 for BC_8.

This introduction is necessary to avoid that two SubCells steps cancel each others when looking at the sequence of operations relating C_0 and C_1 in the reconstruction scenario.

Variants. Other sets of parameters can be chosen. We propose some variants in Table 1. Note that their exact number of rounds (that are the parameters $r_0 = r_1$ and r_{init}), were determined from the security analysis of the cipher, detailed below.

4.2   Design Rationale

Using SKINNY. A forkcipher in IFI paradigm can be instantiated in various ways. We build our forkcipher design reusing the iterative structure of the SPN-based lightweight tweakable block cipher SKINNY. SPNs are very well-researched and allow to apply existing cryptanalysis techniques to the security analysis of our forkcipher. A large number of cryptanalytic results [12,13,49,51,56,57] have further been published against round reduced SKINNY showing that the full version of the cipher have comfortable security margins. Unlike other lightweight block ciphers, such as Midori [16], PRINCE [29], the SKINNYdesign is constructed following the TWEAKEY framework, and in addition supports a number of choices for the tweak size; an important aspect for the choice of SKINNYfor our design. SKINNYis good for lightweight applications on both hardware and software platforms. We also assume that the target application platform does not have AES instruction set available, hence avoiding AES based instantiation.

ForkSkinny Components. In ForkSkinny we have introduced features which aim to serve the forkcipher construction characteristics and security requirements. The 7 bit LFSR introduced in Addconstant avoids the repetition of round constants that could have possibly lead to *slide attack*-like cryptanalyses. The Branch Constant added after forking ensures that in the reconstruction scenario the two non-linear layers positioned around the forking point do not cancel each other. Finally, the required round tweakeys are computed by extending the key schedule of SKINNY by the necessary number of rounds. We chose this particular way of computing the extra tweakeys due to its simplicity, ability to maximally reuse components of SKINNY, and because it was among the most conservative options security-wise.

5   Security Analysis

For most attacks (for instance differential and linear cryptanalysis), the results devised on SKINNY give sufficient arguments to show the resistance of ForkSkinny. First, the series of operations leading M to C_0 correspond exactly to one encryption with SKINNY (up to the round constants) so the existing results transfer

Table 1. The ForkSkinny primitives with their internal parameters for round numbers r_{init}, r_0 and r_1 and their corresponding external parameters of block and tweakey sizes (in bits) for fixed 128 bit keys.

Primitive	block	tweak	tweakey	r_{init}	r_0	r_1
ForkSkinny-64-192	64	64	192	17	23	23
ForkSkinny-128-192	128	64	192	21	27	27
ForkSkinny-128-256	128	128	256	21	27	27
ForkSkinny-128-288	128	160	288	25	31	31
ForkSkinny-128-384	128	256	384	25	31	31

easily in this case. Then, when looking at the relation between M and C_1 we have a version of SKINNY with different round constants and a different tweak after r_{init} rounds. One way to give security arguments here is to look at what happens in the first r_{init} rounds and independently, in the next r_1 rounds to have a (pessimistic) estimation (for instance of the number of active Sboxes). A similar technique can be applied to study the reconstruction path. In both cases, the very large security margins[4] of SKINNY imply that ForkSkinny appears out of reach of the attacks we considered.

Our full security analysis is detailed in the full version [10] of this article. It covers truncated, impossible differential, boomerang, meet-in-the-middle, integral and algebraic attacks. We particularly stress that the boomerang type attack which was shown against ForkAES [17], is not applicable to ForkSkinny. This is due to two reasons: first, the number of rounds after the forking step protects against such attacks by making the boomerang path of very low probability. Second, the branch constant introduced in the right branch protects against such attacks by making the state of two branches different immediately after forking. Note that the attack in [17] against (9 out of 10 rounds) ForkAES in fact uses the property for which there is no difference between the states after forking.

We detail below our analysis of differential and linear attacks.

5.1   Detail of the Evaluation of Differential and Linear Attacks

Arguments in favor of the resistance of ForkSkinny to differential [28] and linear [39] cryptanalysis can easily be deduced from the available analysis on SKINNY. First, we refer to the bounds on the number of active Sboxes provided in the SKINNY specification document (recalled in the full version [10]). These bounds were later refined, and for instance Abdelkhalek et al. [7] showed that in the single key scenario there are no differential characteristics of probability higher than 2^{-128} for 14 rounds or more of SKINNY-128.

The previous results transfer to the case where we look at a trail covering the path from the input message up to C_0. Due to the change in the tweakey schedule we expect different bounds in the related-tweakey for the path from the input message up to C_1. A rough estimate of the minimal number of active Sboxes on this trail can be obtained by summing the bound on r_{init} rounds and the bound on r_1 rounds. For instance for ForkSkinny-128-192 (in TK2 model), 21 rounds activate at least 59 Sboxes. If we consider that the branch starting from the forking point is independent and can start from any internal state difference and tweakey difference (this is the very pessimistic case), only 8 rounds after forking are necessary to go below the characteristic probability of 2^{-128}.

The last case that needs to be evaluated is the reconstruction path scenario. An estimate can be computed following the same idea as before: the number of active Sboxes can be upper bounded by the bound obtained by summing the one for r_0 rounds and the one for r_1 rounds. If we consider that $r_0 = r_1$ as for our concrete instances, we obtain that 16 rounds are required to get more than 64

---

[4] At the time of writing, the best attacks on SKINNY cover at most 55% of the cipher.

active Sboxes. For ForkSkinny-128-192, 30 rounds are required to get more than 64 active Sboxes.

With respect to the parameters we chose, these (optimistic for the attacker) evaluations make us believe that differential attacks pose no threat to our proposal.

Similar arguments lead to the same conclusion for linear attacks. Also, we refer to the FSE 2017 paper [35] by Kranz et al. that looks at the linear hull of a tweakable block cipher and shows that the addition of a tweak does not introduce new linear characteristics, so that no additional precaution should be taken in comparison to a key-only cipher.

6 Tweakable Forkcipher Modes

We demonstrate the applicability of forkciphers by designing provably secure AEAD modes of operation for a tweakable forkcipher. Our AEAD schemes are designed such that (1) they are able to process strings of *arbitrary length* but (2) they are most efficient for data whose total number of blocks (in AD and message) is small, e.g. below four.

We define three forkcipher, nonce-based AEAD modes of operation: PAEF, SAEF and RPAEF. The first mode is fully parallelizable and (quantitatively) optimally secure in the nonce respecting model. The second mode SAEF sequentially encrypts "on-the-fly", has birthday-bounded security, and lends itself to low-overhead implementations. The third mode RPAEF is derived from the first one; it only uses both output blocks of a forkcipher in the final call, allowing to further reduce computational cost even for longer messages. The improved efficiency comes at the price of an n-bit larger tweak, and thus increased HW area footprint.

A Small AE Primitive. While a secure forkcipher does not directly capture integrity, we show in Sect. 6.9 that a secure forkcipher can be used as an AEAD scheme with fixed length messages and AD in the natural way, provably delivering strong AE security guarantees.

6.1 Syntax and Security of AEAD

Our modes following the AEAD syntax proposed by Rogaway [46]. A nonce-based AEAD scheme is a triplet $\Pi = (\mathcal{K}, \mathcal{E}, \mathcal{D})$. The key space $\mathcal{K}$ is a finite set. The deterministic encryption algorithm $\mathcal{E} : \mathcal{K} \times \mathcal{N} \times \mathcal{A} \times \mathcal{M} \to \mathcal{C}$ maps a secret key K, a nonce N, an associated data A and a message M to a ciphertext $C = \mathcal{E}(K, N, A, M)$. The nonce, AD and message domains are all subsets of $\{0,1\}^*$. The deterministic decryption algorithm $\mathcal{D} : \mathcal{K} \times \mathcal{N} \times \mathcal{A} \times \mathcal{C} \to \mathcal{M} \cup \{\perp\}$ takes a tuple (K, N, A, C) and either returns a message $M \in \mathcal{M}$, or a distinguished symbol $\perp$ to indicate an authentication error.

We require that for every $M \in \mathcal{M}$, we have $\{0,1\}^{|M|} \subseteq \mathcal{M}$ (i.e. for any integer m, either all or no strings of length m belong to $\mathcal{M}$) and that for all

$K, N, A, M \in \mathcal{K} \times \mathcal{N} \times \mathcal{A} \times \mathcal{M}$ we have $|\mathcal{E}(K, N, A, M)| = |M| + \tau$ for some non-negative integer τ called the stretch of Π. For correctness of Π, we require that for all $K, N, A, M \in \mathcal{K} \times \mathcal{N} \times \mathcal{A} \times \mathcal{M}$ we have $M = \mathcal{D}(K, N, A, \mathcal{E}(K, N, A, M))$. We let $\mathcal{E}_K(N, A, M) = \mathcal{E}(K, N, A, M)$ and $\mathcal{D}_K(N, A, M) = \mathcal{D}(K, N, A, M)$.

We follow Rogaway's two-requirement definition of AE security. A chosen plaintext attack of an adversary $\mathcal{A}$ against the confidentiality of a nonce-based AE scheme Π is captured with the help of the security games **priv-real** and **priv-real**. In both games, the adversary can make arbitrary chosen plaintext queries to a blackbox encryption oracle, such that each query must have a unique nonce, and such that the queries are replied with the scheme Π using a random secret key (real), or with independent uniform strings of the same length (ideal). The goal of $\mathcal{A}$ is to distinguish the two games. We define the advantage of $\mathcal{A}$ in breaking the confidentiality of Π as $\mathbf{Adv}_{\Pi}^{\mathrm{priv}}(\mathcal{A}) = \Pr[\mathcal{A}^{\mathbf{priv\text{-}real}_{\Pi}} \Rightarrow 1] - \Pr[\mathcal{A}^{\mathbf{priv\text{-}ideal}_{\Pi}} \Rightarrow 1]$.

A chosen ciphertext attack against the integrity of Π is captured with the game **auth**, in which an adversary can make nonce-respecting chosen plaintext and arbitrary chosen ciphertext queries to a black-box instance of Π with the goal of finding a forgery: a tuple that decrypts correctly but is not trivially knwn from the encryption queries. We define the advantage of $\mathcal{A}$ in breaking the integrity of Π as $\mathbf{Adv}_{\Pi}^{\mathrm{priv}}(\mathcal{A}) = \Pr[\mathcal{A}^{\mathbf{auth}_{\Pi}} \text{forges}]$ where "$\mathcal{A}$ forges" denotes a decryption query that returns a value $\neq \perp$. (For convenience, the games are included in full version of this article.)

6.2   Parallel AE from a Forkcipher

The nonce-based AEAD scheme PAEF ("Parallel AE from a Forkcipher") is parameterized by a forkcipher F (Sect. 3) with $\mathcal{T} = \{0,1\}^t$ for a positive t. It is further parameterized by a nonce length $0 < \nu \leq t - 4$. An instance $\mathrm{PAEF}[\mathsf{F}, \nu] = (\mathcal{K}, \mathcal{E}, \mathcal{D})$ has $\mathcal{K} = \{0,1\}^k$ and the encryption (Fig. 6) and decryption algorithms are defined in Fig. 5. Its nonce space is $\mathcal{N} = \{0,1\}^\nu$, and its message and AD space are respectively $\mathcal{M} = \{0,1\}^{\leq n \cdot (2^{(t-\nu-3)}-1)}$, and $\mathcal{A} = \{0,1\}^{\leq n \cdot (2^{(t-\nu-3)}-1)}$ (i.e., AD and message can have at most $2^{(t-\nu-3)} - 1$ blocks). The ciphertext expansion of $\mathrm{PAEF}[\mathsf{F}, \nu]$ is n bits.

In an encryption query, AD and message are partitioned into blocks of n bits. Each block is processed with one call to F using a tweak that is composed of: 1) the nonce; 2) a three-bit flag $f_0 \| f_1 \| f_2$; 3) a $(t - \nu - 3)$-bit encoding of the block index (unique for both AD and message). The nonce-length is a parameter that allows to make a trade-off between the maximal message length and maximal number of queries with the same key. The bit $f_0 = 1$ iff the final block of message is being processed, $f_1 = 1$ iff a block of message is being processed, and $f_2 = 1$ iff the final block of the current input (depending on f_1) is processed and the block is incomplete. The ciphertext blocks are the "left" output blocks of F applied to message blocks, and the right "right" output blocks are xor-summed with the AD output blocks, and the result xored to the final ciphertext block.

The decryption proceeds similarly as the encryption, except that "right" output blocks of the message blocks are reconstructed from ciphertext blocks (using the reconstruction algorithm) to recompute the tag, which is then checked.

6.3   Security of PAEF

We state the formal claim about the nonce-based AE security of PAEF in Theorem 1.

Theorem 1. *Let* F *be a tweakable forkcipher with* $\mathcal{T} = \{0,1\}^t$, *and let* $0 < \nu \leq t - 4$. *Then for any nonce-respecting adversary* $\mathcal{A}$ *whose queries lie in the proper domains of the encryption and decryption algorithms and who makes at most* q_v *decryption queries, we have*

$$\mathbf{Adv}^{\mathrm{priv}}_{\mathrm{PAEF}[\mathsf{F},\nu]}(\mathcal{A}) \leq \mathbf{Adv}^{\mathrm{prtfp}}_{\mathsf{F}}(\mathcal{B}) \quad \text{and} \quad \mathbf{Adv}^{\mathrm{auth}}_{\mathrm{PAEF}[\mathsf{F},\nu]}(\mathcal{A}) \leq \mathbf{Adv}^{\mathrm{prfp}}_{\mathsf{F}}(\mathcal{C}) + \frac{q_v \cdot 2^n}{(2^n - 1)^2}$$

for some adversaries $\mathcal{B}$ *and* $\mathcal{C}$ *who make at most twice as many queries in total as is the total number of blocks in all encryption, respectively all encryption and decryption queries made by* $\mathcal{A}$, *and who run in time given by the running time of* $\mathcal{A}$ *plus an overhead that is linear in the total number of blocks in all* $\mathcal{A}$'s *queries.*

Proof (sketch). The full proof appears in the full version of the paper [10]. For both confidentiality and authenticity, we first replace F with a pair of independent random tweakable permutations. Using a standard hybrid argument we have that $\mathbf{Adv}^{\mathrm{priv}}_{\mathrm{PAEF}[\mathsf{F},\nu]}(\mathcal{A}) \leq \mathbf{Adv}^{\mathrm{prtfp}}_{\mathsf{F}}(\mathcal{B}) + \mathbf{Adv}^{\mathrm{priv}}_{\mathrm{PAEF}[(\pi_0,\pi_1),\nu]}(\mathcal{A})$, and also that $\mathbf{Adv}^{\mathrm{auth}}_{\mathrm{PAEF}[\mathsf{F},\nu]}(\mathcal{A}) \leq \mathbf{Adv}^{\mathrm{prtfp}}_{\mathsf{F}}(\mathcal{C}) + \mathbf{Adv}^{\mathrm{priv}}_{\mathrm{PAEF}[(\pi_0,\pi_1),\nu]}(\mathcal{A})$.

For confidentiality, it is easy to see that in a nonce-respecting attack, every ciphertext block, and each tag is processed using a unique tweak-permutation combination, and all are uniformly distributed. Thus $\mathbf{Adv}^{\mathrm{priv}}_{\mathrm{PAEF}[(\pi_0,\pi_1),\nu]}(\mathcal{A}) = 0$. For authenticity, we analyse the probability of forgery for an adversary $\mathcal{A}'$ that makes a single decryption query against $\mathrm{PAEF}[(\pi_0,\pi_1),\nu]$ by the means of a case analysis, and then use a result of Bellare et al. [21] to obtain $\mathbf{Adv}^{\mathrm{auth}}_{\mathrm{PAEF}[(\pi_0,\pi_1),\nu]}(\mathcal{A}) \leq q_v \cdot \mathbf{Adv}^{\mathrm{auth}}_{\mathrm{PAEF}[(\pi_0,\pi_1),\nu]}(\mathcal{A}')$.

6.4   Sequential AE from a Forkcipher

SAEF (as in "Sequential AE from a Forkcipher," pronounce as "safe") is a nonce-based AEAD scheme parameterized by a tweakable forkcipher F (as defined in Sect. 3) with $\mathcal{T} = \{0,1\}^t$ for a positive $t \leq n$. An instance $\mathrm{SAEF}[\mathsf{F}] = (\mathcal{K}, \mathcal{E}, \mathcal{D})$ has a key space $\mathcal{K} = \{0,1\}^k$, nonce space $\mathcal{N} = \{0,1\}^{t-4}$, and the AD and message spaces are both $\{0,1\}^*$ (although the maximal AD/message length influences the security). The ciphertext expansion of $\mathrm{SAEF}[\mathsf{F}]$ is n bits. The encryption and decryption algorithms are defined in Fig. 7 and the encryption algorithm is illustrated in Fig. 8.

In an encryption query, first AD and then message are processed in blocks of n bits. Each block is processed with exactly one call to F, using a tweak that is composed of: (1) the nonce followed by a 1-bit in the initial F call, and the string $0^{\tau-3}$ otherwise, (2) three-bit flag f. The binary flag f takes different values for processing of different types of blocks in the encryption algorithm.

```
1: function  E(K, N, A, M)                 1: function  D(K, N, A, C)
2:     A₁,...,Aₐ,A∗ ⟵ⁿ A                   2:     A₁,...,Aₐ,A∗ ⟵ⁿ A
3:     M₁,...,Mₘ,M∗ ⟵ⁿ M                   3:     C₁,...,Cₘ,C∗,T ⟵
4:     S ⟵ 0ⁿ; c ⟵ (t − ν − 3)               csplit-bₙ(C)
5:     for i ⟵ 1 to a do                   4:     S ⟵ 0ⁿ; c ⟵ (t − ν − 3)
6:         ◇T ⟵ N‖000‖⟨i⟩_c               5:     for i ⟵ 1 to a do
7:         ∘T ⟵ N‖000‖⟨i⟩_c‖0ⁿ            6:         ◇T ⟵ N‖000‖⟨i⟩_c
8:         S ⟵ S ⊕ F_K^{T,0}(A_i)          7:         ∘T ⟵ N‖000‖⟨i⟩_c‖0ⁿ
9:     end for                             8:         S ⟵ S ⊕ F_K^{T,0}(A_i)
10:    if |A∗| = n then                    9:     end for
11:        ◇T ⟵ N‖001‖⟨a + 1⟩_c          10:    if |A∗| = n then
12:        ∘T ⟵ N‖001‖⟨a + 1⟩_c‖0ⁿ       11:        ◇T ⟵ N‖001‖⟨a + 1⟩_c
13:        S ⟵ S ⊕ F_K^{T,0}(A∗)          12:        ∘T ⟵ N‖001‖⟨a + 1⟩_c‖0ⁿ
14:    else if |A∗| > 0 or |M| = 0         13:        S ⟵ S ⊕ F_K^{T,0}(A∗)
           then                           14:    else if |A∗| > 0 or |T| = 0
15:        ◇T ⟵ N‖011‖⟨a + 1⟩_c               then
16:        ∘T ⟵ N‖011‖⟨a + 1⟩_c‖0ⁿ       15:        ◇T ⟵ N‖011‖⟨a + 1⟩_c
17:        S ⟵ S ⊕ F_K^{T,0}(A∗‖10∗)       16:        ∘T ⟵ N‖011‖⟨a + 1⟩_c‖0ⁿ
18:    end if          ▷ Do nothing if     17:        S ⟵ S ⊕ F_K^{T,0}(A∗‖10∗)
       A = ε, M ≠ ε                        18:    end if          ▷ Do nothing if
19:    for i ⟵ 1 to m do                          A = ε, M ≠ ε
20:        ◇T ⟵ N‖100‖⟨i⟩_c               19:    for i ⟵ 1 to m do
21:        ◇C_i, S' ⟵ F_K^{T,b}(M_i)       20:        ◇T ⟵ N‖100‖⟨i⟩_c
22:        ◇S ⟵ S ⊕ S'                     21:        ◇M_i, S' ⟵ F⁻¹_K^{T,0,b}(C_i)
23:        ∘T ⟵ N‖100‖⟨i⟩_c‖0ⁿ           22:        ◇S ⟵ S ⊕ S'
24:        ∘C_i ⟵ F_K^{T,0}(M_i)           23:        ∘T ⟵ N‖100‖⟨i⟩_c‖0ⁿ
25:        ∘S ⟵ S ⊕ M_i                    24:        ∘M_i ⟵ F⁻¹_K^{T,0,i}(C_i)
26:    end for                            25:        ∘S ⟵ S ⊕ M_i
27:    if |M∗| = n then                    26:    end for
28:        ◇T ⟵ N‖101‖⟨m + 1⟩_c          27:    if |T| = n then
29:        ∘T ⟵ N‖101‖⟨m + 1⟩_c‖S        28:        ◇T ⟵ N‖101‖⟨m + 1⟩_c
30:    else if |M∗| > 0 then                29:        ∘T ⟵ N‖101‖⟨m + 1⟩_c‖S
31:        ◇T ⟵ N‖111‖⟨m + 1⟩_c          30:    else if |T| > 0 then
32:        ∘T ⟵ N‖111‖⟨m + 1⟩_c‖S        31:        ◇T ⟵ N‖111‖⟨m + 1⟩_c
33:    else                               32:        ∘T ⟵ N‖111‖⟨m + 1⟩_c‖S
34:        return S                        33:    else
35:    end if                             34:        if C∗ ≠ S then return ⊥
36:    C∗, T ⟵ F_K^{T,b}(pad10(M∗))       35:        return ε
37:    ◇C∗ ⟵ C∗ ⊕ S                       36:    end if
38:    return                             37:    ◇C∗ ⟵ C∗ ⊕ S
       C₁‖...‖Cₘ‖C∗‖left_{|M∗|}(T)        38:    M∗, T' ⟵ F⁻¹_K^{T,0,b}(C∗ ⊕ S)
39: end function                          39:    T' ⟵ left_{|T|}(T'); P ⟵
                                                 right_{n−|T|}(M∗)
                                          40:    if T' ≠ T return ⊥
                                          41:    if P ≠ left_{n−|T|}(10ⁿ⁻¹) return
                                                 ⊥
                                          42:    return
                                                 M₁‖...‖Mₘ‖left_{|T|}(M∗)
                                          43: end function
```

Fig. 5. The PAEF[F, ν] (unmarked lines and ◇-marked lines) and the RPAEF[F, ν] (unmarked lines and ∘-marked lines) AEAD schemes. Here ⟨i⟩_ℓ is the canonical encoding of an integer i as an ℓ-bit string.

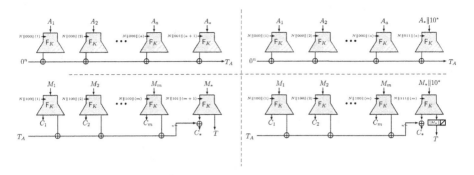

Fig. 6. The encryption algorithm of PAEF[F] mode. The picture illustrates the processing of AD when length of AD is a multiple of n (**top left**) and when the length of AD is not a multiple of n (**top right**), and the processing of the message when length of the message is a multiple of n (**bottom left**) and when the length of message is not a multiple of n (**bottom right**). The white hatching denotes that an output block is not computed.

The values $f = \{000, 010, 011, 110, 111, 001, 100, 101\}$ indicate the processing of respectively: non-final AD block; final complete AD block; final incomplete AD block; final complete AD block to produce tag; final incomplete AD block to produce tag; non-final message block; final complete message block; and final incomplete message block.

One output block of every F call is used as a whitening mask for the following F call, masking either the input (in AD processing) or both the input and the output (in message processing) of this subsequent call. The initial F call in the query is unmasked. The tag is the last "right" output of F produced in the query. The decryption proceeds similarly to the encryption, except that the plaintext blocks and the right-hand outputs of F in the message processing part are computed with the inverse F algorithm.

6.5 Security of SAEF

We state the formal claim about the nonce-based AE security of SAEF in Theorem 2.

Theorem 2. *Let* F *be a tweakable forkcipher with* $\mathcal{T} = \{0,1\}^\tau$. *Then for any nonce-respecting adversary* $\mathcal{A}$ *whose makes at most* q *encryption queries, at most* q_v *decryption queries such that the total number of forkcipher calls induced by all the queries is at most* σ, *with* $\sigma \leq 2^n/2$, *we have*

$$\mathbf{Adv}^{\mathrm{priv}}_{\mathrm{SAEF[F]}}(\mathcal{A}) \leq \mathbf{Adv}^{\mathrm{prtfp}}_{\mathsf{F}}(\mathcal{B}) + 2 \cdot \frac{(\sigma - q)^2}{2^n},$$

$$\mathbf{Adv}^{\mathrm{auth}}_{\mathrm{SAEF[F]}}(\mathcal{A}) \leq \mathbf{Adv}^{\mathrm{prtfp}}_{\mathsf{F}}(\mathcal{C}) + \frac{(\sigma - q + 1)^2}{2^n} + \frac{\sigma(\sigma - q)}{2^n} + \frac{q_v(q + 2)}{2^n}$$

<table>
<tbody></tbody>
</table>

```
 1: function  E(K, N, A, M)                1: function  D(K, N, A, C)
 2:     A_1, ..., A_a, A_* ←ⁿ A            2:     A_1, ..., A_a, A_* ←ⁿ A
 3:     M_1, ..., M_m, M_* ←ⁿ M            3:     C_1, ..., C_m, C_*, T ← csplit-b_n C
 4:     noM ← 0                            4:     noM ← 0
 5:     if |M| = 0 then noM ← 1            5:     if |C| = n then noM ← 1
 6:     Δ ← 0ⁿ; T ← N‖0^{τ−4−ν}‖1          6:     Δ ← 0ⁿ; T ← N‖0^{τ−4−ν}‖1
 7:     for i ← 1 to a do                  7:     for i ← 1 to a do
 8:         T ← T‖000                      8:         T ← T‖000
 9:         Δ ← F_K^{T,0}(A_i ⊕ Δ)         9:         Δ ← F_K^{T,0}(A_i ⊕ Δ)
10:         T ← 0^{τ−3}                   10:         T ← 0^{τ−3}
11:     end for                          11:     end for
12:     if |A_*| = n then                12:     if |A_*| = n then
13:         T ← T‖noM‖10                  13:         T ← T‖noM‖10
14:         Δ ← F_K^{T,0}(A_* ⊕ Δ)        14:         Δ ← F_K^{T,0}(A_* ⊕ Δ)
15:         T ← 0^{τ−3}                   15:         T ← 0^{τ−3}
16:     else if |A_*| > 0 or |M| = 0      16:     else if |A_*| > 0 or |T| = 0
    then                                     then
17:         T ← T‖noM‖11                  17:         T ← T‖noM‖11
18:         Δ ← F_K^{T,0}((A_*‖10*) ⊕ Δ)  18:         Δ ← F_K^{T,0}((A_*‖10*) ⊕ Δ)
19:         T ← 0^{τ−3}                   19:         T ← 0^{τ−3}
20:     end if      ▷ Do nothing if       20:     end if      ▷ Do nothing if
    A = ε, M ≠ ε                             A = ε, M ≠ ε
21:     for i ← 1 to m do                 21:     for i ← 1 to m do
22:         T ← T‖001                     22:         T ← T‖001
23:         C_i, Δ ← F_K^{T,b}(M_i ⊕ Δ) ⊕ 23:         M_i, Δ ← F⁻¹_K^{T,0,b}(C_i ⊕ Δ) ⊕
        (Δ, 0ⁿ)                                  (Δ, 0ⁿ)
24:         T ← 0^{τ−3}                   24:         T ← 0^{τ−3}
25:     end for                          25:     end for
26:     if |M_*| = n then                26:     if |T| = n then
27:         T ← T‖100                     27:         T ← T‖100
28:     else if |M_*| > 0 then            28:     else if |T| > 0 then
29:         T ← T‖101                     29:         T ← T‖101
30:     else                             30:     else
31:         return Δ                     31:         if C_* ≠ Δ then return ⊥
32:     end if                           32:         return ε
33:     C_*, T ← F_K^{T,b}(pad10(M_*) ⊕ Δ) ⊕  33:     end if
        (Δ‖0ⁿ)                           34:     M_*, T' ← F⁻¹_K^{T,0,b}(C_* ⊕ Δ) ⊕
34:     return                                   (Δ, 0ⁿ)
    C_1‖...‖C_m‖C_*‖left_{|M_*|}(T)      35:     T' ← left_{|T|}(T'); P ←
35: end function                             right_{n−|T|}(M_*)
                                         36:     if T' ≠ T return ⊥
                                         37:     if P ≠ left_{n−|T|}(10^{n−1}) return
                                             ⊥
                                         38:     return
                                             M_1‖...‖M_m‖left_{|T|}(M_*)
                                         39: end function
```

Fig. 7. The SAEF[F] AEAD scheme.

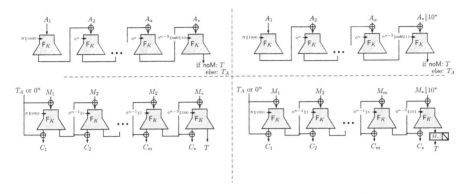

Fig. 8. The encryption algorithm of SAEF[F] mode. The bit noM = 1 iff $|M| = 0$. The picture illustrates the processing of AD when length of AD is a multiple of n (**top left**) and when the length of AD is not a multiple of n (**top right**), and the processing of the message when length of the message is a multiple of n (**bottom left**) and when the length of message is not a multiple of n (**bottom right**). The white hatching denotes that an output block is not computed.

for some adversaries $\mathcal{B}$ and $\mathcal{C}$ who make at most 2σ queries, and who run in time given by the running time of $\mathcal{A}$ plus $\gamma \cdot \sigma$ for some constant γ.

Proof (sketch). The full proof appears in the full version of the paper [10]. As with PAEF, we first replace F with a pair of independent random tweakable permutations, resulting in a similar birthday gap.

For confidentiality, we further replace tweakable permutations by random "tweakable" functions, further increasing the bound by $2 \cdot (\sigma - q)^2/2^{n+1}$ due to an RP-RF switch. Unless there is a non-trivial collision of inputs to f_0 and f_1, confidentiality of SAEF$[(f_0, f_1), \nu]$ is perfect. With such a collision appearing with a probability no greater than $2 \cdot (\sigma - q)^2/2^{n+1}$, we obtain the bound.

In the proof of integrity, we replace certain random permutations (indexed by a specific subset of tweaks) of the underlying tweakable permutations by tweakable functions with the same signature, increasing the bound by $(\sigma - q + 1)^2/2^{n+1}$ due to an RP-RF switch. We then define a variant of the **auth** game (call it **auth′**), which prevents $\mathcal{A}$ to win if a primitive input collision occurs in any of the encryption queries. The transition to the new game increases the bound by $\sigma(\sigma - q)/2^n$. Finally, (using the result of Bellare as for PAEF), we bound the probability of a successful forgery in **auth′** with help of a case analysis by $2 \cdot q_v/(2^n - 1)$.

6.6 Reduced Parallel AE from a Forkcipher

The nonce-based AEAD scheme RPAEF ("Reduced Parallel AE from a Forkcipher") is a derivative of PAEF that only uses the left output block of the underlying forkcipher for most of the message blocks. This allows for *reducing* the computational cost if the unevaluated fork can be switched off (as in ForkSkinny) at the expense of increasing the required tweak size. It is parameterized by a forkcipher F (Sect. 3) with $\mathcal{T} = \{0, 1\}^t$ for a positive $t \geq n + 5$. It is further parameterized by a nonce length $0 < \nu \leq t - (n+4)$. An instance $\text{RPAEF}[\mathsf{F}, \nu] = (\mathcal{K}, \mathcal{E}, \mathcal{D})$ has $\mathcal{K} = \{0,1\}^k$ and the encryption (Fig. 9) and decryption algorithms are defined in Fig. 5. Its nonce space is $\mathcal{N} = \{0,1\}^\nu$, and its message and AD space are respectively $\mathcal{M} = \{0,1\}^{\leq n \cdot (2^{(t-(n+\nu+3))}-1)}$, and $\mathcal{A} = \{0,1\}^{\leq n \cdot (2^{(t-(n+\nu+3))}-1)}$ (i.e. AD and message can have at most $2^{(t-(n+\nu+3))} - 1$ blocks). The ciphertext expansion of $\text{PAEF}[\mathsf{F}, \nu]$ is n bits.

In an encryption query, AD and message are processed in blocks of n bits. Each block is processed with one call to F using a tweak in which the first t bits are the same as in PAEF and the remaining n bits are either equal to a "checksum" of of all AD blocks and all-but-last message blocks, or to n zero bits (all other F calls). For all message blocks except the last one, only the left output block of F is used. The decryption proceeds similarly as the encryption, except that putative message blocks are reconstructed from ciphertext blocks to recompute the "checksum".

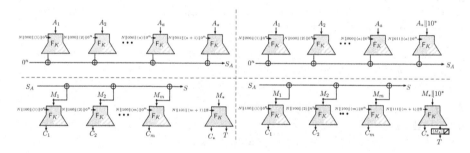

Fig. 9. The encryption algorithm of RPAEF[F] mode. The picture illustrates the processing of AD when length of AD is a multiple of n (**top left**) and when the length of AD is not a multiple of n (**top right**), and the processing of the message when length of the message is a multiple of n (**bottom left**) and when the length of message is not a multiple of n (**bottom right**). The white hatching denotes that an output block is not computed.

6.7   Security of RPAEF

Theorem 3. *Let F be a tweakable forkcipher with $\mathcal{T} = \{0,1\}^t$ and $t \geq n + 5$, and let $0 < \nu \leq t - 4$. Then for any nonce-respecting adversary $\mathcal{A}$ whose queries lie in the proper domains of the encryption and decryption algorithms and who makes at most q_v decryption queries, we have*

$$\mathbf{Adv}_{\text{PAEF}[\mathsf{F},\nu]}^{\text{priv}}(\mathcal{A}) \leq \mathbf{Adv}_{\mathsf{F}}^{\text{prtfp}}(\mathcal{B}) \quad \text{and} \quad \mathbf{Adv}_{\text{PAEF}[\mathsf{F},\nu]}^{\text{auth}}(\mathcal{A}) \leq \mathbf{Adv}_{\mathsf{F}}^{\text{prfp}}(\mathcal{C}) + \frac{2 \cdot q_v}{(2^n - 1)}$$

for some adversaries $\mathcal{B}$ and $\mathcal{C}$ who make at most twice as many queries in total as is the total number of blocks in all encryption, respectively all encryption and decryption queries made by $\mathcal{A}$, and who run in time given by the running time of $\mathcal{A}$ plus an overhead that is linear in the total number of blocks in all $\mathcal{A}$'s queries.

Proof (sketch). The full proof appears in the full version of the paper [10]. For both confidentiality and authenticity, we first replace F with a pair of independent random tweakable permutations, similarly as for PAEF.

For confidentiality, it is easy to see that, exactly as with PAEF, in a nonce-respecting attack every ciphertext block and all tags are uniformly distributed. We have $\mathbf{Adv}^{\mathrm{priv}}_{\mathrm{PAEF}[(\pi_0,\pi_1),\nu]}(\mathcal{A}) = 0$.

For authenticity, we combine a case analysis and the same result of Bellare et al. [21] as used for PAEF to obtain the bound.

6.8 Aggressive RPAEF Instance

We remark that when instantiated with ForkSkinny-128-384 (smaller tweakey would not make sense due to RPAEF's tweak size requirements), one of the three 128-bit tweakey schedule registers is effectively unused for all but last message blocks (it holds the the 0^n tweak component). Based on this observation, we consider a further, more aggressive optimization of RPAEF, which consists in *lowering the numbers of applied rounds* to those from ForkSkinny-128-256 for all but last message blocks, and for all AD blocks. A thorough analysis of this aggressive variant of ForkSkinny with a number of rounds adjusted to the effective tweak size is left as an open question. We do note, however that every tweak will only ever be used with a fixed number of rounds.

6.9 Deterministic MiniAE

In the introduction, we stated that a forkcipher is nearly, but not exactly, an AE primitive: we clarify this statement in the full version of the paper [10]. In short: it is easy to see that the syntax and security goals of a forkcipher, as proposed in Sect. 3, capture neither AE functionality nor AE security goals. Yet, *constructing* a secure PRI (with the same signature) from the forkcipher is trivial: just set $\mathcal{E}(K, N, A, M) = \mathsf{F}_K^{N\|A,b}(M)$ and $\mathcal{D}(K, N, A, C\|T) = \mathsf{F}^{-1}{}_K^{N\|A,0,i}(C)$ iff $T = \mathsf{F}^{-1}{}_K^{N\|A,0,o}(C)$. We prove that when used in this minimalistic "mode" of operation, a secure forkcipher yields a miniature AE scheme for fixed-size messages, which achieves PRI security [48].

7   Hardware Performance

Due to the independent branching of the data flow after the forking point, ForkSkinny comes with inherent data-level parallelism that does not exist in normal (tweakable) blockciphers like SKINNY. We illustrate how round-based hardware implementations amplify the performance boost of our forkcipher modes, well beyond the reduction of blockcipher rounds as argued in Sect. 1. We give a preliminary hardware implementation of all ForkSkinny variants in our three modes of operation, and compare the results with SKINNY-AEAD [19] as the most fairly comparable TBC mode of operation based on SKINNY.

Implementations. Figure 10 presents hardware synthesis results (ASIC) for open cell library NANGATE45NM in typical operating conditions. Messages as small as 8 bytes (64 bits) are considered separately, for which we select M6 as the most suitable SKINNY-AEAD family member. For processing 128-bit blocks, concrete instances are partitioned based matching tweakey lengths. The hardware area is partly based on synthesis results (i.e. the primitive) and partly estimated (i.e. the mode). For details on implementation assumptions, area estimation methodology and synthesis flow, please refer to [10].

For SKINNY-AEAD, we resynthesize the publicly available SKINNY round-based encryption implementations[5] The ForkSkinny implementations are a modification thereof, with a second state register, branch constant logic and extended round constant. We then go on to obtain parallel ForkSkinny implementations, denoted (//), by adding an extra copy of the round function to compute both branches simultaneously. We also implement the aggressive variant of RPAEF with tuned-down number of SKINNY rounds (see Sect. 6.8).

Results Interpretation. When implementations exploit the available primitive-level parallelism, the forkcipher performance boost is substantial. For instance, for messages up to three 128-bit blocks, the speed-up of PAEF and

Implementation (round-based)	Area [GE]	f_{max} [MHz]	Nb. cycles for encrypting $(a + m)$ **64-bit** blocks						General
			$a = 0$			$a = 1$			
			$m=1$	$m=2$	$m=3$	$m=0$	$m=1$	$m=2$	
SK-AEAD M6	6288	1075	96	96	144	48	96	**96**	$48(\lceil\frac{a}{2}\rceil+\lceil\frac{m}{2}\rceil+1)$
PAEF-64-192	**4205**	1265	**63**	126	189	**40**	103	166	$40(a + 1.575m)$
PAEF-64-192 (//)	4811	1265	**40**	**80**	**120**	**40**	**80**	120	$40(a + m)$

Implementation (round-based)	Area [GE]	f_{max} [MHz]	Nb. cycles for encrypting $(a + m)$ **128-bit** blocks						General $(m \geq 1)$
			$a = 0$			$a = 1$			
			$m=1$	$m=2$	$m=3$	$m=0$	$m=1$	$m=2$	
SK-AEAD M5	6778	1075	96	144	192	96	144	192	$48(a + m + 1)$
PAEF-128-256	7189	1053	**75**	150	225	**48**	**123**	198	$48(a + 1.562m)$
PAEF-128-256 (//)	8023	1042	**48**	**96**	**144**	**48**	**96**	**144**	$48(a + m)$
SAEF-128-256 (//)	7064	1042	**48**	**96**	**144**	**48**	**96**	**144**	$48(a + m)$
RPAEF (aggr.)	8203	1052	**87**	**135**	**183**	**48**	**135**	**183**	$48(a+m)+39$
SK-AEAD M1-2	8210	1000	112	168	224	112	168	224	$56(a + m + 1)$
PAEF-128-288	**7989**	971	**87**	174	261	**56**	**143**	230	$56(a + 1.553m)$
PAEF-128-288 (//)	9308	962	**56**	**112**	**168**	**56**	**112**	**168**	$56(a + m)$
RPAEF (cons.)	**8178**	1052	**87**	**143**	**199**	**56**	**143**	**199**	$56(a+m)+31$

Fig. 10. Synthesis results and cycles for encrypting a blocks associated data and m blocks message. Superior performance w.r.t. the baseline (SK-AEAD [19]) is indicated in **bold**. The area is a partly synthesized and partly estimated. RPAEF (conservative) is RPAEF instantiated with ForkSkinny-128-384, and RPAEF (aggressive) is described in Sect. 6.8.

---

[5] Available at https://sites.google.com/site/skinnycipher/implementation.

SAEF (both parallel (//)) ranges from 25% to 50%, where the advantage is largest for the single-block messages. RPAEF shows similar numbers, with a 5% − 22% speed-up for the "aggressive" version. Most notably, for parallel instances (//) the forkcipher invocations are essentially equally fast as block cipher invocations, which results in fewer cycles than SKINNY-AEAD *for all* message sizes. However, this advantage diminishes asymptotically with the message size (cf. the *general* column). For message sizes up to 8 bytes, emphasized by NIST [42], the PAEF-FORKSKINNY-64-192 instances are more than 58% faster (40 vs. 96 cycles) at a considerably smaller implementation size. SAEF has the disadvantage of being a serial mode but it has the smallest area (no block counter and nonce in tweak).

8   Conclusion

The idea of forkcipher opens up numerous interesting open question and research directions. For a detailed discussion we refer to the full version of this article [10].

Ackowledgements. Elena Andreeva was supported in part by the Research Council KU Leuven C1 on Security and Privacy for Cyber-Physical Systems and the Internet of Things with contract number C16/15/058 and by the Research Council KU Leuven, C16/18/004, through the EIT Health RAMSES project, through the IF/C1 on New Block Cipher Structures, and through the NIST project. In addition, this work was supported by the European Commission through the Horizon 2020 research and innovation programme under grant agreement H2020-DS-2014-653497 PANORAMIX and through the grant H2020-DS-SC7-2016-740507 Eunity. The work is supported in part by funding from imec of the Flemish Government. Antoon Purnal is supported by the Horizon 2020 research and innovation programme under Cathedral ERC Advanced Grant 695305. Reza Reyhanitabar's work on this project was initiated when he was with KU Leuven and supported by an EU H2020-MSCA-IF fellowship under grant ID 708815, continued and submitted when he was with Elektrobit Automotive GmbH, and revised while he is now with TE Connectivity. Arnab Roy is supported by the EPSRC grant No. EPSRC EP/N011635/1.

References

1. 3GPP TS 22.261: Service requirements for next generation new services and markets. https://portal.3gpp.org/desktopmodules/Specifications/SpecificationDetails. aspx?specificationId=3107
2. 3GPP TS 36.213: Evolved Universal Terrestrial Radio Access (E-UTRA); Physical layer procedures. https://portal.3gpp.org/desktopmodules/Specifications/ SpecificationDetails.aspx?specificationId=2427
3. CAN FD Standards and Recommendations. https://www.can-cia.org/news/cia-in-action/view/can-fd-standards-and-recommendations/2016/9/30/
4. ISO 11898–1:2015: Road vehicles - Controller area network (CAN) - Part 1: Data link layer and physical signalling. https://www.iso.org/standard/63648.html
5. NB-IoT: Enabling New Business Opportunities. http://www.huawei.com/ minisite/iot/img/nb_iot_whitepaper_en.pdf

6. Specification of Secure Onboard Communication. https://www.autosar. org/fileadmin/user_upload/standards/classic/4-3/AUTOSAR_SWS_ SecureOnboardCommunication.pdf

7. Abdelkhalek, A., Sasaki, Y., Todo, Y., Tolba, M., Youssef, A.M.: MILP modeling for (large) s-boxes to optimize probability of differential characteristics. IACR Trans. Symmetric Cryptol. **2017**(4), 99–129 (2017)

8. Anderson, E., Beaver, C., Draelos, T., Schroeppel, R., Torgerson, M.: ManTiCore: encryption with joint cipher-state authentication. In: Wang, H., Pieprzyk, J., Varadharajan, V. (eds.) ACISP 2004. LNCS, vol. 3108, pp. 440–453. Springer, Heidelberg (2004). https://doi.org/10.1007/978-3-540-27800-9_38

9. Andreeva, E., et al.: COLM v1 (2014). https://competitions.cr.yp.to/round3/ colmv1.pdf

10. Andreeva, E., Lallemand, V., Purnal, A., Reyhanitabar, R., Roy, A., Vizar, D.: Forkcipher: a new primitive for authenticated encryption of very short messages. Cryptology ePrint Archive, Report 2019/1004 (2019). https://eprint.iacr. org/2019/1004

11. Andreeva, E., Neven, G., Preneel, B., Shrimpton, T.: Seven-property-preserving iterated hashing: ROX. In: Kurosawa, K. (ed.) ASIACRYPT 2007. LNCS, vol. 4833, pp. 130–146. Springer, Heidelberg (2007). https://doi.org/10.1007/978-3-540-76900-2_8

12. Ankele, R., Banik, S., Chakraborti, A., List, E., Mendel, F., Sim, S.M., Wang, G.: Related-key impossible-differential attack on reduced-round SKINNY. In: Gollmann, D., Miyaji, A., Kikuchi, H. (eds.) ACNS 2017. LNCS, vol. 10355, pp. 208–228. Springer, Cham (2017). https://doi.org/10.1007/978-3-319-61204-1_11

13. Ankele, R., Kölbl, S.: Mind the gap - a closer look at the security of block ciphers against differential cryptanalysis. In: Cid, C., Jacobson Jr., M. (eds.) SAC 2018. LNCS, vol. 11349, pp. 163–190. Springer, Cham (2018). https://doi.org/10.1007/ 978-3-030-10970-7_8

14. Aumasson, J.P., et al.: CHAE: challenges in authenticated encryption. ECRYPT-CSA D1.1, Revision 1.05, 1 March 2017

15. Avanzi, R.: Method and apparatus to encrypt plaintext data. US patent 9294266B2 (2013). https://patents.google.com/patent/US9294266B2/

16. Banik, S., et al.: Midori: a block cipher for low energy. In: Iwata, T., Cheon, J.H. (eds.) ASIACRYPT 2015. LNCS, vol. 9453, pp. 411–436. Springer, Heidelberg (2015). https://doi.org/10.1007/978-3-662-48800-3_17

17. Banik, S., et al.: Cryptanalysis of forkaes. Cryptology ePrint Archive, Report 2019/289 (2019). https://eprint.iacr.org/2019/289

18. Beierle, C., et al.: The SKINNY family of block ciphers and its low-latency variant MANTIS. In: Robshaw, M., Katz, J. (eds.) CRYPTO 2016. LNCS, vol. 9815, pp. 123–153. Springer, Heidelberg (2016). https://doi.org/10.1007/978-3-662-53008-5_5 .

19. Beierle, C., et al.: Skinny-AEAD and Skinny-Hash. NIST LWC Candidate (2019)

20. Bellare, M.: Practice-oriented provable-security. In: Okamoto, E., Davida, G., Mambo, M. (eds.) ISW 1997. LNCS, vol. 1396, pp. 221–231. Springer, Heidelberg (1998). https://doi.org/10.1007/BFb0030423

21. Bellare, M., Goldreich, O., Mityagin, A.: The power of verification queries in message authentication and authenticated encryption. IACR Cryptology ePrint Archive **2004**, 309 (2004)

22. Bellare, M., Kohno, T., Namprempre, C.: Breaking and provably repairing the SSH authenticated encryption scheme: a case study of the encode-then-encrypt-and-mac paradigm. ACM Trans. Inf. Syst. Secur. **7**(2), 206–241 (2004)

23. Bellare, M., Namprempre, C.: Authenticated encryption: relations among notions and analysis of the generic composition paradigm. In: Okamoto, T. (ed.) ASIACRYPT 2000. LNCS, vol. 1976, pp. 531–545. Springer, Heidelberg (2000). https://doi.org/10.1007/3-540-44448-3_41

24. Bellare, M., Ristenpart, T.: Multi-property-preserving hash domain extension and the EMD transform. In: Lai, X., Chen, K. (eds.) ASIACRYPT 2006. LNCS, vol. 4284, pp. 299–314. Springer, Heidelberg (2006). https://doi.org/10.1007/11935230_20

25. Bellare, M., Rogaway, P.: Encode-then-encipher encryption: how to exploit nonces or redundancy in plaintexts for efficient cryptography. In: Okamoto, T. (ed.) ASIACRYPT 2000. LNCS, vol. 1976, pp. 317–330. Springer, Heidelberg (2000). https://doi.org/10.1007/3-540-44448-3_24

26. Bernstein, D.J.: Cryptographic competitions: CAESAR. http://competitions.cr.yp.to

27. Bertoni, G., Daemen, J., Hoffert, S., Peeters, M., Van Assche, G., Van Keer, R.: Farfalle: parallel permutation-based cryptography. IACR Transactions on Symmetric Cryptology **2017**, (2017). https://tosc.iacr.org/index.php/ToSC/article/view/855

28. Biham, E., Shamir, A.: Differential cryptanalysis of DES-like cryptosystems. J. Cryptol. **4**(1), 3–72 (1991)

29. Borghoff, J., et al.: PRINCE – a low-latency block cipher for pervasive computing applications. In: Wang, X., Sako, K. (eds.) ASIACRYPT 2012. LNCS, vol. 7658, pp. 208–225. Springer, Heidelberg (2012). https://doi.org/10.1007/978-3-642-34961-4_14

30. Dobraunig, C., Eichlseder, M., Mendel, F., Schläffer, M.: ASCON v1.2 (2014). https://competitions.cr.yp.to/round3/asconv12.pdf

31. Hoang, V.T., Krovetz, T., Rogaway, P.: Robust authenticated-encryption AEZ and the problem that it solves. In: Oswald, E., Fischlin, M. (eds.) EUROCRYPT 2015. LNCS, vol. 9056, pp. 15–44. Springer, Heidelberg (2015). https://doi.org/10.1007/978-3-662-46800-5_2

32. Jean, J., Nikolić, I., Peyrin, T., Seurin, Y.: Deoxys v1.41 v1 (2016). https://competitions.cr.yp.to/round3/deoxysv141.pdf

33. Jean, J., Nikolić, I., Peyrin, T.: Tweaks and keys for block ciphers: the TWEAKEY framework. In: Sarkar, P., Iwata, T. (eds.) ASIACRYPT 2014. LNCS, vol. 8874, pp. 274–288. Springer, Heidelberg (2014). https://doi.org/10.1007/978-3-662-45608-8_15

34. Katz, J., Yung, M.: Unforgeable encryption and chosen ciphertext secure modes of operation. In: Goos, G., Hartmanis, J., van Leeuwen, J., Schneier, B. (eds.) FSE 2000. LNCS, vol. 1978, pp. 284–299. Springer, Heidelberg (2001). https://doi.org/10.1007/3-540-44706-7_20

35. Kranz, T., Leander, G., Wiemer, F.: Linear cryptanalysis: key schedules and tweakable block ciphers. IACR Trans. Symmetric Cryptol. **2017**(1), 474–505 (2017)

36. Krovetz, T., Rogaway, P.: OCB v1.1 (2014). https://competitions.cr.yp.to/round3/ocbv11.pdf

37. Krovetz, T., Rogaway, P.: The software performance of authenticated-encryption modes. In: Joux, A. (ed.) FSE 2011. LNCS, vol. 6733, pp. 306–327. Springer, Heidelberg (2011). https://doi.org/10.1007/978-3-642-21702-9_18

38. Liskov, M., Rivest, R.L., Wagner, D.: Tweakable block ciphers. In: Yung, M. (ed.) CRYPTO 2002. LNCS, vol. 2442, pp. 31–46. Springer, Heidelberg (2002). https://doi.org/10.1007/3-540-45708-9_3

39. Matsui, M.: Linear cryptanalysis method for DES cipher. In: Helleseth, T. (ed.) EUROCRYPT 1993. LNCS, vol. 765, pp. 386–397. Springer, Heidelberg (1994). https://doi.org/10.1007/3-540-48285-7_33
40. McGrew, D.A., Viega, J.: The security and performance of the Galois/Counter Mode (GCM) of operation. In: Canteaut, A., Viswanathan, K. (eds.) INDOCRYPT 2004. LNCS, vol. 3348, pp. 343–355. Springer, Heidelberg (2004). https://doi.org/10.1007/978-3-540-30556-9_27
41. Namprempre, C., Rogaway, P., Shrimpton, T.: Reconsidering generic composition. In: Nguyen, P.Q., Oswald, E. (eds.) EUROCRYPT 2014. LNCS, vol. 8441, pp. 257–274. Springer, Heidelberg (2014). https://doi.org/10.1007/978-3-642-55220-5_15
42. NIST: DRAFT Submission Requirements and Evaluation Criteria for the Lightweight Cryptography Standardization Process (2018). https://csrc.nist.gov/Projects/Lightweight-Cryptography
43. Paterson, K.G., Yau, A.K.L.: Cryptography in theory and practice: the case of encryption in ipsec. IACR Cryptology ePrint Archive 2005, 416 (2005). http://eprint.iacr.org/2005/416
44. Paterson, K.G., Yau, A.K.L.: Cryptography in theory and practice: the case of encryption in IPsec. In: Vaudenay, S. (ed.) EUROCRYPT 2006. LNCS, vol. 4004, pp. 12–29. Springer, Heidelberg (2006). https://doi.org/10.1007/11761679_2
45. Reyhanitabar, M.R., Susilo, W., Mu, Y.: Analysis of property-preservation capabilities of the ROX and ESh hash domain extenders. In: Boyd, C., González Nieto, J. (eds.) ACISP 2009. LNCS, vol. 5594, pp. 153–170. Springer, Heidelberg (2009). https://doi.org/10.1007/978-3-642-02620-1_11
46. Rogaway, P.: Authenticated-encryption with associated-data. ACM CCS 2002, 98–107 (2002)
47. Rogaway, P.: Practice-oriented provable security and the social construction of cryptography. IEEE Secur. Priv. 14(6), 10–17 (2016)
48. Rogaway, P., Shrimpton, T.: A provable-security treatment of the key-wrap problem. In: Vaudenay, S. (ed.) EUROCRYPT 2006. LNCS, vol. 4004, pp. 373–390. Springer, Heidelberg (2006). https://doi.org/10.1007/11761679_23
49. Sadeghi, S., Mohammadi, T., Bagheri, N.: Cryptanalysis of reduced round SKINNY block cipher. IACR Trans. Symmetric Cryptol. 2018(3), 124–162 (2018)
50. Sui, H., Wu, W., Zhang, L., Wang, P.: Attacking and fixing the CS mode. In: Qing, S., Zhou, J., Liu, D. (eds.) ICICS 2013. LNCS, vol. 8233, pp. 318–330. Springer, Cham (2013). https://doi.org/10.1007/978-3-319-02726-5_23
51. Tolba, M., Abdelkhalek, A., Youssef, A.M.: Impossible differential cryptanalysis of reduced-round SKINNY. In: Joye, M., Nitaj, A. (eds.) AFRICACRYPT 2017. LNCS, vol. 10239, pp. 117–134. Springer, Cham (2017). https://doi.org/10.1007/978-3-319-57339-7_7
52. Whiting, D., Housley, R., Ferguson, N.: Counter with CBC-MAC (CCM). IETF RFC 3610 (Informational), September 2003. http://www.ietf.org/rfc/rfc3610.txt
53. Wu, H.: ACORN v3 (2014). https://competitions.cr.yp.to/round3/acornv3.pdf
54. Wu, H., Huang, T.: MORUS v2 (2014). https://competitions.cr.yp.to/round3/morusv2.pdf
55. Wu, H., Preneel, B.: AEGIS v1.1 (2014). https://competitions.cr.yp.to/round3/aegisv11.pdf
56. Zhang, P., Zhang, W.: Differential cryptanalysis on block cipher skinny with MILP program. Secur. Commun. Netw. 2018, 3780407:1–3780407:11 (2018)
57. Zhang, W., Rijmen, V.: Division cryptanalysis of block ciphers with a binary diffusion layer. IET Inf. Secur. 13(2), 87–95 (2019)

Anonymous AE

John Chan[(✉)] and Phillip Rogaway

Department of Computer Science, University of California, Davis, USA
jmachan@ucdavis.edu, Rogaway@cs.ucdavis.edu

Abstract. The customary formulation of authenticated encryption (AE) requires the decrypting party to supply the correct nonce with each ciphertext it decrypts. To enable this, the nonce is often sent in the clear alongside the ciphertext. But doing this can forfeit anonymity and degrade usability. Anonymity can also be lost by transmitting associated data (AD) or a session-ID (used to identify the operative key). To address these issues, we introduce anonymous AE, wherein ciphertexts must conceal their origin even when they are understood to encompass everything needed to decrypt (apart from the receiver's secret state). We formalize a type of anonymous AE we call anAE, *anonymous nonce-based AE*, which generalizes and strengthens conventional nonce-based AE, nAE. We provide an efficient construction for anAE, NonceWrap, from an nAE scheme and a blockcipher. We prove NonceWrap secure. While anAE does not address privacy loss through traffic-flow analysis, it does ensure that ciphertexts, now more expansively construed, do not by themselves compromise privacy.

Keywords: Anonymous encryption · Authenticated encryption · Nonces · Privacy · Provable security · Symmetric encryption

1 Introduction

Traditional formulations of authenticated encryption (AE) implicitly assume that auxiliary information is flowed alongside the ciphertext. This information, necessary to decrypt but not normally regarded as part of the ciphertext, may include a nonce, a session-ID (SID), and associated data (AD). But flowing these values in the clear may reveal the sender's identity.

To realize a more private form of encryption, we introduce a primitive we call *anonymous nonce-based AE*, or anAE. Unlike traditional AE [6,10,16,17, 19], anAE treats privacy as a first-class goal. We insist that ciphertexts contain everything the receiver needs to decrypt other than its secret state (including its keys), and ask for privacy even then. We show how to achieve anAE, providing a transform, NonceWrap, that turns a conventional nonce-based AE (nAE) scheme into an anAE scheme. We claim that anAE can improve not only on privacy, but on usability, too.

ⓒ International Association for Cryptologic Research 2019
S. D. Galbraith and S. Moriai (Eds.): ASIACRYPT 2019, LNCS 11922, pp. 183–208, 2019.
https://doi.org/10.1007/978-3-030-34621-8_7

BACKGROUND. The customary formulation for AE, nAE [14, 16, 19], requires the user to provide a nonce not only to encrypt a plaintext, but also to decrypt a ciphertext. Decryption fails if the wrong nonce is provided.

How is the decrypting party supposed to know the right nonce to use? Sometimes it will know it *a priori*, as when communicants speak over a reliable channel and maintain matching counters. But at least as often the nonce is flowed, in the clear, alongside the ciphertext. The *full* ciphertext should be understood as including that nonce, as the decrypting party needs it to decrypt.

Yet transmitting a nonce along with the ciphertext raises both usability and security concerns. Usability is harmed because the ciphertext is no longer self-contained: information beyond it and the operative key are needed to decrypt. At the same time, confidentiality and privacy are harmed because the transmitted nonce *is* information, and information likely correlated to identity. Sending a counter-based nonce, which is the norm, will reveal a message's *ordinality*—its position is the sequence of messages that comprise a session. While the usual definition for nAE effectively *defines* this leakage as harmless, is it always so? A counter-based nonce may be all that is needed to distinguish, say, a high-frequency stock trader (large counters) from a low-frequency stock trader (small counters). With a counter-based nonce, multiple sessions at different points in the sequence can be sorted by point of origin. Perhaps it is nothing but tradition that has led us to accept that nAE schemes, conventionally used, may leak a message's ordinality and the sender's identity.

This paper is about defining and constructing nonce-based AE schemes that are more protective of such metadata. We imagine multiple senders simultaneously communicating with a receiver, as though by broadcast, each session protected by its own key. When a ciphertext arrives, the receiver must decide which session it belongs to. But ciphertexts shouldn't get packaged with a nonce, or even an SID (session identifier) or AD (associated data), any of which would destroy anonymity. Instead, decryption should return these values, along with the underlying plaintext.

A LOUSY APPROACH. One way to conceal the operative nonce and SID would be to encrypt those things under a public key belonging to the receiver. The resulting ciphertext would flow along with an ind\$-secure nAE-encrypted ciphertext (where ind\$ refers to indistinguishability from uniform random bits [17]). While this approach can work, moving to the public-key setting would decimate the trust model, lengthen each ciphertext, and substantially slow each encryption and decryption, augmenting every symmetric-key operation with a public-key one. We prefer an approach that preserves the symmetric trust model and has minimal impact on message lengths and computation time.

CONTRIBUTIONS: DEFINITIONS. We provide a formalization of anonymous AE that we call anAE, *anonymous nonce-based AE*. Our treatment makes anAE encryption identical to encryption under nAE. Either way, encryption is accomplished with a deterministic algorithm $C = \mathcal{E}_K^{N,A}(M)$ operating on the key K, nonce N, associated data A, and plaintext M. As usual, ciphertexts so produced can be decrypted by an algorithm $M = \mathcal{D}_K^{N,A}(C)$. But the receiver employing a privacy-conscious protocol might not *know* what K, N, or A to use, as flowing N or A, or identifying K in any direct way, would damage privacy. So an anAE scheme supplements the decryption algorithm $\mathcal{D}$ with a constellation of alternative algorithms. They let the receiver: initialize a session (`Init`); terminate a session (`Term`); associate an AD with a session, or with all sessions (`Asso`); disassociate an AD with a session, or with all sessions (`Disa`); and decrypt a ciphertext, given nothing else (`Dec`). The last returns not only the plaintext but, also, the nonce, SID, and AD.

After formalizing the syntax for anAE we define security, doing this in the concrete-security, game-based tradition. A single game formalizes confidentiality, privacy, and authenticity, unified as a single notion. It is parameterized by a *nonce policy*, Nx, which defines what nonces a receiver should consider permissible at some point in time. We distinguish this from the nonce or nonces that are *anticipated*, or *likely*, at some point in time, formalized by a different function, Lx. Our treatment of permissible nonces vs. likely nonces may be useful beyond anonymity, and can be used to speed up decryption.

Anonymous AE can be formalized without a user-supplied nonce as an input to encryption, going back to a probabilistic or stateful definition of AE. For this reason, anAE should be understood as one way to treat anonymous AE, not the only way possible. That said, our choice to build on nAE was carefully considered. Maintaining nAE-style encryption, right down to the API, should facilitate backward compatibility and a cleaner migration path from something now quite standard. Beyond this, the reasons for a nonce-based treatment of AE remain valid after privacy becomes a concern. These include minimizing requirements on user-supplied randomness/IVs.

CONTRIBUTIONS: CONSTRUCTIONS. We next investigate how to achieve anAE. Ignoring the AD, an obvious construction is to encipher the nonce using a blockcipher, creating a header Head $= E_{K_1}(N)$. This is sent along with an nAE-encrypted Body $= \mathcal{E}_{K_2}^{N,A}(M)$. But the ciphertext $C =$ Head $\|$ Body so produced would be slow to decrypt, as one would need to trial-decrypt Body under each receiver-known key K_2' until the (apparently) right one is found (according to the nAE scheme's authenticity-check). If the receiver has s active sessions and the message has $|M| = m$ bits, one can expect a decryption time of $\Theta(ms)$.

To do better we put redundancy in the header, replacing it with Head $= E_{K_1}(N \| 0^\rho \| H(AD))$. Look ahead to Fig. 2 for our scheme, NonceWrap. As a concrete example, if the nonce N is 12 bytes [12] and we use the degenerate hash $H(x) = \varepsilon$ (the empty string), then one could encrypt a plaintext M as $C = \text{AES}_{K_1}(N \| 0^{32}) \| \text{GCM}_{K_2}^{N,A}(M)$. Using the header Head to screen candidate keys (only those that produce the right redundancy) and assuming $\rho \geq \lg s$ we can now expect a decryption time of $\Theta(m+s)$ for s blockcipher calls and a single nAE decryption.

In many situations, we can do better, as the receiver will be able to anticipate each nonce for each session. If the receiver is stateful and maintains a dictionary ADT (abstract data type) of all anticipated headers expected to arrive, then a single `lookup` operation replaces the trial decryptions of Head under each prospective key. Using standard data-structure techniques based on hashing or balanced binary trees, the expected run time drops to $\Theta(m+\lg(s))$ for decrypting a length-m string. And one can always fall back to the $\Theta(m + s)$-time process if an unanticipated nonce was used.

Finally, in some situations one can do better still, when all permissible nonces can be anticipated. In such a case the decrypting party need never invert the blockcipher E and the header can be truncated, or some other PRF can be used. In practice, the header could be reduced from 16 bytes to one or two bytes—a savings over a conventional nAE scheme that transmits the nonce.

While NonceWrap encryption is simple, decryption is not; look ahead to Figs. 3 and 4. Even on the encryption side, there are multiple approaches for handling the AD. Among them we have chosen the one that is most bandwidth-efficient and that seems to make the least fuss over the AD.

RELATED WORK. In the CAESAR call for AE algorithms, Bernstein introduced the notion of a *secret message number* (SMN) as a possible alternative to a nonce, which he renamed the *public message number* (PMN) [7]. When the party encrypting a message specifies an SMN, the decrypting party doesn't need to know it. It was an innovative idea, but few CAESAR submissions supported it [2], and none became finalists. Namprempre, Rogaway, and Shrimpton formalized Bernstein's idea by adjusting the nAE syntax and security notion [13]. Their definition didn't capture any privacy properties or advantages of SMNs.

It was also Bernstein who asked (personal communication, 2017) if one could quickly identify which session an AE-encrypted ciphertext belonged to if one was unwilling to explicitly annotate it. NonceWrap does this, assuming a stateful receiver using what we would call a constant-breadth nonce policy.

Coming to the problem from a different angle, Bellare, Ng, and Tackmann contemporaneously investigated the danger of flowing nonces, and recast decryption so that a nonce needn't be provided [5]. Their concern lies in the fact that an encrypting party can't select *any* non-repeating nonce (it shouldn't depend on the plaintext or key), and emphasize that the nAE definition fails to specify which choices are fine.

Our approach to parameterizing an AE's goal using a nonce policy Nx benefits from the evolution of treatments on stateful AE [4,8,11,20]. The introduction of Lx (likely nonces) as something distinct from Nx (permissible nonces) is new.

A privacy goal for semantically secure encryption has been formalized as *key privacy* [3] in the public-key setting and as *which-key concealing encryption* [1] in the shared-key one. But the intent there was narrow: probabilistic encryption (not AE), when the correct key is known, out of band, by the decrypting party.

2 Nonce-Based AE (nAE)

BACKGROUND. An nAE scheme, a nonce-based AE scheme supporting associated data (AD), is determined by a function $\mathcal{E}$, the *encryption algorithm*, with signature $\mathcal{E}: \mathcal{K} \times \mathcal{N} \times \mathcal{A} \times \mathcal{M} \to \mathcal{C}$. We insist that $\mathcal{E}(K, N, A, \cdot)$ be injective for any K, N, A. This ensures that there's a well-defined function $\mathcal{D} = \mathcal{E}^{-1}$ with signature $\mathcal{D}\colon \mathcal{K} \times \mathcal{N} \times \mathcal{A} \times \mathcal{C} \to \mathcal{M} \cup \{\bot\}$ defined by $\mathcal{D}(K,N,A,C) = M$ if $\mathcal{E}(K,N,A,M) = C$ for some (unique) $M \in \mathcal{M}$, while $\mathcal{D}(K,N,A,C) = \bot$ otherwise. The symbol $\bot$ is used to indicate invalidity. We may write $\mathcal{E}_K^{N,A}(M)$ and $\mathcal{D}_K^{N,A}(C)$ for $\mathcal{E}(K,N,A,M)$ and $\mathcal{D}(K,N,A,C)$. We require that the message space $\mathcal{M} \subseteq \{0,1\}^*$ be a set of strings for which $M \in \mathcal{M}$ implies $\{0,1\}^{|M|} \subseteq \mathcal{M}$. Finally, we assume that $|\mathcal{E}_K^{N,A}(M)| = |M| + \tau$ where τ is a constant. We refer to τ as the *expansion* of the scheme.

Let $\mathcal{E}\colon \mathcal{K} \times \mathcal{N} \times \mathcal{A} \times \mathcal{M} \to \mathcal{C}$ be an nAE scheme with expansion τ. A customary way to define nAE security [14,19] associates to an adversary $\mathcal{A}$ the real number $\mathbf{Adv}_{\mathcal{E}}^{\mathrm{nae}}(\mathcal{A}) = \Pr[K \leftarrow \mathcal{K}\colon \mathcal{A}^{E_K(\cdot,\cdot,\cdot),\,D_K(\cdot,\cdot,\cdot)} \Rightarrow 1] - \Pr[\mathcal{A}^{\$(\cdot,\cdot,\cdot),\,\bot(\cdot,\cdot,\cdot)} \Rightarrow 1]$ where the four oracles behave as follows: oracle $E_K(\cdot,\cdot,\cdot)$, on input N,A,M, returns $\mathcal{E}(K,N,A,M)$; oracle $D_K(\cdot,\cdot,\cdot)$, on input N,A,C, returns $\mathcal{D}(K,N,A,C)$; oracle $\$(\cdot,\cdot,\cdot)$, on input N,A,M, returns $|M| + \tau$ uniform random bits; and oracle $\bot(\cdot,\cdot,\cdot)$, on input N,A,C, returns $\bot$. The adversary $\mathcal{A}$ is forbidden from asking its first oracle a query (N,A,M) if it previously asked a query (N,A',M'); nor may it ask its second oracle (N,A,C) if it previously asked its first oracle a query (N,A,M) and received a response of C.

PRIVACY-VIOLATING ASSUMPTIONS OF NAE. The nAE definition quietly embeds a variety of privacy-unfriendly choices. Beginning with syntax, decryption is understood to be performed directly by a function, $\mathcal{D}$, that requires input of K, N, and A. This suggests that the receiver *knows* the right key to use, and that the ciphertext will be delivered within some context that explicitly identifies which session the communication is a part of. But explicitly flowing such information is damaging to privacy. Similarly, the nonce N and AD A are needed by the decrypting party, but flowing either will often prove fatal to anonymity.

Indistinguishability from random bits is routinely understood to buy anonymity: after all, if the encryption of M under keys K and K' are indistinguishable from random bits then they are indistinguishable from each other. But this glosses over the basic problem that the thing that's indistinguishable from random bits isn't everything the adversary will see.

3   Anonymous Nonce-Based AE (anAE)

PRIVACY PRINCIPLE. Our anAE notion can be seen as arising from a basic tenet of secure encryption, which we now make explicit.

> **Privacy principle.** A ciphertext should not by itself compromise the identity of its sender. This should hold even when the term "ciphertext" is understood as the *full* ciphertext—everything the receiver needs to decrypt and that the adversary might see.

The principle implies that it is not OK to just exclude from our understanding of the word *ciphertext* the privacy-violating parts of a transmission that are needed to decrypt. One needs to understand the ciphertext more expansively.

Stated as above, the privacy principle may seem so obvious that it is silly to spell it out. But the fact that nAE blatantly violates this principle, despite being understood as an extremely strong notion of security, suggests otherwise.

While this paper focuses on privacy, attending to the full ciphertext would seem to be the appropriate move when it comes to understanding confidentiality and authenticity as well. Our formulation of anAE does so.

Figuring out *how* to reflect the privacy principle in a definition is non-trivial. We now turn to that task.

SYNTAX. An anAE scheme extends an nAE scheme with five additional algorithms. Formally, an anAE scheme is a six-tuple of deterministic algorithms $\Pi = (\mathsf{Init}, \mathsf{Term}, \mathsf{Asso}, \mathsf{Disa}, \mathsf{Enc}, \mathsf{Dec})$. They create a session, terminate a session, register an AD, deregister an AD, encrypt a plaintext, and decrypt a ciphertext. The encryption algorithm $\mathcal{E} = \mathsf{Enc}$ must be an nAE scheme in its own right. In particular, this means that Enc automatically has an inverse $\mathcal{D} = \mathsf{Enc}^{-1}$, which is *not* what we are denoting Dec. Algorithms $\mathsf{Init}, \mathsf{Term}, \mathsf{Asso}, \mathsf{Disa}$, and Dec are run by the decrypting party (they are, in effect, an alternative to $\mathcal{D} = \mathsf{Enc}^{-1}$) and able to mutate its persistent state $\mathbf{K} \in \mathcal{K}$. Specifically,

- Init, the receiver's *session-initialization algorithm*, takes a key $K \in \mathcal{K}$ and returns a session-ID $\ell \in \mathcal{L}$ that will subsequently be used to name this session. We assume that returned SIDs are always distinct.
- Term, the receiver's *session-termination algorithm*, takes a session-ID $\ell \in \mathcal{L}$ and returns nothing.
- Asso, the receiver's *AD-association algorithm*, on input of either $A \in \mathcal{A}$ or $(A, \ell) \in \mathcal{A} \times \mathcal{L}$, returns nothing.
- Disa, the receiver's *AD-disassociation algorithm*, on input of either $A \in \mathcal{A}$ or $(A, \ell) \in \mathcal{A} \times \mathcal{L}$, returns nothing.
- Dec, the receiver's *decryption algorithm*, takes as input a ciphertext $C \in \mathcal{C}$ and returns either $(\ell, N, A, M) \in \mathcal{L} \times \mathcal{N} \times \mathcal{A} \times \mathcal{M}$ or the symbol $\perp$.

The sets referred to above, all nonempty, are as follows:

- $\mathcal{A}$ is an arbitrary set, the *AD space*.
- $\mathcal{C}$ is a set of strings, the *ciphertext space*.
- $\mathcal{K}$ is a finite set of strings, the *key space*.
- $\mathcal{K}$ is an arbitrary set, the receiver's *persistent state*.
- $\mathcal{L}$ is an arbitrary set, the *session names*.
- $\mathcal{M}$ is a set of strings, the *message space*.
- $\mathcal{N}$ is a finite set, the *nonce space*.

Observe that decryption via Dec is only given the ciphertext C (and, implicitly, the state that the receiver maintains) but is expected, from this alone, to return not only the message but also the operative SID, nonce, and AD. The SID identifies the operative key. For the remainder of the text, we treat the SIDs as natural numbers, that is, we assume as $\mathcal{L} = \mathbb{N}$.

NONCE POLICY. In an AE scheme with stateful decryption [4,8,11,20] the receiver will, at any given point in time, have some set of nonces that it deems acceptable. We allow this set to depend on the nonces already received, but on nothing else. We formalize this by defining a *nonce policy* as a function $Nx\colon \mathcal{N}^{\leq d} \to \mathcal{P}(\mathcal{N})$. By $\mathcal{P}(\mathcal{S})$ we mean the set of all subsets of the set $\mathcal{S}$. The name Nx is meant to suggest the words *next* and *nonce*. The set $Nx(\boldsymbol{N})$ are the *permissible nonces* given the history $\boldsymbol{N}$. The history is a list of previously received nonces. The value $d = \mathrm{depth}(Nx) \in \mathbb{N} \cup \{\infty\}$ is the *depth* of the policy, capturing how many nonces one needs to record in order to know what the next nonce may be. One could reasonably argue that practical nonce policies must have bounded depth, as they would otherwise require the receiver to maintain unlimited state, and decryption would slow as connections grew old. The value $b = \mathrm{breadth}(Nx) = \max_{\boldsymbol{N} \in \mathrm{dom}(Nx)} |Nx(\boldsymbol{N})|$ is the *breadth* of the policy, the maximum number of permissible nonces. For a function $F\colon \mathcal{A} \to \mathcal{B}$ we are writing $\mathrm{dom}(F) = \mathcal{A}$ for its domain. Similarly, we write $\mathrm{range}(F) = \mathcal{B}$ for its range.

We single out two policy extremes. The *permissive* policy $Nx(\Lambda) = \mathcal{N}$ captures what happens in a stateless AE scheme, where repetitions, omissions, and out-of-order delivery are all *permitted*. (The symbol Λ denotes the empty list.) The permissive policy has depth $d = 0$ and breadth $b = |\mathcal{N}|$. Note that while the decryption algorithm *itself* treats all nonces as permissible, there could be some other, higher-level process that restricts this. At the other extreme, assuming a nonce space of $\mathcal{N} = [0..N_{\mathrm{max}}]$, the *strict* policy $Nx(\Lambda) = \{0\}$, $Nx((N)) = \{N+1\}$ (for $N < N_{\mathrm{max}}$), and $Nx((N_{\mathrm{max}})) = \emptyset$ demands an absence of repetitions, omissions, and out-of-order delivery. The nonce starts at zero and must keep incrementing. The depth d and breadth b are both 1. On a reliable channel, this is a natural policy. There is a rich set of policies between these extremes [4,8,11,20].

AD REGISTRATION. A sender may have some data that needs to be authenticated with the ciphertext it sends. Flowing that data in the clear would compromise anonymity. Instead, the receiver will maintain a set of AD values for each session. We can register or remove AD values one-by-one with `Asso` and `Disa`.

There are use cases where an AD value may not be specific to a session. For example, the use of AD in TLS 1.3 does not involve session-specific information; instead, the AD consists of several constants along with the ciphertext length.[1] To accommodate this, we envisage a further set of AD values that are effectively registered to *all* sessions. We refer to this as the set of *global* ADs. These too are added and removed one at a time. When a ciphertext needs to be decrypted, the only AD values that can match it are the global ones and those registered for the session that the ciphertext is seen as belonging to (which the decrypting party will have to determine).

Despite the generality of this treatment, the utility of AD is limited in anAE precisely because AD values *can't* flow in the clear; the only AD values that parties should use are those that can be determined *a priori* by the receiver.

---

[1] While we define anAE to accommodate this use case, it was pointless for TLS to put length of the ciphertext in the AD: nAE ensures that ciphertexts are authenticated, which implies that their length is authenticated. Throwing $|C|$ into the AD contributes nothing to security but does add complexity.

DEFINING SECURITY. Let $\Pi = (\text{Init}, \text{Term}, \text{Asso}, \text{Disa}, \text{Enc}, \text{Dec})$ be an anAE scheme and let Nx be a nonce policy. The anAE security of Π with respect to Nx is captured by the pair of games in Fig. 1. The adversary interacts with either the $\text{Real}_{\Pi,Nx}^{\text{anae}}$ game or the $\text{Ideal}_{\Pi,Nx}^{\text{anae}}$ game and tries to guess which. The advantage of $\mathcal{A}$ attacking Π with respect to Nx is defined as

$$\mathbf{Adv}_{\Pi,Nx}^{\text{anae}}(\mathcal{A}) = \mathbf{Pr}[\, \mathcal{A}^{\text{Real}_{\Pi,Nx}^{\text{anae}}} \rightarrow 1 \,] - \mathbf{Pr}[\, \mathcal{A}^{\text{Ideal}_{\Pi,Nx}^{\text{anae}}} \rightarrow 1 \,],$$

the difference in probability that the adversary outputs "1" in the two games.

In our pseudocode, integers, strings, lists, and associative arrays are silently initialized to 0, ε, Λ, and $\emptyset$. For a nonempty list $\boldsymbol{x} = (x_1, \ldots, x_n)$ we let $\text{tail}(\boldsymbol{x}) = (x_2, \ldots, x_n)$. We write $A \overset{\cup}{\leftarrow} B$, $A \overset{\backslash}{\leftarrow} B$, and $A \overset{\|}{\leftarrow} B$ for $A \leftarrow A \cup B$, $A \leftarrow A \setminus B$, and $A \leftarrow A \parallel B$. When iterating through a string-valued set, we do so in lexicographic order.

We use *associative arrays* (also called *maps* or *dictionaries*) both in our games defining security and in the NonceWrap scheme itself. These are collections of (key, value) pairs with at most one value per key. We write $A[K]$ for doing a lookup in A for the value associated to the key K, returning that value. We write $A[K] \leftarrow X$ to mean adding or reassigning value X to key K. We write A.keys to denote the set of all keys in A. Similarly, A.values denotes the set of all values in A. The last two operations are not always mentioned in abstract treatments of dictionaries, but programming languages like Python do support these methods, and realizations of dictionaries invariably enable them.

EXPLANATION. The "real" anAE game surfaces to the adversary the six procedures of an anAE scheme. Modeling correct use, the Init procedure generates random keys, while calls to Enc may not repeat a nonce within the given session, nor may they employ a fictitious SID or the SID of a terminated session. The game does the needed bookkeeping to keep track of those things, with K_ℓ being the key associated to session ℓ and L recording the set of active session labels and $\text{NE}[\ell]$ being the set of nonces already used for session ℓ.

The "ideal" anAE game provides the same entry points as the "real" one but employs the protocol Π only insofar as INIT returns the same sequence of labels used by Init and, also, the ideal game uses the expansion constant σ from Enc. The sequence of labels returned by INIT could just as well have been fixed as $1, 2, 3, \ldots$. The central idea is that encryption returns uniformly random bits (line 242) regardless of the SID, nonce, AD, or plaintext. This captures both confidentiality and anonymity, and in a strong sense. The same idea is used in the ind\$-form of the nAE definition, but the constraint isn't on the full ciphertext.

As with the all-in-one definition for nAE [19], authenticity is ensured by having the counterpart of the real decryption oracle routinely return $\perp$. When should it *not* return $\perp$? As with nAE, we want the ideal game to return $\perp$ if the ciphertext C was not previously returned from an ENC query (line 250). But we *also* want DEC to return $\perp$ if the relevant session has been torn down, if the relevant nonce is out-of-policy, or if the relevant AD is unregistered. We also want DEC to return $\perp$ if there is more than one in-policy explanation for this ciphertext. All of this is captured in lines 250–253. To express those lines,

Real$_{\Pi,Nx}^{\text{anae}}$	Ideal$_{\Pi,Nx}^{\text{anae}}$
procedure INIT()	**procedure** INIT()
100 $K \twoheadleftarrow \mathcal{K}$	200 $K \twoheadleftarrow \mathcal{K}$
101 $\ell \leftarrow \Pi.\text{Init}(K)$ *Guaranteed new*	201 $\ell \leftarrow \Pi.\text{Init}(K)$
102 $K[\ell] \leftarrow K$; $L \overset{\cup}{\leftarrow} \{\ell\}$; $NE[\ell] \leftarrow \emptyset$	202 $A[\ell] \leftarrow \emptyset$; $L \overset{\cup}{\leftarrow} \{\ell\}$
103 **return** ℓ	203 $NE[\ell] \leftarrow \emptyset$; $ND[\ell] \leftarrow \Lambda$
	204 **return** ℓ
procedure TERM(ℓ)	**procedure** TERM(ℓ)
110 $\Pi.\text{Term}(\ell)$; $L \overset{-}{\leftarrow} \{\ell\}$	210 $L \overset{-}{\leftarrow} \{\ell\}$
procedure ASSO(A)	**procedure** ASSO(A)
120 $\Pi.\text{Asso}(A)$	220 $AD \overset{\cup}{\leftarrow} \{A\}$
procedure ASSO(A, ℓ)	**procedure** ASSO(A, ℓ)
121 $\Pi.\text{Asso}(A, \ell)$	221 $A[\ell] \overset{\cup}{\leftarrow} \{A\}$
procedure DISA(A)	**procedure** DISA(A)
130 $\Pi.\text{Disa}(A)$	230 $AD \overset{-}{\leftarrow} \{A\}$
procedure DISA(A, ℓ)	**procedure** DISA(A, ℓ)
131 $\Pi.\text{Disa}(A, \ell)$	231 $A[\ell] \overset{-}{\leftarrow} \{A\}$
procedure ENC(ℓ, N, A, M)	**procedure** ENC(ℓ, N, A, M)
140 **if** $\ell \notin L$ **or** $N \in NE[\ell]$ **then**	240 **if** $\ell \notin L$ **or** $N \in NE[\ell]$ **then**
141 **return** $\bot$	241 **return** $\bot$
142 $NE[\ell] \overset{\cup}{\leftarrow} \{N\}$	242 $C \twoheadleftarrow \{0,1\}^{|M|+\tau}$
143 **return** $\Pi.\text{Enc}(K[\ell], N, A, M)$	243 $NE[\ell] \overset{\cup}{\leftarrow} \{N\}$
	244 $H[C] \overset{\cup}{\leftarrow} \{(\ell, N, A, M)\}$
	245 **return** C
procedure DEC(C)	**procedure** DEC(C)
150 **return** $\Pi.\text{Dec}(C)$	250 **if** $H[C] = \emptyset$ **then return** $\bot$
	251 **if** $\exists$ unique $(\ell, N, A, M) \in H[C]$ s.t.
	252 $\ell \in L$ **and** $N \in Nx(ND[\ell])$ **and**
	253 $A \in AD \cup A[\ell]$ **then**
	254 $ND[\ell] \overset{\parallel}{\leftarrow} N$
	255 **if** $|ND[\ell]| > d$ **then**
	256 $ND[\ell] \leftarrow \text{tail}(ND[\ell])$
	257 **return** (ℓ, N, A, M)
	258 **return** $\bot$

Fig. 1. Defining anAE security. The games depend on an anAE scheme Π and a nonce policy Nx. The adversary must distinguish the game on the left from the one on the right. Privacy, confidentiality, and authenticity are simultaneously captured.

we need more bookkeeping than the real game did, also recording, in H (for "history"), the (ℓ, N, A, M) value(s) that gave rise to C (line 244); recording in ND$[\ell]$ the sequence of nonces already observed on session ℓ (lines 255–256 truncate the history to only that needed for our decision making); and recording in associative arrays AD and A$[\ell]$ the currently registered AD values.

1AD/SESSION. We anticipate that, in most settings, the user will associate a single AD to a session at any given time. It might be associated to a particular session, or to all sessions, but, once a session has been identified, there is an understood AD for it. A decrypting party that operates in this way is said to be following the *one-AD-per-session restriction*, abbreviated 1AD/session.

STATELESS SCHEMES. Our formalization treats the decrypting party as stateful. Even if there was only one session and one AD, the decrypting party should register K with an Init call, register A with an Asso call, and then call Dec(C). But this sort of use of state is an artifact of the generality of our formulation. To draw out this distinction, we say that an anAE scheme is *stateless* if calls to its Dec algorithm never modify the receiver state.

For stateless anAE, one might provide an alternative API in which keys and AD are provided on each call, as in Dec1(K, A, C). Alternatively, one could initialize a data structure to hold the operative keys and AD values, and this data structure would be provided for decryption, but not side-effected by it. That is what happens in most crypto libraries today, where it is not a string-valued key that is passed to the encryption or decryption algorithms, but an opaque data structure created by a key-preprocessing step.

4 The NonceWrap Scheme

CIPHERTEXT STRUCTURE. Encryption under NonceWrap is illustrated in Fig. 2. The method uses two main primitives: an n-bit blockcipher E and an nAE scheme $\mathcal{E}$. The blockcipher is invoked once for each message encrypted, while the nAE scheme does the bulk of the work. NonceWrap also employs a hash function H, but it is used only for AD processing, outputs only a few bits (we do not seek collision-resistance), and indeed there is no security property from H on which we depend. A poor choice of H (like the constant function) would slow down decryption (in the case of multiple AD values per session), but would have no other adverse effect.

There are two parts to a NonceWrap-produced ciphertext: a header and a body. The header Head would typically be 16 bytes. It not only encodes the nonce N, which would usually be 12 bytes [12], but also some redundancy and a hash of the AD. To create a ciphertext C, the header is generated using a blockcipher E and is prepended to the ciphertext body Body, which is produced using nAE encryption on the nonce, AD, and plaintext. The total length of the ciphertext for M is $|M| + \lambda + \tau$ where λ is the header length—which is, for now, the blocksize $\lambda = n$ of E, and τ is the expansion of the nAE scheme.

When presented with a ciphertext $C = \text{Head} \,\|\, \text{Body}$, a receiver will often be able to determine that it does not belong to a candidate session just by looking

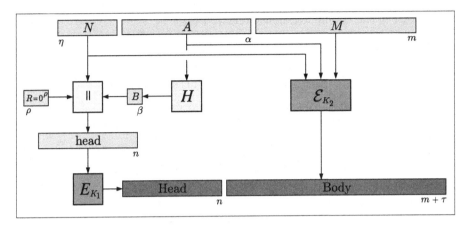

Fig. 2. Scheme illustration. NonceWrap encryption outputs a ciphertext that consists of two parts: a header Head, which is produced from a blockcipher E, and a body Body, which is produced from an nAE scheme $\mathcal{E}$. The hashed AD in the header can be omitted in the customary case where there is one AD per session at any time.

at the prefix Head. It is deciphered with the candidate session key and if the resulting block does not contain the mandated block of zero bits, or the nonce is not within nonce policy, or if the hash field does not contain the hash of a registered AD for this session, then the ciphertext as a whole must be invalid.

The hash of the AD is omitted if 1AD/session is assumed (equivalently, the hash returns the empty string). With a 16-byte header encrypting a 12-byte nonce, there would then be 4 bytes of zeros and roughly a 2^{-32} chance that a header for one session would be considered as a plausible candidate for another. When that does happen, it results in an nAE decryption of a ciphertext Body, not attribution of a ciphertext to an incorrect session. For *that* to happen, the ciphertext body would also have to verify as authentic when decrypted under the incorrect key. It is a little-mentioned property of nAE-secure encryption that a plaintext encrypted under one random key will almost always be deemed inauthentic when decrypted under an independent random key.

ANTICIPATED NONCES. Since the header is computed from the nonce and AD, it may be possible for the receiver to precompute a header before it arrives. This is because the nonce must fall within the protocol's nonce policy and the AD must be registered either specifically to a session or globally across sessions. But even under the 1AD/session assumption, the number of potential headers to precompute would be large if the breadth of the nonce policy is large (as with the permissive policy $Nx(\Lambda) = \mathcal{N}$). To get around this, we introduce a function to name the *anticipated* (or *likely*) nonces, Lx. Given the last few nonces received so far, it returns the nonce or set of nonces that are likely to come next. This is in contrast with Nx, which names the set of nonces that are *permissible* to come next—anything else should be deemed inauthentic. Like the nonce policy Nx, the signature of the anticipated-nonce function is $Lx: \mathcal{N}^{\leq d} \to \mathcal{P}(\mathcal{N})$. We demand that that which is likely is possible: $Lx(\boldsymbol{n}) \subseteq Nx(\boldsymbol{n})$ for all $\boldsymbol{n} \in \mathcal{N}^{\leq d}$.

An anAE scheme that employs Lx and Nx functions is said to be *sharp* if $Lx = Nx$. With a sharp scheme, a ciphertext must be deemed invalid if it employs an unanticipated nonce. Sharpness can aid in efficient decryption.

ALGORITHMIC DETAILS. We now descend more deeply into the structure of NonceWrap. The construction is defined in Fig. 3 and a list of data structures employed is given in Fig. 4.

The NonceWrap scheme maintains a number of dictionaries. The dictionary LNA maps anticipated headers to the set of session, nonce, and AD triples that explain the header. When a session is initialized, the dictionary is populated with headers based on anticipated nonces from an empty nonce history and the set of globally registered ADs. When a session is torn down, all headers belonging to that session are expunged from the dictionary. When a new AD is registered globally, headers are precomputed for each session and their anticipated nonces. If the AD is registered specific to a session, headers are computed for just that session. ADs are also managed in their own associative arrays—one for global and one for session-specific—that map AD hashes to sets of ADs that are preimages of the hash. Deregistering an AD removes it from its respective array and expunges its associated headers from the main dictionary.

NonceWrap decryption comes in three phases. Phase-1 attempts to use the precomputed headers in LNA to quickly determine which session, nonce, and AD are associated to a received ciphertext. As there may be multiple (ℓ, N, A) triples mapped to the header, the receiver tries to decrypt the ciphertext body with each until it arrives at a valid message. If no message is found within this phase, then it falls through to the phase-2, where it attempts to extract the nonce and AD directly by trial-deciphering the header. The receiver tries each session key on the header until it finds a nonce within the session's policy appended with ρ redundant 0-bits. If there are multiple AD values per session, the hash of an AD would be appended. If the AD is properly registered with the receiver, then the receiver has a mapping between the AD hash and its possible preimages. With this, the receiver may now trial-decrypt the ciphertext body. The second phase is repeated until a valid message is found. If none is, then decryption returns $\perp$ and the ciphertext is deemed invalid.

If either phase-1 or phase-2 recovers a valid plaintext, they go into phase-3, where precomputation for the next anticipated header occurs. Entering phase-3 means the receiver knows the (ℓ, N, A, M) for the ciphertext. It can then compute the old set of anticipated nonces prior to receiving N using Lx. It can also compute a new set of anticipated nonces with a nonce history updated with N. With the former, it can expunge all old headers from LNA and, with the latter, it can populate LNA with the next expected headers.

EFFICIENCY. Let s denote the maximum number of active sessions. Let t be the time it takes to compute the E or E^{-1}. Assume an anticipated-nonce policy Lx whose breadth is a small constant. Assume the maximum number of AD values registered either globally or to any one session is a small constant.

$\Pi.\texttt{Init}(K)$

00 $\ell \leftarrow K.\text{ctr}{+}{+}$
01 $K.\text{K1}[\ell] \parallel K.\text{K2}[\ell] \leftarrow K$
02   where $|K.\text{K1}[\ell]| = k_1$
03 $K.L \xleftarrow{\cup} \{\ell\}; \quad K.A[\ell] \leftarrow \emptyset$
04 $K.N[\ell] \leftarrow \Lambda; \quad K1 \leftarrow K.\text{K1}[\ell]$
05 for $N \in Lx(\Lambda)$ do
06   for ADs $\in K.\text{AD.values}$ do
07   for $A \in$ ADs do
08   head $\leftarrow N \parallel 0^\rho \parallel H(A)$
09   Head $\leftarrow E_{K1}(\text{head}))$
0A $K.\text{LNA}[\text{Head}] \xleftarrow{\cup} \{(\ell, N, A)\}$
0B return ℓ

$\Pi.\texttt{Term}(\ell)$

10 for $S \in K.\text{LNA.values}$ do
11    $S \xleftarrow{-} \{\ell\} \times \mathcal{N} \times \mathcal{A}$
12 $K.L \xleftarrow{-} \{\ell\}$

$\Pi.\texttt{Asso}(A)$

20 $B \leftarrow H(A); \quad K.\text{AD}[B] \xleftarrow{\cup} \{A\}$
21 for $\ell \in K.L$ do
22   for $N \in Lx(K.N[\ell])$ do
23   Head $\leftarrow E_{K.\text{K1}[\ell]}(N \parallel 0^\rho \parallel B)$
24      $K.\text{LNA}[\text{Head}] \xleftarrow{\cup} \{(\ell, N, A)\}$

$\Pi.\texttt{Asso}(A, \ell)$

25 $B \leftarrow H(A); \quad K.A[\ell][B] \xleftarrow{\cup} \{A\}$
26 for $N \in Lx(K.N[\ell])$ do
27   Head $\leftarrow E_{K.\text{K1}[\ell]}(N \parallel 0^\rho \parallel B)$
28    $K.\text{LNA}[\text{Head}] \xleftarrow{\cup} \{(\ell, N, A)\}$

$\Pi.\texttt{Disa}(A)$

30 $B \leftarrow H(A); \quad K.\text{AD}[B] \xleftarrow{-} \{A\}$
31 for $S \in K.\text{LNA.values}$ do
32    $S \xleftarrow{-} \mathcal{L} \times \mathcal{N} \times \{A\}$

$\Pi.\texttt{Disa}(A, \ell)$

33 $B \leftarrow H(A); \quad K.A[\ell][B] \xleftarrow{-} \{A\}$
34 for $S \in K.\text{LNA.values}$ do
35    $S \xleftarrow{-} \{\ell\} \times \mathcal{N} \times \{A\}$

$\Pi.\texttt{Enc}(K, N, A, M)$

40 $K_1 \parallel K_2 \leftarrow K$ where $|K_1| = k_1$
41 Head $\leftarrow E_{K_1}(N \parallel 0^\rho \parallel H(A))$
42 Body $\leftarrow \mathcal{E}(K_2, N, A, M)$
43 return $C \leftarrow$ Head $\parallel$ Body

$\Pi.\texttt{Dec}(C)$ *Phase-1 (starting at 51)*

50 Head $\parallel$ Body $\leftarrow C$ where $|\text{Head}| = n$
51 for $(\ell, N, A) \in K.\text{LNA}[\text{Head}]$ do
52    $K_1 \leftarrow K.\text{K1}[\ell]; \quad K_2 \leftarrow K.\text{K2}[\ell]$
53    $M \leftarrow \mathcal{D}(K_2, N, A, \text{Body})$
54   if $M \neq \perp$ then goto 5F

55 for $\ell \in K.L$ do   *Phase-2*
56    $K_1 \leftarrow K.\text{K1}[\ell]; \quad K_2 \leftarrow K.\text{K2}[\ell]$
57    $N \parallel R \parallel B \leftarrow E_{K_1}^{-1}(\text{Head})$
58   where $|N| = \eta$ and $|R| = \rho$
59   if $R \neq 0^\rho$ or $N \notin Nx(K.N[\ell])$ then
5A   continue
5B   for $A \in K.A[\ell][B] \cup K.\text{AD}[B]$ do
5C $M \leftarrow \mathcal{D}(K_2, N, A, \text{Body})$
5D   if $M \neq \perp$ then goto 5F
5E return $\perp$

5F Old $\leftarrow Lx(K.N[\ell])$ *Phase-3*
5G $K.N[\ell] \xleftarrow{\parallel} N$
5H if $|K.N[\ell]| > d$ then
5I $K.N[\ell] \leftarrow \text{tail}(K.N[\ell])$
5J New $\leftarrow Lx(K.N[\ell])$
5K for $N' \in$ Old $\setminus$ New do
5L   for $S \in K.\text{LNA.values}$ do
5M $S \xleftarrow{-} \{\ell\} \times \{N'\} \times \mathcal{A}$
5N for $N' \in$ New $\setminus$ Old do
5O   for $B \in K.A[\ell].\text{keys} \cup K.\text{AD.keys}$ do
5P   Head $\leftarrow E_{K_1}(N' \parallel 0^\rho \parallel B)$
5Q   for $A' \in K.A[\ell][B] \cup K.\text{AD}[B]$ do
5R $K.\text{LNA}[\text{Head}] \xleftarrow{\cup} \{(\ell, N', A')\}$
5S return (ℓ, N, A, M)

Fig. 3. Constructing an anAE scheme. Scheme $\Pi = \text{NonceWrap}[E, H, \mathcal{E}, Lx, Nx]$ depends on a blockcipher $E: \{0,1\}^{k_1} \times \{0,1\}^n \to \{0,1\}^n$, a nonce policy $Nx: \{0,1\}^{\leq d} \to \mathcal{P}(\mathcal{N})$, a hash function $H: \{0,1\}^* \to \{0,1\}^\beta$, an nAE scheme $\mathcal{E}: \mathcal{K} \times \mathcal{N} \times \mathcal{A} \times \mathcal{M} \to \mathcal{C}$ and an anticipated nonce function Lx which always outputs a subset of what policy Nx permits. Data structures employed are described in Fig. 4.

K.L	Set of SIDs
K.K1	Dictionary mapping an SID to a key for the blockcipher E
K.K2	Dictionary mapping an SID to a key for the nAE scheme $\mathcal{E}$
K.N	Dictionary mapping an SID to a list of nonces
K.A	Dictionary mapping an SID to a dict. mapping a hashed AD to a set of ADs
K.AD	Dictionary mapping a hashed AD to a set of ADs
K.LNA	Dictionary mapping a header to a set of (SID, nonce, AD) triples

Fig. 4. Data structures employed for NonceWrap. To achieve good decryption-time efficiency, NonceWrap employs a set ADT and multiple dictionaries, one of which has dictionary-valued entries. Some simplifications are possible for the customary case of 1AD/session.

Assume an amount of redundancy $\rho \in O(\lg s)$ used to create headers. Assume the nAE scheme $\mathcal{E}$ uses time $O(m + a)$ to decrypt a length $m + \tau$ ciphertext with AD A. Assume a nonce can be checked as being in-policy, according to Nx, in constant time. Assume dictionaries are implemented in some customary way, with expected log-time operations. Then the expected time to decrypt a valid ciphertext that used an anticipated nonce will be $O(m + a + t + \lg s)$. The expected time to decrypt an invalid ciphertext, or a valid ciphertext that used an unanticipated nonce, will be $O(m + a + st)$. For a sharp policy we may safely omit phase-2 and get a decryption time of $O(m + a + t + \lg s)$ for any ciphertext.

OPTIMIZATIONS. For a *sharp* scheme, $Nx = Lx$, the anticipated nonces within LNA encompass all valid nonces; a header not stored in the dictionary is necessarily invalid. For such a scheme, phase-2 can be ignored. This improves efficiency and allows for some natural simplifications. In addition, in this case we never compute the inverse E^{-1} of the blockcipher, so there is not longer any *need* for it to be a blockcipher. One can therefore replace the blockcipher $E: \{0,1\}^n \to \{0,1\}^n$ by a PRF $F: \{0,1\}^n \to \{0,1\}^\lambda$ where λ is considerable smaller than n. One or two bytes would typically suffice. After all, all that happens when a collision *does* occur is that one needs to perform a trial decryption of the ciphertext body. A convenient way to construct the PRF F would be by truncating the blockcipher E, setting $F_{K_1}(x) = (E_{K_1}(x))[1..\lambda]$.

At the opposite extreme, when $Lx(n) = \emptyset$ for all n, NonceWrap does not anticipate *any* nonces. In that case *only* the phase-2 is executed, and LNA can be disregarded. This variant is close to standard nAE decryption, and is useful when we need a stateless receiver.

5 NonceWrap Security

ALTERNATIVE CHARACTERIZATION OF NAE. We will find it convenient to use the following alternative formulation of nAE security, which directly models multiple keys and more precisely attends to what is possible for a given expansion. Recall that the expansion of an nAE scheme $\mathcal{E}$ is a constant τ such that $|\mathcal{E}_K^{N,A}(M)| = |M| + \tau$. Let $\mathcal{E}$ and τ be an nAE scheme and its expansion. Let $\mathcal{T}$ be an arbitrary nonempty set. Let $\mathrm{Inj}_\tau^{\mathcal{T}}(\mathcal{M})$ be the set of all functions $f : \mathcal{T} \times \mathcal{M} \rightarrow \{0,1\}^*$ such that $|f(T,M)| = |M| + \tau$ for all $M \in \{0,1\}^*$ and $f(T,\cdot)$ is an injection for all $T \in \mathcal{T}$. For $f \in \mathrm{Inj}_\tau^{\mathcal{T}}$ define $f^{-1} : \mathcal{T} \times \{0,1\}^* \rightarrow \mathcal{M} \cup \{\perp\}$ by $f^{-1}(T,Y) = X$ when $f(T,X) = Y$ for some (unique) $X \in \mathcal{M}$, and $f^{-1}(T,Y) = \perp$ otherwise. Now given an adversary $\mathcal{A}$, define its advantage in attacking the nae-security of $\mathcal{E}$ as the real number

$$\mathbf{Adv}_{\mathcal{E}}^{\mathrm{nae}*}(\mathcal{A}) = \Pr[\text{for } i \in \mathbb{N} \text{ do } K_i \leftarrow \mathcal{K} \colon \mathcal{A}^{E_K(\cdot,\cdot,\cdot,\cdot),D_K(\cdot,\cdot,\cdot,\cdot)} \Rightarrow 1]$$
$$- \Pr[f \leftarrow \mathrm{Inj}_\tau^{\mathbb{N}\times\mathcal{N}\times\mathcal{A}}(\mathcal{M}) \colon \mathcal{A}^{E_f(\cdot,\cdot,\cdot,\cdot),D_f(\cdot,\cdot,\cdot,\cdot)} \Rightarrow 1]$$

where the oracles behave as follows: oracle E_K, on query (i,N,A,M), returns $\mathcal{E}(K_i,N,A,M)$; oracle D_K, on query (i,N,A,C), returns $\mathcal{E}^{-1}(K_i,N,A,C)$; oracle E_f, on query (i,N,A,M), returns $f((i,N,A),M)$; and oracle D_f, on query (i,N,A,C), returns $f^{-1}((i,N,A),C)$. The adversary $\mathcal{A}$ is forbidden from asking its first oracle any query (i,N,A,M) if it previously asked a query (i,N,A',M').

It is a standard exercise, following the PRI-characterization of misuse-resistant AE schemes [18], to show the equivalence of nae (presented in Sect. 2) and nae* security.

MULTI-KEY STRONG-PRP SECURITY. It's also useful for us to define a notion of multi-key strong PRP security, which we denote as prp*-security. In customary strong PRP security, like conventional PRP security, the adversary has access to a forward direction oracle that computes a real or ideal permutation. Strong PRP security adds a backward direction oracle that computes the inverse. To adapt this to the multi-key setting, we treat the PRP as a length-preserving PRI. Define $\mathrm{Inj}^{\mathcal{T}}(\{0,1\}^n) = \mathrm{Inj}_0^{\mathcal{T}}(\{0,1\}^n)$. For an adversary $\mathcal{A}$, we define its advantage in attacking the prp*-security of an n-bit PRP E as the real number

$$\mathbf{Adv}_E^{\mathrm{prp}*}(\mathcal{A}) = \Pr[\text{for } i \in \mathbb{N} \text{ do } K_i \leftarrow \mathcal{K} \colon \mathcal{A}^{F_K(\cdot,\cdot),G_K(\cdot,\cdot)} \Rightarrow 1]$$
$$- \Pr[p \leftarrow \mathrm{Inj}^{\mathbb{N}}(\{0,1\}^n) \colon \mathcal{A}^{F_p(\cdot,\cdot),G_p(\cdot,\cdot)} \Rightarrow 1]$$

where the oracles behave as follows: oracle F_K, on query (i,X), returns $E(K_i,X)$; oracle G_K, on query (i,X), returns $E^{-1}(K_i,X)$; oracle F_p, on query (i,X), returns $p(i,X)$; and oracle G_p, on query (i,X), returns $p^{-1}(i,X)$.

NONCEWRAP SECURITY. To show the security of NonceWrap, we establish that its anae-security is good if E is prp*-secure and $\mathcal{E}$ is nae*-secure.

Theorem 1. *There exists a reduction* R, *explicitly given in the proof of this theorem, as follows: Let* $E \colon \{0,1\}^{k_1} \times \{0,1\}^n \rightarrow \{0,1\}^n$ *be a blockcipher, let* $H \colon \{0,1\}^* \rightarrow \{0,1\}^\beta$ *be a hash function, let* $\mathcal{E} \colon \mathcal{K}_{\mathcal{E}} \times \mathcal{N} \times \mathcal{A} \times \mathcal{M} \rightarrow \mathcal{C}_{\mathcal{E}}$ *be an nAE scheme, and let* Nx$\colon \mathcal{N}^{\leq d} \rightarrow \mathcal{P}(\mathcal{N})$ *be a nonce policy with depth d. Let* Lx *be an anticipated-nonce function with the same signature as* Nx *such that* $Lx(\boldsymbol{n}) \subseteq Nx(\boldsymbol{n})$ *for all* $\boldsymbol{n} \in \mathcal{N}^{\leq d}$. *Let* $\Pi = $ NonceWrap$[E,H,\mathcal{E},Lx,Nx]$ *be a NonceWrap scheme. Let σ be the expansion of Π and τ be the expansion of $\mathcal{E}$.*

Let $\mathcal{A}$ be an adversary that attacks Π. Then R transforms $\mathcal{A}$ into a pair of adversaries $(\mathcal{B}_1,\mathcal{B}_2)$ such that

$$\mathbf{Adv}_{\Pi,Nx}^{\mathrm{anae}}(\mathcal{A}) \leq \mathbf{Adv}_E^{\mathrm{prp*}}(\mathcal{B}_1) + \mathbf{Adv}_{\mathcal{E}}^{\mathrm{nae*}}(\mathcal{B}_2)$$

$$+ \frac{q_e^2}{2^{n+1}} + \frac{q_e^2}{2^{\tau+1}} + \frac{q_e^2 + q_d^2}{2^{\sigma+1}} + \frac{q_e^4}{2^{n+\tau+2}}$$

where q_e and q_d are the number of encryption and decryption queries that $\mathcal{A}$ makes. The resource usage of $\mathcal{B}_1$ and $\mathcal{B}_2$ are similar to that of $\mathcal{A}$.

Proof. We define a sequence of hybrid games that transition the real anae game to the ideal anae game, where the games are using Π and Nx. The first of these hybrids, G_1 replaces the blockcipher E with a random function P from $\mathrm{Inj}^{\mathbb{N}}(\{0,1\}^n)$. Note that $P(i,\cdot)$ is an injection for all $i \in \mathbb{N}$ and is length-preserving, so it is a permutation. We construct an adversary $\mathcal{B}_1$ that attacks the blockcipher E by having it simulate these two games. Whenever $\mathcal{A}$ makes a query, $\mathcal{B}_1$ follows the protocol defined in the real anae game. If the query requires a blockcipher operation, $\mathcal{B}_1$ would query its own forward direction oracle and use that output for the operation instead. It can use its backward direction oracle for inverting the blockcipher. At the end, $\mathcal{B}_1$ outputs the same bit $\mathcal{A}$ returns. The advantage of $\mathcal{B}_1$ is equivalent to $\mathcal{A}$'s advantage in distinguishing the games it simulates as the ciphertexts that the simulated encryption oracles would produce would be identical with the exception of the header, which depends on whether $\mathcal{B}_1$'s oracle is using P or the real blockcipher E. With that, we have:

$$\mathbf{Pr}[\, \mathcal{A}^{\mathrm{Real}_{\Pi}^{\mathrm{anae}}} \,] - \mathbf{Pr}[\, \mathcal{A}^{\mathsf{G}_1} \,] \leq \mathbf{Adv}_E^{\mathrm{prp*}}(\mathcal{B}_1)$$

The next hybrid G_2 replaces NonceWrap's underlying nAE scheme $\mathcal{E}$ with a random function F from $\mathrm{Inj}_{\tau}^{\mathbb{N}\times\mathbb{N}\times\mathcal{A}}(\mathcal{M})$. We construct an adversary $\mathcal{B}_2$ that attacks the nAE scheme by simulating the two hybrid games. Like $\mathcal{B}_1$, adversary $\mathcal{B}_2$ will just follow protocol except it replaces any nAE operations with its oracles. For any blockcipher operations, it simulates P as described in the previous step. It returns the same bit that $\mathcal{A}$ returns. The advantage of $\mathcal{B}_2$ is equivalent to $\mathcal{A}$'s advantage in distinguishing the games it simulates as the only difference between the simulated games is how the ciphertext body is produced, which depends on whether $\mathcal{B}_2$'s oracle is using F or the real nAE scheme $\mathcal{E}$. With that, we have:

$$\mathbf{Pr}[\, \mathcal{A}^{\mathsf{G}_2} \,] - \mathbf{Pr}[\, \mathcal{A}^{\mathsf{G}_3} \,] \leq \mathbf{Adv}_{\mathcal{E}}^{\mathrm{nae*}}(\mathcal{B}_2)$$

At this point we have a real anae game using a NonceWrap scheme built on ideal primitives and we want to measure how well $\mathcal{A}$ can distinguish it from the ideal anae game. For the upcoming parts, we modify the ideal game step-by-step until it is indistinguishable from the real game.

The first hybrid, G_7, makes a simple change to the decryption oracle. Referring to the code in Fig. 1, on line 251, there is a condition that the tuple in the history must be unique. This hybrid simply removes the "unique" condition. Instead, if there are multiple valid tuples that map to a queried ciphertext, the oracle will return the lexicographically first tuple instead of returning $\perp$. Clearly, to distinguish between G_7 and the ideal game, $\mathcal{A}$ would need to call decryption on a ciphertext with multiple valid tuples as the former would return a tuple and the latter would return $\perp$. The probability that this occurs is upper-bounded by the probability that two ciphertexts from encryption are the same as multiple tuples need to be mapped to the same ciphertext in H for there to be multiple valid tuples. Hence, the advantage $\mathcal{A}$ has for distinguishing between these two games is

$$\mathbf{Pr}[\ \mathcal{A}^{\text{Ideal}_{\Pi}^{\text{anae}}}\ ] - \mathbf{Pr}[\ \mathcal{A}^{\mathsf{G}_7}\ ] \leq \frac{q_e^2}{2^{\sigma+1}}$$

The next modification only changes how ciphertexts are generated. Instead of randomly sampling from $\{0,1\}^{|M|+\tau}$ on an encryption query, the encryption oracle will instead use a pair of PRIs to generate a "header" and "body" to create the ciphertext. To do this, we modify the code for the ENC oracle to use the procedure F defined in the top half of Fig. 5. The bottom half of the figure shows the modified encryption oracle. The procedure captures the lazy-sampling of the forward direction of a random function or injection depending on whether the code in grey is executed. Without the grey, the code simulates a function for each tweak T; With the grey, it simulates an injection for each T. Having that, we can use F to capture the pair of PRIs: one from $\text{Inj}_{\tau}^{\mathcal{N}\times\mathcal{N}\times\mathcal{A}}(\mathcal{M})$ for creating the body and one from $\text{Inj}^{\mathbb{N}}(\{0,1\}^n)$ for creating the header.

We can think of G_7 as using two different instances of F, which we label as F_E and $F_{\mathcal{E}}$, without the grey to generate a header and body and concatenating the two results. This is the same as generating a random string of the same length since queries to the encryption oracle can't be repeated, so a random header and body is sampled each time. When we replace the random ciphertext generation with the pair of PRIs, we use F_E and $F_{\mathcal{E}}$ with the grey code. We refer to the game using F for the PRIs as G_6.

To distinguish between G_6 and G_7, $\mathcal{A}$ would need to distinguish the difference between F with and without the grey code. This is the probability that bad gets set to true in F. For now, we don't need to worry about F^{-1} as the adversary has no way of accessing it. On the ith encryption query, the probability that bad gets set to true is at most $(i-1)/2^w$. It follows that the probability bad gets set to true is at most $q_e^2/2^{w+1}$ for q_e encryption queries. The adversary may observe this event in either F_E or $F_{\mathcal{E}}$. Thus, $\mathcal{A}$'s advantage here is

$$\mathbf{Pr}[\ \mathcal{A}^{\mathsf{G}_7}\ ] - \mathbf{Pr}[\ \mathcal{A}^{\mathsf{G}_6}\ ] \leq \frac{q_e^2}{2^{n+1}} + \frac{q_e^2}{2^{\tau+1}}$$

Our next hybrid G_5 changes the decryption oracle and is shown in Fig. 6. The other oracles remain the same. Instead of identifying the SID, nonce, and AD using $\text{H}[C]$ right away, the oracle will search for the tuple by going through all $\ell \in \text{L}$, $N \in Nx(\text{ND}[\ell])$, and $A \in \text{A}[\ell] \cup \text{AD}$. For each of those tuples, it will try to invert the injection on Body to recover M. Now it's possible that the inversion results in an M that wasn't recorded in H since F^{-1} as defined in Fig. 5 can return values that weren't given by the forward oracle. However, we check on line 555 to make sure that the (ℓ, N, A, M) we found is actually mapped to C, which is something required to return a valid tuple in G_6's decryption. The other validity conditions on ℓ, N, and A are already accounted for since we iterate through the sets that validate them. We also iterate through them in lexicographic order, which guarantees that if there are multiple valid tuples, we return the lexicographically first one. Essentially, G_5 does the same as G_6's decryption; it just does it in a roundabout way by searching for the tuple. Hence, G_5 and G_6 are indistinguishable from each other to $\mathcal{A}$.

Instead of looping through the permitted nonces and ADs, we can use the header to figure out the nonce and AD. The header as generated in the previous hybrid's encryption contains the nonce and a hash of the AD. This is just like

---

procedure $F(T, X)$
900 **if** $X \parallel 0^{w-u} \in \text{dom}(f(t, \cdot))$ **then**
901   **return** $f(T, X)$
902 $Y \leftarrow \{0, 1\}^v$
903 **if** $Y \in \text{range}(f(T, \cdot))$ **then**
904   bad $\leftarrow$ true
905      $Y \leftarrow \{0, 1\}^v \setminus \text{range}(f(T, \cdot))$
906 $f(T, X \parallel 0^{w-u}) \leftarrow Y$
907 **return** Y

procedure $F^{-1}(T, Y)$
910 **if** $Y \in \text{range}(f(T, \cdot))$
911      $X' \parallel R \leftarrow f^{-1}(T, Y)$
912   **where** $|X'| = u$
913   **if** $R = 0^{w-u}$ **then return** X'
914   **return** $\bot$
915 $X \leftarrow \{0, 1\}^w$
916 **if** $X \in \text{dom}(f(T, \cdot))$ **then**
917   bad $\leftarrow$ true
918      $X \leftarrow \{0, 1\}^v \setminus \text{dom}(f(T, \cdot))$
919 $f(T, X) \leftarrow Y$
91A $X' \parallel R \leftarrow X$ **where** $|X'| = u$
91B **if** $R = 0^{w-u}$ **then return** X'
91C **return** $\bot$

---

procedure $G_6.\text{ENC}(\ell, N, A, M)$
640 **if** $\ell \notin L$ **or** $N \in \text{NE}[\ell]$ **then return** $\bot$
641 $\text{NE}[\ell] \overset{\cup}{\leftarrow} \{N\}$
642 Head $\leftarrow F_E(\ell, N \parallel 0^\rho \parallel H(A))$
643 Body $\leftarrow F_{\mathcal{E}}((\ell, N, A), M)$
644 $C \leftarrow$ Head $\parallel$ Body
645 $H[C] \overset{\cup}{\leftarrow} \{(\ell, N, A, M)\}$; **return** C

Fig. 5. Top. Lazy-sampling of random functions or injections in the multi-key setting. With the code in grey, the procedures simulate a random injection for each T from u bits to w bits. Without the code in grey, the procedures simulate a random function for each T. **Bottom.** Modified encryption oracle that uses either random functions or random injections to generate the ciphertext. Here, $\rho = n - \eta - \beta$ where η is the length of the nonce. The game using injections is called G_6.

---

procedure $\mathsf{G}_5.\mathrm{DEC}(C)$

550 Head $\|$ Body $\leftarrow C$ **where** $|\mathrm{Head}| = n$

551 **for** $\ell \in \mathrm{L}$ **do**

552   **for** $N \in Nx(\mathrm{ND}[\ell])$ **do**

553   **for** $A \in \mathrm{A}[\ell] \cup \mathrm{AD}$ **do**

554        $M \leftarrow F_{\mathcal{E}}^{-1}((\ell, N, A), \mathrm{Body})$

555   **if** $(\ell, N, A, M) \in \mathrm{H}[C]$ **then**

556          $\mathrm{ND}[\ell] \overset{\|}{\leftarrow} N$

557   **if** $\mathrm{ND}[\ell] \notin \mathrm{dom}(Nx)$ **then** $\mathrm{ND}[\ell] \leftarrow \mathrm{tail}(\mathrm{ND}[\ell])$

558   **return** (ℓ, N, A, M)

559 **return** $\bot$

Fig. 6. G_5's **decryption oracle.** This decryption oracle searches for a (ℓ, N, A) triple to use to recover M. It then validates the resulting quadruple by making sure that it maps to the ciphertext in the history H.

---

procedure $\mathsf{G}_4.\mathrm{ASSO}(A)$

420 $B \leftarrow H(A);\ \mathrm{AD}[B] \overset{\cup}{\leftarrow} \{A\}$

procedure $\mathsf{G}_4.\mathrm{ASSO}(A, \ell)$

421 $B \leftarrow H(A);\ \mathrm{A}[\ell][B] \overset{\cup}{\leftarrow} \{A\}$

procedure $\mathsf{G}_4.\mathrm{DISA}(A)$

430 $B \leftarrow H(A);\ \mathrm{AD}[B] \overset{\frown}{\leftarrow} \{A\}$

procedure $\mathsf{G}_4.\mathrm{DISA}(A, \ell)$

431 $B \leftarrow H(A);\ \mathrm{A}[\ell][B] \overset{\frown}{\leftarrow} \{A\}$

procedure $\mathsf{G}_4.\mathrm{DEC}(C)$ *Resembles phase-2*

450 Head $\|$ Body $\leftarrow C$ **where** $|\mathrm{Head}| = n$

451 **for** $\ell \in \mathrm{L}$ **do**

452    $N \| R \| B \leftarrow F_E^{-1}(\ell, \mathrm{Head})$

453   **where** $|N| = \eta$ **and** $|R| = r$

454   **if** $R \neq 0^\rho$ **or** $N \notin Nx(\mathrm{ND}[\ell])$

455   **then continue**

456   **for** $A \in \mathrm{A}[\ell][B] \cup \mathrm{AD}[B]$ **do**

457      $M \leftarrow f_{\mathcal{E}}^{-1}((\ell, N, A), \mathrm{Body})$

458   **if** $(\ell, N, A, M) \in \mathrm{H}[C]$ **then**

459        $\mathrm{ND}[\ell] \overset{\|}{\leftarrow} N$

45A   **if** $\mathrm{ND}[\ell] \notin \mathrm{dom}(Nx)$ **then**

45B $\mathrm{ND}[\ell] \leftarrow \mathrm{tail}(\mathrm{ND}[\ell])$

45C   **return** (ℓ, N, A, M)

45D **return** $\bot$

Fig. 7. G_4's **decryption oracle.** This decryption oracle resembles phase-2 of NonceWrap. Functionally, it does what the ideal decryption oracle does except instead of looking up a valid tuple in the ciphertext history it iterates through every possibility to search for one.

in NonceWrap encryption. We make modifications to the decryption oracle to do just this. For us to use the AD hash, we also need to modify the ASSO and DISA oracles. The result of these modifications leaves us with hybrid G_4, which is presented in Fig. 7.

Note that decryption now resembles phase-2 of NonceWrap decryption. It's clear that any session it returns is active and any nonce it returns is within the policy as the former is found through iteration and there is an explicit check of the latter. It's also clear that any AD that it returns is registered as $A[\ell][B] \cup AD[B]$ is a subset of all the ℓ's ADs and all the global ADs.

But does G_4 decryption always behave like G_5's decryption? If queried with a C that did not come from the encryption oracle, then both of them return $\perp$ as they both check to make sure $(\ell, N, A, M) \in H[C]$ before returning a tuple. If queried with a C that did, assuming that C was made with an active session key, a nonce under the session's policy, and a properly registered AD, then both decryptions return the same tuple. It's clear that G_5 will find the first lexicographic tuple due to its iteration. If there's only one valid tuple explaining C, then, trivially, the first tuple is returned.

But if there are multiple valid tuples, what happens? If the tuples are under different SIDs, then we arrive at the lexicographically first SID by iteration. If the SIDs are the same, then the header is deciphered and the nonce and AD hash are found. This SID can only have one valid nonce mapped to this header since the header was generated by an injection. Even though G_5 doesn't decipher the header, it still checks the association between nonce and header since it checks whether the tuple is in $H[C]$. This means that G_5, for a fixed session, can only find one nonce—the same nonce as G_4—that is in $H[C]$ even if it iterated through the entirety of the policy. Similarly, the SID can only have one AD hash mapped to this header for the same reason. Even though G_5 iterates through all registered ADs, the ones that it finds that are in $H[C]$ would have their hashes associated to the header. Since G_4 lexicographically iterates through the $A[\ell][B] \cup AD[B]$ subset of registered ADs, it would arrive at the same AD as G_5. Hence, G_5 and G_4 always arrive at the same result for a given ciphertext, making the two indistinguishable.

The next modification adds dictionary LNA from NonceWrap into the game. To start, suppose that we add LNA into the ideal game without actually using it for decryption yet. All other data structures that are needed to support LNA are already exist in our hybrids up to this point; we already manage the active SIDs in the set L and the nonce history of a session in $ND[\ell]$. The structures for ADs were modified from sets into dictionaries in G_4, but we can still derive the set of all valid ADs for a session ℓ from them. The union of all sets in $A[\ell]$.values $\cup$ AD.values is just that. We'll denote this set as $\mathcal{A}_\ell$. All of these data structures are needed to add or remove tuples from LNA. The code for this hybrid G_3 is presented in Fig. 8, but disregard the phase-1 decryption block for now. First, we want to assert a property of LNA.

Lemma 2. *Let* L, ND, A, AD, *and* LNA *be the data structures used in hybrid game* G_3. *Let* $\mathcal{X}$ *be the union of all sets in* LNA.values. *For any SID* ℓ, *let* $\mathcal{A}_\ell$ *be the union of all sets in* $A[\ell]$.values $\cup$ AD.values. *If* $(\ell, N, A) \in \mathcal{X}$ *then* $\ell \in$ L, $N \in Nx(ND[\ell])$, *and* $A \in \mathcal{A}_\ell$.

Proof. Suppose there exists some $(\ell, N, A) \in \mathcal{X}$ such that one of the conditions described in the lemma is false. There are two ways that this can happen: either a value was added into LNA that violated one of the conditions or the condition itself was modified, but LNA was not modified accordingly. We exhaustively check for a case in which this can occur, specifically looking at when we add a tuple or modify the condition.

- Case: $(\ell, N, A) \in \mathcal{X}$ and $\ell \notin L$.
  - When tuple is added in INIT, $\ell \in L$ since INIT adds it to L.
  - When tuple is added in ASSO(A), $\ell \in L$ since the procedure iterates through ℓ to add it.
  - When tuple is added in ASSO(A, ℓ), $\ell \in L$ by assumption.
  - When tuple is added in DEC, $\ell \in L$ since the tuple is added on successful decryption, which happens by iterating through L and finding ℓ.
  - When ℓ is removed from L, all tuples with ℓ as an element are removed from LNA.
- Case: $(\ell, N, A) \in \mathcal{X}$ and $A \notin \mathcal{A}_\ell$.
  - When tuple is added in INIT, $A \in \mathcal{A}_\ell$ since the procedure iterates through AD to get A.
  - When tuple is added in ASSO(A), $A \in \mathcal{A}_\ell$ since the procedure adds A to AD before adding the tuple to LNA.
  - When tuple is added in ASSO(A, ℓ), $A \in \mathcal{A}_\ell$ since the procedure adds A to A[ℓ] before adding the tuple to LNA.
  - When tuple is added in DEC, $A \in \mathcal{A}_\ell$ since the procedure iterates through $\mathcal{A}_\ell$ to add each A.
  - When A is removed in DISA(A), all tuples with A as an element are removed from LNA.
  - When A is removed in DISA(A, ℓ), all tuples with both ℓ and A are removed from LNA. If a tuple containing A is still in $\mathcal{X}$, then it must have a different SID from ℓ.
- Case: $(\ell, N, A) \in \mathcal{X}$ and $N \notin Nx(\text{ND}[\ell])$.
  - When tuple is added in INIT, $N \in Nx(\text{ND}[\ell])$ since ND[ℓ] is initialized to the empty list and the procedure iterates over $Lx(\Lambda)$, which is a subset of $Nx(\Lambda)$.
  - When tuple is added in either ASSO, $N \in Nx(\text{ND}[\ell])$ since the procedure iterates through each nonce in $Lx(\text{ND}[\ell])$, which is a subset of $Nx(\text{ND}[\ell])$.
  - When tuple is added in DEC, ND[ℓ] is appended with a new nonce N' first. Two sets are generated here: $Lx(\text{ND}[\ell])$ and $Lx(\text{ND}[\ell] \parallel N')$. The former is Old and the latter is New in the pseudocode. The procedure iterates over New \ Old, which is a subset of $Nx(\text{ND}[\ell] \parallel N')$ when adding new tuples.
  - When tuple is removed in DEC, the sets Old and New are used again. The procedure iterates over Old \ New and removes tuples containing those nonces from LNA. Hence, any tuple with a nonce not in $Lx(\text{ND}[\ell] \parallel N')$ is removed.

None of these cases provide a situation where $(\ell, N, A) \in \mathcal{X}$ such that $\ell \notin L$, $N \notin Nx(\text{ND}[\ell])$, or $A \notin \mathcal{A}_\ell$. The lemma follows. $\square$

procedure G_3.INIT()

300 $K \xleftarrow{\$} \mathcal{K}$

301 $\ell \leftarrow \Pi.\text{Init}(K)$

302 $K[\ell] \leftarrow K$; $L \xleftarrow{\cup} \{\ell\}$; $NE[\ell] \leftarrow \emptyset$

303 **for** $N \in Lx(\Lambda)$ **do**

304   **for** ADs $\in$ AD.values **do**

305   **for** $A \in$ ADs **do**

306   head $\leftarrow N \parallel 0^\rho \parallel H(A)$

307   Head $\leftarrow F_E(\ell, \text{head})$

308   LNA[Head] $\xleftarrow{\cup} \{(\ell, N, A)\}$

309 **return** ℓ

procedure G_3.TERM(ℓ)

310 **for** $S \in$ LNA.values **do**

311      $S \xleftarrow{\text{--}} \{\ell\} \times \mathcal{N} \times \mathcal{A}$

312 $L \xleftarrow{\text{--}} \{\ell\}$

procedure G_3.ASSO(A)

320 $B \leftarrow H(A)$; AD$[B] \xleftarrow{\cup} \{A\}$

321 **for** $\ell \in L$ **do**

322   **for** $N \in Lx(\text{ND}[\ell])$ **do**

323   Head $\leftarrow F_E(\ell, N \parallel 0^\rho \parallel B)$

324   LNA[Head] $\xleftarrow{\cup} \{(\ell, N, A)\}$

procedure G_3.ASSO(A, ℓ)

325 $B \leftarrow H(A)$; A$[\ell][B] \xleftarrow{\cup} \{A\}$

326 **for** $N \in Lx(\text{ND}[\ell])$ **do**

323   Head $\leftarrow F_E(\ell, N \parallel 0^\rho \parallel B)$

324   LNA[Head] $\xleftarrow{\cup} \{(\ell, N, A)\}$

procedure G_3.DISA(A)

330 $B \leftarrow H(A)$; AD$[B] \xleftarrow{\text{--}} \{A\}$

331 **for** $S \in$ LNA.values **do**

332      $S \xleftarrow{\text{--}} \mathcal{L} \times \mathcal{N} \times \{A\}$

procedure G_3.DISA(A, ℓ)

333 $B \leftarrow H(A)$; A$[\ell][B] \xleftarrow{\text{--}} \{A\}$

334 **for** $S \in$ LNA.values **do**

335      $S \xleftarrow{\text{--}} \{\ell\} \times \mathcal{N} \times \{A\}$

procedure G_3.DEC(C)   *Phase-1*

350 Head $\parallel$ Body $\leftarrow C$ **where** $|\text{Head}| = n$

351 **for** $(\ell, N, A) \in$ LNA[Head] **do**

352      $M \leftarrow F_{\mathcal{E}}^{-1}((\ell, N, A), \text{Body})$

353   **if** $(\ell, N, A, M) \in H[C]$ **then**

354   **goto** 35F

355 **for** $\ell \in L$ **do**   *P-2, same as G_4's*

356      $N \parallel R \parallel B \leftarrow F_E^{-1}(\ell, \text{Head})$

357   **where** $|N| = \eta$ **and** $|R| = r$

358   **if** $R \neq 0^r$ **or** $N \notin Nx(\text{ND}[\ell])$

359   **then continue**

35A   **for** $A \in A[\ell][B] \cup AD[B]$ **do**

35B $M \leftarrow F_{\mathcal{E}}^{-1}((\ell, N, A), \text{Body})$

35C   **if** $(\ell, N, A, M) \in H[C]$ **then**

35D   **goto** 35F

35E **return** $\perp$

35F Old $\leftarrow Lx(\text{ND}[\ell])$ *Phase-3*

35G $\text{ND}[\ell] \xleftarrow{\parallel} N$

35H **if** $|\text{ND}[\ell]| > d$ **then**

35I $\text{ND}[\ell] \leftarrow \text{tail}(\text{ND}[\ell])$

35J New $\leftarrow Lx(\text{ND}[\ell])$

35K **for** $N' \in$ Old $\setminus$ New **do**

35L   **for** $S \in$ LNA.values **do**

35M $S \xleftarrow{\text{--}} \{\ell\} \times \{N'\} \times \mathcal{A}$

35N **for** $N' \in$ New $\setminus$ Old **do**

35O   **for** $B \in A[\ell].\text{keys} \cup AD.\text{keys}$ **do**

35P   Head $\leftarrow F_E(\ell, N' \parallel 0^\rho \parallel B)$

35Q   **for** $A' \in A[\ell][B] \cup AD[B]$ **do**

35R   LNA.[Head] $\xleftarrow{\cup} \{(\ell, N', A')\}$

35S **return** (ℓ, N, A, M)

Fig. 8. Hybrid game resembling NonceWrap. Game G_3 executes procedures similar to those of NonceWrap. For decryption on a ciphertext to succeed, it follows the ideal game. If decryption returns a tuple, then that tuple must have been used to make the queried ciphertext. The encryption oracle is omitted as it is the same as G_5's, which is in Fig. 6.

As per Lemma 2, we have that all tuples recorded in LNA satisfy the validity conditions in ideal decryption. Now when phase-1 decryption is accounted for in G_3 we observe that any successful decryption that occurs must have happened on a tuple in LNA, meeting the validity conditions. Here, success is defined as executing the **goto** instruction on line 354, which instructs the procedure to enter phase-3. The third phase does not modify the tuple being returned in any way; it only does bookkeeping to update the data structures, making sure that they are compliant to the validity conditions. So, whatever tuple was acquired in phase-1 would be returned. If no tuple was found in phase-1, the procedure will enter phase-2 where it iterates through every session as done in G_4's decryption. Whether the valid tuple (ℓ, N, A, M) being returned is found in phase-1 or phase-2, the conditions placed on each component of the tuple remains the same: ℓ must be in L, N must be in $Nx(\mathrm{ND}[\ell])$, A must be in $\mathrm{A}[\ell] \cup \mathrm{AD}$, and the entire tuple must be in $\mathrm{H}[C]$. Thus, G_3 decryption always returns a valid tuple under the same conditions as G_4.

However, in some cases, G_3 does not return the lexicographically first tuple. Suppose that the adversary makes two encryption queries with tuples T_1 and T_2 such that the tuples are different and their parameters are valid for decryption. Suppose it gets back the same ciphertext C both times. Let's say T_1 is the lexicographically first tuple, but its nonce is not within $Lx(\cdot)$. Let's say T_2's nonce *is* within $Lx(\cdot)$. When the adversary queries decryption with C, in G_4, it gets back T_1. On the other hand, it gets back T_2 in G_3 since phase-1 decryption would find T_2 first. The probability this occurs is upper-bounded by the probability of getting the same ciphertext from the encryption oracle, which occurs if the same header and body are outputted by their respective injections. In regards to just the header, the probability that any two headers is the same is $1/2^n$. After q_e encryption queries, any of those pairs of queries can have such a collision. There are about $q_e^2/2$ ways to choose such a pair. Applying the same logic to the ciphertext body, $\mathcal{A}$ gets a collision in both header and body and distinguishes the two hybrids with probability

$$\mathbf{Pr}[\,\mathcal{A}^{G_4}\,] - \mathbf{Pr}[\,\mathcal{A}^{G_3}\,] \leq \frac{q_e^2}{2^{n+1}} \cdot \frac{q_e^2}{2^{\tau+1}} = \frac{q_e^4}{2^{n+\tau+2}} \quad .$$

Observe that G_3 executes almost exactly the same as G_2, which is the real game with ideal primitives does. The only differences in code are the checks for successful decryption. On lines 353 and 35C for G_3, we verify that the tuple was actually used in encryption. On the other hand, in G_2, we move to phase-3 if $M \neq \perp$. This difference can result in the two returning different values. More precisely, if queried with a ciphertext C that was not the result of an encryption query, G_2 may return a tuple while G_3 would never return a tuple. The probability this occurs is upper-bounded by the probability that the function $F_{\mathcal{E}}^{-1}$ on query (T, Y) returns a non-$\perp$ value given that Y was not an output of $F_{\mathcal{E}}$. This is the probability that line 91B in the top half of Fig. 5 returns. That is, the advantage $\mathcal{A}$ has in distinguishing G_3 and G_2 is

$$\mathbf{Pr}[\ \mathcal{A}^{\mathsf{G}_3}\ ] - \mathbf{Pr}[\ \mathcal{A}^{\mathsf{G}_2}\ ] \leq \frac{q_d^2}{2^{\sigma+1}}$$

Summing up all of the bounds computed over the hybrid argument, we get the bound in the theorem statement. □

6 Remarks

COMPLEXITY. While we don't find the anAE definition excessively complex, NonceWrap decryption is quite complicated. One complicating factor is the rich support we have provided for AD values—despite our expectation that implementations will assume the 1AD/Session restriction. Yet we have found that building in the 1AD/Session restriction would only simplify matters modestly. It didn't seem worth it.

We suspect that, no matter what, decryption in anonymous-AE schemes is going to be complicated compared to decryption under conventional nAE. The privacy principle demands that ciphertexts contain everything the receiver needs to decrypt, yet no adversarially worthwhile metadata. The decrypting party must infer this metadata, and it should do so quite efficiently.

TIMING SIDE-CHANNELS. Our anAE definition does not address timing side-channels, and NonceWrap raises several concerns with leaking identity information through decryption times. Timing information might leak how many sessions a header can belong to. In phase-2, nAE decryption is likely to be the operation that takes the longest, and it is possible that an observer might learn information on the number of sessions that produced a valid-looking header. Then there is the timing side-channel that arises from the usage of Lx and Nx. Phase-1 only works on headers in Lx, and is expected to be faster than phase-2, leaking information about whether a nonce was anticipated. We leave the modeling, analysis, and elimination of timing side-channels as an open problem.

THE USAGE PUZZLE. There is an apparent paradox in the use of anonymous AE. If used in an application-layer protocol over something like TCP/IP, then anonymous AE would seem irrelevant because communicated packets already reveal identity. But if used over an anonymity layer like Tor [9], then use of that service would seem to obviate the need for privacy protection. It would seem as though anonymous AE is pointless if the transport provides anonymity, and that pointless if the transport does not provide anonymity.

This reasoning is specious. First, an anonymity layer like Tor only protects a packet while it traverses the Tor network; once it leaves an exit node, the Tor-associated encryption is gone, and end-to-end privacy may still be desired. Second, it simply is not the case that every low-level transport completely leaks identity. For example, while a UDP packet includes a source port, the field need not be used.

To give a concrete example for potential use, consider how NonceWrap (and anAE in general) might fit in with DTLS 1.3 over UDP [15]. Unlike TLS, where session information is presumptively gathered from the underlying transport, DTLS transmits with each record an explicit (sometimes partially explicit) epoch and sequence number (SN). Since UDP itself does not use SNs, the explicit SNs of DTLS are used for replay protection. While DTLS has a mechanism for SN encryption in its latest draft, NonceWrap would seem to improve upon it. The way DTLS associates a key with encrypted records is through the sender's IP and port number at the UDP level. Using NonceWrap, these identifiers could be omitted. If the receiver needs to know source IP and port in order to reply, those values can be moved to the encrypted payload.

Further features of DTLS over UDP might be facilitated by NonceWrap. It provides a mechanism in which an invalid record can often be quickly identified, a feature useful in DTLS. In DTLS, when an SN greater than the next expected one is received, there is an option to either discard the message or keep it in a queue for later. This aligns with NonceWrap's formulation of Lx and Nx.

It is rarely straightforward to deploy encryption in an efficient, privacy-preserving way, and anAE is no panacea. But who's to say how privacy protocols might evolve if one of our most basic tools, AE, is re-envisioned as something more privacy friendly?

Acknowledgments. We thank Dan Bernstein for inspiring this work. Within the CAESAR call, he suggested the use of "secret message numbers" in lieu of nonces; in private communications with the second author, he asked how one might efficiently demultiplex multiple AE communication streams without having marked them in a privacy-compromising manner.

We thank the anonymous ASIACRYPT referees. Their comments brought home that anonymous AE was a concern that transcended our formulation of it. They suggested the name anAE.

This work was supported by NSF CNS 1717542 and NSF CNS 1314855. Many thanks to the NSF for their years of financial support.

References

1. Abadi, M., Rogaway, P.: Reconciling two views of cryptography (the computational soundness of formal encryption). J. Cryptol. **15**(2), 103–127 (2002)
2. Abed, F., Forler, C., Lucks, S.: General classification of the authenticated encryption schemes for the CAESAR competition. Comput. Sci. Rev. **22**, 13–26 (2016)
3. Bellare, M., Boldyreva, A., Desai, A., Pointcheval, D.: Key-privacy in public-key encryption. In: Boyd, C. (ed.) ASIACRYPT 2001. LNCS, vol. 2248, pp. 566–582. Springer, Heidelberg (2001). https://doi.org/10.1007/3-540-45682-1_33
4. Bellare, M., Kohno, T., Namprempre, C.: Breaking and provably repairing the SSH authenticated encryption scheme: a case study of the encode-then-encrypt-and-MAC paradigm. ACM Trans. Inf. Syst. Secur. **7**(2), 206–241 (2004)
5. Bellare, M., Ng, R., Tackmann, B.: Nonces are noticed: AEAD revisited. In: Boldyreva, A., Micciancio, D. (eds.) CRYPTO 2019, Part I. LNCS, vol. 11692, pp. 235–265. Springer, Cham (2019). https://doi.org/10.1007/978-3-030-26948-7_9

6. Bellare, M., Rogaway, P.: Encode-then-encipher encryption: how to exploit nonces or redundancy in plaintexts for efficient cryptography. In: Okamoto, T. (ed.) ASIACRYPT 2000. LNCS, vol. 1976, pp. 317–330. Springer, Heidelberg (2000). https://doi.org/10.1007/3-540-44448-3_24

7. Bernstein, D.: Cryptographic competitions: CAESAR call for submissions. Webpage, January 2014. https://competitions.cr.yp.to/caesar-call.html

8. Boyd, C., Hale, B., Mjølsnes, S.F., Stebila, D.: From Stateless to Stateful: Generic Authentication and Authenticated Encryption Constructions with Application to TLS. In: Sako, K. (ed.) CT-RSA 2016. LNCS, vol. 9610, pp. 55–71. Springer, Cham (2016). https://doi.org/10.1007/978-3-319-29485-8_4

9. Dingledine, R., Mathewson, N., Syverson, P.F.: Tor: The second-generation onion router. In: Blaze, M. (ed.) Proceedings of the 13th USENIX Security Symposium, August 9–13, 2004, San Diego, CA, USA, pp. 303–320. USENIX (2004)

10. Katz, J., Yung, M.: Unforgeable encryption and chosen ciphertext secure modes of operation. In: Goos, G., Hartmanis, J., van Leeuwen, J., Schneier, B. (eds.) FSE 2000. LNCS, vol. 1978, pp. 284–299. Springer, Heidelberg (2001). https://doi.org/10.1007/3-540-44706-7_20

11. Kohno, T., Palacio, A., Black, J.: Building secure cryptographic transforms, or how to encrypt and MAC. IACR Cryptology ePrint Archive 2003:177 (2003)

12. McGrew, D.: An interface and algorithms for authenticated encryption. IETF RFC 5116, January 2018

13. Namprempre, C., Rogaway, P., Shrimpton, T.: AE5 security notions: definitions implicit in the CAESAR call. IACR Cryptology ePrint Archive 2013:242 (2013)

14. Namprempre, C., Rogaway, P., Shrimpton, T.: Reconsidering generic composition. In: Nguyen, P.Q., Oswald, E. (eds.) EUROCRYPT 2014. LNCS, vol. 8441, pp. 257–274. Springer, Heidelberg (2014). https://doi.org/10.1007/978-3-642-55220-5_15

15. Rescorla, E., Tschofenig, H., Modadugu, N.: The datagram transport layer security (DTLS) protocol version 1.3. Internet-Draft draft-ietf-tls-dtls13-31, Internet Engineering Task Force, March 2019. Work in Progress

16. Rogaway, P.: Authenticated-encryption with associated-data. In: Atluri, V. (ed.) ACM CCS 02: 9th Conference on Computer and Communications Security, pp. 98–107. ACM Press, Washington D.C., 18–22 November 2002

17. Rogaway, P., Bellare, M., Black, J., Krovetz, T.: OCB: a block-cipher mode of operation for efficient authenticated encryption. In: ACM CCS 01: 8th Conference on Computer and Communications Security, pp. 196–205. ACM Press, Philadelphia, 5–8 November 2001

18. Rogaway, P., Shrimpton, T.: Deterministic authenticated-encryption: a provable-security treatment of the key-wrap problem. Cryptology ePrint Archive, Report 2006/221 (2006). http://eprint.iacr.org/2006/221

19. Rogaway, P., Shrimpton, T.: A provable-security treatment of the key-wrap problem. In: Vaudenay, S. (ed.) EUROCRYPT 2006. LNCS, vol. 4004, pp. 373–390. Springer, Heidelberg (2006). https://doi.org/10.1007/11761679_23

20. Rogaway, P., Zhang, Y.: Simplifying game-based definitions. In: Shacham, H., Boldyreva, A. (eds.) CRYPTO 2018, Part II. LNCS, vol. 10992, pp. 3–32. Springer, Cham (2018). https://doi.org/10.1007/978-3-319-96881-0_1

Sponges Resist Leakage: The Case of Authenticated Encryption

Jean Paul Degabriele[1(✉)], Christian Janson[2], and Patrick Struck[3]

[1] CNS, Technische Universität Darmstadt, Darmstadt, Germany
jeanpaul.degabriele@crisp-da.de
[2] Cryptoplexity, Technische Universität Darmstadt, Darmstadt, Germany
christian.janson@cryptoplexity.de
[3] CDC, Technische Universität Darmstadt, Darmstadt, Germany
pstruck@cdc.informatik.tu-darmstadt.de

Abstract. In this work we advance the study of leakage-resilient Authenticated Encryption with Associated Data (AEAD) and lay the theoretical groundwork for building such schemes from sponges. Building on the work of Barwell et al. (ASIACRYPT 2017), we reduce the problem of constructing leakage-resilient AEAD schemes to that of building fixed-input-length function families that retain pseudorandomness and unpredictability in the presence of leakage. Notably, neither property is implied by the other in the leakage-resilient setting. We then show that such a function family can be combined with standard primitives, namely a pseudorandom generator and a collision-resistant hash, to yield a nonce-based AEAD scheme. In addition, our construction is quite efficient in that it requires only two calls to this leakage-resilient function per encryption or decryption call. This construction can be instantiated entirely from the T-sponge to yield a concrete AEAD scheme which we call SLAE. We prove this sponge-based instantiation secure in the non-adaptive leakage setting. SLAE bears many similarities and is indeed inspired by ISAP, which was proposed by Dobraunig et al. at FSE 2017. However, while retaining most of the practical advantages of ISAP, SLAE additionally benefits from a formal security treatment.

Keywords: AEAD · Leakage Resilience · Side channels · SLAE · ISAP

1 Introduction

The oldest and most fundamental application of cryptography is concerned with securing the communication between two parties who already share a secret key. The modern cryptographic construct for this application is authenticated encryption with associated data (AEAD), which was the topic of the recent CAESAR competition [6]. Most of the effort in this competition has been directed towards exploring new designs, optimising performance, and offering robust security guarantees. However, there has not been much progress in the development of AEAD constructions that, by design, protect against side-channel attacks.

© International Association for Cryptologic Research 2019
S. D. Galbraith and S. Moriai (Eds.): ASIACRYPT 2019, LNCS 11922, pp. 209–240, 2019.
https://doi.org/10.1007/978-3-030-34621-8_8

This is a challenging problem that is likely to become a primary focus in the area of AEAD design.

Recently, a handful of AEAD designs with this exact goal have emerged. Each of these is based on a different approach with varying trade-offs between complexity, efficiency, and security guarantees. One notable example is the work of Barwell et al. [4], which proposes AEAD constructions with strong security guarantees but pays a relatively high price in terms of complexity and efficiency. Specifically, their constructions achieve security against adaptive leakage but resort to elliptic-curve pairings and secret sharing in order to realise implementations of a leakage-resilient MAC and a leakage-resilient pseudorandom function (employed in a block-wise fashion for encryption) for instantiating their scheme. A more hands-on approach was adopted by Dobraunig et al. in the design of their proposed AEAD scheme ISAP. It was conceived with the intent to protect against Differential Power Analysis (DPA) [10]. ISAP is entirely sponge-based and follows a fairly conventional design, augmented with a rekeying strategy. Arguably, this simpler approach, employing readily-available symmetric primitives, is more likely to lead to a pragmatic solution. However, ISAP's design rationale is predominantly heuristic, lacking any formal security analysis to justify its claims. As such the efficacy of ISAP's approach in resisting side-channel attacks is unclear, both qualitatively and quantitatively, curtailing any objective comparison with the constructions from [4] and others.

In light of the practical advantages that the sponge-based approach offers, we remedy this state of affairs as follows. We propose SLAE, a derivative of ISAP which retains its main structure and benefits but includes certain modifications to admit a formal security proof. We analyse its security in the framework of leakage-resilient cryptography introduced by Dziembowski and Pietrzak [13], adapted to the random transformation model. Specifically, we prove it secure with respect to the leakage-resilient AEAD definition, put forward in [4] by Barwell et al., in the non-adaptive leakage setting. That is, we assume a leakage function that is fixed a priori and whose output is limited to some number of bits λ.

Admittedly, SLAE achieves qualitatively weaker security than the schemes of Barwell et al., since it only achieves non-adaptive leakage resilience. Nevertheless, we contend that SLAE strikes a more pragmatic balance by improving on efficiency and ease of implementation while still benefiting from a provably-secure design. Indeed, several other works [1,12,14,22,24] have settled for and argued that non-adaptive leakage security often suffices in practice. Moreover, as discussed in [24], the syntax of primitives like pseudorandom functions makes adaptive-leakage security impossible to achieve. In fact Barwell et al. achieve security against adaptive leakage by resorting to a specialised implementation of a pseudorandom function which requires an additional random input per invocation. In contrast, SLAE adheres to the standard nonce-based AEAD syntax and requires no source of randomness.

When viewed as sponge-based constructions, SLAE and ISAP look very similar and we do not claim any particular novelty in that respect. Nevertheless, the rationale behind their design is rather different. ISAP was conceived as augmenting a standard sponge-based AEAD design with a rekeying strategy, where the rekeying function is in turn also built from sponges, followed by some optimisations. The rekeying is intended to frustrate Differential Power Analysis (which requires several power traces on the same key but distinct inputs) by running the AEAD scheme with a distinct session key each time its inputs change. In turn, the session key is produced by combining a hash of the inputs and the master key through a rekeying function. Ostensibly, the rekeying function is itself strengthened against DPA by reducing its input data complexity through a low sponge absorption rate. In contrast SLAE is understood through a top-down design where we gradually decompose a leakage-resilient AEAD scheme into smaller components which we then instantiate using sponges. In particular there is no mention of rekeying or session keys. Note that there is more to this distinction than mere renaming. For instance, if we compare the MAC components in ISAP and SLAE we notice that the same value that serves as the MAC session key in ISAP is used directly as the MAC tag in SLAE.

At a more general level, the key premise made in [10] is that sponges offer a promising and practical solution to protect against side-channel attacks. Our work serves to provide formal justification to this claim and allows one to calculate concrete parameters for a desired security level.

1.1   Contribution

Below is an outline of our contributions highlighting how we improve on prior works and some of the challenges we face in our analysis.

A Generic Construction (FGHF′). The composition theorem in [4] reconsiders the N2 construction from [19] in the setting of leakage resilience. Specifically they show that given a MAC that is both *leakage-resilient strongly unforgeable* and a *leakage-resilient pseudorandom function*, together with an encryption scheme that is *leakage-resilient against augmented chosen plaintext attacks*, the N2 construction yields a leakage-resilient AEAD scheme. We extend this result, in the non-adaptive setting, by further decomposing the MAC and the encryption scheme into simpler lower-level primitives, ultimately giving rise to the FGHF′ construction. In turn this constructs a leakage-resilient AEAD scheme from two *fixed-size* leakage-resilient functions $\mathcal{F}$ and $\mathcal{F}'$, a standard pseudorandom generator $\mathcal{G}$, and a collision-resistant vector hash $\mathcal{H}$. The construction requires that both $\mathcal{F}$ and $\mathcal{F}'$ be leakage-resilient pseudorandom functions and that $\mathcal{F}'$ additionally be a leakage-resilient unpredictable function. The latter is a notion that we introduce.

As pointed out in [4], in the adaptive leakage setting any MAC whose verification algorithm recomputes the tag and checks for equality with the candidate tag, simply cannot be strongly unforgeable. They overcome this issue through an ingenious MAC implementation. However this requires three pairing evaluations per verification and a source of randomness. In the FGHF′ construction we show that by settling for non-adaptive leakage security the canonical MAC construction, which recomputes the tag and checks for equality, can be rescued. Specifically, we show that any leakage-resilient unpredictable function gives rise to a canonical MAC which is strongly unforgeable. In contrast to the leakage-free setting, not every pseudorandom function is an unpredictable function. This has to do with the fact that in unpredictability we give the adversary more freedom in what it can query to its oracles, which is in turn a necessary requirement for composition to hold. In addition, we prove that one can combine a collision-resistant hash function with fixed-input-length leakage-resilient pseudorandom and unpredictable functions to obtain corresponding primitives with extended input domains.

For the encryption part, Barwell et al. use Counter Feedback Mode instantiated with a leakage-resilient pseudorandom function and an additional extra call to generate the initial vector from the nonce. Thus multiple calls to the leakage-resilient pseudorandom function are required for each encryption call. In contrast we show that to meet the required security notion, one can do with just one call to the leakage-resilient pseudorandom function and a pseudorandom generator, thereby resulting in a considerably more efficient scheme. Thus, if one is content with non-adaptive leakage security then the FGHF′ construction constitutes a simpler recipe yielding a more efficient AEAD scheme.

All the results needed to prove the security of the FGHF′ construction hold in the general adaptive setting. The limitation to the non-adaptive leakage setting comes from the fact that leakage-resilient unpredictable functions are unattainable in the adaptive-leakage setting if no further restriction is imposed on the set of leakage functions.

Non-Adaptively Leakage-Resilient Functions from Sponges. Having reduced the task of constructing a leakage-resilient AEAD scheme to that of constructing suitable leakage-resilient function families, we turn our attention to the latter problem. We instantiate both $\mathcal{F}$ and $\mathcal{F}'$ with the same sponge-based construction, which we refer to as SLFunc. This construction is essentially the rekeying function employed in Isap [10] instantiated with a random transformation (T-sponge) instead of a random permutation (P-sponge). In [10] this was proposed without proof, instead its security was argued based on its apparent similarity to the GGM construction [15] and the corresponding results in [14,22] for it yielding a leakage-resilient pseudorandom function family. However, there are clear differences between the sponge construction and the GGM construction and we do not see a way to make a direct connection between the security of the two. In fact our proof follows a fairly different strategy from the ones presented in [14,15,22] – which all rely on a hybrid argument whereas ours does not. Moreover, for the overall security of SLAE we need this function family to additionally be leakage-resilient unpredictable, which, as was discussed above, does not follow from it being leakage-resilient pseudorandom. Nevertheless, in both cases we are able to show the intuitive claim made in [10] that λ bits of leakage can be compensated for by increasing the capacity by λ bits.

Another technical challenge that we face here is that we cannot employ the H-coefficient technique which is commonly used to prove the security of various sponge-based constructions. Like most other works on leakage resilience, we resort to arguments based on min-entropy and its chain rule in order to deal with leakage. Unfortunately, such arguments do not combine well with the H-coefficient technique, which precludes us from using it. In turn, this renders the security proof more challenging, as we have to deal with an adversary that may choose its queries (not the leakage function) adaptively. In contrast, the H-coefficient technique would automatically bypass this issue by reducing the security proof to a counting problem.

A Concrete Sponge-Based AEAD Scheme (SLAE). Finally, by instantiating the FGHF' construction with the above sponge-based construction for $\mathcal{F}$ and $\mathcal{F}'$ and matching sponge-based constructions for $\mathcal{G}$ and $\mathcal{H}$ we obtain SLAE. We also present security proofs for the T-sponge instantiations of the pseudorandom generator and the vector hash, which we were unable to readily find in the literature. SLAE is perhaps our most practical contribution – an entirely sponge-based leakage-resilient nonce-based AEAD scheme with provable security guarantees that is simple to implement and reasonably efficient. The efficiency of SLAE could be further optimised using similar techniques to the ones described in [10] for ISAP. Furthermore our security proofs are conducted in the concrete security setting thereby allowing practitioners to easily derive parameter estimates for their desired security level.

1.2 Related Work

To the best of our knowledge, the first authenticated encryption scheme claimed to be leakage-resilient was RCB [3], but it was broken soon after [2].

A series of works [7,8,16,20] have proposed a number of leakage-resilient symmetric encryption schemes, message authentication codes, and authenticated encryption schemes. These constructions assume that a subset of their components (block cipher instances) are leakage-free and that the leakage in the other components is simulatable, an assumption that is somewhat contentious [18,23]. Based on these assumptions, they show that the security of their encryption schemes reduces to the security of a single-block variant of the same scheme. However, the security of the corresponding single-block schemes remains an open question that is implicitly assumed to hold.

Abdalla, Belaïd, and Fouque [1] construct a symmetric encryption scheme that is non-adaptively leakage-resilient against chosen-plaintext attacks. Interestingly, their scheme employs a rekeying function that is not a leakage-resilient pseudorandom function. However their encryption scheme is not nonce-based as it necessitates a source of randomness.

In independent and concurrent work [11] Dobraunig and Mennink analyse the leakage resilience of the duplex sponge construction. While their leakage model is closer to ours, they prove something different. Namely they show that the duplex is indistinguishable from an *adjusted ideal extendible input function* (AIXIF) which is an ideal functionality incorporating leakage. In contrast we show that SLFUNC is both a leakage-resilient PRF (LPRF) and a leakage-resilient unpredictable function (LUF). Furthermore, while the duplex is a more general construction than SLFUNC, we prove better security bounds that allow for a more efficient realisation for the same level of security. Essentially for λ bits of leakage and absorption rate rr, their security bound degrades by $\lambda(rr + 1)$ whereas ours degrades by $\lambda + rr$. In addition, we leverage the leakage resilience of SLFUNC to construct the leakage-resilient AEAD scheme SLAE.

Other independent and concurrent work by Guo et al. [17] proposes an AEAD design, TETSPonge, that combines a sponge construction with two tweakable block cipher instances. While their work and ours share the goal of constructing leakage-resilient AEAD schemes, the two works adopt very different approaches. Both the security definitions and the assumptions on which the security of the schemes rely on are significantly different. One notable difference, is that the leakage resilience of TETSPonge relies crucially on the tweakable block cipher instances being leak-free, presumably due to a hardened implementation, whereas our treatment exploits and exposes the inherent leakage resilience of the sponge construction.

1.3 Organization of the Paper

In Sect. 2 we review the basic concepts and security definitions that we require in the rest of the paper. This is followed by a detailed description of SLAE in Sect. 3. In Sect. 4 we cover the security analysis of the generic FGHF' construction. We conclude with Sect. 5 where we cover the security of the sponge-based primitives used to instantiate FGHF' and thereby obtain SLAE. The full details of the proofs can be found in the full version of this paper. We conclude in Sect. 6 with some remarks on implementing SLAE.

2 Preliminaries

We start by reviewing the basic tools and definitions that we require for our results. We begin by establishing some notation.

2.1 Notation

For any non-negative integer $n \in \mathbb{N}$ we use $[n]$ to denote the set $\{1, \ldots, n\}$, where $[n] = \emptyset$ when $n = 0$. For any two strings s_1 and s_2, $|s_1|$ denotes the size of s_1 and $s_1 \parallel s_2$ denotes their concatenation. For a positive integer $k \leq |s_1|$, we use $\lfloor s_1 \rfloor_k$ to denote the string obtained by truncating s_1 to its leftmost k bits. The empty string is denoted by ε, $\{0,1\}^n$ denotes the set of bit strings of size n, and $\{0,1\}^*$ denotes the set of all strings of finite length. We write $x \leftarrow S$ to denote the process of uniformly sampling a value from the finite set S and assigning it to x.

We make use of the code-based game-playing framework by Bellare and Rogaway [5], where the interaction between a game and the adversary is implicit. In all games, the adversary is given as its input the output of the initialize procedure, it has oracle access to the other procedures described in the game, and its output is fed into the finalize procedure. The output of the finalize procedure is the output of the game. For a game G and an adversary A, $G^A \Rightarrow y$ denotes the event that G outputs y when interacting with A. Similarly, $A^G \Rightarrow x$ denotes the event that A outputs x when interacting with G. By convention all boolean variables Bad are initialized to false, and for any table $p[]$ its entries are all initialized to $\bot$. When lazy-sampling a random function with domain $\mathcal{X}$ and co-domain $\mathcal{Y}$ into a table $p[]$, we use inset(p) and outset(p) to denote respectively the sets of input and output values defined up to that point. That is, inset$(p) = \{X : p[X] \neq \bot \wedge X \in \mathcal{X}\}$ and outset$(p) = \{p[X] : p[X] \neq \bot \wedge X \in \mathcal{X}\}$. If G_1 and G_2 are games and A is an adversary we define the corresponding *adversarial advantage* as

$$\mathbf{Adv}\left(A^{G_1}, A^{G_2}\right) = \Pr[A^{G_1} \Rightarrow 1] - \Pr[A^{G_2} \Rightarrow 1],$$

and the corresponding *game advantage* as

$$\mathbf{Adv}\left(G_1^A, G_2^A\right) = \Pr[G_1^A \Rightarrow \mathsf{true}] - \Pr[G_2^A \Rightarrow \mathsf{true}].$$

We will operate in the random transformation model, where ρ is an idealised random transformation mapping n-bit strings to n-bit strings. For any algorithm $\mathcal{F}$ that uses ρ as a subroutine, we use $Q_{\mathcal{F}}(q, \mu)$ to denote the number of calls to ρ required when evaluating $\mathcal{F}$ q times on a total of μ bits.

2.2 Syntax

Encryption. An *authenticated encryption scheme with associated data* AEAD $= (\mathcal{E}, \mathcal{D})$ is a pair of efficient algorithms such that:

- The deterministic encryption algorithm $\mathcal{E} : \mathcal{K} \times \mathcal{N} \times \mathcal{A} \times \mathcal{M} \to \{0,1\}^*$ takes as input a secret key K, a nonce N, associated data A, and a message M to return a ciphertext C.
- The deterministic decryption algorithm $\mathcal{D} : \mathcal{K} \times \mathcal{N} \times \mathcal{A} \times \{0,1\}^* \to \mathcal{M} \cup \{\bot\}$ takes as input a secret key K, a nonce N, associated data A, and a ciphertext C to return either a message in $\mathcal{M}$ or $\bot$ indicating that the ciphertext is invalid.

Sets $\mathcal{K}$, $\mathcal{N}$, $\mathcal{A}$, and $\mathcal{M}$ denote respectively the key space, the nonce space, the associated data space, and the message space associated to the scheme. We assume throughout that $\mathcal{E}$ and $\mathcal{D}$ are never queried on inputs outside of these sets. An authenticated encryption scheme is required to be *correct* and *tidy*. Correctness requires that for all K, N, A, M if $\mathcal{E}(K, N, A, M) = C$ then $\mathcal{D}(K, N, A, C) = M$. Analogously, tidiness requires that for all K, N, A, C if $\mathcal{D}(K, N, A, C) = M \neq \bot$ then $\mathcal{E}(K, N, A, M) = C$. Furthermore we demand that encryption be length regular, i.e for all K, N, A, M it should hold that $|\mathcal{E}(K, N, A, M)|$ is entirely determined by $|N|$, $|A|$, and $|M|$.

We will use the terms *authenticated encryption scheme* and *symmetric encryption scheme* to refer to the analogously defined encryption scheme which does not admit associated data as part of its input. For such schemes, A is implicitly set to the empty string in the security games.

Message Authentication. A *message authentication code* $\text{MAC} = (T, V)$ is a pair of efficient algorithms with an associated key space $\mathcal{K}$, domain $\mathcal{X}$, and tag length t such that:

- The deterministic tagging algorithm $T : \mathcal{K} \times \mathcal{X} \to \{0, 1\}^t$ takes as input a key K and a value X to return a tag T of size t.
- The deterministic verification algorithm $V : \mathcal{K} \times \mathcal{X} \times \{0, 1\}^t \to \{\top, \bot\}$ takes as input a key K, a value X, and a tag T to return either $\top$ indicating a valid input or $\bot$ otherwise.

We require that for any key $K \in \mathcal{K}$ and any admissible input $X \in \mathcal{X}$, if $T \leftarrow T(K, X)$, then $V(K, X, T) = \top$. When $\mathcal{X} = \{0, 1\}^*$ we end up with the usual MAC definition, however we will also consider MACs over tuples of strings, e.g. $\mathcal{X} = \{0, 1\}^* \times \{0, 1\}^* \times \{0, 1\}^*$. Such MACs where considered in [19] and we follow suit in referring to such MACs as vector MACs.

We say that a MAC is *canonical* if it is implicitly defined by T, where $V(K, X, T)$ consists of running $T' \leftarrow T(K, X)$ and returning $\top$ if $T' = T$ and $\bot$ otherwise.

2.3   The Sponge Construction

The sponge construction is a versatile object that can be used to realise various cryptographic primitives. Several variations of the sponge exist, Fig. 1 illustrates the plain version of the sponge as originally introduced by Bertoni et al. [9]. We give here only a brief overview of its operation and the associated nomenclature that we will use throughout this paper.

The sponge operates iteratively on its inputs through a transformation ρ, and generally includes an *absorbing* phase and a *squeezing* phase. The transformation ρ maps strings of size n to strings of size n. Associated to the sponge are two other values called the rate r and the capacity c, where $n = r + c$. At any given iteration we refer to the output of the transformation as the state, which we denote by S. Furthermore, we denote the leftmost r bits of S by $\bar{S}$ and the remaining c bits by $\hat{S}$. We will at times refer to $\bar{S}$ and $\hat{S}$ as the outer and inner parts of the state, respectively. In the absorbing phase an input M is "absorbed" iteratively r bits at a time. At iteration i input M_i is absorbed by letting $Y_i \leftarrow (M_i \oplus \bar{S}_i) \parallel \hat{S}_i$ and setting $S_{i+1} \leftarrow \rho(Y_i)$. The initial value of S may generally be set to a constant, a concatenation of a secret key and a constant, or by applying the transformation to either of these values. Output is produced from the sponge during the squeezing phase in one or more iterations, r bits at a time.

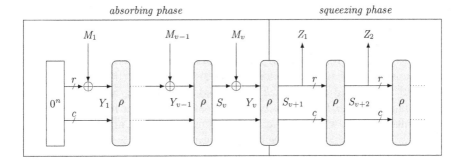

Fig. 1. Illustration of the plain sponge construction.

At iteration i output Z_j is produced by setting $Z_j \leftarrow \bar{S}_i$ and $S_{i+1} \leftarrow \rho(S_i)$. The above variant is normally referred to as the T-sponge, as it employs a fixed-size random transformation. An alternative instantiation, known as the P-sponge, replaces this random transformation with a random permutation.

2.4 The Leakage Model

Our leakage model is based on *leakage resilience* as defined in [13]. This assumes that only computation leaks, and in particular, that only the data that is accessed during computation can leak information. It allows for continuous adaptive leakage, where in each query to a leakage oracle the adversary can specify a leakage function from some predefined set $\mathcal{L}$ that it can chose adaptively based on prior outputs and leakage. Throughout, we restrict ourselves to leakage functions that are deterministic and efficiently computable. While our security definitions are formulated in this general setting, our main results will be in the weaker *granular non-adaptive* leakage setting proposed in [14]. We view the non-adaptive leakage setting as the special case where the leakage set $\mathcal{L}$ is restricted to be a singleton, fixed at the start of the game. In granular leakage, a single time step is with respect to a single computation of some underlying primitive, in our case, the transformation ρ. Correspondingly, in this case the adversary specifies a vector of leakage functions and gets in return the aggregate leakage from the entire evaluation of the higher-level construction. Note that in the granular setting the leakage sets for each iteration can be distinct. Similarly, when studying the leakage resilience of composite constructions we have to consider compositions of leakage functions. For instance, if construction C is composed of primitives A and B with associated leakage sets $\mathcal{L}_A$ and $\mathcal{L}_B$, then we associate to C the Cartesian product of the two leakage sets, i.e. $\mathcal{L}_C = \mathcal{L}_A \times \mathcal{L}_B$. The actual inputs that get fed to the leakage functions are implicitly defined by the construction and its inputs, whereas the combined output is the aggregate output of all function evaluations.

An analysis of sponge-based constructions compels us to consider leakage resilience in the random transformation model. A similar setting, albeit in the random oracle model, was already considered by Standaert et al. in [22]. A central question that arises in idealised settings like this is whether the leakage function should be given access to the ideal primitive. As in [22], we will not give this access to the leakage function. On the one hand, providing the leakage function with unlimited access to the random oracle gives rise to artificial attacks, such as the "future computation attack" discussed in [22], that would not arise in practice. On the other hand, depriving the leakage function from accessing the ideal primitive, means that the leakage function cannot leak any bits of the ideal primitive's output, which may seem overly restrictive. However, for the case of sponge-based constructions this is less problematic because from the adversary's perspective the full output of a transformation call is completely determined by the input to the next transformation call. As such, information about the output of one transformation call can leak as part of the leakage in the next transformation call. Combined with the fact that the only restriction that we will impose on the leakage function is to limit its output length, we think that this leads to a fairly realistic leakage model.

We conclude our discussion on the leakage model by offering our interpretation of the significance of leakage resilience security with respect to practical side channel attacks. One might object that we model leakage by a deterministic function whose output is of a fixed bit-length whereas in practice the leakage is noisy. However through the leakage function we are really trying to capture the maximum amount of information that an adversary may obtain from evaluating the scheme on a single input. Hence, the underlying assumption is that no matter how many times the scheme is run on the same input, in order to even out the noise, the information that the adversary can obtain is limited. Put in more practical terms, this roughly translates to assuming that the scheme's implementation resists Simple Power Analysis (SPA). On the other hand, if the scheme is proven to be leakage-resilient then we are guaranteed that an adversary cannot do much better even if it can observe and accumulate leakage on multiple other (differing) inputs. Thus a proof of leakage resilience can be interpreted as saying that if the scheme's implementation is secure against SPA then, by the inherent properties of the scheme, it is also secure against Differential Power Analysis (DPA). However, a proof of leakage resilience is of course no guarantee that a scheme's implementation will be secure against SPA.

2.5 Authenticated Encryption and Leakage Resilience

Recently, Barwell et al. [4] provided a definitional framework augmenting nonce-based authenticated encryption with leakage. Their security notions capture the leakage resilience setting as defined in [13]. Furthermore, they prove composition theorems analogous to [19] that additionally take leakage into account. Below we reproduce their security definitions and composition result which we will employ in this work, with some minor adaptations. We recast their definitions in a style that admits code-based proofs [5]. Unlike [4] we make no distinction between a scheme and its implementation since we are interested in proving security for the actual scheme. When defining these security notions, we only describe the game and the corresponding adversarial advantage. A scheme is understood to be secure if the adversarial advantage is bounded by a sufficiently small value for all reasonably-resourced adversaries. Our security theorems will then establish a bound on the adversarial advantage in terms of the adversary's resources, without drawing judgement as to what constitutes "small" and "reasonable" since that is a rather subjective matter.

Classifying Adversarial Queries. As usual, the adversary has to be forbidden from making certain queries in order to avoid trivial win conditions. Following the terminology of [4], if an adversary makes a query (N, A, M) to an encryption oracle that returns C, then repeating this query to one of the encryption oracles or querying (N, A, C) to one of the decryption oracles, is considered to be an *equivalent* query. Note that any additional components of a query, such as the leakage function, are ignored for the purpose of determining equivalence between two queries. If an adversary makes equivalent queries across two oracles, it is said to *forward* that query from one oracle to the other. Note that the two oracles do not need to be distinct, and thus forwarded queries include repeated queries to the same oracle.[1]

Let an encryption query refer to any query made to either a challenge encryption oracle or a leakage encryption oracle. Then an adversary against an (authenticated) encryption scheme is said to be *nonce respecting* if it never repeats a nonce in two distinct encryption queries.

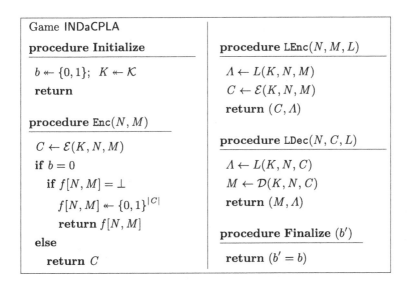

Fig. 2. Game used to define IND-aCPLA security.

---

[1] This is not really required, since contrary to [4] the challenge oracles are not forgetful in our case. Nevertheless we conform to the original definition of forwarded queries.

Chosen-Plaintext Security with Leakage. Barwell et al. introduce an augmented variant of leakage-resilient chosen-plaintext security called IND-aCPLA, that is required by their composition theorem. Here the adversary is given access to three oracles. A challenge oracle that returns either a valid encryption of a message or a random string of appropriate length. A leakage encryption oracle that, upon being queried on a message and a leakage function, returns the corresponding ciphertext and the evaluated leakage. The adversary is not allowed to forward queries between the two encryption oracles. In addition, it has limited access to a leakage decryption oracle which returns the decryption of the queried ciphertext and the leakage corresponding to the queried leakage function. However, it can only query this oracle on inputs forwarded from the leakage encryption oracle. Thus the adversary can obtain decryption leakage, but only on ciphertexts for which it already knows the corresponding message. Below is the formal definition.

Definition 1 (IND-aCPLA Security). *Let* $\mathrm{SE} = (\mathcal{E}, \mathcal{D})$ *be a symmetric encryption scheme and the* INDaCPLA *game be as defined in Fig. 2. Then for any nonce-respecting adversary* $\mathcal{A}$ *that never forwards queries to or from the* Enc *oracle, only makes queries to* LDec *that are forwarded from* LEnc, *and only makes encryption and decryption queries containing leakage functions in the respective sets* $\mathcal{L}_E$ *and* $\mathcal{L}_D$, *its corresponding* IND-aCPLA *advantage is given by:*

$$\mathbf{Adv}_{\mathrm{SE}}^{\mathrm{ind\text{-}acpla}}(\mathcal{A}, \mathcal{L}_E, \mathcal{L}_D) = 2\Pr\left[\mathsf{INDaCPLA}^{\mathcal{A}} \Rightarrow \mathsf{true}\right] - 1.$$

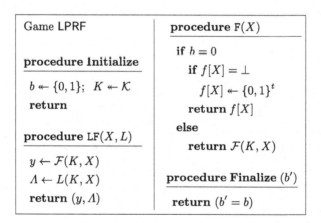

Fig. 3. Game used to define LPRF security.

Leakage-Resilient Function Families. We will distinguish among function families based on their domain $\mathcal{X}$. We will use the terms *fixed-input-length* function when $\mathcal{X} = \{0,1\}^l$ for some $l \in \mathbb{N}$, *variable-input-length* function when

$\mathcal{X} = \{0,1\}^*$, and *vector* function when the domain is a cartesian product of string sets, e.g. $\mathcal{X} = \{0,1\}^* \times \{0,1\}^*$.

For such function families we will consider two security notions: leakage-resilient pseudorandom functions (LPRF) and leakage-resilient unpredictable functions (LUF). While LPRF security is well-established in the literature, LUF security is new. Below are the formal definitions.

Definition 2 (LPRF Security). *Let $\mathcal{F}\colon \mathcal{K} \times \mathcal{X} \to \{0,1\}^t$ be a function family over the domain $\mathcal{X}$ and indexed by $\mathcal{K}$, and the LPRF game be as defined in Fig. 3. Then for any adversary A that never forwards queries to or from the F oracle and only queries leakage functions in the set $\mathcal{L}_F$, its corresponding LPRF advantage is given by:*

$$\mathbf{Adv}_{\mathcal{F}}^{\mathrm{lprf}}\,(A, \mathcal{L}_F) = 2\Pr\left[\mathsf{LPRF}^A \Rightarrow \mathsf{true}\right] - 1\,.$$

Game LUF	procedure Lkg(X, L)
procedure Initialize	$\Lambda \leftarrow L(K, X)$ return Λ
win $\leftarrow$ false; $K \twoheadleftarrow \mathcal{K}$ return	**procedure Guess(X, y')**
procedure F(X)	$y \leftarrow \mathcal{F}(K, X)$ if $X \notin \mathcal{S} \wedge y = y'$
$\mathcal{S} \xleftarrow{\cup} X$ $y \leftarrow \mathcal{F}(K, X)$ return y	win $\leftarrow$ true return $(y = y')$
	procedure Finalize
	return (win)

Fig. 4. Game used to define LUF security.

Definition 3 (LUF Security). *Let $\mathcal{F}\colon \mathcal{K} \times \mathcal{X} \to \{0,1\}^t$ be a function family over the domain $\mathcal{X}$ and indexed by $\mathcal{K}$, and the LUF game be as defined in Fig. 4. Then for any adversary A its corresponding LUF advantage is given by:*

$$\mathbf{Adv}_{\mathcal{F}}^{\mathrm{luf}}\,(A, \mathcal{L}_F) = \Pr\left[\mathsf{LUF}^A \Rightarrow \mathsf{true}\right]\,.$$

Unforgeability in the Presence of Leakage. For message authentication we will require the analogue of strong unforgeability in the leakage setting (SUF-CMLA) put forth in [4]. This is essentially strong unforgeability (SUF-CMA) formulated as a distinguishing game, with a challenge verification oracle and additional tagging and verification oracles that leak. Below is the formal definition.

Game SUFCMLA	procedure LTag(X, L)
	$\Lambda \leftarrow L(K, X)$
procedure Initialize	$T \leftarrow \mathcal{T}(K, X)$
$b \leftarrow \{0, 1\};\ \ K \leftarrow \mathcal{K}$	**return** (T, Λ)
return	
	procedure LVfy(X, T, L)
procedure Vfy(X, T)	$\Lambda \leftarrow L(K, X, T)$
if $b = 0$	$v \leftarrow \mathcal{V}(K, X, T)$
return $\perp$	**return** (v, Λ)
else	
$v \leftarrow \mathcal{V}(K, X, T)$	**procedure Finalize** (b')
return v	**return** $(b' = b)$

Fig. 5. Game used to define SUF-CMLA security.

Definition 4 (SUF-CMLA Security). *Let* MAC $= (\mathcal{T}, \mathcal{V})$ *be a message authentication code and the* SUFCMLA *game be as defined in Fig. 5. For any adversary* $\mathcal{A}$ *that never forwards queries from* LTag *to* Vfy, *and only queries leakage functions to its tagging and verification oracles in the respective sets* $\mathcal{L}_T$ *and* $\mathcal{L}_V$, *its corresponding* SUF-CMLA *advantage is given by:*

$$\mathbf{Adv}^{\text{suf-cmla}}_{\text{MAC}}(\mathcal{A}, \mathcal{L}_T, \mathcal{L}_V) = 2\Pr\left[\text{SUFCMLA}^{\mathcal{A}} \Rightarrow \text{true}\right] - 1.$$

Authenticated Encryption with Leakage. For an authenticated encryption scheme with associated data our target will be LAE security, which is a natural extension of the classical security notion put forth by Rogaway [21] to the leakage setting. This is defined formally below.

Definition 5 (LAE Security). *Let* AEAD $= (\mathcal{E}, \mathcal{D})$ *be an authenticated encryption scheme with associated data and the* LAE *game be as defined in Fig. 6. Then for any adversary* $\mathcal{A}$ *that never forwards queries to or from the* Enc *and* Dec *oracles and only makes encryption and decryption queries containing leakage functions in the respective sets* $\mathcal{L}_{AE}$ *and* $\mathcal{L}_{VD}$, *its corresponding* LAE *advantage is given by:*

$$\mathbf{Adv}^{\text{lae}}_{\text{AEAD}}(\mathcal{A}, \mathcal{L}_{AE}, \mathcal{L}_{VD}) = 2\Pr\left[\text{LAE}^{\mathcal{A}} \Rightarrow \text{true}\right] - 1.$$

Generic Composition in the Leakage Setting. The N2 construction was introduced in [19] and is depicted pictorially in Fig. 7. In [4] Barwell et al. prove a composition theorem for this construction that holds in the leakage setting. We will make use of this theorem and for completeness we reproduce it below, adapted to the random transformation model.

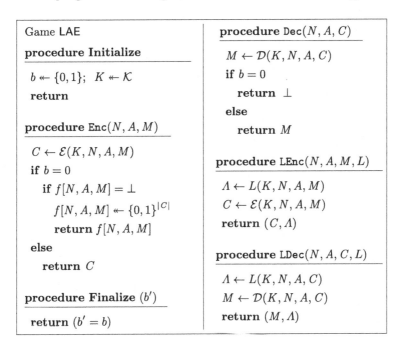

Fig. 6. Game used to define LAE security.

Theorem 1 (LAE Security of the N2 Construction [4]). *Let* $\text{SE} = (\mathcal{E}, \mathcal{D})$ *be a symmetric encryption scheme with associated leakage sets* $(\mathcal{L}_E, \mathcal{L}_D)$ *and* $\text{MAC} = (\mathcal{T}, \mathcal{V})$ *be a MAC with associated leakage sets* $(\mathcal{L}_T, \mathcal{L}_V)$. *Further let* N2 *be the composition of* SE *and* MAC *described in Fig. 7, with associated leakage sets* $(\mathcal{L}_{AE}, \mathcal{L}_{VD})$ *where* $\mathcal{L}_{AE} = \mathcal{L}_E \times \mathcal{L}_T$ *and* $\mathcal{L}_{VD} = \mathcal{L}_D \times \mathcal{L}_V$. *Then for any* LAE *adversary* $\mathcal{A}_{ae}$ *against* N2 *there exist adversaries* $\mathcal{A}_{se}, \mathcal{A}_{prf},$ *and* $\mathcal{A}_{mac}$ *such that:*

$$\mathbf{Adv}_{\text{N2}}^{\text{lae}}(\mathcal{A}_{ae}, \mathcal{L}_{AE}, \mathcal{L}_{VD}) \leq \mathbf{Adv}_{\text{SE}}^{\text{ind-acpla}}(\mathcal{A}_{se}, \mathcal{L}_E, \mathcal{L}_D)$$
$$+ \mathbf{Adv}_{\mathcal{T}}^{\text{lprf}}(\mathcal{A}_{prf}, \mathcal{L}_T) + 2\mathbf{Adv}_{\text{MAC}}^{\text{suf-cmla}}(\mathcal{A}_{mac}, \mathcal{L}_T, \mathcal{L}_V).$$

3   SLAE: A Sponge-Based LAE Construction

SLAE, pronounced "sleigh", is a **S**ponge-based non-adaptive **L**eakage-resilient **AE**AD scheme. It is based on, and is closely related to, a prior sponge-based AEAD scheme called ISAP [10]. ISAP is a nonce-based AEAD scheme intended to inherently resist side-channel attacks while simultaneously fitting the well-established syntax of AEAD schemes. More specifically, it claims security against Differential Power Analysis (DPA) by employing a rekeying mechanism. An important challenge that ISAP overcomes, is to avoid decrypting distinct ciphertexts under the same key without maintaining a state. Furthermore, as noted

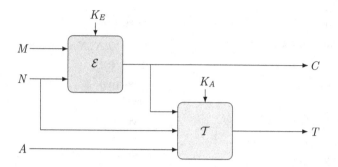

Fig. 7. Graphical representation of the N2 construction.

by ISAP's designers, the sponge construction seems markedly well-suited to protect against side-channels. Typically, the sponge employs a large state that is continually evolving, which intuitively endows it with an intrinsic resilience to information leakage. Thus, in contrast to other designs, ISAP potentially offers a fairly efficient LAE solution that can be instantiated with off-the-shelf primitives. However, as we already noted, ISAP's biggest limitation is that its design is not backed by any formal security analysis, not even in the absence of leakage.

ISAP is composed of a symmetric encryption scheme ISAPENC and a MAC ISAPMAC combined according to the N2 construction. These components were conceived by augmenting established sponge constructs with a rekeying function. In particular the design rationale behind ISAPMAC is to augment a sponge-based suffix MAC with a rekeying function. The rekeying is such that the key fed into the suffix MAC itself depends on the inputs being authenticated and a master authentication key. Similarly ISAPENC is a standard sponge-based encryption scheme whose key is derived from a master encryption key and the nonce. Throughout, the rekeying function is realised from the sponge by setting the absorption rate to be one. Intuitively, ISAP's resistance to DPA comes from the fact that encryption and authentication never use the same key more than once, and the slow absorption rate employed in the rekeying function. Both of these factors limit the so-called data complexity of computations involving secret values, which in turn encumbers DPA attacks. See [10] for more details on ISAP.

SLAE retains the main structure of ISAP, as well as its benefits, but it includes some changes and restrictions that facilitate its security analysis. While the majority of these differences are conceptual, they are substantial enough, however, to *invalidate* any claim that our security proof applies to ISAP. The design of SLAE can be understood across *three* different levels of abstraction. At the highest level, like ISAP, it is the N2 composition of a symmetric encryption scheme SLENC and a MAC SLMAC. At the second abstraction level, SLMAC and SLENC can be viewed in terms of smaller components. Specifically, we view SLMAC as combining a collision-resistant vector hash function $\mathcal{H}$ and a fixed-input-length function $\mathcal{F}'$, and we decompose SLENC into a fixed-input-length function $\mathcal{F}$ and a pseudorandom generator with variable output length $\mathcal{G}$. Indeed this view corresponds to our generic construction of a non-adaptively leakage-resilient AEAD scheme which we refer to as the $FGHF'$ construction.

Note that there is no explicit idea of rekeying in the $FGHF'$ construction. The only leakage-resilient primitives are $\mathcal{F}$ and $\mathcal{F}'$. For security we will require both to be LPRF secure and $\mathcal{F}'$ to additionally be LUF secure. Thus LAE schemes are easy to construct once we have such primitives. Moreover, $\mathcal{F}$ is invoked once for encryption, and likewise $\mathcal{F}'$ is invoked once for authentication, irrespective of the message length. SLAE is obtained by instantiating the four components in the $FGHF'$ construction with T-sponges. This is the third level view. While the design rationale behind the $FGHF'$ construction is quite distinct from that of ISAP, once instantiated, SLAE and ISAP suddenly look very similar.

We now describe SLAE in more detail and then elaborate on the differences between SLAE and ISAP in Sect. 3.4.

3.1 High-Level View of SLAE

As already noted, SLAE $=$ (SLAE-$\mathcal{E}$, SLAE-$\mathcal{D}$) is a nonce-based AEAD scheme composed from a nonce-based symmetric encryption scheme SLENC $=$ (SLENC-$\mathcal{E}$, SLENC-$\mathcal{D}$) and a MAC SLMAC $=$ (SLMAC-$\mathcal{T}$, SLMAC-$\mathcal{V}$). These are combined according to the N2 composition, where the key is split into an encryption key K_E and an authentication key K_A. During encryption, SLENC-$\mathcal{E}$ takes the nonce, message, and key K_E to return a ciphertext which is then fed together with the nonce, associated data, and key K_A, to SLMAC-$\mathcal{T}$ to produce a tag which is then appended onto the ciphertext. Decryption proceeds by reversing these operations in a *verify-then-decrypt* manner, whereby ciphertext decryption using SLENC-$\mathcal{D}$ proceeds *only if* tag verification under SLMAC-$\mathcal{V}$ was successful. The pseudocode for this composition is described in Fig. 8.

3.2 The SLMAC Construction

A pseudocode description of SLMAC $=$ (SLMAC-$\mathcal{T}$, SLMAC-$\mathcal{V}$) can be found in Fig. 9. It is a vector MAC operating on the triple (N, A, C), where verification works by recomputing the tag for the given triple and checking that it is identical to the given tag. As such, the core functionality of SLMAC is captured in the tagging algorithm SLMAC-$\mathcal{T}$, which is additionally depicted in Fig. 10. The tagging algorithm can be understood as being composed of a (sponge-based) vector hash function compressing the triple (N, A, C) into a digest of size w bits, which is then fed to the unpredictable function SLFUNC to produce a tag of size t bits. The nonce N is required to be m bits long, whereas A and C can be of arbitrary length. Accordingly, SLMAC-$\mathcal{T}$ starts by padding both A and C so that their lengths are integer multiples of the sponge rate r. Note that the padding function, lpad, always returns at least a single bit of padding and is always applied, even if the input string is already an integer multiple of r.

$\textsc{Slae-}\mathcal{E}(K, N, A, M)$	$\textsc{Slae-}\mathcal{D}(K, N, A, \overline{C})$
parse K **as** $K_E \parallel K_A$	**parse** K **as** $K_E \parallel K_A$
$C \leftarrow \textsc{SlEnc-}\mathcal{E}(K_E, N, M)$	**parse** $\overline{C}$ **as** $C \parallel T$
$T \leftarrow \textsc{SlMac-}\mathcal{T}(K_A, (N, A, C))$	$v \leftarrow \textsc{SlMac-}\mathcal{V}(K_A, (N, A, C), T)$
$\overline{C} \leftarrow C \parallel T$	**if** $v = \top$
return $\overline{C}$	$\quad M \leftarrow \textsc{SlEnc-}\mathcal{D}(K_E, N, C)$
	$\quad$ **return** M
	else
	$\quad$ **return** $\bot$

Fig. 8. High-level description of $\textsc{Slae}$ in terms of $\textsc{SlMac}$ and $\textsc{SlEnc}$.

$\textsc{SlMac-}\mathcal{T}(K_A, (N, A, C))$	$\textsc{SlMac-}\mathcal{V}(K_A, (N, A, C), T)$
$\boldsymbol{A} \leftarrow A \parallel \mathsf{lpad}(A, r)$	$T' \leftarrow \textsc{SlMac-}\mathcal{T}(K_A, (N, A, C))$
parse $\boldsymbol{A}$ **as** $\boldsymbol{A}_1 \parallel \ldots \parallel \boldsymbol{A}_u$	**if** $T = T'$
$\quad\quad$ **st** $\forall i \; \lvert \boldsymbol{A}_i \rvert = r$	$\quad$ **return** $\top$
$\boldsymbol{C} \leftarrow C \parallel \mathsf{lpad}(C, r)$	**return** $\bot$
parse $\boldsymbol{C}$ **as** $\boldsymbol{C}_1 \parallel \ldots \parallel \boldsymbol{C}_v$	
$\quad\quad$ **st** $\forall i \; \lvert \boldsymbol{C}_i \rvert = r$	
$Y_0 \leftarrow N \parallel IV$	$\textsc{SlFunc}(K_A, H)$
$S_1 \leftarrow \rho(Y_0)$	
// Absorb Associated Data	**parse** H **as** $H_1 \parallel \ldots \parallel H_l$
for i **in** $\{1, \ldots, u\}$	$\quad\quad$ **st** $\forall i \; \lvert H_i \rvert = rr$
$\quad Y_i \leftarrow (\bar{S}_i \oplus \boldsymbol{A}_i) \parallel \hat{S}_i$	$Y_0 \leftarrow K_A \parallel IV$
$\quad S_{i+1} \leftarrow \rho(Y_i)$	$S_1 \leftarrow \rho(Y_0)$
// Separate Inputs	**for** i **in** $\{1, \ldots, l\}$
$S_{u+1} \leftarrow \bar{S}_{u+1} \parallel \left(\hat{S}_{u+1} \oplus (1 \parallel 0^{c-1}) \right)$	$\quad Y_i \leftarrow (\bar{S}_i \oplus H_i) \parallel \hat{S}_i$
// Absorb Ciphertext	$\quad S_{i+1} \leftarrow \rho(Y_i)$
for i **in** $\{u+1, \ldots, u+v\}$	**return** $\lfloor S_{l+1} \rfloor_t$
$\quad Y_i \leftarrow (\bar{S}_i \oplus \boldsymbol{C}_{i-u}) \parallel \hat{S}_i$	
$\quad S_{i+1} \leftarrow \rho(Y_i)$	$\mathsf{lpad}(A, r)$
// Generate Tag	
$H \leftarrow \lfloor S_{u+v+1} \rfloor_w$	$x \leftarrow \lvert A \rvert \bmod r$
$T \leftarrow \lfloor \textsc{SlFunc}(K_A, H) \rfloor_t$	**return** $1 \parallel 0^{r-x-1}$
return T	

Fig. 9. Pseudocode description of $\textsc{SlMac}$ and $\textsc{SlFunc}$.

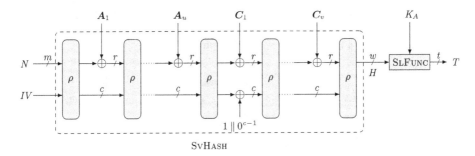

Fig. 10. Graphical illustration of SLMAC-$\mathcal{T}$.

To compute the hash digest H, the internal state is initialised to $\rho(N \parallel IV)$, where IV is a constant string of size $n - m$, and the padded associated data A and the padded ciphertext C are then absorbed block by block. An input separation mechanism is employed in order to demarcate the boundary between A and C. This involves XORing the string $1 \parallel 0^{c-1}$ to the inner part of the state once A has been absorbed, and ensures that distinct pairs $(A, C) \neq (\overline{A}, \overline{C})$ for which $A \parallel C - \overline{A} \parallel \overline{C}$ do not result in the same hash digest.

Once the hash digest is evaluated, it is fed into SLFUNC to compute the final tag. This is also a sponge-based construction for which a graphical representation appears in Fig. 11. Here the state is initialised to $\rho(K_A \parallel IV)$ and the hash digest is then absorbed at a *reduced rate* of rr bits. Once the complete digest has been absorbed the left most t bits of the state are output as the tag.

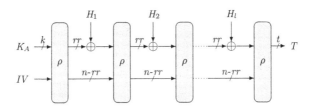

Fig. 11. Graphical illustration of SLFUNC.

3.3 The SLENC Construction

This is the sponge-based symmetric encryption scheme SLENC = (SLENC-$\mathcal{E}$, SLENC-$\mathcal{D}$) described in Fig. 12 and depicted in Fig. 13. It is easy to see that SLENC-$\mathcal{D}(K_E, N, \cdot) =$ SLENC-$\mathcal{E}(K_E, N, \cdot)$, and consequently we only describe the operation of SLENC-$\mathcal{E}$. This algorithm can be viewed as being composed of a pseudorandom function SLFUNC, taking as input the pair (K_E, N), and whose output is then fed into a pseudorandom generator SPRG. The output of SPRG is then used to encrypt the message.

SLEnc-$\mathcal{E}(K_E, N, M)$	SLEnc-$\mathcal{D}(K_E, N, C)$
parse N **as** $N_1 \parallel \ldots \parallel N_l$	**parse** N **as** $N_1 \parallel \ldots \parallel N_l$
st $\forall i \; \lvert N_i \rvert = rr$	**st** $\forall i \; \lvert N_i \rvert = rr$
parse M **as** $M_1 \parallel \ldots \parallel M_v$	**parse** C **as** $C_1 \parallel \ldots \parallel C_v$
st $\forall i < v \; \lvert M_i \rvert = r$ **and** $\lvert M_v \rvert \le r$	**st** $\forall i < v \; \lvert C_i \rvert = r$ **and** $\lvert C_v \rvert \le r$
// First Sponge Iteration	// First Sponge Iteration
$Y_0 \leftarrow K_E \parallel IV$	$Y_0 \leftarrow K_E \parallel IV$
$S_1 \leftarrow \rho(Y_0)$	$S_1 \leftarrow \rho(Y_0)$
// Absorb Nonce	// Absorb Nonce
for i **in** $\{1, \ldots, l\}$	**for** i **in** $\{1, \ldots, l\}$
$Y_i \leftarrow (\bar{S}_i \oplus N_i) \parallel \hat{S}_i$	$Y_i \leftarrow (\bar{S}_i \oplus N_i) \parallel \hat{S}_i$
$S_{i+1} \leftarrow \rho(Y_i)$	$S_{i+1} \leftarrow \rho(Y_i)$
// Squeeze and Encrypt	// Squeeze and Decrypt
for i **in** $\{l+1, \ldots, l+v-1\}$	**for** i **in** $\{l+1, \ldots, l+v-1\}$
$C_{i-l} \leftarrow \bar{S}_i \oplus M_{i-l}$	$M_{i-l} \leftarrow \bar{S}_i \oplus C_{i-l}$
$S_{i+1} \leftarrow \rho(S_i)$	$S_{i+1} \leftarrow \rho(S_i)$
$C_v \leftarrow \lfloor \bar{S}_{l+v} \rfloor_{\lvert M_v \rvert} \oplus M_v$	$M_v \leftarrow \lfloor \bar{S}_{l+v} \rfloor_{\lvert C_v \rvert} \oplus C_v$
return $C_1 \parallel \ldots \parallel C_v$	**return** $M_1 \parallel \ldots \parallel M_v$

Fig. 12. Pseudocode description of the SLEnc encryption scheme.

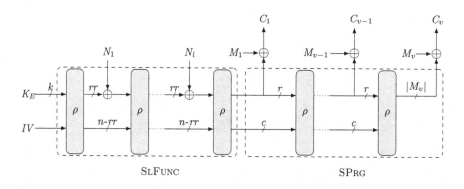

Fig. 13. Graphical illustration of SLEnc-$\mathcal{E}$.

$$\begin{array}{|l|}
\hline
\text{SPRG}(seed, L) \\
\hline
v \leftarrow \lceil \frac{L}{r} \rceil \\
S_1 \leftarrow seed \\
\textbf{for } i \textbf{ in } \{1, \ldots, v-1\} \\
\quad S_{i+1} \leftarrow \rho(S_i) \\
R \leftarrow \bar{S}_1 \parallel \ldots \parallel \bar{S}_v \\
\textbf{return } \lfloor R \rfloor_L \\
\hline
\end{array}$$

Fig. 14. Pseudocode description of SPRG.

The nonce N is required to be m bits long and we do not require any additional padding for the message. The evaluation of SLFUNC proceeds by initialising the internal state to $\rho(K_E \parallel IV)$, with a constant IV of size $n - k$, and then absorbing the nonce at a reduced rate of rr bits. Once the nonce is absorbed, the output state S_{l+1} serves as the seed to the pseudorandom generator SPRG. A separate pseudocode description of SPRG can be found in Fig. 14. The first ciphertext block is generated by XORing the outer part of this state with the first message block. Afterwards the initial state is given as input to the random transformation outputting a new state which is then used to derive the next ciphertext block by simply XORing again the outer state with the next message block. This process is repeated until the whole message has been processed. If the last message block is smaller than r bits, we simply truncate the outer state to the required size and XOR both parts to obtain the last ciphertext block.

3.4 Differences Between SLAE and ISAP

We have already described in passing some of the differences between SLAE and ISAP, but for clarity, we summarise these distinctions below and discuss them in more detail.

The most prominent difference is that SLAE is based on the T-sponge whereas ISAP employs the P-sponge. In particular the security proofs of SLAE rely on treating ρ as a non-invertible transformation. Treating ρ as an invertible random permutation would add another layer of complexity to the security analysis and we chose not to pursue this route at this point.

The description of ISAP actually specifies three distinct permutations, each obtained from the same round function but with a varying number of rounds. These are used in the different components of ISAP as a means of optimisation. Such heuristic optimisations could be employed in SLAE as well, but in our security analysis we instantiate SLAE with the same random transformation throughout. Indeed this is the more conservative assumption, since otherwise we would be treating these variants as being sampled independently at random when in fact they are intimately related.

Another difference between SLAE and ISAP can be seen in their MAC components SLMAC and ISAPMAC. The design of ISAPMAC is based on combining a rekeying function ISAPRK with a sponge-based suffix MAC. In turn, ISAPRK takes as input a hash of the MAC inputs. As a design optimisation, it is then noted that this hash is already being computed as part of the suffix MAC, at which point it is extracted, fed into ISAPRK, and its output (the session key) is fed back into the last permutation of the suffix MAC to yield the MAC tag. In contrast, in SLMAC, the value corresponding to the session key in ISAPMAC is output directly as the MAC tag thereby showing that the last round in ISAPMAC is essentially redundant.

Finally there are some differences in the way we set parameters in SLAE as opposed to ISAP. For instance, ISAP sets the size of the key and the nonce to be equal. On the other hand, our analysis indicates that the limiting factor in the security of SLAE is the key size. As such it makes sense to set the key size k equal to the width of the sponge n while setting the nonce to be much smaller, say between 64 and 128 bits.

4   The Security of FGHF′

In this section we establish the security of the FGHF′ construction which is depicted in Fig. 15. This is an abstraction of SLAE, and proving its security brings us halfway towards proving the security of SLAE. At the same time, we believe the FGHF′ construction to be of independent interest as it serves as a generic blueprint for constructing efficient AEAD schemes that are non-adaptively leakage-resilient.

The FGHF′ construction is a refinement of the N2 construction [19] which builds a nonce-based AEAD scheme from a nonce-based symmetric encryption scheme and a vector MAC. Barwell et al [4] showed that the security of this construction extends to the setting of leakage resilience. Specifically they showed that if the encryption component is IND-aCPLA secure and the vector MAC is both LPRF and SUF-CMLA secure, then the composition is LAE secure. In turn the FGHF′ construction further breaks down the encryption component, denoted by SE[$\mathcal{F}, \mathcal{G}$], and the vector MAC component, denoted by MAC[$\mathcal{H}, \mathcal{F}'$], of N2 into smaller parts. Namely encryption is realised from a fixed-input-length leakage-resilient PRF $\mathcal{F}$ and a standard PRG $\mathcal{G}$, whereas the vector MAC is built from a vector hash function $\mathcal{H}$, and a fixed-input-length function $\mathcal{F}'$ that is both leakage-resilient pseudorandom and leakage-resilient unpredictable.

Since FGHF′ is an instance of N2 we can apply the composition theorem of Barwell et al. [4], which we reproduced and adapted to the random transformation model in Sect. 2.5. Moreover, since we can view non-adaptive leakage as a special case of adaptive leakage where the leakage set is a singleton, the theorem carries over to that setting which is what we are interested in here. Thus to prove that the FGHF′ construction is LAE secure we only need to show that the encryption and MAC components meet the requirements of Theorem 1.

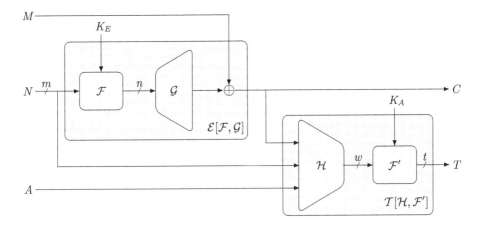

Fig. 15. Graphical representation of the FGHF' construction. It corresponds to the N2 composition of $\mathrm{SE}[\mathcal{F}, \mathcal{G}] = (\mathcal{E}, \mathcal{D})$ and $\mathrm{MAC}[\mathcal{H}, \mathcal{F}'] = (\mathcal{T}, \mathcal{V})$ which are in turn composed of a fixed-input-length LPRF $\mathcal{F}$, a PRG $\mathcal{G}$, a vector hash $\mathcal{H}$, and a fixed-input-length function $\mathcal{F}'$ that is both a LUF and an LPRF.

As it turns out, we can realise an IND-aCPLA secure encryption directly from an LPRF and a variable-output-length PRG. Here the PRG serves only to extend the range of the LPRF in order for the encryption scheme to accommodate variable-length messages. Surprisingly, a standard PRG without any leakage resilience suffices. As for the vector MAC component it needs to be an LPRF over a vector of strings and simultaneously satisfy SUF-CMLA security. Contrary to the leakage-free setting, the latter property is not automatically implied by the former when a MAC is constructed from an LPRF through the canonical construction. This is because the SUFCMLA game is more permissive than the LPRF game with respect to the adversary's queries. Namely, the adversary can forward queries from the LVfy to Vfy, whereas in the LPRF game the adversary is not allowed to forward queries from LF to F. This precludes reducing SUF-CMLA security to LPRF security due to our inability of simulating the verification oracles via the respective LPRF oracles. Note that SUF-CMLA needs to be defined this way for Theorem 1 to hold whereas lifting the restriction in the LPRF game would make it unsatisfiable. We overcome this problem by noting that, in the non-adaptive leakage setting, unpredictability suffices to achieve SUF-CMLA security, and at the same time we can allow the adversary to forward queries between its leakage and challenge oracles while maintaining satisfiability. This leads to our notion of a LUF which we prove to be sufficient to yield SUF-CMLA security. As we will see in the next section we can construct fixed-input-length function families satisfying both notions rather easily from sponges. Given such a function family $\mathcal{F}'$, we can turn it into the required vector MAC by composing it with a collision-resistant vector hash function. Specifically we show that we can extend the domain of LPRFs and LUFs, rather efficiently, by composing them with standard collision-resistant hash functions over appropriate domains.

Combining the results in this section, leads to the LAE security of the FGHF′ construction against non-adaptive leakage. We like this construction as it strikes a practical balance between security and efficiency. By settling for non-adaptive leakage, which seems to suffice for many practical applications, it only requires one call to each of the leakage-resilient primitives, $\mathcal{F}$ and $\mathcal{F}'$, per encryption query. In this work we focused on SLAE which is a specific sponge-based instantiation of FGHF′, but other instantiations, possibly based on different techniques, are of course possible. Thus this construction essentially reduces the problem of designing non-adaptively leakage-resilient AEAD schemes to that of designing function families over small domains that are good LPRFs and LUFs, which conceptually is a much simpler target.

4.1   SE[$\mathcal{F}, \mathcal{G}$] is IND-aCPLA Secure

We begin by proving the security of the encryption component of FGHF′. Note that for this part security holds in more general setting of adaptive leakage. Below is the formal theorem statement and its proof is presented in the full version of this paper.

Theorem 2. *Let* SE[$\mathcal{F}, \mathcal{G}$] *be the encryption scheme depicted in Fig. 15, composed of the function family* $\mathcal{F}: \mathcal{K} \times \{0,1\}^m \to \{0,1\}^n$ *and the PRG* $\mathcal{G}: \{0,1\}^n \to \{0,1\}^*$ *with respective associated leakage sets* $\mathcal{L}_F$ *and* $\mathcal{L}_G$. *Then for any* IND-aCPLA *adversary* $\mathcal{A}_{se}$ *against* SE[$\mathcal{F}, \mathcal{G}$] *and associated leakage sets* $\mathcal{L}_E = \mathcal{L}_D = \mathcal{L}_F \times \mathcal{L}_G$, *there exist an LPRF adversary* $\mathcal{A}_{lprf}$ *against* $\mathcal{F}$ *and a PRG adversary* $\mathcal{A}_{prg}$ *against* $\mathcal{G}$ *such that:*

$$\mathbf{Adv}_{\mathrm{SE}[\mathcal{F},\mathcal{G}]}^{\mathrm{ind\text{-}acpla}} (\mathcal{A}_{se}, \mathcal{L}_E, \mathcal{L}_D) \leq 2\,\mathbf{Adv}_{\mathcal{F}}^{\mathrm{lprf}} (\mathcal{A}_{lprf}, \mathcal{L}_F) + 2\,\mathbf{Adv}_{\mathcal{G}}^{\mathrm{prg}} (\mathcal{A}_{prg}) \ .$$

Let q and μ be such that $\mathcal{A}_{se}$ makes at most q queries totalling μ bits to each of its oracles Enc, LEnc, and LDec, and let q_ρ denote the number of queries it makes to ρ. Then $\mathcal{A}_{lprf}$ makes at most q and $2q$ queries to its oracles F and LF, totalling qm and $2qm$, respectively, and at most $Q_{\mathcal{G}}(2q, 2\mu) + q_\rho$ to ρ. As for $\mathcal{A}_{prg}$, it makes at most q queries to its oracle G totalling μ bits and $Q_{\mathcal{F}}(2q, 2qm) + Q_{\mathcal{G}}(2q, 2\mu) + q_\rho$ queries to ρ.

4.2   MAC[$\mathcal{H}, \mathcal{F}'$] is SUF-CMLA Secure

Next we reduce the SUF-CMLA security of MAC[$\mathcal{H}, \mathcal{F}'$] to the LUF security of $\mathcal{F}'$ and the collision resistance of $\mathcal{H}$. Towards this end, we first show that any LUF $\hat{\mathcal{F}}$ over domain $\mathcal{X}$ yields a SUF-CMLA secure MAC with message space $\mathcal{X}$ via the canonical construction. Then we show that such a function $\hat{\mathcal{F}}$ can be constructed from a fixed-input-length LUF $\mathcal{F}'$ and a collision-resistant hash function with domain $\mathcal{X}$. The formal theorem statements now follow. Their proofs can be found in the full version of this paper.

Theorem 3. *Let* $\hat{\mathcal{F}} \colon \mathcal{K} \times \mathcal{X} \to \{0,1\}^t$ *be a function family with associated leakage set* $\mathcal{L}_{\hat{F}}$, *and let* $\mathrm{MAC}[\hat{\mathcal{F}}]$ *be the corresponding canonical MAC with associated leakage sets* $\mathcal{L}_T$, $\mathcal{L}_V$ *where* $\mathcal{L}_{\hat{F}} = \mathcal{L}_T = \mathcal{L}_V$. *Then for any SUF-CMLA adversary* $\mathcal{A}_{mac}$ *against* $\mathrm{MAC}[\hat{\mathcal{F}}]$, *there exists an adversary* $\mathcal{A}_{luf}$ *against* $\hat{\mathcal{F}}$ *such that:*

$$\mathbf{Adv}_{\mathrm{MAC}[\hat{\mathcal{F}}]}^{\mathrm{suf\text{-}cmla}}\left(\mathcal{A}_{mac}, \mathcal{L}_T, \mathcal{L}_V\right) \leq \mathbf{Adv}_{\hat{\mathcal{F}}}^{\mathrm{luf}}\left(\mathcal{A}_{luf}, \mathcal{L}_{\hat{F}}\right).$$

Let q and μ be such that $\mathcal{A}_{mac}$ makes at most q queries totalling μ bits to each of its oracles Vfy, LTag, and LVfy. Then $\mathcal{A}_{luf}$ makes at most q, $2q$, and $2q$ queries to F, Lkg, and Guess, totalling μ, 2μ, and 2μ bits, respectively.

Theorem 4. *Let* $\mathcal{F}' \colon \mathcal{K} \times \{0,1\}^w \to \{0,1\}^t$ *be a function family with associated leakage set* $\mathcal{L}_{F'}$, *and let* $\mathcal{H} \colon \mathcal{X} \to \{0,1\}^w$ *be a hash function over any domain* $\mathcal{X}$. *Further let their composition* $\hat{\mathcal{F}}$ *be defined as*

$$\hat{\mathcal{F}}(K, X) = \mathcal{F}'(K, \mathcal{H}(X))$$

where $X \in \mathcal{X}$, $K \in \mathcal{K}$, *and* $\mathcal{L}_{\hat{F}} = \mathcal{L}_{F'} \times \mathcal{L}_H$ *for any set of efficiently computable functions* $\mathcal{L}_H$. *Then for any LUF adversary* $\mathcal{A}_{luf}$ *against* $\hat{\mathcal{F}}$, *there exists a corresponding LUF adversary* $\mathcal{A}'_{luf}$ *against* $\mathcal{F}'$ *and an adversary* $\mathcal{A}_{hash}$ *against* $\mathcal{H}$ *such that:*

$$\mathbf{Adv}_{\hat{\mathcal{F}}}^{\mathrm{luf}}\left(\mathcal{A}_{luf}, \mathcal{L}_{\hat{F}}\right) \leq 2\,\mathbf{Adv}_{\mathcal{H}}^{\mathrm{cr}}\left(\mathcal{A}_{hash}\right) + \mathbf{Adv}_{\mathcal{F}'}^{\mathrm{luf}}\left(\mathcal{A}'_{luf}, \mathcal{L}_{F'}\right).$$

Let q and μ be such that $\mathcal{A}_{luf}$ makes at most q queries totalling μ bits to each of its oracles F, Lkg, and Guess, and let q_ρ denote the number of queries it makes to ρ. Then $\mathcal{A}'_{luf}$ makes at most q queries totalling qw bits to each of its oracles F, Lkg, and Guess, and at most $Q_{\mathcal{H}}(3q, 3\mu) + q_\rho$ queries to ρ. As for $\mathcal{A}_{hash}$, it requires at most $Q_{\mathcal{F}'}(3q, 3qw) + Q_{\mathcal{H}}(3q, 3\mu)$ queries to ρ in order to simulate $\mathcal{F}'$ and $\mathcal{H}$.

Combining both theorems, we obtain the following simple corollary reducing the SUF-CMLA security of $\mathrm{MAC}[\mathcal{H}, \mathcal{F}']$ to that of its building blocks $\mathcal{H}$ and $\mathcal{F}'$.

Corollary 1. *Let* $\mathrm{MAC}[\mathcal{H}, \mathcal{F}']$ *be the MAC component depicted in Fig. 15, composed of the hash function* $\mathcal{H}$ *and the function family* $\mathcal{F}'$ *with respective leakage sets* $\mathcal{L}_H$ *and* $\mathcal{L}_{F'}$. *Then for any SUF-CMLA adversary* $\mathcal{A}_{mac}$ *against* $\mathrm{MAC}[\mathcal{H}, \mathcal{F}']$ *with associated leakage sets* $\mathcal{L}_T = \mathcal{L}_V = \mathcal{L}_{F'} \times \mathcal{L}_H$, *there exists a LUF adversary* $\mathcal{A}_{luf}$ *against* $\mathcal{F}'$ *and an adversary* $\mathcal{A}_{hash}$ *against* $\mathcal{H}$ *such that:*

$$\mathbf{Adv}_{\mathrm{MAC}[\mathcal{H}, \mathcal{F}']}^{\mathrm{suf\text{-}cmla}}\left(\mathcal{A}_{mac}, \mathcal{L}_T, \mathcal{L}_V\right) \leq 2\,\mathbf{Adv}_{\mathcal{H}}^{\mathrm{cr}}\left(\mathcal{A}_{hash}\right) + \mathbf{Adv}_{\mathcal{F}'}^{\mathrm{luf}}\left(\mathcal{A}_{luf}, \mathcal{L}_{F'}\right).$$

Suppose $\mathcal{A}_{mac}$ makes at most q queries totalling at most μ bits to each of its oracles Vfy, LTag, and LVfy, and q_ρ to ρ. Then $\mathcal{A}_{luf}$ makes at most $2q$ queries totalling at most $2qw$ bits to each of the oracles in the LUF game, and $Q_{\mathcal{H}}(6q, 6\mu) + q_\rho$ queries to ρ. As for $\mathcal{A}_{hash}$ it needs at most $Q_{\mathcal{F}'}(6q, 6qw) + Q_{\mathcal{H}}(6q, 6\mu)$ queries to ρ to simulate $\mathcal{F}'$ and $\mathcal{H}$.

4.3 $\mathrm{MAC}[\mathcal{H}, \mathcal{F}']$ is LPRF Secure

The final piece needed to apply Theorem 1 is to show that $\mathrm{MAC}[\mathcal{H}, \mathcal{F}']$, or rather its tagging algorithm $\mathcal{T}[\mathcal{H}, \mathcal{F}']$, is a leakage-resilient PRF. Since by assumption $\mathcal{F}'$ is already an LPRF, this result is analogous to Theorem 4 in that it provides us a with simple technique for extending the domain of an LPRF. The proof can be found in the full version of this paper.

Theorem 5. *Let $\mathcal{F}' \colon \mathcal{K} \times \{0,1\}^w \to \{0,1\}^t$ be a function family with associated leakage set $\mathcal{L}_{F'}$, and let $\mathcal{H} \colon \mathcal{X} \to \{0,1\}^w$ be a hash function over the domain $\mathcal{X}$. Further let their composition $\hat{\mathcal{F}}$ be defined as*

$$\hat{\mathcal{F}}(K, X) = \mathcal{F}'(K, \mathcal{H}(X))$$

where $X \in \mathcal{X}$, $K \in \mathcal{K}$, and $\mathcal{L}_{\hat{F}} = \mathcal{L}_{F'} \times \mathcal{L}_H$ for any set of efficiently computable functions $\mathcal{L}_H$. Then for any LPRF adversary $\mathcal{A}_{lprf}$ against $\hat{\mathcal{F}}$, there exists a corresponding LPRF adversary $\mathcal{A}'_{lprf}$ against $\mathcal{F}'$ and an adversary $\mathcal{A}_{hash}$ against $\mathcal{H}$ such that:

$$\mathbf{Adv}^{\mathrm{lprf}}_{\hat{\mathcal{F}}} \left(\mathcal{A}_{lprf}, \mathcal{L}_{\hat{F}} \right) \leq 2\,\mathbf{Adv}^{\mathrm{cr}}_{\mathcal{H}} \left(\mathcal{A}_{hash} \right) + \mathbf{Adv}^{\mathrm{lprf}}_{\mathcal{F}'} \left(\mathcal{A}'_{lprf}, \mathcal{L}_{F'} \right) .$$

Let q and μ be such that $\mathcal{A}_{lprf}$ makes at most q queries totalling μ bits to each of its oracles F and LF, and let q_ρ denote the number of queries it makes to ρ. Then $\mathcal{A}'_{lprf}$ makes at most q queries totalling qw bits to each of its oracles F and LF, and at most $Q_\mathcal{H}(2q, 2\mu) + q_\rho$ queries to ρ. As for $\mathcal{A}_{hash}$, it requires at most $Q_{\mathcal{F}'}(2q, 2qw) + Q_\mathcal{H}(2q, 2\mu)$ queries to ρ in order to simulate $\mathcal{F}'$ and $\mathcal{H}$.

4.4 The FGHF' Composition Theorem

Collecting the results from this section and combining it with the N2 composition theorem we get the following composition theorem for the FGHF' construction.

Theorem 6 (LAE Security of the FGHF' Construction). *Let $\mathcal{F}$ be a fixed-input-length LPRF, $\mathcal{G}$ a PRG, $\mathcal{H}$ a vector hash function, and $\mathcal{F}'$ be a fixed-input-length function that is both an LUF and an LPRF with associated leakage sets $\mathcal{L}_F$, $\mathcal{L}_G$, $\mathcal{L}_H$, and $\mathcal{L}_{F'}$, respectively. Let FGHF' be the composition of $\mathcal{F}$, $\mathcal{G}$, $\mathcal{H}$, and $\mathcal{F}'$ with associated leakage sets $\mathcal{L}_{AE} = \mathcal{L}_{VD} = \mathcal{L}_F \times \mathcal{L}_G \times \mathcal{L}_H \times \mathcal{L}_{F'}$. Then for any LAE adversary $\mathcal{A}_{ae}$ against FGHF' there exist adversaries $\mathcal{A}_{lprf}$, $\mathcal{A}'_{lprf}$, $\mathcal{A}_{prg}$, $\mathcal{A}_{hash}$, and $\mathcal{A}_{luf}$ such that:*

$$\begin{aligned}
\mathbf{Adv}^{\mathrm{lae}}_{\mathrm{FGHF}'} \left(\mathcal{A}_{ae}, \mathcal{L}_{AE}, \mathcal{L}_{VD} \right) \leq\ & 2\,\mathbf{Adv}^{\mathrm{lprf}}_{\mathcal{F}} \left(\mathcal{A}_{lprf}, \mathcal{L}_F \right) + 2\,\mathbf{Adv}^{\mathrm{lprf}}_{\mathcal{F}'} \left(\mathcal{A}'_{lprf}, \mathcal{L}_{F'} \right) \\
& + 2\,\mathbf{Adv}^{\mathrm{prg}}_{\mathcal{G}} \left(\mathcal{A}_{prg} \right) + 6\,\mathbf{Adv}^{\mathrm{cr}}_{\mathcal{H}} \left(\mathcal{A}_{hash} \right) \\
& + 2\,\mathbf{Adv}^{\mathrm{luf}}_{\mathcal{F}'} \left(\mathcal{A}_{luf}, \mathcal{L}_{F'} \right) .
\end{aligned}$$

Now suppose $\mathcal{A}_{ae}$ makes at most q queries totalling at most μ bits to each of its Enc, LEnc, Dec, and LDec oracles, and let q_ρ denote its number of queries to ρ. Then, $\mathcal{A}_{lprf}$ makes at most $2q$ queries totalling $2qm$ bits to each of its oracles F and LF, and at most $2Q_{\mathcal{H}}(2q, 2\mu) + 2Q_{\mathcal{F}'}(2q, 2qw) + Q_{\mathcal{G}}(2q, 2\mu)$ queries to ρ. Similarly, $\mathcal{A}'_{lprf}$ makes at most $2q$ queries totalling $2qw$ bits to each of its oracles F and LF, and at most $2Q_{\mathcal{F}}(q, qm) + 2Q_{\mathcal{G}}(q, \mu) + Q_{\mathcal{H}}(4q, 4\mu)$ queries to ρ. $\mathcal{A}_{luf}$ makes at most $4q$ queries, totalling $4qw$ bits to each of its oracles F and Lkg, and at most $2Q_{\mathcal{F}}(2q, 2qm) + 2Q_{\mathcal{G}}(2q, 2\mu) + Q_{\mathcal{H}}(12q, 12\mu)$ to ρ. As for $\mathcal{A}_{prg}$, it makes at most q queries, totalling μ bits, to its oracle G and at most $2Q_{\mathcal{H}}(2q, 2\mu) + 2Q_{\mathcal{F}'}(2q, 2qw) + Q_{\mathcal{F}}(2q, 2qm) + Q_{\mathcal{G}}(2q, 2\mu)$ queries to ρ. Finally, $\mathcal{A}_{hash}$ requires at most $2Q_{\mathcal{F}}(2q, 2qm) + 2Q_{\mathcal{G}}(2q, 2\mu) + Q_{\mathcal{F}'}(12q, 12qw) + Q_{\mathcal{H}}(12q, 12\mu)$ queries to ρ.

5   Security of Sponge-Based Primitives

We now turn our attention to instantiating the constituent blocks of the FGHF' construction using sponge-based primitives. Specifically we prove the security of the vector hash function SvHASH, the pseudorandom generator SPRG, and the leakage-resilient function family SLFUNC for instantiating both $\mathcal{F}$ and $\mathcal{F}'$. All primitives are based on the T-sponge and this particular instantiation of the FGHF' construction gives rise to SLAE. The most interesting results are Theorems 7 and 8 which substantiate our claim that sponges offer an inherent resistance to non-adaptive leakage. Informally these two theorems state that λ bits of leakage can be compensated for by increasing the capacity, the key, and the output (in the case of LUF security) by λ bits. While this may seem intuitive, and indeed this was already conjectured informally in [10], the actual proofs are fairly involved. While sponge-based hash functions and pseudorandom generators have been studied quite extensively, SvHASH and SPRG are non-standard constructions. Firstly, they are based on a transformation rather than a permutation which is not common in the literature. Secondly, unlike other constructions SPRG treats the whole initial state as the seed, and SvHASH takes a triple of strings as its input. Thus while not particularly novel, we include their security proofs for completeness.

5.1   A Sponge-Based Leakage-Resilient Function Family

Although LPRF and LUF security are incomparable notions, it is still possible to meet both notions simultaneously through a single primitive. Indeed the FGHF' construction requires that such a primitive exist since $\mathcal{F}'$ is required to satisfy both security notions. We now show that the SLFUNC construction is well-suited for this role, and in fact that it can be used to instantiate both the $\mathcal{F}$ and $\mathcal{F}'$ components – as is the case in SLAE. Moreover, the most extensively studied leakage-resilient object is that of a pseudorandom function due to its versatility in several potential applications. SLFUNC yields a practical construction of this primitive against non-adaptive leakage and as such we think it may be of independent interest. The security of SLFUNC is stated formally in the following two theorems. Their proofs can be found in the full version of this paper.

Theorem 7. *Let* SLFUNC *be the function family described in Fig. 9 taking as input strings of size* $(l \cdot rr)$ *bits and returning t-bit strings. Then for any LPRF adversary* $\mathcal{A}$ *against* SLFUNC *and any vector of leakage functions* $[L_0, \ldots, L_l]$ *where each component maps n bits to* λ *bits such that* $\mathcal{L}_\lambda = \{[L_0, \ldots, L_l]\}$*, it holds that:*

$$\mathbf{Adv}^{\mathrm{lprf}}_{\mathrm{SLFUNC}}(\mathcal{A}, \mathcal{L}_\lambda) \leq \frac{q_T(q_T + 2) + (q_{\mathsf{F}} + q_{\mathsf{LF}})q_\rho}{2^{n-rr-1}} + \frac{2q_\rho}{2^{k-\lambda}} + \frac{2l\,q_{\mathsf{F}}\,q_\rho}{2^{n-rr-\lambda}}.$$

In the above q_ρ*,* q_{F}*, and* q_{LF} *denote respectively the number of queries* $\mathcal{A}$ *makes to its oracles* ρ*,* F*, and* LF *and* $q_T = (l+1)(q_{\mathsf{LF}} + q_{\mathsf{F}}) + q_\rho$*. Moreover it is required that* $q_\rho + l(q_{\mathsf{F}} + q_{\mathsf{LF}}) \leq 2^{k-1}$ *and* $(2^{rr})q_\rho + l(q_{\mathsf{F}} + q_{\mathsf{LF}}) \leq 2^{n-1}$*.*

The next Theorem shows that SLFUNC is a good LUF. Its proof bears some similarity to that of Theorem 7 as it uses similar ideas. However one important difference lies in the leakage model that is used in this theorem. Since the Lkg oracle returns only the leakage and no output, we add here an extra leakage function that returns the leakage on the output of SLFUNC. In the LPRF case this was not required since in that game the leakage oracle returns the full output anyway.

Theorem 8. *Let* SLFUNC *be the function family described in Fig. 9 taking as input strings of size* $(l \cdot rr)$ *bits and returning t-bit long strings. Then for any LUF adversary* $\mathcal{A}$ *against* SLFUNC*, and any vector of leakage functions* $[L_0, \ldots, L_{l+1}]$ *where each component maps n bits to* λ *bits such that* $\mathcal{L}_\lambda = \{[L_0, \ldots, L_{l+1}]\}$*, it holds that:*

$$\mathbf{Adv}^{\mathrm{luf}}_{\mathrm{SLFUNC}}(\mathcal{A}, \mathcal{L}_\lambda) \leq \frac{q_T(q_T + 2)}{2^{n-rr}} + \frac{q_\rho}{2^{k-\lambda-1}} + \frac{l\,q_{\mathsf{Lkg}}\,q_\rho}{2^{n-\lambda-1}} + \frac{q_{\mathsf{Guess}}}{2^{t-\lambda-1}}.$$

In the above q_ρ*,* q_{F}*,* q_{Lkg} *and* q_{Guess} *denote respectively the number of queries* $\mathcal{A}$ *makes to its oracles* ρ*,* F*,* Lkg*, and* Guess *and* $q_T = (l+1)(q_{\mathsf{F}} + q_{\mathsf{Lkg}} + q_{\mathsf{Guess}}) + q_\rho$*. Moreover it is required that the following conditions be satisfied* $q_\rho + (l+1)(q_{\mathsf{F}} + q_{\mathsf{Lkg}} + q_{\mathsf{Guess}}) \leq 2^{k-1}$*,* $(2^{rr})q_\rho + (l+1)(q_{\mathsf{F}} + q_{\mathsf{Lkg}} + q_{\mathsf{Guess}}) \leq 2^{n-1}$*, and* $q_{\mathsf{Guess}} + (l+1)(q_{\mathsf{F}} + q_{\mathsf{Lkg}} + q_{\mathsf{Guess}}) \leq 2^{n-1}$*.*

5.2   The Security of SPRG

As explained in Sect. 3.3, SLENC can be decomposed into the cascade of SLFUNC and SPRG, matching the encryption component of the FGHF' construction. A pseudocode description of the variable-output-length pseudorandom generator SPRG is given in Fig. 14. Decomposing SLENC this way requires us to treat all of SPRG's initial state as the seed, which deviates from the more conventional ways of constructing sponge-based pseudorandom generators. Moreover we consider a security definition which allows the adversary to make multiple queries to the PRG, each with differing output lengths.

The security of SPRG is stated formally in Theorem 9. Its proof follows from a standard hybrid argument and can be found in the full version of this paper.

Theorem 9. *Let* SPRG *be the pseudorandom generator described in Fig. 14. Then for any PRG adversary A, it holds that:*

$$\mathbf{Adv}^{\mathrm{prg}}_{\mathrm{SPRG}}(A) \leq \sum_{i=1}^{v_{max}-1} \left(\frac{q_\mathsf{G} q_\rho}{2^c - q_\rho} + \frac{q_\rho q_\mathsf{G} + 2q_\mathsf{G}^2(v_{max} - i)}{2^n} \right) + \frac{q_\mathsf{G}^2}{2^n}.$$

In the above, A can make q_ρ queries to the random transformation ρ and q_G queries to the challenge oracle G of size at most L_{max}, and $v_{max} = \lceil \frac{L_{max}}{r} \rceil$.

5.3   A Sponge-Based Vector Hash Function

The final building block is the sponge-based vector hash function SvHASH which is graphically represented in Fig. 10. It takes as input a triple of strings, namely a nonce, associated data and a ciphertext to return a string digest. A salient feature of this construction is the xoring of $1 \parallel 0^{c-1}$ into the inner state in order to separate the (padded) associated data from the (padded) ciphertext. We prove the security of SvHASH in a modular fashion, by first reducing its security to that of a plain hash function taking a single input and then prove the collision-resistance of this latter construction in the random transformation model. The collision-resistance of SvHASH is stated formally in the following theorem, and the full proof details can be found in the full version of this paper.

Theorem 10. *Let* SvHASH *be the vector hash function described in Fig. 10. Then for any adversary A making q queries to ρ, it holds that:*

$$\mathbf{Adv}^{\mathrm{cr}}_{\mathrm{SvHASH}}(A) \leq \frac{q(q-1)}{2^{w+1}} + \frac{q(q+2)}{2^{c-1}}.$$

5.4   Concrete Security of SLAE

A bound for the security of SLAE is directly obtained by combining Theorem 6 with Theorems 7–10. It only remains to derive concrete bounds for the expressions $Q_\mathcal{F}, Q_\mathcal{G}, Q_\mathcal{H}, Q_{\mathcal{F}'}$ for the specific case of SLAE. Assuming a nonce size of m bits and that the output of $\mathcal{H}$ is w bits long, the following expressions are easily derived from the algorithm definitions. Namely, we have that:

$$Q_\mathcal{F}(q, qm) = q \left\lceil \frac{m+1}{rr} \right\rceil \qquad\qquad Q_{\mathcal{F}'}(q, qw) = q \left\lceil \frac{w+1}{rr} \right\rceil$$

$$Q_\mathcal{G}(q, \mu) = \left\lceil \frac{\mu}{r} \right\rceil \qquad\qquad\qquad Q_\mathcal{H}(q, \mu) = \left\lceil \frac{\mu}{r} \right\rceil + 3q.$$

6 Concluding Remarks

In this work we proposed the $FGHF'$ construction as a template for constructing non-adaptively leakage-resilient AEAD schemes from relatively simpler primitives – requiring only two calls to the leakage-resilient functions per encryption or decryption call. We then presented SLAE as a sponge-based instantiation of this construction, offering good performance and simplicity. Our security analysis shows that λ bits of leakage per transformation call can be compensated for by increasing the sponge capacity by λ bits. However some care is needed in interpreting these results. Like most treatments of leakage resilience we assume that the leakage per evaluation is limited and does not drain the entropy in the secret state. Thus it is implicitly assumed that an implementation is good enough to withstand basic side-channel attacks like Simple Power Analysis (SPA) attacks. The benefit of our leakage-resilience security proof is that resistance to basic attacks automatically translates to resistance against more sophisticated attacks like Differential Power Analysis (DPA).

In the $FGHF'$ construction and SLAE, authenticity is verified by recomputing the MAC tag and testing for equality between the recomputed tag and the one included in the ciphertext. While our leakage model accounts for the leakage that may take place during the tag recomputation, equality testing is assumed to be leak-free. Thus any implementation of SLAE (or any other realisation of the $FGHF'$ construction) needs to ensure that equality testing does not leak, or take additional measures, such as masking, to protect against leakage from this component.

Finally the security of SLAE relies on it being instantiated with a non-invertible transformation rather than a permutation. On the other hand, most practical schemes employ permutations, such as KECCAK-p and XOODOO-p. While in this work we did not specify any concrete transformation, a natural candidate is to use $\rho(x) = p(x) \oplus x$ for $p \in \{\text{KECCAK-}p, \text{XOODOO-}p\}$. Although this construction is known to be differentiable from a random transformation when given access to p, this should not preclude it from being a suitable candidate for instantiating constructions in the random transformation model. Indeed, KECCAK-p and XOODOO-p are also differentiable from a random permutation when given access to their underlying building blocks.

Acknowledgements. We thank Daniel Baur and Christian Schuller for initial discussions during the early stages of this project, and our anonymous reviewers for their constructive comments. Degabriele was supported by the German Federal Ministry of Education and Research (BMBF) as well as by the Hessian State Ministry for Higher Education, Research and Arts (HMWK) within CRISP. Janson was co-funded by the DFG as part of project P2 within the CRC 1119 CROSSING. Struck was funded by the DFG as part of project P1 within the CRC 1119 CROSSING.

References

1. Abdalla, M., Belaïd, S., Fouque, P.-A.: Leakage-resilient symmetric encryption via re-keying. In: Bertoni, G., Coron, J.-S. (eds.) CHES 2013. LNCS, vol. 8086, pp. 471–488. Springer, Heidelberg (2013). https://doi.org/10.1007/978-3-642-40349-1_27
2. Abed, F., Berti, F., Lucks, S.: Insecurity of RCB: leakage-resilient authenticated encryption. Cryptology ePrint Archive, Report 2016/1121 (2016). http://eprint.iacr.org/2016/1121
3. Agrawal, M., et al.: RCB: leakage-resilient authenticated encryption via re-keying. J. Supercomputing **74**(9), 4173–4198 (2018)
4. Barwell, G., Martin, D.P., Oswald, E., Stam, M.: Authenticated encryption in the face of protocol and side channel leakage. In: Takagi, T., Peyrin, T. (eds.) ASIACRYPT 2017. LNCS, vol. 10624, pp. 693–723. Springer, Cham (2017). https://doi.org/10.1007/978-3-319-70694-8_24
5. Bellare, M., Rogaway, P.: The security of triple encryption and a framework for code-based game-playing proofs. In: Vaudenay, S. (ed.) EUROCRYPT 2006. LNCS, vol. 4004, pp. 409–426. Springer, Heidelberg (2006). https://doi.org/10.1007/11761679_25
6. Bernstein, D.J.: CAESAR: Competition for Authenticated Encryption: Security, Applicability, and Robustness (2014)
7. Berti, F., Koeune, F., Pereira, O., Peters, T., Standaert, F.-X.: Leakage-resilient and misuse-resistant authenticated encryption. Cryptology ePrint Archive, Report 2016/996 (2016). http://eprint.iacr.org/2016/996
8. Berti, F., Pereira, O., Peters, T., Standaert, F.-X.: On leakage-resilient authenticated encryption with decryption leakages. IACR Trans. Symm. Cryptol. **2017**(3), 271–293 (2017)
9. Bertoni, G., Daemen, J., Peeters, M., Van Assche, G.: Sponge functions. In: ECRYPT Hash Workshop (2007). https://keccak.team/files/SpongeFunctions.pdf
10. Dobraunig, C., Eichlseder, M., Mangard, S., Mendel, F., Unterluggauer, T.: ISAP - towards side-channel secure authenticated encryption. IACR Trans. Symm. Cryptol. **2017**(1), 80–105 (2017)
11. Dobraunig, C., Mennink, B.: Leakage resilience of the duplex construction. Cryptology ePrint Archive, Report 2019/225 (2019). https://eprint.iacr.org/2019/225
12. Dodis, Y., Pietrzak, K.: Leakage-resilient pseudorandom functions and side-channel attacks on feistel networks. In: Rabin, T. (ed.) CRYPTO 2010. LNCS, vol. 6223, pp. 21–40. Springer, Heidelberg (2010). https://doi.org/10.1007/978-3-642-14623-7_2
13. Dziembowski, S., Pietrzak, K.: Leakage-resilient cryptography. In: 49th FOCS, pp. 293–302. IEEE Computer Society Press, October 2008
14. Faust, S., Pietrzak, K., Schipper, J.: Practical leakage-resilient symmetric cryptography. In: Prouff, E., Schaumont, P. (eds.) CHES 2012. LNCS, vol. 7428, pp. 213–232. Springer, Heidelberg (2012). https://doi.org/10.1007/978-3-642-33027-8_13
15. Goldreich, O., Goldwasser, S., Micali, S.: How to construct random functions. J. ACM **33**(4), 792–807 (1986)
16. Guo, C., Pereira, O., Peters, T., Standaert, F.-X.: Leakage-resilient authenticated encryption with misuse in the leveled leakage setting: definitions, separation results, and constructions. Cryptology ePrint Archive, Report 2018/484 (2018). https://eprint.iacr.org/2018/484

17. Guo, C., Pereira, O., Peters, T., Standaert, F.-X.: Towards lightweight side-channel security and the leakage-resilience of the duplex sponge. Cryptology ePrint Archive, Report 2019/193 (2019). https://eprint.iacr.org/2019/193
18. Longo, J., Martin, D.P., Oswald, E., Page, D., Stam, M., Tunstall, M.J.: Simulatable leakage: analysis, pitfalls, and new constructions. In: Sarkar, P., Iwata, T. (eds.) ASIACRYPT 2014. LNCS, vol. 8873, pp. 223–242. Springer, Heidelberg (2014). https://doi.org/10.1007/978-3-662-45611-8_12
19. Namprempre, C., Rogaway, P., Shrimpton, T.: Reconsidering generic composition. In: Nguyen, P.Q., Oswald, E. (eds.) EUROCRYPT 2014. LNCS, vol. 8441, pp. 257–274. Springer, Heidelberg (2014). https://doi.org/10.1007/978-3-642-55220-5_15
20. Pereira, O., Standaert, F.-X., Vivek, S.: Leakage-resilient authentication and encryption from symmetric cryptographic primitives. In: Ray, I., Li, N., Kruegel, C. (eds.) ACM CCS 2015, pp. 96–108. ACM Press, October 2015
21. Rogaway, P.: Authenticated-encryption with associated-data. In: Atluri, V. (ed.) ACM CCS 2002, pp. 98–107. ACM Press, November 2002
22. Standaert, F., Pereira, O., Yu, Y., Quisquater, J., Yung, M., Oswald, E.: Leakage resilient cryptography in practice. In: Sadeghi, A., Naccache, D. (eds.) Towards Hardware-Intrinsic Security - Foundations and Practice. Information Security and Cryptography, pp. 99–134. Springer, Heidelberg (2010). https://doi.org/10.1007/978-3-642-14452-3_5
23. Standaert, F.-X., Pereira, O., Yu, Y.: Leakage-resilient symmetric cryptography under empirically verifiable assumptions. In: Canetti, R., Garay, J.A. (eds.) CRYPTO 2013. LNCS, vol. 8042, pp. 335–352. Springer, Heidelberg (2013). https://doi.org/10.1007/978-3-642-40041-4_19
24. Yu, Y., Standaert, F.-X., Pereira, O., Yung, M.: Practical leakage-resilient pseudo-random generators. In: Al-Shaer, E., Keromytis, A.D., Shmatikov, V. (eds.) ACM CCS 2010, pp. 141–151. ACM Press, October 2010

Isogenies (2)

Dual Isogenies and Their Application to Public-Key Compression for Isogeny-Based Cryptography

Michael Naehrig[1(✉)] and Joost Renes[2]

[1] Microsoft Research, Redmond, WA, USA
mnaehrig@microsoft.com
[2] Digital Security Group, Radboud University, Nijmegen, The Netherlands
j.renes@cs.ru.nl

Abstract. The isogeny-based protocols SIDH and SIKE have received much attention for being post-quantum key agreement candidates that retain relatively small keys. A recent line of work has proposed and further improved compression of public keys, leading to the inclusion of public-key compression in the SIKE proposal for Round 2 of the NIST Post-Quantum Cryptography Standardization effort. We show how to employ the *dual* isogeny to significantly increase performance of compression techniques, reducing their overhead from 160–182% to 77–86% for Alice's key generation and from 98–104% to 59–61% for Bob's across different SIDH parameter sets. For SIKE, we reduce the overhead of (1) key generation from 140–153% to 61–74%, (2) key encapsulation from 67–90% to 38–57%, and (3) decapsulation from 59–65% to 34–39%. This is mostly achieved by speeding up the pairing computations, which has until now been the main bottleneck, but we also improve (deterministic) basis generation.

Keywords: Post-Quantum Cryptography · Public-key compression · Supersingular elliptic curves · Dual isogenies · Reduced Tate pairings

1 Introduction

Isogeny-based protocols are an alternative to the more mainstream proposals for post-quantum key agreement, such as lattice-based or code-based schemes. Beyond their reliance on a different type of hard computational problem, the main distinguishing characteristics of isogeny-based key exchange and key encapsulation schemes are their small keys and low communication costs. This is for example seen in the supersingular-isogeny Diffie–Hellman (SIDH) scheme first proposed by Jao and De Feo [12,19], its IND-CCA secure key encapsulation variant SIKE [17] and the more recent CSIDH proposal of Castryck, Lange, Martindale, Panny and Renes [5].

J. Renes—Partially supported by the Technology Foundation STW (project 13499 – TYPHOON & ASPASIA), from the Dutch government.

S. D. Galbraith and S. Moriai (Eds.): ASIACRYPT 2019, LNCS 11922, pp. 243–272, 2019.
https://doi.org/10.1007/978-3-030-34621-8_9

The supersingular isogeny key encapsulation scheme SIKE—one of the 17 key encapsulation mechanisms that advanced to the second round of the NIST standardization process for post-quantum cryptography [24]—has the advantage of small public keys and ciphertexts, with sizes in the low hundreds of bytes. For example, the Round 1 submission of SIKE [18] supports public keys of 378 bytes and ciphertexts of 402 bytes at NIST security level 1. In comparison, at the same security level the lattice-based scheme Saber [11] uses 672 byte public keys and 736 byte ciphertexts, while Kyber [4] uses 800 byte public keys and 736 byte ciphertexts. However, such compact keys and ciphertexts in SIKE are contrasted by comparatively high latencies. The recent CSIDH protocol, a non-interactive key exchange scheme and a potential candidate as a drop-in replacement for the standard Diffie-Hellman key exchange, exhibits these characteristics in an even more extreme fashion. It supports even smaller keys, with significantly larger runtimes.

In a similar fashion, techniques for public-key compression for the SIDH protocol also amplify these characteristics. They allow to reduce the communication bandwidth further, but come at the cost of a large computational overhead compared to the uncompressed variant of SIDH. The same techniques apply to SIKE and reduce its public key and ciphertext sizes. Compression has been included in the Round 2 submission of SIKE [17] and, along with the introduction of new parameter sets, has enabled public keys of merely 196 bytes and ciphertexts of only 209 bytes for NIST level 1. We propose several techniques to reduce the large computational overhead, making the use of public-key compression significantly more interesting for practitioners.

Public-Key Compression for SIDH. Let ℓ, m be distinct primes and e_ℓ, e_m be strictly positive integers such that $p = \ell^{e_\ell} \cdot m^{e_m} - 1$ is prime. Let $E/\mathbb{F}_{p^2}$ be a supersingular elliptic curve such that $\#E(\mathbb{F}_{p^2}) = (p+1)^2$, and let $\phi_\ell : E \to E/\langle R \rangle$ be an isogeny of degree ℓ^{e_ℓ} such that $\ker \phi_\ell = \langle R \rangle$ for some point $R \in E[\ell^{e_\ell}]$. Similarly, let $\phi_m : E \to E/\langle S \rangle$ be an isogeny of degree m^{e_m} such that $\ker \phi_m = \langle S \rangle$ for some point $S \in E[m^{e_m}]$. Fix parameters P_ℓ, Q_ℓ such that $\langle P_\ell, Q_\ell \rangle = E[\ell^{e_\ell}]$ and P_m, Q_m such that $\langle P_m, Q_m \rangle = E[m^{e_m}]$. The public keys for SIDH are encoded by the triples of x-coordinates of the form

$$\left[x_{\phi_\ell(P_m)}, x_{\phi_\ell(Q_m)}, x_{\phi_\ell(P_m - Q_m)} \right] , \left[x_{\phi_m(P_\ell)}, x_{\phi_m(Q_\ell)}, x_{\phi_m(P_\ell - Q_\ell)} \right] ,$$

which can be represented with $6 \log_2 p$ bits each.

It is possible to further reduce the size of the public key to $4 \log_2 p$, as first proposed by Azarderakhsh, Jao, Kalach, Koziel, and Leonardi [1], by deterministically generating points U_m, V_m such that $\langle U_m, V_m \rangle = (E/\langle R \rangle)[m^{e_m}]$ and computing $a_0, b_0, a_1, b_1 \in \mathbb{Z}/m^{e_m}\mathbb{Z}$ such that

$$\begin{bmatrix} \phi_\ell(P_m) \\ \phi_\ell(Q_m) \end{bmatrix} = \begin{bmatrix} a_0 & b_0 \\ a_1 & b_1 \end{bmatrix} \begin{bmatrix} U_m \\ V_m \end{bmatrix} .$$

Although initially believed to be orders of magnitude more expensive than the original isogeny computation, the work of Costello, Jao, Longa, Naehrig,

Renes, and Urbanik [7] significantly reduced the cost to a factor 2.4 slowdown, while simultaneously compressing the public keys to $\frac{7}{2}\log_2 p$ bits. Further computational improvements have since been made by Zanon, Simplicio, Pereira, Doliskani and Barreto [30]. An interesting observation by Zanon et al. [30, §2] is that one can equivalently compute $c_0, d_0, c_1, d_1 \in \mathbb{Z}/m^{e_m}\mathbb{Z}$ such that

$$\begin{bmatrix} U_m \\ V_m \end{bmatrix} = \begin{bmatrix} c_0 & d_0 \\ c_1 & d_1 \end{bmatrix} \begin{bmatrix} \phi_\ell(P_m) \\ \phi_\ell(Q_m) \end{bmatrix}, \quad \text{i.e.} \quad \begin{bmatrix} a_0 & b_0 \\ a_1 & b_1 \end{bmatrix} = \begin{bmatrix} c_0 & d_0 \\ c_1 & d_1 \end{bmatrix}^{-1},$$

which they refer to as *reverse* basis decomposition. Its main upside is that the pairing value

$$g_0 := \tau_{m^{e_m}}(\phi_\ell(P_m), \phi_\ell(Q_m)) = \tau_{m^{e_m}}(P_m, Q_m)^{\ell^{e_\ell}},$$

where $\tau_{m^{e_m}}$ is the order-m^{e_m} *reduced Tate pairing* [20], solely depends on public parameters and can therefore be precomputed. The computation of the full compression algorithm is then typically divided into three stages.

1. (*Basis generation*) Compute the basis U_m, V_m.
2. (*Pairing computation*) Compute the pairings

$$g_1 = \tau_{m^{e_m}}(\phi_\ell(P_m), U_m), \qquad g_2 = \tau_{m^{e_m}}(\phi_\ell(P_m), V_m),$$
$$g_3 = \tau_{m^{e_m}}(\phi_\ell(Q_m), U_m), \qquad g_4 = \tau_{m^{e_m}}(\phi_\ell(Q_m), V_m). \tag{1}$$

3. (*Discrete logarithm computation*) Find $c_0, d_0, c_1, d_1 \in \mathbb{Z}/m^{e_m}\mathbb{Z}$ such that

$$g_1 = g_0^{d_0}, \qquad g_2 = g_0^{d_1}, \qquad g_3 = g_0^{-c_0}, \qquad g_4 = g_0^{-c_1}.$$

The first stage is easy for $m = 2$, in which case an *entangled* basis can be computed at extremely low cost [30, §3]. The case of $m = 3$ is more complicated; the work of Costello et al. [7] proposes to use techniques related to explicit 3-descent by Schaefer and Stoll [27] to generate points in $(E/\langle R\rangle) \setminus [3](E/\langle R\rangle)$, while Zanon et al. [30, §3] find that naïve generation of such points via cofactor multiplication yields better results. We observe that the difficulty of applying the 3-descent techniques seems to lie in the fact that there are no known non-trivial 3-torsion points, requiring initial cofactor multiplications to find such points.

The most costly part of the algorithm is the second phase, in which 4 simultaneous pairings are computed. Although optimizations can be made by observing that inputs are shared and by choosing an optimal curve model $E/\langle R\rangle$, the large cost remains.

Finally 4 (simultaneous) discrete logarithms need to be computed, which is feasible due to the smoothness of the group order of points on the elliptic curve. This can be done relatively cheaply with very little requirements on memory [7, §5], and can be sped up through the use of precomputed tables [30, §6].

Contributions. We propose several improvements that together significantly reduce the computational overhead imposed on SIDH and SIKE by public-key compression. The main idea behind our optimizations is to utilize the fact that the pairing values in Eq. (1) satisfy

$$g_1 = \tau_{m^{e_m}}\left(P_m, \widehat{\phi}_\ell(U_m)\right), \qquad\qquad g_2 = \tau_{m^{e_m}}\left(P_m, \widehat{\phi}_\ell(V_m)\right),$$
$$g_3 = \tau_{m^{e_m}}\left(Q_m, \widehat{\phi}_\ell(U_m)\right), \qquad\qquad g_4 = \tau_{m^{e_m}}\left(Q_m, \widehat{\phi}_\ell(V_m)\right),$$

where $\widehat{\phi}_\ell$ denotes the (unique) *dual isogeny* of ϕ_ℓ.

Our first contribution (see Sect. 3) is to propose explicit efficient formulas for computing duals of isogenies of degree 2, 4 and prime $\ell > 2$ between Montgomery curves, as proposed by Costello and Hisil [6] and Renes [25]. Through the use of optimal strategies, these can be used to compute dual isogenies of degrees ℓ^{e_ℓ} for any prime ℓ. However, the crucial observation is that the efficiency of evaluating $\widehat{\phi}_\ell$ increases significantly by re-using values that are obtained during the computation of ϕ_ℓ. In Proposition 3 we describe the kernels of all ℓ-isogenies appearing in the decomposition of $\widehat{\phi}_\ell$, and we give details on the computation and the values to store in Sects. 3.1 and 3.2. Having an efficient way to compute the dual, we gain the flexibility to apply the following idea.

Instead of evaluating ϕ_ℓ on the basis points P_m and Q_m, we pull back the deterministically generated basis points U_m and V_m through $\widehat{\phi}_\ell$ from $E/\langle R\rangle$ to E. This has several advantages, as the starting curve is a fixed system parameter and does not change throughout multiple executions of the protocol. As such, the starting curve can be chosen to have special properties, leading to more efficient operations. For example, precomputing $E(\mathbb{F}_{p^2})[3]$ leads to more efficient basis generation (see Sect. 4.3), while E being defined over $\mathbb{F}_p$ and having the basis points P_m and Q_m (almost) defined over $\mathbb{F}_p$ has significant advantage for the pairing computation (see Sect. 5). More specifically, we summarize our contributions as follows.

- The main contribution is the proposal to use the dual isogeny to pull back computations from $E/\langle R\rangle$ to E. For this purpose we show how to decompose $\widehat{\phi}_\ell$ as a sequence of ℓ-isogenies and how to evaluate them with very little overhead.
- We show how to utilize the dual isogeny in the basis generation phase. First, we adapt the entangled basis construction to an x-only setting. More importantly, we show that pulling back order and independence checking to E gives new interest to 3-descent methods. We analyze these methods in more detail, proving a strong relation to the reduced Tate pairing. Using this connection, we can reduce the cofactor scalar multiplications on E to exponentiations in $\mathbb{F}_{p^2}$ (i. e. pairings), significantly reducing the cost.
- We address the main bottleneck of public-key compression, namely the pairing computation. In the case of $\ell = 2$ we pull back the pairing to the $A = 0$ curve, on which a distortion basis for the m^{e_m}-torsion is available which greatly simplifies the pairing computation. For $\ell > 2$ and $m = 2$ we can pull back to the $A = 6$ curve, for which we can find basis points P_2 and Q_2 such that $[2]P_2$ is $\mathbb{F}_p$-rational and $[2]Q_2 = (x, iy)$, where $x, y \in \mathbb{F}_p$. This constrains many of the field operations to $\mathbb{F}_p$.

- As the m^{e_m}-torsion is a fixed parameter, we propose to use *affine* Weierstrass coordinates for the pairings and to precompute all Miller line functions. This leads to line functions that are very simple to evaluate, at the cost of a precomputed table. However, these tables are only several hundreds of kilobytes large and significantly smaller than those (already) used for discrete logarithms. Therefore, the memory overhead is small.

We have implemented[1] our techniques on top of the C library provided in the Round 2 submission package of SIKE, and compared our implementation to the uncompressed and compressed versions of SIKE as submitted to NIST across all parameter sets SIKEpXXX, where XXX $\in \{434, 503, 610, 751\}$. Our results show that public-key compression for SIDH can be implemented with an induced overhead of 77–86% (resp. 59–61%), compared to the previously best 160–182% (resp. 98–104%) of [17,30] for Alice (resp. Bob) across the different parameter sets (see Table 4a). Moreover, the compression techniques for SIKE induce an overhead of 61–74% (was 140–153%) for key generation, of 38–57% (was 67–90%) for encapsulation and 34–39% for decapsulation (was 59–65%) for the different parameter sets (see Table 4b). Finally, our results show that we speed up the pairing phase by a factor at least 2.97 for $\ell = 2$ and a factor at least 2.70 for $\ell = 3$, while also increasing efficiency of basis generation and decompression for $\ell = 2$. (see Table 2).

Remark 1. As the implementation focuses on $\{\ell, m\} = \{2, 3\}$, which seems to be the optimal parametrization for SIKE, our descriptions often also make this assumption for the sake of simplicity. Everything that is described in this work naturally generalizes to other primes. In that case it should be noted that $\ell = 2$ often exhibits special behavior (e. g. the existence of an entangled basis, or a special case for isogeny formulas [25, Proposition 1]) so we treat it separately, but our contributions work perfectly well by selecting m to be an arbitrary odd prime. Of course, one is also free to choose both ℓ and m to be odd primes without any (theoretical) problems.

Remark 2. The techniques that we describe rely on being able to evaluate the dual isogeny $\widehat{\phi}_\ell$ on a torsion basis $(E/\langle R \rangle)[m^{e_m}]$, which is equivalent to being able to evaluate ϕ_ℓ on $E[m^{e_m}]$. That is, given a point $U \in (E/\langle R \rangle)[m^{e_m}]$ we could (efficiently) solve the two-dimensional discrete logarithm $U = [a]\phi_\ell(P_m) + [b]\phi_\ell(Q_m)$ for some $a, b \in \mathbb{Z}/m^{e_m}\mathbb{Z}$, from which it follows that $\widehat{\phi}_\ell(U) = [\ell^{e_\ell}a]P_m + [\ell^{e_\ell}b]Q_m$. Thus, computing on the points $\widehat{\phi}_\ell(U)$ and $\widehat{\phi}_\ell(V)$ leaks no more information about the secret key than computing on $\phi_\ell(P)$ and $\phi_\ell(Q)$ does.

On the other hand, the evaluation of the dual isogeny itself does rely on secret data, while the intermediate points that are stored are also sensitive (as is the case with the evaluation of ϕ_ℓ). We simply apply the same protections to the dual evaluation as are applied to ϕ_ℓ, which in the implementation of SIKE just means that all algorithms are constant-time (see Sect. 3).

---

[1] The implementation is available as part of the SIDH Library v3.2, https://github.com/microsoft/PQCrypto-SIDH.

2   Preliminaries and Notation

We begin by recalling the basic theory, to remind the reader of typical notions and to establish notation for the rest of this work. As we already discussed public-key compression techniques related to SIDH and SIKE in Sect. 1, we omit those details here.

Elliptic Curves. Let $p > 3$ be prime. An elliptic curve E defined over a field k of characteristic p is a smooth projective curve of genus 1 with specified base point $\mathcal{O}_E$. Although typically defined by the Weierstrass model [29, §III.1], we shall always assume E to be described by the (less general) Montgomery form [22]

$$E : y^2 = x^3 + Ax^2 + x$$

for some $A \in k$ such that $A^2 \neq 4$, and may write E_A to emphasize the curve coefficient. As is the case for the Weierstrass model, the base point $\mathcal{O}_E$ is the unique point at infinity. The points on E form an abelian group with neutral element $\mathcal{O}_E$, and for any $m \in \mathbb{Z}$ we let $[m]P = P + \ldots + P$ be the sum of m copies of P (and a negation if m is negative). For any such non-zero $m \in \mathbb{Z}$, we let

$$E[m] = \{P \in E \mid [m]P = \mathcal{O}_E\}$$

be the m-torsion subgroup and say that E is *supersingular* whenever $\#E[p] = 1$. In that case we have $j(E) \in \mathbb{F}_{p^2}$ [29, Theorem V.3.1], so that E is isomorphic to a curve defined over $\mathbb{F}_{p^2}$. The number of isomorphism classes of supersingular elliptic curves over $\overline{k}$ of characteristic p is seen to be exactly $\lfloor p/12 \rfloor + \varepsilon_p$ [14, Theorem 9.11.11], where

$$\varepsilon_p = \begin{cases} 0 & \text{if } p \equiv 1 \bmod 12\,, \\ 1 & \text{if } p \equiv 5, 7 \bmod 12\,, \\ 2 & \text{if } p \equiv 11 \bmod 12\,. \end{cases}$$

Indeed, in this work we only concern ourselves with Montgomery curves defined over $\mathbb{F}_{p^2}$ for some (large) prime p.

Isogenies and Their Duals. Given any two elliptic curves E and $\overline{E}$ defined over k, an *isogeny* $\phi : E \to \overline{E}$ is a non-constant morphism such that $\phi(\mathcal{O}_E) = \mathcal{O}_{\overline{E}}$. It induces a field embedding $\phi^* : k(\overline{E}) \to k(E)$, and we say that ϕ is *separable* whenever the finite [29, Theorem II.2.4(a)] field extension $k(E)/\phi^*k(\overline{E})$ is separable, in which case we define $\deg \phi = [k(E) : \phi^*k(\overline{E})]$. The map $\phi \mapsto \ker \phi$ defines a correspondence between separable isogenies defined over k emanating from E and subgroups of E that are invariant under the action of $\mathrm{Gal}(\overline{k}/k)$, up to post-composition with an isomorphism [14, Theorem 9.6.19]. Given any isogeny ϕ defined over k, there exists a unique isogeny $\widehat{\phi}$ defined over k of the same degree as ϕ such that $\phi\widehat{\phi} = \widehat{\phi}\phi = [\deg \phi]$. The isogeny $\widehat{\phi}$ is called the *dual isogeny* of ϕ [29, Theorem III.6.1].

Reduced Tate Pairing. Now let $k = \mathbb{F}_q$ be a finite field containing the (cyclic) group of m-th roots of unity μ_m. We denote by

$$\tau_m : E(k)[m] \times E(k)/mE(k) \to \mu_m$$

the *reduced Tate pairing* [20] of order m defined by $\tau_m(S,T) = f_{m,S}(T)^{(q-1)/m}$, where $f_{m,S}$ is a rational function with divisor $m(S) - m(\mathcal{O})$. Interestingly, we have

$$\tau_m(\phi(S), T') = \tau_m(S, \widehat{\phi}(T'))$$

for any isogeny $\phi : E \to E'$ and points $S \in E(k)[m]$, $T' \in E'(k)/mE'(k)$ [3, Theorem IX.9]. Although not generally true, in the cases of our interest we shall always have $E(\mathbb{F}_q)/mE(\mathbb{F}_q) \cong E(\mathbb{F}_q)[m]$, and have the additional property that $\tau_m(S,T) = \tau_m(T,S)^{-1}$ for any $S, T \in E(\mathbb{F}_q)[m]$.

SIDH and SIKE. First we consider the SIDH protocol, proposed in 2011 by Jao and De Feo [19]. Let ℓ, m be distinct primes and e_ℓ, e_m be strictly positive integers such that $p = \ell^{e_\ell} \cdot m^{e_m} - 1$ is prime. Let $E/\mathbb{F}_{p^2}$ be a supersingular elliptic curve such that $\#E(\mathbb{F}_{p^2}) = (p+1)^2$, and let $\phi_\ell : E \to E/\langle R \rangle$ be an isogeny of degree ℓ^{e_ℓ} such that $\ker \phi_\ell = \langle R \rangle$ for some point $R \in E[\ell^{e_\ell}]$. Similarly, let $\phi_m : E \to E/\langle S \rangle$ be an isogeny of degree m^{e_m} such that $\ker \phi_m = \langle S \rangle$ for some point $S \in E[m^{e_m}]$. The shared secret is then (derived from) $j(E/\langle R, S \rangle)$. Notably, this is not feasibly computable from R and $E/\langle S \rangle$ or from S and $E/\langle R \rangle$ respectively.

Instead, we fix public parameters $P_\ell, Q_\ell \in E[\ell^{e_\ell}]$ and $P_m, Q_m \in E[m^{e_m}]$ and derive the points R, S from secret keys $s_0, s_1 \in \mathbb{Z}/\ell^{e_\ell}\mathbb{Z}$ and $t_0, t_1 \in \mathbb{Z}/m^{e_m}\mathbb{Z}$ such that

$$R = [s_0]P_\ell + [s_1]Q_\ell, \quad S = [t_0]P_m + [t_1]Q_m$$

have the desired order. That is, not both s_0 and s_1 (resp. t_0 and t_1) are divisible by ℓ (resp. m). The (naïve) public keys are then $[E/\langle R \rangle, \phi_\ell(P_m), \phi_\ell(Q_m)]$ and $[E/\langle S \rangle, \phi_m(P_\ell), \phi_m(Q_\ell)]$, observing that

$$\begin{aligned}
&(E/\langle R \rangle) / \langle [t_0]\phi_\ell(P_m) + [t_1]\phi_\ell(Q_m) \rangle \\
\cong{}& E/\langle R, S \rangle \\
\cong{}& (E/\langle S \rangle) / \langle [s_0]\phi_m(P_\ell) + [s_1]\phi_m(Q_\ell) \rangle .
\end{aligned}$$

It was noted by Costello, Longa and Naehrig [8, §6] that the public keys can be encoded (up to *simultaneous* sign) by the triples of x-coordinates

$$\left[x_{\phi_\ell(P_m)}, x_{\phi_\ell(Q_m)}, x_{\phi_\ell(P_m - Q_m)} \right], \left[x_{\phi_m(P_\ell)}, x_{\phi_m(Q_\ell)}, x_{\phi_m(P_\ell - Q_\ell)} \right],$$

which can be represented with $6 \log_2 p$ bits each.

Unfortunately, the SIDH key exchange scheme combined with static keys is insecure as the result of an active adaptive attack by Galbraith, Petit, Shani and Ti [15]. Consequently, one must resort to using ephemeral public keys. Alternatively, one can apply standard protocol transformations [13,16] to turn the

IND-CPA key exchange into an IND-CCA key encapsulation mechanism. The resulting scheme is referred to as SIKE [17] and is currently part of the NIST Post-Quantum Cryptography Standardization effort [24]. Although we refer to [17] for more detail on the submission, we remark that the secret keys are chosen such that $s_0 = 1$ and $t_0 = 1$, simplifying some of the treatment.

Field Operations. We denote by $\mathbf{M}$ and $\mathbf{S}$ the cost of an $\mathbb{F}_{p^2}$ field multiplication and squaring respectively, and by $\mathbf{A}$ a field addition or subtraction (which are therefore assumed to have the same cost). We denote by $\mathbf{E}$ the cost of a square root in $\mathbb{F}_{p^2}$. Similarly, we write $\mathbf{m}$ and $\mathbf{s}$ for the cost of an $\mathbb{F}_p$ field multiplication and squaring respectively, while $\mathbf{a}$ denotes an addition or subtraction in $\mathbb{F}_p$. Reflecting the properties of the SIKE implementation, we use $\mathbf{M} = 3\mathbf{m} + 5\mathbf{a}$ and $\mathbf{S} = 2\mathbf{m} + 3\mathbf{a}$.

3   Evaluating Dual Isogenies

In this section we consider the computation of the dual isogeny in the context of SIDH and SIKE. That is, we look towards the case where E is a Montgomery curve defined over some field k with $\mathrm{char}(k) \neq 2$ and $\phi_\ell : E \to E/\langle R \rangle$ a separable isogeny of degree ℓ^{e_ℓ} with kernel $\langle R \rangle$ for some point $R \in E[\ell^{e_\ell}]$. In addition, we could let $p = \ell^{e_\ell} \cdot m^{e_m} - 1$ be a prime, and let E be a supersingular elliptic curve defined over $k = \mathbb{F}_{p^2}$ such that $\#E(\mathbb{F}_{p^2}) = (p+1)^2$. Then R lies in $E(\mathbb{F}_{p^2})$ and, as a result, all arithmetic can be performed over $\mathbb{F}_{p^2}$. The latter is merely a computational advantage and not necessary for the statements below.

The first step to computing $\widehat{\phi}_\ell$ is finding its kernel. For this we note that $\widehat{\phi}_\ell \phi_\ell = [\ell^{e_\ell}]$, hence $\ker(\widehat{\phi}_\ell \phi_\ell) = E[\ell^{e_\ell}] \cong \langle R, S \rangle$ for some point $S \in E[\ell^{e_\ell}]$ of order ℓ^{e_ℓ}. From $\ker \phi_\ell = \langle R \rangle$ it is then immediate that $\ker \widehat{\phi}_\ell = \langle \phi_\ell(S) \rangle$. However, in cryptographic contexts the degree of ϕ_ℓ is too large for ϕ_ℓ to be computed directly, while the same is true for $\widehat{\phi}_\ell$. Instead, ϕ_ℓ is decomposed as

$$\phi_\ell = \phi_\ell^{(e_\ell - 1)} \circ \cdots \circ \phi_\ell^{(0)}$$

as a sequence of ℓ-isogenies. We begin by showing (see Proposition 3) how $\widehat{\phi}_\ell$ can be decomposed in a similar fashion, and describe the kernel of all intermediate ℓ-isogenies.

Proposition 3. *Let E be an elliptic curve defined over a field k and let $\phi_\ell : E \to E/\langle R \rangle$ be an isogeny of degree ℓ^{e_ℓ} with kernel $\langle R \rangle$ for some point $R \in E[\ell^{e_\ell}]$. Let $\phi_\ell = \phi_\ell^{(e_\ell - 1)} \circ \cdots \circ \phi_\ell^{(0)}$, where $\ker \phi_\ell^{(0)} = \langle [\ell^{e_\ell - 1}]R \rangle$ and $\ker \phi_\ell^{(i)} = \langle [\ell^{e_\ell - 1 - i}](\phi_\ell^{(i-1)} \circ \cdots \circ \phi_\ell^{(0)})(R) \rangle$ for $i = 1, \ldots, e_\ell - 1$. Then*

$$\widehat{\phi}_\ell = \widehat{\phi}_\ell^{(0)} \circ \cdots \circ \widehat{\phi}_\ell^{(e_\ell - 1)}, \quad \text{with} \quad \ker \widehat{\phi}_\ell^{(i)} = \langle (\phi_\ell^{(i)} \circ \cdots \circ \phi_\ell^{(0)})([\ell^{e_\ell - 1}]S) \rangle$$

for $i = 0, \ldots, e_\ell - 1$ and any $S \in E[\ell^{e_\ell}]$ such that $\langle R, S \rangle = E[\ell^{e_\ell}]$.

Proof. The first part follows by uniqueness of the dual isogeny, and since

$$\widehat{\phi}_\ell^{(0)} \circ \cdots \circ \widehat{\phi}_\ell^{(e_\ell-1)} \circ \phi_\ell^{(e_\ell-1)} \circ \cdots \circ \phi_\ell^{(0)}$$

$$= \widehat{\phi}_\ell^{(0)} \circ \cdots \circ \widehat{\phi}_\ell^{(e_\ell-2)} \circ \phi_\ell^{(e_\ell-2)} \circ \cdots \circ \phi_\ell^{(0)} \circ [\ell]$$

$$\vdots$$

$$= [\ell^{e_\ell}].$$

Now observe that $E[\ell] = \langle [\ell^{e_\ell-1}]R, [\ell^{e_\ell-1}]S \rangle$, so by using a similar argument as above it follows that $\ker \widehat{\phi}_\ell^{(0)} = \phi_\ell^{(0)}([\ell^{e_\ell-1}]S)$. As any linear relation between the $\ell^{e_\ell-1}$-torsion points $\phi_\ell^{(0)}(R)$ and $\phi_\ell^{(0)}([2]S)$ leads to one between R and S, they form a basis for $(E/\langle[\ell^{e_\ell-1}]R\rangle)[\ell^{e_\ell-1}]$. The statement then follows by proceeding via induction on e_ℓ. $\square$

It is now clear how we can evaluate $\widehat{\phi}_\ell$. We select an arbitrary point S, linearly independent of the kernel point R, and during the computation of ϕ_ℓ we evaluate and store the intermediate evaluations of $[\ell^{e_\ell-1}]S$. These determine the kernels of the ℓ-isogenies appearing in the decomposition of $\widehat{\phi}_\ell$, so it remains to show how to compute the dual of an ℓ-isogeny (i.e. the case $e_\ell = 1$). This of course strongly depends on the choice for ϕ_ℓ, for which we restrict to the parameters of SIKE. That is, we assume E and $E/\langle R\rangle$ to be Montgomery curves and consider the cases where $\ell > 2$ (Sect. 3.1) and where $\ell = 2$ (Sects. 3.2 and 3.3) separately.

Remark 4. This is especially easy in the case of SIKE, where $R = P_\ell + [s_1]Q_\ell$ for some $s_1 \in \mathbb{Z}/\ell^{e_\ell}\mathbb{Z}$. In that case we simply select $S = Q_\ell$ and store intermediate evaluations of $[\ell^{e_\ell-1}]Q_\ell$.

We note that we can write $\phi_\ell = (f_\ell(x), cyf'_\ell(x))$ for some rational function $f_\ell(x)$ in $k(x)$ [14, Theorem 9.7.5] and some $c \in k^*$, where $f'_\ell(x)$ is the formal derivative $df_\ell(x)/dx$ of $f_\ell(x)$. Therefore, the isogeny ϕ_ℓ is determined by $f_\ell(x)$ up to a possible twisting of the y-coordinate by varying c. As the only monomial containing y in Montgomery form is y^2, which has coefficient 1, it follows that $f_\ell(x)$ determines ϕ_ℓ up to composition by $[\pm1]$. Similarly, the dual $\widehat{\phi}_\ell = (\widehat{f}_\ell(x), \widehat{c}yf'_\ell(x))$ is determined by $\widehat{f}_\ell(x)$ up to composition $[\pm1]$. As it suffices for SIDH to compute ϕ_ℓ up to sign, and for our purposes it suffices to compute $\widehat{\phi}_\ell$ up to sign, in what follows we focus on the description of the function $\widehat{f}_\ell$.

3.1   The Case $\ell > 2$

First we consider the case where $\ell > 2$ (which turns out to be the simplest) and let R be a point of order ℓ (i.e. $e_\ell = 1$). In that case, the isogeny $\phi_\ell = (f_\ell(x), cyf'_\ell(x))$ with

$$f_\ell(x) = x \cdot \prod_{T\in\langle R\rangle\setminus\{\mathcal{O}\}} \frac{xx_T - 1}{x - x_T} \tag{2}$$

is an ℓ-isogeny with kernel $\langle R \rangle$, see [6, Theorem 1]. The case $\ell = 3$ is used for computations in the SIKE proposal [17] and in our implementation, but since the more general case follows analogously we also treat it here.

Proposition 5. *Let $E : y^2 = x^3 + Ax^2 + x$ be a Montgomery curve defined over a field k with $\mathrm{char}(k) \neq 2$. Let R and S be two linearly independent points of (prime) order ℓ and let $\phi_\ell = (f_\ell(x), cyf'_\ell(x))$ with*

$$f_\ell(x) = x \cdot \prod_{T \in \langle R \rangle \backslash \{\mathcal{O}\}} \frac{xx_T - 1}{x - x_T}$$

be an ℓ-isogeny of Montgomery curves with dual $\widehat{\phi}_\ell = (\widehat{f}_\ell(x), \widehat{c}y\widehat{f}'_\ell(x))$. Then

$$\widehat{f}_\ell(x) = x \cdot \prod_{T \in \langle \phi_\ell(S) \rangle \backslash \{\mathcal{O}\}} \frac{xx_T - 1}{x - x_T}.$$

Proof. Let $\overline{\phi}_\ell = (\overline{f}_\ell(x), \overline{c}y\overline{f'_\ell})$ be an isogeny with

$$\overline{f}_\ell(x) = x \cdot \prod_{T \in \langle \phi_\ell(S) \rangle \backslash \{\mathcal{O}\}} \frac{xx_T - 1}{x - x_T}.$$

It is clear that $\phi_\ell(S)$ is a point of order ℓ on $E/\langle R \rangle$, so applying [6, Theorem 1] to $\phi_\ell(S)$ shows that $\overline{\phi}_\ell$ is indeed an isogeny such that $\ker \overline{\phi}_\ell \phi_\ell = E[\ell]$. As the kernels are equal, $\overline{\phi}_\ell$ is equal to $\widehat{\phi}_\ell$ up to post-composition with an isomorphism. We finish the proof by showing that the only possible isomorphisms are $[\pm 1]$.

For that purpose we consider the point $(1, \sqrt{A+2})$ of order 4 on E, which satisfies $[\ell](1, \sqrt{A+2}) = (1, \pm\sqrt{A+2})$ depending on the value of $\ell \bmod 4$. No matter which is the case, it follows that $[\ell] = \widehat{\phi}_\ell \phi_\ell$ fixes x-coordinate 1 or, in other words, that $\widehat{f}_\ell f_\ell(1) = 1$. Similarly, considering the point $(0,0)$ of order 2 shows that $\widehat{f}_\ell f_\ell(0) = 0$.

Now note that indeed $\overline{f}_\ell f_\ell(1) = 1$ and $\overline{f}_\ell f_\ell(0) = 0$, so that any isomorphism post-composed with $\overline{\phi}_\ell$ to obtain $\widehat{\phi}_\ell$ must act as the identity on the x-coordinates 0 and 1. By [29, Proposition III.3.1(b)], the only such isomorphisms are $[\pm 1]$. Therefore, $\overline{\phi}_\ell = [\pm 1]\widehat{\phi}_\ell$ and the result follows.   □

Interestingly, Proposition 5 shows that we can compute duals of ℓ-isogenies using the exact same formulas for the isogeny ϕ_ℓ itself. In the case of $\ell = 3$ this (i.e. its projectivized version) can be computed at the cost of $4\mathbf{M} + 2\mathbf{S} + 4\mathbf{A}$ for each first evaluation, and $4\mathbf{M} + 2\mathbf{S} + 2\mathbf{A}$ for each subsequent evaluation [6, Appendix A].

3.2   The Case of 4-Isogenies

Now assume that $\ell = 2$ and that the point R has order 4 (i.e. $e_2 = 2$) such that $[2]R \neq (0,0)$. Again, the isogeny $\phi_2 = (f_2(x), cyf'_2(x))$ can be described by an equation of the form (2), see [25, Proposition 1]. If S is any other point of order 4 linearly independent from R, i.e. $E[4] = \langle R, S \rangle$, then again $\ker \widehat{\phi_2} = \langle \phi_2(S) \rangle$. However, in contrast to the case of $\ell > 2$, applying the formulas from the SIKE proposal [17] (which are essentially those from [25, Proposition 1]) leads to a point $\phi_2(S)$ such that $x_{\phi_2(S)} = 1$ and $[2]\phi_2(S) = (0,0)$. As a result, the dual isogeny can not be described by the formulas of [25, Proposition 1].

Instead, the original work of De Feo–Jao–Plût [12, Eqn. (18)–(21)] describes formulas for a 4-isogeny whose kernel is generated by a point with x-coordinate 1. Unfortunately, unlike before, there is no reason that this isogeny has the correct co-domain. As such, we post-compose with an appropriate isomorphism. One option for such an isomorphism is given in [12, Eqn. (15)], but it is described through the knowledge of a point of order 2. As such a point is not readily (or cheaply) available in our context, one needs to compute a (typically expensive) doubling. We show that this is much cheaper due to the assumption that R has order 4. We summarize this in Proposition 6.

Proposition 6. *Let $E : y^2 = x^3 + Ax^2 + x$ be a Montgomery curve defined over a field k with $\mathrm{char}(k) \neq 2$. Let R be a point of order 4 such that $[2]R \neq (0,0)$, and let $\phi_2 = (f_2(x), cyf'_2(x)) : E \to E/\langle R \rangle : y^2 = x^3 + \widehat{A}x^2 + x$ with*

$$f_2(x) = x \cdot \prod_{T \in \langle R \rangle \setminus \{\mathcal{O}\}} \frac{x x_T - 1}{x - x_T}$$

be a 4-isogeny of Montgomery curves with dual $\widehat{\phi_2} = (\widehat{f_2}(x), \widehat{c}yf'_2(x))$. Then

$$\widehat{f_2}(x) = \frac{(x_R^2 - 1)\overline{X} + (x_R^2 + 1)\overline{Z}}{2x_R \overline{Z}},$$

where

$$\overline{X} = (x + 1)^2 \left((x + 1)^2 - 4(1 - \widehat{A}_{24})x \right), \quad \overline{Z} = 4(1 - \widehat{A}_{24})x(x - 1)^2,$$

and $\widehat{A}_{24} = (\widehat{A} + 2)/4$.

Proof. Let $S = (1, \sqrt{A + 2})$ be a point on E, which has order 4 and is linearly independent from R. As a result, the kernel of the dual of ϕ_2 is generated by $\phi_2(S)$. As $f_2(1) = 1$, the kernel of $\widehat{\phi_2}$ is generated by a point with x-coordinate equal to 1.

The map $\phi_4(x)$ (not to be confused with ϕ_2) computed from [12, Eqn. (18)–(21)] as the concatenation of the maps

$$(x, y) \mapsto \left(\frac{(x - 1)^2}{x}, y\left(1 - \frac{1}{x^2}\right) \right)$$

followed by

$$(x, y) \mapsto \left(\frac{1}{2 - \widehat{A}} \frac{(x + 4)(x + \widehat{A} + 2)}{x}, \frac{y}{2 - \widehat{A}}\left(1 - \frac{4(2 + \widehat{A})}{x^2}\right) \right)$$

is seen to be an isogeny of degree 4 such that the generator of its kernel has x-coordinate 1, and satisfies $\phi_4(x) = \overline{X}/\overline{Z}$. Thus ϕ_4 on $E/\langle R \rangle$ determines an isogeny equal to $\widehat{\phi}_2$ up to post-composition by an isomorphism. As $\ker \widehat{\phi}_4$ is generated by a point of x-coordinate equal to 1 [12, §4.3.2], while $x_R \neq 1$, it follows that $\widehat{\phi}_4 \neq \phi_2$. Taking duals on both sides, we find that $\phi_4 \neq \widehat{\phi}_2$ [29, Theorem III.6.2]. Instead, the isomorphism ψ^{-1}, where

$$\psi : (x, y) \mapsto \left(\frac{x - x_{[2]R}}{x_R - x_{[2]R}}, \frac{y}{x_R - x_{[2]R}} \right)$$

is the map described in [12, Eqn. (15)], maps the kernel of $\widehat{\phi}_4$ to $\langle R \rangle$ and we conclude that $\widehat{\phi}_2 = \psi^{-1}\phi_4$.

At first glance the map ψ requires the use of (the x-coordinate of) $[2]R$, which is generally costly to compute. We show that this simplifies due to R being a point of order 4. Writing $\psi^{-1} = (h(x), yh'(x))$, we note that $h(1) = x_R$ and $h(0) = x_{[2]R}$. Also, let $T = \psi(0,0)$ be a point of order 2 such that $(-1, \sqrt{A-2}) = (1, \sqrt{A+2}) + T$ for an appropriate choice of square roots. Then

$$\psi^{-1}(-1, \sqrt{A-2}) = \psi^{-1}(1, \sqrt{A+2}) + (0,0),$$

implying that $h(-1) = 1/x_R$. Again, by [29, Proposition III.3.1(b)] we have $h(x) = ax + b$ for some $a, b \in k$, for which the above restrictions imply that $b + a = x_R$, $b - a = 1/x_R$ and $b = x_{[2]R}$. It follows that $x_{[2]R} = (x_R + 1/x_R)/2$ and thus

$$h(x) = (x_R - x_{[2]R})x + x_{[2]R} = \frac{(x_R^2 - 1)x + x_R^2 + 1}{2x_R},$$

completing the proof.   □

Projectivizing and writing $x_R = X_R/Z_R$, $x = X/Z$ and $\widehat{A}_{24} = \widehat{a}_{24}/\widehat{c}_{24}$, we can compute $\widehat{f}_2(x)$ as follows. First we compute the coefficients $[K_0, K_1, K_2] = [X_R^2 - Z_R^2, X_R^2 + Z_R^2, 2X_R Z_R]$ through the sequence of operations

$$T_0 \leftarrow X_R^2, \ T_1 \leftarrow Z_R^2, \ K_0 \leftarrow T_0 - T_1, \ K_1 \leftarrow T_0 + T_1,$$
$$K_2 \leftarrow X_R + Z_R, \ K_2 \leftarrow K_2^2, \ K_2 \leftarrow K_2 - K_1,$$

that can be computed at a cost of $3\mathbf{S} + 4\mathbf{A}$. We note that these operations are independent of x and can therefore be shared among multiple evaluations of $\widehat{f}_2(x)$ at distinct points. Moreover, in the context of SIDH and SIKE such an evaluation is always preceded by an evaluation of f_2 in which X_R^2, X_Z^2 and $X_R + Z_R$ are computed. Storing those intermediate values reduces the cost to $1\mathbf{S}+3\mathbf{A}$. We then complete the computation of $\widehat{f}_2(x) = X'/Z'$ via the operations

$$T_0 \leftarrow X + Z, \ T_1 \leftarrow X - Z, \ T_0 \leftarrow T_0^2, \ T_1 \leftarrow T_1^2, \ T_2 \leftarrow T_0 - T_1,$$
$$T_3 \leftarrow \widehat{c}_{24} - \widehat{a}_{24}, \ T_3 \leftarrow T_2 \cdot T_3, \ T_2 \leftarrow \widehat{c}_{24} \cdot T_0, T_2 \leftarrow T_2 - T_3, \ \overline{X} \leftarrow T_2 \cdot T_0,$$
$$\overline{Z} \leftarrow T_3 \cdot T_1, \ X' \leftarrow K_0 \cdot \overline{X}, \ T_0 \leftarrow K_1 \cdot \overline{Z}, \ X' \leftarrow X' + T_0, \ Z' \leftarrow K_2 \cdot \overline{Z},$$

at a cost of $7\mathbf{M}+2\mathbf{S}+6\mathbf{A}$. Summarizing, assuming having stored the intermediate values $[\hat{a}_{24}, \hat{c}_{24}, X_R^2, Z_R^2, 2X_RZ_R]$, the first evaluation of $\hat{f}_2(x)$ can be performed at a cost of $7\mathbf{M} + 3\mathbf{S} + 9\mathbf{A}$. Any subsequent evaluation can be computed at a cost of $7\mathbf{M} + 2\mathbf{S} + 6\mathbf{A}$. For comparison, the evaluation of f_2 in SIKE currently has a cost of $6\mathbf{M} + 2\mathbf{S} + 6\mathbf{A}$. Hence, although the dual is more expensive than the original 4-isogeny, the difference is small.

3.3   The Case of 2-Isogenies

Finally consider $\ell = 2$ and assume that $R \neq (0,0)$ is a point of order 2. We note that 2-isogenies are only employed in the SIKE proposal whenever $e_\ell \neq 0 \bmod 4$, and in that case only a single one is computed. Therefore its cost is negligible to the overall cost of the isogeny. The 2-isogeny is computed as in [25, Proposition 2], and we refer to Proposition 7 for the computation of its dual.

Proposition 7. *Let $E : y^2 = x^3 + Ax^2 + x$ be a Montgomery curve defined over a field k with $\mathrm{char}(k) \neq 2$. Let $R \neq (0,0)$ be a point of order 2 and let $\phi_2 = (f_2(x), cyf_2'(x))$ with*

$$f_2(x) = x \cdot \frac{xx_R - 1}{x - x_R}$$

be a 2-isogeny of Montgomery curves with dual $\hat{\phi}_2 = (\hat{f}_2(x), \hat{c}y\hat{f}_2'(x))$. Then

$$\hat{f}_2(x) = \frac{(x+1)^2}{4x_Rx} .$$

Proof. First we note that $\ker \hat{\phi}_2 = \langle(0,0)\rangle$ by [25, Corollary 1]. An isogeny with such a kernel can be computed by composing the maps

$$(x,y) \mapsto \left(\frac{(x-1)^2}{x}, y\left(1 - \frac{1}{x^2}\right)\right)$$

from [12, Eqn. (19)] followed by the map

$$(x,y) \mapsto \left(\frac{x + \hat{A} + 2}{\sqrt{\hat{A}^2 - 4}}, \frac{y}{\sqrt{\hat{A}^2 - 4}}\right),$$

as observed in [25, Remark 6], where $\hat{A} = 2(1 - 2x_R^2)$ [25, Proposition 2]. After twisting the y-coordinate, this lands on the curve defined by the equation

$$y^2 = x^3 - \frac{2\hat{A}}{\sqrt{\hat{A}^2 - 4}}x^2 + x$$

whose dual is again generated by $(0,0)$. Finally, we post-compose with an isomorphism $\psi(x,y) = (h(x) = ax + b, yh'(x))$. As noted earlier, using the fact that taking duals acts as an involution implies that $h(0) = x_R$ and thus $b = x_R$.

Writing out the curve equation for $\psi(x, y)$ and noting that the coefficient of x^2 is A shows that

$$a = -\sqrt{\widehat{A}^2 - 4(3x_R + A)}/(2\widehat{A}).$$

Composing all these maps leads to the result, for which we omit the details as they are straightforward yet tedious.   □

Letting $x = X/Z$ and $x_R = X_R/Z_R$, the following sequence of operations

$$T_0 \leftarrow X + Z, \ T_0 \leftarrow T_0^2, \ X' \leftarrow Z_R \cdot T_0, \ T_1 \leftarrow X - Z,$$
$$T_1 \leftarrow T_1^2, \ T_1 \leftarrow T_0 - T_1, \ Z' \leftarrow X_R \cdot T_1,$$

computes $\widehat{f}_2(x) = X'/Z'$ at a cost of $2\mathbf{M} + 2\mathbf{S} + 3\mathbf{A}$.

4   Generation of Torsion Bases

As usual we let $p = \ell^{e_\ell} \cdot m^{e_m} - 1$ be a prime, and let $E : y^2 = x^3 + Ax^2 + x$ be a supersingular elliptic curve defined over $k = \mathbb{F}_{p^2}$ such that $\#E(\mathbb{F}_{p^2}) = (p + 1)^2$. Again, we let $\phi_\ell : E \to E/\langle R \rangle$ be a separable isogeny of degree ℓ^{e_ℓ} with kernel $\langle R \rangle$ for some point $R \in E[\ell^{e_\ell}]$. The aim of this section is to describe how to compute $\widehat{\phi}_\ell(U_m)$ and $\widehat{\phi}_\ell(V_m)$ for some deterministically generated basis points U_m and V_m such that $(E/\langle R \rangle)[m^{e_m}] = \langle U_m, V_m \rangle$. This is (naïvely) done in a few steps.[2]

1. Deterministically generate a first point $U \in E/\langle R \rangle$.
2. Repeat 1–2 until $U_m = [\ell^{e_\ell}]U$ has order m^{e_m}.
3. Deterministically generate a second point $V \in E/\langle R \rangle$.
4. Repeat 3–4 until $V_m = [\ell^{e_\ell}]V$ has order m^{e_m} and $(E/\langle R \rangle)[m^{e_m}] = \langle U_m, V_m \rangle$.
5. Compute $\widehat{\phi}_\ell(U_m)$ and $\widehat{\phi}_\ell(V_m)$.

For $\ell = 3$ we do not deviate much from this, yet we remark that it is not necessary to generate the full points U_2 and V_2. Instead, since the dual isogeny computes only on x-coordinates, it suffices to compute x_{U_2} and x_{V_2}. In fact, it is even enough to only obtain x_U and x_V, as the cofactor multiplications naturally factor out during the pairing and discrete logarithm phase [30, §3.1]. However, for the pairing to remain consistent we need to also deterministically compute x_{U-V} (without recovering y_U and y_V). We show how this can be done in Sect. 4.1 and how this applies to the entangled basis generation of [30, §3] in Sect. 4.2.

In the case of $\ell = 2$ we do take an alternative approach. The difference is that checking the order of U and V has to be done through cofactor multiplications $[3^{e_3-1}2^{e_2}]U$ and $[3^{e_3-1}2^{e_2}]V$, both of which are very costly. We propose generating the basis in the following way, recalling that the dual isogeny is defined as $\widehat{\phi}_2 = (\widehat{f}_2(x), \widehat{c}y\widehat{f}_2')$.

---

[2] Note that when considering ϕ_ℓ of degree ℓ^{e_ℓ}, we generate a basis of the m^{e_m}-torsion.

1. Deterministically generate x_U for a point $U \in E/\langle R \rangle$.
2. Compute $\widehat{f}_2(x_U)$ and recover $\widehat{\phi}_2(U)$.
3. Repeat 1–3 until $[2^{e_2}]\widehat{\phi}_2(U)$ has order 3^{e_3}.
4. Deterministically generate x_V for a point $V \in E/\langle R \rangle$.
5. Compute $\widehat{f}_2(x_V)$ and recover $\widehat{\phi}_2(V)$.
6. Repeat 4–6 until $E[3^{e_3}] = \langle [2^{e_2}]\widehat{\phi}_2(U), [2^{e_2}]\widehat{\phi}_2(V) \rangle$.
7. Deterministically generate x_{U-V} and compute $\widehat{f}_2(x_{U-V})$.
8. Modify signs of $\widehat{\phi}_2(U)$, $\widehat{\phi}_2(V)$ so that $x_{\widehat{\phi}_2(U)-\widehat{\phi}_2(V)} = \widehat{f}_2(x_{U-V})$.

We can then obtain $\widehat{\phi}_2(U_3) = [2^{e_2}]\widehat{\phi}_2(U)$ and $\widehat{\phi}_2(V_3) = [2^{e_2}]\widehat{\phi}_2(V)$, but we show in Sect. 4.3 that this is never explicitly necessary. This presents some trade-offs, which we briefly discuss. Firstly, we note that more computation is wasted when a point of the wrong order is generated (i. e. in step 3) or when it is not independent (i. e. in step 6). That is, the evaluation of $\widehat{f}_2$ would unfortunately have been done for nothing. However, since $E[3] \cong \mathbb{Z}/3\mathbb{Z} \times \mathbb{Z}/3\mathbb{Z}$, we expect the points to have full order with probability $8/9$ and to be independent with probability $3/4$. Thus on average we require to perform steps 1–3 only $9/8$ times and step 4–6 only $3/2$ times.

Moreover, we observe that we can check the order of $\widehat{\phi}_2(U)$ and $\widehat{\phi}_2(V)$ on E as opposed to $E/\langle R \rangle$. The main advantage is that E is a fixed public parameter, whereas $E/\langle R \rangle$ varies per choice of R. This allows pre-computation on E, and in particular the generation of 3-torsion points to apply the 3-descent methods of Schaefer and Stoll [27]. We further analyze this in Sect. 4.3 and show how this leads to improved performance.

4.1   Deterministically Generating X-Coordinates

The generation of the points U (and similarly for V) is done in two steps. First, one uses the Elligator 2 map [2] to generate an x-coordinate x_U in $\mathbb{F}_{p^2}$, after which the y-coordinate can be recovered (which is guaranteed to lie in $\mathbb{F}_{p^2}$). Therefore, the generation of the two points U and V requires performing two square roots in $\mathbb{F}_{p^2}$ (although only a single one is needed for the entangled basis, see Sect. 4.2). Evaluating the abscissa as well as the ordinate of $\widehat{\phi}_\ell$ on U and V is also very costly. We show how this can be done much more efficiently.

Instead, we take the approach of SIDH and only ever evaluate $\widehat{f}_\ell$ (i. e. the abscissa of $\widehat{\phi}_\ell$) on U and V, and thus never require their y-coordinates. As usual, we also evaluate $\widehat{f}_\ell$ at their difference (i. e. in step 8) for the computation to remain consistent. This leaves us with the problems of deterministically computing x_{U-V}, and consistently recovering the signs of $\widehat{\phi}_\ell(U)$ and $\widehat{\phi}_\ell(V)$ from $x_{\widehat{\phi}_\ell(U)}$, $x_{\widehat{\phi}_\ell(V)}$, and $x_{\widehat{\phi}_\ell(U-V)}$.

For the generation of x_{U-V} we refer to the techniques applied in the qDSA [26] signature scheme of Renes and Smith. More specifically, in [26, Proposition 3] it is shown that $a \cdot x_{U-V}^2 - 2b \cdot x_{U-V} + c = 0$, where

$$a = (x_U - x_V)^2, \quad c = (x_U x_V - 1)^2,$$
$$b = (x_U x_V + 1)(x_U + x_V) + 2\widehat{A} x_U x_V,$$

and where $\widehat{A}$ is the Montgomery curve coefficient of $E/\langle R \rangle$. It follows that

$$x_{U-V} = \frac{-b \pm \sqrt{b^2 - 4ac}}{2a},$$

allowing to (projectively) compute x_{U-V} at a cost of $1\mathbf{E} + 6\mathbf{M} + 5\mathbf{S} + 15\mathbf{A}$. This is made deterministic by fixing the choice for $\sqrt{b^2 - 4ac}$ in $\mathbb{F}_{p^2}$. As we evaluate $\widehat{f_\ell}$ projectively, there is no need for an inversion to obtain an affine representation. Notably, the computation above does not affect decompression, which uses the points in an x-only three-point ladder.

For the recovery of $\widehat{\phi}_\ell(U)$ and $\widehat{\phi}_\ell(V)$ we refer to [22, §10.3.1]. Writing

$$\widehat{\phi}_\ell(U) = (x_1, y_1), \quad \widehat{\phi}_\ell(V) = (x_2, y_2), \quad \widehat{\phi}_\ell(U - V) = (x_3, y_3),$$

we have

$$x_3 = \frac{(x_2 y_1 + x_1 y_2)^2}{x_1 x_2 (x_1 - x_2)^2} = \frac{x_2^2 y_1^2 + x_1^2 y_2^2 + 2 x_1 x_2 y_1 y_2}{x_1 x_2 (x_1 - x_2)^2}. \tag{3}$$

Using the curve equation for $\widehat{\phi}_\ell(V)$, a simple reorganization shows that

$$y_2 = \frac{x_1 x_2 x_3 (x_1 - x_2)^2 - x_2^2 y_1^2 + x_1^2 (x_2^3 + A x_2^2 + x_2)}{2 x_1 x_2 y_1}. \tag{4}$$

Therefore, it suffices to compute y_1 at the cost of a single square root, after which y_2 is determined. Note that this only recovers $\widehat{\phi}_\ell(U)$ and $\widehat{\phi}_\ell(V)$ up to *simultaneous* sign, determined by the choice of y-coordinate for $\widehat{\phi}_\ell(U)$. As we are only interested in subgroups generated by linear combinations of these two points, this is not an issue. If we only want to verify that our choices of signs are consistent, it suffices to check that Eq. (3) holds. This is what we use in step 8 above, and in Sect. 4.3.

4.2   X-only Entangled Basis Generation for $\ell = 3$

The work of Sect. 4.1 becomes especially simple in the case of $\ell = 3$, where U and V are generated as an *entangled* basis [30, §3]. That is, $U = (x_1, y_1)$, where $x_1 = -\widehat{A}/(1 + t^2)$ is a quadratic non-residue and $t \in \mathbb{F}_{p^2} \setminus \mathbb{F}_p$ such that $t^2 \in \mathbb{F}_{p^2} \setminus \mathbb{F}_p$, and $V = (x_2, y_2)$ where $x_2 = -x_1 - \widehat{A}$ and $y_2 = t \cdot y_1$ [30, Theorem 1]. Writing $U - V = (x_3, y_3)$, we have

$$x_3 = \frac{(y_1 + y_2)^2}{(x_1 - x_2)^2} - \widehat{A} - x_1 - x_2 = \frac{(y_1 + y_2)^2}{(x_1 - x_2)^2} = \frac{(x_1^3 + \widehat{A} x_1^2 + x_1)(1 + t)^2}{(x_1 - x_2)^2},$$

see [22, §10.3.1] again. As done by Zanon et al. [30], we fix $u_0 = 1 + i$ and run over $t = u_0 \cdot r$ for $r = 1, 2, \ldots$ until we succeed. Building tables (r, v) for $r = 1, 2, \ldots$ and $v = 1/(1 + u r^2)$ of quadratric and non-quadratric residues, we can select $x_1 = -\widehat{A} v$, after which the values of x_2 and x_3 can be computed as above. We note that this does not require the computation of y_1, but merely requires checking whether $x_1^3 + \widehat{A} x_1^2 + x_1$ is a square (which has a lower cost).

Having generated x_U, x_V and x_{U-V}, we evaluate the values $\widehat{f}_3(x_U)$, $\widehat{f}_3(x_V)$ and $\widehat{f}_3(x_{U-V})$. After a square root computation to recover $\widehat{\phi}_3(U)$, we use Eqn. (4) to (consistently) recover $\widehat{\phi}_3(V)$.

4.3   Basis Generation with the Reduced Tate Pairing for $\ell = 2$

The situation is more complex for $\ell = 2$, for which there is no (known) analogue of an entangled basis. Instead, checking the order of U and V is done through cofactor multiplications $[3^{e_3-1}2^{e_2}]U$ and $[3^{e_3-1}2^{e_2}]V$. For that purpose, we revisit the 3-descent techniques of Schaefer and Stoll [27].

More precisely, let $T = (x_T, y_T) \in E/\langle R \rangle$ be a point of order 3, necessarily $\mathbb{F}_{p^2}$-rational, and let $g_T(x, y) = y - (\lambda x + \mu)$ be the function defining the tangent line at T. Then Costello et al. [7, §3.3] observe that $U \in [3](E/\langle R \rangle)$ if and only if $g_T(U)$ is a cube in $\mathbb{F}_{p^2}$ for all non-trivial 3-torsion points $T \in (E/\langle R \rangle)[3]$ (and similarly for $g_T(V)$). This method is more complicated due to the fact that 3-torsion points are not readily available on $E/\langle R \rangle$. As such, Costello et al. [7] first find a point of order 3 (and potentially immediately find U), and only afterwards apply the 3-descent techniques. Moreover, since only a single 3-torsion point is found (as opposed to all of $(E/\langle R \rangle)[3]$), a slightly weaker check is performed. It was shown by Zanon et al. [30, §4] that this does not lead to better results than naïve cofactor multiplications.

Explicit 3-Descent and the Reduced Tate Pairing. We begin our analysis by relating the 3-descent techniques to the reduced Tate pairing $\tau_{3^{e_3}}$. That is, we note that for any $T \in (E/\langle R \rangle)[3]$ we have that $g_T(U)$ is a cube in $\mathbb{F}_{p^2}$ if and only if $g_T(U)^{(p^2-1)/3} = 1$. We observe that

$$g_T(U)^{(p^2-1)/3} = \tau_{3^{e_3}}(T, U),$$

which is easily seen by observing that the only non-trivial Miller line function is the first one, which equals $g_T(x, y)$. By properties of the Tate pairing, it follows that $\tau_{3^{e_3}}(T, U) = 1$ if and only if $[3^{e_3-1}2^{e_2}]U \in \langle T \rangle$. In particular, if $U \in [3](E/\langle R \rangle)$, then $[3^{e_3-1}2^{e_2}]U = \mathcal{O}$. Thus all pairings are trivial and we recover the statement from Costello et al. (i.e. $g_T(U)$ is a cube for all 3-torsion points T).

As $\#(E/\langle R \rangle)[3] = 9$, this still leaves many pairings to be computed to test whether $U \in [3](E/\langle R \rangle)$. We can simplify the treatment by fixing a basis for $(E/\langle R \rangle)[3^{e_3}]$. Let $\overline{S}, \overline{T} \in E/\langle R \rangle$ such that $(E/\langle R \rangle)[3^{e_3}] = \langle \overline{S}, \overline{T} \rangle$, and let $S = [3^{e_3-1}]\overline{S}$ and $T = [3^{e_3-1}]\overline{T}$ form a basis for $(E/\langle R \rangle)[3]$. Then it follows by bilinearity of $\tau_{3^{e_3}}$ that

$$U \in [3](E/\langle R \rangle) \iff \tau_{3^{e_3}}(S, U) = \tau_{3^{e_3}}(T, U) = 1,$$

leaving only 2 pairings to be computed. Although this is a good start, we can do a lot more.

For that purpose, we define $h_0 = \tau_{3^{e_3}}(T, \overline{S})$ and note that $h_0 = \tau_{3^{e_3}}(S, \overline{T})^{-1}$. Then there exist (unique) $a, b \in \mathbb{Z}/3^{e_3}\mathbb{Z}$ such that $[2^{e_2}]U = [a]\overline{S} + [b]\overline{T}$, while

$$\tau_{3^{e_3}}(S, U)^{2^{e_2}} = h_0^{-b}, \quad \tau_{3^{e_3}}(T, U)^{2^{e_2}} = h_0^{a}.$$

As h_0 has order 3 and 2^{e_2} is invertible modulo 3, these discrete logarithms can easily be solved to retrieve $a, b \bmod 3$. Hence, we can compute $[3^{e_3-1}2^{e_2}]U = [a \bmod 3]S + [b \bmod 3]T$ at the cost of a single point addition (or, by simply selecting it from a pre-computed table). In practice we can ignore the factor $2^{e_2} \bmod 3$, since it only changes a and b up to a simultaneous factor, while it is enough to compute any generator of $\langle[3^{e_3-1}2^{e_2}]U\rangle$ as opposed to finding $[3^{e_3-1}2^{e_2}]U$ itself.

We can repeat the above by (deterministically) generating $U \in E/\langle R\rangle$ until not both $a = 0$ and $b = 0$, in which case $U \in E/\langle R\rangle \setminus [3](E/\langle R\rangle)$. Once that is done, we repeatedly (and deterministically) generate V until

$$\tau_{3^{e_3}}([3^{e_3-1}2^{e_2}]U, V) \neq 1,$$

which implies that $[3^{e_3-1}2^{e_2}]V \notin \langle[3^{e_3-1}2^{e_2}]U\rangle$, and in turn that $(E/\langle R\rangle)[3^{e_3}] = \langle[2^{e_2}]U, [2^{e_2}]V\rangle$. Under the assumption that $\overline{S}, \overline{T}, S, T$ and h_0 are all precomputed, the cost of generating U is determined by the cost of the 2 pairings, while the generation of V requires a single pairing. As before, the first needs to be repeated 9/8 times on average, while the latter (cheaper) step needs to be repeated 3/2 times on average.

The main drawback of this method is that $\overline{S}, \overline{T}$ form a basis of $(E/\langle R\rangle)[3^{e_3}]$, so to compute a basis we assume to already know one. In fact we do *not* know such a basis on $E/\langle R\rangle$, seemingly making this much less interesting (which is the exact problem that Costello et al. [7] faced). However, by evaluating $\widehat{\phi}_2$ on U and V *before* checking that they are independent (i.e. multiply to independent 3^{e_3}-torsion points) and have the right order, we can apply the above to the public parameter E where we *can* precompute as much as we want. This allows us to check the independence and orders of $\widehat{\phi}_2(U)$ and $\widehat{\phi}_2(V)$ on E much more efficiently than via naïve cofactor multiplication. For completeness, we provide a full description of the method.

1. Deterministically generate x_U for a point $U \in E/\langle R\rangle$.
2. Compute $\widehat{f}_2(x_U)$ and recover $\widehat{\phi}_2(U)$.
3. For $h_0 := \tau_{3^{e_3}}([3^{e_3-1}]P_3, Q_3)$, compute $a, b \in \mathbb{Z}/3\mathbb{Z}$ such that

$$\tau_{3^{e_3}}([3^{e_3-1}]P_3, \widehat{\phi}_2(U)) = h_0^b, \quad \tau_{3^{e_3}}([3^{e_3-1}]Q_3, \widehat{\phi}_2(U)) = h_0^{-a},$$

   and repeat 1–3 until not both a and b are zero.
4. Deterministically generate x_V for a point $V \in E/\langle R\rangle$.
5. Compute $\widehat{f}_2(x_V)$ and recover $\widehat{\phi}_2(V)$.
6. Repeat 4–6 until $\tau_{3^{e_3}}([a \cdot 3^{e_3-1}]P_3 + [b \cdot 3^{e_3-1}]Q_3, \widehat{\phi}_2(V)) \neq 1$.
7. Deterministically generate x_{U-V} and compute $\widehat{f}_2(x_{U-V})$.
8. Modify signs of $\widehat{\phi}_2(U), \widehat{\phi}_2(V)$ so that $x_{\widehat{\phi}_2(U)-\widehat{\phi}_2(V)} = \widehat{f}_2(x_{U-V})$.

As both P_3 and Q_3 are public parameters, the points $[3^{e_3-1}]P_3$ and $[3^{e_3-1}]Q_3$ can be precomputed and the above sequence of steps does not involve any scalar multiplication on E or $E/\langle R \rangle$. Although the improvement is obvious by comparing the number of required field operations, we simply confirm the feasibility of our approach through our implementation in Table 1, leading to a speedup of about 17% across the different parameter sets. Note that by including the iso operation, we also count the overhead generated by evaluating $\widehat{\phi}_2$ as opposed to ϕ_2. That is, Table 1 shows that checking independence and orders on E not only makes up for this slowdown, but even leads to a speedup. This showcases the utility of the methods even before we arrive at the main optimization (i. e. the pairings, see Sect. 5).

Table 1. Performance benchmarks (rounded to 10^3 cycles) on a 3.4 GHz Intel Core i7-6700 (Skylake) processor, for the isogeny (iso) + basis generation (basis) operation for $\ell = 2$. The columns labeled comp denote the results from SIKE, and the columns labeled dual denote our results. Cycle counts are averaged over 10 000 iterations.

	p434		p503		p610		p751	
	comp	dual	comp	dual	comp	dual	comp	dual
iso + basis	9 649	**7 921**	13 332	**11 039**	24 238	**20 269**	37 294	**30 922**

Remark 8. The points U and V (resp. $\widehat{\phi}_2(U)$ and $\widehat{\phi}_2(V)$) that are generated are not a basis for the 3^{e_3}-torsion, as they do not have order 3^{e_3}. Instead, we should use the points $U_3 = [2^{e_2}]U$ and $V_3 = [2^{e_2}]V$ (resp. $[2^{e_2}]\widehat{\phi}_2(U_3)$ and $[2^{e_2}]\widehat{\phi}_2(V_3)$), and by doing so would generate the exact same basis as in the SIKE proposal [17]. However, as noted by Zanon et al. [30, §3.1] in the context of the entangled basis for $\ell = 3$, the cofactors 2^{e_2} naturally factor out during the pairing and discrete logarithm phase and thus do not need to be performed explicitly.

Remark 9. For simplicity the focus of this section is limited to the SIKE parameters where $\ell = 2$ and $m = 3$. However, at no point is any restriction on m made (except not being equal to ℓ), so the above works equally well for any other odd prime m.

5   Pairing Computation

We now turn to the pairing, which is the phase of the compression algorithm on which the use of the dual isogeny has the largest effect. Recall that the reason for computing pairings of the images $\phi_\ell(P_m)$ and $\phi_\ell(Q_m)$ with respect to the deterministically generated basis points U_m and V_m is that in this way, we can transfer the discrete logarithm problems that yield the basis decomposition to the finite field $\mathbb{F}_{p^2}$. They are then solved in the multiplicative group $\mu_{m^{e_m}}$ of m^{e_m}-th roots of unity instead of in the elliptic curve group on the co-domain.

This is more efficient, even including the pairing computation, than solving the discrete logarithm problems on the elliptic curve because field operations are much more efficient and it is possible to precompute large tables of powers of a fixed basis in $\mu_{m^{e_m}}$, as described in [30]. Still, the pairings constitute the main bottleneck of the compression and we discuss how to significantly reduce their computational cost.

5.1   Pulling Back Pairing Arguments

First, recall that we fix a generator g_0 of the group of m^{e_m}-th roots of unity (and the base for the discrete logarithms) as

$$g_0 := \tau_{m^{e_m}}(\phi_\ell(P_m), \phi_\ell(Q_m)) = \tau_{m^{e_m}}(P_m, Q_m)^{\ell^{e_\ell}}.$$

As noted in [30], g_0 can be precomputed via the latter pairing, which only depends on system parameters. We aim to find $c_0, d_0, c_1, d_1 \in \mathbb{Z}/m^{e_m}\mathbb{Z}$ such that

$$g_1 = g_0^{d_0}, \qquad g_2 = g_0^{d_1}, \qquad g_3 = g_0^{-c_0}, \qquad g_4 = g_0^{-c_1},$$

where the g_i are computed as the four pairing values

$$g_1 = \tau_{m^{e_m}}(\phi_\ell(P_m), U_m), \qquad g_2 = \tau_{m^{e_m}}(\phi_\ell(P_m), V_m),$$
$$g_3 = \tau_{m^{e_m}}(\phi_\ell(Q_m), U_m), \qquad g_4 = \tau_{m^{e_m}}(\phi_\ell(Q_m), V_m).$$

Utilizing the dual isogeny $\widehat{\phi}_\ell$ and the torsion basis generation algorithms from the previous section, we compute the pairings with the points $\widehat{\phi}_\ell(U_m)$ and $\widehat{\phi}_\ell(V_m)$ on E instead. That is, the Tate pairing satisfies the property $\tau_m(\phi(S), T) = \tau_m(S, \widehat{\phi}(T))$ as stated in Sect. 2, so that the g_i can be computed as

$$g_1 = \tau_{m^{e_m}}(P_m, \widehat{\phi}_\ell(U_m)), \qquad g_2 = \tau_{m^{e_m}}(P_m, \widehat{\phi}_\ell(V_m)),$$
$$g_3 = \tau_{m^{e_m}}(Q_m, \widehat{\phi}_\ell(U_m)), \qquad g_4 = \tau_{m^{e_m}}(Q_m, \widehat{\phi}_\ell(V_m)).$$

This has the great advantage that the first arguments of all pairings are now fixed torsion basis points on the starting curve.

To see why this is useful, we consider Miller's algorithm [21] for computing pairings, which consists of a loop that carries out a scalar multiplication in a double-and-add fashion of the first pairing argument. On the way, it evaluates and accumulates corresponding line functions at the second argument via a square-and-multiply approach. It was first noted by Scott [28] and further discussed by Costello and Stebila [10] that all information depending on the fixed first argument can be precomputed and stored in a lookup table. This includes all required multiples of the first argument as well as the coefficients of the corresponding line functions. The online phase of the Miller loop consequently only needs to evaluate line functions at the second argument and accumulate them. In particular, this setting thus favors the use of affine coordinates because all inversions for computing the line slopes for point doublings and additions are done as a precomputation and the line functions take a very simple form for affine coordinates. We return to this in Sects. 5.3 and 5.4.

5.2 Special Curves and Torsion Bases for SIKE

From now on we restrict the discussion to the specific setting of SIKE. In particular, we make use of the special starting curve with $A = 6$ that is used in the SIKE proposal.

Let $\ell = 2$ and $m = 3$. Then we are concerned with computing the Tate pairing $\tau_{3^{e_3}}$ with either P_3 or Q_3 as the first argument. This is a special case since there exists a 2-isogeny $\chi : E_0 \to E_6$, while the endomorphism ring of E_0 contains the distortion map $\psi : (x, y) \mapsto (-x, iy)$. As such, there exists a point $P \in E_0(\mathbb{F}_p)[3^{e_3}]$ (any such non-trivial point suffices) such that $E_0[3^{e_3}] = \langle P, \psi(P) \rangle$, i.e. there exists a *distortion basis*, and we set up P_3 and Q_3 such that $P_3 = \chi(P)$ and $Q_3 = \chi\psi(P)$. Finally, by duality of the Tate pairing, we observe that

$$g_1 = \tau_{m^{e_m}}(P, \widehat{\chi\phi_\ell}(U_m)), \qquad g_3 = \tau_{m^{e_m}}(\psi(P), \widehat{\chi\phi_\ell}(U_m)),$$

$$g_2 = \tau_{m^{e_m}}(P, \widehat{\chi\phi_\ell}(V_m)), \qquad g_4 = \tau_{m^{e_m}}(\psi(P), \widehat{\chi\phi_\ell}(V_m)).$$

Hence, by applying an extra (dual of a) 2-isogeny (see Sect. 3.3) we can assume the first arguments to compose a distortion basis. The choice of this basis does not matter, and we simply set $P = [2^{e_2}](x_0, y_0)$ where $x_0 \in \mathbb{F}_p$ is the smallest (positive) integer such that P has order 3^{e_3}. As E_0 is in short Weierstrass form, we can immediately compute with affine Weierstrass coordinates.

The situation is slightly different for $\ell = 3$ and $m = 2$. It is not immediately obvious how to map to E_0 since it is *not* 3-isogenous to E_6. Also, even if we could, it is not possible to pick a distortion basis for $E_0(\mathbb{F}_{p^2})[2^{e_2}]$ according to [8, Lemma 1]. Instead, we map to the Weierstrass curve $E_{a,b} : y^2 = x^3 + ax + b$ over $\mathbb{F}_p$ where $a = -11$ and $b = 14$, which is isomorphic to E_6 via the isomorphism $E_6 \to E_{a,b} : (x, y) \mapsto (x + 2, y)$. Since $E_6(\mathbb{F}_p)[2^{e_2}] \cong \mathbb{Z}/2^{e_2-1}\mathbb{Z} \times \mathbb{Z}/2\mathbb{Z}$, the best we can do is to pick $P_2 \in E_{a,b}(\mathbb{F}_{p^2})[2^{e_2}]$ such that $[2]P_2 \in E_{a,b}(\mathbb{F}_p)$. The second basis point $Q_2 \in E_{a,b}(\mathbb{F}_{p^2})$ can be chosen such that $[2]Q_2 = (x, iy)$, where $x, y \in \mathbb{F}_p$ [9, §3.1].

By setting up the curves and torsion bases this way, the pairings in both the 2^{e_2}- and the 3^{e_3}-torsion groups can be improved by making use of the fact that all operations depending on the first argument are essentially operations in $\mathbb{F}_p$. Furthermore, the distortion basis for the 3^{e_3}-torsion group ensures that pairings with first argument Q_3 can use the same pre-computed table as those with first argument P_3. We explain how this works in detail for both cases.

5.3 Precomputation and the Miller Loop for $\ell = 3$

For $\ell = 3$ we compute order-2^{e_2} pairings of the form $\tau_{2^{e_2}}(P, U)$, meaning that the Miller loop consists of only doubling steps. Recall that for any point $P = (x_1, y_1) \in E_{a,b}(\mathbb{F}_{p^2})$ with $y_1 \neq 0$ its double is given by $[2]P = (x_2, y_2)$, where $x_2 = \lambda_1^2 - 2x_1$, $y_2 = \lambda_1(x_1 - x_2) - y_1$ and $\lambda_1 = (3x_1^2 + a)/(2y_1)$. A Miller doubling step with running point P and pairing value $f \in \mathbb{F}_{p^2}$ then updates f

by computing $f \leftarrow f^2 \cdot g/v$, where the tangent g and vertical line v, evaluated at the second pairing argument $U = (x_U, y_U) \in E_{a,b}(\mathbb{F}_{p^2})$, are given as

$$g = \lambda_1(x_U - x_1) + y_1 - y_U, \quad v = x_U - x_2 .$$

Hence, we can precompute all doublings of the first pairing argument, and store the point coefficients and the slopes used in the doubling formulas. We obtain two tables in the specific setting using the basis points P_2 and Q_2 fixed above as follows.

For P_2 we simply create the table where

$$T_{P_2}[j] = [x_{j+1}, y_{j+1}, \lambda_j] , \quad \text{for } j = 0, \ldots, e_2 - 2 ,$$

denoting $(x_j, y_j, \lambda_j) = (x_{[2^j]P_2}, y_{[2^j]P_2}, (3x_j^2 + a)/(2y_j))$. Since P_2 has order 2^{e_2} and we only carry out $e_2 - 1$ doublings, all doubling operations are well-defined. Note that by the choice of P_2 all point coordinates x_{j+1} and y_{j+1} are in $\mathbb{F}_p$, as are the slopes computed from them, except for the first slope $\lambda_1 \in \mathbb{F}_{p^2} \setminus \mathbb{F}_p$. Therefore, there are exactly $e_2 - 1$ triples and hence $3 \cdot (e_2 - 1)$ field elements in the table. There is an additional $\mathbb{F}_p$ element due to the first slope being an $\mathbb{F}_{p^2}$ element, but the last triple contains a point of order 2 which has y-coordinate 0 and does not have to be stored, keeping the overall element count at $3 \cdot (e_2 - 1)$.

The precomputed table for the point Q_2 is computed similarly. The only difference is that the multiples of Q_2 have the form (x, iy) with $x, y \in \mathbb{F}_p$ instead of being fully defined over $\mathbb{F}_p$. Since all multiples have this form, we can just store x and y and take care of the factor i when computing the line functions in the online phase of the algorithm. The same holds for the slope, as $(3x^2 + a)/(2iy) = -i \cdot (3x^2 + a)/(2y)$. Thus the table T_{Q_2} is defined analogously as

$$T_{Q_2}[j] = [w_{j+1}, z_{j+1}, \kappa_j] , \quad \text{for } j = 0, \ldots, e_2 - 2 ,$$

writing $(w_j, z_j, \kappa_j) = (x_{[2^j]Q_2}, y_{[2^j]Q_2}, (3w_j^2 + a)/(2z_j))$. Again, this table consists of $e_2 - 1$ triples of $\mathbb{F}_p$ elements, except for the first slope, which is in $\mathbb{F}_{p^2}$ and the last y-coordinate, which is 0. So the table stores $3 \cdot (e_2 - 1)$ elements in $\mathbb{F}_p$, and the total table size for storing the precomputed values needed to compute the four $\tau_{2^{e_2}}$ pairings is $6 \cdot (e_2 - 1)$ $\mathbb{F}_p$-elements.

The first Miller iteration for each of the two functions can be computed using $2 \cdot (3\mathbf{M} + \mathbf{S} + 8\mathbf{a}) \approx 2 \cdot (11\mathbf{m} + 26\mathbf{a})$, after which a single Miller iteration can be computed in $2 \cdot (2\mathbf{M} + 1\mathbf{S} + 2\mathbf{m} + 6\mathbf{a}) \approx 2 \cdot (10\mathbf{m} + 19\mathbf{a})$. For all four pairings (we assume $\mathbf{m} \approx \mathbf{s}$), this amounts to $40\mathbf{m} + 76\mathbf{a}$ for each Miller iteration except the first (which is slightly more expensive). For comparison, Zanon et al. [30] state the cost $55\mathbf{m} + 126\mathbf{a}$ for only two pairings, or $110\mathbf{m} + 252\mathbf{a}$ for all four. For a complete description of the algorithms we refer to the full version of the paper [23, Algorithms 1–2].

5.4 Precomputation and the Miller Loop for $\ell = 2$

For $\ell = 2$ we compute order-3^{e_3} pairings of the form $\tau_{3^{e_3}}(P, U)$, meaning that the Miller loop consists of only tripling steps instead. Again, for any point $P = (x_1, y_1) \in E_{a,b}(\mathbb{F}_{p^2})$ with $y_1 \neq 0$ its double is given by $[2]P = (x_2, y_2)$ as before, where $\lambda_1 = 3x_1^2/(2y_1)$ (note that here $a = 0$). If $x_2 \neq x_1$, its triple $[3]P = (x_3, y_3)$ is given by $x_3 = \lambda_2^2 - x_2 - x_1$, $y_3 = \lambda_2(x_1 - x_3) - y_1$ and $\lambda_2 = (y_2 - y_1)/(x_2 - x_1)$. A Miller tripling step with running point P and pairing value $f \in \mathbb{F}_{p^2}$ then updates f by computing $f \leftarrow f^3 \cdot g/v$, where g and v are now quadratic functions evaluated at $U = (x_U, y_U)$ given by

$$
\begin{aligned}
g &= (\lambda_1(x_U - x_1) + y_1 - y_U)(\lambda_2(x_U - x_1) + y_1 - y_U)\,, \\
v &= (x_U - x_2)(x_U - x_3)\,.
\end{aligned}
$$

To compute g, we can precompute λ_1, λ_2, $n_1 = y_1 - \lambda_1 x_1$ and $n_2 = y_1 - \lambda_2 x_1$, so that $g = (\lambda_1 x_U + n_1 - y_U)(\lambda_2 x_U + n_2 - y_U)$. As for the function v, we expand it to $v = x_U^2 - (x_2 + x_3)x_U + x_2 x_3$. Now we precompute $x_{2p3} = x_2 + x_3$ and $x_{23} = x_2 x_3$ and on input of U at the beginning of the loop also $x_{U,2} = x_U^2$. Then $v = x_{U,2} + x_{23} - x_{2p3} x_U$.

Now let $P \in E_0(\mathbb{F}_p)$ be the point of order 3^{e_3} such that $\{P, \psi(P)\}$ is the distortion basis of $E_0[3^{e_3}]$. We denote

$$
(x_2^{(j)}, y_2^{(j)}) = [2 \cdot 3^j]P\,, \quad (x_3^{(j)}, y_3^{(j)}) = [3^{j+1}]P\,, \quad \text{for } j = 0, \ldots e_3 - 2\,,
$$

and define $x_3^{(-1)} = x_1$ and $y_3^{(-1)} = y_1$. Then we define the table T_P by

$$
T_P[j] = \left[\lambda_1^{(j)}, \lambda_2^{(j)}, n_1^{(j)}, n_2^{(j)}, x_{2p3}^{(j)}, x_{23}^{(j)} \right]\,,
$$

where

$$
\begin{aligned}
\lambda_1^{(j)} &= 3(x_3^{(j-1)})^2/(2y_3^{(j-1)})\,, & \lambda_2^{(j)} &= (y_2^{(j)} - y_3^{(j-1)})/(x_2^{(j)} - x_3^{(j-1)})\,, \\
n_1^{(j)} &= y_3^{(j-1)} - \lambda_1 x_3^{(j-1)}\,, & n_2^{(j)} &= y_3^{(j-1)} - \lambda_2 x_3^{(j-1)}\,, \\
x_{2p3}^{(j)} &= x_2^{(j)} + x_3^{(j)}\,, & x_{23}^{(j)} &= x_2^{(j)} \cdot x_3^{(j)}\,.
\end{aligned}
$$

For the last iteration of the Miller loop we append the four extra values $x_3^{(e_3-2)}$, $y_3^{(e_3-2)}$, $\lambda_1^{(e_3-1)}$ and $x_2^{(e_3-1)}$. The second point in the distortion basis has the form $\psi(P) = (x_1, iy_1)$. This means that the functions g and v for pairings with $\psi(P)$ as the first argument are

$$
\begin{aligned}
g &= (-i\lambda_1 x_U + in_1 - y_U)(-i\lambda_2 x_U + in_2 - y_U)\,, \\
v &= x_{U,2} - x_{23} + x_{2p3} x_U\,.
\end{aligned}
$$

As a result, the same precomputed values can be used for those pairings without changes. The different signs and factors of i can be adjusted in the online phase of the pairing. The overall table size for all four $\tau_{3^{e_3}}$ pairings is thus $6 \cdot (e_3 - 1) + 4$ elements in $\mathbb{F}_p$.

A single Miller iteration can be computed in $2 \cdot (8\mathbf{M} + 2\mathbf{s} + 6\mathbf{m} + 18\mathbf{a}) \approx 68\mathbf{m} + 128\mathbf{a}$ for all four pairings (we assume $\mathbf{m} \approx \mathbf{s}$). For comparison, Zanon et al. [30] list $104\mathbf{m} + 2\mathbf{s} + 266\mathbf{a}$ for only two pairings. For a complete description of the algorithm we refer to the full version of the paper [23, Algorithm 3].

The Final Exponentiation. The final exponentiation raises all four pairing values to the power $(p^2 - 1)/m^{em}$. This is done as usual and as described in [7]. It is split up into the easy part, i.e. the power $p - 1$, which is computed by one application of the Frobenius endomorphism and one inversion per pairing value. Here, inversions are pushed down to the subfield $\mathbb{F}_p$ and shared using Montgomery's inversion sharing trick. The hard part of the final exponentiation is raising to the power $(p + 1)/m^{em}$. As $p + 1 = \ell^{e_\ell} \cdot m^{em}$, this is performed through a sequence of e_ℓ cyclotomic powerings by ℓ (e. g. squarings for $\ell = 2$ and cubings for $\ell = 3$).

Remark 10. We obtain significant speedups during the pairing computation as the use of the dual isogeny allows us to fix the first pairing arguments as system parameters, which benefits us for two reasons. Firstly, it allows us to pick the basis points P_m and Q_m of special form, either chosen as a distortion basis for $m = 3$ or as a basis such that the coefficients of (multiples of) $[2]P_m$ are in $\mathbb{F}_p$ and the coefficients of (multiples of) Q_m are of the form (x, iy) for $x, y \in \mathbb{F}_p$ for $m = 2$. In this case, most point doublings, triplings and line functions can be computed with arithmetic in $\mathbb{F}_p$ instead of $\mathbb{F}_{p^2}$. When using a distortion basis, all four pairings (as opposed to only two) share many of these computations.

Secondly, having fixed system parameters enables large precomputations. Although it leads to very significant speedups, it does have an impact on the memory usage. If, instead, one chooses to not use precomputation to keep the memory footprint of the implementation small, the special characteristics of the bases still lead to a reasonable speedup. Simply sharing operations across all four pairings for $m = 3$ and replacing general $\mathbb{F}_{p^2}$ operations by subfield operations can be implemented in the pairing algorithms as they are described by Zanon et al. in [30, §5] using extended Jacobian coordinates and by moving back to the starting curve. Operation counts predict savings of roughly 30% for the four Tate pairings of order 2^{e_2} and about 40% for the Tate pairings of order 3^{e_3}.

6   Implementation Results

We have added all of our techniques to the software library that is part of the SIKE proposal [17], so consider the set of primes $p \in \{\mathsf{p434}, \mathsf{p503}, \mathsf{p610}, \mathsf{p751}\}$, where

$$\mathsf{p434} = 2^{216} \cdot 3^{137} - 1, \qquad \mathsf{p503} = 2^{250} \cdot 3^{159} - 1,$$
$$\mathsf{p610} = 2^{305} \cdot 3^{192} - 1, \qquad \mathsf{p751} = 2^{372} \cdot 3^{239} - 1,$$

targeting the different security levels specified by NIST. The software is compiled with clang version 6.0.1 with the -O flag, and benchmarked on a 3.4 GHz

Intel Core i7-6700 Skylake processor running Ubuntu version 16.04.3 LTS with TurboBoost turned off. This is the exact same setting that was used for the performance numbers of SIKE Round 2 [18, Table 2.1]. Although we rederive their cycle counts for fairness of comparison, there is indeed a negligible difference (see Table 4).

We distinguish between functions in the SIKE library *without* the use of public-key compression techniques (SIKEpXXX), functions in the SIKE library *with* the use of public-key compression (SIKEpXXX_comp), and the functions used in our software (SIKEpXXX_dual). We begin by comparing the functions related to public-key compression to those in the SIKE library in Table 2, showing that we significantly improve the functions that currently bottleneck the computation, and analyze where the remaining bottlenecks are. We consider the impact on the key generation and exchange functions in the IND-CPA secure SIDH protocol in Table 4a and look at the impact on SIKE in Table 4b.

6.1   Cycle Counts for Compression Functions

In this section we discuss the performance of several functions as they are used in public-key compression. For the results we refer to Table 2.

iso. This function takes a secret key as input, and computes the isogeny ϕ_ℓ to obtain the co-domain curve $E/\langle R \rangle$ and potentially the images of basis points. The original compression techniques evaluate ϕ_ℓ at three (x-coordinates of) points, while we do not need to evaluate any points for $\ell = 2$. This leads to a speedup of 18–19% for $\ell = 2$. For $\ell = 3$ we evaluate ϕ_3 at $[2^{e_2-1}]Q_2$ to obtain the intermediate kernels for the dual, leading to a speedup of only 10–11%.

basis. This function starts where iso left off, and outputs U_m and V_m or $\widehat{\phi}_\ell(U_m)$ and $\widehat{\phi}_\ell(V_m)$, respectively. For $\ell = 2$ we apply the techniques described in Sect. 4.3, leading also to a speedup of 13–15%. For $\ell = 3$ the basis generation does not change significantly (as described in Sect. 4.2), while there is the added overhead of applying $\widehat{\phi}_\ell$. This leads to a slowdown for this function of 151–186%. However, since basis generation for $\ell = 3$ contributed only 4% of the total cost, this is much less bad in absolute terms.

pair. We see that the pairing computation significantly bottlenecked compression for $\ell = 3$, while also being the most expensive operation for $\ell = 2$. Applying the results from Sect. 5 leads to a speedup of at least 66% for $\ell = 2$, and a speedup of 62–63% for $\ell = 3$. This has an impressive impact on the efficiency of the full algorithm.

dlog. We have not made any changes to the discrete logarithm computations.

decomp. Decompression is slightly sped up due to simplifications to x-only basis generation in Sect. 4.1 and due to the avoidance of cofactor multiplications of the basis points. We obtain a 14–15% speedup for $\ell = 2$, and a 6–7% speedup for $\ell = 3$.

As a result of the improvements, we note that the pairing phase is no longer a bottleneck for public-key compression. For $\ell = 2$ it is actually significantly cheaper than basis generation, while for $\ell = 3$ it is only moderately more expensive than the basis generation and discrete logarithm phases.

Table 2. Performance benchmarks (rounded to 10^3 cycles) on a 3.4 GHz Intel Core i7-6700 (Skylake) processor, for the compression operations: co-domain generation (iso), basis generation (basis), pairing computation (pair), discrete logarithm computation (dlog) and decompression (decomp). Cycle counts are averaged over 10 000 iterations.

	ℓ	iso	basis	pair	dlog	decomp
SIKEp434_comp	2	5 811	3 838	5 821	923	2 549
SIKEp434_dual	2	4 690	3 231	1 954	923	1 910
SIKEp434_comp	3	6 464	598	4 921	1 222	1 890
SIKEp434_dual	3	5 750	1 618	1 821	1 223	1 741
SIKEp503_comp	2	8 141	5 191	8 033	1 556	3 513
SIKEp503_dual	2	6 594	4 445	2 676	1 554	2 613
SIKEp503_comp	3	9 015	844	6 716	1 532	2 551
SIKEp503_dual	3	7 992	2 219	2 486	1 535	2 380
SIKEp610_comp	2	15 430	8 808	13 458	2 351	5 868
SIKEp610_dual	2	12 778	7 491	4 525	2 349	4 403
SIKEp610_comp	3	15 490	1 340	11 365	2 685	4 365
SIKEp610_dual	3	13 747	3 750	4 214	2 685	4 039
SIKEp751_comp	2	23 133	14 161	21 908	3 529	9 434
SIKEp751_dual	2	18 898	12 024	7 348	3 528	7 135
SIKEp751_comp	3	26 133	2 125	18 224	5 030	6 914
SIKEp751_dual	3	23 316	6 081	6 727	5 055	6 489

6.2   Impact on SIDH and SIKE

Finally, we summarize the impact of improved public-key compression when included in a cryptographic protocol. The schemes that are of interest for this purpose are the passively secure SIDH protocol, and its actively secure variant SIKE. Although one can of course argue about the best metric for comparison, we believe the most interesting from an implementers perspective is the overhead that is caused by including public-key compression. This gives a relatively clear idea of the loss of efficiency that is to be paid for a reduction of the size of the public keys.

In Table 4a we see that, across different SIDH parameter sets, for key generation the overhead is reduced from 160–182% to 77–86% for $\ell = 2$ and from 98–104% to 59–61% for $\ell = 3$, respectively. The overhead for the key exchange phase is reduced by about 10–13% in both cases. For SIKE (see Table 4b), we reduce the overhead of (1) key generation from 140–153% to 61–74%, (2) key encapsulation from 67–90% to 38–57%, and (3) decapsulation from 59–65% to 34–39%. Following the SIKE specification [17, Table 2.1], we also provide the impact on the "total" cost, i. e. on the sum of the costs of encapsulation and decapsulation. This reduces from 62–83% to an overhead of 36–48% across the different parameter sets.

Memory Constraints. Having remarked on the memory usage before, we provide some more detail here. The first notable consequence of our techniques is that we need to build a table containing the kernels of all intermediate ℓ-isogenies appearing in the decomposition of $\widehat{\phi}_\ell$. For $\ell = 2$, and assuming that e_2 is even for simplicity (if not the difference is very minor), we compute a sequence of $e_2/2$ 4-isogenies. For each such isogeny we store 5 elements (see Sect. 3.2), resulting in a table of $5 \cdot e_2/2$ $\mathbb{F}_{p^2}$-elements or simply $5 \cdot e_2$ elements of $\mathbb{F}_p$. For $\ell = 3$, for each 3-isogeny we simply store a generator of the kernel of its dual, requiring 2 elements of $\mathbb{F}_{p^2}$. Hence we store a table containing $4 \cdot e_3$ $\mathbb{F}_p$-elements. Note that these are not precomputed, but need to be temporarily stored on the stack. However, recall from Sect. 5 that we do precompute a table of $6 \cdot (e_2 - 1)$ elements in $\mathbb{F}_p$ for $\ell = 3$ and $6 \cdot (e_3 - 1) + 4)$ for $\ell = 2$ to compute the pairings, i. e. in contrast to the intermediate kernel information, pairing tables are precomputed public parameters.

To aid the discrete logarithm computation, Zanon et al. [30] introduced the use of large precomputed tables. For some fixed window w_3, the discrete logarithms for $\ell = 2$ use a table containing $e_3/w_3 \cdot 3^{w_3}$ or $2 \cdot \lceil e_3/w_3 \rceil \cdot 3^{w_3}$ elements in $\mathbb{F}_{p^2}$ when $w_3 \mid e_3$ or $w_3 \nmid e_3$, respectively. Similarly, the discrete logarithms for $\ell = 3$ use a table of size $e_2/w_2 \cdot 2^{w_2}$ resp. $\lceil e_2/w_2 \rceil \cdot 2^{w_2+1}$ when $w_2 \mid e_2$ resp. $w_2 \nmid e_2$, for some window size w_2. Though small windows of course lead to relatively small tables, for SIKE we always have $4 \le w \le 6$ and the current SIKE submission contains very large tables for the discrete logarithms. We summarize memory requirements in terms of the number of field elements in $\mathbb{F}_p$ for the different parameters in Table 3.

Table 3. Required memory in $\mathbb{F}_p$-elements for storing intermediate information used for computing the dual isogeny (`iso`) and for the precomputed tables for the pairing (`pair`) and discrete logarithm computation (`dlog`).

	ℓ	iso	pair	dlog	ℓ	iso	pair	dlog
SIKEp434_dual		1 080	820	26 244		548	1 290	1 728
SIKEp503_dual	2	1 250	952	30 132	3	636	1 494	3 200
SIKEp610_dual		1 525	1 150	46 656		768	1 824	3 904
SIKEp751_dual		1 860	1 432	45 684		956	2 226	2 976

Section 2.3 of the SIKE specification points out that due to the large tables, the current compression method cannot be used in a straightforward manner on constrained devices. Our methods add another possibility for a time-memory trade-off. As pointed out earlier in Remark 10, the choice of special bases

Table 4. Performance benchmarks (rounded to 10^3 cycles) on a 3.4 GHz Intel Core i7-6700 (Skylake) processor. Cycle counts are averaged over 10 000 iterations. The label oh denotes the cpu overhead over the corresponding uncompressed version of the function.

(a) The SIDH operations: public key generation ($\mathtt{isogen}_2$ and $\mathtt{isogen}_3$) and key exchange ($\mathtt{isoex}_2$ and $\mathtt{isoex}_3$).

	$\mathtt{isogen}_2$		$\mathtt{isoex}_2$		$\mathtt{isogen}_3$		$\mathtt{isoex}_3$	
	cyc	oh	cyc	oh	cyc	oh	cyc	oh
SIKEp434	5 821	–	4 726	–	6 469	–	5 467	–
SIKEp434_comp	16 397	182%	5 425	15%	13 208	104%	6 825	25%
SIKEp434_dual	**10 836**	**86%**	**5 298**	**12%**	**10 412**	**61%**	**6 192**	**13%**
SIKEp503	8 154	–	6 745	–	9 002	–	7 623	–
SIKEp503_comp	22 931	181%	7 582	12%	18 107	101%	9 466	24%
SIKEp503_dual	**15 310**	**88%**	**7 422**	**10%**	**14 270**	**59%**	**8 651**	**13%**
SIKEp610	15 438	–	12 881	–	15 464	–	13 282	–
SIKEp610_comp	40 097	160%	14 458	12%	31 031	101%	16 251	22%
SIKEp610_dual	**27 270**	**77%**	**14 170**	**10%**	**24 527**	**59%**	**14 796**	**11%**
SIKEp751	23 229	–	18 961	–	26 024	–	22 255	–
SIKEp751_comp	62 998	171%	21 517	13%	51 443	98%	27 257	22%
SIKEp751_dual	**41 778**	**80%**	**21 104**	**11%**	**41 298**	**59%**	**24 952**	**12%**

(b) The SIKE operations: public key generation ($\mathtt{KeyGen}$), encapsulation ($\mathtt{Encaps}$), and decapsulation ($\mathtt{Decaps}$).

	Size (B)		KeyGen		Encaps		Decaps	
	pk	ct	cyc	oh	cyc	oh	cyc	oh
SIKEp434	330	346	6 482	–	10 563	–	11 290	–
SIKEp434_comp	196	209	16 397	153 %	20 056	90%	18 622	65%
SIKEp434_dual	196	209	**10 849**	**67%**	**16 600**	**57%**	**15 682**	**39%**
SIKEp503	378	402	9 043	–	14 950	–	15 749	–
SIKEp503_comp	224	248	23 066	155%	27 665	85%	25 646	63%
SIKEp503_dual	224	248	**15 294**	**69%**	**22 875**	**53%**	**21 841**	**39%**
SIKEp610	462	486	15 651	–	28 346	–	28 603	–
SIKEp610_comp	273	297	40 078	156%	47 279	67%	45 536	59%
SIKEp610_dual	273	297	**27 277**	**74%**	**39 238**	**38%**	**38 371**	**34%**
SIKEp751	564	596	26 064	–	42 102	–	45 361	–
SIKEp751_comp	331	363	62 663	140%	78 895	87%	72 924	61%
SIKEp751_dual	331	363	**41 909**	**61%**	**66 096**	**57%**	**62 337**	**37%**

already improves the performance even without precomputation. The precomputed tables can be adjusted in size linearly, where computation of required values can be moved to the online phase. Given that the main bottleneck in both [7] and [30] is clearly the pairing phase, it might be worthwhile to use memory for the pairing tables instead of the discrete logarithm tables and find a more space efficient trade-off than the one currently deployed in the SIKE submission.

Acknowledgements. We thank the anonymous reviewers for their detailed remarks and Paulo S.L.M. Barreto for valuable feedback to improve the paper.

References

1. Azarderakhsh, R., Jao, D., Kalach, K., Koziel, B., Leonardi, C.: Key compression for isogeny-based cryptosystems. In: AsiaPKC 2016, pp. 1–10. ACM (2016)
2. Bernstein, D.J., Hamburg, M., Krasnova, A., Lange, T.: Elligator: elliptic-curve points indistinguishable from uniform random strings. In: ACM SIGSAC 2013, pp. 967–980. ACM (2013)
3. Blake, I., Seroussi, G., Smart, N., Cassels, J.W.S.: Advances in Elliptic Curve Cryptography. Cambridge University Press, Cambridge (2005)
4. Bos, J.W., et al.: CRYSTALS - Kyber: a CCA-secure module-lattice-based KEM. In: EuroS&P 2018, pp. 353–367. IEEE (2018)
5. Castryck, W., Lange, T., Martindale, C., Panny, L., Renes, J.: CSIDH: an efficient post-quantum commutative group action. In: Peyrin, T., Galbraith, S. (eds.) ASIACRYPT 2018. LNCS, vol. 11274, pp. 395–427. Springer, Cham (2018). https://doi.org/10.1007/978-3-030-03332-3_15
6. Costello, C., Hisil, H.: A simple and compact algorithm for SIDH with arbitrary degree isogenies. In: Takagi, T., Peyrin, T. (eds.) ASIACRYPT 2017. LNCS, vol. 10625, pp. 303–329. Springer, Cham (2017). https://doi.org/10.1007/978-3-319-70697-9_11
7. Costello, C., Jao, D., Longa, P., Naehrig, M., Renes, J., Urbanik, D.: Efficient compression of SIDH public keys. In: Coron, J.-S., Nielsen, J.B. (eds.) EUROCRYPT 2017. LNCS, vol. 10210, pp. 679–706. Springer, Cham (2017). https://doi.org/10.1007/978-3-319-56620-7_24
8. Costello, C., Longa, P., Naehrig, M.: Efficient algorithms for supersingular isogeny Diffie-Hellman. In: Robshaw, M., Katz, J. (eds.) CRYPTO 2016. LNCS, vol. 9814, pp. 572–601. Springer, Heidelberg (2016). https://doi.org/10.1007/978-3-662-53018-4_21
9. Costello, C., Longa, P., Naehrig, M., Renes, J., Virdia, F.: Improved classical cryptanalysis of the computational supersingular isogeny problem. Cryptology ePrint Archive, Report 2019/298 (2019). https://eprint.iacr.org/2019/298
10. Costello, C., Stebila, D.: Fixed argument pairings. In: Abdalla, M., Barreto, P.S.L.M. (eds.) LATINCRYPT 2010. LNCS, vol. 6212, pp. 92–108. Springer, Heidelberg (2010). https://doi.org/10.1007/978-3-642-14712-8_6
11. D'Anvers, J.-P., Karmakar, A., Sinha Roy, S., Vercauteren, F.: Saber: module-LWR based key exchange, CPA-secure encryption and CCA-secure KEM. In: Joux, A., Nitaj, A., Rachidi, T. (eds.) AFRICACRYPT 2018. LNCS, vol. 10831, pp. 282–305. Springer, Cham (2018). https://doi.org/10.1007/978-3-319-89339-6_16

12. De Feo, L., Jao, D., Plût, J.: Towards quantum-resistant cryptosystems from supersingular elliptic curve isogenies. J. Math. Cryptol. **8**, 209–247 (2014)
13. Fujisaki, E., Okamoto, T.: Secure integration of asymmetric and symmetric encryption schemes. In: Wiener, M. (ed.) CRYPTO 1999. LNCS, vol. 1666, pp. 537–554. Springer, Heidelberg (1999). https://doi.org/10.1007/3-540-48405-1_34
14. Galbraith, S.D.: Mathematics of Public Key Cryptography. Cambridge University Press, Cambridge (2012)
15. Galbraith, S.D., Petit, C., Shani, B., Ti, Y.B.: On the security of supersingular isogeny cryptosystems. In: Cheon, J.H., Takagi, T. (eds.) ASIACRYPT 2016. LNCS, vol. 10031, pp. 63–91. Springer, Heidelberg (2016). https://doi.org/10.1007/978-3-662-53887-6_3
16. Hofheinz, D., Hövelmanns, K., Kiltz, E.: A modular analysis of the Fujisaki-Okamoto transformation. In: Kalai, Y., Reyzin, L. (eds.) TCC 2017. LNCS, vol. 10677, pp. 341–371. Springer, Cham (2017). https://doi.org/10.1007/978-3-319-70500-2_12
17. Jao, D., et al.: SIKE (2019). Submission to round 2 of [24]. http://sike.org
18. Jao, D., et al.: SIKE (2016). Submission to [24]. http://sike.org
19. Jao, D., De Feo, L.: Towards quantum-resistant cryptosystems from supersingular elliptic curve isogenies. In: Yang, B.-Y. (ed.) PQCrypto 2011. LNCS, vol. 7071, pp. 19–34. Springer, Heidelberg (2011). https://doi.org/10.1007/978-3-642-25405-5_2
20. Lichtenbaum, S.: Duality theorems for curves over P-adic fields. Inventiones Mathematicae **7**, 120–136 (1969)
21. Miller, V.S.: The Weil pairing, and its efficient calculation. J. Cryptol. **17**(4), 235–261 (2004)
22. Montgomery, P.L.: Speeding the pollard and elliptic curve methods of factorization. Math. Comput. **48**(177), 243–264 (1987)
23. Naehrig, M., Renes, J.: Dual isogenies and their application to public-key compression for isogeny-based cryptography. Cryptology ePrint Archive, Report 2019/499 (2019). https://eprint.iacr.org/2019/499
24. National Institute of Standards and Technology: Post-quantum cryptography standardization, December 2016. https://csrc.nist.gov/Projects/Post-Quantum-Cryptography/Post-Quantum-Cryptography-Standardization
25. Renes, J.: Computing isogenies between Montgomery curves using the action of (0, 0). In: Lange, T., Steinwandt, R. (eds.) PQCrypto 2018. LNCS, vol. 10786, pp. 229–247. Springer, Cham (2018). https://doi.org/10.1007/978-3-319-79063-3_11
26. Renes, J., Smith, B.: qDSA: small and secure digital signatures with curve-based Diffie–Hellman key pairs. In: Takagi, T., Peyrin, T. (eds.) ASIACRYPT 2017. LNCS, vol. 10625, pp. 273–302. Springer, Cham (2017). https://doi.org/10.1007/978-3-319-70697-9_10
27. Schaefer, E., Stoll, M.: How to do a p-descent on an elliptic curve. Trans. Am. Math. Soc. **356**(3), 1209–1231 (2004)
28. Scott, M.: Implementing cryptographic pairings. In: Takagi, T., et al. (eds.) Pairing 2007, pp. 177–196. Springer, Heidelberg (2007)
29. Silverman, J.H.: The Arithmetic of Elliptic Curves. GTM, vol. 106. Springer, New York (2009). https://doi.org/10.1007/978-0-387-09494-6
30. Zanon, G.H.M., Simplicio, M.A., Pereira, G.C.C.F., Doliskani, J., Barreto, P.S.L.M.: Faster key compression for isogeny-based cryptosystems. IEEE Trans. Comput. **68**(5), 688–701 (2019)

Optimized Method for Computing Odd-Degree Isogenies on Edwards Curves

Suhri Kim[1], Kisoon Yoon[2], Young-Ho Park[3(✉)], and Seokhie Hong[1]

[1] Center for Information Security Technologies (CIST), Korea University, Seoul, Republic of Korea
suhrikim@gmail.com, shhong@korea.ac.kr
[2] NSHC Inc., Uiwang, Republic of Korea
kisoon.yoon@gmail.com
[3] Sejong Cyber University, Seoul, Republic of Korea
youngho@sjcu.ac.kr

Abstract. In this paper, we present an efficient method to compute arbitrary odd-degree isogenies on Edwards curves. By using the w-coordinate, we optimized the isogeny formula on Edwards curves by Moody and Shumow. We demonstrate that Edwards curves have an additional benefit when recovering the coefficient of the image curve during isogeny computation. For ℓ-degree isogeny where $\ell = 2s + 1$, our isogeny formula on Edwards curves outperforms Montgomery curves when $s \geq 2$. To better represent the performance improvements when w-coordinate is used, we implement CSIDH using our isogeny formula. Our implementation is about 20% faster than the previous implementation. The result of our work opens the door for the usage of Edwards curves in isogeny-based cryptography, especially for CSIDH which requires higher degree isogenies.

Keywords: Isogeny · Post-quantum cryptography · Montgomery curves · Edwards curves · SIDH · CSIDH

1 Introduction

Cryptosystems based on isogenies using supersingular elliptic curves were first proposed by De Feo and Jao [16]. They proposed a Diffie-Hellman type key exchange protocol named Supersingular Isogeny Diffie-Hellman (SIDH). Instead of relying on the discrete logarithm problems where intractability assumption of the problem is broken by Shor's algorithm, the security relies on the problem of finding an isogeny between two given elliptic curves over a finite field. Moreover, since the key sizes are small compared to other post-quantum cryptography (PQC) categories, isogeny-based cryptography has positioned itself as a promising candidate for PQC. Later, SIDH led to the development of the key encapsulation mechanism

This work was supported by the National Research Foundation of Korea (NRF) grant funded by the Korea government (MSIT) (No. NRF-2017R1A2B4011599).

S. D. Galbraith and S. Moriai (Eds.): ASIACRYPT 2019, LNCS 11922, pp. 273–292, 2019.
https://doi.org/10.1007/978-3-030-34621-8_10

called Supersingular Isogeny Key Encapsulation (SIKE), which is a Round 2 candidate in the NIST PQC standardization project [2].

Recently, De Feo *et al.* proposed the improvements to the CRS scheme in [12,23]. The CRS scheme was the first cryptosystem based on isogenies between ordinary curves. However, the scheme was highly inefficient and the use of ordinary curves makes the algorithm suffer from the subexponential attack proposed by [8]. The scheme proposed in [13] optimized the CRS scheme, although several minutes are still required for a single key exchange. Independent from [13], Castryck *et al.* proposed CSIDH (Commutative SIDH), which also adapted the CRS scheme, but applied it to supersingular elliptic curves [7]. Instead of working with supersingular elliptic curves over $\mathbb{F}_{p^2}$ as in SIDH/SIKE, CSIDH works over $\mathbb{F}_p$. CSIDH is a non-interactive key exchange protocol having smaller key sizes than SIDH/SIKE.

Considering the implementation, isogeny-based cryptosystems involve complicating isogeny operations in addition to the standard elliptic curve arithmetic over a finite field. Regarding the isogeny operations, the degree of an isogeny used in the cryptosystem depends on the prime chosen for the scheme. For SIDH or SIKE, p is of the form $p = \ell_A^{e_A} \ell_B^{e_B} f \pm 1$, where ℓ_A and ℓ_B are coprime to each other. The ℓ_A and ℓ_B can be considered as the degree of isogenies dealt in the scheme. Since the complexity of computing isogenies increases as the degree increases, isogenies of degree 3- and 4- were mostly considered for implementing SIDH or SIKE. CSIDH exploits p of the form $p = 4\ell_1\ell_2\cdots\ell_n - 1$, where ℓ_i are odd-primes. Similarly, as ℓ_i are degrees of isogenies used in the scheme, demands for odd-degree isogeny formulas have increased after the proposal of CSIDH. Regarding the elliptic curve arithmetic, it is important to select the form of elliptic curves that can provide efficient curve operations. Until recently, only Montgomery curves were used, as they offer fast computations on both components – *i.e.* isogeny computation and curve arithmetic. The state-of-the-art implementation proposed in [11] is also based on Montgomery curves.

Meanwhile, researches have extended to adopt other forms of elliptic curves that yield efficient arithmetic or isogeny computation. In [9], it was mentioned that due to the birationality between twisted Edwards curves and Montgomery curves, there might exist savings to be gained when twisted Edwards curves are used for SIDH/SIKE. The utilization of elliptic curve arithmetic on twisted Edwards curves was first proposed by Meyer *et al.* [20]. Their method uses twisted Edwards curves for elliptic curve arithmetic and Montgomery curves for isogeny computation. For isogenies on Edwards curves, optimized 3- and 4-isogeny formulas were first proposed in [17], in order to apply Edwards curves in isogeny-based cryptosystems. In [19], they implemented CSIDH by using Montgomery curves for isogenies and twisted Edwards curves for recovering the coefficient of the image curve.

Currently, using Edwards curves for isogeny-based cryptosystems is not so promising. As Bos and Friedberger [5] have demonstrated, working with twisted Edwards curves does not provide faster elliptic curve arithmetic in the setting of SIDH or SIKE. The implementation results in [1,18] also show that Edwards curves do not result in faster performance. In short, Edwards curves for implementing SIDH or SIKE have one critical disadvantage – elliptic curve arithmetic are slower on Edwards curves than on Montgomery curves in SIDH or SIKE settings. When it comes to CSIDH, the most painstaking part is to construct odd-degree isogenies. Although the motivation for the work in [9] is slightly different, the proposed odd-degree isogeny formula can naturally be applied in CSIDH when using Montgomery curves. The only generalized odd-degree isogeny formula on Edwards curves is the formula proposed by Moody and Shumow in [21]. Though, as stated in [19], the coordinate map of the formula is not as simple to compute as in [9].

However, there are still some aspects to optimize the odd-degree isogeny formula on Edwards curves. Until now, the optimization of isogenies on Edwards curves was only done for small degree isogenies. In [17,18], the 3- and 4- isogeny formula on Edwards curves were optimized by substituting the x-coordinate and curve coefficients of Moody and Shumow's formula to y-coordinates using division polynomials and curve equations. As the degree goes higher, optimizing Moody and Shumow's formula by using the method presented in [17,18] is cumbersome. Additional improvements can be achieved on a higher degree isogenies if different approaches are applied for the optimization.

The aim of this work is to construct efficient and generalized odd-degree isogenies on Edwards curves to be suitable for isogeny-based cryptosystems. The following list details the main contributions of this work.

- We exploit the w-coordinate proposed in [14] on Edwards curves. As mentioned above, the main disadvantage of using YZ-coordinates for Edwards curves is that the elliptic curve arithmetic is slower than on Montgomery curves in SIDH or SIKE settings. However, the costs of doubling, tripling, and differential addition using projective w-coordinate are the same as on Montgomery curves, which motivates us to use the w-coordinate system on Edwards curves.
- We present the formula for computing odd-degree isogenies using the w-coordinate. By optimizing the isogeny formula proposed by Moody and Shumow, the computational cost of evaluating an ℓ-isogeny is the same as on Montgomery curves. We also optimized the formula for obtaining the curve coefficient of the image curve. Our formula for computing the curve coefficient does not require additional points and has benefits over Montgomery curves when the degree is higher than 5. Derivations of our isogeny formula and computational cost are presented in Sect. 3, and analysis of our isogeny formula is presented in Sect. 4.
- We present the implementation result of CSIDH using Edwards w-coordinates. The result of our implementation is about 20% faster than the implementation proposed in [7], and 2% faster than the implementation presented in [19]. This result is natural as computing the coefficient of the image curve is more efficient on Edwards w-coordinate than Montgomery x-coordinate. Additionally, when computing elliptic curve arithmetic,

the number of additions and subtractions decreases on w-coordinate Edwards curves compare to x-coordinate Montgomery curves. As the cost of elliptic curve arithmetic is inevitable, the difference in the number of additions is accumulated and resulted in a faster speed than hybrid-CSIDH, proposed in [19].

This paper is organized as follows: In Sect. 2, we review on Edwards curves and their arithmetic using w-coordinates. Also, the description of the SIDH and CSIDH protocol are presented. In Sect. 3, we present our optimization of a generalized odd-degree isogeny formula on Edwards curves. The implementations result of CSIDH using Edwards w-coordinate is presented Sect. 4. We draw our conclusions and future work in Sect. 5.

2   Preliminaries

In this section, we provide the required background that will be used throughout the paper. First, we review the Edwards curves and their arithmetic using the w-coordinate. Then, we introduce the SIDH and CSIDH protocol to illustrate the required degree of an isogeny for each protocol.

2.1   Edwards Curves and Their Arithmetic

Edwards Curves. Edwards elliptic curves over K are defined by the equation,

$$E_d : x^2 + y^2 = 1 + dx^2 y^2, \tag{1}$$

where $d \neq 0, 1$. The E_d has singular points $(1 : 0 : 0)$ and $(0 : 1 : 0)$ at infinity. In Edwards curves, the point $(0, 1)$ is the identity element, and the point $(0, -1)$ has order two. The points $(1, 0)$ and $(-1, 0)$ have order four. The condition that E_d always has a rational point of order four restricts the use of elliptic curves in the Edwards model. Twisted Edwards curves are a generalization of Edwards curves proposed by Bernstein *et al.* in [3], to overcome such deficiency. Twisted Edwards curves are defined by the equation,

$$E_{a,d} : ax^2 + y^2 = 1 + dx^2 y^2, \tag{2}$$

for distinct nonzero elements $a, d \in K$ [3]. Clearly, $E_{a,d}$ is isomorphic to an Edwards curve over $K(\sqrt{a})$. The j-invariant of Edwards curves is defined as $j(E_d) = 16(1 + 14d + d^2)^3 / d(1 - d)^4$. For the same reason as in [11], we use projective curve coefficients on Edwards curves to avoid inversions when recovering the coefficient of the image curves. Let $(C, D) \in \mathbb{P}^1(K)$ where $C \in \bar{K}^\times$ such that $d = D/C$. Then E_d can be expressed as

$$E_{C:D} : Cx^2 + Cy^2 = C + Dx^2 y^2.$$

Arithmetic on Edwards Curves. For points (x_1, y_1) and (x_2, y_2) on Edwards curves E_d, the addition of two points is defined as below, and doubling can be performed with exactly the same formula.

$$(x_1, y_1) + (x_2, y_2) = \left(\frac{x_1 y_2 + y_1 x_2}{1 + d x_1 x_2 y_1 y_2}, \frac{y_1 y_2 - x_1 x_2}{1 - d x_1 x_2 y_1 y_2} \right).$$

Generally, projective coordinates $(X : Y : Z) \in \mathbb{P}^2$ where $x = X/Z$ and $y = Y/Z$ are used for the corresponding affine point (x, y) on E_d to avoid inversions during elliptic curve arithmetic. There are several coordinate systems relating to Edwards curves such as inverted coordinates $(X : Y : Z)$ which represents the point $(Z/X, Z/Y)$ on an Edwards curve or extended coordinates which uses $(X : Y : Z : T)$ with $XY = ZT$, for an efficient computation [4,15].

2.2 w-Coordinate on Edwards Curves

To evaluate the point addition efficiently, Farashahi and Hosseini proposed w-coordinate system on Edwards curves, and we briefly introduce here [14]. In [14], they proposed the rational map w as $w(x, y) = dx^2 y^2$ or $w(x, y) = x^2/y^2$ for points (x, y) on an Edwards curve and presented Montgomery-like formulas for elliptic curve arithmetic on Edwards curves. Although $w(x, y) = dx^2 y^2$ and $w(x, y) = x^2/y^2$ are different rational functions, as they yield identical formula, we shall use the map $w(x, y) = dx^2 y^2$ for the explanation.

Define the rational function w by $w(x, y) = dx^2 y^2$. This function is well defined for all affine points on an Edwards curve. For $P = (x, y)$ on an Edwards curve E_d, $-P = (-x, y)$ so that $w(P) = w(-P)$. Also, $w(O) = 0$. Let $P_1 = (x_1, y_1)$ and $P_2 = (x_2, y_2)$ be the points on E_d. Let $w_0 = w(2P_1)$, $w_3 = w(P_1 + P_2)$, and $w_4 = w(P_1 - P_2)$. The addition formula on Edwards curves gives

$$x_3(1 + dx_1 x_2 y_1 y_2) = x_1 y_2 + x_2 y_1,$$
$$x_4(1 - dx_1 x_2 y_1 y_2) = x_1 y_2 - x_2 y_1,$$
$$y_3(1 - dx_1 x_2 y_1 y_2) = y_1 y_2 - x_1 x_2,$$
$$y_4(1 + dx_1 x_2 y_1 y_2) = y_1 y_2 + x_1 x_2,$$

where $P_1 + P_2 = (x_3, y_3)$ and $P_1 - P_2 = (x_4, y_4)$. By multiplying the above equations and squaring both sides we have,

$$x_3^2 y_3^2 x_4^2 y_4^2 = \frac{(x_1^2 y_2^2 - x_2^2 y_1^2)^2 (y_1^2 y_2^2 - x_1^2 x_2^2)^2}{(1 - d^2 x_1^2 x_2^2 y_1^2 y_2^2)^4}.$$

Multiplying both sides by d^2 of the above equation, we obtained the differential addition formula as presented in [14]. In [14], the doubling and differential addition formulas are defined as,

$$w_0 = \frac{4w_1((w_1 + 1)^2 - ew_1)}{(w_1^2 - 1)^2}, \qquad w_3 w_4 = \frac{(w_1 - w_2)^2}{(w_1 w_2 - 1)^2}.$$

where $e = 4/d$. For the rest of the subsection, we analyze the computational cost of doubling, tripling, and differential additions in the setting of isogeny-based cryptosystems, using projective w-coordinates. The **M** and **S** refers to a field multiplication and squaring, respectively, and **a** and **s** refers to a field addition and subtraction, respectively. In the remainder of this paper, we shall consider WZ-coordinate as projective w-coordinates. As mentioned above, although we define $w(x, y)$ as $w(x, y) = dx^2y^2$, computational costs are identical when $w(x, y)$ is defined as $w(x, y) = x^2/y^2$. Note that these elliptic curve arithmetic form the building blocks when implementing isogeny-based cryptosystems.

Doubling. Let $P = (x, y)$ be a point on an Edwards curve E_d defined as in Eq. (1). Let $d = D/C$, $w = dx^2y^2$, and $w = W/Z$. For $P = (W : Z)$ in projective w-coordinates, the doubling of P gives $[2]P = (W' : Z')$, where W' and Z' are defined as

$$W' = 4WZ(D(W + Z)^2 - 4CWZ),$$
$$Z' = D(W + Z)^2(W - Z)^2.$$

The above equation can be computed as,

$$t_0 = (W + Z)^2, \quad t_1 = (W - Z)^2, \quad t_2 = D \cdot t_0,$$
$$Z' = t_2 \cdot t_1, \quad t_0 = t_0 - t_1, \quad t_1 = C \cdot t_0,$$
$$W' = t_2 - t_1, \quad W' = W' \cdot t_0.$$

The computational cost is **4M+2S**.

Tripling. For $P = (W : Z)$ on an Edwards curve E_d represented in projective coordinates, the tripling of P gives $[3]P = (W' : Z')$, where W' and Z' are defined as

$$W' = W(D(W^2 - Z^2)^2 - Z^2(4D(W + Z)^2 - 16CWZ))^2,$$
$$Z' = Z(-D(W^2 - Z^2)^2 + W^2(4D(W + Z)^2 - 16CWZ))^2.$$

The computational cost is **7M+5S**.

Differential Addition. The differential addition is needed when computing the kernel for SIDH or CSIDH. For example, SIDH starts by computing $R = [m]P + [n]Q$ for chosen basis P and Q and a secret key (m, n). Without loss of generality, we may assume that m is invertible, and compute $R = P + [m^{-1}n]Q$. This can be done by using the Montgomery ladder which requires computing differential additions as a subroutine.

Let $P_1 = (W_1 : Z_1)$ and $P_2 = (W_2 : Z_2)$ be the points on E_d. Let $w_0 = w(P_1 - P_2)$ and $w_3 = w(P_1 + P_2)$. Let $w_0 = W_0/Z_0$ and $w_3 = W_3/Z_3$.

Then,

$$W_3 = Z_0(W_1Z_2 - W_2Z_1)^2,$$
$$Z_3 = W_0(W_1W_2 - Z_1Z_2)^2.$$

The computational cost of differential addition and doubling on Edwards curves is 6M+4S using affine coordinates (SIDH/SIKE settings) and 8M+4S using projective coordinates (CSIDH setting).

2.3 Isogeny-Based Cryptosystems

We recall the SIDH and CSIDH key exchange protocol proposed in [7,16]. For more information, please refer to [7,16] for SIDH and CSIDH, respectively. The notations used in this section will continue to be used throughout the paper.

SIDH Protocol. Fix two coprime numbers ℓ_A and ℓ_B. Let p be a prime of the form $p = \ell_A^{e_A} \ell_B^{e_B} f \pm 1$ for some integer cofactor f, and e_A and e_B be positive integers such that $\ell_A^{e_A} \approx \ell_B^{e_B}$. Then we can easily construct a supersingular elliptic curve E over $\mathbb{F}_{p^2}$ of order $(\ell_A^{e_A} \ell_B^{e_B} f)^2$ [6]. We have full ℓ^e-torsion subgroup on E over $\mathbb{F}_{p^2}$ for $\ell \in \{\ell_A, \ell_B\}$ and $e \in \{e_A, e_B\}$. Choose basis $\{P_A, Q_A\}$ and $\{P_B, Q_B\}$ for the $\ell_A^{e_A}$- and $\ell_B^{e_B}$-torsion subgroups, respectively.

Suppose Alice and Bob want to exchange a secret key. Let $\{P_A, Q_A\}$ be the basis for Alice and $\{P_B, Q_B\}$ be the basis for Bob. For key generation, Alice chooses random elements $m_A, n_A \in \mathbb{Z}/\ell_A^{e_A}\mathbb{Z}$, not both divisible by ℓ_A, and computes the subgroup $\langle R_A \rangle = \langle [m_A]P_A + [n_A]Q_A \rangle$. Then using Velu's formula, Alice computes a curve $E_A = E/\langle R_A \rangle$ and an isogeny $\phi_A : E \to E_A$ of degree $\ell_A^{e_A}$, where $ker\phi_A = \langle R_A \rangle$. Alice computes and sends $(E_A, \phi_A(P_B), \phi_A(Q_B))$ to Bob. Bob repeats the same operation as Alice so that Alice receives $(E_B, \phi_B(P_A), \phi_B(Q_A))$.

For the key establishment, Alice computes the subgroup $\langle R_A' \rangle = \langle [m_A]\phi_B(P_A) + [n_A]\phi_B(Q_A) \rangle$. By using Velu's formula, Alice computes a curve $E_{AB} = E_B/\langle R_A' \rangle$. Bob repeats the same operation as Alice and computes a curve $E_{BA} = E_A/\langle R_B' \rangle$. The shared secret between Alice and Bob is the j-invariant of E_{AB}, i.e. $j(E_{AB}) = j(E_{BA})$.

CSIDH Protocol. CSIDH uses commutative group action on supersingular elliptic curves defined over a finite field $\mathbb{F}_p$. Let $\mathcal{O}$ be an imaginary quadratic order. Let $\mathcal{E}\ell\ell_p(\mathcal{O})$ denote the set of elliptic curves defined over $\mathbb{F}_p$ with the endomorphism ring $\mathcal{O}$. It is well-known that the class group $Cl(\mathcal{O})$ acts freely and transitively on $\mathcal{E}\ell\ell_p(\mathcal{O})$. We call the group action as CM-action and denote the action of an ideal class $[\mathfrak{a}] \in Cl(\mathcal{O})$ on an elliptic curve $E \in \mathcal{E}\ell\ell_p(\mathcal{O})$ by $[\mathfrak{a}]E$.

Let $p = 4\ell_1\ell_2 \cdots \ell_n - 1$ be a prime where $\ell_1, \cdots, \ell_n$ are small distinct odd primes. Let E be a supersingular elliptic curve over $\mathbb{F}_p$ such that $\mathrm{End}_p(E) = \mathbb{Z}[\pi]$, where $\mathrm{End}_p(E)$ is the endomorphism ring of E over $\mathbb{F}_p$. Note that $\mathrm{End}_p(E)$ is a commutative subring of the quaternion order $\mathrm{End}(E)$. Then the trace of

Frobenius is zero, hence $\#E(\mathbb{F}_p) = p + 1$. Since $\pi^2 - 1 = 0 \mod \ell_i$, the ideal $\ell_i\mathcal{O}$ splits as $\ell_i\mathcal{O} = \mathfrak{l}_i\bar{\mathfrak{l}}_i$, where $\mathfrak{l}_i = (\ell_i, \pi - 1)$ and $\bar{\mathfrak{l}}_i = (\ell_i, \pi + 1)$. The group action $[\mathfrak{l}_i]E$ (resp. $[\bar{\mathfrak{l}}_i]E$) is computed via isogeny $\phi_{\mathfrak{l}_i}$ (resp. $\phi_{\bar{\mathfrak{l}}_i}$) over $\mathbb{F}_p$ (resp. $\mathbb{F}_{p^2}$) using Velu's formulas.

Suppose Alice and Bob want to exchange a secret key. Alice chooses a vector $(e_1, \cdots, e_n) \in \mathbb{Z}^n$, where $e_i \in [-m, m]$, for a positive integer m. The vector represents an isogeny associated to the group action by the ideal class $[\mathfrak{a}] = [\mathfrak{l}_1^{e_1} \cdots \mathfrak{l}_n^{e_n}]$, where $\mathfrak{l}_i = (\ell_i, \pi - 1)$. Alice computes the public key $E_A := [\mathfrak{a}]E$ and sends E_A to Bob. Bob repeats the similar operation with his secret ideal $\mathfrak{b}$ and sends the public key $E_B := [\mathfrak{b}]E$ to Alice. Upon receiving Bob's public key, Alice computes $[\mathfrak{a}]E_B$ and Bob computes $[\mathfrak{b}]E_A$. Due to the commutativity, $[\mathfrak{a}]E_B$ and $[\mathfrak{b}]E_A$ are isomorphic to each other so that they can derive a shared secret value from the elliptic curves.

3 Optimized Odd-Degree Isogenies on Edwards Curve

In this section, we present the optimized method for computing odd-degree isogenies on Edwards curves. We used the result of Moody and Shumow as a base formula and optimized it by using w-coordinates. We conclude that the structure of odd-degree isogenies on Edwards curves is similar to the coordinate map on Montgomery curves presented in [9].

3.1 Motivation

After the proposal of CSIDH, demands on a general formula for computing odd-degree isogenies have aroused. The prime p in CSIDH is of the form $p = 4\ell_1\ell_2 \cdots \ell_n - 1$, where ℓ_i are small distinct odd primes. To implement CSIDH, isogeny of degree ℓ_i is required for all i, $1 \leq i \leq n$. The parameter CSIDH-512 presented in [7] uses $n = 74$, meaning that $\ell_1, \ldots, \ell_{73}$ are the 73 smallest odd primes, and ℓ_{74} is a smallest prime distinct from other primes that makes p a prime. Therefore, isogeny formulas of degrees up to at least 587 $(=\ell_{74})$ are required. Although the motivation of the work in [9] is independent of CSIDH scheme, they presented an efficient and generalized odd-degree isogeny formula on Montgomery curves so that the formula can naturally be used for CSIDH. For Edwards curves, optimization of the Moody and Shumow's formula must be performed for the use in CSIDH and other isogeny-based cryptosystems.

Let G be a subgroup of the Edwards curve E_d with odd order $\ell = 2s + 1$, and points $G = \{(0, 1), (\pm\alpha_1, \beta_1), \ldots, (\pm\alpha_s, \beta_s)\}$. Let ϕ be an ℓ-isogeny from E_d with kernel G. The ϕ proposed by Moody and Shumow is given as follows, where $B = \prod_{i=1}^{s} \beta_i$ [21].

$$\phi(x, y) = \left(\frac{x}{B^2} \prod_{i=1}^{s} \frac{\beta_i^2 x^2 - \alpha_i^2 y^2}{1 - d^2\alpha_i^2\beta_i^2 x^2 y^2}, \frac{y}{B^2} \prod_{i=1}^{s} \frac{\beta_i^2 y^2 - \alpha_i^2 x^2}{1 - d^2\alpha_i^2\beta_i^2 x^2 y^2} \right) \tag{3}$$

For optimizing 3-isogeny formula on Edwards curves, Kim et $al.$ used the curve equation and the division polynomial to represent the x-coordinate and the curve coefficient in Eq. (3), in terms of y-coordinate [17]. However, for higher degree isogenies, this optimization method is burdensome. On the other hand, the computational costs of elliptic curve arithmetic are the same for both curves when WZ-coordinate and XZ-coordinate are used for Edwards curves and Montgomery curves, respectively. This motivates us to optimize the odd-degree isogeny on Edwards curves using the w-coordinate. For the rest of the section, we present an odd-degree isogeny formula on Edwards curves expressed in w-coordinate.

3.2 Proposed Odd-Degree Isogeny Formula

We first present the isogeny formula using the w-coordinate, where the rational function w is defined as $w(x, y) = dx^2 y^2$ for points (x, y) on E_d.

Theorem 1. *Let P be a point on the Edwards curve E_d of odd order $\ell = 2s + 1$. Let $\langle P \rangle = \{(0, 1), (\pm \alpha_1, \beta_1), \cdots, (\pm \alpha_s, \beta_s)\}$, where $P = (\alpha_1, \beta_1)$. Let $w_i = d\alpha_i^2 \beta_i^2$ for $1 \leq i \leq s$, and $w = w(Q)$, where $Q = (x, y) \in E_d$. Then for ℓ-isogeny ϕ from E_d to $E_{d'} = E_d / \langle P \rangle$ the evaluation of w, $\phi(w)$, is given by,*

$$w(\phi) = w \prod_{i=1}^{s} \frac{(w - w_i)^2}{(1 - ww_i)^2}. \tag{4}$$

Proof. The proof of Theorem 1 is as follows. From the formula proposed by Moody and Shumow, ϕ is as in Eq. (3), where $d' = B^8 d^\ell$ and $B = \prod_{i=1}^{s} \beta_i$ [21]. In order to use the w-coordinate, we need to express the input and output of an isogeny function in terms of the w-coordinate. The points $(x, y) \in E_d$ and $(\alpha_i, \beta_i) \in E_d$ where $1 \leq i \leq s$, are expressed as $w = dx^2 y^2$ and $w_i = d\alpha_i^2 \beta_i^2$, in w-coordinates, respectively. Let $\phi(x, y) = (X, Y)$ be the image point. Then $w(\phi(x, y)) = d'X^2 Y^2$ so that,

$$d'X^2 Y^2 = B^8 d^\ell \cdot \left(\frac{x}{B^2} \prod_{i=1}^{s} \frac{\beta_i^2 x^2 - \alpha_i^2 y^2}{1 - d^2 \alpha_i^2 \beta_i^2 x^2 y^2} \right)^2 \left(\frac{y}{B^2} \prod_{i=1}^{s} \frac{\beta_i^2 y^2 - \alpha_i^2 x^2}{1 - d^2 \alpha_i^2 \beta_i^2 x^2 y^2} \right)^2.$$

The above equation can be simplified as follows.

$$d'X^2 Y^2 = B^8 d^\ell \cdot \frac{x^2}{B^4} \frac{y^2}{B^4} \left(\prod_{i=1}^{s} \frac{\beta_i^2 x^2 - \alpha_i^2 y^2}{1 - d^2 \alpha_i^2 \beta_i^2 x^2 y^2} \cdot \frac{\beta_i^2 y^2 - \alpha_i^2 x^2}{1 - d^2 \alpha_i^2 \beta_i^2 x^2 y^2} \right)^2$$

$$= dx^2 y^2 \prod_{i=1}^{s} \left(\frac{d(\beta_i^2 x^2 - \alpha_i^2 y^2)(\beta_i^2 y^2 - \alpha_i^2 x^2)}{(1 - d^2 \alpha_i^2 \beta_i^2 x^2 y^2)^2} \right)^2.$$

Since $w_i = d\alpha_i^2 \beta_i^2$ and $w = dx^2 y^2$, the denominator on the inside of the product in the above equation can be simplified as $(1 - ww_i)^4$, which gives,

$$d'X^2 Y^2 = w \prod_{i=1}^{s} \frac{(d(\beta_i^2 x^2 - \alpha_i^2 y^2)(\beta_i^2 y^2 - \alpha_i^2 x^2))^2}{(1 - ww_i)^4}. \tag{5}$$

Now, the numerator on the inside of the product of Eq. (5) can be simplified as follows.

$$
\begin{aligned}
(d(\beta_i^2 x^2 - \alpha_i^2 y^2)(\beta_i^2 y^2 - \alpha_i^2 x^2))^2 &= (d(x^2 y^2 \beta_i^4 - \alpha_i^2 \beta_i^2 x^4 - \alpha_i^2 \beta_i^2 y^4 + x^2 y^2 \alpha_i^4))^2 \\
&= (w(\alpha_i^4 + \beta_i^4) - w_i(x^4 + y^4))^2
\end{aligned}
$$

$$(6)$$

For further simplification of Eq. (6) we use the curve equation. Note that (α_i, β_i) and (x, y) are on the Edwards curve E_d. Then, $\alpha_i^2 + \beta_i^2 = 1 + w_i$ so that

$$
\begin{aligned}
\alpha_i^4 + \beta_i^4 &= (1 + w_i)^2 - 2\alpha_i^2 \beta_i^2 \\
&= (1 + w_i)^2 - 2w_i/d.
\end{aligned}
$$

Similarly for the point (x, y), we have $x^4 + y^4 = (1 + w)^2 - 2w/d$. Substituting the result to Eq. (6), we have,

$$
\begin{aligned}
(d(\beta_i^2 x^2 - \alpha_i^2 y^2)(\beta_i^2 y^2 - \alpha_i^2 x^2))^2 &= \left(w \left((1 + w_i)^2 - \frac{2w_i}{d} \right) - w_i \left((1 + w)^2 - \frac{2w}{d} \right) \right)^2 \\
&= ((w - w_i)(1 - ww_i))^2.
\end{aligned}
$$

Now if we substitute the above equation to Eq. (5), we have

$$
\begin{aligned}
d' X^2 Y^2 &= w \prod_{i=1}^{s} \frac{((w - w_i)(1 - ww_i))^2}{(1 - ww_i)^4} \\
&= w \prod_{i=1}^{s} \frac{(w - w_i)^2}{(1 - ww_i)^2}.
\end{aligned}
$$

which gives the desired result. $\square$

Theorem 1 shows that the evaluation of an isogeny on Edwards curves can be expressed in w-coordinate. Now, it remains to express the coefficient of the image curve in w-coordinates. From the formula proposed by Moody and Shumow, the curve coefficient d' of the image curve $E_{d'}$ is $d' = d^\ell B^8$ where $B = \prod_{i=1}^{s} \beta_i$. Since (α_i, β_i) satisfies the curve equation, $\alpha_i^2 = (1 - \beta_i^2)/(1 - d\beta_i^2)$ so that

$$
\begin{aligned}
w_i &= d\alpha_i^2 \beta_i^2 \\
&= d \left(\frac{1 - \beta_i^2}{1 - d\beta_i^2} \right) \beta_i^2.
\end{aligned}
$$

Solving the above equation for β_i^2, we can express the curve coefficient of the image curve in w-coordinate. However, direct change of d' to w-coordinate is computationally inefficient due to the square root computation. To solve this problem, we refer to the following theorem. Let $P_i = (\alpha_i, \beta_i) \in \langle P \rangle$ for $1 \le i \le s$, where $-P_i = (-\alpha_i, \beta_i)$. We exploit the fact that the set of y-coordinates of $[2]P_i$ where $1 \le i \le s$, is equal to the set of y-coordinates of P_j, where $1 \le j \le s$, up to permutations.

Theorem 2. *The curve coefficient d' of the image curve $E_{d'}$ in Theorem 1 is equal to*

$$d' = d^\ell \prod_{i=1}^{s} \frac{(w_i + 1)^8}{4^4}. \tag{7}$$

Proof. The proof of the Theorem 2 is as follows. From the formula proposed by Moody and Shumow, $d' = d^\ell B^8$ where $B = \prod_{i=1}^{s} \beta_i$. In order to use w-coordinate system for isogeny computations, we also need to express d' in w-coordinate. As denoted above, converting β_i directly to w-coordinate is cumbersome. The idea is that doubling the kernel points also generates the same subgroup since we are only dealing with odd-degree isogenies.

Let $P_i = (\alpha_i, \beta_i)$. Instead of computing the square of the y-coordinate (or x-coordinate) of P_i, we shall compute the square of the y-coordinate (or x-coordinate) of $[2]P_i$. Note that since P is an ℓ-torsion point where $\ell = 2s + 1$, $[2]P_i = \pm P_j$ for some $i, j \in \{1, \ldots, s\}$. Then from the addition formula on Edwards curves, we have

$$[2]P_i = \left(\frac{2\alpha_i \beta_i}{1 + d\alpha_i^2 \beta_i^2}, \frac{\beta_i^2 - \alpha_i^2}{1 - d\alpha_i^2 \beta_i^2} \right).$$

Squaring the x-coordinate of $[2]P_i$, we have

$$\left(\frac{2\alpha_i \beta_i}{1 + d\alpha_i^2 \beta_i^2} \right)^2 = \frac{4\alpha_i^2 \beta_i^2}{(1 + w_i)^2}$$
$$= \frac{4w_i / d}{(1 + w_i)^2}.$$

Since $w_i = d\alpha_i^2 \beta_i^2$, $\beta_i^2 = w_i / d\alpha_i^2$. Hence, by substituting the results, we have

$$d' = d^\ell \prod_{i=1}^{s} \beta_i^8$$
$$= d^\ell \prod_{i=1}^{s} \frac{(w_i + 1)^8}{4^4}$$

which gives the desired result.   □

3.3   Alternate Odd-Degree Isogeny Formula

In this section, we present the isogeny formula by defining the rational function w as $w(x, y) = x^2 / y^2$ for a point (x, y) on E_d. As shown below, the cost of evaluating isogenies is the same as the case when $w(x, y) = dx^2 y^2$. Formulas for computing the coefficient of the image curve are similar in both cases.

Theorem 3. *Let P be a point on the Edwards curve E_d of odd order $\ell = 2s + 1$. Let $\langle P \rangle = \{(0, 1), (\pm\alpha_1, \beta_1), \cdots, (\pm\alpha_s, \beta_s)\}$, where $P = (\alpha_1, \beta_1)$. Let $w_i = \alpha_i^2 / \beta_i^2$*

for $1 \leq i \leq s$ and $w = w(Q)$, where $Q = (x, y) \in E_d$. Then for ℓ-isogeny ϕ from E_d to $E_{d'} = E_d / \langle P \rangle$ the evaluation of w, $\phi(w)$, is given by,

$$w(\phi) = w \prod_{i=1}^{s} \frac{(w - w_i)^2}{(1 - ww_i)^2} \tag{8}$$

Proof. The proof of Theorem 3 is similar to the proof of Theorem 1. From the formula proposed by Moody and Shumow, ϕ is given by Eq. (3). The points $(x, y) \in E_d$ and $(\alpha_i, \beta_i) \in E_d$, where $1 \leq i \leq s$, are expressed as $w = x^2/y^2$ and $w_i = \alpha_i^2/\beta_i^2$ in w-coordinates, respectively. Let $\phi(x, y) = (X, Y)$ be the image point. Then $\phi(x, y)$ can be expressed in w-coordinate as,

$$\phi(w) = \frac{X^2}{Y^2} = \frac{x^2}{y^2} \prod_{i=1}^{s} \frac{(\beta_i^2 x^2 - \alpha_i^2 y^2)^2}{(\beta_i^2 y^2 - \alpha_i^2 x^2)^2}.$$

Simplifying the equation and expressing in w-coordinate, we obtain $\phi(w)$ as in Eq. (8). □

To obtain the coefficient of the image curve, we refer to the following theorem.

Theorem 4. *The curve coefficient d' of the image curve $E_{d'}$ in Theorem 1 is equal to*

$$d' = d^{\ell} \prod_{i=1}^{s} \frac{4^4}{(w_i + 1)^8}. \tag{9}$$

Proof. Let $P_i = (\alpha_i, \beta_i)$ be the point of the kernel. Similar to the proof of the Theorem 2, the Theorem 4 exploits the square of the x-coordinate of $[2]P_i$. From the addition formula on Edwards curves, we have

$$[2]P_i = \left(\frac{2\alpha_i \beta_i}{1 + d\alpha_i^2 \beta_i^2}, \frac{\beta_i^2 - \alpha_i^2}{1 - d\alpha_i^2 \beta_i^2} \right).$$

Squaring the x-coordinate of $[2]P_i$ and dividing both the denominator and numerator by β_i^4, we have,

$$\frac{4\alpha_i^2 \beta_i^2}{(1 + d\alpha_i^2 \beta_i^2)^2} = \frac{4\alpha_i^2 \beta_i^2}{(\alpha_i^2 + \beta_i^2)^2}$$

$$= \frac{4w_i}{(1 + w_i)^2}.$$

Now, since $w_i = \alpha_i^2/\beta_i^2$, $\beta_i^2 = \alpha_i^2/w_i$ so that

$$d' = d^\ell \prod_{i=1}^{s} \beta_i^8$$

$$= d^\ell \prod_{i=1}^{s} \left(\frac{\alpha_i^2}{w_i}\right)^4$$

$$= d^\ell \prod_{i=1}^{s} \frac{4^4}{(w_i+1)^8}$$

which gives the desired result.   □

4   Implementation

In this section, we provide the performance result of our odd-degree isogeny formula by applying to CSIDH. We first compare the computational costs between Montgomery curves and Edwards curves. We then show the performance result of CSIDH when w-coordinate is used.

4.1   Computational Costs

To evaluate the computational costs of the proposed formula, we first projectivize the function into $\mathbb{P}^1$ to avoid inversions. Since both rational maps induce the similar formula, we shall explain this section by defining the rational map as $w(x,y) = x^2/y^2$ for points (x,y) on Edwards curves. Thus, for $(\alpha_i, \beta_i) \in E_d$, $(W_i : Z_i) = (w_i : 1)$ for $i = 1, \ldots, s$ where $w_i = \alpha_i^2/\beta_i^2$. Let ϕ be a degree ℓ isogeny from E_d to $E_{d'}$. For additional input point $(W : Z)$ on the curve E_d, the output is expressed as $(W' : Z')$ where $(W' : Z') = \phi(W : Z)$. Then,

$$W' = W \cdot \prod_{i=1}^{s} (WZ_i - ZW_i)^2,$$

$$Z' = Z \cdot \prod_{i=1}^{s} (WW_i - ZZ_i)^2.$$

Let $F_i = (W - Z)(W_i + Z_i)$ and $G_i = (W + Z)(W_i - Z_i)$. Then the above equation can be rewritten as,

$$W' = W \cdot \prod_{i=1}^{s} (F_i - G_i)^2,$$

$$Z' = Z \cdot \prod_{i=1}^{s} (F_i + G_i)^2.$$

Therefore, computation of $(WZ_i - ZW_i)$ and $(WW_i - ZZ_i)$ cost 2M+6a. For $\ell = 2s + 1$-isogeny, evaluation of an isogeny costs $(4s)$M+2S. To compute the curve coefficients, let $d = D/C$. Then we have,

$$D' = D^\ell \cdot \prod_{i=1}^{s} (2Z_i)^8,$$

$$C' = C^\ell \cdot \prod_{i=1}^{s} (W_i + Z_i)^8,$$

where $d' = D'/C'$. Concluding the section, Table 1 presents the computational costs of evaluation of an isogeny as well as curve coefficient for degree $\ell \in \{3, 5, 7, 9\}$.

As shown in Table 1, the computational costs of evaluating isogenies are identical on both curves. In Table 1, we used the 2-torsion method for Montgomery curves to analyze the computational costs of computing the coefficients. In [9], instead of directly computing the curve coefficients, they exploit the fact that pushing 2-torsion points through an odd-degree isogeny preserves their order on the image curve. When the image of the 2-torsion point is obtained, the curve coefficient of the image curve can be recovered in 2S+5a. For the details of the method, please refer to [10].

Table 1. Computational costs of isogenies of degree 3, 5, 7, and 9 on Montgomery cures and Edwards curves. For computing the curve coefficients on Montgomery curve, the 2-torsion method is used, and the table presents the combined computational cost of evaluating image of the 2-torsion point ($(4s)$M+2S) and recovering curve coefficient (2S)).

	Evaluation		Curve coefficient	
	Montgomery	Edwards (This Work)	Montgomery	Edwards (This Work)
3		4M+2S+6a	2M+3S	4M+6S+8a
5		8M+2S+10a	8M+4S+5a	6M+6S+8a
7		12M+2S+14a	12M+4S+5a	8M+6S+8a
9		16M+2S+18a	16M+4S+5a	10M+6S+8a

Since an additional 2-torsion point is evaluated, the computational cost of recovering the curve coefficient of the image curve is equal to $(4s)$M+4S, where $(4s)$M+2S is for isogeny evaluation and 2S is for recovering from image points. One drawback of the 2-torsion method is that the additional 2-torsion point must be evaluated to recover the curve coefficient. Therefore, the computational cost of obtaining the curve coefficient of the image curve increases as the degree of isogeny increases. Although this is also the case on Edwards curves, an additional 2-torsion point is not required for Edwards curves.

For Montgomery curves, curve coefficients can also be recovered using the x-coordinates of points and the x-coordinate of their differences – *i.e.* x-coordinates of the points P, Q, and $Q - P$ on a Montgomery curve [9]. We shall call this method as `get_a_from_diff` method. Recovering the curve coefficient using this method costs $8\mathbf{M}+5\mathbf{S}+11\mathbf{a}$ and the cost does not increase even if the degree of isogeny increases. In SIDH/SIKE settings, the points P, Q, and $Q - P$ can be seen as a public key $(P_A, Q_A, P_A - Q_A)$ (or $(P_B, Q_B, P_B - Q_B)$ on Bob's side) and are evaluated for each iteration for efficient ladder computations. Therefore, `get_a_from_diff` method are more efficient in SIDH than the 2-torsion method.

Figure 1 depicts the difference in the computational cost of recovering the curve coefficient between Montgomery curves and Edwards curves. The horizontal axis represents the degree of an isogeny and vertical axis represents the number of multiplication used for the computation. The blue line indicates the computational cost on Montgomery curves and the orange line indicates the computational cost on Edwards curves. We considered $1\mathbf{S}$ as $0.8\mathbf{M}$. Note that when WZ-coordinate is used for Edwards curves and XZ-coordinate is used for Montgomery curves, the difference in the performance purely lies on the cost of recovering the coefficients of the image curve, because the costs of all the remaining operations are the same. As shown in Fig. 1(a), when the 2-torsion method is used on Montgomery curves, Edwards curves become more efficient as the degree of isogeny increases. On the other hand, as shown in Fig. 1(b), when `get_a_from_diff` method is used for Montgomery curves, Montgomery curves become more efficient as the degree of isogeny increases. More concretely, Montgomery curves are preferred in SIDH/SIKE settings and are more efficient than Edwards curves for $s \geq 3$. In CSIDH setting, the points P, Q, and $Q - P$ are not evaluated so that the 2-torsion method is used for Montgomery curves. Hence Edwards curves are preferred and are more efficient than Montgomery curves in CSIDH for $s \geq 2$.

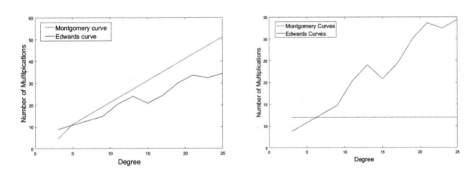

Fig. 1. (a) Computational costs of recovering the curve coefficient of the image curve when the 2-torsion method is used for Montgomery curves. (b) Computational costs of recovering the curve coefficient of the image curve when `get_a_from_diff` method is used for Montgomery curves. (Color figure online)

4.2 Implementation Result of CSIDH Using w-coordinate

To evaluate the performance, the algorithms are implemented in C language. All cycle counts were obtained on one core of an Intel Core i7-6700 (Skylake) at 3.40 GHz, running Ubuntu 16.04 LTS. For compilation, we used GNU GCC version 5.4.0. Before we present the implementation result, we briefly introduce the hybrid-CSIDH proposed by Meyer *et al.*, in order to better explain the results [19].

Hybrid-CSIDH. In [19], Meyer *et al.* proposed hybrid implementation of CSIDH which uses Montgomery curves for elliptic curve arithmetic and isogeny computation, and twisted Edwards curves for computing the coefficients of the image curves. As stated above, computing the image curve is not as straightforward as for the point evaluations on Montgomery curves [9]. However, as presented in [21], computing the image curve is much simpler on twisted Edwards curves. Hence, in [19] by using the fact that conversion between two models costs only two additions, they transformed Montgomery curve to corresponding twisted Edwards curve and computed the image curve and transformed back to Montgomery curve.

Sampling Random Points on Edwards Curves. In order to calculate the class group action, a random point P on a curve is sampled over $\mathbb{F}_p$ or $\mathbb{F}_{p^2} \setminus \mathbb{F}_p$. For Montgomery curve, this can be done by sampling a random $\mathbb{F}_p$-rational x-coordinate, and check whether $x^2 + Ax^2 + x$ is a square or not. For Edwards curves, we sample a random $\mathbb{F}_p$-rational y-coordinate, check whether corresponding x-coordinate is a square or not, and convert to w-coordinate.

Note that for an Edwards curve define as in Eq. (1), $x^2 = (1 - y^2)/(1 - dy^2)$. Thus, we need to check whether $(1 - y^2)/(1 - dy^2)$ is a square or not. This is equivalent to check whether $(1 - y^2)(1 - dy^2)$ is a square or not. After checking the sign, we convert the sampled point on an Edwards curve to projective w-coordinate. For example, when $w = dx^2y^2$ is used for the implementation, the following conversion is required.

$$P = (w : 1) = \left(d \cdot \frac{1 - y^2}{1 - dy^2} \cdot y^2 : 1 \right) = (dy^2(1 - y^2) : 1 - dy^2)$$

This can be done as in Algorithm 1, which costs $3\mathbf{M}+1\mathbf{S}$.

Algorithm 1 Sampling random point on an Edwards curve

Require: An Edwards curve E_d
Ensure: A point $P = (W : Z)$ on E_d in projective w-coordinate for $w = dx^2y^2$
1: Sample a random $y \in \mathbb{F}_p$
2: $Z \leftarrow y^2$ // $Z = y^2$
3: $t_0 \leftarrow d \cdot Z$ // $t_0 = dy^2$
4: $t_1 \leftarrow 1 - Z$ // $t_1 = 1 - y^2$
5: $Z \leftarrow 1 - t_0$ // $Z = 1 - dy^2$
6: $rhs \leftarrow t_1 \cdot Z$ // $rhs = (1 - dy^2)(1 - y^2)$
7: $W \leftarrow t_1 \cdot rhs'$ // $W = dy^2(1 - y^2)$
8: Set $s \leftarrow +1$ if rhs is a square in $\mathbb{F}_p$, else $s \leftarrow -1$
9: **return** P

Remark 1. Another method to sample random points on Edwards curves is to use the idea proposed in [22]. In [22], Moriya et al. proposed a method to sample a random element in $\mathbb{F}_p$, directly in w-coordinate. The idea is to sample a random element in $\mathbb{F}_p$ and consider it as a w-coordinate of $w(P)$. They prove that if $w(2P)$ is a square, then there exist $P' \in E[\pi_p + 1]$ such that $w(P') = w(2P)$. If $w(2P)$ is a non-square, then there exist $P' \in E[\pi_p - 1]$.

Performance of CSIDH Using Edwards Curves. We used prime field $\mathbb{F}_p$ presented in [7], where p is of the form $p = 4\ell_1\ell_2\cdots\ell_{74} - 1$. The $\ell_1,\cdots,\ell_{73}$ are the 73 smallest distinct odd primes and $\ell_{74} = 587$. To compare the performance result with the implementation in [7,19], the field operations implemented in [7] are used for the experiment. We refer to the implementation in [7] as Montgomery-CSIDH and the implementation in [19] as hybrid-CSIDH, for the rest of the paper. Our implementation of CSIDH using Edwards w-coordinate is referred to as Edwards-CSIDH. We used $w = dx^2y^2$ for the implementation.

First, the base field operations were tested in order to visualize the ratio between field operations. Each field operations were repeated 10^8 times.

Table 2. Cycle counts of the field operations over $\mathbb{F}_p$

	Addition	Subtraction	Multiplication
p_{511}	29	24	201

Table 3 illustrates the computational costs of elliptic curve arithmetic and isogeny on Hybrid-CSIDH and Edwards-CSIDH setting. The $[k]P$ represents the computational cost of $[k]P$ on Montgomery curves with respect to the cost on Edwards curves. The additional $3\mathbf{a}$ on Hybrid-CSIDH comes from the curve conversion. Since the number of calls of differential addition when computing $[k]P$ is equal to the bit-length of k, $(\log k \times 4)\mathbf{a}$ are additionally required when using

Montgomery curves compared to Edwards curves. When computing $(2s + 1)$-isogeny, 5 field additions are additionally required for Hybrid-CSIDH for transforming between Montgomery and Edwards curves. However, when computing the image curve, 8 number of field addition is additionally required in Edwards-CSIDH.

Table 3. Computational costs of elliptic curve arithmetic and isogenies on Hybrid-CSIDH and Edwards-CSIDH

	Hybrid-CSIDH	Edwards-CSIDH
Differential addition	8M+4S+7a+4s	8M+4S+3a+4s
Doubling	4M+2S+6a+2s	4M+2S+1a+3s
Addition	4M+2S+3a+3s	4M+2S+3a+3s
$[k]P$	$3\mathbf{a}+(\log k \times 4)\mathbf{a}$	–
$(2s + 1)$-isogeny	$(-3)\mathbf{a}$	–

As shown in Table 4, implementing CSIDH using Edwards w-coordinate is the fastest. When comparing the result between Montgomery-CSIDH and Edwards-CSIDH, the result is not surprising since computing the curve coefficient of the image curve is more efficient on Edwards curves. In order to better compare the result between Hybrid-CSIDH and Edwards-CSIDH, we analyzed the computational cost of each building blocks of CSIDH.

The table below denotes the average number of function calls and differences in the number of field additions of Hybrid-CSIDH with respect to Edwards-CSIDH. The number of additions is omitted as its computational costs are the same for Hybrid-CSIDH and Edwards-CSIDH.

Summing up the result of Tables 2 and 5, although Edwards-CSIDH and Hybrid-CSIDH have the same number of field multiplications and squarings, the efficiency in the number of field additions and subtractions on Edwards-CSIDH lead to the fastest result.

Table 4. Implementation results of CSIDH

	Montgomery [7]	Hybrid [19]	Edwards (This Work)
Alice's keygen	129,165,448 cc	105,438,581 cc	103,239,120 cc
Bob's keygen	128,460,087 cc	105,217,108 cc	103,078,319 cc
Alice's shared key	129,215,839 cc	105,429,541 cc	103,232,321 cc
Bob's shared key	128,426,421 cc	105,204,672 cc	103,084,354 cc

Table 5. Average number of function calls for CSIDH-512 and additional number of field operations for Hybrid-CSIDH with respect to Edwards-CSIDH

	Average number of calls	Hybrid-CSIDH
Doubling	202	+848.4 a
$[k]P$	218	+75,103 a
$(2s + 1)$-isogeny	202	−606 a

5 Conclusion

In this paper, we proposed the optimized method for computing odd-degree isogenies on Edwards curves. By using the w-coordinates, we optimized the isogeny formula proposed by Moody and Shumow. The use of the w-coordinate makes the costs of elliptic curve arithmetic and evaluation of an isogeny identical to that of on Montgomery curves, having efficiency when computing the coefficient of the image curve. For ℓ-degree isogeny where $\ell = 2s + 1$, the proposed formula has benefit over Montgomery curves when $s \geq 2$. We conclude that Montgomery curves are efficient for implementing SIDH or SIKE and Edwards curves are efficient for implementing CSIDH. Additionally, we implemented CSIDH using w-coordinates. Our Edwards-CSIDH is about 20% faster than the Montgomery-CSIDH, and 2% faster than the hybrid-CSIDH. For the future work, we plan to implement constant-time CSIDH using w-coordinate on Edwards curves.

Acknowledgement. We thank the anonymous reviewers for their useful and constructive comments.

References

1. Azarderakhsh, R., Bakos Lang, E., Jao, D., Koziel, B.: EdSIDH: supersingular isogeny Diffie-Hellman key exchange on Edwards curves. In: Chattopadhyay, A., Rebeiro, C., Yarom, Y. (eds.) SPACE 2018. LNCS, vol. 11348, pp. 125–141. Springer, Cham (2018). https://doi.org/10.1007/978-3-030-05072-6_8
2. Azarderakhsh, R., et al.: Supersingular isogeny key encapsulation. Submission to the NIST Post-Quantum Standardization Project (2017)
3. Bernstein, D.J., Birkner, P., Joye, M., Lange, T., Peters, C.: Twisted Edwards curves. In: Vaudenay, S. (ed.) AFRICACRYPT 2008. LNCS, vol. 5023, pp. 389–405. Springer, Heidelberg (2008). https://doi.org/10.1007/978-3-540-68164-9_26
4. Bernstein, D.J., Lange, T.: Inverted Edwards coordinates. In: Boztaş, S., Lu, H.-F.F. (eds.) AAECC 2007. LNCS, vol. 4851, pp. 20–27. Springer, Heidelberg (2007). https://doi.org/10.1007/978-3-540-77224-8_4
5. Bos, J.W., Friedberger, S.J.: Arithmetic considerations for isogeny-based cryptography. IEEE Trans. Comput. **68**(7), 979–990 (2019)
6. Bröker, R.: Constructing supersingular elliptic curves. J. Comb. Number Theory **1**(3), 269–273 (2009)

7. Mendel, F., Nad, T., Schläffer, M.: Finding SHA-2 characteristics: searching through a minefield of contradictions. In: Lee, D.H., Wang, X. (eds.) ASIACRYPT 2011. LNCS, vol. 7073, pp. 288–307. Springer, Heidelberg (2011). https://doi.org/10.1007/978-3-642-25385-0_16

8. Childs, A., Jao, D., Soukharev, V.: Constructing elliptic curve isogenies in quantum subexponential time. J. Math. Cryptol. **8**(1), 1–29 (2014)

9. Costello, C., Hisil, H.: A simple and compact algorithm for SIDH with arbitrary degree isogenies. In: Takagi, T., Peyrin, T. (eds.) ASIACRYPT 2017. LNCS, vol. 10625, pp. 303–329. Springer, Cham (2017). https://doi.org/10.1007/978-3-319-70697-9_11

10. Costello, C., Longa, P., Naehrig, M.: SIDH library (2016–2018). https://github.com/Microsoft/PQCrypto-SIDH

11. Costello, C., Longa, P., Naehrig, M.: Efficient algorithms for supersingular isogeny Diffie-Hellman. In: Robshaw, M., Katz, J. (eds.) CRYPTO 2016. LNCS, vol. 9814, pp. 572–601. Springer, Heidelberg (2016). https://doi.org/10.1007/978-3-662-53018-4_21

12. Couveignes, J.M.: Hard homogeneous spaces (2006). https://eprint.iacr.org/2006/291

13. De Feo, L., Kieffer, J., Smith, B.: Towards practical key exchange from ordinary isogeny graphs. In: Peyrin, T., Galbraith, S. (eds.) ASIACRYPT 2018. LNCS, vol. 11274, pp. 365–394. Springer, Cham (2018). https://doi.org/10.1007/978-3-030-03332-3_14

14. Farashahi, R.R., Hosseini, S.G.: Differential addition on twisted Edwards curves. In: Pieprzyk, J., Suriadi, S. (eds.) ACISP 2017. LNCS, vol. 10343, pp. 366–378. Springer, Cham (2017). https://doi.org/10.1007/978-3-319-59870-3_21

15. Hisil, H., Wong, K.K.-H., Carter, G., Dawson, E.: Twisted Edwards curves revisited. In: Pieprzyk, J. (ed.) ASIACRYPT 2008. LNCS, vol. 5350, pp. 326–343. Springer, Heidelberg (2008). https://doi.org/10.1007/978-3-540-89255-7_20

16. Jao, D., De Feo, L.: Towards quantum-resistant cryptosystems from supersingular elliptic curve isogenies. In: Yang, B.-Y. (ed.) PQCrypto 2011. LNCS, vol. 7071, pp. 19–34. Springer, Heidelberg (2011). https://doi.org/10.1007/978-3-642-25405-5_2

17. Kim, S., Yoon, K., Kwon, J., Hong, S., Park, Y.H.: Efficient isogeny computations on twisted Edwards curves. Secur. Commun. Netw. **2018**, 1–11 (2018)

18. Kim, S., Yoon, K., Kwon, J., Park, Y.H., Hong, S.: New hybrid method for isogeny-based cryptosystems using Edwards curves. IEEE Trans. Inf. Theory (2019). https://doi.org/10.1109/TIT.2019.2938984

19. Meyer, M., Reith, S.: A faster way to the CSIDH. In: Chakraborty, D., Iwata, T. (eds.) INDOCRYPT 2018. LNCS, vol. 11356, pp. 137–152. Springer, Cham (2018). https://doi.org/10.1007/978-3-030-05378-9_8

20. Meyer, M., Reith, S., Campos, F.: On hybrid SIDH schemes using Edwards and Montgomery curve arithmetic (2017). https://eprint.iacr.org/2017/1213

21. Moody, D., Shumow, D.: Analogues of Vélu's formulas for isogenies on alternate models of elliptic curves. Math. Comput. **85**(300), 1929–1951 (2016)

22. Moriya, T., Onuki, H., Takagi, T.: How to construct CSIDH on Edwards curves. Cryptology ePrint Archive, Report 2019/843 (2019). https://eprint.iacr.org/2019/843

23. Stolbunov, A.: Constructing public-key cryptographic schemes based on class group action on a set of isogenous elliptic curves. Adv. Math. Commun. **4**(2), 215–235 (2010)

Hard Isogeny Problems over RSA Moduli
and Groups with Infeasible Inversion

Salim Ali Altuğ[1]([⊠]) and Yilei Chen[2]

[1] Boston University, Boston, USA
saaltug@bu.edu
[2] Visa Research, Palo Alto, USA
yilchen@visa.com

Abstract. We initiate the study of computational problems on elliptic curve isogeny graphs defined over RSA moduli. We conjecture that several variants of the neighbor-search problem over these graphs are hard, and provide a comprehensive list of cryptanalytic attempts on these problems. Moreover, based on the hardness of these problems, we provide a construction of groups with infeasible inversion, where the underlying groups are the ideal class groups of imaginary quadratic orders.

Recall that in a group with infeasible inversion, computing the inverse of a group element is required to be hard, while performing the group operation is easy. Motivated by the potential cryptographic application of building a directed transitive signature scheme, the search for a group with infeasible inversion was initiated in the theses of Hohenberger and Molnar (2003). Later it was also shown to provide a broadcast encryption scheme by Irrer et al. (2004). However, to date the only case of a group with infeasible inversion is implied by the much stronger primitive of self-bilinear map constructed by Yamakawa et al. (2014) based on the hardness of factoring and indistinguishability obfuscation (iO). Our construction gives a candidate without using iO.

1 Introduction

Let $\mathbb{G}$ denote a finite group written multiplicatively. The discrete-log problem asks to find the exponent a given g and $g^a \in \mathbb{G}$. In the groups traditionally used in discrete-log-based cryptosystems, such as $(\mathbb{Z}/q\mathbb{Z})^*$ [11], groups of points on elliptic curves [22,29], and class groups [3,28], computing the inverse $x^{-1} = g^{-a}$ given $x = g^a$ is easy. We say $\mathbb{G}$ is a *group with infeasible inversion* if computing inverses of elements is hard, while performing the group operation is easy (i.e. given g, g^a, g^b, computing g^{a+b} is easy).

The search for a group with infeasible inversion was initiated in the theses of Hohenberger [18] and Molnar [30], motivated with the potential cryptographic application of constructing a directed transitive signature. It was also shown by Irrer et al. [20] to provide a broadcast encryption scheme. The only existing candidate of such a group, however, is implied by the much stronger primitive of self-bilinear maps constructed by Yamakawa et al. [40], assuming the hardness of integer factorization and indistinguishability obfuscation (iO) [2,16].

© International Association for Cryptologic Research 2019
S. D. Galbraith and S. Moriai (Eds.): ASIACRYPT 2019, LNCS 11922, pp. 293–322, 2019.
https://doi.org/10.1007/978-3-030-34621-8_11

In this paper we propose a candidate trapdoor group with infeasible inversion without using iO. The underlying group is isomorphic to the ideal class group of an imaginary quadratic order (henceforth abbreviated as *"the class group"*). In the standard representation of the class group, computing the inverse of a group element is straightforward. The representation we propose uses the volcano-like structure of the isogeny graphs of ordinary elliptic curves. In fact, the initiation of this work was driven by the desire to explore the computational problems on the isogeny graphs defined over RSA moduli.

1.1   Elliptic Curve Isogenies in Cryptography

An isogeny $\varphi : E_1 \rightarrow E_2$ is a morphism of elliptic curves that preserves the identity. Given two isogenous elliptic curves E_1, E_2 over a finite field, finding an explicit rational polynomial that represents an isogeny from E_1 to E_2 is traditionally called the *computational isogeny problem*.

The best way of understanding the nature of the isogeny problem is to look at the *isogeny graphs*. Fix a finite field $\mathbf{k}$ and a prime ℓ different than the characteristic of $\mathbf{k}$. Then the isogeny graph $G_\ell(\mathbf{k})$ is defined as follows: each vertex in $G_\ell(\mathbf{k})$ is a j-invariant of an isomorphism class of curves; two vertices are connected by an edge if there is an isogeny of degree ℓ over $\mathbf{k}$ that maps one curve to another. The structure of the isogeny graph is described in the PhD thesis of Kohel [23]. Roughly speaking, a connected component of an isogeny graph containing ordinary elliptic curves looks like a *volcano* (termed in [15]). The connected component containing supersingular elliptic curves, on the other hand, has a different structure. In this article we will focus on the ordinary case.

A Closer Look at the Algorithms of Computing Isogenies. Let $\mathbf{k}$ be a finite field of q elements, ℓ be an integer such that $\gcd(\ell, q) = 1$. Given the j-invariant of an elliptic curve E, there are at least two different ways to find all the j-invariants of the curves that are ℓ-isogenous to E (or to a twist of E) and to find the corresponding rational polynomials that represent the isogenies:

1. Computing kernel subgroups of E of size ℓ, and then applying Vélu's formulae to obtain explicit isogenies and the j-invariants of the image curves,
2. Calculating the j-invariants of the image curves by solving the ℓth modular polynomial Φ_ℓ over $\mathbf{k}$, and then constructing explicit isogenies from these j-invariants.

Both methods are able to find all the ℓ-isogenous neighbors over $\mathbf{k}$ in time $\mathsf{poly}(\ell, \log(q))$. In other words, *over a finite field*, one can take a stroll around the polynomial-degree isogenous neighbors of a given elliptic curve efficiently.

However, for two random isogenous curves over a sufficiently large field, finding an explicit isogeny between them seems to be hard, even for quantum computers. The conjectured hardness of computing isogenies was used in a key-exchange and a public-key cryptosystem by Couveignes [7] and independently by Rostovtsev and Stolbunov [31]. Moreover, a hash function and a key exchange scheme were proposed based on the hardness of computing isogenies over supersingular curves [4,21]. Isogeny-based cryptography is attracting attention partially due to its conjectured post-quantum security.

1.2 Isogeny Graphs over RSA Moduli

Let p, q be primes and let $N = pq$. In this work we consider computational problems related to elliptic curve isogeny graphs defined over $\mathbb{Z}/N\mathbb{Z}$, where the prime factors p, q of N are unknown. An isogeny graph over $\mathbb{Z}/N\mathbb{Z}$ is defined first by fixing the isogeny graphs over $\mathbb{F}_p$ and $\mathbb{F}_q$, then taking a graph tensor product; obtaining the j-invariants in the vertices of the graph over $\mathbb{Z}/N\mathbb{Z}$ by the Chinese remainder theorem. Working over the ring $\mathbb{Z}/N\mathbb{Z}$ without the factors of N creates new sources of computational hardness from the isogeny problems. Of course, by assuming the hardness of factorization, we immediately lose the post-quantum privilege of the "traditional" isogeny problems. From now on all the discussions of hardness are with respect to the polynomial time classical algorithms.

Basic Neighbor Search Problem over $\mathbb{Z}/N\mathbb{Z}$. When the factorization of N is unknown, it is not clear how to solve the basic problem of finding (even one of) the ℓ-isogenous neighbors of a given elliptic curve. The two algorithms over finite fields we mentioned seem to fail over $\mathbb{Z}/N\mathbb{Z}$ since both of them require solving polynomials over $\mathbb{Z}/N\mathbb{Z}$, which is hard in general when the factorization of N is unknown. In fact, we show that if it is feasible to find all the ℓ-isogenous neighbors of a given elliptic curve over $\mathbb{Z}/N\mathbb{Z}$, then it is feasible to factorize N.

Joint-Neighbor Search Problem over $\mathbb{Z}/N\mathbb{Z}$. Suppose we are given several j-invariants over $\mathbb{Z}/N\mathbb{Z}$ that are connected by polynomial-degree isogenies, we ask whether it is feasible to compute their joint isogenous neighbors. For example, in the isogeny graph on the LHS of Fig. 1, suppose we are given j_0, j_1, j_2, and the degrees ℓ between j_0 and j_1, and m between j_0 and j_2 such that $\gcd(\ell, m) = 1$. Then we can find j_3 which is m-isogenous to j_1 and ℓ-isogenous to j_2, by computing the polynomial $f(x) = \gcd(\Phi_m(j_1, x), \Phi_\ell(j_2, x))$ over $\mathbb{Z}/N\mathbb{Z}$. When $\gcd(\ell, m) = 1$ the polynomial $f(x)$ turns out to be linear with its only root being j_3, hence computing the (ℓ, m) neighbor in this case is feasible.

However, not all the joint-isogenous neighbors are easy to find. As an example, consider the following (ℓ, ℓ^2)-*joint neighbor problem* illustrated on the RHS of Fig. 1. Suppose we are given j_0 and j_1 that are ℓ-isogenous, and asked to find another j-invariant j_{-1} which is ℓ-isogenous to j_0 and ℓ^2-isogenous to j_1. The natural way is to take the gcd of $\Phi_\ell(j_0, x)$ and $\Phi_{\ell^2}(j_1, x)$, but in this case the resulting polynomial is of degree $\ell > 1$ and we are left with the problem of finding a root of it over $\mathbb{Z}/N\mathbb{Z}$, which is believed to be computationally hard without knowing the factors of N.

Currently we do not know if solving this problem is as hard as factoring N. Neither do we know of an efficient algorithm of solving the (ℓ, ℓ^2)-joint neighbor problem. We will list our attempts in solving the (ℓ, ℓ^2)-joint neighbor problem in Sect. 5.2.

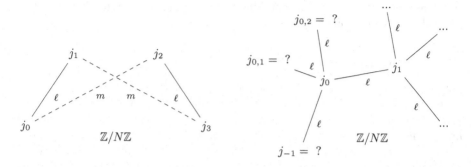

Fig. 1. Left: the (ℓ, m)-isogenous neighbor problem where $\gcd(\ell, m) = 1$. Right: the (ℓ, ℓ^2)-isogenous neighbor problem.

The conjectured computational hardness of the (ℓ, ℓ^2)-joint neighbor problem is fundamental to the infeasibility of inversion in the group we construct.

1.3 Constructing a Trapdoor Group with Infeasible Inversion

To explain the construction of the trapdoor group with infeasible inversion (TGII), it is necessary to recall the connection of the ideal class groups and elliptic curve isogenies. Let $\mathbf{k}$ be a finite field as before and let E be an elliptic curve over $\mathbf{k}$ whose endomorphism ring is isomorphic to an imaginary quadratic order $\mathcal{O}$. The group of invertible $\mathcal{O}$-ideals acts on the set of elliptic curves with endomorphism ring $\mathcal{O}$. The ideal class group $\mathcal{CL}(\mathcal{O})$ acts faithfully and transitively on the set

$$\mathrm{Ell}_{\mathcal{O}}(\mathbf{k}) = \{j(E) : E \text{ with } \mathrm{End}(E) \simeq \mathcal{O}\}.$$

In other words, there is a map

$$\mathcal{CL}(\mathcal{O}) \times \mathrm{Ell}_{\mathcal{O}}(\mathbf{k}) \to \mathrm{Ell}_{\mathcal{O}}(\mathbf{k}), \quad (\mathfrak{a}, j) \mapsto \mathfrak{a} * j$$

such that $\mathfrak{a} * (\mathfrak{b} * j) = (\mathfrak{a}\mathfrak{b}) * j$ for all $\mathfrak{a}, \mathfrak{b} \in \mathcal{CL}(\mathcal{O})$ and $j \in \mathrm{Ell}_{\mathcal{O}}(\mathbf{k})$; for any $j, j' \in \mathrm{Ell}_{\mathcal{O}}(\mathbf{k})$, there is a unique $\mathfrak{a} \in \mathcal{CL}(\mathcal{O})$ such that $j' = \mathfrak{a} * j$. The cardinality of $\mathrm{Ell}_{\mathcal{O}}(\mathbf{k})$ equals to the class number $h(\mathcal{O})$.

We are now ready to provide an overview of the TGII construction with a toy example in Fig. 2.

Parameter Generation. To simplify this overview let us assume that the group $\mathcal{CL}(\mathcal{O})$ is cyclic, in which case the group $\mathbb{G}$ with infeasible inversion is exactly $\mathcal{CL}(\mathcal{O})$ (in the detailed construction we usually choose a cyclic subgroup of $\mathcal{CL}(\mathcal{O})$). To generate the public parameter for the group $\mathcal{CL}(\mathcal{O})$, we choose two primes p, q and curves $E_{0,\mathbb{F}_p}$ over $\mathbb{F}_p$ and $E_{0,\mathbb{F}_q}$ over $\mathbb{F}_q$ such that the endomorphism rings of $E_{0,\mathbb{F}_p}$ and $E_{0,\mathbb{F}_q}$ are both isomorphic to $\mathcal{O}$. Let $N = p \cdot q$. Let E_0 be an elliptic curve over $\mathbb{Z}/N\mathbb{Z}$ as the CRT composition of $E_{0,\mathbb{F}_p}$ and $E_{0,\mathbb{F}_q}$. The j-invariant of E_0, denoted as j_0, equals to the CRT composition of

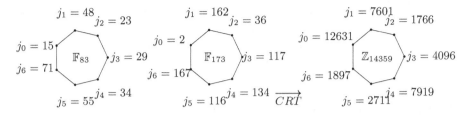

Fig. 2. A representation of $\mathcal{CL}(-251)$ by a 3-isogeny volcano over $\mathbb{Z}_{14359}$ of size $h(-251) = 7$. The $\mathbb{F}_{83}$ part is taken from [31].

the j-invariants of $E_{0,\mathbb{F}_p}$ and $E_{0,\mathbb{F}_q}$. The identity of $\mathcal{CL}(\mathcal{O})$ is represented by j_0. The public parameter of the group is (N, j_0).

In the example of Fig. 2, we set the discriminant D of the imaginary quadratic order $\mathcal{O}$ to be -251. The group order is then the class number $h(\mathcal{O}) = 7$. Choose $p = 83$, $q = 173$, $N = pq = 14359$. Fix a curve E_0 so that $j(E_{0,\mathbb{F}_p}) = 15$, $j(E_{0,\mathbb{F}_q}) = 2$, then $j_0 = \mathsf{CRT}(83, 173; 15, 2) = 12631$. The public parameter is $(14359, 12631)$.

The Encodings. We provide two types of encodings for each group element: the canonical and composable embeddings. The *canonical encoding* of an element is uniquely determined once the public parameter is fixed and it can be directly used in the equivalence test. It, however, does not support efficient group operations. The *composable encoding* of an element, on the other hand, supports efficient group operations with the other composable encodings. Moreover, a composable encoding can be converted to a canonical encoding by an efficient, public extraction algorithm.

An element $x \in \mathcal{CL}(\mathcal{O})$ is canonically represented by the j-invariant of the elliptic curve $x * E_0$ (once again, obtained over $\mathbb{F}_p$ and $\mathbb{F}_q$ then composed by CRT), and we call $j(x * E_0)$ the canonical encoding of x. Note that the canonical encodings of all the elements are fixed once j_0 and N are fixed.

To make things concrete, let $a = \sqrt{-251}$ and consider the toy example above. The ideal class $x = [(3, \frac{a+1}{2})]$ acting on E_0 over $\mathbb{F}_p$ gives $j(x * E_{0,\mathbb{F}_p}) = 48$, over $\mathbb{F}_q$ gives $j(x * E_{0,\mathbb{F}_q}) = 162$. The canonical encoding of x is then $j_1 = \mathsf{CRT}(83, 173; 48, 162) = 7601$. Similarly, the canonical encodings of the ideal classes $[(7, \frac{a-1}{2})]$, $[(5, \frac{a+7}{2})]$, $[(5, \frac{a+3}{2})]$, $[(7, \frac{a+1}{2})]$, $[(3, \frac{a-1}{2})]$ are 1766, 4096, 7919, 2711, 1897.

The Composable Encodings and the Composition Law. To generate a composable encoding of $x \in \mathcal{CL}(\mathcal{O})$, we factorize x as $x = \prod_{x_i \in S} x_i^{e_i}$, where S denotes a generating set, and both the norms $N(x_i)$ and the exponents e_i being polynomial in size. The composable encoding of x then consists of the norms $N(x_i)$ and the j-invariants of $x_i^k * E_0$, for $k \in [e_i]$, for $i \in [|S|]$. The *degree* of a composable encoding is defined to be the product of the norms of the ideals $\prod_{x_i \in S} N(x_i)^{e_i}$. Note that the degree depends on the choice of S and the factorization of x, which is not unique.

As an example let us consider the simplest situation, where the composable encodings are just the canonical encodings themselves together with the norms of the ideals (i.e. the degrees of the isogenies). Set the composable encoding of $x = [(3, \frac{a+1}{2})]$ be $(3, 7601)$, the composable encoding of $y = [(7, \frac{a-1}{2})]$ be $(7, 1766)$.

Let us remark an intricacy of the construction of composable encodings. When the degrees of the composable encodings of x and y are coprime and polynomially large, the composition of x and y can be done simply by concatenating the corresponding encodings. To extract the canonical encoding of $x \circ y$, we take the gcd of the modular polynomials. In the example above, the canonical encoding of $x \circ y$ can be obtained by taking the gcd of $\Phi_7(7601, x)$ and $\Phi_3(1766, x)$ over $\mathbb{Z}/N\mathbb{Z}$. Since the degrees are coprime, the resulting polynomial is linear, with the only root being 4096, which is the canonical encoding of $[(5, \frac{a+7}{2})]$.

Note, however, that if the degrees share prime factors, then the gcd algorithm does not yield a linear polynomial, so the above algorithm for composition does not go through. To give a concrete example to what this means let us go back to our example: if we represent $y = [(7, \frac{a-1}{2})]$ by first factorizing y as $[(3, \frac{a+1}{2})]^2$ we then get the composable encoding of y as $(3, (7601, 1766))$. In this case the gcd of $\Phi_{3^2}(7601, x)$ and $\Phi_3(1766, x)$ over $\mathbb{Z}/N\mathbb{Z}$ yields a degree 3 polynomial, where it is unclear how to extract the roots. Hence, in this case we cannot calculate the canonical embedding of $x \circ y$ simply by looking at the gcd.

Therefore, to facilitate the efficient compositions of the encodings of group elements, we will need to represent them as the product of *pairwise co-prime* ideals with polynomially large norms. This, in particular, means the encoding algorithm will need to keep track on the primes used in the degrees of the composable encodings in the system. In other words, the encoding algorithm is *stateful*.

The Infeasibility of Inversion. The infeasibility of inversion amounts to the hardness of the computation of the canonical embedding of an element $x^{-1} \in \mathbb{G}$ from a composable encoding of x, and it is based on the hardness of the (ℓ, ℓ^2)-isogenous neighbors problem for each ideal of a composable encoding.

Going back to our example, given the composable encoding $(3, 7601)$ of $x = [(3, \frac{a+1}{2})]$, the canonical encoding of $x^{-1} = [(3, \frac{a-1}{2})]$ is a root of $f(x) = \gcd(\Phi_{3^2}(7601, x), \Phi_3(12631, x))$. The degree of f, however, is 3, so that it is not clear how to extract the root efficiently over an RSA modulus.

The Difficulty of Sampling the Class Group Invariants and Its Implications. Let us remark that the actual instantiation of TGII is more involved. A number of challenges arise solely from working with the ideal class groups of imaginary quadratic orders. To give a simple example of the challenges we face, efficiently generating a class group with a known large prime class number is a well-known open problem. Additionally, our construction requires more than the class number (namely, a short basis of the relation lattice of the class group) to support an efficient encoding algorithm.

In our solution, we choose the discriminant D to be of size roughly $\lambda^{O(\log \lambda)}$ and polynomially smooth, so as to make the parameter generation algorithm and the encoding algorithm run in polynomial time. The discriminant D (i.e. the description of the class group $\mathcal{CL}(D)$) has to be hidden to preserve the plausible $\lambda^{O(\log \lambda)}$-security of the TGII. Furthermore, even if D is hidden, there is an $\lambda^{O(\log \lambda)}$ attack by first guessing D or the group order, then solving the discrete-log problem given the polynomially-smooth group order. Extending the working parameters regime seems to require the solutions of several open problems concerning ideal class groups of imaginary quadratic orders.

Summary of the TGII Construction. To summarize, our construction of TGII chooses two sets of j-invariants that correspond to elliptic curves with the same imaginary quadratic order $\mathcal{O}$ over $\mathbb{F}_p$ and $\mathbb{F}_q$, and glues the j-invariants via the CRT composition as the canonical encodings of the group elements in $\mathcal{CL}(\mathcal{O})$. The composable encoding of a group element x is given as several j-invariants that represent the smooth ideals in a factorization of x. The efficiency of solving the (ℓ, m)-joint-neighbor problem over $\mathbb{Z}/N\mathbb{Z}$ facilitates the efficient group operation over coprime degree encodings. The conjectured hardness of the (ℓ, ℓ^2)-joint-neighbor problem over $\mathbb{Z}/N\mathbb{Z}$ is the main reason behind the hardness of inversion, but it also stops us from composing encodings that share prime factors.

The drawbacks of our construction of TGII are as follows.

1. Composition is only feasible for coprime-degree encodings, which means in order to publish arbitrarily polynomially many encodings, the encoding algorithm has to be stateful in order to choose different polynomially large prime factors for the degrees of the encoding (we cannot choose polynomially large prime degrees and hope they are all different).
2. In the definition from [18,30], the composable encodings obtained during the composition are required to be indistinguishable to a freshly sampled encoding. In our construction the encodings keep growing during compositions, until they are extracted to the canonical encoding which is a single j-invariant.
3. In addition to the (ℓ, ℓ^2)-joint-neighbor problem, the security of the TGII construction relies on several other heuristic assumptions. We will list our cryptanalytic attempts in Sect. 5.3. Moreover, even if we have not missed any attacks, the current best attack only requires $\lambda^{O(\log \lambda)}$-time, by first guessing the discriminant or the group order.

The Two Applications of TGII. Let us briefly mention the impact of the limitation of our TGII on the applications of directed transitive signature (DTS) [18,30] and broadcast encryption [20]. For the broadcast encryption from TGII [20], the growth of the composable encodings do not cause a problem. For DTS, in the direct instantiation of DTS from TGII [18,30], the signature is a composable encoding, so the length of the signature keeps growing during the composition, which is an undesirable feature for a non-trivial DTS. So on top of the basic instantiation, we provide an additional compression technique to shrink the composed signature.

Let us also remark that in the directed transitive signature [18, 30], encodings are sampled by the master signer; in the broadcast encryption scheme [20], encodings are sampled by the master encrypter. At least for these two applications, having the master signer/encrypter being stateful is not ideal but acceptable.

Organization. The rest of the paper is organized as follows. Section 2 provides the background of imaginary quadratic fields, elliptic curves and isogenies. Section 3 defines the computational problems for isogeny graphs over composite moduli. Section 4 provides our basic construction of a trapdoor group with infeasible inversion. Section 5 provides a highlight of our cryptanalysis attempts.

2 Preliminaries

Notation and Terminology. Let $\mathbb{C}, \mathbb{R}, \mathbb{Q}, \mathbb{Z}, \mathbb{N}$ denote the set of complex numbers, reals, rationals, integers, and positive integers respectively. For any field K we fix an algebraic closure and denote it by $\bar{K}$. For $n \in \mathbb{N}$, let $[n] := \{1, \ldots, n\}$. For $B \in \mathbb{R}$, an integer n is called B-smooth if all the prime factors of n are less than or equal to B. An n-dimensional vector is written as a bold lower-case letter, e.g. $\mathbf{v} := (v_1, \ldots, v_n)$. For an index $k \in \mathbb{N}$, distinct prime numbers p_i for $i \in [k]$, and $c_i \in \mathbb{Z}/p_i\mathbb{Z}$ we will let $\mathsf{CRT}(p_1, \ldots, p_k; c_1, \ldots, c_k)$ to denote the unique $y \in \mathbb{Z}/(\prod_i^k p_i)\mathbb{Z}$ such that $y \equiv c_i \pmod{p_i}$, for $i \in [k]$. Given a lattice Λ with a basis $\mathbf{B}$, let $\tilde{\mathbf{B}}$ denote the Gram-Schmidt orthogonalization of $\mathbf{B}$.

Let λ denote the security parameter. In theory and by default, an algorithm is called "efficient" if it runs in probabilistic polynomial time over λ.

2.1 Ideal Class Groups of Imaginary Quadratic Orders

We recall the necessary background of ideal class groups from [5, 9, 28].

Let K be an imaginary quadratic field. An *order* $\mathcal{O}$ in K is a subset of K such that (1) $\mathcal{O}$ is a subring of K containing 1; (2) $\mathcal{O}$ is a finitely generated $\mathbb{Z}$-module; (3) $\mathcal{O}$ contains a $\mathbb{Q}$-basis of K. The ring $\mathcal{O}_K$ of integers of K is always an order. For any order $\mathcal{O}$, we have $\mathcal{O} \subseteq \mathcal{O}_K$, in other words $\mathcal{O}_K$ is the maximal order of K with respect to inclusion.

The ideal class group (or class group) of $\mathcal{O}$ is the quotient group $\mathcal{CL}(\mathcal{O}) = I(\mathcal{O})/P(\mathcal{O})$ where $I(\mathcal{O})$ denotes the group of proper (i.e. invertible) fractional $\mathcal{O}$-ideals of, and $P(\mathcal{O})$ is its subgroup of principal $\mathcal{O}$-ideals. Let D be the discriminant of $\mathcal{O}$. Note that since $\mathcal{O}$ is quadratic imaginary we have $D < 0$. Sometimes we will denote the class group $\mathcal{CL}(\mathcal{O})$ as $\mathcal{CL}(D)$, and the class number (the group order of $\mathcal{CL}(\mathcal{O})$) as $h(\mathcal{O})$ or $h(D)$.

Let $D = D_0 \cdot f^2$, where D_0 is the *fundamental discriminant* and f is the *conductor* of $\mathcal{O}$ (or D). The following well-known formula relates the class number of an non-maximal order to that of the maximal one:

$$\frac{h(D)}{w(D)} = \frac{h(D_0)}{w(D_0)} \cdot f \prod_{p \mid f} \left(1 - \frac{\left(\frac{D_0}{p}\right)}{p}\right), \tag{1}$$

where $w(D) = 6$ if $D = -3$, $w(D) = 4$ if $D = -4$, and $w(D) = 2$ if $D < -4$. Let us also remark that the Brauer-Siegel theorem implies that $\ln(h(D)) \sim \ln(\sqrt{|D|})$ as $D \to -\infty$.

Representations. An $\mathcal{O}$-ideal of discriminant D can be represented by its generators, or by its binary quadratic forms. A binary quadratic form of discriminant D is a polynomial $ax^2 + bxy + cy^2$ with $b^2 - 4ac = D$. We denote a binary quadratic form by (a, b, c). The group $SL_2(\mathbb{Z})$ acts on the set of binary quadratic forms and preserves the discriminant. We shall always be assuming that our forms are positive definite, i.e. $a > 0$. Recall that a form (a, b, c) is called *primitive* if $\gcd(a, b, c) = 1$, and a primitive form is called *reduced* if $-a < b \leq a < c$ or $0 \leq b \leq a = c$. Reduced forms satisfy $a \leq \sqrt{|D|/3}$.

A fundamental fact, which goes back to Gauss, is that in each equivalence class, there is a unique reduced form (see Corollary 5.2.6 of [5]). Given a form (a, b, c), denote $[(a, b, c)]$ as its equivalence class. Note that when D is fixed, we can denote a class simply by $[(a, b, \cdot)]$. Efficient algorithms of composing forms and computing the reduced form can be found in [28, Page 9].

2.2 Elliptic Curves and Their Isogenies

In this section we will recall some background on elliptic curves and isogenies. All of this material is well-known and the main references for this section are [14, 23, 34, 35, 37].

Let E be an elliptic curve defined over a finite field $\mathbf{k}$ of characteristic $\neq 2, 3$ with q elements, given by its Weierstrass form $y^2 = x^3 + ax + b$ where $a, b \in \mathbf{k}$. By the Hasse bound we know that the order of the $\mathbf{k}$-rational points $E(\mathbf{k})$ satisfies

$$-2\sqrt{q} \leq \#E(\mathbf{k}) - (q + 1) \leq 2\sqrt{q}.$$

Here, $t = q + 1 - \#E(\mathbf{k})$ is the trace of Frobenius endomorphism $\pi : (x, y) \mapsto (x^q, y^q)$. Let us also recall that Schoof's algorithm [32] takes as inputs E and q, computes t, and hence $\#E(\mathbf{k})$, in time $\mathsf{poly}(\log q)$.

The *j-invariant* of E is defined as $j(E) = 1728 \cdot \frac{4a^3}{4a^3 + 27b^2}$. The values $j = 0$ or 1728 are special and we will choose to avoid these two values throughout the paper. Two elliptic curves are isomorphic over the algebraic closure $\bar{\mathbf{k}}$ if and only if their j-invariants are the same. Note that this isomorphism may not be defined over the base field $\mathbf{k}$, in which case the curves are called twists of each other. It will be convenient for us to use j-invariants to represent isomorphism classes of elliptic curves (including their twists). In many cases, with abuse of notation, a j-invariant will be treated as the same to an elliptic curve over $\mathbf{k}$ in the corresponding isomorphism class.

Isogenies. An *isogeny* $\varphi : E_1 \to E_2$ is a morphism of elliptic curves that preserves the identity. Every nonzero isogeny induces a surjective group homomorphism from $E_1(\bar{\mathbf{k}})$ to $E_2(\bar{\mathbf{k}})$ with a finite kernel. Elliptic curves related by a nonzero isogeny are said to be *isogenous*. By the Tate isogeny theorem [38, pg.139] two elliptic curves E_1 and E_2 are isogenous over $\mathbf{k}$ if and only if $\#E_1(\mathbf{k}) = \#E_2(\mathbf{k})$.

The degree of an isogeny is its degree as a rational map. An isogeny of degree ℓ is called an ℓ-isogeny. When $\mathrm{char}(\mathbf{k}) \nmid \ell$, the kernel of an ℓ-isogeny has cardinality ℓ. Two isogenies ϕ and φ are considered equivalent if $\phi = \iota_1 \circ \varphi \circ \iota_2$ for isomorphisms ι_1 and ι_2. Every ℓ-isogeny $\varphi : E_1 \to E_2$ has a unique dual isogeny $\hat{\varphi} : E_2 \to E_1$ of the same degree such that $\varphi \circ \hat{\varphi} = \hat{\varphi} \circ \varphi = [\ell]$, where $[\ell]$ is the multiplication by ℓ map. The kernel of the multiplication-by-ℓ map is the *ℓ-torsion subgroup*

$$E[\ell] = \{ P \in E(\bar{\mathbf{k}}) : \ell P = 0 \}.$$

When $\ell \nmid \mathrm{char}(\mathbf{k})$ we have $E[\ell] \simeq \mathbb{Z}/\ell\mathbb{Z} \times \mathbb{Z}/\ell\mathbb{Z}$. For a prime $\ell \neq \mathrm{char}(\mathbf{k})$, there are $\ell + 1$ cyclic subgroups in $E[\ell]$ of order ℓ, each corresponding to the kernel of an ℓ-isogeny φ from E. An isogeny from E is defined over $\mathbf{k}$ if and only if its kernel subgroup G is defined over $\mathbf{k}$ (namely, for $P \in G$ and $\sigma \in \mathrm{Gal}(\bar{\mathbf{k}}/\mathbf{k})$, $\sigma(P) \in G$; note that this does not imply $G \subseteq E(\mathbf{k})$). If $\ell \nmid \mathrm{char}(\mathbf{k})$ and $j(E) \neq 0$ or 1728, then up to isomorphism the number of ℓ-isogenies from E defined over $\mathbf{k}$ is $0, 1, 2$, or $\ell + 1$.

Modular Polynomials. Let $\ell \in \mathbb{Z}$, let $\mathbb{H}$ denote the upper half plane $\mathbb{H} := \{\tau \in \mathbb{C} : \mathrm{im}\,\tau > 0\}$ and $\mathbb{H}^* = \mathbb{H} \cup \mathbb{Q} \cup \{\infty\}$. Let $j(\tau)$ be the classical modular function defined on $\mathbb{H}$. For any $\tau \in \mathbb{H}$, the complex numbers $j(\tau)$ and $j(\ell\tau)$ are the j-invariants of elliptic curves defined over $\mathbb{C}$ that are related by an isogeny whose kernel is a cyclic group of order ℓ. The minimal polynomial $\Phi_\ell(y)$ of the function $j(\ell z)$ over the field $\mathbb{C}(j(z))$ has coefficients that are polynomials in $j(z)$ with inter coefficients. Replacing $j(z)$ with a variable x gives the *modular polynomial* $\Phi_\ell(x, y) \in \mathbb{Z}[x, y]$, which is symmetric in x and y. It parameterizes pairs of elliptic curves over $\mathbb{C}$ related by a cyclic ℓ-isogeny (an isogeny is said to be cyclic if its kernel is a cyclic group; when ℓ is a prime every ℓ-isogeny is cyclic). The modular equation $\Phi_\ell(x, y) = 0$ is a canonical equation for the modular curve $Y_0(\ell) = \mathbb{H}/\Gamma_0(\ell)$, where $\Gamma_0(\ell)$ is the congruence subgroup of $\mathrm{SL}_2(\mathbb{Z})$ defined by

$$\Gamma_0(\ell) = \left\{ \begin{pmatrix} a & b \\ c & d \end{pmatrix} \in \mathrm{SL}_2(\mathbb{Z}) \,\middle|\, \begin{pmatrix} a & b \\ c & d \end{pmatrix} \equiv \begin{pmatrix} * & * \\ 0 & * \end{pmatrix} \pmod{\ell} \right\}.$$

The time and space required for computing the modular polynomial Φ_ℓ are polynomial in ℓ, cf. [12, Sect. 3] or [5, Page 386]. In this article we will only use $\{\Phi_\ell \in \mathbb{Z}[x, y]\}_{\ell \in \mathrm{poly}(\lambda)}$, so we might as well assume that the modular polynomials are computed ahead of time.

2.3   Isogeny Volcanoes and the Class Groups

An isogeny from an elliptic curve E to itself is called an *endomorphism*. Over a finite field $\mathbf{k}$, $\mathrm{End}(E)$ is isomorphic to an imaginary quadratic order when E is ordinary, or an order in a definite quaternion algebra when E is supersingular. In this paper we will be focusing on the ordinary case.

Isogeny Graphs. The thesis of [23] describes the graphs that capture the relation of being ℓ-isogenous among elliptic curves over a finite field **k**.

Definition 1 (ℓ-isogeny graph). *Fix a prime ℓ and a finite field **k** such that* char(**k**) $\neq \ell$. *The ℓ-isogeny graph $G_\ell(\mathbf{k})$ has vertex set **k**. Two vertices (j_1, j_2) have a directed edge (from j_1 to j_2) with multiplicity equal to the multiplicity of j_2 as a root of $\Phi_\ell(j_1, Y)$. The vertices of $G_\ell(\mathbf{k})$ are j-invariants and each edge corresponds to an (isomorphism classes of an) ℓ-isogeny.*

For $j_1, j_2 \notin \{0, 1728\}$, an edge (j_1, j_2) occurs with the same multiplicity as (j_2, j_1) and thus the subgraph of $G_\ell(\mathbf{k})$ on $\mathbf{k} \setminus \{0, 1728\}$ can be viewed as an undirected graph. $G_\ell(\mathbf{k})$ has super singular and ordinary components. The ordinary components of $G_\ell(\mathbf{k})$ look like ℓ-*volcanoes*:

Definition 2 (ℓ-volcano). *Fix a prime ℓ. An ℓ-volcano V is a connected undirected graph whose vertices are partitioned into one or more levels V_0, ..., V_d such that the following hold:*

1. *The subgraph on V_0 (the surface, or the crater) is a regular graph of degree at most 2.*
2. *For $i > 0$, each vertex in V_i has exactly one neighbor in level V_{i-1}.*
3. *For $i < d$, each vertex in V_i has degree $\ell + 1$.*

Let $\phi : E_1 \to E_2$ by an ℓ-isogeny of elliptic curves with endomorphism rings $\mathcal{O}_1 = \mathrm{End}(E_1)$ and $\mathcal{O}_2 = \mathrm{End}(E_2)$ respectively. Then, there are three possibilities for $\mathcal{O}_1$ and $\mathcal{O}_2$:

- If $\mathcal{O}_1 = \mathcal{O}_2$, then ϕ is called horizontal,
- If $[\mathcal{O}_1 : \mathcal{O}_2] = \ell$, then ϕ is called descending,
- If $[\mathcal{O}_2 : \mathcal{O}_1] = \ell$, then ϕ is called ascending.

Let E be an elliptic curve over **k** whose endomorphism ring is isomorphic to an imaginary quadratic order $\mathcal{O}$. Then, the set

$$\mathrm{Ell}_{\mathcal{O}}(\mathbf{k}) = \{j(E) \in \mathbf{k} \mid \text{with } \mathrm{End}(E) \simeq \mathcal{O}\}$$

is naturally a $\mathcal{CL}(\mathcal{O})$-torsor as follows: For an invertible $\mathcal{O}$-ideal $\mathfrak{a}$ the $\mathfrak{a}$-torsion subgroup

$$E[\mathfrak{a}] = \{P \in E(\bar{\mathbf{k}}) : \alpha(P) = 0, \forall \alpha \in \mathfrak{a}\}$$

is the kernel of a separable isogeny $\phi_{\mathfrak{a}} : E \to E'$. If the norm $N(\mathfrak{a}) = [\mathcal{O} : \mathfrak{a}]$ is not divisible by char(**k**), then the degree of $\phi_{\mathfrak{a}}$ is $N(\mathfrak{a})$. Moreover, if $\mathfrak{a}$ and $\mathfrak{b}$ are two invertible $\mathcal{O}$-ideals, then $\phi_{\mathfrak{ab}} = \phi_{\mathfrak{a}}\phi_{\mathfrak{b}}$, and if $\mathfrak{a}$ is principal then $\phi_{\mathfrak{a}}$ is an isomorphism. This gives a faithful and transitive action of $\mathcal{CL}(\mathcal{O})$ on $\mathrm{Ell}_{\mathcal{O}}(\mathbf{k})$.

Remark 1 (Linking ideals and horizontal isogenies). When ℓ splits in $\mathcal{O}$ we have $(\ell) = \mathfrak{l} \cdot \bar{\mathfrak{l}}$. Fix an elliptic curve $E(\mathbf{k})$ with $\mathrm{End}(E) \simeq \mathcal{O}$, the two horizontal isogenies $\phi_1 : E \to E_1$ and $\phi_2 : E \to E_2$ can be efficiently associated with the two ideals $\mathfrak{l}$ and $\bar{\mathfrak{l}}$ when $\ell \in \mathrm{poly}(\lambda)$ (cf. [33]). To do so, factorize the characteristic polynomial of Frobenius π as $(x - \mu)(x - \nu) \pmod{\ell}$, where $\mu, \nu \in \mathbb{Z}/\ell\mathbb{Z}$. Given an ℓ-isogeny ϕ from E to E/G, the eigenvalue (say μ) corresponding to the eigenspace G can be verified by picking a point $P \in G$, then check whether $\pi(P) = [\mu]P$ modulo G. If so then μ corresponds to ϕ.

3   Isogeny Graphs over Composite Moduli

Let p, q be distinct primes and set $N = pq$. We will be using elliptic curves over the ring $\mathbb{Z}/N\mathbb{Z}$. We will not be needing a formal treatment of elliptic curves over rings as such a discussion would take us too far afield. Instead, we will be defining objects and quantities over $\mathbb{Z}/N\mathbb{Z}$ by taking the CRT of the corresponding ones over $\mathbb{F}_p$ and $\mathbb{F}_q$, which will suffice for our purposes. This follows the treatment given in [27].

Since the underlying rings will matter, we will denote an elliptic curve over a ring R by $E(R)$. If R is clear from the context we shall omit it from the notation. To begin, let us remark that the number of points $\#(E(\mathbb{Z}/N\mathbb{Z}))$ is equal to $\#(E(\mathbb{F}_p)) \cdot \#(E(\mathbb{F}_q))$, and the j-invariant of $E(\mathbb{Z}/N\mathbb{Z})$ is $\mathsf{CRT}(p, q; j(E(\mathbb{F}_p)), j(E(\mathbb{F}_q)))$.

3.1   Isogeny Graphs over $\mathbb{Z}/N\mathbb{Z}$

Let N be as above. For every prime $\ell \nmid N$ the isogeny graph $G_\ell(\mathbb{Z}/N\mathbb{Z})$ can be defined naturally as the graph tensor product of $G_\ell(\mathbb{F}_p)$ and $G_\ell(\mathbb{F}_q)$.

Definition 3 (ℓ-isogeny graph over $\mathbb{Z}/N\mathbb{Z}$). *Let ℓ, p, and q be distinct primes and let $N = pq$. The ℓ-isogeny graph $G_\ell(\mathbb{Z}/N\mathbb{Z})$ has*

- *The vertex set of $G_\ell(\mathbb{Z}/N\mathbb{Z})$ is $\mathbb{Z}/N\mathbb{Z}$, identified with $\mathbb{Z}/p\mathbb{Z} \times \mathbb{Z}/q\mathbb{Z}$ by* CRT,
- *Two vertices $v_1 = (v_{1,p}, v_{1,q})$ and $v_2 = (v_{2,p}, v_{2,q})$ are connected if and only if $v_{1,p}$ is connected to $v_{2,p}$ in $G_\ell(\mathbb{F}_p)$ and $v_{1,q}$ is connected to $v_{2,q}$ in $G_\ell(\mathbb{F}_q)$.*

Let us make a remark for future consideration. In the construction of groups with infeasible inversion, we will be working with special subgraphs of $G_\ell(\mathbb{Z}/N\mathbb{Z})$, where the vertices over $\mathbb{F}_p$ and $\mathbb{F}_q$ correspond to j-invariants of curves whose endomorphism rings are the same imaginary quadratic order $\mathcal{O}$. Nevertheless, this is a choice we made for convenience, and it does not hurt to define the computational problems over the largest possible graph and to study them first.

3.2   The ℓ-isogenous Neighbors Problem over $\mathbb{Z}/N\mathbb{Z}$

Definition 4 (The ℓ-isogenous neighbors problem). *Let p, q be two distinct primes and let $N = pq$. Let ℓ be a polynomially large prime s.t. $\gcd(\ell, N) = 1$. The input of the ℓ-isogenous neighbor problem is N and an integer $j \in \mathbb{Z}/N\mathbb{Z}$ such that there exists (possibly more than) one integer j' that $\Phi_\ell(j, j') = 0$ over $\mathbb{Z}/N\mathbb{Z}$. The problem asks to find such integer(s) j'.*

The following theorem shows that the problem of finding *all* of the ℓ-isogenous neighbors is at least as hard as factoring N.

Theorem 1. *If there is a probabilistic polynomial time algorithm that finds all the ℓ-isogenous neighbors in Problem 4, then there is a probabilistic polynomial time algorithm that solves the integer factorization problem.*

The idea behind the reduction is as follows. Suppose it is efficient to pick an integer j over $\mathbb{Z}/N\mathbb{Z}$, let $j_p = j \pmod{p}$ and $j_q = j \pmod{q}$, such that j_p has at least two distinct neighbors in $G_\ell(\mathbb{F}_p)$, and j_q has at least two distinct neighbors in $G_\ell(\mathbb{F}_q)$. In this case if we are able to find *all* the integer solutions $j' \in \mathbb{Z}/N\mathbb{Z}$ such that $\Phi_\ell(j, j') = 0$ over $\mathbb{Z}/N\mathbb{Z}$, then there exist two distinct integers j'_1 and j'_2 among the solutions such that $N > \gcd(j'_1 - j'_2, N) > 1$. One can also show that finding *one* of the integer solutions is hard using a probabilistic argument, assuming the underlying algorithm outputs a random solution when there are multiple ones.

In the reduction we pick the elliptic curve E randomly, so we have to make sure that for a non-negligible fraction of the elliptic curves E over $\mathbb{F}_p$, $j(E) \in G_\ell(\mathbb{F}_p)$ has at least two neighbors. The estimate for this relies on the following lemma:

Lemma 1 ([27] (1.9)). *There exists an efficiently computable positive constant c such that for each prime number $p > 3$, for a set of integers $S \subseteq \{s \in \mathbb{Z} \mid |p + 1 - s| < \sqrt{p}\}$, we have*

$$\#' \{E \mid E \text{ is an elliptic curve over } \mathbb{F}_p, \ \#E(\mathbb{F}_p) \in S\}_{/\simeq_{\mathbb{F}_p}} \geq c \, (\#S - 2) \frac{\sqrt{p}}{\log p}.$$

where $\#' \{E\}_{/\simeq_{\mathbb{F}_p}}$ denotes the number of isomorphism classes of elliptic curves over $\mathbb{F}_p$, each counted with weight $(\#\mathrm{Aut}E)^{-1}$.

Theorem 2. *Let p, ℓ be primes such that $6\ell < \sqrt{p}$. The probability that for a random elliptic curve E over $\mathbb{F}_p$ (i.e. a random pair $(a, b) \in \mathbb{F}_p \times \mathbb{F}_p$ such that $4a^3 + 27b^2 \neq 0$) $j(E) \in G_\ell(\mathbb{F}_p)$ having at least two neighbors is $\Omega(\frac{1}{\log p})$.*

Due to the page limitation we refer the readers to the full version for the proof of Theorem 2.

Proof (Proof of Theorem 1). Suppose that there is a probabilistic polynomial time algorithm A that finds all the ℓ-isogenous neighbors in Problem 4 with non-negligible probability η. We will build a probabilistic polynomial time algorithm A' that solves factoring. Given an integer N, A' samples two random integers $a, b \in \mathbb{Z}/N\mathbb{Z}$ such that $4a^3 + 27b^2 \neq 0$, and computes $j = 1728 \cdot \frac{4a^3}{4a^3 + 27b^2}$. With all but negligible probability $\gcd(j, N) = 1$ and $j \neq 0, 1728$; if j happens to satisfy $1 < \gcd(j, N) < N$, then A' outputs $\gcd(j, N)$.

A' then sends N, j_0 to the solver A for Problem 4 for a fixed polynomially large prime ℓ, gets back a set of solutions $\mathcal{J} = \{j_i\}_{i \in [k]}$, where $0 \leq k \leq (\ell + 1)^2$ denotes the number of solutions. With probability $\Omega(\frac{1}{\log^2 N})$, the curve $E : y^2 = x^3 + ax + b$ has at least two ℓ-isogenies over both $\mathbb{F}_p$ and $\mathbb{F}_q$ due to Theorem 2. In that case there exists $j, j' \in \mathcal{J}$ such that $1 < \gcd(j - j', N) < N$, which gives a prime factor of N.

3.3 The (ℓ, M)-Isogenous Neighbors Problem over $\mathbb{Z}/N\mathbb{Z}$

Definition 5 (The (ℓ, m)-isogenous neighbors problem). *Let p and q be two distinct primes. Let $N := p \cdot q$. Let ℓ, m be two polynomially large integers s.t. $\gcd(\ell m, N) = 1$. The input of the (ℓ, m)-isogenous neighbor problem is the j-invariants j_1, j_2 of two elliptic curves E_1, E_2 defined over $\mathbb{Z}/N\mathbb{Z}$. The problem asks to find all the integers j' such that $\Phi_\ell(j(E_1), j') = 0$, and $\Phi_m(j(E_2), j') = 0$ over $\mathbb{Z}/N\mathbb{Z}$.*

When $\gcd(\ell, m) = 1$, applying the Euclidean algorithm on $\Phi_\ell(j_1, x)$ and $\Phi_m(j_2, x)$ gives a linear polynomial over x.

Lemma 2 ([13]). *Let $j_1, j_2 \in \mathrm{Ell}_\mathcal{O}(\mathbb{F}_p)$, and let $\ell, m \neq p$ be distinct primes with $4\ell^2 m^2 < |D|$. Then the degree of $f(x) := \gcd(\Phi_\ell(j_1, x), \Phi_m(j_2, x))$ is less than or equal to 1.*

When $\gcd(\ell, m) = d > 1$, applying the Euclidean algorithm on $\Phi_\ell(j_1, x)$ and $\Phi_m(j_2, x)$ gives a polynomial of degree at least d. We present a proof in the the case where $m = \ell^2$, which has the general idea.

Lemma 3. *Let $p \neq 2, 3$ and $\ell \neq p$ be primes, and let j_0, j_1 be such that $\Phi_\ell(j_0, j_1) = 0 \bmod p$. Let $\Phi_\ell(X, j_0)$ and $\Phi_{\ell^2}(X, j_1)$ be the modular polynomials of levels ℓ and ℓ^2 respectively. Then,*

$$(X - j_1) \cdot \gcd(\Phi_\ell(X, j_0), \Phi_{\ell^2}(X, j_1)) = \Phi_\ell(X, j_0)$$

in $\mathbb{F}_p[X]$. In particular,

$$\deg(\gcd(\Phi_\ell(X, j_0), \Phi_{\ell^2}(X, j_1))) = \ell$$

Proof. Without loss of generality we can, and we do, assume that $\Phi_\ell(X, j_0)$, $\Phi_\ell(X, j_1)$, and $\Phi_{\ell^2}(X, j_1)$ split over $\mathbb{F}_p$ (otherwise we can base change to an extension $k'/\mathbb{F}_p$, where the full ℓ^2-torsion is defined, this does not affect the degree of the gcd).

Assume that the degree of the gcd is N_{gcd}. We have,

$$\deg(\Phi_\ell(X, j_0)) = \ell + 1, \qquad \deg(\Phi_{\ell^2}(X, j_1)) = \ell(\ell + 1). \qquad (2)$$

Let E_0, E_1 denote the (isomorphism classes of) elliptic curves with j-invariants j_0 and j_1 respectively, and $\varphi_\ell : E_0 \to E_1$ be the corresponding isogeny. We count the number N_{ℓ^2} of cyclic ℓ^2-isogenies from E_1 two ways. First, N_{ℓ^2} is the number of roots of $\Phi_{\ell^2}(X, j_1)$, which, by (2) and the assumption that $\ell^2 + \ell < p$, is $\ell^2 + \ell$.

Next, recall (cf. Corollary 6.11 of [36]) that every isogeny of degree ℓ^2 can be decomposed as a composition of two degree ℓ isogenies (which are necessarily cyclic). Using this N_{ℓ^2} is bounded above by $N_{\mathrm{gcd}} + \ell^2$, where the first factor counts the number of ℓ^2-isogenies $E_1 \to E$ that are compositions $E_1 \xrightarrow{\hat{\varphi}_\ell} E_0 \to E$, and the second factor counts the isogenies that are compositions $E_1 \to E' \to E$, where $E' \not\cong E_1$. Note that we are not counting compositions $E_1 \xrightarrow{\phi} \tilde{E} \xrightarrow{\hat{\phi}} E_1$ since these do not give rise to cyclic isogenies.

This shows that $\ell^2 + \ell \leq \ell^2 + N_{\ell^2} \Rightarrow N_{\text{gcd}} \geq \ell$. On the other hand, by (2) $N_{\text{gcd}} \leq \ell$ since $\Phi_\ell(X, j_0)/(X - j_0)$ has degree ℓ and each root except for j_1 gives a (possibly cyclic) ℓ^2-isogeny by composition with $\hat{\varphi}_\ell$. This implies that $N_{\text{gcd}} = \ell$ and that all the ℓ^2-isogenies obtained this way are cyclic. In particular, we get that the gcd is $\Phi_\ell(X, j_0)/(X - j_1)$.

Discussions. Let us remark that we do not know if solving the (ℓ, ℓ^2)-isogenous neighbors problem is as hard as factoring. To adapt the same reduction in the proof of Theorem 1, we need the feasibility of sampling two integers j_1, j_2 such that $\Phi_\ell(j_1, j_2) = 0 \pmod{N}$, and j_1 or j_2 has to have another isogenous neighbor over $\mathbb{F}_p$ or $\mathbb{F}_q$. However the feasibility is unclear to us in general.

From the cryptanalytic point of view, a significant difference of the (ℓ, ℓ^2)-isogenous neighbors problem and the ℓ-isogenous neighbors problem is the following. Let ℓ be an odd prime. Recall that an isogeny $\phi : E_1 \to E_2$ of degree ℓ can be represented by a rational polynomial

$$\phi : E_1 \to E_2, \quad (x, y) \mapsto \left(\frac{f(x)}{h(x)^2}, \frac{g(x, y)}{h(x)^3} \right),$$

where $h(x)$ is its *kernel polynomial* of degree $\frac{\ell-1}{2}$. The roots of $h(x)$ are the x-coordinates of the kernel subgroup $G \subset E_1[\ell]$ such that $\phi : E_1 \to E_1/G$.

Given a single j-invariant j' over $\mathbb{Z}/N\mathbb{Z}$, it is infeasible to find a rational polynomial ϕ of degree ℓ that maps from a curve E with j-invariant j' to another curve j'', since otherwise j'' is a solution to the ℓ-isogenous neighbors problem. However, if we are given two j-invariants $j_1, j_2 \in \mathbb{Z}/N\mathbb{Z}$ such that $\Phi_\ell(j_1, j_2) = 0$ $\pmod{N}$, as in the (ℓ, ℓ^2)-isogenous neighbors problem; then it is feasible to compute a pair of curves E_1, E_2 such that $j(E_1) = j_1$, $j(E_2) = j_2$, together with an explicit rational polynomial of an ℓ-isogeny from E_1 to E_2. This is because the arithmetic operations involved in computing the kernel polynomial $h(x)$ mentioned in [8,12,33] works over $\mathbb{Z}/N\mathbb{Z}$ by reduction mod N, and does not require the factorization of N.

Proposition 1. *Given $\ell, N \in \mathbb{Z}$ such that $\gcd(\ell, N) = 1$, and two integers $j_1, j_2 \in \mathbb{Z}/N\mathbb{Z}$ such that $\Phi_\ell(j_1, j_2) = 0$ over $\mathbb{Z}/N\mathbb{Z}$, the elliptic curves E_1, E_2, and the kernel polynomial $h(x)$ of an isogeny ϕ from E_1, E_2 can be computed in time polynomial in $\ell, \log(N)$. From the kernel polynomial $h(x)$ of an isogeny ϕ, computing $f(x), g(x, y)$, hence the entire rational polynomial of ϕ, is feasible over $\mathbb{Z}/N\mathbb{Z}$ via Vélu's formulae [39].*

However, it is unclear how to utilize the rational polynomial to solve the (ℓ, ℓ^2)-joint neighbors problem. We postpone further discussions on the hardness and cryptanalysis to Sect. 5.

4   Trapdoor Group with Infeasible Inversion

In this section we present the construction of the trapdoor group with infeasible inversion. As the general construction is somewhat technical we will present it in two steps: first we will go over the basic algorithms that feature a simple encoding and composition rule, which suffices for the instantiations of the applications; we will then move to the general algorithms that offer potential optimization and flexibility.

4.1   Definitions

Let us first provide the definition of a TGII, adapted from the original definition in [18,30] to match our construction. The main differences are:

1. The trapdoor in the definition of [18,30] is only used to invert an encoded group element, whereas we assume the trapdoor can be use to encode and decode (which implies the ability of inverting).
2. We classify the encodings of the group elements as *canonical encodings* and *composable encodings*, whereas the definition from [18,30] does not. In our definition, the canonical encoding of an element is uniquely determined once the public parameter is fixed. It can be directly used in the equivalence test, but it does not support efficient group operations. Composable encodings of group elements support efficient group operations. A composable encoding, moreover, can be converted into a canonical encoding by an efficient, public extraction algorithm.

Definition 6. *Let* $\mathbb{G} = (\circ, 1_{\mathbb{G}})$ *be a finite multiplicative group where* $\circ$ *denotes the group operator, and* $1_{\mathbb{G}}$ *denotes the identity. For* $x \in \mathbb{G}$, *denote its inverse by* x^{-1}. $\mathbb{G}$ *is associated with the following efficient algorithms:*

Parameter generation. $\mathsf{Gen}(1^{\lambda})$ *takes as input the security parameter* 1^{λ}, *outputs the public parameter* PP *and the trapdoor* τ.

Private sampling. $\mathsf{TrapSam}(\mathsf{PP}, \tau, x)$ *takes as inputs the public parameter* PP, *the trapdoor* τ, *and a plaintext group element* $x \in \mathbb{G}$, *outputs a composable encoding* $\mathsf{enc}(x)$.

Composition. $\mathsf{Compose}(\mathsf{PP}, \mathsf{enc}(x), \mathsf{enc}(y))$ *takes as inputs the public parameter* PP, *two composable encodings* $\mathsf{enc}(x), \mathsf{enc}(y)$, *outputs* $\mathsf{enc}(x \circ y)$. *We often use the notation* $\mathsf{enc}(x) \circ \mathsf{enc}(y)$ *for* $\mathsf{Compose}(\mathsf{PP}, \mathsf{enc}(x), \mathsf{enc}(y))$.

Extraction. $\mathsf{Ext}(\mathsf{PP}, \mathsf{enc}(x))$ *takes as inputs the public parameter* PP, *a composable encoding* $\mathsf{enc}(x)$ *of* x, *outputs the canonical encoding of* x *as* $\mathsf{enc}^{*}(x)$.

The hardness of inversion requires that it is infeasible for any efficient algorithm to produce the canonical encoding of x^{-1} *given a composable encoding of* $x \in \mathbb{G}$.

Hardness of Inversion. *For any p.p.t. algorithm* A,

$$\Pr[z = \mathsf{enc}^{*}(x^{-1}) \mid z \leftarrow A(\mathsf{PP}, \mathsf{enc}(x))] < \mathsf{negl}(\lambda),$$

where the probability is taken over the randomness in the generation of PP, x, $\mathsf{enc}(x)$, *and the adversary* A.

4.2   Construction Details: Basic

In this section we provide the formal construction of the TGII with the basic setting of algorithms. The basic setting assumes that in the application of TGII, the encoding sampling algorithm can be stateful, and it is easy to determine which encodings have to be pairwise composable, and which are not. Under these assumptions, we show that we can always sample composable encodings so that the composition always succeeds. That is, the degrees of the any two encodings are chosen to be coprime if they will be composed in the application, and not coprime if they will not be composed. The reader may be wondering why we are distinguishing pairs that are composable and those that are not, as opposed to simply assuming that every pairs of encoding are composable. The reason is for security, meanly due to the parallelogram attack in Sect. 5.3.

The basic setting suffices for instantiating the directed transitive signature [18,30] and the broadcast encryption schemes [20], where the master signer and the master encrypter are stateful. We will explain how to determine which encodings are pairwise composable in these two applications, so as to determine the prime degrees of the encodings (the rest of the parameters are not application-specific and follow the universal solution from this section).

For convenience of the reader and for further reference, we provide in Fig. 3 a summary of the parameters, with the basic constraints they should satisfy, and whether they are public or hidden. The correctness and efficiency reasons behind these constraints will be detailed in the coming paragraphs, whereas the security reasons will be explained in Sect. 5.

Parameter Generation. The parameter generation algorithm $\mathsf{Gen}(1^\lambda)$ takes the security parameter 1^λ as input, first chooses a non-maximal order $\mathcal{O}$ of an imaginary quadratic field as follows:

1. Select a square-free negative integer $D_0 \equiv 1 \bmod 4$ as the fundamental discriminant, such that D_0 is polynomially large and $h(D_0)$ is a prime.
2. Choose $k = O(\log(\lambda))$, and a set of distinct polynomially large prime numbers $\{f_i\}_{i\in[k]}$ such that the odd-part of $\left(f_i - \left(\frac{D_0}{f_i}\right)\right)$ is square-free and not divisible by $h(D_0)$. Let $f = \prod_{i\in[k]} f_i$.
3. Set $D = f^2 D_0$. Recall from Eqn. (1) that

$$h(D) = 2 \cdot \frac{h(D_0)}{w(D_0)} \prod_{i\in[k]} \left(f_i - \left(\frac{D_0}{f_i}\right)\right) \tag{3}$$

Let $\mathcal{CL}(\mathcal{O})_{\mathrm{odd}}$ be the odd part of $\mathcal{CL}(\mathcal{O})$, $h(D)_{\mathrm{odd}}$ be largest odd factor of $h(D)$. Note that due to the choices of D_0 and $\{f_i\}$, $\mathcal{CL}(\mathcal{O})_{\mathrm{odd}}$ is cyclic, and we have $|D|, h(D)_{\mathrm{odd}} \in \lambda^{O(\log \lambda)}$. The group with infeasible inversion $\mathbb{G}$ is then $\mathcal{CL}(\mathcal{O})_{\mathrm{odd}}$ with group order $h(D)_{\mathrm{odd}}$.

We then sample the public parameters as follows:

1. Choose two primes p, q, and elliptic curves $E_{0,\mathbb{F}_p}$, $E_{0,\mathbb{F}_q}$ with discriminant D, using the CM method (cf. [25] and more).

Parameters	Basic constraints	Public?
The modulus N	$N = pq$, p, q are primes, $|p|, |q| \in \mathsf{poly}(\lambda)$	Yes
The identity $j(E_0)$	$\mathrm{End}(E_0(\mathbb{F}_p)) \simeq \mathrm{End}(E_0(\mathbb{F}_q)) \simeq \mathcal{O}$	Yes
$\#(E_0(\mathbb{F}_p))$, $\#(E_0(\mathbb{F}_q))$	not polynomially smooth	No
The discriminant D of $\mathcal{O}$	$D = D_0 \cdot f^2$, $D \approx \lambda^{O(\log \lambda)}$, D is poly smooth	No
The class number $h(D)$	follows the choice of D	No
A set S in an encoding:	$S = \{C_i = [(p_i, b_i, \cdot)]\}_{i \in [w]}$ generates $\mathcal{CL}(D)_{\mathrm{odd}}$	See below
The number w of ideals	$w \in O(\log \lambda)$	Yes
The degree p_i of isogenies	$p_i \in \mathsf{poly}(\lambda)$	Yes
The basis $\mathbf{B}$ of Λ_S	$\|\tilde{\mathbf{B}}\| \in \mathsf{poly}(\lambda)$	No

Fig. 3. Summary of the choices of parameters in the basic setting.

2. Check whether p and q are safe RSA primes (if not, then back to the previous step and restart). Also, check whether the number of points $\#(E_0(\mathbb{F}_p))$, $\#(E_0(\mathbb{F}_q))$, $\#(\tilde{E}_0(\mathbb{F}_p))$, $\#(\tilde{E}_0(\mathbb{F}_q))$ (where $\tilde{E}$ denotes the quadratic twist of E) are polynomially smooth (if yes, then back to the previous step and restart). p, q and the number of points should be hidden for security.
3. Set the modulus N as $N := p \cdot q$ and let $j_0 = \mathsf{CRT}(p, q; j(E_{0,\mathbb{F}_p}), j(E_{0,\mathbb{F}_q}))$. Let j_0 represent the identity of $\mathbb{G}$.

Output (N, j_0) as the public parameter PP. Keep (D, p, q) as the trapdoor τ (D and the group order of $\mathbb{G}$ should be hidden for security).

The Sampling Algorithm and the Group Operation of the Composable Encodings. Next we provide the definitions and the algorithms for the composable encoding.

Definition 7 (Composable encoding). *Given a factorization of x as $\prod_{i=1}^{w} C_i^{e_i}$, where $w \in O(\log \lambda)$; $C_i = [(p_i, b_i, \cdot)] \in \mathbb{G}$, $e_i \in \mathbb{N}$, for $i \in [w]$. A composable encoding of $x \in \mathbb{G}$ is represented by*

$$\mathsf{enc}(x) = (L; T_1, ..., T_w) = ((p_1, ..., p_w); (j_{1,1}, ..., j_{1,e_1}), ..., (j_{w,1}, ..., j_{w,e_w})),$$

where all the primes in the list $L = (p_1, ..., p_w)$ are distinct; for each $i \in [w]$, $T_i \in (\mathbb{Z}/N\mathbb{Z})^{e_i}$ is a list of the j-invariants such that $j_{i,k} = C_i^k * j_0$, for $k \in [e_i]$. The degree of an encoding $\mathsf{enc}(x)$ is defined to be $d(\mathsf{enc}(x)) := \prod_{i=1}^{w} p_i^{e_i}$.

Notice that the factorization of $x = \prod_{i=1}^{w} C_i^{e_i}$ has to satisfy $e_i \in \mathsf{poly}(\lambda)$, for all $i \in [w]$, so as to ensure the length of $\mathsf{enc}(x)$ is polynomial. Looking ahead, we also require each p_i, the degree of the isogeny that represents the C_i-action, to be polynomially large so as to ensure Algorithm 3 in the encoding sampling algorithm and Algorithm 6 in the extraction algorithm run in polynomial time.

The composable encoding sampling algorithm requires the following subroutine:

Algorithm 3. $\mathsf{act}(\tau, j, C)$ *takes as input the trapdoor $\tau = (D, p, q)$, a j-invariant $j \in \mathbb{Z}/N\mathbb{Z}$, and an ideal class $C \in \mathcal{CL}(\mathcal{O})$, proceeds as follows:*

1. *Let $j_p = j \mod p$, $j_q = j \mod q$.*
2. *Compute $j'_p := C * j_p \in \mathbb{F}_p$, $j'_q := C * j_q \in \mathbb{F}_q$.*
3. *Output $j' := \mathsf{CRT}(p, q; j'_p, j'_q)$.*

Algorithm 4. (Sample a composable encoding) *Given as input the public parameter* $\mathsf{PP} = (N, j_0)$, *the trapdoor* $\tau = (D, p, q)$, *and* $x \in \mathbb{G}$, $\mathsf{TrapSam}(\mathsf{PP}, \tau, x)$ *produces a composable encoding of x is sampled as follows:*

1. *Choose $w \in O(\log \lambda)$ and a generation set $S = \{C_i = [(p_i, b_i, \cdot)]\}_{i \in [w]} \subset \mathbb{G}$.*
2. *Sample a short basis $\mathbf{B}$ (in the sense that $\|\tilde{\mathbf{B}}\| \in \mathsf{poly}(\lambda)$) for the relation lattice Λ_S:*

$$\Lambda_S := \left\{ \mathbf{y} \mid \mathbf{y} \in \mathbb{Z}^w, \prod_{i \in [w]} C_i^{y_i} = 1_{\mathbb{G}} \right\}. \tag{4}$$

3. *Given x, S, $\mathbf{B}$, sample a short vector $\mathbf{e} \in \{\mathsf{poly}(\lambda) \cap \mathbb{N}\}^w$ such that $x = \prod_{i \in [w]} C_i^{e_i}$.*
4. *For all $i \in [w]$:*
   (a) Let $j_{i,0} := j_0$.
   (b) For $k = 1$ to e_i: compute $j_{i,k} := \mathsf{act}(\tau, j_{i,k-1}, C_i)$.
   (c) Let $T_i := (j_{i,1}, \ldots, j_{i,e_i})$.
5. *Let $L \in \mathbb{N}^w$ be a list where the ith entry of L is p_i.*
6. *Output the composable encoding of x as*

$$\mathsf{enc}(x) = (L; T_1, \ldots, T_w) = ((p_1, \ldots, p_w); (j_{1,1}, \ldots, j_{1,e_1}), \ldots, (j_{w,1}, \ldots, j_{w,e_w})).$$

Remark 2 (Thinking of each adjacent pair of j-invariants as an isogeny). In each T_i, each adjacent pair of the j-invariants can be thought of representing an isogeny ϕ that corresponds to the ideal class $C_i = [(p_i, b_i, \cdot)]$. Over the finite field, C_i can be explicitly recovered from an adjacent pair of the j-invariants and p_i (cf. Remark 1). Over $\mathbb{Z}/N\mathbb{Z}$, the rational polynomial of the isogeny ϕ can be recovered from the adjacent pair of the j-invariants and p_i (cf. Proposition 1), but it is not clear how to recover b_i in the binary quadratic form representation of C_i.

Remark 3 (The only stateful step in the sampling algorithm). Recall that the basic setting assumes the encoding algorithm is stateful, where the state records the prime factors of the degrees used in the existing composable encodings. The state is only used in the first step to choose the $\{p_i\}$ of the ideals in the generation set $S = \{C_i = [(p_i, b_i, \cdot)]\}_{i \in [w]}$.

Group Operations. Given two composable encodings, the group operation is done by simply concatenating the encodings if their degrees are coprime, or otherwise outputting "failure".

Algorithm 5. *The encoding composition algorithm* $\mathsf{Compose}(\mathsf{PP}, \mathsf{enc}(x), \mathsf{enc}(y))$ *parses* $\mathsf{enc}(x) = (L_x; T_{x,1}, \ldots, T_{x,w_x})$, $\mathsf{enc}(y) = (L_y; T_{y,1}, \ldots, T_{y,w_y})$, *produces the composable encoding of $x \circ y$ as follows:*

– *If* $\gcd(d(\mathrm{enc}(x)), d(\mathrm{enc}(y))) = 1$, *then output the composable encoding of* $x \circ y$ *as*

$$\mathrm{enc}(x \circ y) = (L_x \| L_y; T_{x,1}, \ldots, T_{x,w_x}, T_{y,1}, \ldots, T_{y,w_y}).$$

– *If* $\gcd(d(\mathrm{enc}(x)), d(\mathrm{enc}(y))) > 1$, *output "failure".*

The canonical encoding and the extraction algorithm.

Definition 8 (Canonical Encoding). *The canonical encoding of* $x \in \mathbb{G}$ *is* $x * j_0 \in \mathbb{Z}/N\mathbb{Z}$.

The canonical encoding of x can be computed by first obtaining a composable encoding of x, and then converting the composable encoding into the canonical encoding using the extraction algorithm. The extraction algorithm requires the following subroutine.

Algorithm 6 (The "gcd" operation). *The algorithm* $\mathrm{gcd.op}(\mathsf{PP}, \ell_1, \ell_2; j_1, j_2)$ *takes as input the public parameter* PP, *two degrees* ℓ_1, ℓ_2 *and two* j-*invariants* j_1, j_2, *proceeds as follows:*

– *If* $\gcd(\ell_1, \ell_2) = 1$, *then it computes* $f(x) = \gcd(\Phi_{\ell_2}(j_1, x), \Phi_{\ell_1}(j_2, x))$ *over* $\mathbb{Z}/N\mathbb{Z}$, *and outputs the only root of* $f(x)$;
– *If* $\gcd(\ell_1, \ell_2) > 1$, *it outputs* $\perp$.

Algorithm 7. $\mathrm{Ext}(\mathsf{PP}, \mathrm{enc}(x))$ *converts the composable encoding* $\mathrm{enc}(x)$ *into the canonical encoding* $\mathrm{enc}^*(x)$. *The algorithm maintains a pair of lists* (U, V), *where* U *stores a list of* j-*invariants* $(j_1, \ldots, j_{|U|})$, V *stores a list of degrees where the* i^{th} *entry of* V *is the degree of isogeny between* j_i *and* j_{i-1} *(when* $i = 1$, j_{i-1} *is the* j_0 *in the public parameter). The lengths of* U *and* V *are always equal during the execution of the algorithm.*

The algorithm parses $\mathrm{enc}(x) = (L; T_1, \ldots, T_w)$, *proceeds as follows:*

1. *Initialization: Let* $U := T_1$, $V := (L_1, \ldots, L_1)$ *of length* $|T_1|$ *(i.e. copy* L_1 *for* $|T_1|$ *times).*
2. *For* $i = 2$ *to* w:
   (a) *Set* $u_{\mathsf{temp}} := |U|$.
   (b) *For* $k = 1$ *to* $|T_i|$:
       i. *Let* $t_{i,k,0}$ *be the* k^{th} j-*invariant in* T_i, *i.e.* $j_{i,k}$;
       ii. *For* $h = 1$ *to* u_{temp}:
           – *If* $k = 1$, *compute* $t_{i,k,h} := \mathrm{gcd.op}(\mathsf{PP}, L_i, V_h; t_{i,k,h-1}, U_h)$;
           – *If* $k > 1$, *compute* $t_{i,k,h} := \mathrm{gcd.op}(\mathsf{PP}, L_i, V_h; t_{i,k,h-1}, t_{i,k-1,h})$;
       iii. *Append* $t_{i,k,u_{\mathsf{temp}}}$ *to the list* U, *append* L_i *to the list* V.
3. *Return the last entry of* U.

Example 1. Let us give a simple example for the composition and the extraction algorithms. Let ℓ, m, n be three distinct polynomially large primes. Let the composable encoding of an element y be $\mathrm{enc}(y) = ((\ell); (j_{1,1}, j_{1,2}, j_{1,3}))$,

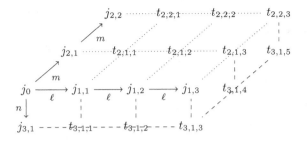

Fig. 4. An example for the composable encoding and the extraction algorithm.

based on the factorization of $y = C_1^{e_1} = [(\ell, b_\ell, \cdot)]^3$. Let the composable encoding of an element z be $\mathsf{enc}(z) = ((m, n); (j_{2,1}, j_{2,2}), (j_{3,1}))$, based on the factorization of $z = C_2^{e_2} \cdot C_3^{e_3} = [(m, b_m, \cdot)]^2 \cdot [(n, b_n, \cdot)]^1$. Then the composable encoding of $x = y \circ z$ obtained from Algorithm 5 is $\mathsf{enc}(x) = ((\ell, m, n); (j_{1,1}, j_{1,2}, j_{1,3}), (j_{2,1}, j_{2,2}), (j_{3,1}))$.

Next we explain how to extract the canonical encoding of x from $\mathsf{enc}(x)$. In Fig. 4, the j-invariants in $\mathsf{enc}(x)$ are placed on the solid arrows (their positions do not follow the relative positions on the volcano). We can think of each gcd operation in Algorithm 6 as fulfilling a missing vertex of a parallelogram defined by three existing vertices.

When running $\mathsf{Ext}(\mathsf{PP}, \mathsf{enc}(x))$, the list U is initialized as $(j_{1,1}, j_{1,2}, j_{1,3})$, the list V is initialized as (ℓ, ℓ, ℓ). Let us go through the algorithm for $i = 2$ and $i = 3$ in the second step.

- When $i = 2$, u_{temp} equals to $|U| = 3$. The j-invariants $\{t_{2,k,h}\}_{k \in [|T_2|], h \in [u_{\mathsf{temp}}]}$ are placed on the dotted lines, computed in the order of $t_{2,1,1}$, $t_{2,1,2}$, $t_{2,1,3}$, $t_{2,2,1}$, $t_{2,2,2}$, $t_{2,2,3}$. The list U is updated to $(j_{1,1}, j_{1,2}, j_{1,3}, t_{2,1,3}, t_{2,2,3})$, the list V is updated to (ℓ, ℓ, ℓ, m, m)
- When $i = 3$, u_{temp} equals to $|U| = 5$. The j-invariants $\{t_{3,1,h}\}_{h \in [u_{\mathsf{temp}}]}$ are placed on the dashed lines, computed in the order of $t_{3,1,1}, \ldots, t_{3,1,5}$. In the end, $t_{3,1,5}$ is appended to U, n is appended to V.

The canonical encoding of x is then $t_{3,1,5}$.

On Correctness and Efficiency. We now verify the correctness and efficiency of the parameter generation, encoding sampling, composition, and the extraction algorithms.

To begin with, we verify that the canonical encoding correctly and uniquely determines the group element in $\mathcal{CL}(\mathcal{O})$. It follows from the choices of the elliptic curves $E_0(\mathbb{F}_p)$ and $E_0(\mathbb{F}_q)$ with $\mathrm{End}(E_0(\mathbb{F}_p)) \simeq \mathrm{End}(E_0(\mathbb{F}_q)) \simeq \mathcal{O}$, and the following bijection once we fix E_0:

$$\mathcal{CL}(\mathcal{O}) \to \mathrm{Ell}_{\mathcal{O}}(\mathbf{k}), \quad x \mapsto x * j(E_0(\mathbf{k})), \text{ for } \mathbf{k} \in \{\mathbb{F}_p, \mathbb{F}_q\}$$

Next, we will show that generating the parameters, i.e. the curves $E_{0, \mathbb{F}_p}$, $E_{0, \mathbb{F}_q}$ with a given fundamental discriminant D_0 and a conductor $f = \prod_i^k f_i$, is efficient when $|D_0|$ and all the factors of f are of polynomial size. Let u be an integer such that $f \mid u$. Choose a p and t_p such that $t_p^2 - 4p = u^2 D_0$. Then, compute the Hilbert class polynomial H_{D_0} over $\mathbb{F}_p$ and find one of its roots j. From j, descending on the volcanoes $G_{f_i}(\mathbb{F}_p)$ for every f_i gives the j-invariant for the curve with desired discriminant. The same construction works verbatim for q.

We then show that sampling the composable encodings can be done in polynomial time heuristically:

1. Given a logarithmically large set $S = \{C_i = [(\ell_i, b_i, \cdot)] \in \mathcal{CL}(\mathcal{O})\}_{i \in [w]}$, a possibly big basis of the relation lattice Λ_S can be obtained by solving the discrete-log problem over $\mathcal{CL}(\mathcal{O})$, which can be done in polynomial time since the group order is polynomially smooth.

2. Suppose that the lattice Λ_S satisfies the Gaussian heuristic (this is the only heuristic we assume). That is, for all $1 \leq i \leq w$, the ith successive minimum of Λ_S, denoted as λ_i, satisfies $\lambda_i \approx \sqrt{w} \cdot h(\mathcal{O})^{1/w} \in \mathsf{poly}(\lambda)$. Since $w = O(\log(\lambda))$, the short basis $\mathbf{B}$ of Λ_S, produced by the LLL algorithm, satisfies $\|\mathbf{B}\| \leq 2^{\frac{w}{2}} \cdot \lambda_w \in \mathsf{poly}(\lambda)$.

3. Given a target group element $x \in \mathcal{CL}(\mathcal{O})$, the polynomially short basis $\mathbf{B}$, we can sample a vector $\mathbf{e} \in \mathbb{N}^w$ such that $\prod_{i=1}^m C_i^{e_i} = x$ and $\|\mathbf{e}\|_1 \in \mathsf{poly}(\lambda)$ in polynomial time using e.g. Babai's algorithm [1]. (In Sect. 5.3, we will explain that the GPV sampler [17] is preferred for the security purpose.)

4. The unit operation $\mathsf{act}(\tau, j, C)$ is efficient when the ideal class C corresponds to a polynomial degree isogeny, since it is efficient to compute polynomial degree isogenies over the finite fields.

5. The length of the final output $\mathsf{enc}(x)$ is $(w + \|\mathbf{e}\|_1) \cdot \mathsf{poly}(\lambda) \in \mathsf{poly}(\lambda)$.

The algorithm $\mathsf{Compose}(\mathsf{PP}, \mathsf{enc}(x), \mathsf{enc}(y))$ concatenates $\mathsf{enc}(x)$, $\mathsf{enc}(y)$, so it is efficient as long as $\mathsf{enc}(x)$, $\mathsf{enc}(y)$ are of polynomial size.

The correctness of the unit operation $\mathsf{gcd.op}$ follows the commutativity of the endomorphism ring $\mathcal{O}$. The operation $\mathsf{gcd.op}(\mathsf{PP}, \ell_1, \ell_2; j_1, j_2)$ is efficient when $\gcd(\ell_1, \ell_2) = 1$, $\ell_1, \ell_2 \in \mathsf{poly}(\lambda)$, given that solving the (ℓ_1, ℓ_2) isogenous neighbor problem over $\mathbb{Z}/N\mathbb{Z}$ is efficient under these conditions.

When applying $\mathsf{Ext}()$ (Algorithm 7) on a composable encoding $\mathsf{enc}(x) = (L_x; T_{x,1}, \ldots, T_{x,w_x})$, it runs $\mathsf{gcd.op}$ for $\max_{i=1}^{w_x} |T_{x_i}| \cdot (\sum_{i=1}^{w_x} |T_{x_i}|)$ times. So obtaining the canonical encoding is efficient as long as all the primes in L_x are polynomially large, and $|T_{x,i}| \in \mathsf{poly}(\lambda)$ for all $i \in [w_x]$.

5   Cryptanalysis

We provide a highlight of the cryptanalytic attempts we have made and discuss the impacts and the countermeasures. The details of our cryptanalysis attempts can be found in the full version.

The security of our cryptosystem relies on the conjectured hardness of solving various problems over $\mathbb{Z}/N\mathbb{Z}$ without knowing the factors of N. So we start from the feasibility of performing several individual computational tasks over $\mathbb{Z}/N\mathbb{Z}$;

then focus on the (ℓ, ℓ^2)-isogenous neighbor problem over $\mathbb{Z}/N\mathbb{Z}$, whose hardness is necessary for the security of our candidate TGII; finally address all the other attacks in the TGII construction.

5.1 The (in)feasibility of Performing Computations over $\mathbb{Z}/N\mathbb{Z}$

Factoring Polynomials over $\mathbb{Z}/N\mathbb{Z}$. The task of finding roots of polynomials of degree $d \geq 2$ over $\mathbb{Z}/N\mathbb{Z}$ sits in the subroutines of many potential algorithms we need to consider, so let us begin with a clarification on the status of this problem. No polynomial time algorithm is known to solve this problem in general. In a few special cases, finding at least one root is feasible. For example, if a root of a polynomial over $\mathbb{Z}/N\mathbb{Z}$ is known to be the same as the root over $\mathbb{Q}$, then we can use LLL [26]; or if a root is known to be smaller than roughly $O(N^{1/d})$, then Coppersmith-type algorithms can be used to find such a root [6]. However, these families of polynomials only form a negligible portion of all the polynomials with polynomially bounded degrees.

Feasible Information from a Single j-invariant. From any $j \in \mathbb{Z}/N\mathbb{Z}, j \neq 0, 1728$, we can find the coefficients a and b of the Weierstrass form of an elliptic curve $E(\mathbb{Z}/N\mathbb{Z})$ with $j(E) = j$ by computing $a = 3j(1728-j), b = 2j(1728-j)^2$. But choosing a curve over $\mathbb{Z}/N\mathbb{Z}$ with a given j-invariant together with a point on the curve seems tricky. Nevertheless, it is always feasible to choose a curve together with the x-coordinate of a point on it, since a random $x \in \mathbb{Z}/N\mathbb{Z}$ is the x-coordinate of some point on the curve with probability roughly $\frac{1}{2}$. It is also known that computing the multiples of a point P over $E(\mathbb{Z}/N\mathbb{Z})$ is feasible solely using the x-coordinate of P (cf. [10]). The implication of this is that we should at the very least not give out the group orders of the curves involved in the scheme. More precisely, we should avoid the j-invariants corresponding to curves (or their twists) with polynomially smooth cardinalities over either $\mathbb{F}_p$ or $\mathbb{F}_q$. Otherwise Lenstra's algorithm [27] can be used to factorize N.

In our application we also assume that the endomorphism rings of $E(\mathbb{F}_p)$ and $E(\mathbb{F}_q)$ are isomorphic and not given out (the reason will be explained later). Computing the discriminant of $\mathcal{O} \simeq \mathrm{End}(E(\mathbb{F}_p)) \simeq \mathrm{End}(E(\mathbb{F}_q))$ or the number of points of E over $\mathbb{Z}/N\mathbb{Z}$ seems to be hard given only N and a j-invariant. In fact Kunihiro and Koyama (and others) have reduced factorizing N to computing the number of points of general elliptic curves over $\mathbb{Z}/N\mathbb{Z}$ [24]. However, these reductions are not efficient in the special case, where the endomorphism rings of $E(\mathbb{F}_p)$ and $E(\mathbb{F}_q)$ are required to be isomorphic. So, the result of [24] can be viewed as evidence that the polynomial time algorithms for counting points on elliptic curves over finite fields may fail over $\mathbb{Z}/N\mathbb{Z}$ without making use of the fact that the endomorphism rings of $E(\mathbb{F}_p)$ and $E(\mathbb{F}_q)$ are isomorphic.

Let ℓ be a prime. We will be concerned with degree ℓ isogenies. If we are only given a single j-invariant $j_1 \in \mathbb{Z}/N\mathbb{Z}$, then finding an integer j_2 such that $\Phi_\ell(j_1, j_2) = 0 \pmod{N}$ seems hard. Nevertheless, we remark that Theorem 1 does not guarantee that finding j_2 is as hard as factoring when the endomorphism rings of $E(\mathbb{F}_p)$ and $E(\mathbb{F}_q)$ are isomorphic. However, as of now, we do not know how to make use of the condition that the endomorphism rings are isomorphic to mount an attack on the problem.

Feasible Information from More j-invariants. In the construction of a TGII we are not only given a single j-invariant, but many j-invariants with each neighboring pair of them satisfying the ℓth modular polynomial, a polynomial degree $\ell + 1$. We will study what other information can be extracted from these neighboring j-invariants.

In Proposition 1, we have explained that given two integers $j_1, j_2 \in \mathbb{Z}/N\mathbb{Z}$ such that $\Phi_\ell(j_1, j_2) = 0$ over $\mathbb{Z}/N\mathbb{Z}$, the elliptic curves E_1, E_2, and the kernel polynomial $h(x)$ of an isogeny ϕ from E_1, E_2 can be computed in time polynomial in $\ell, \log(N)$. However, it is not clear how to use the explicit expression of ϕ to break factoring or solve the inversion problem.

A natural next step is to recover a point in the kernel of ϕ, but it is also not clear how to recover even the x-coordinate of a point in the kernel when $\ell \geq 5$. For $\ell = 3$, on the other hand, the kernel polynomial does reveal the x-coordinate of a point P in the kernel $G \subset E_1[3]$ (note that $h(\cdot)$ is of degree 1 in this particular case). But revealing the x-coordinate of a point $P \in E_1[3]$ does not immediately break factoring, since $3P$ is O over both $\mathbb{F}_p$ and $\mathbb{F}_q$. At this moment we do not know of a full attack from a point in $\ker(\phi)$. Nevertheless, we still choose to take an additional safeguard by avoiding the use of 3-isogenies since it reveals the x-coordinate of a point in $E_1[3]$, and many operations on elliptic curves are feasible given the x-coordinate of a point.

5.2   Tackling the (ℓ, ℓ^2)-Isogenous Neighbor Problem over $\mathbb{Z}/N\mathbb{Z}$

The (ℓ, ℓ^2)-isogenous neighbor problem is essential to the hardness of inversion in our TGII construction. In addition to Definition 5, we assume that the endomorphism rings of the curves in the problem are isomorphic to an imaginary quadratic order $\mathcal{O}$.

The Hilbert Class Polynomial Attack. We first note that the discriminant D of the underlying endomorphism ring $\mathcal{O}$ cannot be polynomially large, otherwise we can compute the Hilbert class polynomial H_D in polynomial time and therefore solve the (ℓ, ℓ^2)-isogenous neighbor problem. Given j_0, j_1 such that $\Phi_\ell(j_0, j_1) = 0$, compute the polynomial $\gamma(x)$,

$$\gamma(x) := \gcd(\Phi_\ell(j_0, x), \Phi_{\ell^2}(j_1, x), H_D(x)) \in (\mathbb{Z}/N\mathbb{Z})[x].$$

The gcd of $\Phi_\ell(j_0, x)$ and $\Phi_{\ell^2}(j_1, x)$ gives a polynomial of degree ℓ. The potential root they share with $H_D(x)$ is the only one with the same endomorphism ring with j_0 and j_1, which is j_{-1}. So $\gamma(x)$ is a linear function.

Survey of the Ionica-Joux Algorithm. Among the potential solutions to the (ℓ, ℓ^2)-isogenous neighbor problem, finding the one corresponding to the image of a horizontal isogeny would break our candidate group with infeasible inversion, so it is worth investigating algorithms which find isogenies with specific directions. However, the only known such algorithm over the finite fields, that of Ionica and Joux [19], does not seem to work over $\mathbb{Z}/N\mathbb{Z}$. In the full version we provide a detailed survey of this algorithm.

More About Modular Curves and Characteristic Zero Attacks. Given j, solving $\Phi_\ell(j, x)$ is not the only way to find the j-invariants of the ℓ-isogenous curves. Alternative complex analytic (i.e. characteristic zero) methods have been discussed, for instance, in [12, Section 3]. However, these methods all involve solving polynomials of degree ≥ 2 to get started.

As mentioned in Sect. 2.2, the curve $\mathbb{H}/\Gamma_0(\ell)$ parameterizes pairs of elliptic curves over $\mathbb{C}$ related by a cyclic ℓ-isogeny. The (ℓ, ℓ^2)-isogenous neighbor problem, on the other hand, concerns curves that are horizontally ℓ-isogenous, i.e. ℓ-isogenous and have the same endomorphism ring. To avoid an attack through characteristic zero techniques, we make sure that there is no immediate quotient of $\mathbb{H}$ that parametrizes curves which are related with an ℓ-isogeny and have the same endomorphism ring. Below, we first go over the well-known moduli description of modular curves to make sure that they don't lead to an immediate attack, and then show that there is indeed no quotient of $\mathbb{H}$ between $\mathbb{H}/SL_2(\mathbb{Z})$ and $\mathbb{H}/\Gamma_0(\ell)$, so we don't have to worry about possible attacks on that end.

Let $\Gamma := SL_2(\mathbb{Z})$, and let $\Gamma(\ell)$ and $\Gamma_1(\ell)$ denote the congruence subgroups,

$$\Gamma(\ell) := \left\{ \begin{pmatrix} a & b \\ c & d \end{pmatrix} \in SL_2(\mathbb{Z}) \middle| \begin{pmatrix} a & b \\ c & d \end{pmatrix} \equiv \begin{pmatrix} 1 & 0 \\ 0 & 1 \end{pmatrix} \pmod{\ell} \right\},$$

$$\Gamma_1(\ell) := \left\{ \begin{pmatrix} a & b \\ c & d \end{pmatrix} \in SL_2(\mathbb{Z}) \middle| \begin{pmatrix} a & b \\ c & d \end{pmatrix} \equiv \begin{pmatrix} 1 & * \\ 0 & 1 \end{pmatrix} \pmod{\ell} \right\}.$$

It is well-known that the curves $\mathbb{H}/\Gamma_1(\ell)$ and $\mathbb{H}/\Gamma(\ell)$ parametrize elliptic curves with extra data on their ℓ-torsion (cf. [23]). $\mathbb{H}/\Gamma_1(\ell)$ parametrizes (E, P), where P is a point on E having order exactly ℓ, and $\mathbb{H}/\Gamma(\ell)$ parametrizes triples (E, P, Q), where $E[\ell] = \langle P, Q \rangle$ and they have a fixed Weil pairing. These curves carry more information than the ℓ-isogenous relation and they are not immediately helpful for solving the (ℓ, ℓ^2)-isogenous neighbor problem.

As for the quotients between $\mathbb{H}/SL_2(\mathbb{Z})$ and $\mathbb{H}/\Gamma_0(\ell)$, the following lemma shows that there are indeed none.

Lemma 4. *Let ℓ be a prime. If $H \leq \Gamma$ is such that $\Gamma_0(\ell) \leq H \leq \Gamma$, then either $H = \Gamma_0(\ell)$ or $H = \Gamma$.*

Proof. Let $\sigma_1 = \left(\begin{smallmatrix} 1 & 1 \\ 0 & 1 \end{smallmatrix} \right)$, $\sigma_2 = \left(\begin{smallmatrix} 1 & 0 \\ 1 & 1 \end{smallmatrix} \right)$, $\sigma_3 = \sigma_1 \sigma_2^{-1}$, and recall that $SL_2(\mathbb{Z}/\ell\mathbb{Z}) = \langle \sigma_1, \sigma_2 \rangle = \langle \sigma_1, \sigma_3 \rangle$. Recall that the natural projection $\pi : \Gamma \to SL_2(\mathbb{Z}/\ell\mathbb{Z})$ is surjective. Assume that $H \neq \Gamma_0(\ell)$. This implies that $\pi(H) = SL_2(\mathbb{Z}/\ell\mathbb{Z})$ (we shall give a proof below). Assuming this claim for the moment let $g \in \Gamma \backslash H$.

Since $\pi(\Gamma) = \pi(H)$ there exists $h \in H$ such that $\pi(g) = \pi(h)$. Therefore, $gh^{-1} \in \ker(\pi) = \Gamma(\ell) \subset H$. Therefore, $g \in H$ and $\Gamma = H$.

To see that $\pi(\Gamma) = \pi(H)$, first note that since $\Gamma_0(\ell) \subset H$ we have all the upper triangular matrices in $\pi(H)$. Next, let $h = \left(\begin{smallmatrix} h_1 & h_2 \\ h_3 & h_4 \end{smallmatrix}\right) \in H\backslash\Gamma_0(\ell)$ such that $\pi(h) = \left(\begin{smallmatrix} \bar{h}_1 & \bar{h}_2 \\ \bar{h}_3 & \bar{h}_4 \end{smallmatrix}\right) \in \pi(H)\backslash\pi(\Gamma_0(\ell))$ (note that this difference is non-empty since otherwise $\Gamma_0(\ell) = H$).

We have two cases depending on $\bar{h}_1 = 0$ or not. If $\bar{h}_1 = 0$ then $\bar{h}_3 \neq 0$ and $\sigma_3 = \left(\begin{smallmatrix} \bar{h}_3^{-1} & \bar{h}_4 \\ 0 & \bar{h}_3 \end{smallmatrix}\right)\bar{h}^{-1} \in \pi(H)$. On the other hand, if $\bar{h}_1 \neq 0$ multiplying on the right by $\left(\begin{smallmatrix} \bar{h}_1^{-1} & -\bar{h}_2 \\ 0 & \bar{h}_1 \end{smallmatrix}\right) \in \pi(H)$ we see that $\left(\begin{smallmatrix} 1 & 0 \\ \bar{h}_3\bar{h}_1^{-1} & 1 \end{smallmatrix}\right) \in \pi(H)$. For any integer m, the m'th power of this matrix is $\left(\begin{smallmatrix} 1 & 0 \\ m\bar{h}_3\bar{h}_1^{-1} & 1 \end{smallmatrix}\right) \in \pi(H)$. Taking $m \equiv \bar{h}_1\bar{h}_3^{-1}$ shows that $\sigma_2 \in \pi(H)$. This shows that $\pi(H) = SL_2(\mathbb{Z}/\ell\mathbb{Z})$.

5.3   Cryptanalysis of the Candidate Group with Infeasible Inversion

We now cryptanalyze the concrete candidate TGII. Recall the format of an encoding of a group element x from Definition 7:

$$\mathsf{enc}(x) = (L_x; T_{x,1}, \ldots, T_{x,w_x})$$
$$= ((p_{x,1}, \ldots, p_{x,w_x}); (j_{x,1,1}, \ldots, j_{x,1,e_{x,1}}), \ldots, (j_{x,w_x,1}, \ldots, j_{x,w_x,e_{x,w_x}})).$$

The "exponent vector" $\mathbf{e}_x \in \mathbb{Z}^{w_x}$ can be read from the encoding as $\mathbf{e}_x = (|T_{x,1}|, \ldots, |T_{x,w_x}|)$.

We assume polynomially many composable encodings are published in the applications of a TGII. In down-to-earth terms it means the adversary is presented with polynomially many j-invariants on the crater of a volcano, and the explicit isogenies between each pair of the neighboring j-invariants (due to Proposition 1).

We will be considering the following model on the adversary's attacking strategy.

Definition 9 (The GCD attack model). *In the GCD attack model, the adversary is allowed to try to find the inverse of a target group element only by executing the unit gcd operation* $\mathsf{gcd.op}(\mathsf{PP}, \ell_1, \ell_2; j_1, j_2)$ *given in Algorithm 6 for polynomially many steps, where* $\ell_1, \ell_2; j_1, j_2$ *are from the published encodings or obtained from the previous executions of the gcd evaluations.*

We do not know how to prove the construction of TGII is secure even if the adversary is restricted to attack in the GCD model. Our cryptanalysis attempts can be classified as showing (1) how to prevent the attacks that obey the GCD evaluation law; (2) how to prevent the other attack approaches (by e.g. guessing the class group invariants).

Preventing the Trivial Leakage of Inverses. In applications we are often required to publish the encodings of elements that are related in some way.

A typical case is the following: for $x, y \in \mathcal{CL}(\mathcal{O})$, the scheme may require publishing the encodings of x and $z = y \circ x^{-1}$ without revealing a valid encoding of x^{-1}. As a toy example, let $x = [(p_x, b_x, \cdot)]$, $y = [(p_y, b_y, \cdot)]$, where p_x and p_y are distinct primes. Let j_0, the j-invariant of a curve E_0, represent the identity element in the public parameter. Let $((p_x); (j_x))$ be a composable encoding of x and $((p_y); (j_y))$ be a composable encoding of y.

Naively, a composable encoding of $z = y \circ x^{-1}$ could be $((p_x, p_y); (j_{x^{-1}}), (j_y))$, where $j_{x^{-1}}$ is the j-invariant of $E_{x^{-1}} = x^{-1} E_0$. Note, however, that $((p_x); (j_{x^{-1}}))$ is a valid encoding of x^{-1}. In other words such an encoding of $y \circ x^{-1}$ trivially reveals the encoding of x^{-1}.

One way of generating an encoding of $z = y \circ x^{-1}$ without trivially revealing $j_{x^{-1}}$ is to first pick a generator set of ideals where the norms of the ideals are coprime to p_x and p_y, then solve the discrete-log of z over these generators to compute the composable encoding. This is the approach we take in this paper.

Parallelogram Attack. In the applications we are often required to publish the composable encodings of group elements a, b, c such that $a \circ b = c$. If the degrees of the three encodings are coprime, then we can recover the encodings of a^{-1}, b^{-1}, and c^{-1} using the following "parallelogram attack". This is a non-trivial attack which obeys the gcd evaluation law in Definition 9.

Let us illustrate the attack via the examples in Fig. 5, where the solid arrows represent the isogenies that are given as the inputs (the j-invariants of the target curves are written at the head of the arrows, their positions do not follow the relative positions on the volcano; the degree of the isogeny is written on the arrow); the dashed lines and the j-invariants on those lines are obtained from the gcd evaluation law.

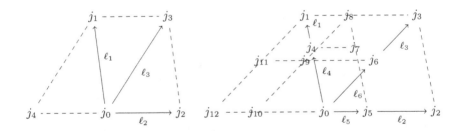

Fig. 5. The parallelogram attack.

For simplicity let us first look at the example in the left figure. Let composable encodings of a, b, c be given by $(\ell_1; (j_1))$, $(\ell_2; (j_2))$, $(\ell_3; (j_3))$, where ℓ_1, ℓ_2, ℓ_3 are polynomial and pairwise coprime. A composable encoding of b^{-1} then can be written as $(\ell_2; (j_4))$, where j_4 is the root of the linear equation $f(x) = \gcd(\Phi_{\ell_3}(j_1, x), \Phi_{\ell_2}(j_0, x))$. This is due to the relation $c \circ b^{-1} = a$, which, in particular implies that j_1 and j_4 are connected by an isogeny of degree ℓ_3.

The simple attack above uses the fact that the degrees of the entire encodings of a, b and c are polynomial. Let us use the example in the right figure to illustrate that even if the encodings are composed of many polynomial degree isogenies (so that the total degrees may be super-polynomial), the attack is still effective. The idea is to view the composition law as filling the missing edges of a parallelogram given the j-invariants on a pair of adjacent edges. The final goal is to find the missing corner j_{12} in the parallelogram $j_0 - j_3 - j_1 - j_{12}$. To arrive there we need the j-invariants on a pair of adjacent edges to begin with, so we first have to fill the j-invariants on, for instance, the edge $j_1 - j_3$. Therefore, as the first step, we consider the parallelogram $j_0 - j_2 - j_3 - j_1$. To fill the j-invariants on the edge $j_1 - j_3$, we first compute j_7 as the root of $f_7(x) = \gcd(\Phi_{\ell_4}(j_5, x), \Phi_{\ell_5}(j_4, x))$, then compute j_8 as the root of $f_8(x) = \gcd(\Phi_{\ell_1}(j_7, x), \Phi_{\ell_5}(j_1, x))$ (the polynomials f_7, f_8 are linear since the degrees of $\mathsf{enc}(a)$ and $\mathsf{enc}(b)$ are coprime). In the second step, we consider the parallelogram $j_0 - j_3 - j_1 - j_{12}$. To find j_{12} we use the gcd evaluation law to find j_9, j_{10}, j_{11}, j_{12} one-by-one (using the condition that the degrees of $\mathsf{enc}(c)$ and $\mathsf{enc}(b)$ are coprime).

The parallelogram attack is very powerful, in the sense that it is not preventable when application requires to publish the composable encodings of a, b, c such that $a \circ b = c$, and $\mathsf{enc}(a)$, $\mathsf{enc}(b)$, $\mathsf{enc}(c)$ to be pairwise composable. However, the parallelogram attack does not seem to work when 2 out of the 3 pairs of $\mathsf{enc}(a)$, $\mathsf{enc}(b)$ and $\mathsf{enc}(c)$ are not composable. In the applications of directed transitive signature and broadcast encryption, there are encodings of a, b, c such that $a \circ b = c$. Luckily, only one pair of the encodings among the three has to be composable to provide the necessary functionalities of these applications.

Hiding the Class Group Invariants. In the applications of TGII, it is reasonable to assume that the closure of the gcd-compositions of the published j-invariants covers all the $h(D)$ j-invariants. So inverting a group element can be done by solving the discrete-log problem over $\mathcal{CL}(D)$. However, the class number $h(D)$ is polynomially smooth, so the discrete-log problem over $\mathcal{CL}(D)$ can be solved in polynomial time once $h(D)$ is given, and $h(D)$ can be recovered from D or any basis of a relation lattice of $\mathcal{CL}(D)$. So we do need to hide the discriminant D, the class number $h(D)$, and any lattice Λ defined above. In the full version, we describe the details of how to hide these class group invariants.

Let us remark that if self-composition of an encoding is feasible, then one can efficiently guess all the polynomially smooth factors of $h(D)$. However for our construction self-composition is infeasible, due to the hardness of the (ℓ, ℓ^2)-isogenous neighbor problem. Nevertheless, one can still attack by first guessing D or $h(D)$, which takes $\lambda^{O(\log \lambda)}$ time according to the current setting of parameter.

Acknowledgments. The research of Salim Ali Altuğ is supported by the grant DMS-1702176. The research of Yilei Chen was conducted at Boston University supported by the NSF MACS project and NSF grant CNS-1422965.

References

1. Babai, L.: On lovász' lattice reduction and the nearest lattice point problem. Combinatorica **6**(1), 1–13 (1986)
2. Barak, B., et al.: On the (im)possibility of obfuscating programs. In: Kilian, J. (ed.) CRYPTO 2001. LNCS, vol. 2139, pp. 1–18. Springer, Heidelberg (2001). https://doi.org/10.1007/3-540-44647-8_1
3. Buchmann, J.A., Williams, H.C.: A key-exchange system based on imaginary quadratic fields. J. Cryptology **1**(2), 107–118 (1988)
4. Charles, D.X., Lauter, K.E., Goren, E.Z.: Cryptographic hash functions from expander graphs. J. Cryptology **22**(1), 93–113 (2009)
5. Cohen, H.: A Course in Computational Algebraic Number Theory. Graduate Texts in Mathematics. Springer, Heidelberg (1995). https://doi.org/10.1007/978-3-662-02945-9
6. Coppersmith, D.: Small solutions to polynomial equations, and low exponent RSA vulnerabilities. J. Cryptology **10**(4), 233–260 (1997)
7. Couveignes, J.-M.: Hard homogeneous spaces. Cryptology ePrint Archive, Report 2006/291 (2006)
8. Couveignes, J.-M., Morain, F.: Schoof's algorithm and isogeny cycles. In: Adleman, L.M., Huang, M.-D. (eds.) ANTS 1994. LNCS, vol. 877, pp. 43–58. Springer, Heidelberg (1994). https://doi.org/10.1007/3-540-58691-1_42
9. Cox, D.A.: Primes of the Form $x^2 + ny^2$: Fermat, Class Field Theory, and Complex Multiplication, vol. 34. Wiley, Hoboken (2011)
10. Demytko, N.: A new elliptic curve based analogue of RSA. In: Helleseth, T. (ed.) EUROCRYPT 1993. LNCS, vol. 765, pp. 40–49. Springer, Heidelberg (1994). https://doi.org/10.1007/3-540-48285-7_4
11. Diffie, W., Hellman, M.E.: New directions in cryptography. IEEE Trans. Inf. Theory **22**(6), 644–654 (1976)
12. Elkies, N.D., et al.: Elliptic and modular curves over finite fields and related computational issues. AMS IP Stud. Adv. Math. **7**, 21–76 (1998)
13. Enge, A., Sutherland, A.V.: Class invariants by the CRT method. In: Hanrot, G., Morain, F., Thomé, E. (eds.) ANTS 2010. LNCS, vol. 6197, pp. 142–156. Springer, Heidelberg (2010). https://doi.org/10.1007/978-3-642-14518-6_14
14. De Feo, L.: Mathematics of isogeny based cryptography. arXiv preprint arXiv:1711.04062 (2017)
15. Fouquet, M., Morain, F.: Isogeny volcanoes and the SEA algorithm. In: Fieker, C., Kohel, D.R. (eds.) ANTS 2002. LNCS, vol. 2369, pp. 276–291. Springer, Heidelberg (2002). https://doi.org/10.1007/3-540-45455-1_23
16. Garg, S., Gentry, C., Halevi, S., Raykova, M., Sahai, A., Waters, B.: Candidate indistinguishability obfuscation and functional encryption for all circuits. In: FOCS, pp. 40–49. IEEE Computer Society (2013)
17. Gentry, C., Peikert, C., Vaikuntanathan, V.: Trapdoors for hard lattices and new cryptographic constructions. In: STOC, pp. 197–206 (2008)
18. Hohenberger, S.R.: The cryptographic impact of groups with infeasible inversion. Master's thesis, Massachusetts Institute of Technology (2003)
19. Ionica, S., Joux, A.: Pairing the volcano. Math. Comput. **82**(281), 581–603 (2013)
20. Irrer, J., Lokam, S., Opyrchal, L., Prakash, A.: Infeasible group inversion and broadcast encryption. University of Michigan Electrical Engineering and Computer Science Tech Note CSE-TR-485-04 (2004)

21. Jao, D., De Feo, L.: Towards quantum-resistant cryptosystems from supersingular elliptic curve isogenies. In: Yang, B.-Y. (ed.) PQCrypto 2011. LNCS, vol. 7071, pp. 19–34. Springer, Heidelberg (2011). https://doi.org/10.1007/978-3-642-25405-5_2
22. Koblitz, N.: Elliptic curve cryptosystems. Math. Comput. **48**(177), 203–209 (1987)
23. Kohel, D.R.: Endomorphism rings of elliptic curves over finite fields. PhD thesis, University of California, Berkeley (1996)
24. Kunihiro, N., Koyama, K.: Equivalence of counting the number of points on elliptic curve over the ring Zn and factoring n. In: Nyberg, K. (ed.) EUROCRYPT 1998. LNCS, vol. 1403, pp. 47–58. Springer, Heidelberg (1998). https://doi.org/10.1007/BFb0054116
25. Lay, G.-J., Zimmer, H.G.: Constructing elliptic curves with given group order over large finite fields. In: Adleman, L.M., Huang, M.-D. (eds.) ANTS 1994. LNCS, vol. 877, pp. 250–263. Springer, Heidelberg (1994). https://doi.org/10.1007/3-540-58691-1_64
26. Lenstra, A.K., Lenstra, H.W., Lovász, L.: Factoring polynomials with rational coefficients. Math. Ann. **261**(4), 515–534 (1982)
27. Lenstra, H.W.: Factoring integers with elliptic curves. Ann. Math. **126**(3), 649–673 (1987)
28. McCurley, K.S.: Cryptographic key distribution and computation in class groups. IBM Thomas J. Watson Research Division (1988)
29. Miller, V.S.: Use of elliptic curves in cryptography. In: Williams, H.C. (ed.) CRYPTO 1985. LNCS, vol. 218, pp. 417–426. Springer, Heidelberg (1986). https://doi.org/10.1007/3-540-39799-X_31
30. Molnar, D.: Homomorphic signature schemes. B.s. thesis, Harvard College (2003)
31. Alexander Rostovtsev and Anton Stolbunov. Public-key cryptosystem based on isogenies. Cryptology ePrint Archive, Report 2006/145 (2006)
32. Schoof, R.: Elliptic curves over finite fields and the computation of square roots mod p. Math. Comput. **44**(170), 483–494 (1985)
33. Schoof, R.: Counting points on elliptic curves over finite fields. J. Théor. Nombres Bordeaux **7**(1), 219–254 (1995)
34. Silverman, J.H.: The Arithmetic of Elliptic Curves. GTM, vol. 106. Springer, New York (2009). https://doi.org/10.1007/978-0-387-09494-6
35. Silverman, J.H.: Advanced Topics in the Arithmetic of Elliptic Curves. Graduate Texts in Mathematics, vol. 151. Springer, New York (2013). https://doi.org/10.1007/978-1-4612-0851-8
36. Sutherland, A.V.: Isogeny kernels and division polynomials. https://ocw.mit.edu/courses/mathematics/18-783-elliptic-curves-spring-2017/lecture-notes/MIT18_783S17_lec6.pdf. Accessed 03 Sept 2018
37. Sutherland, A.V.: Isogeny volcanoes. Open Book Ser. **1**(1), 507–530 (2013)
38. Tate, J.: Endomorphisms of abelian varieties over finite fields. Inventiones Math. **2**(2), 134–144 (1966)
39. Vélu, J.: Isogénies entre courbes elliptiques. Comptes Rendus de l'Académie des Sciences de Paris **273**, 238–241 (1971)
40. Yamakawa, T., Yamada, S., Hanaoka, G., Kunihiro, N.: Self-bilinear map on unknown order groups from indistinguishability obfuscation and its applications. In: Garay, J.A., Gennaro, R. (eds.) CRYPTO 2014. LNCS, vol. 8617, pp. 90–107. Springer, Heidelberg (2014). https://doi.org/10.1007/978-3-662-44381-1_6

Multilinear Maps

On Kilian's Randomization of Multilinear Map Encodings

Jean-Sébastien Coron$^{(\boxtimes)}$ and Hilder V. L. Pereira

University of Luxembourg, Luxembourg City, Luxembourg

jean-sebastien.coron@uni.lu

Abstract. Indistinguishability obfuscation constructions based on matrix branching programs generally proceed in two steps: first apply Kilian's randomization of the matrix product computation, and then encode the matrices using a multilinear map scheme. In this paper we observe that by applying Kilian's randomization *after* encoding, the complexity of the best attacks is significantly increased for CLT13 multilinear maps. This implies that much smaller parameters can be used, which improves the efficiency of the constructions by several orders of magnitude.

As an application, we describe the first concrete implementation of multiparty non-interactive Diffie-Hellman key exchange secure against existing attacks. Key exchange was originally the most straightforward application of multilinear maps; however it was quickly broken for the three known families of multilinear maps (GGH13, CLT13 and GGH15). Here we describe the first implementation of key exchange that is resistant against known attacks, based on CLT13 multilinear maps. For $N = 4$ users and a medium level of security, our implementation requires 18 GB of public parameters, and a few minutes for the derivation of a shared key.

1 Introduction

Multilinear Maps and Indistinguishability Obfuscation. Since the breakthrough construction of Garg, Gentry and Halevi [GGH13a], cryptographic multilinear maps have shown amazingly powerful applications in cryptography, most notably the first plausible construction of program obfuscation [GGH+13b]. A multilinear map scheme encodes plaintext values $\{a_i\}$ into encodings $\{[a_i]\}$ such that the a_i's are hidden; only a restricted class of polynomials can then be evaluated over these encoded values; eventually one can determine whether the evaluation is zero or not, using the zero testing procedure of the multilinear map scheme.

The goal of program obfuscation is to hide secrets in arbitrary running programs. The first plausible construction of general program obfuscation was described by Garg, Gentry, Halevi, Raykova, Sahai and Waters (GGHRSW) in [GGH+13b], based on multilinear maps; the construction has opened many new research directions, because the notion of indistinguishability obfuscation (iO) has tremendous applications in cryptography [SW14]. Since the publication of the GGHRSW construction, many variants of GGHRSW have been described

© International Association for Cryptologic Research 2019
S. D. Galbraith and S. Moriai (Eds.): ASIACRYPT 2019, LNCS 11922, pp. 325–355, 2019.
https://doi.org/10.1007/978-3-030-34621-8_12

[MSW14, AGIS14, PST14, BGK+14, BMSZ16]. Currently there are essentially only three known candidate constructions of multilinear maps:

- **GGH13.** The first candidate construction of multilinear maps is based on ideal lattices [GGH13a]. Its security relies on the difficulty of the NTRU problem and the principal ideal problem (PIP) in certain number fields.
- **CLT13.** An analogous construction but over the integers was described in [CLT13], based on the DGHV fully homomorphic encryption scheme [DGHV10].
- **GGH15.** Gentry, Gorbunov and Halevi described another multilinear maps scheme [GGH15], based on the Learning With Errors (LWE) problem with encoding over matrices, and defined with respect to a directed acyclic graph.

However the security of multilinear maps is still poorly understood. The most important attacks against multilinear maps are "zeroizing attacks", which consist in using linear algebra to recover the secrets of the scheme from the encodings of zero. At Eurocrypt 2015, Cheon *et al.* described a devastating zeroizing attack against CLT13; when CLT13 is used to implement non-interactive multipartite Diffie-Hellman key exchange, the attack completely breaks the protocol [CHL+15]. The attack was also extended to encodings variants, where encodings of zero are not directly available [CGH+15]. The key-exchange protocol based on GGH13 was also broken by a zeroizing attack in [HJ16]. Finally, the Diffie-Hellman key exchange protocol under GGH15 was broken in [CLLT16], using an extension of the Cheon *et al.* zeroizing attack.

However, not all attacks against the above multilinear map schemes can be applied to indistinguishability obfuscation. While multipartite key exchange based on any of the three families of multilinear map schemes is broken, iO is not necessarily broken by zeroizing attacks, because of the particular structure that iO constructions induce on the computation of multilinear map encoded values. Namely, in iO constructions, no low-level encodings of zeroes are available, and the obfuscation of a matrix branching program can only produce zeroes at the last level, moreover when evaluated in a very specific way. However some partial attacks against iO constructions have already been described. In [CGH+15] it was shown how to break the GGHRSW branching-program obfuscator when instantiated using CLT13, when the branching program to be obfuscated has a very simple structure (input partition). For GGH13, Miles, Sahai and Zhandry introduced "annihilation attacks" [MSZ16] that can break many obfuscation schemes based on GGH13; however, the attack does not apply to the GGHRSW construction, because in GGHRSW the matrix program is embedded in a larger matrix with random entries (diagonal padding). In [CGH17], the authors showed how to break iO constructions under GGH13, using a variant of the input partitioning attack; the attack applies against the GGHRSW construction with diagonal padding. A new tensoring technique was introduced in [CLLT17] to break iO constructions for branching programs without the input partition structure. Finally, an attack against iO over GGH15 was described in [CVW18] based on computing the rank of a certain matrix.

Obfuscating Matrix Branching Programs. The GGHRSW construction and its variants consist of a "core component" for obfuscating matrix branching programs, and a bootstrapping procedure to obfuscate arbitrary programs based on the core component, using fully homomorphic encryption and proofs of correct computation. The core component relies on multilinear maps for evaluating a product of encoded matrices corresponding to a branching program, without revealing the underlying value of those matrices.

More precisely, the core component of the GGHRSW construction and its variants proceeds in two steps: first apply Kilian's randomization of the matrix product computation, and then encode the matrices using a multilinear map scheme. In this paper, our main observation is that for CLT13 multilinear maps, the complexity of the best attacks is significantly increased when Kilian's randomization is also applied *after* encoding. We note that applying Kilian's randomization "on the encoding side" was already used in GGH15 multilinear maps as an additional safeguard [GGH15, §5.1]. For CLT13 this implies that one can use much smaller parameters (noise and encoding size), which improves the efficiency of the constructions by several orders of magnitude.

More precisely, a matrix branching program BP of length n is evaluated on input $x \in \{0,1\}^\ell$ by computing:

$$C(x) = b_0 \times \prod_{i=1}^{n} B_{i,x_{\mathsf{inp}(i)}} \times b_{n+1} \tag{1}$$

where $\{B_{i,b}\}_{1 \le i \le n, b \in \{0,1\}}$ are square matrices and b_0 and b_{n+1} are bookend vectors; then $\mathsf{BP}(x) = 0$ if $C(x) = 0$, and $\mathsf{BP}(x) = 1$ otherwise. The function $\mathsf{inp}(i)$ indicates which bit of x is read at step i of the product matrix computation. To obfuscate a matrix branching program, the GGHRSW construction proceeds in two steps. First one randomizes the matrices $B_{i,b}$ as in Kilian's protocol [Kil88]: choose $n+1$ random invertible matrices $\{R_i\}_{i=0}^{n}$ and set $\tilde{B}_{i,b} = R_{i-1} B_{i,b} R_i^{-1}$, with also $\tilde{b}_0 = b_0 R_0^{-1}$ and $\tilde{b}_{n+1} = R_n b_{n+1}$. The randomized matrix branching program can then be evaluated by computing $C(x) = \tilde{b}_0 \times \prod_{i=1}^{n} \tilde{B}_{i,x_{\mathsf{inp}(i)}} \times \tilde{b}_{n+1}$. Namely the successive randomization matrices R_i cancel each other; therefore the matrix product computation evaluates to the same result as in (1).

The second step in the GGHRSW construction is to encode the entries of the matrices $\tilde{B}_{i,b}$ using a multilinear map scheme. Every entry of a given matrix is encoded separately; the bookend vectors $\tilde{b}_0$ and $\tilde{b}_n$ are also encoded similarly. Therefore one defines the matrices and vectors $\hat{B}_{i,b} = \mathsf{Encode}_{\{i+1\}}(\tilde{B}_{i,b}), \hat{b}_0 = \mathsf{Encode}_{\{1\}}(\tilde{b}_0), \hat{b}_n = \mathsf{Encode}_{\{n+2\}}(\tilde{b}_{n+2})$. The matrix branching program from (1) can then be evaluated over the encoded matrices:

$$\hat{C}(x) = \hat{b}_0 \times \prod_{i=1}^{n} \hat{B}_{i,x_{\mathsf{inp}(i)}} \times \hat{b}_{n+1} \tag{2}$$

Eventually one obtains an encoded $\hat{C}(x)$ over the universe set $S = \{1, \ldots, n+2\}$, and one can use the zero-testing procedure of the multilinear map scheme to check if $C(x) = 0$, thereby learning the output of the branching program $\mathsf{BP}(x)$, without revealing the values of the matrices $\boldsymbol{B}_{i,b}$.

(In)efficiency of iO. However, even with some efficiency improvements (as in [AGIS14]), the main issue is that indistinguishability obfuscation is currently not feasible to implement in practice. The first obstacle is that when converting the input circuit to a matrix branching program using Barrington's theorem [Bar86], one induces an enormous cost in performance, as the length of the branching program grows exponentially with the depth of the circuit being evaluated. The second obstacle is that the multilinear map noise and parameters grow with the degree of the polynomial being computed over encoded elements, which corresponds to the length of the matrix branching program.

In this paper, we consider both issues. For the second one, we show that for CLT13 multilinear maps, when applying Kilian's randomization "on the encoding side", one can significantly reduce the noise and encoding size while keeping the same level of security; this leads to major improvements of performance. For the first issue, we craft a sequence of matrix products that only performs a multipartite DH key-exchange, rather than generating one from a circuit through Barrington's theorem, so that its degree becomes much more manageable. We can then describe the first concrete implementation of multipartite DH key-exchange based on multilinear maps that is resistant against existing attacks.

Kilian's Randomization on the Encoding Side. As already observed in [GGH15], Kilian's randomization can also be applied over the encoding space, as an additional safeguard. Namely starting from the encoded matrices $\hat{\boldsymbol{B}}_{i,b}$ used to compute $\hat{C}(x)$ as in Eq. (2), one can again choose $n + 1$ random invertible matrices $\{\hat{\boldsymbol{R}}_i\}_{i=0}^n$ and then randomize the matrices $\hat{\boldsymbol{B}}_{i,b}$ with:

$$\bar{\boldsymbol{B}}_{i,b} = \hat{\boldsymbol{R}}_{i-1}\hat{\boldsymbol{B}}_{i,b}\hat{\boldsymbol{R}}_i^{-1}$$

with also $\bar{\boldsymbol{b}}_0 = \hat{\boldsymbol{b}}_0\hat{\boldsymbol{R}}_0^{-1}$ and $\bar{\boldsymbol{b}}_{n+1} = \hat{\boldsymbol{R}}_n\hat{\boldsymbol{b}}_{n+1}$. Since the matrices $\hat{\boldsymbol{R}}_i$ cancel each other in the matrix product computation, the evaluation proceeds exactly as in (2), with $\hat{C}(x) = \bar{\boldsymbol{b}}_0 \times \prod_{i=1}^n \bar{\boldsymbol{B}}_{i,x_{\mathrm{inp}(i)}} \times \bar{\boldsymbol{b}}_{n+1}$, and therefore the same zero-testing procedure can be applied to $\hat{C}(x)$. Note that the $\hat{\boldsymbol{R}}_i$ matrices are applied on the encoding side, that is on the encoded matrices $\hat{\boldsymbol{B}}_{i,b}$, instead of the plaintext matrices $\boldsymbol{B}_{i,b}$ as previously; obviously both randomizations (before and after encoding) can be applied independently.

In this paper we focus on Kilian's randomization on the encoding side in the context of the CLT13 multilinear maps. In CLT13 the encoding space is the set of integers modulo x_0, where $x_0 = \prod_{j=1}^n p_j$; therefore the matrices $\{\hat{\boldsymbol{R}}_i\}_{i=0}^n$ are random invertible matrices modulo x_0. We show that the complexity of the best attacks against CLT13 is significantly increased thanks to Kilian's randomization of the encodings. One can therefore use much smaller parameters (noise size and encoding size), which can improve the efficiency of a construction by several orders of magnitude.

More precisely, the security of CLT13 is based on the hardness of the multi-prime Approximate-GCD problem. Given $x_0 = \prod_{i=1}^{n} p_i$ for random primes p_i, and polynomially many integers c_j such that

$$c_j \equiv r_{ij} \pmod{p_i} \tag{3}$$

for small integers r_{ij}'s, the goal is to recover the secret primes p_i's. The multi-prime Approximate-GCD problem is an extension of the single-prime problem, with a single prime p to be recovered from encodings $c_j = q_j \cdot p + r_j$ and $x_0 = q_0 \cdot p$, for small integers r_j. The two main approaches for solving the Approximate-GCD problem are the orthogonal lattice attacks and the GCD attacks.

First Contribution: Solving the Multi-prime Approximate-GCD Problem. For the single-prime Approximate-GCD problem, the classical orthogonal lattice attack has complexity $2^{\Omega(\gamma/\eta^2)}$, where γ is the size of x_0 and η is the size of the prime p; see [DGHV10, §5.2]. However, extending the attack to the multi-prime case as in CLT13 is actually not straightforward; in particular, we argue that the approach described in [CLT13] is incomplete and does not recover the primes p_i's, except for small values of n; we note that solving the multi-prime case was actually considered as an open problem in [GGM16].

Our first contribution is to solve this open problem with an algorithm that proceeds in two steps. The first step is the classical orthogonal lattice attack; it recovers a basis of the lattice generated by the vectors $r_i = c \bmod p_i$, where $c = (c_1, \ldots, c_t)$. However, the vectors r_i cannot be recovered directly; namely by applying LLL or BKZ one recovers a basis of moderately short vectors, and not necessarily the r_i's which are the shortest vectors in the lattice. Therefore the approach described in [CLT13] does not work, except in low dimension. In the second step of our algorithm, using the lattice basis obtained from the first step, we show that by computing the eigenvalues of a well chosen matrix, we can recover the primes p_i's, as in the Cheon et al. attack [CHL+15]. The asymptotic complexity of the full attack is the same as in the single-prime case; using $\gamma = \eta \cdot n$ for the size of x_0 as previously, where n is the number of primes p_i, the complexity is $2^{\Omega(n/\eta)}$. Therefore, as in [CLT13], one must take $n = \omega(\eta \log \lambda)$ to prevent the lattice attack, where λ is the security parameter.

Second Contribution: Extension to the Vector Approximate-GCD Problem. When working with matrix branching programs and Kilian's randomization on the encoding side, we must actually consider a vector variant of the Approximate-GCD problem, in which we have access to randomized vectors of encodings instead of scalar values as in (3). Therefore, our second contribution is to extend the orthogonal lattice attack to the Vector Approximate-GCD problem, and to show that the extended attack has complexity $2^{\Omega(m \cdot n/\eta)}$, for vectors of dimension m. This implies that the new condition on the number n of primes p_i in CLT13 becomes:

$$n = \omega \left(\frac{\eta}{m} \log \lambda \right)$$

Compared to the previous condition, the number of primes n in CLT13 can therefore be divided by a factor m, for the same level of security, where m is the matrix dimension. This implies that the encoding size γ can also be divided by a factor m, which provides a significant improvement in efficiency.

Third Contribution: GCD Attacks Against the Vector Approximate-GCD Problem. The naive GCD attack against the Approximate-GCD problem with $c_1 = q_1 \cdot p + r_1$ and $x_0 = q_0 \cdot p$ consists in computing $\gcd(c_1 - r_1, x_0)$ for all possible r_1 and has complexity $\mathcal{O}(2^\rho)$, where ρ is the bitsize of r_1. At Eurocrypt 2012, Chen and Nguyen [CN12] described an improved attack based on multipoint polynomial evaluation, with complexity $\tilde{\mathcal{O}}(2^{\rho/2})$. The Chen-Nguyen attack was later extended by Lee and Seo at Crypto 2014 [LS14], when the c_i's are multiplicatively masked by a random secret z modulo x_0, as it is the case in the CLT13 scheme; their attack has the same complexity $\tilde{\mathcal{O}}(2^{\rho/2})$.

As previously, when working with matrix branching programs and Kilian's randomization on the encoding side, we must consider the vector variant of the Approximate-GCD problem. Our third contribution is therefore to extend the Lee-Seo attack to this vector variant; we obtain a complexity $\tilde{\mathcal{O}}(2^{m \cdot \rho/2})$ instead of $\tilde{\mathcal{O}}(2^{\rho/2})$, where m is the vector dimension. Assuming that this is the best possible attack, one can therefore divide the noise size ρ by a factor m. Similarly, when Kilian's randomization is applied to a $m \times m$ matrix, we show that the attack complexity becomes $\tilde{\mathcal{O}}(2^{m^2 \cdot \rho/2})$, and therefore the noise size ρ used to encode those matrices in CLT13 can be divided by m^2. Combined with the previous improvement, this improves the efficiency of CLT13 based constructions by several orders of magnitude.

Fourth Contribution: Non-interactive DH Key Exchange from Multilinear Maps. In principle the most straightforward application of multilinear maps is non-interactive multipartite Diffie-Hellman (DH) key exchange with N users, a natural generalization of the DH protocol for 3 users based on the bilinear pairing. This was originally described for GGH13, CLT13 and GGH15, but was quickly broken for the three families of multilinear maps; in particular, key exchange based on CLT13 was broken by the Cheon et al. attack [CHL+15]. The main question is therefore:

Can we construct a practical N-way non-interactive key-exchange protocol from candidate multilinear maps constructions?

In this paper we provide a first step in that direction. Namely our fourth contribution is to describe the first implementation of N-way DH key exchange resistant against known attacks. Our construction is based on CLT13 multilinear maps and is secure against the Cheon et al. attack and its variants. Our construction contains many ingredients from the GGHRSW and other similar constructions. Namely we express the session key as the result of a matrix product computation, and we embed the matrices into larger randomized matrices before encoding, together with some special "bookend" components at the start and end of the computation, as in [GGH+13b]. We use the "multiplicative bundling" technique from [GGH+13b] to prevent the adversary from combining the matrices in arbitrary ways. As explained previously, we use Kilian's randomization on the encoding side. With no additional cost, we can also use the straddling set systems from [BGK+14] to further constrain the attacker, and Kilian's randomization at the plaintext level. Finally, we use k repetitions in order to prevent the Cheon et al. attack against CLT13, when considering input partitioning attacks as in [CGH+15], and its extension with the tensoring attack [CLLT17]. We argue that the extended Cheon et al. attack has complexity $\Omega(m^{2k-1})$ in our scheme, where m is the matrix dimension and k the number of repetitions.

For $N = 4$ users and a medium (62 bits) level of security, our implementation requires 18 GB of public parameters, and a few minutes for the derivation of a shared key. We note that without Kilian's randomization of encodings our construction would be completely unpractical, as it would require more than 100 TB of public parameters.

Related Work. In [MZ18], Ma and Zhandry described a multilinear map scheme built on top of CLT13 that is provably resistant against zeroizing attack, and which can be used to directly construct a non-interactive DH key-exchange. More precisely, the authors develop a new weak multilinear map model for CLT13 to capture all known attack strategies against CLT13. The authors then construct a new multilinear map scheme on top of CLT13 that is secure in this model. The construction is based on multiplying matrices of CLT13 encodings as in iO schemes. To prevent zeroizing attacks, the same input is read multiple times, as in iO constructions. The input consistency is ensured by a clever use of "enforcing" matrices based on some permutation invariant property. Finally, the authors construct a non-interactive DH key-exchange scheme based on their new multilinear map scheme. However, the authors do not provide implementation results nor concrete parameters (except for multilinear map degree and number of public encodings), so it is difficult to assess the practicality of their construction. The authors still provide the following parameters for a 4-party DH key exchange with 80 bits of security; see Table 1. We provide our corresponding parameters for comparison (see more details in Sect. 7).

Table 1. Comparison of parameters for 4-party DH key exchange, with 80 bits of security.

Scheme	MMap degree	Public encodings	Public-key size
Boneh et al. [BISW17]	4150	2^{44}	
Ma-Zhandry (setting 1)	52	2^{62}	
Ma-Zhandry (setting 2)	160	2^{33}	
Ma-Zhandry (setting 3)	1040	2^{19}	
Ma-Zhandry (setting 4)	2000	2^{14}	
Our construction	266	2^{20}	1848 GB

The main advantage of the Ma-Zhandry construction is that it has a proof of security in a weak multilinear map model, whereas our construction has heuristic security only. It seems from Table 1 that our construction would require a smaller multilinear map degree for the same number of public encodings. We stress however that providing concrete parameters is actually a complex optimization problem (see Sect. 7), so Table 1 should be handled with care. In any case, the Ma-Zhandry construction can certainly benefit from our analysis, since Kilian's randomization on the encoding side can also be applied "for free" in their construction.

Source Code. We provide the source code of our construction, and the source code of the attacks, in [CP19].

2   Preliminaries

We denote by $[a]_n$ or $a \bmod n$ the unique integer $x \in (-\frac{n}{2}, \frac{n}{2}]$ which is congruent to a modulo n. The set $\{1, 2, \dots, n\}$ is denoted by $[n]$.

2.1   The CLT13 Multilinear Map

We briefly recall the (asymmetric) CLT13 multilinear map scheme; we refer to [CLT13] for a full description. For large secret primes p_i's, let $x_0 = \prod_{k=1}^n p_i$, where n is the number of primes. We denote by η the bitsize of the p_i's, and by γ the bitsize of x_0; therefore $\gamma \simeq n \cdot \eta$. The plaintext space of CLT13 is $\mathbb{Z}_{g_1} \times \mathbb{Z}_{g_2} \times \cdots \times \mathbb{Z}_{g_n}$ for secret prime integers g_i's of α bits.

The CLT13 scheme is based on CRT representations. We denote by $\mathsf{CRT}(a_1, \dots, a_n)$ or $\mathsf{CRT}(a_i)_i$ the number $a \in \mathbb{Z}_{x_0}$ such that $a \equiv u_i \pmod{p_i}$ for all $i \in [n]$. An encoding of a vector $\mathbf{m} = (m_1, \dots, m_n)$ at level set $S = \{j\}$ is an integer $c \in \mathbb{Z}_{x_0}$ such that $c = [\mathsf{CRT}(m_1 + g_1 r_1, \dots, m_n + g_n r_n)/z_j]_{x_0}$ for integers r_i of size ρ bits, where z_j is a secret mask in $\mathbb{Z}_{x_0}$ uniformly chosen during the parameters generation procedure of the multilinear map. This gives:

$$c \equiv \frac{m_i + g_i r_i}{z_j} \pmod{p_i} \tag{4}$$

for all $1 \le i \le n$. To support a ℓ-level multilinearity, one uses ℓ distinct z_j's.

It is clear that encodings from the same level can be added via addition modulo x_0. Similarly multiplication between encodings can be done by modular multiplication in $\mathbb{Z}_{x_0}$, but the encodings must be of disjoint level sets; the resulting encoding level set is then the union of the input level sets. At the top level set $S = \{1, \dots, \ell\}$, one can zero-test an encoding by multiplication with the zero-test parameter $p_{zt} = \left(\prod_{j=1}^\ell z_j \right) \cdot \mathsf{CRT}(p_i^* h_i g_i^{-1})_i \bmod x_0$, where $p_i^* = x_0/p_i$ and the h_i's are random β-bit integers. Namely given a top-level encoding c with

$c = \frac{\mathsf{CRT}(m_i + g_i r_i)_i}{\prod_{j=1}^{\ell} z_j}$ mod x_0, we obtain after multiplication by p_{zt}:

$$c \cdot p_{zt} = \mathsf{CRT}(h_i p_i^*(m_i g_i^{-1} + r_i))_i = \sum_{i=1}^{n} h_i p_i^*(m_i g_i^{-1} + r_i) \pmod{x_0} \qquad (5)$$

and therefore if $m_i = 0$ for all $1 \le i \le n$ then the result will be small compared to x_0. From the previous equation the high-order bits of $c \cdot p_{zt}$ mod x_0 only depend on the m_i's; therefore from the zero-testing procedure one can extract a value that only depends on the m_i's.

2.2 The Approximate-GCD Problem and Its Variant

The security of the CLT13 multilinear map scheme is based on the Approximate-GCD problem. For a specific η-bit prime integer p, we use the following distribution over γ-bit integers:

$$\mathcal{D}_{\gamma,\rho}(p) = \left\{ \text{Choose } q \leftarrow \mathbb{Z} \cap [0, 2^{\gamma}/p), \ r \leftarrow \mathbb{Z} \cap (-2^{\rho}, 2^{\rho}) : \ \text{Output } x = q \cdot p + r \right\}$$

We also consider a noise-free $x_0 = q_0 \cdot p$ where q_0 is a random $(\gamma - \eta)$-bit prime integer (alternatively the product of $\gamma/\eta - 1$ primes of size η bits each).

Definition 1 (Approximate-GCD problem with noise-free x_0). *For a random η-bit prime integer p, given $x_0 = q_0 \cdot p$ and polynomially many samples from $\mathcal{D}_{\gamma,\rho}(p)$, output p.*

We also consider the following variant, in which instead of being given elements from $\mathcal{D}_{\gamma,\rho}(p)$, we get vectors of elements multiplied by a secret random invertible matrix $\boldsymbol{K}$ modulo x_0.

Definition 2 (Vector Approximate-GCD problem with noise-free x_0). *For a random η-bit prime integer p, generate $x_0 = q_0 \cdot p$ and a random invertible $m \times m$ matrix $\boldsymbol{K}$ modulo x_0. Given x_0 and polynomially many samples $\tilde{\boldsymbol{v}} = \boldsymbol{v} \cdot \boldsymbol{K}$ mod x_0 where $\boldsymbol{v} \leftarrow (\mathcal{D}_{\gamma,\rho}(p))^m$, output p.*

The vector variant of the Approximate-GCD problem cannot be easier than the original problem, since any algorithm solving the vector variant can be used to solve the Approximate-GCD problem, simply by generating vectors $\tilde{\boldsymbol{v}} = \boldsymbol{v} \cdot \boldsymbol{K}$ (mod x_0) for some random matrix $\boldsymbol{K}$. However, the vector variant could be harder to solve, so that smaller parameters could be used when dealing with the Vector Approximate-GCD problem. We show in the next sections that the generalizations of the attacks to the vector variant indeed have higher complexity.

In the context of the CLT13 scheme, one actually works with multiple primes p_i's. Therefore we consider the multi-prime variant of the Approximate-GCD problem.

Definition 3 (Multi-prime Approximate-GCD problem). *For n random η-bit prime integers p_i, let $x_0 = \prod_{i=1}^{n} p_i$. Given x_0 and polynomially many integers $c_j = \mathsf{CRT}(r_{ij})_i$ where $r_{ij} \leftarrow \mathbb{Z} \cap (-2^{\rho}, 2^{\rho})$, output the primes p_i.*

Finally, we consider the vector variant of the multi-prime Approximate-GCD problem.

Definition 4 (Vector multi-prime Approximate-GCD problem). *For n random η-bit prime integers p_i, let $x_0 = \prod_{i=1}^{n} p_i$. Let $\boldsymbol{K}$ be a random invertible $m \times m$ matrix modulo x_0. Given x_0 and polynomially many vectors $\tilde{\boldsymbol{v}} = \boldsymbol{v} \cdot \boldsymbol{K}$ mod x_0, where $\boldsymbol{v} = (v_1, \ldots, v_m)$ and $v_j = \mathsf{CRT}(r_{ij})_i$ where $r_{ij} \leftarrow \mathbb{Z} \cap (-2^\rho, 2^\rho)$, output the primes p_i.*

The two main approaches for solving the Approximate-GCD problem are the orthogonal lattice attacks and the GCD attacks. We consider the orthogonal lattice attacks in Sect. 3, and the GCD attacks in Sect. 4.

3   Lattice Attack Against the Approximate-GCD Problem

We first recall the lattice attack against the single-prime Approximate-GCD problem [DGHV10, §B.1], based on the Nguyen-Stern orthogonal lattice attack [NS01]. As mentioned in introduction, extending the attack to the multi-prime case is actually not straightforward; in particular, we argue that the approach described in [CLT13] is incomplete and does not recover the primes p_i's, except for small values of n. Therefore, we describe a new algorithm for solving the multi-prime Approximate-GCD problem, using a variant of the Cheon *et al.* attack against CLT13. We then extend the algorithm to the vector variant of the Approximate-GCD problem. Finally, we run our attacks against both the multi-prime Approximate-GCD problem and the vector variant, in order to derive concrete parameters for our construction. We provide the source code of our attacks in [CP19].

3.1   The Orthogonal Lattice

We first recall the definition of the orthogonal lattice, following [NS97]. Let L be a lattice in $\mathbb{Z}^m$. The orthogonal lattice $L^\perp$ is defined as the set of elements in $\mathbb{Z}^m$ which are orthogonal to all the lattice points of L, for the usual dot product. We define the lattice $\bar{L} = (L^\perp)^\perp$; it is the intersection of $\mathbb{Z}^m$ with the $\mathbb{Q}$-vector space generated by L; we have that $L \subset \bar{L}$ and the determinant of $\bar{L}$ divides the determinant of L. We have that $\dim(L) + \dim(L^\perp) = m$ and $\det(L^\perp) = \det(\bar{L})$.

From Minkowski's bound, we expect that a reduced basis of a "random" lattice L has short vectors of norm $\simeq (\det L)^{1/\dim L}$. For a "random" lattice L, we also expect that $\det(L) \simeq \det(\bar{L}) = \det(L^\perp)$. Moreover, for a lattice L generated by a set of d "random" vectors $\boldsymbol{b}_i \in \mathbb{Z}^m$, from Hadamard inequality we expect that $\det L \simeq \prod_{i=1}^{d} \|\boldsymbol{b}_i\|$. In that case, we therefore expect the short vectors of $L^\perp$ to have norm $\simeq (\det L^\perp)^{1/(m-d)} \simeq (\det L)^{1/(m-d)} \simeq (\prod_{i=1}^{d} \|\boldsymbol{b}_i\|)^{1/(m-d)}$.

3.2   The Classical Orthogonal Lattice Attack Against the Single-Prime Approximate-GCD Problem

In this section we recall the lattice attack against the Approximate-GCD problem, based on the Nguyen-Stern orthogonal lattice attack [NS01]; see also the analysis in [DGHV10, §B.1]. We consider a set of t integers $x_i = p \cdot q_i + r_i$ and $x_0 = p \cdot q_0$, for $r_i \in (-2^\rho, 2^\rho) \cap \mathbb{Z}$. We consider the lattice L of vectors $\boldsymbol{u}$ that are orthogonal to $\boldsymbol{x}$ modulo x_0, where $\boldsymbol{x} = (x_1, \ldots, x_t)$:

$$L = \{\boldsymbol{u} \in \mathbb{Z}^t \mid \boldsymbol{u} \cdot \boldsymbol{x} \equiv 0 \pmod{x_0} \}$$

The lattice L is of full rank t since it contains $x_0 \mathbb{Z}^t$. Moreover, we have $\det L = [\mathbb{Z}^t : L] = x_0 / \gcd(x_0, x_1, \ldots, x_t) = x_0$. Therefore, applying lattice reduction should yield a reduced basis $(\boldsymbol{u}_1, \ldots, \boldsymbol{u}_t)$ with vectors of length

$$\|\boldsymbol{u}_k\| \leq 2^{\iota t} \cdot (\det L)^{1/t} \approx 2^{\iota t + \gamma/t} \tag{6}$$

where γ is the size of x_0, for some constant $\iota > 0$ depending on the lattice reduction algorithm, where $2^{\iota t}$ is the Hermite factor.

Now given a vector $\boldsymbol{u} \in L$, we have $\boldsymbol{u} \cdot \boldsymbol{x} \equiv 0 \pmod{x_0}$, which implies that $\boldsymbol{u} \cdot \boldsymbol{r} \equiv 0 \pmod{p}$ where $\boldsymbol{r} = (r_1, \ldots, r_t)$. The main observation is that if $\boldsymbol{u}$ is short enough, the equality will hold over $\mathbb{Z}$. More precisely, if $\|\boldsymbol{u}\| \cdot \|\boldsymbol{r}\| < p$, we get $\boldsymbol{u} \cdot \boldsymbol{r} = 0$ in $\mathbb{Z}$. From (6), this happens under the condition:

$$2^{\iota t + \gamma/t} \cdot 2^\rho < 2^\eta. \tag{7}$$

In that case, the vectors $(\boldsymbol{u}_1, \ldots, \boldsymbol{u}_{t-1})$ from the previous lattice reduction step should be orthogonal to the vector $\boldsymbol{r}$. One can therefore recover $\pm \boldsymbol{r}$ by computing the rank 1 lattice orthogonal to those vectors. From $\boldsymbol{r}$ one can recover p by computing $p = \gcd(x_0, x_1 - r_1)$.

Asymptotic Complexity. We derive a heuristic lower bound for the complexity of the attack, as in [DGHV10, §5.2]. From condition (7) the attack requires a minimal lattice dimension $t > \gamma/\eta$; therefore from the same condition we must have $\iota < \eta^2/\gamma$. Achieving an Hermite factor of $2^{\iota t}$ heuristically requires at least $2^{\Omega(1/\iota)}$ time, by using BKZ reduction with block-size $\beta = \omega(1/\iota)$ [HPS11]. Therefore, the orthogonal lattice attack has time complexity at least $2^{\Omega(\gamma/\eta^2)}$.

3.3   Lattice Attack Against Multi-prime Approximate GCD

We consider the setting of CLT13, that is we are given a modulus $x_0 = \prod_{i=1}^n p_i$ and a set of integers $x_j \in \mathbb{Z}_{x_0}$ such that $x_j \bmod p_i = r_{ij}$ for $r_{ij} \in (-2^\rho, 2^\rho) \cap \mathbb{Z}$, and the goal is to recover the secret primes p_i.

First Step: Orthogonal Lattice Attack. As previously we consider the integer vector $\boldsymbol{x}$ formed by the first t integers x_j, and we consider the lattice L of vectors $\boldsymbol{u}$ that are orthogonal to $\boldsymbol{x}$ modulo x_0:

$$L = \{\boldsymbol{u} \in \mathbb{Z}^t \mid \boldsymbol{u} \cdot \boldsymbol{x} \equiv 0 \pmod{x_0} \}$$

Note that the lattice L is of full rank t since it contains $x_0 \mathbb{Z}^t$. For $1 \leq i \leq n$, let $\boldsymbol{r}_i = \boldsymbol{x} \bmod p_i$. For any $\boldsymbol{u} \in \mathbb{Z}^t$, if $\boldsymbol{u} \cdot \boldsymbol{r}_i = 0$ in $\mathbb{Z}$ for all $1 \leq i \leq n$, then $\boldsymbol{u} \cdot \boldsymbol{x} \equiv 0$

(mod x_0). Therefore, denoting by L_r the lattice generated by the vectors $\boldsymbol{r}_i$, the lattice L contains the sublattice $L_r^\perp$ of the vectors orthogonal in $\mathbb{Z}$ to the n vectors $\boldsymbol{r}_i$'s. Assuming that the n vectors $\boldsymbol{r}_i$'s are linearly independent, we have $\dim L_r^\perp = t - n$, and we expect a reduced basis of $L_r^\perp$ to have vectors of norm $(\prod_{i=1}^n \|\boldsymbol{r}_i\|)^{1/(t-n)} \simeq 2^{\rho \cdot n/(t-n)}$.

Given a vector $\boldsymbol{u} \in L$, we have $\boldsymbol{u} \cdot \boldsymbol{x} \equiv 0 \pmod{x_0}$, which implies that $\boldsymbol{u} \cdot \boldsymbol{r}_i \equiv 0 \pmod{p_i}$ for all $1 \leq i \leq n$. As previously, if $\boldsymbol{u}$ is short enough, the equalities will hold over $\mathbb{Z}$. More precisely, if $\|\boldsymbol{u}\| \cdot \|\boldsymbol{r}_i\| < p_i$ for all $1 \leq i \leq n$, we get $\boldsymbol{u} \cdot \boldsymbol{r}_i = 0$ in $\mathbb{Z}$ for all i; therefore we must have $\boldsymbol{u} \in L_r^\perp$ under the condition $\|\boldsymbol{u}\| < (\min p_i)/(\max \|\boldsymbol{r}_i\|) \simeq 2^{\eta - \rho}$. Hence, when applying lattice reduction to the lattice L, we expect to recover the vectors from the sublattice $L_r^\perp$ if there is a gap of at least $2^{\iota \cdot t}$ between the short vectors in $L_r^\perp$ and the other vectors in $L \setminus L_r^\perp$, where $2^{\iota \cdot t}$ is the Hermite factor. Since the vectors in $L \setminus L_r^\perp$ must have norm at least approximately $2^{\eta - \rho}$, this gives the condition:

$$2^{\rho \cdot n/(t-n)} \cdot 2^{\iota t} < 2^{\eta - \rho}, \tag{8}$$

In that case, applying lattice reduction to L should yield a reduced basis $(\boldsymbol{u}_1, \ldots, \boldsymbol{u}_t)$ where the first $t - n$ vectors belong to the sublattice $L_r^\perp$. By computing the rank n lattice orthogonal to those vectors, one recovers a basis $B = (\boldsymbol{b}_1, \ldots, \boldsymbol{b}_n)$ of the lattice $\bar{L}_r = (L_r^\perp)^\perp$, where L_r is the lattice generated by the n vectors $\boldsymbol{r}_i$, However this does not necessarily reveal the original vectors $\boldsymbol{r}_i$. Namely even by applying LLL or BKZ on the basis B, we do not necessarily recover the short vectors $\boldsymbol{r}_i$'s, except possibly in low dimension; therefore the approach described in [CLT13] only works when n is small.

However, the main observation is that since each vector $\boldsymbol{b}_j$ of the basis B is a linear combination of the vectors $\boldsymbol{r}_i$, it can play the same role as a zero-tested value in the CLT13 scheme. More precisely, since the vectors $\boldsymbol{b}_1, \ldots, \boldsymbol{b}_n$ form a basis of $\bar{L}_r$, we can write for all $1 \leq j \leq n$:

$$\boldsymbol{b}_j = \sum_{i=1}^n \lambda_{ji} \boldsymbol{r}_i$$

for unknown coefficients $\lambda_{ji} \in \mathbb{Q}$. The above equation is analogous to Eq. (5) on the zero-tested value $c \cdot p_{zt}$, which is a linear combination of the r_i's over $\mathbb{Z}$ when all m_i's are zero. Therefore, we can apply a variant of the Cheon et al. attack to recover the primes p_i's, by computing the eigenvalues of a well chosen matrix. Since we have n vectors $\boldsymbol{b}_j$ instead of a single p_{zt} value, we only need to work with equations of degree 2 in the x_j's, instead of degree 3 as in [CHL+15].

Second Step: Algebraic Attack. The second step of the attack is similar to the Cheon al. attack. Recall that we receive as input $x_0 = \prod_{i=1}^n p_i$ and a set of integers $x_j \in \mathbb{Z}_{x_0}$ such that $x_j \bmod p_i = r_{ij}$ for $r_{ij} \in (-2^\rho, 2^\rho) \cap \mathbb{Z}$. Since we must work with an equation of degree 2 in the inputs, we consider an additional integer $y \in \mathbb{Z}_{x_0}$ with $y \bmod p_i = s_i$ with $s_i \in (-2^\rho, 2^\rho) \cap \mathbb{Z}$ for all $1 \leq i \leq n$.

We define the column vector $\boldsymbol{x} = \begin{bmatrix} x_1 \ldots x_n \end{bmatrix}^T$. Instead of running the orthogonal lattice attack with $\boldsymbol{x}$, we run the orthogonal lattice attack from the previous step with the column vector $\boldsymbol{z}$ of dimension $t = 2n$ defined as follows:

$$\boldsymbol{z} = \begin{bmatrix} \boldsymbol{x} \\ y \cdot \boldsymbol{x} \end{bmatrix}$$

Letting $\boldsymbol{r}_i = \boldsymbol{x} \bmod p_i$, this gives the column vectors for $1 \le i \le n$:

$$\boldsymbol{z} \bmod p_i = \begin{bmatrix} \boldsymbol{r}_i \\ s_i \cdot \boldsymbol{r}_i \end{bmatrix}$$

We denote by $\boldsymbol{Z}$ the $2n \times n$ matrix of column vectors $\boldsymbol{z} \bmod p_i$:

$$\boldsymbol{Z} = \begin{bmatrix} \boldsymbol{r}_1 & \cdots & \boldsymbol{r}_n \\ s_1 \cdot \boldsymbol{r}_1 & \cdots & s_n \cdot \boldsymbol{r}_n \end{bmatrix} = \begin{bmatrix} \boldsymbol{R} \\ \boldsymbol{R} \cdot \boldsymbol{U} \end{bmatrix}$$

where $\boldsymbol{R}$ is the $n \times n$ matrix of column vectors $\boldsymbol{r}_i$, and $\boldsymbol{U} := \mathsf{diag}(s_1, \ldots, s_n)$.

By applying the orthogonal lattice attack of the first step on the known vector $\boldsymbol{z}$, we obtain a basis of the lattice intersection of $\mathbb{Z}^{2n}$ with the $\mathbb{Q}$-vector space generated by the n vectors $\boldsymbol{z} \bmod p_i$, which corresponds to the columns of the matrix $\boldsymbol{Z}$. Therefore we obtain two matrices $\boldsymbol{W_0}$ and $\boldsymbol{W_1}$ such that:

$$\boldsymbol{W_0} = \boldsymbol{R} \cdot \boldsymbol{A}$$
$$\boldsymbol{W_1} = \boldsymbol{R} \cdot \boldsymbol{U} \cdot \boldsymbol{A}$$

for some unknown matrix $\boldsymbol{A} \in \mathbb{Q}^{n \times n}$. Therefore, as in the Cheon *et al.* attack, we compute the matrix:

$$\boldsymbol{W} = \boldsymbol{W_1} \cdot \boldsymbol{W_0}^{-1} = \boldsymbol{R} \cdot \boldsymbol{U} \cdot \boldsymbol{R}^{-1}$$

and by computing the eigenvalues of $\boldsymbol{W}$, one recovers the components s_i of the diagonal matrix $\boldsymbol{U}$, from which we recover the p_i's by taking gcd's. We provide the source code of the attack in [CP19].

Asymptotic Complexity. As previously, we derive a heuristic lower bound for the complexity of the attack. The attack requires a lattice dimension $t = 2n$, and moreover the vectors $\boldsymbol{r}_i$ have norm $\simeq 2^{2\rho}$ instead of 2^ρ; therefore condition (8) gives $4\rho + 2\iota n < \eta$ which implies the condition $\iota < \frac{\eta}{2n}$. Achieving an Hermite factor of $2^{\iota t}$ heuristically requires $2^{\Omega(1/\iota)}$ time, by using BKZ reduction with block-size $\beta = \omega(1/\iota)$ [HPS11]. Therefore, the orthogonal lattice attack has time complexity at least $2^{\Omega(n/\eta)}$. Note that with $\gamma = \eta \cdot n$, we get the same time complexity lower bound $2^{\Omega(\gamma/\eta^2)}$ as for the single-prime Approximate-GCD problem. Finally, as shown in [CLT13], to prevent the orthogonal lattice attack, one must take:

$$n = \omega(\eta \log \lambda) \tag{9}$$

Namely, in that case there exists a function $c(\lambda)$ such that $n(\lambda) = c(\lambda)\eta(\lambda) \log_2 \lambda$ with $c(\lambda) \to \infty$ for $\lambda \to \infty$. With a time complexity at least $2^{k \cdot n/\eta}$ for some $k > 0$, the time complexity is therefore at least $2^{k \cdot c(\lambda) \log_2 \lambda} = \lambda^{k \cdot c(\lambda)}$. This implies that the attack is not polynomial time under Condition 9.

3.4 Lattice Attack Against the Vector Approximate-GCD Problem

In this section we extend the previous orthogonal lattice attack to the vector variant of the Approximate-GCD problem with multiple primes p_i's. We still consider a modulus $x_0 = \prod_{i=1}^{n} p_i$, but instead of scalar values x_j, we consider t row vectors $\boldsymbol{v}_j$, each with m components $(\boldsymbol{v}_j)_k$, such that:

$$(\boldsymbol{v}_j)_k = r_{ijk} \pmod{p_i}$$

for all components $1 \leq k \leq m$ and all $1 \leq i \leq n$, where $r_{ijk} \in (-2^\rho, 2^\rho) \cap \mathbb{Z}$. We consider the $t \times m$ matrix $\boldsymbol{V}$ of row vectors $\boldsymbol{v}_j$. We don't publish the matrix $\boldsymbol{V}$ directly; instead we first generate a random secret $m \times m$ invertible matrix $\boldsymbol{K}$ modulo x_0 and publish the $t \times m$ matrix:

$$\tilde{\boldsymbol{V}} = \boldsymbol{V} \cdot \boldsymbol{K} \pmod{x_0}$$

The goal is to recover the primes p_i's as in the previous attack.

Actually, we cannot solve the original multi-prime vector Approximate-GCD problem directly, since the algebraic step of the attack requires degree 2 equations in the inputs. Instead, we assume that we can additionally obtain two $m \times m$ matrices:

$$\tilde{\boldsymbol{C}}_0 = \boldsymbol{K}^{-1} \cdot \boldsymbol{C}_0 \cdot \boldsymbol{K}' \pmod{x_0}$$
$$\tilde{\boldsymbol{C}}_1 = \boldsymbol{K}^{-1} \cdot \boldsymbol{C}_1 \cdot \boldsymbol{K}' \pmod{x_0}$$

for some random invertible matrix $\boldsymbol{K}'$ modulo x_0, where the components of the matrices $\boldsymbol{C}_0, \boldsymbol{C}_1 \in \mathbb{Z}_{x_0}^{m \times m}$ are small modulo each p_i. This assumption is verified in our construction of Sect. 5.

First Step: Orthogonal Lattice Attack. In our extended attack we consider the lattice L of vectors $\boldsymbol{u}$ that are orthogonal to all columns of $\tilde{\boldsymbol{V}}$ modulo x_0:

$$L = \{\boldsymbol{u} \in \mathbb{Z}^t \mid \boldsymbol{u} \cdot \tilde{\boldsymbol{V}} \equiv 0 \pmod{x_0}\}$$

Since the matrix $\boldsymbol{K}$ is invertible, we obtain:

$$L = \{\boldsymbol{u} \in \mathbb{Z}^t \mid \boldsymbol{u} \cdot \boldsymbol{V} \equiv 0 \pmod{x_0}\} \tag{10}$$

The lattice L is of full rank t since it contains $x_0 \mathbb{Z}^t$. Let $\boldsymbol{R}_i = \boldsymbol{V} \bmod p_i$. As previously, the lattice L contains the sublattice L' of dimension $t - m \cdot n$ of the vectors orthogonal in $\mathbb{Z}$ to the $m \cdot n$ column vectors in $\boldsymbol{R}_i$ for $1 \leq i \leq n$. We expect a reduced basis of L' to have vectors of norm $\simeq 2^{\rho \cdot m \cdot n/(t - m \cdot n)}$. Therefore, applying lattice reduction to L should yield a reduced basis $(\boldsymbol{u}_1, \ldots, \boldsymbol{u}_t)$ where the first $t - m \cdot n$ vectors belong to the sublattice L', under the modified condition:

$$2^{t + \rho \cdot m \cdot n/(t - m \cdot n)} < 2^{\eta - \rho} \tag{11}$$

As previously, by computing the rank $n \cdot m$ lattice orthogonal to the vectors $(\boldsymbol{u}_1, \ldots, \boldsymbol{u}_{t-m \cdot n})$, we obtain a basis of the lattice intersection of $\mathbb{Z}^t$ with the $\mathbb{Q}$-vector space generated by the column vectors of the $\boldsymbol{R}_i$'s.

Second Step: Algebraic Attack. The second step is similar to the second step of the attack from Sect. 3.3 and is described in the full version of this paper [CP18], with a lattice dimension $t = 2mn$.

Asymptotic Complexity. As previously, we derive a heuristic lower bound for the complexity of the attack. Since the attack requires a lattice dimension $t = 2mn$, condition (11) with noise size 2ρ instead of ρ gives $4\rho + 2\iota mn < \eta$ which gives the new condition $\iota < \frac{\eta}{2mn}$. Therefore, the orthogonal lattice attack has time complexity at least $2^{\Omega(n \cdot m/\eta)}$. This implies that to prevent the orthogonal lattice attack, we must have:

$$n = \omega\left(\frac{\eta}{m}\log\lambda\right)$$

Compared to the original condition of [CLT13] recalled by (9), the value of n can therefore be divided by m. This implies that the encoding size $\gamma = \eta \cdot n$ can also be divided by m. We show in Sect. 7 that this brings a significant improvement in practice.

3.5 Practical Experiments and Concrete Parameters

Practical Experiments. We have run our two attacks from Sects. 3.3 and 3.4 against the multi-prime approximate-GCD problem and its vector variant; we provide the source code in [CP19]. We summarize the running times for various values of n in Tables 2 and 3. We see that the running time of the lattice step in the vector variant is roughly the same as in the non-vector variant, when the number of primes n is divided by m in the vector variant. This confirms the asymptotic analysis of the previous section.

For the algebraic step of the non-vector problem, it is significantly more efficient to compute the matrix kernel and eigenvalues modulo some arbitrary prime integer q of size η, instead of over the rationals. However we have not found a similar optimization for the vector variant; we see in Table 3 that for larger n the cost of the algebraic step becomes prohibitive (but still polynomial time) for the vector variant. In this paper we conservatively fix our concrete parameters by considering the lattice step only. We leave as an open problem the derivation of a "practical" algebraic step for the vector variant.

LLL and BKZ Practical Complexity. To derive concrete parameters for our construction from Sect. 5, we have run more experiments with LLL and BKZ lattice reduction algorithms applied to a lattice similar to the lattice L of the previous section. Recall that we must apply lattice reduction on the lattice:

$$L = \{\boldsymbol{u} \in \mathbb{Z}^t \mid \boldsymbol{u} \cdot \tilde{\boldsymbol{V}} \equiv 0 \pmod{x_0}\}$$

with $t = 2nm$. We write $\boldsymbol{u} = [\boldsymbol{u_1}, \boldsymbol{u_2}]$ with $\boldsymbol{u_1} \in \mathbb{Z}^{t-m}$ and $\boldsymbol{u_2} \in \mathbb{Z}^m$. Similarly we write $\tilde{\boldsymbol{V}} = \begin{bmatrix} \boldsymbol{A} \\ \boldsymbol{W} \end{bmatrix}$ where $\boldsymbol{W}$ is a $m \times m$ matrix. With high probability $\boldsymbol{W}$ is

Table 2. Running time of the LLL step and the algebraic step for solving the multi-prime approximate-GCD problem, on a 3.2 GHz Intel Core i5.

n	η	ρ	lat. dim.	Time LLL	Time alg.
20	335	80	40	1.5 s	0.6 s
30	335	80	60	9 s	0.7 s
40	335	80	80	37 s	1.5 s
60	335	80	120	4 min	4 s
80	335	80	160	20 min	8 s

Table 3. Running time of the LLL step and the algebraic step for solving the vector multi-prime approximate-GCD problem, on a 3.2 GHz Intel Core i5.

n	m	η	ρ	lat. dim.	Time LLL	Time alg.
4	5	335	80	40	1.4 s	2.3 s
6	5	335	80	60	9 s	20 s
8	5	335	80	80	32 s	27 min
12	5	335	80	120	6 min	–
16	5	335	80	160	12 min	–

invertible modulo x_0, otherwise we can partially factor x_0. We obtain

$$\boldsymbol{u} \in L \iff \boldsymbol{u_1}\boldsymbol{A} + \boldsymbol{u_2}\boldsymbol{W} \equiv 0 \pmod{x_0}$$
$$\iff \boldsymbol{u_1}\boldsymbol{A}\boldsymbol{W}^{-1} + \boldsymbol{u_2} \equiv 0 \pmod{x_0}$$

Therefore, a basis of L is given by the matrix of row vectors:

$$L = \begin{bmatrix} \boldsymbol{I}_{t-m} & -\boldsymbol{A}\boldsymbol{W}^{-1} \\ & x_0\boldsymbol{I}_m \end{bmatrix}$$

For simplicity, we have performed our experiments on a simpler lattice:

$$L' = \begin{bmatrix} \boldsymbol{I}_{t-m} & \boldsymbol{A'} \\ & x_0\boldsymbol{I}_m \end{bmatrix}$$

where the components of $\boldsymbol{A'}$ are randomly generated modulo x_0. We expect to obtain a reduced basis $(\boldsymbol{u_1}, \dots, \boldsymbol{u_t})$ with vectors of norm:

$$\|\boldsymbol{u_k}\| \simeq 2^{\iota \cdot t}(\det L)^{1/t} \simeq 2^{\iota \cdot t + m \cdot \gamma/t}$$

where $2^{\iota \cdot t}$ is the Hermite factor, and γ the size of x_0. Experimentally, we observed the following running time (expressed in number of clock cycles) for the LLL lattice reduction algorithm in the Sage implementation:

$$T_{LLL}(t, \gamma, m) \simeq 2 \cdot t^{3.3} \cdot \gamma \cdot m \tag{12}$$

The Sage implementation also includes an implementation of BKZ 2.0 [CN11]. Experimentally we observed the following running-times (in number of clock cycles):

$$T_{BKZ}(t, \beta) \simeq b(\beta) \cdot t^{4.3} \tag{13}$$

where the observed constant $b(\beta)$ and the Hermite factor are given in Table 4. However we were not able to obtain experimental results for block-sizes $\beta > 60$, so for BKZ-80 and BKZ-100 we used extrapolated values, assuming that the cost of BKZ sieving with blocksize β is $\mathsf{poly}(t) \cdot 2^{0.292\beta + o(\beta)}$ (see [BDGL16]). The Hermite factors for BKZ-80 and BKZ-100 are from [CN11].

Table 4. Experimental values of running time and Hermite factor for LLL and BKZ as a function of the blocksize β. The parameters for $\beta = 80, 100$ are extrapolated.

	LLL	BKZ-60	BKZ-80	BKZ-100
(Hermite factor)$^{1/t} = 2^\iota$	1.021	1.011	1.01	1.009
Running time parameter $b(\beta)$	–	10^3	$6 \cdot 10^4$	$3 \cdot 10^6$

Setting Concrete Parameters. When applying LLL or BKZ with blocksize β on the original lattice L, we obtain an orthogonal vector $\boldsymbol{u}$ under the condition (11), which gives with $t = 2nm$ and vectors with noise size 2ρ instead of ρ:

$$\iota \cdot 2nm + 4\rho < \eta \tag{14}$$

Therefore we must run LLL or BKZ-β with a large enough blocksize β so that ι is small enough for condition (14) to hold. For security parameter λ, we require that $T_{lat}(t, \gamma) \geq 2^\lambda$, with $t = 2nm$, where the running time (in number of clock cycles) $T_{lat}(t, \gamma)$ is given by (12) or (13), for $\gamma = \eta \cdot n$. We use that condition to provide concrete parameters for our scheme in Sect. 7.

4   GCD Attacks Against the Approximate-GCD Problem and Its Variants

4.1   The Naive GCD Attack

For simplicity we first consider the single prime variant of the Approximate-GCD problem. More precisely, we consider $x_0 = q_0 \cdot p$ and an encoding c with $c \equiv r \pmod{p}$, where r is a small integer of size ρ bits. The naive GCD attack, which has complexity $\mathcal{O}(2^\rho)$, consists in performing an exhaustive search of r and computing $\gcd(c - r, x_0)$ to obtain the factor p.

4.2   The Chen-Nguyen Attack

At Eurocrypt 2012, Chen and Nguyen described an improved attack based on multipoint polynomial evaluation [CN12], with complexity $\tilde{\mathcal{O}}(2^{\rho/2})$. One starts from the equation:

$$p = \gcd\left(x_0, \prod_{i=0}^{2^\rho - 1} (c - i) \pmod{x_0} \right) \tag{15}$$

The main observation is that the above product modulo x_0 can be written as the product of $2^{\rho/2}$ evaluations of a single polynomial of degree $2^{\rho/2}$. Using a tree structure, it is possible to evaluate a polynomial of degree $2^{\rho/2}$ at $2^{\rho/2}$ points in $\tilde{\mathcal{O}}(2^{\rho/2})$ time and memory, instead of $\mathcal{O}(2^\rho)$.

More precisely, one can define the following polynomial $f(x)$ of degree $2^{\rho/2}$, with coefficients modulo x_0; we assume for simplicity that ρ is even:

$$f(x) = \prod_{i=0}^{2^{\rho/2} - 1} (c - (x + i)) \bmod x_0$$

One can then rewrite (15) as the product of $2^{\rho/2}$ evaluations of the polynomial $f(x)$:

$$p = \gcd\left(x_0, \prod_{k=0}^{2^{\rho/2} - 1} f(2^{\rho/2} k) \pmod{x_0} \right)$$

There are classical algorithms which can evaluate a polynomial $f(x)$ of degree d at d points, using at most $\tilde{\mathcal{O}}(d)$ operations in the coefficient ring; see for example [Ber03]. Therefore, the Chen-Nguyen Attack has time and memory complexity $\tilde{\mathcal{O}}(2^{\rho/2})$. We provide in [CP19] an implementation of the Chen-Nguyen attack in Sage; our running time is similar to [CN12, Table 1]; see Table 5 below for practical experiments. In practice, the running time in number of clock cycles of the Chen-Nguyen attack with a γ-bit x_0 is well approximated by:

$$T_{CN}(\rho, \gamma) = 0.3 \cdot \rho^2 \cdot 2^{\rho/2} \cdot \gamma \cdot \log^2 \gamma \tag{16}$$

4.3   The Lee-Seo Attack

The Chen-Nguyen attack was later extended by Lee and Seo at Crypto 2014 [LS14], when the encodings are multiplicatively masked by a random secret z modulo x_0, as it is the case in the CLT13 scheme; their attack has the same complexity $\tilde{\mathcal{O}}(2^{\rho/2})$. Namely in the asymmetric CLT13 scheme recalled in Sect. 2.1, an encoding c at level set $\{i_0\}$ is such that:

$$c \equiv \frac{r_i \cdot g_i + m_i}{z_{i_0}} \pmod{p_i}$$

for some random secret z_{i_0} modulo x_0. Therefore, we consider the following variant of the Approximate-GCD problem. Instead of being given encodings c_i with $c_i \equiv r_i \pmod{p}$ for small r_i's, we are given encodings c_i with:

$$c_i \equiv r_i \cdot z \pmod{p}$$

for some random integer z modulo x_0, where the r_i's are still ρ-bit integers. Since $c_1/c_2 \equiv r_1/r_2 \pmod{p}$, the naive GCD attack consists in guessing r_1 and r_2 and computing $p = \gcd(c_1/c_2 - r_1/r_2 \bmod x_0, x_0)$, with complexity $\mathcal{O}(2^{2\rho})$.

The Lee-Seo attack with complexity $\tilde{\mathcal{O}}(2^{\rho/2})$ is as follows. First, one generates two lists L_1 and L_2 of such encodings, and we look for a collision modulo p between those two lists; such collision will appear with good probability when the size of the two lists is at least $2^{\rho/2}$. More precisely, let c_i be the elements of L_1 and d_j be the elements of L_2, with $c_i \equiv r_i \cdot z \pmod{p}$ and $d_j = s_j \cdot z \pmod{p}$. If $r_i = s_j$ for some pair (i, j), then $c_i \equiv d_j \pmod{p}$ and therefore:

$$p = \gcd \left(\prod_{i,j}(c_i - d_j) \bmod x_0, x_0 \right)$$

where the product is over all $c_i \in L_1$ and $d_j \in L_2$. A naive computation of this product would take time $|L_1| \cdot |L_2| = 2^\rho$; however, as in the Chen-Nguyen attack, this product can be computed in time and memory $\tilde{\mathcal{O}}(2^{\rho/2})$. Namely one can define the polynomial $f(x) = \prod_i (c_i - x) \bmod x_0$ of degree $|L_1| = 2^{\rho/2}$ and the previous equation can be rewritten:

$$p = \gcd \left(\prod_j f(d_j) \bmod x_0, x_0 \right)$$

This corresponds to the multipoint evaluation of the degree $2^{\rho/2}$ polynomial $f(x)$ at the $2^{\rho/2}$ points of the list L_2; therefore, this can be computed in time and memory $\tilde{\mathcal{O}}(2^{\rho/2})$.

As observed in [LS14], if only a small set of elements c_i is available (much less than $2^{\rho/2}$), one can still generate exponentially more c_i's by using small linear integer combinations of the original c_i's, and the above attack still applies, with only a slight increase in the noise ρ. We provide in [CP19] an implementation of the Lee-Seo attack in Sage. Its running time is roughly the same as Chen-Nguyen, except that the attack is probabilistic only; its success probability can be increased by taking slightly larger lists L_1 and L_2 to improve the collision probability.

4.4   GCD Attack Against the Vector Approximate GCD Problem

We now consider the Vector Approximate-GCD problem (Definition 2). We consider a set of row vectors v_i of dimension m, such that for each vector v_i, all components $(v_i)_j$ of v_i are small modulo p:

$$(v_i)_j = r_{ij} \pmod{p}$$

However, we only obtain the randomized vectors:

$$\tilde{v}_i = v_i \cdot K \pmod{x_0}$$

for some random invertible matrix K modulo x_0. The goal is still to recover the prime p.

Our attack is similar to the Lee-Seo attack recalled previously. We only consider the first component $c_i = (\tilde{v}_i)_1$ of each vector $\tilde{v}_i$. We have:

$$c_i = (\tilde{v}_i)_1 = \sum_{j=1}^{m} (v_i)_j \cdot K_{j1} = \sum_{j=1}^{m} r_{ij} \cdot K_{j1} \pmod{p}$$

We build the two lists L_1 and L_2 from the c_i's as in the Lee-Seo attack. Since each c_i is a linear combination modulo p of m random values r_{ij}'s (where the coefficients are initially generated at random modulo p), it has $m \cdot \rho$ bits of entropy modulo p, instead of ρ in the Lee-Seo attack. Therefore a collision between the two lists will occur with good probability when the lists have size at least $2^{m \cdot \rho/2}$. This implies that the attack has time and memory complexity $\tilde{\mathcal{O}}(2^{m \cdot \rho/2})$. Note that the entropy of each c_i modulo p is actually upper-bounded by the bitsize η of p. If $m \cdot \rho > \eta$, the attack complexity becomes $\tilde{\mathcal{O}}(2^{\eta/2})$, which corresponds to the complexity of the Pollard's rho factoring algorithm. We provide in [CP19] an implementation of the attack in Sage; see Table 5 below for practical experiments.

With an attack complexity $\tilde{\mathcal{O}}(2^{m\rho/2})$ instead of $\tilde{\mathcal{O}}(2^{\rho/2})$, one can therefore divide the size of the noise ρ by a factor m compared to the original CLT13, which is a significant improvement. For example, it is recommended in [CLT13] to take $\rho = 89$ bits for $\lambda = 80$ bits of security; with a vector dimension $m = 10$, one can now take $\rho = 9$ for the same level of security. Note that we can take $m \cdot \rho/2 < \lambda$ because we only require that the running time in number of clock-cycles is at least 2^λ. More precisely, the running time can be approximated by $T_{CN}(m\rho, \gamma)$ for a γ-bit x_0, where $T_{CN}(\rho, \gamma)$ is given by (16), and we require $T_{CN}(m\rho, \gamma) \geq 2^\lambda$.

With Matrices. The previous GCD attack can be generalized to $m \times m$ matrices V_i instead of m-dimensional vectors v_i. More precisely, we consider a set of matrices V_i of dimension $m \times m$ with small components modulo p, that is:

$$(V_i)_{jk} = r_{ijk} \pmod{p} \tag{17}$$

for ρ-bit integers r_{ijk}. As previously, instead of publishing the matrices V_i, we publish the randomized matrices

$$\tilde{V}_i = K \cdot V_i \cdot K' \pmod{x_0} \tag{18}$$

for two random invertible $m \times m$ matrices K and K' modulo x_0. In that case, each component of $\tilde{V}_i$ depends on the m^2 elements of the matrix V_i. This implies that the entropy of each component of $\tilde{V}_i$ is now $m^2 \cdot \rho$ and therefore the GCD attack has complexity $\tilde{\mathcal{O}}(2^{m^2 \cdot \rho/2})$.

Formally, using the Kronecker product, we can rewrite (18) as $\mathsf{vec}(\tilde{V}_i) = (K'^T \otimes K)\mathsf{vec}(V_i)$, where $\mathsf{vec}(V_i)$ denotes the column vector of dimension m^2 formed by stacking the columns of V_i on top of one another, and similarly for $\mathsf{vec}(\tilde{V}_i)$. We can therefore apply the previous attack with vectors of dimension m^2 instead of m; the attack complexity is therefore $\tilde{\mathcal{O}}(2^{m^2 \cdot \rho/2})$. This implies that we can divide the noise size ρ by a factor m^2 compared to [CLT13], where m is the matrix dimension. We provide in [CP19] an implementation of the attack in Sage; see Table 5 below for practical experiments.

With Multiple Primes $p_i's$. Instead of considering an encoding c that is small modulo a single prime p, we consider as in CLT13 a modulus $x_0 = \prod_{i=1}^{n} p_i$ and an integer $c \in \mathbb{Z}_{x_0}$ such that $c \bmod p_i = r_i$ for ρ-bit integers r_i. With good probability, we have $|r_i| \leq 2^\rho/n$ for some i but not all i, and Eq. (15) from the Chen-Nguyen attack can be rewritten:

$$p_i \Big| \gcd \left(x_0, \prod_{j=0}^{\lfloor 2^\rho/n \rfloor} (c - j) \pmod{x_0} \right)$$

where the gcd is not equal to x_0; therefore a sub-product of the p_i's is revealed. Since the number of terms in the product is divided by n, the complexity of the Chen-Nguyen attack for recovering a single p_i (or a sub-product of the p_i's) is divided by $\sqrt{n}$. By repeating the same attack n times in different intervals of the r_i's, one can recover all the p_i's; the running time of the Chen-Nguyen attack is then increased by a factor $\sqrt{n}$.

Similarly, in the Lee-Seo attack with multiple primes p_i's, the collision probability for recovering a single p_i is multiplied by n, and therefore the attack complexity is divided by $\sqrt{n}$ for recovering a single p_i. The same applies to our variant attack against the Vector Approximate GCD problem and to the matrix variant. In the later case, with noise size ρ_m, the running time of the attack in number of clock cycles can therefore be approximated by

$$T_{GCD}(m, \gamma, \rho_m, n) = T_{CN}(\rho, \gamma)/\sqrt{n} \tag{19}$$

with $\rho = m^2 \rho_m$. We will use that approximation to provide concrete parameters for our scheme in Sect. 7.

Practical Experiments. We provide in Table 5 the result of practical experiments against the Approximate-GCD problem and its vector variant with a single prime p. We see that our attack against the vector variant with dimension m and noise size ρ_v has roughly the same running time as the Chen-Nguyen attack on the original problem with noise $\rho = m \cdot \rho_v$; similarly, the running time of our attack against $m \times m$ matrices with noise ρ_m has roughly the same running time as Chen-Nguyen with noise $\rho = m^2 \cdot \rho_m$; this confirms the above analysis. We provide the source code in [CP19].

5   Our Construction

5.1   Non-interactive Multipartite Diffie-Hellman Key Exchange

A multipartite key exchange protocol aims to derive a shared value between N parties. This is achieved via a procedure in which the parties broadcast some values and then use some secret information together with the values broadcasted by the other parties to set up the shared key. In a non-interactive protocol, the parties broadcast their public values only once and at the same time (or equivalently, the values broadcasted by each party do not depend on the values broadcasted by the others). Following the notation of [BS03], such protocol can be described with three randomized probabilistic polynomial-time algorithms as follows.

Table 5. Running time of the Chen-Nguyen attack against the Approximate-GCD problem and our attack against the vector variant and matrix variants with $\eta = 100$ and x_0 of size $\gamma = 16\,000$, on a 3.2 GHz Intel Core i5.

AGCD: Chen-Nguyen	ρ	12	16	20	24
	Time (s)	0.3	2.5	15	94
m-vector AGCD: our attack ($m = 4$)	ρ_v	3	4	5	6
	Time (s)	1.5	9.3	53	301
$m \times m$-matrix AGCD: our attack ($m = 2$)	ρ_m	3	4	5	6
	Time (s)	1.5	10	54	300

- Setup($1^\lambda, N$): This algorithm runs in polynomial time in the security parameter $\lambda \in \mathbb{N}$ and in the number of parties N, and outputs the public parameters params.
- Publish(params, u): Given a party $u \in [N]$, this algorithm generates a pair of keys $(\mathsf{sk}_u, \mathsf{pk}_u)$. Party u broadcasts pk_u and keeps sk_u secret.
- KeyGen(params, v, sk_v, $\{\mathsf{pk}_u\}_{u \neq v}$): Party $v \in [N]$ uses its secret sk_v and all the values pk_u broadcasted by other parties to generate a session key s_v.

We say that the protocol is correct if $s = s_1 = s_2 = \cdots = s_N$, i.e., if all the parties share the same value at the end. We say that the protocol is secure if no probabilistic polynomial-time adversary can distinguish the shared value s from a random string given the public parameters params and the broadcasted values $\mathsf{pk}_1, \ldots, \mathsf{pk}_N$.

5.2   Our Construction

We describe our N-party one-round key exchange protocol. We start with the Setup procedure, which is run a single time by a trusted authority to generate the public parameters. As illustrated in Table 6, Setup generates for each party v two sequences of matrices $(\boldsymbol{C}_{i,b}^{(v)})_{i=1,\ldots,\ell}$ for $b \in \{0, 1\}$. In the KeyGen procedure, each party v will use the product of the matrices $\boldsymbol{C}_{i,b}^{(v)}$ on his row v to generate the session-key. The product is computed according to the secret-key sk_v of Party v and the secret-keys sk_u of the other parties. Therefore, in the Publish procedure, each party u will compute and publish the partial sub-products corresponding to his sk_u on the other rows $v \neq u$, to be used by each party v on his row v.

Table 6. Public matrices for $N = 3$ generated during the Setup procedure.

Party 1	$\boldsymbol{C}_{1,0}^{(1)}$	$\boldsymbol{C}_{2,0}^{(1)}$	$\cdots$	$\boldsymbol{C}_{\ell,0}^{(1)}$
	$\boldsymbol{C}_{1,1}^{(1)}$	$\boldsymbol{C}_{2,1}^{(1)}$	$\cdots$	$\boldsymbol{C}_{\ell,1}^{(1)}$
Party 2	$\boldsymbol{C}_{1,0}^{(2)}$	$\boldsymbol{C}_{2,0}^{(2)}$	$\cdots$	$\boldsymbol{C}_{\ell,0}^{(2)}$
	$\boldsymbol{C}_{1,1}^{(2)}$	$\boldsymbol{C}_{2,1}^{(2)}$	$\cdots$	$\boldsymbol{C}_{\ell,1}^{(2)}$
Party 3	$\boldsymbol{C}_{1,0}^{(3)}$	$\boldsymbol{C}_{2,0}^{(3)}$	$\cdots$	$\boldsymbol{C}_{\ell,0}^{(3)}$
	$\boldsymbol{C}_{1,1}^{(3)}$	$\boldsymbol{C}_{2,1}^{(3)}$	$\cdots$	$\boldsymbol{C}_{\ell,1}^{(3)}$

Setup($1^\lambda, N$): given a security parameter λ and the number of participants N, we set the length μ of each parties' secret, the number of repetitions k, and the dimension m of the matrices, with $m \equiv 0 \pmod 3$. We then instantiate the CLT13 multilinear map with degree of multilinearity $\ell + 2$ with $\ell := \mu N k$. Let $g = \prod_{i=1}^{n} g_i$ be the integer defining the message space $\mathbb{Z}_g$. Let ν be the number of high-order bits that can be extracted from a zero-tested value.

To ensure that all users $1 \leq u \leq N$ compute the same session-key, we define $\boldsymbol{A}_{i,b}^{(u)}$ as a larger matrix embedding a matrix $\boldsymbol{B}_{i,b}$ that is the same for all users, with some random block padding in the diagonal and the multiplicative bundling scalars $\alpha_{i,b}^{(u)}$ to prevent the adversary from switching the corresponding bits b_i's between the k repetitions of the secret keys:

$$\boldsymbol{A}_{i,b}^{(u)} \sim \begin{pmatrix} \$ & \cdots & \$ & & & & \\ \vdots & \ddots & \vdots & & & & \\ \$ & \cdots & \$ & & & & \\ & & & \$ & \cdots & \$ & \\ & & & \vdots & \ddots & \vdots & \\ & & & \$ & \cdots & \$ & \\ & & & & & & \alpha_{i,b}^{(u)} \cdot \boldsymbol{B}_{i,b} \end{pmatrix} \tag{20}$$

More precisely, we first sample 2ℓ random invertible matrices $\boldsymbol{B}_{i,b}$ in $\mathbb{Z}_g^{m' \times m'}$ where $m' = m/3$, for $1 \leq i \leq \ell$ and $b \in \{0,1\}$. For each $u \in [N]$, we additionally sample $2\ell N$ scalars $\alpha_{i,b}^{(u)}$ in $\mathbb{Z}_g^\star$ and $4\ell N$ random invertible matrices $\boldsymbol{S}_{i,b}^{(u)}$ and $\boldsymbol{T}_{i,b}^{(u)}$ in $\mathbb{Z}_g^{m' \times m'}$, for $1 \leq i \leq \ell$ and $b \in \{0,1\}$. As illustrated in (20), we let

$$\boldsymbol{A}_{i,b}^{(u)} := \mathsf{diag}(\boldsymbol{S}_{i,b}^{(u)}, \ \boldsymbol{T}_{i,b}^{(u)}, \ \alpha_{i,b}^{(u)} \cdot \boldsymbol{B}_{i,b}) \tag{21}$$

The scalars $\alpha_{i,b}^{(u)}$ must satisfy the following condition:

$$\forall u,v \in [N], \forall i \in [N\mu], \forall b \in \{0,1\}, \quad \prod_{j=0}^{k-1} \alpha_{j \cdot N \cdot \mu + i - 1, b}^{(u)} = \prod_{j=0}^{k-1} \alpha_{j \cdot N \cdot \mu + i - 1, b}^{(v)} \quad (\text{mod } g)$$

In addition, we sample the vectors s^*, t^* uniformly from $\mathbb{Z}_g^{m'}$, and for each $u \in [N]$ we define a left bookend vector

$$s^{(u)} := (0, \ldots, 0, \$, \ldots, \$, s^*) \in \mathbb{Z}_g^m$$

where the block of 0's and the block of randoms have the same length $m' = m/3$ as s^*, and similarly a right bookend vector $t^{(u)} := (\$, \ldots, \$, 0, \ldots, 0, t^*) \in \mathbb{Z}_g^m$.

We let $\tilde{A}_{i,b}^{(u)} \in \mathbb{Z}_{x_0}^{m \times m}$ be the matrix obtained by encoding each entry of $A_{i,b}^{(u)}$ independently. Similarly we encode $s^{(u)}$ and $t^{(u)}$ entry-wise, obtaining $\tilde{s}^{(u)}$ and $\tilde{t}^{(u)}$. For each $u \in [N]$, we sample uniformly random invertible matrices $K_i^{(u)} \in \mathbb{Z}_{x_0}^{m \times m}$ for $0 \le i \le \ell$. We then use Kilian's randomization "on the encoding side" and define:

$$C_{i,b}^{(u)} := K_{i-1}^{(u)} \tilde{A}_{i,b}^{(u)} \left(K_i^{(u)} \right)^{-1} \quad (\text{mod } x_0)$$

Similarly, we define $\bar{s}^{(u)} := \tilde{s}^{(u)} \left(K_0^{(u)} \right)^{-1}$ (mod x_0) and $\bar{t}^{(u)} := K_\ell^{(u)} \tilde{t}^{(u)} p_{zt}$ (mod x_0). Note that thanks to Kilian's randomization "on the encoding side", the matrices $A_{i,b}^{(u)}$ can be encoded with denominator $z_j = 1$ in (4) for all levels j; namely we obtain the same distribution in the final $C_{i,b}^{(u)}$ as with random z_j's. Finally we output params, which is defined as the set containing all the matrices $C_{i,b}^{(u)}$, the bookend vectors $\bar{s}^{(u)}$ and $\bar{t}^{(u)}$, and the scalars $\mu, k, N, \ell, x_0, \nu$ and m.

Publish(params, u): Party u samples a bit string $\mathsf{sk}^{(u)} \in \{0,1\}^\mu$ and for each $v \in [N]$ such that $u \ne v$, Party u computes k products using matrices from the row of party v. This ensures that from the extraction procedure of the multilinear map scheme, each user u can derive the session key from his own $\mathsf{sk}^{(u)}$ by computing on his row u the partial products corresponding to his $\mathsf{sk}^{(u)}$, combined with the published partial matrix products from the other users. More precisely, Party u computes and broadcasts the following products:

$$D_r^{(u \to v)} := \prod_{i=0}^{\mu-1} C_{(r-1)N\mu + (u-1)\mu + i, \mathsf{sk}^{(u)}[i]}^{(v)} \quad (\text{mod } x_0) \tag{22}$$

for each $v \ne u$ and $r \in [k]$. The notation $u \to v$ stands for "computed by u to be used by v". We let $\mathsf{pk}_u = \{D_r^{(u \to v)} : v \in [N], v \ne u, r \in [k]\}$.

KeyGen(params, v, $\mathsf{sk}^{(v)}$, $\{\mathsf{pk}_u\}_{u \ne v}$): Using secret $\mathsf{sk}^{(v)}$, party v computes the products $D_r^{(v \to v)}$ for all $r \in [k]$ using (22), and then the product

$$z^{(v)} := \bar{s}^{(v)} \left(\prod_{r=1}^k \left(\prod_{u=1}^N D_r^{(u \to v)} \right) \right) \bar{t}^{(v)} \quad (\text{mod } x_0). \tag{23}$$

Eventually the shared key is obtained by applying a strong randomness extractor to the ν most-significant bits of $z^{(v)}$. This terminates the description of our construction.

Correctness. It is easy to verify the correctness of our construction. Namely defining sk as the concatenated secret-keys with the k repetitions:

$$\mathsf{sk} = (\underbrace{\mathsf{sk}^{(1)}, \ldots, \mathsf{sk}^{(N)}}_{\text{First repetition}}, \ldots, \underbrace{\mathsf{sk}^{(1)}, \ldots, \mathsf{sk}^{(N)}}_{k\text{-th repetition}}) \tag{24}$$

we obtain from (22) and (23), and then from the cancellation of Kilian's randomization on the encoding side:

$$z^{(v)} = \bar{s}^{(v)} \left(\prod_{i=1}^{\ell} C_{i,\mathsf{sk}[i]}^{(v)} \right) \bar{t}^{(v)} = \tilde{s}^{(v)} \left(\prod_{i=1}^{\ell} \tilde{A}_{i,\mathsf{sk}[i]}^{(v)} \right) \tilde{t}^{(v)} p_{zt} \pmod{x_0}.$$

This corresponds to a zero-tested encoding of:

$$v_v = s^{(v)} \cdot \left(\prod_{i=1}^{\ell} A_{i,\mathsf{sk}[i]}^{(v)} \right) \cdot t^{(v)} = s^* \cdot \left(\prod_{i=1}^{\ell} \alpha_{i,\mathsf{sk}[i]}^{(v)} \right) \cdot \left(\prod_{i=1}^{\ell} B_{i,\mathsf{sk}[i]} \right) \cdot t^* \pmod{g}$$

From the condition satisfied by the $\alpha_{i,b}^{(v)}$'s, the products $\prod_{i=1}^{\ell} \alpha_{i,\mathsf{sk}[i]}^{(v)}$ are independent from v. Therefore, each party v will extract from $z^{(v)}$ the same session-key, as required.

5.3 Additional Safeguard: Straddling Sets

As an additional safeguard one can use the straddling set systems from [BGK+14]. Like the multiplicative bundling scalars $\alpha_{i,b}^{(u)}$, this prevents the adversary from switching the secret-key bits between the k repetitions. Additionally, the straddling set system prevents the adversary from mixing the matrices $\tilde{A}_{i,0}^{(u)}$ and $\tilde{A}_{i,1}^{(u)}$, since in that case the matrices are encoded at a different level set.

6 The Cheon et al. Attack and Its Generalization Using Tensor Products

At Eurocrypt 2015, Cheon *et al.* described in [CHL+15] a total break of the basic key-exchange protocol of CLT13. The attack was then extended and applied to several constructions based on CLT13. In the full version of this paper [CP18], we argue that the complexity of the Cheon *et al.* attack against our construction is $\Omega(m^{2k-1})$, where m is the matrix dimension and k the number of repetitions. Therefore, the Cheon *et al.* attack is prevented by using a large enough k.

7 Optimizations and Implementation

In this section we describe a few optimizations in order to obtain a concrete implementation of our construction from Sect. 5.

7.1 Encoding of Elements

For the bookend vectors, the components are CLT13-encoded with random noise of size ρ_b bits. Letting α be the size of the g_i's, for simplicity we take $\rho_b = \alpha$. Therefore the encoded bookend vectors have $\alpha \cdot (2m/3) + \rho_b \cdot m = 5\alpha m/3$ bits of entropy on each slot. For the matrices, we can use a much smaller encoding noise thanks to the analysis from Sect. 4.4. On a single slot, the matrices $A_{i,b}^{(u)}$ have entropy $\simeq \alpha \cdot m^2/3$, and when CLT13-encoded with noise ρ_m, the matrices $\tilde{A}_{i,b}^{(u)}$ have entropy $\simeq \alpha \cdot m^2/3 + \rho_m \cdot m^2$ on each slot; the GCD attack complexity is therefore $\tilde{\mathcal{O}}(2^{m^2 \cdot (\rho_m + \alpha/3)/2})$. For the parameters from Table 7 below, it suffices to take $\rho_m = 2$ to prevent GCD attacks.

7.2 Number of Matrices per Level

Instead of taking only two matrices $A_{i,0}^{(u)}$, $A_{i,1}^{(u)}$ for each $1 \leq i \leq \ell$, we can take 2^τ matrices for each i. In that case, the secret key of each user has μ words of τ bits, where each word selects one of the 2^τ matrices; the size of the secret-key is therefore $\mu \cdot \tau$ bits. For the same secret-key size, one can therefore divide the total degree ℓ by a factor τ, but the number of encoded matrices is multiplied by a factor $2^\tau/\tau$. In order to minimize the size of the public parameters, we use $\tau = 3$.

7.3 Other Attacks

Orthogonal Lattice Attack on Zero-Tested Values. There is an orthogonal lattice attack against the values obtained by subtracting two zero-tested last-level encodings from two different rows. The attack is analogous to the attack described in Sect. 3.3, and is prevented under the condition $n = \omega(\frac{\nu^2}{\eta - \nu} \log \lambda)$, where ν is the number of extracted bits in the zero-tested values.

Meet-in-the-Middle Attack. Given the matrix products $D_r^{(u \to v)}$ published by each party u corresponding to his secret $\mathsf{sk}^{(u)}$, there is a meet-in-the-middle attack that can recover $\mathsf{sk}^{(u)}$. Since each $\mathsf{sk}^{(u)}$ has length $\mu \cdot \tau$ bits, the attack's complexity is $\mathcal{O}(2^{\mu \cdot \tau/2})$. More precisely, the attack complexity is at least $M(m, \gamma) \cdot 2^{\mu \cdot \tau/2}$, where $M(m, \gamma)$ is the time it takes to multiply $m \times m$ matrices with entries of size γ. We ensure $M(m, \gamma) \cdot 2^{\mu \cdot \tau/2} \geq 2^\lambda$.

7.4   Concrete Parameters and Implementation Results

In this section we propose concrete parameters for our key-exchange construction with $N = 4$ parties. These parameters are generated so that all known attacks have running time $\geq 2^\lambda$ clock cycles. In the construction the total number of encoded matrices is $2^\tau \cdot \ell \cdot N$ with $\tau = 3$, with a total degree $\ell = \mu \cdot k \cdot N$. Therefore, the total number of CLT13 encodings is $N_{CLT13} \simeq 2^\tau \cdot \ell \cdot N \cdot m^2$. The size of the secret key is $\tau\mu = 3\mu$ bits. The size η of the primes p_i is adjusted so that we extract $\nu = \lambda$ bits. During the publish phase, each party must broadcast $k \cdot (N - 1)$ matrices of dimension $m \times m$ and γ-bit entries. The size of those broadcasted values along with the other parameters are shown in Table 7.

Table 7. Concrete parameters for a 4-party key-exchange.

	λ	η	m	n	μ	α	k	$\gamma = n \cdot \eta$	ℓ	N_{CLT13}	params	Broadcast
Small	52	1759	6	160	15	11	2	$281 \cdot 10^3$	120	$1.4 \cdot 10^5$	4.8 GB	7.6 MB
Medium	62	2602	6	294	21	12	2	$764 \cdot 10^3$	168	$1.9 \cdot 10^5$	18.5 GB	20 MB
Large	72	3761	6	1349	27	14	2	$5073 \cdot 10^3$	216	$2.5 \cdot 10^5$	157.8 GB	137 MB
High	82	5159	9	4188	33	16	2	$21605 \cdot 10^3$	264	$6.8 \cdot 10^5$	1848.0 GB	1312 MB

The main difference with the original (insecure) key-exchange protocol from [CLT13] is that we get a much larger public parameter size; for $\lambda = 62$ bits of security, we need 18 GB of public parameters, instead of 70 MB originally. However our construction would be completely unpractical without Kilian's randomization on the encoding side. Namely for $\lambda = 62$ and a degree $\ell = 168$, one would need primes p_i of size $\eta \simeq (\alpha + \rho) \cdot \ell \simeq 2.4 \cdot 10^4$ with $\alpha = 80$ and $\rho = 62$ as in [CLT13]. Since $\gamma = \omega(\eta^2 \log \lambda)$ in [CLT13], one would need $\gamma \simeq 4 \cdot 10^9$. With $N_{CLT13} = 1.9 \cdot 10^5$, that would require 100 TB of public parameter size. Hence Kilian's randomization on the encoding side provides a reduction of the public parameter size by a factor $\simeq 10^4$.

We have implemented the key-exchange protocol in SAGE [S+17] and executed it on a machine with processor Intel Core i5-8600K CPU (3.60 GHz), 32 GB of RAM, and Ubuntu 18.04.2 LTS. The execution times are shown in Table 8. We could not run the Large and High instantiations ($\lambda = 72$ and $\lambda = 82$) because of the huge parameter size. While the Setup time is significant, since we need to sample all the random values and perform expensive operations like CRT and inverting matrices, the Publish and KeyGen times remain reasonable. In fact, each user just has to multiply $m \times m$ matrices $\mu \cdot k \cdot (N - 1)$ times to publish their values and $k \cdot (\mu + N)$ times to derive the shared key. We provide the source code of the key-exchange in [CP19].

Table 8. Timings for a 4-party key-exchange.

	Setup (once)	Publish (per party)	KeyGen (per party)
Small	2 h 20 min	45 s	19 s
Medium	12 h 23 min	3 min 35 s	1 min 24 s

8 Conclusion

We have shown that Kilian's randomization "on the encoding side" can bring orders of magnitude efficiency improvements for iO based constructions when instantiated with CLT13 multilinear maps. As an application, we have described the first concrete implementation of multipartite DH key exchange secure against existing attacks. The main advantage of Kilian's randomization is that it can be applied essentially for free in any existing implementation; for example it could be easily integrated in the 5Gen framework [LMA+16] for experimenting with program obfuscation constructions.

References

[AGIS14] Ananth, P.V., Gupta, D., Ishai, Y., Sahai, A.: Optimizing obfuscation: avoiding Barrington's theorem. In: ACM CCS. ACM (2014)

[Bar86] Barrington, D.A.M.: Bounded-width polynomial-size branching programs recognize exactly those languages in NC^1. In: Proceedings of the 18th Annual ACM Symposium on Theory of Computing, Berkeley, California, USA, 28–30 May 1986 (1986)

[BDGL16] Becker, A., Ducas, L., Gama, N., Laarhoven, T.: New directions in nearest neighbor searching with applications to lattice sieving. In: Proceedings of the Twenty-Seventh Annual ACM-SIAM Symposium on Discrete Algorithms, SODA 2016, Arlington, VA, USA, 10–12 January 2016 (2016)

[Ber03] Bernstein, D.J.: Fast multiplication and its applications. Algorithmic Number Theory **44**, 325–384 (2003)

[BGK+14] Barak, B., Garg, S., Kalai, Y.T., Paneth, O., Sahai, A.: Protecting obfuscation against algebraic attacks. In: Nguyen, P.Q., Oswald, E. (eds.) EUROCRYPT 2014. LNCS, vol. 8441, pp. 221–238. Springer, Heidelberg (2014). https://doi.org/10.1007/978-3-642-55220-5_13

[BISW17] Boneh, D., Ishai, Y., Sahai, A., Wu, D.J.: Lattice-based SNARGs and their application to more efficient obfuscation. In: Coron, J.-S., Nielsen, J.B. (eds.) EUROCRYPT 2017. LNCS, vol. 10212, pp. 247–277. Springer, Cham (2017). https://doi.org/10.1007/978-3-319-56617-7_9

[BMSZ16] Badrinarayanan, S., Miles, E., Sahai, A., Zhandry, M.: Post-zeroizing obfuscation: new mathematical tools, and the case of evasive circuits. In: Fischlin, M., Coron, J.-S. (eds.) EUROCRYPT 2016. LNCS, vol. 9666, pp. 764–791. Springer, Heidelberg (2016). https://doi.org/10.1007/978-3-662-49896-5_27

[BS03] Boneh, D., Silverberg, A.: Applications of multilinear forms to cryptography. Contemp. Math. **324**, 71–90 (2003)

[CGH+15] Coron, J.-S., et al.: Zeroizing without low-level zeroes: new MMAP attacks and their limitations. In: Gennaro, R., Robshaw, M. (eds.) CRYPTO 2015. LNCS, vol. 9215, pp. 247–266. Springer, Heidelberg (2015). https://doi.org/10.1007/978-3-662-47989-6_12

[CGH17] Chen, Y., Gentry, C., Halevi, S.: Cryptanalyses of candidate branching program obfuscators. In: Coron, J.-S., Nielsen, J.B. (eds.) EUROCRYPT 2017. LNCS, vol. 10212, pp. 278–307. Springer, Cham (2017). https://doi.org/10.1007/978-3-319-56617-7_10

[CHL+15] Cheon, J.H., Han, K., Lee, C., Ryu, H., Stehlé, D.: Cryptanalysis of the multilinear map over the integers. In: Oswald, E., Fischlin, M. (eds.) EUROCRYPT 2015. LNCS, vol. 9056, pp. 3–12. Springer, Heidelberg (2015). https://doi.org/10.1007/978-3-662-46800-5_1

[CLLT16] Coron, J.-S., Lee, M.S., Lepoint, T., Tibouchi, M.: Cryptanalysis of GGH15 multilinear maps. In: Robshaw, M., Katz, J. (eds.) CRYPTO 2016. LNCS, vol. 9815, pp. 607–628. Springer, Heidelberg (2016). https://doi.org/10.1007/978-3-662-53008-5_21

[CLLT17] Coron, J.-S., Lee, M.S., Lepoint, T., Tibouchi, M.: Zeroizing attacks on indistinguishability obfuscation over CLT13. In: Fehr, S. (ed.) PKC 2017. LNCS, vol. 10174, pp. 41–58. Springer, Heidelberg (2017). https://doi.org/10.1007/978-3-662-54365-8_3

[CLT13] Coron, J.-S., Lepoint, T., Tibouchi, M.: Practical multilinear maps over the integers. In: Canetti, R., Garay, J.A. (eds.) CRYPTO 2013. LNCS, vol. 8042, pp. 476–493. Springer, Heidelberg (2013). https://doi.org/10.1007/978-3-642-40041-4_26

[CN11] Chen, Y., Nguyen, P.Q.: BKZ 2.0: better lattice security estimates. In: Lee, D.H., Wang, X. (eds.) ASIACRYPT 2011. LNCS, vol. 7073, pp. 1–20. Springer, Heidelberg (2011). https://doi.org/10.1007/978-3-642-25385-0_1

[CN12] Chen, Y., Nguyen, P.Q.: Faster algorithms for approximate common divisors: breaking fully-homomorphic-encryption challenges over the integers. In: Pointcheval, D., Johansson, T. (eds.) EUROCRYPT 2012. LNCS, vol. 7237, pp. 502–519. Springer, Heidelberg (2012). https://doi.org/10.1007/978-3-642-29011-4_30

[CP18] Coron, J.-S., Pereira, H.V.L.: On kilian's randomization of multilinear map encodings. Cryptology ePrint Archive, Report 2018/1129 (2018). https://eprint.iacr.org/2018/1129

[CP19] Coron, J.-S., Pereira, H.V.L.: Implementation of key-exchange based on CLT13 multilinear maps (2019). https://github.com/coron/cltexchangeimpl

[CVW18] Chen, Y., Vaikuntanathan, V., Wee, H.: GGH15 beyond permutation branching programs: proofs, attacks, and candidates. In: Shacham, H., Boldyreva, A. (eds.) CRYPTO 2018. LNCS, vol. 10992, pp. 577–607. Springer, Cham (2018). https://doi.org/10.1007/978-3-319-96881-0_20

[DGHV10] van Dijk, M., Gentry, C., Halevi, S., Vaikuntanathan, V.: Fully homomorphic encryption over the integers. In: Gilbert, H. (ed.) EUROCRYPT 2010. LNCS, vol. 6110, pp. 24–43. Springer, Heidelberg (2010). https://doi.org/10.1007/978-3-642-13190-5_2

[GGH13a] Garg, S., Gentry, C., Halevi, S.: Candidate multilinear maps from ideal lattices. In: Johansson, T., Nguyen, P.Q. (eds.) EUROCRYPT 2013. LNCS, vol. 7881, pp. 1–17. Springer, Heidelberg (2013). https://doi.org/10.1007/978-3-642-38348-9_1

[GGH+13b] Garg, S., Gentry, C., Halevi, S., Raykova, M., Sahai, A., Waters, B.: Candidate indistinguishability obfuscation and functional encryption for all circuits. In: FOCS. IEEE Computer Society (2013)

[GGH15] Gentry, C., Gorbunov, S., Halevi, S.: Graph-induced multilinear maps from lattices. In: Dodis, Y., Nielsen, J.B. (eds.) TCC 2015. LNCS, vol. 9015, pp. 498–527. Springer, Heidelberg (2015). https://doi.org/10.1007/978-3-662-46497-7_20

[GGM16] Galbraith, S.D., Gebregiyorgis, S.W., Murphy, S.: Algorithms for the approximate common divisor problem. LMS J. Comput. Math. 19(A), 58–72 (2016)

[HJ16] Hu, Y., Jia, H.: Cryptanalysis of GGH map. In: Fischlin, M., Coron, J.-S. (eds.) EUROCRYPT 2016. LNCS, vol. 9665, pp. 537–565. Springer, Heidelberg (2016). https://doi.org/10.1007/978-3-662-49890-3_21

[HPS11] Hanrot, G., Pujol, X., Stehlé, D.: Analyzing blockwise lattice algorithms using dynamical systems. In: Rogaway, P. (ed.) CRYPTO 2011. LNCS, vol. 6841, pp. 447–464. Springer, Heidelberg (2011). https://doi.org/10.1007/978-3-642-22792-9_25

[Kil88] Kilian, J.: Founding cryptography on oblivious transfer. In: Proceedings of the 20th Annual ACM Symposium on Theory of Computing, Chicago, Illinois, USA, 2–4 May 1988 (1988)

[LMA+16] Lewi, K., et al.: 5Gen: a framework for prototyping applications using multilinear maps and matrix branching programs. In: Proceedings of the 2016 ACM SIGSAC Conference on Computer and Communications Security, CCS 2016 (2016)

[LS14] Lee, H.T., Seo, J.H.: Security analysis of multilinear maps over the integers. In: Garay, J.A., Gennaro, R. (eds.) CRYPTO 2014. LNCS, vol. 8616, pp. 224–240. Springer, Heidelberg (2014). https://doi.org/10.1007/978-3-662-44371-2_13

[MSW14] Miles, E., Sahai, A., Weiss, M.: Protecting obfuscation against arithmetic attacks. Cryptology ePrint Archive, Report 2014/878 (2014). https://eprint.iacr.org/2014/878

[MSZ16] Miles, E., Sahai, A., Zhandry, M.: Annihilation attacks for multilinear maps: cryptanalysis of indistinguishability obfuscation over GGH13. In: Robshaw, M., Katz, J. (eds.) CRYPTO 2016. LNCS, vol. 9815, pp. 629–658. Springer, Heidelberg (2016). https://doi.org/10.1007/978-3-662-53008-5_22

[MZ18] Ma, F., Zhandry, M.: The MMap strikes back: obfuscation and new multilinear maps immune to CLT13 zeroizing attacks. In: Beimel, A., Dziembowski, S. (eds.) TCC 2018. LNCS, vol. 11240, pp. 513–543. Springer, Cham (2018). https://doi.org/10.1007/978-3-030-03810-6_19

[NS97] Nguyen, P., Stern, J.: Merkle-Hellman revisited: a cryptanalysis of the Qu-Vanstone cryptosystem based on group factorizations. In: Kaliski, B.S. (ed.) CRYPTO 1997. LNCS, vol. 1294, pp. 198–212. Springer, Heidelberg (1997). https://doi.org/10.1007/BFb0052236

[NS01] Nguyen, P.Q., Stern, J.: The two faces of lattices in cryptology. In: Silverman, J.H. (ed.) CaLC 2001. LNCS, vol. 2146, pp. 146–180. Springer, Heidelberg (2001). https://doi.org/10.1007/3-540-44670-2_12

[PST14] Pass, R., Seth, K., Telang, S.: Indistinguishability obfuscation from semantically-secure multilinear encodings. In: Garay, J.A., Gennaro, R. (eds.) CRYPTO 2014. LNCS, vol. 8616, pp. 500–517. Springer, Heidelberg (2014). https://doi.org/10.1007/978-3-662-44371-2_28

[S+17] Stein, W.A., et al.: Sage Mathematics Software (Version 8.0). The Sage Development Team (2017). http://www.sagemath.org

[SW14] Sahai, A., Waters, B.: How to use indistinguishability obfuscation: deniable encryption, and more. In: Proceedings of the Forty-Sixth Annual ACM Symposium on Theory of Computing, STOC 2014, New York, NY, USA. ACM (2014)

Cryptanalysis of CLT13 Multilinear Maps with Independent Slots

Jean-Sébastien Coron$^{(\boxtimes)}$ and Luca Notarnicola

University of Luxembourg, Luxembourg City, Luxembourg
{jean-sebastien.coron,luca.notarnicola}@uni.lu

Abstract. Many constructions based on multilinear maps require independent slots in the plaintext, so that multiple computations can be performed in parallel over the slots. Such constructions are usually based on CLT13 multilinear maps, since CLT13 inherently provides a composite encoding space, with a plaintext ring $\bigoplus_{i=1}^{n} \mathbb{Z}/g_i\mathbb{Z}$ for small primes g_i's. However, a vulnerability was identified at Crypto 2014 by Gentry, Lewko and Waters, with a lattice-based attack in dimension 2, and the authors have suggested a simple countermeasure. In this paper, we identify an attack based on higher dimension lattice reduction that breaks the author's countermeasure for a wide range of parameters. Combined with the Cheon *et al.* attack from Eurocrypt 2015, this leads to the recovery of all the secret parameters of CLT13, assuming that low-level encodings of almost zero plaintexts are available. We show how to apply our attack against various constructions based on composite-order CLT13. For the [FRS17] construction, our attack enables to recover the secret CLT13 plaintext ring for a certain range of parameters; however, breaking the indistinguishability of the branching program remains an open problem.

1 Introduction

Multilinear Maps. In 2013, Garg, Gentry and Halevi described the first plausible construction of cryptographic multilinear maps based on ideal lattices [GGH13a]. Since then many amazing applications of multilinear maps have been found in cryptography, including program obfuscation [GGH+13b]. Shortly after the publication of GGH13, an analogous construction over the integers was described in [CLT13], based on the DGHV fully homomorphic encryption scheme [DGHV10]. The GGH15 scheme is the third known family of multilinear maps, based on the LWE problem with encoding over matrices [GGH15].

In the last few years, many attacks have appeared against multilinear maps, and the security of multilinear maps is still poorly understood. An important class of attacks against multilinear maps are "zeroizing attacks", which can recover the secret parameters from encodings of zero, using linear algebra. For the non-interactive multipartite Diffie-Hellman key exchange, the zeroizing attack from Cheon *et al.* [CHL+15] recovers all secret parameters from CLT13; the attack can also be extended to encoding variants where encodings of zero are not directly available [CGH+15]. The zeroizing attack from [HJ16] also breaks

S. D. Galbraith and S. Moriai (Eds.): ASIACRYPT 2019, LNCS 11922, pp. 356–385, 2019.
https://doi.org/10.1007/978-3-030-34621-8_13

the Diffie-Hellman key-exchange over GGH13. Finally, the key exchange over GGH15 was also broken in [CLLT16], using an extension of the Cheon *et al.* zeroizing attack.

Even though direct multipartite key exchange protocols are broken for the three known families of multilinear maps, more complex constructions based on multilinear maps are not necessarily broken, in particular indistinguishability obfuscation (iO); namely low-level encodings of zero are generally not available in iO constructions. However the Cheon *et al.* attack against CLT13 was extended in [CGH+15] to matrix branching programs where the input can be partitioned into three independent sets. The attack was further extended in [CLLT17] to branching programs without a simple input partition structure, using a tensoring technique. For GGH13 based obfuscation, Miles, Sahai and Zhandry introduced "annihilation attacks" that can break a certain class of matrix branching programs [MSZ16]; the attack was later extended in [CGH17] to break the [GGH+13b] obfuscation under GGH13, using a variant of the input partitioning attack. Finally, Chen, Vaikuntanathan and Wee described in [CVW18] an attack against iO over GGH15, based on computing the rank of a well chosen matrix. In general, the above attacks only apply against branching programs with a simple structure, and breaking more complex constructions (such as dual-input branching programs) is currently infeasible.

Multilinear Maps with Independent Slots. Many constructions based on multilinear maps require independent slots in the plaintext, so that multiple computations can be performed in parallel over the slots when evaluating the multilinear map. For example, [GLW14] and [GLSW15] use independent slots to obtain improved security reductions for witness encryption and obfuscation. Multilinear maps with independent slots were also used in the circuit based constructions of [AB15, Zim15]. The construction from [FRS17], which gives a powerful technique for preventing zeroizing attacks against iO, is also based on multilinear maps with independent slots.

The CLT13 multilinear map scheme inherently supports a composite integer encoding space, with a plaintext ring $\mathbb{Z}/G\mathbb{Z} \simeq \bigoplus_{i=1}^{n} \mathbb{Z}/g_i\mathbb{Z}$ for small secret primes g_i's and $G = g_1 \cdots g_n$. For example, in the construction from [FRS17], every branching program works independently modulo each g_i. In that case, the main difference with the original CLT13 is that the attacker can obtain encodings of subring elements which are zero modulo all g_i's except one; for example, in [FRS17] this would be done by carefully choosing the input so that all branching programs would evaluate to zero except one. Whereas in the original CLT13 construction, one never provides encodings of subring elements; instead one uses an "all-or-nothing" approach: either the plaintext element is zero modulo all g_i's, or it is non-zero modulo all g_i's (with high probability).

The Attack and Countermeasure from [GLW14]. At Crypto 2014, Gentry, Lewko and Waters observed that using CLT13 with independent slots leads to a simple lattice attack in dimension 2, which efficiently recovers the (secret) plaintext ring $\bigoplus_{i=1}^{n} \mathbb{Z}/g_i\mathbb{Z}$ [GLW14, Appendix B]. Namely, when using CLT13 with independent slots, the attacker can obtain encodings where all slots are zero modulo g_i except one. For example, for a matrix branching program evaluation as in [FRS17], the result of the program evaluation could have the form:

$$A(x) \equiv \sum_{i=1}^{n} h_i \cdot (r_i + m_i \cdot (g_i^{-1} \bmod p_i)) \cdot \frac{x_0}{p_i} \quad (\bmod \ x_0)$$

where $m_i = 0$ for all i except $m_j \neq 0$ for some $1 \leq j \leq n$. This implies:

$$g_j \cdot A(x) \equiv h_j(r_j g_j + m_j)\frac{x_0}{p_j} + \sum_{i \neq j} g_j h_i r_i \frac{x_0}{p_i} \quad (\bmod \ x_0)$$

and therefore $g_j \cdot A(x) \bmod x_0$ is "small" (significantly smaller than x_0). Since g_j is very small, we can then recover g_j using lattice reduction in dimension 2, while normally the g_i's are secret in CLT13. Moreover, once we know g_j, we can simply multiply the evaluation by g_j to obtain a "small" result, even if the evaluation of the branching program is non-zero modulo g_j; in particular, this cancels the effect of the protection against input partitioning from [FRS17].

The countermeasure considered in [GLW14, Appendix B] is to give many "buddies" to each g_i, so that we do not have a plaintext element which is non-zero modulo a single isolated g_i. Then, either an encoding is 0 modulo g_i and all its prime buddies g_j, or it is (with high probability) non-zero modulo all of them. In other words, instead of using individual g_i's to define the plaintext slots, every slot is defined modulo a product of θ prime g_i's, for some $1 \leq \theta < n$. Therefore, we obtain a total of $\lfloor n/\theta \rfloor$ plaintext slots (instead of n). While the above attack can be extended by multiplying $A(x)$ by the θ corresponding g_i's, for large enough θ the right-hand side of the equation is not "small" anymore and the attack is thwarted.

Our Contributions. In this paper we identify an attack based on higher dimension lattice reduction that breaks the countermeasure from [GLW14, Appendix B] for a wide range of parameters, with significant impact on the security of CLT13 multilinear maps with independent slots. More precisely, our contributions are as follows:

1. **Analysis of the attack from** [GLW14]. Our first contribution is to provide a theoretical study of the above attack, in order to derive a precise bound on θ as a function of the CLT13 parameters (there was no explicit bound in [GLW14]), where θ is the number of primes g_i's for each plaintext slot. We argue that, when ν denotes the number of bits that can be extracted from zero-testing in CLT13, the 2-dimensional lattice attack requires:

$$\alpha\theta < \frac{\nu}{2} \tag{1}$$

where α is the bit size of the g_i's.

2. **Breaking the countermeasure from** [GLW14]. Our main contribution is to extend the 2-dimensional attack to break the countermeasure for larger values of θ. Our attack is based on higher dimension lattice reduction, by using a similar orthogonal lattice attack as in [NS99] for solving the hidden subset sum problem. In this extension, we use ℓ encodings $\{c_j : 1 \leq j \leq \ell\}$ where the corresponding plaintexts have only θ non-zero components modulo the g_i's (instead of $\ell = 1$ in the previous attack). Using a lattice attack in dimension $\ell + 1$, we show that our attack requires the approximate condition $\left(1 + \frac{1}{\ell}\right)\alpha\theta < \nu$ for the parameters. Therefore, for moderately large values of ℓ, we get the simpler condition:

$$\alpha\theta < \nu$$

which improves (1) by a factor 2.

In the same vein, we show how to further improve this condition by considering products of encodings of the form $c_j \cdot d_k$ for $1 \leq j \leq \ell$ and $1 \leq k \leq d$, where as previously, the plaintexts of the c_j's have only θ non-zero components modulo the g_i's. In that case, using a variant of the previous lattice attack (this time in dimension $\ell + d$), the bound improves to:

$$\alpha\theta = \mathcal{O}(\nu^2)$$

The above bound also applies when a vector of zero-testing elements is available, instead of a single p_{zt}. While the original attack from [GLW14] recovers the secret plaintext ring of CLT13, we additionally recover the plaintext messages $\{m_j : 1 \leq j \leq \ell\}$ for the encodings $\{c_j : 1 \leq j \leq \ell\}$, up to a scaling factor.

We provide in Sect. 4.5 the result of practical experiments. For the original parameters of [CLT13], our attack takes a few seconds for $\theta = 40$, and a few hours for θ as large as 160, while the original attack from [GLW14] only works for $\theta = 1$. In summary, our attack is more powerful than the attack in [GLW14], as it additionally recovers secret information about the plaintext messages, moreover for much larger values of θ. Finally, we suggest a set of secure parameters for CLT13 multilinear maps that prevents our extended attack. For $\lambda = 80$ bits of security, we recommend to take $\theta \geq 1789$.

3. **Recovering all the secret parameters of CLT13.** For the range of parameters derived previously, we show how to combine our attack with the Cheon et al. attack from [CHL+15], in order to recover all secret parameters of CLT13. More precisely, when intermediate-level encodings of partially zero messages are available, our approach consists in applying the lattice attack to generate intermediate-level encodings of zero; then the Cheon et al. attack is applied on these newly-created encodings of zero, to recover all secret parameters.

4. **Application to CLT13-based constructions.** Finally we show how our attack affects the parameter selection of several schemes based on CLT13 multilinear maps with independent slots, namely the constructions from [GLW14, GLSW15, Zim15] and [FRS17]. For the [FRS17] construction, our

attack enables to recover the secret CLT13 plaintext ring for a certain range of parameters; however, breaking the indistinguishability of the branching program remains an open problem.

Source Code. We provide in https://pastebin.com/7WEMHBE9 the source code of our attacks in Sage [S+17].

2   The CLT13 Multilinear Map Scheme

We first recall the CLT13 multilinear map scheme over the integers [CLT13]. For $n \in \mathbb{Z}_{\geq 1}$, the instance generation of CLT13 generates n distinct secret "large" primes $p_1, \ldots, p_n$ of size η bits, and publishes the modulus $x_0 = \prod_{i=1}^n p_i$. We let γ denote the bit size of x_0; therefore $\gamma \simeq n \cdot \eta$. One also generates n distinct secret "small" prime numbers $g_1, \ldots, g_n$ of size α bits. The plaintext ring is composite, *i.e.* a plaintext is an element $\boldsymbol{m} = (m_1, \ldots, m_n)$ of the ring $\mathbb{Z}/G\mathbb{Z} \simeq \bigoplus_{i=1}^n \mathbb{Z}/g_i\mathbb{Z}$ where $G = \prod_{i=1}^n g_i$. Let $\kappa \in \mathbb{Z}_{\geq 1}$ be the multilinearity parameter. For $k \in \{1, \ldots, \kappa\}$, an encoding at level k of the plaintext $\boldsymbol{m}$ is an integer $c \in \mathbb{Z}$ such that

$$c \equiv \frac{r_i g_i + m_i}{z^k} \pmod{p_i}, \text{ for all } 1 \leq i \leq n \tag{2}$$

for "small" random integers r_i of bit size ρ. The random mask $z \in (\mathbb{Z}/x_0\mathbb{Z})^\times$ is the same for all encodings. It is clear that two encodings at the same level can be added, and the underlying plaintexts get added in the ring $\mathbb{Z}/G\mathbb{Z}$. Similarly, the product of two encodings at level i and j gives an encoding of the product plaintexts at level $i + j$, as long as the numerators in (2) do not grow too large, *i.e.* they must remain smaller than each p_i.

For an encoding at the last level κ, one defines the following zero-testing procedure. The instance generation publishes the zero-testing parameter p_{zt}, defined by

$$p_{zt} = \sum_{i=1}^n h_i z^\kappa (g_i^{-1} \bmod p_i) \frac{x_0}{p_i} \bmod x_0, \tag{3}$$

where $h_i \in \mathbb{Z}$ are "small" random integers of size n_h bits. Given an encoding c at the last level κ, we compute the integer:

$$\omega := p_{zt} \cdot c \bmod x_0 \equiv \sum_{i=1}^n h_i(r_i + m_i(g_i^{-1} \bmod p_i)) \frac{x_0}{p_i} \pmod{x_0} \tag{4}$$

and we consider that c encodes the zero message if ω is "small" compared to x_0. Namely, if $m_i = 0$ for all i, we obtain $\omega \equiv \sum_{i=1}^n h_i r_i \frac{x_0}{p_i} \pmod{x_0}$, and since the integers h_i and r_i are "small", the resulting ω will be "small" compared to x_0.

More precisely, let ρ_f be the maximum bit size of the noise r_i in the encodings. Then the integers $h_i r_i x_0/p_i$ have size roughly $\gamma - \eta + n_h + \rho_f$, and therefore letting

$$\nu = \eta - n_h - \rho_f, \tag{5}$$

the integers $h_i r_i x_0 / p_i$ have size roughly $\gamma - \nu$ bits. Therefore, when $m_i = 0$ for all i, the integer ω has size roughly $\gamma - \nu$ bits; whereas when $m_i \neq 0$ for some i, we expect that ω is of full size modulo x_0, that is γ bits. The parameter ν in (5) corresponds to the number of bits that can be extracted from zero-testing; namely from (4), the ν most significant bits of ω only depend on the plaintext messages m_i, and not on the noise r_i. Note that to get a proper zero-testing procedure, one needs to use a *vector* of n elements p_{zt}; namely with a single p_{zt} there exist encodings c with $m_i \neq 0$ while $p_{zt} \cdot c$ is "small" modulo x_0. In the rest of the paper, for simplicity, we mainly consider a single p_{zt}, as it is usually the case in constructions over CLT13 multilinear maps. We refer to [CLT13, Section 3.1] for the setting of the parameters.

3 Basic Attack Against CLT13 with Independent Slots

Many constructions based on multilinear maps require independent slots in the plaintext, so that multiple computations can be performed in parallel over the slots when evaluating the multilinear map; see for example [GLW14, GLSW15] and [AB15, Zim15, FRS17]. The CLT13 multilinear maps inherently provide independent slots, as the plaintext ring is $\bigoplus_{i=1}^{n} \mathbb{Z}/g_i\mathbb{Z}$ for small secret primes $g_1, \ldots, g_n$. Therefore we can have independent computations performed over the n plaintext slots modulo g_i; for example, in the construction from [FRS17], every branching program works independently modulo each g_i.

The Basic Attack from [GLW14]. When using CLT13 with independent slots, the attacker can obtain encodings of plaintext elements where all slots are zero modulo g_i except one. For example, in the [FRS17] construction where each branching program works modulo g_i, the attacker can choose the input so that the resulting evaluation is 0 modulo all g_i's except one, say g_1, without loss of generality. Let c be a level-κ encoding of a plaintext $\boldsymbol{m} = (m_1, \ldots, m_n)$ where $m_i = 0$ for all $2 \leq i \leq n$. From Eq. (4) we obtain the following zero-testing evaluation:

$$\omega \equiv h_1 \cdot m_1 \cdot (g_1^{-1} \bmod p_1) \cdot \frac{x_0}{p_1} + \sum_{i=1}^{n} h_i \cdot r_i \cdot \frac{x_0}{p_i} \pmod{x_0}$$

This implies:

$$g_1 \cdot \omega \equiv h_1 \cdot m_1 \cdot \frac{x_0}{p_1} + \sum_{i=1}^{n} g_1 \cdot h_i \cdot r_i \cdot \frac{x_0}{p_i} \pmod{x_0}$$

and therefore $g_1 \cdot \omega \bmod x_0$ is significantly smaller than x_0, as the integers h_i and r_i are "small". This implies that we can recover g_1, and similarly the other g_i's using lattice reduction in dimension 2, while normally the g_i's are secret in CLT13. This eventually recovers the plaintext ring. We analyze the attack below.

The Countermeasure from [GLW14]. The following countermeasure was therefore suggested by the authors: instead of using individual g_i's to define the plaintext slots, every slot is defined modulo a product of θ prime g_i's, where $2 \leq \theta < n$. Therefore, a plaintext element cannot be non-zero modulo a single prime g_i; it has to be non-zero modulo at least θ primes g_i's. This gives a total of n/θ plaintext slots (instead of n); for simplicity we assume that θ divides n.

Therefore, the original plaintext ring $R = \mathbb{Z}/g_1\mathbb{Z} \times \cdots \times \mathbb{Z}/g_n\mathbb{Z}$ can be rewritten as $R = \bigoplus_{j=1}^{n/\theta} R_j$, where for all $1 \leq j \leq n/\theta$, the subrings R_j are such that $R_j \simeq \bigoplus_{i=1}^{\theta} \mathbb{Z}/g_{(j-1)\theta+i}\mathbb{Z}$. We can assume that the attacker can obtain encodings of random subring plaintexts in R_j for any $1 \leq j \leq n/\theta$. In that case, the attacker obtains an encoding c of $\boldsymbol{m} = (m_1, \ldots, m_n) \in R$ where $m_i \equiv 0$ (mod g_i) for all $i \in \{1, \ldots, n\} \setminus \{(j-1)\theta+1, \ldots, j\theta\}$. In that case we will say that $\boldsymbol{m}$ has non-zero support of length θ.

Analysis of the Basic Attack. In this section we analyze in more details the attack from [GLW14], and we derive an explicit bound on the parameter θ, as a function of the other CLT13 parameters. Given an integer $1 \leq \theta < n$ (the above attack is obtained for $\theta = 1$), we consider a message having non-zero support of length θ; that is, (without loss of generality) of the form $\boldsymbol{m} = (m_1, \ldots, m_n) \in \mathbb{Z}^n$ with $0 \leq m_i < g_i$ such that $m_i = 0$ for $\theta + 1 \leq i \leq n$, i.e. we assume that the non-zero support of $\boldsymbol{m}$ is located in the first slot. We consider a top level κ encoding c of $\boldsymbol{m}$, that is:

$$c \equiv \frac{r_i g_i + m_i}{z^\kappa} \quad (\text{mod } p_i), \quad 1 \leq i \leq n$$

with integers r_i of bit size ρ_f. From zero-testing, we obtain from (4):

$$\omega \equiv p_{zt} \cdot c \equiv \sum_{i=1}^{\theta} h_i(g_i^{-1} \bmod p_i) m_i \frac{x_0}{p_i} + \sum_{i=1}^{n} h_i r_i \frac{x_0}{p_i} \quad (\text{mod } x_0)$$

By multiplying out by $g := \prod_{i=1}^{\theta} g_i$ we obtain

$$g\omega \equiv \sum_{i=1}^{\theta} h_i m_i \frac{g}{g_i} \frac{x_0}{p_i} + \sum_{i=1}^{n} g h_i r_i \frac{x_0}{p_i} \quad (\text{mod } x_0),$$

$$g\omega \equiv U \quad (\text{mod } x_0) \tag{6}$$

where $U = \sum_{i=1}^{\theta} h_i m_i (g/g_i)(x_0/p_i) + \sum_{i=1}^{n} g h_i r_i (x_0/p_i)$. Since the integers h_i and r_i are "small" in order to ensure correct zero-testing, the integer U is "small" in comparison to x_0. More precisely, the proposition below shows that if $g \cdot U$ is a bit smaller than x_0, then we can recover g and U by lattice reduction in dimension 2.

Proposition 1. Let $g, \omega, U \in \mathbb{Z}_{\geq 1}$ and $x_0 \in \mathbb{Z}_{\geq 1}$ be such that $g\omega \equiv U$ (mod x_0), $\omega \in (\mathbb{Z}/x_0\mathbb{Z})^\times$ and $\gcd(U, g) = 1$. Assume that $g \cdot U < x_0/10$. Given ω and x_0 as input, one can recover g and U in polynomial time.

Proof. Without loss of generality we can assume $g \leq U$, since otherwise we can apply the algorithm with $U\omega^{-1} \equiv g \pmod{x_0}$. Let $B \in \mathbb{Z}_{\geq 1}$ such that $U \leq Bg \leq 2U$. When the bit size of g and U is unknown, such a B can be found by exhaustive search in polynomial time. We consider the lattice $L \subseteq \mathbb{Z}^2$ of vectors (Bx, y) such that $x\omega \equiv y \pmod{x_0}$. From $g\omega \equiv U \pmod{x_0}$ it follows that L contains the vector $\boldsymbol{v} = (Bg, U)$. We show that $\boldsymbol{v}$ is a shortest non-zero vector in L.

By Minkowski's Theorem, we have $\lambda_1(L) \leq \sqrt{2 \det(L)}$. From Hadamard's Inequality, with $\det(L) = Bx_0$, we obtain:

$$\lambda_2(L) \geq \frac{\det(L)}{\lambda_1(L)} \geq \frac{\sqrt{\det(L)}}{\sqrt{2}} = \frac{\sqrt{Bx_0}}{\sqrt{2}} > \sqrt{5BgU} \geq \sqrt{5}U.$$

Moreover, we have:

$$\|\boldsymbol{v}\| = ((Bg)^2 + U^2)^{1/2} \leq \sqrt{5}U.$$

This implies that $\|\boldsymbol{v}\| < \lambda_2(L)$ and therefore $\boldsymbol{v}$ is a multiple of a shortest non-zero vector in L: we write $\boldsymbol{v} = k\boldsymbol{u}$ with $\|\boldsymbol{u}\| = \lambda_1(L)$, and $k \in \mathbb{Z}\backslash\{0\}$. Letting $\boldsymbol{u} = (Bu_1, u_2)$, we have $g = ku_1$ and $U = ku_2$. Hence k divides both g and U. Since $\gcd(g, U) = 1$ one has $k = \pm 1$. This shows that $\boldsymbol{v}$ is a shortest non-zero vector of L.

By running Lagrange-Gauss reduction on the matrix of row vectors:

$$\begin{bmatrix} B & \omega \\ 0 & x_0 \end{bmatrix}$$

one obtains in polynomial time a length-ordered basis $(\boldsymbol{b}_1, \boldsymbol{b}_2)$ of L satisfying $\|\boldsymbol{b}_1\| = \lambda_1(L)$ and $\|\boldsymbol{b}_2\| = \lambda_2(L)$, which enables to recover g and U. □

Using the same notations as in Sect. 2, the integer $g = \prod_{i=1}^{\theta} g_i$ has approximate bit size $\theta \cdot \alpha$, while the integer U has an approximate bit size $\gamma - \eta + n_h + \rho_f + \theta\alpha$. From the condition $g \cdot U < x_0/10$ of Proposition 1, we obtain by dropping the term $\log_2(10)$, the simplified condition

$$\gamma - \eta + n_h + \rho_f + \theta \cdot \alpha + \theta \cdot \alpha < \gamma.$$

Writing as previously $\nu = \eta - n_h - \rho_f$ for the number of bits that can be extracted during zero testing, the attack works under the condition:

$$2\alpha\theta < \nu \tag{7}$$

where α is the bit size of the g_i's. In the next section we describe a high-dimensional lattice reduction attack with an improved bound on θ.

4   An Extended Attack Against CLT13 with Independent Slots

Outline of Our New Attack. Our new attack improves the bound on θ compared to the attack recalled in Sect. 3; it also enables to recover multiples of the underlying plaintext messages, instead of only the CLT13 plaintext ring. The main difference is that we work with several messages instead of a single one, using high-dimensional lattice reduction instead of dimension 2.

Let $\ell \geq 1$ be an integer. Assume that we have ℓ level-κ encodings c_j of plaintext elements $\boldsymbol{m}_j = (m_{j1}, \ldots, m_{jn})$ for $1 \leq j \leq \ell$, where each message has non-zero support of length θ. Without loss of generality, we can assume that $m_{ji} = 0$ for all $\theta + 1 \leq i \leq n$ and all $1 \leq j \leq \ell$. We consider the zero-testing evaluations $\omega_j = p_{zt} \cdot c_j \bmod x_0$ of these encodings, which gives as previously:

$$\omega_j \equiv \sum_{i=1}^{\theta} h_i(r_{ji} + m_{ji}(g_i^{-1} \bmod p_i))\frac{x_0}{p_i} + \sum_{i=\theta+1}^{n} h_i r_{ji}\frac{x_0}{p_i} \pmod{x_0}, \quad 1 \leq j \leq \ell$$

for integers r_{ji}. We can rewrite the above equation as:

$$\omega_j \equiv \sum_{i=1}^{\theta} \alpha_i \cdot m_{ji} + R_j \pmod{x_0}, \quad 1 \leq j \leq \ell \tag{8}$$

for some integers α_i, where for each evaluation ω_j, the integer R_j is significantly smaller than x_0.

We can see Eq. (8) as an instance of a "noisy" hidden subset sum problem. Namely in [NS99], the authors consider the following hidden subset sum problem. Given a positive integer M, and a vector $\boldsymbol{b} = (b_1, \ldots, b_\ell) \in \mathbb{Z}^\ell$ with entries in $[0, M-1]$, find integers $\alpha_1, \ldots, \alpha_n \in [0, M-1]$ such that there exist vectors $\boldsymbol{x}_1, \ldots, \boldsymbol{x}_n \in \mathbb{Z}^\ell$ with entries in $\{0, 1\}$ satisfying:

$$\boldsymbol{b} \equiv \alpha_1 \boldsymbol{x}_1 + \alpha_2 \boldsymbol{x}_2 + \cdots + \alpha_n \boldsymbol{x}_n \pmod{M}$$

In our case, the weights $\alpha_1, \ldots, \alpha_n$ are hidden as in [NS99], but for each equation we have an additional hidden noisy term R_j. Moreover, the weights $\alpha_i = h_i \cdot (g_i^{-1} \bmod p_i) \cdot x_0/p_i$ have a special structure, instead of being random in [NS99]. Thanks to this special structure, using a variant of the orthogonal lattice approach from [NS99], we can recover the secret product $g = g_1 \cdots g_\theta$ and the plaintext elements m_{ji} up to a scaling factor.

4.1   Preliminaries on Lattices

Let L be a lattice in $\mathbb{R}^d$ of rank $0 < n \leq d$. We recall that Hadamard's Inequality gives the following upper bound on the determinant of L, for every basis $\mathcal{B}$ of L:

$$\det(L) \leq \prod_{b \in \mathcal{B}} \|\boldsymbol{b}\|$$

Based on Hadamard's Inequality, we prove the following simple lemma.

Lemma 2. *Let* $1 \leq n \leq d$ *be integers and let* $L \subseteq \mathbb{Z}^d$ *be a lattice of rank* n. *Let* $\boldsymbol{x}_1, \ldots, \boldsymbol{x}_{n-1} \in L$ *be linearly independent. Then for every vector* $\boldsymbol{y} \in L$ *not in the linear span of* $\boldsymbol{x}_1, \ldots, \boldsymbol{x}_{n-1}$, *one has* $\|\boldsymbol{y}\| \geq \det(L) / \prod_{i=1}^{n-1} \|\boldsymbol{x}_i\|$.

Proof. Since $\boldsymbol{x}_1, \ldots, \boldsymbol{x}_{n-1}, \boldsymbol{y} \in L$ are linearly independent, they generate a rank-n sublattice L' of L and hence $\det(L) \leq \det(L')$ as $\det(L)$ divides $\det(L')$. By Hadamard's Inequality, $\det(L) \leq \det(L') \leq \|\boldsymbol{y}\| \cdot \prod_{i=1}^{n-1} \|\boldsymbol{x}_i\|$. The bound follows. □

We recall that the LLL algorithm [LLL82], given an input basis of L, produces a reduced basis of L with respect to the choice of a parameter $\delta \in (1/4, 1)$; we call such a basis δ-*reduced*. More precisely, we will use the following theorem.

Theorem 3. *Let* $1 \leq n \leq d$ *be integers and let* $L \subseteq \mathbb{Z}^d$ *be a lattice of rank* n. *Let* $\{\boldsymbol{b}_i : 1 \leq i \leq n\}$ *be a basis of* L. *Let* $B \in \mathbb{Z}_{\geq 1}$ *be such that* $\|\boldsymbol{b}_i\|^2 \leq B$ *for* $1 \leq i \leq n$. *Let* $\delta \in (1/4, 1)$. *Then the* LLL *algorithm with reduction parameter* δ *outputs a* δ-*reduced basis* $\{\boldsymbol{b}_i' : 1 \leq i \leq n\}$ *after* $\mathcal{O}(n^5 d \log^3 B)$ *operations. Moreover, the first vector in such a basis satisfies:*

$$\|\boldsymbol{b}_1'\| \leq c^{(n-1)/2} \|\boldsymbol{x}\|$$

for every non-zero $\boldsymbol{x} \in L$, *and where* $c = 1/(\delta - 1/4)$.

4.2   Our First Lattice-Based Attack

Setting. In this section, we describe our first attack based on a variant of the hidden subset-sum problem. We consider plaintext elements $\boldsymbol{m}_1, \ldots, \boldsymbol{m}_\ell \in \mathbb{Z}^n$ and write m_{ji} for the i-th entry of the j-th message, where $0 \leq m_{ji} < g_i$ for all $1 \leq i \leq n$ and $1 \leq j \leq \ell$. As previously, we assume that $m_{ji} = 0$ for all $\theta + 1 \leq i \leq n$. We write M for the matrix of row vectors $\boldsymbol{m}_j$ for $1 \leq j \leq \ell$; and we will denote its columns by $\hat{\boldsymbol{m}}_i$ for $1 \leq i \leq n$, that is, $M = \begin{bmatrix} \hat{\boldsymbol{m}}_1 \mid \cdots \mid \hat{\boldsymbol{m}}_n \end{bmatrix} \in \text{Mat}_{\ell \times n}(\mathbb{Z})$. By construction, the vectors $\hat{\boldsymbol{m}}_i$ for $\theta + 1 \leq i \leq n$ are all zero. We also assume that for all $1 \leq i \leq \theta$, $\hat{\boldsymbol{m}}_i \not\equiv \boldsymbol{0} \pmod{g_i}$. For $1 \leq j \leq \ell$, we let c_j denote an encoding of $\boldsymbol{m}_j$ at the last level κ:

$$c_j \equiv \frac{r_{ji} g_i + m_{ji}}{z^\kappa} \pmod{p_i}, \quad 1 \leq i \leq n$$

where $r_{ji} \in \mathbb{Z}$ are ρ_f-bit integers. Letting $\boldsymbol{c} = (c_j)_{1 \leq j \leq \ell}$, this gives a vector equation over $\mathbb{Z}^\ell$:

$$\boldsymbol{c} \equiv z^{-\kappa} (g_i \boldsymbol{r}_i + \hat{\boldsymbol{m}}_i) \pmod{p_i}, \quad 1 \leq i \leq n \tag{9}$$

for $\boldsymbol{r}_i = (r_{ji})_{1 \leq j \leq \ell}$. Let p_{zt} be the zero-testing parameter, as defined in (3). From zero-testing we obtain the following equations:

$$\omega_j \equiv c_j \cdot p_{zt} \equiv \sum_{i=1}^{\theta} h_i m_{ji} (g_i^{-1} \bmod p_i) \frac{x_0}{p_i} + \sum_{i=1}^{n} h_i r_{ji} \frac{x_0}{p_i} \pmod{x_0}, \ 1 \leq j \leq \ell$$

which can be rewritten as $\omega_j \equiv \sum_{i=1}^{\theta} \alpha_i m_{ji} + R_j \pmod{x_0}$, where we use the shorthand notations:

$$\alpha_i := h_i(g_i^{-1} \bmod p_i)\frac{x_0}{p_i}, \quad 1 \leq i \leq \theta \tag{10}$$

and $R_j := \sum_{i=1}^{n} h_i r_{ji}\frac{x_0}{p_i}$ for $1 \leq j \leq \ell$. As a vector equation, this reads:

$$\boldsymbol{\omega} \equiv p_{zt} \cdot \boldsymbol{c} \equiv \sum_{i=1}^{\theta} \alpha_i \hat{\boldsymbol{m}}_i + \boldsymbol{R} \pmod{x_0} \tag{11}$$

with $\boldsymbol{\omega} = (\omega_j)_{1 \leq j \leq \ell}$; for $1 \leq i \leq \theta$ the vectors $\hat{\boldsymbol{m}}_i$ are as above and $\boldsymbol{R} = (R_j)_{1 \leq j \leq \ell} = \sum_{i=1}^{n} h_i \frac{x_0}{p_i}\boldsymbol{r}_i$.

In the above equation, the components of $\boldsymbol{R}$ have approximate bit size $\rho_R = \gamma - \eta + n_h + \rho_f$. Using, as previously, $\nu = \eta - n_h - \rho_f$ as the number of bits that can be extracted, we have therefore $\rho_R = \gamma - \nu$. As explained above, Eq. (11) is similar to an instance of the hidden subset sum problem, so we describe a variant of the orthogonal lattice attack from [NS99], which recovers the secret CLT13 plaintext ring and the hidden plaintexts $\{\hat{\boldsymbol{m}}_i : 1 \leq i \leq \theta\}$, up to a scaling factor. For the sequel, we assume that the prime numbers $g_1, \ldots, g_\theta$ are distinct, and that for every $1 \leq i \leq \theta$, we have $\gcd(g_i, h_i x_0/p_i) = 1$.

The Orthogonal Lattice L. We consider the lattice L of vectors $(B\boldsymbol{u}, v) \in \mathbb{Z}^{\ell+1}$, with $\boldsymbol{u} \in \mathbb{Z}^\ell$ and $v \in \mathbb{Z}$, such that $(\boldsymbol{u}, v)$ is orthogonal to $(\boldsymbol{\omega}, 1)$ modulo x_0, where $B \in \mathbb{Z}_{\geq 1}$ is a scaling factor that will be determined later. Since L contains the sublattice $x_0 \mathbb{Z}^{\ell+1}$, it has full-rank $\ell + 1$. We note that this lattice is known (i.e. we can construct a basis for it) since $\boldsymbol{\omega}$ and x_0 are given. Our attack is based on the fact that L contains a rank-ℓ sublattice L', generated by reasonably short vectors $\{(B\boldsymbol{u}_i, v_i) : 1 \leq i \leq \ell\}$ of L, which can be used to reveal the secret product $g - \prod_{i=1}^{\theta} g_i$.

More precisely, for every $(B\boldsymbol{u}, v) \in L$, we obtain from (11):

$$\langle \boldsymbol{u}, \boldsymbol{\omega} \rangle + v \equiv \sum_{i=1}^{\theta} \alpha_i \langle \boldsymbol{u}, \hat{\boldsymbol{m}}_i \rangle + \langle \boldsymbol{u}, \boldsymbol{R} \rangle + v \equiv 0 \pmod{x_0}$$

and therefore, the vector $(\langle \boldsymbol{u}, \hat{\boldsymbol{m}}_1 \rangle, \ldots, \langle \boldsymbol{u}, \hat{\boldsymbol{m}}_\theta \rangle, \langle \boldsymbol{u}, \boldsymbol{R} \rangle + v)$ is orthogonal modulo x_0 to the vector $\boldsymbol{a} = (\alpha_1, \ldots, \alpha_\theta, 1)$. To obtain balanced components, we use another scaling factor $C \in \mathbb{Z}_{\geq 1}$ and we consider the vector:

$$\boldsymbol{p}_{\boldsymbol{u},v} := (C\langle \boldsymbol{u}, \hat{\boldsymbol{m}}_1 \rangle, \ldots, C\langle \boldsymbol{u}, \hat{\boldsymbol{m}}_\theta \rangle, \langle \boldsymbol{u}, \boldsymbol{R} \rangle + v)$$

Following the original orthogonal lattice attack from [NS99], if a vector $(B\boldsymbol{u}, v) \in L$ is short enough, then the associated vector $\boldsymbol{p}_{\boldsymbol{u},v} = (C\boldsymbol{x}, y)$ will also be short, and if $(\boldsymbol{x}, y)$ becomes shorter than a shortest non-zero vector orthogonal to $\boldsymbol{a}$ modulo x_0, we must have $\boldsymbol{p}_{\boldsymbol{u},v} = 0$, which implies $\langle \boldsymbol{u}, \hat{\boldsymbol{m}}_i \rangle = 0$ for all $1 \leq i \leq \theta$. We will see that in our setting, because of the specific structure of the coefficients α_i's, we only get $\langle \boldsymbol{u}, \hat{\boldsymbol{m}}_i \rangle \equiv 0 \pmod{g_i}$ for all $1 \leq i \leq \theta$.

Therefore, by applying lattice reduction to L, we expect to recover the lattice $\Lambda^\perp$ of vectors $\boldsymbol{u}$ which are orthogonal to all $\hat{\boldsymbol{m}}_i$ modulo g_i; since by assumption $\hat{\boldsymbol{m}}_i \not\equiv \boldsymbol{0} \pmod{g_i}$ for all $1 \leq i \leq \theta$, the lattice $\Lambda_i^\perp = \{\boldsymbol{u} \in \mathbb{Z}^\ell : \langle \boldsymbol{u}, \hat{\boldsymbol{m}}_i \rangle \equiv 0 \pmod{g_i}\}$ has determinant g_i, and since $g_1, \ldots, g_\theta$ are distinct primes, the lattice $\Lambda^\perp = \cap_{i=1}^\theta \Lambda_i^\perp$ has determinant equal to $g = \prod_{i=1}^\theta g_i$. In particular, any basis for this lattice reveals g by computing its determinant.

The Lattice $A^\perp$. Henceforth, we must study the short vectors in the lattice of vectors orthogonal to $\boldsymbol{a}$ modulo x_0. More precisely, we consider the lattice $A^\perp$ of vectors $(C\boldsymbol{x}, y) \in \mathbb{Z}^{\theta+1}$, such that $(\boldsymbol{x}, y)$ is orthogonal to $\boldsymbol{a} = (\alpha_1, \ldots, \alpha_\theta, 1)$ modulo x_0; therefore $\boldsymbol{p}_{u,v} \in A^\perp$. The lattice $A^\perp$ has full-rank $\theta + 1$ and we have $\det(A^\perp) = C^\theta x_0$. Namely, we have an abstract group isomorphism $A^\perp \simeq (C\mathbb{Z})^\theta \oplus x_0\mathbb{Z}$, sending $(C\boldsymbol{x}, y)$ to $(C\boldsymbol{x}, \langle \boldsymbol{x}, \boldsymbol{a} \rangle + y)$.

As mentioned previously, the coefficients α_i's in the vector $\boldsymbol{a}$ have a particular structure. Namely, we have $\alpha_i = (g_i^{-1} \bmod p_i)h_i x_0/p_i$, and therefore

$$g_i \cdot \alpha_i \equiv h_i \cdot \frac{x_0}{p_i} \pmod{x_0}$$

for all $1 \leq i \leq \theta$. Therefore the lattice $A^\perp$ contains the θ linearly independent short vectors $\boldsymbol{q}_i = (0, \ldots, 0, Cg_i, 0, \ldots, 0, -s_i)$, where $s_i = h_i \cdot x_0/p_i$. Using $C := 2^{\rho_R - \alpha}$, we get $\|\boldsymbol{q}_i\| \simeq C \cdot 2^\alpha$.

We now derive a condition on $\|\boldsymbol{p}_{u,v}\|$ so that the vector $\boldsymbol{p}_{u,v}$ belongs to the sublattice of $A^\perp$ generated by the short vectors $\{\boldsymbol{q}_i : 1 \leq i \leq \theta\}$. From Lemma 2, if $\|\boldsymbol{p}_{u,v}\| < \det(A^\perp)/\prod_{i=1}^\theta \|\boldsymbol{q}_i\|$, then $\boldsymbol{p}_{u,v}$ must belong to the linear span generated by the $\boldsymbol{q}_i$'s; since by assumption, the g_i's are distinct primes and $\gcd(s_i, g_i) = 1$ for all $1 \leq i \leq \theta$, this implies that it must belong to the sublattice generated by the $\boldsymbol{q}_i$'s. In that case, we have:

$$\langle \boldsymbol{u}, \hat{\boldsymbol{m}}_i \rangle \equiv 0 \pmod{g_i}, \quad 1 \leq i \leq \theta \tag{12}$$

From $\det(A^\perp) = C^\theta \cdot x_0$ and $\|\boldsymbol{q}_i\| \simeq C \cdot 2^\alpha$, the previous condition $\|\boldsymbol{p}_{u,v}\| < \det(A^\perp)/\prod_{i=1}^\theta \|\boldsymbol{q}_i\|$ gives the approximate condition:

$$\|\boldsymbol{p}_{u,v}\| < 2^{\gamma - \alpha \cdot \theta} \tag{13}$$

Short Vectors in L. We now study the short vectors of L; more precisely, we explain that L contains ℓ linearly independent short vectors of norm roughly $2^{\rho_R + \alpha\theta/\ell}$. We show that these vectors can be derived from the lattice $\Lambda^\perp$ of vectors $\boldsymbol{u} \in \mathbb{Z}^\ell$ satisfying (12), i.e. that are orthogonal to $\hat{\boldsymbol{m}}_i$ modulo g_i for every $1 \leq i \leq \theta$. This is a full-rank lattice of dimension ℓ and determinant $g = \prod_{i=1}^\theta g_i$, with $g \simeq 2^{\alpha\theta}$. Therefore, we heuristically expect that the lattice $\Lambda^\perp$ contains ℓ linearly independent vectors of norm roughly $(\det \Lambda^\perp)^{1/\ell} \simeq 2^{\alpha\theta/\ell}$. We show that from any short $\boldsymbol{u} \in \Lambda^\perp$, we can generate a vector $(\boldsymbol{u}, v)$ with small v, and orthogonal to $(\boldsymbol{\omega}, 1)$ modulo x_0, and consequently a short vector

$(Bu, v) \in L$. For this, we write $\langle u, \hat{m}_i \rangle = k_i g_i$ with $k_i \in \mathbb{Z}$, and we have:

$$\langle u, \omega \rangle + v \equiv \sum_{i=1}^{\theta} \alpha_i \langle u, \hat{m}_i \rangle + \langle u, R \rangle + v \equiv \sum_{i=1}^{\theta} k_i \cdot g_i \cdot \alpha_i + \langle u, R \rangle + v \pmod{x_0}$$

$$\equiv \sum_{i=1}^{\theta} k_i \cdot s_i + \langle u, R \rangle + v \pmod{x_0}$$

Therefore, it suffices to let $v := -\langle u, R \rangle - \sum_{i=1}^{\theta} k_i \cdot s_i$ to obtain $\langle u, \omega \rangle + v \equiv 0$ (mod x_0); the vector (u, v) is then orthogonal to $(\omega, 1)$ modulo x_0, and thus $(Bu, v) \in L$. We obtain $|v| \simeq \|u\| \cdot 2^{\rho_R}$; therefore letting $B := 2^{\rho_R}$, we get $\|(Bu, v)\| \simeq 2^{\rho_R} \|u\|$. In summary, the lattice L contains a sublattice L' of rank ℓ, generated by ℓ vectors of norm roughly $2^{\rho_R + \alpha\theta/\ell}$. That the recovered vectors are indeed linearly independent is the content of the following lemma, which we prove in Appendix A.1.

Lemma 4. *Let $\{(Bu_j, v_j) : 1 \le j \le \ell+1\}$ be a basis of the lattice L and assume that the vectors $\{p_{u_j, v_j} : 1 \le j \le \ell\}$ belong to the sublattice of $A^\perp$ generated by the vectors $\{q_i : 1 \le i \le \theta\}$. Then the vectors $\{u_j : 1 \le j \le \ell\}$ are $\mathbb{R}$-linearly independent.*

Recovering $g = \prod_{i=1}^{\theta} g_i$. By applying lattice reduction to the lattice L, we expect that the first ℓ vectors $\{(Bu_j, v_j) : 1 \le j \le \ell\}$ of a reduced basis belong to the above sublattice L' and have norm roughly:

$$\|(Bu_j, v_j)\| \simeq 2^{\rho_R + \alpha\theta/\ell} \cdot 2^{\iota(\ell+1)}, \quad 1 \le j \le \ell \tag{14}$$

where $2^{\iota(\ell+1)}$ is the Hermite factor for some positive constant ι depending on the lattice reduction algorithm. With $C = 2^{\rho_R - \alpha}$, we have $\|p_{u_i, v_i}\| \simeq \|(Bu_i, v_i)\|$ for all $1 \le i \le \ell$. From the condition given by (13), we have that $u_i \in A^\perp$ if $\|p_{u_i, v_i}\| < 2^{\gamma - \alpha \cdot \theta}$; therefore combining with (14) we get the approximate condition:

$$\rho_R + \frac{\alpha\theta}{\ell} + \iota(\ell+1) < \gamma - \alpha\theta$$

Using $\rho_R = \gamma - \nu$ where ν is the number of bits that can be extracted from zero-testing, this condition becomes

$$\alpha\theta \left(1 + \frac{1}{\ell}\right) + \iota(\ell+1) < \nu. \tag{15}$$

In summary, when Condition (15) is satisfied, we expect to recover a basis $\{u_i : 1 \le i \le \ell\}$ of the lattice $A^\perp$; then since $\det(A^\perp) = g = \prod_{i=1}^{\theta} g_i$, the absolute value of the determinant of the basis matrix reveals g.

From Eq. (15), we observe that the parameter ℓ can be kept relatively small (say $\ell \simeq 10$), as larger values of ℓ would not significantly improve the bound; this implies that the lattice dimension $\ell+1$ on which LLL is applied can be kept relatively small. Moreover for LLL, experiments show that $2^\iota \simeq 1.021$ so that ι is approximately 0.03, and therefore for such small values of ℓ, the term $\iota \cdot (\ell + 1)$ is negligible. Thus we can use the simpler approximate bound for our attack:

$$\alpha\theta < \nu \tag{16}$$

This gives a factor 2 improvement compared to the previous bound given by (7), following the attack of [GLW14]. In the next subsection we will see how to get a much more significant improvement, with $\alpha\theta = \mathcal{O}(\nu^2)$.

A Proven Variant. The above algorithm is heuristic only. Below we describe a proven variant that can recover a vector $\boldsymbol{u}$ such that $\langle \boldsymbol{u}, \hat{\boldsymbol{m}}_i \rangle \equiv 0 \pmod{g_i}$ for all $1 \leq i \leq \theta$, using the LLL reduction algorithm. Although we only recover a single vector $\boldsymbol{u}$ instead of a lattice basis, this will be enough when combined with the Cheon *et al.* attack to recover all secret parameters of CLT13 (see Sect. 5). We provide the proof of Proposition 5 in Appendix A.2.

Proposition 5. *Let $\ell, \theta \in \mathbb{Z}_{\geq 1}$, $x_0 \in \mathbb{Z}_{\geq 1}$ and let $g_i \in \mathbb{Z}_{\geq 2}$ be distinct α-bit prime numbers for $1 \leq i \leq \theta$ and some $\alpha \in \mathbb{Z}_{\geq 1}$. For $1 \leq i \leq \theta$, let $\alpha_i \in \mathbb{Z}$ such that $g_i \cdot \alpha_i \equiv s_i \pmod{x_0}$, for $s_i \in \mathbb{Z}$ satisfying $|s_i| \leq 2^{\rho_R}$, for some $\rho_R \in \mathbb{Z}_{\geq 1}$ and assume that $\gcd(g_i, s_i) = 1$. For $1 \leq i \leq \theta$, let $\hat{\boldsymbol{m}}_i \in \mathbb{Z}^\ell$ be vectors with entries in $[0, g_i) \cap \mathbb{Z}$ such that $\hat{\boldsymbol{m}}_i \not\equiv \boldsymbol{0} \pmod{g_i}$, and let $\boldsymbol{R} \in \mathbb{Z}^\ell$ such that $\|\boldsymbol{R}\|_\infty \leq 2^{\rho_R}$. Let $\boldsymbol{\omega} \in \mathbb{Z}^\ell$ such that $\boldsymbol{\omega} \equiv \sum_{i=1}^\theta \alpha_i \hat{\boldsymbol{m}}_i + \boldsymbol{R} \pmod{x_0}$. Assume that*

$$\alpha\theta\left(1 + \frac{1}{\ell}\right) + \frac{\ell + \theta}{2} + \log_2(\ell\sqrt{\ell+1} \cdot \theta) + 4 < \log_2(x_0) - \rho_R. \tag{17}$$

Given the integers $\ell, \theta, \rho_R, x_0$ and the vector $\boldsymbol{\omega}$, one can recover in polynomial time a vector $\boldsymbol{u} \in \mathbb{Z}^\ell$ such that $\langle \boldsymbol{u}, \hat{\boldsymbol{m}}_i \rangle \equiv 0 \pmod{g_i}$ for all $1 \leq i \leq \theta$, satisfying $\|\boldsymbol{u}\| \leq 2^{\ell/2}\sqrt{\ell(\ell+1)}(\prod_{i=1}^\theta g_i)^{1/\ell}$.

We remark that by replacing $\log_2(x_0) - \rho_R$ by $\gamma - \rho_R = \nu$, we recover, up to additional logarithmic terms, the approximate bound established in (15).

4.3   Extended Orthogonal Lattice Attack

In this section we describe an extended attack that significantly improves the bound on θ established in (16). Let $\ell, d \geq 1$ be integers. As previously, we assume that we have encodings c_j of plaintext elements $\boldsymbol{m}_j = (m_{j1}, \ldots, m_{jn})$ for $1 \leq j \leq \ell$, where only the first θ components of each $\boldsymbol{m}_j$ are non-zero, that is, $m_{ji} = 0$ for $\theta + 1 \leq i \leq n$. However, we assume that these encodings are at level $\kappa - 1$, and that we also have an additional set of d level-1 encodings $\{c'_k : 1 \leq k \leq d\}$ of plaintext elements $\boldsymbol{x}_k = (x_{k1}, \ldots, x_{kn})$ for $1 \leq k \leq d$. By computing the top-level κ product encodings, we can therefore obtain the following zero-testing evaluations:

$$\omega_{jk} \equiv (c_j \cdot c'_k) \cdot p_{zt} \equiv \sum_{i=1}^\theta h_i m_{ji} x_{ki}(g_i^{-1} \bmod p_i)\frac{x_0}{p_i} + \sum_{i=1}^n h_i r_{jki}\frac{x_0}{p_i} \pmod{x_0} \tag{18}$$

for some integers r_{jki}. Since every encoding c_j encodes a message with non-zero support of length θ, the product encodings $c_j c'_k$ maintain their zero slots. Note that the same remains valid if the encodings c_j are at even lower levels, because they can be raised to level $\kappa - 1$ without removing their zero slots. As previously, we rewrite Eq. (18) as:

$$\omega_{jk} \equiv \sum_{i=1}^{\theta} \alpha_{ik} m_{ji} + R_{jk} \quad (\text{mod } x_0)$$

where we let

$$\alpha_{ik} = h_i x_{ki} (g_i^{-1} \bmod p_i) \frac{x_0}{p_i}, \quad 1 \leq i \leq \theta, \ 1 \leq k \leq d$$

and $R_{jk} = \sum_{i=1}^{n} h_i r_{jki} x_0 / p_i$ for all $1 \leq j \leq \ell$ and $1 \leq k \leq d$. As before, for $1 \leq i \leq \theta$, we denote by $\hat{m}_i \in \mathbb{Z}^{\ell}$ the vector with components m_{ji} for $1 \leq j \leq \ell$, and similarly ω_k and R_k the corresponding vectors in $\mathbb{Z}^{\ell}$. We assume that $\hat{m}_i \not\equiv 0 \ (\text{mod } g_i)$ for all i. The previous equation can then be rewritten as:

$$\omega_k \equiv \sum_{i=1}^{\theta} \alpha_{ik} \hat{m}_i + R_k \quad (\text{mod } x_0) \tag{19}$$

The difference with Eq. (11) from our first lattice attack is that the vectors $\{\hat{m}_i : 1 \leq i \leq \theta\}$ now satisfy d equations for $1 \leq k \leq d$, instead of a single equation, as in Subsect. 4.2. With more constraints on the vectors $\hat{m}_i$, we can therefore break the countermeasure from [GLW14] for much higher values of θ. In order to derive a condition on the parameters, we proceed as previously. Namely, the lattices that we considered in Subsect. 4.2 now admit natural higher-dimensional analogues.

The Orthogonal Lattice L. As previously, for a scaling factor $B \in \mathbb{Z}_{\geq 1}$, we consider the lattice L of vectors $(B\boldsymbol{u}, \boldsymbol{v}) \in \mathbb{Z}^{\ell+d}$, with $\boldsymbol{u} \in \mathbb{Z}^{\ell}$ and $\boldsymbol{v} \in \mathbb{Z}^d$, such that $(\boldsymbol{u}, \boldsymbol{v})$ is orthogonal to the d vectors $\{(\omega_k, e_k) : 1 \leq k \leq d\}$ modulo x_0, where $e_k \in \mathbb{Z}^d$ is the kth unit vector for $1 \leq k \leq d$. This gives for all $1 \leq k \leq d$ and all $(B\boldsymbol{u}, \boldsymbol{v}) \in L$, writing $\boldsymbol{v} = (v_1, \dots, v_d)$:

$$\langle \boldsymbol{u}, \omega_k \rangle + v_k \equiv \sum_{i=1}^{\theta} \alpha_{ik} \langle \boldsymbol{u}, \hat{m}_i \rangle + \langle \boldsymbol{u}, R_k \rangle + v_k \equiv 0 \quad (\text{mod } x_0)$$

and therefore the vector $(\langle \boldsymbol{u}, \hat{m}_1 \rangle, \dots, \langle \boldsymbol{u}, \hat{m}_\theta \rangle, \langle \boldsymbol{u}, R_1 \rangle + v_1, \dots, \langle \boldsymbol{u}, R_d \rangle + v_d)$ is orthogonal modulo x_0 to the d vectors $\boldsymbol{a}_k = (\alpha_{1k}, \dots, \alpha_{\theta k}, e_k)$, for $1 \leq k \leq d$. Again, using a scaling factor $C \in \mathbb{Z}_{\geq 1}$, we let

$$\boldsymbol{p}_{\boldsymbol{u}, \boldsymbol{v}} = (C\langle \boldsymbol{u}, \hat{m}_1 \rangle, \dots, C\langle \boldsymbol{u}, \hat{m}_\theta \rangle, \langle \boldsymbol{u}, R_1 \rangle + v_1, \dots, \langle \boldsymbol{u}, R_d \rangle + v_d).$$

The Lattice $A^{\perp}$. In order to bound the norm of the vector $\boldsymbol{p}_{\boldsymbol{u}, \boldsymbol{v}}$, we must study the short vectors in the lattice of vectors orthogonal to the vectors $\boldsymbol{a}_k$

modulo x_0 (instead of single vector $\boldsymbol{a}$). As previously, we consider the lattice $A^\perp$ of vectors $(C\boldsymbol{x}, \boldsymbol{y}) \in \mathbb{Z}^{\theta+d}$ such that $(\boldsymbol{x}, \boldsymbol{y})$ is orthogonal to the d vectors $\{\boldsymbol{a}_k : 1 \le k \le d\}$ modulo x_0; therefore $\boldsymbol{p}_{u,v} \in A^\perp$. The lattice $A^\perp$ has full-rank $\theta + d$ and determinant $C^\theta x_0^d$. As previously, the coefficients α_{ik} in the vectors $\boldsymbol{a}_k$ have a special structure, since they satisfy the congruence relations

$$g_i \cdot \alpha_{ik} \equiv h_i \cdot x_{ik} \cdot \frac{x_0}{p_i} \pmod{x_0}$$

for all $1 \le i \le \theta$ and $1 \le k \le d$. Therefore letting $s_{ik} = h_i \cdot x_{ik} \cdot x_0/p_i$, the lattice $A^\perp$ contains the θ short vectors $\boldsymbol{q}_i = (0, \ldots, 0, Cg_i, 0, \ldots, 0, -s_{i1}, \ldots, -s_{id})$ for $1 \le i \le \theta$. Using $C = 2^{\rho_R - \alpha}$, we get as previously $\|\boldsymbol{q}_i\| \simeq C \cdot 2^\alpha$.

We now derive a bound on $\|\boldsymbol{p}_{u,v}\|$ so that $\boldsymbol{p}_{u,v}$ belongs to the sublattice generated by the θ vectors $\{\boldsymbol{q}_i : 1 \le i \le \theta\}$. We expect a reduced basis of $A^\perp$ to have the first θ vectors with approximately the same norm as the vectors $\{\boldsymbol{q}_i : 1 \le i \le \theta\}$, and to have the last d vectors with norm U satisfying $(C \cdot 2^\alpha)^\theta \cdot U^d \simeq \det(A^\perp)$. Using $\det(A^\perp) = C^\theta x_0^d$, this gives $U \simeq x_0/2^{\alpha\theta/d}$. This implies that, heuristically, if $\|\boldsymbol{p}_{u,v}\| < U$, then $\boldsymbol{p}_{u,v}$ must belong to the sublattice generated by the θ vectors $\{\boldsymbol{q}_i : 1 \le i \le \theta\}$. As previously, in that case we have that for all $1 \le i \le \theta$:

$$\langle \boldsymbol{u}, \hat{\boldsymbol{m}}_i \rangle \equiv 0 \pmod{g_i}. \tag{20}$$

Short Vectors in L. We now study the short vectors of L; as previously, we show that L contains ℓ linearly independent short vectors of norm roughly $2^{\rho_R + \alpha\theta/\ell}$, which can be derived from the lattice $\Lambda^\perp$ of vectors $\boldsymbol{u} \in \mathbb{Z}^\ell$ satisfying (20). Since, as previously, $\Lambda^\perp$ heuristically contains ℓ linearly independent vectors of norm roughly $(\det \Lambda^\perp)^{1/\ell} \simeq 2^{\alpha\theta/\ell}$, the lattice L contains ℓ linearly independent vectors of norm roughly $2^{\rho_R + \alpha\theta/\ell}$. Therefore, by applying lattice reduction to the lattice L, we expect that the first ℓ vectors $\{(B\boldsymbol{u}_i, \boldsymbol{v}_i) : 1 \le i \le \ell\}$ of the basis have norm roughly:

$$\|(B\boldsymbol{u}_i, \boldsymbol{v}_i)\| \simeq B \cdot 2^{\alpha\theta/\ell} \cdot 2^{\iota(\ell+d)}$$

where $2^{\iota(\ell+d)}$ is the Hermite factor. With $B = 2^{\rho_R}$ and $C = 2^{\rho_R - \alpha}$, we have $\|\boldsymbol{p}_{u_i,v_i}\| \simeq \|(B\boldsymbol{u}_i, \boldsymbol{v}_i)\|$. From the condition $\|\boldsymbol{p}_{u_i,v_i}\| < U$, we get the condition:

$$\rho_R + \frac{\alpha\theta}{\ell} + \iota(\ell+d) < \gamma - \frac{\alpha\theta}{d}$$

which gives using $\rho_R = \gamma - \nu$:

$$\alpha\theta \cdot \left(\frac{1}{\ell} + \frac{1}{d}\right) + \iota(\ell+d) < \nu \tag{21}$$

Remark that with $d = 1$ the previous bound gives Eq. (15). Since (21) is concave and symmetric in both ℓ and d, the optimum is to take $\ell = d$. This gives the bound:

$$\frac{2\alpha\theta}{\ell} + 2\iota\ell < \nu \tag{22}$$

Recovering $g = \prod_{i=1}^{\theta} g_i$. When the above condition is satisfied, as previously we expect to recover a basis $\{u_i : 1 \leq i \leq \ell\}$ of the lattice $\Lambda^{\perp}$. Then since $\det(\Lambda^{\perp}) = g = \prod_{i=1}^{\theta} g_i$, the absolute value of the determinant of the basis matrix reveals g. In particular, it follows that the attack requires $\ell > 2\alpha\theta/\nu$, and we must have:

$$\iota < \frac{\nu^2}{4\alpha\theta}$$

Heuristically, achieving a Hermite factor of $2^{\iota 2\ell}$ requires $2^{\Omega(1/\iota)}$ using BKZ reduction with block-size $\beta = \omega(1/\iota)$, [HPS11]. The attack has therefore complexity $2^{\Omega(\alpha\theta/\nu^2)}$; the attack has therefore (heuristic) polynomial-time complexity under the condition:

$$\alpha\theta = \mathcal{O}(\nu^2)$$

which significantly improves our previous bound given by (16). Conversely, one expects that the attack is prevented under the condition:

$$\theta = \omega\left(\frac{\nu^2}{\alpha} \log \lambda\right) \tag{23}$$

In Sect. 4.5 we provide concrete parameters for CLT13 multilinear maps with independent slots. We will see that Condition (23) requires a much higher value for θ than the condition $2\theta\alpha \geq \nu$ for preventing the [GLW14] attack. Namely for $\lambda = 80$ bits of security, the bound $2\theta\alpha \geq \nu$ already holds for $\theta = 2$, while a concrete application of Condition (23) requires $\theta \geq 1789$.

Analogy of the Attacks. We remark that our extended attacks share similarities with the 2-dimensional attack from Sect. 3. For $\ell, d \in \mathbb{Z}_{\geq 1}$, our extended lattice attack works by reducing the $(\ell + d)$-dimensional lattice

$$L_{(\ell,d)} = \{(B\boldsymbol{u}, \boldsymbol{v}) \in \mathbb{Z}^{\ell} \times \mathbb{Z}^d : \langle(\boldsymbol{u}, \boldsymbol{v}), (\boldsymbol{\omega}_k, \boldsymbol{c}_k)\rangle \equiv 0 \pmod{x_0}, 1 \leq k \leq d\},$$

where $B \in \mathbb{Z}_{\geq 1}$ is fixed. With this notation, the three attacks work by reducing the lattices $L_{(1,1)}$, $L_{(\ell,1)}$ and $L_{(\ell,d)}$, respectively. Note that $L_{(1,1)}$ is the lattice $\{(Bu, v) \in \mathbb{Z}^2 : u\omega + v \equiv 0 \pmod{x_0}\}$. For the extended attacks, the $\ell \times \ell$ top-left submatrix of a reduced basis of $L_{(\ell,d)}$ (divided by B) has determinant $\pm g$. Note that this coincides with the 2-dimensional case $\ell = d = 1$: the first entry (divided by B) of the first vector in a reduced basis equals $\pm g$ (i.e. a "1×1 submatrix" of determinant $\pm g$). As such, our higher-dimensional attacks are consistent generalizations of the 2-dimensional attack.

Summary. We have described a lattice-based attack, which under the condition $\alpha\theta = \mathcal{O}(\nu^2)$, and given as input a collection of encodings (or products of encodings) of messages with non-zero support of length θ, outputs the secret plaintext ring of CLT13. More precisely, our extended lattice attack with the improved bound $\alpha\theta = \mathcal{O}(\nu^2)$ can be described in the following three steps, with parameters $\ell, d \geq 1$. We provide in https://pastebin.com/7WEMHBE9 the source code in Sage [S+17].

Input: Sets of level-κ encodings $\{c_j \cdot c'_k \bmod x_0 : 1 \leq j \leq \ell, 1 \leq k \leq d\}$ where c_j encodes a message of non-zero support of length θ.

Output: $g = \prod_{i=1}^{\theta} g_i$

1. For $1 \leq k \leq d$, compute the vectors $\boldsymbol{\omega}_k \in \mathbb{Z}^{\ell}$ with $(\boldsymbol{\omega}_k)_j = c_j \cdot c'_k \cdot p_{zt} \bmod x_0$.
2. Let $B = 2^{\rho_R}$ and compute a LLL-reduced basis of the lattice $L_{(\ell,d)} \subseteq \mathbb{Z}^{\ell+d}$ of vectors $\{(B\boldsymbol{u}, \boldsymbol{v}) \in \mathbb{Z}^{\ell} \times \mathbb{Z}^d : \langle (\boldsymbol{u}, \boldsymbol{v}), (\boldsymbol{\omega}_k, \boldsymbol{e}_k) \rangle \equiv 0 \pmod{x_0}, 1 \leq k \leq d\}$, where $\boldsymbol{e}_k \in \mathbb{Z}^d$ is the kth unit vector for $1 \leq k \leq d$. Denote by $\{(B\boldsymbol{u}_j, \boldsymbol{v}_j) : 1 \leq j \leq \ell + d\}$ the LLL-reduced basis.
3. Form the $\ell \times \ell$ matrix $\boldsymbol{P}$ of vectors $\{\boldsymbol{u}_j : 1 \leq j \leq \ell\}$ and compute $|\det(\boldsymbol{P})| = g = \prod_{i=1}^{\theta} g_i$.

Variant with Multiple p_{zt}. In many concrete constructions based on composite order multilinear maps, intermediate-level encodings of almost zero plaintexts are not necessarily available. We refer to Sect. 6 for the application of our attacks to concrete constructions. In order to get around this assumption, we consider a variant of the above attack, where we have multiple zero-testing elements p_{zt} instead of a single one. Namely, as described in [CLT13], in order to get a proper zero-testing procedure, one needs to use a vector of n elements p_{zt}. We denote by $p_{zt,k}$ for $1 \leq k \leq n$ those zero-testing elements:

$$p_{zt,k} = \sum_{i=1}^{n} h_{ik} z^{\kappa} (g_i^{-1} \bmod p_i) \frac{x_0}{p_i} \bmod x_0$$

for corresponding integers h_{ik}. As previously, we assume that we have encodings c_j of plaintext elements $\boldsymbol{m}_j = (m_{j1}, \ldots, m_{jn})$ for $1 \leq j \leq \ell$, where only the first θ components of each $\boldsymbol{m}_j$ are non-zero, that is, $m_{ji} = 0$ for $\theta + 1 \leq i \leq n$. We can now assume that these encodings are at the last level κ. Thanks to the multiple zero-testing elements, we can therefore obtain the following zero-testing evaluations:

$$\omega_{jk} \equiv c_j \cdot p_{zt,k} \equiv \sum_{i=1}^{\theta} h_{ik} m_{ji} (g_i^{-1} \bmod p_i) \frac{x_0}{p_i} + \sum_{i=1}^{n} h_{ik} r_{jki} \frac{x_0}{p_i} \pmod{x_0}$$

for some integers r_{jki}, which is similar to (18) with $h_{ik} = h_i \cdot x_{ki}$. Therefore the same attack applies and the secret $g = \prod_{i=1}^{\theta} g_i$ can be recovered in (heuristic) polynomial-time under the condition $\alpha\theta = \mathcal{O}(\nu^2)$.

4.4 Revealing Information About the Plaintext Elements

We show that our attack not only reveals the secret CLT13 plaintext ring, but also information about the secret plaintext elements $\{\hat{m}_i : 1 \leq i \leq \theta\}$. Namely, the orthogonal lattice attack not only recovers $g = \prod_{i=1}^{\theta} g_i$, but also constructs a matrix U of rows $\{\boldsymbol{u}_j : 1 \leq j \leq \ell\}$ orthogonal to the vectors $\{\hat{\boldsymbol{m}}_i : 1 \leq i \leq \theta\}$ modulo g_i (*i.e.* a basis of the lattice $\Lambda^{\perp}$, following the previous notation) and we can use this matrix U in order to recover scalar multiples of the plaintext vectors $\{\hat{m}_i : 1 \leq i \leq \theta\}$.

More precisely, we show that for each $1 \leq i \leq \theta$, we can recover the one-dimensional linear space generated by $\hat{m}_i$ modulo g_i. The first step is to factor $g = \prod_{i=1}^{\theta} g_i$ to recover the primes g_i's; this is feasible if the g_i's are small enough.[1] Since we have a basis matrix U of the lattice of vectors u with $\langle u, \hat{m}_i \rangle \equiv 0$ (mod g_i) for all $1 \leq i \leq \theta$, it suffices to compute the $\mathbb{Z}/g_i\mathbb{Z}$-kernel of the $\ell \times \ell$ matrix $U_{g_i} = U \bmod g_i$; assuming that $\hat{m}_i \not\equiv 0$ (mod g_i), we have that $\ker(U_{g_i})$ has dimension 1 over $\mathbb{Z}/g_i\mathbb{Z}$ and therefore, we recover a non-trivial multiple $\lambda_i \hat{m}_i$ of the original messages $\hat{m}_i$ modulo g_i, for $1 \leq i \leq \theta$. With the ECM [Len87] the factorization of $g = \prod_{i=1}^{\theta} g_i$ can be computed in time $\exp(c\sqrt{\alpha \ln \alpha})$ for some positive constant c and where α is the bit size of the g_i's, which gives a sub-exponential time attack.

Alternatively, to avoid the factorization of g, we can compute the integer right kernel of the matrix $[U \mid g I_\ell]$, where I_ℓ denotes the identity matrix in dimension ℓ. The following proposition shows that we can recover in polynomial time a non-trivial multiple of the vector $\hat{m}$, such that $\hat{m} \equiv \hat{m}_i$ (mod g_i) for all $1 \leq i \leq \theta$.

Proposition 6. *Let $\ell, \theta \in \mathbb{Z}_{\geq 1}$. Let $g_1, \ldots, g_\theta$ be distinct prime numbers. For $1 \leq i \leq \theta$, let $\hat{m}_i \in \mathbb{Z}^\ell \cap [0, g_i)^\ell$ be vectors such that $\hat{m}_i \not\equiv \mathbf{0}$ (mod g_i). Let $\{u_j : 1 \leq j \leq \ell\}$ be a basis of the lattice of vectors $u \in \mathbb{Z}^\ell$ such that $\langle u, \hat{m}_i \rangle \equiv 0$ (mod g_i) for all $1 \leq i \leq \theta$. Then, given $g = \prod_{i=1}^{\theta} g_i$ and the vectors $\{u_j : 1 \leq j \leq \ell\}$, one can recover in polynomial time a vector $\lambda \cdot \hat{m} \in \mathbb{Z}^\ell \cap [0, g)^\ell$ with $\gcd(\lambda, g) = 1$, such that $\hat{m} \equiv \hat{m}_i$ (mod g_i) for all $1 \leq i \leq \theta$.*

Proof. By the Chinese Remainder Theorem, there exists a unique vector $\hat{m} \in \mathbb{Z}^\ell \cap [0, g)^\ell$ satisfying $\hat{m} \equiv \hat{m}_i$ (mod g_i) for all $1 \leq i \leq \theta$. Consider the composition of maps $\mathbb{Z}^\ell \xrightarrow{\pi} (\mathbb{Z}/g\mathbb{Z})^\ell \xrightarrow{\phi} \mathbb{Z}/g\mathbb{Z}$, where π is reduction modulo g and ϕ sends u to $\langle u, \hat{m} \rangle$. By the Chinese Remainder Theorem, the map ϕ corresponds to a vector of maps $\phi = (\phi_1, \ldots, \phi_\theta) : (\prod_i \mathbb{Z}/g_i\mathbb{Z})^\ell \to \prod_i \mathbb{Z}/g_i\mathbb{Z}$ with components $\phi_i : (\mathbb{Z}/g_i\mathbb{Z})^\ell \to \mathbb{Z}/g_i\mathbb{Z}$ for $1 \leq i \leq \theta$. Let $1 \leq i \leq \theta$; since $\hat{m}_i \not\equiv \mathbf{0}$ (mod g_i), the map ϕ_i is surjective with kernel $\ker(\phi_i) = \operatorname{im}(U_{g_i})$ where $U_{g_i} = U \bmod g_i$. Since g_i is prime, $\ker(\phi_i) = \operatorname{im}(U_{g_i})$ is a $\mathbb{Z}/g_i\mathbb{Z}$-vector space of dimension $\ell - 1$. It follows that the kernel of U_{g_i} has dimension 1 over $\mathbb{Z}/g_i\mathbb{Z}$. This holds for all $1 \leq i \leq \theta$, so by the Chinese Remainder Theorem, $\ker(U_g)$ (where U_g is the matrix U modulo g) is a free $\mathbb{Z}/g\mathbb{Z}$-module of rank 1, generated by $\hat{m}$. In particular, there exists $k \in \mathbb{Z}^\ell$ such that $(\hat{m}, k)$ belongs to the $\mathbb{Z}$-kernel of the matrix $[U \mid g I_\ell]$. The integer kernel of this matrix can be computed in polynomial time from g and U and the left $\ell \times \ell$ submatrix of the Hermite normal form of the basis of the $\mathbb{Z}$-kernel gives in the first row a vector $\lambda \hat{m}$ with $\lambda \in (\mathbb{Z}/g\mathbb{Z})^\times$. $\square$

---

[1] For the concrete parameters provided in [CLT13], the g_i's are 80-bit primes; therefore the factorization is straightforward.

4.5   Concrete Parameters and Practical Experiments

Concrete Parameters. We provide concrete parameters for CLT13 multilinear maps with independent slots, for various values of the security parameter λ. We start from the same concrete parameters as provided in [CLT13]; we assume that the encoding noise is set so that the number of extracted bits is $\nu = 2\lambda + 12$; we take $\alpha = \lambda$. We then provide the minimum value of θ that ensures the same level of security against lattice attacks; see Table 1. As in [CLT13], the goal is to ensure that the best attack takes at least 2^λ clock cycles.

While in Table 1 the number of independent slots $n_{\text{slots}} = \lfloor n/\theta \rfloor$ appears to be relatively small, it is always possible to increase the number of independent slots by increasing the value of n.

Table 1. Concrete parameters for CLT13 multilinear maps with independent slots, for security parameter λ.

Instantiation	λ	n	η	$\gamma = n \cdot \eta$	ν	θ	n_{slots}
Small	52	1080	1981	$2.1 \cdot 10^6$	116	540	2
Medium	62	2364	2055	$4.9 \cdot 10^6$	136	1182	2
Large	72	8250	2261	$18.7 \cdot 10^6$	156	1472	5
Extra	80	26115	2438	$63.7 \cdot 10^6$	172	1789	14

Practical Experiments. We have run our extended attack from Sect. 4.3 with the "Extra" parameters of CLT13 from Table 1, for increasing values of θ. Note that for such parameters, the original attack from [GLW14] only applies for $\theta = 1$. To improve efficiency, we give as input to LLL a truncated matrix basis, where we keep only the ν most significant bits. Table 2 shows that our attack works in practice for much larger values of θ than the original attack from [GLW14], which can only work for $\theta = 1$. We provide in https://pastebin.com/7WEMHBE9 the source code in Sage [S+17].

Table 2. Running time of our LLL-based attack, as a function of the parameter θ, for the "Extra" parameters of CLT13. The lattice dimension is $\ell + d = 2\ell$.

	θ	α	ν	$\ell = d$	Lat. dim.	Running time
Basic attack [GLW14]	1	80	172	1	2	ε
Extended attack (Sect. 4.3)	2	80	172	2	4	ε
Extended attack (Sect. 4.3)	40	80	172	39	78	10 s
Extended attack (Sect. 4.3)	100	80	172	100	200	11 min
Extended attack (Sect. 4.3)	160	80	172	163	326	11 h

5   Application to the Cheon *et al.* Attack

In 2015, Cheon *et al.* published in [CHL+15] a polynomial time attack against CLT13 resulting in a total break of the multipartite Diffie-Hellman key exchange protocol. The attack relies on the availability of low-level encodings of zero. In this section, we show how to adapt the Cheon *et al.* attack to the setting of CLT13 with independent slots: we assume that no encodings of zero are available to the attacker (otherwise the Cheon *et al.* attack would apply immediately), but as previously, the attacker can obtain low-level encodings where only θ components of the plaintext are non-zero. In particular, this contributes to a cryptanalysis of CLT13 multilinear maps where no encodings of zero are available beforehand; this was considered as an open problem in [CLR15, Section 4].

5.1   The Original Cheon *et al.* Attack with Encodings of Zero

We first recall the basic Cheon *et al.* attack against CLT13. For simplicity, we take $\kappa = 3$; the attack is easily extended to $\kappa > 3$. Consider a set $\mathcal{A} = \{a_j : 1 \leq j \leq n\}$ of encodings of zero at level one, a pair $\mathcal{B} = \{b_0, b_1\}$ of encodings at level one, and a set $\mathcal{C} = \{c_k : 1 \leq k \leq n\}$ of encodings at level one. We write $a_j \equiv a_{ji}/z \pmod{p_i}$, $b_t \equiv b_{ti}/z \pmod{p_i}$, $c_k \equiv c_{ki}/z \pmod{p_i}$, with integers $a_{ji} \equiv 0 \pmod{g_i}$, for all $1 \leq j, i, k \leq n$ and $t \in \{0, 1\}$. We obtain the zero-testing evaluations:

$$\omega_{jk}^{(t)} = a_j b_t c_k p_{zt} \bmod x_0 = \sum_{i=1}^{n} \frac{h_i}{g_i} a_{ji} b_{ti} c_{ki} \frac{x_0}{p_i}$$

where the equality holds over $\mathbb{Z}$ because the products $a_j b_t c_k$ are level-3 encodings of 0. This can be written in matrix form as

$$\omega_{jk}^{(t)} = \begin{bmatrix} a_{j1} \cdots a_{jn} \end{bmatrix} \begin{bmatrix} b_{t1} p_{zt,1} & & \\ & \ddots & \\ & & b_{tn} p_{zt,n} \end{bmatrix} \begin{bmatrix} c_{k1} \\ \vdots \\ c_{kn} \end{bmatrix}.$$

where $p_{zt,i} = (h_i/g_i) \cdot x_0/p_i$ for all $1 \leq i \leq n$. Writing out the matrices $\boldsymbol{W}_t = (\omega_{jk}^{(t)})_{1 \leq j,k \leq n}$ for $t \in \{0, 1\}$, one obtains the integer matrix equalities $\boldsymbol{W}_t = \boldsymbol{A} \Delta_t \boldsymbol{C}$ for $t \in \{0, 1\}$, where the rows of $\boldsymbol{A}$ are the vectors $(a_{j1}, \cdots, a_{jn})_j$, the columns of $\boldsymbol{C}$ are the vectors $(c_{k1}, \cdots, c_{kn})_k$, and Δ_t is the diagonal matrix $\mathrm{diag}(b_{t1} p_{zt,1}, \ldots, b_{tn} p_{zt,n})$.

Provided that at least one of $\boldsymbol{W}_0, \boldsymbol{W}_1$ is invertible over $\mathbb{Q}$ (say $\boldsymbol{W}_1$), one then evaluates over $\mathbb{Q}$ the matrix product:

$$\boldsymbol{W}_0 \cdot \boldsymbol{W}_1^{-1} = \boldsymbol{A}(\Delta_0 \Delta_1^{-1})\boldsymbol{A}^{-1}$$

The attacker can thus compute the eigenvalues of $\boldsymbol{W}_0 \boldsymbol{W}_1^{-1}$, by factoring the characteristic polynomial (over $\mathbb{Q}$). By similarity of these matrices, these eigenvalues coincide with those of $\Delta_0 \Delta_1^{-1} = \mathrm{diag}(b_{01}/b_{11}, \ldots, b_{0n}/b_{1n})$, which are $\{b_{0i}/b_{1i} : 1 \leq i \leq n\}$. These ratios are now enough to factor x_0. Namely, writing the quotients $b_{0i}/b_{1i} = x_i/y_i$ for coprime integers x_i, y_i and using that $b_t \equiv b_{ti}/z$ (mod p_i), we obtain:

$$x_i b_1 - y_i b_0 \equiv (x_i b_{1i} - y_i b_{0i})/z \equiv 0 \pmod{p_i}$$

and therefore $\gcd(x_i b_1 - y_i b_0, x_0) = p_i$ with good probability. In summary, the Cheon et al. attack recovers all secret p_i's in polynomial time given the low-level encodings of zero $\{a_j : 1 \leq j \leq n\}$.

5.2   Adaptation of the Cheon *et al.* Attack to Our Cryptanalysis

We now show how to adapt the Cheon et al. attack when no encodings of zero are available, but the attacker can obtain low-level encodings where only θ components of the underlying plaintexts are non-zero. The attack is divided in two steps: first the attacker generates encodings of zero using the orthogonal lattice attack from Sect. 4, and then applies the original Cheon et al. attack to reveal the primes $\{p_i : 1 \leq i \leq n\}$.

We consider the following setting with $\kappa = 4$. Let $\ell \geq 1$; we consider a set $\mathcal{Y} = \{y_j : 1 \leq j \leq \ell\}$ of level-one encodings of messages $\boldsymbol{m}_1, \ldots, \boldsymbol{m}_\ell$ where only the first θ components of each $\boldsymbol{m}_j$ are non-zero. Moreover, we consider as in the previous section three sets $\mathcal{A} = \{a_j : 1 \leq j \leq n\}$, $\mathcal{B} = \{b_0, b_1\}$ and $\mathcal{C} = \{c_k : 1 \leq k \leq n\}$ of level-one encodings of non-zero messages.

First Step: Orthogonal Lattice Attack. We show that the orthogonal lattice attack from Sect. 4.2 can compute a short vector $\boldsymbol{u} \in \mathbb{Z}^\ell$ such that $y' = \langle \boldsymbol{u}, \boldsymbol{y} \rangle$ is a level-1 encoding of zero, where $\boldsymbol{y} = (y_1, \ldots, y_\ell)$. We write for all $1 \leq j \leq \ell$:

$$y_j \equiv \frac{r_{ji} \cdot g_i + m_{ji}}{z} \pmod{p_i}, \quad 1 \leq i \leq n,$$

with the usual CLT13 notations, where $m_{ji} = 0$ for $\theta + 1 \leq i \leq n$. Note that our orthogonal lattice attack from Sect. 4.2 uses level-κ encodings; therefore it can be applied on level-κ encodings of the form:

$$e_j = y_j \cdot a_1 \cdot b_0 \cdot c_1 \bmod x_0$$

for level-one encodings a_1, b_0, c_1; we obtain:

$$e_j \equiv \frac{r'_{ji} \cdot g_i + m_{ji} \cdot x_i}{z^\kappa} \pmod{p_i}, \quad 1 \leq i \leq n$$

for some $r'_{ji} \in \mathbb{Z}$ and where x_i is the i-th component of the plaintext corresponding to the encoding $a_1 \cdot b_0 \cdot c_1$. Clearly, since the messages $\{\boldsymbol{m}_j : 1 \leq j \leq \ell\}$ have non-zero support of length θ, the messages $\{(m_{ji} \cdot x_i)_{1 \leq i \leq n} : 1 \leq j \leq \ell\}$ have non-zero support of length at most θ. Therefore, applying the orthogonal lattice attack from Sect. 4.2 on the encodings e_j (i.e. on the vector $\boldsymbol{\omega} = p_{zt} \cdot (e_j)_{1 \leq j \leq \ell} \bmod x_0$), we obtain a vector $\boldsymbol{u} \in \mathbb{Z}^\ell$ such that $\langle \boldsymbol{u}, \hat{\boldsymbol{m}}_i \cdot x_i \rangle \equiv 0$

$(\bmod \, g_i)$ for all $1 \le i \le \theta$, where the $\hat{\boldsymbol{m}}_i$'s are the vectors $(m_{1i}, \ldots, m_{\ell i})$ for $1 \le i \le \theta$. Provided that $x_i \not\equiv 0 \pmod{g_i}$, this implies $\langle \boldsymbol{u}, \hat{\boldsymbol{m}}_i \rangle \equiv 0 \pmod{g_i}$ for all $1 \le i \le \theta$. Therefore, for all $1 \le i \le n$, we can write $\sum_{j=1}^{\ell} u_j m_{ji} = k_i g_i$ for integers k_i (and $k_i = 0$ for $\theta + 1 \le i \le n$). This gives:

$$y' = \sum_{j=1}^{\ell} u_j y_j \equiv g_i \left(\sum_{j=1}^{\ell} u_j r_{ji} + k_i \right) \cdot z^{-1} \pmod{p_i}, \quad 1 \le i \le n$$

and therefore y' is a level-1 encoding of zero, moreover with small noise since the vector $\boldsymbol{u}$ is short. Note that we only need a single vector $\boldsymbol{u}$; therefore the first step of the attack is proven by Proposition 5.

Second Step: Cheon *et al.* **attack.** The second step consists in applying the Cheon *et al.* attack with the three sets $\mathcal{A}' = \{y' \cdot a_j : 1 \le j \le n\}$, $\mathcal{B} = \{b_0, b_1\}$ and $\mathcal{C} = \{c_k : 1 \le k \le n\}$. Since y' is an encoding of zero, all encodings in $\mathcal{A}'$ are encodings of zero, and we can apply the Cheon *et al.* attack on the three sets $\mathcal{A}'$, $\mathcal{B}$ and $\mathcal{C}$ to recover all secret primes p_i.

Since the orthogonal lattice attack more generally provides a set of ℓ vectors $\boldsymbol{u}_j \in \mathbb{Z}^{\ell}$ (instead of a single $\boldsymbol{u}$; and all satisfying $\langle \boldsymbol{u}_j, \hat{\boldsymbol{m}}_i \rangle \equiv 0 \pmod{g_i}$ for all i), a variant of the above attack with $\kappa = 3$ consists in starting from a set $\mathcal{A} = \{a_j : 1 \le j \le n\}$ of $\ell = n$ encodings where only the first θ components of the underlying plaintexts are non-zero, and then generating a set $\mathcal{A}' = \{\langle \boldsymbol{u}_j, \boldsymbol{a} \rangle : 1 \le j \le n\}$ of encodings of zero, with the vector of encodings $\boldsymbol{a} = (a_1, \ldots, a_n)$. One can then apply the Cheon *et al.* attack as previously on the three sets $\mathcal{A}'$, $\mathcal{B}$ and $\mathcal{C}$.

Note that the first step of the attack above (*i.e.* the generation of encodings of zero) uses the orthogonal lattice attack from Sect. 4.2 with the bound $\alpha\theta < \nu$. The attack from Sect. 4.3 is easily adapted to reach the improved bound $\alpha\theta = \mathcal{O}(\nu^2)$. In this case the attacker can obtain $\ell \cdot d$ level-two encodings of zero given by $\{\langle \boldsymbol{u}_j, \boldsymbol{c}_k \rangle : 1 \le j \le \ell, 1 \le k \le d\}$ where $\boldsymbol{c}_k$ is the vector of encodings $(c_j \cdot c_k')_{1 \le j \le \ell}$ with the encodings $c_j \cdot c_k'$ considered in Sect. 4.3.

6 Application to Constructions Based on CLT13 with Independent Slots

In this section we show that our orthogonal lattice attack from Sect. 4 can be applied to various constructions over CLT13 multilinear maps with independent slots.

6.1 The Multilinear Subgroup Elimination Assumption from [GLW14, GLSW15]

The multilinear subgroup elimination assumption is used in [GLW14] for witness encryption and in [GLSW15] for constructing program obfuscation, based on a single assumption, independent of the particular circuit to be obfuscated.

The multilinear subgroup elimination assumption is stated for a generic model of composite-order multilinear maps. Below, we show that our attacks break this assumption over CLT13 composite-order multilinear maps. We note that since the GLW14 scheme also includes encodings of zeroes, it could also be broken more directly by the Cheon *et al.* attack. We recall the definition from [GLSW15].

Definition 1 ((μ, ν)-multilinear subgroup elimination assumption). *Let G be a group of order $N = a_1 \cdots a_\mu b_1 \cdots b_\nu c$ where $a_1, \ldots, a_\mu, b_1, \ldots, b_\nu, c$ are $\mu + \nu + 1$ distinct primes. We give out generators $x_{a_1}, \ldots, x_{a_\mu}, x_{b_1}, \ldots, x_{b_\nu}$ for each prime order subgroup except for the subgroup of order c. For each $1 \le i \le \mu$, we also give out a group element h_i sampled uniformly at random from the subgroup of order $ca_1 \cdots a_{i-1} a_{i+1} \cdots a_\mu$. The challenge term is a group element $T \in G$ that is either sampled uniformly at random from the subgroup of order $ca_1 \cdots a_\mu$ or uniformly at random from the subgroup of order $a_1 \cdots a_\mu$. The task is to distinguish between these two distributions of T.*

For simplicity, we consider the assumption with $\mu = 1$ and $\nu = 0$; the generalization of our attack to any (μ, ν) is straightforward. Therefore G is a group of order $a_1 c$. The challenge $T \in G$ is either generated at random from the subgroup of order $a_1 c$, or from the subgroup of order a_1. In the context of a CLT13 instantiation, we assume that $a_1 = \prod_{i=1}^{\theta} g_i$ and $c = \prod_{i=\theta+1}^{n} g_i$. In that case, a_1 and c are not primes, but the assumption can still be considered for composite a_i's, b_i's and c. The encoding T is then either generated from a random plaintext $m \in \bigoplus_{i=1}^{n} \mathbb{Z}/g_i\mathbb{Z}$, or from a random plaintext with only the θ first components non-zero, that is $m \equiv 0 \pmod{g_i}$ for $\theta + 1 \le i \le n$. It is easy to see that our attacks from Sects. 4.2 and 4.3 apply in this setting. Namely, when only the first θ components of the plaintext m corresponding to the challenge T are non-zero, our attacks recover the product $a_1 = \prod_{i=1}^{\theta} g_i$, whereas the attacks will fail when m is a random plaintext. Therefore the challenge T is easily distinguished unless θ is large enough; more precisely, θ must satisfy the bound given by (23) to prevent the attack.

6.2 The Zimmerman Circuit Obfuscation Scheme

At Eurocrypt 2015, Zimmerman described a technique to obfuscate programs without matrix branching programs, based on composite-order multilinear maps [Zim15]. A plaintext m belongs to $\mathbb{Z}/N\mathbb{Z}$ for a composite modulus $N = N_{\text{ev}} \cdot N_{\text{chk}}$, and the ring $\mathbb{Z}/N\mathbb{Z}$ is viewed as a direct product of an "evaluation" ring $\mathbb{Z}/N_{\text{ev}}\mathbb{Z}$ to evaluate the circuit, and a "checksum" ring $\mathbb{Z}/N_{\text{chk}}\mathbb{Z}$ to prevent the adversary from evaluating a different circuit; those two evaluations are performed in parallel. Using the CLT13 notations from Sect. 2, one can let $N_{\text{ev}} = \prod_{i=1}^{\theta} g_i$ and $N_{\text{chk}} = \prod_{i=\theta+1}^{n} g_i$. In that case, the parameter θ must satisfy the bound given by (23) to prevent our lattice attack.

6.3   The FRS17 Construction for Preventing Input Partitioning Attacks

At Asiacrypt 2017, Fernando, Rasmussen and Sahai described three constructions of "stamping functions" for preventing input-partitioning attacks on matrix branching programs [FRS17]. Their third construction is based on permutation hash functions and is instantiated over CLT13 multilinear maps with independent slots. More precisely, the permutation hash function is written as a matrix branching program, and multiple such permutation hash functions h_i are evaluated in parallel along with the main matrix branching program; this is to ensure that only inputs of the form $x \| h(x)$ can be evaluated, where $h(x) = h_1(x) \| \cdots \| h_t(x)$, which prevents input partitioning attacks.

Matrix Branching Programs. We first recall the construction of [GGH+13b] to obfuscate matrix branching programs. A matrix branching program BP of length n_p on ℓ-bit inputs $x \in \{0, 1\}^\ell$ is evaluated by computing:

$$C(x) = b_0 \cdot \prod_{i=1}^{n_p} B_{i, x_{\text{inp}(i)}} \cdot b_{n_p+1} \tag{24}$$

where $\{B_{i,b} : 1 \le i \le n_p, b \in \{0,1\}\}$ are $2n_p$ square matrices and b_0 and b_{n_p+1} are bookend vectors; then $\mathsf{BP}(x) = 0$ if $C(x) = 0$, and $\mathsf{BP}(x) = 1$ otherwise. The integer $\mathsf{inp}(i) \in \{1, \dots, \ell\}$ indicates which bit of x is read at step i of the product matrix computation. The matrices $B_{i,b}$ are first randomized by choosing $n_p + 1$ random invertible matrices $\{R_i : 0 \le i \le n_p\}$ and letting $\tilde{B}_{i,b} = R_{i-1} B_{i,b} R_i^{-1}$ for $1 \le i \le n_p$, with also $\tilde{b}_0 = b_0 R_0^{-1}$ and $\tilde{b}_{n_p+1} = R_{n_p} b_{n_p+1}$. We obtain a randomized matrix branching program with the same result since the randomization matrices R_i cancel each other: $C(x) = \tilde{b}_0 \cdot \prod_{i=1}^{n_p} \tilde{B}_{i, x_{\text{inp}(i)}} \cdot \tilde{b}_{n_p+1}$.

The entries of the matrices $\tilde{B}_{i,b}$ are then independently encoded, as well as the bookend vectors $\tilde{b}_0$ and $\tilde{b}_{n_p}$. We obtain the matrices and vectors $\hat{B}_{i,b} = \mathsf{Encode}_{\{i+1\}}(\tilde{B}_{i,b})$, $\hat{b}_0 = \mathsf{Encode}_{\{1\}}(\tilde{b}_0)$ and $\hat{b}_{n_p+1} = \mathsf{Encode}_{\{n_p+2\}}(\tilde{b}_{n_p+1})$. Here $\mathsf{Encode}_{\{i\}}(\cdot)$ denotes an encoding relative to the singleton i. The matrix branching program from (24) can then be evaluated over the encoded matrices:

$$\hat{C}(x) = \hat{b}_0 \cdot \prod_{i=1}^{n_p} \hat{B}_{i, x_{\text{inp}(i)}} \cdot \hat{b}_{n_p+1} \tag{25}$$

The resulting $\hat{C}(x)$ is then a last-level encoding that can be zero-tested to check if $C(x) = 0$, which reveals the output of the branching program $\mathsf{BP}(x)$, without revealing the matrices $B_{i,b}$.

Application to the FRS17 Construction. The [FRS17] scheme constructs a modified matrix branching program BP' that receives as input $u \| v_1 \dots v_t$ and checks whether $v_i = h_i(u)$ for all $1 \le i \le t$, where the h_i's are permutation hash functions; in that case, BP' returns $\mathsf{BP}(u)$ where BP is the original branching program; otherwise, it returns some non-zero value. As explained in [FRS17], multiple branching programs can be evaluated in parallel with composite order multilinear maps; with the countermeasure from [GLW14] over CLT13, each branching program is then evaluated modulo a product of θ of the primes g_i's, instead of a single g_i in [FRS17].

It is easy to generate an input $u\|v_1\ldots v_t$ such that $\mathsf{BP}(u) = 0$ and $v_i = h_i(u)$ for all $1 \le i \le t$ except for some $i = i^\star$; in that case, only one of the $t + 1$ parallel matrix branching program will evaluate to a non-zero value. The orthogonal lattice attack from Sect. 4.2 can therefore recover the secret plaintext ring $\bigoplus_{i=1}^n \mathbb{Z}/g_i\mathbb{Z}$ of CLT13, under the condition $\alpha\theta < \nu$. Alternatively, if multiple p_{zt}'s are available, the extended attack from Sect. 4.3 applies under the condition $\alpha\theta = \mathcal{O}(\nu^2)$, as described at the end of Sect. 4.3.

We note however that in both cases, our attack against [FRS17] only recovers the secret plaintext ring $\bigoplus_{i=1}^n \mathbb{Z}/g_i\mathbb{Z}$ of CLT13, and not all secret parameters of CLT13; we leave that as an open problem.

Acknowledgments. We would like to thank the Asiacrypt 2019 referees for their numerous helpful comments. The second author is supported by the Luxembourg National Research Fund through grant PRIDE15/10621687/SPsquared.

A   Proofs

A.1   Proof of Lemma 4

Let $\mathcal{B} = \{(B\boldsymbol{u}_j, v_j) : 1 \le j \le \ell + 1\}$ be a basis of L. We show that the vectors $\{\boldsymbol{u}_j : 1 \le j \le \ell\}$ corresponding to the first ℓ vectors, must necessarily be linearly independent over $\mathbb{R}$. For the sake of contradiction, we assume they are linearly dependent. For every vector $(B\boldsymbol{u}_j, v_j)$ (with $1 \le j \le \ell$), we consider the associated vector $\boldsymbol{p}_{\boldsymbol{u}_j, v_j}$. By assumption, the vectors $\{\boldsymbol{p}_{\boldsymbol{u}_j, v_j} : 1 \le j \le \ell\}$ belong to the lattice generated by the vectors $\{\boldsymbol{q}_i : 1 \le i \le \theta\}$, so there are integers $\beta_{ij} \in \mathbb{Z}$ such that $\boldsymbol{p}_{\boldsymbol{u}_j, v_j} = \sum_{i=1}^{\theta} \beta_{ij}\boldsymbol{q}_i$ for every $1 \le j \le \ell$. The definition of the vectors $\{\boldsymbol{q}_i : 1 \le i \le \theta\}$ gives $\boldsymbol{p}_{\boldsymbol{u}_j, v_j} = (C\beta_{1j}g_1, \ldots, C\beta_{\theta j}g_\theta, -\sum_{i=1}^{\theta} \beta_{ij}s_i)$ for every $1 \le j \le \ell$; and from the definition of the vector $\boldsymbol{p}_{\boldsymbol{u}_j, v_j}$, we conclude by equalizing the components, the relations

$$\beta_{ij}g_i = \langle \boldsymbol{u}_j, \hat{\boldsymbol{m}}_i \rangle \tag{26}$$

and

$$-\sum_{i=1}^{\theta} \beta_{ij}s_i = \langle \boldsymbol{u}_j, \boldsymbol{R} \rangle + v_j \tag{27}$$

for every $1 \le j \le \ell, 1 \le i \le \theta$. Combining Eqs. (26) and (27) gives

$$v_j = -\sum_{i=1}^{\theta} \frac{s_i}{g_i}\langle \boldsymbol{u}_j, \hat{\boldsymbol{m}}_i \rangle - \langle \boldsymbol{u}_j, \boldsymbol{R} \rangle, \; 1 \le j \le \ell$$

This implies that if the vectors $\{u_j : 1 \leq j \leq \ell\}$ are linearly dependent over $\mathbb{R}$, then also the vectors $\{(Bu_j, v_j) : 1 \leq j \leq \ell\}$ are linearly dependent over $\mathbb{R}$, which contradicts the fact that $\mathcal{B}$ is a basis of L. $\square$

A.2   Proof of Proposition 5

Let $a = (\alpha_1, \ldots, \alpha_\theta, 1) \in \mathbb{Z}^{\theta+1}$. We let $C = 2^{\rho_R - \alpha + 1}$ and consider the lattice $A^\perp$ of vectors $(Cx, y) \in \mathbb{Z}^\theta \times \mathbb{Z}$ such that (x, y) is orthogonal to a modulo x_0. Further, we let $B = \theta 2^{\rho_R + 2}$ and let $L \subseteq \mathbb{Z}^{\ell+1}$ denote the lattice of vectors $(Bu, v) \in \mathbb{Z}^\ell \times \mathbb{Z}$ such that the vector (u, v) is orthogonal to the vector $(\omega, 1)$ modulo x_0.

Let $\Lambda^\perp$ be the lattice of vectors $u \in \mathbb{Z}^\ell$ such that $\langle u, \hat{m}_i \rangle \equiv 0 \pmod{g_i}$ for all $1 \leq i \leq \theta$. We denote by u_0 a shortest non-zero vector of $\Lambda^\perp$. We write $\langle u_0, \hat{m}_i \rangle = k_i g_i$ with $k_i \in \mathbb{Z}$. To u_0 we thus associate the vector $F(u_0) = (Bu_0, -\sum_{i=1}^\theta k_i s_i - \langle u_0, R \rangle)$. From the definition of ω and the congruence relations $g_i \alpha_i \equiv s_i \pmod{x_0}$, we have that $(u_0, -\sum_{i=1}^\theta k_i s_i - \langle u_0, R \rangle)$ is orthogonal to $(\omega, 1)$ modulo x_0, and therefore $F(u_0) \in L$.

Letting $g = \prod_{i=1}^\theta g_i$, we now show that $F(u_0)$ has square norm upper bounded by

$$\|F(u_0)\|^2 \leq (\ell+1)B^2\|u_0\|^2 \leq \ell(\ell+1)B^2 g^{2/\ell}. \tag{28}$$

Indeed, we write $\|F(u_0)\|^2 \leq B^2\|u_0\|^2 + (\sum_{i=1}^\theta |k_i s_i| + \|u_0\|\|R\|)^2$. From $\|\hat{m}_i\| \leq \sqrt{\ell} 2^\alpha$, we obtain $2^{\alpha-1}|k_i| \leq |k_i| g_i \leq \|u_0\|\|\hat{m}_i\| \leq \sqrt{\ell} 2^\alpha \|u_0\|$; i.e. $|k_i| \leq 2\sqrt{\ell}\|u_0\|$ for all i. Combined with $\|R\| \leq \sqrt{\ell}\|R\|_\infty \leq \sqrt{\ell} 2^{\rho_R}$, this gives

$$\sum_{i=1}^\theta |k_i s_i| + \|u_0\|\|R\| \leq \sqrt{\ell}\|u_0\| \cdot (2^{\rho_R+1}\theta + 2^{\rho_R}) \leq \sqrt{\ell}\|u_0\|(2 \cdot 2^{\rho_R+1}\theta) = \sqrt{\ell}B\|u_0\|$$

Therefore, $\|F(u_0)\|^2 \leq B^2\|u_0\|^2 + \ell B^2\|u_0\|^2 = (\ell+1)B^2\|u_0\|^2$. Now, since u_0 has length $\lambda_1(\Lambda^\perp)$, it follows from Minkowski's Theorem that $\|u_0\| \leq \sqrt{\ell} g^{1/\ell}$ where $g = \det(\Lambda^\perp)$, and (28) easily follows.

Let $x_1 = (Bu_1, v_1)$ be the first vector in a $(3/4)$-reduced basis of the lattice L, obtained from LLL. By Theorem 3, it satisfies $\|x_1\| \leq 2^{\ell/2}\|F(u_0)\|$, that is, combined with (28), $\|x_1\| \leq 2^{\ell/2}\sqrt{\ell(\ell+1)}Bg^{1/\ell}$. In particular, we obtain the bounds

$$\|u_1\| \leq 2^{\ell/2}\sqrt{\ell(\ell+1)} \cdot g^{1/\ell} \tag{29}$$

$$|v_1| \leq 2^{\ell/2}B\sqrt{\ell(\ell+1)} \cdot g^{1/\ell}. \tag{30}$$

For simplicity we write $K = 2^{\ell/2}\sqrt{\ell(\ell+1)}g^{1/\ell}$. Now, to the vector $x_1 \in L$, we associate, for C as above, the vector $f(x_1) = (C\langle u_1, \hat{m}_1 \rangle, \ldots, C\langle u_1, \hat{m}_\theta \rangle, \langle u_1, R \rangle + v_1) \in A^\perp$. Because $(Bu_1, v_1) \in L$, it is a direct check that $f(x_1) \in A^\perp$. Its square norm is upper bounded by

$$\|f(x_1)\|^2 \leq C^2 \sum_{i=1}^\theta \|u_1\|^2\|\hat{m}_i\|^2 + (\|u_1\|\|R\| + v_1)^2.$$

Using once again that $\|\hat{\boldsymbol{m}}_i\| \leq 2^\alpha \sqrt{\ell}$ and $\|\boldsymbol{R}\| \leq 2^{\rho_R} \sqrt{\ell}$, and combining with (29) and (30), we obtain

$$\|f(\boldsymbol{x}_1)\|^2 \leq C^2 K^2 \cdot \theta \ell 2^{2\alpha} + (K\sqrt{\ell}2^{\rho_R} + KB)^2 \leq C^2 K^2 \cdot \theta \ell 2^{2\alpha} + (2K\sqrt{\ell}B)^2$$
$$= K^2 \ell (C^2 \theta 2^{2\alpha} + 4B^2)$$

so that, using $C^2 \theta 2^{2\alpha} \leq B^2 = 16\theta^2 2^{2\rho_R}$, this gives

$$\|f(\boldsymbol{x}_1)\| \leq 4\sqrt{5} \cdot \sqrt{\ell} \cdot \theta \cdot K \cdot 2^{\rho_R}. \tag{31}$$

We now consider the vectors $\{\boldsymbol{q}_i : 1 \leq i \leq \theta\}$ defined by $\boldsymbol{q}_i = (0, \ldots 0, Cg_i, 0, \ldots, 0, -s_i) \in \mathbb{Z}^{\theta+1}$. They are linearly independent; moreover, from the congruence relations $g_i \alpha_i \equiv s_i \pmod{x_0}$ for $1 \leq i \leq \theta$ we deduce that for all i, $\langle \boldsymbol{q}_i, \boldsymbol{a} \rangle \equiv 0 \pmod{x_0}$; i.e. $\boldsymbol{q}_i \in A^\perp$. Further, as $|s_i| \leq 2^{\rho_R}$, their norm is upper bounded by $\|\boldsymbol{q}_i\|^2 \leq C^2 g_i^2 + 2^{2\rho_R} \leq C^2 g_i^2 + Cg_i^2 \leq 2C^2 g_i^2$ because $Cg_i \geq 2^{\rho_R - \alpha + 1} \cdot 2^{\alpha - 1} = 2^{\rho_R}$. Consequently,

$$\prod_{i=1}^\theta \|\boldsymbol{q}_i\| \leq 2^{\theta/2} C^\theta \prod_{i=1}^\theta g_i = 2^{\theta/2} C^\theta g. \tag{32}$$

Now, (17) together with $g \leq 2^{\alpha\theta}$, implies $(1 + 1/\ell)\log_2(g) + (\ell + \theta)/2 + \log_2(4\sqrt{5}\sqrt{\ell+1}\theta\ell) < \log_2(x_0) - \rho_R$ and, by raising to the power of 2, we obtain $g^{1+1/\ell} \cdot 2^{\ell/2} \cdot 2^{\theta/2} \cdot 4\sqrt{5}\sqrt{\ell+1}\theta\ell < x_0/2^{\rho_R}$. This is equivalent to

$$g^{1/\ell} \cdot 2^{\ell/2} \cdot 2^{\rho_R} \cdot 4\sqrt{5}\sqrt{\ell+1} \cdot \theta\ell < \frac{C^\theta x_0}{C^\theta 2^{\theta/2} g}. \tag{33}$$

The left hand side is lower bounded by $\|f(\boldsymbol{x}_1)\|$ by (31), and the right hand side is upper bounded by $\det(A^\perp)/\prod_{i=1}^\theta \|\boldsymbol{q}_i\|$, by (32) together with $\det(A^\perp) = C^\theta x_0$. Therefore (33) implies $\|f(\boldsymbol{x}_1)\| < \det(A^\perp)/\prod_{i=1}^\theta \|\boldsymbol{q}_i\|$. It follows from Lemma 2 that $f(\boldsymbol{x}_1)$ is in the linear span generated by the vectors $\{\boldsymbol{q}_i : 1 \leq i \leq \theta\}$. Since g_i are distinct prime numbers and $\gcd(s_i, g_i) = 1$ for $1 \leq i \leq \theta$, we conclude that $f(\boldsymbol{x}_1)$ is in the sublattice generated by the vectors $\{\boldsymbol{q}_i : 1 \leq i \leq \theta\}$. Consequently, for all $1 \leq i \leq \theta$, one has $\langle \boldsymbol{u}_1, \hat{\boldsymbol{m}}_i \rangle \equiv 0 \pmod{g_i}$.

The rows $\{\boldsymbol{b}_j : 1 \leq j \leq \ell + 1\}$ of the matrix

$$\begin{bmatrix} B\boldsymbol{I}_\ell & -\boldsymbol{\omega}^T \\ 0 & x_0 \end{bmatrix},$$

where $\boldsymbol{I}_\ell$ denotes the $\ell \times \ell$ identity matrix, form a $\mathbb{Z}$-basis of L. Hence, by running LLL on this matrix with $\delta = 3/4$, we obtain a vector $\boldsymbol{x}_1$ of which the first ℓ entries, divided by B, produce a vector $\boldsymbol{u} = \boldsymbol{u}_1$ satisfying $\langle \boldsymbol{u}_1, \hat{\boldsymbol{m}}_i \rangle \equiv 0 \pmod{g_i}$ for all i. By Theorem 3, the algorithm terminates in polynomial time. $\qquad\square$

References

[AB15]   Applebaum, B., Brakerski, Z.: Obfuscating circuits via composite-order graded encoding. In: Dodis, Y., Nielsen, J.B. (eds.) TCC 2015. LNCS, vol. 9015, pp. 528–556. Springer, Heidelberg (2015). https://doi.org/10.1007/978-3-662-46497-7_21

[CGH+15]   Coron, J.-S., et al.: Zeroizing without low-level zeroes: new MMAP attacks and their limitations. In: Gennaro, R., Robshaw, M. (eds.) CRYPTO 2015, Part I. LNCS, vol. 9215, pp. 247–266. Springer, Heidelberg (2015). https://doi.org/10.1007/978-3-662-47989-6_12

[CGH17]   Chen, Y., Gentry, C., Halevi, S.: Cryptanalyses of candidate branching program obfuscators. In: Coron, J.-S., Nielsen, J.B. (eds.) EUROCRYPT 2017, Part III. LNCS, vol. 10212, pp. 278–307. Springer, Cham (2017). https://doi.org/10.1007/978-3-319-56617-7_10

[CHL+15]   Cheon, J.H., Han, K., Lee, C., Ryu, H., Stehlé, D.: Cryptanalysis of the multilinear map over the integers. In: Oswald, E., Fischlin, M. (eds.) EUROCRYPT 2015, Part I. LNCS, vol. 9056, pp. 3–12. Springer, Heidelberg (2015). https://doi.org/10.1007/978-3-662-46800-5_1

[CLLT16]   Coron, J.-S., Lee, M.S., Lepoint, T., Tibouchi, M.: Cryptanalysis of GGH15 multilinear maps. In: Robshaw, M., Katz, J. (eds.) CRYPTO 2016, Part II. LNCS, vol. 9815, pp. 607–628. Springer, Heidelberg (2016). https://doi.org/10.1007/978-3-662-53008-5_21

[CLLT17]   Coron, J.-S., Lee, M.S., Lepoint, T., Tibouchi, M.: Zeroizing attacks on indistinguishability obfuscation over CLT13. In: Fehr, S. (ed.) PKC 2017, Part I. LNCS, vol. 10174, pp. 41–58. Springer, Heidelberg (2017). https://doi.org/10.1007/978-3-662-54365-8_3

[CLR15]   Cheon, J.H., Lee, C., Ryu, H.: Cryptanalysis of the new CLT multilinear maps. IACR Cryptology ePrint Archive, 2015:934 (2015)

[CLT13]   Coron, J.-S., Lepoint, T., Tibouchi, M.: Practical multilinear maps over the integers. In: Canetti, R., Garay, J.A. (eds.) CRYPTO 2013. LNCS, vol. 8042, pp. 476–493. Springer, Heidelberg (2013). https://doi.org/10.1007/978-3-642-40041-4_26

[CVW18]   Chen, Y., Vaikuntanathan, V., Wee, H.: GGH15 beyond permutation branching programs: proofs, attacks, and candidates. In: Shacham, H., Boldyreva, A. (eds.) CRYPTO 2018, Part II. LNCS, vol. 10992, pp. 577–607. Springer, Cham (2018). https://doi.org/10.1007/978-3-319-96881-0_20

[DGHV10]   van Dijk, M., Gentry, C., Halevi, S., Vaikuntanathan, V.: Fully homomorphic encryption over the integers. In: Gilbert, H. (ed.) EUROCRYPT 2010. LNCS, vol. 6110, pp. 24–43. Springer, Heidelberg (2010). https://doi.org/10.1007/978-3-642-13190-5_2

[FRS17]   Fernando, R., Rasmussen, P.M.R., Sahai, A.: Preventing CLT attacks on obfuscation with linear overhead. In: Takagi, T., Peyrin, T. (eds.) ASIACRYPT 2017, Part III. LNCS, vol. 10626, pp. 242–271. Springer, Cham (2017). https://doi.org/10.1007/978-3-319-70700-6_9

[GGH13a]   Garg, S., Gentry, C., Halevi, S.: Candidate multilinear maps from ideal lattices. In: Johansson, T., Nguyen, P.Q. (eds.) EUROCRYPT 2013. LNCS, vol. 7881, pp. 1–17. Springer, Heidelberg (2013). https://doi.org/10.1007/978-3-642-38348-9_1

[GGH+13b] Garg, S., Gentry, C., Halevi, S., Raykova, M., Sahai, A., Waters, B.: Candidate indistinguishability obfuscation and functional encryption for all circuits. In: FOCS, pp. 40–49. IEEE Computer Society (2013)

[GGH15] Gentry, C., Gorbunov, S., Halevi, S.: Graph-induced multilinear maps from lattices. In: Dodis, Y., Nielsen, J.B. (eds.) TCC 2015, Part II. LNCS, vol. 9015, pp. 498–527. Springer, Heidelberg (2015). https://doi.org/10.1007/978-3-662-46497-7_20

[GLSW15] Gentry, C., Lewko, A.B., Sahai, A., Waters, B.: Indistinguishability obfuscation from the multilinear subgroup elimination assumption. In: IEEE 56th Annual Symposium on Foundations of Computer Science, FOCS 2015, Berkeley, CA, USA, 17–20 October, 2015, pp. 151–170 (2015)

[GLW14] Gentry, C., Lewko, A., Waters, B.: Witness encryption from instance independent assumptions. In: Garay, J.A., Gennaro, R. (eds.) CRYPTO 2014, Part I. LNCS, vol. 8616, pp. 426–443. Springer, Heidelberg (2014). https://doi.org/10.1007/978-3-662-44371-2_24

[HJ16] Hu, Y., Jia, H.: Cryptanalysis of GGH map. In: Fischlin, M., Coron, J.-S. (eds.) EUROCRYPT 2016, Part I. LNCS, vol. 9665, pp. 537–565. Springer, Heidelberg (2016). https://doi.org/10.1007/978-3-662-49890-3_21

[HPS11] Hanrot, G., Pujol, X., Stehlé, D.: Analyzing blockwise lattice algorithms using dynamical systems. In: Rogaway, P. (ed.) CRYPTO 2011. LNCS, vol. 6841, pp. 447–464. Springer, Heidelberg (2011). https://doi.org/10.1007/978-3-642-22792-9_25

[Len87] Lenstra Jr., H.W.: Factoring integers with elliptic curves. Ann. Math. (2) **126**(3), 649–673 (1987)

[LLL82] Lenstra, A.K., Lenstra, H.W., Lovasz, L.: Factoring polynomials with rational coefficients. Math. Ann. **261**, 515–534 (1982)

[MSZ16] Miles, E., Sahai, A., Zhandry, M.: Annihilation attacks for multilinear maps: cryptanalysis of indistinguishability obfuscation over GGH13. In: Robshaw, M., Katz, J. (eds.) CRYPTO 2016, Part II. LNCS, vol. 9815, pp. 629–658. Springer, Heidelberg (2016). https://doi.org/10.1007/978-3-662-53008-5_22

[NS99] Nguyen, P., Stern, J.: The hardness of the hidden subset sum problem and its cryptographic implications. In: Wiener, M. (ed.) CRYPTO 1999. LNCS, vol. 1666, pp. 31–46. Springer, Heidelberg (1999). https://doi.org/10.1007/3-540-48405-1_3

[S+17] Stein, W.A., et al.: Sage Mathematics Software (Version 8.0). The Sage Development Team (2017). http://www.sagemath.org

[Zim15] Zimmerman, J.: How to obfuscate programs directly. In: Oswald, E., Fischlin, M. (eds.) EUROCRYPT 2015, Part II. LNCS, vol. 9057, pp. 439–467. Springer, Heidelberg (2015). https://doi.org/10.1007/978-3-662-46803-6_15

Algebraic XOR-RKA-Secure Pseudorandom Functions from Post-Zeroizing Multilinear Maps

Michel Abdalla[1,2(✉)] [ID], Fabrice Benhamouda[3] [ID], and Alain Passelègue[4]

[1] DIENS, École normale supérieure, CNRS, PSL Research University, Paris, France
michel.abdalla@ens.fr
[2] Inria, Paris, France
[3] Algorand Foundation, NewYork, NY, USA
fabrice.benhamouda@normalesup.org
[4] Inria, ENS Lyon, Lyon, France
alain.passelegue@inria.fr

Abstract. Due to the vast number of successful related-key attacks against existing block-ciphers, related-key security has become a common design goal for such primitives. In these attacks, the adversary is not only capable of seeing the output of a function on inputs of its choice, but also on related keys. At Crypto 2010, Bellare and Cash proposed the first construction of a pseudorandom function that could provably withstand such attacks based on standard assumptions. Their construction, as well as several others that appeared more recently, have in common the fact that they only consider linear or polynomial functions of the secret key over complex groups. In reality, however, most related-key attacks have a simpler form, such as the XOR of the key with a known value. To address this problem, we propose the first construction of RKA-secure pseudorandom function for XOR relations. Our construction relies on multilinear maps and, hence, can only be seen as a feasibility result. Nevertheless, we remark that it can be instantiated under two of the existing multilinear-map candidates since it does not reveal any encodings of zero. To achieve this goal, we rely on several techniques that were used in the context of program obfuscation, but we also introduce new ones to address challenges that are specific to the related-key-security setting.

Keywords: Pseudorandom functions · Related-key security · Multilinear maps · Post-zeroizing constructions

1 Introduction

Context. Most of the security models used to prove the security of cryptographic schemes usually assume that an adversary has only a black-box access to the cryptosystem. In particular, the adversary has no information about the

S. D. Galbraith and S. Moriai (Eds.): ASIACRYPT 2019, LNCS 11922, pp. 386–412, 2019.
https://doi.org/10.1007/978-3-030-34621-8_14

secret key, nor can it modify the latter. Unfortunately, it has been shown that this is not always true in practice. For instance, an adversary may learn information from physical measures such as the running time of the protocol or its energy consumption, or may also be able to inject faults in the cryptosystem. In the specific case of fault attacks, in addition to possibly being able to learn partial information about the key, the adversary may be able to force the cryptosystem to run with a different, but related, secret key. Then, by observing the outcomes of the cryptosystem with this new related key, an adversary may be able to break it. Such an attack is called a related-key attack (RKA) and has often been used against concrete blockciphers [Knu93, Bih94, BDK05, BDK+10].

Formalization of RKA Security. Following the seminal cryptanalytic works by Biham and Knudsen, theoreticians have defined new security models in order to capture such attacks. In 2003, Bellare and Kohno formalized the foundations for RKA security [BK03]. Specifically, let $F: \mathcal{K} \times \mathcal{D} \to \mathcal{R}$ be a pseudorandom function and let $\Phi \subseteq \mathsf{Fun}(\mathcal{K}, \mathcal{K})$ be a set of functions on the key space $\mathcal{K}$, called a class of related-key deriving (RKD) functions. We say that F is Φ-RKA secure if it is hard to distinguish an oracle which, on input a pair $(\phi, x) \in \Phi \times \mathcal{D}$, outputs $F(\phi(K), x)$, from an oracle which, on the same input pair, outputs $G(\phi(K), x)$, where $K \in \mathcal{K}$ is a random target key and $G: \mathcal{K} \times \mathcal{D} \to \mathcal{R}$ is a random function.

Existing Constructions. Building provably RKA-secure pseudorandom functions in non-idealized model has been a long-standing open question until the work of Bellare and Cash in 2010 [BC10]. Their construction is adapted from the Naor-Reingold PRF [NR97] and is obtained by applying a generic framework to it. This generic framework has led to other constructions based under different assumptions [LMR14] and been recently extended to circumvent its limitations [ABPP14, ABP15, ABPP18]. However, despite being simple and elegant, the existing frameworks crucially rely on the algebraic structure of the pseudorandom functions and thus only allow to build RKA-secure pseudorandom functions for algebraic classes of RKD functions. More precisely, existing constructions use a key whose components are elements in $\mathbb{Z}_p$, and are proven secure against an attacker that is given the capability to perform operations (additions, multiplications, or even polynomial evaluation) modulo p, with p super-polynomial in the security parameter for instantiations over cyclic or multilinear groups [BC10, LMR14, ABPP14, ABP15], or p polynomial but still much larger than 2, for instantiations based on lattices [LMR14, BP14].

Unfortunately, the algebraic classes of RKD functions above are not very natural as they seem difficult to implement from the perspective of an attacker[1]. Moreover, they also do not seem to match attacks against concrete blockciphers such as AES (e.g., [BDK05, BDK+10]). To address these shortcomings, we focus

---

[1] In practice, laser fault attacks can be used to switch bits inside a chip, but it seems unlikely that such a fault attack can be used to apply an operation modulo some prime number on a register.

in this paper on the XOR class of RKD functions, which seems more relevant for practice, as suggested by Bellare and Kohno [BK03]. In this class, which corresponds to the class of functions $\Phi_\oplus = \{\phi_s\colon K \in \{0,1\}^k \mapsto K \oplus s \in \{0,1\}^k \mid s \in \{0,1\}^k\}$ with $\mathcal{K} = \{0,1\}^k$ being the keyspace, the adversary is allowed to flip bits of the secret key.

In the context of pseudorandom functions, there have been a few proposals for protecting against weak versions of XOR-related-key attacks. In [AW14], for instance, Applebaum and Widder proposed several schemes that can provably resist XOR attacks by restricting the capabilities of the adversary. In particular, they consider models where the adversary only uses a bounded number of related keys or where it only performs random XOR operations. In [JW15], Jafargholi and Wichs proposed constructions of pseudorandom functions based on continuous non-malleable codes that can resist several forms of related-key attacks, including XOR-related-key attacks. Their solutions, however, only guarantee pseudorandomness under the original key and not under both the original and related keys as in standard notions of related-key security. Another advantage of our construction compared to theirs is that the key in our case is just a bit string.

In the random-oracle model, there is a straightforward construction of a pseudorandom function secure against XOR-related-key attacks: $F(K, x) = H(K\|x)$, with H being a hash function modeled as a random oracle, and $\|$ being the string concatenation operator. However, from a theoretical perspective, having a construction in the random-oracle model does not provide any guarantee that the primitive can be realized under non-interactive (and falsifiable) assumptions. To the best of our knowledge, building a pseudorandom function that provably resists XOR-related-key attacks in non-idealized models still remains a major open problem. Furthermore, we would like to point out that we do not know of any construction of a pseudorandom function secure against XOR-related-key attacks in the generic multilinear group model.

Our Contributions. In this paper, we provide the first provably secure construction of an RKA-secure pseudorandom function for XOR relations. To achieve this goal, our construction departs significantly from previous RKA-secure pseudorandom functions in that the secret key no longer lies in an algebraic group. As in prior constructions (e.g., [BC10, ABPP14, ABP15, LMR14]), our new scheme also requires public parameters, such as generators of a cyclic group, which cannot be tampered with. However, unlike these constructions, the secret key in our scheme is just a bit string, whose values are used to select a subset of the common parameters which will be used in the evaluation of the pseudorandom function.[2] The evaluation is performed using an asymmetric multilinear map [GGH13a, CLT13, GGH15].

---

[2] Please note that these common parameters are public and can therefore be stored in a non-secure read-only part of the memory which can potentially more easily be made tamper-proof.

In particular, we prove our construction under two non-interactive assumptions. We further prove that it is hard to come up with encodings of zero (in the generic multilinear map model) given only the encodings revealed by these assumptions. Therefore these assumptions are plausible under some current instantiations of multilinear maps. To the best of our knowledge, this is the first construction using multilinear maps with such a level of security. In [BMSZ16], Badrinarayanan et al. were the first to take into account zeroizing attacks (i.e., attacks using encodings of zero) against current multilinear maps constructions into their constructions and to propose a scheme which is proven not to reveal encodings of zero, in the generic multilinear map model. But, contrary to us, they did not provide a proof of their scheme under a non-interactive assumption.

Overview of Our Techniques. As mentioned above, our construction is based on asymmetric multilinear maps [GGH13a, CLT13, GGH15]. Informally speaking, a multilinear map is a generalization of bilinear maps. It allows to "encode" scalars a (in some finite field $\mathbb{Z}_p$) into some encodings $[a]_{\mathcal{S}}$ with respect to an index set (also called an index) $\mathcal{S} \subseteq \mathcal{U}$, where $\mathcal{S}$ indicates the level of the encoding $[a]_{\mathcal{S}}$ and $\mathcal{U}$ denotes the top-level index set. We can add two elements $[a]_{\mathcal{S}}$ and $[b]_{\mathcal{S}}$ belonging to the same index set $\mathcal{S}$ to obtain $[a+b]_{\mathcal{S}}$. We can also multiply elements $[a]_{\mathcal{S}_1}$ and $[b]_{\mathcal{S}_2}$ to compute $[a \cdot b]_{\mathcal{S}_1 \cup \mathcal{S}_2}$ to obtain the encoding of $a \cdot b$ with respect to the index set $\mathcal{S}_1 \cup \mathcal{S}_2$, as long as the two index sets are disjoint ($\mathcal{S}_1 \cap \mathcal{S}_2 = \emptyset$). Finally, it is possible to test whether an element at level $\mathcal{U}$ is an encoding of 0.

Let k and n be the lengths of the secret key K and input x, respectively, and let $\mathcal{U} = \{1, \ldots, k+n\}$. Next, let $\{a_{i,b}\}_{i \in [k], b \in \{0,1\}}$ and $\{c_{j,b}\}_{j \in [n], b \in \{0,1\}}$ be random scalars in $\mathbb{Z}_p$ and let $\hat{a}_{i,b} = [a_{i,b}]_{\{i\}}$ be the encoding of $a_{i,b}$ at index level $\{i\}$ and $\hat{c}_{j,b} = [c_{j,b}]_{\{j+k\}}$ be the encoding of $c_{j,b}$ at index level $\{j+k\}$. The starting point of our construction is the function

$$F_{\mathsf{pp}}(K, x) = \left[\prod_{i=1}^{k} a_{i,K_i} \prod_{j=1}^{n} c_{j,x_j} \right]_{\mathcal{U}},$$

where the public parameters pp include $\{\hat{a}_{i,b}\}_{i \in [k], b \in \{0,1\}}$ and $\{\hat{c}_{j,b}\}_{j \in [n], b \in \{0,1\}}$ as well as the public parameters of the multilinear map.

Since the encodings of the scalars $a_{i,b}$ and $c_{j,b}$ are included in pp, it is not hard to see that the user in possession of the secret K can efficiently evaluate this function at any point x by computing the multilinear map function with the help of the encodings $\hat{a}_{i,K_i}$ and $\hat{c}_{j,x_j}$. To prove it secure and be able to instantiate the scheme with existing multilinear map candidates, however, is not straightforward as one needs to show that the adversary cannot mix too many existing encodings and create an encoding of zero.

While a proof in the generic multilinear map model [Sho97, GGH+13b, BR14, BGK+14, Zim15] is possible, we would like to rely on non-interactive complexity assumptions over multilinear maps which are likely to hold in existing multilinear map constructions. To achieve this goal, we change the way in which the index sets are defined using techniques from program obfuscation [GGH+13b, BGK+14, Zim15]. More precisely, we make use of the notion of strong straddling sets [BGK+14, MSW14, Zim15, BMSZ16], which informally allows to partition a set into two disjoint sets of subsets so that they cannot be mixed. As in [MSW14, Zim15, BMSZ16], we first construct strong straddling set systems over sets $\mathscr{S}_i$ of k^2 fresh symbols, for $i \in \{0, \ldots, k\}$ and use $\mathscr{S}_i$ for $i \geq 1$ to prevent the adversary from mixing an exponential number of inputs. In addition to that, and unlike [MSW14, Zim15, BMSZ16], we use $\mathscr{S}_0$ in the proof to prevent the adversary from mixing an internal private representation of the key (used by the reduction) with the parameters.

As we show in Sect. 4, the resulting scheme can be proven secure in the generic multilinear map model. In particular, we are actually able to prove that no polynomial-time adversary can construct a (non-trivial) encoding of zero.[3] In addition to that, as we show in Sect. 5, one benefit of using (strong) straddling sets is that it allows us to prove the security of our construction under non-interactive assumptions and avoid the use of idealized models.[4] Finally, we also prove the plausibility of these new assumptions in the full version [ABP17] by showing that they hold in the generic multilinear map model.

We would like to stress that the security proof in Sect. 5 is a much stronger result qualitatively than a direct proof in the generic multilinear map model since the latter model is only used to show that our (non-interactive) assumptions are plausible.

Concrete Instantiations. The security of our scheme relies on two new non-interactive assumptions on multilinear maps. However, contrary to most classical assumptions (such as the Decisional Diffie-Hellman assumption for low-level group elements) or the multilinear subgroup assumption used to construct witness encryption and indistinguishable obfuscation in [GLW14, GLSW15], our new assumptions do not reveal any encoding of zero. More precisely, similarly to what Badrinarayanan et al. did in [BMSZ16], we show in the ideal multilinear map model, that given an instance for one of our assumptions, the adversary cannot construct any encoding of zero (even at the top level). In particular, this implies that our assumptions holds in the hybrid graded encoding model [GMM+16], as this model just restricts the shape of the encodings of zero that the adversary can construct, while we prove that the adversary cannot construct any such encoding.

Our assumptions are therefore not broken by most of the attacks against multilinear maps [GGH13a, CHL+15, CGH+15, HJ16, CLLT16] including the annihilation attacks [MSZ16, CGH17] and recent attacks such as [CLLT17, CVW18], Hence, while the GGH13 and CLT15 multilinear map candidates [GGH13a,

---

[3] Note that, to prove that two games are indistinguishable in the generic multilinear map model, it suffices to prove that the adversary cannot generate encodings of zeros in either game as it would only obtain random handles which carry no information.

[4] The security result in Sect. 4 is therefore implied by the security result in Sect. 5. Section 4 should be seen as a warm-up for Sect. 5.

CLT15] might not be used for our construction because of recent attacks without encodings of zero [CFL+16, ABD16, CJL16], our assumptions are plausible when implemented with the GGH15[5] or the CLT13 multilinear map candidates [GGH15, CLT13].

Additional Related Work. In addition to the work mentioned above, a few other constructions of pseudorandom functions against related-key attacks for linear and polynomial functions have been proposed in [BLMR13, LMR14]. While their RKA-secure pseudorandom functions also require multilinear maps, their security proofs are based on assumptions which reveal encodings of zero and hence are subject to the multiple attacks mentioned above. Moreover, these works do not consider XOR-related-key attacks.

Related-key security with respect to XOR relations has also been considered in the context of Even-Mansour ciphers [CS15, FP15, Men16, WLZZ16]. Unlike our work, which aims to prove security under well specified non-interactive complexity assumptions, all these works rely on idealized models, which we want to avoid.

In [FX15], the authors proposed efficient constructions of identity-based encryption and key encapsulation schemes that remain secure against related-key attacks for a large class of functions, which include XOR relations. While their results are interesting, we remark that achieving RKA security for randomized primitives appears to be significantly easier than for deterministic ones, as already noted by Bellare and Cash [BC10].

Finally, it is worth mentioning here that, due to the results of Bellare, Cash, and Miller [BCM11], RKA security for pseudorandom functions can be transferred to several other primitives, including identity-based encryption, signatures, and chosen-ciphertext-secure public-key encryption.

Organization. The rest of the paper is organized as follows. Section 2 presents standard definitions of (related-key) pseudorandom functions and multilinear maps. It also introduces the generic multilinear map model and the notion of straddling set system. Section 3 explains our construction of XOR-RKA PRF. Section 4 provides a first security proof of our new scheme in the generic multilinear map model. It also shows that is not feasible for an adversary to generate

---

[5] The GGH15 multilinear map is defined over a graph. Index sets (and in particular straddling set indexes) are defined independently of the graph structure of GGH15, hence they can be implemented as on GGH13. Specifically, index sets are based on the z_i's, as for GGH13, and not on the graph structure. More precisely, the construction relies on distinct z_i's for every index and an encoding with index set S depends on $\prod_{i \in S} z_i$, exactly as on GGH13. This is completely independent of the graph structure and then we can use a simple chain as a graph, as in the case of indistinguishable obfuscation candidate. A clean exposition of how to implement straddling sets for GGH15 is given in [Hal15, Sect. 4.2]. Note also that we consider multilinear maps with plaintext space corresponding to a prime-order finite field. This can be obtained using [Hal15, Sect. 4.1]. Attacks such as [CHKL18] do not apply in this setting.

non-trivial encodings of zero. Finally, Sect. 5 describes our main result, which is a proof of security for our new construction under two new non-interactive assumptions. The formal proof of security, as well as the proofs of our assumptions are detailed in the appendix. In particular, we prove that our assumptions are not only secure in the generic multilinear map model, but also that it is not feasible for an adversary to generate (non-trivial) encodings of zero. Hence, our assumptions are plausible given some current instantiations of multilinear maps.

2   Definitions

2.1   Notation and Games

Notation. We denote by κ the security parameter. Let $F \colon \mathcal{K} \times \mathcal{D} \to \mathcal{R}$ be a function that takes a key $K \in \mathcal{K}$ and an input $x \in \mathcal{D}$ and returns an output $F(K, x) \in \mathcal{R}$. The set of all functions $F \colon \mathcal{K} \times \mathcal{D} \to \mathcal{R}$ is then denoted by $\mathsf{Fun}(\mathcal{K}, \mathcal{D}, \mathcal{R})$. Likewise, $\mathsf{Fun}(\mathcal{D}, \mathcal{R})$ denotes the set of all functions mapping $\mathcal{D}$ to $\mathcal{R}$. If S is a set, then we denote by $s \xleftarrow{\$} S$ the operation of picking at random s in S. If $\vec{x}$ is a vector then we denote by $|\vec{x}|$ its length, so $\vec{x} = (x_1, \ldots, x_{|\vec{x}|})$. For a binary string x, we denote its length by $|x|$, x_i its i-th bit, so $x \in \{0, 1\}^{|x|}$ and $x = x_1 \| \ldots \| x_n$.

Games [BR06]. Most of our definitions and proofs use the code-based game-playing framework, in which a game has an **Initialize** procedure, procedures to respond to adversary oracle queries, and a **Finalize** procedure. To execute a game G with an adversary $\mathcal{A}$, we proceed as follows. First, **Initialize** is executed and its outputs become the input of $\mathcal{A}$. When $\mathcal{A}$ executes, its oracle queries are answered by the corresponding procedures of G. When $\mathcal{A}$ terminates, its outputs become the input of **Finalize**. The output of the latter, denoted $G^{\mathcal{A}}$ is called the output of the game, and we let "$G^{\mathcal{A}} \Rightarrow 1$" denote the event that this game output takes the value 1. The running time of an adversary by convention is the worst case time for the execution of the adversary with any of the games defining its security, so that the time of the called game procedures is included.

2.2   Pseudorandom Functions

Our definitions of pseudorandom functions and related-key secure pseudorandom functions include a **Setup** algorithm that is used to generate public parameters. For classical PRFs, the public parameters could actually just be included in the key. However, to our knowledge, all the known proven RKA-secure PRFs [BC10, ABPP14, LMR14, ABP15], use public parameters. Contrary to the key itself, the public parameters cannot be modified by the related-key deriving function. In our case, the **Setup** algorithm is specified explicitly in the construction for clarity.

PRFs [GGM86, BC10]. Consider a pseudorandom function $F_{\mathsf{pp}} \colon \mathcal{K} \times \mathcal{D} \to \mathcal{R}$ with public parameters pp. The advantage of an adversary $\mathcal{A}$ in attacking the standard PRF security of a function F_{pp} is defined via

$$\mathbf{Adv}_{F_{\mathsf{pp}}}^{\mathsf{prf}}(\mathcal{A}) = \Pr\left[\mathrm{PRFReal}_{F_{\mathsf{pp}}}^{\mathcal{A}} \Rightarrow 1\right] - \Pr\left[\mathrm{PRFRand}_{F_{\mathsf{pp}}}^{\mathcal{A}} \Rightarrow 1\right].$$

Game $\text{PRFReal}_{F_{\text{pp}}}$ first runs the Setup algorithm to generate the public parameters pp which it outputs. It then picks $K \xleftarrow{\$} \mathcal{K}$ at random, and responds to oracle query $\mathbf{Fn}(x)$ via $F_{\text{pp}}(K, x)$. Game $\text{PRFRand}_{F_{\text{pp}}}$ runs Setup and outputs the public parameters pp. It then picks $f \xleftarrow{\$} \text{Fun}(\mathcal{D}, \mathcal{R})$ and responds to oracle query $\mathbf{Fn}(x)$ via $f(x)$.

RKA-PRFs [BK03, BC10]. Consider a pseudorandom function $F_{\text{pp}}: \mathcal{K} \times \mathcal{D} \to \mathcal{R}$ with public parameters pp. Let $\Phi \subseteq \text{Fun}(\mathcal{K}, \mathcal{K})$; the members of Φ are called RKD (Related-Key Deriving) functions. An adversary is said to be Φ-restricted if its oracle queries (ϕ, x) satisfy $\phi \in \Phi$. The advantage of a Φ-restricted adversary $\mathscr{A}$ in attacking the RKA-PRF security of F_{pp} is defined via

$$\mathbf{Adv}^{\text{prf-rka}}_{\Phi, F_{\text{pp}}}(\mathscr{A}) = \Pr\left[\text{RKPRFReal}^{\mathscr{A}}_{F_{\text{pp}}} \Rightarrow 1\right] - \Pr\left[\text{RKPRFRand}^{\mathscr{A}}_{F_{\text{pp}}} \Rightarrow 1\right].$$

Game $\text{RKPRFReal}_{F_{\text{pp}}}$ first runs the Setup algorithm to generate the public parameters which it outputs. It then picks a key $K \xleftarrow{\$} \mathcal{K}$ at random, and responds to oracle query $\mathbf{RKFn}(\phi, x)$ via $F_{\text{pp}}(\phi(K), x)$. Game $\text{RKPRFRand}_{F_{\text{pp}}}$ runs Setup to generate the public parameters pp and outputs them. It then picks $G \xleftarrow{\$} \text{Fun}(\mathcal{K}, \mathcal{D}, \mathcal{R})$ and $K \xleftarrow{\$} \mathcal{K}$ at random, and responds to oracle query $\mathbf{RKFn}(\phi, x)$ via $G(\phi(K), x)$. We say that F_{pp} is a Φ-RKA-secure PRF if for any Φ-restricted adversary, its advantage in attacking the RKA-PRF security is negligible.

XOR-RKA-PRFs. Let $F_{\text{pp}}: \mathcal{K} \times \mathcal{D} \to \mathcal{R}$ be a pseudorandom function with public parameters pp and $\mathcal{K} = \{0, 1\}^k$ for some integer $k \geq \kappa$. We say that F_{pp} is XOR-RKA-secure if it is a $\Phi_\oplus$-RKA-secure PRF according to the above definition, where $\Phi_\oplus = \{\phi_s: K \in \{0, 1\}^k \mapsto K \oplus s \in \{0, 1\}^k \mid s \in \{0, 1\}^k\}$.

2.3 Multilinear Maps

We informally introduced multilinear maps in the introduction. Let us now introduce formal definitions, following the notations of [Zim15].

Definition 1 (Formal Symbol). *A formal symbol is a bitstring in $\{0, 1\}^*$. Distinct variables denote distinct bitstrings, and we call a fresh formal symbol any bitstring in $\{0, 1\}^*$ that has not already been assigned to a formal symbol.*

Definition 2 (Index Sets). *An index set (also called index) is a set of formal symbols.*

Definition 3 (Multilinear Map). *A multilinear map is defined by a tuple of six algorithms (MM.Setup, MM.Encode, MM.Add, MM.Mult, MM.ZeroTest, MM.Extract) with the following properties:*

- MM.Setup *takes as inputs the security parameter κ in unary and an index set $\mathscr{U}$, termed the* top-level index set, *and generates public parameters mm.pp, secret parameters mm.sp, and a prime number p;*

- MM.Encode *takes as inputs secret parameters* mm.sp, *a scalar* $x \in \mathbb{Z}_p$, *and an index set* $\mathcal{S} \subseteq \mathcal{U}$ *and outputs:*

$$\text{MM.Encode}(\text{mm.sp}, x, \mathcal{S}) \rightarrow [x]_{\mathcal{S}} \, ;$$

For the index set $\mathcal{S} = \emptyset$, $[x]_{\emptyset}$ *is simply the scalar* $x \in \mathbb{Z}_p$.
- MM.Add *takes as inputs public parameters* mm.pp *and two encodings with same index set* $\mathcal{S} \subseteq \mathcal{U}$ *and outputs:*

$$\text{MM.Add}(\text{mm.pp}, [x]_{\mathcal{S}}, [y]_{\mathcal{S}}) \rightarrow [x+y]_{\mathcal{S}} \, ;$$

- MM.Mult *takes as inputs public parameters* mm.pp *and two encodings with index sets* $\mathcal{S}_1, \mathcal{S}_2 \subseteq \mathcal{U}$ *respectively and outputs:*

$$\text{MM.Mult}(\text{mm.pp}, [x]_{\mathcal{S}_1}, [y]_{\mathcal{S}_2}) \rightarrow \begin{cases} [xy]_{\mathcal{S}_1 \cup \mathcal{S}_2} & \textit{if } \mathcal{S}_1 \cap \mathcal{S}_2 = \emptyset \\ \bot & \textit{otherwise} \end{cases} \, ;$$

- MM.ZeroTest *takes as inputs public parameters* mm.pp *and a top-level encoding (with index set* $\mathcal{U}$ *) and outputs:*

$$\text{MM.ZeroTest}(\text{mm.pp}, [x]_{\mathcal{S}}) \rightarrow \begin{cases} \textit{"zero"} & \textit{if } \mathcal{S} = \mathcal{U} \textit{ and } x = 0 \\ \textit{"non-zero"} & \textit{otherwise} \end{cases} \, ;$$

- MM.Extract *takes public parameters* mm.pp *and a top-level encoding* $[x]_{\mathcal{U}}$ *as inputs and outputs a canonical representation of* $[x]_{\mathcal{U}}$.

Remark 4. The MM.Extract algorithm is needed for our pseudorandom function to be deterministic with all currently known instantiations of multilinear maps [GGH13a, CLT13, GGH15, CLT15]. Indeed, in these instantiations, the same group element has many different representations, and the extraction procedure enables to extract a unique representation from any top-level group element (i.e., of index $\mathcal{U}$).

This extraction is necessary for our proof under non-interactive assumptions in Sect. 5 to work. For our proof in the generic multilinear map model, this is not required. For this reason, our generic multilinear map model does not support extraction for the sake of simplicity. Actually, this only strengthens the result, as before extraction, the adversary still has to possibility to add top-level group elements while extracted values are not necessarily homomorphic.

Conventions. In order to ease the reading, we adopt the following conventions in the rest of the paper:

- Scalars are noted with lowercase letter, e.g. $a, b, \ldots$
- Encodings are noted either as their encoding at index set $\mathcal{S}$, $[a]_{\mathcal{S}}$ or simply with a hat, when the index set is clear from the context, e.g. $\hat{a}, \hat{b}, \ldots$ In particular, $\hat{a}$ is an encoding of the scalar a.

- Index sets as well as formal variables are noted with uppercase letters, e.g. $X, S, \mathscr{S}, \ldots$
- We denote by $S_1 \cdot S_2$ or $S_1 S_2$ the union of sets S_1 and S_2. This notation implicitly assumes that the two sets are disjoint. If S_1 is an element, then $S_1 \cdot S_2$ stands for $\{S_1\} \cdot S_2$.
- The top-level index set is referred as $\mathscr{U}$.

We also naturally extend these notations when clear from the context, so for instance $\hat{a} + \hat{b} = \mathsf{MM.Add(mm.pp, \hat{a}, \hat{b})}$ and $\hat{a} \cdot \hat{b} = \mathsf{MM.Mult(mm.pp, \hat{a}, \hat{b})}$.

2.4 Generic Multilinear Map Model

Our construction is proven secure under two non-interactive assumptions. One is very classical and is a variant of DDH. The other is relatively simple but is not classical and we prove it in the generic multilinear map model to show its plausibility. That is why we need to introduce the generic multilinear map model, in addition to the fact that we also prove in this model that our construction does not enable the adversary to produce encodings of zero.

The generic multilinear map model is similar to the generic group model [Sho97]. Roughly speaking, the adversary has only the capability to apply operations (add, multiply, and zero-test) of the multilinear map to encodings. A scheme is secure in the generic multilinear map model if for any adversary breaking the real scheme, there is a generic adversary that breaks a modified scheme in which encodings are replaced by fresh nonces, called *handles*, that it can supply to a stateful oracle $\mathscr{M}$, defined as follows:

Definition 5 (Generic Multilinear Map Oracle). *A generic multilinear map oracle is a stateful oracle $\mathscr{M}$ that responds to queries as follows:*

- *On a query* $\mathsf{MM.Setup}(1^\kappa, \mathscr{U})$, *$\mathscr{M}$ generates a prime number p and parameters* $\mathsf{mm.pp, mm.sp}$ *as fresh nonces chosen uniformly at random from $\{0,1\}^\kappa$. It also initializes an internal table $T \leftarrow []$ that it uses to store queries and handles. It finally returns* $(\mathsf{mm.pp, mm.sp}, p)$ *and set internal state so that subsequent* $\mathsf{MM.Setup}$ *queries fail.*
- *On a query* $\mathsf{MM.Encode}(z, x, \mathcal{S})$, *with $z \in \{0,1\}^\kappa$ and $x \in \mathbb{Z}_p$, it checks that $z = \mathsf{mm.sp}$ and $\mathcal{S} \subseteq \mathscr{U}$ and outputs $\perp$ if the check fails, otherwise it generates a fresh handle $h \xleftarrow{\$} \{0,1\}^\kappa$, adds $h \mapsto (x, \mathcal{S})$ to T, and returns h.*
- *On a query* $\mathsf{MM.Add}(z, h_1, h_2)$, *with $z, h_1, h_2 \in \{0,1\}^\kappa$, it checks that $z = \mathsf{mm.pp}$, that h_1 and h_2 are handles in T which are mapped to values $(x_1, \mathcal{S}_1)$ and $(x_2, \mathcal{S}_2)$ such that $\mathcal{S}_1 = \mathcal{S}_2 = \mathcal{S} \subseteq \mathscr{U}$, and returns $\perp$ if the check fails. If it passes, it generates a fresh handle $h \xleftarrow{\$} \{0,1\}^\kappa$, adds $h \mapsto (x_1 + x_2, \mathcal{S})$ to T, and returns h.*
- *On a query* $\mathsf{MM.Mult}(z, h_1, h_2)$, *with $z, h_1, h_2 \in \{0,1\}^\kappa$, it checks $z = \mathsf{mm.pp}$, that h_1 and h_2 are handles in T which are mapped to values $(x_1, \mathcal{S}_1)$ and $(x_2, \mathcal{S}_2)$ such that $\mathcal{S}_1 \cup \mathcal{S}_2 \subseteq \mathscr{U}$ and $\mathcal{S}_1 \cap \mathcal{S}_2 = \emptyset$, and returns $\perp$ if the check fails. If it passes, it generates a fresh handle $h \xleftarrow{\$} \{0,1\}^\kappa$, adds $h \mapsto (x_1 x_2, \mathcal{S}_1 \cup \mathcal{S}_2)$ to T, and returns h.*

– *On a query* MM.ZeroTest(z, h), *with* $z, h \in \{0, 1\}^\kappa$, *it checks* $z = $ mm.pp, *that h is a handle in T such that it is mapped to a value $(x, \mathscr{U})$, and returns $\perp$ if the check fails. If it passes, it returns "zero" if $x = 0$ and "non-zero" otherwise.*

Remark 6. In order to ease the reading, we actually use a slightly different and more intuitive characterization of the generic multilinear map oracle in our proofs. Informally, instead of considering encodings as nonces, we consider these as formal polynomials (that can be computed easily), whose formal variables are substituted with their join value distribution from the real game. In our construction, formal variables are $\hat{a}_{i,b}, \hat{c}_{j,b}, \hat{z}_{i_1,i_2,b_1,b_2}$—please refer to the construction in Sect. 3 for details. This variant characterization follows the formalization from [Zim15, Appendix B], please refer to this section for more formal definitions.

2.5   Actual Instantiations

While Definition 3 is a very natural definition and is what we actually would like as a multilinear map, up to now, we still do not know any such construction. Known constructions [GGH13a, CLT13, GGH15, CLT15] of multilinear maps are actually "noisy" variants of our formal definition. That is, each encoding includes a random error term, and similarly to what happens for lattice-based constructions, this error term grows when performing operations (addition or multiplication). Eventually, this error term becomes too big and the MM.ZeroTest can no longer recover the correct answer. This noise implicitly restricts the number of operations that can be performed. Intuitively, in current constructions, the errors are added when performing an addition and multiplied when performing a multiplication. However, the fact that current instantiations are noisy does not pose any problem regarding our construction, as the number of operations for evaluating our pseudorandom function is fixed and independent from the instantiation of the multilinear map.

2.6   Straddling Sets

Our construction and its proofs use strong straddling sets [BGK+14, MSW14, Zim15, BMSZ16], in order to prevent the adversary from mixing too many encodings and creating encodings of zero. We recall their definition below. We first recall that, for a set $\mathscr{S}$, we say that $\{S_1, \ldots, S_k\}$, for some integer k, is a partition of $\mathscr{S}$, if and only if $\cup_{i=1}^{k} S_i = \mathscr{S}$, $S_i \neq \emptyset$ and $S_i \cap S_j = \emptyset$, for any $1 \leq i, j \leq k$, $i \neq j$.

Definition 7 ((Strong) Straddling Set System). *For $k \in \mathbb{N}$, a k-straddling set system over a set $\mathscr{S}$ consists of two partitions $S_0 = \{S_{0,1}, \ldots, S_{0,k}\}$ and $S_1 = \{S_{1,1}, \ldots, S_{1,k}\}$ of $\mathscr{S}$ such that the following holds: for any $\mathscr{T} \subseteq \mathscr{S}$, if T_0, T_1 are distinct subsequences of $S_{0,1}, \ldots, S_{0,k}, S_{1,1}, \ldots, S_{1,k}$ such that T_0 and T_1 are partitions of $\mathscr{T}$, then $\mathscr{T} = \mathscr{S}$ and $T_0 = S_b$ and $T_1 = S_{1-b}$ for some $b \in \{0, 1\}$.*

Moreover, we say that S_0, S_1 is a strong k-straddling set system if for any $1 \leq i, j \leq k$, $S_{0,i} \cap S_{1,j} \neq \emptyset$.

A strong k-straddling set system is clearly also a k-straddling set system. Intuitively, a k-straddling set system ensures that the only two solutions to build a partition of $\mathcal{S}$ from combining sets in S_0 or S_1 are to use either every element in S_0 or every element in S_1.

We are only using strong straddling set systems in this paper, for the sake of simplicity. However, we only rely on the straddling set property in all our proofs, except the proof of one of our non-interactive assumptions, namely the Sel-Prod assumption. Let us now recall the construction of strong straddling set systems from [MSW14]. For the sake of completeness, we also recall the construction of straddling set systems from [BGK+14] in the full version [ABP17].

Construction 8 (Strong Straddling Set Systems [MSW14]). Let k be a fixed integer and let $\mathcal{S} = \{1, \ldots, k^2\}$. Then the following partitions $S_b = (S_{b,1}, \ldots, S_{b,k})$, for $b \in \{0, 1\}$, form a strong k-straddling set system over $\mathcal{S}$:

$$S_{0,i} = \{k(i-1)+1, k(i-1)+2, \ldots, ki\}$$
$$S_{1,i} = \{i, k+i, 2k+i, \ldots, k(k-1)+i\}.$$

This construction naturally extends to any set with k^2 elements (see Fig. 1 for an illustration of the case $k = 5$).

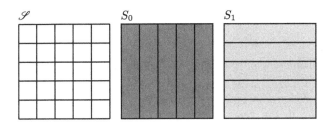

Fig. 1. Construction of strong 5-straddling set systems

Strong straddling set systems also satisfy the following lemma. Please refer to [BMSZ16] for proofs of constructions and lemma.

Lemma 9. *Let S_0, S_1 be a strong k-straddling set system over a set $\mathcal{U}$. Then for any $\mathcal{T} \subsetneq \mathcal{U}$ that can be written as a disjoint union of sets from S_0, S_1, there is a unique $b \in \{0, 1\}$ such that $\mathcal{T} = \cup_{i \in I} S_{b,i}$ for some $I \subseteq \{1, \ldots, k\}$.*

3 Our Construction

Let us now describe our construction of an XOR-RKA-secure pseudorandom function, for security parameter κ, key set $\mathcal{K} = \{0, 1\}^k$, with $k = 2\kappa$, and domain $\mathcal{D} = \{0, 1\}^n$ for some integer n.

3.1   Intuition

Construction Overview. The starting point of our construction is the Naor-Reingold pseudorandom function, defined as $\mathsf{NR} : (\vec{a}, x) \in \mathbb{Z}_p^{2n} \times \{0,1\}^n \mapsto g^{\prod_{i=1}^n a_{i,x_i}}$, where $\vec{a}$ is the secret key. As we are interested in XOR relations, we want the key to be a bitstring. A simple solution is to tweak this construction by considering the function $f_{\vec{a},\vec{c}} : (K, x) \in \{0,1\}^k \times \{0,1\}^n \mapsto g^{\prod_{i=1}^k a_{i,K_i} \cdot \prod_{j=1}^n c_{i,x_i}}$, with $\vec{a} \in \mathbb{Z}_p^k$ and $\vec{c} \in \mathbb{Z}_p^n$. It is easy to see that without knowing $\vec{a}$ nor $\vec{c}$, the outputs of this function are computationally indistinguishable from random (they correspond to NR evaluations with key $(\vec{a}, \vec{c})$ and on input (K, x)). However, given a key $K \in \{0,1\}^k$, one needs the values $\vec{a}, \vec{c}$ in order to be able to evaluate this function, so these values need to be made public. Then, it becomes very easy, even without knowing K, to distinguish this function from a random one.

That is why we use a multilinear map: this allows us to publicly reveal low-level encodings of elements in $\vec{a}$ and $\vec{c}$. These encodings let anyone evaluate the function on any key K and any input x, while keeping the outputs of the function computationally indistinguishable from random to an adversary that does not know the secret key K. Formally, we let $\mathscr{U} = \{1, \ldots, k+n\}$ be the set of indices for a multilinear map, $(a_{i,b})_{i \in \{1,\ldots,k\}, b \in \{0,1\}}$ and $(c_{j,b})_{j \in \{1,\ldots,n\}, b \in \{0,1\}}$ be random scalars in $\mathbb{Z}_p$ and $\hat{a}_{i,b} = [a_{i,b}]_{\{i\}}$ be an encoding of $a_{i,b}$ at index index i and $\hat{c}_{j,b} = [c_{j,b}]_{\{j+k\}}$ be an encoding of $c_{j,b}$ at index level $j + k$. We then consider the function:

$$F_{\mathsf{pp}}(K, x) = \left[\prod_{i=1}^k a_{i,K_i} \prod_{j=1}^n c_{j,x_j} \right]_{\mathscr{U}} = \prod_{i=1}^k \hat{a}_{i,K_i} \cdot \prod_{j=1}^n \hat{c}_{j,x_j},$$

with public parameters pp including the public parameters of the multilinear map as well as $\{\hat{a}_{i,b}\}_{i \in \{1,\ldots,k\}, b \in \{0,1\}}$ and $(\hat{c}_{j,b})_{j \in \{1,\ldots,n\}, b \in \{0,1\}}$.

This construction can be easily proven to be an XOR-RKA secure pseudorandom function in the generic multilinear map model, and it is also easy to show that it does not let an adversary create encodings of zero. However, it seems very hard to prove that this construction is secure under a non-interactive assumption[6]. Hence, we modify this construction by using a more complex set of indices and straddling sets. While this makes the proof in the generic multilinear map model a bit harder, this allows us to prove the security of our construction under non-interactive assumptions, whose hardness seems plausible even with current instantiations of multilinear maps. In particular, we prove in the full version [ABP17] that these assumptions are secure in the generic multilinear map model and do not let an adversary generate (non-trivial) encodings of zero.

---

[6] In particular, the strategy we would like to use (replacing $\hat{a}_{i,b}$ by independent random elements $\hat{\gamma}_i$ when evaluating the function) does not work, as the adversary would be able to distinguish between the two games by comparing $\hat{a}_i, 0$ and $\hat{a}_i, 1$ to $\hat{\gamma}_i$ (by computing the differences and applying $\mathsf{MM.ZeroTest}$).

Proof Overview. In the proof, we need to show that an oracle $(s, x) \mapsto F_{pp}(K \oplus s, x)$ (where K is chosen secretly uniformly at random) looks indistinguishable from a random oracle. We first remark that we can write $F_{pp}(K \oplus s, x)$ as:

$$F_{pp}(K \oplus s, x) = \prod_{i=1}^{k} \hat{\gamma}_{i,s_i} \cdot \prod_{j=1}^{n} \hat{c}_{j,x_j},$$

where $\hat{\gamma}_{i,b} = [a_{i,K_i \oplus b}]_{\{i\}}$ is an encoding of $a_{i,K_i \oplus b}$ for $i \in \{1, \ldots, k\}$. Thus, instead of using $K \in \{0,1\}^k$ as the key, we can use an alternative private representation: $(\hat{\gamma}_{i,b})_{i \in \{1,\ldots,k\}}$.

The main idea of our reduction is to be able to replace this private representation of the key, by completely random group elements, independent of the public parameters. We remark that the function

$$((\hat{\gamma}_{i,b})_{i \in \{1,\ldots,k\}}, s) \mapsto \prod_{i=1}^{k} \hat{\gamma}_{i,s_i},$$

is a slight variant of the Naor-Reingold pseudorandom function with $(\hat{\gamma}_{i,b})_{i \in \{1,\ldots,k\}}$ being the key and $s \in \{0,1\}^k$ the input. It is actually possible to prove it is a pseudorandom function under a DDH-like assumption, and from that to prove that our construction is XOR-RKA-secure.

Unfortunately, it is obviously not true that given the public parameters $(\hat{a}_{i,b})_{i \in \{1,\ldots,k\}, b \in \{0,1\}}$ (which are encodings of $a_{i,b}$), the real group elements $(\hat{\gamma}_{i,b})_{i \in \{1,\ldots,k\}}$ (which are encodings $a_{i,K_i \oplus b}$) are indistinguishable from encodings of independent uniform values: it is straightforward to check that $\hat{a}_{i,b}$ corresponds to the same scalar as $\hat{\gamma}_{i,b}$ or $\hat{\gamma}_{i,1-b}$ (depending whether $K_i = 0$ or 1), using MM.ZeroTest (and after multiplying these group elements by the same group elements to get a top-level encodings). Our solution is to use more complex index sets based on strong straddling sets to solve this issue.

The first step consists in using a first strong k-straddling set $\mathscr{S}_0$: we use the indices of the first partition for $\hat{a}_{i,0}$ and $\hat{a}_{i,1}$ (for each i, $\hat{a}_{i,0}$ and $\hat{a}_{i,1}$ get the same index), and the indices of the second partition for $\hat{\gamma}_{i,0}$ and $\hat{\gamma}_{i,1}$. This prevents the adversary from comparing one group element $\hat{a}_{i,b}$ with a group element $\hat{\gamma}_{i,b'}$ directly. But this is not sufficient, as the adversary still could check whether:

$$\prod_{i=1}^{k} (\hat{a}_{i,0} + \hat{a}_{i,1}) = \prod_{i=1}^{k} (\hat{\gamma}_{i,0} + \hat{\gamma}_{i,1}),$$

for example. When $(\hat{\gamma}_{i,b})_{i \in \{1,\ldots,k\}}$ are correctly generated, the equality is satisfied, while otherwise, it is not with overwhelming probability. More generally, the adversary can generate expression which contains an exponential number of monomials when expanded. We do not know how to prove anything reasonable on these expressions, so instead, we are using k additional strong k-straddling sets to prevent this from happening, in a similar way as they are used in [Zim15].

3.2   Actual Construction

Index Set. First, similarly to [Zim15], for each $i \in \{0, \ldots, k\}$, we construct a strong k-straddling set system over a set $\mathscr{S}_i$ of $2k - 1$ fresh formal symbols. We denote by $S_{i,b}$ the two partitions forming each of this straddling set, for $b \in \{0, 1\}$, and by $S_{i,b,j}$ their elements, for $1 \leq j \leq k$. We also define:

$$\mathsf{BitCommit}_{i,b} = S_{i,b,i} \qquad \mathsf{BitFill}_{i_1,i_2,b_1,b_2} = S_{i_1,b_1,i_2} \cdot S_{i_2,b_2,i_1}$$

for any $i, i_1, i_2 \in \{1, \ldots, k\}$ and $b, b_1, b_2 \in \{0, 1\}$. Intuitively, the straddling set systems $\mathscr{S}_i$ for $i \geq 1$ play the same role as in [Zim15] (preventing the adversary from mixing an exponential number of inputs), while $\mathscr{S}_0$ is used in the proof to prevent the adversary from mixing the private representation of the key with the public parameters.

Let X_j be fresh formal symbols for $j \in \{1, \ldots, n\}$. We then define the top-level index set as follows:

$$\mathscr{U} = \prod_{I=0}^{k} \mathscr{S}_i \prod_{j=1}^{n} X_j.$$

Setup. The algorithm Setup first generates the parameters $(\mathsf{mm.pp}, \mathsf{mm.sp}, p)$ for the multilinear map by running $\mathsf{MM.Setup}(1^\kappa, \mathscr{U})$. Then it generates the following elements:

$$
\begin{array}{ll}
a_{i,b} \xleftarrow{\$} \mathbb{Z}_p & \text{for } i \in \{1, \ldots, k\} \text{ and } b \in \{0, 1\} \\[4pt]
\hat{a}_{i,b} \leftarrow [a_{i,b}]_{S_{0,0,i} \mathsf{BitCommit}_{i,b}} & \text{for } i \in \{1, \ldots, k\} \text{ and } b \in \{0, 1\} \\[4pt]
c_{j,b} \xleftarrow{\$} \mathbb{Z}_p & \text{for } j \in \{1, \ldots, n\} \text{ and } b \in \{0, 1\} \\[4pt]
\hat{c}_{j,b} \leftarrow [c_{j,b}]_{X_j} & \text{for } j \in \{1, \ldots, n\} \text{ and } b \in \{0, 1\} \\[4pt]
\hat{z}_{i_1,i_2,b_1,b_2} \leftarrow [1]_{\mathsf{BitFill}_{i_1,i_2,b_1,b_2}} & \text{for } i_1, i_2 \in \{1, \ldots, k\} \text{ and } b_1, b_2 \in \{0, 1\} \\
& \text{with } i_1 < i_2,
\end{array}
$$

and outputs the following parameters:

$$\mathsf{pp} = \left(\mathsf{mm.pp}, (\hat{a}_{i,b})_{i,b}, (\hat{c}_{j,b})_{j,b}, (\hat{z}_{i_1,i_2,b_1,b_2})_{i_1,i_2,b_1,b_2} \right).$$

Intuitively, $\mathsf{BitCommit}_{i,b} = S_{i,b,i}$ is associated to the (public) encoding used to evaluate the function if the i-th bit of the key is b. By definition of a straddling set, the only way to reach the top-level $\mathscr{U}$, which contains $\mathscr{S}_i$, once we have used an encoding with index $S_{I,b,i}$ is to use every index $S_{i,b,j}$ with $j \neq i$. These is done by multiplying the terms $\hat{z}_{i,j,K_i,K_j}$. Therefore, using $K_i = b$ is like "committing" to the partition $S_{i,b}$ of $\mathscr{S}_i$, and terms $\hat{z}_{i,j,K_i,K_j}$ are then used to "fill" this partition.

Remark 10. For the sake of simplicity, we set $\hat{z}_{i_1,i_2,b_1,b_2}$ to be encodings of 1, but one could also simply set it to encodings of a random value, as long as they are all encodings of the same value.

Evaluation. The output of the PRF on a key $K \in \{0,1\}^k$ and an input $x \in \{0,1\}^n$ is

$$F_{\mathsf{pp}}(K, x) = \mathsf{MM.Extract}\left(\prod_{i=1}^{k} \hat{a}_{i,K_i} \prod_{j=1}^{n} \hat{c}_{j,x_j} \prod_{i_1=1}^{k} \prod_{i_2=i_1+1}^{k} \hat{z}_{i_1,i_2,K_{i_1},K_{i_2}}\right).$$

We can re-write it as:

$$F_{\mathsf{pp}}(K, x) = \mathsf{MM.Extract}\left(\left[\prod_{i=1}^{k} a_{i,K_i} \prod_{j=1}^{n} c_{j,x_i}\right]_{\mathscr{U}}\right).$$

Extraction. As explained in Remark 4, the role of extraction (MM.Extract) is dual. First, it ensures the correctness of the PRF, as in currently known instantiations of multilinear maps [GGH13a, CLT13, GGH15, CLT15], a scalar has many different encodings. Second, it is used in our proof of security under non-interactive assumptions in Sect. 5, as in the security proof we change the way the group element

$$\prod_{i=1}^{k} \hat{a}_{i,K_i} \prod_{j=1}^{n} \hat{c}_{j,x_j} \prod_{i_1=1}^{k} \prod_{i_2=i_1+1}^{k} \hat{z}_{i_1,i_2,K_{i_1},K_{i_2}}$$

is computed. We indeed recall that due to the fact that a scalar has many different encodings, any group element (as the above one) leaks information on the exact computation used to obtain it, instead of just depending on its discrete logarithm. The usual way to solve this issue is to randomize the resulting group element using encodings of zero. However, in this paper, we do not want to use any encoding of zeros, hence the requirement for this extraction. For the proof in the generic multilinear map model in Sect. 4, this is not an issue, and we just ignore the extraction (see Remark 4).

4   Security in the Generic Multilinear Map Model

In this section, we prove the security of our construction in the generic multilinear map model. As already explained at the end of Sect. 3, we suppose in this section that no extraction is performed. We actually even prove that no polynomial-time adversary can construct a (non-trivial) encoding of zero, in any of the two experiments RKPRFReal and RKPRFRand, with non-negligible probability. This implies, in particular, that these two experiments cannot be distinguished by a polynomial-time adversary, in the generic multilinear map model, as an adversary only sees handles which only leak information when two of them correspond to the same (top-level) element.

This section is mainly a warm-up to familiarize with the model, since the fact that we prove our construction under some assumptions that are proven secure in the generic multilinear map model and proven to not let an adversary generate encodings of zero also implies the results below. However, it is very simple to modify the proof below in order to prove the security of the simplified construction proposed in Sect. 3.1, which is of independent interest.

We first need to formally define the notion of (non-trivial) encoding of zero. We follow the definition of Badrinarayanan et al. [BMSZ16].

Definition 11 ((Non-trivial) encoding of zero). *An adversary $\mathcal{A}$ in the generic multilinear map model with multilinear map oracle $\mathcal{M}$ returns a (non-trivial) encoding of zero if it returns a handle h (output by $\mathcal{M}$) such that h corresponds to the element 0 in $\mathcal{M}$'s table and the polynomial corresponding to the handle is not identically null.*

Theorem 12 (Impossibility of constructing encodings of zero). *In the generic multilinear map model with oracle $\mathcal{M}$, for any adversary $\mathcal{A}$ making at most $q_{\mathcal{M}}$ queries to the oracle $\mathcal{M}$ and q_{RKFn} queries to the oracle $\mathbf{RKFn}$, we have:*

$$\Pr\left[\, \mathrm{PRFReal}_{F_{pp}}^{\mathcal{A}} \Rightarrow an\ encoding\ of\ 0 \right] \leq q_{\mathcal{M}} \left(\frac{q_{RKFn}}{2^k} + \frac{k+n}{p} \right)$$

and

$$\Pr\left[\, \mathrm{PRFRand}_{F_{pp}}^{\mathcal{A}} \Rightarrow an\ encoding\ of\ 0 \right] \leq q_{\mathcal{M}} \frac{k+n}{p}.$$

Proof (Theorem 12). We first introduce a technical lemma, whose proof is given in the full version [ABP17].

Lemma 13. *Let k and n be two positive integers. Let $\mathcal{U}$ be the index defined in Sect. 3. Let $\hat{z}_{i_1,i_2,b_1,b_2} = [1]_{\mathsf{BitFill}_{i_1,i_2,b_1,b_2}}$ for $1 \leq i_1 < i_2 \leq k$ and $b_1, b_2 \in \{0,1\}$. Let Z_1 and Z_2 be two subsets of $\{(i_1,i_2) \mid 1 \leq i_1 < i_2 \leq k\} \times \{0,1\}^2$. If $t_1 = \prod_{(i_1,i_2,b_1,b_2) \in Z_1} \hat{z}_{i_1,i_2,b_1,b_2}$ and $t_2 = \prod_{(i_1,i_2,b_1,b_2) \in Z_2} \hat{z}_{i_1,i_2,b_1,b_2}$ have the same index set, then $Z_1 = Z_2$.*

We need to show that the adversary cannot generate a non-trivial encoding of zero.

$\mathrm{RKPRFRand}_{F_{pp}}^{\mathcal{A}}$. We start by proving it in the game $\mathrm{RKPRFRand}_{F_{pp}}^{\mathcal{A}}$. In this game, except for $\hat{z}_{i_1,i_2,b_1,b_2}$, all the handles the adversary sees correspond to fresh new formal variables, as the oracle $\mathbf{RKFn}$ only returns fresh new formal variables (of index $\mathcal{U}$). The only polynomials the adversary can generate are therefore of the form:

$$P = \sum_{\ell=1}^{L} Q_\ell \prod_{(i_1,i_2,b_1,b_2) \in Z_\ell} \hat{z}_{i_1,i_2,b_1,b_2},$$

where Q_ℓ are polynomials over all the elements except $\hat{z}_{i_1,i_2,b_1,b_2}$, and Z_ℓ are distinct subsets of $\{(i_1,i_2) \mid 1 \leq i_1 < i_2 \leq k\} \times \{0,1\}^2$ (L might be exponential in $q_{\mathcal{M}}$, but that does not matter for what follows).

Let us now show that if P is not the zero polynomial, then when replacing $\hat{z}_{i_1,i_2,b_1,b_2}$ by 1, the resulting polynomial is still a non-zero polynomial. From Lemma 13, one can assume that elements $\prod_{(i_1,i_2,b_1,b_2)\in Z_\ell} \hat{z}_{i_1,i_2,b_1,b_2}$ all have distinct indices. Therefore, the polynomials Q_ℓ all have distinct indices too. No monomial in two different Q_ℓ of the sum $\sum_\ell Q_\ell$ (when forgetting the indices) can therefore cancel out, otherwise this would mean that the adversary can construct two equal monomials (without $\hat{z}_{i,b}$) with two different indices. This is impossible as, except for $\hat{z}_{i,b}$, two distinct handles correspond to two fresh variables (or in other words, all the handles except $\hat{z}_{i,b}$ are encodings of scalars chosen uniformly and independently at random).

We therefore simulate the oracle $\mathcal{M}$ as follows: we do everything as normal, but we make the zero-testing oracle always output "non-zero" except when its input corresponds to the zero polynomial. The Schwarz-Zippel lemma ensures that any non-zero polynomial of degree at most $k + n$ and whose variables are fixed to uniformly random values in $\mathbb{Z}_p$ does not evaluate to zero, except with probability at most $(k + n)/p$. In other words, the zero-testing oracle outputs "zero" on a non-zero polynomial with probability at most $(k + n)/p$, as this polynomial remains non-zero and has degree at most $(k + n)$, when we replace $\hat{z}_{i_1,i_2,b_1,b_2}$ by 1. As we can suppose that the zero-testing oracle is queried with the output of the adversary without loss of generality, using at most $q_\mathcal{M}$ hybrid games (replacing one-by-one every output of the zero-testing oracle with "non-zero") we get that:

$$\Pr\left[\text{PRFRand}^{\mathscr{A}}_{F_{\mathsf{pp}}} \Rightarrow \text{an encoding of } 0\right] \leq q_\mathcal{M} \frac{k + n}{p}.$$

$\text{RKPRFReal}^{\mathscr{A}}_{F_{\mathsf{pp}}}$. Let us now look at the game $\text{RKPRFReal}^{\mathscr{A}}_{F_{\mathsf{pp}}}$. The analysis is more complicated as the adversary has access to new formal variables

$$\hat{y}_{s,x} = F_{\mathsf{pp}}(K \oplus s, x)$$

returned by queries $\mathbf{RKFn}(\phi_s, x)$.

We use the same simulator as in the previous case. We need to show that if a polynomial P produced by the adversary is not zero, it remains non-zero when $\hat{z}_{i_1,i_2,b_1,b_2}$ is replaced by 1 and $\hat{y}_{s,x}$ is replaced by its value.

We first consider the case where P is not a top-level polynomial. In this case, P cannot contain these new variables $\hat{y}_{s,x}$ as these variables are top-level. Then, as in $\text{RKPRFRand}^{\mathscr{A}}_{F_{\mathsf{pp}}}$, the zero-testing oracle outputs "non-zero" except with probability at most $(k + n)/p$.

Let us now suppose that P is a top-level polynomial. This polynomial has the form:

$$P = \sum_{\ell=1}^{L} Q_\ell \prod_{i=1}^{k} \hat{a}_{i,K'_{\ell,i}} \prod_{i_1=1}^{k} \prod_{i_2=i_1+1}^{k} \hat{z}_{i_1,i_2,K'_{\ell,i_1},K'_{\ell,i_2}} + \sum_{j=1}^{q'} \lambda_j \hat{y}_{s_j,x_j},$$

with L being a non-negative integer (possibly exponential in $q_\mathcal{M}$), Q_ℓ being non-zero polynomials in the formal variables $\hat{c}_{j,b}$, K'_ℓ being distinct bitstrings in $\{0,1\}^k$ (chosen by the adversary), q' an integer less or equal to $q_{\mathbf{RKFn}}$, $(s_1, x_1), \ldots, (s_{q'}, x_{q'})$ queries to $\mathbf{RKFn}$, and λ_j some scalar in $\mathbb{Z}_p$. Indeed, the adversary can ask for an encoding of any polynomial of the form $Q_\ell \prod_{i=1}^k \hat{a}_{i,K'_{\ell,i}}$, and by definition of straddling set systems, the unique way to obtain a top-level encoding from such a polynomial is by multiplying it with an encoding of $\prod_{i_1=1}^k \prod_{i_2=i_1+1}^k \hat{z}_{i_1,i_2,K'_{\ell,i_1},K'_{\ell,i_2}}$.

Let us suppose that P is not zero but becomes zero when $\hat{z}_{i_1,i_2,b_1,b_2}$ is replaced by 1 and $\hat{y}_{s,x}$ is replaced by its value. In this case, in particular, the first monomial (for any order) of the term

$$Q_1 \prod_{i=1}^k \hat{a}_{i,K'_{1,i}} \prod_{i_1=1}^k \prod_{i_2=i_1+1}^k \hat{z}_{i_1,i_2,K'_{1,i_1},K'_{1,i_2}}$$

necessarily needs to be canceled out by some $\hat{y}_{s_j,x_j}$. The probability over $K \xleftarrow{\$} \{0,1\}^k$ that this happens is at most:

$$\Pr\left[\exists j' \in \{1, \ldots, q'\},\ K'_1 = K \oplus s_{j'}\right] \leq \frac{q'}{2^k} \leq \frac{q_{\mathbf{RKFn}}}{2^k}.$$

As before, thanks to the Schwarz-Zippel lemma, we get that the zero-testing oracle outputs "zero", on input a non-zero polynomial, with probability at most:

$$\frac{q_{\mathbf{RKFn}}}{2^k} + \frac{k+n}{p}.$$

This concludes the proof of Theorem 12. $\qquad\qquad\qquad\qquad\qquad\qquad\qquad\square$

Remark 14. We never use the properties of the straddling set system $\mathscr{S}_0$ in this proof. These properties are only used in our proof under non-interactive assumptions in Sect. 5.

We obtain the following immediate corollary.

Corollary 15 (Security in the generic multilinear map model). *Let $\mathscr{A}$ be an adversary in the generic multilinear map model with oracle $\mathscr{M}$ against the XOR-RKA security of the PRF F defined in Sect. 3. If $\mathscr{A}$ makes at most $q_{\mathscr{M}}$ queries to the oracle $\mathscr{M}$ and $q_{\mathbf{RKFn}}$ queries to the oracle $\mathbf{RKFn}$, then:*

$$\mathbf{Adv}^{\text{prf-rka}}_{\Phi_\oplus, F_{\text{pp}}}(\mathscr{A}) \leq \frac{q_{\mathscr{M}}\, q_{\mathbf{RKFn}}}{2^k} + \frac{2q_{\mathscr{M}}(k+n)}{p}.$$

Proof (Corollary 15). We just consider an intermediate game where we simulate everything as before except the zero-testing oracle which always outputs "non-zero" unless its input is zero, as a polynomial. This game is indistinguishable from both $\text{RKPRFReal}^{\mathscr{A}}_{F_{\text{pp}}}$ and $\text{RKPRFRand}^{\mathscr{A}}_{F_{\text{pp}}}$ according to Theorem 12 (up to the bounds in this lemma). Corollary 15 follows. $\qquad\qquad\qquad\square$

5   Security Under Non-interactive Assumptions

In this section, we show that our construction is an XOR-RKA PRF under two non-interactive assumptions defined below.

5.1   Assumptions

We use two assumptions that we call the (k, n, X, Y)-XY-DDH assumption, which is roughly a generalization of the standard DDH assumption, and the (k, n)-Sel-Prod assumption. We show in the full version [ABP17] that both these assumptions are secure in the generic multilinear map model, and even that an adversary against these assumptions cannot generate encodings of zero. As explained in the "concrete instantiations" paragraph of Sect. 1, contrary to most assumptions considered on multilinear maps (e.g., classical DDH-like assumptions and the multilinear subgroup assumption [GLW14,GLSW15]), these assumptions are therefore plausible at least with two current instantiations of multilinear maps [CLT13,GGH15].

To ensure the impossibility of generating encodings of zero, in these two assumptions, we restrict the adversary's capabilities as follows: it is only provided parameters mm.pp, so it can only run MM.Add, MM.Mult and MM.ZeroTest (and of course use the elements generated by the assumption), but we do not allow the adversary to generate new encodings of a chosen scalar. In particular, this forces us to let the assumption contain the group elements $\hat{z}_{i_1, i_2, b_1, b_2}$. It is straightforward to get rid of these additional elements by allowing the adversary to generate any element of the multilinear map, at the cost of getting an implausible assumption under current instantiations of multilinear maps.

Finally, our assumption implicitly contains a list $\mathcal{L}$ of a polynomial number of encodings of independent uniform random values at non-zero index, index being implicit parameters of the assumption. We could avoid this artifact with the previous proposition as well, or by giving a sufficient number of encodings of 0 and 1, but once again, in that case, the assumption would most likely not hold with currently known multilinear maps instantiations. We believe this is a small price to pay to get plausible assumptions, as the resulting assumptions are still non-interactive.

We insist on the fact that the encodings in $\mathcal{L}$ are encodings of independent uniformly random scalars. At least in the generic multilinear group model, our assumptions hold whatever the list of indices of these encodings is. We do not have any constraint on this list of indices.

Definition 16 ((k, n, X, Y)-XY-DDH). *Let k and n be two positive integers. Let X and Y be two non-empty and disjoint indices in the index set $\mathcal{U}$ of our construction in Sect. 3. The advantage of an adversary $\mathcal{D}$ against the (k, n, X, Y)-XY-DDH problem is:*

$$\mathbf{Adv}^{(k,n,X,Y)\text{-XY-DDH}}(\mathcal{D}) = \Pr\left[(k, n, X, Y)\text{-XY-DDH-L}^{\mathcal{D}} \Rightarrow 1\right]$$
$$-\Pr\left[(k, n, X, Y)\text{-XY-DDH-R}^{\mathcal{D}} \Rightarrow 1\right],$$

where the games (k, n, X, Y)-XY-DDH-L$^{\mathscr{D}}$ *and* (k, n, X, Y)-XY-DDH-R$^{\mathscr{D}}$ *are defined in Fig. 2. The* (n, k, X, Y)-XY-DDH *assumption holds when this advantage is negligible for any polynomial-time adversary* $\mathscr{D}$.

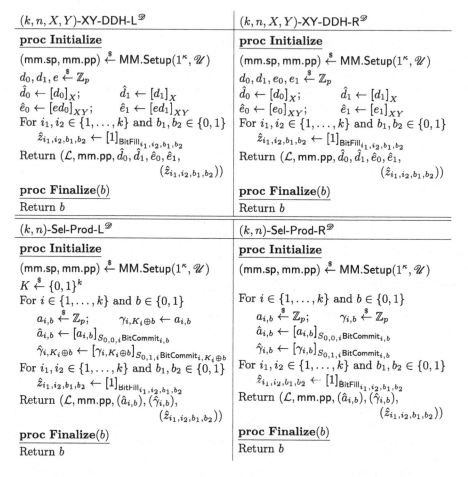

Fig. 2. Games defining the advantage of an adversary $\mathscr{D}$ against the XY-DDH and Sel-Prod problems.

This assumption is very close to the classical DDH assumption with indices, with two main differences: the presence of elements $\hat{z}_{i_1,i_2,b_1,b_2}$ which are necessary to prove our construction and the implicit presence of encodings of random values at non-zero indices (list $\mathcal{L}$ described previously) instead of a polynomial number of encodings of 0 and 1. Without the elements $\hat{z}_{i_1,i_2,b_1,b_2}$, the proof of this assumption in the generic multilinear map model would be completely straightforward. The difficulty of the proof is to deal with these elements.

In the security proof of our construction, this assumption is used in a similar way as the DDH assumption in the proof of the Naor-Reingold PRF.

Definition 17 $((k,n)$-Sel-Prod$)$. *Let* k *and* n *be two positive integers. The advantage of an adversary* $\mathscr{D}$ *against the* (k,n,X,Y)-Sel-Prod *problem is:*

$$\mathbf{Adv}^{(k,n)\text{-Sel-Prod}}(\mathscr{D}) = \Pr\left[\,(k,n)\text{-Sel-Prod-L}^{\mathscr{D}} \Rightarrow 1\,\right]$$
$$-\Pr\left[\,(k,n)\text{-Sel-Prod-R}^{\mathscr{D}} \Rightarrow 1\,\right],$$

where the games (k,n)-Sel-Prod-L$^{\mathscr{D}}$ *and* (k,n)-Sel-Prod-R$^{\mathscr{D}}$ *are defined in Fig. 2. The* (n,k)-Sel-Prod *assumption holds when this advantage is negligible for any polynomial-time adversary* $\mathscr{D}$.

Intuitively, this assumption states that, given a low-level encodings of $a_{i,0}$ and $a_{i,1}$ at indices $S_{0,i}$ from the first partition of the straddling set $\mathscr{S}$, where $i \in \{1,\ldots,k\}$, then it is hard to distinguish low-level encodings of $\gamma_{i,0} = a_{i,K_i}$ and $\gamma_{i,1} = a_{i,1-K_i}$ (where K_i is a random bit) at indices $S_{1,i}$ from encodings of fresh random values (at the same index). The hardness of this assumption crucially relies on two facts: First, one can only compare top-level encodings, and thus, the only way to compare elements whose index is in the first partition of $\mathscr{S}$ with elements whose index is in the second partition is to combine k such elements to reach the same index. Second, the latter can be hard only if k, the size of the partitions, is big. Indeed, assume $k = 2$, then one can just guess K_1 and K_2 and check if the relation carries on between encodings of $a_{i,b}$ and encodings of $\gamma_{i,b}$. Therefore, we show that if k is big enough, this assumption holds in the generic multilinear map model.

As explained in Sect. 3.1, this assumption is used to switch from the key K to a private independent key represented by the encodings $\gamma_{i,b}$. More precisely, under this assumption, we can replace the encodings $\hat{a}_{i,K_i \oplus s_i}$ at index from the first partition of the straddling set $\mathscr{S}_0$, used in the computation of the output with relation s, to encodings of uniformly random scalars at index from the second partition of $\mathscr{S}_0$. In particular, doing this change, we no longer need to know the key K to simulate correctly the output, but only the relations s for each query.

Remark 18. For the sake of simplicity, we do not explicitly specify the noise level in our assumptions. It can easily be made to work with our proof.

5.2 Security of Our Construction

In this whole section, we set $\mathscr{S} = \prod_{i=0}^{k} \mathscr{S}_i$ and $\mathscr{S}' = \prod_{i=1}^{k} \mathscr{S}_i$, so $\mathscr{S} = \mathscr{S}_0 \cdot \mathscr{S}'$. Please refer to Sect. 3 for notation.

Theorem 19 (Security under non-interactive assumptions). *Let* $\mathscr{A}$ *be a polynomial-time adversary against the XOR-RKA security of the PRF F defined in Sect. 3. We suppose that* $\mathscr{A}$ *makes at most* $q_{\mathbf{RKFn}}$ *queries to the oracle* $\mathbf{RKFn}$.

We can define an adversary $\mathscr{D}$ against the (k,n)-Sel-Prod problem, $(k-1)$ adversaries $\mathscr{B}_{i'}$ against the $(k,n,\mathscr{S}'\prod_{i=1}^{i'-1} S_{0,1,i}, S_{0,1,i'})$-XY-DDH problem for $i' \in \{2,\ldots,k\}$, and n adversaries $\mathscr{C}_{j'}$ against the $(k,n,\mathscr{S}\prod_{j=1}^{j'-1} X_j, X_{j'})$-XY-DDH problem for $j' \in \{1,\ldots,n\}$, such that:

$$\mathbf{Adv}_{\Phi_\oplus,F_{pp}}^{\text{prf-rka}}(\mathscr{A}) \leq \mathbf{Adv}^{(k,n)\text{-Sel-Prod}}(\mathscr{D})$$

$$+ \sum_{i'=2}^{k} q_{RKFn} \cdot \mathbf{Adv}^{(k,n,\mathscr{S}'\prod_{i=1}^{i'-1} S_{0,1,i}, S_{0,1,i'})\text{-XY-DDH}}(\mathscr{B}_{i'})$$

$$+ \sum_{j'=1}^{n} q_{RKFn} \cdot \mathbf{Adv}^{(k,n,\mathscr{S}\prod_{j=1}^{j'-1} X_j, X_{j'})\text{-XY-DDH}}(\mathscr{C}_{j'}).$$

Furthermore, all these adversaries run in polynomial time (their running time is approximately the same as $\mathscr{A}$).

Below, we provide a sketch of the proof. The full proof is given in the full version [ABP17].

Sketch of Proof. The proof follows a sequence of hybrid games. The first hybrid corresponds exactly to RKPRFReal$_F^{\mathscr{A}}$, while the last game corresponds to RKPRFRand$_F^{\mathscr{A}}$. Here is how we proceed. First, instead of computing the output using encodings $\hat{a}_{i,b}$ of $a_{i,b}$ with index $S_{0,0,i}$BitCommit$_{i,b}$, we use encodings $\hat{\gamma}_{i,b}$ of $a_{i,K_i\oplus b}$ with index $S_{0,1,i}$BitCommit$_{i,K_i\oplus b}$. That is, we use the second partition $S_{0,1}$ of the straddling set $\mathscr{S}_0$ instead of the first one $(S_{0,0})$ to reach top-level index (which contains $\mathscr{S}_0$). Also, we now compute the output using only the relation s instead of the key K. More precisely, the output on a query (s,x) is computed as:

$$\mathsf{MM.Extract}\left(\prod_{i=1}^{k} \hat{\gamma}_{i,s_i} \prod_{j=1}^{n} \hat{c}_{j,x_j} \prod_{i_1=1}^{k} \prod_{i_2=i_1+1}^{k} \hat{z}_{i_1,i_2,K_{i_1},K_{i_2}}\right),$$

which can be computed without knowing K. This does not change anything regarding the output (thanks to the extraction), so these two games are indistinguishable.

However, using the (k,n)-Sel-Prod assumption, we can now switch the encodings $\hat{\gamma}_{i,b}$ to encodings of fresh random scalars $\gamma_{i,b} \in \mathbb{Z}_p$. The rest of the proof is very similar to the proof of the Naor-Reingold pseudorandom function. We do $k+n$ hybrid games, where in the j-th hybrid, we just switch products of encodings $\prod_{i=1}^{j} \hat{\gamma}_{i,s_i}$ to encodings of uniformly fresh random values using the XY-DDH assumption with the proper indices. These modifications are done in a lazy fashion to obtain a polynomial-time reduction.

Acknowledgements. The first author was supported in part by the European Research Council under the European Union's Seventh Framework Programme

(FP7/2007–2013 Grant Agreement 339563 – CryptoCloud). The second author was supported in part by the Defense Advanced Research Projects Agency (DARPA) and Army Research Office (ARO) under Contract No. W911NF-15-C-0236. Part of this work was done while the second author was at École normale supérieure, Paris, France, and at IBM Research, Yorktown Heights, NY, USA. Part of this work was done while the third author was at École normale supérieure, Paris, France, and at UCLA, Los Angeles, CA, USA.

References

[ABD16] Albrecht, M., Bai, S., Ducas, L.: A subfield lattice attack on overstretched NTRU assumptions. In: Robshaw, M., Katz, J. (eds.) CRYPTO 2016. LNCS, vol. 9814, pp. 153–178. Springer, Heidelberg (2016). https://doi.org/10.1007/978-3-662-53018-4_6

[ABP15] Abdalla, M., Benhamouda, F., Passelègue, A.: An algebraic framework for pseudorandom functions and applications to related-key security. In: Gennaro, R., Robshaw, M. (eds.) CRYPTO 2015. LNCS, vol. 9215, pp. 388–409. Springer, Heidelberg (2015). https://doi.org/10.1007/978-3-662-47989-6_19

[ABP17] Abdalla, M., Benhamouda, F., Passelègue, A.: Algebraic XOR-RKA-secure pseudorandom functions from post-zeroizing multilinear maps. Cryptology ePrint Archive, Report 2017/500 (2017). http://eprint.iacr.org/2017/500

[ABPP14] Abdalla, M., Benhamouda, F., Passelègue, A., Paterson, K.G.: Related-key security for pseudorandom functions beyond the linear barrier. In: Garay, J.A., Gennaro, R. (eds.) CRYPTO 2014. LNCS, vol. 8616, pp. 77–94. Springer, Heidelberg (2014). https://doi.org/10.1007/978-3-662-44371-2_5

[ABPP18] Abdalla, M., Benhamouda, F., Passelègue, A., Paterson, K.G.: Related-key security for pseudorandom functions beyond the linear barrier. J. Cryptol. 31(4), 917–964 (2018)

[AW14] Applebaum, B., Widder, E.: Related-key secure pseudorandom functions: the case of additive attacks. Cryptology ePrint Archive, Report 2014/478 (2014). http://eprint.iacr.org/2014/478

[BC10] Bellare, M., Cash, D.: Pseudorandom functions and permutations provably secure against related-key attacks. In: Rabin, T. (ed.) CRYPTO 2010. LNCS, vol. 6223, pp. 666–684. Springer, Heidelberg (2010). https://doi.org/10.1007/978-3-642-14623-7_36

[BCM11] Bellare, M., Cash, D., Miller, R.: Cryptography secure against related-key attacks and tampering. In: Lee, D.H., Wang, X. (eds.) ASIACRYPT 2011. LNCS, vol. 7073, pp. 486–503. Springer, Heidelberg (2011). https://doi.org/10.1007/978-3-642-25385-0_26

[BDK05] Biham, E., Dunkelman, O., Keller, N.: Related-key boomerang and rectangle attacks. In: Cramer, R. (ed.) EUROCRYPT 2005. LNCS, vol. 3494, pp. 507–525. Springer, Heidelberg (2005). https://doi.org/10.1007/11426639_30

[BDK+10] Biryukov, A., Dunkelman, O., Keller, N., Khovratovich, D., Shamir, A.: Key recovery attacks of practical complexity on AES-256 variants with up to 10 Rounds. In: Gilbert, H. (ed.) EUROCRYPT 2010. LNCS, vol. 6110, pp. 299–319. Springer, Heidelberg (2010). https://doi.org/10.1007/978-3-642-13190-5_15

[BGK+14] Barak, B., Garg, S., Kalai, Y.T., Paneth, O., Sahai, A.: Protecting obfuscation against algebraic attacks. In: Nguyen, P.Q., Oswald, E. (eds.) EUROCRYPT 2014. LNCS, vol. 8441, pp. 221–238. Springer, Heidelberg (2014). https://doi.org/10.1007/978-3-642-55220-5_13

[Bih94] Biham, E.: New types of cryptanalytic attacks using related keys. In: Helleseth, T. (ed.) EUROCRYPT 1993. LNCS, vol. 765, pp. 398–409. Springer, Heidelberg (1994). https://doi.org/10.1007/3-540-48285-7_34

[BK03] Bellare, M., Kohno, T.: A theoretical treatment of related-key attacks: RKA-PRPs, RKA-PRFs, and applications. In: Biham, E. (ed.) EUROCRYPT 2003. LNCS, vol. 2656, pp. 491–506. Springer, Heidelberg (2003). https://doi.org/10.1007/3-540-39200-9_31

[BLMR13] Boneh, D., Lewi, K., Montgomery, H., Raghunathan, A.: Key homomorphic PRFs and their applications. In: Canetti, R., Garay, J.A. (eds.) CRYPTO 2013. LNCS, vol. 8042, pp. 410–428. Springer, Heidelberg (2013). https://doi.org/10.1007/978-3-642-40041-4_23

[BMSZ16] Badrinarayanan, S., Miles, E., Sahai, A., Zhandry, M.: Post-zeroizing obfuscation: new mathematical tools, and the case of evasive circuits. In: Fischlin, M., Coron, J.S. (eds.) EUROCRYPT 2016. LNCS, vol. 9666, pp. 764–791. Springer, Heidelberg (2016). https://doi.org/10.1007/978-3-662-49896-5_27

[BP14] Banerjee, A., Peikert, C.: New and improved key-homomorphic pseudorandom functions. In: Garay, J.A., Gennaro, R. (eds.) CRYPTO 2014. LNCS, vol. 8616, pp. 353–370. Springer, Heidelberg (2014). https://doi.org/10.1007/978-3-662-44371-2_20

[BR06] Bellare, M., Rogaway, P.: The security of triple encryption and a framework for code-based game-playing proofs. In: Vaudenay, S. (ed.) EUROCRYPT 2006. LNCS, vol. 4004, pp. 409–426. Springer, Heidelberg (2006). https://doi.org/10.1007/11761679_25

[BR14] Brakerski, Z., Rothblum, G.N.: Virtual black-box obfuscation for all circuits via generic graded encoding. In: Lindell, Y. (ed.) TCC 2014. LNCS, vol. 8349, pp. 1–25. Springer, Heidelberg (2014). https://doi.org/10.1007/978-3-642-54242-8_1

[CFL+16] Cheon, J.H., Fouque, P.A., Lee, C., Minaud, B., Ryu, H.: Cryptanalysis of the new CLT multilinear map over the integers. In: Fischlin, M., Coron, J.S. (eds.) EUROCRYPT 2016. LNCS, vol. 9665, pp. 509–536. Springer, Heidelberg (2016). https://doi.org/10.1007/978-3-662-49890-3_20

[CGH+15] Coron, J.S., et al.: Zeroizing without low-level zeroes: new MMAP attacks and their limitations. In: Gennaro, R., Robshaw, M. (eds.) CRYPTO 2015. LNCS, vol. 9215, pp. 247–266. Springer, Heidelberg (2015). https://doi.org/10.1007/978-3-662-47989-6_12

[CGH17] Chen, Y., Gentry, C., Halevi, S.: Cryptanalyses of candidate branching program obfuscators. In: Coron, J.S., Nielsen, J.B. (eds.) EUROCRYPT 2017. LNCS, vol. 10212, pp. 278–307. Springer, Cham (2017). https://doi.org/10.1007/978-3-319-56617-7_10

[CHKL18] Cheon, J.H., Hhan, M., Kim, J., Lee, C.: Cryptanalysis on the HHSS obfuscation arising from absence of safeguards. IEEE Access 6, 40096–40104 (2018)

[CHL+15] Cheon, J.H., Han, K., Lee, C., Ryu, H., Stehlé, D.: Cryptanalysis of the multilinear map over the integers. In: Oswald, E., Fischlin, M. (eds.) EUROCRYPT 2015. LNCS, vol. 9056, pp. 3–12. Springer, Heidelberg (2015). https://doi.org/10.1007/978-3-662-46800-5_1

[CJL16] Cheon, J.H., Jeong, J., Lee, C.: An algorithm for NTRU problems and cryptanalysis of the GGH multilinear map without a low-level encoding of zero. LMS J. Comput. Math. 19A, 255–266 (2016)

[CLLT16] Coron, J.S., Lee, M.S., Lepoint, T., Tibouchi, M.: Cryptanalysis of GGH15 multilinear maps. In: Robshaw, M., Katz, J. (eds.) CRYPTO 2016. LNCS, vol. 9815, pp. 607–628. Springer, Heidelberg (2016). https://doi.org/10.1007/978-3-662-53008-5_21

[CLLT17] Coron, J.S., Lee, M.S., Lepoint, T., Tibouchi, M.: Zeroizing attacks on indistinguishability obfuscation over CLT13. In: Fehr, S. (ed.) PKC 2017. LNCS, vol. 10174, pp. 41–58. Springer, Heidelberg (2017). https://doi.org/10.1007/978-3-662-54365-8_3

[CLT13] Coron, J.S., Lepoint, T., Tibouchi, M.: Practical multilinear maps over the integers. In: Canetti, R., Garay, J.A. (eds.) CRYPTO 2013. LNCS, vol. 8042, pp. 476–493. Springer, Heidelberg (2013). https://doi.org/10.1007/978-3-642-40041-4_26

[CLT15] Coron, J.S., Lepoint, T., Tibouchi, M.: New multilinear maps over the integers. In: Gennaro, R., Robshaw, M. (eds.) CRYPTO 2015. LNCS, vol. 9215, pp. 267–286. Springer, Heidelberg (2015). https://doi.org/10.1007/978-3-662-47989-6_13

[CS15] Cogliati, B., Seurin, Y.: On the provable security of the iterated even-mansour cipher against related-key and chosen-key attacks. In: Oswald, E., Fischlin, M. (eds.) EUROCRYPT 2015. LNCS, vol. 9056, pp. 584–613. Springer, Heidelberg (2015). https://doi.org/10.1007/978-3-662-46800-5_23

[CVW18] Chen, Y., Vaikuntanathan, V., Wee, H.: GGH15 beyond permutation branching programs: proofs, attacks, and candidates. In: Shacham, H., Boldyreva, A. (eds.) CRYPTO 2018. LNCS, vol. 10992, pp. 577–607. Springer, Cham (2018). https://doi.org/10.1007/978-3-319-96881-0_20

[FP15] Farshim, P., Procter, G.: The related-key security of iterated Even-Mansour ciphers. In: Leander, G. (ed.) FSE 2015. LNCS, vol. 9054, pp. 342–363. Springer, Heidelberg (2015). https://doi.org/10.1007/978-3-662-48116-5_17

[FX15] Fujisaki, E., Xagawa, K.: Efficient RKA-secure KEM and IBE schemes against invertible functions. In: Lauter, K., Rodríguez-Henríquez, F. (eds.) LATINCRYPT 2015. LNCS, vol. 9230, pp. 3–20. Springer, Cham (2015). https://doi.org/10.1007/978-3-319-22174-8_1

[GGH13a] Garg, S., Gentry, C., Halevi, S.: Candidate multilinear maps from ideal lattices. In: Johansson, T., Nguyen, P.Q. (eds.) EUROCRYPT 2013. LNCS, vol. 7881, pp. 1–17. Springer, Heidelberg (2013). https://doi.org/10.1007/978-3-642-38348-9_1

[GGH+13b] Garg, S., Gentry, C., Halevi, S., Raykova, M., Sahai, A., Waters, B.: Candidate indistinguishability obfuscation and functional encryption for all circuits. In: 54th Annual Symposium on Foundations of Computer Science, pp. 40–49. IEEE Computer Society Press, October 2013

[GGH15] Gentry, C., Gorbunov, S., Halevi, S.: Graph-induced multilinear maps from lattices. In: Dodis, Y., Nielsen, J.B. (eds.) TCC 2015. LNCS, vol. 9015, pp. 498–527. Springer, Heidelberg (2015). https://doi.org/10.1007/978-3-662-46497-7_20

[GGM86] Goldreich, O., Goldwasser, S., Micali, S.: How to construct random functions. J. ACM 33(4), 792–807 (1986)

[GLSW15] Gentry, C., Lewko, A.B., Sahai, A., Waters, B.: Indistinguishability obfuscation from the multilinear subgroup elimination assumption. In: Guruswami, V. (ed.) 56th Annual Symposium on Foundations of Computer Science, pp. 151–170. IEEE Computer Society Press, October 2015

[GLW14] Gentry, C., Lewko, A., Waters, B.: Witness encryption from instance independent assumptions. In: Garay, J.A., Gennaro, R. (eds.) CRYPTO 2014. LNCS, vol. 8616, pp. 426–443. Springer, Heidelberg (2014). https://doi.org/10.1007/978-3-662-44371-2_24

[GMM+16] Garg, S., Miles, E., Mukherjee, P., Sahai, A., Srinivasan, A., Zhandry, M.: Secure obfuscation in a weak multilinear map model. In: Hirt, M., Smith, A. (eds.) TCC 2016. LNCS, vol. 9986, pp. 241–268. Springer, Heidelberg (2016). https://doi.org/10.1007/978-3-662-53644-5_10

[Hal15] Halevi, S.: Graded encoding, variations on a scheme. Cryptology ePrint Archive, Report 2015/866 (2015). http://eprint.iacr.org/2015/866

[HJ16] Hu, Y., Jia, H.: Cryptanalysis of GGH map. In: Fischlin, M., Coron, J.S. (eds.) EUROCRYPT 2016. LNCS, vol. 9665, pp. 537–565. Springer, Heidelberg (2016). https://doi.org/10.1007/978-3-662-49890-3_21

[JW15] Jafargholi, Z., Wichs, D.: Tamper detection and continuous non-malleable codes. In: Dodis, Y., Nielsen, J.B. (eds.) TCC 2015. LNCS, vol. 9014, pp. 451–480. Springer, Heidelberg (2015). https://doi.org/10.1007/978-3-662-46494-6_19

[Knu93] Knudsen, L.R.: Cryptanalysis of LOKI 91. In: Seberry, J., Zheng, Y. (eds.) AUSCRYPT 1992. LNCS, vol. 718, pp. 196–208. Springer, Heidelberg (1993). https://doi.org/10.1007/3-540-57220-1_62

[LMR14] Lewi, K., Montgomery, H., Raghunathan, A.: Improved constructions of PRFs secure against related-key attacks. In: Boureanu, I., Owesarski, P., Vaudenay, S. (eds.) ACNS 2014. LNCS, vol. 8479, pp. 44–61. Springer, Cham (2014). https://doi.org/10.1007/978-3-319-07536-5_4

[Men16] Mennink, B.: XPX: generalized tweakable Even-Mansour with improved security guarantees. In: Robshaw, M., Katz, J. (eds.) CRYPTO 2016. LNCS, vol. 9814, pp. 64–94. Springer, Heidelberg (2016). https://doi.org/10.1007/978-3-662-53018-4_3

[MSW14] Miles, E., Sahai, A., Weiss, M.: Protecting obfuscation against arithmetic attacks. Cryptology ePrint Archive, Report 2014/878 (2014). http://eprint.iacr.org/2014/878

[MSZ16] Miles, E., Sahai, A., Zhandry, M.: Annihilation attacks for multilinear maps: cryptanalysis of indistinguishability obfuscation over GGH13. In: Robshaw, M., Katz, J. (eds.) CRYPTO 2016. LNCS, vol. 9815, pp. 629–658. Springer, Heidelberg (2016). https://doi.org/10.1007/978-3-662-53008-5_22

[NR97] Naor, M., Reingold, O.: Number-theoretic constructions of efficient pseudorandom functions. In: 38th Annual Symposium on Foundations of Computer Science, pp. 458–467. IEEE Computer Society Press, October 1997

[Sho97] Shoup, V.: Lower bounds for discrete logarithms and related problems. In: Fumy, W. (ed.) EUROCRYPT 1997. LNCS, vol. 1233, pp. 256–266. Springer, Heidelberg (1997). https://doi.org/10.1007/3-540-69053-0_18

[WLZZ16] Wang, P., Li, Y., Zhang, L., Zheng, K.: Related-key almost universal hash functions: definitions, constructions and applications. In: Peyrin, T. (ed.) FSE 2016. LNCS, vol. 9783, pp. 514–532. Springer, Heidelberg (2016). https://doi.org/10.1007/978-3-662-52993-5_26

[Zim15] Zimmerman, J.: How to obfuscate programs directly. In: Oswald, E., Fischlin, M. (eds.) EUROCRYPT 2015. LNCS, vol. 9057, pp. 439–467. Springer, Heidelberg (2015). https://doi.org/10.1007/978-3-662-46803-6_15

Homomorphic Encryption

Numerical Method for Comparison
on Homomorphically Encrypted Numbers

Jung Hee Cheon$^{(\boxtimes)}$, Dongwoo Kim, Duhyeong Kim, Hun Hee Lee,
and Keewoo Lee

Department of Mathematical Sciences, Seoul National University, Seoul, South Korea
{jhcheon,dwkim606,doodoo1204,hunheelee,activecondor}@snu.ac.kr

Abstract. We propose a new method to compare numbers which are
encrypted by Homomorphic Encryption (HE). Previously, comparison
and min/max functions were evaluated using Boolean functions where
input numbers are encrypted bit-wise. However, the bit-wise encryption
methods require relatively expensive computations for basic arithmetic
operations such as addition and multiplication.

In this paper, we introduce iterative algorithms that approximately
compute the min/max and comparison operations of several numbers
which are encrypted word-wise. From the concrete error analyses, we
show that our min/max and comparison algorithms have $\Theta(\alpha)$ and
$\Theta(\alpha \log \alpha)$ computational complexity to obtain approximate values within
an error rate $2^{-\alpha}$, while the previous minimax polynomial approxima-
tion method requires the exponential complexity $\Theta(2^{\alpha/2})$ and $\Theta(\sqrt{\alpha} \cdot
2^{\alpha/2})$, respectively. Our algorithms achieve (quasi-)optimality in terms
of asymptotic computational complexity among polynomial approximations for min/max and comparison operations. The comparison algorithm
is extended to several applications such as computing the top-k elements
and counting numbers over the threshold in encrypted state.

Our method enables word-wise HEs to enjoy comparable performance
in practice with bit-wise HEs for comparison operations while showing much better performance on polynomial operations. Computing an
approximate maximum value of any two ℓ-bit integers encrypted by
HEAAN, up to error $2^{\ell-10}$, takes only 1.14 ms in amortized running
time, which is comparable to the result based on bit-wise HEs.

Keywords: Homomorphic Encryption · Comparison · Min/Max ·
Iterative method

1 Introduction

Homomorphic Encryption (HE) is a cryptographic primitive which allows arithmetic operations over encrypted data without any decryption process. From this
distinctive property, HE has received lots of attention in many privacy preserving
applications. The HE schemes can be classified as word-wise HEs [8,13,26,29]
and bit-wise HEs [18,23] according to the basic operations provided by them.
Basic operations of word-wise HEs are component-wise addition and multiplication of an encrypted array over $\mathbb{Z}_p$ for a positive integer $p > 2$ [8,26] or the

© International Association for Cryptologic Research 2019
S. D. Galbraith and S. Moriai (Eds.): ASIACRYPT 2019, LNCS 11922, pp. 415–445, 2019.
https://doi.org/10.1007/978-3-030-34621-8_15

field $\mathbb{C}$ of complex numbers [13], and all other operations are built upon two basic operations. Contrary to word-wise HEs, basic operations of bit-wise HEs are logical gates such as NAND gate [23] and look-up table based operations [18, 19].

When input numbers are encrypted word-wise, polynomial operations consisting of additions and multiplications are quite natural, but it is rather hard to carry out non-polynomial operations such as comparison and min/max functions. On the other hand, when each bit of ℓ-bit integers is encrypted separately (e.g., $a = \sum_{i=0}^{\ell-1} a_i 2^i$ is encrypted as $\text{Enc}(a_0), \text{Enc}(a_1), ..., \text{Enc}(a_{\ell-1})$), comparing two ℓ-bit integers can be done by evaluating a Boolean function in $\Theta(\ell)$ homomorphic multiplications with depth $\log \ell$ [16]. However, this bit-wise encryption method is rather inefficient for homomorphic addition and multiplication since it requires sequential computation of each carry bit transferred from lower-bit operations.

In this paper, we propose an efficient numerical method for comparison and min/max functions, which can be efficiently exploited by word-wise HEs. Instead of evaluating a Boolean function over bit-wise encrypted inputs, we homomorphically evaluate *iterative algorithms* to obtain approximate min/max values and the comparison result over word-wise encrypted inputs.

Our method is especially effective in real-world applications which require several min/max or comparison operations between a large amount of polynomial operations. The statement is experimentally evidenced by a very recent work [15] on privacy-preserving clustering analysis over word-wise encrypted data which utilizes our comparison algorithm as one of the core building blocks. Their HE solution shows more than 400 times faster performance than the previously best known result [34] which encrypts data bit-wise.

1.1   Our Idea

To perform non-polynomial operations over word-wise HEs, previous works [12, 30, 36] utilized general polynomial approximation methods (e.g., Taylor, least square, minimax). To obtain the desired error bound in the given interval, they choose an appropriate degree of an approximate polynomial. As the degree grows, the lower error is guaranteed; however, the higher computational cost is required which is very critical part in HE.

To obtain an approximate value within $2^{-\alpha}$ relative error through general polynomial approximations, the approximate polynomial should have the degree at least $\Theta(2^\alpha)$ (see Sect. 6). However, the evaluation of a general polynomial of degree $\Theta(2^\alpha)$ requires at least exponential computational complexity $\Theta(2^{\alpha/2})$ [39]. In this respect, the general polynomial approximation methods, which mainly consider the optimality of polynomial degree rather than computational complexity, may not be the best solution for HE applications.

This observation leads us to utilize some *well-structured* polynomials which can be evaluated much more efficiently than general polynomials. In particular, we aim to structure approximate polynomials as *compositions* of some constant-degree polynomials observing that the utilization of a composite function has a substantial advantage in computational complexity: When a polynomial f of degree $\Theta(2^\alpha)$ is expressed as $g \circ g \circ \cdots \circ g$ for some constant-degree polynomial g, then f can be computed in a linear complexity $\Theta(\alpha)$, not $\Theta(2^{\alpha/2})$. In algorithmic perspective, the composite polynomial $g \circ g \circ \cdots \circ g$ essentially corresponds to an iterative algorithm which repeatedly computes g. As a result, our goal becomes to find *iterative algorithms* to compute min/max and comparison operations.

Our new iterative algorithms of min/max and comparison operations are constructed in two steps. We first observe that min/max and comparison operations can be expressed by *square root* and *inverse* operations. To be precise, for computing the maximum value between two numbers, we use the following identity

$$\max(a, b) = \frac{a+b}{2} + \frac{|a-b|}{2} = \frac{a+b}{2} + \frac{\sqrt{(a-b)^2}}{2},$$

and this identity can be utilized to obtain the maximum value among several numbers. To obtain the comparison result of several distinct positive numbers as well as the maximum value, we devise another identity

$$\lim_{k \to \infty} \frac{a_i^k}{a_1^k + \cdots + a_n^k} = \begin{cases} 1 & \text{if } a_i \text{ is maximal, and} \\ 0 & \text{otherwise.} \end{cases}$$

For $k = 2$, the equation can be interpreted as a sigmoid approximation of the step function which corresponds to the comparison operation (see Sect. 5). Our second observation is that there exist efficient iterative algorithms for square root and inverse operations and they can be utilized as core building blocks of min/max and comparison operations. From these observations and several optimization techniques to reduce the computational complexity, we finally devise new iterative algorithms for min/max and comparison operations.

In our algorithms, the size of intermediate values such as a_i^k grow exponentially as k increases, so they are not easy to be computed only with additions and multiplications in the bounded plaintext space. Instead, we remark that several most significant bits of a_i^k are sufficient for the approximate computation of our algorithms, and they can be obtained by an efficient bit-extraction [28,32] or the rounding-off operation [13] which is supported by the approximate HE scheme HEAAN almost for free.

1.2 Our Result

We introduce new iterative algorithms for min/max and comparison with numerical approaches, which are much more efficient than general polynomial approximation methods such as Taylor, least square and minimax approximations. Through the rigorous analysis on the error compared to the true value, we compute the minimal depth and computational complexity of our algorithms, and provide the strategies to choose the number of iterations.

Both theoretical and experimental results evidence the efficiency of our algorithms. In theoretical aspect, our algorithms achieve (quasi-)*optimal* asymptotic computational complexity among all possible polynomial approximations for min/max and comparison operations. In experimental aspect, our algorithms based on word-wise HE scheme HEAAN enjoy comparable performance with the previous algorithms based on bit-wise HE in amortized running time sense. Specific results on our algorithms are summarized as follows:

First, for min/max algorithm,

- To obtain an approximate min/max value of two ℓ-bit integers a and b up to error $2^{\ell-\alpha}$ for $\alpha > 0$, our max algorithm denoted by Max requires $\Theta(\alpha)$ depth and complexity.
- Under the condition $|a - b| \geq c$ for some small $c > 0$, the required depth and complexity are reduced to $\Theta(\log \alpha + 2 \log(1/c))$.
- The homomorphic evaluation of Max on 2^{16} pairs of 32-bit integers preserving top-10 most significant bits takes 75 s (1.14 ms as the amortized running time).

Second, for comparison algorithm,

- To obtain an approximate value of $\text{comp}(a, b) = (a > b?)$ with error bounded by $2^{-\alpha}$ where $\max(a, b)/\min(a, b) \geq c$ for some fixed $c > 1$, our comparison algorithm denoted by Comp requires $\Theta(\log(\alpha/\log c) \cdot \log(\alpha + \log(\alpha/\log c)))$ depth and complexity.
- The homomorphic evaluation of Comp on 2^{16} pairs of 32-bit integers with 14-bit precisions takes about 230 s (3.5 ms as the amortized running time).

We additionally provide some implementation results on several applications of the comparison algorithm. For example, we can compute the index of the maximum element among 16 encrypted 7-bit integers (where the maximum is at least twice larger than the others) with 7-bit precisions with amortized running time of about 75.9 ms. We also propose an efficient solution to the so-called threshold counting problem, which aims to count the number of data exceeding a certain value. For any 32 encrypted 7-bit integers, the amortized running time of our solution is 135 ms.

1.3   Related Work

There are a lot of work that consider comparison-related operations in HE schemes [5,6,10,16,19,21,24,37,43]. Most of the work deal with min/max, equality test, and sorting based on the bit-wise encryption approach. In other words, they encrypt each bit of numbers separately to provide bit-wise access.

Chillotti et al. [19] calculate the maximum of two numbers of which each bit is encrypted into a distinct ciphertext by a bit-wise HE scheme [18,19]. They express the max function by controlled Mux gates via weighted finite automata approach, and the implementation of their max algorithm on 8-bit integers took approximately a millisecond. Some other works [16,21,37,43] implemented a Boolean function corresponding to the comparison operation, where input numbers are still encrypted bit-wise. Cheon et al. [16] calculate a comparison operation over two 10-bit integers in 307 ms using the plaintext space $\mathbb{Z}_{2^{14}}$. More recent work of Crawford et al. [21] takes a few seconds to compute a comparison result of 8-bit integers. Since the comparison operation can be simultaneously done in 1800 plaintext slots, the amortized running time becomes just a few milliseconds. These bit-wise encryption methods show very nice performance on comparison operations as described above, but polynomial operations including addition and multiplication of large numbers are significantly inefficient compared to word-wise encryption methods.

On the other hand, Boura et al. [5] compute absolute function and sign function, which correspond to min/max and comparison respectively, over word-wise encrypted numbers by approximating the functions via Fourier series over a target interval. This method has an advantage on numerical stability compared to general polynomial approximation methods: Since Fourier series is a periodic function, the approximate function does not diverge to ∞ outside of the interval, while approximate polynomials obtained by polynomial approximation methods diverge. The homomorphic evaluation of the sign function over wide-wise encrypted inputs is also described in [6], which implemented the evaluation phase of discretized neural network based on HE. It utilizes the bootstrapping technique of [18] to homomorphically extract the sign value of the input number and bootstrap the corresponding ciphertext in the same time. Recently, there have been proposed a method to approximate the sign function over $x \in [-0.25, 0.25]$ by a hyperbolic tangent function $\tanh(kx) = \frac{e^{kx} - e^{-kx}}{e^{kx} + e^{-kx}}$ for sufficiently large $k > 0$ [17]. To efficiently compute $\tanh(kx)$, they first approximate $\tanh(x)$ to x and then repeatedly apply the double-angle formula $\tanh(2x) = \frac{2\tanh(x)}{1+\tanh^2(x)}$ where the inverse operation was substituted by a low-degree (e.g., 1 or 3) minimax approximation polynomial. Due to the low degree of the polynomial, their method is efficient to obtain an approximate value of the sign function with low precision.

When applying min/max and comparison functions on real-world applications such as machine learning, there have been some attempts to detour these functions by substituting them with other HE-friendly operations. For example, Gilad-Bachrah et al. [30] expressed the maximum of positive numbers $a_1, ..., a_n$ as $\lim_{k \to \infty} (\sum_{i=1}^{n} a_i^k)^{1/k}$; however, they substituted the max function by the simple summation $\sum_{i=1}^{n} a_i$ due to the hardness of evaluating $x^{1/k}$ for large k in HE.

2 Preliminaries

2.1 Notations

All logarithms are base 2 unless otherwise indicated. $\mathbb{Z}$, $\mathbb{R}$ and $\mathbb{C}$ denote the integer ring, the real number field and complex number field, respectively. For a real-valued function f defined over $\mathbb{R}$ and a domain $I \subset \mathbb{R}$, we denote the infinite norm of f over the domain I by $||f||_{\infty, I} := \max_{x \in I} |f(x)|$. If $I = \mathbb{R}$, then we omit the second term of the subscript. For a power-of-two integer N, we define a polynomial ring $R := \mathbb{Z}[X]/(X^N + 1)$. For an integer $q \geq 0$, a quotient polynomial ring R/qR is denoted by R_q. A positive integer d denotes the number of iterations in inverse and square root algorithms, and d' and t denote the numbers of iterations in the comparison algorithm.

2.2   Homomorphic Encryption

Homomorphic Encryption (denoted as HE afterwards) is a cryptographic primitive which allows arithmetic operations such as additions and multiplications over encrypted data without decryption process. HE is regarded as a promising solution which prevents private information leakage during analyses on sensitive data such as biomedical data and financial data. A number of HE schemes [4,7,8,13,18,20,22,23,26,29] have been suggested following Gentry's blueprint [27], and are achieving successes in various applications [5,11,14,30,35].

An HE scheme consists of the following algorithms:

- KeyGen(params). For parameters params determined by a level parameter L and a security parameter λ, output a public key pk, a secret key sk, and an evaluation key evk.
- $\mathrm{Enc}_{\mathrm{pk}}(m)$. For a message m, output a ciphertext ct of m.
- $\mathrm{Dec}_{\mathrm{sk}}(\mathrm{ct})$. For a ciphertext ct of m, output the message m.
- $\mathrm{Add}_{\mathrm{evk}}(\mathrm{ct}_1, \mathrm{ct}_2)$. For ciphertexts ct_1 and ct_2 of m_1 and m_2, output the ciphertext $\mathrm{ct}_{\mathrm{add}}$ of $m_1 + m_2$.
- $\mathrm{Mult}_{\mathrm{evk}}(\mathrm{ct}_1, \mathrm{ct}_2)$. For ciphertexts ct_1 and ct_2 of m_1 and m_2, output the ciphertext $\mathrm{ct}_{\mathrm{mult}}$ of $m_1 \cdot m_2$.

3   Iterative Algorithms for Inverse and Square Root

In this section, we introduce approximate algorithms computing the inverse and the square root of a real number through additions and multiplications, so that they can be efficiently computed based on word-wise HEs. We additionally analyze the error rate of each algorithm to measure the quality of the approximation.

3.1   Inverse Algorithm

One of the most popular algorithms to compute the inverse of a (positive) real number is Goldschmidt's division algorithm [31]. For $x \in (0, 2)$, the main idea of Goldschmidt's algorithm $\mathrm{Inv}(x; d)$ is

$$\frac{1}{x} = \frac{1}{1 - (1 - x)} = \prod_{i=0}^{\infty} \left(1 + (1 - x)^{2^i}\right) \approx \prod_{i=0}^{d} \left(1 + (1 - x)^{2^i}\right).$$

The value $1 + (1 - x)^{2^i}$ converges to 1 as $i \to \infty$, so the approximation holds for sufficiently large $d > 0$.

Lemma 1. *For $x \in (0, 2)$ and a positive integer d, the error rate of the output of* $\mathrm{Inv}(x; d)$ *compared to $1/x$ is bounded by $(1 - x)^{2^{d+1}}$. In fact, the error is always negative, i.e., the output of $\mathrm{Inv}(x; d)$ is always smaller than $1/x$.*

Algorithm 1. $\text{Inv}(x; d)$

Input: $0 < x < 2$, $d \in \mathbb{N}$
Output: an approximate value of $1/x$ (refer Lemma 1)
1: $a_0 \leftarrow 2 - x$
2: $b_0 \leftarrow 1 - x$
3: **for** $n \leftarrow 0$ **to** $d - 1$ **do**
4: $\quad b_{n+1} \leftarrow b_n^2$
5: $\quad a_{n+1} \leftarrow a_n \cdot (1 + b_{n+1})$
6: **end for**
7: **return** a_d

Proof. We can simply compute $|\frac{a_d - 1/x}{1/x}| = 1 - x \cdot a_d = (1 - x)^{2^{d+1}}$. $\quad\square$

Remark 1. Lemma 1 implies that if we have tighter lower/upper bound of x, then it guarantees an exponential convergence in the number of iteration d. For example, assuming that $x \in [2^{-n}, 1)$ for some $n \in \mathbb{N}$, the error rate of $\text{Inv}(x; d)$ is bounded by $(1 - 2^{-n})^{2^{d+1}}$ which implies that only $d = \Theta(\log \alpha + n)$ number of iterations suffice for Algorithm 1 to achieve the error bound $2^{-\alpha}$.

3.2 Square Root Algorithm

In order to compute the square root of a positive real number, we exploit a two-variable iterative method proposed by Wilkes in 1951 [44]. The algorithm consists of simple addition and multiplication operations for each iteration, and it has an exponential convergence rate depending on the input value.

Algorithm 2. $\text{Sqrt}(x; d)$

Input: $0 \leq x \leq 1$, $d \in \mathbb{N}$
Output: an approximate value of $\sqrt{x}$ (refer Lemma 2)
1: $a_0 \leftarrow x$
2: $b_0 \leftarrow x - 1$
3: **for** $n \leftarrow 0$ **to** $d - 1$ **do**
4: $\quad a_{n+1} \leftarrow a_n \left(1 - \frac{b_n}{2}\right)$
5: $\quad b_{n+1} \leftarrow b_n^2 \left(\frac{b_n - 3}{4}\right)$
6: **end for**
7: **return** a_d

Lemma 2. *For $x \in (0, 1)$ and a positive integer d, the error rate of the output of $\text{Sqrt}(x; d)$ compared to $\sqrt{x}$ is bounded by $(1 - \frac{x}{4})^{2^{d+1}}$. In fact, the error is always negative, i.e., the output of $\text{Sqrt}(x; d)$ is always smaller than $\sqrt{x}$.*

Proof. Since $-1 \le b_0 \le 0$, we can easily check that $-1 \le b_n \le 0$ for all $n \in \mathbb{N}$. Then, $|b_{n+1}| = |b_n| \cdot |\frac{b_n(b_n-3)}{4}| \le |b_n|$ gives $|b_{n+1}| \le |b_n|^2 \cdot (1 - \frac{x}{4})$, and it holds that $|b_d| \le |b_0|^{2^d} \cdot (1 - \frac{x}{4})^{2^d-1} < (1 - \frac{x}{4})^{2^{d+1}}$.

From the definition of a_n and b_n, the equality $x(1+b_n) = a_n^2$ can be obtained by a simple induction. Hence, the error rate is

$$\left| \frac{a_n - \sqrt{x}}{\sqrt{x}} \right| = 1 - \sqrt{1 + b_n} < |b_n|,$$

which implies the result of the lemma.   □

Remark 2. Similarly to Remark 1, Lemma 2 implies that if we have tighter lower/upper bound of x, it guarantees an exponential convergence rate, e.g., if $x \in [2^{-n}, 1)$, then $d = \Theta(\log \alpha + n)$ iterations are sufficient for Algorithm 2 to achieve the error bound $2^{-\alpha}$.

Absolute Value. By observing $|x| = \sqrt{x^2}$, we can also compute the absolute value of $-1 \le x \le 1$ by $\mathtt{Sqrt}(x^2; d)$ for some sufficiently large $d > 0$. By Lemma 2, the error rate compared to the true value $|x|$ is bounded by $\left(1 - \frac{x^2}{4}\right)^{2^{d+1}}$.

4   Approximate Min/Max Algorithms

In this section, we describe approximate algorithms for min/max operations applying the square root algorithm described in the previous section. Our main goal is to obtain the min/max value and the comparison result between ℓ-bit positive integers (or ℓ-bit precision positive real numbers) for some given integer $\ell > 0$. Since our inverse and square root algorithms require input value to be contained in a prefixed interval (e.g., $[0, 1]$), we need to scale down the large input values into small range. For this reason, when two inputs $\bar{a}, \bar{b} \in [0, 2^\ell)$ are given, we first scale down

$$(a, b) \leftarrow \left(\frac{\bar{a}}{2^\ell}, \frac{\bar{b}}{2^\ell} \right)$$

so that $a, b \in [0, 1)$. After running the algorithms we desired, we will scale up the output value by the factor 2^ℓ. For example, after we obtain an approximate value x of $\max(a, b)$, then we can compute $2^\ell \cdot x \approx \max(\bar{a}, \bar{b})$. Note that this scaling procedure preserves the error rate compared to the true value.

4.1   Min/Max Algorithm for Two Numbers

In this subsection, we describe the Min and Max algorithms which approximately compute the minimum and maximum values of given two inputs contained in $[0, 1)$, respectively. The approximate min/max algorithms, which we denote by

`Min` and `Max`, respectively, can be directly obtained from the following observations:

$$\min(a, b) = \frac{a + b}{2} - \frac{\sqrt{(a - b)^2}}{2}, \quad \max(a, b) = \frac{a + b}{2} + \frac{\sqrt{(a - b)^2}}{2}.$$

For the square root part of the formula we will use the square root algorithm described in Sect. 3.2 as a subroutine, which leads us to the algorithms:

$$\texttt{Min}(a, b; d) = \frac{a + b}{2} - \frac{\texttt{Sqrt}((a - b)^2; d)}{2}, \quad \text{and}$$

$$\texttt{Max}(a, b; d) = \frac{a + b}{2} + \frac{\texttt{Sqrt}((a - b)^2; d)}{2}.$$

Algorithm 3. $\texttt{Min}(a, b; d)$, $\texttt{Max}(a, b; d)$

Input: $a, b \in [0, 1)$, $d \in \mathbb{N}$
Output: an approximate value of $\min(a, b)$ and $\max(a, b)$ (refer Theorem 1,2)
1: $x = \frac{a+b}{2}$ and $y = \frac{a-b}{2}$
2: $z \leftarrow \texttt{Sqrt}(y^2; d)$
3: **return** $x - z$ for $\texttt{Min}(a, b; d)$
      $x + z$ for $\texttt{Max}(a, b; d)$

Assume that one would like to obtain a good enough approximate value of min/max of $a, b \in [0, 1)$. Roughly speaking, we can obtain an approximate min/max value with an error up to $2^{-\alpha}$ in about 2α iterations.

Theorem 1. *If $d \geq 2\alpha - 3$ for some $\alpha > 0$, then the error of $\texttt{Max}(a, b; d)$ (resp. $\texttt{Min}(a, b; d)$) from the true value $\max(a, b)$ (resp. $\min(a, b)$) is bounded by $2^{-\alpha}$ for any $a, b \in [0, 1)$.*

Proof. By Lemma 2, we obtain $\left| \texttt{Sqrt}((a - b)^2; d) - |a - b| \right| < \left(1 - \frac{(a-b)^2}{4} \right)^{2^{d+1}} \cdot |a - b|$. Therefore, the error of $\texttt{Max}(a, b; d)$ (resp. $\texttt{Min}(a, b; d)$) from $\max(a, b)$ (resp. $\min(a, b)$) is bounded by $\frac{1}{2} \cdot \left(1 - \frac{(a-b)^2}{4} \right)^{2^{d+1}} \cdot |a - b|$.

Considering $|a - b|$ as a variable x, let us find the maximal value of $f(x) = (1 - \frac{x^2}{4})^{2^{d+1}} \cdot x$ for $x \in [0, 1)$. By a simple computation, one can check that $f'(x) = (1 - \frac{x^2}{4})^{2^{d+1}-1} \cdot \left(1 - \left(\frac{1}{4} + 2^d \right) x^2 \right) = 0$ has a unique solution $x_0 = 1/\sqrt{2^d + \frac{1}{4}}$ in $[0, 1)$ so that x_0 is the maximal point of $f(x)$. Hence, we obtain the following inequality

$$\left(1 - \frac{(a - b)^2}{4} \right)^{2^{d+1}} \cdot |a - b| \leq \left(1 - \frac{1}{2^{d+2} + 1} \right)^{2^{d+1}} \cdot \frac{1}{\sqrt{2^d + \frac{1}{4}}}$$

$$< \frac{1}{\left(1 + \frac{1}{2^{d+2}} \right)^{2^{d+1}}} \cdot 2^{-\frac{d}{2}} < 2^{-\frac{d+1}{2}},$$

using the fact that $(1 + x)^{1/x} \geq 2$ for $x \in [0, 1)$. Therefore, under the condition $d > 2\alpha - 3$, the error of $\texttt{Max}(a, b; d)$ (and $\texttt{Min}(a, b; d)$) is upper bounded by $2^{-\alpha}$. $\square$

By Theorem 1, we can select an appropriate parameter d depending on α, i.e., the quality of the approximation. For example, let $\ell = 64$ so that $\bar{a}$ and $\bar{b}$ are 64-bit positive integers. If one aims to obtain exact maximum value between $\bar{a}$ and $\bar{b}$, then one can set $d = 2 \cdot 64 - 3 = 125$. But if one only aims to obtain an approximate value within an error less than 2^{48}, i.e., obtain the top 16 bits of the maximum value in 64-bit representation, one can set much smaller d as $d = 2 \cdot 16 - 3 = 29$. In this case, the output would be a 64-bit integer of which top-16 bits coincide with those of the true maximum value.

Parameter Reduction over the Restricted Domain. We can improve the condition on the parameter d in Theorem 1 from $\Theta(\alpha)$ to $\Theta(\log \alpha)$ by adding some conditions on a and b: $|a - b| \geq c$ for some constant $0 < c < 1$. In other words, $d = \Theta(\log \alpha)$ provides appropriate min/max results with probability $(1 - c)^2$ for uniform randomly chosen a and b from $[0, 1)$.

Theorem 2. *If $d \geq \log \alpha + 2 \log(1/c) + 1$ for some $\alpha > 0$ and $0 < c < 1$, then the error of $\texttt{Max}(a, b; d)$ (resp. $\texttt{Min}(a, b; d)$) from the true value $\max(a, b)$ (resp. $\min(a, b)$) is bounded by $2^{-\alpha}$ for any $a, b \in [0, 1)$ satisfying $|a - b| \geq c$.*

Proof. We resume at the upper bound $\frac{1}{2} \cdot \left(1 - \frac{(a-b)^2}{4}\right)^{2^{d+1}} \cdot |a - b|$ of the error of $\texttt{Max}(a, b; d)$ (resp. $\texttt{Min}(a, b; d)$) from $\max(a, b)$ (resp. $\min(a, b)$) as in the proof of Theorem 1.

Since $|a - b| \geq c$, we obtain

$$\frac{1}{2} \cdot \left(1 - \frac{(a - b)^2}{4}\right)^{2^{d+1}} \cdot |a - b| \leq \left(1 - \frac{c^2}{4}\right)^{2^{d+1}}.$$

Since $(1 - x)^{1/x} < \frac{1}{e} < \frac{1}{2}$ for $0 < x < 1$, if $d \geq \log \alpha + 2 \log(1/c) + 1$, it holds that

$$\left(1 - \frac{c^2}{4}\right)^{2^{d+1}} = \left(\left(1 - \frac{c^2}{4}\right)^{4/c^2}\right)^{2^{(d+2\log c-1)}} < 2^{-2^{(d+2\log c-1)}} \leq 2^{-\alpha},$$

which is the conclusion we wanted. $\square$

Note that the area of the bad region $\{(a, b) \in [0, 1) \times [0, 1) : |a - b| \leq c\}$, where the theorem does not hold, is $1 - (1 - c)^2$ ($\approx 2c$ if c is very small). Consider a, b as a uniform random variable in $[0, 1)$, and assume that we want to obtain an appropriate output of $\texttt{Max}(a, b; d)$ and $\texttt{Min}(a, b; d)$ with probability $1 - \epsilon$ for $0 < \epsilon < 1$. Then by combining the results from Theorems 1 and 2, it suffices to set $d \approx \min(2\alpha - 3, \log \alpha + 2 \log(1/c) + 1)$.

Depth and Complexity of $\texttt{Min}/\texttt{Max}$ Algorithms. Since the depth of the $\texttt{Sqrt}(\cdot; d)$ algorithm is $2d + 1$, the depth of $\texttt{Min}(\cdot, \cdot; d)$ and $\texttt{Max}(\cdot, \cdot; d)$ algorithms is also $2d + 1$. Since the algorithm is iterative, the complexity is indeed $\Theta(d)$.

4.2   Min/Max Algorithm for Several Numbers

With the basic min/max algorithm for two numbers in Sect. 4.1, we are able to construct a min/max algorithm for several numbers. Let $a_{1,0}, a_{2,0}, ..., a_{n,0}$ be given numbers contained in $[0,1)$, and our aim is to obtain an approximate value of the maximum value among them. For convenience of analysis, assume that n is a power-of-two integer. For some positive integer $d > 0$, we first run $\mathtt{Max}(a_{2i-1,0}, a_{2i,0}; d)$ for $1 \leq i \leq n/2$ and denote the outputs by $a_{i,1}$, respectively. Repeatedly, we obtain the outputs $a_{i,2}$ of $\mathtt{Max}(a_{2i-1,1}, a_{2i,1})$ for $1 \leq i \leq n/4$. Then, we can inductively construct a binary tree structure $\{a_{i,j}\}_{0 \leq j \leq \log n, 1 \leq i \leq n/2^j}$, and $a_{1,\log n}$ would be the desired approximate maximum value. The same argument can be applied to the case of $\mathtt{Min}$ algorithm.

Algorithm 4. $\mathtt{ArrayMax}(a_1, a_2, ..., a_n; d)$

Input: $a_1, a_2, ..., a_n \in [0,1)$, $d \in \mathbb{N}$
Output: an approximate value of $\max(a_1, a_2, ..., a_n; d)$ (refer Theorem 3)
 1: $(a_{1,0}, a_{2,0}, ..., a_{n,0}) \leftarrow (a_1, a_2, ..., a_n)$
 2: $d \leftarrow n$
 3: **for** $j \leftarrow 0$ **to** $\lfloor \log n \rfloor$ **do**
 4:   **if** d is odd **then**
 5:     $a_{\lceil d/2 \rceil, j+1} \leftarrow a_{d,j}$
 6:   **end if**
 7:   $d \leftarrow \lfloor n/2 \rfloor$
 8:   **for** $i \leftarrow 1$ **to** d **do**
 9:     $a_{i,j+1} \leftarrow \mathtt{Max}(a_{2i-1,j}, a_{2i,j}; d)$
10:   **end for**
11: **end for**
12: **return** $a_{1, \lceil \log n \rceil}$

Theorem 3. *Let n be a power-of-two integer. The numbers $a_1, a_2, ..., a_n \in [0,1)$ satisfying $|a_i - a_j| \geq c > 0$ for any $1 \leq i < j \leq n$ are given. When $d \geq \log(\alpha + \log \log n) + 2\log(1/c) + 1$, the error of the output of $\mathtt{ArrayMax}(a_1, a_2, ..., a_n; d)$ (resp. $\mathtt{ArrayMin}(a_1, a_2, ..., a_n; d)$) from the true value $\max(a_1, a_2, ..., a_n)$ (resp. $\min(a_1, a_2, ..., a_n)$) is bounded by $2^{-\alpha}$. Note that the error is always negative, i.e., the output value is always smaller than the true value.*

Proof. Refer to Appendix A. □

Theorem 2 was applied in this theorem for the good region $\{(a_i)_{1 \leq i \leq n} \in [0,1)^n : |a_i - a_j| \geq c$ for any $1 \leq i < j \leq n$ and some $c > 0\}$. Note that we can also apply Theorem 1 to obtain the worst-case analysis: In this case, d should be set as $d = 2(\alpha + \log \log n) - 3$. The area of the good region, is exactly $(1 - (n-1)c)^n$ ($\approx 1 - n(n-1)c$ when c is very small) referring to [9]. Therefore, if one want to obtain an output of $\mathtt{ArrayMax}$ or $\mathtt{ArrayMin}$ within error $2^{-\alpha}$ with probability $1 - \epsilon$ for $0 < \epsilon < 1$, then by Theorem 3 it suffices to set $d \approx \min(2(\alpha + \log \log n) - 3, \log(\alpha + \log \log n) + 2\log(1/c) + 1)$.

Remark 3. We set n be a power-of-two integer for convenience of the error analysis, but the theorem still holds for a non-power-of-two integer n.

Depth and Complexity of `ArrayMin`/`ArrayMax` Algorithms. Since we constructed a binary tree of depth $\log n$ with the number of nodes n, the depth is $\log n \cdot (2d + 1)$ and the complexity is $\Theta(nd)$.

5   Approximate Comparison Algorithms

In this section, we propose approximate comparison algorithms for various purposes. The core idea of algorithms starts with a simple fact that the comparison result of two numbers a and b can be evaluated as $\mathrm{comp}(a,b) := \chi_{(0,\infty)}(a - b)$ where $\chi_{(0,\infty)}$ is a step function over $\mathbb{R}$ defined as $\chi_{(0,\infty)}(x) := \begin{cases} 1 & \text{if } x > 0 \\ 0 & \text{otherwise} \end{cases}$.

However, it is challenging to evaluate discontinuous functions such as $\chi_{(0,\infty)}$ in word-wise HE. To overcome this problem, we first approximate the step function by a globally smooth function called sigmoid $\sigma(x) = 1/(1 + e^{-x})$. The error between the sigmoid and $\chi_{(0,\infty)}$ can be controlled by scaling the sigmoid as $\sigma_k(x) := \sigma(kx)$. Following the notation, it holds that

$$\lim_{k \to \infty} ||\chi_{(0,\infty)} - \sigma_k||_{\infty, \mathbb{R}-[-\epsilon, \epsilon]} = 0$$

for any $\epsilon > 0$. In other words, we can approximately evaluate the step function $\chi_{(0,\infty)}$ through the scaled sigmoid function σ_k for sufficiently large k (Fig. 1).

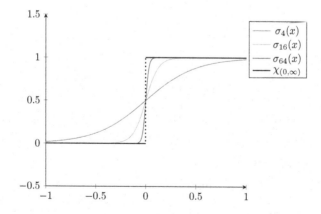

Fig. 1. Approximation of the step function $\chi_{(0,\infty)}$ by scaled sigmoid functions

Though a scaled sigmoid function is a continuous function contrary to $\chi_{(0,\infty)}$, $\sigma_k(a - b) = e^{ka}/(e^{ka} + e^{kb})$ still requires exponential function evaluations which cannot be easily done in HE. This obstacle can be simply overcome by taking logarithm on each input of comparison. Since the log function is a strictly increasing function, it does not reverse the order, i.e., $\log a > \log b$ if and only if $a > b$. Therefore, the evaluation of $\chi_{(0,\infty)}$ on $x = \log a - \log b$ also outputs the correct comparison result of a and b. As a result, we obtain the following approximation formula:

$$\text{comp}(a, b) \approx \sigma_k(\log a - \log b) = \frac{e^{k \log a}}{e^{k \log a} + e^{k \log b}} = \frac{a^k}{a^k + b^k}.$$

5.1   Comparison Between Two Numbers

In this subsection, we discuss how to efficiently evaluate the approximate comparison equation $a^k/(a^k + b^k) \approx \text{comp}(a, b)$ with basic operations such as addition and multiplication. For given two ℓ-bit positive integers $\bar{a}$ and $\bar{b}$, we first scale them down to $a, b \in \left[\frac{1}{2}, \frac{3}{2}\right)$ via the mapping $\bar{x} \mapsto x := \frac{1}{2} + \frac{x}{2^\ell}$ which is *order-preserving*, i.e., $x > y$ if and only if $\bar{x} > \bar{y}$. We may scale those ℓ-bit integers to $[0, 1)$ as in min/max algorithms, but note that the range $\left[\frac{1}{2}, \frac{3}{2}\right)$ is more suitable than $[0, 1)$ to exploit Inv algorithm.

From the observation in the beginning of Sect. 5, the followings hold:

$$\lim_{k \to \infty} \frac{\max(a, b)^k}{a^k + b^k} = 1, \text{ and } \lim_{k \to \infty} \frac{\min(a, b)^k}{a^k + b^k} = 0 \text{ if } a \neq b, \tag{1}$$

so that we obtained the approximate values if we set sufficiently large $k > 0$. Our comparison algorithm denoted by Comp is described as Algorithm 5.

Algorithm 5. Comp$(a, b; d, d', t, m)$

Input: distinct numbers $a, b \in \left[\frac{1}{2}, \frac{3}{2}\right)$, $d, d', t, m \in \mathbb{N}$
Output: an approximate value of $\text{comp}(a, b)$ (refer Theorem 4)
1: $a_0 \leftarrow \frac{a}{2} \cdot \text{Inv}\left(\frac{a+b}{2}; d'\right)$
2: $b_0 \leftarrow 1 - a_0$
3: **for** $n \leftarrow 0$ to $t - 1$ **do**
4:     $inv \leftarrow \text{Inv}(a_n^m + b_n^m; d)$
5:     $a_{n+1} \leftarrow a_n^m \cdot inv$
6:     $b_{n+1} \leftarrow 1 - a_{n+1}$
7: **end for**
8: **return** a_t

The first preparatory stage of the algorithm is to (1-norm) normalize the given input into the new pair (a, b) with $a, b \in [0, 1]$ satisfying $a + b = 1$. This normalization provides lower and upper bounds $1/2^{k-1} \leq a^k + b^k \leq 1$ so that $a^k + b^k$ can be an appropriate input of Inv algorithm. The next step is to approximate the value of $a^k/(a^k + b^k)$. One naive approach could be to compute $a^k \cdot \text{Inv}(a^k + b^k; d)$ for some positive integer $d > 0$. However, since the $a^k + b^k$ could be as small as $1/2^{k-1}$, it requires too large parameter d for sufficiently nice approximation of $1/(a^k + b^k)$ with Inv algorithm (see Remark 1).

In order to overcome this bottleneck we approximate the value of $a^k/(a^k + b^k)$ by performing the operation $a^m \cdot \text{Inv}(a^m + b^m; d)$ repeatedly for small m. The additional parameter m, which we normally choose as a power-of-two integer, satisfies $m^t = k$. As an illustration, let us take the two steps of the iteration. We first compute $(a_1, b_1) = (\frac{a^m}{a^m + b^m}, \frac{b^m}{a^m + b^m})$ applying $\text{Inv}(a^m + b^m; d)$, and then compute $(a_2, b_2) = (\frac{a_1^m}{a_1^m + b_1^m}, \frac{b_1^m}{a_1^m + b_1^m}) = (\frac{a^{2m}}{a^{2m} + b^{2m}}, \frac{b^{2m}}{a^{2m} + b^{2m}})$ again using $\text{Inv}(a_1^m + b_1^m; d)$. Then, in t steps we arrive at $\frac{a^{m^t}}{a^{m^t} + b^{m^t}} = \frac{a^k}{a^k + b^k}$.

This modification requires more Inv algorithms to be used, but it allows us to set much smaller d for Inv algorithm, because $a^m + b^m$ at each steps is in the range $[1/2^{m-1}, 1]$ while $a^n + b^n$ is in the range $[1/2^{n-1}, 1]$. Therefore, it makes a trade-off between the number of iterations t and the parameter d.

Theorem 4. *Let $a, b \in [\frac{1}{2}, \frac{3}{2})$ satisfying $\max(a, b)/\min(a, b) \geq c$ for some fixed $1 < c < 3$. When $t \geq \frac{1}{\log m}[\log(\alpha + 1) - \log \log c]$, $d \geq \log(\alpha + t + 2) + m - 2$, and $d' \geq \log(\alpha + 2) - 1$, the error of (the vector) $\text{Comp}(a, b; d, d', t, m)$ compared to the true value $\text{comp}(a, b)$ is bounded by $2^{-\alpha}$. Note that the error is always toward $1/2$, i.e., the output value is always in between $1/2$ and the true value.*

Proof. Without loss of generality we may assume that $a > b$. Note that the step 1 and 2 of our algorithm scales a, b to non-negative numbers a_0, b_0 with $a_0 + b_0 = 1$. Let us execute the first round of iteration. Note that

$$\left| a_0^m \text{Inv}(a_0^m + b_0^m; d) - \frac{a_0^m}{a_0^m + b_0^m} \right| = a_0^m \cdot |\text{Inv}(a_0^m + b_0^m; d) - (a_0^m + b_0^m)^{-1}|$$

$$\leq (1 - (a_0^m + b_0^m)^{-1})^{2^{d+1}} \cdot \frac{a_0^m}{a_0^m + b_0^m}.$$

Since $(1 - (a_0^m + b_0^m)^{-1})^{2^{d+1}} < e^{-2^{d+1}/2^{m-1}} < 2^{-2^{d-m+2}}$ from the lower bound estimate $a_0^m + b_0^m \geq 2^{-m+1}$, we can conclude that the error rate for one iteration is bounded by $K = 2^{-2^{d-m+2}}$. Thus, the error rate for t iterations is bounded by $1 - (1 - K)^t \leq tK < 2^t K$. Since we want this bound to be smaller than $2^{-\alpha-2}$ we get the desired lower bound for d, namely $d \geq \log(\alpha + t + 2) + m - 2$.

Now we wish to bound the difference

$$\left| 1 - \frac{a^{m^t}}{a^{m^t} + b^{m^t}} \right| = 1 - \frac{1}{1 + (b/a)^{m^t}} \leq \left(\frac{b}{a} \right)^{m^t} \leq c^{-m^t}$$

by $2^{-\alpha-1}$, which leads us to the condition $t \geq \frac{1}{\log m}[\log(\alpha + 1) - \log \log c]$.

Finally, we examine the step 1 and 2 of our algorithm, whose error rate is bounded by $2^{-2^{d'+1}}$. If we require this bound to be smaller than $2^{-\alpha-2}$, we get the condition $d' \geq \log(\alpha + 2) - 1$, which is implied by our assumption on d'.

Summing up all the error rates, we get the conclusion we wanted. ☐

Remark 4. We note that introducing the condition on the ratio of inputs with the constant c is not unrealistic or harsh. In the case of n-bit integers, setting the lower bound $c = a/b \geq \left(\frac{1}{2} + \frac{2^n - 1}{2^n}\right) / \left(\frac{1}{2} + \frac{2^n - 2}{2^n}\right)$ allows us to compare *any* two n-bit integers. Similar argument also applies to the case of real numbers, if we consider finite precision and input bounds. To sum up, an appropriate c generally exists in real-world applications.

Depth and Complexity of `Comp` Algorithm. The depth and complexity of `Comp` is $d' + 1 + t(d + \log m + 2)$ and $\Theta(d' + t(d + \log m))$ respectively. When we set $m = 2$ which roughly gives $t = \log(\alpha/\log c)$ and $d = \log(\alpha + \log(\alpha/\log c))$, those depth and complexity are optimized as $\Theta(\log(\alpha/\log c) \cdot \log(\alpha + \log(\alpha/\log c)))$. For $c = 1 + 2^{-\alpha}$, it is simplified as $\Theta(\alpha \log \alpha)$.

5.2 Max Index of Several Numbers

Given several distinct numbers $a_1, a_2, ..., a_n \in \left[\frac{1}{2}, \frac{3}{2}\right)$, assume that we want to obtain the index of the maximum value. This problem can be easily solved by observing Eq. (1) with another point of view. As the exponent k increases, then the gap between $\max(a, b)^k$ and $\min(a, b)^k$ becomes larger so that $\max(a, b)^k$ becomes a dominant term of $a^k + b^k$. This observation is also applicable to the comparison of several numbers, i.e., $\max(a_1, a_2, ..., a_n)^k$ is a dominant term of $\sum_{i=1}^{n} a_i^k$ when k is large enough. As a result, Eq. (1) can be generalized as followings:

$$\lim_{k \to \infty} \frac{a_j^k}{a_1^k + a_2^k + \cdots + a_n^k} = 1 \iff a_j = \max(a_1, ..., a_n),$$

$$\lim_{k \to \infty} \frac{a_j^k}{a_1^k + a_2^k + \cdots + a_n^k} = 0 \iff a_j \neq \max(a_1, ..., a_n).$$

From these properties, we construct the algorithm `MaxIdx` of which the output indicates the index of the maximum value, as a simple generalization of the comparison algorithm `Comp` in the previous section.

Theorem 5. *Let $a_1, a_2, \ldots, a_n \in \left[\frac{1}{2}, \frac{3}{2}\right)$ be n distinct elements, and the ratio of maximum value over the second maximum value be $1 < c < 3$. If $t \geq \frac{1}{\log m}[\log(\alpha + \log n + 1) - \log \log c]$ and $\min(d, d') \geq \log(\alpha + t + 2) + (m - 1)\log n - 1$, the error of the output of `MaxIdx`$(a_1, ..., a_n; d, d', m, t)$ compared to the true value is (component-wise) bounded by $2^{-\alpha}$. Note that the error is always toward $1/2$, i.e., the output value is always in between $1/2$ and the true value.*

Proof. Refer to Appendix A. □

Depth and Complexity of `MaxIdx` Algorithm. The depth and complexity of `MaxIdx` is $d' + 1 + t(d + \log m + 2)$ and $\Theta(n + d' + t(d + n \log m))$ respectively, as that of `Comp`, and is again optimized when $m = 2$ roughly giving $t = \log((\alpha + \log n)/\log c)$, $d = \log(\alpha + \log((\alpha + \log n)/\log c)) + \log n$. Note that when $\log n \leq \alpha$, depth of `MaxIdx` (asymptotically) does not exceed the depth of `Comp`.

Algorithm 6. MaxIdx$(a_1, a_2, ..., a_n; d, d', m, t)$

Input: n distinct numbers $(a_1, a_2, ..., a_n)$ with $a_i \in \left[\frac{1}{2}, \frac{3}{2}\right)$, $d, d', m, t \in \mathbb{N}$

Output: $(b_1, b_2, ..., b_n)$ where b_i is close to 1 if a_i is the largest among a_j's and is close to 0 otherwise (refer Theorem 5)

1: $inv \leftarrow \mathtt{Inv}(\sum_{j=1}^{n} a_j/n; d')$
2: **for** $j \leftarrow 1$ **to** $n-1$ **do**
3:    $b_j \leftarrow a_j/n \cdot inv$ // Initial 1-norm normalization
4: **end for**
5: $b_n \leftarrow 1 - \sum_{k=1}^{n-1} b_j$
6: **for** $i \leftarrow 1$ **to** t **do**
7:    $inv \leftarrow \mathtt{Inv}(\sum_{j=1}^{n} b_j^m; d)$
8:    **for** $j \leftarrow 0$ **to** $n-1$ **do**
9:        $b_j \leftarrow b_j^m \cdot inv$
10:   **end for**
11:   $b_n \leftarrow 1 - \sum_{k=1}^{n-1} b_j$
12: **end for**
13: **return** $(b_1, b_2, ..., b_n)$

Remark 5. Under the same condition on d, d', m and t with Theorem 5, we can obtain an approximate maximal value among n distinct numbers $a_1, a_2, ..., a_n$ by computing $\sum_{i=1}^{n} b_i a_i$ for $(b_1, b_2, ..., b_n) \leftarrow \mathtt{MaxIdx}(a_1, .., a_n; d, d', m, t)$. This idea is basically derived from the equality

$$\lim_{k \to \infty} \frac{a_1^{k+1} + a_2^{k+1} + \cdots + a_n^{k+1}}{a_1^k + a_2^k + \cdots + a_n^k} = \max(a_1, a_2, ..., a_n).$$

Let a_1 be the unique maximum element without loss of generality, then $1 - 2^{-\alpha} \leq b_1 \leq 1$ and $0 \leq b_i \leq 2^{-\alpha}$ for $2 \leq i \leq n$. Then, the error of $\sum_{i=1}^{n} b_i a_i$ compared to the true value $\max(a_1, ..., a_n)$ is bounded by $2^{-\alpha} \cdot \max(a_1, \sum_{i=2}^{n} a_i) \leq \frac{3n}{2} \cdot 2^{-\alpha}$.

6 Asymptotic Optimality of Our Methods

In this section, we compare the efficiency of our min/max and comparison algorithms with general polynomial approximation methods, in terms of computational complexity. As the result, we prove the (quasi-)optimality of our algorithms in terms of asymptotic computational complexity among polynomial evaluations to obtain approximate min/max and comparison results.

There have been various approaches on dealing with non-polynomial homomorphic operations in many applications of word-wise HE [12, 30, 36], and those works commonly use polynomial approximation. Since our algorithms are based on addition and multiplication, they can be also viewed as polynomial evaluations. However, the main difference is that our polynomial evaluations are represented as recursive algorithms so that the complexity is significantly lower than that of general polynomial evaluation of the same degree.

As described in Theorems 1–5, we estimated an approximation error of our methods (Algorithms 3–6) through the infinite norm, i.e., the maximal error over the domain. Therefore, the *minimax polynomial approximation* [40] which targets the (degree-)optimal polynomial approximation with respect to the error measured by the infinite norm should be compared with our methods. The upper bound of the error of minimax polynomial approximation is given by Jackson's inequality [41] which is a well-known result in approximation theory. The inequality originally covers both algebraic and trigonometric polynomial approximation of general functions, but it can be simplified fitting into our case as following [38]. If a function f defined on $[-1, 1]$ satisfies L-Lipschitz condition, i.e., $|f(x_1) - f(x_2)| \le L \cdot |x_1 - x_2|$ for any $x_1, x_2 \in [-1, 1]$, then it holds that

$$||f - p_k||_{\infty, [-1,1]} \le \frac{L\pi}{2(k+1)} \tag{2}$$

where p_k is the degree-k minimax polynomial of f over the interval $[-1, 1]$. Namely, the maximal error between the degree-k minimax polynomial and the original Lipschitz function is $O(1/k)$.

6.1   Min/Max from Minimax Approximation

As described in Sect. 4, the min/max functions can be simply described with the absolute function as

$$\min(a, b) = \frac{a + b}{2} - \frac{|a - b|}{2}, \quad \max(a, b) = \frac{a + b}{2} + \frac{|a - b|}{2}.$$

Since the absolute function can also be expressed as $|x| = x - 2 \cdot \min(x, 0) = 2 \cdot \max(x, 0) - x$, the evaluation of min and max functions are actually equivalent to the evaluation of the absolute function with some additional linear factors. Hence it suffices to consider the minimax polynomial approximation of the absolute function $f(x) = |x|$. We assume that a and b are scaled numbers in $[0, 1)$.

In the case of $f(x) = |x|$, it is proved that the error upper bound $O(1/k)$ of Jackson's inequality is quite *tight* in terms of asymptotic complexity:

$$\lim_{k \to \infty} k \cdot |||x| - p_k||_{\infty, [-1,1]} = \beta$$

for some constant $\beta \approx 0.28$ [3]. For more details of experimental results on the equation above, we refer the readers to [38, p. 19]. As a result, to obtain an approximation error at most $2^{-\alpha}$ for $f(x) = |x|$, it requires the degree of the minimax polynomial to be at least $\Theta(2^\alpha)$. Since general polynomial of degree n requires at least $\sqrt{n}$ multiplications [39], the evaluation of the minimax polynomial requires at least $\Theta(2^{\alpha/2})$ multiplications. In contrast, our min/max algorithms require only $\Theta(\alpha)$ complexity by Theorem 1. Note that the depths of minimax polynomial evaluation and our min/max algorithms are $\alpha + O(1)$ and $4\alpha - 6$, respectively, both of which are $\Theta(\alpha)$.

Even without asymptotic point of view, our method outperforms the minimax approximation in terms of the required number of multiplications when α

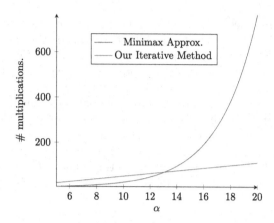

Fig. 2. The actual number of multiplications in minimax approximation and our iterative method for `Max`

is larger than 13. Easy computations show that the required number of multiplications in our iterative method and the minimax approximation method to achieve certain error bound $2^{-\alpha}$ are $3 \cdot (2\alpha - 3) = 6\alpha - 9$ and (approximately) $\sqrt{2\beta} \cdot 2^{\alpha/2}$, respectively (refer Fig. 2). Here $2\alpha - 3$ is the minimal number of iterations in `Min/Max`, and 3 is the number of multiplications in each iteration.

6.2   Comparison from Minimax Approximation

Since the comparison equation is expressed as $\mathrm{comp}(a, b) = \chi_{(0,\infty)}(a - b)$, one needs to find a minimax polynomial of the step function $\chi_{(0,\infty)}$. Note that the evaluations of comp and $\chi_{(0,\infty)}$ are equivalent since the step function can also be expressed as $\chi_{(0,\infty)}(x) = \mathrm{Comp}(x, 0)$. Let a and b be scaled numbers contained in $\left[\frac{1}{2}, \frac{3}{2}\right)$ as discussed in Sect. 5. Then the range of $(a - b)$ is $(-1, 1)$, so we can still consider the approximation over the interval $[-1, 1]$.

Contrary to the absolute function $|x|$, the minimax polynomial approximation of $\chi_{(0,\infty)}$ over an interval $[-1, 1]$, which contains 0, *never* gives a nice error bound $||\chi_{(0,\infty)} - p_k||_{\infty, [-1,1]}$ since the step function is discontinuous on $x = 0$. Therefore, it is inevitable to abandon a good polynomial approximation of $\chi_{(0,\infty)}$ over an interval $(-\epsilon, \epsilon)$ for some small $\epsilon > 0$, and our goal should be reduced to find an approximate polynomial p of $\chi_{(0,\infty)}$ which minimizes $||\chi_{(0,\infty)} - p||_{\infty, [-1,-\epsilon]\cup[\epsilon,1]}$. Namely, we should aim to obtain a nice approximate result of comparison on a and b satisfying $|a - b| \geq \epsilon$, not for all $a, b \in \left[\frac{1}{2}, \frac{3}{2}\right)$.

Let us denote by $q_{k,\epsilon}$ the degree-k approximate polynomial which minimizes $||\chi_{(0,\infty)} - p||_{\infty, [-1,-\epsilon]\cup[\epsilon,1]}$. For the step function $\chi_{(0,\infty)}$, there exists a tighter upper bound on the approximation error than Jackson's inequality as following:

$$\lim_{k \to \infty} \sqrt{\frac{k-1}{2}} \cdot \left(\frac{1+\epsilon}{1-\epsilon}\right)^{\frac{k-1}{2}} \cdot ||\chi_{(0,\infty)} - q_{k,\epsilon}||_{\infty, [-1,-\epsilon]\cup[\epsilon,1]} = \frac{1-\epsilon}{2\sqrt{\pi\epsilon}},$$

which was proved by Eremenko and Yuditskii [25]. Assume that k is large enough so that $\sqrt{\frac{k-1}{2}} \cdot \left(\frac{1+\epsilon}{1-\epsilon}\right)^{\frac{k-1}{2}} \cdot ||\chi_{(0,\infty)} - q_{k,\epsilon}||_{\infty,[-1,-\epsilon]\cup[\epsilon,1]}$ is sufficiently close to the limit value. To obtain an approximation error at most $2^{-\alpha}$ for $\chi_{(0,\infty)}$ over $[-1,-\epsilon] \cup [\epsilon, 1]$, the degree k should be chosen to satisfy

$$\sqrt{\frac{k-1}{2}} \cdot \left(\frac{1+\epsilon}{1-\epsilon}\right)^{\frac{k-1}{2}} \cdot \frac{2\sqrt{\pi\epsilon}}{1-\epsilon} > 2^{\alpha}.$$

Let us consider two cases: $\epsilon = \omega(1)$ and $\epsilon = 2^{-\alpha}$. In the case of $\epsilon = \omega(1)$, i.e., ϵ is a constant with respect to α, the polynomial degree k should be at least $\Theta(\alpha)$. Therefore, the required depth and computational complexity of q_k evaluation considering the Paterson-Stockmeyer method are $\Theta(\log\alpha)$ and $\Theta(\sqrt{\alpha})$, respectively. In the case of $\epsilon = 2^{-\alpha}$, the polynomial degree k should be at least $\Theta(\alpha \cdot 2^{\alpha})$, needing $\Theta(\alpha)$ depth and $\Theta(\sqrt{\alpha} \cdot 2^{\alpha/2})$ multiplications with the Paterson-Stockmeyer method.

For a fair (conservative) comparison between the above polynomial approximation and our comparison method, we set $c = \frac{3}{3-2\epsilon}$ where $1 < c < 3$ is a constant defined in Theorem 4 so that the domain $D_1 := \{(a,b) \in \left[\frac{1}{2}, \frac{3}{2}\right]^2 : |a-b| \geq \epsilon\}$ for the above polynomial approximation is completely contained in the domain $D_2 := \{(a,b) \in \left[\frac{1}{2}, \frac{3}{2}\right]^2 : \max(a,b)/\min(a,b) \geq c\}$ for our method. In this setting, the depth and complexity $\Theta(\log(\alpha/\log c) \cdot \log(\alpha + \log(\alpha/\log c)))$ of our Comp algorithm becomes $\Theta(\log^2\alpha)$ if $\epsilon = \omega(1)$ and $\Theta(\alpha\log\alpha)$ if $\epsilon = 2^{-\alpha}$ (Fig. 3).

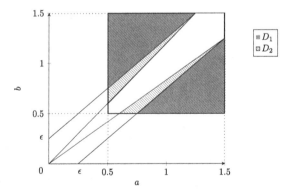

Fig. 3. Regions $D_1 \subset D_2$ for $\epsilon = \frac{3}{2} \cdot \left(1 - \frac{1}{c}\right)$

The comparison results on the complexity of our methods and minimax polynomial approximation are summarized in Table 1. As discussed above, we set two cases $\epsilon = \omega(1)$ and $\epsilon = 2^{-\alpha}$ for the comparison operation.

(Quasi-)optimality of Our Methods. The comparison of computational complexity on our method and minimax approximation method implies the

Table 1. Complexity of our methods and minimax approximation method

		Minimax approx.	Our method
min/max		$\Theta(2^{\alpha/2})$	$\Theta(\alpha)$
Comparison	$\epsilon = \omega(1)$	$\Theta(\sqrt{\alpha})$	$\Theta\left(\log^2 \alpha\right)$
	$\epsilon = 2^{-\alpha}$	$\Theta\left(\sqrt{\alpha} \cdot 2^{\alpha/2}\right)$	$\Theta\left(\alpha \log \alpha\right)$

(quasi-)optimality of our Min/Max and Comp algorithms in terms of asymptotic computational complexity. What Jackson's inequality implies is that *any* polynomial evaluation to obtain an absolute value (hence a min/max result) within $2^{-\alpha}$ error requires $\omega(2^{\alpha})$ degree. Regardless of how the polynomial of degree $\omega(2^{\alpha})$ is well-structured, the complexity of the polynomial evaluation should be at least the depth $\omega(\alpha)$. In this respective, our Min/Max algorithm is optimal in asymptotic complexity among the polynomial evaluations to obtain an approximate min/max result. In the same manner, any polynomial evaluation to obtain a comparison result within $2^{-\alpha}$ error requires at least $\omega(\log \alpha)$ and $\omega(\alpha)$ complexity for the cases $\epsilon = \omega(1)$ and $\epsilon = 2^{-\alpha}$, respectively. Therefore, our Comp algorithm achieves a kind of quasi-optimal asymptotic complexity with an additional factor $\log \alpha$.

Remark 6. In [5], Boura, Gama and Georgieva proposed a different approach for evaluating the absolute function and the step function which use Fourier approximation, and the evaluations can be efficiently done in HEAAN which supports operations of complex numbers. For the fair comparison with our method, we look into the theoretical upper bound of errors in Fourier approximation. By Jackson's inequality for Fourier approximation [33], the upper bound for error of the Fourier approximation of an Lipschitz function f is given as

$$\|f - S_k f\|_\infty \leq K \cdot \frac{\log k}{k}$$

for some $K > 0$ where $S_k f(x) := \sum_{n=-k}^{k} \hat{f}(n) \cdot e^{inx}$ is the k-th Fourier approximation of f, which can be viewed as a polynomial of e^{ix} and e^{-ix}.

We note that the upper bound of the Fourier approximation error for the absolute function can be reduced to $\Theta(1/k)$. As a result, to make the error upper bound less than $2^{-\alpha}$ following theoretical results, one needs at least $\Theta(2^{\alpha})$-th (resp. $\Theta(\alpha \cdot 2^{\alpha})$-th) Fourier approximation for the absolute function (resp. step function). Moreover, exponential functions e^{ix} and e^{-ix} should be also approximately evaluated which derives an additional error. Therefore, this Fourier approximation approach still requires exponential computational complexity with respect to α. To sum up, in asymptotic complexity sense, the Fourier approximation approach in [5] requires more computations than our method to obtain the result within a certain level of error.

7   Applications of Comparison Algorithms

In this section, we exploit our comparison algorithms proposed in Sect. 5 for several applications: Threshold Counting and Top-k Max.

7.1   Threshold Counting

In this subsection, we give a solution to the problem asked at the very beginning of HE. In 1978, Rivest et al. [42] first proposed the concept of HE and listed some problems to be solved with HE:

$\cdots$ *This organization permits the loan company to utilize the storage facilities of the time—sharing service, but generally makes it difficult to utilize the computational facilities without compromising the privacy of the stored data. The loan company, however, wishes to be able to answer such questions as:*

- *What is the size of the average loan outstanding?*
- *How much income from loan payments is expected next month?*
- *How many loans over $5,000 have been granted?*

While the first two problems can be answered with simple arithmetic operations, the last problem requires comparison-like operation intrinsically. We propose a solution to the third problem with our Comp algorithm. First, we abstract the problem to "Threshold Counting" problem. The goal of threshold counting problem is to find the number of a_i's larger than b for given $(a_1, a_2, ..., a_n)$ and b. The algorithm is rather simple. We compare a_i's with b and sum up the values $\mathrm{comp}(a_i, b)$. We can use usual packing method of HE to compare several elements in a single operation. We remark that if $a_i = b$ then a_i is counted as $1/2$, not 0 or 1, but in real-world applications this error may be ignored or adjusted by subtracting a very small constant to the threshold b.

Algorithm 7. $\mathtt{Threshold}(a_1, a_2, .., a_n; b; d, d', t, m)$

Input: n numbers $(a_1, a_2, ..., a_n)$ with $a_i \in [0, 1)$, $b \in [0, 1)$, $d, d', m, t \in \mathbb{N}$
Output: an approximate value of the number of a_i's larger than b
1: **for** $i \leftarrow 1$ **to** n **do**
2:    $c_i \leftarrow \mathtt{Comp}(a_i, b; d, d', m, t)$ // Can be done in a SIMD manner via HE.
3: **end for**
4: $sum \leftarrow 0$
5: **for** $j \leftarrow 1$ **to** k **do**
6:    $sum \leftarrow sum + c_i$
7: **end for**
8: **return** sum

7.2 Top-k Max

Applying the MaxIdx algorithm in Sect. 5.2 recursively, we can obtain top-k maximum values which we call top-k max algorithm. For given distinct numbers $a_1, a_2, ..., a_n \in \left[\frac{1}{2}, \frac{3}{2}\right)$ and some positive integers $d, d', m, t \geq 0$, let $(b_1, b_2, ..., b_n) \leftarrow$ MaxIdx$(a_1, a_2, ..., a_n; d, d', m, t)$. Then as noted in Remark 5, $\sum_{i=1}^{n} b_i a_i$ is an approximate maximum value of $a_1, ..., a_n$ since $b_i \approx 1$ if and only if a_i is the maximum. Now, to compute the second maximum value, let a_j be the (unique) maximum value, and define $c_i := (1 - b_i)a_i$ for $1 \leq i \leq n$. Then $c_i = (1 - b_i)a_i \approx a_i$ for all $i \neq j$ and $c_j = (1 - b_j)a_j \approx 0$. Since we assume that a_i's are positive numbers, the output of MaxIdx$(c_1, c_2, ..., c_n; d, d', m, t)$ indeed indicates the index of the second maximum value. This algorithm can be generalized as following.

Algorithm 8. Top-k-Max$(a_1, a_2, .., a_n; d, d', m, t)$

Input: n distinct numbers $(a_1, a_2, ..., a_n)$ with $a_i \in [0, 1)$, $d, d', m, t \in \mathbb{N}$
Output: $(m_1, m_2, ..., m_k)$ where m_i denotes an approximate value of the ith largest number among $\{a_1, a_2, ..., a_n\}$
1: **for** $i \leftarrow 1$ **to** n **do**
2:     $c_i \leftarrow a_i$
3: **end for**
4: **for** $j \leftarrow 1$ **to** k **do**
5:     $(b_1, b_2, ..., b_n) \leftarrow$ MaxIdx$(c_1, c_2, ..., c_n; d, d', m, t)$
6:     $m_j \leftarrow \sum_{i=1}^{n} b_i c_i$
7:     $(c_1, c_2, ..., c_n) \leftarrow ((1 - b_1)c_1, (1 - b_2)c_2, ..., (1 - b_n)c_n)$
8: **end for**
9: **return** $(m_1, m_2, ..., m_k)$

Theorem 6. *Let $a_1, a_2, ..., a_n \in [1/2, 3/2]$ be n distinct elements, and let the ratio of i-th maximum value over the $(i + 1)$-th maximum value $\frac{\max_i}{\max_{i+1}} > c_i$ for $1 \leq i \leq k$. For some $c > 1$ and $\alpha > 0$ satisfying $2^\alpha \cdot (1 - 2^{-\alpha})^{\frac{k(k-1)}{2}} > c^k$, assume that $c_i = c/(1 - 2^{-\alpha})^{i-1}$ and $\frac{(1-2^{-\alpha})^k \max_{k+1}}{2^{-\alpha} \max_1} > c$. If t, d and d' satisfy the same conditions in Theorem 5, the output $(m_1, ..., m_k)$ of Top-k-Max$(a_1, ..., a_n; d, d', m, t)$ satisfies $(1 - 2^{-\alpha})^j \max_j \leq m_j \leq \max_j$ for $1 \leq j \leq k$.*

Proof. Refer to Appendix A.   □

8   Experimental Results

This section illustrates some implementation results of the algorithms we described in the previous sections based on the approximate HE scheme called HEAAN [13]. We also propose some reasonable parameters, and show that the algorithms can be carried out with HEAAN very well.

We first show the performance of Max algorithm for several setups based on HEAAN. We also implement Comp algorithm based on HEAAN and show that it can be exploited to solve the threshold counting problem efficiently. Lastly, we show the performance of our MaxIdx algorithm.

8.1 Approximate HE Scheme HEAAN

Cheon et al. [13] proposed an HE scheme HEAAN which supports approximate computations of real/complex numbers. By abandoning the exact computation, HEAAN achieves big advantages in ciphertext/plaintext ratio and speed. Since many real-world applications require real number computations, HEAAN has a strength in various real-world problems [11,14,15,35,36], which usually deal with approximate computation of real numbers, compared to the other HE schemes. For an efficiently computable (field) isomorphism $\tau : \mathbb{R}[X]/(X^N + 1) \to \mathbb{C}^{N/2}$, the basic algorithms are following:

- KeyGen($L, 1^\lambda$).
  - Given the level parameter L and the security parameter λ, select power-of-two integers N and set $q_\ell = 2^\ell$ for $1 \le \ell \le L$.
  - Set the secret and error distributions $\chi_{\text{key}}, \chi_{\text{err}}, \chi_{\text{enc}}$ over R.
  - Sample $s \leftarrow \chi_{\text{key}}$. Set the secret key as $\text{sk} \leftarrow (1, s)$.
  - Sample $a \leftarrow U(R_{q_L})$ and $e \leftarrow \chi_{\text{err}}$. Set the public key as $\text{pk} \leftarrow (b, a) \in R_{q_L}^2$ where $b \leftarrow [-a \cdot s + e]_{q_L}$.
  - Sample $a' \leftarrow U(R_{q_L^2})$ and $e' \leftarrow \chi_{\text{err}}$. Set the evaluation key as $\text{evk} \leftarrow (b', a') \in R_{q_L^2}^2$ where $b' \leftarrow [-a's + e' + q_L \cdot s^2]_{q_L^2}$.
- Enc$_{\text{pk}}(\boldsymbol{m})$.
  - For a plaintext $\boldsymbol{m} = (m_0, ..., m_{N/2-1})$ in $\mathbb{C}^{N/2}$ and a scaling bit $p > 0$, compute a polynomial $\mathtt{m} \leftarrow \lfloor 2^p \cdot \tau^{-1}(\boldsymbol{m}) \rceil \in R$
  - Sample $v \leftarrow \chi_{\text{enc}}$ and $e_0, e_1 \leftarrow \chi_{\text{err}}$. Output $\text{ct} = [v \cdot \text{pk} + (\mathtt{m} + e_0, e_1)]_{q_L}$.
- Dec$_{\text{sk}}(\text{ct})$.
  - For a ciphertext $\text{ct} = (c_0, c_1) \in R_{q_\ell}^2$, compute $\mathtt{m}' = [c_0 + c_1 \cdot s]_{q_\ell}$.
  - Output a plaintext vector $\boldsymbol{m}' = 2^{-p} \cdot \tau(\mathtt{m}') \in \mathbb{C}^{N/2}$.
- Add(ct, ct'). For $\text{ct}, \text{ct}' \in R_{q_\ell}^2$, output $\text{ct}_{\text{add}} \leftarrow [\text{ct} + \text{ct}']_{q_\ell}$.
- Sub(ct, ct'). For $\text{ct}, \text{ct}' \in R_{q_\ell}^2$, output $\text{ct}_{\text{sub}} \leftarrow [\text{ct} - \text{ct}']_{q_\ell}$.
- Mult$_{\text{evk}}$(ct, ct'). For $\text{ct} = (c_0, c_1), \text{ct}' = (c_0', c_1') \in \mathcal{R}_{q_\ell}^2$, let $(d_0, d_1, d_2) = (c_0 c_0', c_0 c_1' + c_1 c_0', c_1 c_1')$. Compute $\text{ct}'_{\text{mult}} \leftarrow [(d_0, d_1) + \lfloor q_L^{-1} \cdot d_2 \cdot \text{evk} \rceil]_{q_\ell}$, and output $\text{ct}_{\text{mult}} \leftarrow [\lfloor (1/p) \cdot \text{ct}'_{\text{mult}} \rceil]_{q_{\ell-1}}$.

For details on the correctness and security of the scheme, we refer the readers to [13].

Table 2. HEAAN implementation of `Max` algorithm for several precision bits. HEAAN parameters $(\log N, Q, \lambda)$ were chosen as [a]$(17, 930, 192.2)$, [b]$(17, 1170, 147.0)$, [c]$(17, 1410, 131.5)$, and [d]$(17, 1890, 107.7)$.

Algorithm	# precision bits	# iterations	Running time	
	α	d	Total (s)	Amortized (ms)
Max	8	11	$48^{[a]}$	0.73
	10	14	$75^{[b]}$	1.14
	12	17	$127^{[c]}$	1.94
	16	23	$237^{[d]}$	3.62

8.2 Implementations of Various Non-polynomial Operations

All experiments on our method were implemented in C++ on Linux with Intel Xeon CPU E5-2620 v4 at 2.10 GHz processor with multi-threading (8 threads) turned on for speed acceleration. Note that we checked the security level of HEAAN parameters we used in our implementation through a security estimator constructed by Albrecht [1,2]. More precisely, we set the level parameter L to be the minimum required considering the depth of algorithms (without bootstrapping), the dimension N to be the minimum ensuring the security parameter $\lambda \geq 128$, and the scaling bit p to be 40 or around.

In the rest of the section, we present both the actual running time and the amortized running time considering the plaintext batching technique of HEAAN. We note that the amortized running time is important as much as the actual running time in various applications which require a number of same operations. For example, even a basic task such as threshold counting can be performed simultaneously with only a single homomorphic comparison. More seriously, k-nearest neighbor algorithm for classification and k-means algorithm for clustering requires substantial numbers of min/max and comparison, which can also be parallelized in the same manner with the above threshold counting.

Max of Two Integers. We first show the performance of Algorithm 3 (`Max`) which outputs an approximate value of the maximum value given two large integers. Since HEAAN supports at most $N/2$ operations simultaneously in a SIMD manner, the actual experiment is to compute $\max(a_i, b_i)$ for $1 \leq i \leq N/2$. In Table 2, minimal iteration d required for `Max` to achieve each bit precision α is provided. The number of iterations are empirically chosen considering the worst case, which is smaller than the theoretical expectation of Theorem 1. For example, when $\alpha = 10$, then $d = 14$ suffices while theoretical requirement is $d \geq 17$. The amortized running time is measured by dividing total running time by the number of plaintext slots.

Table 3. Implementation of `Comp` for several precision bits. HEAAN parameters $(\log N, Q, \lambda)$ were chosen as [a]$(17, 1600, 121, 6)$, and [b]$(17, 1870, 108.9)$.

Algorithm	# precision bits	# iterations	Running time	
	α	(d', d, t)	Total (s)	Amortized (ms)
Comp (exact)	7	$(5, 5, 5)$	225[a]	3.43
	8	$(5, 5, 6)$	310[b]	4.72
Comp ($c = 1.01$)	14	$(5, 5, 5)$	230[a]	3.50
Comp ($c = 1.05$)	24	$(5, 5, 5)$	259[a]	3.94

We remark that our performance only depends on the precision α, not on the input bitsize ℓ. It provides us much flexibility when we need only approximate maximum value. For example, our implementation shows that we can obtain an approximate maximum value of any two 32-bit integers with an error up to 2^{22} in 1.14 ms (with amortized time sense).

The performance of our `Max` algorithm is comparable, in amortized running time sense, to the previous results of which input numbers are encrypted bit-wise. For example, the max algorithm from [19] based on a bit-wise HE, which expressed the max function by a number of logical gates via weighted finite automata, takes about 1 ms to compute the maximum of two 8-bit integers.

Comparison of Two Integers. We also implemented our `Comp` algorithm for various setups on the number of precision bits α and the lower bound c of the ratio $\frac{\max(a,b)}{\min(a,b)}$. As in the previous subsection, we put integers in full $N/2$ plaintext slots of HEAAN ciphertext so that the `Comp` algorithm supports $N/2$ simultaneous comparison operations. For each setup, we empirically chose optimal parameters $m = 4$, d, d' and t. Refer to Algorithm 5 for definitions of the parameters.

In Table 3, `Comp` (exact) denotes the comparison experiment considering the worst case, i.e., comparing *any* of two α-bit integers scaled into $[\frac{1}{2}, \frac{3}{2})$ with α-bit precision, which corresponds to $c = \left(\frac{1}{2} + \frac{2^\alpha - 1}{2^\alpha}\right) / \left(\frac{1}{2} + \frac{2^\alpha - 2}{2^\alpha}\right)$. For the cases $c = 1.01$ and $c = 1.05$, we took 32-bit integers satisfying the ratio lower bound as input.

As same as `Max`, our empirically chosen parameters d, d' and t and are smaller than the theoretical expectation from Theorem 4. For example, for 7-bit precision of `Comp` (exact), it was expected to be $d, d' > 5.9$ and $t > 5.5$ from the theorem, but we found that a bit smaller parameters were sufficient.

The result shows that when we do not need exact comparison, i.e., when we are given that two inputs has enough difference, we can set the parameters as more efficient ones. For example, the same iteration $(d', d, t) = (5, 5, 5)$ guarantees 14, or 24 bit precision when c is 1.01 or 1.05, respectively, while it only guarantees 7-bit precision if we need exact comparison. When c is 1.05, only $(d', d, t) = (5, 4, 4)$ iteration suffices for 8-bit precision. Note that each result

Table 4. Implementation of MaxIdx and Threshold for 2^4 and 2^5 encrypted 7-bit integers, respectively. HEAAN parameters $(\log N, Q, \lambda)$ were chosen as $(17, 1800, 111.3)$.

Algorithm	# precision bits	# iterations	Running time	
	α	(d', d, t)	Total (s)	Amortized (ms)
MaxIdx	7	(3, 11, 3)	311	75.9
Threshold	6	(3, 5, 5)	278	135

shows high performance of Comp showing less than 5 ms of amortized running time considering 2^{16} number of plaintext slots in one ciphertext.

In [21], Crawford et al. reported some recent implementation results on the comparison operation based on HElib, where the input integers were bit-wise encrypted. We referred their comparison experiment on 8-bit integers which uses the 15709-th cyclotomic polynomial, and it took about a second with 8 threads. Considering ciphertexts over 15709-th cyclotomic polynomial have 682 plaintext slots, the amortized running time is around 1.5 ms. This shows that the performance of our word-wise comparison is comparable, in amortized running time, to that of a bit-wise comparison which has been regarded to be one of the most natural approaches to compare numbers.

Max Index for Several Numbers. We present an experimental evaluation of the MaxIdx algorithm. For experiment, we compute max index of 16 encrypted 7-bit integers. We assume that the maximum integer has non-zero most significant bit, while other integers have most and 2nd-most significant bits zero. This condition corresponds to the lower bound $c = \left(\frac{1}{2} + \frac{2^6}{2^7}\right) / \left(\frac{1}{2} + \frac{2^5-1}{2^7}\right) = \frac{128}{95}$.

The parameter chosen by considering worst-case is a little better than the theoretical estimation (Theorem 5) which suggests t and d to satisfy $t > 2$ and $d > 14$. Total running time is about 311 s, and we can run $2^{16}/2^4 = 2^{12}$ number of Max index algorithms with one ciphertext resulting amortized running time to be only about 75 ms.

Threshold Counting. For Threshold algorithm, we assume that the threshold b is encrypted. This is because in some scenarios the threshold could be private information. If b is not secret, the algorithm shows a better performance since a constant multiplication is faster than a ciphertext multiplication in HE.

For a power-of-two integer $k \leq N/2$, HEAAN supports a packing method which packs k real numbers in a single ciphertext, enabling us to perform parallel computations over encryption. As mentioned in the Sect. 7.1, we utilize this packing method to solve threshold counting with exactly one Comp query and then use RotateSum to sum up the results of the Comp.

For experimental results, we assume that given 2^5 number of 7-bit integers, we want to calculate the number of elements bigger than an encrypted 7-bit threshold. Then, we can take the lower bound $c = \left(\frac{1}{2} + \frac{2^6-1}{2^6}\right) / \left(\frac{1}{2} + \frac{2^6-2}{2^6}\right) = \frac{191}{190}$, and it suffices to bound error size to be smaller than $2^{-\alpha} = 2^{-6}$ for each result of comparison, since we evaluate the addition of 2^5 comparison results, whose true value is an integer. In Table 4, we can see that it takes about 278 s to get the number of elements bigger than the given threshold. Since we can pack at most 2^{16} numbers in one ciphertext, we can manage 2^{11} threshold counting problems for 2^5 numbers with only a single ciphertext, resulting about 135 ms of amortized running time. If we allow some errors in the final result, or we are given that the gap between threshold and other numbers are large, we can get more efficient result than above.

Acknowledgement. We thank Minki Hhan for suggesting a new interpretation on the efficiency of our algorithms, and Yongsoo Song for several valuable comments. We also thank to anonymous reviewers of ASIACRYPT 2019. This work was supported by the National Research Foundation of Korea (NRF) Grant funded by the Korean Government (MSIT) (No. 2017R1A5A1015626).

A Proofs

Proof of Theorem 3. By Theorem 2, the error of $\mathtt{Max}(\cdot, \cdot; d)$ algorithm from the true value is bounded by $2^{(-\alpha-\log\log n)} = 2^{-\alpha}/\log n$. Note from the proof of Lemma 2 that the output of the square root algorithm $\mathtt{Sqrt}(x; d)$ is always smaller than the true value $\sqrt{x}$, so that the same holds for the max algorithm $\mathtt{Max}(\cdot, \cdot; d)$. This means that $a_{i,1} = \mathtt{Max}(a_{2i-1,0}, a_{2i,0}; d)$ can be written $a_{i,1} = \max(a_{2i-1,0}, a_{2i,0}) - \epsilon_i$ for $1 \leq i \leq n/2$ with $0 \leq \epsilon_i \leq 2^{-\alpha}/\log n$. Now we have

$$\max(a_{2i-1,1}, a_{2i,1}) = \max(\max(a_{4i-3,0}, a_{4i-2,0}) - \epsilon_{2i-1}, \max(a_{4i-1,0}, a_{4i,0}) - \epsilon_{2i})$$
$$\geq \max(a_{4i-3,0}, a_{4i-2,0}, a_{4i-1,0}, a_{4i,0}) - \max(\epsilon_{2i-1}, \epsilon_{2i})$$
$$\geq \max(a_{4i-3,0}, a_{4i-2,0}, a_{4i-1,0}, a_{4i,0}) - 2^{-\alpha}/\log n,$$

which implies that the error of $a_{i,2} = \mathtt{Max}(a_{2i-1,1}, a_{2i,1}; d)$ from $\max(a_{2i-1,1}, a_{2i,1})$ is bounded by $2 \cdot 2^{-\alpha}/\log n$ for $1 \leq i \leq n/4$. We can repeat the above procedure to get the conclusion that the error of $a_{1,\log n}$ from $\max(a_1, ..a_n)$ is bounded by $\log n \cdot 2^{-\alpha}/\log n = 2^{-\alpha}$.

For the case of min algorithm we note that the approximate values are larger than the true values and we can apply a similar approach to the above with reversed inequalities. □

Proof of Theorem 5. Note that $\mathtt{MaxIdx}$ is a natural generalization of $\mathtt{Comp}$. Without loss of generality, we assume that a_1 is the unique maximum element, and we only consider the error between the output b_1 of $\mathtt{MaxIdx}$ and the real value 1. At Step 1–4, $(a_i)_{i=1}^n$ is scaled to $(b_i)_{i=1}^n$ whose sum is 1. Moreover, every input of $\mathtt{Inv}$ is bounded by $\frac{n}{2^m}$ since $\sum_{k=1}^n b_j$ is always set to be 1 before the $\mathtt{Inv}$

algorithm. Note that each b_j from the iterations is nothing but $a_j^{m^t}/\sum_{i=1}^n a_i^{m^t}$ with t being increased by one as the iteration go. The error of MaxIdx algorithm is also composed of three parts as Theorem 4; an error from the convergence of $\lim_{m\to\infty} a_1^m/\sum_{i=1}^n a_i^m = 1$, and an error from the approximation of $1/(\sum_{i=1}^n b_i^m)$ by our Inv algorithm and an error coming from Steps 1–4.

Now, the error analysis is almost the same as the proof of Theorem 4 with minor differences in the values of errors. The first part of the error is bounded by $n \cdot (1/c)^{m^t}$ since $1 - \frac{a_1^N}{\sum_{i=1}^n b_i^N} = 1 - \frac{1}{1+\sum_{i=2}^n (b_i/a_1)^N} \le n/c^N$. The second part of the error (from the Inv algorithm) is bounded by $(1 - n^{-(m-1)})^{2^{d+1}}$ since $n^{-(m-1)}$ is the lower bound of the denominators $\sum_{i=1}^n b_i^m$ by Cauchy-Schwartz inequality. As a result, we can conclude that the conditions $t \ge \frac{1}{\log m}[\log(\alpha + \log n + 1) - \log \log c]$ and $d, d' \ge \log(\alpha + t + 1) + (m-1)\log n - 1$ suffice to make the total error of MaxIdx less than $2^{-\alpha}$ by a similar argument as in Theorem 4.   □

Proof of Theorem 6. Without loss of generality, let a_i be the ith maximum value $\max_i$ for $1 \le i \le n$.

For $1 \le i < k$, since $(1 - 2^{-\alpha})^i a_{i+1} > (1 - 2^{-\alpha})^k a_{k+1}$, we first obtain $\frac{(1-2^{-\alpha})^i a_{i+1}}{2^{-\alpha} a_1} > c$. For $j = 1$, the statement holds directly by Theorem 5. After obtaining m_1, the algorithm takes $(\epsilon_1 a_1, (1 - \epsilon_2)a_2, ..., (1 - \epsilon_n)a_n)$ as an input of MaxIdx$(\cdots ; d, d', m, t)$, where $0 \le \epsilon_i \le 2^{-\alpha}$. Since the following inequalities

$$(1 - \epsilon_2)a_2 \ge (1 - 2^{-\alpha}) \cdot \frac{2^{-\alpha}}{1 - 2^{-\alpha}} \cdot ca_1 \ge c \cdot \epsilon_1 a_1, \text{ and}$$

$$(1 - \epsilon_2)a_2 > (1 - \epsilon_2)c_2 a_3 \ge ca_3 \ge c \cdot (1 - \epsilon_j)a_j \text{ for } 3 \le j \le n$$

hold, the output m_2 satisfies $(1 - 2^{-\alpha})^2 a_2 \le m_2 \le a_2$ by Theorem 5.

Inductively, assume that we have obtained $m_1, m_2, ..., m_{j-1}$ satisfying the statement condition. After obtaining an approximate value m_{j-1} of the $(j-1)^{\text{th}}$ maximum value a_{j-1}, the next input of MaxIdx algorithm is $(\delta_1 a_1, \delta_2 a_2, ..., \delta_n a_n)$ where $0 \le \delta_i \le 2^{-\alpha}$ for $i < j$ and $(1 - 2^{-\alpha})^j \le \delta_i \le 1$ for otherwise. From the following inequalities

$$\delta_j a_j \ge (1 - 2^{-\alpha})^j \cdot \frac{2^{-\alpha}}{(1 - 2^{-\alpha})^j} \cdot ca_1 \ge c \cdot \delta_i a_i \text{ for } 1 \le i < j, \text{ and}$$

$$\delta_j a_j > \delta_j c_j a_{j+1} \ge ca_{j+1} \ge c \cdot \delta_i a_i \text{ for } i > j,$$

by Theorem 5 the output m_{j+1} satisfies $(1 - 2^{-\alpha})\delta_j a_j \le m_j \le \delta_j a_j$ so that the statement also holds for j. Therefore, the theorem is proved by induction.   □

References

1. Albrecht, M.R.: A sage module for estimating the concrete security of learning with errors instances (2017). https://bitbucket.org/malb/lwe-estimator
2. Albrecht, M.R., Player, R., Scott, S.: On the concrete hardness of learning with errors. J. Math. Cryptol. **9**(3), 169–203 (2015)

3. Bernstein, S.: Sur la meilleure approximation de |x| par des polynomes de degrés donnés. Acta Math. **37**(1), 1–57 (1914)

4. Bos, J.W., Lauter, K., Loftus, J., Naehrig, M.: Improved security for a ring-based fully homomorphic encryption scheme. In: Stam, M. (ed.) IMACC 2013. LNCS, vol. 8308, pp. 45–64. Springer, Heidelberg (2013). https://doi.org/10.1007/978-3-642-45239-0_4

5. Boura, C., Gama, N., Georgieva, M.: Chimera: a unified framework for B/FV, TFHE and HEAAN fully homomorphic encryption and predictions for deep learning. Cryptology ePrint Archive, Report 2018/758 (2018). https://eprint.iacr.org/2018/758

6. Bourse, F., Minelli, M., Minihold, M., Paillier, P.: Fast homomorphic evaluation of deep discretized neural networks. In: Shacham, H., Boldyreva, A. (eds.) CRYPTO 2018. LNCS, vol. 10993, pp. 483–512. Springer, Cham (2018). https://doi.org/10.1007/978-3-319-96878-0_17

7. Brakerski, Z.: Fully homomorphic encryption without modulus switching from classical GapSVP. In: Safavi-Naini, R., Canetti, R. (eds.) CRYPTO 2012. LNCS, vol. 7417, pp. 868–886. Springer, Heidelberg (2012). https://doi.org/10.1007/978-3-642-32009-5_50

8. Brakerski, Z., Gentry, C., Vaikuntanathan, V.: (Leveled) fully homomorphic encryption without bootstrapping. In: Proceedings of ITCS, pp. 309–325. ACM (2012)

9. Brown, K.: Probability of intersecting intervals. https://www.mathpages.com/home/kmath580/kmath580.htm

10. Chatterjee, A., SenGupta, I.: Sorting of fully homomorphic encrypted cloud data: can partitioning be effective? IEEE Trans. Serv. Comput. (2017)

11. Cheon, J.H., et al.: Toward a secure drone system: flying with real-time homomorphic authenticated encryption. IEEE Access **6**, 24325–24339 (2018)

12. Cheon, J.H., Jeong, J., Lee, J., Lee, K.: Privacy-preserving computations of predictive medical models with minimax approximation and non-adjacent form. In: Brenner, M., et al. (eds.) FC 2017. LNCS, vol. 10323, pp. 53–74. Springer, Cham (2017). https://doi.org/10.1007/978-3-319-70278-0_4

13. Cheon, J.H., Kim, A., Kim, M., Song, Y.: Homomorphic encryption for arithmetic of approximate numbers. In: Takagi, T., Peyrin, T. (eds.) ASIACRYPT 2017. LNCS, vol. 10624, pp. 409–437. Springer, Cham (2017). https://doi.org/10.1007/978-3-319-70694-8_15

14. Cheon, J.H., Kim, D., Kim, Y., Song, Y.: Ensemble method for privacy-preserving logistic regression based on homomorphic encryption. IEEE Access **6**, 46938–46948 (2018)

15. Cheon, J.H., Kim, D., Park, J.H.: Towards a practical clustering analysis over encrypted data. Cryptology ePrint Archive, Report 2019/465 (2019). https://eprint.iacr.org/2019/465

16. Cheon, J.H., Kim, M., Kim, M.: Search-and-compute on encrypted data. In: Brenner, M., Christin, N., Johnson, B., Rohloff, K. (eds.) FC 2015. LNCS, vol. 8976, pp. 142–159. Springer, Heidelberg (2015). https://doi.org/10.1007/978-3-662-48051-9_11

17. Chialva, D., Dooms, A.: Conditionals in homomorphic encryption and machine learning applications. Cryptology ePrint Archive, Report 2018/1032 (2018). https://eprint.iacr.org/2018/1032

18. Chillotti, I., Gama, N., Georgieva, M., Izabachène, M.: Faster fully homomorphic encryption: bootstrapping in less than 0.1 seconds. In: Cheon, J.H., Takagi, T. (eds.) ASIACRYPT 2016. LNCS, vol. 10031, pp. 3–33. Springer, Heidelberg (2016). https://doi.org/10.1007/978-3-662-53887-6_1

19. Chillotti, I., Gama, N., Georgieva, M., Izabachène, M.: Faster packed homomorphic operations and efficient circuit bootstrapping for TFHE. In: Takagi, T., Peyrin, T. (eds.) ASIACRYPT 2017. LNCS, vol. 10624, pp. 377–408. Springer, Cham (2017). https://doi.org/10.1007/978-3-319-70694-8_14

20. Costache, A., Smart, N.P.: Which ring based somewhat homomorphic encryption scheme is best? In: Sako, K. (ed.) CT-RSA 2016. LNCS, vol. 9610, pp. 325–340. Springer, Cham (2016). https://doi.org/10.1007/978-3-319-29485-8_19

21. Crawford, J.L., Gentry, C., Halevi, S., Platt, D., Shoup, V.: Doing real work with FHE: the case of logistic regression. In: Proceedings of the 6th Workshop on Encrypted Computing and Applied Homomorphic Cryptography, pp. 1–12. ACM (2018)

22. van Dijk, M., Gentry, C., Halevi, S., Vaikuntanathan, V.: Fully homomorphic encryption over the integers. In: Gilbert, H. (ed.) EUROCRYPT 2010. LNCS, vol. 6110, pp. 24–43. Springer, Heidelberg (2010). https://doi.org/10.1007/978-3-642-13190-5_2

23. Ducas, L., Micciancio, D.: FHEW: bootstrapping homomorphic encryption in less than a second. In: Oswald, E., Fischlin, M. (eds.) EUROCRYPT 2015. LNCS, vol. 9056, pp. 617–640. Springer, Heidelberg (2015). https://doi.org/10.1007/978-3-662-46800-5_24

24. Emmadi, N., Gauravaram, P., Narumanchi, H., Syed, H.: Updates on sorting of fully homomorphic encrypted data. In: 2015 International Conference on Cloud Computing Research and Innovation (ICCCRI), pp. 19–24. IEEE (2015)

25. Eremenko, A., Yuditskii, P.: Uniform approximation of sgn(x) by polynomials and entire functions. J. d'Analyse Mathématique 101(1), 313–324 (2007)

26. Fan, J., Vercauteren, F.: Somewhat practical fully homomorphic encryption. IACR Cryptology ePrint Archive, 2012:144 (2012)

27. Gentry, C.: A fully homomorphic encryption scheme. Ph.D. thesis, Stanford University (2009). http://crypto.stanford.edu/craig

28. Gentry, C., Halevi, S., Smart, N.P.: Better bootstrapping in fully homomorphic encryption. In: Fischlin, M., Buchmann, J., Manulis, M. (eds.) PKC 2012. LNCS, vol. 7293, pp. 1–16. Springer, Heidelberg (2012). https://doi.org/10.1007/978-3-642-30057-8_1

29. Gentry, C., Sahai, A., Waters, B.: Homomorphic encryption from learning with errors: conceptually-simpler, asymptotically-faster, attribute-based. In: Canetti, R., Garay, J.A. (eds.) CRYPTO 2013. LNCS, vol. 8042, pp. 75–92. Springer, Heidelberg (2013). https://doi.org/10.1007/978-3-642-40041-4_5

30. Gilad-Bachrach, R., Dowlin, N., Laine, K., Lauter, K., Naehrig, M., Wernsing, J.: Cryptonets: applying neural networks to encrypted data with high throughput and accuracy. In: International Conference on Machine Learning (2016)

31. Goldschmidt, R.E.: Applications of division by convergence. Ph.D. thesis, Massachusetts Institute of Technology (1964)

32. Halevi, S., Shoup, V.: Bootstrapping for HElib. In: Oswald, E., Fischlin, M. (eds.) EUROCRYPT 2015. LNCS, vol. 9056, pp. 641–670. Springer, Heidelberg (2015). https://doi.org/10.1007/978-3-662-46800-5_25

33. Jackson, D.: The Theory of Approximation, vol. 11. American Mathematical Society (1930)

34. Jäschke, A., Armknecht, F.: Unsupervised machine learning on encrypted data. In: Cid, C., Jacobson Jr., M. (eds.) SAC 2018. LNCS, pp. 453–478. Springer, Cham (2018). https://doi.org/10.1007/978-3-030-10970-7_21

35. Kim, A., Song, Y., Kim, M., Lee, K., Cheon, J.H.: Logistic regression model training based on the approximate homomorphic encryption. BMC Med. Genomics **11**(4), 83 (2018)

36. Kim, M., Song, Y., Wang, S., Xia, Y., Jiang, X.: Secure logistic regression based on homomorphic encryption: design and evaluation. JMIR Med. Inform. **6**(2), e19 (2018)

37. Kocabas, O., Soyata, T.: Utilizing homomorphic encryption to implement secure and private medical cloud computing. In: 2015 IEEE 8th International Conference on Cloud Computing (CLOUD), pp. 540–547. IEEE (2015)

38. Pachón, R., Trefethen, L.N.: Barycentric-Remez algorithms for best polynomial approximation in the chebfun system. BIT Numer. Math. **49**(4), 721 (2009)

39. Paterson, M.S., Stockmeyer, L.J.: On the number of nonscalar multiplications necessary to evaluate polynomials. SIAM J. Comput. **2**(1), 60–66 (1973)

40. Phillips, G.M.: Best approximation. In: Phillips, G.M. (ed.) Interpolation and Approximation by Polynomials. CBM, pp. 49–118. Springer, New York (2003). https://doi.org/10.1007/0-387-21682-0_2

41. Powell, M.J.D.: Approximation Theory and Methods. Cambridge University Press, Cambridge (1981)

42. Rivest, R.L., Adleman, L., Dertouzos, M.L.: On data banks and privacy homomorphisms. Found. Secur. Comput. **4**(11), 169–180 (1978)

43. Togan, M., Morogan, L., Plesca, C.: Comparison-based applications for fully homomorphic encrypted data. In: Proceedings of the Romanian Academy-Series A: Mathematics, Physics, Technical Sciences, Information Science, vol. 16, p. 329 (2015)

44. Wilkes, M.V.: The Preparation of Programs for an Electronic Digital Computer: with Special Reference to the EDSAC and the Use of a Library of Subroutines. Addison-Wesley Press (1951)

Multi-Key Homomorphic Encryption from TFHE

Hao Chen[1], Ilaria Chillotti[2], and Yongsoo Song[1(✉)]

[1] Microsoft Research, Redmond, USA
{haoche,yongsoo.song}@microsoft.com
[2] imec-COSIC, KU Leuven, Leuven, Belgium
ilaria.chillotti@kuleuven.be

Abstract. In this paper, we propose a Multi-Key Homomorphic Encryption (MKHE) scheme by generalizing the low-latency homomorphic encryption by Chillotti et al. (ASIACRYPT 2016). Our scheme can evaluate a binary gate on ciphertexts encrypted under different keys followed by a bootstrapping.

The biggest challenge to meeting the goal is to design a multiplication between a bootstrapping key of a single party and a multi-key RLWE ciphertext. We propose two different algorithms for this hybrid product. Our first method improves the ciphertext extension by Mukherjee and Wichs (EUROCRYPT 2016) to provide better performance. The other one is a whole new approach which has advantages in storage, complexity, and noise growth.

Compared to previous work, our construction is more efficient in terms of both asymptotic and concrete complexity. The length of ciphertexts and the computational costs of a binary gate grow linearly and quadratically on the number of parties, respectively. We provide experimental results demonstrating the running time of a homomorphic NAND gate with bootstrapping. To the best of our knowledge, this is the first attempt in the literature to implement an MKHE scheme.

Keywords: Multi-Key Homomorphic Encryption · Bootstrapping

1 Introduction

Cryptographic primitives for secure computation have been actively studied in recent years. Homomorphic Encryption (HE) and Multi-Party Computation (MPC) are the most promising solutions with different models and performance trade-offs. HE is useful for outsourcing the storage and computation to a public cloud, but all data providers should agree on the same public key generated by a secret key owner. In MPC, multiple parties can build an interactive protocol to evaluate a circuit without revealing an auxilarity information beyond the computation result, but it usually suffers from a high communication and round complexity.

© International Association for Cryptologic Research 2019
S. D. Galbraith and S. Moriai (Eds.): ASIACRYPT 2019, LNCS 11922, pp. 446–472, 2019.
https://doi.org/10.1007/978-3-030-34621-8_16

López-Alt et al. [28] proposed the notion of Multi-Key Homomorphic Encryption (MKHE) which is a variant of HE supporting computation on ciphertexts encrypted under different keys. This attractive primitive can address the aforementioned issues of HE and MPC, and it has many applications such as round-efficient MPC (e.g. [2,18,24,30,33]) and spooky encryption [19]. There have been several researches (e.g. [7,10,17,30,31]) on MKHE. However, all the previous works were purely abstract and far from practical. In particular, the efficiency of MKHE remained an open question for years because there has been no study to implement or compare the MKHE schemes empirically.

Chillotti et al. [13] proposed an HE scheme (called TFHE) based on the learning with errors (LWE) [32] assumption and its ring variant (RLWE) [29]. This HE scheme can evaluate an arbitrary binary gate on encrypted bits followed by a bootstrapping. TFHE has advantages in running time and usability compared to other HE schemes. Its bootstrapping has a low-latency and it makes simpler the task of implementing a binary circuit without background knowledge on HE (every gate of the plaintext circuit can be automatically replaced by its bootstrapped homomorphic version).

The TFHE scheme supports an operation called *external product*, which multiplies a Ring GSW (RGSW) ciphertext to an RLWE ciphertext and returns an RLWE ciphertext. A bootstrapping key consists of several RGSW encryptions, and each of them is recursively multiplied to an RLWE ciphertext to refresh it. In the multi-key case, the main difference is that we take a multi-key ciphertext as the input of bootstrapping. Hence we should be able to multiply a bootstrapping key, which is generated by a single party, to a multi-key ciphertext, which is associated with multiple parties.

We propose an RGSW-like cryptosystem and present two methods to multiply a single-key encryption to a multi-key RLWE ciphertext. The first algorithm consists of two phases: generation of a multi-key RGSW ciphertext and multi-key external product. It is similar to previous ciphertext extension method (firstly proposed by Clear and McGoldrick [17] and simplified by Mukherjee and Wichs [30]), but our scheme is simpler, lighter and faster. Our second algorithm for hybrid product is a completely different approach to achieve the same functionality. A single-key ciphertext directly acts on a multi-key RLWE ciphertext without any expensive multi-key RGSW operation. It achieves even better asymptotic complexity and less noise growth, and thereby improves the overall performance.

In summary, the length of ciphertext and the computational costs of a single binary gate grow linearly and quadratically on the number of involved parties, respectively (see Table 1 for comparison). Furthermore, our scheme is easy to implement and compatible with existing techniques for advanced functionalities such as the threshold decryption [3,26], circuit bootstrapping [14], and plaintext packing [4,8].

Finally, we provide a proof-of-concept implementation with concrete parameter sets. For example, it took about 0.27, 1.45 and 7.16 s to evaluate a bootstrapped NAND gate when the number of parties is 2, 4 and 8, respectively, on a personal computer.

Overview of Our Scheme. We adapt the formalization of (R)LWE over the real torus $\mathbb{T} = \mathbb{R}$ (mod 1) from Chillotti et al. [13]. We generalize TFHE to support the homomorphic computation on ciphertexts encrypted under independently generated keys. Let $R = \mathbb{Z}[X]/(X^N + 1)$ and $T = \mathbb{T}[X]/(X^N + 1)$ for a power-of-two integer N. We use a gadget vector $\mathbf{g} = (B^{-1}, \ldots, B^{-d}) \in \mathbb{Z}^d$ for some base B and degree d.

Each party (of index i) independently generates the LWE secret $\mathbf{s}_i \in \{0,1\}^n$ and the RLWE secret $z_i \in R$. A multi-key encryption of $m \in \{0,1\}$ is a vector of the form $\overline{\mathsf{ct}} = (b, \mathbf{a}_1, \ldots, \mathbf{a}_k) \in \mathbb{T}^{kn+1}$ such that $b + \langle \mathbf{a}_1, \mathbf{s}_1 \rangle + \cdots + \langle \mathbf{a}_k, \mathbf{s}_k \rangle \approx \frac{1}{4}m$ (mod 1) where k denotes the number of involved parties and $\mathbf{s}_i = (s_{i,j})_{1 \leq j \leq n}$ are their LWE secrets. The homomorphic evaluation of a NAND gate consists in an initial linear combination followed by a bootstrapping that takes care of the non-linear part of the gate together with the noise reduction. In particular, the noise reduction is performed by homomorphically computing the decryption formula (on the exponent of X) and by selecting the correct output of the bootstrapping encoded in a fixed test polynomial. The following three steps describe in more detail the NAND evaluation idea: for more details on the original TFHE bootstrapping we refer to [15].

First, we evaluate the linear combination for the NAND gate $m = m_1 \barwedge m_2$ on encrypted bits m_1, m_2 and return a ciphertext $\overline{\mathsf{ct}}' = (b', \mathbf{a}'_1, \ldots, \mathbf{a}'_k)$ satisfying $b' + \sum_{i=1}^k \langle \mathbf{a}'_i, \mathbf{s}_i \rangle \approx \frac{1}{2}m$ (mod 1). The evaluation is done after arranging the entries and extending the dimension of input ciphertexts to share the same secret.

In the second step, we extract the most significant bits $\tilde{b} = \lfloor 2N \cdot b' \rceil$ and $\tilde{\mathbf{a}}_i = \lfloor 2N \cdot \mathbf{a}'_i \rceil$, and initialize the accumulator $\overline{\mathbf{c}} = (-\frac{1}{8}X^{\tilde{b}} \cdot h(X), \mathbf{0}) \in T^{k+1}$ for the testing polynomial $h(X) = \sum_{-N/2 < d < N/2} X^d$ which is a multi-key RLWE encryption with respect to the concatenated RLWE secret $\overline{\mathbf{z}} = (1, z_1, \ldots, z_k) \in R^{k+1}$. Then, we evaluate Mux gates (data selector) recursively to obtain an RLWE encryption of $-\frac{1}{8}X^{\tilde{b}+\sum_{i=1}^k \langle \tilde{\mathbf{a}}_i, \mathbf{s}_i \rangle} \cdot h(X)$ using encryptions of $s_{i,j} \in \{0,1\}$.

Finally, from the output of accumulator, we extract an LWE encryption $\overline{\mathsf{ct}}^* = (b^*, \mathbf{a}_1^*, \ldots, \mathbf{a}_k^*)$ such that $b^* + \sum_{i=1}^k \langle \mathbf{a}_i^*, \mathbf{z}_i^* \rangle \approx \frac{1}{4}m$ (mod 1) where $\mathbf{z}_i^* \in \mathbb{Z}^N$ is a permuted coefficient vector of z_i. Finally, we perform the multi-key-switching procedure from $(\mathbf{z}_1^*, \ldots, \mathbf{z}_k^*)$ to $(\mathbf{s}_1, \ldots, \mathbf{s}_k)$ by repeating the ordinary key-switching procedure from $\mathbf{z}_i^*$ to $\mathbf{s}_i$.

The main difference between TFHE and our multi-key variant is in the second step. In the multi-key case, the i-th bootstrapping key (encryptions of $s_{i,j}$ for $1 \leq j \leq n$) is generated by a single party but we should multiply it to a multi-key RLWE ciphertext. We propose an RLWE-based scheme (called uni-encryption) which supports this hybrid product. In the key generation phase, each party takes a Common Reference String (CRS) $\mathbf{a} \in T^d$ and set a public key $\mathbf{b}_i \approx -z_i \cdot \mathbf{a}$ (mod 1). A party i can uni-encrypt a plaintext $\mu_i \in R$ into a ciphertext $(\mathbf{d}_i, \mathbf{F}_i = [\mathbf{f}_{i,0}|\mathbf{f}_{i,1}]) \in T^d \times T^{d \times 2}$ such that $\mathbf{d}_i \approx r_i \cdot \mathbf{a} + \mu_i \cdot \mathbf{g}$ (mod 1) and $\mathbf{f}_0 + z_i \cdot \mathbf{f}_1 \approx r_i \cdot \mathbf{g}$ (mod 1). Our hybrid product function multiplies a

Table 1. Memory (bit-size) and computational costs (number of scalar operations) of MKHE schemes. k denotes the number of parties and n is the dimension of the (R)LWE assumption. PK and EVK denote the public and evaluation (or bootstrapping) keys, respectively.

Scheme	Space		Time		Bootstrap
	Type	Complexity	Type	Complexity	
CZW17 [10]	EvalKey	$\tilde{O}(k^3 n)$	EvalKey Gen	$\tilde{O}(k^3 n)$	No
	Ciphertext	$\tilde{O}(kn)$	Hom Mult	$\tilde{O}(k^3 n)$	
PS16 #2 [31]	PK	$\tilde{O}(kn^4)$	Hom Mult	$\tilde{O}(k^{2.37} n^{2.37})$	No
	Ciphertext	$\tilde{O}(k^2 n^2)$			
BP16 [7]	PK	$\tilde{O}(kn^3)$	Hom NAND	$\text{poly}(k, n)$	Yes
	Ciphertext	$\tilde{O}(kn)$			
This work (Method 1)	Eval Key	$\tilde{O}(k^2 n^2)$	Eval Key Gen	$\tilde{O}(k^2 n^2)$	Yes
	Ciphertext	$\tilde{O}(kn)$	Hom NAND	$\tilde{O}(k^2 n^2)$	
This work (Method 2)	Ciphertext	$\tilde{O}(kn)$	Hom NAND	$\tilde{O}(k^2 n^2)$	Yes

uni-encryption of μ_i to a multi-key RLWE encryption $\overline{c} \in T^{k+1}$ and returns a multi-key RLWE ciphertext, i.e., the output $\overline{c}' \in T^{k+1}$ satisfies that $\langle \overline{c}', \overline{z} \rangle \approx \mu_i \cdot \langle \overline{c}, \overline{z} \rangle$ (mod 1). We propose two different algorithms to achieve this functionality.

Our first hybrid product algorithm is an improvement of the GSW extension algorithm in previous work [7,17,30]. It aims to transform a uni-encryption of μ_i into a multi-key RGSW encryption $\overline{D}_i \in T^{d(k+1) \times (k+1)}$ of the same message under the concatenated key $\overline{z} \in R^{k+1}$ satisfying $\overline{D}_i \overline{z} \approx \mu_i \cdot (\mathbf{I}_{k+1} \otimes \mathbf{g})$ in $T^{d(k+1)}$. Then, we can perform the multi-key external product between $\overline{c}$ and $\overline{D}_i$ to multiply them. Compared to previous algorithm, we reduce the dimension of ciphertexts from $2k$ down to $(k+1)$ by merging duplicated components $(1, z_1, \ldots, 1, z_k)$ into $\overline{z} = (1, z_1, \ldots, z_k)$. In addition, we observe that the uni-encryption is not used for encrypting real messages, but only for generating a bootstrapping key. Hence we propose a symmetric key encryption to reduce the size of ciphertexts and complexity of extension algorithm. However, the first method does not change the asymptotic complexity $O(kd^2 \cdot N \log N)$ of extension process (see Sect. 3.2 for details).

We propose a new framework in our second algorithm for hybrid product. The previous GSW extension is done independently from the input multi-key RLWE ciphertext $\overline{c}$. Instead, we work on $\overline{c}$ directly to avoid expensive multi-key RGSW operations. There are two main advantages of this approach: its complexity $O(kd \cdot N \log N)$ is asymptotically better and the noise variance is reduced by a factor of $O(d \cdot B^2)$. For these reasons, we used the second algorithm in our implementation.

Related Works. López-Alt et al. [28] firstly proposed an MKHE scheme based on the NTRU assumption. Clear and McGoldrick [17] introduced an LWE-based construction, and it was significantly simplified by Mukherjee and Wichs [30]. These schemes are *single-hop* for keys where the list of parties has to be known before the computation starts. This work was improved in concurrent researches by Peikert-Shiehian [31] and Brakerski-Perlman [7] which design multi-hop (dynamic for keys) MKHEs. Chen, Zhang and Wang [10] constructed a scheme which can encrypt a ring element compared to a single bit of prior works. Unfortunately, there have been no research with implementation results because all previous schemes were impractical.

We summarize the performance of relatively efficient MKHE schemes in Table 1. We only consider the second (main) one between two schemes described in [31]. All existing schemes except [7] and [10] use variants of the GSW scheme to encrypt plaintexts. Therefore, the size of ciphertexts grows at least quadractically on the number k of parties in the computation.

Similar to our scheme, [7] encrypts a bit in a single LWE ciphertext. However, they proposed a purely abstract bootstrapping based on the evaluation of a huge branching program of length $L = \text{poly}(k, n)$ representing the NAND gate followed by LWE decryption. A memory-complexity tradeoff was proposed to keep a linear storage requirement on k, but even the asymptotic complexity of bootstrapping is not analyzed in the paper.

The construction of a batched MKHE scheme is an orthogonal research issue. Chen et al. [10] proposed a multi-key variant of BGV [6] with a larger plaintext space. However, it is a leveled scheme so a large constant (depending on the maximum level of a circuit to be evaluated) is hidden in the $\tilde{O}(\cdot)$ notation. Moreover, the space and time complexity of homomorphic multiplication grow rapidly as the number of parties increases. Its complexity is quasi-linear on the security parameter, however, our scheme can be implemented using a smaller parameter.

Since [7] and [10] use the GSW extension to generate evaluation (bootstrapping) keys, our improved (compact and symmetric) method can be directly applied to these schemes for better performance.

2 Background

2.1 Notation

All logarithms are in base two unless otherwise indicated. We denote vectors in bold, e.g. $\mathbf{a}$, and matrices in upper-case bold, e.g. $\mathbf{A}$. We denote by $\langle \cdot, \cdot \rangle$ the usual dot product of two vectors. For a real number r, $\lfloor r \rceil$ denotes the nearest integer to r, rounding upwards in case of a tie. We use $x \leftarrow D$ to denote the sampling x according to distribution D. For a finite set S, $U(S)$ denotes the uniform distribution on S. For a real $\alpha > 0$, D_α denotes the Gaussian distribution of variance α^2. We let λ denote the security parameter throughout the paper: all known valid attacks against the cryptographic scheme under scope should take

$\Omega(2^\lambda)$ bit operations. For a positive integer k, $[k] = \{1, 2, \ldots, k\}$ denotes the index set.

2.2   Multi-Key Homomorphic Encryption

A multi-key homomorphic encryption MKHE consists of five PPT algorithms Setup, KeyGen, Enc, Dec, and NAND.

- $pp \leftarrow$ MKHE.Setup(1^λ): Given the security parameter λ, returns a public parameter pp.
- $(sk, pk) \leftarrow$ MKHE.KeyGen(pp): Generates its secret and public keys. We assume that each party has its own ID (index) mapped to the keys.
- ct $\leftarrow$ MKHE.Enc($m; pk$): Given a bit $m \in \{0, 1\}$, returns a ciphertext ct $\in \{0, 1\}^*$. We assume that every ciphertext contains IDs of relevant parties.
- $m \leftarrow$ MKHE.Dec($\overline{\text{ct}}; \{sk_i\}_{i \in [k]}$): Given a ciphertext $\overline{\text{ct}}$, let $\{sk_i\}_{i \in [k]}$ be the sequence of secret keys of relevant parties. Decrypts the ciphertext into a bit $m \in \{0, 1\}$.
- $\overline{\text{ct}}' \leftarrow$ MKHE.NAND($\overline{\text{ct}}_1, \overline{\text{ct}}_2, \{pk_i\}_{i \in [k]}$): Given ciphertexts $\overline{\text{ct}}_1$ and $\overline{\text{ct}}_2$, let k be the number of parties relevant to either $\overline{\text{ct}}_1$ or $\overline{\text{ct}}_2$, and $\{pk_i\}_{i \in [k]}$ be the sequence of their public keys. Evaluates the NAND gate and returns a ciphertext $\overline{\text{ct}}'$. The output ciphertext implicitly includes k indices of related parties.

An MKHE scheme is called secure if its encryption is semantically secure. The output $\overline{\text{ct}}' \leftarrow$ MKHE.NAND($\overline{\text{ct}}_1, \overline{\text{ct}}_2, \{pk_i\}_{i \in [k]}$) of homomorphic NAND should satisfy MKHE.Dec($\overline{\text{ct}}', \{sk_i\}_{i \in [k]}$) $= m_1 \overline{\wedge} m_2$ with an overwhelming probability if $\overline{\text{ct}}_1$ and $\overline{\text{ct}}_2$ are encryptions of m_1 and m_2, respectively.

2.3   TLWE and TRLWE

The TFHE scheme, presented for the first time in [13], is based on the TLWE (resp. TRLWE) problem, which is the torus variant of the LWE (resp. RLWE) problem. Instead of working over $\mathbb{Z}/q\mathbb{Z}$, or over the ring $\mathbb{Z}[X]/(X^N + 1)$ modulo q in the ring variant, in TFHE we work over the real Torus $\mathbb{T} = \mathbb{R} \bmod 1$ and over $T = \mathbb{T}[X]/(X^N + 1)$, the set of cyclotomic polynomials over $\mathbb{T}$ for a power-of-two integer N. In this section and in the following one we present an overview of the TFHE scheme: for more details we refer to [15].

We denote by $R = \mathbb{Z}[X]/(X^N + 1)$ the set of cyclotomic polynomials over $\mathbb{Z}$. Then, we observe that $\mathbb{T}$ and T are modules over $\mathbb{Z}$ and R, respectively. This means that they are groups with respect to the addition and they are provided with an external product by an integer or an integer polynomial.

A TLWE sample is a pair $(b, \mathbf{a}) \in \mathbb{T}^{n+1}$, where $\mathbf{a}$ is sampled uniformly over $\mathbb{T}^n$ and $b = \langle \mathbf{a}, \mathbf{s} \rangle + e$. The secret key $\mathbf{s}$ and error e are sampled from a key distribution χ on $\mathbb{Z}^n$ and a Gaussian with standard deviation $\alpha > 0$.

By following the same path, a TRLWE sample is a pair of polynomials $(b, a) \in T^2$, where a is sampled uniformly from T and $b = a \cdot z + e \pmod 1$ for an error e. The secret key z is an integer polynomial of degree N sampled from a key distribution ψ on R and the error polynomial e is sampled from a Gaussian distribution with standard deviation β. We will set ψ as the uniform distribution on the set of polynomials of R with binary coefficients in $\{0,1\}$. For $a, b \in R$ (resp. T), we denote by $a \approx b \pmod 1$ if $a = b + e \pmod 1$ for a small error $e \in R$ (resp. $\mathbb{R}[X]/(X^N + 1)$).

We can then define two problems for both TLWE and TRLWE:

- Decision problem: for a fixed TLWE secret $\mathbf{s}$ (resp. TRLWE secret z), distinguish the uniform distribution over $\mathbb{T}^{n+1}$ (resp. T^2) from the TLWE (resp. TRLWE) samples.
- Search problem: given arbitrarily many samples from the TLWE (resp. TRLWE) distribution, find the secret $\mathbf{s}$ (resp. z).

TLWE samples can be used to encrypt Torus messages. By fixing the message space as a discrete subset $\mathcal{M} \subseteq \mathbb{T}$, a message $\mu \in \mathcal{M}$ can be encrypted by adding the trivial TLWE sample $(\mu, \mathbf{0})$ to a TLWE sample generated as described in previous paragraphs. Then, the corresponding ciphertext ct is a pair $(b, \mathbf{a}) \in \mathbb{T}^{n+1}$, with $b = -\langle \mathbf{a}, \mathbf{s} \rangle + e + \mu$. In order to decrypt, we compute the phase $\varphi_\mathbf{s}$ of the ciphertext ct, which is equal to $\varphi_\mathbf{s}(\mathsf{ct}) = b + \langle \mathbf{a}, \mathbf{s} \rangle$, and we approximate it to the nearest message possible in $\mathcal{M}$ to retrieve μ. By following the same footstep, we can use TRLWE samples to encrypt torus polynomial messages in T.

Thanks to the $\mathbb{Z}$-module structure of the torus and to the R-module structure of T, the TLWE and TRLWE samples have additive homomorphic properties. The external integer homomorphic multiplication can be performed thanks to the TRGSW ciphertexts we define in the next section.

2.4   TRGSW and External Product

For a base integer $B \geq 2$ and a degree d, we call $\mathbf{g} = (B^{-1}, \ldots, B^{-d})$ the *gadget vector*. For an integer $k \geq 1$, the gadget matrix is defined by

$$\mathbf{G}_k = \mathbf{I}_k \otimes \mathbf{g} = \begin{bmatrix} \mathbf{g} & 0 & \ldots & 0 \\ 0 & \mathbf{g} & \ldots & 0 \\ \vdots & \vdots & \ddots & \vdots \\ 0 & 0 & \ldots & \mathbf{g} \end{bmatrix} \in \mathbb{T}^{dk \times k}.$$

For any $\mathbf{u} \in \mathbb{T}^k$, we define its base decomposition by a dk-dimensional vector $\mathbf{v} = \mathbf{G}_k^{-1}(\mathbf{u})$ with coefficients in $\mathbb{Z} \cap (-B/2, B/2]$ which minimizes $\|\mathbf{v}^T \cdot \mathbf{G}_k - \mathbf{u}^T\|_\infty$. The decomposition error $\|\mathbf{v}^T \cdot \mathbf{G}_k - \mathbf{u}^T\|_\infty$ is bounded by $\frac{1}{2}B^{-1}$.

We identify an arbitrary element of T to the vector of its coefficients in $\mathbb{T}^N$, and naturally extend the base decomposition $\mathbf{G}_k^{-1}(\cdot)$ to a function $T^k \to R^{dk}$ by applying the basic decomposition function coefficient wisely.

Then, we can define the TRGSW samples as the torus variant of RGSW samples, in the same way as we did in previous section[1]. For a fixed TRLWE

---

[1] We define only the Ring version TRGSW, since this is the only sample we need in this paper. TGSW can be defined in the same way. For more details we refer to [15].

secret $s \in R$, we define a TRGSW sample as $\mathbf{C} = \mathbf{Z} + \mu \cdot \mathbf{G}_2$, where each line of the matrix $\mathbf{Z} \in T^{d \times 2}$ is a TRLWE encryption of 0, $\mathbf{G}_2$ is the gadget matrix and the message $\mu \in R$ is an integer polynomial.

TRGSW samples are homomorphic with respect to the addition and to an internal multiplication. Furthermore, an external product, noted $\boxdot$, with TRLWE can be defined as $\mathbf{A} \boxdot \mathbf{b} = \mathbf{G}_2^{-1}(\mathbf{b}) \cdot \mathbf{A}$, for all TRLWE samples $\mathbf{b}$ and TRGSW samples $\mathbf{A}$ encrypted with the same secret key. In the following sections, we define a variant of the TRGSW samples and an adapted external product. The internal product between two TRGSW samples $\mathbf{A}$ and $\mathbf{B}$ encrypted with the same secret key can be defined as a list of independent external products between the cipher $\mathbf{A}$ and the lines composing the cipher $\mathbf{B}$.

The scheme TFHE has been implemented and is publicly available at [16]. In Sect. 5 we present some experimental results we obtained by implementing our Multi-Key scheme on top of the TFHE library.

In the rest of the paper, in order to lighten the notations, we will abandon the 'T' notation in front of LWE, RLWE and RGSW.

3   Basic Schemes

In this section, we present the LWE [32] and RGSW [20,23] schemes and describe some extended algorithms that will be used in our MKHE scheme.

3.1   Multi-Key-Switching on LWE Ciphertexts

We first describe the standard LWE-based scheme and generalize its key-switching algorithm to the multi-key case.

- LWE.Setup(1^λ): It takes the security parameter as input and generates the LWE dimension n, key distribution χ, error parameter α. Set the decomposition base B' and degree d' for gadget vector $\mathbf{g}' = (B'^{-1}, \ldots, B'^{-d'})$. Return the public parameter $pp^{\text{LWE}} = (n, \chi, \alpha, B', d')$.

An LWE secret $\mathbf{s}$ is sampled from the distribution χ. We use the key-switching gadget vector $\mathbf{g}' = (B'^{-1}, \ldots, B'^{-d'})$. Recall that the base decomposition algorithm with respect to $\mathbf{g}'$ transforms an element $a \in \mathbb{T}$ into the d'-dimensional vector $\mathbf{g}'^{-1}(a)$ with coefficients in $\mathbb{Z}_{B'}$ which minimizes $|a - \langle \mathbf{g}'^{-1}(a), \mathbf{g}' \rangle|$.

We assume that the following LWE algorithms implicitly takes pp^{LWE} as an input.

- LWE.KeyGen(): Sample the LWE secret $\mathbf{s} \leftarrow \chi$.
- LWE.Enc($m, \mathbf{s}$): This is a standard LWE encryption which takes a bit $m \in \{0, 1\}$ as an input. It samples $a \leftarrow U(\mathbb{T}^n)$ and $e \leftarrow D_\alpha$, and returns the ciphertext $\mathsf{ct} = (b, \mathbf{a}) \in \mathbb{T}^{n+1}$ where $b = -\langle \mathbf{a}, \mathbf{s} \rangle + \frac{1}{4}m + e \pmod{1}$.

Note that the scaling factor is $1/4$, as in FHEW [20] or TFHE [13]. We described a symmetric encryption for simplicity, but this algorithm can be replaced by any LWE-style encryption schemes such as public key encryption [27]. The only requirement is that the output ciphertext should be a vector $\mathsf{ct} = (b, \mathbf{a}) \in \mathbb{T}^{n+1}$ satisfying $b + \langle \mathbf{a}, \mathbf{s} \rangle \approx \frac{1}{4} m \pmod{1}$.

- $\mathsf{LWE.KSGen}(\mathbf{t}, \mathbf{s})$: Given LWE secrets $\mathbf{t} \in \mathbb{Z}^N$ and $\mathbf{s} \in \mathbb{Z}^n$, it returns the key-switching key $\mathsf{KS} = \{\mathbf{K}_j\}_{j \in [N]} \in (\mathbb{T}^{d' \times (n+1)})^N$ from $\mathbf{t}$ to $\mathbf{s}$. For each $j \in [N]$, the j-th entry is generated by sampling $\mathbf{A}_j \leftarrow U(\mathbb{T}^{d' \times n})$ and $\mathbf{e}_j \leftarrow D_\beta^{d'}$, and returning $\mathbf{K}_j = [\mathbf{b}_j | \mathbf{A}_j]$ where $\mathbf{b}_j = -\mathbf{A}_j \mathbf{s} + \mathbf{e}_j + t_j \cdot \mathbf{g}' \pmod{1}$.

We can transform an LWE ciphertext corresponding to $\mathbf{t}$ into another LWE encryption of the same message under the secret $\mathbf{s}$ using a key-switching key $\mathsf{KS} \leftarrow \mathsf{LWE.KSGen}(\mathbf{t}, \mathbf{s})$.

We consider the notion of extended LWE encryption and the multi-key-switching procedure. For k LWE secrets $\mathbf{s}_1, \ldots, \mathbf{s}_k \in \mathbb{Z}^n$, an extended cipher-text $\overline{\mathsf{ct}} = (b, \mathbf{a}_1, \ldots, \mathbf{a}_k) \in \mathbb{T}^{kn+1}$ will be called an encryption of $m \in \{0,1\}$ with respect to the concatenated secret $\overline{\mathbf{s}} = (\mathbf{s}_1, \ldots, \mathbf{s}_k)$ if $\langle \overline{\mathsf{ct}}, (1, \overline{\mathbf{s}}) \rangle = b + \sum_{i=1}^k \langle \mathbf{a}_i, \mathbf{s}_i \rangle \approx \frac{1}{2} m \pmod{1}$.

- $\mathsf{LWE.MKSwitch}(\overline{\mathsf{ct}}, \{\mathsf{KS}_i\}_{i \in [k]})$: Given a ciphertext $\mathsf{ct} = (b, \mathbf{a}_1, \ldots, \mathbf{a}_k) \in \mathbb{T}^{kN+1}$ and a sequence of the key-switching keys $\mathsf{KS}_i = \{\mathbf{K}_{i,j}\}_{j \in [N]}$, compute $(b_i', \mathbf{a}_i') = \sum_{j=1}^N \mathbf{g}'^{-1}(a_{i,j}) \cdot \mathbf{K}_{i,j} \pmod{1}$ for all $i \in [k]$ and let $b' = b + \sum_{i=1}^k b_i'$ $\pmod{1}$. Return the ciphertext $\overline{\mathsf{ct}}' = (b', \mathbf{a}_1', \ldots, \mathbf{a}_k') \in \mathbb{T}^{kn+1}$.

This multi-key-switching algorithm takes as the input an extended ciphertext $\overline{\mathsf{ct}} \in \mathbb{T}^{kN+1}$ corresponding to $\overline{\mathbf{t}} = (\mathbf{t}_1, \ldots, \mathbf{t}_k)$ and a sequence of key-switching keys from $\mathbf{t}_i$ to $\mathbf{s}_i$ and returns an encryption of the same message under $\overline{\mathbf{s}} = (\mathbf{s}_1, \ldots, \mathbf{s}_k)$.

Security. The j-th component $\mathbf{K}_j$ of a key-switching key $\mathsf{KS} = \{\mathbf{K}_j\}_{j \in [N]}$ from $\mathbf{t} \in \mathbb{Z}^N$ to $\mathbf{s} \in \mathbb{Z}^n$ is generated by adding $t_j \cdot \mathbf{g}'$ to the first column of a matrix in $\mathbb{T}^{d' \times (n+1)}$ whose rows are LWE instances under the secret $\mathbf{s}$. Therefore, $\mathsf{KS} \leftarrow \mathsf{LWE.KSGen}(\mathbf{t}, \mathbf{s})$ is computationally indistinguishable from the uniform distribution over $(\mathbb{T}^{d' \times (n+1)})^N$ under the LWE assumption with parameter (n, χ, β) if $\mathbf{s}$ is sampled according to χ.

Correctness. We show that if $\mathsf{ct} = (b, \mathbf{a}_1, \ldots, \mathbf{a}_k)$ is an LWE ciphertext encrypted by $\overline{\mathbf{t}} = (\mathbf{t}_1, \ldots, \mathbf{t}_k)$ and $\{\mathsf{KS}_i\}_{i \in [k]}$ are key-switching keys from $\mathbf{t}_i \in \mathbb{Z}^N$ to $\mathbf{s}_i \in \mathbb{Z}^n$, then the output ciphertext encrypts the same message under the concatenated secret $\overline{\mathbf{s}} = (\mathbf{s}_1, \ldots, \mathbf{s}_k)$. The correctness of this algorithm is simply shown by the following equation:

$$\langle \overline{\mathsf{ct}}', (1, \overline{\mathbf{s}}) \rangle = b + \sum_{i=1}^{k} (b_i' + \langle \mathbf{a}_i', \mathbf{s}_i \rangle)$$

$$\approx b + \sum_{i=1}^{k} \sum_{j=1}^{N} \langle \mathbf{g}'^{-1}(a_{i,j}), t_{i,j} \cdot \mathbf{g}' \rangle \approx \langle \overline{\mathsf{ct}}, (1, \overline{\mathbf{t}}) \rangle \pmod 1.$$

Therefore, $\overline{\mathsf{KS}} = \{\mathsf{KS}_i\}_{i \in [k]}$ can be considered as a key-switching key from $\overline{\mathbf{t}} \in \mathbb{Z}^{kN}$ to $\overline{\mathbf{s}} \in \mathbb{Z}^{kn}$.

3.2   Multi-key RLWE and Hybrid Product

In this section, we present a ring-based scheme supporting two algorithms UniEnc and Prod. First, UniEnc is a *single-key* symmetric encryption which can encrypt a ring element. We can multiply a uni-encryption to a *multi-key* RLWE ciphertext using the hybrid product algorithm Prod. In fact, we provide two different methods to perform this operation. We will explain their performance and (dis)advantages later.

- RLWE.Setup(1^λ): It takes input the secret parameter λ.
  1. Set the RLWE dimension N which is a power of two.
  2. Set the key distribution ψ over R and choose the error parameter α.
  3. Set the base integer $B \geq 2$ and the decomposition degree d for the gadget vector $\mathbf{g} = (B^{-1}, \ldots, B^{-d})$.
  4. Generate a random vector $\mathbf{a} \leftarrow U(T^d)$.
  Return the public parameter $pp^{\mathtt{RLWE}} = (N, \psi, \alpha, B, d, \mathbf{a})$.

Our RLWE-based scheme is based on the CRS model since the public parameter $pp^{\mathtt{RLWE}}$ contains a CRS $\mathbf{a} \in T^d$. The parameter should be chosen appropriately so that the RLWE problem with parameter (N, ψ, α) achieves at least λ-bit security level. We assume that the following RLWE algorithms implicitly takes $pp^{\mathtt{RLWE}}$ as an input.

- RLWE.KeyGen(): Sample the secret $z \leftarrow \psi$ and set $\mathbf{z} = (1, z)$. Sample an error vector $\mathbf{e} \leftarrow D_\alpha^d$ and set the public key as $\mathbf{b} = -z \cdot \mathbf{a} + \mathbf{e} \pmod 1$. Return $(z, \mathbf{b}) \in R \times T^d$.
- RLWE.UniEnc(μ, z): For an input plaintext $\mu \in R$ and a secret key z, it generates and returns the ciphertexts $(\mathbf{d}, \mathbf{F}) \in T^d \times T^{d \times 2}$ as follows:
  1. Sample $r \leftarrow \psi$ and an error $\mathbf{e}_1 \leftarrow D_\alpha^d$. Output the vector $\mathbf{d} = r \cdot \mathbf{a} + \mu \cdot \mathbf{g} + \mathbf{e}_1 \in T^d$.
  2. Sample $\mathbf{f}_1 \leftarrow U(T^d)$ and $\mathbf{e}_2 \leftarrow D_\alpha^d$. Output the ciphertext $\mathbf{F} = [\mathbf{f}_0 | \mathbf{f}_1] \in T^{d \times 2}$ where $\mathbf{f}_0 = -z \cdot \mathbf{f}_1 + r \cdot \mathbf{g} + \mathbf{e}_2 \pmod 1$.

A uni-encryption consists of three polynomial vectors of dimension d, three-fourth the size of ordinary RGSW ciphertexts in $T^{2d \times 2}$. The first component $\mathbf{d}$ and the CRS $\mathbf{a}$ together form an encryption of μ under the randomness r. We can consider $\mathbf{F}$ as an encryption of r under the secret z. In the following, we describe two different algorithms for hybrid product.

- $\mathtt{RLWE.Prod}(\overline{\mathbf{c}}(\mathbf{d}_i, \mathbf{F}_i), \{\mathbf{b}_j\}_{j \in [k]})$: Given a multi-key RLWE ciphertext $\overline{\mathbf{c}} \in T^{k+1}$ and the public keys of k parties associated to $\overline{\mathbf{c}}$, multiply a uni-encryption $(\mathbf{d}_i, \mathbf{F}_i)$ encrypted by the i-th party to $\overline{\mathbf{c}}$ as follows. We use the notation $z_0 = 1$ and $\mathbf{b}_0 = -\mathbf{a}$ in this algorithm (in the context of $\mathbf{b}_j \approx -z_j \cdot \mathbf{a}$).

Method 1. The first method consists of two steps. We first generate an extended RGSW ciphertext $\overline{\mathbf{D}}_i \in T^{d(k+1) \times (k+1)}$ by combining a uni-encryption $(\mathbf{d}_i, \mathbf{F}_i)$ and the set of public keys $\{\mathbf{b}_j\}_{j \in [k]}$, then multiply it to $\overline{\mathbf{c}}$ using the multi-key external product.

Step 1. Ciphertext Extension. $\overline{\mathbf{D}}_i \leftarrow \mathtt{RLWE.Extend}((\mathbf{d}_i, \mathbf{F}_i), \{\mathbf{b}_j\}_{j \in [k]})$: For $0 \le j \le k$, compute the vectors $\mathbf{x}_j, \mathbf{y}_j \in R_Q^d$ by $\mathbf{x}_j[\ell] = \langle \mathbf{g}^{-1}(\mathbf{b}_j[\ell]), \mathbf{f}_0 \rangle$ and $\mathbf{y}_j[\ell] = \langle \mathbf{g}^{-1}(\mathbf{b}_j[\ell]), \mathbf{f}_1 \rangle$ for all $\ell \in [d]$, i.e., $[\mathbf{x}_j | \mathbf{y}_j] = \mathbf{M}_j \mathbf{F}_i \in T^{d \times 2}$ where $\mathbf{M}_j \in R^{d \times d}$ is the matrix of which ℓ-th row vector is $\mathbf{g}^{-1}(\mathbf{b}_j[\ell])$.

Return the expanded ciphertext

$$\overline{\mathbf{D}}_i = \begin{bmatrix} \mathbf{d}_i + \mathbf{x}_0 & 0 & \cdots & \mathbf{y}_0 & \cdots & 0 \\ \mathbf{x}_1 & \mathbf{d}_i & \cdots & \mathbf{y}_1 & \cdots & 0 \\ \vdots & \vdots & \ddots & \vdots & \ddots & \vdots \\ \mathbf{x}_i & 0 & \cdots & \mathbf{d}_i + \mathbf{y}_i & \cdots & 0 \\ \vdots & \vdots & \ddots & \vdots & \ddots & \vdots \\ \mathbf{x}_k & 0 & \cdots & \mathbf{y}_k & \cdots & \mathbf{d}_i \end{bmatrix} \in T^{d(k+1) \times (k+1)}.$$

Step 2. Multi-key GSW External Product. $\overline{\mathbf{c}}' \leftarrow \overline{\mathbf{c}} \boxdot \overline{\mathbf{D}}_i$: For a multi-key RLWE ciphertext $\overline{\mathbf{c}} \in T^{k+1}$ and a multi-key RGSW ciphertext $\overline{\mathbf{D}}_i \in T^{d(k+1) \times (k+1)}$, return the ciphertext $\overline{\mathbf{c}}' = \mathbf{G}_{k+1}^{-1}(\overline{\mathbf{c}}) \cdot \overline{\mathbf{D}}_i \pmod{1}$.

Method 2. Given a multi-key RLWE ciphertext $\overline{\mathbf{c}} = (c_0, c_1, \ldots, c_k) \in T^{k+1}$, we first compute the following polynomials for all $0 \le j \le k$:

$$u_j = \langle \mathbf{g}^{-1}(c_j), \mathbf{d}_i \rangle,$$
$$v_j = \langle \mathbf{g}^{-1}(c_j), \mathbf{b}_j \rangle,$$
$$w_{j,0} = \langle \mathbf{g}^{-1}(v_j), \mathbf{f}_{i,0} \rangle,$$
$$w_{j,1} = \langle \mathbf{g}^{-1}(v_j), \mathbf{f}_{i,1} \rangle.$$

Then, we return the multi-key RLWE ciphertext $\bar{c}' = (c_0', \ldots, c_k') \in T^{k+1}$ where $c_0' = u_0 + \sum_{j=0}^{k} w_{j,0} \pmod 1$, $c_i' = u_i + \sum_{j=0}^{k} w_{j,1} \pmod 1$, and $c_j' = u_j$ for $j \in [k] \backslash \{i\}$.

Security. We claim that the distribution

$$\mathcal{D}_0 = \{(\mathbf{a}, \mathbf{b}, \mathbf{d}, \mathbf{F}) : pp^{\text{RLWE}} \leftarrow \text{RLWE.Setup}(1^\lambda),$$
$$(z, \mathbf{b}) \leftarrow \text{RLWE.KeyGen}()(\mathbf{d}, \mathbf{F}) \leftarrow \text{RLWE.UniEnc}(\mu, z)\}$$

is computationally indistinguishable from the uniform distribution over $T^{d \times 5}$ for any $\mu \in R$ under the RLWE assumption with parameter (N, ψ, α). We consider the following distributions: First, we can switch $(\mathbf{b}, \mathbf{a})$ and $\mathbf{F} = [\mathbf{f}_0 | \mathbf{f}_1]$ into independent uniform distributions on $T^{d \times 2}$ using the RLWE assumption of the secret z. Hence $\mathcal{D}_0$ is computationally indistinguishable from

$$\mathcal{D}_1 = \{(\mathbf{a}, \mathbf{b}, \mathbf{d}, \mathbf{F}) : \mathbf{a}, \mathbf{b} \leftarrow U(T^d), \mathbf{F} \leftarrow U(T^{d \times 2}),$$
$$r \leftarrow \psi, \mathbf{e}_1 \leftarrow D_\alpha^d, \mathbf{d} = r \cdot \mathbf{a} + \mu \cdot \mathbf{g} + \mathbf{e}_1 \pmod 1\}.$$

Then, $\mathbf{d}$ is changed to a uniform distribution using the RLWE assumption of secret r again. Therefore, $\mathcal{D}_1$ is indistinguishable from the distribution

$$\mathcal{D}_2 = \{(\mathbf{a}, \mathbf{b}, \mathbf{d}, \mathbf{F}) : \mathbf{a}, \mathbf{b}, \mathbf{d} \leftarrow U(T^d), \mathbf{F} \leftarrow U(T^{d \times 2})\}.$$

Since $\mathcal{D}_2$ is independent from μ, our RLWE scheme is semantically secure.

Correctness of Method 1. Let $(z_j, \mathbf{b}_j)$ be the RGSW key of the j-th party for $j \in [k]$. Suppose that $(\mathbf{d}_i, \mathbf{F}_i = [\mathbf{f}_{i,0} | \mathbf{f}_{i,1}])$ is a uni-encryption of $\mu_i \in R$ of the i-th party, i.e., $\mathbf{d}_i \approx r_i \cdot \mathbf{a} + \mu_i \cdot \mathbf{g} \pmod 1$ and $\mathbf{f}_{i,0} + z_i \cdot \mathbf{f}_{i,1} \approx r_i \cdot \mathbf{g} \pmod 1$ for some $r_i \leftarrow \psi$.

We call $\overline{\mathbf{D}} \in T^{d(k+1) \times (k+1)}$ a multi-key RGSW encryption of $\mu \in R$ under the concatenated secret $\bar{z} = (1, z_1, \ldots, z_k) \in R^{k+1}$ if $\overline{\mathbf{D}} \cdot \bar{z} \approx \mu \cdot \mathbf{G}_{k+1} \bar{z} \pmod 1$. We first claim that $\overline{\mathbf{D}}_i \leftarrow \text{RLWE.Extend}((\mathbf{d}_i, \mathbf{F}_i), \{\mathbf{b}_j\}_{j \in [k]})$ is a valid RGSW encryption of μ_i corresponding to $\bar{z}$. It suffices to show that $\mathbf{x}_j + z_i \cdot \mathbf{y}_j + z_j \cdot \mathbf{d}_i \approx \mu_i z_j \cdot \mathbf{g} \pmod 1$ for all $0 \le j \le k$. We can combine the following equations to obtain the desired result:

$$\mathbf{x}_j + z_i \cdot \mathbf{y}_j = \mathbf{M}_j \mathbf{F}_i z_i \approx \mathbf{M}_j (r \cdot \mathbf{g}) \approx r_i \cdot \mathbf{b}_j \pmod 1,$$
$$z_j \cdot \mathbf{d}_i \approx r_i z_j \cdot \mathbf{a} + \mu_i z_j \cdot \mathbf{g} \approx -r_i \cdot \mathbf{b}_j + \mu_i z_j \cdot \mathbf{g} \pmod 1.$$

The multi-key external product $\boxdot$ is a natural generalization of the ordinary external product between RLWE and RGSW ciphertexts to the multi-key setting [7,13]. Let us suppose that $\bar{c} \in T^{k+1}$ is an RLWE ciphertext and $\overline{\mathbf{D}}_i \in T^{d(k+1) \times (k+1)}$ is an RGSW encryption of μ_i with respect to the secret $\bar{z} \in R^{k+1}$, i.e., $\overline{\mathbf{D}}_i \bar{z} \approx \mu_i \cdot \mathbf{G}_{k+1} \bar{z} \pmod 1$. Then their external product $\bar{c}' = \bar{c} \boxdot \overline{\mathbf{D}}_i$ satisfies that $\langle \bar{c}', \bar{z} \rangle = \mathbf{G}_{k+1}^{-1}(\bar{c}) \cdot \overline{\mathbf{D}}_i \bar{z} \approx \mathbf{G}_{k+1}^{-1}(\bar{c}) \cdot \mu_i \mathbf{G}_{k+1} \bar{z} \approx \mu_i \cdot \langle \bar{c}, \bar{z} \rangle \pmod 1$, as desired.

Correctness of Method 2. We note that $\overline{\mathbf{c}}'$ is generated by adding $\sum_{j=0}^{k} w_{j,0}$ and $\sum_{j=0}^{k} w_{j,1}$ to the zeroth and i-th components of $(u_0, \ldots, u_k)$. Hence we have $\langle \overline{\mathbf{c}}', \overline{\mathbf{z}} \rangle = \sum_{j=0}^{k} u_j \cdot z_j + \sum_{j=0}^{k} (w_{j,0} + w_{j,1} \cdot z_i) \pmod 1$.

From the definition of u_j, v_j, $w_{j,0}$ and $w_{j,1}$, we obtain

$$\sum_{j=0}^{k} u_j \cdot z_j \approx \sum_{j=0}^{k} \langle \mathbf{g}^{-1}(c_j), r_i \cdot \mathbf{a} + \mu_i \cdot \mathbf{g} \rangle \cdot z_j \approx \mu_i \cdot \langle \overline{\mathbf{c}}, \overline{\mathbf{z}} \rangle - r_i \cdot \sum_{j=0}^{k} v_j \pmod 1,$$

$$\sum_{j=0}^{k} (w_{j,0} + w_{j,1} z_j) = \sum_{j=0}^{k} \langle \mathbf{g}^{-1}(v_j), \mathbf{f}_{i,0} + z_i \cdot \mathbf{f}_{i,1} \rangle \approx r_i \cdot \sum_{j=0}^{k} v_j \pmod 1,$$

and conclude that $\langle \overline{\mathbf{c}}', \overline{\mathbf{z}} \rangle \approx \mu_i \cdot \langle \overline{\mathbf{c}}, \overline{\mathbf{z}} \rangle$, as desired.

Performance. In the first method, the Extend algorithm transforms a uni-encryption $(\mathbf{d}_i, \mathbf{F}_i)$ generated by the i-th party into a valid multi-key RGSW ciphertext $\overline{\mathbf{D}}_i$ encrypting the same message under the concatenated secret $\overline{\mathbf{z}} = (1, z_1, \ldots, z_k) \in R^{k+1}$. It can be viewed as a variant of RGSW extension of previous work [10,17,30]. However, we improve its performance by reducing the dimension of extended ciphertexts by almost half from $2k$ down to $(k+1)$. Moreover, we proposed a symmetric encryption since uni-encryption is not used for plaintext encryption but only for generating the bootstrapping keys of our MKHE scheme. Therefore, our uni-encryption, extension and thereby external product algorithms are better in terms of size, complexity, and noise growth. Our solution can be directly applied to [10] to improve its key-switching key generation in the same context. We note that Extend requires $2(k+1)d^2 = O(kd^2)$ polynomial multiplications to generate $\mathbf{x}_j$, $\mathbf{x}_j$ for $0 \leq j \leq k$ and we can store an extended ciphertext using $(2k+3)d = O(kd)$ polynomials due to its sparsity. In addition, the external product (Step 2) takes $(3k+1)d = O(kd)$ polynomial multiplications since $\overline{\mathbf{D}}_i$ is a sparse matrix generated by the extension algorithm.

Meanwhile, our second method updates the input multi-key RLWE ciphertext $\overline{\mathbf{c}}$ without generating any multi-key GSW ciphertext. It requires only $4(k+1)d = O(kd)$ polynomial multiplications, and also enjoys a comparative advantage in terms of noise growth. Roughly speaking, it computes the same function at the plaintext level, but uses a different circuit representation.[2] The next paragraph will explain more about it in detail.

Comparison. The first method is asymptotically slower than the second method, however, we note that the ciphertext extension depends only on $(\mathbf{d}_i, \mathbf{F}_i)$ and $\{\mathbf{b}_j\}_{j \in [k]}$. The extended ciphertext $\overline{\mathbf{D}}_i$ can be pre-computed and reused in

---

[2] For the reader who is familiar with the GSW scheme, let us cite a similar example. For GSW ciphertexts C_i, we denote by $\boxtimes$ the multiplication between GSW ciphertexts. Both $C_1 \boxtimes (C_2 \boxtimes C_3)$ and $(C_1 \boxtimes C_2) \boxtimes C_3$ are computing the same function (product of three plaintexts) but latter one introduces a much smaller error.

the multiplications with different multi-key RLWE ciphertexts. In particular, Step 2 requires less number of polynomial multiplication than Method 2 (about 3/4 times) so one can take an advantage of complexity from this pre-processing.

On the other hand, we point out that the second method introduces a much smaller noise. We denote by $V_B \approx B^2/12$ the variance of a uniform distribution on $\mathbb{Z}_B$. We show in Appendix A that the first method outputs a ciphertext satisfying

$$\langle \overline{\mathbf{c}}', \overline{\mathbf{z}} \rangle = \mu_i \cdot \langle \overline{\mathbf{c}}, \overline{\mathbf{z}} \rangle + e \pmod 1$$

for some error e of variance $V_1 \approx (k+1)d^2 \cdot N^2 \cdot V_B^2 \cdot \beta^2$ while the second method holds the same equation but with different error variance $V_2 \approx \frac{1}{2}(kd + k + 1) \cdot N^2 \cdot V_B \cdot \beta^2 \leq (d \cdot V_B)^{-1} \cdot V_1$.

In summary, the first method with pre-processing can have an advantage in complexity (by a factor of about 3/4) by making a trade-off between storage and complexity. Meanwhile, the second method has a smaller noise growth. In other words, one may use a smaller parameter while achieving the same level of noise. For these reasons, the second method is more practical than the first one in almost all aspects.

The controlled selector gate (called CMux in [15]) is one direct application of hybrid product. It aims to securely choose $\overline{\mathbf{c}}_\mu = (1 - \mu) \cdot \overline{\mathbf{c}}_0 + \mu \cdot \overline{\mathbf{c}}_1$ between two multi-key RLWE ciphertexts $\overline{\mathbf{c}}_0$ and $\overline{\mathbf{c}}_1$ using an encrypted bit $\mu \in \{0, 1\}$. The CMux gate is a core operation in the bootstrapping of our scheme.

- $\underline{\texttt{RLWE.CMux}(\overline{\mathbf{c}}_0, \overline{\mathbf{c}}_1(\mathbf{d}_i, \mathbf{F}_i), \{\mathbf{b}_j\}_{j \in [k]})}$: Given multi-key ciphertexts $\overline{\mathbf{c}}_0, \overline{\mathbf{c}}_1 \in T^{k+1}$, a uni-encryption $(\mathbf{d}_i, \mathbf{F}_i)$ (encrypting a bit $\mu_i \in \{0, 1\}$) and the set of public keys $\{\mathbf{b}_j\}_{1 \leq j \leq k}$, compute and return $\overline{\mathbf{c}}' \leftarrow \overline{\mathbf{c}}_0 + \texttt{RLWE.Prod}(\overline{\mathbf{c}}_1 - \overline{\mathbf{c}}_0(\mathbf{d}_i, \mathbf{F}_i), \{\mathbf{b}_j\}_{j \in [k]})$.

4 Multi-key Variant of TFHE

4.1 Description

In this section, we explicitly describe an MKHE scheme based on the LWE and RGSW schemes. Our scheme can bootstrap a ciphertext after the evaluation of a binary gate as in TFHE [13], but it requires to pre-compute the bootstrapping key corresponding to the set of parties involved in a computation.

- $\underline{\texttt{MKHE.Setup}(1^\lambda)}$:
    - Run $\texttt{LWE.Setup}(1^\lambda)$ to generate the parameter $pp^{\texttt{LWE}} = (n, \chi, \alpha, B', d')$.
    - Run $\texttt{RLWE.Setup}(1^\lambda)$ to generate the parameter $pp^{\texttt{RLWE}} = (N, \psi, \beta, B, d, \mathbf{a})$.
    - Return the generated public parameters $pp^{\texttt{MKHE}} = (pp^{\texttt{LWE}}, pp^{\texttt{RLWE}})$.

We assume that all other algorithms of MKHE implicitly take $pp^{\texttt{MKHE}}$ as an input.

- $\underline{\texttt{MKHE.KeyGen}()}$: Each party i independently generates its keys as follows.

- Sample the LWE secret by $s_i \leftarrow$ LWE.KeyGen().
- Run $(z_i, b_i) \leftarrow$ RLWE.KeyGen() and set the public key as $PK_i = b_i$. We write $z_i^* = (z_{i,0}, -z_{i,N-1}, \ldots, -z_{i,1}) \in \mathbb{Z}^N$ for $z_i = z_{i,0} + z_{i,1}X + \cdots + z_{i,N-1}X^{N-1}$.
- Generate $(d_{i,j}, F_{i,j}) \leftarrow$ RLWE.UniEnc$(s_{i,j}, z_i)$ for $j \in [n]$ and set the bootstrapping key as $BK_i = \{(d_{i,j}, F_{i,j})\}_{j \in [n]}$.
- Generate the key-switching key $KS \leftarrow$ LWE.KSGen(z_i^*, s_i).
- Return the secret key s_i. Publish the triple (PK_i, BK_i, KS_i) of public, bootstrapping, and key-switching keys.

We remark that for any $a = a_0 + a_1 X + \cdots + a_{N-1}X^{N-1} \in T$ and the vector of its coefficients $(a_0, \ldots, a_{N-1}) \in \mathbb{T}^N$, the constant term of $a \cdot z \in T$ is equal to $\langle a, z^* \rangle$ modulo 1.

• MKHE.Enc(m): For an input bit $m \in \{0, 1\}$, run LWE.Enc(m) and return an LWE encryption with the scaling factor $1/4$. The output ciphertext $ct = (b, a) \in \mathbb{T}^{n+1}$ satisfies $b + \langle a, s \rangle \approx \frac{1}{4}m \pmod{1}$.

The dimension of a ciphertext increases after homomorphic computations. The indices of related parties should be stored together with a ciphertext for the correct decryption and homomorphic operations.

- MKHE.Dec$(\overline{ct}, \{s_i\}_{i \in [k]})$: For a ciphertext $\overline{ct} = (b, a_1, \ldots, a_k) \in \mathbb{T}^{kn+1}$ and a tuple of secrets $(s_1, \ldots, s_k)$, return the bit $m \in \{0, 1\}$ which minimizes $|b + \sum_{i=1}^{k} \langle a_i, s_i \rangle - \frac{1}{4}m|$.
- MKHE.NAND$(\overline{ct}_1, \overline{ct}_2, \{(PK_i, BK_i, KS_i)\}_{i \in [k]})$: Given two LWE ciphertexts $\overline{ct}_1 \in \mathbb{T}^{k_1 n+1}$ and $\overline{ct}_2 \in \mathbb{T}^{k_2 n+1}$, let k be the number of parties that are associated with either $\overline{ct}_1$ or $\overline{ct}_2$. For $i \in [k]$, $PK_i = b_i$, $BK_i = \{(d_{i,j}, F_{i,j})\}_{j \in [n]}$ and KS_i are the public key, bootstrapping key and key-switching key of the j-th party, respectively.

This algorithm consists of three phases. The first step expands the input LWE ciphertexts and evaluate the NAND gate $m = m_1 \bar{\wedge} m_2$ homomorphically on encrypted bits.

1-1. Extend $\overline{ct}_1$ and $\overline{ct}_2$ to the ciphertexts $\overline{ct}_1', \overline{ct}_2' \in \mathbb{T}^{kn+1}$ which encrypt the same messages under the concatenated secret key $\bar{s} = (s_1, \ldots, s_k) \in \mathbb{Z}^{kn}$. It is simply done by rearranging the components and putting zeros in the empty slots.

1-2. Compute $\overline{ct}' = (\frac{5}{8}, 0, \ldots, 0) - \overline{ct}_1' - \overline{ct}_2' \pmod{1}$.

To be precise, if an input ciphertext $\overline{ct}_i = (b_i, a_{i,1}, \ldots, a_{i,k_i})$ is an encryption corresponding to a tuple $(j_1, \ldots, j_{k_i}) \in [k]^{k_1}$ of indices, then (1-1) returns

$$\overline{ct}_i' = (b_i, a_{i,1}', \ldots, a_{i,k}') \text{ where } a_{i,j}' = \begin{cases} a_{i,\ell} & \text{if } j = j_\ell \text{ for some } \ell \in [k_i], \\ 0 & \text{otherwise;} \end{cases} \text{ for } j \in$$

$[k]$. It is clear from the definition that $\langle \overline{ct}_i(1, s_{j_1}, \ldots, s_{j_{k_i}}) \rangle = \langle \overline{ct}_i', (1, \bar{s}) \rangle$ for $\bar{s} = (s_1, \ldots, s_k)$.

If $\langle \overline{ct}_i', (1, \bar{s}) \rangle = \frac{1}{4}m_i + e_i \pmod{1}$ for some errors $e_i \in \mathbb{R}$, then the output ciphertext satisfies that $\langle \overline{ct}', (1, \bar{s}) \rangle = \frac{1}{2}m + e' \pmod{1}$ for $m = m_1 \bar{\wedge} m_2$ and

$e' = \pm\frac{1}{8} - e_1 - e_2$ which is bounded by $\frac{1}{4}$ when $|e_i| \leq \frac{1}{16}$. The next step, called homomorphic accumulator [20], is to evaluate the decryption circuit of an extended LWE ciphertext using the external product of RGSW scheme for bootstrapping.

2-1. Let $\overline{\mathsf{ct}}' = (b', \mathbf{a}'_1, \ldots, \mathbf{a}'_k) \in \mathbb{T}^{kn+1}$. Compute $\tilde{b} = \lfloor 2N \cdot b' \rceil$ and $\tilde{\mathbf{a}}_i = \lfloor 2N \cdot \mathbf{a}'_i \rceil$. Initialize the RLWE ciphertext as $\overline{\mathbf{c}} = (-\frac{1}{8}h(X) \cdot X^{\tilde{b}}, \mathbf{0}) \in T^{k+1}$ where $h = \sum_{-\frac{N}{2} < j < \frac{N}{2}} X^j = 1 + X + \cdots + X^{\frac{N}{2}-1} - X^{\frac{N}{2}+1} - \cdots - X^{N-1}$.
2-2. Let $\tilde{\mathbf{a}}_i = (\tilde{a}_{i,j})_{j \in [n]}$ for $i \in [k]$. Compute

$$\overline{\mathbf{c}} \leftarrow \mathtt{RLWE.CMux}(\overline{\mathbf{c}}, X^{\tilde{a}_{i,j}} \cdot \overline{\mathbf{c}}, (\mathbf{d}_{i,j}, \mathbf{F}_{i,j}), \{\mathbf{b}_\ell\}_{\ell \in [k]})$$

recursively for all $i \in [k]$ and $j \in [n]$.
2-3. Return $\overline{\mathbf{c}} \leftarrow (\frac{1}{8}, \mathbf{0}) + \overline{\mathbf{c}} \pmod 1$.

The accumulator $\overline{\mathbf{c}}$ is initialized in (2-1) as the trivial RLWE encryption of $-\frac{1}{8}h(X) \cdot X^{\tilde{b}}$. The main computation is done in (2-2) using the Mux gate. In each step, it homomorphically selects one of $\overline{\mathbf{c}}$ and $X^{\tilde{a}_{i,\ell}} \cdot \overline{\mathbf{c}}$ using the encryption $(\mathbf{d}_{i,j}, \mathbf{F}_{i,j})$ of $s_{i,j} \in \{0,1\}$. The output is a multi-key RLWE ciphertext satisfying

$$\langle \overline{\mathbf{c}}, \overline{\mathbf{z}} \rangle \approx -\frac{1}{8}h(X) \cdot X^{\tilde{b}+\sum_{i=1}^{k}\langle \tilde{\mathbf{a}}_i, \mathbf{s}_i \rangle}$$

$$= -\frac{1}{8}\left(\sum_{-\frac{N}{2} < j < \frac{N}{2}} X^j\right) \cdot X^{\tilde{b}+\sum_{i=1}^{k}\langle \tilde{\mathbf{a}}_i, \mathbf{s}_i \rangle} \pmod 1.$$

Since $\tilde{b} + \sum_{i=1}^{k}\langle \tilde{\mathbf{a}}_i, \mathbf{s}_i \rangle \approx (2N) \cdot \langle \overline{\mathsf{ct}}', (1, \overline{\mathbf{s}}) \rangle \approx N \cdot m \pmod{2N}$, the constant term of $\langle \overline{\mathbf{c}}, \overline{\mathbf{z}} \rangle$ is approximately equal to either $-\frac{1}{8}$ (if $m = 0$) or $\frac{1}{8}$ (otherwise; $m = 1$), which is $\frac{1}{4}m - \frac{1}{8}$. Finally, the term $\frac{1}{8}$ is cancelled out in (2–3).

We stress that we proposed two different algorithms for the underlying hybrid product algorithm of CMux.

3-1. For $\overline{\mathbf{c}} = (c_0, c_1, \ldots, c_k) \in T^{k+1}$, let b^* be the constant term of c_0 and $\mathbf{a}^*_i$ be the coefficient vector of c_i for $i \in [k]$. Construct the LWE ciphertext $\overline{\mathsf{ct}}^* = (b^*, \mathbf{a}^*_1, \ldots, \mathbf{a}^*_k) \in \mathbb{T}^{kN+1}$.
3-2. Let $\overline{\mathsf{KS}} = \{\mathsf{KS}_i\}_{i \in [k]}$. Run the multi-key-switching algorithm and return the ciphertext $\overline{\mathsf{ct}}'' \leftarrow \mathtt{LWE.MKSwitch}(\overline{\mathsf{ct}}^*, \overline{\mathsf{KS}})$.

In the last step, we transform $\overline{\mathbf{c}}$ into an LWE ciphertext and run the multi-key-switching algorithm. As we noted above, $\langle \mathbf{a}^*_i, \mathbf{z}^*_i \rangle \pmod 1$ is equal to the constant term of $c_i \cdot z_i$ for $i \in [k]$. Hence, (3-1) returns an LWE ciphertext $\overline{\mathsf{ct}}^*$ satisfying $\langle \overline{\mathsf{ct}}^*, (1, \overline{\mathbf{z}}^*) \rangle \approx \frac{1}{4}m \pmod 1$ for $\overline{\mathbf{z}}^* = (\mathbf{z}^*_1, \ldots, \mathbf{z}^*_k)$. Finally, (3-2) switches the LWE key into $\overline{\mathbf{s}}$ so the output LWE ciphertext satisfies that $\langle \overline{\mathsf{ct}}, (1, \overline{\mathbf{s}}) \rangle \approx \frac{1}{4}m \pmod 1$, as desired.

Security. Our scheme is semantically secure under the (R)LWE assumption described in the previous section, so the parameters $pp^{\texttt{LWE}}$ and $pp^{\texttt{RLWE}}$ should be chosen properly to achieve at least λ-bit of security level.

We note that each party publishes uni-encryptions of $s_1, \ldots, s_n$ encrypted by z as well as a key-switching key from $\mathbf{z}^* = (z_0, -z_{N-1}, \ldots, -z_1)$ to $\mathbf{s}$. Similar to TFHE [13] and all other bootstrappable (fully) HE schemes [22] such as [9,11,20,25], our scheme requires an additional circular security assumption.

Correctness Conditions. Our scheme should satisfy the following requirements to guarantee its correctness:

- In (2-1), the quantized ciphertext $(\tilde{b}, \tilde{\mathbf{a}}_1, \ldots, \tilde{\mathbf{a}}_k) \in \mathbb{Z}_{2N}^{kn+1}$ should satisfy $\tilde{b} + \sum_{j=1}^{k} \langle \tilde{\mathbf{a}}_j, \mathbf{s}_j \rangle = N \cdot m + \tilde{e}$ for some $\tilde{e} \in \mathbb{Z}$ with $|\tilde{e}| < N/2$. This noise $\tilde{e}$ consists of two parts $\tilde{e} = 2N \cdot e' + e''$ for $e' = \pm\frac{1}{8} - e_1 - e_2$ from the step (1-2) and a rounding error $e'' = (\tilde{b} - 2N \cdot b') + \sum_{j=1}^{k} \langle \tilde{\mathbf{a}}_j - 2N \cdot \mathbf{a}'_j, \mathbf{s}_j \rangle$.
- The error $e \in \mathbb{R}$ of an output LWE ciphertext $\overline{\mathsf{ct}}$ should be small enough for the correct decryption and further computations. It is the sum of the constant term of an RLWE error which is accumulated from the external products during (2–3), and the multi-key-switching error from (3-2).

We provide a rigorous noise estimation in Appendix A. We refer the reader to Sect. 5 for a recommended parameter set.

Performance. The accumulation step (2-2) is the most expensive part of the whole pipeline and all other algorithms including multi-key-switching are asymptotically faster. We run the CMux algorithm $k \cdot n$ times, each of which has almost the same complexity as the hybrid product $\texttt{RLWE.Prod}$ described in Sect. 3.2. The computing server can choose one of two proposed algorithms and inherit their (dis)advantages. The performance of gate bootstrapping would be $k \cdot n$ times of the chosen hybrid product algorithm.

If the first method is chosen, we can pre-compute the extended RGSW ciphertexts $\overline{\mathbf{D}}_{i,j} \leftarrow \texttt{RLWE.Extend}((\mathbf{d}_{i,j}, \mathbf{F}_{i,j}), \{\mathbf{b}_\ell\}_{\ell \in [k]})$ and set $\overline{\mathsf{BK}} := \{\overline{\mathbf{D}}_{i,j}\}_{i \in [k], j \in [n]}$ as a shared bootstrapping key which can be reused in the evaluation of an arbitrary Boolean gate on the same set of k parties. However, it requires more space ($O(k^2 nd)$ polynomials) to store $\overline{\mathsf{BK}}$.

4.2 Discussion

We presented a multi-key variant of the TFHE scheme. However, we can simply design some variants of this basic scheme with better functionality and versatility.

More Bootstrapped Gates. We described only the Multi-Key bootstrapped NAND gate in the previous section, but any arbitrary binary bootstrapped gate (such as AND, OR, XOR, etc.) can be evaluated in the same way, as it is done in TFHE: it is sufficient to modify the initial linear combination before bootstrapping.

Time-Space Trade-Off. Brakerski and Perlman [7] suggested a method to reduce down the memory requirement by generating a temporary evaluation key in each step. Since our first method generates an expanded bootstrapping key $\overline{\mathsf{BK}}$ whose size grows quadratically with the number of parties, we can apply this idea to have a linearly-growing space complexity. In this case, we lose the reusability of a expanded bootstrapping key which is the only advantage of first method compared to the second solution. Therefore, we do *not* have any motivation to adapt this optimization technique.

Distributed Decryption. HE has some attractive applications in the construction of advanced cryptographic primitives such as round-efficient MPC [2,18,24,30,33]. In particular, the distributed property of threshold HE [3,26] makes an important role to achieve this functionality. Any secure multi-party protocol can be built between key owners to evaluate the decryption circuit, but we introduce a simple example in this paragraph.

Since our MKHE scheme is based on the standard LWE encryption, the techniques for threshold decryption such as noise smudging (a.k.a. noise flooding) [2] can be directly applied to our scheme. The noise distribution, parametrized by a constant $\gamma > 0$, should have a medium size which is smaller than 1 but sufficiently larger than the error of an input ciphertext to prevent the leakage of extra information beyond the decrypted value. See [2] for parameter choice and security proof.

- $\mathsf{MKHE.PartDec}(\overline{\mathsf{ct}}, \mathbf{s}_i)$: For a ciphertext $\overline{\mathsf{ct}} = (b, \mathbf{a}_1, \dots, \mathbf{a}_k) \in \mathbb{T}^{kn+1}$ and the i-th secret $\mathbf{s}_i$, sample an error $e_i \leftarrow D_\gamma$ and return the value $p_i = \langle \mathbf{a}_i, \mathbf{s}_i \rangle + e_i$ (mod 1).
- $\mathsf{MKHE.Merge}(b, \{p_j\}_{j \in [k]})$: For the first entry b of an input ciphertext and the partial decryptions $\{p_j\}_{j \in [k]}$, output the bit $m \in \{0, 1\}$ which minimizes $|b + \sum_{j=1}^{k} p_j - \frac{1}{4}m|$.

Faster Evaluation of a Look-Up Table (LUT). There have been some progresses in TFHE-type schemes to accelerate the evaluation of a LUT. For example, Chillotti et al. [14] suggested a *vertical* packing method for TRLWE combined with a circuit bootstrapping algorithm which gives a speed-up compared to the gate-by-gate bootstrapping, while Bonnoron et al. [4] (see also [8]) suggested a method to encrypt more than one bit in a single ciphertext. It is easy to see that these techniques are directly applicable to our MKHE scheme.

5 Experimental Results

We present a proof-of-concept implementation to convince the reader that our scheme is practical. The implementation took a few days of coding and it is based on the TFHE library [16]. Our source code is publicly available at https://github. com/ilachill/MK-TFHE.

Table 2. Recommended parameter sets.

Set	LWE				RLWE (RGSW)			
	n	α	B'	d'	N	β	B	d
I	560	$3.05 \cdot 10^{-5}$	2^2	8	1024	$3.72 \cdot 10^{-9}$	2^9	3
II							2^8	4
III							2^6	5

In Table 2, we present three candidate parameter sets. We increase the dimensions of LWE and RLWE to have a more conservative parameter.[3] Our parameters achieve at least 110-bit security level according to the LWE Estimator [1], which is a common reference in the domain.[4]

As mentioned before, we used the second hybrid product algorithm in implementation. We set the LWE/RLWE secret distributions χ and ψ as the uniform distributions over the set of binary vectors in $\mathbb{Z}^n$ and over the polynomials in R with binary coefficients, respectively.

We show in Appendix A that the standard deviation of bootstrapping error grows linearly on the number of parties. Hence the growth of parameter with respect to the maximal number of involved parties is very slow. We control the noise by changing the decomposition degree and exponent which do not affect the security level.

We adapt a space-time trade-off technique in [13,20] which reduces the complexity of key-switching procedure by publishing all LWE encryptions of $a \cdot B'^i \cdot t_j$ for $i \in [d']$, $j \in [N]$, and $a \in \mathbb{Z}_{B'}$, compared to the encryptions of $B'^i \cdot t_j$ in the scheme description. Hence our implementation of multi-key-switching is purely represented by a summation of LWE vectors. It does not make any change in asymptotic complexity.

Our experimental results are summarized in Table 3. All experiments are performed on a Intel Core i7-4910MQ at 2.90 GHz laptop, running on a single thread, which takes 13ms to execute a gate bootstrapping of the TFHE library. On the left sides of table, we describe the *local* complexity of our scheme such as key generation timing of each party. This part is independent from k. The other side presents the *global* performance corresponding to the multi-key operation. The parameter sets I, II and III support homomorphic computation on any number of parties up to 2, 4 and 8, respectively. A smaller parameter has a

---

[3] In [15], the authors recommend to take more conservative parameters for the original TFHE scheme as well. This new parameter set will affect their gate bootstrapping timing by making it increase of a few milliseconds with respect to the original given execution timing of about 13 ms.

[4] https://bitbucket.org/malb/lwe-estimator/src/master/.

Table 3. Performance of our implementation. k denotes the number of parties in computation.

Set	KG	BK	KS	ct	k	$\overline{ct}$	NAND
I	1.1 s	0.62 MB	70.1 MB	2.19 KB	2	4.38 KB	0.27 s
II	1.2 s	0.82 MB	70.1 MB	2.19 KB	2	4.38 KB	0.43 s
					4	8.77 KB	1.45 s
III	1.3 s	1.03 MB	70.1 MB	2.19 KB	2	4.38 KB	0.50 s
					4	8.77 KB	1.90 s
					8	17.32 KB	7.16 s

better performance but a larger one makes the scheme more flexible because more parties can join the computation dynamically. We believe that the code has space for optimization. This, with a more accurate choice of the parameters could produce better execution timings.

6 Conclusion

We designed a practical MKHE scheme by generalizing the gate bootstrapping of TFHE to the multi-key case. Our main technical contribution is to establish a new hybrid product between single-key and multi-key ciphertexts which provides better storage, computational cost and noise growth. We implemented our scheme to present its concrete performance.

As we discussed in Sect. 4.2, one future direction is to implement advanced functionalities of TFHE in the multi-key setting. Another direction is to design a practical MKHE scheme from another HE system (e.g. BFV [5,21], CKKS [12]) which has advantages in amortized complexity. Finally, one primary open problem in this area is how to construct an MKHE scheme without the CRS model.

Acknowledgments. The second author (I.C.) has been supported in part by ERC Advanced Grant ERC-2015-AdG-IMPaCT and by the FWO under an Odysseus project GOH9718N. Any opinions, findings and conclusions or recommendations expressed in this material are those of the author(s) and do not necessarily reflect the views of the ERC or FWO.

A   Noise Estimation

For the decomposition base B and degree d, let $\epsilon^2 = 1/(12B^{2d})$ be the variance of uniform distribution over the interval $(-\frac{1}{2}B^{-d}, \frac{1}{2}B^{-d}]$. We denote by $V_B =$
$\begin{cases} \frac{1}{12}(B^2 - 1) & \text{if } B \text{ is odd,} \\ \frac{1}{12}(B^2 + 2) & \text{if } B \text{ is even;} \end{cases}$ the mean square of a uniform distribution over $\mathbb{Z} \cap$

$(-B/2, B/2]$. We similarly define ϵ'^2 and $V_{B'}$ based on the parameter B' and d' for the key-switching algorithm. We set the RGSW and LWE secret distributions χ, ψ as uniform distributions over $\{0,1\}^N$ and $\{0,1\}^n$, respectively.

The variance of a random variable e over $\mathbb{R}$ is denoted by $\mathsf{Var}(e)$. For a random variable e over $\mathbb{R}[X]/(X^N + 1)$, it denotes the variance of a coefficient when all coefficients have the same variance. If $\mathbf{e}$ is a vector of random variables, $\mathsf{Var}(\mathbf{e})$ denotes the maximum of its entries' variances.

We mainly compute the variance of a noise. Our average-case analysis is based on the heuristic assumption that a noise behaves like a Gaussian distribution, which has been empirically shown in the previous work (Fig. 10, [15]).

Hybrid Product Method 1

Step 1: Ciphertext extension. Let us suppose that

$$\mathbf{b}_j = -z_j \cdot \mathbf{a} + \mathbf{e}_j \quad (\text{mod } 1) \text{ for } j \in [k],$$
$$\mathbf{d}_i = r_i \cdot \mathbf{a} + \mu_i \cdot \mathbf{g} + \mathbf{e}_{i,1} \quad (\text{mod } 1),$$
$$\mathbf{f}_{i,0} + z_i \cdot \mathbf{f}_{i,1} = r_i \cdot \mathbf{g} + \mathbf{e}_{i,2} \quad (\text{mod } 1) \text{ for some } \mu \in R \text{ and}$$
$$\overline{\mathbf{D}}_i \leftarrow \mathsf{RLWE.Extend}\left((\mathbf{d}_i, \mathbf{F}_i), \{\mathbf{b}_j\}_{j \in [k]}\right).$$

We let $\mathbf{e}_0 = \mathbf{0}$ for simplicity. Then for any $0 \le j \le k$, the j-th row of $\overline{\mathbf{D}}_i$ satisfies that

$$\mathbf{x}_j + z_i \cdot \mathbf{y}_j = \mathbf{M}_j \cdot (r_i \cdot \mathbf{g} + \mathbf{e}_{i,2}) = r_i \cdot \mathbf{b}_j + (r_i \cdot \mathbf{e}'_j + \mathbf{M}_j \cdot \mathbf{e}_{i,2}) \quad (\text{mod } 1)$$

for the decomposition error $\mathbf{e}'_j \in \mathbb{R}^d$ such that $\mathbf{e}'[\ell] = \langle \mathbf{g}^{-1}(\mathbf{b}_j[\ell]), \mathbf{g} \rangle - \mathbf{b}_j[\ell]$ for $\ell \in [d]$, and

$$z_j \cdot \mathbf{d}_i = r_i z_j \cdot \mathbf{a} + \mu_i z_j \cdot \mathbf{g} + z_j \cdot \mathbf{e}_{i,1}$$
$$= -r_i \cdot \mathbf{b}_j + \mu_i z_j \cdot \mathbf{g} + (r_i \cdot \mathbf{e}_j + z_j \cdot \mathbf{e}_{i,1}) \quad (\text{mod } 1).$$

Therefore, the j-th row is decrypted into $\mu_i z_j \cdot \mathbf{g} + \mathbf{e}_{i,j}$ for the GSW extension error $\mathbf{e}_{i,j} = r_i \cdot (\mathbf{e}_j + \mathbf{e}'_j) + z_j \cdot \mathbf{e}_{i,1} + \mathbf{M}_j \cdot \mathbf{e}_{i,2}$.

Its variance is bounded by

$$V_{exp} \le (N/2)\epsilon^2 + (1 + d \cdot V_B) \cdot N\beta^2$$

since $\mathsf{Var}(r_i) = \mathsf{Var}(z_j) = 1/2$, $\mathsf{Var}(\mathbf{e}_j) \le \mathsf{Var}(\mathbf{e}_{i,1}) = \mathsf{Var}(\mathbf{e}_{i,2}) = \beta^2$, $\mathsf{Var}(\mathbf{e}'_j) = \epsilon^2$ and $\mathsf{Var}(\mathbf{M}_j \cdot \mathbf{e}_{i,2}) = dN \cdot V_B \cdot \beta^2$.

Step 2: Multi-key GSW external product. Let $\bar{\mathbf{c}}$ and $\overline{\mathbf{D}}$ be multi-key RLWE and RGSW ciphertexts. Suppose that $\overline{\mathbf{D}}$ satisfies $\overline{\mathbf{D}}\bar{\mathbf{z}} = \mu \cdot \mathbf{G}_{k+1}\bar{\mathbf{z}} + \bar{\mathbf{e}}$ for a plaintext $\mu \in R$ and an error vector $\bar{\mathbf{e}}$. We denote by $\mathsf{VarErr}(\overline{\mathbf{D}}) = \mathsf{Var}(\bar{\mathbf{e}})$. The external product outputs an RLWE ciphertext $\bar{\mathbf{c}}'$ satisfying

$$\langle \bar{\mathbf{c}}', \bar{\mathbf{z}} \rangle = \mathbf{G}_{k+1}^{-1}(\bar{\mathbf{c}}) \cdot \overline{\mathbf{D}}\bar{\mathbf{z}} \quad (\text{mod } 1)$$
$$= \mathbf{G}_{k+1}^{-1}(\bar{\mathbf{c}}) \cdot (\mu \cdot \mathbf{G}_{k+1}\bar{\mathbf{z}} + \bar{\mathbf{e}}) \quad (\text{mod } 1)$$
$$= \mu \cdot \langle \bar{\mathbf{c}}, \bar{\mathbf{z}} \rangle + \left(\mu \cdot \langle \bar{\mathbf{e}}', \bar{\mathbf{z}} \rangle + \mathbf{G}_{k+1}^{-1}(\bar{\mathbf{c}}) \cdot \bar{\mathbf{e}}\right) \quad (\text{mod } 1)$$

for the decomposition error $\overline{\mathbf{e}}' = \mathbf{G}_{k+1}^{-1}(\overline{\mathbf{c}}) \cdot \mathbf{G}_{k+1} - \overline{\mathbf{c}}$. Therefore, the variance of external product error $e_{ep} = \mu \cdot \langle \overline{\mathbf{e}}', \overline{\mathbf{z}} \rangle + \mathbf{G}_{k+1}^{-1}(\overline{\mathbf{c}}) \cdot \overline{\mathbf{e}}$ is

$$V_{ep} = \mu^2 \cdot \epsilon^2 (1 + kN/2) + (k+1)dN \cdot V_B \cdot \mathsf{VarErr}(\overline{\mathbf{D}})$$

since $\mathsf{Var}(\overline{\mathbf{e}}') = \epsilon^2$ and $\mathsf{Var}(\mathbf{G}_{k+1}^{-1}(\overline{\mathbf{c}})) = V_B$.

In our case, $\overline{\mathbf{D}} = \overline{\mathbf{D}}_i$ is an extended RGSW ciphertext whose error variance is $V_{exp} \leq (N/2)\epsilon^2 + (1 + d \cdot V_B) \cdot N\beta^2$. As a result, our first method returns a ciphertext whose noise variance is

$$V_1 = \mu_i^2 \cdot \epsilon^2 (1 + kN/2) + (k+1)dN \cdot V_B \cdot V_{exp}.$$

In our MKHE scheme, the decomposition error ϵ^2 can be easily controlled. Hence the extension error is mainly dominated by $V_{exp} \approx dN \cdot V_B \cdot \beta^2$. Similarly, the noise of hybrid product is dominated by $V_1 \approx (k+1)dN \cdot V_B \cdot V_{exp} \approx (k+1)d^2 \cdot N^2 \cdot V_B^2 \cdot \beta^2$.

Hybrid Product Method 2. As shown earlier, the output $\overline{\mathbf{c}}'$ of the second multiplication algorithm satisfies $\langle \overline{\mathbf{c}}', \overline{\mathbf{z}} \rangle = \sum_{j=0}^{k} u_j \cdot z_j + \sum_{j=0}^{k} (w_{j,0} + w_{j,1} \cdot z_i)$. The first term is

$$\sum_{j=0}^{k} u_j \cdot z_j = \sum_{j=0}^{k} \langle \mathbf{g}^{-1}(c_j), r_i \cdot \mathbf{a} + \mu_i \cdot \mathbf{g} + \mathbf{e}_{i,1} \rangle \cdot z_j \pmod 1$$

$$= \mu_i \cdot \langle \overline{\mathbf{c}}, \overline{\mathbf{z}} \rangle + \mu_i \cdot e' + r_i \cdot \sum_{j=0}^{k} \langle \mathbf{g}^{-1}(c_j), z_j \cdot \mathbf{a} \rangle + \sum_{j=0}^{k} \langle \mathbf{g}^{-1}(c_j), \mathbf{e}_{i,1} \rangle \cdot z_j \pmod 1$$

$$= \mu_i \cdot \langle \overline{\mathbf{c}}, \overline{\mathbf{z}} \rangle - r_i \cdot \sum_{j=0}^{k} v_j + \mu_i \cdot e' + r_i \cdot \sum_{j=0}^{k} \langle \mathbf{g}^{-1}(c_j), \mathbf{e}_j \rangle + \left\langle \sum_{j=0}^{k} z_j \cdot \mathbf{g}^{-1}(c_j), \mathbf{e}_{i,1} \right\rangle$$

for the decomposition error $e' = \sum_{j=0}^{k} (\langle \mathbf{g}^{-1}(c_j), \mathbf{g} \rangle - c_j) \cdot z_j$, while the second term is

$$\sum_{j=0}^{k} (w_{j,0} + w_{j,1}z_i) = \sum_{j=0}^{k} \langle \mathbf{g}^{-1}(v_j), \mathbf{f}_{i,0} + z_i \cdot \mathbf{f}_{i,1} \rangle \pmod 1$$

$$= \sum_{j=0}^{k} \langle \mathbf{g}^{-1}(v_j), r_i \cdot \mathbf{g} + \mathbf{e}_{i,2} \rangle = r_i \cdot \sum_{j=0}^{k} v_j + r_i \cdot e'' + \left\langle \sum_{j=0}^{k} \mathbf{g}^{-1}(v_j), \mathbf{e}_{i,2} \right\rangle \pmod 1$$

for $e'' = \sum_{j=0}^{k} (\langle \mathbf{g}^{-1}(v_j), \mathbf{g} \rangle - v_j)$. Note that $\mathsf{Var}(e') = \epsilon^2 (1 + kN/2)$ and $\mathsf{Var}(e'') = \epsilon^2 (k+1)$.

Therefore, the noise of $\overline{\mathbf{c}}'$ is

$$\mu_i \cdot e' + r_i \cdot \sum_{j=0}^{k} \langle \mathbf{g}^{-1}(c_j), \mathbf{e}_j \rangle + \left\langle \sum_{j=0}^{k} z_j \cdot \mathbf{g}^{-1}(c_j), \mathbf{e}_{i,1} \right\rangle + r_i \cdot e'' + \left\langle \sum_{j=0}^{k} \mathbf{g}^{-1}(v_j), \mathbf{e}_{i,2} \right\rangle,$$

and its variance

$$V_2 = \mu_i^2 N \epsilon^2 (1 + kN/2) + (N^2/2)(k+1)V_B\beta^2 + dN(1 + kN/2)V_B\beta^2$$
$$+ (N/2)\epsilon^2(k+1) + (k+1)NV_B\beta^2,$$

is dominated by $V_2 \approx \frac{1}{2}(kd + k + 1) \cdot N^2 \cdot V_B \cdot \beta^2$.

Rounding Error. In (2-2), we compute $\tilde{b} = \lfloor 2N \cdot b' \rceil$ and $\tilde{a}_i = \lfloor 2N \cdot a_i' \rceil$. We assume that each of the rounding errors behaves like a uniform random variable on the interval $\mathbb{R} \pmod 1 = (-0.5, 0.5]$. Therefore, the total rounding error $(\tilde{b} - \lfloor 2N \cdot b' \rceil) + \sum_{j=1}^{k} \langle \tilde{a}_j - \lfloor 2N \cdot a_j' \rceil, s_j \rangle$ has the variance of $\frac{1}{12}(1 + kn/2)$.

Mux Gate. Suppose that $\overline{c}_0, \overline{c}_1$ are RLWE ciphertexts and $\overline{C}$ is an RGSW encryption of $\mu \in \{0, 1\}$ with error $\overline{e}$. The mux gate is to compute $\overline{c} = \overline{c}_0 + \text{RLWE.Prod}(\overline{c}_1 - \overline{c}_0, \overline{C})$ to choose $\overline{c}_\mu$ homomorphically:

$$\langle \overline{c}, \overline{z} \rangle = \langle \overline{c}_0, \overline{z} \rangle + G_{k+1}^{-1}(\overline{c}_1 - \overline{c}_0) \cdot (\mu \cdot G_{k+1}\overline{z} + \overline{e}) \pmod 1$$
$$= (1 - \mu) \cdot \langle \overline{c}_0, \overline{z} \rangle + \mu \cdot \langle \overline{c}_1, \overline{z} \rangle + (\mu \cdot \langle \overline{e}', \overline{z} \rangle + G_{k+1}^{-1}(\overline{c}_1 - \overline{c}_0) \cdot \overline{e}) \pmod 1,$$

for the decomposition error $\overline{e}' = G_{k+1}^{-1}(\overline{c}_1 - \overline{c}_0) \cdot G_{k+1} - (\overline{c}_1 - \overline{c}_0)$. The noise has the variance of $\mu^2 \cdot \epsilon^2(1 + kN/2) + (k+1)dN \cdot V_B \cdot \text{VarErr}(\overline{C})$, exactly the same as external product.

Accumulation. The initial RLWE ciphertext has no noise. All bootstrapping keys $\overline{C}_{i,\ell}$ have the same variance of noise $\text{VarErr}(\overline{C}_{i,\ell}) = (N/2)\epsilon^2 + (1 + N + dNV_B)\beta^2$ from the expansion algorithm. We recursively evaluate the mux gate $k \cdot n$ times and an encrypted secret $s_{i,\ell}$ is sampled uniformly from $\{0, 1\}$. Therefore, the output of accumulator has an error of variance

$$\frac{1}{2}kn \cdot \epsilon^2(1 + kN/2) + (k+1)kdnN \cdot V_B \cdot \left((N/2)\epsilon^2 + (1 + N + dNV_B)\beta^2\right). \quad (1)$$

Multi-key Switching. Let $\overline{ct} = (b, a_1, \ldots, a_k)$ be an input LWE ciphertext and $\overline{ct}' = (b', a_1', \ldots, a_k')$ be the output of multi-key-switching algorithm. Then, we have

$$\langle \overline{ct}', (1, \overline{s}) \rangle = b + \sum_{i=1}^{k}(b_i' + \langle a_i', s_i \rangle) \pmod 1$$

$$= b + \sum_{i=1}^{k}\sum_{j=1}^{N} \langle g'^{-1}(a_{i,j}), t_{i,j} \cdot g' + e_{i,j} \rangle \pmod 1$$

$$= \langle \overline{ct}, (1, \overline{t}) \rangle + \sum_{i=1}^{k}\sum_{j=1}^{N} (t_{i,j} \cdot e_{i,j}' + \langle g'^{-1}(a_{i,j}), e_{i,j} \rangle) \pmod 1$$

for the decomposition error $e'_{i,j} = \langle \mathbf{g}'^{-1}(a_{i,j}), \mathbf{g} \rangle - a_{i,j}$. As a result, the variance of a multi-key-switching error $e_{ks} = \sum_{i=1}^{k} \sum_{j=1}^{N} \left(t_{i,j} \cdot e'_{i,j} + \langle \mathbf{g}'^{-1}(a_{i,j}), \mathbf{e}_{i,j} \rangle \right)$ is obtained by

$$\mathsf{Var}(e_{ks}) = kN \left(\frac{1}{2} \epsilon'^2 + d' \cdot V_{B'} \cdot \alpha^2 \right). \tag{2}$$

We note that this term does not include the error of input LWE ciphertext. If $\langle \mathsf{ct}', (1, \overline{\mathbf{t}}) \rangle = \frac{1}{4}m + e \pmod 1$ for a bit $m \in \{0, 1\}$ and an error $e \in \mathbb{R}$, then ct' will be an encryption of the same message m with error $e' = e + e_{ks}$.

Multi-key Switching (Modified). Different from the previous algorithm, the key-switching key of the i-th party consists of LWE encryptions of $a \cdot B'^{\ell} \cdot t_{i,j}$ for $1 \leq j \leq N$, $0 \leq \ell < d'$ and $a \in \mathbb{Z}_{B'}$ encrypted under the secret $\mathbf{s}_i$. For an input LWE ciphertext $\overline{\mathsf{ct}} = (b, \mathbf{a}_1, \ldots, \mathbf{a}_k)$, the (modified) multi-key switching algorithm computes $\mathbf{g}'^{-1}(a_{i,j}) = (a_{i,j,\ell})_{0 \leq \ell < d'}$ for each $1 \leq i \leq k$ and $1 \leq j \leq N$, and then compute the summation of LWE encryptions of $a_{i,j,\ell} \cdot B'^{\ell} \cdot t_{i,j}$ for $1 \leq i \leq k$, $1 \leq j \leq N$ and $0 \leq \ell < d'$. Therefore, the output ciphertext $\overline{\mathsf{ct}}'$ satisfies that

$$\langle \overline{\mathsf{ct}}', (1, \overline{\mathbf{s}}) \rangle = b + \sum_{i=1}^{k} \sum_{j=1}^{N} \sum_{\ell=0}^{d'-1} \mathbf{g}'^{-1}(a_{i,j})[\ell] \cdot B'^{\ell} \cdot t_{i,j} + e_{i,j,a_{i,j,\ell}} \pmod 1$$

$$= b + \sum_{i=1}^{k} \sum_{j=1}^{N} (a_{i,j} + e'_{i,j}) \cdot t_{i,j} + \sum_{i=1}^{k} \sum_{j=1}^{N} \sum_{\ell=0}^{d'-1} e_{i,j,a_{i,j,\ell}} \pmod 1$$

$$= \langle \overline{\mathsf{ct}}, (1, \overline{\mathbf{t}}) \rangle + \left(\sum_{i=1}^{k} \sum_{j=1}^{N} t_{i,j} \cdot e'_{i,j} + \sum_{i=1}^{k} \sum_{j=1}^{N} \sum_{\ell=0}^{d'-1} e_{i,j,a_{i,j,\ell}} \right) \pmod 1,$$

for the decomposition error $e'_{i,j} = \langle \mathbf{g}'^{-1}(a_{i,j}), \mathbf{g}' \rangle - a_{i,j}$. As a result, the variance of a multi-key-switching error $e_{ks} = \sum_{i=1}^{k} \sum_{j=1}^{N} t_{i,j} \cdot e'_{i,j} + \sum_{i=1}^{k} \sum_{j=1}^{N} \sum_{\ell=0}^{d'-1} e_{i,j,a_{i,j,\ell}}$ is obtained by

$$\mathsf{Var}(e_{ks}) = kN \left(\frac{1}{2} \epsilon_{\mathbf{K}}^2 + d'\alpha^2 \right), \tag{3}$$

which is smaller than that of standard key-switching error (2).

Bootstrapping. The bootstrapping noise is simply the sum of the accumulation and multi-key-switching errors so that it has the variance of $(1) + (3)$.

References

1. Albrecht, M.R., Player, R., Scott, S.: On the concrete hardness of learning with errors. J. Math. Cryptol. **9**(3), 169–203 (2015)

2. Asharov, G., Jain, A., López-Alt, A., Tromer, E., Vaikuntanathan, V., Wichs, D.: Multiparty computation with low communication, computation and interaction via threshold FHE. In: Pointcheval, D., Johansson, T. (eds.) EUROCRYPT 2012. LNCS, vol. 7237, pp. 483–501. Springer, Heidelberg (2012). https://doi.org/10.1007/978-3-642-29011-4_29

3. Boneh, D., et al.: Threshold cryptosystems from threshold fully homomorphic encryption. In: Shacham, H., Boldyreva, A. (eds.) CRYPTO 2018. LNCS, vol. 10991, pp. 565–596. Springer, Cham (2018). https://doi.org/10.1007/978-3-319-96884-1_19

4. Bonnoron, G., Ducas, L., Fillinger, M.: Large FHE gates from tensored homomorphic accumulator. In: Joux, A., Nitaj, A., Rachidi, T. (eds.) AFRICACRYPT 2018. LNCS, vol. 10831, pp. 217–251. Springer, Cham (2018). https://doi.org/10.1007/978-3-319-89339-6_13

5. Brakerski, Z.: Fully homomorphic encryption without modulus switching from classical GapSVP. In: Safavi-Naini, R., Canetti, R. (eds.) CRYPTO 2012. LNCS, vol. 7417, pp. 868–886. Springer, Heidelberg (2012). https://doi.org/10.1007/978-3-642-32009-5_50

6. Brakerski, Z., Gentry, C., Vaikuntanathan, V.: (Leveled) fully homomorphic encryption without bootstrapping. In: Proceedings of ITCS, pp. 309–325. ACM (2012)

7. Brakerski, Z., Perlman, R.: Lattice-based fully dynamic multi-key FHE with short ciphertexts. In: Robshaw, M., Katz, J. (eds.) CRYPTO 2016. LNCS, vol. 9814, pp. 190–213. Springer, Heidelberg (2016). https://doi.org/10.1007/978-3-662-53018-4_8

8. Carpov, S., Izabachène, M., Mollimard, V.: New techniques for multi-value homomorphic evaluation and applications. IACR Cryptology ePrint Archive, 2018:622 (2018)

9. Chen, H., Han, K.: Homomorphic lower digits removal and improved FHE bootstrapping. In: Nielsen, J.B., Rijmen, V. (eds.) EUROCRYPT 2018. LNCS, vol. 10820, pp. 315–337. Springer, Cham (2018). https://doi.org/10.1007/978-3-319-78381-9_12

10. Chen, L., Zhang, Z., Wang, X.: Batched multi-hop multi-key FHE from ring-LWE with compact ciphertext extension. In: Kalai, Y., Reyzin, L. (eds.) TCC 2017. LNCS, vol. 10678, pp. 597–627. Springer, Cham (2017). https://doi.org/10.1007/978-3-319-70503-3_20

11. Cheon, J.H., Han, K., Kim, A., Kim, M., Song, Y.: Bootstrapping for approximate homomorphic encryption. In: Nielsen, J.B., Rijmen, V. (eds.) EUROCRYPT 2018. LNCS, vol. 10820, pp. 360–384. Springer, Cham (2018). https://doi.org/10.1007/978-3-319-78381-9_14

12. Cheon, J.H., Kim, A., Kim, M., Song, Y.: Homomorphic encryption for arithmetic of approximate numbers. In: Takagi, T., Peyrin, T. (eds.) ASIACRYPT 2017. LNCS, vol. 10624, pp. 409–437. Springer, Cham (2017). https://doi.org/10.1007/978-3-319-70694-8_15

13. Chillotti, I., Gama, N., Georgieva, M., Izabachène, M.: Faster fully homomorphic encryption: bootstrapping in less than 0.1 seconds. In: Cheon, J.H., Takagi, T. (eds.) ASIACRYPT 2016. LNCS, vol. 10031, pp. 3–33. Springer, Heidelberg (2016). https://doi.org/10.1007/978-3-662-53887-6_1

14. Chillotti, I., Gama, N., Georgieva, M., Izabachène, M.: Faster packed homomorphic operations and efficient circuit bootstrapping for TFHE. In: Takagi, T., Peyrin, T. (eds.) ASIACRYPT 2017. LNCS, vol. 10624, pp. 377–408. Springer, Cham (2017). https://doi.org/10.1007/978-3-319-70694-8_14

15. Chillotti, I., Gama, N., Georgieva, M., Izabachène, M.: TFHE: fast fully homomorphic encryption over the torus. J. Cryptol. (2019)
16. Chillotti, I., Gama, N., Georgieva, M., Izabachène, M.: TFHE: fast fully homomorphic encryption library, August 2016. https://tfhe.github.io/tfhe/
17. Clear, M., McGoldrick, C.: Multi-identity and multi-key leveled FHE from learning with errors. In: Gennaro, R., Robshaw, M. (eds.) CRYPTO 2015. LNCS, vol. 9216, pp. 630–656. Springer, Heidelberg (2015). https://doi.org/10.1007/978-3-662-48000-7_31
18. Cramer, R., Damgård, I., Nielsen, J.B.: Multiparty computation from threshold homomorphic encryption. In: Pfitzmann, B. (ed.) EUROCRYPT 2001. LNCS, vol. 2045, pp. 280–300. Springer, Heidelberg (2001). https://doi.org/10.1007/3-540-44987-6_18
19. Dodis, Y., Halevi, S., Rothblum, R.D., Wichs, D.: Spooky encryption and its applications. In: Robshaw, M., Katz, J. (eds.) CRYPTO 2016. LNCS, vol. 9816, pp. 93–122. Springer, Heidelberg (2016). https://doi.org/10.1007/978-3-662-53015-3_4
20. Ducas, L., Micciancio, D.: FHEW: bootstrapping homomorphic encryption in less than a second. In: Oswald, E., Fischlin, M. (eds.) EUROCRYPT 2015. LNCS, vol. 9056, pp. 617–640. Springer, Heidelberg (2015). https://doi.org/10.1007/978-3-662-46800-5_24
21. Fan, J., Vercauteren, F.: Somewhat practical fully homomorphic encryption. IACR Cryptology ePrint Archive, 2012:144 (2012)
22. Gentry, C.: Fully homomorphic encryption using ideal lattices. In: Proceedings of the Forty-first Annual ACM Symposium on Theory of Computing, STOC 2009, pp. 169–178. ACM (2009)
23. Gentry, C., Sahai, A., Waters, B.: Homomorphic encryption from learning with errors: conceptually-simpler, asymptotically-faster, attribute-based. In: Canetti, R., Garay, J.A. (eds.) CRYPTO 2013. LNCS, vol. 8042, pp. 75–92. Springer, Heidelberg (2013). https://doi.org/10.1007/978-3-642-40041-4_5
24. Dov Gordon, S., Liu, F.-H., Shi, E.: Constant-round MPC with fairness and guarantee of output delivery. In: Gennaro, R., Robshaw, M. (eds.) CRYPTO 2015. LNCS, vol. 9216, pp. 63–82. Springer, Heidelberg (2015). https://doi.org/10.1007/978-3-662-48000-7_4
25. Halevi, S., Shoup, V.: Bootstrapping for HElib. In: Oswald, E., Fischlin, M. (eds.) EUROCRYPT 2015. LNCS, vol. 9056, pp. 641–670. Springer, Heidelberg (2015). https://doi.org/10.1007/978-3-662-46800-5_25
26. Jain, A., Rasmussen, P.M.R., Sahai, A.: Threshold fully homomorphic encryption. Cryptology ePrint Archive, Report 2017/257 (2017). https://eprint.iacr.org/2017/257
27. Lindner, R., Peikert, C.: Better key sizes (and attacks) for LWE-based encryption. In: Kiayias, A. (ed.) CT-RSA 2011. LNCS, vol. 6558, pp. 319–339. Springer, Heidelberg (2011). https://doi.org/10.1007/978-3-642-19074-2_21
28. López-Alt, A., Tromer, E., Vaikuntanathan, V.: On-the-fly multiparty computation on the cloud via multikey fully homomorphic encryption. In: Proceedings of the Forty-fourth Annual ACM Symposium on Theory of Computing, pp. 1219–1234. ACM (2012)
29. Lyubashevsky, V., Peikert, C., Regev, O.: On ideal lattices and learning with errors over rings. In: Gilbert, H. (ed.) EUROCRYPT 2010. LNCS, vol. 6110, pp. 1–23. Springer, Heidelberg (2010). https://doi.org/10.1007/978-3-642-13190-5_1

30. Mukherjee, P., Wichs, D.: Two round multiparty computation via multi-key FHE. In: Fischlin, M., Coron, J.-S. (eds.) EUROCRYPT 2016. LNCS, vol. 9666, pp. 735–763. Springer, Heidelberg (2016). https://doi.org/10.1007/978-3-662-49896-5_26
31. Peikert, C., Shiehian, S.: Multi-key FHE from LWE, revisited. In: Hirt, M., Smith, A. (eds.) TCC 2016. LNCS, vol. 9986, pp. 217–238. Springer, Heidelberg (2016). https://doi.org/10.1007/978-3-662-53644-5_9
32. Regev, O.: On lattices, learning with errors, random linear codes, and cryptography. In: Proceedings of the Thirty-seventh Annual ACM Symposium on Theory of Computing, STOC 2005, pp. 84–93. ACM, New York (2005)
33. Schoenmakers, B., Veeningen, M.: Universally verifiable multiparty computation from threshold homomorphic cryptosystems. In: Malkin, T., Kolesnikov, V., Lewko, A.B., Polychronakis, M. (eds.) ACNS 2015. LNCS, vol. 9092, pp. 3–22. Springer, Cham (2015). https://doi.org/10.1007/978-3-319-28166-7_1

Homomorphic Encryption
for Finite Automata

Nicholas Genise[1]($\boxtimes$), Craig Gentry[2], Shai Halevi[2], Baiyu Li[3],
and Daniele Micciancio[3]

[1] Rutgers University, Piscataway, NJ, USA
nicholasgenise@gmail.com
[2] Algorand Foundation, New York, NY, USA
craigbgentry@gmail.com, shaih@alum.mit.edu
[3] University of California, San Diego, La Jolla, CA, USA
{baiyu,daniele}@cs.ucsd.edu

Abstract. We describe a somewhat homomorphic GSW-like encryption scheme, natively encrypting matrices rather than just single elements. This scheme offers much better performance than existing homomorphic encryption schemes for evaluating encrypted (nondeterministic) finite automata (NFAs). Differently from GSW, we do not know how to reduce the security of this scheme from LWE, instead we reduce it from a stronger assumption, that can be thought of as an inhomogeneous variant of the NTRU assumption. This assumption (that we term iNTRU) may be useful and interesting in its own right, and we examine a few of its properties. We also examine methods to encode regular expressions as NFAs, and in particular explore a new optimization problem, motivated by our application to encrypted NFA evaluation. In this problem, we seek to minimize the number of states in an NFA for a given expression, subject to the constraint on the ambiguity of the NFA.

Keywords: Finite automata · Inhomogeneous NTRU · Homomorphic encryption · Regular expressions

1 Introduction

Homomorphic encryption (HE) [40] enables computation on encrypted data even without knowing the secret key. Ten years after Gentry described the first scheme capable of supporting arbitrary computations [19], we now have an arsenal of several different schemes and variations, with various capabilities and tradeoffs (see, e.g., [9,10,14,17,22,32,43] for a few examples).

Our original motivation for the current work is the simple example of encrypted virus scan: consider a center that deploys many remote systems, operating in many different environments, and wants to protect them against viruses that it knows about. The center would like to periodically send updated virus

---

N. Genise—This work was done when the author was at UCSD.
C. Gentry and S. Halevi—This work was done when the authors were in IBM Research.

S. D. Galbraith and S. Moriai (Eds.): ASIACRYPT 2019, LNCS 11922, pp. 473–502, 2019.
https://doi.org/10.1007/978-3-030-34621-8_17

signatures to all its systems, and have them scan their systems to check for infections. The virus signatures, however, could be sensitive, perhaps because some of them are not yet widely known and exposing the signatures could tip the hand of the center as it develops countermeasures.

A plausible solution would have the center encrypt the virus signatures, the remote systems could then perform the virus scan on the encrypted signatures, and report the (encrypted) results to the center. The center could then decrypt, and take appropriate actions when infections are detected. As virus signatures usually take the form of many small regular expressions[1], this application calls for a homomorphic encryption scheme that can quickly test for a match against many small regular expressions. Equivalently, it should quickly evaluate (many, encrypted) non-deterministic finite automata (NFAs) on a given cleartext file. Notice that this is quite different from, and incomparable to, the DFA computation problem studied in previous works on homomorphic encryption, like [15,16,18]. Specifically, nondeterminism aside, the crucial difference is that those works consider the evaluation of a plaintext automaton on an encrypted file. In other words, the roles of the input and the program are reversed. In our motivating application, the problem studied in [15,16,18] would correspond to searching for arbitrary (possibly nonregular) patterns, on files described by regular languages, a very unlikely scenario.

Evaluating an encrypted NFA on a cleartext string $w = w_1 \cdots w_k$ can be done by computing a product of a single vector $\mathbf{v}$ (representing the initial state of the NFA) by many matrices $\mathbf{M}_{w_i}$ (representing the transition matrices of the NFA associated to each input symbol w_i). Namely the operation that we want to support is computing

$$\mathbf{u} := \left(\prod_{i=k}^{1} \mathbf{M}_{w_i} \right) \times \mathbf{v},$$

(with operations over the integers), where the matrices $\mathbf{M}_{w_i}$ and the vector $\mathbf{v}$ are encrypted.[2] Most of the HE schemes from above can be used to carry out this computation, but none of them is ideal for the job. For practical purposes, the homomorphic schemes that offer the best performance are either the BGV-type schemes (scale-invariant or not), or GSW-type schemes.

BGV-Type Schemes. These schemes have an advantage that they can use *packed ciphertexts*, where each ciphertext encrypts not just one plaintext element but a vector of them, and each ciphertext operation affects all the elements of the

---

[1] For example, many ClamAV virus signatures (https://www.clamav.net/downloads) are regular expressions of the form $\Sigma^* K_1 \cdots \Sigma^* K_n \cdot \Sigma^*$ with no more than 1 K symbols, where Σ is the alphabet and each K_i is a set of a few hex strings.

[2] The initial vector $\mathbf{v}$ is not required to be encrypted, as it reveals no information about the automaton. However, the intermediate vectors obtained after each matrix-vector multiplication should be kept secret. So, we will need a scheme supporting matrix-vector multiplication where both the matrix and the vector are encrypted.

vector simultaneously, cf. [42]. Moreover, they can even be made to support efficient matrix-vector operations, as was demonstrated in [23].[3]

However, for BGV-type schemes it is crucial to keep the computation multiplicative depth to a minimum, which in our case means using a binary multiplication tree. But this means that we have to use matrix-matrix multiplication[4] (rather than the matrix-vector products that are computed in the sequential procedure). This increases the total work (and hence the computation time) by a factor equal to the dimension of these matrices—which must be substantial for security reasons.

GSW-Type Schemes. A major advantage of GSW-like schemes is the asymmetric noise growth, that makes it possible to handle sequential processing of products [12]. For our purposes, it lets us evaluate the product while performing only matrix-vector multiplications.

While "textbook GSW" can only encrypt individual elements, it is possible to adapt the ciphertext-packing techniques from [42] also to GSW, as long as we have a priori bound on the size of the plaintext vectors that occur in the computation. However porting the matrix-multiplication optimizations from [23] is far from simple, and we expect significant overhead when trying to implement it in practice.

In [25], Hiromasa, Abe, and Okamoto proposed a GSW-like FHE scheme that is capable of encrypting square matrices and doing homomorphic matrix addition and multiplication. The HAO15 FHE scheme can be viewed as a matrix extension of the standard GSW-FHE scheme, where the secret key $S = [I | -S']$ consists of a random secret matrix S'. Like in GSW [22], the decryption invariant for a ciphertext C encrypting a message M relative to the secret key S is

$$S \times C = M \times S \times G + E \pmod{q},$$

where E is a low-norm error and G is the "gadget matrix" from [36]. Notice that M and S are both matrices in the matrix-FHE case, whereas in the GSW scheme M is a scalar and S is a vector. The GSW security reduction [22] from the learning-with-errors (LWE) problem still applies to the HAO15 scheme, except that an additional circular security assumption is required. Being able to encrypt matrices in an atomic operation and support homomorphic matrix operations makes the HAO15 scheme an interesting candidate to use in our application of homomorphic NFA evaluation. Moreover, as we will show in Sect. 3.1, the HAO15 scheme with some modification can also encrypt vectors and homomorphically multiply an encrypted matrix by an encrypted vector. However, the HAO15 scheme is not optimal due to overhead in the size of keys and ciphertexts. So we seek to find a better solution that would allow us to scan longer strings with faster execution times in practice.

---

[3] The techniques in [23] only handle multiplication of plaintext matrices by encrypted vectors, but many of these tools can be adapted to the case of encrypted matrices.

[4] Technically, the nodes on the rightmost path of the tree can use matrix-vector multiplications, but this makes hardly any difference on the efficiency of the overall computation.

1.1   Our New HE Scheme

In this work we introduce a new scheme, that can be viewed as another GSW-type encryption for matrices but with a different hardness assumption. (Alternatively, it can be viewed as a variant of the GGH15 graded encoding [20], but with no zero-test parameter.) In addition, our scheme can also encrypt vectors and natively support homomorphic matrix-vector multiplication. Similar to the HAO15 scheme, the decryption invariant in our scheme for a ciphertext $\mathbf{C} \leftarrow \mathsf{Enc}_\mathbf{S}(\mathbf{M})$ encrypting a matrix $\mathbf{M}$ is also $\mathbf{S} \times \mathbf{C} = \mathbf{MSG} + \mathbf{E} \pmod{q}$, where $\mathbf{E}$ is a low-norm error matrix.[5] Differently from the HAO15 scheme, in our construction we assume that the key $\mathbf{S}$ is a square invertible matrix, and so we can express the ciphertext as $\mathbf{C} := \mathbf{S}^{-1}(\mathbf{M} \times \mathbf{S} \times \mathbf{G} + \mathbf{E}) \bmod q$. As a result, both keys and ciphertexts are smaller in our scheme.

The operations of the scheme, and the analysis of the noise development are identical to the GSW scheme, except that here we typically cannot ensure that the plaintext size never grows, and instead must use properties of the application to reason about the plaintext size.

When it comes to security, however, we can no longer use the GSW reduction [22] from the LWE problem. That reduction relies heavily on the scalar $\mathbf{M}$ commuting with the vector $\mathbf{S}$, which no longer holds in our case. Instead, we reduce the security of this scheme from a stonger assumption, that can be viewed as an inhomogeneous version of NTRU (or alternatively as an LWE instance with an additional hint).

1.2   The iNTRU Hardness Assumption

Recall that in LWE[6], we are given two matrices $\mathbf{A}, \mathbf{B} \in \mathbb{Z}_q^{n \times m}$ $(m > n)$, with $\mathbf{A}$ a uniformly random matrix, and need to decide if $\mathbf{B}$ is also a uniformly random matrix, or it is chosen as $\mathbf{B} = \mathbf{SA} + \mathbf{E}$ with a uniform $\mathbf{S} \in \mathbb{Z}_q^{n \times n}$ and a low-norm $\mathbf{E} \in \mathbb{Z}_q^{n \times m}$.

It is easy to see that this problem becomes easy if we are also given a trapdoor for the matrix $\mathbf{A}$, in this case it is even easy to recover the secret matrix $\mathbf{S}$ when $\mathbf{B} = \mathbf{SA} + \mathbf{E}$. But what if we are given a trapdoor for the matrix $\mathbf{B}$ instead? In this case we do not know of any effective distinguisher, so we assume that the decision problem is still hard and show a hardness reduction from this version of LWE to our hardness assumption, iNTRU, in Sect. 4. We remark that this "LWE with a trapdoor for $\mathbf{B}$" assumption is not standard and it deserves further study.

Once we know a trapdoor for $\mathbf{B}$, we might as well consider the case where $\mathbf{B}$ is the gadget matrix $\mathbf{G}$ (for which everyone knows a trapdoor). Namely we assume that the following decision problem is hard:

---

[5] As we describe later, we use a slightly different variant to encrypt the vector $\mathbf{v}$.

[6] Here we refer to the multiple-secret variant of LWE, which can be reduced from the normal LWE.

iNTRU. As in LWE, we have the parameters n, m, q, with $m > n \log q$ and $q > m$. The input is a matrix $\mathbf{A} \in \mathbb{Z}_q^{n \times m}$, which is either uniform in $\mathbb{Z}_q^{n \times m}$, or is set as $\mathbf{A} := \mathbf{S}^{-1}(\mathbf{G} - \mathbf{E}) \bmod q$ (with $\mathbf{S} \in \mathbb{Z}_q^{n \times n}$ a random invertible matrix, $\mathbf{G}$ the gadget matrix, and $\mathbf{E}$ a low-norm matrix). The goal is to decide which is the case.

One can think of the above problem as an inhomogeneous version of NTRU, over matrices, as follows. Recall that in the NTRU cryptosystem [26], the secret key is given by two polynomials (or ring elements) with small coefficients f, g, and the corresponding public key is the product $h = f^{-1} \cdot g$. The NTRU cryptosystem can be proved secure under the assumption that this public key h is pseudorandom, i.e., indistinguishable from a uniformly random polynomial (or ring element) with arbitrary coefficients. We extend this assumption as follows. First, we replace g with a sequence of vectors $g_1, \ldots, g_k$, chosen independently at random, with small coefficients. Then, the assumption is that $f^{-1}g_1, f^{-1}g_2, \ldots, f^{-1}g_k$ is pseudorandom. This is a simple syntactic extension of NTRU (that would allow, for example, the encryption of longer messages), akin to changing some parameter, and not a qualitative change in the security assumption. Next, we add a (known, constant) "shift", replacing each g_i with $(2^{i-1} - g_i)$, and still requiring $f^{-1}(1 - g_1), f^{-1}(2 - g_2), \ldots, f^{-1}(2^{k-1} - g_k)$ to be indistinguishable from uniform. We call this the "inhomogeneous" NTRU assumption. Finally, instead of working over a ring of polynomials of degree n, we replace each $f, g_1, \ldots, g_k$ with a square $n \times n$ random matrix with small entries. Intuitively, moving from polynomial rings (which are commutative) to the ring of matrices, should only make the assumption weaker, though we do not know how to prove a formal relation between the two problems. This last problem is essentially equivalent to the pseudorandomness of $\mathbf{A} = \mathbf{S}^{-1}(\mathbf{G} - \mathbf{E})$, where $\mathbf{E} = [\mathbf{E}_0 | \ldots | \mathbf{E}_k]$ is a random matrix with small entries, and $\mathbf{G} = [\mathbf{0} | \mathbf{I} | 2\mathbf{I} | \ldots | 2^{k-1}\mathbf{I}]$ is a constant known matrix. In fact, putting $\mathbf{A}$ in Hermite Normal Form [35] "cancels out" the $\mathbf{S}$ matrix, and gives a sequence of square matrices $-\mathbf{E}_0^{-1}(2\mathbf{I}^{i-1} - \mathbf{E}_i)$, corresponding to the matrix version of our inhomogeneous NTRU problem[7] with $f = -\mathbf{E}_0$ and $g_i = \mathbf{E}_i$.

1.3   From Regular Expression to NFAs

While our scheme directly supports the evaluation of (encrypted) NFAs, patterns (e.g., virus signatures) are typically, and most conveniently, represented by regular expressions. Since the noise growth of our homomorphic encryption scheme depends on the details of the NFA being evaluated and its computations, the conversion of regular expressions to NFA is a critical part of our application. In Sect. 5 we describe a specific conversion following the method of [3,13] based on the use of *partial derivatives* of regular expressions, which is both very elegant and efficient. Derivatives of regular expressions [13] are themselves regular expressions and they are defined similarly to formal derivatives of arithmetic

---

[7] Matrix-NTRU has been used in lattice-based signatures [5], though the most efficient versions of these lattice signatures use the standard, algebraic NTRU assumption.

expressions, e.g., $d_a(e_0 + e_1) = d_a(e_0) + d_a(e_1)$ for the sum (set union) operation, and $d_a(e^*) = d_a(e)e^*$ for exponentiation (Kleene star). Informally, when parsing an input string according to regular expression e, the derivative $d_a(e)$ represents the part of the input to be expected after reading a first symbol "a". A regular expression e can be converted into an automaton with states labeled by derivatives (modulo a natural equivalence relation on regular expressions), and transitions of the form $e \xrightarrow{a} d_a(e)$. A classical result of Brzozowski [13] shows that this produces an automaton with a finite number of states, and, in fact, the minimal DFA of the regular expression. As our homomorphic encryption scheme supports the evaluation of nondeterministic automata, we are interested in the conversion of regular expressions to NFAs, which are potentially much smaller than the equivalent minimal DFA. However, optimizing NFAs in our application is far from trivial. To start with, in stark contrast to the DFA case, minimizing the number of states of an NFA is a PSPACE-complete problem. Moreover, due to noise growth, minimizing the number of states may not even be the right goal for our homomorphic encryption application. We address the first issue by using the *partial derivative* construction of [3], where a partial derivative $\partial_a(e)$ maps an expression e to a *set* of regular expressions (representing possible nondeterministic choices), and in particular $\partial_a(e_0 + e_1) = \partial_a(e_0) \cup \partial_a(e_1)$. This construction results in NFAs that, while not necessarily minimal, have a very small number of states, bounded by the number of alphabet symbols in the input regular expression. In order to bound the noise growth, we show that a simple optimization of the homomorphic NFA evaluation procedure[8] allows to relate the noise growth to the *degree of ambiguity* of the NFA, a standard quantity studied in automata theory, which can be evaluated in polynomial time [45]. We reduce the problem of finding an optimal noise to a variant of NFA minimization problem with bounded ambiguity. Although solving this optimization problem is hard in general, we use techniques of determining ambiguity in Sect. 5 to explore some tradeoffs between automata size and degree of ambiguity/noise growth.

1.4   Implementation and Performance

We implemented our scheme in C++ using the Number Theory Library (NTL) and describe its details in Sect. 6. Despite being a simple implementation without optimizations, the on-line pattern matching was exceptionally fast. For example, we could homomorphically match a 65536 bit string in 394 s on an encrypted NFA with 1024 states of size 66 Mb. Using the same set of parameters, we estimate that an HAO15 implementation can only match up to 16000 bits with a slower execution time and a bigger program size. More performance details and comparisons can be found in Sect. 6.

---

[8] Namely, one can let the initial state vector $\mathbf{v}$ be an "errorless" encryption, because the initial state does not reveal any information about the rest of the automaton.

1.5 Related Work

As already mentioned, the problem of homomorphically evaluating finite automata or branching programs has been considered before [12,15,16,18], but in a very different context, where the branching program or automaton are publicly known, and the computation is performed homomorphically on an encrypted input string. This is motivated, for example, by applications to FHE bootstrapping, where the program is specified by the publicly known decryption/refreshing procedure, and the input in the (encrypted) secret key. In our setting, the role of the program and input are reversed, and we want the computation to be homomorphic on the automaton, rather than the input string. In the case of general computation, program and input are easily interchanged using a universal Turing machine. But in the case of restricted models of computation, like finite automata, swapping the program and the input results in a completely different problem.

On the Relation with Other Matrix-FHE Schemes. As we mentioned earlier, the HAO15 [25] FHE scheme is also capable of encrypting square matrices and doing homomorphic matrix addition and multiplication on ciphertexts. In the private-key version of their scheme, the secrete key is $\mathbf{S} = [\mathbf{I}_r | - \mathbf{S}']$ for a secret matrix $\mathbf{S}'$, and a matrix $\mathbf{M} \in \mathbb{Z}^{r \times r}$ is encrypted as

$$C = \left(\frac{\mathbf{S}'\mathbf{A} + \mathbf{E}}{\mathbf{A}} \right) + \left(\frac{\mathbf{MS}}{\mathbf{0}} \right) \times \mathbf{G} \bmod q,$$

where $\mathbf{A} \leftarrow \mathbb{Z}_q^{n \times N}$, $\mathbf{E} \leftarrow \chi^{r \times N}$ for $N = (n + r) \lceil \log q \rceil$.

It may be tempting to claim that our scheme is the same as the HAO15 scheme due to having the same decryption invariant $\mathbf{SC} = \mathbf{MSG} + \mathbf{E}$. However, these two schemes are not quite identical. The relation between them is very similar to the relation between NTRU and RLWE Regev-like schemes[9], where the difference is that the secret key $\mathbf{S}$ is a small square matrix for NTRU (representing multiply-by-s in the ring), whereas the secret key is $\mathbf{S} = [\mathbf{I}|\mathbf{S}']$ in RLWE (where $\mathbf{S}'$ represents multiply-by-s' in the ring). Notice that, instead of the Regev invariant, both the HAO15 scheme and our scheme use the GSW-like invariant $\mathbf{SC} = \mathbf{MSG} + \mathbf{E}$ for a small noise matrix $\mathbf{E}$.

More specifically, in our scheme the secret key $\mathbf{S}$ is a small square matrix that must be invertible, while in HAO15 we have $\mathbf{S} = [\mathbf{I} | - \mathbf{S}']$ where $\mathbf{S}'$ can be any random matrix. Consider the "leveled versions" of the HAO15 scheme and our scheme, in which the secret key matrices $\mathbf{S}_0, \mathbf{S}_1, \ldots, \mathbf{S}_L$ are generated such that $\mathbf{S}_i$ is used to encrypt the matrices in level i of the computation. In both schemes it holds that

$$\mathbf{S}_i \mathbf{C}_i = \mathbf{MS}_{i+1} \mathbf{G} + \mathbf{E}_i.$$

---

[9] Consider writing both NTRU and RLWE-Regev in matrix form, representing ring elements by their matrices: In both NTRU and RLWE-Regev we have a ciphertext matrix $\mathbf{C}$ encrypting a plaintext matrix $\mathbf{M}$ relative to the secret matrix $\mathbf{S}$ (and plaintext space mod p) if $\mathbf{SC} = \mathbf{M} + p\mathbf{E} \bmod q$.

The security of the HAO15 scheme can be reduced from the standard LWE assumption, while our scheme relies on the NTRU-like assumption that we introduce. On the other hand, our scheme is more efficient: we encrypt a matrix $\mathbf{M} \in \mathbb{Z}_q^{r \times r}$ in a ciphertext matrix of dimension $\max(r, \lambda)$, whereas the HAO15 scheme requires a dimension $r + \lambda$ ciphertext matrix. One can view our scheme as an NTRU-like variant of the HAO15 scheme (or perhaps an NTRU-like variant of the GSW scheme). From that viewpoint, we introduce in this work the assumption that lets us adapt NTRU to get a GSW-like scheme.

When applied to homomorphically evaluating NFAs, the efficiency advantage of our scheme is more significant. Note that the HAO15 scheme can be used to do homomorphic matrix-vector multiplication as well. But, since we rely on an NTRU-like assumption, the noise bound in our scheme is smaller than the noise bound in the HAO15 scheme, which allows us to homomorphically evaluate longer strings with the same lattice parameters. In terms of the complexity of the homomorphic computation on encrypted NFAs, our scheme runs faster than the HAO15 scheme in practice due to smaller ciphertexts. For more detailed performance comparison, we refer the readers to Sect. 6 and Appendix C.

Recently, Wang et al. [44] proposed another matrix-FHE scheme, similar to [9], that has smaller ciphertexts than the HAO15 scheme and can be reduced from the standard LWE assumption. We note that it is possible to perform homomorphic matrix-vector multiplication in their scheme. However, their scheme relies heavily on tensor product to perform homomorphic multiplication, so the security and the complexity of applying their scheme to homomorphic NFA computation is at least on the same level as the HAO15 scheme.

2   Preliminaries

We denote vectors by lower-case bold letters (e.g., $\mathbf{v}$), and we assume they are always in column form. We denote matrices by upper-case bold letters (e.g., $\mathbf{M}$). A distribution $\mathcal{D}$ over a finite set X is ϵ-uniform if its statistical distance from the uniform distribution over X is at most ϵ, where the statistical difference between two distributions $\mathcal{D}_1, \mathcal{D}_2$ over a finite domain X is $\frac{1}{2} \sum_{x \in X} |\mathcal{D}_1(x) - \mathcal{D}_2(x)|$. We denote by $x \leftarrow \mathcal{D}$ drawing x from the distribution $\mathcal{D}$, and for a set X we denote by $x \leftarrow X$ drawing x uniformly at random from X.

2.1   Leftover Hash Lemma

A distribution $\mathcal{D}$ over X has min-entropy k if $\max_{x \in X} \mathcal{D}(x) = 2^{-k}$. A family $\mathcal{H}$ of hash functions from X to Y (with Y a finite set) is said to be 2-universal if for all distinct $x, x' \in X$, $\Pr_{h \leftarrow \mathcal{H}}[h(x) = h(x')] = 1/|Y|$.

Lemma 1. *(Leftover Hash Lemma [24]). Let $\mathcal{H}$ be a family of 2-universal hash functions from X to Y, and let $\mathcal{D}$ be a distribution over X with min-entropy k. Suppose that $h \leftarrow \mathcal{H}$ and $x \leftarrow \mathcal{D}$ are chosen independently, then, $(h, h(x))$ is $(\frac{1}{2}\sqrt{|Y|/2^k})$-uniform over $\mathcal{H} \times Y$.*

In this work we apply Lemma 1 to the hashing family $\mathcal{H} : \mathbb{Z}_q^m \to \mathbb{Z}_q^n$ defined by

$$\mathcal{H} = \{h_A(\mathbf{v}) = \mathbf{A}\mathbf{v} \bmod q\}_{\mathbf{A} \in \mathbb{Z}_q^{n \times m}},$$

(which is clearly 2-universal). In particular we use the following corollary:

Corollary 1. *Fix the integers k, n, m, m', q, and let $\mathcal{D}_1, \mathcal{D}_2, \ldots, \mathcal{D}_m$ be independent distributions over $\mathbb{Z}_q^m$, all with min-entropy at least k. Let $\mathcal{D}$ be a distribution over matrices $\mathbf{R} \in \mathbb{Z}_q^{m \times m'}$, where the i'th column is drawn from $\mathcal{D}_i$. Then the distribution*

$$\{(\mathbf{A}, \mathbf{A}\mathbf{R} \bmod q) : \mathbf{A} \leftarrow \mathbb{Z}_q^{n \times m}, \mathbf{R} \leftarrow \mathcal{D}\}$$

is $(\frac{m'}{2}\sqrt{q^n/2^k})$-uniform over $\mathbb{Z}_q^{n \times m} \times \mathbb{Z}_q^{n \times m'}$.

2.2   Gadget Lattice Sampling

Definitions. We consider the norm of a matrix as the length of its longest column in the l_2 norm. A lattice Λ is a discrete subgroup of $\mathbb{R}^n$ (we only consider full-rank, integer lattices). It can be represented by a basis $\mathbf{B} \in \mathbb{Z}^{n \times n}$ where the lattice is the set of all integer combinations of $\mathbf{B}$'s columns. Let $\mathbf{G} = [\mathbf{I}|2\mathbf{I}|\cdots|2^{\ell-1}\mathbf{I}] \in \mathbb{Z}_q^{n \times n\ell}$ where $\ell = \lceil \log_2(q) \rceil$. The G-lattice for a fixed modulus q is $\Lambda_q^\perp(\mathbf{G}) = \{\mathbf{x} \in \mathbb{Z}^{n\ell} : \mathbf{G}\mathbf{x} \bmod q = \mathbf{0}\}$. The distribution sampled over $\Lambda_q^\perp(\mathbf{G})$ and its integer cosets is the discrete gaussian, a gaussian distribution conditioned on being in the lattice. The probability a sample equals some lattice coset vector $\mathbf{y}$ is proportional to $\exp(-\pi\|\mathbf{y}\|^2/s^2)$ where $s > 0$ is the width of the gaussian (we are only concerned with $\mathbf{0}$-centered distributions). Denote a discrete gaussian of width s on a lattice coset $\Lambda + \mathbf{c}$ as $\mathcal{D}_{\Lambda+\mathbf{c},s}$. We can efficiently sample from $\mathcal{D}_{\Lambda_q^\perp(\mathbf{G})+\mathbf{v},s}$ for any $q \geq 2$ and $s \geq \sqrt{5\ln(2n\ell + 4)/\pi}$ (Theorem 4.1 [36] and Lemma 2.3 [11]). We denote $\mathbf{G}^{-1}(\mathbf{v})$ as a discrete gaussian vector $\mathbf{y}$ such that $\mathbf{G}\mathbf{y} = \mathbf{v} \bmod q$. Further, we assume the width is set just above twice the smoothing parameter (defined below) of the G-lattice.

Concentration and Min-entropy. The smoothing parameter [37] of a lattice is needed for our purposes, and it is denoted as $\eta_\epsilon(\Lambda)$ for an $\epsilon > 0$. Informally, this is the smallest width for which a discrete gaussian shares many properties of the continuous gaussian distribution. If $\mathbf{B}$ is a basis with minimum Gram-Schmidt norm $\|\widetilde{\mathbf{B}}\|$, we can bound the smoothing parameter $\eta_\epsilon(\Lambda) \leq \|\widetilde{\mathbf{B}}\|\omega(\sqrt{\log n})$ for negligible $\epsilon(n) = n^{-\omega(1)}$ [21]. Discrete gaussian samples' l_2 norms are bounded by their width as follows.

Lemma 2. *(Lemma 1.5 [6]) Let $\Lambda \subset \mathbb{R}^n$ be a lattice, $r \geq \eta_\epsilon(\Lambda)$ for some $\epsilon \in (0, 1)$, and $\mathbf{c} \in \mathbb{R}^n$. Then,*

$$\Pr(\|\mathcal{D}_{\Lambda+\mathbf{c},r} \geq r\sqrt{n}\|) \leq 2^{-n} \cdot \left(\frac{1+\epsilon}{1-\epsilon}\right).$$

Therefore, we can efficiently sample a discrete gaussian $\mathbf{G}^{-1}(\cdot)$ with length less than $\widetilde{O}(\sqrt{n \log q})^{10}$ with overwhelming probability, and assume $\mathbf{G}^{-1}(\cdot)$'s support is $\mathbb{Z}_q^{n\ell}$. Since we will be using the leftover hash lemma on discrete gaussian input, we will use the following lemma on the min-entropy of a discrete gaussian. Further, the proof of Lemma 3 is identical to the proof of [38, Lemma 2.11].

Lemma 3. *(Lemma 2.11 [38]) Let $\Lambda + \mathbf{v} \subset \mathbb{R}^n$ be a lattice coset, $c > 0$, and $s \geq 2^{1+c} \eta_\epsilon(\Lambda)$ for $\epsilon \in (0,1)$. Then for any $\mathbf{y} \in \Lambda + \mathbf{v}$ and for $\mathbf{x} \leftarrow \mathcal{D}_{\Lambda+\mathbf{v},s}$,*

$$\Pr(\mathbf{x} = \mathbf{y}) \leq 2^{-n(1+c)} \left(\frac{1+\epsilon}{1-\epsilon} \right).$$

Leftover Hash Lemma with $\mathbf{G}^{-1}(\cdot)$. Let $m = n\ell$, now we can replace the distributions $\mathcal{D}_i$ in Corollary 1 with independent discrete gaussian samples $\mathbf{G}^{-1}(\mathbf{v})$ (with potential repeats in the coset vector $\mathbf{v}$). Let $\mathbf{R} \leftarrow \mathbf{G}^{-1}(\mathbf{X})$ in Corollary 1 for some $\mathbf{X} \in \mathbb{Z}_q^{n \times m'}$ with $\mathbf{R}$'s columns sampled independently. Then by the lemmas above, the min-entropy a column of $\mathbf{R}$ is at least $n(1+c) \log q - 2$ whenever $\mathbf{G}^{-1}(\cdot)$'s width is just above twice $\eta_\epsilon(\Lambda_q^\perp(\mathbf{G}))$ for any $\epsilon \in (0, 1/2]$. Say we let $c = \log_q(2)$ in Lemma 3. This implies the distribution

$$\{(\mathbf{A}, \mathbf{AR} \bmod q) : \mathbf{A} \leftarrow \mathbb{Z}_q^{n \times m}, \mathbf{R} \leftarrow \mathbf{G}^{-1}(\mathbf{X})\}$$

is $O(m' 2^{-n/2})$-uniform for any $\mathbf{X} \in \mathbb{Z}_q^{m \times m'}$.

3   The Schemes

Given an NFA $\mathcal{M}$ of r states over a finite alphabet Σ, we denote by $\mathbf{M}_\sigma \in \{0,1\}^{r \times r}$ the transition matrix of $\mathcal{M}$ for each input symbol $\sigma \in \Sigma$, where $(\mathbf{M}_\sigma)_{j,i} = 1$ if and only if there is a transition from state i to state j on σ. Let $\mathbf{v} \in \{0,1\}^r$ be the vector representing the initial states. To check if a string $w = w_1 \cdots w_k \in \Sigma^*$ is accepted by $\mathcal{M}$, we simply check whether there are any non-zero entries in the vector $(\prod_{i=k}^1 \mathbf{M}_{w_i}) \times \mathbf{v}$ that correspond to final states. So we need a scheme that can compute matrix-vector multiplication homomorphically over encrypted matrices and vectors.

3.1   The HAO15 Matrix-FHE Scheme [25]

The FHE scheme from [25] can be extended to support homomorphic matrix-vector multiplication. We first recall the private-key version of the HAO15 scheme, and we then slightly extend it for vector encryption and homomorphic matrix-vector multiplication. For a given security parameter λ, choose lattice parameters n, m, q and a noise distribution χ over $\mathbb{Z}_q$. Let $\ell = \lceil \log q \rceil$, $m = (n + r) \log q$, and $N = (n + r)\ell$. Here we describe a leveled version of the HAO15 scheme that supports multiplication depth up to $k \geq 1$. We abuse notation and have $\mathbf{G} = [\mathbf{0}|\mathbf{I}|2\mathbf{I}|\cdots|2^{\ell-1}\mathbf{I}]$ in this subsection.

---

[10] $\widetilde{O}(\cdot)$ hides poly-logarithmic factors in n.

Key Generation. Same as in HAO15, the secret key for level $i \geq 0$ is set to $\mathsf{sk}_i := \mathbf{S}_i = [\mathbf{I}_r | - \mathbf{S}'_i]$, where $\mathbf{S}'_i \leftarrow \chi^{r \times n}$.

Matrix Encryption. Given a plaintext matrix $\mathbf{M} \in \{0,1\}^{r \times r}$ and a level $i \geq 0$, to encrypt it for the i'th level of computation, the HAO15 scheme outputs

$$\mathbf{C} := \mathsf{HAO.MatEnc}_{\mathsf{sk}_i}(M) = \begin{pmatrix} \mathbf{S}'_i \mathbf{A}' + \mathbf{E} \\ \mathbf{A}' \end{pmatrix} + \begin{pmatrix} \mathbf{M} \mathbf{S}_{i-1} \\ \mathbf{0}_{n \times (n+r)} \end{pmatrix} \mathbf{G} \bmod q,$$

where $\mathbf{A}' \leftarrow \mathbb{Z}_q^{n \times N}$ and $\mathbf{E} \leftarrow \chi^{r \times N}$. For $i = 0$, we consider $\mathbf{S}_{-1} = [\mathbf{I}_r | \mathbf{0}_{r \times n}]$. Notice that $\mathbf{C} \in \mathbb{Z}_q^{(r+n) \times N}$. The decryption procedure is exactly the same as in [25], but we skip it as it is not needed in our application.

Vector Encryption and Decryption. For a vector $\mathbf{v} \in \mathbb{Z}_q^r$, we can follow the same idea as in the matrix encryption procedure, except that we do not multiply $\mathbf{v}$ by $\mathbf{S}$ nor $\mathbf{G}$. Since we only need to encrypt the initial state vector to evaluate an NFA, we always encrypt a vector using the secret key for the first level:

$$\mathbf{c} := \mathsf{HAO.VecEnc}_{\mathsf{sk}_0}(\mathbf{v}) = \begin{pmatrix} \mathbf{S}'_0 \mathbf{a} + \mathbf{e} \\ \mathbf{a} \end{pmatrix} + \begin{pmatrix} \mathbf{v} \\ \mathbf{0}_n \end{pmatrix} \bmod q,$$

where $\mathbf{a} \leftarrow \mathbb{Z}_q^n$ and $\mathbf{e} \leftarrow \chi^r$. Note that $\mathbf{c}$ has dimension $r + n$. To decrypt a ciphertext vector $\mathbf{c}$ from the i'th level of a computation, output the vector

$$\mathbf{v}' := \mathsf{HAO.VecDec}_{\mathsf{sk}_i}(\mathbf{c}) = \lceil \mathbf{S}_i \mathbf{c} \rfloor_2.$$

Homomorphic Operations. To add and multiply two ciphertext matrices $\mathbf{C}_1$ and $\mathbf{C}_2$, we follow [25]: $\mathsf{HAO.Add}(\mathbf{C}_1, \mathbf{C}_2) = \mathbf{C}_1 + \mathbf{C}_2$, and $\mathsf{HAO.Mul}(\mathbf{C}_1, \mathbf{C}_2) = \mathbf{C}_1 \times \mathbf{G}^{-1}(\mathbf{C}_2)$. To multiply a ciphertext matrix $\mathbf{C}$ by an encrypted vector $\mathbf{c}$, output

$$\mathsf{HAO.Mul}(\mathbf{C}, \mathbf{c}) := \mathbf{C} \times \mathbf{G}^{-1}(\mathbf{c}).$$

The security of this extended scheme can be proved in the same way as in [25], reducing from the standard $\mathsf{DLWE}_{n,m,q,\chi}$ hardness assumption. It is easy to check that, if $\mathbf{C}$ is an encryption of $\mathbf{M} \in \{0,1\}^{r \times r}$ for level i and $\mathbf{c}$ is an encryption of $\mathbf{v}$ of level $i - 1$, then $\mathbf{S}_i \times (\mathbf{C} \times \mathbf{G}^{-1}(\mathbf{c})) = \mathbf{M}\mathbf{v} + \mathbf{e}'$ for some low norm error vector $\mathbf{e}'$. More generally, for any $\mathbf{M}_i \in \{0,1\}^{r \times r}$ for $i = 1, \dots, k$ and $\mathbf{v} \in \mathbb{Z}_q^r$, if $\mathbf{C}_i \leftarrow \mathsf{HAO.MatEnc}_{\mathsf{sk}_i}(\mathbf{M}_i)$ with an error matrix $\mathbf{E}_i$ for each i, $\mathbf{c}_0 \leftarrow \mathsf{HAO.VecEnc}_{\mathsf{sk}_0}(\mathbf{v})$ with an error vector $\mathbf{e}$, and $\mathbf{c}_i \leftarrow \mathsf{HAO.Mul}(\mathbf{C}_i, \mathbf{c}_{i-1})$ for $i = 1, \dots, k$, then $\mathbf{S}_k \times \mathbf{c}_k = (\prod_{j=k}^{1} \mathbf{M}_j)\mathbf{v} + \mathbf{e}_k$ where

$$\mathbf{e}_k = \mathbf{E}_k \mathbf{G}^{-1}(\mathbf{c}_{k-1}) + \sum_{i=2}^{k} (\prod_{j=k}^{i} \mathbf{M}_j)\mathbf{E}_{i-1}\mathbf{G}^{-1}(\mathbf{c}_{i-2}) + (\prod_{j=k}^{1} \mathbf{M}_j)\mathbf{e}.$$

The l_∞ norm of $\mathbf{e}_k$ can be bounded by

$$\|\mathbf{e}_k\|_\infty \leq \chi N (1 + k \max_{1 \leq i \leq k} \| \prod_{j=k}^{i} \mathbf{M}_j\|_\infty). \tag{1}$$

To successfully decrypt $\mathbf{c}_k$, we require $\|\mathbf{e}_k\|_\infty \leq q/8$ as in [25].

3.2   Our New Matrix-HE Scheme

To achieve sufficient level of security and a desired capability of homomorphic
NFA evaluation, we may need to use a large lattice dimension n in practice.
The above extension of the HAO15 scheme seems suboptimal with an overhead n in ciphertext dimension. In this section we describe a new matrix homomorphic encryption scheme that supports atomic matrix and vector encryption and matrix-vector multiplication. Our scheme is more efficient in practical
applications.

Fix integer parameters n, m, q (to be determined later) and an error distribution χ over $\mathbb{Z}_q$ that outputs with high probability integers of magnitude $\ll q$.
Given any NFA with $r \leq n$ states, we pad its transition matrices $\mathbf{M}_\sigma$ with 0
entries such that $\mathbf{M}_\sigma \in \{0,1\}^{n \times n}$ for all $\sigma \in \Sigma$. For our application we use two
variants of (private-key) encryption, one for matrices and the other for vectors.
Both variants share a noise-sampling procedure, that takes as input the secret
key and another vector (that comes from the plaintext) and outputs a noise
vector for use in the encryption (which may be different than just sampling from
χ). We denote this procedure by $\mathbf{e} \leftarrow \mathsf{NoiseSamp}(\mathsf{sk}, \mathbf{v})$, and will describe it later
in this section.

Key Generation. We draw two matrices using χ, a square matrix $\mathbf{S} \leftarrow \chi^{n \times n}$ and
a rectangular $\mathbf{E} \leftarrow \chi^{n \times m}$ (which is only used in the $\mathsf{NoiseSamp}$ procedure). We
insist that $\mathbf{S}$ is invertible, and re-sample if it is not (which happens with a small
probability $\approx 1/q$). The secret key is $\mathsf{sk} := (\mathbf{S}, \mathbf{E})$.

The $\mathsf{NoiseSamp}$ Procedure. To prove semantic security of our encryption method,
we need a somewhat convoluted procedure for sampling the noise. Specifically,
the procedure $\mathsf{NoiseSamp}((\mathbf{S}, \mathbf{E}), \mathbf{v})$ begins by sampling $\mathbf{r} \leftarrow \mathbf{G}^{-1}(\mathbf{v})$, then outputs $\mathbf{e} := \mathbf{E} \times \mathbf{r} \bmod q$.

Basic "Encryption" Transformation. Underlying both the vector and matrix
encryption procedure, is the following "encryption" procedure (in quotes, since it
does not have a matching decryption procedure). Given the secret key $\mathsf{sk} = (\mathbf{S}, \mathbf{E})$
and a vector $\mathbf{v} \in \mathbb{Z}_q^n$, we draw a noise vector $\mathbf{e} \leftarrow \mathsf{NoiseSamp}(\mathsf{sk}, \mathbf{v})$, then output
the "ciphertext"

$$\mathbf{c} := \mathsf{Enc}^*_{\mathsf{sk}}(\mathbf{v}) = \mathbf{S}^{-1}(\mathbf{v} + \mathbf{e}).$$

We remark that the low-order bits of $\mathbf{v}$ are lost in this transformation, due the
added noise. Still, the "ciphertext" satisfies the property that $\mathbf{S}\mathbf{c} \approx \mathbf{v}$, up to the
low-norm noise vector $\mathbf{e}$.

We provide in Sect. 4 a detailed proof that the procedure above provides
semantic security for $\mathbf{v}$, under the inhomogeneous NTRU hardness assumption.

Vector Encryption and Decryption. As with Regev encryption [39], to convert
the above to real encryption we just need to multiply $\mathbf{v}$ by a large enough scalar
β so that $\|\mathbf{e}\|_\infty < \beta$ with high probability. Let b be an upper bound on the l_∞

norm of vectors that can be dealt with (which depends on the application), we assume that $b \ll q$ and set $\beta := \lfloor q/b \rfloor$.

To encrypt a vector $\mathbf{v} \in \mathbb{Z}_b^n$ we just set $\mathbf{c} := \mathsf{VecEnc}_{\mathsf{sk}}(\mathbf{v}) = \mathsf{Enc}_{\mathsf{sk}}^*(\beta \cdot \mathbf{v})$. To decrypt we set $\mathbf{u} := \mathbf{S} \times \mathbf{c} = \beta \cdot \mathbf{v} + \mathbf{e} \pmod{q}$, then decode each entry of $\mathbf{u}$ to the nearest multiple of β. Namely, we decrypt as

$$\mathbf{v} := \mathsf{VecDec}_{\mathsf{sk}}(\mathbf{c}) = \left\lceil \frac{b \cdot (\mathbf{S} \times \mathbf{c} \bmod q)}{q} \right\rfloor.$$

Matrix Encryption and Decryption. Matrix encryption is similar, except that instead of just multiplying by a large scalar, we use the GSW technique of redundant encoding using $\mathbf{G}$.

The "native plaintext space" consists of square matrices $\mathbf{M} \in \mathbb{Z}_q^{n \times n}$. To encrypt $\mathbf{M}$ we first compute $\mathbf{M}' = \mathbf{M} \times \mathbf{G} \pmod{q}$ and let $\mathbf{m}'_j$ be the j'th column of $\mathbf{M}'$ ($j = 1, \ldots, m$). Then we set

$$\mathbf{c}_j := \mathsf{Enc}_{\mathsf{sk}}^*(\mathbf{m}'_j), \text{ and } \mathbf{C} := \mathsf{MatEnc}_{\mathsf{sk}}(\mathbf{M}) = [\mathbf{c}_1|\mathbf{c}_2|\ldots|\mathbf{c}_m].$$

Note that the ciphertext $\mathbf{C}$ has the form $\mathbf{C} = \mathbf{S}^{-1} \times (\mathbf{MG} + \mathbf{E}')$, where $\mathbf{E}'$ is the low-norm matrix consisting of all the noise vectors that were drawn inside of $\mathsf{Enc}_{\mathsf{sk}}^*$. In other words, the property that this ciphertext satisfies is $\mathbf{S} \times \mathbf{C} \approx \mathbf{M} \times \mathbf{G}$, up to the low-norm error matrix $\mathbf{E}'$.

In our application we never need to decrypt matrices, but note that we could compute $\mathbf{U} := \mathbf{S} \times \mathbf{C} = \mathbf{MG} + \mathbf{E}' \pmod{q}$, and then recover $\mathbf{M}$ from $\mathbf{U}$ (since $\mathbf{E}'$ is low norm and $\mathbf{G}$ is the gadget matrix that has a known trapdoor).

3.3   A Leveled NFA-Homomorphic Scheme

Computing a Single Product Chain. To enable homomorphic computation of a product of k matrices by a vector, $(\prod_{i=k}^{1} \mathbf{M}_i) \times \mathbf{v}$, we choose $k+1$ secret keys as above, $\mathsf{sk}_i = (\mathbf{S}_i, \mathbf{E}_i)$, for $i = 0, 1, \ldots, k$. We then encrypt the vector $\mathbf{v}$ under the first key sk_0, and for $1 \leq i \leq k$ we use sk_i to encrypt the matrix $\mathbf{M}'_i = \mathbf{M}_i \times \mathbf{S}_{i-1}$. In other words, we prepare the ciphertexts

$$\mathbf{c} = \mathbf{S}_0^{-1} \times (\beta\mathbf{v} + \mathbf{e}) \bmod q,$$

and

$$\mathbf{C}_i = \mathbf{S}_i^{-1} \times (\mathbf{M}_i \mathbf{S}_{i-1} \mathbf{G} + \mathbf{E}'_i) \bmod q, \text{ for } i = 1, \ldots, k,$$

where the noise vectors/matrices are all low-norm. To perform the homomorphic computation, we initialize $\mathbf{c}_0 := \mathbf{c}$, and then repeatedly set

$$\mathbf{c}_i := \mathbf{C}_i \times \mathbf{G}^{-1}(\mathbf{c}_{i-1}) \bmod q,$$

outputting the final vector ciphertext $\mathbf{c}_k$. We now show (by induction) that for every i, the vector ciphertext $\mathbf{c}_i$ is a valid encryption of the plaintext vector $\mathbf{v}_i = (\prod_{j=i}^{1} \mathbf{M}_j) \times \mathbf{v}$ under the key sk_i. This holds by definition for $\mathbf{v}_0 = \mathbf{v}$,

so we now assume that it holds for $i \geq 0$ and show for $i + 1$. By assumption we have
$$c_i = S_i^{-1} \times (\beta v_i + e_i),$$
for some low-norm noise vector e_i. Hence we get

$$
\begin{aligned}
c_{i+1} = C_{i+1} \times G^{-1}(c_i) &= S_{i+1}^{-1} \times (M_{i+1} S_i G + E'_{i+1}) \times G^{-1}(c_i) \\
&= S_{i+1}^{-1} \times (M_{i+1} S_i \times c_i + E'_{i+1} \times G^{-1}(c_i)) \\
&= S_{i+1}^{-1} \times (M_{i+1} S_i \times S_i^{-1} \times (\beta v_i + e_i) + E'_{i+1} \times G^{-1}(c_i)) \\
&= S_{i+1}^{-1} \times (\beta \underbrace{M_{i+1} v_i}_{v_{i+1}} + \underbrace{M_{i+1} e_i + E'_{i+1} \times G^{-1}(c_i)}_{e_{i+1}}).
\end{aligned}
$$

Since e_i, E'_{i+1}, and $G^{-1}(c_i)$ are all low norm, the noise term e_{i+1} will be low norm as long as M_{i+1} is. We conclude that $c_k = S_k^{-1}(\beta v_k + e_k) \pmod q$, where the noise term is

$$
e_k = \Big(\prod_{j=k}^{1} M_j\Big) e + \sum_{i=2}^{k} \Big(\prod_{j=k}^{i} M_j\Big) E'_{i-1} G^{-1}(c_{i-2}) + E'_k G^{-1}(c_{k-1}) \pmod q.
\tag{2}
$$

Hence as long as all the products $\prod_{j=k}^{i} M_j$ have low norm, the final noise term e_k will also have low norm. We will present a detailed analysis on the bounds of the noise terms in relation with NFAs in Sect. 5.

Encrypting and Evaluating an NFA. To be able to evaluate this NFA on strings of up to k symbols, we set the parameters so that $\beta = \lfloor q/b \rfloor$ is sufficiently larger than $\max_{w \in \Sigma^{\leq k}} \| \prod_{i=|w|}^{1} M_{w_i} \|_\infty$, then choose $k + 1$ secret keys sk_i for $i = 0, \ldots, k$. We encrypt the initial state vector v under sk_0, and encrypt each of the matrices M_σ for $\sigma \in \Sigma$ under all the other keys. Namely we set

$$c = \mathsf{VecEnc}_{\mathsf{sk}_0}(v), \text{ and } C_{\sigma,i} = \mathsf{MatEnc}_{\mathsf{sk}_i}(M_\sigma S_{i-1}) \text{ for } i = 1, \ldots, k.$$

Clearly this method provides semantic security for the NFA, so long as the basic "encryption" transformation from above is semantically secure.

To evaluate the encrypted NFA on a k-symbol string $w_1 w_2 \ldots w_k$, we apply the chain-product procedure from above to evaluate homomorphically the product $(\prod_{i=k}^{1} M_{w_i}) \times v$. Namely we set $c'_0 = c$ and then $c'_i = C_{w_i,i} \times G^{-1}(c'_{i-1})$ for $i = 1, \ldots, k$. At the end of the evaluation, we decrypt the final ciphertext c'_k to $u = \mathsf{VecDec}_{\mathsf{sk}_k}(c'_k)$ and check if the computation is accepting.

Circular Security for Better Efficiency. As usual, we can improve efficiency by assuming circular security of the encryption. Namely, instead of choosing all the secret keys independently, we choose just a single secret key and use it everywhere. This means that we only need the ciphertexts

$$c = S^{-1} \times (\beta v + e), \text{ and } C_\sigma = S^{-1} \times (M_\sigma S G + E_\sigma) \text{ for each } \sigma \in \Sigma.$$

3.4   The Parameters

To determine the parameters that are needed for certain NFA (or a class of NFAs) on k-symbol strings, we first need an upper bound on the size of the plaintext, specifically

$$B_{\mathsf{ptxt}} \geq \max_{w \in \Sigma^{\leq k}} \| \prod_{i=|w|}^{1} \mathbf{M}_{w_i} \|_\infty.$$

(See Sect. 5 for methods of converting regular expressions to NFAs while keeping this bound small.) Once we have the bound B_{ptxt}, we use it on Eq. 2 to compute a high probability bound on the expression

$$B^* \geq \| B_{\mathsf{ptxt}} \cdot \mathbf{e} + k \cdot B_{\mathsf{ptxt}} \cdot \mathbf{E} \times \mathbf{G}^{-1}(\mathbf{c}) \|,$$

where $\mathbf{e}, \mathbf{E}$ are noise terms that are output by the NoiseSamp procedure. This value B^* bounds with high probability the size of the noise that we can get when evaluating the NFA, and so we need to choose $q > B^* \cdot B_{\mathsf{ptxt}}$ (since our plaintext can be as large as B_{ptxt}).

At the same time, we need to set n large enough relative to q to ensure the required security level (say $q < 2^{n/\lambda}$), and $m > O(n \log q)$ (since we rely on the leftover hash lemma). As usual with lattice-based systems, there is a weak circular dependence between these constraints, but it is not hard to find values that satisfy them all.

4   Security Analysis

Below we define (two variants of) the inhomogeneous NTRU problem, one over a ring and one over integer matrices. We describe some properties of this problem, and show that hardness of the matrix variant implies the security of our encryption scheme.

4.1   Inhomogeneous NTRU

We begin with the ring variant of our hardness assumption. Fix a ring R, a modulus q, and an error distribution χ over R, producing with overwhelming probability elements with norm $\ll q$ and $-\chi = \chi$. Denoting $\ell = \lceil \log q \rceil$, the iNTRU distribution with these parameters is defined as follows:

$$\mathsf{iNTRU} = \left\{ \begin{array}{l} \text{draw } s \leftarrow R/qR, \text{ and } e_i \leftarrow \chi, \text{ for } i = 0, \ldots, \ell, \\ \text{set } a_0 := e_0/s \bmod q, \\ \text{and } a_i := (2^{i-1} - e_i)/s \bmod q \text{ for } i = 1, \ldots, \ell, \\ \hspace{4cm} \text{output } (a_0, \ldots, a_{\ell-1}) \end{array} \right\} . \quad (3)$$

The inhomogeneous NTRU problem is to distinguish between this distribution and the uniform distribution over $(R/qR)^\ell$.

In the matrix variant of this assumption, the ring elements s, e_i are replaced by n-by-n integer matrices, and the a_i's are similarly replaced with matrices $\mathbf{A}_0 := -\mathbf{S}^{-1} \times \mathbf{E}_0$, $\mathbf{A}_i := \mathbf{S}^{-1} \times (2^i \mathbf{I} - \mathbf{E}_i)$. In matrix notation, let $m' = n(\ell+1)$ and $\mathbf{G}'$ be the gadget matrix[11] $\mathbf{G}' = [\mathbf{0}|\mathbf{I}|2\mathbf{I}|4\mathbf{I}|\ldots|2^{\ell-1}\mathbf{I}] \in \mathbb{Z}^{n \times m'}$, and let χ be a distribution over $\mathbb{Z}$, producing with overwhelming probability integers of magnitude $\ll q$. The matrix-iNTRU distribution (MiNTRU) with these parameters is defined as follows:

$$\text{MiNTRU} = \left\{ \begin{array}{c} \text{draw } \mathbf{S} \leftarrow \mathbb{Z}_q^{n \times n}, \text{ and } \mathbf{E}' \leftarrow \chi^{n \times m'}, \\ \text{output } \mathbf{A}' := \mathbf{S}^{-1} \times (\mathbf{G}' - \mathbf{E}') \bmod q \end{array} \right\}. \tag{4}$$

As before, the hardness assumption says that MiNTRU is pseudorandom, namely that the matrix $\mathbf{A}'$ is indistinguishable from a matrix uniform in $\mathbb{Z}_q^{n \times m'}$.

Small-Secret Inhomogeneous NTRU. Similarly to LWE, here too we can prove that the inhomogeneous NTRU problem remains hard even when the secret is chosen from the error distribution. We lose a little on parameters in the conversion, specifically the extra block at the beginning of $\mathbf{G}'$. With the parameters n, m', q, χ as above, let $m = n\lceil \log q \rceil = m' - n$, and $\mathbf{G} = [\mathbf{I}|2\mathbf{I}|4\mathbf{I}|\ldots|2^{\ell-1}\mathbf{I}] \in \mathbb{Z}^{n \times m}$. The matrix-iNTRU distribution with small secret (MiNTRU^s) is as follows:

$$\text{MiNTRU}^s = \left\{ \begin{array}{c} \text{draw } \mathbf{S} \leftarrow \chi^{n \times n}, \text{ and } \mathbf{E} \leftarrow \chi^{n \times m}, \\ \text{output } \mathbf{A} := \mathbf{S}^{-1} \times (\mathbf{G} - \mathbf{E}) \bmod q \end{array} \right\}. \tag{5}$$

Lemma 4. *For the parameters n, m, m', q, χ as above, if* MiNTRU *is pseudorandom in $\mathbb{Z}_q^{n \times m'}$, then* MiNTRU^s *is pseudorandom in $\mathbb{Z}_q^{n \times m}$.*

Proof. We show that if we could distinguish MiNTRU^s from uniformly random n-by-m matrices over $\mathbb{Z}_q$ then we could also distinguish MiNTRU from uniformly random n-by-m' matrices over $\mathbb{Z}_q$. Given a MiNTRU instance that we want to distinguish, $\mathbf{A}' = [\mathbf{A}'_0|\mathbf{A}'_1|\ldots|\mathbf{A}'_\ell]$ (with $\mathbf{A}'_i \in \mathbb{Z}_q^{n \times n}$), we set

$$\mathbf{A}_i = \mathbf{A}'_0^{-1} \times \mathbf{A}'_i \bmod q, \text{ for } i = 1, \ldots, \ell,$$

(aborting if $\mathbf{A}'_0$ is not invertible), then run the MiNTRU^s distinguisher on $\mathbf{A} = [\mathbf{A}_1|\mathbf{A}_2|\ldots|\mathbf{A}_\ell]$. Observe that if $\mathbf{A}'$ is uniformly random then so is $\mathbf{A}$, and if $\mathbf{A}'$ is chosen from the MiNTRU distribution then

$$\mathbf{A}_i = \mathbf{A}'_0^{-1} \times \mathbf{A}'_i = -\mathbf{E}'_0^{-1} \times \mathbf{S} \times \mathbf{S}^{-1} \times (2^{i-1}\mathbf{I} - \mathbf{E}'_i) = -\mathbf{E}'_0^{-1} \times (2^{i-1}\mathbf{I} - \mathbf{E}'_i),$$

for $i = 1, \ldots, \ell$, and hence $\mathbf{A}$ follows the MiNTRU^s distribution as needed. $\square$

---

[11] We use a slightly larger gadget matrix than usual, with an extra first block. The reason will become clear when we prove Lemma 4 below.

4.2 Security Reduction

We next show that pseudorandomness of MiNTRU^s (or equivalently MiNTRU) with some error distribution χ, implies the semantic security of our scheme with a related error distribution (but not quite the same). Specifically, let n, m, q, χ be the parameters of the MiNTRU^s distribution above. For a fixed pair of matrices $\mathbf{E}, \mathbf{Y} \in \mathbb{Z}_q^{n \times m}$, consider the distribution

$$\psi[\mathbf{E}, \mathbf{Y}] = \{\mathbf{R} \leftarrow \mathbf{G}^{-1}(\mathbf{Y}), \text{ output } \mathbf{E} \times \mathbf{R} \bmod q\}.$$

In the provable version of our scheme, the secret key includes the square invertible matrix $\mathbf{S} \leftarrow \chi^{n \times n}$, and in addition a fixed error matrix $\mathbf{E} \leftarrow \chi^{n \times m}$, and we use the error distribution $\psi[\mathbf{E}, \mathbf{M} \times \mathbf{G}]$ when encrypting a matrix $\mathbf{M} \in \mathbb{Z}_q^{n \times n}$. Namely we draw a sample $\mathbf{R} \leftarrow \mathbf{G}^{-1}(\mathbf{MG}) \in \mathbb{Z}_q^{m \times m}$, then output the ciphertext $\mathbf{C} := \mathbf{S}^{-1} \times (\mathbf{MG} - \mathbf{ER}) \bmod q$. Note that given a MiNTRU^s sample $\mathbf{S}^{-1} \times (\mathbf{G} - \mathbf{E})$, one can efficiently generate samples of the form $\mathbf{S}^{-1} \times (\mathbf{M}_i \mathbf{G} - \mathbf{ER})$. This means Proposition 1 is a reduction from CPA security to distinguishing a single MiNTRU^s sample.

Proposition 1. *If MiNTRU^s is pseudorandom, then our encryption scheme using the error distribution $\psi[\mathbf{E}, \mathbf{M} \times \mathbf{G}]$ is semantically secure.*

Proof. We use the "real-or-random" formulation of semantic security for secret-key encryption [7]. Namely, we have a challenger that chooses a secret key $\mathsf{sk} = (\mathbf{S}, \mathbf{E})$, where $\mathbf{S} \leftarrow \chi^{n \times n}, \mathbf{E} \leftarrow \chi^{n \times m}$, and a bit $\sigma \leftarrow \{0, 1\}$, then the adversary repeatedly chooses messages $\mathbf{M}_i \in \mathbb{Z}_q^{n \times n}$ for $i = 1, \ldots, k$ and sends them to the challenger, who replies either with uniformly random matrices $\mathbf{C}_i \in \mathbb{Z}_q^{n \times m}$ if $\sigma = 0$, or with ciphertexts $\mathbf{C}_i := \mathsf{MatEnc}_{\mathsf{sk}}(\mathbf{M}_i) = \mathbf{S}^{-1} \times (\mathbf{M}_i \mathbf{G} + \mathbf{E}_i)$ if $\sigma = 1$, where $\mathbf{E}_i \leftarrow \psi[\mathbf{E}, \mathbf{M}_i \mathbf{G}]$, for $i = 1, \ldots, k$. The adversary eventually outputs a guess σ' for σ, and is considered successful if $\sigma' = \sigma$ with probability significantly larger than $1/2$.

We show that an adversary Adv with a noticeable advantage ϵ can be transformed into a distinguisher between MiNTRU^s and the uniform distribution over $\mathbb{Z}_q^{n \times m}$, with an advantage close to ϵ. The distinguisher D receives as input $\mathbf{A} \in \mathbb{Z}_q^{n \times m}$ that is either an instance of MiNTRU^s or a uniformly random matrix, and it interacts with the adversary Adv as follows:

When receiving a matrix $\mathbf{M}_i$ from Adv, the distinguisher D draws a sample $\mathbf{R}_i \leftarrow \mathbf{G}^{-1}(\mathbf{M}_i \mathbf{G})$, and replies with the "ciphertext" $\mathbf{C}_i := \mathbf{AR}_i \bmod q$. When Adv eventually outputs a guess σ', the distinguisher D outputs the same guess. We next show that the distinguishing advantage of D is very close to ϵ.

If $\mathbf{A}$ is a uniformly random matrix in $\mathbb{Z}_q^{n \times m}$ then, by the leftover hash lemma, each $\mathbf{C}_i = \mathbf{A} \times \mathbf{G}^{-1}(\text{something}) \bmod q$ is statistically close to uniformly random matrices in $\mathbb{Z}_q^{n \times m}$ and independent of $\mathbf{A}$. On the other hand, if $\mathbf{A} = \mathbf{S}^{-1} \times (\mathbf{G} - \mathbf{E})$ is an instance of MiNTRU^s, then we have

$$\mathbf{C}_i = \mathbf{A} \times \mathbf{G}^{-1}(\mathbf{M}_i \mathbf{G}) = \mathbf{S}^{-1} \times \left(\mathbf{G} \times \mathbf{G}^{-1}(\mathbf{M}_i \mathbf{G}) - \mathbf{E} \times \mathbf{G}^{-1}(\mathbf{M}_i \mathbf{G}) \right)$$
$$= \mathbf{S}^{-1} \times \left(\mathbf{M}_i \mathbf{G} - \mathbf{E} \times \mathbf{G}^{-1}(\mathbf{M}_i \mathbf{G}) \right),$$

which is identical to the distribution produced by our encryption procedure. □

4.3 Hardness of MiNTRU from LWE with a Trapdoor

Here we prove the reduction alluded to in Sect. 1.2. We define a trapdoor oracle for an arbitrary matrix $\mathbf{B} \in \mathbb{Z}_q^{n \times m}$ as an oracle which takes as input $\mathbf{B}$, a vector $\mathbf{v} \in \mathbb{Z}_q^n$, and outputs a discrete Gaussian integer vector $\mathbf{x} \in \mathbb{Z}^m$ conditioned on $\mathbf{Bx} \bmod q = \mathbf{v}$. Repeated calls to the oracle are assumed to use independent random coins. Further, we assume the oracle's distribution samples above the smoothing parameter of

$$\Lambda_q^{\perp}(\mathbf{B}) = \{\mathbf{x} \in \mathbb{Z}^m : \mathbf{Bx} = \mathbf{0} \bmod q\}$$

for a *uniformly random* $\mathbf{B}$, for some negligible function $\epsilon(n)$. In general, the smoothing parameter of $\Lambda_q^{\perp}(\mathbf{B})$ is just above the smoothing parameter of $\mathbb{Z}^m$, for some negligible $\epsilon(n)$, when $m > n \log q$, [36, Lemma 2.4].

Let n-secret LWE define the distribution

$$\{(\mathbf{A}, \mathbf{B} = \mathbf{SA} + \mathbf{E}) : \mathbf{A} \leftarrow \mathbb{Z}_q^{n \times m}, \mathbf{S} \leftarrow \mathbb{Z}_q^{n \times n}, \mathbf{E} \leftarrow \chi^{n \times m}\}$$

for some distribution χ. Next, we show the pseudorandomness of MiNTRU follows from the n-secret LWE distribution with a trapdoor oracle for $\mathbf{B}$. Let $\mathbf{G} \in \mathbb{Z}_q^{n \times m'}$ be any formulation of the gadget matrix. ($\mathbf{G} = [\mathbf{0}|\mathbf{I}|2\mathbf{I}|\cdots|2^{\log q-1}\mathbf{I}] \in \mathbb{Z}_q^{n \times n(\log q+1)}$ in the MiNTRU definition.)

Proposition 2. *Let* $n \in \mathbb{N}$, $q < 2^{poly(n)}$, χ *be a distribution over* $\mathbb{Z}_q$, $m \geq n \log q$, *and* m' *be the number of columns in the* $\mathbf{G}$*-matrix. Further, let* $q = \omega(\sqrt{m})$. *Then, the pseudorandomness of* MiNTRU *with error distribution* $\chi^{n \times m} \cdot \mathbf{B}^{-1}(\mathbf{G})$ *follows from the pseudorandomness of* n*-secret LWE with a trapdoor oracle for* $\mathbf{B}$.

Proof. We show a reduction from the n-secret LWE with a trapdoor oracle for $\mathbf{B}$ to MiNTRU with error distribution $\chi^{n \times m} \cdot \mathbf{B}^{-1}(\mathbf{G})$. Given as input a pair of matrices $(\mathbf{A}, \mathbf{B})$, we call m' times the trapdoor oracle for $\mathbf{B}$ to get $\mathbf{X} \leftarrow \mathbf{B}^{-1}(\mathbf{G})$. Then the reduction outputs $\mathbf{A} \times \mathbf{X} \bmod q$. Notice when $(\mathbf{A}, \mathbf{B})$ is generated uniformly and independently, then $\mathbf{AX} \bmod q$ is negligibly close to uniformly random by leftover hash lemma, along with Lemmas 2 and 3. Conversely, we have $\mathbf{S}^{-1} \in \mathbb{Z}_q^{n \times n}$ exists with high probability and $\mathbf{A} = \mathbf{S}^{-1} \times (\mathbf{B} - \mathbf{E}) \bmod q$ when $(\mathbf{A}, \mathbf{B})$ is sampled from the n-secret LWE distribution. Therefore,

$$\mathbf{A} \times \mathbf{B}^{-1}(\mathbf{G}) = \mathbf{S}^{-1} \times (\mathbf{G} - \mathbf{EB}^{-1}(\mathbf{G})) = \mathbf{S}^{-1} \times (\mathbf{G} - \mathbf{E}') \bmod q.$$

So $\mathbf{AX} \bmod q$ is an instance of MiNTRU with the desired error distribution. □

Remark 1. There is an identical reduction from n-secret LWE with a trapdoor for $\mathbf{B}$ with small secrets to MiNTRU^s.

5 Converting Regular Expressions to Automata

In real world applications, regular languages or finite automata are often represented by regular expressions, which have a very compact form and are convenient to store. So it is important for our scheme to be useful when NFAs are specified using regular expressions. In this section we present an efficient method to convert regular expressions to NFAs of relatively small sizes, and we discuss how to find a suitable NFA to bound the noise growth. We assume the reader has some familiarity with regular languages, regular expressions, and finite automata. See Appendix A for basic notation and definitions.

Partial Derivatives and NFAs. Let Σ be a finite alphabet, and RE be the set of all regular expressions over Σ. We consider the basic operations such as union ("+"), concatenation ("·"), and Kleene star ("*") on regular expressions. For any regular expression e, the *language* of e is denoted by $\mathcal{L}(e)$. To convert a regular expression to an NFA, we start with Antimirov's partial derivative construction [3], which is an elegant extension of Brzozowski's derivative construction [13] to NFAs. For any symbol $a \in \Sigma$, the *partial derivative* of e w.r.t. a, denoted as $\partial_a(e)$, is a set of regular expressions defined inductively as

$$\partial_a(\epsilon) = \emptyset, \qquad \partial_a(e_0 + e_1) = \partial_a(e_0) \cup \partial_a(e_1), \qquad \partial_a(e^*) = \partial_a(e)e^*$$

$$\partial_a(a_i) = \begin{cases} \{\epsilon\} & \text{if } a_i = a \\ \emptyset & \text{otherwise} \end{cases} \qquad \partial_a(e_0 \cdot e_1) = \begin{cases} \partial_a(e_0)e_1 \cup \partial_a(e_1) & \text{if } \epsilon \in \mathcal{L}(e_0) \\ \partial_a(e_0)e_1 & \text{otherwise} \end{cases}$$

where e, e_0, e_1 range over RE. The partial derivative of e w.r.t. any string is $\partial_\epsilon(e) = \{e\}$ and $\partial_{ua}(e) = \bigcup\{\partial_a(f) \mid f \in \partial_u(e)\}$ where $u \in \Sigma^*$ and $a \in \Sigma$. A regular expression e' is a *partial derivative term* of e if e' is an element of $\partial_w(e)$ for some $w \in \Sigma^*$, and $\partial(e)$ is the set of all partial derivative terms of e.

Definition 1 (Partial derivative NFA). *For any regular expression e, the partial derivative NFA of e is $\mathcal{M}_{\mathcal{PD}}(e) = (Q, \Sigma, \delta, Q_I, Q_F)$, where $Q = \partial(e)$, $Q_I = \{e\}$, $Q_F = \{e' \in \partial(e) \mid \epsilon \in \mathcal{L}(e')\}$, and for any $e' \in Q$ and $a \in \Sigma$, $\delta(e', a) = \partial_a(e')$.*

Remark 2. It was shown in [3] that $\partial(e)$ is a finite set (with respect to syntactic equality on regular expressions). In fact, $|\partial(e)| \leq r + 1$ where r is the number of occurrences of alphabet symbols in e.

The language of e satisfies $\mathcal{L}(e) = \bigcup_{a \in \Sigma} a \cdot \partial_a(e)$. It follows that the language accepted by $\mathcal{M}_{\mathcal{PD}}(e)$ is exactly $\mathcal{L}(e)$.

Ambiguity Measure. As will be shown later, when evaluating an encrypted NFA, the noise growth is closely related to the amount of nondeterministic choices of the NFA. Here we describe some notions that characterize this quantity. Let $\mathcal{M} = (Q, \Sigma, \delta, Q_I, Q_F)$ be an NFA. For any string $w = w_1 \cdots w_k$ where $w_1, \ldots, w_k \in \Sigma$, a *path of w from state s to state t* is a finite sequence of states $s = s_{i_0}, s_{i_1}, \ldots, s_{i_k} = t$ such that $s_{i_j} \in \delta(s_{i_{j-1}}, w_j)$ for all $1 \leq j \leq k$. A path is accepting if $s \in Q_I$ and $t \in Q_F$. The *degree of ambiguity of $\mathcal{M}$*, denoted as $\mathrm{da}(\mathcal{M}, k)$, is the maximal number of accepting paths for a string of length k. If $\mathrm{da}(\mathcal{M}, k) \leq 1$ for all $k > 0$, then we say $\mathcal{M}$ is *unambiguous*.[12] We say that $\mathcal{M}$ is

---

[12] Notice that a DFA $\mathcal{M}$ has $\mathrm{da}(\mathcal{M}, k) \leq 1$ for all $k \geq 0$, but the converse is not necessarily true. An NFA can have multiple nondeterministic choices at every state but still satisfies $\mathrm{da}(\mathcal{M}, k) \leq 1$, in such cases at most one of these choices could lead to a final state.

finitely ambiguous if $\sup\{\mathrm{da}(\mathcal{M}, k) \mid k \geq 0\} < \infty$, and $\mathcal{M}$ is infinitely ambiguous otherwise. Clearly $\mathrm{da}(\mathcal{M}, k) \leq |Q|^{k+1}$ for any NFA. To upper bound the quantity $\mathrm{da}(\mathcal{M}, k)$ using a function of k, we can define the *degree of growth of ambiguity of* $\mathcal{M}$, denoted as $\deg(\mathcal{M})$, to be the minimal *degree* of a polynomial $h(\cdot)$ such that $\mathrm{da}(\mathcal{M}, k) \leq h(k)$ for all $k \geq 0$. If no such polynomial exists, we simply set $\deg(\mathcal{M}) = \infty$. Note that $\mathcal{M}$ is finitely ambiguous if and only if $\deg(\mathcal{M}) = 0$. It was shown in [45] that $\deg(\mathcal{M})$ can be computed in time $O(r^6|\Sigma|)$ for any NFA $\mathcal{M}$ with r states.

On Optimizing NFA. For our application of evaluating encrypted NFA, an optimal NFA should be such that its encryption can be correctly evaluated on as many strings as possible. Concretely, we want to find an NFA such that the noise term at the end of evaluation is small enough for a successful decryption. Recall that (n, q) is the lattice parameter in our scheme, b is the maximum l_∞ norm on plaintext vectors, and χ is an error distribution from which we sample noise terms. As we assume the first state will be the only initial state in all our NFAs, we can encrypt the initial state vector with no noise. As a result, we obtain the following bounds on the noise due to homomorphic evaluation of NFAs, which can be bounded using the ambiguity measures of $\mathcal{M}$.

Proposition 3. *For any $n \geq 1$, if $\mathcal{M}$ is an NFA with $r \leq n$ states, and w a string of length k, the noise vector $\mathbf{e}^{(k)}$ at the end of homomorphic evaluation of encrypted $\mathcal{M}$ on w satisfies the following bounds:*

- *If $\mathcal{M}$ is unambiguous, then $\|\mathbf{e}^{(k)}\|_\infty \leq bnk\chi \log_b q$.*
- *If $\mathcal{M}$ is finitely ambiguous, then $\|\mathbf{e}^{(k)}\|_\infty \leq bnrk\chi \log_b q$.*
- *If $\mathcal{M}$ is infinitely ambiguous, then $\|\mathbf{e}^{(k)}\|_\infty \leq bnk^{\deg(\mathcal{M})+1}\chi \log_b q$.*

Notice that both the number of states and the degree of ambiguity contribute to the bound on the noise growth. To find a small noise growth for the general case of processing an arbitrary long input string, we can try to solve the following optimization problem on NFA minimization with bounded ambiguity.

Definition 2 (NFA Minimization with Bounded Ambiguity Problem). *For a given NFA of r states and a function $B : \mathbb{N} \to \mathbb{N}$, find an equivalent NFA $\mathcal{M}$ with a minimal number of states such that $\mathrm{da}(\mathcal{M}, k) \leq B(k)$ for all $k \geq 1$.*

A closely related problem is to find a minimal NFA $\mathcal{M}$ with a given bound on $\deg(\mathcal{M})$. Conversely, we can consider a similar minimization problem of finding an NFA $\mathcal{M}$ with minimal $\deg(\mathcal{M})$ when given a regular expression and a bound on the number of states. These problems seem to be hard in general as evidenced by several exponential separation results in automata theory, and we briefly mention a few. It was shown in [30] that, for each $r > 0$, there exists an NFA of r states such that the minimal equivalent NFA $\mathcal{M}'$ of bounded $\deg(\mathcal{M}')$ have $2^r - 1$ states.[13] With a more strict bound on the ambiguity, it was known [28] that

---

[13] Note that $\deg(\mathcal{M}')$ is bounded if and only if $\mathrm{da}(\mathcal{M}', k)$ is at most a polynomial in k for all $k > 0$.

there exist NFAs of r states such that the equivalent finitely ambiguous NFAs have at least $2^{\Omega(r^{1/3})}$ states. A more tractable problem of finding a minimal unambiguous NFA is NP-complete [8, 29].

On the other hand, unambiguous NFAs can have much smaller size than equivalent DFAs. A well-known example is the language $L_r = (0+1)^*0(0+1)^{r-2}$ for any $r \geq 2$: its partial derivative NFA has r states and is unambiguous, but its minimal equivalent DFA requires 2^{r-1} states [34]. The exponential upper bound 2^r can actually be met: it was shown in [31] that there exists a series $\{\mathcal{M}_r\}_{r \geq 1}$ of unambiguous NFAs such that $\mathcal{M}_r$ has r states but the minimal equivalent DFA of $\mathcal{M}_r$ has 2^r states. Notice that, if the size of the given regular expression is small, the bound on the size of the noise is dominated by the degree of ambiguity, which is same for unambiguous NFAs and DFAs. So we can exploit the fact that our scheme supports homomorphic encryption of NFAs and try to find a small unambiguous NFA, which can be much more efficient than encrypting DFAs.

Some particular useful classes of regular languages are the pattern matching languages L such that $L = \Sigma^* K \Sigma^*$, $L = K \Sigma^*$, or $L = \Sigma^* K$ where K is a finite set of strings. One can check using the criterion in [45] that the partial derivative NFA for such a language is unambiguous, but its minimal equivalent DFA may have exponentially many states. Even if K can be specified using a DFA of m states, the minimal equivalent DFA of L may still have $2^{m-2} + 1$ states. As our scheme supports encryption of NFAs, pattern matching on encrypted patterns can be much more efficient than previous approaches via DFAs.

6 Implementation and Performance

This section describes a proof of concept implementation of our scheme[14] and compares its performance with the HAO15 matrix-FHE scheme [25] when applied to homomorphic evaluation of encrypted NFAs.

Implementation. We implemented our scheme in C++ using the NTL library (version 10.5.0) for a power of two modulus, q, and we performed experiments on an Intel i7-2600 3.4 GHz CPU. The implementation is naive in that it only uses NTL's native functionality with no further optimizations. It can be done in a few hundred lines of code and a few days' programming effort. There are many opportunities for optimization since the code was written for simplicity and not efficiency. Despite this, we noticed exceptionally fast evaluation times as listed in Table 1.

In our experiments, we set lattice parameters to $n = 1024$ and $q = 2^{42}$. We kept the modulus both as a power of two and as a power of the maximum l_∞ norm b on plaintext vectors in order to take advantage of bit-shifting instead of multiplications and divisions modulo q. The noise matrices $\mathbf{E}_i \leftarrow \chi_q^{n \times m}$ and the secret keys $\mathbf{S} \leftarrow \chi_q^{n \times n}$ were chosen as uniformly random binary matrices with

---

[14] The source code of our proof-of-concept implementation can be accessed at https://www.dropbox.com/s/10g2nocx3pmyu4t/henfa.zip.

Table 1. Running times for each function along with memory for a 1024-state NFA accepting the language $(0+1)^*0(0+1)^{r-1}$ for $r = 11$. "NFA Enc. Time" is the time to encrypt the NFA, "Matching" is the time to evaluate an encrypted NFA on an input of k symbols, "Enc. NFA" is the memory storage for the encrypted NFA, and the last column measures the total RAM used during encryption, evaluation, and decryption. Total RAM usage was measured with the "sys/resource.h" library in unix.

Input Length ($4k$)	NFA Enc. Time	Matching	Enc. NFA	RAM used
256 bit S.L.	16.35 s	1.53 s	66 Mb	172 Mb
512 bit S.L.	16.66 s	3.34 s	66 Mb	172 Mb
1024 bit S.L.	16.53 s	6.63 s	66 Mb	172 Mb
16384 bit S.L.	16.76 s	98.97 s	66 Mb	172 Mb
65536 bit S.L.	16.42 s	394.47 s	66 Mb	172 Mb

the latter being invertible modulo q. We used NTL's pseudorandom number generator "Random_ZZ" for all random matrices.

Notice that MiNTRU^s can be cryptanalyzed by NTRU attacks like dimension reduction [33] and the hybrid attack [27] for key recovery. Therefore, we use the uSVP attack to estimate the time for a key recovery attack as in [1] and set the LWE noise parameter as $\alpha = \sqrt{2n}/q$ in the on-line LWE bit security estimator[15]. Rough estimates show that our scheme achieves 100 bits of security with these parameters.

We conducted tests on r-state partial derivative NFAs accepting the pattern-matching languages $(0+1)^*0(0+1)^{r-1}$ with finite ambiguity, for some r smaller than the lattice dimension n. Notice that the equivalent minimal DFA's have 2^{r-1} states. In the experiments, we pad the transition matrices to n-dimensional matrices by adding transitions from nonreachable states to final states to increase ambiguity, and hence we effectively obtain n-state NFAs. The strings scanned were randomly generated. At the end of each scan, our code checked for any decryption errors. We observed no decryption errors nor noise overflow. The experiment results for $r = 11$ are listed in Table 1, where time was measured using C++'s "time.h" library.

Consider the worst case where the NFA has infinite ambiguity, but bounded degree of growth of ambiguity. Then the final noise term $\mathbf{e}^{(k)}$ has norm $\|\mathbf{e}^{(k)}\|_\infty \leq bnk^{\deg(\mathcal{M})+1}\chi \log_b q$ as discussed in the previous section. By setting the modulus just above the error growth, we see that the bit length of the modulus is linear in $\deg(\mathcal{M}) + 1$. Now as we view total memory for the encrypted NFA, $n^2|\Sigma| \log_2(q) \log_b(q)$ bits, we see that efficiency is quadratic in NFA's number of states and quadratic in the degree of growth of ambiguity (though we have some control over $\log_b(q)$ by choosing a large base b). This gives us an exact relation between the number of states, the NFA's ambiguity, and performance.

---

[15] https://bitbucket.org/malb/lwe-estimator.

Table 2. Maximal lengths of strings can be scanned on any n-state NFA in both schemes without decryption error. In all cases, the noise parameter is set to $\alpha = \sqrt{2n}/q$.

Lattice parameters	$n = 1024, q = 2^{42}$		$n = 4096, q = 2^{111}$		$n = 32768, q = 2^{883}$	
	Ours	HAO15	Ours	HAO15	Ours	HAO15
Unambiguous	564918	141229	1.577e25	3.943e24	2.176e255	5.441e254
Finitely ambiguous	551	137	3.850e21	9.626e20	6.642e250	1.660e250
Infinitely ambiguous	82	65	250782489	199046193	1.295e85	1.028e85

Performance Improvement over HAO15. Now we compare the performance of our scheme with the HAO15 matrix-FHE scheme for homomorphic evaluation of encrypted NFAs. Let $\mathcal{M}$ be an NFA of $r \le n$ states, where n is the lattice dimension, and let k be the length of the string to be scanned on $\mathcal{M}$. For the HAO15 scheme, applying the NFA ambiguity analysis technique as in Proposition 3, we can rewrite Eq. 1 to obtain the following bound on the l_∞ norm of the final noise vector $\mathbf{e}_k$:

$$\|\mathbf{e}_k\|_\infty \le \chi(n + r) \log q + \chi(n + r) \log q \sum_{l=2}^{k} da(\mathcal{M}, l) + \chi da(\mathcal{M}, k), \quad (6)$$

which must be bounded away from $q/4$ for successful decryption of the final ciphertext vector.

Using Proposition 3 and the bound in Eq. 6, one can determine each scheme's capability of homomorphic NFA evaluation. For concrete results, we consider three cases of the ambiguity of $\mathcal{M}$:

1. $\mathcal{M}$ is unambiguous, so $da(\mathcal{M}, l) \le 1$;
2. $\mathcal{M}$ is finitely ambiguous, so $da(\mathcal{M}, l) \le r$; and
3. $\mathcal{M}$ is infinitely ambiguous and its degree of growth of ambiguity is $\deg(\mathcal{M}) = 2$, so $da(\mathcal{M}, l) \le l^2$.

Furthermore, we consider three sets of lattice parameters for at least 100 bits of security, and hence three different maximal sizes r for $\mathcal{M}$. We list in Table 2 the maximal lengths of strings can be scanned without decryption error using both schemes on any n-state NFA. The results show that we can almost always evaluate twice long strings using our scheme.

For the running time, the computational complexity of k homomorphic matrix multiplications in the HAO15 scheme, assuming naive matrix-vector multiplication of complexity $O(n^2)$, is $O(k(r + n)^2 \log q)$. On the other hand, the complexity of our homomorphic evaluation procedure is $O(kn^2 \log q)$. So using the same parameter and matrix multiplication algorithm, we expect our scheme runs three times faster than an implementation of the HAO15 scheme.

Potential Optimizations. One potential optimization is parallelization through the unused states. Say we must evaluate a long string (10000 bits) but only use a 100 state NFA. Then, we can evaluate ten such NFAs in parallel by setting the transition matrix for symbol $a \in \Sigma$ as the block diagonal matrix with the blocks as the smaller transition matrices in the small parameter setting. The total number of states must stay above a few hundred for this corresponds to the lattice dimension of the underlying lattice problem.

Let $\mathbf{G} = \mathbf{I}_n \otimes \mathbf{g}^t$ for $\mathbf{g}^t = (1, b, \cdots, b^{\log_b(q)-1})$ as in [36]. We expect to see smaller noise growth via a randomized bit decomposition for the decomposition of the encrypted state vector, as used in [2]. This can be done with a simple tweak to Babai's nearest plane algorithm [4] on the G-matrix's null lattice $\Lambda_q^{\perp}(\mathbf{G}) = \{\mathbf{x} \in \mathbb{Z}^m : \mathbf{Gx} = \mathbf{0} \bmod q\}$ and its cosets.

A   Definitions on Regular Expressions and NFA

We recall some standard definitions about regular languages and finite automata [46]. Let Σ be a finite alphabet, and Σ^* the free monoid generated by Σ. A *string* w is an element of Σ^*, which can be written as a finite sequence of symbols $w = w_1 w_2 \cdots w_k$ where $w_1, \ldots, w_k \in \Sigma$, and its *length* is $|w| = k$. The *empty string* is denoted by ϵ, which is the neutral element of Σ^*. The *concatenation* of two strings $u = u_1 \cdots u_m$ and $v = v_1 \cdots v_n$ is a string $uv = u_1 \cdots u_m v_1 \cdots v_n$. A *language* over Σ is a subset of Σ^*. For any languages L and K, we consider the following regular operations: (union) $L \cup K$, (product) $LK = \{uv \mid u \in L, v \in K\}$, and (Kleene star) $L^* = \cup_{i \geq 0} L^i$, where $L^0 = \{\epsilon\}$, and $L^i = LL^{i-1}$ for $i > 0$. *Regular languages* are the smallest class of languages containing the basic languages $\emptyset$, $\{\epsilon\}$, and $\{a_i\}$ for all $a_i \in \Sigma$ that are closed under regular operations.

A *nondeterministic finite automaton (NFA)* over Σ is a quintuple $M = (Q, \Sigma, \delta, Q_I, Q_F)$, where $Q = \{s_1, \ldots, s_n\}$ is a finite set of states, $\delta : Q \times \Sigma \to \wp(Q)$ is a transition function, $Q_I \subseteq Q$ is the set of initial states, and $Q_F \subseteq Q$ is the set of final states. We can extend δ to a function $\delta : Q \times \Sigma^* \to \wp(Q)$ over strings in the natural way. Without loss of generality, we assume that all our NFAs have a single initial state s_1. A string $w \in \Sigma^*$ is *accepted* by an NFA M if $\delta(s_1, w) \cap Q_F \neq \emptyset$. The set of all the strings accepted by an NFA M is called the *language of* M, and it is denoted by $\mathcal{L}(M)$. A *deterministic finite automaton (DFA)* is an NFA such that $\delta(s, a_i)$ is a singleton set for all $s \in Q$ and $a_i \in \Sigma$, and $|Q_I| = 1$.

A *regular expression* over Σ is a formal expression generated by the following grammar rules:

$$\mathsf{RE} \to \epsilon \mid a_i \mid (\mathsf{RE} + \mathsf{RE}) \mid (\mathsf{RE} \cdot \mathsf{RE}) \mid (\mathsf{RE})^*,$$

where a_i ranges over Σ. The operator $*$ takes the highest precedence, followed by $\cdot$, and then by $+$. The parentheses can be omitted when there is no ambiguity. The operator $\cdot$ is usually omitted as well, and concatenations can be written as juxtapositions of regular expressions. For a regular expression e, its *language* $\mathcal{L}(e)$ can be defined inductively as follows:

$$\mathcal{L}(\epsilon) = \{\epsilon\}, \qquad\qquad\qquad \mathcal{L}(a_i) = \{a_i\},$$
$$\mathcal{L}(e_0 + e_1) = \mathcal{L}(e_0) \cup \mathcal{L}(e_1), \qquad \mathcal{L}(e_0 \cdot e_1) = \{uv \mid u \in \mathcal{L}(e_0), v \in \mathcal{L}(e_1)\},$$
$$\mathcal{L}(e^*) = \cup_{i \geq 0} \mathcal{L}(e)^i,$$

where a_i ranges over Σ, and e_0, e_1 are regular expressions. For any set R of regular expressions, let $\mathcal{L}(R) = \cup_{e \in R} \mathcal{L}(e)$. It is well known that the languages defined by regular expressions are exactly the regular languages, which are exactly the languages accepted by finite automata.

For any sets R, T of regular expressions, we write RT for the set of regular expressions

$$RT = \{e \cdot f \mid e \in R, f \in T\},$$

and we write $Re = \{f \cdot e \mid f \in R\}$ and $eR = \{e \cdot f \mid f \in R\}$; in particular, $\emptyset T = R\emptyset = \emptyset e = e\emptyset = \emptyset$.

B   Proofs

In this section we present proofs that are omitted in the main paper.

Proposition 3. *For any $n \geq 1$, if $\mathcal{M}$ is an NFA with $r \leq n$ states, and w a string of length k, the noise vector $\mathbf{e}^{(k)}$ at the end of homomorphic evaluation of encrypted $\mathcal{M}$ on w satisfies the following bounds:*

- *If $\mathcal{M}$ is unambiguous, then $\|\mathbf{e}^{(k)}\|_\infty \leq bnk\chi \log_b q$.*
- *If $\mathcal{M}$ is finitely ambiguous, then $\|\mathbf{e}^{(k)}\|_\infty \leq bnrk\chi \log_b q$.*
- *If $\mathcal{M}$ is infinitely ambiguous, then $\|\mathbf{e}^{(k)}\|_\infty \leq bnk^{\deg(\mathcal{M})+1}\chi \log_b q$.*

Proof. Let $\mathcal{M} = (Q, \Sigma, \delta, \{s_1\}, Q_F)$ be an NFA with r states $s_1, \ldots, s_r$, and for each input symbol $\sigma \in \Sigma$, denote by $\mathbf{M}_\sigma \in \{0,1\}^{n \times n}$ the transition matrix of $\mathcal{M}$ on σ (padded with 0s in the extra columns and rows), where $(\mathbf{M}_\sigma)_{t,s} = 1$ if $t \in \delta(s, \sigma)$, and $(\mathbf{M}_\sigma)_{t,s} = 0$ otehwise. For any $t \in Q$ let $\mathcal{M}_t = (Q, \Sigma, \delta, Q, \{t\})$ be the NFA obtained from $\mathcal{M}$ by setting all states to be initial and t the only final state. Notice that $\mathrm{da}(\mathcal{M}_t, l)$ is an upper bound on the total number of paths in $\mathcal{M}$ on a string of length l from any state to t.

Let $w = w_1 \cdots w_k$ be the string to be scanned on $\mathcal{M}$. For all $1 \leq i \leq k$, the encrypted state vector $\mathbf{q}^{(i)}$ after reading w_i is:

$$\mathbf{q}^{(i)} = \sum_{j=0}^{\log_b q} C_{w_i,j} \mathbf{q}_j^{(i-1)} = \beta \mathbf{S}^{-1} \mathbf{M}_{w_i} \cdots \mathbf{M}_{w_1} \mathbf{v} + \mathbf{S}^{-1}(\mathbf{M}_{w_i} \mathbf{e}^{(i-1)} + \sum_{j=0}^{\log_b q} \mathbf{E}_{w_i,j} \mathbf{q}_j^{(i-1)}),$$

where $\mathbf{e}^{(i-1)}$ is the noise term after reading the previous symbol w_{i-1}. As in our assumption, s_1 is always the sole initial state in $\mathcal{M}$, we can set the initial noise $\mathbf{e}^{(0)} = \mathbf{0}$ without leaking any additional information about the NFA $\mathcal{M}$. By expanding all the noise terms, we get

$$\mathbf{e}^{(k)} = \sum_{l=2}^{k} \mathbf{M}_{w_k} \cdots \mathbf{M}_{w_l} \sum_{j=0}^{\log_b q} \mathbf{E}_{w_{l-1},j} \mathbf{q}_j^{(l-2)} + \sum_{j=0}^{\log_b q} \mathbf{E}_{w_k,j} \mathbf{q}_j^{(k-1)}. \qquad (7)$$

Notice that, for any symbol $a \in \Sigma$, the (t, s)'th entry of $\mathbf{M}_a$ is 1 if $t \in \delta(s, a)$ and it is 0 otherwise. So the (t, s)'th entry of the product $\mathbf{M}_{w_k} \cdots \mathbf{M}_{w_l}$ counts the number of paths from s to t on the string $w_l \cdots w_k$, where $1 \leq l \leq k$. Let $\mathbf{1}$ be the vector whose entries are all 1. Then the t'th entry of the vector $\mathbf{M}_{w_k} \cdots \mathbf{M}_{w_l} \mathbf{1}$ counts the total number of paths from an arbitrary state to t on this string, which is at most $\mathrm{da}(\mathcal{M}_t, k - l + 1)$. Thus we have

$$\| \mathbf{M}_{w_k} \cdots \mathbf{M}_{w_l} \sum_{j=0}^{\log_b q} \mathbf{E}_{w_{l-1},j} \mathbf{q}_j^{(l-2)} \|_\infty \leq bn\chi \log_b q \cdot \max_{t \in Q} \{ \mathrm{da}(\mathcal{M}_t, k - l + 1) \}.$$

It follows that the final noise vector $\mathbf{e}^{(k)}$ can be bounded by

$$\| \mathbf{e}^{(k)} \|_\infty \leq bn\chi \log_b q \cdot \sum_{l=1}^{k-1} \max_{t \in Q} \{ \mathrm{da}(\mathcal{M}_t, l) \} + bn\chi \log_b q \qquad (8)$$

If $\mathcal{M}$ is unambiguous, then $\mathrm{da}(\mathcal{M}_t, l) \leq 1$ for all $t \in Q$ and $l \geq 0$, so

$$\| \mathbf{e}^{(k)} \|_\infty \leq bkn\chi \log_b q.$$

If $\mathcal{M}$ is finitely ambiguous, then for all $s, t \in Q$, the number of paths of w from s to t is at most 1 [45]. So $\mathrm{da}(\mathcal{M}_t, l) \leq r$ for all $t \in Q$ and $l \geq 0$, and $\mathbf{e}^{(k)}$ can be bounded by

$$\| \mathbf{e}^{(k)} \|_\infty \leq bknr\chi \log_b q.$$

For the case where $\mathcal{M}$ is infinitely ambiguous, notice that $\mathrm{da}(\mathcal{M}_t, l) \leq l^{\deg(\mathcal{M})}$ for all $l \geq 1$, and we have

$$\| \mathbf{e}^{(k)} \|_\infty \leq b\chi \log_b q \sum_{l=1}^{k-1} l^{\deg(\mathcal{M})} + b\chi \log_b q$$

$$\leq bnk^{\deg(\mathcal{M})+1} \chi \log_b q. \qquad \square$$

C   Performance Comparisons with HAO15

In this section we present a brief analysis of applying the matrix-FHE scheme of HAO15 [25] to the case of homomorphic evaluation of NFA.

Fix an NFA $\mathcal{M}$ of r states and with an alphabet Σ, and let $\mathbf{M}_\sigma \in \{0,1\}^{r \times r}$ for $\sigma \in \Sigma$ be its transition matrices on symbol σ. Recall the "leveled version" of the HAO15 scheme as described in Sect. 3.1. To encrypt $\mathcal{M}$ for homomorphic evaluation on any string of length at most k, we sample $k+1$ secret keys sk_i for $i = 0, 1, \ldots, k$, and for each $\sigma \in \Sigma$, we encrypt $\mathbf{M}_\sigma$ with all keys sk_i to get $\mathbf{C}_{\sigma,i} \leftarrow \mathsf{HAO.MatEnc}_{\mathsf{sk}_i}(\mathbf{M}_\sigma)$. We also encrypt the initial state vector $\mathbf{v} = (1, 0, \ldots, 0)^t$ in a ciphertext $\mathbf{c} = \mathsf{HAO.VecEnc}_{\mathsf{sk}_0}(\mathbf{v})$.

To scan $w = w_1 \cdots w_k$ on $\mathcal{M}$, set $\mathbf{c}_0 = \mathbf{c}$ and $\mathbf{c}_i = \mathsf{HAO.Mul}(\mathbf{C}_{w_i,i}, \mathbf{c}_{i-1}) = \mathbf{C}_{w_i,i} \times \mathbf{G}^{-1}(\mathbf{c}_{i-1})$. Then each ciphertext $\mathbf{c}_i$ satisfies $\mathbf{S}_i \mathbf{c}_i = (\prod_{j=i}^1 \mathbf{M}_{w_j}) \times \mathbf{v} + \mathbf{e}_i$ for some noise vector $\mathbf{e}_i$. By Eq. 1, the l_∞ norm of $\mathbf{e}_k$ can be bounded by

$$\|\mathbf{e}_k\|_\infty \leq \chi N + \chi N \sum_{l=2}^k \mathrm{da}(\mathcal{M}, l) + \chi \mathrm{da}(\mathcal{M}, k),$$

which must be bounded away from $q/4$.

For performance comparison, consider two cases of the ambiguity measures of $\mathcal{M}$:

- $\mathcal{M}$ is finitely ambiguous: We have $\mathrm{da}(\mathcal{M}, l) \leq r$ for all $1 \leq l \leq k$, so w.h.p.

$$\|\mathbf{e}_k\|_\infty \leq \alpha q(n+r)(kr+1) \log q,$$

where $\alpha = \sqrt{2n}/q$ is the LWE noise parameter. Thus, in the HAO15 scheme we can homomorphically evaluate $\mathcal{M}$ on strings of length $k \leq \frac{1}{\alpha(n+r)r \log q}$. For example, assuming at least 100 bit of security is needed, for an NFA of up to 1024 states on strings of length up to 275, we need $n = 1024$ and $q = 2^{42}$. On the other hand, using our scheme we can evaluate $\mathcal{M}$ on strings of length $k \leq \frac{q}{b^2 n \chi r \log_b q}$. So, using our scheme with the above sets of parameters, we can homomorphically evaluate an NFA of up to 1024 states on strings of length up to 551.
- $\mathcal{M}$ is infinitely ambiguous: We have $\mathrm{da}(\mathcal{M}, l) \leq l^{\deg(\mathcal{M})}$, so w.h.p.

$$\|\mathbf{e}_k\|_\infty \leq \alpha q(n+r) \log q \cdot \left(\sum_{l=1}^k l^{\deg(\mathcal{M})} + 1 \right) \leq \alpha q(n+r) \log q k^{\deg(\mathcal{M})+1}$$

Using the same parameters as the above to achieve at least 100 bit of security, and assuming that $\deg(\mathcal{M}) = 2$ for the NFA $\mathcal{M}$, we can homomorphically evaluate $\mathcal{M}$ on strings of length up to 65 in the HAO15 scheme, whereas we can homomorphically evaluate $\mathcal{M}$ on strings of length up to 82 in our scheme.

Moreover, the computational complexity of k homomorphic matrix multiplications, assuming naive matrix-vector multiplication of complexity $O(n^2)$, is $O(k(r+n)^2 \log q)$. On the other hand, the complexity of our homomorphic evaluation procedure is $O(kn^2 \log q)$.

References

1. Albrecht, M.R., et al.: Estimate all the {LWE, NTRU} schemes! IACR Cryptology ePrint Archive 2018:331 (2018)
2. Alperin-Sheriff, J., Peikert, C.: Faster bootstrapping with polynomial error. In: Garay, J.A., Gennaro, R. (eds.) CRYPTO 2014, Part I. LNCS, vol. 8616, pp. 297–314. Springer, Heidelberg (2014). https://doi.org/10.1007/978-3-662-44371-2_17
3. Antimirov, V.M.: Partial derivatives of regular expressions and finite automaton constructions. Theor. Comput. Sci. 155(2), 291–319 (1996)
4. Babai, L.: On Lovász' lattice reduction and the nearest lattice point problem. Combinatorica 6(1), 1–13 (1986)
5. Bai, S., Galbraith, S.D.: An improved compression technique for signatures based on learning with errors. In: Topics in Cryptology - CT-RSA 2014 - The Cryptographer's Track at the RSA Conference 2014, San Francisco, CA, USA, February 25–28, 2014. Proceedings, pp. 28–47 (2014)
6. Banaszczyk, W.: New bounds in some transference theorems in the geometry of numbers. Math. Ann. 296(1), 625–635 (1993)
7. Bellare, M., Desai, A., Jokipii, E., Rogaway, P.: A concrete security treatment of symmetric encryption: analysis of the DES modes of operation. In: Proceedings of 38th Annual Symposium on Foundations of Computer Science (FOCS 1997), pp. 394–403. IEEE Press (1997)
8. Björklund, H., Martens, W.: The tractability frontier for NFA minimization. J. Comput. Syst. Sci. 78(1), 198–210 (2012)
9. Brakerski, Z.: Fully homomorphic encryption without modulus switching from classical GapSVP. In: Safavi-Naini, R., Canetti, R. (eds.) CRYPTO 2012. LNCS, vol. 7417, pp. 868–886. Springer, Heidelberg (2012). https://doi.org/10.1007/978-3-642-32009-5_50
10. Brakerski, Z., Gentry, C., Vaikuntanathan, V.: (Leveled) fully homomorphic encryption without bootstrapping. ACM Trans. Comput. Theory 6(3), 13 (2014)
11. Brakerski, Z., Langlois, A., Peikert, C., Regev, O., Stehlé, D.: Classical hardness of learning with errors. In: Boneh, D., Roughgarden, T., Feigenbaum, J. (eds.) Symposium on Theory of Computing Conference, STOC 2013, Palo Alto, CA, USA, June 1–4, 2013, pp. 575–584. ACM (2013)
12. Brakerski, Z., Vaikuntanathan.: Lattice-based FHE as secure as PKE. In: Naor, M. (ed.) Innovations in Theoretical Computer Science, ITCS 2014, pp. 1–12. ACM (2014)
13. Brzozowski, J.A.: Derivatives of regular expressions. J. ACM 11(4), 481–494 (1964)
14. Cheon, J.H., Kim, A., Kim, M., Song, Y.: Homomorphic encryption for arithmetic of approximate numbers. In: Takagi, T., Peyrin, T. (eds.) ASIACRYPT 2017, Part I. LNCS, vol. 10624, pp. 409–437. Springer, Cham (2017). https://doi.org/10.1007/978-3-319-70694-8_15
15. Chillotti, I., Gama, N., Georgieva, M., Izabachène, M.: Faster fully homomorphic encryption: bootstrapping in less than 0.1 seconds. In: Cheon, J.H., Takagi, T. (eds.) ASIACRYPT 2016, Part I. LNCS, vol. 10031, pp. 3–33. Springer, Heidelberg (2016). https://doi.org/10.1007/978-3-662-53887-6_1
16. Chillotti, I., Gama, N., Georgieva, M., Izabachène, M.: Faster packed homomorphic operations and efficient circuit bootstrapping for TFHE. In: Takagi, T., Peyrin, T. (eds.) ASIACRYPT 2017, Part I. LNCS, vol. 10624, pp. 377–408. Springer, Cham (2017). https://doi.org/10.1007/978-3-319-70694-8_14

17. Fan, J., Vercauteren, F.: Somewhat practical fully homomorphic encryption. IACR Cryptol. ePrint Arch. **2012**, 144 (2012)
18. Gama, N., Izabachène, M., Nguyen, P.Q., Xie, X.: Structural lattice reduction: generalized worst-case to average-case reductions and homomorphic cryptosystems. In: Fischlin, M., Coron, J.-S. (eds.) EUROCRYPT 2016, Part II. LNCS, vol. 9666, pp. 528–558. Springer, Heidelberg (2016). https://doi.org/10.1007/978-3-662-49896-5_19
19. Gentry, C.: Fully homomorphic encryption using ideal lattices. In: Proceedings of the 41st ACM Symposium on Theory of Computing - STOC 2009, pp. 169–178. ACM (2009)
20. Gentry, C., Gorbunov, S., Halevi, S.: Graph-induced multilinear maps from lattices. In: Dodis, Y., Nielsen, J.B. (eds.) TCC 2015, Part II. LNCS, vol. 9015, pp. 498–527. Springer, Heidelberg (2015). https://doi.org/10.1007/978-3-662-46497-7_20
21. Gentry, C., Peikert, C., Vaikuntanathan, V.: Trapdoors for hard lattices and new cryptographic constructions. In: Dwork, C. (ed.) Proceedings of the 40th Annual ACM Symposium on Theory of Computing, Victoria, British Columbia, Canada, 17–20 May 2008, pp. 197–206. ACM (2008)
22. Gentry, C., Sahai, A., Waters, B.: Homomorphic encryption from learning with errors: conceptually-simpler, asymptotically-faster, attribute-based. In: Canetti, R., Garay, J.A. (eds.) CRYPTO 2013, Part I. LNCS, vol. 8042, pp. 75–92. Springer, Heidelberg (2013). https://doi.org/10.1007/978-3-642-40041-4_5
23. Halevi, S., Shoup, V.: Faster homomorphic linear transformations in HElib. IACR Cryptol. ePrint Arch. **2018**, 244 (2018)
24. Håstad, J., Impagliazzo, R., Levin, L.A., Luby, M.: A pseudorandom generator from any one-way function. SIAM J. Comput. **28**(4), 1364–1396 (1999)
25. Hiromasa, R., Abe, M., Okamoto, T.: Packing messages and optimizing bootstrapping in GSW-FHE. In: Katz, J. (ed.) PKC 2015. LNCS, vol. 9020, pp. 699–715. Springer, Heidelberg (2015). https://doi.org/10.1007/978-3-662-46447-2_31
26. Hoffstein, J., Pipher, J., Silverman, J.H.: NTRU: a ring-based public key cryptosystem. In: Buhler, J.P. (ed.) ANTS 1998. LNCS, vol. 1423, pp. 267–288. Springer, Heidelberg (1998). https://doi.org/10.1007/BFb0054868
27. Howgrave-Graham, N.: A hybrid lattice-reduction and meet-in-the-middle attack against NTRU. In: Menezes, A. (ed.) CRYPTO 2007. LNCS, vol. 4622, pp. 150–169. Springer, Heidelberg (2007). https://doi.org/10.1007/978-3-540-74143-5_9
28. Hromkovic, J., Schnitger, G.: Ambiguity and communication. Theory Comput. Syst. **48**(3), 517–534 (2011)
29. Jiang, T., Ravikumar, B.: Minimal NFA problems are hard. SIAM J. Comput. **22**(6), 1117–1141 (1993)
30. Leung, H.: Separating exponentially ambiguous finite automata from polynomially ambiguous finite automata. SIAM J. Comput. **27**(4), 1073–1082 (1998)
31. Leung, H.: Descriptional complexity of NFA of different ambiguity. Int. J. Found. Comput. Sci. **16**(5), 975–984 (2005)
32. López-Alt, A., Tromer, E., Vaikuntanathan, V.: On-the-fly multiparty computation on the cloud via multikey fully homomorphic encryption. In: STOC, pp. 1219–1234 (2012)
33. May, A., Silverman, J.H.: Dimension reduction methods for convolution modular lattices. In: Silverman [41], pp. 110–125
34. Meyer, A.R., Fischer, M.J.: Economy of description by automata, grammars, and formal systems. In: 12th Annual Symposium on Switching and Automata Theory, East Lansing, Michigan, USA, 13–15 October 1971, pp. 188–191 (1971)

35. Micciancio, D.: Improving lattice based cryptosystems using the hermite normal form. In: Silverman [41], pp. 126–145
36. Micciancio, D., Peikert, C.: Trapdoors for lattices: simpler, tighter, faster, smaller. In: Pointcheval, D., Johansson, T. (eds.) EUROCRYPT 2012. LNCS, vol. 7237, pp. 700–718. Springer, Heidelberg (2012). https://doi.org/10.1007/978-3-642-29011-4_41
37. Micciancio, D., Regev, O.: Worst-case to average-case reductions based on gaussian measures. In: 45th Symposium on Foundations of Computer Science (FOCS 2004), 17–19 October 2004, Rome, Italy, Proceedings, pp. 372–381. IEEE Computer Society (2004)
38. Peikert, C., Rosen, A.: Efficient collision-resistant hashing from worst-case assumptions on cyclic lattices. In: Halevi, S., Rabin, T. (eds.) TCC 2006. LNCS, vol. 3876, pp. 145–166. Springer, Heidelberg (2006). https://doi.org/10.1007/11681878_8
39. Regev, O.: On lattices, learning with errors, random linear codes, and cryptography. J. ACM 56(6), 34 (2009)
40. Rivest, R., Adleman, L., Dertouzos, M.: On data banks and privacy homomorphisms. In: Foundations of Secure Computation, pp. 169–177. Academic Press (1978)
41. Silverman, J.H. (ed.): CaLC 2001. LNCS, vol. 2146. Springer, Heidelberg (2001). https://doi.org/10.1007/3-540-44670-2
42. Smart, N.P., Vercauteren, F.: Fully homomorphic SIMD operations. Des. Codes Cryptography 71(1), 57–81 (2014). Early verion at http://eprint.iacr.org/2011/133
43. van Dijk, M., Gentry, C., Halevi, S., Vaikuntanathan, V.: Fully homomorphic encryption over the integers. In: Gilbert, H. (ed.) EUROCRYPT 2010. LNCS, vol. 6110, pp. 24–43. Springer, Heidelberg (2010). https://doi.org/10.1007/978-3-642-13190-5_2
44. Wang, B., Wang, X., Xue, R., Huang, X.: Matrix FHE and its application in optimizing bootstrapping. Comput. J. 61(12), 1845–1861 (2018)
45. Weber, A., Seidl, H.: On the degree of ambiguity of finite automata. Theor. Comput. Sci. 88(2), 325–349 (1991)
46. Yu, S.: Regular languages. In: Rozenberg, G., Salomaa, A. (eds.) Handbook of Formal Languages, vol. 1, pp. 41–110. Springer, Heidelberg (1997). https://doi.org/10.1007/978-3-642-59136-5_2

Combinatorial Cryptography

Efficient Explicit Constructions of Multipartite Secret Sharing Schemes

Qi Chen[1](✉), Chunming Tang[2], and Zhiqiang Lin[2]

[1] Advanced Institute of Engineering Science for Intelligent Manufacturing, Guangzhou University, Guangzhou 510006, China
chenqi.math@gmail.com
[2] College of Mathematics and Information Science, Guangzhou University, Guangzhou 510006, China
ctang@gzhu.edu.cn, linzhiqiang0824@163.com

Abstract. Multipartite secret sharing schemes are those having a multipartite access structure, in which the set of participants is divided into several parts and all participants in the same part play an equivalent role. Secret sharing schemes for multipartite access structures have received considerable attention due to the fact that multipartite secret sharing can be seen as a natural and useful generalization of threshold secret sharing.

This work deals with efficient and explicit constructions of ideal multipartite secret sharing schemes, while most of the known constructions are either inefficient or randomized. Most ideal multipartite secret sharing schemes in the literature can be classified as either hierarchical or compartmented. The main results are the constructions for ideal hierarchical access structures, a family that contains every ideal hierarchical access structure as a particular case such as the disjunctive hierarchical threshold access structure and the conjunctive hierarchical threshold access structure, and the constructions for compartmented access structures with upper bounds and compartmented access structures with lower bounds, two families of compartmented access structures.

On the basis of the relationship between multipartite secret sharing schemes, polymatroids, and matroids, the problem of how to construct a scheme realizing a multipartite access structure can be transformed to the problem of how to find a representation of a matroid from a presentation of its associated polymatroid. In this paper, we give efficient algorithms to find representations of the matroids associated to the three families of multipartite access structures. More precisely, based on know results about integer polymatroids, for each of the three families of access structures, we give an efficient method to find a representation of the integer polymatroid over some finite field, and then over some finite extension of that field, we give an efficient method to find a presentation of the matroid associated to the integer polymatroid. Finally, we construct ideal linear schemes realizing the three families of multipartite access structures by efficient methods.

Keywords: Secret sharing schemes · Multipartite access structures · Matroids · Polymatroids

ⓒ International Association for Cryptologic Research 2019
S. D. Galbraith and S. Moriai (Eds.): ASIACRYPT 2019, LNCS 11922, pp. 505–536, 2019.
https://doi.org/10.1007/978-3-030-34621-8_18

1   Introduction

Secret sharing is an important cryptographic primitive, by means of which a secret value is distributed into shares among a number of participants in such a way that only the qualified sets of participants can recover the secret value from their shares. A scheme is *perfect* if the unqualified subsets do not obtain any information about the secret. The first proposed secret sharing schemes [8,31] realized *threshold access structures*, in which the qualified subsets are those having at least a given number of participants. In addition, these schemes are *ideal* and *linear*. A scheme is ideal if the share of every participant has the same length as the secret, and it is linear if the linear combination of the shares of different secrets results in shares for the same linear combination of the secret values. Even though there exists a linear secret sharing scheme for every access structure [6,24], the known general constructions are not impractical because the length of the shares grows exponentially with the number of participants. Actually, the optimization of secret sharing schemes for general access structures has appeared to be an extremely difficult problem and not much is known about it. Nevertheless, secret sharing schemes have found numerous applications in cryptography and distributed computing, such as threshold cryptography [16], secure multiparty computations [5,11,14,15], and oblivious transfer [32,36]. In many of the applications mentioned above, we hope to use practical schemes, namely, the linear schemes in which the size of the share of each participant is a polynomial of the size of the secret. In particular, we want to use the ideal schemes since they are the most space-efficient.

Due to the difficulty of constructing an ideal liner scheme for every given access structure, it is worthwhile to find families of access structures that admit ideal linear schemes and have useful properties for the applications of secret sharing. Several such families are formed by multipartite access structures, in which the set of participants is divided into different parts and all participants in the same part play an equivalent role. Weighted threshold access structures [4,31], hierarchical access structures [18,34,35], and compartmented access structures [9,22,37] are typical examples of such multipartite access structures. Readers can refer to [19] for comprehensive survey on multipartite access structures. A great deal of the ongoing research in this area is devoted to the properties of multipartite access structures and to secret sharing schemes (especially ideal and linear ones) that realize them.

The first class of multipartite access structures is weighted threshold access structures which appeared in the seminal paper by Shamir [31]. Weighted threshold access structures do not admit an ideal secret sharing scheme in general. Ideal multipartite secret sharing and their access structures were initially studied by Kothari [25] and by Simmons [34]. Kothari [25] presented some ideas to construct ideal linear schemes with hierarchical properties. Simmons [34] introduced the multilevel access structures (also called disjunctive hierarchical threshold access structures (DHTASs) in [35]) and compartmented access structures, and constructed ideal linear schemes for some of them by geometric method [8], but the method is inefficient. The efficient method to construct ideal linear schemes for DHTASs was presented by Brickell [9] based on primitive polynomials over finite fields. He also presented a more general family, that is the so-called compartmented access structures with lower bounds (LCASs) as a generalization of Simmons' compartmented access structures and offered a method to construct ideal linear schemes realizing LCASs too. This method is efficient to construct schemes realizing Simmons' compartmented access structures but is inefficient to construct the schemes realizing LCASs in general because it is required to check (possible exponentially) many matrices for non-singularity. Tassa [35] presented conjunctive hierarchical threshold access structures (CHTASs) and offered a method to construct ideal linear schemes realizing them based on Birkhoff interpolation. In the case of random allocation of participant identities, this method is probabilistic. A method is probabilistic if it produces a scheme for the given access structure with high probability. In the probabilistic method, it is still required to check many matrices for non-singularity. In general, we hope to construct schemes by efficient methods. By allocating participant identities in a monotone way, Tassa gave an efficient method to construct ideal linear schemes for CHTASs over a sufficiently large prime field based on Birkhoff interpolation. Tassa and Dyn [37] presented compartmented access structures with upper bounds (UCASs) and offered probabilistic methods to construct ideal linear schemes for UCASs, LCASs and CHTASs based on bivariate interpolation.

Another related line of work deals with the characterization of the ideal multipartite secret sharing schemes and their access structures. This line of research was initiated by Brickell [9] and by Brickell and Davenport [10]. They introduced the relationship between secret sharing schemes and matroids, and characterized the ideal secret sharing schemes by matroids. Beimel et al. [4] characterized ideal weighted threshold secret sharing schemes by matroids. The bipartite access structures were characterized in [29] and some partial results about the tripartite case were presented in [13] and [22]. On the basis of the works in [9,10], Farràs et al. [17–19] introduced integer polymatroids for the study of ideal multipartite secret sharing schemes. They studied the connection of multipartite secret sharing schemes, matroids and polymatroids, and presented many new families of multipartite access structures such as ideal hierarchical access structures (IHASs) and compartmented access structures with upper and lower bounds. Their work implies the problem of how to construct a scheme realizing a multipartite access structure can be transformed to the problem of how to find a representation of a matroid from a presentation of its associated polymatroid. Nevertheless, Farràs et al. [17,19] pointed out it remains open whether or not exist efficient algorithms to obtain representations of matroids from representations of their associated polymatroids in general.

1.1   Related Work

Efficient Explicit Constructions of Ideal Multipartite Secret Sharing. The most of the known constructions of ideal multipartite secret sharing schemes are either inefficient or randomized in the literature. Efficient methods to construct ideal hierarchical secret sharing schemes were given by Brickell [9] and by

Tassa [35]. Brickell's construction provides a representation of a matroid associated to DHTASs over finite fields of the form $\mathbb{F}_{q^\lambda}$ with $\lambda \geq mk^2$, where q is a prime power, m is the number of parts that the set of participants is divided into, and k is the rank of the matroid. An irreducible polynomial of degree λ over $\mathbb{F}_q$ has to be found, but this can be done in time polynomial in q and λ by using the algorithm given by Shoup [33]. Therefore, a representation can be found in time polynomial in the size of the ground set. Accordingly, ideal linear schemes realizing DHTASs can be obtained by an efficient method. Tassa [35] offered a representation of a matroid associated to CHTASs over prime fields $\mathbb{F}_p$ with p larger than $2^{-k+2}(k-1)^{(k-1)/2}(k-1)!N^{(k-1)(k-2)/2}$, where k is the rank of the matroid and N is the maximum identity assigning to participants. A matrix M is the representation if some of its submatrices are nonsingular. The non-singularity of these submatrices depends on the Birkhoff interpolation. There is an efficient algorithm to solve this kind of interpolation over the prime fields $\mathbb{F}_p$, and consequently, ideal linear schemes for CHTASs can be obtained by an efficient method. Ball et al. [1] extended the methods in [9,35] and obtained two different kinds of representations of biuniform matroids, one by using a primitive element of an extension field and another one by using a large prime field. The schemes for some bipartite access structures can be obtained based on these representations. In addition, efficient methods to construct schemes for some multilevel access structures with two levels and three levels were presented in [7] and [21], respectively.

Multipartite Secret Sharing, Polymatroids and Matroids. On the basis of the connection of multipartite secret sharing schemes, matroids and polymatroids, Farràs et al. [17–19] introduced a unified method based on polymatroid techniques, which simplifies the task of determining whether a given multipartite access structures is ideal or not. In particular, they characterized ideal secret sharing schemes for hierarchical access structures in [18] by the unified method. They defined the accurate form of IHASs and showed that every ideal hierarchical access structure is of this form or it can be obtained from a structure of this form by removing some participants. Moreover, they presented a general method to construct ideal linear schemes realizing multipartite access structures. Specially, to construct a secret sharing scheme realizing a given multipartite access structure, first find an integer polymatroid associated to the access structure, then find a representation of the integer polymatroid over some finite field, and third find a representation of the matroid associated the access structure over some finite extension of the finite field based on the representation of the integer polymatroid. The result in [17] implies the matroid can be used to construct an ideal linear scheme realizing the access structure.

1.2   Our Results

In this paper, we study how to construct secret sharing schemes realizing multipartite access structures. The main results are the constructions for IHASs, a family that contains all ideal hierarchical access structure as a particular case such as DHTASs and CHTASs, and the constructions for UCASs and LCASs, two special families of compartmented access structures. We give efficient methods to explicitly construct ideal linear schemes realizing these access structures combining the general polymatroid-based method in [17] and Brickell's method to construct ideal linear schemes for DHTASs in [9]. The ideal of our construction is described as follows.

Our method to construct multipartite schemes is closely related to the representations of the multipartite matroid associated to the given multipartite access structure. The problem of how to obtain a representation of the multipartite matroid can be transformed to find a matrix M such that its some special submatrices are nonsingular. Thus, our main goal is that providing a polynomial time algorithms to construct such a matrix M such that all the determinants of those special submatrices are nonzero over some finite fields. More precisely, we construct the matrix M with special form such that every determinant of those submatrices can be viewed as a nonzero polynomial on x of degree at most t over some finite field $\mathbb{F}_q$. Based on such a matrix M, over $\mathbb{F}_{q^\lambda}$ with $\lambda > t$, the algorithm given by Shoup [33] implies a representation of the matroid associated the given access structure can be found in time polynomial in the size of the ground set.

The idea of finding a matrix M such that the determinants of some of its submatrices are denoted by a nonzero polynomial on x comes from Brickell [9]. This is the key to find a representation of the matroid. This is related to the determinant function of matrix. To solve this question, we introduce approaches to calculate two class of matrices with special form, one can be applied to the constructions for IHASs and another one can be applied to the constructions for UCASs and LCASs.

Specifically, based on the integer polymatroids associated to the three families of multipartite access structures presented in [17–19], for each of the three families of access structures, we give an efficient method to find a representation of the integer polymatroid over some finite field, and then over some finite extension of that field, we give an efficient method to find a presentation of the matroid associated to the integer polymatroid. Accordingly, we construct ideal linear schemes for these access structures. First, we construct a $\mathbb{F}_q$-representation of an integer polymatroid that is as simple as possible. In the constructions for IHASs, the representation is constructed based on unit matrix, and in the constructions for UCASs and LCASs, the representations are constructed based on Vandermonde matrix. Second, based on the special representation for some access structure, we construct the matrix M satisfied the required conditions such that every determinant of some of its submatrices can be viewed as a nonzero polynomial on x over $\mathbb{F}_q$. Thus, a representation of the matroid associated the given access structure can be found in time polynomial in the size of the ground set by the algorithm in [33].

In addition, we compare our results with the efficient methods to construct multipartite secret sharing schemes from [9,35] in Sect. 4.3. In particular, we point out that our construction for DHTASs is the same as the one in [9], but we improve the bound for the size of the ground set.

1.3   Organization of the Paper

Section 2 introduces some knowledge about access structures, secret sharing schemes, polymatroids, matroids, and the methods to construct secret sharing schemes by matroids and polymatroids. Section 3 introduces the approaches to calculate the determinant functions of two classes of matrices with special form. Section 4 gives two classes of constructions for ideal linear secret sharing schemes realizing IHASs. Section 5 construct ideal linear secret sharing schemes realizing UCASs and LCASs.

2   Preliminaries

We introduce here some notation that will be used all through the paper. In particular, we recall the compact and useful representation of multipartite access structures as in [17–19].

We use $\mathbb{Z}_+$ to denote the set of the non-negative integers. For every positive integer i we use the notation $[i] := \{1, \ldots, i\}$ and for every $i, j \in \mathbb{Z}_+$ we use the notation $[i, j] := \{i, \ldots, j\}$ with $i < j$. Consider a finite set J and given two vectors $\boldsymbol{u} = (u_i)_{i \in J}$ and $\boldsymbol{v} = (v_i)_{i \in J}$ in $\mathbb{Z}_+^J$, we write $\boldsymbol{u} \leq \boldsymbol{v}$ if $u_i \leq v_i$ for every $i \in J$. The *modulus* $|\boldsymbol{u}|$ of a vector $\boldsymbol{u} \in \mathbb{Z}_+^J$ is defined by $|\boldsymbol{u}| = \sum_{i \in J} u_i$. For every subset $X \subseteq J$, we notate $\boldsymbol{u}(X) = (u_i)_{i \in X} \in \mathbb{Z}_+^X$. For every positive integer m, we notate $J_m = \{1, \ldots, m\}$ and $J'_m = \{0, 1, \ldots, m\}$. Of course the vector notation that has been introduced here applies as well to $\mathbb{Z}_+^m = \mathbb{Z}_+^{J_m}$.

2.1   Access Structures and Secret Sharing Schemes

Let $P = \{p_1, \ldots, p_n\}$ denote the set of participants and its power set be denoted by $\mathcal{P}(P) = \{\mathcal{V} : \mathcal{V} \subseteq P\}$ which contains all the subsets of P. A collection $\Gamma \subseteq \mathcal{P}(P)$ is monotone if $\mathcal{V} \in \Gamma$ and $\mathcal{V} \subseteq \mathcal{W}$ imply that $\mathcal{W} \in \Gamma$. An *access structure* is a monotone collection $\Gamma \subseteq \mathcal{P}(P)$ of nonempty subsets of P. Sets in Γ are called *authorized*, and sets not in Γ are called *unauthorized*. An authorized set $\mathcal{V} \in \Gamma$ is called a *minimal authorized set* if for every $\mathcal{W} \subsetneq \mathcal{V}$, the set $\mathcal{W}$ is unauthorized. An unauthorized set $\mathcal{V} \notin \Gamma$ is called a *maximal unauthorized set* if for every $\mathcal{W} \supsetneq \mathcal{V}$, the set $\mathcal{W}$ is authorized. The set $\Gamma^* = \{\mathcal{V} : \mathcal{V}^c \notin \Gamma\}$ is called the *dual* access structure to Γ. It is easy to see that Γ^* is monotone too. In particular, an access structure is said to be *connected* if all participants are in at least one minimal authorized subset.

A family $\Pi = (\Pi_i)_{i \in J_m}$ of subsets of P is called here a *partition* of P if $P = \bigcup_{i \in J_m} \Pi_i$ and $\Pi_i \cap \Pi_j = \emptyset$ whenever $i \neq j$. For a partition Π of a set P, we consider the mapping $\Pi : \mathcal{P}(P) \to \mathbb{Z}_+^m$ defined by $\Pi(\mathcal{V}) = (|\mathcal{V} \cap \Pi_i|)_{i \in J_m}$. We write $\mathbf{P} = \Pi(\mathcal{P}(P)) = \{\boldsymbol{u} \in \mathbb{Z}_+^m : \boldsymbol{u} \leq \Pi(P)\}$. For a partition Π of a set P, a Π-*permutation* is a permutation σ on P such that $\sigma(\Pi_i) = \Pi_i$ for every part Π_i

of Π. An access structure on P is said to be Π-*partite* if every Π-permutation is an automorphism of it.

As in [17–19], we describe a multipartite access structure in a compact way by taking into account that its members are determined by the number of elements they have in each part. If an access structure Γ on P is Π-partite, then $\mathcal{V} \in \Gamma$ if and only if $\Pi(\mathcal{V}) \in \Pi(\Gamma)$. That is, Γ is completely determined by the partition Π and set of vectors $\Pi(\Gamma) \subseteq \mathbf{P} \subseteq \mathbb{Z}_+^m$. Moreover, the set $\Pi(\Gamma) \subseteq \mathbf{P}$ is monotone increasing, that is, if $\boldsymbol{u} \in \Pi(\Gamma)$ and $\boldsymbol{v} \in \mathbf{P}$ is such that $\boldsymbol{u} \leq \boldsymbol{v}$, then $\boldsymbol{v} \in \Pi(\Gamma)$. Therefore, $\Pi(\Gamma)$ is univocally determined by $\min \Pi(\Gamma)$, the family of its minimal vectors, that is, those representing the minimal qualified subsets of Γ. By an abuse of notation, we will use Γ to denote both a Π-partite access structure on P and the corresponding set $\Pi(\Gamma)$ of points in $\mathbf{P}$, and the same applies to $\min \Gamma$.

Now, we introduce some families of multipartite access structures.

Definition 1 (Ideal hierarchical access structures). *Take $\hat{\boldsymbol{k}}$, $\boldsymbol{k} \in \mathbb{Z}_+^m$ such that $\hat{k}_1 = 0$ and $\hat{k}_i \leq \hat{k}_{i+1} < k_i \leq k_{i+1}$ for $i \in [m-1]$. The following access structures are called* ideal hierarchical access structures (IHASs)

$$\Gamma = \{\boldsymbol{u} \in \mathbf{P} : |\boldsymbol{u}([\ell])| \geq k_\ell \text{ for some } \ell \in J_m \text{ and } |\boldsymbol{u}([i])| \geq \hat{k}_{i+1} \text{ for all } i \in [\ell-1]\}. \tag{1}$$

In particular, if $\hat{k}_i = 0$ for every $i \in J_m$ and $0 < k_1 < \cdots < k_m = k$, then IHASs is the *disjunctive hierarchical threshold access structures (DHTASs)*, which can be denoted by

$$\Gamma = \{\boldsymbol{u} \in \mathbf{P} : |\boldsymbol{u}([i])| \geq k_i \text{ for some } i \in J_m\}, \tag{2}$$

and if $0 = \hat{k}_1 < \cdots < \hat{k}_m$ and $k_1 = \cdots = k_m = k$ then IHASs is the *conjunctive hierarchical threshold access structures (CHTASs)*, which can be denoted by

$$\Gamma = \{\boldsymbol{u} \in \mathbf{P} : |\boldsymbol{u}([i])| \geq \tilde{k}_i \text{ for all } i \in J_m\}, \tag{3}$$

where $\tilde{k}_i = \hat{k}_{i+1}$ for $i \in [m-1]$ and $\tilde{k}_m = k_m$.

Definition 2 (Compartmented access structures). *Take $\boldsymbol{t} \in \mathbb{Z}_+^m$ and $k \in \mathbb{N}$ such that $k \geq |\boldsymbol{t}|$. The following access structures are called* compartmented access structures with lower bounds (LCASs)

$$\min \Gamma = \{\boldsymbol{u} \in \mathbf{P} : |\boldsymbol{u}| = k \text{ and } \boldsymbol{u} \geq \boldsymbol{t}\}. \tag{4}$$

Take $\boldsymbol{r} \in \mathbb{Z}_+^m$ such that $\boldsymbol{r} \leq \Pi(P)$ and $r_i \leq k \leq |\boldsymbol{r}|$ for every $i \in J_m$. The following access structures are called compartmented access structure with upper bound (UCASs)

$$\min \Gamma = \{\boldsymbol{u} \in \mathbf{P} : |\boldsymbol{u}| = k \text{ and } \boldsymbol{u} \leq \boldsymbol{r}\}. \tag{5}$$

We next present the definition of *unconditionally secure perfect secret sharing scheme* as given in [3,12]. For more information about this definition and secret sharing in general, see [2].

Definition 3 (Secret sharing schemes). *Let $P = \{p_1, \ldots, p_n\}$ be a set of participants. A distribution scheme $\Sigma = (\Phi, \mu)$ with domain of secrets S is a pair, where μ is a probability distribution on some finite set $\mathcal{R}$ called the set of random strings and Φ is a mapping from $S \times \mathcal{R}$ to a set of n-tuples $S_1 \times S_2 \times \cdots \times S_n$, where S_i is called the domain of shares of p_i. A dealer distributes a secret $s \in S$ according to Σ by first sampling a random string $r \in \mathcal{R}$ according μ, computing a vector of shares $\Phi(s, r) = (s_1, \ldots, s_n)$, and privately communicating each share s_i to participant p_i. For a set $V \subseteq P$, we denote $\Phi_V(s, r)$ as the restriction of $\Phi(s, r)$ to its V-entries (i.e., the shares of the participants in V).*

Let S be a finite set of secrets, where $S \geq 2$. A distribution scheme $\Sigma = (\Phi, \mu)$ with domain of secrets S is a secret sharing scheme realizing an access structure $\Gamma \subseteq \mathcal{P}(P)$ if the following two requirements hold:

CORRECTNESS. *The secret s can be reconstructed by any authorized set of participants. That is, for any authorized set $V \in \Gamma$ (where $V = \{p_{i_1}, \ldots, p_{i_{|V|}}\}$), there exists a reconstruction function $Recon_V : S_{i_1} \times \cdots \times S_{i_{|V|}} \to S$ such that for every $s \in S$ and every random string $r \in \mathcal{R}$,*

$$Recon_V\big(\Phi_V(s, r)\big) = s.$$

PRIVACY. *Every unauthorized set can learn nothing about the secret (in the information theoretic sense) from their shares. Formally, for any unauthorized set $W \notin \Gamma$, every two secrets $s, s' \in S$, and every possible $|W|$-tuple of shares $(s_i)_{u_i \in W}$,*

$$Pr\big[\Phi_W(s, r) = (s_i)_{u_i \in W}\big] = Pr\big[\Phi_W(s', r) = (s_i)_{u_i \in W}\big]$$

when the probability is over the choice of r from $\mathcal{R}$ at random according to μ.

Definition 4 (Ideal linear secret sharing schemes). *Let $P = \{p_1, \ldots, p_n\}$ be a set of participants. Let $\Sigma = (\Phi, \mu)$ be a secret sharing scheme with domain of secrets S, where μ is a probability distribution on a set $\mathcal{R}$ and Φ is a mapping from $S \times \mathcal{R}$ to $S_1 \times S_2 \times \cdots \times S_n$, where S_i is called the domain of shares of p_i. We say that Σ is an ideal linear secret sharing scheme over a finite field $\mathbb{K}$ if $S = S_1 = \cdots = S_n = \mathbb{K}$, $\mathcal{R}$ is a $\mathbb{K}$-vector space, Φ is a $\mathbb{K}$-linear mapping, and μ is the uniform probability distribution.*

This paper deals with *unconditionally secure perfect ideal linear secret sharing schemes*.

2.2   Polymatroids and Matroids

In this section we introduce the definitions and some properties of polymatroids and matroids. Most results of this section are from [17–19]. For more background on matroids and polymatroids, see [23, 28, 30, 38].

Definition 5. *A polymatroid S is defined by a pair (J, h), where J is the finite ground set and $h : \mathcal{P}(J) \to \mathbb{R}$ is the rank function that satisfies*

(1) $h(\emptyset) = 0$;
(2) h *is* monotone increasing: *if* $X \subseteq Y \subseteq J$, *then* $h(X) \leq h(Y)$;
(3) h *is* submodular: *if* $X, Y \subseteq J$, *then* $h(X \cup Y) + h(X \cap Y) \leq h(X) + h(Y)$.

An integer polymatroid $\mathcal{Z}$ *is a polymatroid with an integer-valued rank function* h. *An integer polymatroid such that* $h(X) \leq |X|$ *for any* $X \subseteq J$ *is called a* matroid.

While matroids abstract some properties related to linear dependency of collections of vectors in a vector space, integer polymatroids do the same with collections of subspaces. Suppose $(V_i)_{i \in J}$ is a finite collection of subspaces of a $\mathbb{K}$-vector space V, where $\mathbb{K}$ is a finite field. The mapping $h(X) : \mathcal{P}(J) \to \mathbb{Z}$ defined by $h(X) = \dim(\sum_{i \in X} V_i)$ is the rank function of an integer polymatroid with ground set J. Integer polymatroids and, in particular, matroids that can be defined in this way are said to be $\mathbb{K}$-*representable*.

Following the analogy with vector spaces we make the following definitions. For an integer polymatroid $\mathcal{Z}$, the set of *integer independent vectors* of $\mathcal{Z}$ is

$$\mathcal{D} = \{ \boldsymbol{u} \in \mathbb{Z}_+^J : |\boldsymbol{u}(X)| \leq h(X) \; for \; every \; X \subseteq J \},$$

in which the maximal integer independent vectors are called the *integer bases* of $\mathcal{Z}$. Let $\mathcal{B}$ or $\mathcal{B}(\mathcal{Z})$ denote the collection of all integer bases of $\mathcal{Z}$. Then all the elements of $\mathcal{B}(\mathcal{Z})$ have the identical modulus. In fact, every integer polymatroid $\mathcal{Z}$ is univocally determined by $\mathcal{B}(\mathcal{Z})$ since h is determined by $h(X) = \max\{|\boldsymbol{u}(X)| : \boldsymbol{u} \in \mathcal{B}(\mathcal{Z})\}$.

Given an integer polymatroid $\mathcal{Z} = (J, h)$ and a subset $X \subseteq J$, let $\mathcal{Z}|X = (X, h)$ denote a new integer polymatroid restricted $\mathcal{Z}$ on X, and $\mathcal{B}(\mathcal{Z}, X) = \{ \boldsymbol{u} \in \mathcal{D} : \text{supp}(\boldsymbol{u}) \subseteq X \text{ and } |\boldsymbol{u}| = h(X) \}$ where $\text{supp}(\boldsymbol{u}) = \{ i \in J : u_i \neq 0 \}$. Then there is a natural bijection between $\mathcal{B}(\mathcal{Z}, X)$ and $\mathcal{B}(\mathcal{Z}|X)$.

We next introduce a class of polymatroids as follows.

Definition 6 (Boolean polymatroids). *Let* S *be a finite set and consider a family* $(S_i)_{i \in J}$ *of subsets of* S. *The mapping* $h : \mathcal{P}(J) \to \mathbb{Z}$ *defined by* $h(X) = |\bigcup_{i \in X} S_i|$ *is clearly the rank function of an integer polymatroid. Integer polymatroids that can be defined in this way are called Boolean polymatroids.*

Boolean polymatroids are very simple integer polymatroids that are representable over every finite field $\mathbb{K}$. If $|S| = t$, we can assume that S is a basis of the vector space $V = \mathbb{K}^t$. For every $i \in J$, consider the vector subspace $V_i = \langle S_i \rangle$. Obviously, these subspaces form a $\mathbb{K}$-representation of $\mathcal{Z}$.

2.3   Secret Sharing Schemes, Matroids and Polymatroids

In this section we review the methods to construct ideal linear secret sharing schemes for multipartite access structures by matroids and polymatroids. Most results of this section are from [17–19]. We first introduce the method to construct ideal linear schemes by matroids.

Let $P = \{p_1, \ldots, p_n\}$ be a set of participants and $p_0 \notin P$ be the dealer. Suppose $\mathcal{M}$ is a matroid on the finite set $P' = P \cup \{p_0\}$, and let
$$\Gamma_{p_0}(\mathcal{M}) = \{A \subseteq P : h(A \cup \{p_0\}) = h(A)\}.$$
Then $\Gamma_{p_0}(\mathcal{M})$ is an access structure on P because monotonicity property is satisfied, which is called *the port of the matroid $\mathcal{M}$ at the point p_0*.

Matroid ports play a very important role in secret sharing. Brickell [9] proved that the ports of representable matroids admit ideal secret sharing schemes and provided a method to construct ideal schemes for ports of $\mathbb{K}$-representable matroids. These schemes are called a $\mathbb{K}$-*vector space secret sharing schemes*. This method was described by Massey [26,27] in terms of linear codes. Suppose M is a $k \times (n+1)$ matrix over $\mathbb{K}$. Then the columns of M determine a $\mathbb{K}$-representable matroid $\mathcal{M}$ with ground set P' such that the columns of M are in one-to-one correspondence with the elements in P'. In this situation, the matrix M is called a $\mathbb{K}$-*representation* of the matroid $\mathcal{M}$. Moreover, M is a generator matrix of some $(n+1, k)$ linear code C over $\mathbb{K}$, that is, a matrix whose rows span C. A code C of length $n+1$ and dimension k is called an $(n+1, k)$ linear code over $\mathbb{K}$ which is a k-dimensional subspace of $\mathbb{K}^{n+1}$. A secret sharing scheme can be constructed by the matrix M based the code C as follows.

Let $s \in \mathbb{K}$ be a secret value. Secret a codeword $\boldsymbol{c} = (c_0, c_1, \ldots, c_n) \in C$ uniformly at random such that $c_0 = s$, and define the share-vector as $(c_1, \ldots, c_n)$, that is c_i is the share of the participant p_i for $i \in [n]$. Let $LSSS(M)$ denote this secret sharing scheme.

Theorem 1 ([26]). *$LSSS(M)$ is a perfect ideal linear scheme such that a set $\mathcal{V} \subset P$ is qualified if and only if the first column in M is a linear combination of the columns with indices in $\mathcal{V}$.*

The *dual code* $C^\perp$ for a code C consists of all vectors $\boldsymbol{c}^\perp \in \mathbb{K}^{n+1}$ such that $\langle \boldsymbol{c}^\perp, \boldsymbol{c} \rangle = 0$ for all $\boldsymbol{c} \in C$, where $\langle \cdot, \cdot \rangle$ denotes the standard inner product. Suppose M and M^* are generator matrices of some $(n+1, k)$ linear code C and its dual $C^\perp$ over $\mathbb{K}$, respectively. Then $LSSS(M)$ and $LSSS(M^*)$ realize Γ and Γ^*, respectively. Sometimes it is not easy to construct an ideal linear scheme for a given access structure Γ directly. In this case we can first construct a scheme for Γ^* and then translate the scheme into an ideal linear scheme for Γ^* using the explicit transformation of [20]. In Sect. 5.2, we will present the construction for LCASs (4) by this method.

This paper deals with unconditionally secure perfect ideal linear secret sharing schemes. Brickell's method can be applied to construct such schemes. Nevertheless, it is difficult to determine whether a given access structure admits an ideal linear secret sharing scheme or not. Moreover, even for access structures that admit such schemes, it may not be easy to construct them. Some strategies based on matroids and polymatroids were presented in [17,19] to attack those problems for multipartite access structures.

The relationship between ideal multipartite access structures and integer polymatroids is summarized as follows.

Theorem 2 ([17]). *Let* $\Pi = (\Pi_i)_{i \in J_m}$ *be a partition of the set* P, *and* $\mathcal{Z}' = (J'_m, h)$ *is an integer polymatroid such that* $h(\{0\}) = 1$ *and* $h(\{i\}) \leq |\Pi_i|$ *for every* $i \in J_m$. *Take* $\Gamma_0(\mathcal{Z}') = \{X \subseteq J_m : h(X \cup \{0\}) = h(X)\}$ *and*

$$\Gamma_0(\mathcal{Z}', \Pi) = \{u \in \mathbf{P} : \text{there exist } X \in \Gamma_0(\mathcal{Z}') \text{ and } v \in \mathcal{B}(\mathcal{Z}'|J_m, X) \text{ such that } v \leq u\}.$$

Then $\Gamma = \Gamma_0(\mathcal{Z}', \Pi)$ *is a* Π-*partite access structure on* P *and a matroid port. Moreover, if* $\mathcal{Z}'$ *is* $\mathbb{K}$-*representable, then* Γ *can be realized by some* $\mathbb{L}$-*vector space secret sharing scheme over every large enough finite extension* $\mathbb{L}$ *of* $\mathbb{K}$. *In addition,* $\mathcal{Z}'$ *is univocally determined by* Γ *if it is connected.*

The general method presented by Farràs et al. [17] to construct ideal schemes for the multipartite access structures satisfying the conditions in Theorem 2 is summarized as follows.

Let $\Pi_0 = \{p_0\}$ and $\Pi' = (\Pi_i)_{i \in J'_m}$ be a partition of the set $P' = P \cup \{p_0\}$ such that $|\Pi_i| = n_i$. Given a connected Π-partite access structure Γ satisfying the conditions in Theorem 2.

Step 1. Find an integer polymatroid $\mathcal{Z}'$ such that $\Gamma = \Gamma_0(\mathcal{Z}', \Pi)$;
Step 2. Find a representation $(V_i)_{i \in J'_m}$ of $\mathcal{Z}'$ over some finite field $\mathbb{K}$;
Step 3. Over some finite extension of $\mathbb{K}$, find a representation of the matroid $\mathcal{M}$ such that Γ is a port of $\mathcal{M}$. More precisely, construct a $k \times (n+1)$ matrix $M = (M_0|M_1|\cdots|M_m)$ with the following properties:
   1. $k = h(J'_m)$ and $n = \sum_{i=1}^m n_i$;
   2. M_i is a $k \times n_i$ matrix whose columns are vectors in V_i;
   3. M_u is nonsingular for any $u \in \mathcal{B}(\mathcal{Z}')$, where M_u is the $k \times k$ submatrix of M formed by any u_i columns in every M_i.

Farràs et al. [17–19] proved that all the multipartite access structures introduced in Sect. 2.1 are connected matroid ports. Moreover, they presented the associated integer polymatroids and proved that they are representable. Therefore, the results in [17–19] solve Step 1. In this paper, we will give an efficient method to explicitly solve Steps 2 and 3, and hence to construct ideal linear schemes for those families of access structures. Our method is based on the properties of determinant functions.

3   Some Properties of Determinant Functions

In this section, we study determinant functions of two classes of matrices with special form, which will be applied to the constructions of representations of matroids associated to multipartite access structures.

3.1   The First Class of Matrices

In this Section, we introduce the approach to calculate the determinant of a class of matrices with special form. This approach is very useful to calculate the determinant of the matrices with some zero blocks. This class of matrices will be applied to the construction of representable matroid associated to IHASs. We will use the well known Laplace Expansion Theorem of determinant.

Theorem 3 (The Laplace Expansion Theorem). *Take a $n \times n$ matrix A. Let $\mathbf{r} = (r_1, \ldots, r_k)$ be a list of k column indices for A such that $1 \leq r_1 < \cdots < r_k < n$ where $1 \leq k < n$ and $\mathbf{t} = (t_1, \ldots, t_k)$ be a list of k row indices for A such that $1 \leq t_1 < \cdots < t_k < n$ where $1 \leq k < n$. The submatrix obtained by keeping the entries in the intersection of any column and row that are in the lists is denoted by $S(A : \mathbf{r}, \mathbf{t})$. The submatrix obtained by removing the entries in the columns and rows that are in the list is denoted by $S'(A : \mathbf{r}, \mathbf{t})$. Then the determinant of A is*

$$\det(A) = (-1)^{|\mathbf{r}|} \sum_{\mathbf{t} \in \mathcal{T}} (-1)^{|\mathbf{t}|} \det\big(S(A : \mathbf{r}, \mathbf{t})\big) \det\big(S'(A : \mathbf{r}, \mathbf{t})\big),$$

where $\mathcal{T}$ denotes the set of all k-tuples $\mathbf{t} = (t_1, \ldots, t_k)$ for which $1 \leq t_1 < \cdots < t_k < n$.

Example 1. Take a 7×7 matrix $A = (A_1 | A_2 | A_3)$ where A_1 and A_2 are 7×2 blocks, and A_3 is a 7×3 block. Then the determinant of A can be calculated as follows.

Take $\mathbf{r_1} = (r_{1,1}, r_{1,2}) = (1, 2)$ and $\mathbf{t_1} = (t_{1,1}, t_{1,2})$. Then from Theorem 3,

$$\det(A) = (-1)^{|\mathbf{r_1}|} \sum_{\mathbf{t_1} \in \mathcal{T}_1} (-1)^{|\mathbf{t_1}|} \det\big(S(A : \mathbf{r_1}, \mathbf{t_1})\big) \det\big(S'(A : \mathbf{r_1}, \mathbf{t_1})\big),$$

where $\mathcal{T}_1$ denotes the set of all 2-tuples $\mathbf{t_1} = (t_{1,1}, t_{1,2})$ for which $1 \leq t_{1,1} < t_{1,2} \leq 7$. We proceed to calculate $\det(S'(A : \mathbf{r_1}, \mathbf{t_1}))$ by Theorem 3. Take $\mathbf{r_2} = (r_{2,1}, r_{2,2}) = (3, 4)$, $\mathbf{r} = (\mathbf{r_1}, \mathbf{r_2}) = (r_{1,1}, r_{1,2}, r_{2,1}, r_{2,2})$, $\mathbf{t_2} = (t_{2,1}, t_{2,2})$, $\mathbf{t} = (\mathbf{t_1}, \mathbf{t_2}) = (t_{1,1}, t_{1,2}, t_{2,1}, t_{2,2})$, and let $\mathcal{T}_2$ denote the set of all 2-tuples $\mathbf{t_2} = (t_{2,1}, t_{2,2})$ for which $1 \leq t_{2,1} < t_{2,2} \leq 7$. For a given $\mathbf{t_1} = (t_{1,1}, t_{1,2})$, let

$$\mathcal{T}_2(\mathbf{t_1}) = \mathcal{T}_2 \backslash \{(t_{2,1}, t_{2,2}) : t_{2,1} \in \{t_{1,1}, t_{1,2}\} \text{ or } t_{2,2} \in \{t_{1,1}, t_{1,2}\}\}.$$

Then

$$\det(S'(A : \mathbf{r_1}, \mathbf{t_1})) = 1)^{|\mathbf{r_2}|} \sum_{\mathbf{t_2} \in \mathcal{T}_2(\mathbf{t_1})} (-1)^{|\mathbf{t_2}|} \det(S(A : \mathbf{r_2}, \mathbf{t_2})) \det(S'(A : \mathbf{r}, \mathbf{t})).$$

Hence the determinant of A can also be denoted by

$$\det(A) = (-1)^{|\mathbf{r}|} \sum_{\mathbf{t_1} \in \mathcal{T}_1} \sum_{\mathbf{t_2} \in \mathcal{T}_2(\mathbf{t_1})} (-1)^{|\mathbf{t}|} \det\big(S(A : \mathbf{r_1}, \mathbf{t_1})\big) \det\big(S(A : \mathbf{r_2}, \mathbf{t_2})\big) \det\big(S'(A : \mathbf{r}, \mathbf{t})\big).$$

In general, we have the following result.

Proposition 1. *Take a $n \times n$ matrix $A = (A_1 | \cdots | A_m)$ where A_i is a $n \times n_i$ matrix, and take $n_0 = 0$. For every $i \in J_m$, let $\mathbf{r_i} = (r_{i,1}, \ldots, r_{i,n_i}) = (\sum_{j=0}^{i-1} n_j + 1, \ldots, \sum_{j=0}^{i} n_j)$, and $\mathbf{t_i} = (t_{i,1}, \ldots, t_{i,n_i})$ be a list of n_i row indices for A_i such that $1 \leq t_{i,1} < \cdots < t_{i,n_i} \leq n$. Take $\mathbf{r} = (\mathbf{r_1}, \ldots, \mathbf{r_m})$ and $\mathbf{t} = (\mathbf{t_1}, \ldots, \mathbf{t_m})$. Let $\mathcal{T}_i$ denote the set of all n_i-tuples $\mathbf{t_i} = (t_{i,1}, \ldots, t_{1,n_i})$ for which $1 \leq t_{i,1} < \cdots < t_{1,n_i} \leq n$. For a given $\mathbf{t_i} = (t_{i,1}, \ldots, t_{1,n_i})$, take $S(\mathbf{t_i}) = \{t_{i,1}, \ldots, t_{i,n_i}\}$, and for given $\mathbf{t_{i'}} = (t_{i',1}, \ldots, t_{i',n_{i'}})$ with $i' \in [i-1]$, take*

$$\mathcal{T}_i(\mathbf{t_{i'}}, i' \in [i-1]) = \mathcal{T}_i \backslash \Big\{(t_{i,1}, \ldots, t_{i,n_i}) : t_{i,j} \in \bigcup_{i'=1}^{i-1} S(\mathbf{t_{i'}}) \text{ for some } j \in [n_i]\Big\}.$$

Then

$$\det(A) = (-1)^{|r|} \sum_{t_1 \in T_1} \sum_{t_2 \in T_2(t_1)} \cdots \sum_{\substack{t_{m-1} \in T_{m-1}(t_{i'}, \\ i' \in [m-2])}} (-1)^{|t|} \prod_{i=1}^{m-1} \det\left(S(A : r_i, t_i)\right) \det\left(S'(A : r, t)\right).$$

Proof. Theorem 3 implies

$$\det(A) = (-1)^{|r_1|} \sum_{t_1 \in T_1} (-1)^{|t_1|} \det(S(A : r_1, t_1)) \det(S'(A : r_1, t_1)).$$

We proceed to calculate $\det(S'(A : r_1, t_1))$ by Theorem 3 and the following result can be obtained

$$\det(S'(A : r_1, t_1)) = (-1)^{|r_2|} \sum_{t_2 \in T_2(t_1)} (-1)^{|t_2|} \det(S(A : r_2, t_2)) \det(S'(A : (r_1, r_2), (t_1, t_2)).$$

Accordingly, $\det(S'(A : (r_1, \ldots, r_i), (t_1, \ldots, t_i)))$ can be obtained by Theorem 3 for $i \in [2, m-1]$, and the result follows.   □

Example 2. Take

$$A = \begin{pmatrix} a_{1,1} & a_{1,2} & 0 & 0 & 0 & 0 & 0 \\ a_{2,1} & a_{2,2} & a_{2,3} & a_{2,4} & 0 & 0 & 0 \\ a_{3,1} & a_{3,2} & a_{3,3} & a_{3,4} & a_{3,5} & a_{3,6} & a_{3,7} \\ 0 & 0 & a_{4,3} & a_{4,4} & a_{4,5} & a_{4,6} & a_{4,7} \\ 0 & 0 & a_{5,3} & a_{5,4} & a_{5,5} & a_{5,6} & a_{5,7} \\ 0 & 0 & 0 & 0 & a_{6,5} & a_{6,6} & a_{6,7} \\ 0 & 0 & 0 & 0 & a_{7,5} & a_{7,6} & a_{7,7} \end{pmatrix}.$$

Then from Example 1,

$$\det(A) = (-1)^{|r|} \sum_{t_1 \in T_1} \sum_{t_2 \in T_2(t_1)} (-1)^{|t|} \det(S(A : r_1, t_1)) \det(S(A : r_2, t_2)) \det(S'(A : r, t)).$$

Note that the T_1 and T_2 are different from the ones in Example 1. Here, there are some zero blocks in A. In this case, T_1 denotes the set of all 2-tuples $t_1 = (t_{1,1}, t_{1,2})$ for which $1 \le t_{1,1} < t_{1,2} \le 3$ and T_2 denotes the set of all 2-tuples $t_2 = (t_{2,1}, t_{2,2})$ for which $2 \le t_{2,1} < t_{2,2} \le 5$.

This example implies that Proposition 1 is more suitable for calculating the determinant function of the matrix which has more zero blocks in its submatrices consist of some columns.

3.2   The Second Class of Matrices

In this section, we introduce the calculation approach to the determinant function of another class of matrices with special form. These matrices will be applied to the construction of representable matroid associated to UCASs and LCASs. Recall that the determinant function is linear in the columns of a matrix as follows.

Proposition 2. *If a and b are scalars, $\bar{\alpha}$ and $\bar{\beta}$ are columns vectors, and B is some matrix, then $\det\left((a\bar{\alpha} + b\bar{\beta} \,|B)\right) = a\det\left((\bar{\alpha}\,|B)\right) + b\det\left((\bar{\beta}\,|B)\right)$.*

Example 3. Let $A_i = (a_{u,v})_{2\times3}$ and $B_i = (b_{u,v})_{3\times2}$ be a 2×3 matrix and a 3×2 matrix, respectively. Then $AB = \left(\sum_{i_1=1}^{3} b_{i_1,1}\bar{a}_{i_1} \,\middle|\, \sum_{i_2=1}^{3} b_{i_2,2}\bar{a}_{i_2}\right)$ is a 2×2 matrix, where $\bar{a}_i$ denotes the ith column of A. Hence, from Proposition 2,

$$\det(AB) = \sum_{i_1=1}^{3} b_{i_1,1}\det\left(\left(\bar{a}_{i_1}\,\middle|\, \sum_{i_2=1}^{3} b_{i_2,2}\bar{a}_{i_2}\right)\right)$$

$$= \sum_{i_1=1}^{3}\sum_{i_2=1}^{3} b_{i_1,1}b_{i_2,2}\det\left((\bar{a}_{i_1}|\bar{a}_{i_2})\right)$$

$$= b_{1,1}b_{2,2}\det\left((\bar{a}_1|\bar{a}_2)\right) + b_{1,1}b_{3,2}\det\left((\bar{a}_1|\bar{a}_3)\right) + b_{2,1}b_{1,2}\det\left((\bar{a}_2|\bar{a}_1)\right)$$

$$+ b_{2,1}b_{3,2}\det\left((\bar{a}_2|\bar{a}_3)\right) + b_{3,1}b_{1,2}\det\left((\bar{a}_3|\bar{a}_1)\right) + b_{3,1}b_{2,2}\det\left((\bar{a}_3|\bar{a}_2)\right)$$

$$= \sum_{1\le j_1<j_2\le3} \det\begin{pmatrix} b_{j_1,1} & b_{j_1,2} \\ b_{j_2,1} & b_{j_2,2} \end{pmatrix}\det\left((\bar{a}_{j_1}|\bar{a}_{j_2})\right).$$

In general, we have the following proposition.

Proposition 3. *Take a $k\times k$ matrix $(AB|D)$ where $A = (a_{u,v})$ is a $k\times r$ matrix, $B = (b_{u,v})$ is a $r\times l$ matrix, and $k\ge r\ge l$, and take $\boldsymbol{j} = (j_1,\dots,j_l)$ such that $1\le j_1 < \cdots < j_l \le r$. Let $A(\boldsymbol{j})$ and $B(\boldsymbol{j})$ denote the $k\times l$ submatrix formed by the j_1th column, ..., j_lth column of A and the $l\times l$ submatrix formed by the j_1th row, ..., j_lth row of B, respectively. Then*

$$\det\left((AB|D)\right) = \sum_{\boldsymbol{j}\in\mathcal{J}} \det\left(B(\boldsymbol{j})\right)\det\left((A(\boldsymbol{j})|D)\right),$$

where $\mathcal{J}$ denotes the set of all l-tuples $\boldsymbol{j} = (j_1,\dots,j_l)$ for which $1\le j_1 < \cdots < j_l \le r$.

Proof. If there are two identical columns in a square matrix, then its determinant equals 0. Therefore, from this and Proposition 2,

$$\det\left((AB|D)\right) = \det\left(\left(\sum_{i_1=1}^{r} b_{i_1,1}\bar{a}_{i_1}\,\middle|\cdots\middle|\, \sum_{i_l=1}^{r} b_{i_l,l}\bar{a}_{i_l}\,\middle|\, D\right)\right)$$

$$= \sum_{i_v\in[r],v\in[l]} \left(\prod_{v\in[l]} b_{i_v,v}\right)\det\left((\bar{a}_{i_1}|\cdots|\bar{a}_{i_l}|D)\right)$$

$$= \sum_{i} \left(\prod_{v\in[l]} b_{i_v,v}\right)\det\left((\bar{a}_{i_1}|\cdots|\bar{a}_{i_l}|D)\right),$$

where the summation is over all l-tuples $\boldsymbol{i} = (i_1, \ldots, i_l)$ for which $i_v \in [r]$ and $i_v \neq i_{v'}$, $v \neq v' \in [l]$.

For a given $\boldsymbol{j} = (j_1, \ldots, j_l)$ such that $1 \leq j_1 < \cdots < j_l \leq r$, let $S(\boldsymbol{j})$ denote the set of all the permutations on the set $\{j_1, \ldots, j_l\}$. we claim that

$$\sum_{\boldsymbol{i} \in S(\boldsymbol{j})} \left(\prod_{v \in [l]} b_{i_v, v} \right) \det \left((\bar{\boldsymbol{a}}_{i_1} | \cdots | \bar{\boldsymbol{a}}_{i_l} | D) \right) = \det \left(B(\boldsymbol{j}) \right) \det \left((A(\boldsymbol{j}) | D) \right)$$

Without loss of generality, we may assume that $\boldsymbol{j} = (1, \ldots, l)$, that is $j_v = v$ with $v \in [l]$. Then

$$\left(\prod_{v \in [l]} b_{i_v, v} \right) \det \left((\bar{\boldsymbol{a}}_{i_1} | \cdots | \bar{\boldsymbol{a}}_{i_l} | D) \right) = \mathrm{sgn}(\boldsymbol{i}) \left(\prod_{v \in [l]} b_{i_v, v} \right) \det \left((\bar{\boldsymbol{a}}_1 | \cdots | \bar{\boldsymbol{a}}_l | D) \right),$$

where $\mathrm{sgn}(\boldsymbol{i})$ denotes the sign of $\boldsymbol{i}$. Note that for $\boldsymbol{j} = (1, \ldots, l)$,

$$\sum_{\boldsymbol{i} \in S(\boldsymbol{j})} \mathrm{sgn}(\boldsymbol{i}) \left(\prod_{v \in [l]} b_{i_v, v} \right) = \det \left(B(\boldsymbol{j}) \right).$$

This implies the claim, and the result follows.   □

We next give a formula to calculate the determinant function of a matrix with special form which will be used to the scheme for UCASs and LCASs.

Proposition 4. *Let $G = (A_1 B_1 | \cdots | A_m B_m)$ be a $k \times k$ matrix such that A_i is a $k \times r_i$ block and B_i is a $r_i \times l_i$ block, where $l_i \leq r_i \leq k$ and $\sum_{i=1}^{m} l_i = k$. For any $\boldsymbol{j}_i = (j_{i,1}, \ldots, j_{i,l_i})$ with $i \in J_m$ such that $1 \leq j_{i,1} < \cdots < j_{i,l_i} \leq r_i$, let $A_i(\boldsymbol{j}_i)$ and $B_i(\boldsymbol{j}_i)$ denote the $k \times l_i$ submatrix formed by the $j_{i,1}$th column, $\ldots$, j_{i,l_i}th column of A_i and the $l_i \times l_i$ submatrix formed by the $j_{i,1}$th row, $\ldots$, j_{i,l_i}th row of B_i, respectively. Then*

$$\det(G) = \sum_{\boldsymbol{j}_i, i \in [m]} \left(\prod_{i=1}^{m} \det \left(B_i(\boldsymbol{j}_i) \right) \right) \det \left((A_1(\boldsymbol{j}_1) | \cdots | A_m(\boldsymbol{j}_m)) \right),$$

where the summation is over all l_i-tuples $\boldsymbol{j}_i = (j_{i,1}, \ldots, j_{i,l_i})$ with $i \in J_m$, for which $1 \leq j_{i,1} < \cdots < j_{i,l_i} \leq r_i$.

Proof. Let $A_i := (a_{u,v}^{(i)})$ with $u \in [k]$ and $v \in [r_i]$, $B_i := (b_{u,v}^{(i)})$ with $u \in [r_i]$ and $v \in [l_i]$, and $\bar{\boldsymbol{a}}_j^{(i)}$ denote the jth column of matrix A_i. From Proposition 3,

$$\det(G) = \det \left(\left(\sum_{i_{1,1}=1}^{r_1} b_{i_{1,1},1}^{(1)} \bar{\boldsymbol{a}}_{i_{1,1}}^{(1)} \Big| \cdots \Big| \sum_{i_{1,l_1}=1}^{r_1} b_{i_{1,l_1},l_1}^{(1)} \bar{\boldsymbol{a}}_{i_{1,l_1}}^{(1)} \Big| A_2 B_2 \Big| \cdots \Big| A_m B_m \right) \right)$$

$$= \sum_{\boldsymbol{j}_1} \det \left(B_1(\boldsymbol{j}_1) \right) \det \left((A_1(\boldsymbol{j}_1) | A_2 B_2 | \cdots | A_m B_m) \right),$$

where the summation is over all l_1-tuples $\boldsymbol{j}_1 = (j_{1,1}, \ldots, j_{1,l_1})$, for which $1 \leq j_{1,1} < \cdots < j_{1,l_1} \leq r_1$. The conclusion can be obtained by computing $A_i B_i$ for $i \in [2, m]$ using the similar method to $A_1 B_1$.   □

4 Secret Sharing Schemes for Ideal Hierarchical Access Structures

In this section, we construct ideal linear secret sharing schemes realizing IHASs by an efficient method. We will present two classes of constructions based on the same representation of an integer polymatroid. We first present an integer polymatroid $\mathcal{Z}'$ satisfying Theorem 2 such that the IHASs (1) are of the form $\Gamma_0(\mathcal{Z}', \Pi)$.

Given two vectors $\hat{\boldsymbol{k}}, \boldsymbol{k} \in \mathbb{Z}_+^{J_m'}$ such that $\hat{k}_0 = \hat{k}_1 = 0$, $k_0 = 1$, $k_m = k$, and $\hat{k}_i \leq \hat{k}_{i+1} < k_i \leq k_{i+1}$ for $i \in [0, m-1]$, consider the subsets $S_i = [\hat{k}_i + 1, k_i]$ of the set $S = [k]$ and the Boolean polymatroid $\mathcal{Z}' = \mathcal{Z}'(\hat{\boldsymbol{k}}, \boldsymbol{k})$ with ground J_m' defined from them. The following result was presented in Section IX of [18].

Lemma 1. *Let $\Pi = (\Pi_i)_{i \in J_m}$ be a partition of the set P with $|\Pi_i| \geq h(\{i\}) = k_i - \hat{k}_i$. Then the IHASs (1) are of the form $\Gamma_0(\mathcal{Z}', \Pi)$.*

Now we introduce a linear representation of the polymatroid defined in Lemma 1, that is a collection $(V_i)_{i \in J_m'}$ of subspaces of some vector space. Recalled that Boolean polymatroids are representable over every finite field. Here, we give a simple representation of $\mathcal{Z}'$ based on the unit matrix as follows.

Take a $k \times k$ unit matrix I_k, and for every $i \in J_m'$, let E_i denote the submatrix formed by the $(\hat{k}_i + 1)$th column to the k_ith column of I_k. Consider the $\mathbb{F}_q$-vector subspace $V_i \subseteq \mathbb{F}_q^k$ spanned by all the columns of E_i. Let the integer polymatroid $\mathcal{Z}' = (J_m', h)$ such that $h(X) = \dim\left(\sum_{i \in X} V_i\right)$ for every $X \subseteq J_m'$. We have the following result.

Proposition 5. *For the integer polymatroid $\mathcal{Z}'$ defined above, the IHASs (1) are of the form $\Gamma_0(\mathcal{Z}', \Pi)$ and $\mathcal{B}(\mathcal{Z}') = \mathcal{B}_1 \cup \mathcal{B}_2$, where*

$$\mathcal{B}_1 = \{\boldsymbol{u} \in \mathbb{Z}_+^{J_m'} : |\boldsymbol{u}| = k, u_0 = 0 \text{ and } \hat{k}_{i+1} \leq |\boldsymbol{u}([i])| \leq k_i \text{ for all } i \in [m-1]\}, \tag{6}$$

$$\mathcal{B}_2 = \{\boldsymbol{u} \in \mathbb{Z}_+^{J_m'} : |\boldsymbol{u}| = k, u_0 = 1 \text{ and } \hat{k}_{i+1} - 1 \leq |\boldsymbol{u}([i])| \leq k_i - 1 \text{ for all } i \in [m-1]\}.$$

Proof. Suppose the set $S = [k]$ and the subsets $S_i = [\hat{k}_i + 1, k_i]$ for every $i \in J_m'$. Then for every $X \subseteq J_m'$, $h(X) = \dim\left(\sum_{i \in X} V_i\right) = |\cup_{i \in X} S_i|$. This implies $\mathcal{Z}'$ is a linear representation of the polymatroid $\mathcal{Z}'(\hat{\boldsymbol{k}}, \boldsymbol{k})$, and the first claim follows. In addition, since I_k is nonsingular and E_i is an submatrix of I_k for every $i \in J_m'$, it follows that any k distinct columns vectors from E_i with $i \in J_m'$ are linearly independent, and the second claim follows. $\quad\square$

This proposition implies that the collection $(V_i)_{i \in J_m'}$ is a linear representation of the integer polymatroid $\mathcal{Z}'$ associated to the IHASs (1). We will present two class of constructions for ideal linear schemes realizing IHASs by representable matroids obtained based on $\mathcal{Z}'$.

4.1 Construction for Ideal Hierarchical Access Structures

In this section, we represent a class of ideal linear scheme for IHASs, which can be obtained by a representation of the matroid associated to IHASs.

Suppose $\Pi_0 = \{p_0\}$ and let $\Pi' = (\Pi_i)_{i \in J'_m}$ and $\Pi = (\Pi_i)_{i \in J_m}$ be the partition of $P' = P \cup \{p_0\}$ and P, respectively, such that $|\Pi_i| = n_i$. For every $i \in J_m$, take different elements $\beta_{i,v} \in \mathbb{F} \backslash \{0\}$ with $v \in [n_i]$ and define a $(k_i - \hat{k}_i) \times n_i$ matrix

$$B_i = \left((\beta_{i,v} x^{m-i})^{u-1} \right) \qquad u \in [k_i - \hat{k}_i], \; v \in [n_i].$$

Let a $k \times (n+1)$ matrix be defined as

$$M = (M_0 | M_1 | \cdots | M_m), \tag{7}$$

where $M_0 = (1, 0, \ldots, 0)^T$ is a k-dimensional column vector and $M_i = E_i B_i$ for every $i \in J_m$. Then the secret sharing scheme $LSSS(M)$ is as follows:

Secret Sharing Scheme.

1. Let $s \in \mathbb{K}$ be a secret value. The dealer chooses randomly a k-dimensional vector $\boldsymbol{a}$ such that $\boldsymbol{a} M_0 = s$;
2. The share of each participant $p_{i,j}$ from compartment Π_i is $\boldsymbol{a} \boldsymbol{b}_{i,j}^T$, where $\boldsymbol{b}_{i,j}^T$ denotes the jth column of M_i with $i \in J_m$ and $j \in [n_i]$.

We proceed to show that $LSSS(M)$ is a perfect ideal linear scheme realizing IHASs. This can be done by proving M is a representation of the matroid associated the IHASs (1). Obviously, M satisfies the first two conditions in Step 3 of Sect. 2.3. We will prove that it satisfies the third condition too. We first give the following lemmas.

Lemma 2. *For any $\boldsymbol{u} \in \mathcal{B}_1$, (6), $\det(M_{\boldsymbol{u}})$ is a nonzero polynomial on x of degree at most K where*

$$K = \frac{1}{2} \sum_{i=1}^{m-1} k_i(k_i - 1) - \sum_{i=2}^{m-1} (m-i)(k_i - k_{i-1})\hat{k}_i.$$

Proof. For every $i \in J_m$, take

$$B'_i = \left(\beta_{i,v}^{u-1} \right) \qquad u \in [k_i - \hat{k}_i], \; v \in [n_i],$$

and for any $\boldsymbol{u} \in \mathcal{B}_1$, (6), let $B_i(u_i)$ and $B'_i(u_i)$ denote the submatrices formed by any u_i columns in B_i and B'_i, respectively.

Let us exemplify how such an event may occur. Assume, for example, that $m = 3$, $\boldsymbol{k} = (k_1, k_2, k_3) = (3, 5, 7)$, $\hat{\boldsymbol{k}} = (\hat{k}_1, \hat{k}_2, \hat{k}_3) = (0, 1, 2)$. Take $\boldsymbol{u} = (u_1, u_2, u_3) = (2, 2, 3)$ and the corresponding matrix $M_{\boldsymbol{u}}$ has the following form:

$$M_{\boldsymbol{u}} = \begin{pmatrix} 1 & 1 & 0 & 0 & 0 & 0 & 0 \\ \beta_{1,1}x^2 & \beta_{1,2}x^2 & 1 & 1 & 0 & 0 & 0 \\ (\beta_{1,1}x^2)^2 & (\beta_{1,2}x^2)^2 & \beta_{2,1}x & \beta_{2,2}x & 1 & 1 & 1 \\ 0 & 0 & (\beta_{2,1}x)^2 & (\beta_{2,2}x)^2 & \beta_{3,1} & \beta_{3,2} & \beta_{3,3} \\ 0 & 0 & (\beta_{2,1}x)^3 & (\beta_{2,2}x)^3 & \beta_{3,1}^2 & \beta_{3,2}^2 & \beta_{3,3}^2 \\ 0 & 0 & 0 & 0 & \beta_{3,1}^3 & \beta_{3,2}^3 & \beta_{3,3}^3 \\ 0 & 0 & 0 & 0 & \beta_{3,1}^4 & \beta_{3,2}^4 & \beta_{3,3}^4 \end{pmatrix}.$$

Suppose $1 \leq t_{1,1} < t_{1,2} \leq 3$, $2 \leq t_{2,1} < t_{2,2} \leq 5$, $3 \leq t_{3,1} < t_{3,2} < t_{3,3} \leq 7$, and $\{t_{1,1}, t_{1,2}, t_{2,1}, t_{2,2}, t_{3,1}, t_{3,2}, t_{3,3}\} = [7]$. Let $\hat{B}_1$ and $\hat{B}'_1$ be the blocks formed by the $t_{1,1}$th and $t_{1,2}$th rows of $B_1(u_1)$ and $B'_1(u_1)$, respectively, $\hat{B}_2$ and $\hat{B}'_2$ be the blocks formed by the $t_{2,1}$th and $t_{2,2}$th rows of $B_2(u_2)$ and $B'_2(u_2)$, respectively, and $\hat{B}_3$ and $\hat{B}'_3$ be the blocks formed by the $t_{3,1}$th, $t_{3,2}$th and $t_{3,3}$th rows of $B_3(u_3)$ and $B'_3(u_3)$, respectively. Then Proposition 1 implies that the summation in $\det(M_u)$ can be denoted by

$$|a_t x^t| := \det(\hat{B}_1)\det(\hat{B}_2)\det(\hat{B}_3) = \det(\hat{B}'_1)\det(\hat{B}'_2)\det(\hat{B}'_3)x^t$$

where $t = 2(t_{1,1}-1)+2(t_{1,2}-1)+(t_{2,1}-2)+(t_{2,2}-2)$. Therefore, when $t_{1,1} = 1$, $t_{1,2} = 2$, $t_{2,1} = 3$ and $t_{2,2} = 4$, t is minimal. In this case $t = 5$ and $\hat{B}'_i$ with $i \in [3]$ are all nonsingular. This implies $a_5 \neq 0$.

In addition, take $\boldsymbol{u}' = (u'_1, u'_2, u'_3)$ such that $\boldsymbol{u}'([i]) = k_i$ for every $i \in [3]$. Then $\boldsymbol{u}' \in \mathcal{B}_1$. In this case let $t'_{1,1} = 1$, $t'_{1,2} = 2$, $t'_{1,3} = 3$, $t'_{2,1} = 4$, $t'_{2,2} = 5$, $t'_{3,1} = 6$ and $t'_{3,2} = 7$, then $t \leq 2\sum_{i'=1}^{3}(t'_{1,i'} - 1) + \sum_{i'=1}^{2}(t'_{2,i'} - 2) = 11$. Therefore, $\det(M_u)$ is a nonzero polynomial on x of degree at most 11. In fact, by computing, we have $t < 11$.

In general, for any $\boldsymbol{u} \in \mathcal{B}_1$, let $\hat{B}_i$ and $\hat{B}'_i$ be the blocks formed by all the $t_{i,i'}$th rows of $B_i(u_i)$ and $B'_i(u_i)$, respectively, where $i' \in [u_i]$ such that

$$\hat{k}_i + 1 \leq t_{i,1} < \cdots < t_{i,u_i} \leq k_i \text{ and } \bigcup_{i=1}^{m} \{t_{i,i'} : i' \in [u_i]\} = [k].$$

Then Proposition 1 implies that the summation in $\det(M_u)$ can be denoted by

$$|a_t x^t| = \prod_{i=1}^{m} \det(\hat{B}_i) = \prod_{i=1}^{m} \det(\hat{B}'_i)x^t$$

where

$$t = \sum_{i=1}^{m-1} \left((m-i)\sum_{i'=1}^{u_i}(t_{i,i'} - \hat{k}_i - 1)\right) = \sum_{j=1}^{m-1}\left(\sum_{i=1}^{j}\left(\sum_{i'=1}^{u_i}(t_{i,i'} - \hat{k}_i - 1)\right)\right). \quad (8)$$

For every $j \in [m-1]$, take $T_j = \sum_{i=1}^{j}\left(\sum_{i'=1}^{u_i}(t_{i,i'} - \hat{k}_i - 1)\right)$. We have that T_{m-1} is minimal if $\bigcup_{i=1}^{m-1}\{t_{i,i'} : i' \in [u_i]\} = [|\boldsymbol{u}([m-1])|]$. In this case T_{m-2} is minimal if $\bigcup_{i=1}^{m-2}\{t_{i,i'} : i' \in [u_i]\} = [|\boldsymbol{u}([m-2])|]$. Therefore, t is minimal if $\bigcup_{i=1}^{j}\{t_{i,i'} : i' \in [u_i]\} = [|\boldsymbol{u}([j])|]$ for all $j \in [m-1]$. This implies that $t_{1,i'} = i'$ and $t_{i,i'} = |\boldsymbol{u}([i-1])| + i'$ for $i \in [2, m-1]$. Hence,

$$t \geq (m-1)\sum_{i'=1}^{u_1}(i'-1) + \sum_{i=2}^{m-1}\left((m-i)\sum_{i'=1}^{u_i}(|\boldsymbol{u}([i-1])| + i' - \hat{k}_i - 1)\right) = t_0.$$

In this case each $\hat{B}'_i$ is nonsingular since it is the square submatrix formed by the successive u_i rows of $B'_i(u_i)$. This implies that $a_{t_0} \neq 0$.

In addition, take a vector $\boldsymbol{u}' \in \mathbb{Z}_+^m$ such that $|\boldsymbol{u}([i])| = k_i$ for every $i \in J_m$. Then $\boldsymbol{u}' \in \mathcal{B}_1$. In this case $t_{1,i'} = i'$ with $i' \in [k_1]$ and $t'_{i,i'} = k_{i-1} + i'$ with $i \in [2, m-1]$ and $i' \in [k_i - k_{i-1}]$. Then

$$t \le (m-1) \sum_{i'=1}^{k_1} (i'-1) + \sum_{i=2}^{m-1} \left((m-i) \sum_{i'=1}^{k_i-k_{i-1}} (k_{i-1} + i' - \hat{k}_i - 1) \right)$$

$$= (m-1) \sum_{i'=1}^{k_1} (i'-1) + \sum_{i=2}^{m-1} \left((m-i) \sum_{i'=1}^{k_i-k_{i-1}} (k_{i-1} + i' - 1) \right) - \sum_{i=2}^{m-1} (m-i) \sum_{i'=1}^{k_i-k_{i-1}} \hat{k}_i$$

$$= \sum_{i=1}^{m-1} (1 + 2 + \cdots + (k_i - 1)) - \sum_{i=2}^{m-1} (m-i)(k_i - k_{i-1})\hat{k}_i$$

$$= \frac{1}{2} \sum_{i=1}^{m-1} k_i(k_i - 1) - \sum_{i=2}^{m-1} (m-i)(k_i - k_{i-1})\hat{k}_i.$$

$$(9)$$

This implies the conclusion.

Lemma 3. *For any $\boldsymbol{u} \in \mathcal{B}_2$, (6), $\det(M_{\boldsymbol{u}})$ is a nonzero polynomial on x of degree at most K.*

Proof. Let M' denote the submatrix obtained by removing the first row and the first column of M and take $\boldsymbol{k}', \hat{\boldsymbol{k}}' \in \mathbb{Z}_+^m$ such that for every $i \in J_m$, $k'_i = k_i - 1$, and $\hat{k}'_i = \hat{k}_i$ if $\hat{k}_i = 0$ and $\hat{k}'_i = \hat{k}_i - 1$ if $\hat{k}_i > 0$. For every $i \in J_m$, let E'_i denote the submatrix formed by the $(\hat{k}'_i + 1)$th column to the k'_ith column of I_{k-1}. Let D_1 and D'_1 denote the submatrices formed by the last k'_1 rows of B_1 and B'_1, respectively. For every $i \in [2, m]$, if $\hat{k}_i = 0$, let D_i and D'_i denote the submatrices formed by the last $k'_i - 1$ rows of B_i and B'_i, respectively, and if $\hat{k}_i > 0$, let $D_i = B_i$ and $D'_i = B'_i$. Then

$$M' = (M'_1 | \cdots | M'_m)$$

where $M'_i = E'_i D_i$ and for any $\boldsymbol{u} \in \mathcal{B}_2$, (6), $\det(M_{\boldsymbol{u}}) = \det\left(M'_{\boldsymbol{u}(J_m)} \right)$. In particular, for any $\boldsymbol{u} \in \mathcal{B}_2$, (6), $\hat{k}'_{i+1} \le |\boldsymbol{u}([i])| \le k'_i$ for all $i \in [m-1]$ and $|\boldsymbol{u}| = k - 1$. Therefore, this claim can be proved by the same method in the proof of Lemma 2.

For any $\boldsymbol{u} \in \mathcal{B}_2$, (6), let $D'_i(u_i)$ denote the block formed by any u_i columns in D'_i, and let $\hat{D}'_i$ be the block formed by all the $t_{i,i'}$th rows of $D'_i(u_i)$. Here, $i' \in [u_i]$ such that $\hat{k}'_i + 1 \le t_{i,1} < \cdots < t_{i,u_i} \le k'_i$ and $\bigcup_{i=1}^m \{t_{i,i'} : i' \in [u_i]\} = [k-1]$. Then the summation in $\det\left(M'_{\boldsymbol{u}(J_m)} \right)$ can be denoted by $|b_{t'} x^{t'}| = \prod_{i=1}^m \det(\hat{D}'_i) x^{t'}$. Similar to (8),

$$t' = \sum_{i=1}^{m-1} \left((m-i) \sum_{i'=1}^{u_i} (t_{i,i'} - \hat{k}'_i - y_i) \right)$$

where $y_i = 0$ if $\hat{k}'_i = 0$ and $y_i = 1$ if $\hat{k}'_i > 0$. From $\hat{k}'_i = \hat{k}_i$ if $\hat{k}_i = 0$ and $\hat{k}'_i = \hat{k}_i - 1$ if $\hat{k}_i > 0$, we have

$$t' = \sum_{i=1}^{m-1} \left((m-i) \sum_{i'=1}^{u_i} (t_{i,i'} - \hat{k}_i) \right).$$

Similar to the proof in Lemma 2, we can obtain t' is minimal if $t_{1,i'} = i'$ and $t_{i,i'} = |\boldsymbol{u}([i-1])| + i'$ for $i \in [2, m-1]$, and in this case each $\hat{D}'_i$ is nonsingular, thus $\det\left(M'_{\boldsymbol{u}(J_m)}\right)$ is a nonzero polynomial on x. In addition, take a vector $\boldsymbol{u}' \in \mathbb{Z}_+^m$ such that $|\boldsymbol{u}([i])| = k'_i$ for every $i \in J_m$. Then $\hat{k}'_{i+1} \le |\boldsymbol{u}'([i])| \le k'_i$ for all $i \in [m-1]$ and $|\boldsymbol{u}'| = k-1$. In this case $t_{1,i'} = i'$ with $i' \in [k'_1]$ and $t'_{i,i'} = k'_{i-1} + i'$ with $i \in [2, m-1]$ and $i' \in [k'_i - k'_{i-1}]$. Similar to (9),

$$t' \le (m-1) \sum_{i'=1}^{k'_1} i' + \sum_{i=2}^{m-1} \left((m-i) \sum_{i'=1}^{k'_i - k'_{i-1}} (k'_{i-1} + i' - \hat{k}_i) \right)$$

$$= (m-1) \sum_{i'=1}^{k_1} (i'-1) + \sum_{i=2}^{m-1} \left((m-i) \sum_{i'=1}^{k_i - k_{i-1}} (k_{i-1} + i' - \hat{k}_i - 1) \right) = K$$

since $k'_i = k_i - 1$ for every $i \in J_m$. This implies $\det\left(M'_{\boldsymbol{u}(J_m)}\right)$ is a nonzero polynomial on x of degree at most K, and the claim follows. □

The following result was proved by Shoup [33].

Theorem 4 ([33]). *Take a finite field $\mathbb{F}_{q^\lambda}$ where q is a prime power and λ is a positive integer. Then there exists an element $x \in \mathbb{F}_{q^\lambda}$ such that its minimal polynomial over $\mathbb{F}_q$ is of degree λ which can be found in time $O(q, \lambda)$.*

Now, take a finite field $\mathbb{F}_{q^\lambda}$, where $q > \max_{i \in J_m}\{n_i\}$ is a prime power and $\lambda > K$. Take all $\beta_{i,v}$ in the matrix (7) from $\mathbb{F}_q \backslash \{0\}$ and take $x \in \mathbb{F}_{q^\lambda}$ such that its minimal polynomial over $\mathbb{F}_q$ is of degree λ. We have the following result.

Theorem 5. *The matrix (7) is a representation of the matroid associated to IHASs (1) over $\mathbb{F}_{q^\lambda}$ for some prime power $q > \max_{i \in J_m}\{n_i\}$ and some $\lambda > K$. Moreover, such a representation can be obtained in time $O(q, \lambda)$.*

Proof. Since all the entries in the matrix (7), except the powers of x, are in $\mathbb{F}_q$, and Theorem 4 implies that such an element x can be found in time $O(q, \lambda)$, it follows that for any $\boldsymbol{u} \in \mathcal{B}(\mathcal{Z}')$, (6), $\det(M_{\boldsymbol{u}})$ must be a nonzero $\mathbb{F}_q$-polynomial on x with degree smaller than λ, and consequently, the matrix $M_{\boldsymbol{u}}$ is nonsingular. This implies the claim. □

Proposition 6. *Suppose M is the matrix (7). Then $LSSS(M)$ realizes the IHASs (1) over $\mathbb{F}_{q^\lambda}$ defined as in Theorem 5. Moreover, such a scheme can be obtained in time $O(q, \lambda)$.*

Proof. Theorem 1 implies that proving this claim is equivalent to proving that $\boldsymbol{v}(J_m) \in \Gamma$ if and only M_0 is a linear combination of all the columns in $M_{\boldsymbol{v}(J_m)}$.

Let $\boldsymbol{v}(J_m) \in \min \Gamma$, (1), namely, $\boldsymbol{v}(J_m) = (v_1, v_2, \dots, v_\ell, 0, \dots, 0)$ for some $\ell \in J_m$ such that $\hat{k}_{i+1} \le |\boldsymbol{v}([i])| < k_i$ for all $i \in [\ell-1]$ and $|\boldsymbol{v}([\ell])| = k_\ell$. Then there must exist a vector $\boldsymbol{u} \in \mathcal{B}_1$, (6), such that $\boldsymbol{u} \ge \boldsymbol{v}$ and $u_i = v_i$ for every

$i \in [\ell]$. Note that the last $k - k_\ell$ rows of $M_{v(J_m)}$ are all zero rows, it follows that $M_{u(J_m)}$ has the following form

$$M_{u(J_m)} = \begin{pmatrix} \hat{M}_{v(J_m)} & A_1 \\ O & A_2 \end{pmatrix}$$

where $\hat{M}_{v(J_m)}$ is the square block formed by the first k_ℓ rows of $M_{v(J_m)}$, A_1 is a $(k-k_\ell) \times k_\ell$ block and A_2 is a $(k-k_\ell) \times (k-k_\ell)$ block. From Theorem 5, $M_{u(J_m)}$ is nonsingular. This with $\det(M_{u(J_m)}) = \det(\hat{M}_{v(J_m)}) \cdot \det(A_2)$ implies that $\hat{M}_{v(J_m)}$ is nonsingular. In this case, the k_ℓ-dimensional column vector formed by the first k_ℓ elements of M_0 can be spanned by the columns of $\hat{M}_{v(J_m)}$. Accordingly, M_0 can be spanned by the columns in $M_{v(J_m)}$ as the last $k - k_\ell$ elements of M_0 are all zero. Hence, M_0 can be spanned by the columns in $M_{v(J_m)}$ for any $v(J_m) \in \Gamma$.

Assume that $v(J_m) \notin \Gamma$. We know every unauthorized subset may be completed into an authorized subset (though not necessarily minimal) by adding to it at most k participants. Without loss of generality, we may assume that there exists a vector $v'(J_m) \in \Gamma$ such that $v'(J_m) \geq v(J_m)$ and $|v'(J_m)| = |v(J_m)|+1$.

First, assume that $v(J_m) = (v_1, v_2, \ldots, v_\ell, 0, \ldots, 0)$ for some $\ell \in J_m$ such that $\hat{k}_{i+1} - 1 \leq |v([i])| \leq k_i - 1$ for all $i \in [\ell-1]$ and $|v([\ell])| = k_\ell - 1$. Then for the vector $v(J'_m)$ with $u_0 = 1$, namely, $v(J'_m) = (1, v_1, v_2, \ldots, v_\ell, 0, \ldots, 0)$, there must exist a vector $u(J'_m) \in \mathcal{B}_2$, (6), such that $u(J'_m) \geq v(J'_m)$ and $u_i = v_i$ for every $i \in [0, \ell]$. From Theorem 5, $M_{u(J'_m)}$ is nonsingular. This with $v(J_m) \leq u(J_m)$ implies that M_0 can't be spanned by all the columns in $M_{v(J_m)}$.

Second, assume that $v(J_m) = (v_1, v_2, \ldots, v_m)$ with $|v(J_m)| \geq k$ such that for some $\ell \in J_m$, $|v([\ell])| = \hat{k}_{\ell+1} - 1$, $\hat{k}_{i+1} - 1 \leq |v([i])| < k_i$ for every $i \in [\ell-1]$, and $v_i = n_i$ for every $i \in [\ell+1, m]$. Then M_0 can't be spanned by the columns in $M_{v'(J_m)}$ for any $v'(J_m) \leq v(J_m)$ if M_0 can't be spanned by the columns in $M_{v(J_m)}$. We claim that every column in $M_{v(J_m)}$ can be spanned by the columns in $M_{u(J_m)}$ for any $u(J_m) \leq v(J_m)$ with $|u(J_m)| = k - 1$ such that $|u([i])| = |v([i])|$ for every $i \in [l]$ and $\hat{k}_{i+1} - 1 \leq |u([i])| < k_i$ for every $i \in [\ell+1, m-1]$.

For such a vector $u(J_m)$, if $u_0 = 1$, then $u(J'_m) \in \mathcal{B}_2$, (6). This implies M_0 can't be spanned by the columns in $M_{u(J_m)}$. Furthermore, M_0 can't be spanned by the columns in $M_{v(J_m)}$ if the claim is true.

We proceed to prove the claim. Note that

$$M_{u(J'_m)} = \left(M_{u([0,\ell])} \middle| M_{u([\ell+1,m])} \right) = \begin{pmatrix} D_1 & O \\ D_2 & \bar{M}_{u([\ell+1,m])} \end{pmatrix}$$

where $\bar{M}_{u([\ell+1,m])}$ is the square block formed by the last $k - \hat{k}_{\ell+1}$ rows of $M_{u([\ell+1,m])}$. As $M_{u(J'_m)}$ is nonsingular, thus $\bar{M}_{u([\ell+1,m])}$ is nonsingular. On the other hand, $M_{v(J_m)} = \left(M_{v([\ell])} \middle| M_{v([\ell+1,m])} \right)$, where

$$M_{v([\ell+1,m])} = \begin{pmatrix} O \\ \bar{M}_{v([\ell+1,m])} \end{pmatrix}$$

for which $\bar{M}_{v([\ell+1,m])}$ is the block formed by the last $k - \hat{k}_{\ell+1}$ rows of $M_{v([\ell+1,m])}$. Since $\bar{M}_{u([\ell+1,m])}$ is a submatrix of $\bar{M}_{v([\ell+1,m])}$ and $\bar{M}_{u([\ell+1,m])}$ is nonsingular, it follows that any column in $\bar{M}_{v([\ell+1,m])}$ can be spanned by the columns in $\bar{M}_{u([\ell+1,m])}$. Accordingly, any column in $M_{v([\ell+1,m])}$ is a linear combination of the columns in $M_{u([\ell+1,m])}$. This with $M_{v([\ell])} = M_{u([\ell])}$ implies the claim.   □

4.2 Another Construction for Ideal Hierarchical Access Structures

In this section, we give another construction of ideal linear secret sharing schemes for IHASs using the similar technique in Sect. 4.1. The parameters of this construction may be better than the construction in Sect. 4.1 in some cases.

For every $i \in J_m$, take n_i different elements $\beta_{i,v} \in \mathbb{F}\backslash\{0\}$ and let the $(k_i - \hat{k}_i) \times n_i$ matrix B_i be defined as follows:

$$B_i = \left((\beta_{i,v} x^{i-1})^{k_i - \hat{k}_i - u}\right) \quad u \in [k_i - \hat{k}_i], \; v \in [n_i].$$

Take a k-dimensional column vector $M_0 = (1, 0, \ldots, 0)^T$ and $M_i = E_i B_i$ for every $i \in J_m$. Define a $k \times (n+1)$ matrix as

$$M = (M_0 | M_1 | \cdots | M_m). \tag{10}$$

Similar to the case in Sect. 4.1, we will prove that $LSSS(M)$ realizes IHASs. First, we give an example to explain this construction as follows.

Example 4. As in Lemma 2, assume that $m = 3$, $\boldsymbol{k} = (k_1, k_2, k_3) = (3, 5, 7)$, and $\hat{\boldsymbol{k}} = (\hat{k}_1, \hat{k}_2, \hat{k}_3) = (0, 1, 2)$. Take $\boldsymbol{u} = (u_1, u_2, u_3) = (2, 2, 3)$ and the matrix M_u has the following form:

$$M_u = \begin{pmatrix} \beta_{1,1}^2 & \beta_{1,2}^2 & 0 & 0 & 0 & 0 & 0 \\ \beta_{1,1} & \beta_{1,2} & (\beta_{2,1}x)^3 & (\beta_{2,2}x)^3 & 0 & 0 & 0 \\ 1 & 1 & (\beta_{2,1}x)^2 & (\beta_{2,2}x)^2 & (\beta_{3,1}x^2)^4 & (\beta_{3,2}x^2)^4 & (\beta_{3,3}x^2)^4 \\ 0 & 0 & \beta_{2,1}x & \beta_{2,2}x & (\beta_{3,1}x^2)^3 & (\beta_{3,2}x^2)^3 & (\beta_{3,3}x^2)^3 \\ 0 & 0 & 1 & 1 & (\beta_{3,1}x^2)^2 & (\beta_{3,2}x^2)^2 & (\beta_{3,3}x^2)^2 \\ 0 & 0 & 0 & 0 & \beta_{3,1}x^2 & \beta_{3,2}x^2 & \beta_{3,3}x^2 \\ 0 & 0 & 0 & 0 & 1 & 1 & 1 \end{pmatrix}.$$

Note that M_u can be transformed to the following form by exchanging rows and columns

$$\tilde{M}_u = \begin{pmatrix} 1 & 1 & 1 & 0 & 0 & 0 & 0 \\ \beta_{3,1}x^2 & \beta_{3,2}x^2 & \beta_{3,3}x^2 & 0 & 0 & 0 & 0 \\ (\beta_{3,1}x^2)^2 & (\beta_{3,2}x^2)^2 & (\beta_{3,3}x^2)^2 & 1 & 1 & 0 & 0 \\ (\beta_{3,1}x^2)^3 & (\beta_{3,2}x^2)^3 & (\beta_{3,3}x^2)^3 & \beta_{2,1}x & \beta_{2,2}x & 0 & 0 \\ (\beta_{3,1}x^2)^4 & (\beta_{3,2}x^2)^4 & (\beta_{3,3}x^2)^4 & (\beta_{2,1}x)^2 & (\beta_{2,2}x)^2 & 1 & 1 \\ 0 & 0 & 0 & (\beta_{2,1}x)^3 & (\beta_{2,2}x)^3 & \beta_{1,1} & \beta_{1,2} \\ 0 & 0 & 0 & 0 & 0 & \beta_{1,1}^2 & \beta_{1,2}^2 \end{pmatrix},$$

Therefore, $|\det(M_u)| = |\det(\tilde{M}_u)|$.

Take $\boldsymbol{\kappa} = (\kappa_1, \kappa_2, \kappa_3) = (k - \hat{k}_3, k - \hat{k}_2, k - \hat{k}_1) = (5, 6, 7)$, and $\hat{\boldsymbol{\kappa}} = (\hat{\kappa}_1, \hat{\kappa}_2, \hat{\kappa}_3) = (k - k_3, k - k_2, k - k_1) = (0, 2, 4)$. Then Lemma 2 implies that $\det(\tilde{M}_u)$ is a nonzero polynomial on x of degree at most L with

$$L = \frac{1}{2} \sum_{i=1}^{2} \kappa_i(\kappa_i - 1) - (\kappa_2 - \kappa_1)\hat{\kappa}_2 = 23.$$

Accordingly, $\det(M_u)$ is a nonzero polynomial on x of degree at most L.

In general, we have the following lemma.

Lemma 4. *For any* $\boldsymbol{u} \in \mathcal{B}(\mathcal{Z}')$, *(6),* $\det(M_u)$ *is a nonzero polynomial on* x *of degree at most* L *where*

$$L = \frac{1}{2} \sum_{i=2}^{m} (k - \hat{k}_i)(k - \hat{k}_i - 1) - \sum_{i=2}^{m-1} (i - 1)(\hat{k}_{i+1} - \hat{k}_i)(k - k_i).$$

Proof. For every $i \in J_m$, take

$$\tilde{B}_i = \left((\beta_{m-i+1,v} x^{m-i})^{u-1} \right) \qquad u \in [k_{m-i+1} - \hat{k}_{m-i+1}], \ v \in [n_{m-i+1}]$$

and let $\tilde{E}_i$ be the submatrix formed by the $(k - k_{m-i+1} + 1)$th column to the $(k - \hat{k}_{m-i+1})$th column of I_k. Let

$$\tilde{M} = (\tilde{M}_0 | \tilde{M}_2 | \cdots | \tilde{M}_m),$$

where $\tilde{M}_0 = (0, 0, \ldots, 0, 1)^T$ is a k-dimensional column vector and $\tilde{M}_i = \tilde{E}_i \tilde{B}_i$ for every $i \in J_m$. Take $\tilde{\Pi}_0 = \Pi_0$ and $\tilde{\Pi}_i = \Pi_{m-i+1}$ for every $i \in J_m$. Then $\tilde{\Pi} = (\tilde{\Pi}_i)_{i \in J_m'}$ is a partition of $P' = P \cup \{p_0\}$ too. Moreover, take $\boldsymbol{\kappa}, \hat{\boldsymbol{\kappa}} \in \mathbb{Z}_+^{J_m'}$ such that $\kappa_0 = k$, $\hat{\kappa}_0 = k - 1$, and for every $i \in J_m$, $\kappa_i = k - \hat{k}_{m-i+1}$ and $\hat{\kappa}_i = k - k_{m-i+1}$. Then $\hat{\kappa}_i \leq \hat{\kappa}_{i+1} < \kappa_i \leq \kappa_{i+1}$ for $i \in [m - 1]$.

If $\boldsymbol{u} \in \mathcal{B}_1$, (6), then for any matrix M_u, as in Example 4, by exchanging rows and columns we can obtain the matrix $\tilde{M}_u$ such that $|\det(M_u)| = |\det(\tilde{M}_u)|$. As $\hat{k}_{m-i+1} \leq |\boldsymbol{u}([m - i])| \leq k_{m-i}$ for every $i \in [m - 1]$,

$$\hat{\kappa}_{i+1} = k - k_{m-i} \leq |\boldsymbol{u}([m - i + 1, m])| \leq k - \hat{k}_{m-i+1} = \kappa_i$$

for every $i \in [m - 1]$. From Lemma 2, $\det(\tilde{M}_u)$ is a nonzero polynomial on x of degree at most L where

$$L = \frac{1}{2} \sum_{i=1}^{m-1} \kappa_i(\kappa_i - 1) - \sum_{i=2}^{m-1} (m - i)(\kappa_i - \kappa_{i-1})\hat{\kappa}_i$$

$$= \frac{1}{2} \sum_{i=2}^{m} (k - \hat{k}_i)(k - \hat{k}_i - 1) - \sum_{i=2}^{m-1} (i - 1)(\hat{k}_{i+1} - \hat{k}_i)(k - k_i).$$

If $\boldsymbol{u} \in \mathcal{B}_2$, (6), then for any matrix M_u, we can obtain a matrix $\tilde{M}_u$ such that $|\det(M_u)| = |\det(\tilde{M}_u)| = |\det(\tilde{M}_u')|$, where $\tilde{M}_u'$ is the submatrix obtained by removing the first column and the last row of $\tilde{M}_u$. In this case $\hat{k}_{m-i+1} - 1 \le |\boldsymbol{u}([m-i])| \le k_{m-i} - 1$ for every $i \in [m-1]$, hence

$$\hat{\kappa}_{i+1} = (k-1) - (k_{m-i} - 1) \le |\boldsymbol{u}([m-i+1,m])| \le (k-1) - (\hat{k}_{m-i+1} - 1) = \kappa_i$$

for every $i \in [m-1]$. Lemma 2 implies that $\det(\tilde{M}_u')$ is a nonzero polynomial on x of degree at most L too, and the claim follows. $\qquad\square$

Now, take a finite field $\mathbb{F}_{q^\lambda}$, where $q > \max_{i \in J_m}\{n_i\}$ is a prime power and $\lambda > L$. Take all $\beta_{i,v}$ in the matrix (10) from $\mathbb{F}_q \backslash \{0\}$ and take $x \in \mathbb{F}_{q^\lambda}$ such that its minimal polynomial over $\mathbb{F}_q$ is of degree λ. Using the similar method to prove Theorem 5 and Proposition 6, we can obtain the following results.

Theorem 6. *The matrix (10) is a representation of the matroid associated to IHASs (1) over $\mathbb{F}_{q^\lambda}$ for some prime power $q > \max_{i \in J_m}\{n_i\}$ and some $\lambda > L$. Moreover, such a representation can be obtained in time $O(q, \lambda)$.*

Proposition 7. *Suppose M is the matrix (10). Then $LSSS(M)$ realizes the IHASs (1) over $\mathbb{F}_{q^\lambda}$ defined as in Theorem 6. Moreover, such a scheme can be obtained in time $O(q, \lambda)$.*

Remark 1. In some cases, Proposition 7 can give schemes for IHASs superior to the ones given by Proposition 6. For example, Proposition 6 can give the scheme for the DHTASs (2) over $\mathbb{F}_{q^\lambda}$ with $\lambda > K = \frac{1}{2}\sum_{i=1}^{m-1} k_i(k_i - 1)$ since $\hat{k}_1 = \cdots = \hat{k}_m = 0$ and the scheme for the CHTASs (3) over $\mathbb{F}_{q^\lambda}$ with $\lambda > K = \frac{1}{2}\sum_{i=1}^{m-1} k_i(k_i - 1) = \frac{1}{2}(m-1)k(k-1)$ since $0 = \hat{k}_1 < \cdots < \hat{k}_m$ and $k_1 = \cdots = k_m = k$.

On the other hand, Proposition 7 give the scheme for the DHTASs (2) over $\mathbb{F}_{q^\lambda}$ with $\lambda > L = \frac{1}{2}\sum_{i=2}^m (k - \hat{k}_i)(k - \hat{k}_i - 1) = \frac{1}{2}(m-1)k(k-1)$ and the scheme for the DHTASs (3) over $\mathbb{F}_{q^\lambda}$ with $\lambda > L = \frac{1}{2}\sum_{i=1}^{m-1}(k - \tilde{k}_i)(k - \tilde{k}_i - 1)$.

Therefore, Proposition 6 gives the scheme for DHTASs superior to the one given by Proposition 7. Nevertheless, Proposition 7 gives the scheme for CHTASs superior to the one given by Proposition 6.

4.3   Comparisons

Comparison to the Construction of Brickell. Brickell [9] presented an efficient method to construct the ideal linear scheme for the DHTASs (2) over $\mathbb{F}_{q^{\lambda'}}$ with $q > \max_{i \in J_m}\{n_i\}$ and $\lambda' \ge mk^2$. Proposition 6 gives a scheme for the DHTASs (2) too. In fact, our scheme is the same as Brickell's scheme. Nevertheless, Proposition 6 implies the scheme for the DHTASs (2) can be obtained over $\mathbb{F}_{q^\lambda}$ with $\lambda > K = \frac{1}{2}\sum_{i=1}^{m-1} k_i(k_i - 1)$. Therefore, we improve the bound for the field size since

$$\frac{1}{2}\sum_{i=1}^{m-1} k_i(k_i - 1) + 1 \le \frac{1}{2}(m-1)k_{m-1}(k_{m-1} - 1) + 1 < \frac{1}{2}(m-1)k_{m-1}^2 < mk^2.$$

The reason for the improvement is that we give a relatively precise description of $\det(M_u)$ by the formula provided in Proposition 1.

Comparison to the Construction of Tassa. Tassa [35] presented an efficient method to construct the ideal linear scheme for the CHTAS (3) over $\mathbb{F}_p$ where

$$p > 2^{-k+2}(k-1)^{(k-1)/2}(k-1)!N^{(k-1)(k-2)/2} \tag{11}$$

is a prime and N is the maximum identity assigning to participants. Proposition 7 gives a scheme for the CHTAS (3) over $\mathbb{F}_{q^\lambda}$ with $q > \max_{i \in J_m}\{n_i\}$ and $\lambda > L = \frac{1}{2}\sum_{i=1}^{m-1}(k-\tilde{k}_i)(k-\tilde{k}_i-1)$.

Since $(k-1)! \geq 2^{k-2}$ when $k \geq 2$, it follows that the right hand of (11) is great than or equal to $(k-1)^{(k-1)/2}N^{(k-1)(k-2)/2}$. From this with $N \geq n \geq \max_{i \in J_m}\{n_i\}$, we have

$$q^L \leq N^{(k-1)(k-2)/2} < 2^{-k+2}(k-1)^{(k-1)/2}(k-1)!N^{(k-1)(k-2)/2}$$

if $L \leq \frac{1}{2}(k-1)(k-2)$. In fact, $\max_{i \in J_m}\{n_i\} \ll N$ in general. This implies in this case $2^{-k+2}(k-1)^{(k-1)/2}(k-1)!N^{(k-1)(k-2)/2} \gg q^L$, and consequently, our result is superior to Tassa's result. In the case of $L > \frac{1}{2}(k-1)(k-2)$, it is very possible that q^L is smaller than the right hand of (11). In particular, our efficient methods can also work for non-prime fields.

5 Secret Sharing Schemes for Compartmented Access Structures

In this section, we study ideal linear secret sharing schemes for two families of compartmented access structures by efficient methods.

5.1 Construction for Compartmented Access Structures with Upper Bounds

In this section, we construct ideal linear secret sharing schemes realizing UCASs. We first present a representation of the integer polymatroid $\mathcal{Z}'$ satisfying Theorem 2 such that the UCASs (5) are of the form $\Gamma_0(\mathcal{Z}', \Pi)$.

Take $\Pi = (\Pi_i)_{i \in J_m}$ be a partition of the set P such that $|\Pi_i| = n_i$. Let $r \in \mathbb{Z}_+^{J'_m}$ and $k \in \mathbb{N}$ such that $r_0 = 1$, $r(J_m) \leq \Pi(P)$ and $r_i \leq k \leq |r(J_m)|$ for every $i \in J_m$. The following result was presented in Section 8.2 of [17].

Lemma 5. *Suppose $\mathcal{Z}' = (J'_m, h)$ is an integer polymatroid such that $h(X) = \min\{k, |r(X)|\}$ for every $X \subseteq J'_m$. Then the UCASs (5) are of the form $\Gamma_0(\mathcal{Z}', \Pi)$.*

Now, we introduce a linear representation of the polymatroid defined in Lemma 5. Take different elements $\alpha_{i,j} \in \mathbb{F}_q$ with $i \in J'_m$ and $j \in [r_i]$, where $q \geq \max_{i \in J_m}\{n_i, |r(J_m)| + 1\}$ is a prime power. For every $i \in J'_m$, let

$$A_i = (\alpha_{i,v}^{u-1}) \qquad u \in [k], \ v \in [r_i]$$

and consider the $\mathbb{F}_q$-vector subspace $V_i \subseteq \mathbb{F}_q^k$ spanned by all the columns of A_i. Let the integer polymatroid $\mathcal{Z}' = (J'_m, h)$ such that $h(X) = \dim\left(\sum_{i \in X} V_i\right)$ for every $X \subseteq J'_m$. We have the following result.

Proposition 8. *For the integer polymatroid* $\mathcal{Z}'$ *defined above, the UCASs (5) are of the form* $\Gamma_0(\mathcal{Z}', \Pi)$ *and*

$$\mathcal{B}(\mathcal{Z}') = \{\boldsymbol{u} \in \mathbb{Z}_+^{J'_m} : |\boldsymbol{u}| = k \text{ and } \boldsymbol{u} \leq \boldsymbol{r}\}. \tag{12}$$

Proof. Let $A = (A_0|A_1|\cdots|A_m)$. Then it is a $k \times (|\boldsymbol{r}(J_m)| + 1)$ Vandermonde matrix. Therefore, any $k \times k$ submatrix of A is nonsingular. This with $\dim(V_i) = r_i$ for every $i \in J'_m$ implies the second claim. In addition, $\left| \bigcup_{i \in X} \{\boldsymbol{a}_{i,v} : v \in [r_i]\} \right| = |\boldsymbol{r}(X)|$ for every $X \subseteq J'_m$ where $\boldsymbol{a}_{i,v}$ denotes the vth columns of A_i. Hence, $h(X) = \min\{k, |\boldsymbol{r}(X)|\}$ for every $X \subseteq J'_m$, and the first claim follows. $\square$

This proposition implies that the collection $(V_i)_{i \in J'_m}$ is a linear representation of the integer polymatroid $\mathcal{Z}'$ associated to the UCASs (5). We next present a matrix M based on $\mathcal{Z}'$, which is a representation of a matroid $\mathcal{M}$ such that the UCASs (5) are of the form $\Gamma_{p_0}(\mathcal{M})$.

Let $\Pi_0 = \{p_0\}$ and let $\Pi' = (\Pi_i)_{i \in J'_m}$ and $\Pi = (\Pi_i)_{i \in J_m}$ be the partition of $P' = P \cup \{p_0\}$ and P, respectively, such that $|\Pi_i| = n_i$. For every $i \in J'_m$, take n_i different elements $\beta_{i,v} \in \mathbb{F}_q$ with $v \in [n_i]$ and let

$$B_i = \left((\beta_{i,v} x)^{u-1} \right) \qquad u \in [r_i], \ v \in [n_i].$$

Let a $k \times (n+1)$ matrix be defined as

$$M = (M_0|M_1|\cdots|M_m) \tag{13}$$

where $M_i = A_i B_i$. We have the following result.

Lemma 6. *For any* $\boldsymbol{u} \in \mathcal{B}(\mathcal{Z}')$, *(12),* $\det(M_{\boldsymbol{u}})$ *is a nonzero polynomial on* x *of degree at most* $k(r-1)$, *where* $r = \max_{i \in J_m}\{r_i\}$.

Proof. Without loss of generality, we may assume that $M_{\boldsymbol{u}}$ is the $k \times k$ submatrix of M formed by the first u_i columns in every M_i. For every $i \in J'_m$, take $\bar{B}_i = \left(\beta_{i,v}^{u-1} \right)$ with $u \in [r_i]$ and $v \in [n_i]$, and let B'_i and $\bar{B}'_i$ denote the submatrices formed by the first u_i columns in B_i and $\bar{B}_i$, respectively. In addition, for any $\boldsymbol{j}_i = (j_{i,1}, \ldots, j_{i,u_i})$ with $i \in J_m$ such that $1 \leq j_{i,1} < \cdots < j_{i,u_i} \leq r_i$, let $B'_i(\boldsymbol{j}_i)$ and $\bar{B}'_i(\boldsymbol{j}_i)$ denote the $u_i \times u_i$ submatrices formed by the $j_{i,1}$th row, $\ldots$, j_{i,u_i}th row of B'_i and $\bar{B}'_i$, respectively, and let $A_i(\boldsymbol{j}_i)$ denote the submatrix formed by the first u_i columns in A_i. Then

$$\det\left(B'_i(\boldsymbol{j}_i)\right) = \det\left(\bar{B}'_i(\boldsymbol{j}_i)\right) x^{\sum_{v=1}^{u_i}(j_{i,v}-1)}.$$

If $j_{i,v} = r_i - u_i + v$ for $v \in [u_i]$, then the exponent of x in $\det(B'_i(\boldsymbol{j}_i))$ is maximum, that is

$$\sum_{v=1}^{u_i}(j_{i,v} - 1) = \sum_{v=1}^{u_i}(r_i - u_i + v - 1) = u_i(r_i - u_i) + \sum_{v=1}^{u_i-1} v = \frac{1}{2}u_i(2r_i - u_i - 1). \tag{14}$$

Note that in this case $\bar{B}'_i(\boldsymbol{j}_i)$ is the submatrix formed by of the last u_i rows of $\bar{B}'_i$, it follows $\det(\bar{B}'_i(\boldsymbol{j}_i)) \neq 0$. Hence, Proposition 4 implies that $\det(M_{\boldsymbol{u}})$ can be viewed as a polynomial on x and the summation with maximum exponent of x in it is

$$\left(\prod_{i=1}^{m} \det\left(\bar{B}_i'(\boldsymbol{j}_i)\right)\right) \det\left(\left(A_0(\boldsymbol{j}_0)|A_1(\boldsymbol{j}_1)|\cdots|A_m(\boldsymbol{j}_m)\right)\right)x^t, \qquad (15)$$

where for $i \in J_m$ and $v \in [u_i]$, $j_{i,v} = r_i - u_i + v$. As $\sum_{i=1}^{m} u_i^2 \geq \sum_{i=1}^{m} u_i$ and $\sum_{i=1}^{m} u_i = k$ or $k - 1$, from (14), we have

$$t = \frac{1}{2}\sum_{i=1}^{m} u_i(2r_i - u_i - 1) = \sum_{i=1}^{m} u_i r_i - \frac{1}{2}\sum_{i=1}^{m}(u_i^2 + u_i) \leq k(r-1). \qquad (16)$$

In addition, the matrix $\left(A_0(\boldsymbol{j}_0)|A_1(\boldsymbol{j}_1)|\cdots|A_m(\boldsymbol{j}_m)\right)$ is nonsingular, thus $\det(M_u)$ is a nonzero polynomial on x of degree t. Using the same method, we can prove this claim for any $\boldsymbol{u} \in \mathcal{B}(\mathcal{Z}')$, (12). □

Now, take a finite field $\mathbb{F}_{q^\lambda}$, where $q \geq \max_{i \in J_m}\{n_i, |\boldsymbol{r}(J_m)| + 1\}$ is a prime power and $\lambda > k(r-1)$. Take $\alpha_{i,v}$ and $\beta_{i,v}$ in the matrix (13) from $\mathbb{F}_q$ and take $x \in \mathbb{F}_{q^\lambda}$ such that its minimal polynomial over $\mathbb{F}_q$ is of degree λ. Then similar to Theorem 5 and Proposition 6, from this lemma, we can obtain the following result.

Theorem 7. *The matrix (13) is a representation of the matroid associated to UCASs (5) over $\mathbb{F}_{q^\lambda}$ for some prime power $q \geq \max_{i \in J_m}\{n_i, |\boldsymbol{r}(J_m)| + 1\}$ and some $\lambda > k(r-1)$. Moreover, such a representation can be obtained in time $O(q, \lambda)$.*

Proposition 9. *Suppose M is the matrix (13). Then $LSSS(M)$ realizes the UCASs (5) over $\mathbb{F}_{q^\lambda}$ defined as in Theorem 7. Moreover, such a scheme can be obtained in time $O(q, \lambda)$.*

Proof. If $\boldsymbol{u}(J_m) \in \min \Gamma$ and $u_0 = 0$, then $\boldsymbol{u}(J_m') \in \mathcal{B}(\mathcal{Z}')$, (12). Theorem 7 implies $M_{\boldsymbol{u}(J_m)}$ is nonsingular. Accordingly, M_0 can be spanned by the columns in $M_{\boldsymbol{u}(J_m)}$ for any $\boldsymbol{u}(J_m) \in \Gamma$. Assume that $\boldsymbol{u}(J) \notin \Gamma$. As $h(\{(i)\}) = r_i$ for every $i \in J_m$, thus without loss of generality, we may assume that $\boldsymbol{u}(J_m) \leq \boldsymbol{r}(J_m)$. Furthermore, we may assume that $|\boldsymbol{u}(J_m)| = k - 1$, since if $|\boldsymbol{u}(J_m)| < k - 1$, we may find a vector $\boldsymbol{u}'(J_m) \geq \boldsymbol{u}(J_m)$ such that $\boldsymbol{u}'(J_m) \leq \boldsymbol{r}(J_m)$ and $|\boldsymbol{u}'(J_m)| = k - 1$. In this case if $u_0 = 1$, then $\boldsymbol{u}(J_m') \in \mathcal{B}(\mathcal{Z}')$. Theorem 7 implies $M_{\boldsymbol{u}(J_m')}$ is nonsingular, and the claim follows. □

5.2 Construction for Compartmented Access Structures with Lower Bounds

In this section, we describe ideal linear secret sharing schemes realizing LCASs based on the schemes for the dual access structures of LCASs.

The dual access structures of LCASs (4) presented in [37] are defined as

$$\Gamma^* = \{\boldsymbol{u} \in \mathbf{P} : |\boldsymbol{u}| \geq l \text{ or } u_i \geq \tau_i \text{ for some } i \in J_m\} \qquad (17)$$

where $l = |P| - k + 1$, $\tau_i = |\Pi_i| - t_i + 1$ for $i \in J$, and $|\tau| \geq l + m - 1$.

We first present a representation of the integer polymatroid $\mathcal{Z}'$ satisfying Theorem 2 such that the access structures (17) are of the form $\Gamma_0(\mathcal{Z}', \Pi)$.

Take $\Pi = (\Pi_i)_{i \in J_m}$ be a partition of the set P such that $|\Pi_i| = n_i$. Let $\boldsymbol{\tau} \in \mathbb{Z}_+^{J'_m}$ and $l \in \mathbb{N}$ such that $\tau_0 = 1$, $\boldsymbol{\tau}(J_m) \leq \Pi(P)$ and $|\boldsymbol{\tau}(J_m)| \geq l + m - 1$. Take $\boldsymbol{\tau}' \in \mathbb{Z}_+^{J'_m}$ such that $\tau'_0 = 1$ and $\tau'_i = \tau_i - 1$ for every $i \in J_m$. The following result was presented in Section IV-D of [19].

Lemma 7. *Suppose $\mathcal{Z}' = (J'_m, h)$ is an integer polymatroid with h satisfying*

(1) $h(\{0\}) = 1$;
(2) $h(X) = \min\{l, 1 + |\boldsymbol{\tau}'(X)|\}$ *for every* $X \subseteq J_m$;
(3) $h(X \cup \{0\}) = h(X)$ *for every* $X \subseteq J_m$.

Then the access structures (17) are of the form $\Gamma_0(\mathcal{Z}', \Pi)$.

We next introduce a linear representation of the polymatroid defined in Lemma 7. Take elements $\alpha_{i,j} \in \mathbb{F}_q$ with $i \in J'_m$ and $j \in [\tau_i]$ where $q > \max_{i \in J_m} \{n_i, |\boldsymbol{\tau}'(J_m)|\}$ is a prime power such that

- $\alpha_{i,1} = \alpha_0$ for all $i \in J'_m$ and
- the elements α_0 and $\alpha_{i,j}$ with $i \in J_m$ and $j \in [2, \tau_i]$ are pairwise distinct.

For every $i \in J'_m$, let

$$A_i = \left(\alpha_{i,v}^{u-1}\right) \qquad u \in [l], \ v \in [\tau_i]$$

and consider the $\mathbb{F}_q$-vector subspace $V_i \subseteq \mathbb{F}_q^k$ spanned by all the columns of A_i. Let the integer polymatroid $\mathcal{Z}' = (J'_m, h)$ such that $h(X) = \dim\left(\sum_{i \in X} V_i\right)$ for every $X \subseteq J'_m$.

Proposition 10. *For the integer polymatroid $\mathcal{Z}'$ defined above, the access structures (17) are of the form $\Gamma_0(\mathcal{Z}', \Pi)$ and $\mathcal{B}(\mathcal{Z}') = \mathcal{B}_1 \cup \mathcal{B}_2$, where*

$$\mathcal{B}_1 = \{\boldsymbol{u} \in \mathbb{Z}_+^{J'_m} : |\boldsymbol{u}| = l, \ u_0 = 0, \ u_{i'} \leq \tau_{i'} \text{ for some } i' \in J_m$$
$$\text{and } u_i \leq \tau'_i \text{ for all } i \in J_m \setminus \{i'\}\}, \tag{18}$$

$$\mathcal{B}_2 = \{\boldsymbol{u} \in \mathbb{Z}_+^{J'_m} : |\boldsymbol{u}| = l, \ u_0 = 1 \text{ and } \boldsymbol{u}(J_m) \leq \boldsymbol{\tau}'(J_m)\}.$$

Proof. Proving the first claim is equivalent to proving that h satisfies the three conditions in Lemma 7. First, $h(\{0\}) = 1$ as $\dim(V_0) = 1$. Let A be the matrix formed by the column A_0 and the last τ'_i columns of A_i for every $i \in J_m$. Then it is a $l \times (1 + |\boldsymbol{\tau}'(J_m)|)$ Vandermonde matrix. Accordingly, any $l \times l$ submatrix of A is nonsingular. Since $\left|\bigcup_{i \in X}\{a_{i,v} : v \in [\tau_i]\}\right| = 1 + |\boldsymbol{\tau}'(X)|$ for every $X \subseteq J_m$ where $a_{i,v}$ denotes the vth columns of A_i, it follows that $h(X) = \min\{l, 1 + |\boldsymbol{\tau}'(X)|\}$ for every $X \subseteq J_m$. Moreover, $V_0 \subseteq V_i$ for every $X \subseteq J_m$, Therefore, $h(X \cup \{0\}) = h(X)$ for every $X \subseteq J_m$.

In addition, since any $l \times l$ submatrix of A is nonsingular, on the one hand, any l distinct columns from A_i with $i \in J_m$ are linearly independent, and on the other hand, A_0 and any $l - 1$ columns from the last τ'_i columns of A_i with $i \in J_m$ are linearly independent too. This implies the second claim. $\square$

We next present a matrix M which is a representation of a matroid $\mathcal{M}$ such that the access structures (17) are of the form $\Gamma_{p_0}(\mathcal{M})$.

Suppose $\Pi_0 = \{p_0\}$ and let $\Pi' = (\Pi_i)_{i \in J'_m}$ and $\Pi = (\Pi_i)_{i \in J_m}$ be the partition of $P' = P \cup \{p_0\}$ and P, respectively, such that $|\Pi_i| = n_i$. Take $\beta_{0,1} = 0$ and for every $i \in J_m$, take n_i different elements $\beta_{i,v} \in \mathbb{F}_q$ with $v \in [n_i]$ such that $\beta_{i,v} \neq 0$. For every $i \in J'_m$, take

$$B_i = \left((\beta_{i,v} x)^{u-1} \right) \qquad u \in [\tau_i], \ v \in [n_i]$$

and $M_i = A_i B_i$. Define a $l \times (n+1)$ matrix as

$$M = (M_0 | M_1 | \cdots | M_m). \tag{19}$$

Lemma 8. *For any $\boldsymbol{u} \in \mathcal{B}(\mathcal{Z}')$, (18), $\det(M_{\boldsymbol{u}})$ is a nonzero polynomial on x of degree at most $l(\tau - 1)$, where $\tau = \max_{i \in J_m}\{\tau_i\}$.*

Proof. Without loss of generality, we may assume that $M_{\boldsymbol{u}}$ is the $l \times l$ submatrix of M formed by the first u_i columns in every M_i. For every $i \in J'_m$, take $\bar{B}_i = \left(\beta_{i,v}^{u-1} \right)$ with $u \in [\tau_i]$ and $v \in [n_i]$, and let $\bar{B}'_i$ denote the submatrix formed by the first u_i columns in $\bar{B}_i$. Proposition 4 implies that $\det(M_{\boldsymbol{u}})$ can be viewed as a polynomial on x.

In the case of $\boldsymbol{u} \in \mathcal{B}_1$, let the summation with maximum exponent of x in $\det(M_{\boldsymbol{u}})$ be denoted by $a_{t_1} x^{t_1}$. Then similar to (15),

$$a_{t_1} x^{t_1} = \left(\prod_{i=1}^{m} \det\left(\bar{B}'_i(\boldsymbol{j}_i) \right) \right) \det\left((A_1(\boldsymbol{j_1})| \cdots | A_m(\boldsymbol{j_m})) \right) x^{t_1},$$

where $\boldsymbol{j}_i = (j_{i,1}, \ldots, j_{i,u_i})$ with $i \in J_m$ such that $j_{i,v} = \tau_i - u_i + v$ for $v \in [u_i]$. In this case the matrix $\left(A_1(\boldsymbol{j_1})| \cdots | A_m(\boldsymbol{j_m}) \right)$ is nonsingular since its all columns are pairwise distinct. From this and each $\bar{B}'_i(\boldsymbol{j}_i)$ is nonsingular, we have that $a_{t_1} \neq 0$. In addition, as $u_i \leq \tau_i$ for every $i \in J_m$, the inequality (16) implies $t_1 \leq l(\tau - 1)$.

In the case of $\boldsymbol{u} \in \mathcal{B}_2$, let the summation with maximum exponent of x in $\det(M_{\boldsymbol{u}})$ be denoted by $a_{t_2} x^{t_2}$. Then

$$a_{t_2} x^{t_2} = \left(\prod_{i=1}^{m} \det\left(\bar{B}'_i(\boldsymbol{j}_i) \right) \right) \det\left((A_0|A_1(\boldsymbol{j_1})| \ldots | A_m(\boldsymbol{j_m})) \right) x^{t_2},$$

where $\boldsymbol{j}_i = (j_{i,1}, \ldots, j_{i,u_i})$ with $i \in J_m$ such that $j_{i,v} = \tau_i - u_i + v$ for $v \in [u_i]$. In this case $u_i \leq \tau_i - 1$ for every $i \in J_m$. Therefore, from the inequality (16), $t_2 \leq l(\tau - 1)$. Moreover, $a_{t_2} \neq 0$ as $\bar{B}'_i(\boldsymbol{j}_i)$ with $i \in J'_m$ and $\left(A_0(\boldsymbol{j_0})| \cdots | A_m(\boldsymbol{j_m}) \right)$ are all nonsingular. $\qquad\square$

Now, take a finite field $\mathbb{F}_{q^\lambda}$ with $q > \max_{i \in J_m}\{n_i, |\tau'(J_m)|\}$ is a prime power and $\lambda > l(\tau - 1)$. Take $\alpha_{i,v}$ and $\beta_{i,v}$ in the matrix (19) from $\mathbb{F}_q \backslash \{0\}$ and take $x \in \mathbb{F}_{q^\lambda}$ such that its minimal polynomial over $\mathbb{F}_q$ is of degree λ. Similar to Theorem 7, we can obtain the following result.

Theorem 8. *The matrix (19) is a representation of the matroid associated to access structures (17) over $\mathbb{F}_{q^\lambda}$ for some prime power $q > \max_{i \in J_m}\{n_i, |\tau'(J_m)|\}$ and some $\lambda > l(\tau - 1)$. Moreover, such a representation can be obtained in time $O(q, \lambda)$.*

Proposition 11. *Suppose M is the matrix (19). Then $LSSS(M)$ realizes the access structures (17) over $\mathbb{F}_{q^\lambda}$ defined as in Theorem 8. Moreover, such a scheme can be obtained in time $O(q, \lambda)$.*

Proof. Let $u(J_m) \in \Gamma^*$, (17), be a minimal set, then $|u(J_m)| = l$ and $u(J_m) \leq \tau'(J_m)$, or $u_i = \tau_i$ for some $i \in J_m$. In the first case, Theorem 8 implies M_0 is can be spanned by all the columns in $M_{u(J_m)}$. Moreover, Theorem 8 implies any τ_i columns of M_i are linearly independent. From this with $h(\{0, i\}) = h(\{i\}) = \tau_i$ for every $i \in J_m$, M_0 is a linear combination of any τ_i columns in M_i. Hence, in the second case M_0 can be spanned by all the columns in $M_{u(J_m)}$ too.

Assume that $u(J_m) \notin \Gamma^*$, (17). Then $u(J_m) \leq \tau'(J_m)$ and $|u(J_m)| \leq l - 1$. Without loss of generality, we may assume that $|u(J_m)| = l-1$, since if $|u(J_m)| < l - 1$, we may find a vector $u'(J_m) \geq u(J_m)$ such that $u'(J_m) \leq \tau'(J_m)$ and $|u'(J_m)| = l-1$. As $l \leq |\tau'(J_m)|+1$, the above-described procedure is possible. In this case if $u_0 = 1$, then $u(J'_m) \in \mathcal{B}_2$. Theorem 8 implies $M_{u(J'_m)}$ is nonsingular, and the claim follows.   □

Remark 2. From the dual relationship of the access structures (17) and the LCASs (4), we can translate the scheme in Proposition 11 into an ideal linear scheme for the LCASs (4) using the explicit transformation of [20]. Specially, the efficient construction of ideal linear schemes realizing LCASs (4) can be obtained over $\mathbb{F}_{q^\lambda}$ in time $O(q, \lambda)$ for some prime power $q > \max_{i \in J_m}\{n_i, \sum_{i=1}^m (n_i - t_i)\}$ and some $\lambda > (n - k + 1)t$, where $t = \max_{i \in J_m}\{n_i - t_i\}$.

Acknowledgements. The authors would like to thank the reviewers for their helpful comments and suggestions. This research was supported in part by the Foundation of National Natural Science of China (No. 61772147, 61702124), Guangdong Province Natural Science Foundation of major basic research and Cultivation project (No. 2015A030308016), Project of Ordinary University Innovation Team Construction of Guangdong Province (No. 2015KCXTD014), Collaborative Innovation Major Projects of Bureau of Education of Guangzhou City (No. 1201610005) and National Cryptography Development Fund (No. MMJJ20170117).

References

1. Ball, S., Padró, C., Weiner, Z., Xing, C.: On the representability of the biuniform matroid. SIAM J. Discrete Math. **27**(3), 1482–1491 (2013)
2. Beimel, A.: Secret-sharing schemes: a survey. In: Chee, Y.M., et al. (eds.) IWCC 2011. LNCS, vol. 6639, pp. 11–46. Springer, Heidelberg (2011). https://doi.org/10.1007/978-3-642-20901-7_2
3. Beimel, A., Chor, B.: Universally ideal secret sharing schemes. IEEE Trans. Inf. Theory **40**(3), 786–794 (1994)

4. Beimel, A., Tassa, T., Weinreb, E.: Characterizing ideal weighted threshold secret sharing. SIAM J. Discrete Math. **22**(1), 360–397 (2008)
5. Ben-Or, M., Goldwasser, S., Wigderson, A.: Completeness theorems for noncryptographic fault-tolerant distributed computations. In: Proceedings of the 20th ACM Symposium on the Theory of Computing, pp. 1–10 (1988)
6. Benaloh, J., Leichter, J.: Generalized secret sharing and monotone functions. In: Goldwasser, S. (ed.) CRYPTO 1988. LNCS, vol. 403, pp. 27–35. Springer, New York (1990). https://doi.org/10.1007/0-387-34799-2_3
7. Beutelspacher, A., Wettl, F.: On 2-level secret sharing. Des. Codes Cryptogr. **3**(2), 127–134 (1993)
8. Blakley, G.R.: Safeguarding cryptographic keys. In: Proceedings of the National Computer Conference 1979, AFIPS Proceedings, vol. 48, pp. 313–317 (1979)
9. Brickell, E.F.: Some ideal secret sharing schemes. J. Combin. Maths. Combin. Comp. **9**, 105–113 (1989)
10. Brickell, E.F., Davenport, D.M.: On the classification of ideal secret sharing schemes. J. Cryptol. **4**, 123–134 (1991)
11. Chaum, D., Crépeau, C., Damgård, I.: Multiparty unconditionally secure protocols. In: Proceedings of the 20th ACM Symposium on the Theory of Computing, pp. 11–19 (1988)
12. Chor, B., Kushilevitz, E.: Secret sharing over infinite domains. J. Cryptol. **6**(2), 87–96 (1993)
13. Collins, M.J.: A note on ideal tripartite access structures. Cryptology ePrint Archive, Report 2002/193. http://eprint.iacr.org/2002/193
14. Cramer, R., Damgård, I., Maurer, U.: General secure multi-party computation from any linear secret-sharing scheme. In: Preneel, B. (ed.) EUROCRYPT 2000. LNCS, vol. 1807, pp. 316–334. Springer, Heidelberg (2000). https://doi.org/10.1007/3-540-45539-6_22
15. Cramer, R., et al.: On codes, matroids and secure multi-party computation from linear secret sharing schemes. In: Shoup, V. (ed.) CRYPTO 2005. LNCS, vol. 3621, pp. 327–343. Springer, Heidelberg (2005). https://doi.org/10.1007/11535218_20
16. Desmedt, Y., Frankel, Y.: Threshold cryptosystems. In: Brassard, G. (ed.) CRYPTO 1989. LNCS, vol. 435, pp. 307–315. Springer, New York (1990). https://doi.org/10.1007/0-387-34805-0_28
17. Farràs, O., Martí-Farré, J., Padró, C.: Ideal multipartite secret sharing schemes. J. Cryptol. **25**(3), 434–463 (2012)
18. Farràs, O., Padró, C.: Ideal hierarchical secret sharing schemes. IEEE Trans. Inf. Theory **58**(5), 3273–3286 (2012)
19. Farràs, O., Padró, C., Xing, C., Yang, A.: Natural generalizations of threshold secret sharing. IEEE Trans. Inf. Theory **60**(3), 1652–1664 (2014)
20. Fehr, S.: Efficient construction of the dual span program. Manuscript, May (1999)
21. Giulietti, M., Vincenti, R.: Three-level secret sharing schemes from the twisted cubic. Discrete Math. **310**(22), 3236–3240 (2010)
22. Herranz, J., Sáez, G.: New results on multipartite access structures. IEE Proc. Inf. Secur. **153**(4), 153–162 (2006)
23. Herzog, J., Hibi, T.: Discrete polymatroids. J. Algebraic Combinat. **16**(3), 239–268 (2002)
24. Ito, M., Saito, A., Nishizeki, T.: Secret sharing schemes realizing general access structure. In: Proceedings of the IEEE Global Telecommunication Conference, Globecom 1987, pp. 99–102 (1987)

25. Kothari, S.C.: Generalized linear threshold scheme. In: Blakley, G.R., Chaum, D. (eds.) CRYPTO 1984. LNCS, vol. 196, pp. 231–241. Springer, Heidelberg (1985). https://doi.org/10.1007/3-540-39568-7_19
26. Massey, J.L.: Minimal codewords and secret sharing. In: Proceedings of the 6th Joint Swedish-Russian Workshop on Information Theory, pp. 276–279 (1993)
27. Massey, J.L.: Some applications of coding theory in cryptography. Codes and Ciphers: Cryptography and Coding IV, pp. 33–47 (1995)
28. Oxley, J.G.: Matroid Theory. Oxford Science Publications. The Clarendon Press, Oxford University Press, New York (1992)
29. Padró, C., Sáez, G.: Secret sharing schemes with bipartite access structure. IEEE Trans. Inf. Theory **46**(7), 2596–2604 (2000)
30. Schrijver, A.: Combinatorial Optimization. Polyhedra and Efficiency. Springer, Berlin (2003)
31. Shamir, A.: How to share a secret. Commun. ACM **22**, 612–613 (1979)
32. Shankar, B., Srinathan, K., Rangan, C.P.: Alternative protocols for generalized oblivious transfer. In: Rao, S., Chatterjee, M., Jayanti, P., Murthy, C.S.R., Saha, S.K. (eds.) ICDCN 2008. LNCS, vol. 4904, pp. 304–309. Springer, Heidelberg (2007). https://doi.org/10.1007/978-3-540-77444-0_31
33. Shoup, V.: New algorithm for finding irreducible polynomials over finite fields. Math. Comput. **54**, 435–447 (1990)
34. Simmons, G.J.: How to (really) share a secret. In: Goldwasser, S. (ed.) CRYPTO 1988. LNCS, vol. 403, pp. 390–448. Springer, New York (1990). https://doi.org/10.1007/0-387-34799-2_30
35. Tassa, T.: Hierarchical threshold secret sharing. J. Cryptol. **20**(2), 237–264 (2007)
36. Tassa, T.: Generalized oblivious transfer by secret sharing. Des. Codes Cryptol. **58**(1), 11–21 (2011)
37. Tassa, T., Dyn, N.: Multipartite secret sharing by bivariate interpolation. J. Cryptol. **22**(2), 227–258 (2009)
38. Welsh, D.J.A.: Matroid Theory. Academic Press, London (1976)

Perfectly Secure Oblivious RAM with Sublinear Bandwidth Overhead

Michael Raskin[1(✉)] and Mark Simkin[2]

[1] Technical University of Munich, Munich, Germany
raskin@mccme.ru
[2] Aarhus University, Aarhus, Denmark
simkin@cs.au.dk

Abstract. Oblivious RAM (ORAM) has established itself as a fundamental cryptographic building block. Understanding which bandwidth overheads are possible under which assumptions has been the topic of a vast amount of previous works. In this work, we focus on perfectly secure ORAM and we present the first construction with sublinear bandwidth overhead *in the worst-case*. All prior constructions with perfect security require linear communication overhead in the worst-case and only achieve sublinear bandwidth overheads in the amortized sense. We present a fundamentally new approach for constructing ORAM and our results significantly advance our understanding of what is possible with perfect security.

Our main construction, Lookahead ORAM, is perfectly secure, has a worst-case bandwidth overhead of $\mathcal{O}\left(\sqrt{N}\right)$, and a total storage cost of $\mathcal{O}(N)$ on the server-side, where N is the maximum number of stored data elements. In terms of concrete server-side storage costs, our construction has the smallest storage overhead among all perfectly and statistically secure ORAMs and is only a factor 3 worse than the most storage efficient computationally secure ORAM. Assuming a client-side position map, our construction is the first, among all ORAMs with worst-case sublinear overhead, that allows for a $\mathcal{O}(1)$ online bandwidth overhead without server-side computation. Along the way, we construct a conceptually extremely simple statistically secure ORAM with a worst-case bandwidth overhead of $\mathcal{O}\left(\sqrt{N}\frac{\log N}{\log \log N}\right)$, which may be of independent interest.

1 Introduction

More and more sensitive data is stored online. A basic attempt to keep data private while storing it online on an untrusted server is to simply encrypt each data entry. Unfortunately, this is not always sufficient. For instance, Islam et al. [IKK12] showed that by observing the access patterns induced by encrypted search queries over an encrypted database, an honest-but-curious server storing the database, could learn significant amounts of information about the queries' contents. For example, it might be undesirable to let the server know whether the same block was accessed twice, or two different blocks have been accessed. A more viable approach is to not only encrypt the data, but also hide the access patterns. Oblivious RAM (ORAM) is a cryptographic primitive that allows a

© International Association for Cryptologic Research 2019
S. D. Galbraith and S. Moriai (Eds.): ASIACRYPT 2019, LNCS 11922, pp. 537–563, 2019.
https://doi.org/10.1007/978-3-030-34621-8_19

client to do exactly this, at the cost of some bandwidth and storage overhead. It enables a client to outsource his data to an untrusted server inside of an ORAM data structure, and then read and write to his dataset without revealing the position that was accessed or the operation that was performed. Goldreich and Ostrovsky [Gol87, GO96] first introduced the notion of ORAM, presented the first constructions thereof, and proved the first lower bound of $\Omega(\log n)$ on the bandwidth overhead for a certain type of constructions, where n the maximum number of data elements stored in the data structure. Boyle and Naor [BN16] revisit this lower bound proof and highlight that it only holds for statistically secure ORAMS that behave in a "balls-and-bins" fashion. The same lower bound without these restriction was recently proven by Larsen and Nielsen [LN18].

Understanding which upper bounds can be achieved in which setting has been the topic of numerous works [OS97, GMOT11, SCSL11, DMN11, SSS12, SvS+13, CNS18, PPRY18, AKL+18]. Most commonly, these works measure bandwidth overhead in one of two ways. Either they consider the worst-case overhead, meaning the largest overhead any one operation on the ORAM data structure can incur, or they consider the amortized overhead, meaning the average overhead per operation in a longer sequence of operations. The best known upper bound for worst-case bandwidth overhead is due to Stefanov et al. [SvS+13], who present a statistically secure construction with $\mathcal{O}\left(\log^2 N\right)$ overhead[1]. The best known upper bound for amortized bandwidth overhead is due to Asharov et al. [AKL+18], who present a computationally secure construction with bandwidth overhead of $\mathcal{O}(\log N)$, which matches the lower of Larsen and Nielsen. The first perfectly secure ORAM construction is due to Damgård et al. [DMN11] and has an amortized bandwidth overhead of $\mathcal{O}\left(\log^3 N\right)$ and a multiplicative storage overhead of $\mathcal{O}(\log N)$. This was recently improved upon by Chan et al. [CNS18], who present a perfectly secure construction with the same amortized bandwidth overhead and a storage overhead of $\mathcal{O}(1)$. Given the conceptual complexity of both these constructions it is not clear whether the de-amortization tricks of Goodrich et al. [GMOT11] can be applied. In the multi-server setting, where the ORAM data structure is distributed among several non-colluding servers, Chan et al. [CKN+18] present a perfectly secure construction with worst-case bandwidth overhead of $\mathcal{O}\left(\log^2 N\right)$. Since all existing perfectly secure single-server constructions require linear worst-case bandwidth overhead, we pose the natural question:

Can we construct a perfectly secure single-server ORAM with sublinear worst-case bandwidth overhead?

We believe this is an important theoretical question. It is well known that randomization and computational assumptions are powerful tools in algorithm and protocol design. It is a common theme in research to investigate the power of these tools for specific problems by understanding the upper bounds that we can achieve with and without them. In this work we make a significant

---

[1] For a larger data block size of $\mathcal{O}\left(\log^2 N\right)$ they even achieve an overhead of $\mathcal{O}(\log N)$ data blocks.

step towards understanding what ORAM bandwidth overhead can be achieved in the worst-case with perfect security and without relying on randomization[2] or computational assumptions. See [CNS18] for a further discussion about the importance of perfectly secure ORAM.

1.1 Our Contribution

We present the first construction of perfectly secure Oblivious RAM with sublinear worst-case bandwidth overhead and, furthermore, we also make the following contributions:

Novel Approach to Constructing ORAM. We present a fundamentally new approach for constructing ORAM. Somewhat surprisingly, and despite the large amount of research interest that ORAM has received, all existing constructions are based on a handful of conceptually different approaches. We believe it is of theoretical and practical interest to explore new ways to construct this primitive. In this work, we present two new constructions.

Our first construction is conceptually extremely simple. It is statistically secure, meaning that even a computationally unbounded adversary cannot break the obliviousness guarantees and all operations on the data structure succeed with an overwhelming probability. It has a worst-case bandwidth overhead of $\mathcal{O}\left(\sqrt{N}\frac{c+\log t+\log N}{\log\left(c+\log t+\log N\right)}\right)$, where N is the maximum number of data blocks to be stored, t is an upper bound on the number of accesses, and c is the correctness parameter that provides an upper bound of 2^{-c} on the failure probability. To the best of our knowledge, it is one of the conceptually simplest known ORAM constructions to date. The underlying logic is easy to implement and the proof of security is straightforward. Our main construction, called Lookahead ORAM, is loosely based on our first construction. It has a worst-case bandwidth overhead of $\mathcal{O}\left(\sqrt{N}\right)$ and is *perfectly* secure, in the sense that every operation on the ORAM data structure succeeds with probability 1 and obliviousness is guaranteed against an unbounded adversary. The hidden constants behind the big-O notation are small and our construction is significantly faster in the worst-case than the fastest perfectly secure single-server ORAM construction of Chan et al. [CNS18] in the amortized case for any practical parameter range. For instance, for $N = 2^{20}$ our construction has a worst-case bandwidth overhead of less than 7 000 data blocks, whereas their construction has a amortized bandwidth overhead of around 160 000 data blocks[3].

---

[2] Our main construction is using randomness exclusively for the sake of security, but not for efficiency. We believe this is unavoidable.

[3] Our estimate of Chan et al.'s construction is computed by instantiating it with Batcher's Bitonic sort [Bat68] and a hidden constant of 1. For our construction we took the concrete parameters one obtains assuming a server-side position map.

ORAM	Client storage		Server storage	Bandwidth overhead	Online overhead	Section
	persistent	temporary				
Matrix	$\mathcal{O}(N)$	$\mathcal{O}(1)$	$\approx 18N + \frac{N \log N}{2}$	$\mathcal{O}\left(\sqrt{N}\log N\right)$	–	3
Matrix	$\mathcal{O}(1)$	$\mathcal{O}(1)$	$\approx 142N + 2N \log N$	$\mathcal{O}\left(\sqrt{N}\log N\right)$	–	5
Lookahead	$\mathcal{O}(N)$	$\mathcal{O}\left(\sqrt{N}\right)$	$N + 2\sqrt{N}$	$\mathcal{O}\left(\sqrt{N}\right)$	$\mathcal{O}(1)$	4
Lookahead	$\mathcal{O}(1)$	$\mathcal{O}\left(\sqrt{N}\right)$	$6N + 12\sqrt{N}$	$\mathcal{O}\left(\sqrt{N}\right)$	–	5
Lookahead	$\mathcal{O}(1)$	$\mathcal{O}(1)$	$6N + 12\sqrt{N}$	$\mathcal{O}\left(\sqrt{N}\log\sqrt{N}\right)$	–	5.1

Fig. 1. Overview of the different parameter settings for Matrix and Lookahead ORAM. For Matrix ORAM we crudely estimate the parameters for $c = 20$ and $t = 15$. All overheads are stated in data blocks and assume block size $\Theta(\log N)$. Asymptotically line 3 with a client-side storage of $\mathcal{O}(N)$ makes little sense. From a practical perspective, however, the $\mathcal{O}(1)$ online overhead is a powerful feature and the concrete client-side storage is significantly smaller than the concrete amount of data stored on the server-side. The asymptotical behaviour of the position map size also improves in case of faster growth of block size.

Small Concrete Storage Overhead. Assuming a client-side position map, Lookahead ORAM has the smallest concrete storage overhead among all existing ORAM constructions with sublinear worst-case bandwidth overhead. Our construction only incurs an additive server-side storage overhead of $\sqrt{2N}$ data blocks. A small storage overhead is particularly beneficial in outsourced storage settings where data owners have to pay their storage provider for the storage they consume. At the cost of a slightly increased total server-side storage, namely $6N + 12\sqrt{N}$, we can reduce the client-side storage of Lookahead ORAM to $\mathcal{O}(1)$. In this case, our construction has the smallest storage overhead among all statistically and perfectly secure ORAMs and is only a multiplicative factor of around 3 larger than the most storage efficient computationally secure ORAMs [GMOT11] with sublinear worst-case bandwidth overhead. For a more detailed discussion of the concrete server-side storage costs of Lookahead ORAM see Sects. 4.5 and 5. Lookahead ORAM, by default, requires the client to have a temporary client-side storage of $\mathcal{O}\left(\sqrt{N}\right)$ during each access. We show how to reduce the temporary client-side storage to $\mathcal{O}(1)$ at the cost of increasing the worst-case bandwidth overhead to $\mathcal{O}\left(\sqrt{N}\log\sqrt{N}\right)$ in Sect. 5.1. We illustrate the different parameter options and their efficiency in Fig. 1.

Constant Online Bandwidth Overhead. One approach to circumvent the $\Omega(\log N)$ lower bound on the bandwidth overhead was introduced by Boneh et al. [BMP11] and then improved upon in [DSS14, RFK+15]. Their main idea was to split the total bandwidth overhead into two parts. The first part, the so called online overhead, is the amount of data that needs to be transmitted between the client and the server to retrieve a desired data element obliviously. The second part, the offline overhead, is the amount of data that needs to be transmitted between the two parties to ensure obliviousness of future accesses.

One can think of the offline overhead as background work that, usually, moves around encrypted data elements in the ORAM data structure to ensure the desired obliviousness guarantees. Splitting the total bandwidth overhead this way and then minimizing the online overhead has practical advantages. It allows the client to efficiently retrieve data from the server without much latency during bursts of requests and then do the background work during quieter phases.

In [BMP11], Boneh et al. presented a computationally secure construction, for a primitive, which is strongly related to ORAM and has $\mathcal{O}(\log N)$ online and $\mathcal{O}\left(\sqrt{N}\log N\right)^4$ worst-case overhead. In [DSS14, RFK+15] this idea of splitting the total overhead has been further refined, and computationally secure constructions that achieve an online bandwidth overhead of $\mathcal{O}(1)$ are presented. However, these constructions require some server-side computation during the online phase, which renders these solutions not applicable for "raw" storage providers that do not support these ORAM constructions explicitly. That is, our construction works in combination with arbitrary storage providers like Dropbox or Google Drive, whereas the constructions from [DSS14, RFK+15] only work with storage providers that explicitly implement their given scheme.

If the client stores the position map and a small client-side storage that can hold up to $\sqrt{2N}$ data blocks locally, then Lookahead ORAM allows the client to obliviously retrieve arbitrary many elements from the ORAM data structure with no bandwidth overhead in the online phase and no server-side computation in the online or offline phase. That is, in the online phase, the client can directly download the desired elements from the server. Our construction is the first to provide such a feature among all computationally, statistically, and perfectly secure ORAMs with sublinear worst-case bandwidth overhead.

To provide a better feeling for how expensive it is to store the stash and the position map locally, consider a 1 GB database with a block size of 1 KB. To be able to make use of our online overhead feature, the client would need to store a roughly 2 MB stash and a 8 MB position map locally. As mentioned before, if the client chooses to not use the minimizing online overhead feature, it can reduce its persistent storage to $\mathcal{O}(1)$.

Attack on [GMSS16]. We identify a flaw in the ORAM construction of [GMSS16] and outline an attack that breaks the claimed obliviousness guarantees in Sect. A. We have contacted the authors and they have acknowledged our attack.

1.2   Other Related Work

A vast amount of works have contributed to our current understanding of ORAM. In this section we merely provide a high-level overview of the works that are directly related to our work.

---

[4] This worst-case complexity is slightly different from the original paper. The paper has a superlinear worst-case overhead due to an expensive reshuffling phase, but when splitting shuffling over $\sqrt{N}$ accesses, one can achieve the stated complexity.

In order to achieve practical efficiency and overcome the $\Omega(\log N)$ lower bound, several works have looked at different refinements of the classical ORAM notion in the client server model. Path-PIR [MBC14] uses server-side computations to achieve a practically very small, yet still poly-logarithmic bandwidth overhead. In [AKST14], Apon et al. formally define the notion of Verifiable Oblivious Storage, which generalizes the notion of ORAM by allowing the server to perform computations, and show that the ORAM lower bound does not apply to their setting by providing a scheme with constant overhead per access based on Fully Homomorphic Encryption. In [DDF+16] a scheme, called Onion ORAM, is presented that breaks the lower bound, but only relies on additively homomorphic encryption. In this work we will only focus on the classical notion of ORAM that does *not* allow server-side computation.

Demertzis et al. [DPP18] present a *computationally* secure ORAM construction with worst-case bandwidth overhead $\mathcal{O}\left(N^{1/3}\right)$ and perfect correctness. Several recent works cite Demertzis et al. and claim that their construction is perfectly secure. This is *not* correct and this claim is not made by the authors of that paper either. Their construction is a modification of the square-root ORAM construction and requires the client to store a random permutation, which represents the position map. This position map can only be stored succinctly by using a pseudorandom function. Therefore, their construction is either only computationally secure, or requires linear client-side storage, or requires linear bandwidth overhead.

Lastly, a work by Gordon et al. [GMSS16] presents an ORAM construction that may seem *superficially* similar to ours. However, our work significantly differs from theirs in terms of performance guarantees we achieve, underlying ideas we present, and security we obtain. Their construction, called Matrix-ORAM, arranges the data elements in a fixed number of rows. The size of each row linearly depends on the size of the total database and each row has its own stash. Accesses to their data structure are performed in a conceptually and concretely different manner to ours. The authors claim a logarithmic bandwidth overhead. In contrast, our main construction has a rectangular shape[5], and has a bandwidth overhead of $\mathcal{O}\left(\sqrt{N}\right)$.

We discovered a flaw in their construction and present a concrete attack on their scheme. This flaw is discussed in detail in Sect. A.

2 Preliminaries

On a high level, the ORAM security definition assumes a honest-but-curious server and says that for any two data access sequences, the corresponding access sequences to the ORAM data structure should be indistinguishable. The security definition is taken from [SSS12].

Definition 1 *(Security Definition). Let*

$$\vec{y} := ((\mathsf{op}_1, \mathsf{a}_1, \mathsf{data}_1), \ldots, (\mathsf{op}_M, \mathsf{a}_M, \mathsf{data}_M))$$

---

[5] One may even say they look matrix shaped.

be a data request sequence of length M, where each op_ℓ is either a read at position a_ℓ or a write of data_ℓ at position a_ℓ. Let $\mathsf{oram}(\overrightarrow{y})$ denote the (possibly randomized) sequence of accesses to the remote storage given the sequence of data requests $\overrightarrow{y}$. An ORAM construction is said to be secure if for any two data request sequences $\overrightarrow{y}$ and $\overrightarrow{z}$ of the same length, their access patterns $\mathsf{oram}(\overrightarrow{y})$ and $\mathsf{oram}(\overrightarrow{z})$ are indistinguishable and the construction is correct in the sense that it returns on input $\overrightarrow{y}$ data that is consistent with $\overrightarrow{y}$ with probability at least $1 - 2^{-c}$. We call c the correctness parameter. We call an ORAM perfectly, respectively statistically, secure if the two distributions above are perfectly, respectively statistically indistinguishable.

Note that in ORAM schemes, the server holding the encoded data does not perform any computations.

Position Map. All known ORAM schemes need to maintain a position map of size $\mathcal{O}(N)$ that keeps track of the ordering of elements inside the ORAM data structure on the server. For the sake of simplicity we will assume that the client stores the full position map locally. From a practical point of view, this seems to be a reasonable assumption in many client-server settings. For example, the position map of a 1 GB database containing 1 KB blocks is only around 8 MB large. From a theoretical point of view, to reduce the client's persistent storage to $\mathcal{O}(1)$, both of our constructions can be combined with the well-known approach of recursively storing the position map in a sequence of smaller ORAMs, which was first introduced in [SCSL11]. Recursively storing the position map on the server increases the number of round-trips per access to $\mathcal{O}(\log N)$, but it does not change the asymptotic bandwidth overheads of our constructions. We explain how to combine our main construction with the recursive approach in detail in Sect. 5.

Block Size. If we want to use the recursive ORAM approach mentioned above to store the position map on the server-side, then the data blocks need to be $\Omega(\log N)$ large. In the setting, where the client stores the position map locally, we do not make any assumptions about the data block size. However, for the construction to be useful, the data blocks on the server should in total be larger than the position map that the client stores locally.

Integrity. The ORAM security definition assumes the server to be honest-but-curious. Similar to previous works [SvS+13], our construction can, at the cost of giving up perfect for computational security, be extended to prevent tampering of an actively malicious server by using a Merkle Tree on top of our ORAM data structures.

3   A Simple Matrix Bucket ORAM

In this section, we will present a very simple oblivious RAM construction with reasonable efficiency and a simple proof of security. To the best of our knowledge this is one of the, arguably, simplest ORAM constructions known to date. Apart from being interesting on its own, it will also serve a stepping stone towards our main construction by introducing some of the ideas behind our main approach.

Initially we are given an array A of length N of data elements. To initialize our scheme, we create a empty $\sqrt{N} \times \sqrt{N}$ matrix C, in which each matrix cell is a bucket of size w. We randomly (and independently) assign each element from A to a bucket in C. Once all elements from A are distributed among buckets in C, we encrypt each bucket separately, and store the matrix on the server. For the sake of simplicity, we assume that the client stores a position map σ that maps indices of elements from A to columns of C locally.

Init(A)	Access(ℓ, x)
1 $C \leftarrow$ initBucketMatrix()	1 $c \leftarrow$ readColumn($C, \sigma(\ell)$)
2 $\sigma \leftarrow$ initPosMap()	2 $(i,j) \leftarrow$ pickRandBucket()
3 for $\ell = 1 \ldots N$ do	3 $r \leftarrow$ readRow(C, i)
4 $\quad (i,j) \leftarrow$ pickRandBucket()	4 $r_{\mathsf{dec}} \leftarrow$ decrypt(r)
5 $\quad C[i,j]$.put($A[\ell] \| \ell$)	5 $c_{\mathsf{dec}} \leftarrow$ decrypt(c)
6 $\quad \sigma(\ell) = j$	6 $data \leftarrow$ popData(ℓ, c_{dec})
7 $C \leftarrow$ encryptBuckets(C)	7 putData($x \| \ell, r_{\mathsf{dec}}, j$)
	8 $\sigma(\ell) = j$
	9 $c^* \leftarrow$ encrypt(c_{dec})
	10 $r^* \leftarrow$ encrypt(r_{dec})
	11 writeBack(c^*, r^*)
	12 return $data$

Fig. 2. Pseudocode of simple matrix ORAM construction

To obliviously access some element with index ℓ in A, we need to access column $\sigma(\ell)$ in C. In addition, we need to pick a uniformly random bucket (i,j) in the matrix and obtain row i. We find the element with index ℓ in the retrieved column and perform our desired operation (read or write). We then remove element ℓ from its current bucket, put it into bucket (i,j), re-encrypt all retrieved buckets, and write back the retrieved row and column. Lastly, we update the position map to point to the new column that stores ℓ, i.e., set $\sigma(\ell) = j$. The pseudocode implementing this construction is given in Fig. 2

3.1 Security

We prove the following theorem

Theorem 1. *Let $\mathcal{E} = (\mathsf{gen}, \mathsf{encrypt}, \mathsf{decrypt})$ be an IND-CPA secure encryption scheme. Then the construction in Fig. 2 is a statistically secure ORAM scheme with $\mathcal{O}\left(\sqrt{N} \frac{c + \log t + \log N}{\log(c + \log t + \log N)}\right)$ bandwidth overhead and a total storage cost of $\mathcal{O}(N \log N)$ data blocks, where N is the number of data elements, c the correctness parameter, and t is the upper bound on the number of accesses.*

Proof. The key idea of why the proposed scheme is oblivious stems from the basic observation that every column intersects with every row. Intuitively, this means that if we obliviously write an element into some uniformly random position in a row, then, from an adversarial point of view, every column is equally likely a potential candidate for reading that element in a future access. In our scheme, whenever we read an element through a column access, we move it to a new uniformly random bucket and, in particular, a new uniformly random column, through a row access. Importantly, the movement of each element is completely independent of the access history and the other elements residing in the matrix. From these observations it is straightforward to see that the proposed scheme is oblivious.

What remains to show is the relation between the bucket size and the correctness parameter, i.e. we want to pick our buckets sufficiently large such that a bucket overflows with negligible (in the correctness parameter) probability. Towards this goal, we make an observation that simplifies our analysis. Let $\mathsf{Exp}_{\mathsf{move}}^{N,t}$ be the experiment of first throwing N balls into N buckets once and then picking up a random ball from a random bucket and moving it to a new random bucket t times. This experiment expresses the actual movement of data during t many accesses in our oblivious ram construction. Let $\mathsf{Exp}_{\mathsf{throw}}^{N}$ be the experiment of throwing N balls into N initially empty buckets of capacity w. Let $\mathsf{Load}_i^{>w}$ denote the event that bucket with index i at some point in time has more than w many elements in it and $\mathsf{Load}^{>w}$ the event that this happens to any of the buckets. We will analyze the probability of the event of one bucket overflowing in $\mathsf{Exp}_{\mathsf{throw}}^{N}$ and use the following lemma to put $\mathsf{Exp}_{\mathsf{throw}}^{N}$ and $\mathsf{Exp}_{\mathsf{move}}^{N,t}$ into relation.

Lemma 1. *Let $t > 0$, then*

$$\Pr[\mathsf{Load}_i^{>w}|\mathsf{Exp}_{\mathsf{move}}^{N,t}] \leq t \cdot \Pr[\mathsf{Load}_i^{>w}|\mathsf{Exp}_{\mathsf{throw}}^{N}]$$

Proof. Given N balls and N bins, there are N^N different possibilities to distribute the balls among the bins. Let us call each way to distribute the balls a constellation. Let X be one arbitrary but fixed constellation among them and, since all of them are equally probable, we have $\Pr[X|\mathsf{Exp}_{\mathsf{throw}}^{N}] = \frac{1}{N^N}$. Let us now consider constellations, which are one ball move away from X. There are exactly $N^2 - N$ such constellations, because we can select any of the N balls, and pick any of $N - 1$ buckets distinct from the current bucket of the selected ball. Selecting a random ball uniformly and moving it to a random bucket yields the original constellation with probability $\frac{1}{N^2}$, and each of the neighbouring constellations with probability $\frac{1}{N^2}$. As all ball moves are reversible, each constellation can be obtained from $N^2 - N$ other constellations. The probability of obtaining a constellation after a uniform selection of constellation and a single random ball move is therefore equal to $\frac{N^2-N}{N^2}\frac{1}{N^N} + \frac{1}{N}\frac{1}{N^N} = \frac{1}{N^N}$. The lemma follows by induction over t and then applying the union bound.

Using this lemma it is sufficient to upper bound the probability of a bucket overflowing in the experiment $\mathsf{Exp}_{\mathsf{throw}}^{N}$ and then apply the union bound over all buckets. Let us first look at the probability of some single bucket i overflowing by one element after $\mathsf{Exp}_{\mathsf{throw}}^N$, i.e. the probability of a bucket containing (exactly) $z = w + 1$ balls after throwing N balls into N buckets at random once. In the analysis we assume N to be sufficiently large, i.e. N should be large enough for our bucket size w to be at least 8, so that our inequalities work out. In the following calculation we will use two inequalities. First, $\forall x \geq 0$ it holds that $\left(1 - \frac{1}{x}\right)^x \leq e^{-1}$ and, secondly, $\forall 0 \leq k \leq n$ it holds that $\left(\frac{n}{k}\right)^k \leq \binom{n}{k} \leq \left(\frac{en}{k}\right)^k$.

$$
\begin{aligned}
\Pr[\mathsf{Load}_i^z | \mathsf{Exp}_{\mathsf{throw}}^N] &= \binom{N}{z} \left(\frac{1}{N}\right)^z \left(1 - \frac{1}{N}\right)^{N-z} \\
&\leq \left(\frac{eN}{z}\right)^z N^{-z} \left(\left(1 - \frac{1}{N}\right)^N\right)^{1-\frac{z}{N}} \\
&\leq \left(\frac{eN}{z}\right)^z N^{-z} e^{\frac{z}{N}-1} \\
&= e^z z^{-z} e^{\frac{z}{N}} e^{-1} \\
&= 2^{z(\log e - \log z) + \log e\left(\frac{z}{N}-1\right)} \\
&\leq 2^{z(\log e - \log z)} \\
&\leq 2^{-\frac{1}{2}z \log z}
\end{aligned}
$$

We can provide an upper bound on the event of a single bucket having more than w balls after throwing N balls into N buckets using geometric series as follows

$$
\begin{aligned}
\Pr[\mathsf{Load}_i^{>w} | \mathsf{Exp}_{\mathsf{throw}}^N] &\leq \sum_{z=w+1}^{N} 2^{-\frac{1}{2}z \log z} \\
&= \sum_{z=1}^{N-w} 2^{-\frac{1}{2}(w+z) \log (w+z)} \\
&= \sum_{z=1}^{N-w} 2^{-\frac{1}{2}(w \log (w+z) + z \log (w+z))} \\
&\leq \sum_{z=1}^{N-w} 2^{-\frac{1}{2}(w \log w + z \log z)} \\
&= 2^{-\frac{1}{2}w \log w} \sum_{z=1}^{N-w} 2^{-\frac{1}{2}z \log z} \\
&\leq 2^{-\frac{1}{2}w \log w + 1}
\end{aligned}
$$

Applying the union bound over all buckets and using Lemma 1 we obtain

$$
\Pr[\mathsf{Load}^{>w} | \mathsf{Exp}_{\mathsf{move}}^{N,t}] \leq 2^{-\frac{1}{2}w \log w + 1} tN
$$

We want to bound this probability of a bad event happening by some correctness parameter c, i.e. we want this probability to be smaller than 2^{-c}.

$$2^{-\frac{1}{2}w \log w + 1} t N \leq 2^{-c}$$

$$\Leftrightarrow -\frac{1}{2} w \log w + 1 + \log t + \log N \leq -c$$

$$\Leftrightarrow w \log w \geq 2(c + \log t + \log N + 1)$$

Hence, the bucket size $w \in \mathcal{O}\left(\frac{c + \log t + \log N}{\log (c + \log t + \log N)}\right)$ and therefore the total bandwidth cost in our construction is $\mathcal{O}\left(\sqrt{N} \frac{c + \log t + \log N}{\log (c + \log t + \log N)}\right)$.

4 Main Construction

In this section we are going to present our main Lookahead ORAM construction. The first difference between our Matrix Bucket ORAM and Lookahead ORAM is that we replace all buckets by cells that can only hold single elements. As a first try to construct a more efficient ORAM we could do the following: Initially, randomly shuffle the data, distribute the data elements among matrix cells, and encrypt each cell separately. To access an element, we retrieve the column corresponding to that element and a row corresponding to a uniformly random cell. After accessing the desired element, we swap the accessed element with the uniformly random cell, re-encrypt both row and column, and write them back into the matrix.

On an intuitive level, one could hope for this to be a secure ORAM construction, since every element will be swapped into a new uniformly random column at every access. Unfortunately this is not the case. The problem is that the distribution of columns into which elements are swapped is not uniformly random when conditioned on the observed row accesses. In particular, the difference with the simple matrix construction is that, here, the accessed element will change the position of another element, i.e the swap partner. It turns out that the server can infer information about the positions of accessed elements whenever we access the same row twice.

Figure 3 illustrates why the straightforward approach of directly swapping the accessed element with an element from a uniformly random cell fails. In this figure, the root node depicts a 2×2 matrix holding four encrypted entries. Initially the server has no knowledge about the arrangement of data elements in the matrix. Let us assume we access two different data elements. With probability non-negligible in the security parameter the following events will occur. On the first access the server observes the second column and second row being accessed. Edges from the root to the first layer show the possible swaps that could have happened, given the observed row and column accesses. $a \leftrightarrow b$ means element a was accessed and swapped with element b. On the second access the server observes the first column and, again, the second row being accessed. The leaf nodes of this tree represent all possible arrangements of elements in the matrix, given the observed access pattern. Dashed boxes indicate the case, where the first accessed element has changed its column. Solid boxes indicate that the first accessed element is in its original column. Counting the leaf nodes it can be seen that the first accessed element will more likely than not have switched columns after the two accesses. Hence, from the server's point of view, the elements are not distributed uniformly at random and the approach does not provide the desired obliviousness guarantees.

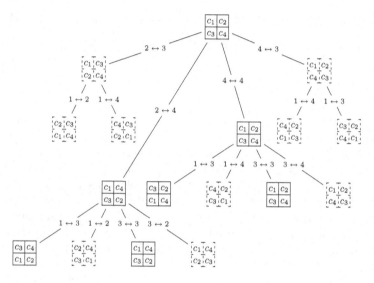

Fig. 3. Illustration of why the naive approach of swapping two random cells via a column and a row access fails. The root node depicts a encrypted 2×2 matrix. The leaf nodes depict all possible arrangements of data elements that are possible after the observed access pattern. Dashed boxes indicate the case, where the first accessed element has changed its column. Solid boxes indicate that the first accessed element is in its original column.

4.1   Intuition for Lookahead ORAM

The main issue with this first approach is that the row accesses reveal too much information. Ideally, we would like to have a swap procedure that allows us to directly access the desired element instead of the whole column and then swap that accessed cell with a new cell without revealing the column or row of that new cell. Observe, that to perform a swap, we have to perform two tasks. We have to remove the accessed element from its cell and put it into the cell of its swap partner. Symmetrically, we have to remove the swap partner from its cell and put it into the cell of the accessed element.

To realize such a swap procedure, we introduce two auxiliary stashes stash_{acc} and stash_{swap}, where stash_{acc} will be storing accessed elements and stash stash_{swap} will store pre-selected swap partners. From a high-level perspective, these stashes will help us to pretend that we immediately swap accessed elements to an unknown new location in the matrix. From the server's point of view, the client will always read both full stashes, and a uniformly random cell in the matrix, since the client behaves as if accessed elements are immediately magically swapped to their new locations in the matrix. In reality, accessed elements will go to stash_{acc} from where they will be eventually evicted into the cell of their respective swap partner obliviously. Swap partners will be readily waiting in stash_{swap} and upon accessing some element in the matrix, the swap partner will be swapped from stash_{swap} into the accessed cell directly. As an invariant we have that each element is either at its expected location in the matrix or in one of the stashes.

Two issues that need to be addressed are, how do we get swap partners into stash_{swap} before they are used and how to get accessed elements from stash_{acc} into their new cells in the matrix. To solve both these issues, we introduce a (stateful) round-robin column access that will iterate through the columns. Using the round-robin column access, we perform two "maintenance" tasks.

To empty stash_{acc}, we evict all elements from it, whose destination is somewhere in the column of the current round-robin column access. Note that for a matrix of size $\sqrt{N} \times \sqrt{N}$, the round-robin access will have accessed every cell of the matrix in $\sqrt{N}$ steps. This means that no element in stash_{acc} will wait for more than $\sqrt{N}$ steps to be evicted. Since, we add at most one element to the stash per access, this means that stash_{acc} will never contain more than $\sqrt{N}$ elements.

The second task is to ensure that, whenever we use a swap partner from stash_{swap}, the content of the swap partner's cell must already be available in that stash. Observe, that the swap partner's cell is a uniformly random cell in the matrix, which does not depend on the access pattern and can be selected upfront. Assume stash_{swap} contains $\sqrt{N}$ many preselected swap partners in a queue. These swap partners are sufficient for the next $\sqrt{N}$ accesses. Now upon performing an access, one swap partner from the stash will be used, and we pre-select a uniformly random cell that will be the swap partner once all other swap partners from stash_{swap} are used. At this point in time, the content of the pre-selected swap partner is (likely) not in the stash. However, since we have $\sqrt{N} - 1$ many more accesses before it will be used, we can be certain that the round-robin column access will fetch it in time before it will be used. Since our stashes accommodate accessed elements waiting to be evicted and swap partners waiting to be used, the total stash size is $2\sqrt{N}$.

One detail that we have swept under the rug so far, is the case, where the element we want to access is not in the matrix, but somewhere in the stashes. To get an intuitive feeling for the handling of these cases, it is helpful to keep in mind that at every access, we are basically (virtually) swapping two cells in the matrix. The stashes are just auxiliary data structures that make this process happen.

Globals

1 C, σ
2 stash $= (\text{stash}_{swap}, \text{stash}_{acc})$
3 ind_{col}

Init(A)

1 $C :=$
   $\text{initEmptyMatrix}(\sqrt{N} \times \sqrt{N})$
2 $(A', \sigma') \leftarrow \text{shuffle}(A)$
3 $\sigma \leftarrow \text{fillMatrix}(A', \sigma', C)$
4 $(\text{stash}_{swap}, \text{stash}_{acc}) :=$
   $\text{initStash}()$
5 **for** $i := 1 \ldots \sqrt{N}$ **do**
6 $\quad (i, j) \leftarrow \text{pickRandCell}()$
7 $\quad \text{enqueue}(((i, j), \bot), \text{stash}_{swap})$
8 $\text{ind}_{col} := 0$
9 **for** $i := 1 \ldots \sqrt{N}$ **do**
10 $\quad \text{Background}();$

Access(ℓ, x)

1 $v.pos := \sigma(\ell)$
2 $(v.val, where) :=$
   $\text{ReadVirtual}(v.pos)$
3 $s := \text{dequeue}(\text{stash}_{swap})$
4 **if** $x \neq \bot$ **then**
5 $\quad v.val := x$
6 $\text{SwapVirtual}(v, s, where)$
7 $\text{swapInPosMap}((i, j), s.pos, \sigma)$
8 $(i', j') \leftarrow \text{pickRandCell}()$
9 $\text{enqueue}(((i', j'), \bot), \text{stash}_{swap})$
10 $\text{Background}()$
11 **return** v

SwapVirtual(v, s, where)

1 **switch** *where* **do**
2 $\quad$ **case** *matrix* **do**
3 $\qquad C[v.pos] := s.val$
4 $\quad$ **case** *accstash* **do**
5 $\qquad C[v.pos] := s.val$
6 $\qquad \text{remove}(v.pos, v.val, \text{stash}_{acc})$
7 $\quad$ **case** *swapstash* **do**
8 $\qquad C[v.pos] := \bot$
9 $\qquad t := \text{find}(v.pos, \text{stash}_{swap})$
10 $\qquad \text{replaceVal}(t, s.val, \text{stash}_{swap})$

11 $\text{put}(s.pos, v.val, \text{stash}_{acc})$

Background()

1 $c = \text{readColumn}(\text{ind}_{col}, C)$
2 **for** $t \in \text{stash}_{acc}$ **do**
3 $\quad$ **if** $t.pos.j = \text{ind}_{col}$ **then**
4 $\qquad c[t.pos.i] := t.value$
5 $\qquad \text{remove}((i, j), \text{stash}_{acc})$

6 **for** $t \in \text{stash}_{swap}$ **do**
7 $\quad$ **if** $t.pos.j = \text{ind}_{col}$ **then**
8 $\qquad$ **if** $c[t.pos.i] \neq \bot$ **then**
9 $\qquad\quad t.value := c[t.pos.i]$
10 $\qquad\quad c[t.pos.i] := \bot$

11 $\text{writeColumn}(\text{ind}_{col}, c, C)$
12 $\text{ind}_{col} := \text{ind}_{col} + 1 \mod \sqrt{N}$

ReadVirtual((i, j))

1 $v := C[i, j]$
2 **if** $v \neq \bot$ **then**
3 $\quad$ **return** $(v, matrix)$
4 $v := \text{find}((i, j), \text{stash}_{acc})$
5 **if** $v \neq \bot$ **then**
6 $\quad$ **return** $(v, accstash)$
7 $v := \text{find}((i, j), \text{stash}_{swap})$
8 **return** $(v, swapstash)$

Fig. 4. Pseudocode of lookahead ORAM

Let (i, j) be the cell in the matrix that is expected to contain the element we are accessing and let the next swap partner from stash$_{swap}$ be some data element v originating from some cell (a, b). If the desired data element is not at position (i, j) in the matrix, but rather in stash$_{acc}$ (waiting to be evicted to cell (i, j)), then take the swap partner from stash$_{swap}$ and place its value v into the matrix at position (i, j). From now on the element in stash$_{acc}$ that was expected to be at (i, j) will be waiting to be evicted to (a, b). If the accessed element, expected at location (i, j), is in stash$_{swap}$, this means that (i, j) is pre-selected as a swap partner for some future access. Therefore the contents of (i, j) are not supposed to be in the matrix, but rather in the stash$_{swap}$. In this case, we find (i, j) in the stash$_{swap}$, put the value of (i, j) into stash$_{acc}$ to be evicted into (a, b), and replace the value of (i, j) in stash$_{swap}$ with v.

4.2   Formal Description

Given this intuition about how our construction work, we are now ready to formally present our construction in Fig. 4. Let C be the matrix containing the encrypted data entries and σ be the position map that maps array indices to matrix positions in C. We implement stash$_{swap}$ as a queue and stash$_{acc}$ as an map from positions to values. Both matrix C and the stashes are stored on the server-side. Init(A) initializes the ORAM data structure, by permuting the elements and storing them in an encrypted matrix C. Initially, both stashes are created empty. stash$_{swap}$, is filled up with random elements from the encrypted matrix. To write value x or just read at position ℓ in array A inside the ORAM data structure, we use Access ($\ell, \times$), which makes use of the ReadVirtual, WriteVirtual, and Background subroutines. ReadVirtual (i, j) reads the stash and the matrix cell $C[i, j]$ to find the data element that is expected to be at position (i, j) inside the matrix C. SwapVirtual(v, s, where) simulates a swap of accessed value $v.val$ at position $v.pos$ with pre-selected swap partner s with value $s.val$ from position $s.pos$. Background() implements the round-robin column access, which takes care of flushing elements out of stash$_{acc}$ and fetching elements into stash$_{swap}$.

4.3   Security

Theorem 2. *Let $\mathcal{E} = $ (gen, encrypt, decrypt) be an IND-CPA secure encryption scheme. Then the construction in Fig. 4 is a perfectly secure ORAM scheme with $\mathcal{O}\left(\sqrt{N}\right)$ bandwidth overhead and a total storage cost of $N + \mathcal{O}\left(\sqrt{N}\right)$ data blocks, where N is the number of data elements.*

Proof. Instead of directly arguing about the security of our proposed construction, we will rather argue about the security of an idealized version, which leaks the same amount of information about the access pattern, but is easier to analyze. As previously explained, from a high-level perspective, our construction directly accesses the desired element in the matrix and then swaps it with a random cell. The swap is immediately applied to the position map and we always directly access the cell in the matrix, which should contain a desired element according to the position map. The stash and the round-robin column accesses are there to enable us to (virtually) swap the accessed element into a new cell without leaking anything about that new location.

IdealizedAccess$_1(\sigma, \ell)$	IdealizedAccess$_2(\sigma, \ell)$
1 $(i,j) \leftarrow \sigma(\ell)$	1 $(i,j) \leftarrow \sigma(\ell)$
2 $(i',j') \leftarrow$ pickRandCell(C)	2 $(i',j') \leftarrow$ pickRandCell(C)
3 $v \leftarrow$ readCell(i,j,C)	3 $v \leftarrow$ readCell(i,j,C)
4 $C \leftarrow$ readMatrix(C)	4 $C \leftarrow$ readMatrix(C)
5 swapInPosMap$((i,j),(i',j'),\sigma)$	5 swapInPosMap$((i,j),(i',j'),\sigma)$
6 swapInPosMap$((i,j),(i',j'),C)$	6 $(C,\sigma) \leftarrow$ fullReshuffle(C,σ);
7 return v, σ	return v, σ

Fig. 5. Idealized access procedures

Since both the stash and the round-robin column access are always executed independently of the access pattern, they leak no information. Hence, instead of analyzing our construction directly, we can now analyze a construction with the idealized access procedure IdealizedAccess$_1$ depicted in Fig. 5 on the left. The initialization procedure corresponding to IdealizedAccess$_1$ is a straightforward adaption of our main construction and is not stated explicitly. In IdealizedAccess$_1$, we directly access the cell that contains our data element and, next, we retrieve the full matrix to perform the swap operation locally. From an efficiency point of view this is clearly a useless construction, but w.r.t. obliviousness both our main construction and this idealized version thereof leak the same amount of information about the access pattern. More formally, the success probability of any distinguisher D, distinguishing two data access sequences, is the same in our main construction and in the construction with IdealizedAccess$_1$.

Lemma 2. *Let* oram *be the main construction from Fig. 4 and* oram$_1^*$ *the construction using the access procedure* IdealizedAccess$_1$. *Then, for any distinguisher* D, *for any two data request sequences* $\overrightarrow{y}$ *and* $\overrightarrow{z}$ *we have*

$$| \Pr[\mathsf{D}(\mathsf{oram}(\overrightarrow{y})] - \Pr[\mathsf{D}(\mathsf{oram}(\overrightarrow{z}))]|$$
$$= | \Pr[\mathsf{D}(\mathsf{oram}_1^*(\overrightarrow{y}))] - \Pr[\mathsf{D}(\mathsf{oram}_1^*(\overrightarrow{z}))]|$$

There are two components that are observable by the server in both our real construction and the idealized access that can leak information about the access pattern. The first component is the swap logic that moves accessed elements to a new position. The second component is the direct accesses to the desired elements, i.e. we need to show that conditioned on previously observed accesses, each new access will fetch a uniformly random cell in the matrix.

Towards showing the first part, let v_k for $1 \leq k \leq N$ be some arbitrary data elements. Let $\mathsf{Exp}_{\mathsf{swap}}^{N,t}$ be the experiment of, initially, distributing the N data elements v_k in a matrix C with N cells uniformly at random and then for t steps repeatedly swapping the contents of two uniformly random cells.

Lemma 3. *Let C be a matrix of size $\sqrt{N} \times \sqrt{N}$ and let $v_k \in \mathcal{V}$ for $1 \leq k \leq N$ be arbitrary values from some value space $\mathcal{V}$. Then C, after running experiment $\mathsf{Exp}_{\mathsf{swap}}^{N,t}$, is a uniformly random permutation of the data elements v_k.*

Proof. Initially distributing the data elements v_k uniformly at random in the matrix C corresponds to a uniformly random permutation. For $\mathsf{Exp}_{\mathsf{swap}}^{N,1}$, i.e. distributing the elements and then swapping two uniformly random cells once, the statement holds, since we apply a random permutation of two elements to a uniformly random permutation. The statement for $t > 1$ follows by induction over t.

To conclude the security proof it remains to show that even conditioned on the previously observed accesses the distribution of data elements in C is uniformly random. Assume the server observes an access to position (i, j) in the matrix to fetch some data element v_k. The accessed element is going to be swapped into every position in the matrix with equal probability. Since each element is equally likely, with a probability of $\frac{1}{N}$, to be selected as a swap partner, every element is equally likely to end up in (i, j). From the accessed cells, no other information about the access pattern is leaked. Hence, from the server's perspective all distributions of access patterns are equally likely, no matter what the actual data access sequence is.

For the sake of clarity, let us look at the slightly modified access procedure IdealizedAccess$_2$ depicted in Fig. 5 on the right. In IdealizedAccess$_2$ we do not just swap two cells locally, but we fully reshuffle the whole matrix. Due to the full reshuffle, each access is completely independent of the previously observed access pattern. It is straightforward to see that IdealizedAccess$_2$ is a secure ORAM construction. Since in both IdealizedAccess$_1$ and IdealizedAccess$_2$ the access patterns are distributed uniformly at random, IdealizedAccess$_2$ leaks as much information as IdealizedAccess$_1$.

4.4   Online Overhead

For some practical applications it may be of interest to split the total bandwidth overhead into an online and an offline overhead. The point of this is to minimize the online bandwidth, which represents the amount of data that needs to be transmitted, when a client requests an element, and then do some background work, the offline overhead, to ensure the security of the ORAM, when no data requests are actively pending. This way we can minimize the practical latency of user requests despite the inherent lower bound on the overhead shown in [Gol87, GO96].

Looking at our main construction, it is straightforward to split the total bandwidth overhead into online and offline overhead. In the online phase, assuming we download the stash once, we can directly access the desired elements in the matrix on the server without any overhead. In the offline phase we need to do the remaining work, i.e. perform the round-robin column access, fill up the stash of pre-selected swap partners and flush out elements that need to go back into the matrix.

While storing the stash locally can theoretically compromise data integrity in case of a client device failure, the only part of the stash that cannot be randomly reinitialized is the cache of recently accessed elements. If the server is significantly more reliable than the client device, recently accessed elements can be written to a separate server-side buffer of size $\sqrt{N}$ in a round-robin manner. Each element will be stored in this buffer for $\sqrt{N}$ operations which is enough for the background operations to write it to its long-term server-side storage position.

It is also possible to allow multiple online accesses in a single burst. The simplest way to do it increases local storage requirements by twice the size of the maximum allowed burst length. Note that our implementation faithfully simulates the oram_1^* construction as long as the background work is performed at least $\sqrt{N}$ times between committing to a swap partner and its use, and background work is also performed at least $\sqrt{N}$ times between accessing the element and evicting it from the recently accessed element stash. If there is additional stash space of size b for the potential swap partners and the same amount of space for the recently accessed elements, we only need to have at least $\sqrt{N}$ background work operations performed during every interval when $b + \sqrt{N}$ access operations are performed. Note that it is possible to perform some part of the background work, handle an additional burst and then continue the background work as long as enough background work is done to prevent exhaustion of the swap partners or overflow of the recently accessed elements stash.

The same analysis shows that there is a trade off between stash size and bandwidth overhead: for example, if we have a stash twice as large as needed, we can afford doing only half the background processing step after each access.

4.5   Trading Off Bandwidth and Storage Overhead

For our main construction, the server's storage overhead comes from the two auxiliary stashes that it needs to store in addition to the encrypted data elements. For a matrix C, where the number of columns equals the number of rows, this results in an additive storage overhead of $2\sqrt{N}$. More generally, by considering an arbitrary rectangle C with H rows and W columns, we can trade off the concrete storage and bandwidth overhead costs of our construction. The number of columns W affects the time it takes the round-robin column access to iterate over the whole rectangle C and thus it also affects the stashes which have to be of size W each. The number of rows H, affects the size of the column that we need to download at each access. For example, by setting $H = 2\sqrt{N}$ and $W = \frac{1}{2}\sqrt{N}$, we can, in comparison to a quadratically shaped matrix C, directly reduce the additive storage overhead to $\sqrt{N}$ and maintain a bandwidth overhead of $3\sqrt{N} + 1$. By setting $H = \sqrt{2N}$ and $W = \sqrt{\frac{N}{2}}$, we get a bandwidth overhead of $2\sqrt{2}\sqrt{N} + 1$ and a storage overhead of $\sqrt{2N}$.

5   Constant Client-Side Storage

So far we have assumed that the full position map is stored explicitly on the client side. Following the approach of [SCSL11], we show that our construction can be modified to only require $\mathcal{O}(1)$ client-side storage at the cost of $\mathcal{O}(\log N)$ rounds of interaction per access between the client and the server. To store the position map on the server side, we will create a sequence of ORAM data structures that increase in size, which represent the position map. This means that now the server stores a sequence of position map ORAMs and an ORAM data structure that contains the actual data. To access an element at a certain index, the client will use the position map ORAMs to determine, which index it should query in the ORAM data structure that stores the actual data.

Server-side storage of the position map also requires specifying some details of background processing of the stashes and the position map. We describe the necessary changes.

Theorem 3. *Consider an arbitrary ORAM construction with a client-side position map used one time per access. Assume there exist such constants K and C that for N entries with a block size of $D \geqslant 4 \log N$, the ORAM construction has a multiplicative bandwidth overhead of $C\sqrt{N}$ and total storage of $(N + K\sqrt{N})D$ bits.*

Such an ORAM can be converted into an ORAM with a server-side position map with a multiplicative bandwidth overhead of $2C\sqrt{N} + 4$ and total storage $ND + 2KD\sqrt{N} + 2N \log N$ bits.

Proof. The proof goes by induction over N. In the base case, for size $N \leqslant 4$, we use a linear ORAM, which simply reads and writes all the blocks for each access. This ORAM has a bandwidth overhead of 4 and no storage overhead.

For an arbitrary N, we use the induction hypothesis to replace an ORAM with client-side position map with an ORAM that stores the position map on the server-side. We encode 4 blocks addresses in the position map into one storage block inside the ORAM data structure. Using a block size of $4 \log N$ results in an ORAM data structure that stores $\frac{N}{4}$ blocks, storing 4 addresses each. Now we apply the induction hypothesis.

Since we want to store $\frac{N}{4}$ blocks with $4 \log N$ bits per block, by induction hypothesis our position map ORAM has a total storage cost of $N \log N + 2K(4 \log N)\sqrt{\frac{N}{4}} + 2\frac{N}{4} \log N$ bits. Adding the storage costs of our main ORAM, the total storage cost in bits will be

$$N \log N + 2K(4 \log N)\sqrt{\frac{N}{4}} + 2\frac{N}{4} \log N + ND + KD\sqrt{N}$$
$$= ND + K(4 \log N)\sqrt{N} + KD\sqrt{N} + \frac{3}{2}N \log N$$
$$< ND + 2KD\sqrt{N} + 2N \log N$$

In our construction, every access to the main ORAM requires one access to the position map. Note that we do not need any additional position map accesses for the background work that moves data out of the stash. The position map access adds its bandwidth cost of $2C\sqrt{\frac{N}{4}} + 4$ to the main ORAM bandwidth cost of $C\sqrt{N}$. The total overhead is $C\sqrt{N} + 2C\sqrt{\frac{N}{4}} + 4 = 2C\sqrt{N} + 4$.

This completes the proof of the inductive step.

By setting the parameters as described in Sect. 4.5, we get

Corollary 1. *Lookahead ORAM with a client-side position map, block size $D = 4\log N$, multiplicative bandwidth overhead $\mathcal{O}\left(\sqrt{N}\right)$, and total storage of $(N + \sqrt{2N})4\log N$ bits can be converted into an ORAM with server-side position map and multiplicative bandwidth overhead $\mathcal{O}\left(\sqrt{N}\right)$ and total storage of $\log N(6N + 12\sqrt{N})$ bits.*

Proof. We need to show that we can implement the Lookahead ORAM with a single position map access per main ORAM access. As a straightforward implementation of `swapInPosMap` would use two accesses, we describe the necessary modification.

Each cell in the ORAM stores its array index in addition to data. This also applies to the stashes (so each entry keeps matrix position, array index, and data). The client downloads the entire content of stashes before reading any data from the matrix, and if one of the stashes contains the required index, the stash overrides the position map. If an operation changes the position map, we also update the index entries in the stashes if necessary. When we perform an access, the client reads the position map entry for the index being accessed, and the updated position for this index is written to the position map. For the swap partner, we delay the update to the position map just like we delay it for the data block itself. During the background work stage, in addition to a single column of the data matrix, we also scan a single line of the position map matrix and the position map stash and perform the pending updates from the top-level stash if applicable.

5.1 Constant Client-Side Temporary Storage

Using a recursive position map, we reduce the persistent client-side storage to $\mathcal{O}(1)$. However, during any operation on the ORAM data structure, the client still needs to temporarily store $\mathcal{O}\left(\sqrt{N}\right)$ data blocks. If desired, this can be reduced to $\mathcal{O}(1)$ data blocks at the cost of increasing the bandwidth overhead by a multiplicative factor of $\mathcal{O}\left(\log\sqrt{N}\right)$. We outline our solution here and leave the details to the interested reader.

Recall that during each ORAM operation, we have to access one matrix cell, one matrix column j, and the two stashes. For the sake of simplicity lets assume that the data matrix has height and width $\sqrt{N}$. Upon each ORAM operation we first process the stashes stash_{acc} and stash_{swap} into a temporary stashes $\mathsf{stash}^{acc}_{final}$ and $\mathsf{stash}^{swap}_{final}$, which have the following property: If either stash_{swap} or stash_{acc} contains a cell (i, j') with $j' = j$, then this cell will be in the i-th position of $\mathsf{stash}^{swap}_{final}$ or $\mathsf{stash}^{acc}_{final}$ respectively. For example, if stash_{swap} contains a cell that is associated with the second cell from the top in the current column j, then it will be in the second position in $\mathsf{stash}^{swap}_{final}$. Once we have such stashes, we can iterate over the current column j and both final stashes one cell each at a time and perform the necessary operations. We explain the generation of $\mathsf{stash}^{acc}_{final}$. The generation of $\mathsf{stash}^{swap}_{final}$ is completely analogous:

1. Create an empty stash $\mathsf{stash}^{acc}_{final}$
2. Iterate over stash_{acc} one cell at a time. If the current cell (i, j) is associated with the current column, we append it to $\mathsf{stash}^{acc}_{final}$ with a priority i. If it is not relevant for the current column, then we append it to $\mathsf{stash}^{acc}_{final}$ with a priority of ∞.
3. We append $\sqrt{N}$ dummy elements to $\mathsf{stash}^{acc}_{final}$, where dummy element i has priority $i + \frac{1}{2}$. At this point $\mathsf{stash}^{acc}_{final}$ contains $2\sqrt{N}$ cells.
4. We use an oblivious sorting algorithm [Bat68, AKS83] to sort $\mathsf{stash}^{acc}_{final}$ according to its priorities from smallest to largest.
5. We iterate over $\mathsf{stash}^{acc}_{final}$ one data element at a time. Whenever we read a real cell i, we set the priority of the dummy cell right after it to ∞.
6. We again obliviously sort $\mathsf{stash}^{acc}_{final}$.

A visual illustration of this final stash generation is depicted in Fig. 6.

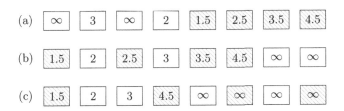

Fig. 6. Illustration of the generation of $\mathsf{stash}^{acc}_{final}$ with $\sqrt{N} = 4$. Each row represents $\mathsf{stash}^{acc}_{final}$ at a certain stage in the stash generation algorithm. The squares represent data cells and their labels represent their assigned priorities. Shaded rectangles represent dummy elements. (a) depicts $\mathsf{stash}^{acc}_{final}$ at the end of step 3. (b) depicts the stash at the end of step 4. (c) depicts it after step 5.

The post-processing of the two final stashes is straightforward. Again we only explain the post-processing of $\mathsf{stash}^{acc}_{final}$, since the post-processing of $\mathsf{stash}^{swap}_{final}$ is completely analogous. We iterate over $\mathsf{stash}^{acc}_{final}$ and assign each real data cell priority $-\infty$ and each dummy element priority ∞. We use oblivious sort to sort $\mathsf{stash}^{acc}_{final}$ from small to large. We interpret the first $\sqrt{N}$ as stash_{acc} and delete the remaining last $\sqrt{N}$ (dummy) elements.

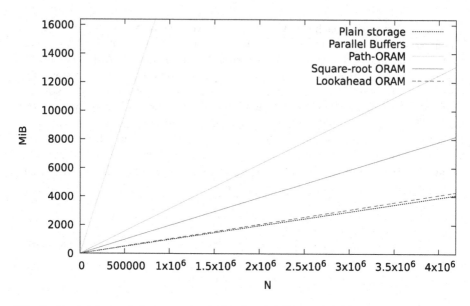

Fig. 7. Comparison of the storage overheads of different ORAM constructions. The x-axis shows different amounts of data blocks. The data block size is fixed to 1024 bytes. The y-axis plots the total required storage on the server-side in MiB. For the related works we compare ourselves to, the values are computed based on the concrete formulas and constants that are reported in the respective papers.

Regarding the efficiency of our procedure we observe that we perform a constant amount of oblivious sorts of $\sqrt{N}$ elements per ORAM operation. This can be done with bandwidth overhead of $\mathcal{O}\left(\sqrt{N}\log\sqrt{N}\right)$ data elements with $\mathcal{O}(1)$ temporary storage on the client side [AKS83].

6   Evaluation

To provide a rough idea of the practical performance of our Lookahead ORAM construction, we implemented a prototype with a client-side position map and evaluated it in terms of concrete bandwidth and server-side storage overhead.

We assume that each encrypted block has an extra 40-byte encryption/MAC overhead, and every stash entry has an additional status 20-byte header with status and position information. We assume 4-byte words are also used for denoting the request types.

When encrypting N data blocks of size B each, the server's total storage is $N \times (B + 40)$. The corresponding position map is $8N$, and the stash $2\sqrt{N} \times (B + 20)$ bytes large. Whenever the client performs an operation on the ORAM data structure, it needs to download $40 + (B + 40) \times (\sqrt{N} + 1)$ and upload $80 + (B + 40) \times (\sqrt{N} + 1)$ bytes.

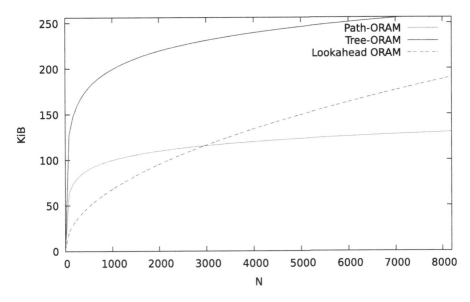

Fig. 8. Comparison of the bandwidth overhead of Lookahead ORAM and Path-ORAM. The x-axis shows different amounts of data blocks. The data block size is fixed to 1024 bytes. The y-axis shows the total amount of transmitted data per access in MiB.

During initialization of our ORAM, we fill the storage with zeros and then fill the stash with pre-selected swap partners. During this initalization, we upload $44 + (B + 40) \times N$ bytes to the server.

All of the above formulas have been calculated based on our theoretical construction and validated empirically using our prototype implementation. In the following, the data block size is fixed to 1024 bytes.

Storage Overhead. We compare our storage overhead on the server-side to the overheads of the most storage efficient related works with sublinear worst-case bandwidth overheads. For our comparison, we measure the total storage requirements on the server-side for varying N. For the related works we compare ourselves to, we computed the data points based on the formulas (including constants) given in the respective papers. For Square-root ORAM we take the storage value of $2N + 4\sqrt{N}$. It should be noted that in [SvS+13] the authors obtain provable security for Path-ORAM with a storage overhead of $20N$, but evaluate their scheme on smaller parameter settings. Since we are interested in provable security and correctness guarantees, we compare our scheme to theirs with a storage overhead of $20N$. Apart from comparing ourselves to the most well known constructions, we also provide a comparison to the parallel buffers construction of Stefanov et al. [SSS12], which is claimed to be practically efficient. We use the $3.2N$ storage estimate for this scheme. The results are depicted in Fig. 7. As expected, assuming a client-side position map, Lookahead ORAM has the smallest storage overhead among all for the tested setting. Storing the position map on the server-side would increase the storage overhead of Lookahead ORAM by a factor of roughly 6.

Bandwidth Overhead. We compare ourselves to Path-ORAM (with a client-side position map), which is known to be the most efficient construction in terms of asymptotic and practical worst-case bandwidth overhead. For our comparison we use their self-reported bandwidth overhead of $10 \log N$. For comparison we also show the Tree ORAM bandwidth overhead of $20 \log N$. We have not included the Parallel Buffers ORAM because all the concrete constants are only provided for specific (and very large) numbers of blocks. We have not included Square-Root ORAM since it is slower than all other depicted ORAMs in terms of bandwidth overhead. The results of our comparison are depicted in Fig. 8. As expected from the asymptotic behaviour of Path-ORAM and Lookahead ORAM, we can see that Path-ORAM becomes more efficient for large values of N. However, for values of $N < 3000$, Lookahead ORAM is more efficient in terms of concrete bandwidth overhead.

Acknowledgements. Michael Raskin was supported by funding from the European Research Council (ERC) under the European Union's Horizon 2020 research and innovation programme under grant agreement No 787367 (PaVeS). Mark Simkin was supported by funding from the European Research Council (ERC) under the European Unions's Horizon 2020 research and innovation program under grant agreement No 669255 (MPCPRO) and No 731583 (SODA).

A   Attack on [GMSS16]

In a work by Gordon et al. [GMSS16] the authors present an ORAM, called M-ORAM. Their construction has a flaw and does not provide obliviousness. In the following, we give a high-level overview of their scheme and sketch our attack that breaks their obliviousness claims.

The construction partitions the server-side storage into a fixed number of rows and a number of columns that depends on the dataset's size. Every cell in their rectangular storage layout holds one data element. Additionally, every row has its own separate constant-sized stash.

Initially, all data elements are present in the storage rectangle in a randomly permuted order and the stashes are empty. Simply speaking an access is performed by accessing one element in each row of their data structure. In one of the rows the desired element is accessed and in all other rows a uniformly random cell is selected. More precisely, the authors claim that to achieve obliviousness not all "dummy" cells are selected uniformly at random, instead some of them are random cells from the previous access. After retrieving one cell from each row, the client shuffles the cells and puts one cell into each stash. The client picks one random block from each stash and sends it back to the server as the new content of the retrieved cells.

Let $x_1, \ldots, x_N$ be some data elements stored in the ORAM data structure, the access sequences $(\mathsf{read}(x_1), \mathsf{read}(x_2), \mathsf{read}(x_1))$ and $(\mathsf{read}(x_1), \mathsf{read}(x_2),$ $\mathsf{read}(x_3))$ can be distinguished with a success probability that is non-negligible in the security parameter. From a high-level perspective, every access selects a subset of cells from the data structure and every two subsets corresponding to two consecutive accesses intersect at some random cells. For three accesses the proposed approach breaks down. Looking at our first access sequence, the proposed construction has a slightly higher bias of the first and third access subset intersecting, since we are accessing the same element.

We have contacted the authors and they have acknowledged our attack.

References

AKL+18. Asharov, G., Komargodski, I., Lin, W.-K., Nayak, K., Peserico, E., Shi, E.: Optorama: Optimal oblivious ram. Cryptology ePrint Archive, Report 2018/892 (2018). https://eprint.iacr.org/2018/892

AKS83. Ajtai, M., Komlós, J., Szemerédi, E.: An 0(n log n) sorting network. In: Proceedings of the Fifteenth Annual ACM Symposium on Theory of Computing, STOC 1983, New York, NY, USA, pp. 1–9. ACM (1983)

AKST14. Apon, D., Katz, J., Shi, E., Thiruvengadam, A.: Verifiable oblivious storage. In: Krawczyk, H. (ed.) PKC 2014. LNCS, vol. 8383, pp. 131–148. Springer, Heidelberg (2014). https://doi.org/10.1007/978-3-642-54631-0_8

Bat68. Batcher, K.E.: Sorting networks and their applications. In: Proceedings of the April 30-May 2, 1968, Spring Joint Computer Conference, AFIPS 1968 (Spring), New York, NY, USA, pp. 307–314. ACM (1968)

BMP11. Boneh, D., Mazieres, D., Popa, R.A.: Making oblivious ram practical, Remote oblivious storage (2011)

BN16. Boyle, E., Naor, M.: Is there an oblivious RAM lower bound? In: Sudan, M. (ed.) ITCS 2016: 7th Conference on Innovations in Theoretical Computer Science, Cambridge, MA, USA, 14–16 January 2016, pp. 357–368. Association for Computing Machinery (2016)

CKN+18. Chan, T.-H.H., Katz, J., Nayak, K., Polychroniadou, A., Shi, E.: More is less: perfectly secure oblivious algorithms in the multi-server setting. In: Peyrin, T., Galbraith, S. (eds.) ASIACRYPT 2018, Part III. LNCS, vol. 11274, pp. 158–188. Springer, Cham (2018). https://doi.org/10.1007/978-3-030-03332-3_7

CNS18. Chan, T.-H.H., Nayak, K., Shi, E.: Perfectly secure oblivious parallel RAM. In: Beimel, A., Dziembowski, S. (eds.) TCC 2018, Part II. LNCS, vol. 11240, pp. 636–668. Springer, Cham (2018). https://doi.org/10.1007/978-3-030-03810-6_23

DDF+16. Devadas, S., van Dijk, M., Fletcher, C.W., Ren, L., Shi, E., Wichs, D.: Onion ORAM: a constant bandwidth blowup oblivious RAM. In: Kushilevitz, E., Malkin, T. (eds.) TCC 2016, Part II. LNCS, vol. 9563, pp. 145–174. Springer, Heidelberg (2016). https://doi.org/10.1007/978-3-662-49099-0_6

DMN11. Damgård, I., Meldgaard, S., Nielsen, J.B.: Perfectly secure oblivious RAM without random oracles. In: Ishai, Y. (ed.) TCC 2011. LNCS, vol. 6597, pp. 144–163. Springer, Heidelberg (2011). https://doi.org/10.1007/978-3-642-19571-6_10

DPP18. Demertzis, I., Papadopoulos, D., Papamanthou, C.: Searchable encryption with optimal locality: achieving sublogarithmic read efficiency. In: Shacham, H., Boldyreva, A. (eds.) CRYPTO 2018, Part I. LNCS, vol. 10991, pp. 371–406. Springer, Cham (2018). https://doi.org/10.1007/978-3-319-96884-1_13

DSS14. Dautrich, J., Stefanov, E., Shi, E.: Burst oram: minimizing oram response times for bursty access patterns. In: 23rd USENIX Security Symposium (USENIX Security 14), pp. 749–764 (2014)

GMOT11. Goodrich, M.T., Mitzenmacher, M., Ohrimenko, O., Tamassia, R.: Oblivious ram simulation with efficient worst-case access overhead. In: Proceedings of the 3rd ACM Workshop on Cloud Computing Security Workshop, pp. 95–100. ACM (2011)

GMSS16. Gordon, S., Miyaji, A., Su, C., Sumongkayothin, K.: M-ORAM: a matrix ORAM with log N bandwidth cost. In: Kim, H., Choi, D. (eds.) WISA 2015. LNCS, vol. 9503, pp. 3–15. Springer, Cham (2016). https://doi.org/10.1007/978-3-319-31875-2_1

GO96. Goldreich, O., Ostrovsky, R.: Software protection and simulation on oblivious rams. J. ACM (JACM) **43**(3), 431–473 (1996)

Gol87. Goldreich, O.: Towards a theory of software protection and simulation by oblivious RAMs. In: Aho, A., (ed.) 19th Annual ACM Symposium on Theory of Computing, New York City, NY, USA, 25–27 May 1987, pp. 182–194. ACM Press (1987)

IKK12. Islam, M.S., Kuzu, M., Kantarcioglu, M.: Access pattern disclosure on searchable encryption: ramification, attack and mitigation. In: ISOC Network and Distributed System Security Symposium - NDSS 2012, San Diego, CA, USA, 5–8 February 2012. The Internet Society (2012)

LN18. Larsen, K.G., Nielsen, J.B.: Yes, there is an oblivious RAM lower bound!. In: Shacham, H., Boldyreva, A. (eds.) CRYPTO 2018, Part II. LNCS, vol. 10992, pp. 523–542. Springer, Cham (2018). https://doi.org/10.1007/978-3-319-96881-0_18

MBC14. Mayberry, T., Blass, E.-O., Chan, A.H.: Efficient private file retrieval by combining ORAM and PIR. In: ISOC Network and Distributed System Security Symposium - NDSS 2014, San Diego, CA, USA, 23–26 February 2014. The Internet Society (2014)

OS97. Ostrovsky, R., Shoup, V.: Private information storage (extended abstract). In: 29th Annual ACM Symposium on Theory of Computing, El Paso, TX, USA, 4–6 May 1997, pp. 294–303. ACM Press (1997)

PPRY18. Patel, S., Persiano, G., Raykova, M., Yeo, K.: PanORAMa: oblivious RAM with logarithmic overhead. In: Thorup, M. (ed.) 59th Annual Symposium on Foundations of Computer Science, Paris, France, 7–9 October 2018, pp. 871–882. IEEE Computer Society Press (2018)

RFK+15. Ren, L., et al.: Constants count: practical improvements to oblivious ram. In: 24th USENIX Security Symposium (USENIX Security 15), Washington, D.C., pp. 415–430. USENIX Association (2015)

SCSL11. Shi, E., Chan, T.-H.H., Stefanov, E., Li, M.: Oblivious RAM with $O((\log N)^3)$ worst-case cost. In: Lee, D.H., Wang, X. (eds.) ASIACRYPT 2011. LNCS, vol. 7073, pp. 197–214. Springer, Heidelberg (2011). https://doi.org/10.1007/978-3-642-25385-0_11

SSS12. Stefanov, E., Shi, E., Song, D.X.: Towards practical oblivious RAM. In: ISOC Network and Distributed System Security Symposium - NDSS 2012, San Diego, CA, USA, February 5–8, 2012. The Internet Society (2012)

SvS+13. Stefanov, E., et al.: Path ORAM: an extremely simple oblivious RAM protocol. In: Sadeghi, A.-R., Gligor, V.D., Yung, M., (eds.) ACM CCS 13: 20th Conference on Computer and Communications Security, Berlin, Germany, 4–8 November 2013, pp. 299–310. ACM Press (2013)

How to Correct
Errors in Multi-server PIR

Kaoru Kurosawa[(⊠)]

Ibaraki University, Hitachi, Japan
kaoru.kurosawa.kk@vc.ibaraki.ac.jp

Abstract. Suppose that there exist a user and ℓ servers $S_1, \ldots, S_\ell$. Each server S_j holds a copy of a database $\mathbf{x} = (x_1, \ldots, x_n) \in \{0, 1\}^n$, and the user holds a secret index $i_0 \in \{1, \ldots, n\}$. A b error correcting ℓ server PIR (Private Information Retrieval) scheme allows a user to retrieve x_{i_0} correctly even if and b or less servers return false answers while each server learns no information on i_0 in the information theoretic sense. Although there exists such a scheme with the total communication cost $O(n^{1/(2k-1)} \times k\ell \log \ell)$ where $k = \ell - 2b$, the decoding algorithm is very inefficient.

In this paper, we show an efficient decoding algorithm for this b error correcting ℓ server PIR scheme. It runs in time $O(\ell^3)$.

Keywords: Private information retrieval · Information theoretic · Error correcting

1 Introduction

Private information retrieval (PIR) was introduced by Chor, Kushilevitz, Goldreich and Sudan [8]. In this model, a server S holds a database $\mathbf{x} = (x_1, \ldots, x_n) \in \{0, 1\}^n$, and a user holds a secret index $i_0 \in \{1, \ldots, n\}$. The user should be able to retrieve x_{i_0} without revealing no information on i_0 to the server S. A trivial solution is that S sends the entire $\mathbf{x}$ to the user. Can the user obtain x_{i_0} with less than n bits of communication?

Unfortunately, Chor et al. [8] showed that n bits are required in the information theoretic setting. (In what follows, we consider information theoretic setting.) To get around this, they considered an ℓ server PIR scheme such that each server S_j has a copy of the database $\mathbf{x}$, where the ℓ servers do not communicate each other. In particular, they showed a two server protocol whose total communication cost is $O(n^{1/3})$.[1] The ℓ server PIR schemes have been improved further by [1,3,4,6,10,12,16,22].

Beimel and Stahl [5] considered what can be done if some of the servers break down. In a (k, ℓ) robust PIR schemes, the user can retrieve x_{i_0} if k out of ℓ servers respond. Woodruff and Yekhanin [21] showed a (k, ℓ) robust PIR scheme whose total communication cost is

$$O(n^{1/(2k-1)} \times k\ell \log \ell).$$

---

[1] I.e., the total number of bits communicated between the user and the servers.

© International Association for Cryptologic Research 2019
S. D. Galbraith and S. Moriai (Eds.): ASIACRYPT 2019, LNCS 11922, pp. 564–574, 2019.
https://doi.org/10.1007/978-3-030-34621-8_20

Currently this is the best known (k, ℓ) robust PIR scheme.

Beimel and Stahl [5] also considered what can be done if some of the servers return false answers. A b-error correcting ℓ server PIR scheme is an (ℓ, ℓ) robust PIR scheme with the additional property such that the user can compute x_{i_0} correctly even if b (or less) servers return false answers. They [5] showed that a (k, ℓ) robust PIR scheme can be used as a b error correcting ℓ server PIR scheme if

$$\ell \geq k + 2b.$$

However, their generic decoding algorithm is very inefficient as they mentioned in [5, page 314].

To summarize, although there exists a b error correcting ℓ server PIR scheme with the total communication cost $O(n^{1/(2k-1)} \times k\ell \log \ell)$ [5,21], where $k = \ell - 2b$, the decoding algorithm [5] is very inefficient.

In this paper, we show an efficient decoding algorithm for the above b error correcting ℓ server PIR scheme. The running time is $O(\ell^3)$. We achieve this by extending Berlekamp-Welch decoding algorithm [23] for Reed-Solomon codes to our problem. While a codeword is defined by using a polynomial $f(x)$ in a Reed-Solomon code, it is defined by using $(f(x), f'(x))$ in the b error correcting ℓ server PIR scheme. This is the difficulty which we must overcome.

A ℓ server PIR scheme is said to be t-private if any coalition of t servers learn no information on i_0. Woodruff and Yekhanin [21] showed a t-private (k, k) robust PIR scheme with the total communication cost $O(n^{\lfloor (2k-1)/t \rfloor} \times k\ell/t \log \ell)$. It is easily generalized to a t-private (k, ℓ) robust PIR scheme, and the latter can be used as a t-private b error correcting ℓ server PIR scheme if $\ell \geq k + 2b$ [5]. Our decoding algorithm can be applied to this scheme too.

1.1 Related Works

In the above model, the user wants one bit. What if the data is partitioned into blocks of m bits each and the user wants an entire block. The user could invoke a PIR scheme m times. Chor et al. [8] showed a more efficient protocol than this. Goldberg [14] and Devet et al. [11] considered b error correcting PIR schemes in this model.

Sun et al. [19,20] and Banawan et al. [2] considered the case where the size of x_i is very large, and hence only the download cost is of interest (but not the upload cost).

In the computational setting, PIR has been studied by [7,9,15,17,18]. [13] is a good survey.

2 Preliminaries

2.1 PIR

In the model of (k, ℓ) robust PIR schemes, there exist ℓ servers $S_1, \ldots, S_\ell$ such that each server S_j has a copy of a database $\mathbf{x} = (x_1, \ldots, x_n) \in \{0, 1\}^n$.

The user should be able to retrieve x_{i_0} if k servers respond while any server S_j should learn no information on i_0 in the information theoretic sense.

Definition 1. A (k, ℓ) robust PIR scheme consists of three algorithms $(\mathcal{Q}, \mathcal{A}, \mathcal{C})$ as follows.

1. The user U runs $\mathcal{Q}(n, i_0)$ to generate ℓ queries $(q_1, \ldots, q_\ell)$ together with an auxiliary information aux.
2. He sends q_j to server S_j for $j = 1, \ldots, \ell$.
3. Each server S_j returns $a_j = \mathcal{R}_0(j, \mathbf{x}, q_j)$ to U, where $\mathbf{x} = (x_1, \ldots, x_n) \in \{0, 1\}^n$ is a copy of a database.
4. Upon receiving (at least) k answers $a_{j_1}, \ldots, a_{j_k}$ from servers, U runs

$$\mathcal{C}((j_1, a_{j_1}), \ldots, (j_k, a_{j_k}), aux)$$

   to compute x_{i_0}. (See step 1 for aux.)

It must satisfy the following requirements.

– Correctness F
  For any n, $\mathbf{x} \in \{0, 1\}^n$, $i_0 \in \{1, \ldots, n\}$ and $\{j_1, \ldots, j_k\} \subset \{1, \ldots, \ell\}$, it holds that

$$\mathcal{C}((j_1, a_{j_1}), \ldots, (j_k, a_{j_k}), aux) = x_{i_0}$$

  if $(q_1, \ldots, q_\ell)$ and $(a_1, \ldots, a_\ell)$ are computed from n, $\mathbf{x} \in \{0, 1\}^n$ and $i_0 \in \{1, \ldots, n\}$.
– Privacy F
  Any server learns no information on i_0. Formally, for any $i_1, i_2 \in \{1, \ldots, n\}$, q_j generated by $\mathcal{Q}(n, i_1)$ and q_j generated by $\mathcal{Q}(n, i_2)$ are identically distributed for $j = 1, \ldots, \ell$.

Definition 2. A b-error correcting ℓ server PIR scheme is an (ℓ, ℓ) robust PIR scheme with the additional property such that the user can compute x_{i_0} correctly even if b (or less) answers among $(a_1, \ldots, a_\ell)$ are false.

Definition 3. The total communication cost of a (k, ℓ) robust PIR scheme is the number of bits communicated between the user U and the ℓ servers $S_1, \ldots, S_\ell$.
　　The total communication cost of a b-error correcting ℓ server PIR scheme is defined similarly.

2.2　Technical Lemma

Woodruff and Yekhanin [21] proved the following lemma.

Lemma 1. *Suppose that (y_i, u_i) are given for $i = 1, \ldots, s$, where $y_i \in \mathbb{F}_p$ and $u_i \in \mathbb{F}_p$. Then there exists at most one polynomial $f(\lambda)$ over $\mathbb{F}_p$ of degree $\leq 2s - 1$ such that $f(i) = y_i$ and $f'(i) = u_i$ for $i = 1, \ldots, s$.*

3   Robust PIR of Woodruff and Yekhanin

Let

$$\mathbf{x} = (x_1, \ldots, x_n) \in \{0,1\}^n$$

be a database. Woodruff and Yekhanin [21] showed a (k, ℓ) robust PIR scheme such that the total communication cost is

$$O(n^{1/(2k-1)} \times k\ell \log \ell).$$

In their (k, k)-robust PIR scheme, the user somehow obtains $(f(i), f'(i))$ from a server S_i for $i = 1, \ldots, k$, where $f(\lambda)$ is a polynomial of degree $2k - 1$ such that $f(0) = x_{i_0}$. He then reconstruct $f(\lambda)$ from

$$(f(1), f'(1)), \ldots, (f(k), f'(k)).$$

3.1 (k, k)-Robust PIR Scheme

For a given (n, k), consider m such that

$$\binom{m}{2k-1} \geq n. \tag{1}$$

There exists such m which also satisfies [21]

$$m = O(kn^{1/(2k-1)}). \tag{2}$$

Then we can consider an injection

$$E : \{1, \ldots, n\} \to \{0,1\}^m$$

such that each $E(i)$ has the Hamming weight $2k - 1$.

Let p be a prime such that $k < p \leq 2k$. For a database $\mathbf{x} = (x_1, \ldots, x_n) \in \{0,1\}^n$, define a function $F : \mathbb{F}_p^m \to \mathbb{F}_p$ by

$$F(z_1, \ldots, z_m) = x_1 \cdot \Big(\prod_{E(1)_j = 1} z_j \Big) + \ldots + x_n \cdot \Big(\prod_{E(n)_j = 1} z_j \Big) \tag{3}$$

where $E(i)_j$ is the jth coordinate of $E(i) \in \{0,1\}^m$.

For example, let $n = m = 4$ and $2k - 1 = 3$. Define E as

$$E(1) = (1,1,1,0), E(2) = (1,1,0,1), E(3) = (1,0,1,1), E(4) = (0,1,1,1).$$

Then

$$F(z_1, \ldots, z_4) = x_1(z_1 z_2 z_3) + x_2(z_1 z_2 z_4) + x_3(z_1 z_3 z_4) + x_4(z_2 z_3 z_4).$$

(A1) The degree of $F(z_1, \ldots, z_m)$ is $2k - 1$ because each $E(i)$ has the Hamming weight $2k - 1$.

(A2) For each i, it holds that $F(E(i)) = x_i$.

Their (k, k)-robust PIR scheme is as follows.

1. The user chooses $\mathbf{V} = (v_1, \ldots, v_m) \in \mathbb{F}_p^m$ randomly.
2. For $i = 1, \ldots, k$, he sends

$$\mathbf{Q}_i = E(i_0) + i \cdot \mathbf{V} \in \mathbb{F}_p^m$$

   to a server S_i, where i_0 is the secret index of the user.
3. For $i = 1, \ldots, k$, S_i returns $y_i \in \mathbb{F}_p$ and $\mathbf{B}_i \in \mathbb{F}_p^m$ such that

$$y_i = F(\mathbf{Q}_i)$$
$$\mathbf{B}_i = (F_{z_1}(\mathbf{Q}_i), \ldots, F_{z_m}(\mathbf{Q}_i))$$

   to the user, where F is defined by Eq. (3) and F_z is the partial derivative of F by z.

Now define

$$f(\lambda) = F(E(i_0) + \lambda \mathbf{V}). \tag{4}$$

Then the degree of $f(\lambda)$ is $2k - 1$ from (A1). Therefore $f(\lambda)$ is written as

$$f(\lambda) = a_0 + a_1 \lambda + \ldots + a_{2k-1} \lambda^{2k-1}. \tag{5}$$

Further it holds that

$$f(i) = y_i, \tag{6}$$
$$f'(i) = \mathbf{B}_i \cdot \mathbf{V}^T \tag{7}$$

for $i = 1, \ldots, k$. (Eq. (7) is obtained by using the chain rule.) The above equations give $2k$ linear equation in $(a_0, \ldots, a_{2k-1})$.

The user computes $(a_0, \ldots, a_{2k-1})$ by solving this set of equations. Finally the user obtains x_{i_0} from

$$x_{i_0} = F(E(i_0)) = f(0) = a_0.$$

See (A2).

(Privacy). For any i, $\mathbf{Q}_i = E(i_0) + i \cdot \mathbf{V}$ is random because $\mathbf{V}$ is randomly chosen. Therefore any sever S_i learns no information on i_0.

(Communication Cost). The user sends $\mathbf{Q}_i \in \mathbb{F}_p^m$ to each sever S_i, and S_i returns $(y_i, \mathbf{B}_i) \in \mathbb{F}_p^{m+1}$. Since $m = O(kn^{1/(2k-1)})$ and $p \le 2k$, the total communication cost is given by

$$O(n^{1/(2k-1)} \times k^2 \log k).$$

3.2 (k, ℓ)-Robust PIR

Let p be a prime such that $\ell < p \le 2\ell$. Then the above scheme is easily generalized to a (k, ℓ)-robust PIR scheme. In steps 2 and 3, just replace "$i = 1, \ldots, k$" with "$i = 1, \ldots, \ell$".

The total communication cost is given by

$$O(n^{1/(2k-1)} \times k\ell \log \ell).$$

4 Error Correcting PIR of Beimel and Stahl

Beimel and Stahl [5] showed that a robust PIR scheme can be used as an error correcting PIR.

Proposition 1. *A (k, ℓ) robust PIR scheme is also a b error correcting ℓ server PIR if*

$$\ell \geq k + 2b.$$

Their generic decoding algorithm is as follows.

1. For each subset B of servers such that $|B| = k$, compute x_{i_0} by running the (k, ℓ) robust PIR scheme.
2. Find the largest A such that for every $B \subset A$ such that $|B| = k$, the user reconstructs the same value of x_{i_0}.
3. Output this value as the value of x_{i_0}.

This algorithm is, however, very inefficient because $\binom{\ell}{k}$ is very large in general, as Beimel and Stahl mentioned in [5, page 314].

From Proposition 1 [5], the (k, ℓ) robust PIR scheme of Woodruff and Yekhanin [21] is also a b error correcting ℓ server PIR scheme if $\ell \geq k + 2b$. However, the decoding algorithm is very inefficient as shown above.

For this b error correcting ℓ server PIR scheme, we can consider a variant of the decoding algorithm as follows.

1. For each subset **BAD** of servers such that $|\mathbf{BAD}| = b$, check if the user reconstructs the same value of x_{i_0} for every $B \subset A \setminus \mathbf{BAD}$ such that $|B| = k$.
2. If the check succeeds, then output this value as the value of x_{i_0}.

Still it is very inefficient because $\binom{\ell}{b}$ is very large in general.

To summarize, although there exists a b error correcting ℓ server PIR scheme with the total communication cost $O(n^{1/(2k-1)} \times k\ell \log \ell)$ [5, 21], where $k = \ell - 2b$, the decoding algorithm [5] is very inefficient.

5 Proposed Decoding Algorithm

In this section, we show an efficient decoding algorithm for the above b error correcting ℓ server PIR scheme. The running time is $O(\ell^3)$.

We achieve this by extending Berlekamp-Welch decoding algorithm [23, 24] for Reed-Solomon codes to our problem. While a codeword is defined by using a polynomial $f(x)$ in a Reed-Solomon code, it is defined by using $(f(x), f'(x))$ in the b error correcting ℓ server PIR scheme. This is the difficulty which we must overcome.

5.1   Berlekamp-Welch Algorithm

Consider a Reed Solomon code of length ℓ with dimension k over $\mathbb{F}_p$. A codeword is given by

$$\mathbf{c} = (f(1), \ldots, f(\ell))$$

for some polynomial $f(\lambda)$ of degree at most $k - 1$. Let

$$\mathbf{r} = (r_1, \ldots, r_\ell)$$

be the received vector which includes at most b errors, where

$$\ell \geq 2b + k. \tag{8}$$

Note that $r_i = f(i)$ if r_i has no error.

Now Berlekamp-Welch decoding algorithm [23] works as follows. Since the number of errors is at most b, there exists a monic polynomial $R_1(\lambda)$ of degree b such that $R_1(i) = 0$ if $r_i \neq f(i)$. Then it holds that

$$R_1(i)f(i) = R_1(i)r_i$$

for $i = 1, \ldots, \ell$. Let $R_0(\lambda) = R_1(\lambda)f(\lambda)$. Then we have

$$R_0(i) = R_1(i)r_i \tag{9}$$

for $i = 1, \ldots, \ell$. $R_0(\lambda)$ has $b + k$ unknown coefficients and $R_1(\lambda)$ has b unknown coefficients. Hence there are $(b+k)+b = k+2b$ unknowns in total. On the other hand, Eq. (9) gives ℓ linear equation in these unknowns.

Therefore we can obtain $R_0(\lambda)$ and $R_1(\lambda)$ by solving this set of linear equations, and can find $f(\lambda) = R_0(\lambda)/R_1(\lambda)$.

5.2   Proposed Decoding Algorithm

We show an efficient decoding algorithm for the b error correcting ℓ server PIR scheme. Fix (b, ℓ) and k such that

$$\ell \geq k + 2b. \tag{10}$$

See Proposition 1 for Eq. (10).

Consider the (k, ℓ) robust PIR scheme of Woodruff and Yekhanin [21]. If all servers are honest, then the user obtains

$$\mathbf{c} = (c_1, \ldots, c_\ell)$$

such that

$$c_i = (f(i), f'(i))$$

for $i = 1, \ldots, \ell$ from Eqs. (6) and (7), where

$$\deg f(\lambda) = 2k - 1. \tag{11}$$

See Sect.3.1.

Suppose that b or less servers return false answers. Then the user obtains

$$\mathbf{c}' = (c_1', \ldots, c_\ell')$$

which includes b or less errors. Let

$$c_i' = (\hat{y}_i, \hat{u}_i)$$

for $i = 1, \ldots, \ell$. Note that

$$(\hat{y}_i, \hat{u}_i) = (f(i), f'(i))$$

if c_i' has no error.

Now consider two polynomials $R_0(\lambda)$ and $R_1(\lambda)$ over $\mathbb{F}_p$ with the following properties:

(P1) $\deg R_0(\lambda) \le 2k - 1 + 2b$.
(P2) $R_1(\lambda)$ is a monic polynomial with $\deg R_1(\lambda) = 2b$.
(P3) $R_0(i) - \hat{y}_i R_1(i) = 0$ for $i = 1, \ldots, \ell$.
(P4) $R_0'(i) - \hat{u}_i R_1(i) - \hat{y}_i R_1'(i) = 0$ for $i = 1, \ldots, \ell$.

Theorem 1. *There exist such polynomials $R_0(\lambda)$ and $R_1(\lambda)$.*

Proof. Define

$$\mathbf{BAD} = \{i \mid (\hat{y}_i, \hat{u}_i) \ne (f(i), f'(i))\}.$$

Then $c = |\mathbf{BAD}| \le b$. Let

$$B(z) = z^{b-c} \prod_{i \in \mathbf{BAD}} (z - i).$$

Let

$$R_1(\lambda) = B(\lambda)^2,$$
$$R_0(\lambda) = f(\lambda)R_1(\lambda) = f(\lambda)B(\lambda)^2.$$

Then it is easy to see that (P1) and (P2) are satisfied. Further

$$\begin{aligned}
R_0(i) - \hat{y}_i R_1(i) &= f(i)B(i)^2 - \hat{y}_i B(i)^2 \\
&= (f(i) - \hat{y}_i)B(i)^2 \\
&= 0
\end{aligned}$$

because $B(i) = 0$ if $f(i) \ne \hat{y}_i$. Also

$$\begin{aligned}
&R_0'(i) - \hat{u}_i R_1(i) - \hat{y}_i R_1'(i) \\
&= f'(i)R_1(i) + f(i)R_1'(i) - \hat{u}_i R_1(i) - \hat{y}_i R_1'(i) \\
&= (f'(i) - \hat{u}_i)R_1(i) + (f(i) - \hat{y}_i)R_1'(i) \\
&= (f'(i) - \hat{u}_i)B(i)^2 + 2(f(i) - \hat{y}_i)B(i)B'(i) \\
&= 0
\end{aligned}$$

because $B(i) = 0$ if $(f(i), f'(i)) \ne (\hat{y}_i, \hat{u}_i)$. Therefore (P3) and (P4) are satisfied. □

Theorem 2. *We can find $R_0(\lambda)$ and $R_1(\lambda)$ which satisfy (P1) $\sim$ (P4) in time $O(\ell^3)$.*

Proof. From (P1) and (P2), the number of unknown coefficients of $R_0(\lambda)$ and $R_1(\lambda)$ are given by

$$2k + 2b + 2b = 2(k + 2b).$$

On the other hand, (P3) and (P4) give

$$2\ell \geq 2(k + 2b)$$

linear equations involving them. (See Eq. (10).) Further there exists a solution for this set of linear equations from Theorem 1. Hence we can find a solution in time $O(\ell^3)$.

Consequently we can find $R_0(\lambda)$ and $R_1(\lambda)$ which satisfy (P1) $\sim$ (P4) in time $O(\ell^3)$.

□

Theorem 3. *It holds that*

$$f(\lambda) = R_0(\lambda)/R_1(\lambda)$$

for any $R_0(\lambda)$ and $R_1(\lambda)$ which satisfy (P1) $\sim$ (P4),

Proof. Let

$$Q(\lambda) = R_0(\lambda) - f(\lambda)R_1(\lambda).$$

Then

$$Q'(\lambda) = R_0'(\lambda) - f'(\lambda)R_1(\lambda) - f(\lambda)R_1'(\lambda).$$

Since there are at most b errors, there exist

$$\ell - b \geq k + 2b - b = k + b(= s)$$

points such that $\hat{y}_i = f(i)$ and $\hat{u}_i = f'(i)$. For these $k + b$ points, we have

$$\begin{aligned} Q(i) &= R_0(i) - f(i)R_1(i) \\ &= R_0(i) - \hat{y}_i R_1(i) \\ &= 0 \end{aligned}$$

and

$$\begin{aligned} Q'(i) &= R_0'(i) - f'(i)R_1(i) - f(i)R_1'(i) \\ &= R_0'(i) - \hat{u}_i R_1(i) - \hat{y}_i R_1'(i) \\ &= 0 \end{aligned}$$

from (P3) and (P4). On the other hand,

$$\begin{aligned} \deg Q(\lambda) &\leq \max(\deg R_0(\lambda), \deg f(\lambda) + \deg R_1(\lambda)) \\ &= 2(k + b) - 1(= 2s - 1) \end{aligned}$$

This means that $Q(\lambda) = 0$ from Lemma 1. Therefore we have $f(\lambda) = R_0(\lambda)/R_1(\lambda)$.

□

Our decoding algorithm of the user is given as follows.

1. The user obtains $(\hat{y}_i, \hat{u}_i)$ from the answer of a server S_i for $i = 1, \ldots, \ell$.
2. He computes two polynomials $R_0(\lambda)$ and $R_1(\lambda)$ which satisfy (P1) $\sim$ (P4) in time $O(\ell^3)$. See Theorem 2.
3. He computes $f(\lambda) = R_0(\lambda)/R_1(\lambda)$. See Theorem 3.
4. Finally he computes $x_{i_0} = f(0)$.

It runs in time $O(\ell^3)$.

6   Extension to t-Private PIR Scheme

A ℓ server PIR scheme is said to be t-private if any coalition of t servers learn no information on i_0. Woodruff and Yekhanin [21] showed a t-private (k, k) robust PIR scheme with the total communication cost $O(n^{\lfloor (2k-1)/t \rfloor} \times k\ell/t \, \log \ell)$ such as follows.

Let $d = \lfloor (2k-1)/t \rfloor$. For a given n, consider m such that

$$\binom{m}{d} \geq n. \tag{12}$$

There exists such m which also satisfies [21]

$$m = O(dn^{1/d}). \tag{13}$$

1. The user chooses $\mathbf{V}_1, \ldots, \mathbf{V}_t \in \mathbb{F}_p^m$ randomly.
2. For $i = 1, \ldots, k$, the user sends

$$Q_i = E(i_0) + i \cdot \mathbf{V}_1 + \ldots + i^t \cdot \mathbf{V}_t$$

to the server S_i.

The rest is the same as in Sect. 3.1. A t-private (k, ℓ) robust PIR scheme is obtained similarly.

Beimel and Stahl [5] showed that a t-private (k, ℓ) robust PIR scheme can be used as a t-private b error correcting ℓ server PIR scheme if $\ell \geq k + 2b$. Now it is easy to see that our decoding algorithm can also be applied to this scheme.

References

1. Ambainis, A.: Upper bound on the communication complexity of private information retrieval. In: Degano, P., Gorrieri, R., Marchetti-Spaccamela, A. (eds.) ICALP 1997. LNCS, vol. 1256, pp. 401–407. Springer, Heidelberg (1997). https://doi.org/10.1007/3-540-63165-8_196
2. Banawan, K., Ulukus, S.: Private information retrieval from Byzantine and colluding databases, Allerton, pp. 1091–1098 (2017)

3. Beimel, A., Ishai, Y.: Information-theoretic private information retrieval: a unified construction. In: Orejas, F., Spirakis, P.G., van Leeuwen, J. (eds.) ICALP 2001. LNCS, vol. 2076, pp. 912–926. Springer, Heidelberg (2001). https://doi.org/10. 1007/3-540-48224-5_74

4. Beimel, A., Ishai, Y., Kushilevitz, E., Raymond, J.F.: Breaking the O(n1/(2k]1)) barrier for information-theoretic private information retrieval. In: FOCS f02, pp. 261–270 (2002)

5. Beimel, A., Stahl, Y.: Robust information-theoretic private information retrieval. J. Cryptol. **20**(3), 295–321 (2007)

6. Chee, Y.M., Feng, T., Ling, S., Wang, H., Zhang, L.F.: Query-efficient locally decodable codes of subexponential length. Comput. Complex. **22**(1), 159–189 (2013)

7. Chor, B., Gilboa, N.: Comput. Private Information Retrieval, STOC (1997)

8. Chor, B., Goldreich, O., Kushilevitz, E., Sudan, M.: Private information retrieval. J. ACM **45**(6), 965–981 (1998)

9. Cachin, C., Micali, S., Stadler, M.: Computationally private information retrieval with polylogarithmic communication. In: Stern, J. (ed.) EUROCRYPT 1999. LNCS, vol. 1592, pp. 402–414. Springer, Heidelberg (1999). https://doi.org/10. 1007/3-540-48910-X_28

10. Dvir, Z., Gopi, S.: 2-server PIR with subpolynomial communication. J. ACM **63**(4), 39 (2016)

11. Devet, C., Goldberg, I., Heninger, N.: Optimally Robust Private Information Retrieval. In: USENIX Security Symposium, pp. 269–283 (2012)

12. Efremenko, K.: 3-query locally decodable codes of subexponential length. SIAM J. Comput. **41**(6), 1694–1703 (2012)

13. Gasarch, W.: A survey on private information retrieval. http://citeseerx.ist.psu. edu/viewdoc/summary?doi=10.1.1.9.8246

14. Goldberg, I.: Improving the robustness of private information retrieval. IEEE Symp. Secur. Priv. **131–148**, 131–148 (2007)

15. Gentry, C., Ramzan, Z.: Single-database private information retrieval with constant communication rate. In: Caires, L., Italiano, G.F., Monteiro, L., Palamidessi, C., Yung, M. (eds.) ICALP 2005. LNCS, vol. 3580, pp. 803–815. Springer, Heidelberg (2005). https://doi.org/10.1007/11523468_65

16. Itoh, T., Suzuki, Y.: Improved constructions for query-efficient locally decodable codes of subexponential length. IEICE Trans. **93**(2), 263–270 (2010)

17. Kushilevitz, E., Ostrovsky, R.: Replication is not needed: single database, computationally private information retrieval. In: FOCS (1997)

18. Lipmaa, H.: An oblivious transfer protocol with log-squared communication. In: Zhou, J., Lopez, J., Deng, R.H., Bao, F. (eds.) ISC 2005. LNCS, vol. 3650, pp. 314–328. Springer, Heidelberg (2005). https://doi.org/10.1007/11556992_23

19. Sun, H., Jafar, S.A.: The capacity of private information retrieval. IEEE Trans. Information Theory **63**(7), 4075–4088 (2017)

20. Sun, H., Jafar, S.A.: The capacity of robust private information retrieval with colluding databases. IEEE Trans. Information Theory **64**(4), 2361–2370 (2018)

21. Woodruff, D., Yekhanin, S.: A geometric approach to information-theoretic private information retrieval. SIAM J. Comput. **37**(4), 1046–1056 (2007)

22. Yekhanin, S.: Towards 3-query locally decodable codes of subexponential length. J. ACM **55**, 1 (2008)

23. Berlekamp-Welch algorithm. https://en.wikipedia.org/wiki/Berlekamp%E2%80 %93Welch_algorithm

24. Lecture 10 Reed Solomon Codes Decoding: Berlekamp-Welch. http://people.ece. umn.edu/~arya/EE5583/lecture10.pdf

Multiparty Computation (2)

UC-Secure Multiparty Computation from One-Way Functions Using Stateless Tokens

Saikrishna Badrinarayanan[1]([✉]), Abhishek Jain[2], Rafail Ostrovsky[1], and Ivan Visconti[3]

[1] UCLA, Los Angeles, USA
{saikrishna,rafail}@cs.ucla.edu
[2] JHU, Baltimore, USA
abhishek@cs.jhu.edu
[3] University of Salerno, Fisciano, Italy
visconti@unisa.it

Abstract. We revisit the problem of universally composable (UC) secure multiparty computation in the stateless hardware token model.
- We construct a three round multi-party computation protocol for general functions based on one-way functions where each party sends two tokens to every other party. Relaxing to the two-party case, we also construct a two round protocol based on one-way functions where each party sends a single token to the other party, and at the end of the protocol, both parties learn the output.
- One of the key components in the above constructions is a new two-round oblivious transfer protocol based on one-way functions using only one token, which can be reused an unbounded polynomial number of times.

All prior constructions required either stronger complexity assumptions, or larger number of rounds, or a larger number of tokens.

Keywords: Secure computation · Hardware tokens.

1 Introduction

Hardware Token Model. The seminal work of Katz [Kat07] initiated the study of Universally Composable (UC) [Can01] protocols using tamper-proof

S. Badrinarayanan—Research supported in part by the IBM PhD Fellowship.

A. Jain—Research supported in part by NSF SaTC grant 1814919 and Darpa Safeware grant W911NF-15-C-0213.

R. Ostrovsky—Research supported in part by NSF-BSF Grant 1619348, DARPA/SPAWAR N66001-15-C-4065, ODNI/IARPA 2019-1902070008 US-Israel BSF grant 2012366, JP Morgan Faculty Award, OKAWA Foundation Research Award, IBM Faculty Research Award, Xerox Faculty Research Award, B. John Garrick Foundation Award, Teradata Research Award, and Lockheed-Martin Corporation Research Award. The views expressed are those of the authors and do not reflect position of the Department of Defense or the U.S. Government.

I. Visconti—Research supported in part by the European Union's Horizon 2020 research and innovation programme under grant agreement No 780477 (project PRIViLEDGE).

© International Association for Cryptologic Research 2019
S. D. Galbraith and S. Moriai (Eds.): ASIACRYPT 2019, LNCS 11922, pp. 577–605, 2019.
https://doi.org/10.1007/978-3-030-34621-8_21

hardware tokens. In this model, each party can create hardware tokens that compute functions of its choice such that an adversary that has access to these tokens does not learn anything more than their input/output behavior. The main appeal of this model is that its security relies on a physical assumption and does not require all the players to trust a common entity. Instead each player can construct its own tokens or rely on its own token manufacturer.

Over the years, two different versions of the hardware token model have been studied: stateful tokens, and stateless (a.k.a. resettable) tokens. The latter model is more realistic, and, in practice, places weaker requirements on the token manufacturer. This makes it appealing from both theoretical and practical viewpoints. In this work, we focus on the *stateless* hardware token model.

Minimizing Complexity. There are three main parameters in the study of UC secure multiparty computation (MPC) in the stateless hardware token model: complexity assumption, number of rounds in the protocol, and the number of tokens. Since the introduction of the stateless hardware token model [CGS08], several works [CGS08, GIS+10, CKS+14, DKMN15a, HPV16] have investigated various trade-offs between these three parameters (see Sect. 1.2 for details). However, in the *multiparty* setting, the best known protocols based on the minimal assumption of one-way functions[1] require $O(d)$ rounds [HPV16], where d is the depth of the circuit being computed. This leaves open the following question (w.r.t. any polynomial number of tokens):

Does there exist a constant round UC secure multiparty computation protocol for general functions based on one-way functions in the stateless hardware token model?

Token Reusability. Since tamper-proof hardware tokens can be expensive to manufacture, it is very desirable to allow *reuse* of tokens across multiple sessions. Indeed, for this reason, the reusable token model was put forth by [CKS+14], where a set of tokens can be reused across multiple protocol executions (for different function evaluations, on different set of inputs) between the same set of parties, a.k.a. concurrent *self composition*. While the ability to reuse a setup for concurrent self composition typically comes for free in setup models such as the common reference string model, it is not the case for the hardware token model. As such, it was put forth as an explicit goal by [CKS+14].

In the setting of *two-party* computation, [HPV16] constructed round-optimal (i.e., two-round) protocols based on one-way functions, with unlimited token reusability (even in the stronger Global UC model [CDPW07, CJS14]). However, their protocol requires a polynomial number of tokens. The concurrent work of [DKMN15a] requires only one token, but does not support unlimited token reuse. This leaves open the following question:

---

[1] One-way function is a necessary assumption in the stateless hardware token model since an unbounded adversary can simply "learn" a stateless token [GIS+10].

Does there exist a two round UC secure two-party computation protocol for general functions based on one-way functions in the reusable stateless token model?

In this work, we resolve both of the aforementioned questions in the affirmative.

1.1 Our Results

We continue the study of UC secure computation in the stateless hardware token model.

Multiparty Computation. Our first result is a three-round UC-secure multiparty computation protocol based on the minimal assumption of one-way functions. Our protocol requires each party to send two tokens to every other party.

Theorem 1 (Informal). *Assuming one-way functions, there exists a three-round UC multiparty computation protocol in the stateless hardware token model.*

If we restrict our attention to the case of two parties, where the parties communicate over simultaneous broadcast channels[2], we can further reduce the round-complexity of our protocol to *two-rounds*. We emphasize that at the end of the protocol, *both* parties learn the output. Our protocol requires each party to send only one token to the other party. Prior to our work, no such two-party computation protocol was known in the literature.

Theorem 2 (Informal). *Assuming one-way functions, there exists a two-round UC two-party computation protocol over simultaneous broadcast channels, in the stateless hardware token model.*

We emphasize that the protocols in Theorems 1 and 2 allow for unlimited token reuse across multiple sessions between the same set of parties.

Oblivious Transfer. A key component in our constructions is a new two-round UC oblivious transfer protocol in the stateless hardware token model based on one-way functions, and relying upon a single token. Crucially, unlike [DKMN15a] who support an a priori bounded number of uses of the token, our protocol supports *unlimited* token reuse.

Theorem 3 (Informal). *Assuming one-way functions, there exists a two-round UC oblivious transfer protocol in the reusable stateless hardware token model, using a single token.*

---

[2] This is the standard model for multiparty computation, where in each round, every party simultaneously broadcasts a message to the other parties. However, a rushing adversary may wait to receive the honest party's message in any round before deciding its own message.

By combining the above theorem with the work of Ishai et al. [IKO+11], we can obtain a two-round secure two-party computation protocol in the unidirectional-message model based on one-way functions with one reusable token. Unlike Theorem 2, however, only one of the two parties learns the output at the end of the protocol.

Discussion and Future Work. The work of Hazay et al. [HPV16] puts forth GUC security as a more desirable notion of security for protocols in the hardware token model. Our protocols do not achieve GUC security, and it is an interesting open problem to extend our results to the stronger model of [HPV16].

Further, unlike the work of [HPV16], who construct black-box protocols, our protocols make non-black-box use of one-way functions due to the use of ZK arguments. It is an interesting open problem to construct optimal black-box protocols in the hardware token model.

1.2   Related Work

Katz established the first feasibility results for UC secure multiparty computation (MPC) using stateful hardware tokens. Subsequently, this model was extensively explored in several directions with the purpose of improving upon the complexity assumptions, round-complexity of protocols and the number of required tokens [MS08, GKR08, Kol10, DKM11, DKM12].

The study of UC-secure protocols in the stateless hardware token model was initiated by Chandran et al. [CGS08]. They constructed a polynomial round protocol for multi-commitment functionality where each party exchanges one token with the other party, based on enhanced trapdoor permutations. Subsequent to their work, Goyal et al. [GIS+10] constructed constant-round protocols assuming collision-resistant hash functions (CRHFs). However, these improvements were achieved at the cost of requiring a polynomial number of tokens. Choi et al. [CKS+14] subsequently improved upon their result by decreasing the number of required tokens to only one, while still using only constant rounds and CRHFs.

Recently, Hazay et al. [HPV16] constructed two rounds two-party computation protocols based on one-way functions, and three-round MPC protocols based on oblivious transfer in the Global UC model. They also construct a multiparty protocol from one-way functions where the round complexity is linear in the depth of the circuit being computed. All of their protocols require a polynomial number of tokens. In a concurrent work, Döttling et al. [DKMN15a] construct two-round oblivious transfer from stateless tokens based on one-way functions, but their protocol does not support unbounded token reuse. Badrinarayanan et al. [BJOV18] constructed a non-interactive UC-secure two party computation protocol in the stateless hardware token model based on one way functions.

Döttling et al. [DKMN15b] construct information-theoretic UC-secure protocols in a model where the tokens can be reset only a bounded number of times. In a different work, Döttling et al. [DMMN13] construct UC-secure protocols for resettable functionalities using stateless tokens. In contrast, we focus on securely computing general functionalities using stateless tokens in this work.

The work of Agrawal et al. [AAG+14] proves lower bounds on the number of token queries necessary for secure computation in the stateless hardware token model. We do not seek to optimize the query complexity of tokens in this work.

Niles [Nil15] and Mechler et al. [MMN16] provide a new formulation for tamper-proof hardware tokens that can be reused across different protocol executions. Their security definition is different from GUC security studied in [HPV16].

Recently, Hazay et al. [HPV17] constructed constant round adaptively secure protocols in the stateless token model. In this work, we focus on static corruptions.

2   Technical Overview

We first describe the techniques used in our new two round oblivious transfer protocol in the next subsection. In the subsequent subsection, we describe the techniques for the two party computation protocol. We then build upon these techniques to construct the MPC protocol and discuss this in the final subsection.

2.1   Two-Round Oblivious Transfer (OT)

We design a new two-round OT protocol based on one-way functions where the sender $\mathcal{S}$ sends a single token $\mathbf{T}$ to the receiver $\mathcal{R}$. Our protocol combines multiple ideas from prior works to address some standard issues that arise when dealing with *stateless* tokens, together with our new ideas for improving upon the parameters achieved in prior works. Below, we discuss our approach for the case where the token is only used for a single execution. However, our approach easily extends to allow for resuability of token.

Our starting approach is to divide the computation into two parts: in the first part, the receiver $\mathcal{R}$ performs a *random* OT execution with the token $\mathbf{T}$. In the second part, $\mathcal{R}$ interacts with the sender $\mathcal{S}$ to perform standard OT using the random OT instance. In more detail, the sender embeds two random strings (r_0, r_1) in the token $\mathbf{T}$ and sends it to $\mathcal{R}$. The receiver secret-shares its input bit b into two parts (s, z) s.t. $\mathsf{s} \oplus \mathsf{z} = \mathsf{b}$, and then uses z to learn $\mathsf{r_z}$ from $\mathbf{T}$. At the same time, $\mathcal{R}$ sends $\mathsf{s} = \mathsf{b} \oplus \mathsf{z}$ to the sender $\mathcal{S}$ to obtain (M_0, M_1) s.t. $M_0 = (\mathsf{m_0} \oplus \mathsf{r_0})$, $M_1 = (\mathsf{m_1} \oplus \mathsf{r_1})$ if $\mathsf{s} = 0$ and $M_0 = (\mathsf{m_0} \oplus \mathsf{r_1})$, $M_1 = (\mathsf{m_1} \oplus \mathsf{r_0})$ otherwise. Using the mask $\mathsf{r_z}$ learned from the token, $\mathcal{R}$ appropriately unmasks one of the two values (M_0, M_1) to learn $\mathsf{m_b}$.

The immediate problem with this naive approach is that an adversarial receiver can simply reset the stateless token and run it two times on different inputs to learn both $\mathsf{r_0}$ and $\mathsf{r_1}$. Using these masks, the receiver can then recover both $(\mathsf{m_0}, \mathsf{m_1})$, completely breaking the security.

We address this issue in a similar manner as many prior works such as [GIS+10, CKS+14, DKMN15a]. The basic idea is to require $\mathcal{S}$ to *authenticate* $\mathcal{R}$'s query to the token. Namely, $\mathcal{R}$ commits to its query z and then obtains a signature σ on the commitment from $\mathcal{S}$.[3] In order to query $\mathbf{T}$ on z, the receiver must provide σ and the appropriate decommitment information. The unforgeability of the signature scheme ensures that an adversarial receiver cannot query the token more than once.

Input-Dependent Aborts. Unfortunately, this modification introduces a subtle problem: a malicious sender can subliminally communicate s to the token by embedding bit s into the signature value σ. This allows the token to learn the receiver's input bit b. It can now decide whether or not to abort based on this input bit, which effectively signals the bit b back to the sender, breaking the security of the protocol.

Similar to [DKMN15a], we address this problem by hiding the signature from the token. Specifically, instead of sending σ to $\mathbf{T}$, $\mathcal{R}$ proves knowledge of σ via a zero-knowledge argument of knowledge. Since $\mathbf{T}$ is stateless, we require this argument of knowledge to be resettably sound [BGGL01]. Recent works have constructed such argument systems based on one-way functions [CPS13, BP13, COPV13, COP+14, BP15, CPS16].

While this modification prevents subliminal communication from the sender to the token, unfortunately, the protocol still remains susceptible to input-dependent aborts. In particular, an adversarial token can simply decide to abort or not based on the random bit z. This effectively signals z back to the sender, who combines it with s to learn the receiver's input b.

A natural idea to prevent such an attack is to secret-share z into two parts and query the token on each part separately. The hope here would be that the adversarial token can only signal back one of the two secret shares of z back to the sender, which does not suffice for learning b. Unfortunately, this idea immediately fails since an adversarial token may be *stateful*, and therefore have a joint view of all the queries made by the receiver.

Leakage-Resilient Secret-Sharing. Our first step to address the problem of input-dependent aborts is to employ leakage-resilient secret-sharing sharing. Roughly, $\mathcal{R}$ secret-shares its input b into $2n$ random bits $\mathsf{b}_1, \ldots, \mathsf{b}_{2n}$ s.t. b is the inner product of $(\mathsf{b}_1, \ldots, \mathsf{b}_n)$ with $(\mathsf{b}_{n+1}, \ldots, \mathsf{b}_{2n})$. Each bit b_i is further secret shared into $(\mathsf{s}_i, \mathsf{z}_i)$ s.t. $\mathsf{b}_i = (\mathsf{s}_i \oplus \mathsf{z}_i)$.

The main idea is that due to the leakage-resilient properties of inner product, even given all of the bits $(\mathsf{z}_1, \ldots, \mathsf{z}_{2n})$, an adversarial token cannot signal back any one bit of information to $\mathcal{S}$ that is sufficient for learning b.

Unfortunately, however, it is not immediately clear how to integrate the above secret-sharing scheme with the rest of the protocol. In particular, while our strategy of performing OT via a random OT is compatible with the XOR-based secret sharing, it does not seem to be compatible with inner-product based secret sharing.

---

[3] For simplicity, here we assume a non-interactive commitment scheme. In order to use a two-round commitment scheme based on one-way functions, we use the token $\mathbf{T}$ to generate the first commitment message.

Delegation of Computation. Towards building a solution, let us first assume that we have a trusted third party that computes the following function G: it takes as input the bits $b_1, \ldots, b_{2n}$, recomputes b and then outputs one of the two hardcoded values (m_0, m_1), depending upon b.

Clearly, given access to such a third party, performing OT is straightforward. Our main idea is to implement such a party via *garbled circuits*. Namely, we augment the functionality of the token $\mathbf{T}$ to compute a garbled circuit for G and send it to the receiver $\mathcal{R}$ so that it can evaluate it on its own. In other words, the token delegates the computation of G to $\mathcal{R}$.

Note, however, that to evaluate the garbled circuit, the receiver needs to obtain input wire labels via OT. Thus, it may seem that we come back in a full circle and not made any progress at all.

The crucial observation is that the input wire labels for the garbled circuit can be obtained in the same manner as earlier, without leaking any information about b. In particular, the receiver uses the same process as described earlier to obtain one of the two wire labels for each bit b_i. Namely, it first queries the token on a random bit z_i to learn a random mask r_i. At the same time, it obtains the masked input wire labels for the ith input from $\mathcal{S}$. It then uses the mask r_i to recover the wire label corresponding to bit b_i. This process is repeated in parallel for every position i.

Since the garbled circuit gets a full view of the input of the receiver, we require the token to prove its well-formedness via a resettable zero-knowledge argument of knowledge [CGGM00, COPV13]. This ensures that the garbled circuit cannot do an input-dependent abort and signal the bit b back to $\mathcal{S}$. Note that a similar proof could not have been given by the sender $\mathcal{S}$ about the physical token.

Trapdoor Mechanism. A crucial issue that arises while proving UC security of the protocol is the following: when $\mathcal{R}$ is corrupted, the proof of well-formedness of the garbled circuit given by the token $\mathbf{T}$ must be simulated in order to "force" the correct output. However, the UC simulator cannot rewind the adversary, nor does it have access to its code! To get around this issue, we implement the following trapdoor generation mechanism that allows the simulator to recover a trapdoor that can then be used to perform straight-line simulation. In the first round, along with the other messages, $\mathcal{R}$ also sends a random string x. The sender $\mathcal{S}$ then sends $y = \mathrm{OWF}(x)$, where OWF is a one-way function, and a signature on y along with the other messages in the second round. Upon being queried with y, the token proves, via a resettable witness-indistinguishable argument of knowledge[4] (RWIAOK), that either the garbled circuit is well-formed or it knows an inverse of y. Here, we crucially rely on the asymmetry between the simulator and the adversary: since an honest sender's token is implemented by the simulator, it already knows the trapdoor x that was sent by the adversarial receiver during the protocol. In contrast, once an adversarial sender learns x in

---

[4] Such argument systems can be constructed from one-way functions [COPV13].

the protocol execution, it has no way of signaling it to its token. At this juncture, we remark that this trapdoor mechanism is crucially used in all our other secure computation protocols as well.

We refer the reader to the technical sections for further details about our OT protocol.

2.2 Two-Party Computation

A Cloning Strategy. Consider parties P_1 and P_2 with inputs x_1 and x_2 respectively who wish to securely evaluate a function f in two rounds such that both parties learn the output. The main idea at the heart of our protocol is the following: instead of running a two-party computation protocol between remote players that would require several rounds of interaction, we ask a player to construct a clone of itself in the form of a stateless token that can then be remotely activated by its creator to perform the actual two-party computation. We explain this in more detail below.

P_1 creates a clone of itself in the form of a stateless token T_1 and sends it to P_2. P_2 does the same thing by sending a token T_2 to P_1. Then, P_2 can simply execute a secure two-party computation protocol Π for f *locally* with T_1, while P_1 can do the same with T_2. An immediate problem with this idea is that since the tokens are stateless, an adversarial P_2, for example, can simply reset the token T_1 during the execution of Π, which may completely break its security.

Input Authentication. We solve this issue by allowing the sender to remotely activate the token only for one input of the other party, in a similar manner as in our OT protocol. We describe the strategy for P_1. Upon being activated, P_2's token first outputs a commitment to P_2's input, and proves the knowledge of the committed value using a resettable zero-knowledge argument of knowledge (RZKAOK). Party P_1 then signs this commitment and sends it to P_2. In the course of the two-party computation with token T_1, P_2 proves that its behavior is correct with respect to the input inside the signed commitment.

Implementing Two-Party Computation. We implement the actual two-party computation between the token T_1 and P_2 via garbled circuits and OT. (The two-party computation between T_2 and P_1 is implemented in a symmetric manner.) In more detail, the token T_1 prepares and outputs a garbled circuit for the functionality $f(x_1, \cdot)$ and proves its well-formedness via a RZKAOK. In order to evaluate this garbled circuit, P_2 needs the wire labels corresponding to its input x_2, which in turn requires the use of OT.

Instead of performing OTs with the token T_1, P_2 runs multiple parallel executions of OT with P_1, where P_2 plays the receiver and P_1 plays the sender. The role of the OT token in this protocol is played by T_1. By using our new two-round OT protocol, we are able to ensure that the communication between P_1 and P_2 only requires two rounds. Barring several details concerning the security proof (see below), this already yields a two-round two-party computation protocol.

We now briefly discuss some of the steps in the security proofs of the above protocol. Consider a malicious P_2^*. The simulator can extract its input using the non-black-box extractor of RZKAOK given by token T_2. Here, the extractor requires the code of the token T_2 which is not an issue since T_2 is disconnected from its creator P_2^* and the environment. Once it obtains the output y_2 from the ideal functionality, it can simulate the two-party computation protocol between token T_1 and P_2^*. Here, we will need to simulate the proof of the well-formedness of the garbled circuit and we will rely on the trapdoor generation mechanism used in the OT protocol to achieve this task.

Simulating Input Commitment. Another issue that arises is how does the simulator generate the proofs for the input commitment? That is, in the ideal world, consider the setting where the party P_1 is corrupted. Now, the simulator's token T_2 on behalf of honest party P_2, while interacting with P_1 will have to prove via a RZKAOK that it indeed knows the honest party's input inside the commitment. In the ideal world, P_2 does not know the honest party's input and so the commitment will be just to some random string. However, we can not simulate the RZKAOK argument given by the token as we don't have the code of the environment that it is interacting with in the setting of UC security. Neither can we use the trapdoor mechanism as the trapdoor is generated only much later after interacting with the adversary's token. We overcome this issue by noting that we actually don't need the full power of zero knowledge here and instead, all we require is a resettable strong witness indistinguishable argument of knowledge. That is, we just require that the input commitment being used is changed honestly to an indistinguishable one (a commitment to a random string) and simultaneously can change the proof to prove knowledge of this committed value. As a result, we do the following: in the reduction, we first simulate the RZKAOK argument. Here, we crucially use the fact that this happens only inside the security reduction and the *final UC simulator does not need the environment's code*. We then switch the input commitment to be a commitment of a random string by relying on the hiding property and finally, switch the proof back to honestly prove knowledge of the committed value.

The above discussion ignores several subtleties that arise in the proof. A more detailed explanation of our protocol and proof can be found in Sect. 5.

2.3   Multiparty Computation

We now describe the techniques used in our MPC protocol. At a high level, we follow the same recipe as in the two party case: that is, each party creates a clone of itself that can then be remotely activated by the creator to perform the actual MPC. As in the two party setting, we will also use the trapdoor mechanism described in the OT section to help simulate the resettable arguments in the security proof. However, there are several additional challenges that arise in the context of MPC and we describe them below.

First, lets describe the approach in more detail. Consider a set of n parties $P_1, \ldots, P_n$. Now, in order to offload the heavy computation of an actual MPC protocol to remote players, we would require each party P_i to interact with a set of $(n-1)$ tokens - one each from every other party, in an actual MPC protocol. Unlike the two party setting where we essentially performed a two party computation between a party and a token, here, the tokens can not talk to each other! Therefore, each party P_i has to facilitate as the channel through which the messages are exchanged between all these tokens taking party in the MPC.

The next question is what sort of MPC protocol do we run amongst P_i and the $(n-1)$ tokens? Recall that our goal is to base the security of our entire protocol only on the existence of one way functions. In the two party setting, we overcame this issue by running a semi-honest two party protocol based on garbled circuits and oblivious transfer (OT) and composing it with appropriately resettable arguments. We then used our new two round OT protocol to compute the OTs required by the semi-honest construction. Taking a similar approach, we would need to run a semi-honest secure MPC protocol that can be based on just OT and one way functions. While there are several such protocols in literature, a crucial issue that arises is that we would need to instantiate it with an MPC protocol where all the OT executions can be made in parallel once before the execution of the rest of the protocol. We know of protocols in the OT hybrid model [Bea96, Kil88, IPS08, IKO+11] assuming just one way functions that satisfy this structure and we use such an MPC protocol and use our two round OT protocol to run the OT executions. As in the two party setting, we perform the input authentication and trapdoor generation before running the MPC protocol and this help facilitates the proof.

Extra Round. The description so far seems to suggest that the protocol runs in only two rounds. However, unlike the two party setting, we need an extra round for the following reason. Let's recall how the actual MPC is computed. Consider party P_i. In order to run the underlying MPC protocol, the $(n-1)$ tokens in possession of P_i do act as the OT receiver in some executions of the initial parallel OT calls. As a result, the tokens need to know the output of the OT invocations before proceeding with the rest of the computation. However, its not at all clear how to deliver this output to the tokens. To illustrate the issue more clearly with an example, consider two tokens T_1 and T_2 in the presence of party P_i. Let's suppose that in some OT invocation, T_1 is the sender and T_2 is the receiver. Now, clearly, the OT has to be performed "externally" via their token creators as the respective sender and receiver and not amongst the tokens themselves using P_i as the channel because our OT protocol is not resettably secure. Therefore, lets suppose we perform the OT protocol amongst their respective creators P_1 and P_2. At the end of this OT, the party P_2 only learns the output. However, we need to transmit this to its token T_2. To solve this, in the third round, we have P_2 send the OT outputs in an encrypted (and signed) form which is then relayed to T_2 via the party P_i.

At this point, we stop and reflect why this was not an issue in the two party setting. There, recall that the only OT to be performed involved the party P_i as the receiver and the corresponding token (of the other party) as the sender. Therefore, by just running the two round OT protocol, P_i learns the output and we avoid this extra round.

Finally, to ease the exposition and simplify the proof, unlike in the two party setting, we treat the token that computes the MPC different from the one that takes part in the OT protocol. Hence, we require every party to send two tokens to every other party. We refer the reader to the technical section for more details.

3 Preliminaries

UC Secure Computation. The UC framework, introduced by [Can01] offers advanced security guarantees since it deals with the security of protocols that may be arbitrarily composed. We include the formal definitions in the full version.

OT. Ideal 2-choose-1 oblivious transfer (OT) is a two-party functionality that takes two inputs m_0, m_1 from a sender and a bit b from a receiver. It outputs m_b to the receiver and $\perp$ to the sender. We use $\mathcal{F}_{ot}$ to denote this functionality. The ideal oblivious transfer(OT) functionality $\mathcal{F}_{ot}$ is formally defined in the supplementary material. Given UC oblivious transfer, it is possible to obtain UC secure two-party computation of any functionality [IPS08, IKO+11].

Token Functionality. We model a tamper-proof hardware token as an ideal functionality $\mathcal{F}_{WRAP}$ in the UC framework, following Katz [Kat07]. A formal definition of this functionality can be found in the full version. Note that our ideal functionality models stateful tokens. Although all our protocols use stateless tokens, an adversarially generated token may be stateful.

Cryptographic Primitives. We use the following primitives all of which can be constructed from one way functions: pseudorandom functions, digital signatures, commitments, garbled circuits [GGM86, Yao86, Rom90, Nao91]. Additionally, we use the following advanced primitives recently constructed based on one way functions: resettable zero knowledge argument of knowledge, resettably sound zero knowledge argument of knowledge, resettable witness indistinguishable argument of knowledge and resettably sound witness indistinguishable argument of knowledge [CGGM00, BGGL01, CPS13, BP13, COPV13, COP+14, BP15, CPS16].

Interactive Proofs for a "Stateless" Player. We consider the notion of an interactive proof system for a "stateless" prover/verifier. By "stateless", we mean that the verifier has no extra memory that can be used to remember the transcript of the proof so far. Consider a stateless verifier. To get around the issue of not knowing the transcript, the verifier signs the transcript at each step and sends it back to the prover. In the next round, the prover is required to send this signed transcript back to the verifier and the verifier first checks the signature and then uses the transcript to continue with the protocol execution. Without loss of generality, we can also include the statement to be proved as part of the transcript. It is easy to see that such a scenario arises in our setting if the stateless token acts as the verifier in an interactive proof with another party.

4   Oblivious Transfer

In this section, we construct a two round UC oblivious transfer protocol with unbounded reusability based on one-way functions using only one stateless hardware token. The token is sent by the OT sender to the OT receiver in an initial token transfer phase.

We first describe our protocol for the case where the token sent by the OT sender can only be used for a single OT protocol execution. We then describe a modification to make the token *reusable*, such that it can be used to perform an unbounded polynomial number of OT executions between the same pair of parties, with different pairs of inputs.

Formally, we show the following theorem:

Theorem 4. *Assuming one-way functions exist, there exists a two round UC secure unbounded OT protocol in the stateless hardware token model.*

Combining this with the result of Ishai et al. [IKO+11], we obtain the following corollary:

Corollary 5. *Assuming one-way functions exist, there exists a two round UC secure two-party computation protocol using one stateless hardware token where only one party learns the output.*

4.1   Overview

Consider a sender S with inputs (m_0, m_1) and a receiver R with choice bit b who wish to run an OT protocol.

Token Transfer Phase. Initially, as part of the token transfer phase, S creates a stateless token T that has a prf key k_S and a signing key and verification key pair (sk, vk) for a signature scheme hardwired into it. Additionally, S chooses two random strings (r_0, r_1) and creates a circuit C that, given input $b_1, \ldots, b_{2n}$, outputs r_b where $b = \langle (b_1, \ldots, b_n), (b_{n+1}, \ldots, b_{2n}) \rangle$. (Here, $\langle x, y \rangle$ denotes the inner product of x and y.) The sender creates a garbled version of this circuit $\tilde{C}$ and hardwires it into the token, together with the randomness used to create the garbled circuit. S sends the token T to R.

Round 1. R picks a key k_R for a pseudorandom function and sends c which is a commitment to this key. Also, R picks $2n$ bits $b_1, \ldots, b_{2n}$ uniformly at random such that $\langle B_1, B_2 \rangle = b$ where $B_1 = (b_1, \ldots, b_n)$ and $B_2 = (b_{n+1}, \ldots, b_{2n})$. Then, for each $i \in [2n]$, $\mathcal{R}$ sends $s_i = (b_i \oplus z_i)$ where $z_i = \mathsf{PRF}(k_\mathcal{R}, i)$.

Round 2. $\mathcal{S}$ computes a signature $\sigma = \mathsf{Sign}_{sk}(c)$. Also, for each $i \in [2n]$, $\mathcal{S}$ computes $A_{i,0} = \mathsf{PRF}(k_\mathcal{S}, i, 0)$ and $A_{i,1} = \mathsf{PRF}(k_\mathcal{S}, i, 1)$. Looking ahead, $A_{i,0}$ and $A_{i,1}$ will be the token's output when queried with $z_i = 0$ or $z_i = 1$ respectively. Let the pair of labels for the ith input wire to the garbled circuit be $L_{i,0}$ and $L_{i,1}$. If $s_i = 0$, $\mathcal{S}$ computes $Z_{i,0} = (L_{i,0} \oplus A_{i,0})$ and $Z_{i,1} = (L_{i,1} \oplus A_{i,1})$. On the other hand, if $s_i = 1$, $\mathcal{S}$ computes $Z_{i,0} = (L_{i,1} \oplus A_{i,0})$ and $Z_{i,1} = (L_{i,0} \oplus A_{i,1})$. Also, $\mathcal{S}$ computes $\alpha_0 = (m_0 \oplus r_0)$ and $\alpha_1 = (m_1 \oplus r_1)$. $\mathcal{S}$ sends $(Z_{i,0}, Z_{i,1})$ for each $i \in [2n]$ along with $(\alpha_0, \alpha_1, \sigma)$.

Output Computation. First, $\mathcal{R}$ aborts if $\mathsf{Verify}_{vk}(c, \sigma) = 0$. Now, for each $i \in [2n]$, $\mathcal{R}$ queries the token $\mathbf{T}$ using input (z_i, c, i) along with a resettably sound zero-knowledge argument of knowledge (RSZKAOK) for the following NP statement:

$$\text{There exists } (k_\mathcal{R}, \sigma) \text{ such that } c = \mathsf{Commit}(k_\mathcal{R}), \mathsf{Verify}_{vk}(c, \sigma) = 1 \text{ and}$$
$$z_i = \mathsf{PRF}(k_\mathcal{R}, i).$$

$\mathbf{T}$ first verifies the proof. It aborts if the proof doesn't verify. Then, $\mathbf{T}$ outputs $A_{i,z_i} = \mathsf{PRF}(k_\mathcal{S}, i, z_i)$ and $\sigma_i = \mathsf{Sign}_{sk}(i)$. Now, for each i, $\mathcal{R}$ computes the label value as $L_{i,b_i} = Z_{i,z_i} \oplus A_{i,z_i}$.

After this, $\mathcal{R}$ queries the token with the $2n$ signatures - $\sigma_1, \ldots, \sigma_{2n}$ and receives a garbled circuit $\tilde{C}$ in response along with a resettable zero knowledge (RZK) argument that it was generated correctly. In order to facilitate simulation of this proof, we actually implement it via a resettable witness indistinguishable argument of knowledge (RWIAOK) which can be proven by using a "trapdoor witness" that is generated as follows: $\mathcal{R}$, in the first round of the protocol, picks a random $\bar{x}$ and sends it to $\mathcal{S}$. In the second round, $\mathcal{S}$ computes $y = \mathsf{OWF}(\bar{x})$ and a signature $\sigma_y = \mathsf{Sign}_{sk}(y)$. Now, when $\mathcal{R}$ queries the token to get the garbled circuit, he also sends y and gives a RSZKAOK that he knows a signature on y with respect to the verification key vk. The token, via the RWIAOK proves that either the garbled circuit was correctly generated or that it knows a pre-image of y. Using the corresponding label values, $\mathcal{R}$ evaluates the garbled circuit to recover its output r_b. $\mathcal{R}$ then uses this value along with α_b to recover m_b.

The correctness of the protocol follows by inspection. Below, we provide a brief overview of the security proofs against malicious receivers and malicious senders.

Security Against a Malicious Receiver. Consider a malicious receiver $\mathcal{R}^*$. For each i, let's suppose $\mathcal{R}^*$ queries $\mathbf{T}$ with (z_i, c, i) and a valid RSZKAOK argument. First, from the security of the pseudorandom function, observe that the output of $\mathbf{T}$ for a query containing index i' gives no information at all about its output for index $i \neq i'$. Therefore, we now need to argue that $\mathcal{R}^*$ can not query the token with $(1 - z_i, c', i)$ and receive a valid output. If $\mathcal{R}^*$ produces a different (c') that would break the security of the signature scheme. Fixing $(c') = (c)$, we observe that, from the statistical binding property of the commitment, there is a unique $k_{\mathcal{R}}$ and hence a unique value of $z_i = \mathsf{PRF}(k_{\mathcal{R}}, i)$. Therefore, if $\mathcal{R}^*$ produces a valid argument for $z_i' \neq z_i$, then that would violate the soundness of the RSZKAOK system.

The simulation strategy in the ideal world is as follows: the simulator first retrieves all the z_i values by observing the queries to **T**. It then extracts the receiver's input b from the set of z_i and s_i values. The simulator forwards b to the ideal OT functionality to receive m_b. It then computes a simulated garbled circuit as output of the token. Note that by using additional signatures on each output of the token, we force the receiver to query for the garbled circuit from the token only after it gets all the label keys and messages from the sender. This ensures that the simulator has enough time to extract the adversary's input and produce a simulated garbled circuit. Further, the simulator observes the query $\bar{x}$ from the receiver in the first round and uses that as the witness in the RWIAOK given by the token.

Security Against a Malicious Sender. To prove security against a malicious sender $\mathcal{S}^*$, the simulator, which receives the token's code M from the ideal functionality when the token is created, runs the code M on both $z_i = 0$ and $z_i = 1$ for every i by producing simulated RSZKAOK arguments as input. Note that in order to produce simulated RSZKAOK arguments, the simulator requires the code of the verifier which in our case is the token. Observe that this does not violate UC security since the simulator only needs the code of the token (which it does receive as per the model) and not the environment's code. In its interaction with the sender $\mathcal{S}^*$, the simulator picks σ_i uniformly at random and not as the output of a PRF. Using the sender's responses $A_{i,0}, A_{i,1}$ along with the outputs from the token - $(Z_{i,0}, Z_{i,1})$ on both $z_i = 0$ and $z_i = 1$, the simulator can compute both the label values for each input wire to the garbled circuit. Further, the simulator sends a random y as input to receive the garbled circuit $\tilde{C}$ and produces a simulated RSZKAOK of the signature. Then, from the soundness of the RWIAOK, the simulator is guaranteed that $\tilde{C}$ was indeed garbled correctly using two messages (r_0, r_1). Finally, the simulator can extract both m_0 and m_1 using the garbled circuit $\tilde{C}$, all the labels and the messages α_0, α_1.[5]

Further, note that due to the "leakage resilience" of the inner product, $\mathcal{S}^*$ doesn't learn anything about b even if the malicious token selectively aborts. That is, $\mathcal{S}^*$ can't learn b unless it learns all the b_i values. For this, the token has to signal information about each z_i by selectively aborting to help $\mathcal{S}^*$ recover the respective b_i and this can happen only with negligible probability since the z_i's are essentially picked uniformly at random. That is, in the proof, the situation where the simulator fails to extract both messages while the honest party doesn't abort happens only with negligible probability.

In the above protocol description, we treated the RSZKAOK and RWIAOK argument systems as non-interactive protocols, but in reality they are interactive

---

[5] An alternate proof strategy is for the simulator to directly extract the values r_0 and r_1 using the extractor of the RWIAOK but we won't delve further into that.

proofs. This doesn't increase the round complexity of our protocol since these protocols are only executed between the receiver and the token. However, since the token is stateless, it can't "remember" anything about the proof. We fix this by simply having the token sign the statement and transcript along with its message in every round.

4.2   Construction

Notation. Let n denote the security parameter. Let $\mathsf{OWF} : \{0,1\}^n \to \{0,1\}^{2n}$ be a one-way function, $\mathsf{PRF} : \{0,1\}^n \times \{0,1\}^{n+1} \to \{0,1\}^n$ and $\mathsf{PRF}_1 : \{0,1\}^n \times \{0,1\}^n \to \{0,1\}$ be two pseudorandom functions, $, = (\mathsf{Commit}, \mathsf{Decommit})$ be a non-interactive [6]computationally hiding and statistically binding commitment scheme that uses n bits of randomness to commit to one bit, let $(\mathsf{Gen}, \mathsf{Sign}, \mathsf{Verify})$ be a signature scheme, $(\mathsf{RSZKAOK.Prove}, \mathsf{RSZKAOK.Verify})$ be a resettably-sound zero-knowledge(RSZKAOK) argument of knowledge system for a "state-less verifier" and $(\mathsf{RWIAOK.Prove}, \mathsf{RWIAOK.Verify})$ be a resettable witness indistinguishable (RWIAOK) argument of knowledge system for a "stateless prover" as defined in Sect. 3. Let $(\mathsf{Garble}, \mathsf{Eval})$ be a garbling scheme for poly sized circuits that take inputs of length $(2n)$ bits and produces an output of length n bits. Let's say the sender $\mathcal{S}$ has private inputs $(\mathsf{m}_0, \mathsf{m}_1) \in \{0,1\}^{2n}$ and receiver $\mathcal{R}$ has private input $\mathsf{b} \in \{0,1\}$.

Note that all these primitives can be constructed assuming the existence of one-way functions.

NP languages. We will use the following NP languages in our OT protocol.

1. NP language L_1^{OT} characterized by the following relation R_1^{OT}.
   Statement : $\mathsf{st} = (\mathsf{z}, i, \mathsf{c})$
   Witness : $\mathsf{w} = (\mathsf{k}_\mathcal{R}, \mathsf{r}, \sigma)$
   $R_1^{OT}(\mathsf{st}, \mathsf{w}) = 1$ if and only if :
   – $\mathsf{z} = \mathsf{PRF}_1(\mathsf{k}_\mathcal{R}, i)$ AND
   – $\mathsf{Verify}_{\mathsf{vk}}(\mathsf{c}, \sigma) = 1$
   – $\mathsf{c} = \mathsf{Commit}(\mathsf{k}_\mathcal{R}; \mathsf{r})$
2. NP language L_2^{OT} characterized by the following relation R_2^{OT}.
   Statement : $\mathsf{st} = (\mathsf{y})$
   Witness : $\mathsf{w} = (\sigma_\mathsf{y})$
   $R_2^{OT}(\mathsf{st}, \mathsf{w}) = 1$ if and only if :
   – $\mathsf{Verify}_{\mathsf{vk}}(\mathsf{y}, \sigma_\mathsf{y}) = 1$
3. NP language L_3^{OT} characterized by the following relation R_3^{OT}.
   Statement : $\mathsf{st} = (\tilde{C}, \mathsf{y})$
   Witness : $\mathsf{w} = (C, \mathsf{k}, \mathsf{r}_0, \mathsf{r}_1 \bar{\mathsf{x}})$
   $R_3^{OT}(\mathsf{st}, \mathsf{w}) = 1$ if and only if :

---

[6] To ease the exposition, we use non-interactive commitments that are based on injective one-way functions. We describe later how the protocol can be modified to use a two-round commitment scheme that relies only on one-way functions without increasing the round complexity of the protocol.

– Either
  - $\tilde{\mathcal{C}} = \mathsf{Garble}(\mathcal{C}, \mathsf{k})$ (AND)
  - circuit $\mathcal{C}$ on input $(\mathsf{b}_1, \ldots, \mathsf{b}_{2n})$, outputs r_b where
    $\mathsf{b} = \langle(\mathsf{b}_1, \ldots, \mathsf{b}_n), (\mathsf{b}_{n+1}, \ldots, \mathsf{b}_{2n})\rangle$.
    (OR)
– $\mathsf{y} = \mathsf{OWF}(\overline{\mathsf{x}})$.

OT Protocol. We now describe our two round OT protocol π^{OT}.

– **Token Exchange Phase:**
  $\mathcal{S}$ picks two random keys $\{\mathsf{k}_\mathcal{S}, \mathsf{k}_\mathcal{V}\} \xleftarrow{\$} \{0,1\}^{2n}$ for the function PRF and computes $(\mathsf{sk}, \mathsf{vk}) \leftarrow (\mathsf{Gen}(n))$. Then, $\mathcal{S}$ creates a single token $\mathbf{T}_\mathcal{S}$ containing the codes in Figs. 1 and 2. $\mathcal{S}$ picks two random values $\mathsf{r}_0, \mathsf{r}_1$. Consider a circuit $\mathcal{C}$ that, given input $\mathsf{b}_1, \ldots, \mathsf{b}_{2n}$, outputs r_b where $\mathsf{b} = \langle(\mathsf{b}_1, \ldots, \mathsf{b}_n), (\mathsf{b}_{n+1}, \ldots, \mathsf{b}_{2n})\rangle$. $\mathcal{S}$ creates a garbled version of this circuit - $\tilde{\mathcal{C}}$ using keys $\{\mathsf{L}_{i,\mathsf{b}}\}$ for all $i \in [2n]$ and $\mathsf{b} \in \{0,1\}$. This is hardwired into the token. $\mathcal{S}$ sends vk and $\mathbf{T}_\mathcal{S}$ to the receiver $\mathcal{R}$.

– **Oblivious Transfer Phase:**
  1. **Round 1:** $\mathcal{R}$ does the following:
     - Choose a random key $\mathsf{k}_\mathcal{R} \xleftarrow{\$} \{0,1\}^n$ for the function PRF_1 and a random string $\overline{\mathsf{x}} \xleftarrow{\$} \{0,1\}^n$. Compute $\mathsf{y} = \mathsf{OWF}(\overline{\mathsf{x}})$.
     - Compute $\mathsf{c} = \mathsf{Commit}(\mathsf{k}_\mathcal{R}; \mathsf{r})$ using a random string $\mathsf{r} \xleftarrow{\$} \{0,1\}^{n^2}$.
     - Pick $2n$ bits $\mathsf{b}_1, \ldots, \mathsf{b}_{2n}$ uniformly at random such that $\langle \mathsf{B}_1, \mathsf{B}_2 \rangle = \mathsf{b}$ where $\mathsf{B}_1 = (\mathsf{b}_1, \ldots, \mathsf{b}_n)$ and $\mathsf{B}_2 = (\mathsf{b}_{n+1}, \ldots, \mathsf{b}_{2n})$. Then, for each $i \in [2n]$, compute $\mathsf{s}_i = (\mathsf{b}_i \oplus \mathsf{z}_i)$ where $\mathsf{z}_i = \mathsf{PRF}_1(\mathsf{k}_\mathcal{R}, i)$.
     - Send $\mathsf{c}, \overline{\mathsf{x}}$ and $\{\mathsf{s}_i\}_{i=1}^{2n}$ to $\mathcal{S}$.
  2. **Round 2:** $\mathcal{S}$ does the following:
     - Compute $\mathsf{y} = \mathsf{OWF}(\overline{\mathsf{x}})$, $\sigma = \mathsf{Sign}_\mathsf{sk}(\mathsf{c}; \mathsf{r}')$ and $\sigma_\mathsf{y} = \mathsf{Sign}_\mathsf{sk}(\mathsf{y})$.
     - For each $i \in [2n]$, compute $\mathsf{A}_{i,\mathsf{s}_i} = \mathsf{PRF}(\mathsf{k}_\mathcal{S}, i, \mathsf{s}_i)$.
     - Compute $\alpha_0 = (\mathsf{m}_0 \uplus \mathsf{r}_0)$ and $\alpha_1 = (\mathsf{m}_1 \oplus \mathsf{r}_1)$.
     - Send $(\{\mathsf{A}_{i,\mathsf{s}_i}\}_{i=1}^{2n}, \sigma, \sigma_\mathsf{y})$ along with (α_0, α_1) to $\mathcal{R}$.

– **Output Computation Phase:** $\mathcal{R}$ does the following:
  - Abort if $\mathsf{Verify}_\mathsf{vk}(\mathsf{c}, \sigma) = 0$ or $\mathsf{Verify}_\mathsf{vk}(\mathsf{y}, \sigma_\mathsf{y}) = 0$.
  - For each $i \in [2n]$, query $\mathbf{T}_\mathcal{S}$ with input $(\mathsf{z}_i, i, \mathsf{c}, \text{"prove"})$. Using the prover algorithm (RSZKAOK.Prove), engage in an execution of an RSZKAOK argument with $\mathbf{T}_\mathcal{S}$ (who acts as the verifier) for the statement $\mathsf{st}_1 = (\mathsf{z}_i, i, \mathsf{c}) \in L_1^{OT}$ using witness $\mathsf{w}_1 = (\mathsf{k}_\mathcal{R}, \mathsf{r}, \sigma)$. That is, as part of the RSZKAOK, if the next message of the prover is msg, query $\mathbf{T}_\mathcal{S}$ with input $(\mathsf{z}_i, i, \mathsf{c}, \mathsf{msg})$ in that round.
  - Let $\{(\mathsf{Z}_{i,\mathsf{z}_i}, \sigma_i)\}_{i=1}^{2n}$ be the outputs received from $\mathbf{T}_\mathcal{S}$. For each i, compute $\mathsf{L}_{i,\mathsf{b}_i} = (\mathsf{Z}_{i,\mathsf{s}_i} \oplus \mathsf{A}_{i,\mathsf{s}_i})$.
  - Query $\mathbf{T}_\mathcal{S}$ with input $(\sigma_1, \ldots, \sigma_{2n}, \mathsf{y}, \text{"prove}_1\text{"})$. Using the prover algorithm (RSZKAOK.Prove), engage in an execution of an RSZKAOK argument with $\mathbf{T}_\mathcal{S}$ (who acts as the verifier) for statement $(\mathsf{st}_2 = \mathsf{y}) \in L_2^{OT}$ using witness $\mathsf{w}_2 = (\sigma_\mathsf{y})$. That is, as part of the RSZKAOK, if the prover's next message is msg, query $\mathbf{T}_\mathcal{S}$ with input $(\sigma_1, \ldots, \sigma_{2n}, \mathsf{y}, \mathsf{msg})$ in that round.

- Let $(\tilde{\mathcal{C}}, \sigma_{\tilde{\mathcal{C}}, \mathsf{y}})$ be the output of $\mathbf{T}_{\mathcal{S}}$. Using the algorithm (RWIAOK.Verify), engage in an execution of a RWIAOK with $\mathbf{T}_{\mathcal{S}}$ (who acts as the prover) for the statement $\mathsf{st}_3 = (\tilde{\mathcal{C}}, \mathsf{y}) \in L_3^{QT}$. As part of the RWIAOK, if the next message of the verifier is msg, query $\mathbf{T}_{\mathcal{S}}$ with input $(\tilde{\mathcal{C}}, \mathsf{y}, \sigma_{\tilde{\mathcal{C}}, \mathsf{y}}, \mathsf{msg})$ in that round. Initially, query with $(\tilde{\mathcal{C}}, \mathsf{y}, \sigma_{\tilde{\mathcal{C}}, \mathsf{y}}, \text{"prove"})$. Abort if the argument doesn't verify.
- Using the keys $\{\mathsf{L}_{i, \mathsf{b}_i}\}_{i=1}^{2n}$ and the garbled circuit $\tilde{\mathcal{C}}$, run the algorithm Eval to recover the value r_b.
- Then, compute $\mathsf{m}_\mathsf{b} = (\alpha_\mathsf{b} \oplus \mathsf{r}_\mathsf{b})$

Remarks:

1. To be more precise, we use a 2-round commitment scheme where the first message is actually sent by the token (acting on behalf of the receiver of the commitment) independent of the value being committed to. This has been abstracted out as part of the commitment scheme.
2. The verification key vk can be output by the token itself instead of being sent by $\mathcal{S}$ along with the token. This would then strictly imply that the token exchange phase has no communication messages.

4.3   Token Reusability

Observe that the sender's input messages $(\mathsf{m}_0, \mathsf{m}_1)$ don't appear in the token at all. For each execution, the token just evaluates a garbled circuit $\tilde{\mathcal{C}}$ generated using a circuit C that contains two random strings $(\mathsf{r}_0, \mathsf{r}_1)$. In the current construction, the strings $(\mathsf{r}_0, \mathsf{r}_1)$ and the garbled circuit $\tilde{\mathcal{C}}$ were hardwired into the token. Instead, we can just hardwire two PRF keys - k_r and $\mathsf{k}_{\tilde{\mathcal{C}}}$ in the token. Then, the token can use the first key k_r to generate the pair of random strings $(\mathsf{r}_0, \mathsf{r}_1)$ and thereby the circuit C for each execution. Similarly, the second key $\mathsf{k}_{\tilde{\mathcal{C}}}$ can be used to generate the randomness required to garble the circuit for that execution. Thus, the same token can be re-used to run an unbounded number of oblivious transfer executions between the same pair of sender and receiver.

4.4   Security

We defer the formal proof of security to the full version of the paper.

5   Two Round Two-Party Computation

In this section, we study two-party computation in the simultaneous broadcast channel using stateless hardware tokens. We first construct a two round UC secure two-party computation protocol for general functions in this model based on one-way functions using two tokens. Specifically, each party sends a single token to the other party in a token exchange phase prior to the protocol communication phase. Formally, we show the following theorem:

Constants: $(k_V, k_S, vk, sk, \tilde{C}, \{L_{i,b}\})$
Case 1: If Input $=(z_i, i, c, msg)$:

- If msg $=$ "prove", the token does the following:
  1. Consider a random tape defined by $\mathsf{PRF}(k_V, i, 0)$.
  2. Using the above randomness and the verifier algorithm ($\mathsf{RSZKAOK.Verify}$), initiate an execution of a RSZK argument with the querying party playing the role of the prover for the statement $st_1 = (z_i, i, c) \in L_1^{OT}$. Output the first message of the verifier.
- If msg $\neq$ "prove", the token does the following:
  1. Consider a random tape defined by $\mathsf{PRF}(k_V, i, 0)$.
  2. Using the above randomness and msg as the message sent by the prover, run the verifier algorithm ($\mathsf{RSZKAOK.Verify}$) to continue an execution of a RSZK argument with the querying party playing the prover's role for the statement $st_1 = (z_i, i, c) \in L_1^{OT}$.
  3. Compute the next message msg$'$ of the verifier.
  4. If msg$' \notin \{accept, reject\}$, output msg$'$. If msg$' = reject$, abort.
  5. If msg$' = accept$:
     If $z_i = 0$, compute $Z_{i,0} = (L_{i,0} \oplus A_{i,0})$ and $Z_{i,1} = (L_{i,1} \oplus A_{i,1})$. If $z_i = 1$, compute $Z_{i,0} = (L_{i,1} \oplus A_{i,0})$ and $Z_{i,1} = (L_{i,0} \oplus A_{i,1})$. Output $(Z_{i,0}, Z_{i,1}, \sigma_i = \mathsf{Sign}_{sk}(i))$.

Case 2: If Input $=(\sigma_1, \ldots, \sigma_{2n}, y, msg)$:

- Abort if $\mathsf{Verify}_{vk}(i, \sigma_i) = 0$ for any $i \in [2n]$.
- If msg $=$ "prove", the token does the following:
  1. Consider a random tape defined by $\mathsf{PRF}(k_V, 1^{n+1})$.
  2. Using the above randomness and the verifier algorithm ($\mathsf{RSZKAOK.Verify}$), initiate an execution of a RSZKAOK with the querying party playing the role of the prover for the statement $st_2 = y \in L_2^{OT}$. Output the first message of the verifier.
- If msg $\neq$ "prove", the token does the following:
  1. Consider a random tape defined by $\mathsf{PRF}(k_V, 1^{n+1})$.
  2. Using the above randomness and msg as the message sent by the prover, run the verifier algorithm ($\mathsf{RSZKAOK.Verify}$) to continue an execution of a RSZKAOK with the querying party playing the prover's role for the statement $st_2 = y \in L_2^{OT}$.
  3. Compute the next message msg$'$ of the verifier.
  4. If msg$' \notin \{accept, reject\}$, output msg$'$. If msg$' = reject$, abort.
  5. If msg$' = accept$, output $(\tilde{C}, \sigma_{\tilde{C},y} = \mathsf{Sign}_{sk}(\tilde{C}, y))$.

Continues in Figure 2.

Fig. 1. Code of token $\mathbf{T}_S$

Theorem 6. *Assuming one-way functions exist, there exists a two round UC-secure two-party computation protocol over simultaneous broadcast channels for any functionality f in the stateless hardware token model where each party sends one token.*

Continuing from Figure 1.

Case 3: If Input $= (\tilde{\mathcal{C}}, y, \sigma_{\tilde{\mathcal{C}}, y}, \mathsf{msg})$:

- Abort if $\mathsf{Verify}_{vk}(\tilde{\mathcal{C}}, y, \sigma_{\tilde{\mathcal{C}}, y}) = 0$.
- If $\mathsf{msg} = $ "prove", the token does the following:
    1. Consider a random tape defined by $\mathsf{PRF}(k_{\mathcal{V}}, 0^{n+1})$.
    2. Using the above randomness and the prover algorithm ($\mathsf{RWIAOK.Prove}$), initiate an execution of a RWIAOK with the querying party playing the role of the verifier for the statement $\mathsf{st}_3 = (\tilde{\mathcal{C}}, y) \in L_3^{OT}$ using witness $\mathsf{w}_3 = (\mathcal{C}, \{L_{i,0}, L_{i,1}\}, r_0, r_1, \bot)$ where $i \in [2n], b \in \{0,1\}$. Output the first message of the prover.
- If $\mathsf{msg} \neq $ "prove", the token does the following:
    1. Consider a random tape defined by $\mathsf{PRF}(k_{\mathcal{V}}, 0^{n+1})$.
    2. Using the above randomness and msg as the message sent by the prover, run the prover algorithm ($\mathsf{RWIAOK.Prove}$) to continue an execution of a RWIAOK with the querying party playing the verifier's role for the statement $\mathsf{st}_3 \in L_3^{OT}$.

Fig. 2. Continuing code of token $\mathbf{T}_{\mathcal{S}}$

5.1   Construction

Let f be any two-party functionality. Consider two parties P_1 and P_2 with inputs $x_1 \in \{0,1\}^n$ and $x_2 \in \{0,1\}^n$ respectively who wish to compute f on their joint inputs. Below, we describe a two round protocol Π^{2pc} for securely computing f.

Notation. Let n denote the security parameter and $\mathsf{OWF} : \{0,1\}^n \to \{0,1\}^{\mathsf{poly}(n)}$ be a one-way function. Let $\mathsf{PRF} : \{0,1\}^n \times \{0,1\}^{n+1} \to \{0,1\}^n$ be a pseudorandom function , $= (\mathsf{Commit}, \mathsf{Decommit})$ be a non-interactive[7] statistically binding commitment scheme that uses n bits of randomness to commit to one bit and $(\mathsf{Gen}, \mathsf{Sign}, \mathsf{Verify})$ be a signature scheme. Let $\mathsf{RZKAOK} = (\mathsf{RZKAOK.Prove}, \mathsf{RZKAOK.Verify})$ be a resettable zero-knowledge argument of knowledge for a "stateless prover", $\mathsf{RWIAOK} = (\mathsf{RWIAOK.Prove}, \mathsf{RWIAOK.Verify})$ be a resettable witness indistinguishable argument of knowledge for a "stateless prover" and $\mathsf{RSZKAOK} = (\mathsf{RSZKAOK.Prove}, \mathsf{RSZKAOK.Verify})$ be a resettably-sound zero-knowledge argument of knowledge for a "stateless verifier" as defined in Sect. 3. Let $(\mathsf{Garble}, \mathsf{Eval})$ be a garbling scheme for poly sized circuits that take inputs of length (n) bits and produces outputs of length n bits.

Let $(\mathsf{OT}_1, \mathsf{OT}_2, \mathsf{OT}_3)$ denote the 2-message oblivious transfer protocol from Sect. 4. That is, the algorithm OT_1 is used by the receiver to compute the first message ot_1. The algorithm OT_2 is used by the sender to compute the second

---

[7] To ease the exposition, we use non-interactive commitments that are based on injective one-way functions. We describe later how the protocol can be modified to use a two-round commitment scheme that relies only on one-way functions without increasing the round complexity of the protocol.

message ot_2 and the algorithm OT_3 is used by the receiver to compute the output.

NP languages. We will use the following NP languages in our protocol.

- Language L_1 characterized by the following relation R_1. Statement : st = (b, c)
  Witness : $w = (a, x, r)$
  $R_1(st, w) = 1$ if and only if :
  - $b = OWF(a)$ AND
  - $c = Commit(x; r)$
- Language L_2 characterized by the following relation R_2:
  Statement : st $= (\tilde{C}, b, c)$
  $w = (x, r_c, C, k, a)$
  $R_2(st, w) = 1$ if and only if :
  - Either
    * $c = Commit(x; r_c)$ AND
    * $\tilde{C} = Garble(C, k)$ AND
    * circuit C, on any input α, outputs $f(x, \alpha)$.
      (OR)
  - $b = OWF(a)$
- NP language L_3 characterized by the following relation R_3.
  Statement : st $= (c, vk_1, b, vk_2)$
  Witness : $w = (\sigma_c, \sigma_b)$
  $R_3(st, w) = 1$ if and only if :
  - $Verify_{vk}(c, \sigma_c) = 1$ AND
  - $Verify_{vk}(b, \sigma_b) = 1$

The Protocol. We now describe protocol Π^{2pc} for UC secure two-party computation in the stateless hardware token model. Let party P_1 have input x_1 and P_2 have input $\bar{x}_2$. Recall that function to be computed is denoted by f.

- **Token Exchange Phase:**
  P_1 does the following:
  - Compute $(sk_{c_2}, vk_{c_2}) \leftarrow Gen(n)$, $(sk_{b_2}, vk_{b_2}) \leftarrow Gen(n)$, $(sk_1, vk_1) \leftarrow Gen(n)$. Pick keys $\{k_1^{rzk}, k_1^{rszk}, k_1^{rwi}\} \xleftarrow{\$} \{0, 1\}^{3n}$ for the PRF.
  - Choose a random string a_1 and compute $b_1 = OWF(a_1)$. Also, compute $c_1 = Commit(x_1; r_{c_1})$ using a random string r_{c_1}.
  - Consider a circuit C^1 that, given an n-bit input (α), outputs $f(x_1, \alpha)$. Create a garbled version of this circuit - $\tilde{C}^1$ using keys $\{L_{i,b}^1\}$ for all $i \in [n]$ and $b \in \{0, 1\}$. This is hardwired into the token. Compute $\sigma_{\tilde{C}^1} = Sign_{sk_1}(\tilde{C}^1; r_{\tilde{C}^1})$ using a random string $r_{\tilde{C}^1}$.
  - Create a token $\mathbf{T}_1^{2pc}$ containing the codes in Figs. 3 and 4. Note that this involves performing steps carried out in the token exchange phase of the OT protocol in Sect. 4.
  - P_1 sends the token $\mathbf{T}_1^{2pc}$ to P_2.
  The protocol is symmetric. That is, P_2 sends $\mathbf{T}_2^{2pc}$ to P_1.

– **Communication Phase:**
  1. **Round 1:**
     P_1 does the following:
     - For each $i \in [n]$, compute $\mathsf{ot}_{1,i}^{1 \to 2} = \mathsf{OT}_1(x_{1,i})$ where $x_{1,i}$ denotes the ith bit of x_i. Here the superscript denotes that its sent from P_1 to P_2.
     - Send $(b_1, c_1, \mathsf{vk}_{b_2}, \mathsf{vk}_{c_2}, \{\mathsf{ot}_{1,i}^{1 \to 2}\}_{i=1}^n)$ to P_2 where b_1, c_1 were computed in the token exchange phase.

     P_2 performs the same operations symmetrically and sends $(b_2, c_2, \mathsf{vk}_{b_1}, \mathsf{vk}_{c_1}, \{\mathsf{ot}_{1,i}^{2 \to 1}\}_{i=1}^n)$ to P_1.
  2. **Round 2:**
     P_1 does the following:
     - Using the verifier algorithm ($\mathsf{RZKAOK.Verify}$), engage in an execution of a RZKAOK with the token $\mathbf{T}_2^{2pc}$ (who acts as the prover) for the statement that $(\mathsf{st}_1) \in L_1$ where $\mathsf{st}_2 = (b_2, c_2)$. This is done by querying token $\mathbf{T}_2^{2pc}$ with input ("activate"). As part of the RZKAOK, if the next message of the verifier is msg, query the token with input (msg) in that round.
     - Abort if the above argument doesn't verify.
     - Compute $\sigma_{c_2} = \mathsf{Sign}_{\mathsf{sk}_{c_2}}(c_2; r_{c_2})$ and $\sigma_{b_2} = \mathsf{Sign}_{\mathsf{sk}_{b_2}}(b_2; r_{b_2})$ using random strings r_{c_2} and r_{b_2}.
     - For each $i \in [n]$, compute $\mathsf{ot}_{2,i}^{1 \to 2} = \mathsf{OT}_2(L_{i,0}^1, L_{i,1}^1, \mathsf{ot}_{1,i}^{2 \to 1})$ where $L_{i,0}^1$ and $L_{i,1}^1$ are the labels of the garbled circuit $\tilde{C}^1$.
     - Send $(\sigma_{c_2}, \sigma_{b_2}, \mathsf{ot}_{2,1}^{1 \to 2}, \ldots, \mathsf{ot}_{2,n}^{1 \to 2})$ to P_2.

     P_2 symmetrically sends $(\sigma_{c_1}, \sigma_{b_1}, \mathsf{ot}_{2,1}^{2 \to 1}, \ldots, \mathsf{ot}_{2,n}^{2 \to 1})$ to P_1.
– **Output Computation:**
  P_1, does the following :
  - For each $i \in [n]$, run the "Output computation phase" of the OT protocol using input $x_{1,i}$ and $\mathsf{ot}_{2,i}^{2 \to 1}$ as the messages from the sender. For any query msg to be made to the token in the OT protocol, query token $\mathbf{T}_2^{2pc}$ using input ("OT", msg). Compute output $L_{i,x_{1,i}}^2$ for each $i \in [n]$.
  - Query $\mathbf{T}_2^{2pc}$ using input $(c_1, b_1, \text{"2pc"})$. Using the prover algorithm ($\mathsf{RSZKAOK.Prove}$), engage in an execution of an RSZKAOK argument with $\mathbf{T}_2^{2pc}$ (who acts as the verifier) for statement $\mathsf{st}_3 = (c_1, b_1, \mathsf{vk}_{c_1}, \mathsf{vk}_{b_1}) \in L_3$ using witness $w_3 = (\sigma_{c_1}, \sigma_{b_1})$. That is, as part of the RSZKAOK, if the next message of the prover is msg, query $\mathbf{T}_2^{2pc}$ with input (c_1, b_1, msg) in that round.
  - Let $(\tilde{C}^2, \sigma_{\tilde{C}^2})$ be the output of $\mathbf{T}_2^{2pc}$. Then, using the verifier algorithm ($\mathsf{RWIAOK.Verify}$), engage in an execution of a RWIAOK with $\mathbf{T}_2^{2pc}$ (who acts as the prover) for the statement $\mathsf{st}_2 = (\tilde{C}^2, b_1, c_1) \in L_2$. As part of the RWIAOK, if the verifier's next message is msg, query $\mathbf{T}_2^{2pc}$ with input $(\tilde{C}^2, \sigma_{\tilde{C}^2}, \mathsf{msg})$ in that round. Initially, query with $(\tilde{C}^2, \sigma_{\tilde{C}^2}, \text{"prove"})$. Abort if the argument doesn't verify.
  - Using the keys $\{L_{i,x_{1,i}}^2\}_{i=1}^n$, and the garbled circuit $\tilde{C}^2$, run the algorithm Eval to recover the output y_1.

  P_2 performs the same operations symmetrically to receive output y_2.

Note: For better understanding of the rest of the protocol, this figure actually describes the code of token $\mathbf{T}_2^{2pc}$ created by P_2. The code of $\mathbf{T}_1^{2pc}$ is symmetrical.
Constants: $(\tilde{C}^2, \sigma_{\tilde{C}^2}, \{L_{i,0}^2, L_{i,1}^2\}_{i=1}^n, x_2, c_2, r_{x_2}, a_2, b_2, k_2^{rzk}, k_2^{rszk}, k_2^{rwi}, \mathsf{PRF}$
$(\mathsf{sk}_{c_1}, \mathsf{vk}_{c_1}), (\mathsf{sk}_{b_1}, \mathsf{vk}_{b_1}), (\mathsf{sk}_2, \mathsf{vk}_2))$

1. If input $= (\text{"OT"}, \mathsf{msg})$, respond as done by the token in Section 4 using input as msg.
2. If input $= (\text{"activate"})$, do the following: using a random tape defined by $\mathsf{PRF}(k_2^{rzk}, 0^{n+1})$ and the prover algorithm (RZKAOK.Prove), initiate an execution of a RZKAOK with the querying party playing the role of the verifier for the statement $(\mathsf{st}_1) \in L_1$ where $\mathsf{st}_1 = (b_2, c_2)$ using witness (a_2, x_2, r_{x_2}). Output the first message of the prover.
3. If input $= (\mathsf{msg})$, do the following: using a random tape defined by $\mathsf{PRF}(k_2^{rzk}, 0^{n+1})$ and the prover algorithm (RZKAOK.Prove), continue an execution of a RZKAOK with the querying party playing the role of the verifier for the statement $(\mathsf{st}_1) \in L_1$ where $\mathsf{st}_2 = (b_2, c_2)$ using witness (a_2, x_2, r_{x_2}). Output the next message of the prover.

Continues in Figure 4.

Fig. 3. Code of token $\mathbf{T}_2^{2pc}$

Remark: In the above description, we were assuming non-interactive commitments (which require injective one way functions) to ease the exposition. In order to rely on just one way functions, we switch our commitment protocol to a two-round one where the receiver sends the first message. Now, we tweak our protocol as follows: after receiving the token, P_1 receives the first round of the commitment from the token $\mathbf{T}_2$ and uses that to compute c_1. P_2 does the same thing symmetrically after interacting with $\mathbf{T}_1$.

Reusability: If we want to allow our tokens to be reused an unbounded number of times for performing multiple two party computation protocols between the same pair of parties, we can tweak the protocol as follows: instead of hardwiring P_1's input x_1 and the garbled circuit $\tilde{C}_1$ inside the token $\mathbf{T}_1^{2pc}$, we can just hardwire an encryption key ek_1 for a symmetric encryption scheme. Now, in this setting, the tokens are exchanged apriori in an initial token exchange phase. Then, in the first round, when P_1 sends $c_1 = \mathsf{Commit}(x_1)$ and message b_1 to party P_2, it also sends $\mathsf{ct}_1 = \mathsf{enc}_{\mathsf{ek}_1}(x_1)$ and $\sigma_{\mathsf{ct}_1} = \mathsf{Sign}(\mathsf{ct}_1)$. That is, it sends an encryption of its input and a signature on this encryption. Party P_2 is now expected to also query the token with $(\mathsf{ct}_1, \sigma_{\mathsf{ct}_1})$ along with c_1, b_1. The token $\mathbf{T}_1^{2pc}$ verifies the signature, decrypts the message to learn the input x_1 and then proceeds to use it for the rest of the computation as before. A similar procedure is also performed with respect to P_2's initial messages and $\mathbf{T}_2^{2pc}$'s token responses.

Continuing from Figure 3.

4. If input = $(c_1, b_1, \text{"2pc"})$, do the following:
   - If msg = "prove", the token does the following: Using a random tape defined by $\mathsf{PRF}(k_2^{\mathsf{RSZK}}, 1^{n+1})$ and the verifier algorithm $(\mathsf{RSZKAOK.Verify})$, initiate an execution of a RSZKAOK with the querying party playing the role of the prover for the statement $st_3 = (c_1, b_1, vk_{c_1}, vk_{b_1}) \in L_3$. Output the first message of the verifier.
   - If msg $\neq$ "prove", the token does the following:
     (a) Using a random tape defined by $\mathsf{PRF}(k_2^{\mathsf{RSZK}}, 1^{n+1})$ and msg as the message sent by the prover, run the verifier algorithm $(\mathsf{RSZKAOK.Verify})$ to continue an execution of a RSZKAOK with the querying party playing the prover's role for the statement $st_3 = (c_1, b_1) \in L_3$.
     (b) Compute the next message msg' of the verifier.
     (c) If msg' $\notin \{\mathsf{accept}, \mathsf{reject}\}$, output msg'. If msg' = reject, abort.
     (d) If msg' = accept, output $(\tilde{\mathcal{C}}^2, \sigma_{\tilde{c}^2})$.
5. if input = $(\tilde{\mathcal{C}}^2, \sigma_{\tilde{c}^2}, \text{"prove"})$
   - Abort if the signature $\sigma_{\tilde{c}^2}$ doesn't verify.
   - using a random tape defined by $\mathsf{PRF}(k_2^{rwi}, 0^{n+1})$ and the prover algorithm $(\mathsf{RWIAOK.Prove})$, initiate an execution of a RWIAOK with the querying party playing the role of the verifier for the statement $st_2 = (\tilde{\mathcal{C}}^2, b_1, c_1) \in L_2$ using witness $(x_2, r_{c_2}, \mathcal{C}^2, \{L_{i,0}^2, L_{i,1}^2\}_{i=1}^n, \bot)$.
   - Output the first message of the prover.
6. If input = $(\tilde{\mathcal{C}}^2, \sigma_{\tilde{c}^2}, \mathsf{msg})$, do:
   - Abort if the signature $\sigma_{\tilde{c}^2}$ doesn't verify.
   - Using a random tape defined by $\mathsf{PRF}(k_2^{rwi}, 0^{n+1})$ and the prover algorithm $(\mathsf{RWIAOK.Prove})$, continue an execution of a RWIAOK with the querying party playing the role of the verifier for the statement $st_2 = (\tilde{\mathcal{C}}^2, b_1, c_1) \in L_2$ using witness $(x_2, r_{c_2}, \mathcal{C}^2, \{L_{i,0}^2, L_{i,1}^2\}_{i=1}^n, \bot)$.
   - Output the next message of the prover.

Fig. 4. Code of token $\mathbf{T}_2^{2pc}$

5.2 Security

We formally prove security in the full version of the paper.

6 Three Round MPC

In this section, we use our unbounded reusable OT protocol to construct a three round UC secure MPC protocol for general functions in the stateless hardware token model amongst n parties based on one-way functions. In this protocol, each party sends two tokens to every other party in a token exchange phase prior to the protocol communication phase. Formally, we show the following theorem:

Theorem 7. *Assuming one-way functions exist, there exists a three round protocol that UC-securely realizes any n-party functionality f in the stateless hardware token model where each party sends two tokens to every other party.*

6.1   Construction

Let f be any functionality. Consider n parties $P_1, \ldots, P_n$ with inputs $\mathsf{inp}_1, \ldots,$ inp_n respectively who wish to compute f on their joint inputs. Below, we describe a three round protocol Π^{mpc} for securely computing f.

Notation. We first list some notation and the primitives used.

- Let λ denote the security parameter.
- Let OWF : $\{0,1\}^\lambda \to \{0,1\}^{\mathsf{poly}(\lambda)}$ be a one-way function. Let PRF : $\{0,1\}^\lambda \times \{0,1\}^{\lambda+1} \to \{0,1\}^\lambda$ be a pseudorandom function, Com = (Commit, Decommit) be a non-interactive statistically binding commitment scheme that uses λ bits of randomness to commit to one bit, (Gen, Sign, Verify) be a signature scheme and (setup, enc, dec) be a private key encryption scheme.
- Let RWIAOK = (RWIAOK.Prove, RWIAOK.Verify) be a resettable witness indistinguishable argument of knowledge for a "stateless prover" and RZKAOK = (RZKAOK.Prove, RZKAOK.Verify) be a resettable zero-knowledge argument of knowledge for a "stateless prover" as defined in Sect. 3.
- Let $(\mathsf{OT}_1, \mathsf{OT}_2, \mathsf{OT}_3)$ denote the 2-message oblivious transfer protocol from Sect. 4. That is, the algorithm OT_1 is used by the receiver to compute the first message ot_1. The algorithm OT_2 is used by the sender to compute the second message ot_2 and the algorithm OT_3 is used by the receiver to compute the output.
- Let π denote a semi-malicious secure MPC protocol in the correlated randomness model where the correlated randomness is the following: between every pair of parties, there exists an OT channel. That is: between every pair of parties P_i, P_j, there exists a set of tuples $\{s_{0,k}, s_{1,k}, b_k\}_{k \in [p(\lambda)]}$ for some fixed polynomial p such that P_i knows $\{s_{0,k}, s_{1,k}\}_{k \in [\mathsf{poly}(\lambda)]}$ and P_j knows $\{b_k\}_{k \in [\mathsf{poly}(\lambda)]}$. We know that such a protocol can be constructed assuming just the existence of one way functions [Bea96, Kil88, IPS08, IKO+11]. Lets say its an ℓ round protocol. Let $\pi.\mathsf{Round}_i$ denote the algorithm used by any party to generate the message in round i and let $\pi.\mathsf{Out}$ denote the algorithm used to compute the final output. Let Sim^π denote the simulator for this protocol. We require that Sim^π can generate simulated correlated randomness without knowing the output of the protocol or the input and randomness of the corrupted parties.

NP languages. We will use the following NP languages in our protocol.

- Language L_1 characterized by the following relation R_1. Statement : st = (b, c)
  Witness : $\mathsf{w} = (\mathsf{a}, \mathsf{x}, \mathsf{r})$
  $R_1(\mathsf{st}, \mathsf{w}) = 1$ if and only if :
  - $\mathsf{b} = \mathsf{OWF}(\mathsf{a})$ AND
  - $\mathsf{c} = \mathsf{Commit}(\mathsf{x}; \mathsf{r})$

- Language L_2 characterized by the following relation R_2:

  Statement : $\mathsf{st} = (\mathsf{b}, \mathsf{c}, \mathsf{Trans}, \mathsf{msg}, i)$

  $w = (\mathsf{x}, r_c, \mathsf{cor.rand}, \mathsf{a})$

  $R_2(\mathsf{st}, w) = 1$ if and only if :
  - Either
    * $\mathsf{c} = \mathsf{Commit}(\mathsf{x}; r_c)$ AND
    * $\mathsf{msg} = \pi.\mathsf{Round}_i(\mathsf{x}, \mathsf{Trans}, \mathsf{cor.rand})$

      (OR)
  - $\mathsf{b} = \mathsf{OWF}(\mathsf{a})$

The Protocol. We now describe protocol Π^{mpc} in the stateless hardware token model. Recall that each party P_i has input inp_i and the function to be computed is denoted by f.

- **Token Exchange Phase:**

  Each party P_i does the following:
  1. For each party P_j, create token $\mathbf{T}_{ot}^{i \to j}$ as done in Sect. 4.
  2. Compute $(\mathsf{sk}_i, \mathsf{vk}_i) \leftarrow \mathsf{Gen}(\lambda)$, $\mathsf{ek}_i \leftarrow \mathsf{setup}(\lambda)$. Pick keys $\{k_i^{rzk}, k_i^{rwi}\} \xleftarrow{\$} \{0,1\}^{2n}$ for the PRF.
  3. Choose a random string a_i and compute $b_i = \mathsf{OWF}(a_i)$.
  4. Pick a random string r_i to run the MPC protocol π. Set $x_i = (\mathsf{inp}_i \| r_i)$. Compute $c_i = \mathsf{Commit}(x_i; r_{c_i})$ using a random string r_{c_i}.
  5. For each party P_k, create a token $\mathbf{T}_{mpc}^{i \to k}$ containing the code in Fig. 5.
  6. P_i broadcasts all the tokens created above.

- **OT Phase:**
  1. For each $k \in [n]$, every pair of parties P_i and P_j with $i > j$ perform a set of $p(\lambda)$ executions of the Oblivious Transfer protocol from Sect. 4 using the token $\mathbf{T}_{ot}^{i \to j}$. Here, P_i picks random inputs (s_0, s_1) for each execution independently and P_j picks a random bit b in each execution independently. This process takes two rounds.
  2. In round 3, every party P_i does the following: For each $k \in [n]$, for each $j \in [n]$ and each OT execution t with party P_j, do:
     (a) If $i > j$, compute $\mathsf{ct} = \mathsf{enc}(\mathsf{ek}, \{s_{0,t}, s_{1,t}\})$ and $\sigma_{\mathsf{ct}} = \mathsf{Sign}_{\mathsf{sk}}(\mathsf{ct})$. Output $(\mathsf{ct}, \sigma_{\mathsf{ct}})$.
     (b) If $i < j$, compute $\mathsf{ct} = \mathsf{enc}(\mathsf{ek}, \{b, s_{b,t}\})$ and $\sigma_{\mathsf{ct}} = \mathsf{Sign}_{\mathsf{sk}}(\mathsf{ct})$. Output $(\mathsf{ct}, \sigma_{\mathsf{ct}})$.

- **Input Commitment Phase:**
  1. **Round 1:**

     Each party P_i broadcasts (b_i, c_i) where b_i, c_i were computed in the token exchange phase.
  2. **Round 2:**

     Each party P_i does the following:
     - For each $j \in [n]$, using the verifier algorithm (RZKAOK.Verify), engage in an execution of a RZKAOK with the token $\mathbf{T}_{mpc}^{j \to i}$ (who acts as the prover) for the statement that $(\mathsf{st}_j) \in L_1$ where $\mathsf{st}_j = (b_j, c_j)$. As part of the RZKAOK, if the next message of the verifier is msg, query the token with input ("RZKAOK",msg) in that round.

- Abort if the above argument doesn't verify.
- For each $j \in [n]$, compute and broadcast $\sigma_{c_j} = \mathsf{Sign}_{\mathsf{sk}}(c_j)$ and $\sigma_{b_j} = \mathsf{Sign}_{\mathsf{sk}}(b_j)$.

- **Computation Phase:**

  Each party P_i does the following :

  1. Run an execution of the MPC protocol π amongst itself and the $(n-1)$ "MPC" tokens it received. That is, protocol π is executed amongst the n parties $\mathbf{T}_{mpc}^{1 \to i}, \ldots, \mathbf{T}_{mpc}^{(i-1) \to i}, P_i, \mathbf{T}_{mpc}^{(i+1) \to i}, \ldots, \mathbf{T}_{mpc}^{n \to i}$ for the functionality f where the kth party uses input inp_i, randomness r_i and correlated randomness as the decrypted values in the set of authenticated ciphertexts ct broadcast by party P_k in the OT phase.[8] Initiate the protocol by sending "MPC" to each token.

  2. Here, P_i acts as the channel and sends the messages broadcast by any party (aka token) to all the other parties (aka tokens). Along with each message, to each token $\mathbf{T}_{mpc}^{j \to i}$, P_i also sends the following:
     - The set of $\mathsf{ct}, \sigma_{\mathsf{ct}}$ broadcast by party P_j in the OT phase. This is the encryption of the correlated randomness for the token $\mathbf{T}_{mpc}^{j \to i}$ in the protocol π.
     - For each $k \in [n]$, $(b_k, c_k, \sigma_{b_k}, \sigma_{c_k})$ which are the authenticated input commitments.

  3. Whenever a token $\mathbf{T}_{mpc}^{j \to i}$ sends a message msg_j in round t, additionally it also acts as a prover in an execution of a RWIAOK argument with every other token $\mathbf{T}_{mpc}^{k \to i}$ as the verifier for the statement $\mathsf{st}_{j,t} = (b_k, c_j, \mathsf{Trans}, \mathsf{msg}_j, t) \in L_2$ using witness $w_{j,t} = (x_j, r_{c_j}, \mathsf{cor.rand}, \perp)$. Here, Trans denotes the transcript of the protocol upto round $(t-1)$ and $\mathsf{cor.rand}$ is the decryption of all the ct it receives. Once again P_i acts as the channel.

  4. Finally, compute and output $\mathsf{out} = \pi.\mathsf{Out}(x_i, \mathsf{Trans})$ where Trans denotes the transcript of the protocol.

Remark: In the above description, we were assuming non-interactive commitments (which require injective one way functions) to ease the exposition. In order to rely on just one way functions, we switch our commitment protocol to a two-round one where the receiver sends the first message.

Reusability: If we want to allow our tokens to be reused an unbounded number of times for performing multiple MPC protocols between the same set of parties, we can tweak the protocol as follows: instead of hardwiring P_i's input $x_i = (\mathsf{inp}_i, r_i)$ inside the tokens $\mathbf{T}_{mpc}^{i \to j}$ sent by P_i, we can just hardwire an encryption key ek_i for a symmetric encryption scheme. Now, in this setting, the tokens are exchanged apriori in an initial token exchange phase. Then, in the first round, when P_i sends $c_i = \mathsf{Commit}(x_i)$ and message b_i, it also sends $\mathsf{ct}_i = \mathsf{enc}_{\mathsf{ek}_i}(x_i)$ and

---

[8] To ease the exposition, we assume that x_k and r_k are hardwired inside each token. Instead, we can have each party broadcast encrypted signed versions of them which are sent to the respective token along with the other messages.

Note: For better understanding of the rest of the protocol, this figure actually describes the code of token $\mathbf{T}_{mpc}^{j \to i}$ created by party P_j and sent to P_i. The code of $\mathbf{T}_{mpc}^{i \to j}$ is symmetrical.

Constants: $(x_j, c_j, r_{x_j}, a_j, b_j, k_j^{rzk}, k_j^{rwi}, \mathsf{PRF}, (\mathsf{sk}, \mathsf{vk}), \mathsf{ek})$

1. If input = ("RZKAOK", msg): using a random tape defined by $\mathsf{PRF}(k_j^{rzk}, 0^{n+1})$ and the prover algorithm (RZKAOK.Prove), engage in an execution of a RZKAOK with the querying party playing the role of the verifier for the statement $(\mathsf{st}_j) \in L_1$ where $\mathsf{st}_j = (b_j, c_j)$ using witness (a_j, x_j, r_{c_j}) where msg is the next message of the verifier in the protocol. Output the next message of the prover.
2. If input = ("MPC", $\{\mathsf{ct}, \sigma_{\mathsf{ct}}\}$, $\{b_k, c_k, \sigma_{b_k}, \sigma_{c_k}\}$), do the following:
   (a) If the signatures verify, engage in an execution of the MPC protocol π with $(n-1)$ other parties for the functionality f using input inp_j, randomness r_j and correlated randomness as the set of decryptions of $\{\mathsf{ct}\}$. Here, the querying party acts as the channel.
   (b) In round t, if party P_k sends a message msg, also engage in an execution of a RWIAOK argument acting as the verifier with party P_k as the prover for the statement $\mathsf{st}_{k,t} = (b_j, c_k, \mathsf{Trans}, \mathsf{msg}, t) \in L_2$ where Trans denotes the transcript of the protocol upto round $(t-1)$.
   (c) In round t, after sending a message msg, for every other party P_k, engage in an execution of a RWIAOK argument using the prover algorithm with party P_k acting as the verifier for the statement $\mathsf{st}_{j,t} = (b_k, c_j, \mathsf{Trans}, \mathsf{msg}, t) \in L_2$ using witness $w_{j,t} = (x_j, r_{c_j}, \mathsf{cor.rand}, \perp)$ where Trans denotes the transcript of the protocol upto round $(t-1)$ and cor.rand is the decryption of the set of ct using the secret key ek.

Fig. 5. Code of token $\mathbf{T}_{mpc}^{j \to i}$

$\sigma_{\mathsf{ct}_i} = \mathsf{Sign}(\mathsf{ct}_i)$. That is, it sends an encryption of its input and a signature on this encryption. Every party P_j is now expected to also query the token $\mathbf{T}_{mpc}^{i \to j}$ with $(\mathsf{ct}_i, \sigma_{\mathsf{ct}_i})$ along with c_i, b_i in every query. The token verifies the signature, decrypts the message to learn the input x_i and then proceeds to use it for the rest of the computation.

6.2 Security

We formally prove security in the full version of the paper.

References

[AAG+14] Agrawal, S., Ananth, P., Goyal, V., Prabhakaran, M., Rosen, A.: Lower bounds in the hardware token model. In: TCC (2014)

[Bea96] Beaver, D.: Correlated pseudorandomness and the complexity of private computations. In: STOC (1996)

[BGGL01] Barak, B., Goldreich, O., Goldwasser, S., Lindell, Y.: Resettably-sound zero-knowledge and its applications. In: FOCS (2001)

[BJOV18] Badrinarayanan, S., Jain, A., Ostrovsky, R., Visconti, I.: Non-interactive secure computation from one-way functions. In: Peyrin, T., Galbraith, S. (eds.) ASIACRYPT 2018. LNCS, vol. 11274, pp. 118–138. Springer, Cham (2018). https://doi.org/10.1007/978-3-030-03332-3_5

[BP13] Bitansky, N., Paneth, O.: On the impossibility of approximate obfuscation and applications to resettable cryptography. In: STOC (2013)

[BP15] Bitansky, N., Paneth, O.: On non-black-box simulation and the impossibility of approximate obfuscation. SIAM J. Comput. 44, 1325–1383 (2015)

[Can01] Canetti, R.: Universally composable security: a new paradigm for cryptographic protocols. In: FOCS (2001)

[CDPW07] Canetti, R., Dodis, Y., Pass, R., Walfish, S.: Universally composable security with global setup. In: TCC (2007)

[CGGM00] Canetti, R., Goldreich, O., Goldwasser, S., Micali, S.: Resettable zero-knowledge (extended abstract). In: STOC (2000)

[CGS08] Chandran, N., Goyal, V., Sahai, A.: New constructions for UC secure computation using tamper-proof hardware. In: EUROCRYPT (2008)

[CJS14] Canetti, R., Jain, A., Scafuro, A.: Practical UC security with a global random oracle. In: CCS (2014)

[CKS+14] Choi, S.G., Katz, J., Schröder, D., Yerukhimovich, A., Zhou, H.-S.: (Efficient) universally composable oblivious transfer using a minimal number of stateless tokens. In: TCC (2014)

[COP+14] Chung, K.-M., Ostrovsky, R., Pass, R., Venkitasubramaniam, M., Visconti, I.: 4-round resettably-sound zero knowledge. In: TCC (2014)

[COPV13] Chung, K.-M., Ostrovsky, R., Pass, R., Visconti, I.: Simultaneous resettability from one-way functions. In: FOCS (2013)

[CPS13] Chung, K.-M., Pass, R., Seth, K.: Non-black-box simulation from one-way functions and applications to resettable security. In: STOC (2013)

[CPS16] Chung, K.-M., Pass, R., Seth, K.: Non-black-box simulation from one-way functions and applications to resettable security. SIAM J. Comput. 45(2), 415–458 (2016)

[DKM11] Döttling, N., Kraschewski, D., Müller-Quade, J.: Unconditional and composable security using a single stateful tamper-proof hardware token. In: TCC (2011)

[DKM12] Döttling, N., Kraschewski, D., Müller-Quade, J.: Statistically secure linear-rate dimension extension for oblivious affine function evaluation. In: ICITS (2012)

[DKMN15a] Döttling, N., Kraschewski, D., Müller-Quade, J., Nilges, T.: From stateful hardware to resettable hardware using symmetric assumptions. In: ProvSec (2015)

[DKMN15b] Döttling, N., Kraschewski, D., Müller-Quade, J., Nilges, T.: General statistically secure computation with bounded-resettable hardware tokens. In: TCC (2015)

[DMMN13] Döttling, N., Mie, T., Müller-Quade, J., Nilges, T.: Implementing resettable UC-functionalities with untrusted tamper-proof hardware-tokens. In: TCC (2013)

[GGM86] Goldreich, O., Goldwasser, S., Micali, S.: How to construct random functions. J. ACM 33, 792–807 (1986)

[GIS+10] Goyal, V., Ishai, Y., Sahai, A., Venkatesan, R., Wadia, A.: Founding cryptography on tamper-proof hardware tokens. In: TCC (2010)

[GKR08] Goldwasser, S., Kalai, Y.T., Rothblum, G.N.: One-time programs, In: CRYPTO (2008)

[HPV16] Hazay, C., Polychroniadou, A., Venkitasubramaniam, M.: Composable security in the tamper-proof hardware model under minimal complexity. In: Hirt, M., Smith, A. (eds.) TCC 2016. LNCS, vol. 9985, pp. 367–399. Springer, Heidelberg (2016). https://doi.org/10.1007/978-3-662-53641-4_15

[HPV17] Hazay, C., Polychroniadou, A., Venkitasubramaniam, M.: Constant round adaptively secure protocols in the tamper-proof hardware model. In: PKC (2017)

[IKO+11] Yuval, I., Kushilevitz, E., Prabhakaran, M., Sahai, A.: Efficient non-interactive secure computation. In: EUROCRYPT (2011)

[IPS08] Yuval, I., Prabhakaran, M., Sahai, A.: Founding cryptography on oblivious transfer - efficiently. In: CRYPTO (2008)

[Kat07] Katz, J.: Universally composable multi-party computation using tamper-proof hardware. In: EUROCRYPT (2007)

[Kil88] Kilian, J.: Founding cryptography on oblivious transfer. In: STOC (1988)

[Kol10] Kolesnikov, V.: Truly efficient string oblivious transfer using resettable tamper-proof tokens. In: TCC (2010)

[MMN16] Mechler, J., Müller-Quade, J., Nilges, T.: Universally composable (non-interactive) two-party computation from untrusted reusable hardware tokens. IACR Cryptol. ePrint Archive **2016**, 615 (2016)

[MS08] Moran, T., Segev, G.: David and goliath commitments: UC computation for asymmetric parties using tamper-proof hardware. In: EUROCRYPT (2008)

[Nao91] Naor, M.: Bit commitment using pseudorandomness. J. Cryptol. **4**, 151–158 (1991)

[Nil15] Nilges, T.: The Cryptographic Strength of Tamper-Proof Hardware. PhD thesis, Karlsruhe Institute of Technology (2015)

[Rom90] Rompel, J.: One-way functions are necessary and sufficient for secure signatures. In: Proceedings of the Twenty-Second Annual ACM Symposium on Theory of Computing, pp. 387–394. ACM (1990)

[Yao86] Yao, A.C.C.: How to generate and exchange secrets (extended abstract). In: FOCS (1986)

Efficient UC Commitment Extension
with Homomorphism for Free
(and Applications)

Ignacio Cascudo[1]([✉]), Ivan Damgård[2], Bernardo David[3], Nico Döttling[4],
Rafael Dowsley[5], and Irene Giacomelli[6]

[1] IMDEA Software Institute, Madrid, Spain
ignacio.cascudo@imdea.org
[2] Aarhus University, Aarhus, Denmark
ivan@cs.au.dk
[3] IT University of Copenhagen, Copenhagen, Denmark
bernardo@bmdavid.com
[4] CISPA Helmholtz Center for Information Security, Saarbrücken, Germany
nico.doettling@gmail.com
[5] Bar Ilan University, Tel Aviv, Israel
rafael@dowsley.net
[6] Protocol Labs, Inc., Basel, Switzerland
irene@protocol.ai

Abstract. Homomorphic universally composable (UC) commitments allow for the sender to reveal the result of additions and multiplications of values contained in commitments without revealing the values themselves while assuring the receiver of the correctness of such computation on committed values. In this work, we construct essentially optimal additively homomorphic UC commitments from any (not necessarily UC or homomorphic) extractable commitment, while the previous best constructions require oblivious transfer. We obtain amortized linear computational complexity in the length of the input messages and rate 1. Next, we show how to extend our scheme to also obtain multiplicative homomorphism at the cost of asymptotic optimality but retaining low concrete complexity for practical parameters. Moreover, our techniques yield public coin protocols, which are compatible with the Fiat-Shamir heuristic. These results come at the cost of realizing a restricted version of the homomorphic commitment functionality where the sender is allowed to perform any number of commitments and operations on committed messages but is only allowed to perform a single batch opening

I. Cascudo—This work was done while Ignacio Cascudo was with Department of Mathematics, Aalborg University, Denmark.
I. Damgård and R. Dowsley—This project has received funding from the European Research Council (ERC) under the European Unions' Horizon 2020 research and innovation programme under grant agreement No 669255 (MPCPRO).
B. David—This work was partially supported by DFF grant number 9040-00399B (TrA²C).
I. Giacomelli—This work was done while Irene Giacomelli was with the ISI Foundation (Turin, Italy) and supported by Intesa Sanapolo Innovation Center.

S. D. Galbraith and S. Moriai (Eds.): ASIACRYPT 2019, LNCS 11922, pp. 606–635, 2019.
https://doi.org/10.1007/978-3-030-34621-8_22

of a number of commitments. Although this functionality seems restrictive, we show that it can be used as a building block for more efficient instantiations of recent protocols for secure multiparty computation and zero knowledge non-interactive arguments of knowledge.

1   Introduction

A commitment scheme is the digital equivalent of a locked box containing a committed message chosen by a prover. Once the prover gives away the box to a verifier, the content cannot be changed, the commitment is binding. On the other hand, the verifier cannot look into the box so the message is hidden until the prover gives away the key to the box. Commitments are perhaps the most fundamental building block in cryptographic protocols and despite the conceptual simplicity of the primitive, it has far-reaching consequences and many applications, e.g., to coin-flipping, zero-knowledge proofs and many other things.

The simplest form of commitment that only have the basic binding and hiding properties follow from one-way functions. On the other hand, one may wish for many other properties, such as non-malleability, security under composition etc. The strongest form of commitments, namely UC secure commitments, has all these properties, but on the other hand can only be implemented under setup assumptions, such as the common reference string model. In this model, UC commitments imply secure key exchange, so since some sort of public-key technology seems to be required, it was believed for a long time that even if UC commitments are the gold standard for security, they must be much less efficient than the weaker type that only requires symmetric primitives.

However, in [19] and independently in [9,24], this was shown to be false: one can push the use of public-key technology into a preprocessing phase that is only needed once and for all and the cost of which does not depend on the number of commitments to be done later. Notably, the actual commitment and opening protocols only requires simple finite field algebra and a pseudorandom generator. After this, a long line of research optimized this approach [17,22], culminating in [16] where it was shown that after doing $O(k + s)$ string OTs in the setup phase (where s is the statistical security parameter and k is the message length) one can commit at rate approaching 1, that is, the communication required is $k + o(1)$ bits, furthermore the computational complexity is linear in k^1. Finally, the commitments are additively homomorphic, i.e., one commits to vectors over a finite field $\mathbb{F}$, and if $\boldsymbol{a}, \boldsymbol{b} \in \mathbb{F}^k$ have been committed, prover and verifier can compute a commitment to $\boldsymbol{a} + \boldsymbol{b}$ which, if opened, would reveal only the sum.

The first construction from this line of work [19] had also a multiplicatively homomorphic property, namely the prover can send the verifier a single message, and this allows the verifier to compute a commitment to $\boldsymbol{a} * \boldsymbol{b}$, the coordinate-wise (Schur) product of the vectors. However, subsequent constructions did not have this property.

---

[1] All this holds in an amortized sense, assuming we make enough commitments so that the cost of the setup phase is dwarfed.

So, while this line of research has resulted in constructions that are optimal in several respects, it still leaves some important and natural questions unanswered:

Is it overkill to use OT in the setup phase? All efficient earlier schemes [16,17,19,22,24] use OT in the preprocessing phase, but this is in general a stronger primitive than commitment. Even UC commitments do not always imply OT, this depends on the setup assumption. It is therefore natural to ask if we can make do with only commitment in the preprocessing, thus obtaining a proper "commitment extension" result.

Can we make an efficient multi-verifier scheme? The commitments from [16], and in fact all constructions from this line of work, can only work with one verifier because security against a corrupt prover depends on the verifier's private choice of selections bits in the initial OT's. Thus, if a prover needs to commit towards several verifiers, the only known solution is to run many instances of the scheme, one for each verifier and then on top of this have the prover convince the verifiers that (s)he committed to the same message. This seems quite far from an ideal solution.

Can we also get multiplicatively homomorphic schemes? The most efficient constructions are not multiplicative, but one earlier scheme was in fact "fully homomorphic" [19]. So it is natural to ask if we can solve the above problems and also get multiplication at the same time.

1.1   Our Contributions

In this paper, we come up with positive answers to all of the above questions. We present a protocol for UC secure commitments that has the well known structure consisting of a preprocessing phase and a phase where the actual commitments are built, computed on and opened. In addition to achieving the same asymptotic efficiency as the former best scheme [16] in the single-verifier additive case, our protocol supports multiple verifiers and multiplicative homomorphism.

In contrast to previous work, however, the preprocessing only makes use of a commitment scheme (and not OT)[2]. Notably, however, this commitment scheme does not need to be homomorphic, and in fact it does not even need to be UC secure. It just needs to be extractable and hiding - here, extractable means that the simulator can extract the committed value from a corrupt prover. For UC full security one usually needs also equivocation (when the prover is honest, the simulator can fake a commitment and later open it to any value). The commitment scheme we build uses only a PRG and finite field arithmetic after the preprocessing. It has rate 1, it is additively homomorphic, and linear time. Security does not depend on any secret choices of the verifier, so the scheme easily extends to multiple verifiers with no essential loss of security. Finally, we show how to make the scheme multiplicative, the scheme is then only quasilinear, and we get constant rate instead of rate 1.

---

[2] The scheme of [9] can be constructed from an extractable commitment and an equivocal commitment. However, it is intrinsically incompatible with homomorphic operations.

All these results come at the cost that what we implement is a slightly weaker commitment functionality than the standard one. Namely, it allows opening of committed values only in a final stage and after this the functionality stops working. Equivalently, one can think of this as a functionality one can use exactly as the standard one, except that when opening a value the prover simply tells the verifier what the committed value is. Of course a corrupt prover can lie, but there is a final verification stage where the prover will be caught if he lied.

We show that despite this limitation there are a wide range of applications for the scheme. While we describe these in more detail below, it is already intuitively clear that our functionality is sufficient for ZK proofs, for instance: the verifier needs to decide to accept or reject only at the end of the protocol so it is sufficient that a cheating prover is caught at that point. As a simple example of the power of our construction, consider that UC secure commitments are easy to implement in the (global) random oracle model [11]: one simply inputs the message concatenated with some randomness to the oracle and uses the output as the commitment. Of course, a random oracle based scheme has no homomorphic properties: a random oracle "by definition" has no such structure. But nevertheless, we can use it as commitment scheme in our preprocessing and get a homomorphic scheme. In general, one can think of our protocol as a "commitment extension" result. It is similar to the well known OT extension protocols, but incomparable because we get extra homomorphic properties (and perhaps UC security) for free, but we realize a slightly weaker functionality.

Techniques. On the technical side, our approach is best described by referring to previous work such as [16]: the main idea there was that the prover commits to a vector a by encoding it using a linear code C. He then additively secret shares each coordinate in the codeword $C(a)$ to get two shares for each position. Using the OT's from the preprocessing, the verifier will learn one out of the two shares for each position, however, the prover does not know which shares the verifier has. To open, the prover must reveal $C(a)$ and all shares, and the verifier can now check that the prover sent a codeword and that the shares are consistent with $C(a)$ and with the shares the verifier knows.

Intuitively, since the verifier has only one share of each coordinate, $C(a)$ is unknown to him at commit time. On the other hand, if the prover wants to open a different value, he must change to a different codeword. However, if C has large minimum distance, this means the prover must change many coordinates and therefore must lie about many of the shares. Since he does not know which shares he can change without being detected, this can only be done with negligible success probability[3].

In order to avoid having to do an OT for each codeword position and each commitment, instead the prover chooses seeds $s_{i,j}$ for a PRG, where i points to a codeword position and $j = 0, 1$. The shares for all the commitments are then

---

[3] This argument works, even if the prover did not choose a codeword at commit time. If we also want to have additive homomorphism, we need to check that the prover chose something that it at least close to a codeword. This can be done using, e.g., the interactive proximity testing from [16].

constructed by running the PRG on all these seeds and for each i an OT is done that transfers either $s_{i,0}$ or $s_{i,1}$ to the verifier.

Our key observation now is that it is actually sufficient if the prover simply commits to the seeds in the preprocessing phase, if we are careful later. Namely, we run the same protocol as we would have done had the OTs been used, but at the end of the protocol, the verifier will ask the prover to reveal either $s_{i,0}$ or $s_{i,1}$ for each i. Note that, as long as a corrupt prover cannot predict which seeds he will be asked for, he is in the exactly same position as in the original protocol. The verifier will receive the same information as before, but cannot verify it until the end, so hence openings can only be done, or at least can only be verified, at the end. A corrupt verifier clearly has no advantage compared to the OT based protocol: he learns the same information, only later.

A very nice "side effect" of this is that we can now easily have several verifiers. They just need to receive the prover's initial commitments (assuming, of course that the initial commitments support this). Then at the end, they can decide, e.g., by coin flipping which seeds to ask for.

We also extend the commitment scheme to allow for proving multiplicative relations on committed values. For this purpose, we require the code C to have the property that its square C^{*2} is also a good code, with large minimum distance. Here C^{*2} is defined to be the span of all pairwise Schur-products of words from C. Moreover, we replace the 2-party additive secret sharing by 3-party linear secret sharing which is multiplicative: the Schur-product of sets of shares of $u, v \in \mathbb{F}$ is (essentially) an additive secret sharing of uv. The effect of all this is that if we multiply two commitments to a, b by multiplying corresponding components of them, we obtain a commitment to $a * b$ of essentially the same form as in the original protocol, except that underlying code is now C^{*2}. See more details within. The new demands we place on C imply that we can only get constant rate and not rate 1 and also that complexity will be quasilinear rather than linear. The main motivation for this construction is that we get the multiplicative property and at the same time have multiple verifiers and use only commitment for preprocessing. An earlier scheme that achieves multiplicative homomorphism was constructed in [19] via building first an elaborate VSS (verifiable secret sharing) scheme. Our construction obtains similar asymptotic complexity, but it requires less conditions on the underlying linear code. Indeed, our multiplicatively homomorphic scheme can be constructed from any linear code whose minimum distance and squares minimum distance are large enough. In contrast, [19] requires in addition a code whose duals minimum distance is large enough (i.e., equivalent to multiplicative secret sharing scheme). Thanks to this, for fixed security parameters we can give an explicit bound for the rate of our multiplicative commitment based on recent results on squares of cyclic codes (details in Sect. 4).

1.2   Applications

Efficient Zero-Knowledge Arguments. A recent line of research is concerned with the construction of practically efficient succinct non-interactive zero-knowledge arguments of knowledge (e.g. [1,10,31]) with a particular focus on optimizing the efficiency of the prover while keeping verification complexity sub-linear.

One such approach, originally dating back to [5], compiles a public coins interactive proof system for a language $\mathcal{L}$ into a zero-knowledge proof system for the same language. This transformation is conceptually simple: Instead of sending its messages to the verifier in the clear, the prover provides only commitments of his messages to the verifier. At the end of the protocol, the prover provides a zero-knowledge proof to the verifier which asserts that the verifier of the original proof system would accept the committed transcript. This transformation has received renewed interest in the light of efficient P-delegation schemes [25,29].

Wahby et al. [31] observed that this approach can be implemented in a particularly efficient way if the verifier of the interactive proof system is algebraic: In this case the zero-knowledge proof in the transformation of [5] can be implemented very efficiently via homomorphic commitments.

We show that using our homomorphic commitment scheme, this transformation can be performed at a very low overhead, i.e. we can convert any public coin interactive proof system with algebraic verifier into an honest-verifier zero-knowledge proof system such that the communication complexity of the protocol grows only by a small factor and both prover and verifier incur only a small constant factor overhead. Using the Fiat Shamir transform [20], we can convert such a proof system into a succinct non-interactive zero-knowledge argument.

Committed MPC. The so called "Committed MPC" protocol [21] requires a multiparty additively homomorphic commitment protocol that supports additions of commitments generated by different senders. While a generic approach for constructing such schemes from any two-party additively homomoprhic commitments was proposed in [21], their generic construction for t parties requires t^2 calls to the underlying commitment scheme. If instantiated with the previously best two-party additively homomorphic commitment protocol of [16] using a $[n, k, s]$ code, this construction would require nt^2 OTs plus extra communication in the order of $O(nmt^2)$ to commit to m messages of length k. We provide a new generic construction from multi-receiver additively homomorphic commitments which can be instantiated with our new protocols, requiring only nt non-homomorphic commitments (*e.g.* random oracle commitments) plus extra communication in the order of $O(smt)$ to achieve the same.

Insured MPC. The topic of MPC with financial penalties has attracted increasing attention recently [2,4,6,7,26]. The main idea is to combine MPC techniques with cryptocurrencies in order to provide monetary incentives for the participants to act honestly during the protocol execution. Insured MPC [4], the most efficient solution to date, uses a publicly verifiable additively homomorphic multi-receiver commitment as an important component to build the protocol. However, the employed commitment scheme is a bottleneck in that construction as its complexity grows quadratically in the number of participants. Using our new techniques together with an authenticated bulletin board (which is also used in the previous construction), it is possible to dramatically improve the performance of publicly verifiable additively homomorphic multi-receiver commitment. We can obtain extremely efficient instantiations, for instance, by using the canonical random oracle commitment scheme. The improvement in computational and communication complexity achieved for this application is very similar to that of the Committed MPC case, since the previously best protocol for publicly verifiable additively homomorphic multi-receiver commitments [4] has a very similar structure to the multi-sender protocol of [21]. Thus, we basically go from quadratic to linear in the number of players.

2   Preliminaries

In this section we establish notation and introduce notions that will be used throughout the paper. We borrow much of the notation from [16].

Notation. The set of the n first positive integers is denoted $[n] = \{1, 2, \ldots, n\}$. Given a finite set D, sampling a uniformly random element from D is denoted $r \xleftarrow{\$} D$. Vectors of elements of some field are denoted by bold lower-case letters, while matrices are denoted by bold upper-case letters. We denote finite fields by $\mathbb{F}$ and write $\mathbb{F}_q$ for the finite field of size q. For $z \in \mathbb{F}^k$, $z[i]$ denotes the i'th entry of the vector, where $z[1]$ is the first element of z. The coordinate-wise (Schur) product of two vectors is denoted by $*$, i.e. if $a, b \in \mathbb{F}^n$, then $a * b \in \mathbb{F}^n$ and $(a * b)[i] = a[i]b[i]$. If $A \subseteq [n]$, we will use π_A to denote the projection that outputs the coordinates with index in A of a vector. For a matrix $\mathbf{M} \in \mathbb{F}^{n \times k}$, we let $\mathbf{M}[\cdot, j]$ denote the j'th column of $\mathbf{M}$ and $\mathbf{M}[i, \cdot]$ denote the i'th row. The row support of $\mathbf{M}$ is the set of indices $I \subseteq \{1, \ldots, n\}$ such that $\mathbf{M}[i, \cdot] \neq \mathbf{0}$.

We say that a function ϵ is negligible in n if for every positive polynomial p there exists a constant c such that $\epsilon(n) < \frac{1}{p(n)}$ when $n > c$. Two ensembles $X = \{X_{\kappa,z}\}_{\kappa \in \mathbb{N}, z \in \{0,1\}^*}$ and $Y = \{Y_{\kappa,z}\}_{\kappa \in \mathbb{N}, z \in \{0,1\}^*}$ of binary random variables are said to be *statistically indistinguishable*, denoted by $X \approx_s Y$, if for all z it holds that $|\Pr[\mathcal{D}(X_{\kappa,z}) = 1] - \Pr[\mathcal{D}(Y_{\kappa,z}) = 1]|$ is negligible in κ for every probabilistic algorithm (distinguisher) $\mathcal{D}$. In case this only holds for computationally bounded (non-uniform probabilistic polynomial-time (*PPT*)) distinguishers we say that X and Y are *computationally indistinguishable* and denote it by $\approx_c$.

2.1   Coding Theory

For a vector $x \in \mathbb{F}^n$, we denote the Hamming-weight of x by $\|x\|_0 = |\{i \in [n] : x[i] \neq 0\}|$. Let $\mathsf{C} \subset \mathbb{F}^n$ be a linear subspace of $\mathbb{F}^n$. We say that C is an $\mathbb{F}$-linear $[n, k, d]$ code, if C has dimension k and it holds for every nonzero $x \in \mathsf{C}$ that $\|x\|_0 \geq d$, i.e., the minimum distance of C, denoted $\mathrm{dist}(\mathsf{C})$, is at least d. The distance $\mathrm{dist}(\mathsf{C}, x)$ between C and a vector $x \in \mathbb{F}^n$ is the minimum of $\|c - x\|_0$ when $c \in \mathsf{C}$. The rate of an $\mathbb{F}$-linear $[n, k, d]$ code is $\frac{k}{n}$ and its relative minimum distance is $\frac{d}{n}$.

A matrix $\mathbf{G} \in \mathbb{F}^{n \times k}$ is a generator matrix of C if $\mathsf{C} = \{\mathbf{G}x : x \in \mathbb{F}^k\}$, and we write $\mathsf{C}(x) = \mathbf{G}x$. The code C is systematic if it has a generator matrix $\mathbf{G}$ such that the submatrix given by the top k rows of $\mathbf{G}$ is the identity matrix $\mathbf{I} \in \mathbb{F}^{k \times k}$.

For an $\mathbb{F}$-linear $[n, k, d]$ code C, we denote by $\mathsf{C}^{\odot m}$ the *m-interleaved product* of C, which is defined by $\mathsf{C}^{\odot m} = \{\mathbf{C} \in \mathbb{F}^{n \times m} : \forall i \in [m] : \mathbf{C}[\cdot, i] \in \mathsf{C}\}$.

In other words, $C^{\odot m}$ consists of all $\mathbb{F}^{n \times m}$ matrices for which all columns are in C. We can think of $C^{\odot m}$ as a linear code with symbol alphabet $\mathbb{F}^m$, where we obtain codewords by taking m arbitrary codewords of C and bundling together the components of these codewords into symbols from $\mathbb{F}^m$. For a matrix $\mathbf{E} \in \mathbb{F}^{n \times m}$, $\|\mathbf{E}\|_0$ is the number of nonzero rows of $\mathbf{E}$, and the code $C^{\odot m}$ has minimum distance at least d' if all nonzero $\mathbf{C} \in C^{\odot m}$ satisfy $\|\mathbf{C}\|_0 \geq d'$. With this definition, it is easy to see that $\mathsf{dist}(C^{\odot m}) = \mathsf{dist}(C)$.

For an $\mathbb{F}$-linear $[n, k, d]$ code C, we denote by C^{*2} the *Schur square* of C, which is defined as the linear subspace of $\mathbb{F}^n$ generated by all the possible vectors of the form $\boldsymbol{v} * \boldsymbol{w}$ with $\boldsymbol{v}, \boldsymbol{w} \in C$. This is an $[n, \hat{k}, \hat{d}]$ code where $\hat{k} \geq k$ and $\hat{d} \leq d$.

2.2 Interactive Proximity Testing and Linear Time Building Blocks

We will use the interactive proximity testing technique and corresponding linear time building blocks introduced in [16]. As stated in [16], this technique consists in the following argument: suppose we sample a function $\mathbf{H}$ from an almost universal family of linear hash functions (from $\mathbb{F}^m$ to $\mathbb{F}^\ell$), and we apply this to each of the rows of a matrix $\mathbf{X} \in \mathbb{F}^{n \times m}$, obtaining another matrix $\mathbf{X}' \in \mathbb{F}^{n \times \ell}$; because of linearity, if $\mathbf{X}$ belonged to an interleaved code $C^{\odot m}$, then $\mathbf{X}'$ belongs to the interleaved code $C^{\odot \ell}$. Theorem 1 states that we can test whether $\mathbf{X}$ is close to $C^{\odot m}$ by testing instead if $\mathbf{X}'$ is close to $C^{\odot \ell}$ (with high probability over the choice of the hash function) and moreover, if these elements are close to the respective codes, the set of rows that have to be modified in each of the matrices in order to correct them to codewords are the same.

Definition 1 (Almost Universal Linear Hashing [16]). *We say that a family* $\mathcal{H}$ *of linear functions* $\mathbb{F}^n \rightarrow \mathbb{F}^s$ *is* ϵ-*almost universal, if it holds for every non-zero* $\mathbf{x} \in \mathbb{F}^\mathbf{n}$ *that*

$$\Pr_{\mathbf{H} \xleftarrow{s} \mathcal{H}} [\mathbf{H}(\mathbf{x}) = 0] \leq \epsilon,$$

where $\mathbf{H}$ *is chosen uniformly at random from the family* $\mathcal{H}$. *We say that* $\mathcal{H}$ *is universal, if it is* $|\mathbb{F}^{-s}|$-*almost universal. We will identify functions* $H \in \mathcal{H}$ *with their transformation matrix and write* $\mathbf{H}(\mathbf{x}) = \mathbf{H} \cdot \mathbf{x}$.

Theorem 1 (Theorem 1 in [16]). *Let* $\mathcal{H} : \mathbb{F}^m \rightarrow \mathbb{F}^{2s+t}$ *be a family of* $|\mathbb{F}|^{-2s}$-*almost universal* $\mathbb{F}$-*linear hash functions. Further let* C *be an* $\mathbb{F}$-*linear* $[n, k, s]$ *code. Then for every* $\mathbf{X} \in \mathbb{F}^{n \times m}$ *at least one of the following statements holds, except with probability* $|\mathbb{F}|^{-s}$ *over the choice of* $\mathbf{H} \xleftarrow{s} \mathcal{H}$:

1. $\mathbf{X}\mathbf{H}^\top$ *has distance at least* s *from* $C^{\odot(2s+t)}$
2. *For every* $\mathbf{C}' \in C^{\odot(2s+t)}$ *there exists a* $\mathbf{C} \in C^{\odot m}$ *such that* $\mathbf{X}\mathbf{H}^\top - \mathbf{C}'$ *and* $\mathbf{X} - \mathbf{C}$ *have the same row support*

Remark 1 ([16]). If the first item in the statement of the Theorem does not hold, the second one must hold. Then we can efficiently recover a codeword $\mathbf{C}$ with distance at most $s - 1$ from $\mathbf{X}$ using erasure correction, given a codeword $\mathbf{C}' \in C^{\odot(2s+t)}$ with distance at most $s - 1$ from $\mathbf{X}\mathbf{H}^\top$. More specifically, we compute the row support of $\mathbf{X}\mathbf{H}^\top - \mathbf{C}'$, erase the corresponding rows of $\mathbf{X}$ and recover $\mathbf{C}$ from $\mathbf{X}$ using erasure correction[4]. The last step is possible as the distance between $\mathbf{X}$ and $\mathbf{C}$ is at most $s - 1$.

In order to achieve linear time and optimal rate (*i.e.*, rate-1) in our constructions, we will need to instantiate interactive proximity testing with a family of linear time almost universal linear hash functions and a linear time encodable error correcting code that achieves rate 1. Theorems 3 and 6 from [16] guarantee that explicit constructions of such building blocks exist. The following theorem is a strengthening of Theorem 3 of [16] in that the output of the hash functions is guaranteed to be uniformly random given that its first l inputs are uniformly random. The full proof is given in the full version of this paper [18].

Theorem 2. *Fix a finite field $\mathbb{F}$ of constant size, let $s \in \mathbb{N}$ be a statistical security parameter, let $n \in \mathbb{N}$ and let $l = s + O(\log(n))$. Then there exists an explicit family $\mathcal{H} : \mathbb{F}^{l+n} \to \mathbb{F}^l$ of $|\mathbb{F}|^{-s}$-universal hash functions that can be represented by $O(s^2)$ bits and computed in time $O(n)$. Moreover, it holds for any function $H \in \mathcal{H}$ that if $\mathbf{x} = (x_1, \ldots, x_l, \ldots x_{l+n})$ is such that the $x_1, \ldots, x_l$ are independently uniform and $x_{l+1}, \ldots, x_{l+n}$ are independent of $x_1, \ldots, x_l$, then $H(\mathbf{x})$ is distributed uniformly random.*

2.3   Universal Composability

The protocols presented in this paper are proven secure in the Universal Composability (UC) framework introduced by Canetti in [12]. We refer the reader to the full version of this paper [18] and [12] for further details.

Adversarial Model: Our protocols will be proven secure against static and active adversaries: the adversary may deviate from the protocol in any arbitrary way but only corrupt parties before the protocol execution starts.

Setup Assumption: Since UC commitment protocols cannot be obtained in the plain model [13], they need a setup assumption, *i.e.*, a resource available to all parties before the protocol starts. In this work, our goal is to prove security in the $\mathcal{F}_{\text{COM}}$-hybrid model [12,14], where the parties have access to an ideal (non-homomorphic) commitment functionality (our constructions are described in the $\mathcal{F}_{\text{COM}}$-hybrid model for the sake of clarity, but they actually only need the underlying commitments to be extractable). Functionality $\mathcal{F}_{\text{COM}}$ is described in Fig. 1. Notice that we describe a version of $\mathcal{F}_{\text{COM}}$ that operates with a set V of multiple receivers instead of a single receiver. However, $\mathcal{F}_{\text{COM}}$ can operate as a standard two-party commitment functionality with a single receiver by setting $V = \{V_1\}$, in which case it can be realized in the CRS model under different assumptions with security against static malicious adversaries by a number of protocols such as [8,13,27].

---

[4] Recall that erasure correction for linear codes can be performed efficiently via gaussian elimination.

A recent result by Camenisch *et al.* [11] shows that the "canonical" random oracle commitment realizes this functionality in the Global Random Oracle model without extra computational assumptions achieving security against static malicious adversaries. We observe that the protocol in [11] supports multiple receivers. In this protocol, the sender commits to a message m with randomness r by sending to the receiver the output c of the global random oracle when queried on (r, m) and opens by revealing (r, m), which allows the receiver to verify by querying the global random oracle with the pair (r, m) received as opening and checking that the response is equal to c. Given that the random oracle functionality in this model is global, any number of receivers who have received the commitment and the opening can trivially obtain the same result in the verification.

Ideal Functionalities: In Sect. 3, we construct an additively homomorphic string commitment protocol that UC-realizes functionality $\mathcal{F}_{\mathrm{AHCOM}}$, described in Fig. 2. Similarly to a functionality of [16], $\mathcal{F}_{\mathrm{AHCOM}}$ augments the standard multiple commitments functionality $\mathcal{F}_{\mathrm{MCOM}}$ from [14] by introducing a command for adding two previously stored commitments and an abort command in the Commit Phase. Moreover, $\mathcal{F}_{\mathrm{AHCOM}}$ gives an honest sender commitments to random messages instead of letting it submit a message as input, which can be straightforwardly used to commit to arbitrary messages with additive homomorphism as shown in [16]. In order to model corruptions, functionality $\mathcal{F}_{\mathrm{AHCOM}}$ lets a corrupted sender choose the messages it wants to commit to. The abort is necessary to deal with inconsistent commitments that could be sent by a corrupted party. However, differently from [16] or [14], this functionality can operate with a set V of multiple receivers but only allows for a single opening of a batch of commitments, after which it halts, not allowing further commitments, additions or openings. Notice that this functionality can operate as a two-party commitment functionality with a single receiver by setting $V = \{V_1\}$. Section 4 shows how to modify the construction of Sect. 3 to obtain a protocol that UC-realizes the augmented functionality $\mathcal{F}_{\mathrm{MHCOM}}$ (Fig. 3), which also allows for multiplication of committed values.

Functionality $\mathcal{F}_{\mathrm{COM}}$

$\mathcal{F}_{\mathrm{COM}}$ is parameterized by commitment length λ. $\mathcal{F}_{\mathrm{COM}}$ interacts with a sender P, a set of receivers $V = \{V_1, \ldots, V_t\}$ and an adversary $\mathcal{S}$ and proceeds as follows:

- **Commit Phase:** Upon receiving a message (commit, $sid, ssid, P, V, m$) from P where $m \in \{0, 1\}^{\lambda}$, record the tuple $(ssid, P, V, m)$ and send (receipt, $sid, ssid, P, V$) to every receiver $V_i \in V$ and $\mathcal{S}$. Ignore subsequent commit messages with the same $ssid$.
- **Open Phase:** Upon receiving a message (reveal, $sid, ssid$) from P, if a tuple $(ssid, P, V, m)$ was previously recorded, then send (reveal, $sid, ssid, P, V, m$) to every receiver $V_i \in V$ and $\mathcal{S}$. Otherwise, ignore.

Fig. 1. Functionality $\mathcal{F}_{\mathrm{COM}}$.

Functionality $\mathcal{F}_{\text{AHCOM}}$

$\mathcal{F}_{\text{AHCOM}}$ interacts with a sender P, a set of receivers $V = \{V_1, \ldots, V_t\}$ and an adversary $\mathcal{S}$ and proceeds as follows:
- **Commit Phase**: The length of the committed messages λ is fixed and known to all parties.
  - If P is honest, upon receiving a message (commit, $sid, ssid, P, V$) from P, sample a random $\boldsymbol{m} \leftarrow \{0,1\}^\lambda$, record the tuple $(ssid, P, V, \boldsymbol{m})$, send the message (commit, $sid, ssid, P, V, \boldsymbol{m}$) to P and send the message (receipt, $sid, ssid, P, V$) to every receiver $V_i \in V$ and $\mathcal{S}$. Ignore any future commit messages with the same $ssid$ from P to V.
  - If P is corrupted, upon receiving a message (commit, $sid, ssid, P, V, \boldsymbol{m}$) from P, where $\boldsymbol{m} \in \{0,1\}^\lambda$, record the tuple $(ssid, P, V, \boldsymbol{m})$ and send the message (receipt, $sid, ssid, P, V$) to every receiver $V_i \in V$ and $\mathcal{S}$. Ignore any future commit messages with the same $ssid$ from P to V.
  - If a message (abort, $sid, ssid$) is received from $\mathcal{S}$, the functionality halts.
- **Addition**: Upon receiving a message (add, $sid, ssid_1, ssid_2, ssid_3, P, V$) from P: If tuples $(ssid_1, P, V, \boldsymbol{m}_1)$, $(ssid_2, P, V, \boldsymbol{m}_2)$ were previously recorded and $ssid_3$ is unused, record $(ssid_3, P, V, \boldsymbol{m}_1 + \boldsymbol{m}_2)$ and send the message (add, $sid, ssid_1, ssid_2, ssid_3, P, V$, success) to P, every receiver $V_i \in V$ and $\mathcal{S}$.
- **Open Phase**: Upon receiving a message (reveal, $sid, ssid_1, \ldots, ssid_o$) from P, for every $ssid \in \{ssid_1, \ldots, ssid_o\}$, if a tuple $(ssid, P, V, \boldsymbol{m})$ was previously recorded, then send (reveal, $sid, ssid, P, V, \boldsymbol{m}$) to every receiver $V_i \in V$ and $\mathcal{S}$, if not, send nothing. Finally, halt.

Fig. 2. Functionality $\mathcal{F}_{\text{AHCOM}}$

Functionality $\mathcal{F}_{\text{MHCOM}}$

Augment the functionality $\mathcal{F}_{\text{AHCOM}}$ (Figure 2) with the step:
- **Multiplication**: Upon receiving a message (mult, $sid, ssid_1, ssid_2, ssid_3, P, V$) from P: If tuples $(ssid_1, P, V, \boldsymbol{m}_1)$, $(ssid_2, P, V, \boldsymbol{m}_2)$ were previously recorded and $ssid_3$ is unused, record $(ssid_3, P, V, \boldsymbol{m}_1 * \boldsymbol{m}_2)$ and send the message (mult, $sid, ssid_1, ssid_2, ssid_3, P, V$, success) to P, every receiver $V_i \in V$ and $\mathcal{S}$.

Fig. 3. Functionality $\mathcal{F}_{\text{MHCOM}}$

3 Rate-1 Linear Time Additively Homomorphic Commitments

In this section, we construct a linear time additively homomorphic commitment protocol that achieves amortized rate-1 and linear time in the length of committed messages assuming an extractable (not homomorphic) commitment and a PRG as building blocks. Protocol Π_{AHCOM} realizes $\mathcal{F}_{\text{AHCOM}}$, which only allows for commitments to random messages. Interestingly, in this case we can achieve sublinear communication complexity in the commitment phase while maintaining rate-1 in the opening phase. Even though committing to random messages is useful for a number of applications (*e.g.* [23]) that $\mathcal{F}_{\text{AHCOM}}$ is sufficient for building a protocol Π_{ARBHCOM} that commits to arbitrary messages achieving rate-1 and running in linear time as discussed in [16]. Essentially, Protocol Π_{AHCOM} achieves the same asymptotic efficiency as the former best UC commitment scheme [16], while supporting multiple verifiers and without requiring OT in the preprocessing phase, resulting in better concrete efficiency.

The main idea is to use a "delayed watchlist" mechanism where the sender first commits to seeds that will be stretched by a PRG to instantiate the watchlist but only allows the receivers to learn the watch bits in a later point, at which the receivers choose a random subset of the seed commitments to be opened. Basically, the watchlist is viewed as a matrix $\mathbf{R} = \mathbf{R}_0 + \mathbf{R}_1$ such that, for each row of $\mathbf{R}$, the receiver learns only a row from either $\mathbf{R}_0$ or $\mathbf{R}_1$ without revealing to the sender which one. Instead of using a number of 1-out-of-2 random OTs to obtain seeds that are stretched to generate each line of $\mathbf{R}_0$ or $\mathbf{R}_1$ in the beginning of the protocol as in previous works, the receiver relies on simple commitments to each seed sent by the sender. This scheme achieves rate-1 using similar techniques as [16]: first having the sender adjust the bottom bits of the watchlist matrix $\mathbf{R}$ so that its columns are codewords of random strings (in the top bits of $\mathbf{R}$) and then using interactive proximity testing to convince the receiver that these columns are indeed "very close" to codewords. In order to "open" a commitment, the sender reveals the columns from both $\mathbf{R}_0$ and $\mathbf{R}_1$ corresponding to that commitment, allowing a receiver who knows rows from each of these matrices to check that the revealed column vector corresponds to the watchlist with high probability. However, in our new scheme, the receiver only chooses which commitments to seeds will be revealed after the sender has sent this opening information. Otherwise, the sender would learn which rows of $\mathbf{R}_0$ or $\mathbf{R}_1$ the receiver would check, being able to open commitments to arbitrary messages. Protocol Π_{AHCOM} is described in Figs. 4 and 5.

In comparison to the protocol of [16], our scheme realizes a functionality with a caveat that only one opening of a batch of commitments is allowed (after which it terminates). However, this limited functionality is sufficient for a number of applications that we discuss in later sections. Moreover, our protocol has two important properties that the scheme of [16] lacks: it is public coin and supports multiple receivers. Notice that the watch bits of the receiver (represented by a row from either $\mathbf{R}_0$ or $\mathbf{R}_1$) are chosen at random but in public by the receiver. Hence, given an underlying commitment that support multiple receivers (*e.g.*, the canonical random oracle commitment scheme), it is sufficient to have the receivers run a simple commit-then-open coin tossing protocol to choose the watch bits they will learn, then have the sender publicly open his seed commitments. Interestingly, having the receivers broadcast their coin tossing commitments at the beginning of the protocol (before the sender broadcasts opening information), allows the simulator to both equivocate and extract commitments solely by extracting the underlying commitments. Notice that the simulator can equivocate a commitment by knowing in advance the watch bits to be learned by the receivers and extract a commitment by learning the whole watchlist, which are fixed in the sender's seed commitments. In order to eliminate interaction with the receivers, the random watch bits to be opened can be selected with the help of a random oracle following the Fiat-Shamir heuristic.

Efficiency. We achieve the same asymptotic complexity as [16] but with a preprocessing phase that can be instantiated with lower concrete complexity since it only requires extractable commitments. All phases of the Π_{AHCOM} run in linear time (requiring a constant number of operations per committed bit) when we use a linear time PRG (*i.e.*, with a constant number of operations per generated bit [30]) a linear time encodable code C (*e.g.* the one from [16]) and a linear time linear almost universal hash function $\mathbf{H}$ (*e.g.* the one from [16]). The cost of the calls to $\mathcal{F}_{\text{AHCOM}}$ is amortized over the number of commitments, which does not need to be very large if $\mathcal{F}_{\text{AHCOM}}$ is instantiated with cheap random oracle based commitments. The commitment phase achieves sublinear communication complexity when committing to random messages, since a rate-1 $[n, k, s]$-code C is used and only $\mathbf{W}, \mathbf{T_0}, \mathbf{T_1}$ (of size $O(1)$) are exchanged. Even if the trick from [16] is used to commit to arbitrary messages, only k extra bits need to be sent per message. In this case, our protocol achieves rate-1, meaning that the amortized overhead per committed bit is $o(1)$ for a sufficiently large number of commitments. The opening phase as described in Fig. 5 does not achieve rate-1, since the sender has to send both $\mathbf{A_0}[\cdot, j]\ \mathbf{A_1}[\cdot, j]$. However, it can be modified to achieve rate-1 using the same technique from [16], where a batch of commitments are opened by performing interactive proximity testing on a matrix $\mathbf{A}'$ containing the columns of $\mathbf{A}$ corresponding to the commitments to be opened. The receivers can use another coin-tossing to select a hash function $\mathbf{H}$, then the sender sends $\mathbf{A}'$, $\mathbf{T_0}' = \mathbf{A_0}'\mathbf{H}$ and $\mathbf{T_1}' = \mathbf{A_1}'\mathbf{H}$. The receivers check that $\mathbf{A}'\mathbf{H} = \mathbf{T_0}' + \mathbf{T_1}'$, that all columns in $\mathbf{A}'$ are in C and that $\mathbf{\Delta T_0}' + (\mathbf{I} - \mathbf{\Delta})\mathbf{T_1}' = \mathbf{B}'\mathbf{H}$, where $\mathbf{B}'$ contains the columns from $\mathbf{B}$ corresponding to the commitments being checked. This technique can be proven secure with the same techniques used for the case of a corrupt sender.

Security Analysis. For the sake of clarity, we will prove Protocol Π_{AHCOM}'s security in the $\mathcal{F}_{\text{COM}}$-hybrid model, *i.e.* assuming access to an ideal functionality for commitments. The proof of security for Protocol Π_{AHCOM} is very similar to that of the scheme of [16], with the exception that all information the simulator needs to extract and equivocate commitments will be obtained from $\mathcal{F}_{\text{COM}}$ instead of an OT functionality. However, our simulator will only rely on the fact that it can extract the messages sent by the adversary to $\mathcal{F}_{\text{COM}}$ before it opens its commitments. Essentially, our simulators only need an underlying commitment scheme that is extractable, not a full blown UC commitment scheme (which would also allow the simulator to open the underlying commitments to arbitrary messages). The security of Protocol Π_{AHCOM} is formally stated in Theorem 3.

Theorem 3. *Protocol Π_{AHCOM} UC-realizes $\mathcal{F}_{\text{AHCOM}}$ in the $\mathcal{F}_{\text{COM}}$-hybrid model with computational security against a static adversary. Formally, there exists a simulator S such that for every static adversary $\mathcal{A}$, and any environment $\mathcal{Z}$, the environment cannot distinguish Π_{AHCOM} composed with $\mathcal{F}_{\text{COM}}$ and $\mathcal{A}$ from S composed with $\mathcal{F}_{\text{AHCOM}}$. That is, $\mathsf{IDEAL}_{\mathcal{F}_{\text{AHCOM}}, S, \mathcal{Z}} \approx_c \mathsf{HYBRID}_{\Pi_{\text{AHCOM}}, \mathcal{A}, \mathcal{Z}}^{\mathcal{F}_{\text{COM}}}$.*

Protocol Π_{AHCOM}

Let C be a systematic binary linear $[n, k, s]$ code, where s is the statistical security parameter and n is $k + O(s)$. Let $\mathcal{H}$ be a family of linear almost universal hash functions $\mathbf{H} : \{0,1\}^m \to \{0,1\}^l$. Let $\mathsf{PRG} : \{0,1\}^\ell \to \{0,1\}^{m+l}$ be a pseudorandom generator. Protocol Π_{AHCOM} is run by a sender P and a set of receivers $V = \{V_1, \ldots, V_t\}$, who interact with $\mathcal{F}_{\mathrm{COM}}$ and proceed as follows:

Commitment Phase

1. On input $(\mathsf{commit}, sid, ssid_1, \ldots, ssid_m, P, V)$, P proceeds as follows:
   (a) For $i \in [n]$ and $j \in \{0,1\}$, sample $\boldsymbol{s}_{i,j} \xleftarrow{\$} \{0,1\}^\ell$ and send $(\mathsf{commit}, sid, ssid_{i,j}, P, V, \boldsymbol{s}_{i,j})$ to $\mathcal{F}_{\mathrm{COM}}$.
   (b) Compute $\mathbf{R_j}[i, \cdot] = \mathsf{PRG}(\boldsymbol{s}_{i,j})$ and set $\mathbf{R} = \mathbf{R_0} + \mathbf{R_1}$ so that $\mathbf{R_0}, \mathbf{R_1}$ forms an additive secret sharing of $\mathbf{R}$.
   (c) Adjust the bottom $n - k$ rows of $\mathbf{R}$ so that all columns are codewords in C by constructing a matrix $\mathbf{W}$ with dimensions as $\mathbf{R}$ and 0s in the top k rows, such that $\mathbf{A} := \mathbf{R} + \mathbf{W} \in \mathsf{C}^{\odot m+l}$ (recall that C is systematic). Set $\mathbf{A_0} = \mathbf{R_0}, \mathbf{A_1} = \mathbf{R_1} + \mathbf{W}$ and broadcast $(sid, ssid_1, \ldots, ssid_m, \mathbf{W})$ (only sending the bottom $n - k = O(s)$ rows).
2. Upon receiving all messages $(\mathsf{receipt}, sid, ssid_{i,j}, P, V)$ from $\mathcal{F}_{\mathrm{COM}}$ and $(sid, ssid_1, \ldots, ssid_m, \mathbf{W})$ from P, every receiver $V_i \in V$ proceeds as follows:
   (a) Sample $\boldsymbol{r}_i \xleftarrow{\$} \{0,1\}^n$ and $\boldsymbol{r}_i' \xleftarrow{\$} \{0,1\}^\ell$, and send $(\mathsf{commit}, sid, ssid, V_i, V', \boldsymbol{r}_i)$ and $(\mathsf{commit}, sid, ssid', V_i, V', \boldsymbol{r}_i')$ to $\mathcal{F}_{\mathrm{COM}}{}^a$, where $V' = P \cup V \setminus V_i$.
   (b) Upon receiving $(\mathsf{receipt}, sid, ssid, V_j, V')$ and $(\mathsf{receipt}, sid, ssid', V_j, V')$ from $\mathcal{F}_{\mathrm{COM}}$ for all $V_j \in V \setminus V_i$, send $(\mathsf{reveal}, sid, ssid')$ to $\mathcal{F}_{\mathrm{COM}}$.
   (c) Upon receiving $(\mathsf{reveal}, sid, ssid', V_j, V', \boldsymbol{r}_j')$ from $\mathcal{F}_{\mathrm{COM}}$ for all $V_j \in V \setminus V_i$, set $\boldsymbol{r}' = \boldsymbol{r}_1' \oplus \ldots \oplus \boldsymbol{r}_t'$.
3. Upon receiving $(\mathsf{commit}, sid, ssid, V_i, V')$ and $(\mathsf{reveal}, sid, ssid', V_j, V', \boldsymbol{r}_j')$ from $\mathcal{F}_{\mathrm{COM}}$ for all $V_j \in V$, P proceeds as follows:
   (a) Use $\boldsymbol{r}' = \boldsymbol{r}_1' \oplus \ldots \oplus \boldsymbol{r}_t'$ as a seed for a random function $\mathbf{H} \in \mathcal{H}$ (note that we identify the function with its matrix and all functions in $\mathcal{H}$ are linear).
   (b) Set matrices $\mathbf{P}, \mathbf{P_0}$ and $\mathbf{P_1}$ as the first l columns of $\mathbf{A}, \mathbf{A_0}$ and $\mathbf{A_1}$, respectively, and remove these columns from $\mathbf{A}, \mathbf{A_0}$ and $\mathbf{A_1}$. Renumber the remaining columns of $\mathbf{A}, \mathbf{A_0}$ and $\mathbf{A_1}$ from 1 and associate each $ssid_i$ (commitment id from step 1) with a different column index in these matrices. Notice that $\mathbf{P} = \mathbf{P_0} + \mathbf{P_1}$.
   (c) For $i \in \{0,1\}$, compute $\mathbf{T}_i = \mathbf{A}_i \mathbf{H} + \mathbf{P}_i$ and broadcast $(sid, ssid_1, \ldots, ssid_m, \mathbf{T_0}, \mathbf{T_1})$. Note that $\mathbf{A}\mathbf{H} + \mathbf{P} = \mathbf{A_0}\mathbf{H} + \mathbf{P_0} + \mathbf{A_1}\mathbf{H} + \mathbf{P_1} = \mathbf{T_0} + \mathbf{T_1}$, and $\mathbf{A}\mathbf{H} + \mathbf{P} \in \mathsf{C}^{\odot l}$.

---

[a]We abuse notation and assume that each receiver V_i in Π_{AHCOM} has access to an instance of $\mathcal{F}_{\mathrm{COM}}$ that takes as message with the appropriate length where it acts as sender and where all other receivers plus sender P act as receivers.

Fig. 4. Commit phase for the protocol Π_{AHCOM}.

Proof. We give the proof in the full version of this paper [18]. Note that constructing a simulator for the case where all parties are honest is trivial. Hence,

---

Protocol Π_{AHCOM}

Addition of Commitments

1. On input $(\text{add}, sid, ssid_1, ssid_2, ssid_3, P, V)$, P finds indexes i and j corresponding to $ssid_1$ and $ssid_2$ respectively and check that $ssid_3$ is unused. P appends the column $\mathbf{A}[\cdot, i] + \mathbf{A}[\cdot, j]$ to $\mathbf{A}$, likewise appends to $\mathbf{A_0}$ and $\mathbf{A_1}$ the sum of their i-th and j-th columns, and associates $ssid_3$ with the new column index. P broadcasts $(\text{add}, sid, ssid_1, ssid_2, ssid_3)$. Note that this maintains the properties $\mathbf{A} = \mathbf{A_0} + \mathbf{A_1}$ and $\mathbf{A} \in C^{\odot m'}$, where m' is the current number of columns (after appending columns for addition results).
2. Upon receiving $(\text{add}, sid, ssid_1, ssid_2, ssid_3)$, every receiver $V_i \in V$ stores the message.

Opening

1. On input $(\text{reveal}, sid, ssid_1, \ldots, ssid_o)$, P finds the set $J = \{j_1, \ldots, j_o\}$ of indexes associated to $ssid_1, \ldots, ssid_o$ and broadcasts $(sid, ssid_1, \ldots, ssid_o, (\mathbf{A_0}[\cdot, j], \mathbf{A_1}[\cdot, j])_{j \in J})$.
2. Upon receiving message $(sid, ssid_1, \ldots, ssid_o, (\mathbf{A_0}[\cdot, j], \mathbf{A_1}[\cdot, j])_{j \in J})$, every $V_i \in V$ sends $(\text{reveal}, sid, ssid)$ to $\mathcal{F}_{\text{COM}}$ and waits for $(\text{reveal}, sid, ssid, V_j, V', r_j)$ from $\mathcal{F}_{\text{COM}}$ for all $V_j \in V \setminus V_i$. V_i sets $\mathbf{r} = \mathbf{r_1} \oplus \cdots \oplus \mathbf{r_t}$ and sets the diagonal matrix $\mathbf{\Delta}$ such that it contains $r[1], \ldots, r[n]$ in the diagonal.
3. Upon receiving $(\text{reveal}, sid, ssid, V_j, V', r_j)$ from $\mathcal{F}_{\text{COM}}$ for all $V_j \in V$, P sets $\mathbf{r} = \mathbf{r_1} \oplus \ldots \oplus \mathbf{r_t}$, sends $(\text{reveal}, sid, ssid_{i, r[i]})$ to $\mathcal{F}_{\text{COM}}$ for $i \in [n]$ and halts.
4. Upon receiving $(\text{reveal}, sid, ssid_{i, r[i]}, P, V, \mathbf{s}_{i, r[i]})$ from $\mathcal{F}_{\text{COM}}$ for $i \in [n]$, every receiver $V_j \in V$ proceeds as follows:
   (a) Compute $\mathbf{S}[i, \cdot] = \text{PRG}(\mathbf{s}_{i, r[i]})$, obtaining a matrix $\mathbf{S}$. Note that each row of $\mathbf{S}$ is a row from either $\mathbf{R_0}$ or $\mathbf{R_1}$, which form an additive secret sharing of $\mathbf{R}$ held by P. Set $\mathbf{B} = \mathbf{\Delta W} + \mathbf{S}$. Define the matrix $\mathbf{Q}$ as the first l columns of $\mathbf{B}$ and remove these columns from $\mathbf{B}$, renumbering the remaining columns from 1. Note that, for $\mathbf{A}$ from the commitment phase, $\mathbf{A} = \mathbf{A_0} + \mathbf{A_1}$, $\mathbf{B} = \mathbf{\Delta A_1} + (\mathbf{I} - \mathbf{\Delta})\mathbf{A_0}$, $\mathbf{A} \in C^{\odot m}$, *i.e.*, $\mathbf{A}$ initially held by P is additively shared and for each row index, V knows either a row from $\mathbf{A_0}$ or from $\mathbf{A_1}$.
   (b) Check that $\mathbf{\Delta T_1} + (\mathbf{I} - \mathbf{\Delta})\mathbf{T_0} = \mathbf{BH} + \mathbf{Q}$ and that $\mathbf{T_0} + \mathbf{T_1} \in C^{\odot l}$. If any check fails, abort. Notice that $\mathbf{T_0}, \mathbf{T_1}$ form an additive sharing of $\mathbf{AH} + \mathbf{P}$, where V knows some of the shares, namely the rows of $\mathbf{BH} + \mathbf{Q}$.
   (c) For every message $(\text{add}, sid, ssid_1, ssid_2, ssid_3)$ received from P, append $\mathbf{B}[\cdot, j] + \mathbf{B}[\cdot, i]$ to $\mathbf{B}$, where i and j are the index corresponding to $ssid_1$ and $ssid_2$ respectively and associate $ssid_3$ with the new column index. Note that this maintains the property $\mathbf{B} = \mathbf{\Delta A_1} + (\mathbf{I} - \mathbf{\Delta})\mathbf{A_0}$.
   (d) For every $j \in J$, check that $\mathbf{A_0}[\cdot, j] + \mathbf{A_1}[\cdot, j] \in C$ and that, for $i \in [n]$, it holds that $\mathbf{B}[i, j] = \mathbf{A}_{r[i]}[i, j]$ (recall that $r[i]$ is the i-th entry on the diagonal of $\mathbf{\Delta}$). If all checks succeed, for every $j \in J$, output the first k positions in $\mathbf{A_0}[\cdot, j] + \mathbf{A_1}[\cdot, j]$ as the opened string and halt. Otherwise, abort by outputting $(sid, ssid_j, \perp)$.

Fig. 5. Addition of commitments and opening phase for the protocol Π_{AHCOM}.

the theorem follows by establishing security against an adversary that corrupts P and all but one receiver in V or an adversary who corrupts all receivers in V. See the full version for each of these.

4 Achieving Multiplicative Homomorphism

In this section, we modify our additively homomorphic commitment protocol described Sect. 3 (protocol Π_{AHCOM}) so that it is also homomorphic for (coordinatewise) multiplication of messages. That is, if we denote the scheme from Sect. 3 by com, our goal is that given commitments $\mathrm{com}(\boldsymbol{a})$, $\mathrm{com}(\boldsymbol{b})$ the prover can construct a commitment $\mathrm{com}(\boldsymbol{a} * \boldsymbol{b})$. In order to do this we need to introduce a second auxiliary commitment scheme prodcom, also described below.

Both com and prodcom can be obtained by changing the instantiation of two of the building blocks of protocol Π_{AHCOM}. Namely, at the core of the construction of the commitment scheme in Sect. 3 (as well as in the ones from [16,17,22]) there is a linear error correcting code C, which is used to encode the message and which needs to have a large enough minimum distance; and there is the 2-out-of-2 additive secret sharing scheme Add_2, which is applied to each coordinate of the encoding. Our modifications are as follows: first, we need a linear code C such that also its *(Schur) square* C^{*2} has a large enough minimum distance. We will use C as the linear code in com and C^{*2} as the linear code in prodcom (with a certain caveat described below). As for the secret sharing schemes, we will use the *replicated secret sharing scheme* RSS_3 (described below) for com and the additive 3-out-of-3 Add_3 secret sharing scheme for prodcom. RSS_3 is the secret sharing scheme where the secret $s \in \{0, 1\}$ is additively split into three parts, i.e., $s = r_0 + r_1 + r_2$ where r_0, r_1 are uniformly random and independent, and the shares are defined to be the pairs $s_0 = (r_0, r_1)$, $s_1 = (r_1, r_2)$, $s_2 = (r_2, r_0)$. RSS_3 is a multiplicative secret sharing scheme, which means that shares of s, s' can locally be transformed into shares by Add_3 of the product $s \cdot s'$. More precisely, $s \cdot s' = t_0 + t_1 + t_2$, where $t_i = r_i r'_i + r_i r'_{i+1} + r'_i r_{i+1}$ (where sums in the indices are modulo 3) and note that all this information is contained in the i-th shares s_i, s'_i of s and s'. The rationale for the choices of codes and secret sharing schemes is then that from the watchlists of $\mathrm{com}(\boldsymbol{a}), \mathrm{com}(\boldsymbol{b})$ a verifier can compute a watchlist to a commitment $\mathrm{prodcom}(\boldsymbol{a} * \boldsymbol{b})$. Indeed, given the j-th share (in RSS_3) of the i-th coordinates $(\mathsf{C}(\boldsymbol{a}))_i, (\mathsf{C}(\boldsymbol{b}))_i$ the verifier can determine the j-th share (in Add_3) of $(\mathsf{C}(\boldsymbol{a}) * \mathsf{C}(\boldsymbol{b}))_i$, and note $\mathsf{C}(\boldsymbol{a}) * \mathsf{C}(\boldsymbol{b})$ is a codeword in C^{*2} having $\boldsymbol{a} * \boldsymbol{b}$ as the vector of its first k coordinates.

But our goal is to construct $\mathrm{com}(\boldsymbol{a} * \boldsymbol{b})$ rather than $\mathrm{prodcom}(\boldsymbol{a} * \boldsymbol{b})$. We do that as follows: the prover constructs commitments $\mathrm{com}(\boldsymbol{y})$, $\mathrm{prodcom}(\boldsymbol{y})$ of a random vector $\boldsymbol{y}$ with both commitment schemes, where for every coordinate i, the verifier will later request to open the share with the same index r_i in $\mathrm{prodcom}(\boldsymbol{y})$ as he does for $\mathrm{com}(\boldsymbol{a}), \mathrm{com}(\boldsymbol{b}), \mathrm{com}(\boldsymbol{y})$ (note that for com that means the additive shares indexed by r_i and $r_i + 1$). The sender needs to prove that $\mathrm{com}(\boldsymbol{y})$, $\mathrm{prodcom}(\boldsymbol{y})$ are indeed commitments to the same vector, which will be detailed later. From $\mathrm{com}(\boldsymbol{a}), \mathrm{com}(\boldsymbol{b})$ the prover constructs all the shares in $\mathrm{prodcom}(\boldsymbol{a} * \boldsymbol{b})$ as mentioned above, and then announces all three additive shares of $a * b - y$. For each coordinate i, the receiver can determine the r_i-th share of this vector from the watchlists of $\mathrm{com}(a), \mathrm{com}(b), \mathrm{prodcom}(y)$ and contrast this with the information that the prover opens. Now assuming the verifier does not abort, the prover and verifier can simply construct $\mathrm{com}(a * b)$ by adding $a * b - y$ to $\mathrm{com}(y)$.[5] We need to address some technical details: commitments with $\mathrm{prodcom}$ are to messages of length k' (the dimension of C^{*2}) rather than messages of length k and in general it can happen that $k' > k$, so when we say $\mathrm{prodcom}(y)$ we mean that the commitment is to a vector $y||z$ where z is of length $k' - k$. Moreover, initially we cannot choose the random vectors we commit to since these are generated pseudorandomly from the seeds, so the prover will need to send some correction information in order to commit to the same value in the two schemes. In order to do that, and simultaneously prepare to prove that $\mathrm{com}(y)$ and $\mathrm{prodcom}(y)$ are commitments to the same vector y, we define the linear code $\widetilde{\mathsf{C}}$ defined as the concatenation of C and C^{*2}. More precisely,

$$\widetilde{\mathsf{C}} = \{(y, c, y, c') : (y, c) \in \mathsf{C}, (y, c') \in \mathsf{C}^{*2}\}. \tag{1}$$

The prover, having used the PRGs to construct pairs of random vectors r, r' in $\{0, 1\}^n$ and additive splittings of them, will concatenate the two vectors and send correction information $z \in \{0, 1\}^{2n}$ so that $(r||r') - z \in \widetilde{\mathsf{C}}$ (as before, the first k bits of z can be taken to be 0, so the prover needs to send only $2n - k$ bits). Now given a batch of supposed codewords of this form the interactive proximity testing technique is applied so that the sender proves they are indeed codewords in $\widetilde{\mathsf{C}}$, and therefore they are associated to commitments $(\mathrm{com}(y), \mathrm{prodcom}(y))$. Note that since the first n coordinates of the codewords in $\widetilde{\mathsf{C}}$ are codewords in C, this test also guarantees all properties of the interactive proximity test for the additive case, so we do not need to perform that one separately.

We note that $\widetilde{d} = \mathrm{dist}(\widetilde{\mathsf{C}}) \geq \mathrm{dist}(\mathsf{C}^{*2})$,[6] so we need a lower bound on $\mathrm{dist}(\mathsf{C}^{*2})$ to obtain the same guarantees as in the additive case. Furthermore, a difference with the proof for the additive-only commitment scheme is that now the verifier sees 2 out of 3 additive shares of the first n coordinates and 1 out of 3 coordinates of the last n, which affects the cheating probabilities of a corrupt prover: we will show that it is enough to assume that $\mathrm{dist}(\mathsf{C}^{*2}) > \beta s$, where $\beta = 1/(\log_2 3 - 1) = 1.709...$ (which satisfies $(2/3)^\beta = 1/2$), in order to guarantee that the cheating prover can succeed with probability at most 2^{-s}. Protocol Π_{MHCOM} is described in Figs. 6, 7 and 8. Notice that for consistency with the notation of Sect. 3, we describe our fully homomorphic commitment protocol for random messages. However, a commitment to chosen messages m can be created using the protocol Π_{MHCOM} simply sending $c = m - a$, where $a = \pi_{[k]}(\mathbf{A}[\cdot, i])$ is one of the random messages that the prover gets in the commit phase of Π_{MHCOM}. Now, in order

---

[5] More precisely, the last share of each coordinate of $\mathsf{C}(y)$ is added with the corresponding (now public) coordinate of $\mathsf{C}(a * b - y)$.

[6] One may think that the tighter lower bound $\mathrm{dist}(\widetilde{\mathsf{C}}) \geq \mathrm{dist}(\mathsf{C}) + \mathrm{dist}(\mathsf{C}^{*2})$ holds, but this is not necessarily true if the dimension k' of C^{*2} is larger than k, as in that case there will be codewords of the form $(0^k, 0^{n-k}, 0^k, c')$ where $c' \neq 0^{n-k}$. Indeed take $(0^k, c')$ to be the encoding by C^{*2} of $(0^k||z)$ for a nonzero $z \in \{0, 1\}^{k'-k}$.

to allow multiplication of commitments to chosen messages it is enough that all the players locally adjust the shares of the random messages used as OTP keys (e.g., the prover P adds $C(c)$ to $\mathbf{A}_2[\cdot, i]$ and every receiver in V adds $\Delta C(c)$ and $\Delta' C(c)$ to $\mathbf{B}[\cdot, i]$ and $\mathbf{B}'[\cdot, i]$, respectively) and then execute the multiplication step as detailed in Fig. 7.

Finally notice that for the sake of simplicity, in the commit phase of Protocol Π_{MHCOM} we use the same notation and the same construction both for random messages that are actually input to commitments (or used to construct a commitment to a chosen message as explained above) and for the auxiliary random messages that are needed in the multiplication step (i.e., $\boldsymbol{y}$ in the notation used in the introduction of this section), so that all those messages are encoded in columns of the big matrix $\tilde{\mathbf{A}}$. However, committing with prodcom, and hence creating and manipulating the last n rows of the matrix $\hat{\mathbf{A}}$ (what we call $\hat{\mathbf{A}}$), is only necessary for the random messages used in the multiplication step, and could be saved for the remaining random messages. On the other hand, the current structure of the commit phase, where we do not distinguish between the two roles for the random messages, allows us to use only a single interactive proximity test instead of two (i.e., one for C as in protocol Π_{AHCOM} to guarantee the additive property and another one for $\tilde{C}$ and the auxiliary random messages to guarantee that the same value $\boldsymbol{y}$ is encoded using C and C^{*2}).

Security Analysis. The proof of security for Protocol Π_{MHCOM} is similar to that of Π_{AHCOM}. Indeed, the following Theorem 4 can be proved by adapting the description of the simulators for the security proof of the additive-only construction to the new watchlist setting (i.e., three additive shares instead of two, of which the verifier knows either two - in the base commitment given by matrix $\mathbf{A}$ - or one - in the product commitment given by $\hat{\mathbf{A}}$) and adding to both simulators the step to simulate the multiplication command. More details are given in the full version [18].

Theorem 4. *Protocol* Π_{MHCOM} *UC realizes* $\mathcal{F}_{\mathrm{MHCOM}}$ *in the* $\mathcal{F}_{\mathrm{COM}}$-*hybrid model with computational security against a static adversary. Formally, there exists a simulator $\mathcal{S}$ such that for every static adversary $\mathcal{A}$, and any environment $\mathcal{Z}$, the environment cannot distinguish Π_{MHCOM} composed with $\mathcal{F}_{\mathrm{COM}}$ and $\mathcal{A}$ from $\mathcal{S}$ composed with $\mathcal{F}_{\mathrm{MHCOM}}$. That is,* $\mathsf{IDEAL}_{\mathcal{F}_{\mathrm{MHCOM}}, \mathcal{S}, \mathcal{Z}} \approx_c$ $\mathsf{HYBRID}_{\Pi_{\mathrm{MHCOM}}, \mathcal{A}, \mathcal{Z}}^{\mathcal{F}_{\mathrm{COM}}}$.

Efficiency. Since we commit to every random message with both com and prodcom, the total length of the commitment will be $2n - k + o(k)$ bits per message of k bits. For chosen messages we need to add an extra k bits per message for a total of $2n$ bits. If C has rate R, our commitments have then rate $R/2$. Moreover, for multiplying two commitments the prover needs to have created an additional commitment of a random message with both com and prodcom (hence communicating $2n$ bits), and then communicate all shares of a related commitment with prodcom (the $\boldsymbol{w}_i$'s in the protocol), which amounts to $3n$ bits. So the communication of this step is $5n$ bits. The question is then what rates we can have under our new requirements on $\mathsf{dist}(C^{*2})$.

Families of binary codes $\{C_n\}$ with constant rate (of C_n) and constant relative minimum distance of C_n^{*2} exist based on algebraic geometry [28]. For fixed values of the security parameter s the families of cyclic codes constructed in [15] give better rates. As an example, for $s = 60$, where our protocol needs $\mathsf{dist}(C^{*2}) \geq 103$, Table 2 in [15] gives a $[4095, 338]$ cyclic code with $\mathsf{dist}(C^{*2}) \geq 135$, which has rate around 0.08. Hence the commitments will have rate 0.04.[7] We need to send $25k$ bits per k-bit message we commit to, and $62.5k$ bits to construct a commitment to the product of two messages.

5   Applications to Efficient Zero-Knowledge Arguments

In this section, we outline how to use a variant of the homomorphic commitments constructed in Sect. 3 and 4 to compile a certain class of public coin interactive proof system into public coin honest-verifier zero-knowledge proof systems. Using the Fiat-Shamir heuristic, we can convert such a zero-knowledge proof system into a non-interactive zero-knowledge proof system. As an application, we can improve a recent construction of zkSNARKs [31] in a certain parameter regime. Specifically, the zkSNARK construction of [31] uses additively homomorphic vector commitments[8] to transform a public coin interactive proof system into a zero-knowledge protocol. The commitments in [31] are instantiated using number-theoretic assumptions. One of the core ideas of [31] is that general algebraic relations between commitments can be reduced to linear relations between vector-commitments in a way that only induces a constant additional overhead for low-degree relations. The construction of [31] is general enough that it can be instantiated with homomorphic commitment schemes with some additional properties. We remark though that [31] utilizes an additional optimization which relies on *compressing* homomorphic commitments, which is not available in our setting.

Our main observation is that for this application the unveil of the commitments in the protocol of [31] can be delayed until the very end of the protocol, which makes this protocol compatible with our commitment scheme.

The notion of interactive proof system we focus on will be resettably sound public coin interactive proofs with *algebraic verifier*. Such a proof system proceeds in t rounds, where in each round i the prover sends a message p_i, upon which the verifier answers with a uniformly random message v_i. We require all the messages p_i and v_i to be vectors over a field $\mathbb{F}$. After the conversation is over, the verifier evaluates a system of low degree polynomials $F_1, \ldots, F_s$ in the p_i and v_i and accepts if all F_i evaluate to 0, otherwise it rejects. At the heart of this kind of protocol is the sum-check protocol, which lets a prover prove statements of the form $\sum_{x \in \{0,1\}^n} P(x) = L$, where $P \in \mathbb{F}[X_1, \ldots, X_n]$ is a low-degree polynomial and $L \in \mathbb{F}$.

---

[7] And naturally from this one can also obtain a $[4095 \cdot \ell, 338 \cdot \ell]$-code with the same minimum distance of its square, by simply applying the $[4095, 338]$ to each block of 338 bits of the message.

[8] In [31] they are referred to as *multi-commitments*.

Protocol Π_{MHCOM}

Let C be a systematic binary linear $[n, k]$ code, such that C^{*2} is also systematic and satisfies $\text{dist}(C^{*2}) \geq \beta s$, where $\beta = 1/(\log_2 3 - 1)$ and s is the statistical security parameter. Let $\widetilde{C}$ be the code defined in (1). Let $\mathcal{H}$ be a family of linear almost universal hash functions $\mathbf{H} : \{0, 1\}^m \to \{0, 1\}^l$. Let $\text{PRG} : \{0, 1\}^\ell \to \{0, 1\}^{m+l}$ be a pseudorandom generator. Protocol Π_{MHCOM} is run by a sender P and a set of receivers $V = \{V_1, \ldots, V_t\}$, who interact with $\mathcal{F}_{\text{COM}}$ as follows:

Commitment Phase

1. On input $(\text{commit}, sid, ssid_1, \ldots, ssid_m, P, V)$, P proceeds as follows:
   (a) For $i \in [n]$ and $j \in \{0, 1, 2\}$, sample $\boldsymbol{s}_{i,j} \xleftarrow{\$} \{0, 1\}^\ell$, $\widehat{\boldsymbol{s}}_{i,j} \xleftarrow{\$} \{0, 1\}^\ell$ and send $(\text{commit}, sid, ssid_{i,j}, P, V, \boldsymbol{s}_{i,j})$, $(\text{commit}, sid, \widehat{ssid}_{i,j}, P, V, \widehat{\boldsymbol{s}}_{i,j})$ to $\mathcal{F}_{\text{COM}}$.
   (b) Compute $\mathbf{R}_{\mathbf{j}}[i, \cdot] = \text{PRG}(\boldsymbol{s}_{i,j})$ and $\widehat{\mathbf{R}}_{\mathbf{j}}[i, \cdot] = \text{PRG}(\widehat{\boldsymbol{s}}_{i,j})$ and set $\mathbf{R} = \mathbf{R}_{\mathbf{0}} + \mathbf{R}_{\mathbf{1}} + \mathbf{R}_{\mathbf{2}}$ and $\widehat{\mathbf{R}} = \widehat{\mathbf{R}}_{\mathbf{0}} + \widehat{\mathbf{R}}_{\mathbf{1}} + \widehat{\mathbf{R}}_{\mathbf{2}}$.
   (c) Adjust the bottom $n - k$ rows of $\mathbf{R}$ so that all columns are codewords in C by constructing a matrix $\mathbf{W}$ with dimensions as $\mathbf{R}$ and 0s in the top k rows, such that $\mathbf{A} := \mathbf{R} + \mathbf{W} \in C^{\odot m+l}$ (recall that C is systematic). Set $\mathbf{A}_{\mathbf{0}} = \mathbf{R}_{\mathbf{0}}, \mathbf{A}_{\mathbf{1}} = \mathbf{R}_{\mathbf{1}}, \mathbf{A}_{\mathbf{2}} = \mathbf{R}_{\mathbf{2}} + \mathbf{W}$.
   (d) Adjust $\widehat{\mathbf{R}}$ so that all columns are codewords in C^{*2} and the first k rows are the same as in $\mathbf{A}$ by constructing a matrix $\widehat{\mathbf{W}}$ with dimensions as $\widehat{\mathbf{R}}$ such that $\widehat{\mathbf{A}} := \widehat{\mathbf{R}} + \widehat{\mathbf{W}} \in (C^{*2})^{\odot m+l}$ and $\widehat{\mathbf{A}}[i, \cdot] = \mathbf{A}[i, \cdot]$ for all $i \in [k]$. Set $\widehat{\mathbf{A}}_{\mathbf{0}} = \widehat{\mathbf{R}}_{\mathbf{0}}, \widehat{\mathbf{A}}_{\mathbf{1}} = \widehat{\mathbf{R}}_{\mathbf{1}}, \widehat{\mathbf{A}}_{\mathbf{2}} = \widehat{\mathbf{R}}_{\mathbf{2}} + \widehat{\mathbf{W}}$ and broadcast $(sid, ssid_1, \ldots, ssid_m, \mathbf{W}, \widehat{\mathbf{W}})$ (sending the bottom $n - k$ rows of $\mathbf{W}$ and the entire matrix $\widehat{\mathbf{W}}$).
2. Upon receiving all $(\text{receipt}, sid, ssid_{i,j}, P, V)$ from $\mathcal{F}_{\text{COM}}$ and $(sid, ssid_1, \ldots, ssid_m, \mathbf{W}, \widehat{\mathbf{W}})$ from P, every $V_i \in V$ proceeds as follows:
   (a) Sample $\boldsymbol{r}_i \xleftarrow{\$} \mathbb{Z}_3^n$, $\boldsymbol{r}_i' \xleftarrow{\$} \{0, 1\}^\ell$ and send $(\text{commit}, sid, ssid, V_i, V', \boldsymbol{r}_i)$ and $(\text{commit}, sid, ssid', V_i, V', \boldsymbol{r}_i')$ to $\mathcal{F}_{\text{COM}}$, where $V' = P \cup V \setminus V_i$.
   (b) and (c) as is the commit phase of Π_{AHCOM} (Figure 4).
3. Upon receiving $(\text{commit}, sid, ssid, V_i, V')$ and $(\text{reveal}, sid, ssid', V_j, V', \boldsymbol{r}_j')$ from $\mathcal{F}_{\text{COM}}$ for all $V_j \in V$, P proceeds as follows:
   (a) Use $\boldsymbol{r}' = \boldsymbol{r}_1' \oplus \ldots \oplus \boldsymbol{r}_t'$ as a seed for a random function $\mathbf{H} \in \mathcal{H}$.
   (b) Define the matrices $\widetilde{\mathbf{A}} = \begin{pmatrix} \mathbf{A} \\ \widehat{\mathbf{A}} \end{pmatrix}$ and $\widetilde{\mathbf{A}}_{\mathbf{i}} = \begin{pmatrix} \mathbf{A}_i \\ \widehat{\mathbf{A}}_i \end{pmatrix}$ for $i \in \{0, 1, 2\}$. Note that $\widetilde{\mathbf{A}} \in \widetilde{C}^{\odot m+l}$ and $\widetilde{\mathbf{A}} = \widetilde{\mathbf{A}}_{\mathbf{0}} + \widetilde{\mathbf{A}}_{\mathbf{1}} + \widetilde{\mathbf{A}}_{\mathbf{2}}$. Set the matrices $\widetilde{\mathbf{P}}$ and $\widetilde{\mathbf{P}}_i$ as the first l columns of $\widetilde{\mathbf{A}}$ and $\widetilde{\mathbf{A}}_i$, respectively, and remove these columns from $\widetilde{\mathbf{A}}, \widetilde{\mathbf{A}}_i, \mathbf{A}, \mathbf{A}_i, \widehat{\mathbf{A}}, \widehat{\mathbf{A}}_i$ for $i \in \{0, 1, 2\}$. Renumber the remaining columns from 1 and associate each commitment $ssid_i$ (commitment id from step 1) with a different column in these matrices. Notice that $\widetilde{\mathbf{P}} = \widetilde{\mathbf{P}}_0 + \widetilde{\mathbf{P}}_1 + \widetilde{\mathbf{P}}_2$.
   (c) For $i \in \{0, 1, 2\}$, compute the matrix $\widetilde{\mathbf{T}}_i = \widetilde{\mathbf{A}}_i \mathbf{H} + \widetilde{\mathbf{P}}_i$ and broadcast $(sid, ssid_1, \ldots, ssid_m, \widetilde{\mathbf{T}}_{\mathbf{0}}, \widetilde{\mathbf{T}}_{\mathbf{1}}, \widetilde{\mathbf{T}}_{\mathbf{2}})$. Note that $\widetilde{\mathbf{A}} \mathbf{H} + \widetilde{\mathbf{P}} = \widetilde{\mathbf{T}}_0 + \widetilde{\mathbf{T}}_1 + \widetilde{\mathbf{T}}_2$, and $\mathbf{A} \mathbf{H} + \mathbf{P} \in C^{\odot l}$.

Fig. 6. Commit phase for the protocol Π_{MHCOM}.

Protocol Π_{MHCOM}

Addition of Commitments

1. On input $(\mathsf{add}, sid, ssid_1, ssid_2, ssid_3, P, V)$, P finds indexes i and j corresponding to $ssid_1$ and $ssid_2$ respectively and check that $ssid_3$ is unused. P appends the column $\mathbf{A}[\cdot, i] + \mathbf{A}[\cdot, j]$ to $\mathbf{A}$, likewise appends to $\mathbf{A_0}, \mathbf{A_1}, \mathbf{A_2}$ the sum of their i-th and j-th columns, and associates $ssid_3$ with the new column index. P broadcasts $(\mathsf{add}, sid, ssid_1, ssid_2, ssid_3)$ to V.
2. Upon receiving $(\mathsf{add}, sid, ssid_1, ssid_2, ssid_3)$, every $V_i \in V$ stores the message.

Multiplication of Commitments

1. On input $(\mathsf{mult}, sid, ssid_1, ssid_2, ssid_3, P, V)$, P finds indexes i and j corresponding to $ssid_1$ and $ssid_2$ respectively and check that $ssid_3$ is unused. Then, P proceeds as follows:
   (a) For $l \in \{0, 1, 2\}$, compute $\boldsymbol{v}_l = \mathbf{A}_l[\cdot, i] * \mathbf{A}_l[\cdot, j] + \mathbf{A}_l[\cdot, i] * \mathbf{A}_{l+1}[\cdot, j] + \mathbf{A}_{l+1}[\cdot, i] * \mathbf{A}_l[\cdot, j]$. Note that $\boldsymbol{v}_0, \boldsymbol{v}_1, \boldsymbol{v}_2$ are shares of $\mathbf{A}[\cdot, i] * \mathbf{A}[\cdot, j]$ in the scheme Add_3 and known to P only. Let h be the index of the first unused column from $\mathbf{A}$ and $\widehat{\mathbf{A}}$, compute $\boldsymbol{w}_l = \boldsymbol{v}_l - \widehat{\mathbf{A}}_l[\cdot, h]$ for $l = 0, 1, 2$ and broadcast $(sid, ssid, h, \boldsymbol{w}_0, \boldsymbol{w}_1, \boldsymbol{w}_2)$ to V. Note that $\boldsymbol{w}_0, \boldsymbol{w}_1, \boldsymbol{w}_2$ are shares of $\mathbf{A}[\cdot, i] * \mathbf{A}[\cdot, j] - \widehat{\mathbf{A}}[\cdot, h]$ in the scheme Add_3 and are known to $P \cup V$.
   (b) Let $\boldsymbol{u} = \pi_{[k]}(\boldsymbol{w}_0 + \boldsymbol{w}_1 + \boldsymbol{w}_2)$ (i.e., $\boldsymbol{u}$ consists of the first k components of $\mathbf{A}[\cdot, i] * \mathbf{A}[\cdot, j] - \widehat{\mathbf{A}}[\cdot, h]$), append the columns $\mathbf{A}[\cdot, h] + \mathsf{C}(\boldsymbol{u})$ and $\mathbf{A_2}[\cdot, h] + \mathsf{C}(\boldsymbol{u})$ to $\mathbf{A}$ and $\mathbf{A_2}$, respectively. Append the column $\mathbf{A}_i[\cdot, h]$ to $\mathbf{A}_i$ for $i = 0, 1$ and associate $ssid_3$ with the new column index. Note that since $\pi_{[k]}(\widehat{\mathbf{A}}[\cdot, h]) + \pi_{[k]}(\mathbf{A}[\cdot, h])$, for $l \in \{1, \ldots, k\}$ the l-th component of the newly appended column in $\mathbf{A}$ is equal to $\mathbf{A}[l, i] * \mathbf{A}[l, j]$. Broadcast $(\mathsf{add}, sid, ssid_1, ssid_2, ssid_3)$ to V.
2. Upon receiving $(\mathsf{mult}, sid, ssid_1, ssid_2, ssid_3)$, every $V_i \in V$ stores the message.

Note that this maintains the properties $\mathbf{A} = \mathbf{A_0} + \mathbf{A_1} + \mathbf{A_2}$ and $\mathbf{A} \in \mathsf{C}^{\odot m'}$, where m' is the current number of columns.

Opening (Part 1)

1. On input $(\mathsf{reveal}, sid, ssid_1, \ldots, ssid_o)$, P finds the set $J = \{j_1, \ldots, j_o\}$ of indexes associated to $ssid_1, \ldots, ssid_o$ and broadcasts $(sid, ssid_1, \ldots, ssid_o, (\mathbf{A_0}[\cdot, j], \mathbf{A_1}[\cdot, j], \mathbf{A_2}[\cdot, j])_{j \in J})$.
2. Upon receiving message $(sid, ssid_1, \ldots, ssid_o, (\mathbf{A_0}[\cdot, j], \mathbf{A_1}[\cdot, j], \mathbf{A_2}[\cdot, j])_{j \in J})$, every receiver $V_i \in V$ sends $(\mathsf{reveal}, sid, ssid)$ to $\mathcal{F}_{\mathrm{COM}}$ and waits for $(\mathsf{reveal}, sid, ssid, V_j, V', r_j)$ from $\mathcal{F}_{\mathrm{COM}}$ for all $V_j \in V \setminus V_i$. V_i sets $\boldsymbol{r} = \boldsymbol{r_1} + \cdots + \boldsymbol{r_t}$ (where the sum is in $\mathbb{Z}_3^n$) and sets the diagonal matrices $\boldsymbol{\Delta}, \boldsymbol{\Delta}'$ such that the i-th element in $\boldsymbol{\Delta}$ (resp. $\boldsymbol{\Delta}'$) is 1 if $\boldsymbol{r}[i] = 2$ (resp. $\boldsymbol{r}[i] = 1$) and 0 otherwise.
3. Upon receiving $(\mathsf{reveal}, sid, ssid, V_j, V', r_j)$ from $\mathcal{F}_{\mathrm{COM}}$ for all $V_j \in V$, P sets $\boldsymbol{r} = \boldsymbol{r_1} + \ldots + \boldsymbol{r_t}$, sends $(\mathsf{reveal}, sid, ssid_{i,r[i]})$, $(\mathsf{reveal}, sid, ssid_{i,r[i]+1})$ and $(\mathsf{reveal}, sid, \widehat{ssid}_{i,r[i]})$ to $\mathcal{F}_{\mathrm{COM}}$ for $i = 1, \ldots, n$ and halts.

Fig. 7. Addition and multiplication steps, and opening phase for the protocol Π_{MHCOM}.

Protocol Π_{MHCOM}

Opening (Part 2)

4. Upon receiving the messages $(\mathsf{reveal}, sid, ssid_{i,r[i]}, P, V, \boldsymbol{s}_{i,r[i]})$, $(\mathsf{reveal}, sid, ssid_{i,r[i]+1}, P, V, \boldsymbol{s}_{i,r[i]+1})$ and $(\mathsf{reveal}, sid, \widehat{ssid}_{i,r[i]}, P, V, \widehat{\boldsymbol{s}}_{i,r[i]})$ from $\mathcal{F}_{\text{COM}}$ for $i \in \{1, \ldots, n\}$, every receiver $V_j \in V$ proceeds as follows:

   (a) Compute $\mathbf{S}[i, \cdot] = \mathsf{PRG}(\boldsymbol{s}_{i,r[i]})$, $\mathbf{S}'[i, \cdot] = \mathsf{PRG}(\boldsymbol{s}_{i,r[i]+1})$ and $\widehat{\mathbf{S}}[i, \cdot] = \pi_{\mu+l}(\mathsf{PRG}(\widehat{\boldsymbol{s}}_{i,r[i]}))$ obtaining matrices $\mathbf{S}$, $\mathbf{S}'$ and $\widehat{\mathbf{S}}$. Note for each i, the i-th row of $\mathbf{S}$, $\mathbf{S}'$, $\widehat{\mathbf{S}}$ will equal the i-th row of $\mathbf{R}_{r[i]}$, $\mathbf{R}_{r[i]+1}$, $\widehat{\mathbf{R}}_{r[i]}$ respectively. Set $\mathbf{B} = \boldsymbol{\Delta}\mathbf{W} + \mathbf{S}$, $\mathbf{B}' = \boldsymbol{\Delta}'\mathbf{W} + \mathbf{S}'$ and $\widehat{\mathbf{B}} = \boldsymbol{\Delta}\widehat{\mathbf{W}} + \widehat{\mathbf{S}}$. Define the matrices[a] $\mathbf{Q}$, $\mathbf{Q}'$, $\widehat{\mathbf{Q}}$ as the first l columns of $\mathbf{B}$, $\mathbf{B}'$, $\widehat{\mathbf{B}}$ and remove these columns from the latter matrices, renumbering the remaining columns from 1.

   (b) Notice that $\widetilde{\mathbf{T}}_0, \widetilde{\mathbf{T}}_1, \widetilde{\mathbf{T}}_2$ form an additive sharing of $\widetilde{\mathbf{A}}\mathbf{H} + \widetilde{\mathbf{P}}$, and the verifiers know some of the shares, namely the rows of $\mathbf{B}\mathbf{H} + \mathbf{Q}$ and $\mathbf{B}'\mathbf{H} + \mathbf{Q}'$ (shares for the first n rows of $\widetilde{\mathbf{A}}\mathbf{H} + \widetilde{\mathbf{P}}$) and the rows of $\widehat{\mathbf{B}}\mathbf{H} + \widehat{\mathbf{Q}}$ (shares for the last n rows). For $i \in \{0, 1, 2\}$, parse $\widetilde{\mathbf{T}}_i$ as $\widetilde{\mathbf{T}}_i = \begin{pmatrix} \mathbf{T}_i \\ \widehat{\mathbf{T}}_i \end{pmatrix}$. Check that $\mathbf{B}\mathbf{H} + \mathbf{Q} = \boldsymbol{\Delta}\mathbf{T}_2 + \boldsymbol{\Delta}'\mathbf{T}_1 + (1 - \boldsymbol{\Delta} - \boldsymbol{\Delta}')\mathbf{T}_0$, $\mathbf{B}'\mathbf{H} + \mathbf{Q}' = \boldsymbol{\Delta}\mathbf{T}_0 + \boldsymbol{\Delta}'\mathbf{T}_2 + (1 - \boldsymbol{\Delta} - \boldsymbol{\Delta}')\mathbf{T}_1$ and $\widehat{\mathbf{B}}\mathbf{H} + \widehat{\mathbf{Q}} = \boldsymbol{\Delta}\widehat{\mathbf{T}}_2 + \boldsymbol{\Delta}'\widehat{\mathbf{T}}_1 + (1 - \boldsymbol{\Delta} - \boldsymbol{\Delta}')\widehat{\mathbf{T}}_0$, and that $\mathbf{T}_0 + \mathbf{T}_1 + \mathbf{T}_2 \in \mathsf{C}^{\odot\ell}$. If any check fails, abort.

   (c) For every $(\mathsf{add}, sid, ssid_1, ssid_2, ssid_3)$ received from P, append $\mathbf{B}[\cdot, a] + \mathbf{B}[\cdot, b]$ to $\mathbf{B}$ and append $\mathbf{B}'[\cdot, a] + \mathbf{B}'[\cdot, b]$ to $\mathbf{B}'$ (a, b are the index corresponding to $ssid_1$, $ssid_2$ respectively and $ssid_3$ is associated with the new column index). For every $(\mathsf{mult}, sid, ssid_1, ssid_2, ssid_3)$ received from P:
   - given $(sid, ssid, h, \boldsymbol{w}_0, \boldsymbol{w}_1, \boldsymbol{w}_2)$, check that $\boldsymbol{w}_0 + \boldsymbol{w}_1 + \boldsymbol{w}_2 \in \mathsf{C}^{*2}$ and $\boldsymbol{w}_{r[i]} = \mathbf{B}[\cdot, a] * \mathbf{B}[\cdot, b] + \mathbf{B}[\cdot, a] * \mathbf{B}'[\cdot, b] + \mathbf{B}'[\cdot, a] * \mathbf{B}[\cdot, b] + \widehat{\mathbf{B}}[\cdot, h]$;
   - let $\boldsymbol{u} = \pi_{[k]}(\boldsymbol{w}_0 + \boldsymbol{w}_1 + \boldsymbol{w}_2)$, append the columns $\mathbf{B}[\cdot, h] + \boldsymbol{\Delta}\mathsf{C}(\boldsymbol{u})$ and $\mathbf{B}'[\cdot, h] + \boldsymbol{\Delta}'\mathsf{C}(\boldsymbol{u})$ to $\mathbf{B}$ and $\mathbf{B}'$, respectively.

   Note that the properties detailed in footnote[a] are maintained.

   (d) For every $j \in J$, check that $\mathbf{A}_0[\cdot, j] + \mathbf{A}_1[\cdot, j] + \mathbf{A}_2[\cdot, j] \in \mathsf{C}$ and that, for $i = 1, \ldots, n$, it holds that $\mathbf{B}[i, j] = \mathbf{A}_{r[i]}[i, j]$ and $\mathbf{B}'[i, j] = \mathbf{A}_{r[i]+1}[i, j]$. If all checks succeed, for every $j \in J$, output the first k positions in $\mathbf{A}_0[\cdot, j] + \mathbf{A}_1[\cdot, j] + \mathbf{A}_2[\cdot, j]$ as the opened string and halts. Otherwise, abort by outputting $(sid, ssid_j, \perp)$.

---

[a] Note that we have $\mathbf{A} = \mathbf{A}_0 + \mathbf{A}_1 + \mathbf{A}_2$, $\mathbf{B} = \boldsymbol{\Delta}\mathbf{A}_2 + \boldsymbol{\Delta}'\mathbf{A}_1 + (1 - \boldsymbol{\Delta} - \boldsymbol{\Delta}')\mathbf{A}_0$ and $\mathbf{B}' = \boldsymbol{\Delta}\mathbf{A}_0 + \boldsymbol{\Delta}'\mathbf{A}_2 + (1 - \boldsymbol{\Delta} - \boldsymbol{\Delta}')\mathbf{A}_1$. This means that $\mathbf{A}$ held by P is shared in the replicated secret sharing scheme RSS_3 and for each row index, V knows one share (i.e., V knows the corresponding rows from exactly two of the matrices $\mathbf{A}_0$, $\mathbf{A}_1$, $\mathbf{A}_2$). Moreover, $\widehat{\mathbf{A}} = \widehat{\mathbf{A}}_0 + \widehat{\mathbf{A}}_1 + \widehat{\mathbf{A}}_2$ and $\widehat{\mathbf{B}} = \boldsymbol{\Delta}\widehat{\mathbf{A}}_2 + \boldsymbol{\Delta}'\widehat{\mathbf{A}}_1 + (\mathbf{I} - \boldsymbol{\Delta} - \boldsymbol{\Delta}')\widehat{\mathbf{A}}_0$ i.e., $\widehat{\mathbf{A}}$ held by P is shared in the additive secret sharing scheme Add_3 and for each row index, V knows one share (V knows the corresponding row from exactly one of the matrices $\widehat{\mathbf{A}}_0$, $\widehat{\mathbf{A}}_1$, $\widehat{\mathbf{A}}_2$).

Fig. 8. Opening phase (continued) for the protocol Π_{MHCOM}.

While it can be shown that any constant round proof system can be immediately compiled into a non-interactive argument system via the Fiat-Shamir heuristic [20], super-constant round protocols need to fulfil a stronger soundness property called *resettable soundness* for the Fiat-Shamir transform to result in a sound protocol.

We will now outline how to compile any resettable sound public coin interactive proof system into an honest-verifier zero-knowledge proof systems in a way that only slightly increases the communication complexity and only affects the efficiency of prover and verifier by a small constant factor.

The basic idea of the transformation is simple and follows the paradigm of *committed conversations* [5]. The prover and verifier run the interactive proof system with the modification that instead of sending its messages in the plain, the prover sends commitments to its messages. After the protocol is over the prover convinces the verifier that the commitment values pass the verification equations $F_1, \ldots, F_s$. The homomorphic property of the commitments will be used to implement this check efficiently. While our protocol Π_{MHCOM} does support evaluation of low degree polynomials, we will focus on linear/affine verification equations $F_1, \ldots, F_s$ and will therefore rely on the additively homomorphic commitment scheme Π_{AHCOM}, with several modifications which are discussed in the full version [18].

Instantiation. We will now discuss instantiating the hyrax protocol of [31] with the modified version of the commitment scheme Π_{AHCOM}.

To prove satisfiability of an algebraic circuit of depth d, width G and input/witness size $|w|$, the hyrax protocol has proof size $(10d \log(G) + \sqrt{|w|}) \cdot \kappa$ assuming that a group element in a DLOG-hard group $\mathbb{G}$ has size κ. The verifier runtime is $O(\sqrt{|w|} + d \cdot \log(G))$ whereas the prover runtime is linear in the size of the circuit C.

Replacing the DLOG-based homomorphic commitment in the hyrax protocol with our commitment protocol Π_{AHCOM} as outlined above, the main optimization which is not available is compression of the witness w. Consequently, in our instantiation proof size will depend linearly on the size of the witness $|w|$.

One of the key ideas in the hyrax protocol is to reduce all algebraic relations between commitments to linear relations between vector commitments, an idea also used in bulletproofs [10]. In this way, general algebraic relations can be proven using a protocol which just supports linear relations between vectors. This transformation only incurs a small constant factor additional overhead. Omitting details, there are three main steps. In the first step reduce multiplicative relations to linear relations, in the second step show that many linear relations can be compressed into a single linear relation, and in. the third step step reduce linear relations between commitments to linear relations between vector commitments. All three steps are implemented using a Schnorr-style protocol. In [31] these transformations are provided for the concrete case of DLOG-based commitments, but these ideas can be implemented using arbitrary homomorphic vector commitments.

The main improvement of our protocol over [31] is that we only rely on simple private key primitives. On the turn side, our vector-commitments are not compressing, which leads to the proof-size to depend linearly on the witness-size $|w|$ instead of $\sqrt{|w|}$. However, the proof size does not depend multiplicatively on the computational security parameter κ, but rather on $|\mathbb{F}|$, which is a statistical security parameter an can therefore be chosen much smaller. Consequently, we get an advantage in terms of proof-size whenever the proof-size is dominated by d rather than $|w|$.

6 Applications to Secure Multiparty Computation

6.1 Committed MPC

A recent work by Frederiksen *et al.* [21] has shown that additively homomorphic commitments can be leveraged to construct efficient preprocessed MPC. However, their "Committed MPC" protocol requires a multiparty commitment functionality that allows for multiple senders and for computing linear combinations between commitments generated by different senders. We will show a generic construction of such a protocol from functionality $\mathcal{F}_{\text{AHCOM}}$ that can be instantiated with Protocol Π_{AHCOM}, achieving significantly better efficiency than the construction of [21].

Functionality $\mathcal{F}_{\text{MSAHCOM}}$. Our protocol will realize the multiparty additively homomorphic commitment functionality from [21] with the difference that it will only allow for a single batch verification of opened commitments. While it allows for openings before verification, the validity of those will not be ensured by $\mathcal{F}_{\text{MSAHCOM}}$, which will let the adversary choose any value to be provided as an opening. $\mathcal{F}_{\text{MSAHCOM}}$ will allow for a single verification phase where all parties check whether the openings they have received are valid, after which the functionality halts. This functionality is sufficient for realizing the "Committed MPC" protocol of [21], since the parties can use the intermediate (non-verified) openings to compute the protocol and in the end verify that the result is correct. Other small differences is that we omit the Partial Open interface used to open a commitment to a single receiver and provide an interface for single addition operations. Notice that our procedures for opening a commitment for all receivers can be trivially adapted to opening towards a specific receiver by sending the corresponding messages only to that receiver and that single additions of commitments can be trivially used for computing linear combinations as in the functionality of [21]. We present Functionality $\mathcal{F}_{\text{MSAHCOM}}$ in Fig. 9.

Protocol Π_{MSAHCOM}. While a generic construction of such a protocol from any two-party additively homomorphic commitment scheme is presented in [21], we can significantly simplify and improve the efficiency of this construction departing from a multi-receiver scheme as defined in $\mathcal{F}_{\text{AHCOM}}$. We construct a protocol where every party acts both as sender and receiver of all commitments. In this protocol, each party first uses $\mathcal{F}_{\text{AHCOM}}$ to commit to random values towards the others. A joint random commitment in the new multi-sender protocol is defined as the commitment to the sum of all random messages contained in

---

Functionality $\mathcal{F}_{\text{MSAHCOM}}$

$\mathcal{F}_{\text{MSAHCOM}}$ is parameterized by $n \in \mathbb{N}$. $\mathcal{F}_{\text{MSAHCOM}}$ interacts with a set of parties $P = \{P_1, \ldots, P_t\}$ and an adversary $\mathcal{S}$ (who may abort at any time):

- **Init** Upon receiving (init, sid) from all parties in P, forward the message to $\mathcal{S}$ and initialize empty lists raw and actual.
- **Commit:** Upon receiving $(\text{commit}, sid, \mathcal{I})$ from all parties in P where $\mathcal{I}$ is a set of unused identifiers, for every $ssid \in \mathcal{I}$, sample a random $\boldsymbol{x}_{ssid} \xleftarrow{\$} \mathbb{F}^k$, set $\text{raw}[ssid] = \boldsymbol{x}_{ssid}$ and send $(\text{commit} - \text{recorded}, sid, \mathcal{I})$ to all parties P and $\mathcal{S}$.
- **Input:** Upon receiving $(\text{input}, sid, ssid, P_i, \boldsymbol{y})$ from $P_i \in P$ and $(\text{input}, sid, ssid, P_i)$ from all other parties in P, if $\text{raw}[ssid] = \boldsymbol{x}_{ssid} \neq \perp$, set $\text{raw}[ssid] = \perp$, set $\text{actual}[ssid] = \boldsymbol{y}$ and send $(\text{input} - \text{recorded}, sid, ssid, P_i)$ to all parties in P and $\mathcal{S}$.
- **Random:** Upon receiving $(\text{random}, sid, ssid)$ from all parties in P, if $\text{raw}[ssid] = \boldsymbol{x}_{ssid} \neq \perp$, set $\text{actual}[ssid] = \boldsymbol{x}_{ssid}$, set $\text{raw}[ssid] = \perp$ and send $(\text{random} - \text{recorded}, sid, ssid)$ to all parties P and $\mathcal{S}$.
- **Addition:** Upon receiving a message $(\text{add}, sid, ssid_1, ssid_2, ssid_3)$ from all parties in P: if $\text{actual}[ssid] = \boldsymbol{x}_{ssid} \neq \perp$ for $ssid \in \{ssid_1, ssid_2\}$ and $\text{raw}[ssid_3] = \text{actual}[ssid_3] = \perp$, set $\text{actual}[ssid_3] = \text{actual}[ssid_1] + \text{actual}[ssid_2]$ and send the message $(\text{add} - \text{recorded}, sid, ssid_1, ssid_2, ssid_3)$ to all P and $\mathcal{S}$.
- **Open:** Upon receiving $(\text{open}, sid, ssid)$ from all parties P, if $\text{actual}[ssid] = \boldsymbol{x}_{ssid} \neq \perp$, send $(\text{open}, sid, ssid, \boldsymbol{x}_{ssid})$ to $\mathcal{S}$. If $\mathcal{S}$ answers with $(\text{open}, sid, ssid, \boldsymbol{x}'_{ssid})$, send $(\text{open}, sid, ssid, \boldsymbol{x}'_{ssid})$ to all parties in P.
- **Verify:** Upon receiving a message (verify, sid) from all parties in P, let $ssid_1, \ldots, ssid_o$ be the $ssids$ of opened commitments (i.e. for which $(\text{open}, sid, ssid, \boldsymbol{x}'_{ssid})$ messages were sent). For $ssid \in \{ssid_1, \ldots, ssid_o\}$, set $b = 1$ if $\text{actual}[ssid] = \boldsymbol{x}'_{ssid}$ or $b = 0$ if not, and send $(\text{verify}, sid, ssid, b)$ to every party in P.

Fig. 9. Functionality for additively homomorphic commitments with multiple senders.

the individual commitments by each party. Linear combinations between joint commitments can be computed by having each party (acting as a sender in the underlying multi-receiver commitment scheme) compute the same linear combination on its own "shares" of the joint commitment. Opening a joint commitment works by having each party open their individual commitments, allowing everybody to compute the joint commitment as the sum of the opened messages. Using standard tricks, these joint random commitments can be easily turned into commitments to arbitrary messages (Fig. 10).

Security Analysis. To verify correctness, notice that Π_{MSAHCOM} computes a random commitment identified by $ssid$ as a commitment to $\sum_{i \in [t]} \text{raw}^i[ssid]$, where $\text{raw}^i[ssid]$ is supposed to be the value obtained by P_i from $\mathcal{F}^i_{\text{AHCOM}}$. In the verification procedure, all parties obtain $\boldsymbol{x}_j$ for $j \in [t]$ directly from $\mathcal{F}^j_{\text{AHCOM}}$, being able to verify that the previously opened commitments are indeed valid. If a commitment identified by $ssid$ is set to an arbitrary message $\boldsymbol{y}$, the sender P_j holding $\boldsymbol{y}$ broadcasts $\boldsymbol{w} = \boldsymbol{y} - \sum_{i \in [t]} \text{raw}^i[ssid]$, which also allows all parties

---

Protocol Π_{MSAHCOM}

Given a set of parties $P = \{P_1, \ldots, P_t\}$, for each party $P_i \in P$, Π_{MSAHCOM} uses an instance of $\mathcal{F}_{\text{AHCOM}}$ denoted as $\mathcal{F}^i_{\text{AHCOM}}$ where P_i is the sender with a set of receivers $V_i = P \setminus P_i$. Parties in $P = \{P_1, \ldots, P_t\}$ interact with each other and with $\mathcal{F}^1_{\text{AHCOM}}, \ldots, \mathcal{F}^t_{\text{AHCOM}}$, proceeding as follows:

1. **Commit** On input $(\text{commit}, sid, ssid, \mathcal{I})$ where $\mathcal{I} = \{ssid_1, \ldots, ssid_\gamma\}$ each party $P_i \in P$, for $ssid \in \mathcal{I}$, sends $(\text{commit}, sid, ssid, P_i, V_i)$ to $\mathcal{F}^i_{\text{AHCOM}}$, receiving as answer $(\text{receipt}, sid, ssid, P_i, V_i, \boldsymbol{x}_{ssid})$ and setting $\text{raw}^i[ssid] = \boldsymbol{x}_{ssid}$ and $\text{actual}^i[ssid] = \perp$.

2. **Input** On input $(\text{input}, sid, ssid, \boldsymbol{y})$ for P_i and input $(\text{input}, sid, ssid, P_j)$ for every P_j for $j \neq i$, parties P proceed as follows:
   (a) For every $j \in [t], j \neq i$, P_j aborts if $\text{actual}^j[ssid] \neq \perp$. Otherwise, P_j sends $(sid, ssid, \text{raw}^j[ssid])$ to P_i.
   (b) Upon receiving $(sid, ssid, \text{raw}^j[ssid])$ from P_j for every $j \in [t], j \neq i$, P_i sets $\boldsymbol{x} = \sum_{j \in [t]} \text{raw}^j[ssid]$, $\boldsymbol{w} = \boldsymbol{y} - \boldsymbol{x}$, $\text{actual}^i[ssid] = \boldsymbol{w}$ and broadcasts $(sid, ssid, P_i, \boldsymbol{w})$.
   (c) Upon receiving $(sid, ssid, P_i, \boldsymbol{w})$, every party $P_j \in P$ sets $\text{actual}^j[ssid] = \boldsymbol{w}$.

3. **Random:** On input $(\text{random}, sid, ssid)$, if $\text{actual}^i[ssid] = \perp$, each party $P_i \in P$ sets $\text{actual}^i[ssid] = \boldsymbol{0}^k$.

4. **Addition:** On input $(\text{add}, sid, ssid_1, ssid_2, ssid_3)$, if $\text{actual}^i[ssid_1] \neq \perp$, $\text{actual}^i[ssid_2] \neq \perp$ and $\text{actual}^i[ssid_3] = \perp$, every party $P_i \in P$ sets $\text{actual}^i[ssid_3] = \text{actual}^i[ssid_1] + \text{actual}^i[ssid_2]$ and sends $(\text{add}, sid, ssid_1, ssid_2, ssid_3, P_i, V_i)$ to $\mathcal{F}^i_{\text{AHCOM}}$. All parties proceed after receiving $(\text{add}, sid, ssid_1, ssid_2, ssid_3, P_i, V_i, \text{success})$ from $\mathcal{F}^i_{\text{AHCOM}}$.

5. **Open:** On input $(\text{open}, sid, ssid)$, each $P_i \in P$ broadcasts $(sid, ssid, \text{raw}^i[ssid])$. Upon receiving $(sid, ssid, \text{raw}^j[ssid])$ for $j \in [t], j \neq i$, each party $P_i \in P$ computes $\boldsymbol{x}' = \text{actual}^i[ssid] + \sum_{j \in [t]} \text{raw}^j[ssid]$ and outputs $(sid, ssid, \boldsymbol{x}')$.

6. **Verify:** On input (verify, sid), let $ssid_1, \ldots, ssid_o$ be the $ssid$s of opened commitments (*i.e.* for which $(\text{open}, sid, ssid)$ inputs were received), every $P_i \in P$ sends $(\text{reveal}, sid, ssid_1, \ldots, ssid_o)$ to $\mathcal{F}^i_{\text{AHCOM}}$. For every $ssid \in \{ssid_1, \ldots, ssid_o\}$, upon receiving $(\text{reveal}, sid, ssid, P_j, V_j, \boldsymbol{x}_j)$ for $j \in [t], j \neq i$, each party $P_i \in P$ sets $\boldsymbol{x}_i = \text{raw}^i[ssid]$, computes $\boldsymbol{x} = \text{actual}^i[ssid] + \sum_{j \in [t]} \boldsymbol{x}_j$, sets $b = 1$ if $\boldsymbol{x}' = \boldsymbol{x}$ (where $\boldsymbol{x}'$ is the value previously opened) or $b = 0$ if not, and outputs $(\text{verify}, sid, ssid, b)$.

Fig. 10. Protocol Π_{MSAHCOM}

to retrieve $\boldsymbol{y}$ when values $\text{raw}^i[ssid]$ are released and to verify the correctness of this opening when $\boldsymbol{x}_j$ (corresponding to $\text{raw}^j[ssid]$) are revealed. Notice that addition are simply computed by adding the $\text{actual}^i[ssid]$ vectors and, since all of these vectors are linear combinations of themselves, opening and verification of a result addition works the same way as for the other commitments.

Theorem 5. *Protocol Π_{MSAHCOM} UC realizes $\mathcal{F}_{\text{MSAHCOM}}$ in the $\mathcal{F}_{\text{AHCOM}}$-hybrid model with statistical security against a static adversary. Formally, there exists a simulator $\mathcal{S}$ such that for every static adversary $\mathcal{A}$, and any environment $\mathcal{Z}$ the following holds: $\text{IDEAL}_{\mathcal{F}_{\text{MSAHCOM}}, \mathcal{S}, \mathcal{Z}} \approx_s \text{HYBRID}^{\mathcal{F}_{\text{AHCOM}}}_{\Pi_{\text{MSAHCOM}}, \mathcal{A}, \mathcal{Z}}$.*

Proof (Sketch). Notice that Π_{MSAHCOM} only performs operations with random values obtained from $\mathcal{F}_{\text{AHCOM}}$. Hence, upon learning the opening of any commitment from $\mathcal{F}_{\text{MSAHCOM}}$, the simulator can simply cheat in the openings of random values from the emulated $\mathcal{F}^i_{\text{AHCOM}}$ in order to equivocate a commitment. Similarly, if it needs to extract any commitment done in Π_{MSAHCOM}, the simulator can compute it from the messages sent by the adversary in the protocol and the messages the adversary obtains from the emulated $\mathcal{F}^i_{\text{AHCOM}}$.

Efficiency. Notice that our construction of Π_{MSAHCOM} using $\mathcal{F}_{\text{AHCOM}}$ as a black box actually communicates more bits than necessary. In Π_{MSAHCOM}'s opening phase, all parties broadcast the messages in commitments generated by $\mathcal{F}_{\text{AHCOM}}$ and, later on, verify these openings by opening the commitments through $\mathcal{F}_{\text{AHCOM}}$, sending the same messages again. If instantiated with Π_{AHCOM}, our construction can be made more efficient by having the parties broadcast columns $\mathbf{A}_0[\cdot, j], \mathbf{A}_1[\cdot, j]$ (Step 1 of Π_{AHCOM}'s opening phase) during the opening phase of Π_{MSAHCOM}. Later on, for verification, the parties only need to execute the remaining steps of the opening phase of Π_{AHCOM} in order to verify that the columns they have previously obtained are actually valid. In a setting with t parties, our protocol only requires t individual multi-receiver commitments, where the construction of [21] requires t^2 two-party commitments. Their constructions also require extra communication in the order of $O(skt^2)$ for generating a batch of m commitments, where s is the security parameter and k is the message length. Moreover, instantiating the construction of [21] with the previously best two-party additively homomorphic commitments [16] implies a high cost of nt^2 OTs for the setup phase (with an underlying $[n, k, s]$ code) and extra communication in the order of $O(nmt^2)$ bits for generating a batch of m commitments to random messages. On the other hand, our construction instantiated with protocol Π_{AHCOM} can do the same with nt calls to $\mathcal{F}_{\text{COM}}$ (which can be instantiated much cheaper than an OT by calling a random oracle and sending its output) and extra communication in the order of $O(smt)$ bits. In the opening phase, the construction of [21] requires communication in the order of $O(nt^2)$ bits, while our construction only requires communication in the order of $O(nt)$ bits, assuming broadcast channels.

6.2   Insured MPC

Recently, Andrychowicz et al. [2] started a line of work [4,6,7,26] that deals with the problem of fairness in multiparty computation by combining MPC protocols with cryptocurrencies. The main idea is to provide financial incentives for the parties to act honestly. In a nutshell, each party provides a security deposit before the protocol execution or right before the outputs are revealed. After that, the protocol is executed and if no problem happens, then the security deposits are reimbursed. On the other hand, if some problem happens, the security deposit of the parties who misbehaved/aborted is used to compensate the remaining parties. This combination of MPC and cryptocurrency techniques also allows to have both inputs and outputs consisting of both data and monetary assets and distribute the funds according to the output of the computation.

The most efficient solution to date, due to Baum et al. [4], uses a publicly verifiable additively homomorphic multi-receiver commitment scheme as a central building block. By combining such commitment scheme with a smart contract, an authenticated bulletin board, and a MPC scheme that output verifiably secret shared outputs, they obtained an efficient MPC protocol with public detection of cheating behavior that financially punishes misbehaving parties. Nevertheless, the main bottleneck of their protocol is the multi-party commitment scheme, as its complexity grows quadratically in the number of parties. With our techniques it is possible to greatly improve the performance of publicly verifiable additively homomorphic multi-receiver commitments.

The functionality for publicly verifiable additively homomorphic commitment $\mathcal{F}_{\mathrm{PVHCOM}}$ is described in the full version [18] and the set of external verifiers U is allowed to be dynamic by adding procedures for registering and deregistering parties following the approach of Badertscher et al. [3]. Assuming that the underlying commitment protocol Π_{COM} used as a building block is publicly verifiable, Protocol Π_{AHCOM} is trivially publicly verifiable when all the messages are posted to an authenticated bulletin board, straightforwardly realizing functionality $\mathcal{F}_{\mathrm{PVHCOM}}$. The "canonical" random oracle commitment scheme (that realizes $\mathcal{F}_{\mathrm{COM}}$ in the programmable Global Random Oracle model without extra computational assumptions according to a recent result by Camenisch *et al.* [11]) is a clear example of a scheme that is publicly verifiable when the messages are posted to an authenticated bulletin board, and Π_{AHCOM} instantiated using that commitment scheme can be used to remarkably improve the performance of publicly verifiable additively homomorphic commitments and consequently of the Insured MPC protocol of Baum et al. [4]. The efficiency improvements achieved in this application are similar to those of the Committed MPC case, since the previously best publicly verifiable multi-receiver additively homomorphic commitment protocol of [4] has a very similar structure to the commitment protocol of [21].

References

1. Ames, S., Hazay, C., Ishai, Y., Venkitasubramaniam, M.: Ligero: lightweight sublinear arguments without a trusted setup. In: Thuraisingham, B.M., Evans, D., Malkin, T., Xu, D. (eds.) ACM CCS 2017, pp. 2087–2104. ACM Press, October/November (2017)
2. Andrychowicz, M., Dziembowski, S., Malinowski, D., Mazurek, L.: Secure multiparty computations on bitcoin. In: 2014 IEEE Symposium on Security and Privacy, pp. 443–458. IEEE Computer Society Press, May (2014)
3. Badertscher, C., Maurer, U., Tschudi, D., Zikas, V.: Bitcoin as a transaction ledger: a composable treatment. In: Katz, J., Shacham, H. (eds.) CRYPTO 2017. LNCS, vol. 10401, pp. 324–356. Springer, Cham (2017). https://doi.org/10.1007/978-3-319-63688-7_11
4. Baum, C., David, B., Dowsley, R.: Insured mpc: efficient secure multiparty computation with punishable abort. Cryptology ePrint Archive, Report 2018/942 (2018). https://eprint.iacr.org/2018/942

634   I. Cascudo et al.

5. Ben-Or, M., Goldreich, O., Goldwasser, S., Håstad, J., Kilian, J., Micali, S., Rogaway, P.: Everything provable is provable in zero-knowledge. In: Goldwasser, S. (ed.) CRYPTO 1988. LNCS, vol. 403, pp. 37–56. Springer, New York (1990). https://doi.org/10.1007/0-387-34799-2_4

6. Bentov, I., Kumaresan, R.: How to use bitcoin to design fair protocols. In: Garay, J.A., Gennaro, R. (eds.) CRYPTO 2014. LNCS, vol. 8617, pp. 421–439. Springer, Heidelberg (2014). https://doi.org/10.1007/978-3-662-44381-1_24

7. Bentov, I., Kumaresan, R., Miller, A.: instantaneous decentralized poker. In: Takagi, T., Peyrin, T. (eds.) ASIACRYPT 2017. LNCS, vol. 10625, pp. 410–440. Springer, Cham (2017). https://doi.org/10.1007/978-3-319-70697-9_15

8. Blazy, O., Chevalier, C., Pointcheval, D., Vergnaud, D.: Analysis and Improvement of Lindell?s UC-Secure Commitment Schemes. In: Jacobson, M., Locasto, M., Mohassel, P., Safavi-Naini, R. (eds.) ACNS 2013. LNCS, vol. 7954, pp. 534–551. Springer, Heidelberg (2013). https://doi.org/10.1007/978-3-642-38980-1_34

9. Brandão, L.T.A.N.: Very-efficient simulatable flipping of many coins into a well. In: Cheng, C.-M., Chung, K.-M., Persiano, G., Yang, B.-Y. (eds.) PKC 2016. LNCS, vol. 9615, pp. 297–326. Springer, Heidelberg (2016). https://doi.org/10.1007/978-3-662-49387-8_12

10. Bünz, B., Bootle, J., Boneh, D., Poelstra, A., Wuille, P., Maxwell, G.: Bulletproofs: short proofs for confidential transactions and more. In: 2018 IEEE Symposium on Security and Privacy, pp. 315–334. IEEE Computer Society Press, May (2018)

11. Camenisch, J., Drijvers, M., Gagliardoni, T., Lehmann, A., Neven, G.: The wonderful world of global random oracles. In: Nielsen, J.B., Rijmen, V. (eds.) EUROCRYPT 2018. LNCS, vol. 10820, pp. 280–312. Springer, Cham (2018). https://doi.org/10.1007/978-3-319-78381-9_11

12. Canetti, R.: Universally composable security: a new paradigm for cryptographic protocols. In: 42nd FOCS, pp. 136–145. IEEE Computer Society Press, October (2001)

13. Canetti, R., Fischlin, M.: Universally composable commitments. In: Kilian, J. (ed.) CRYPTO 2001. LNCS, vol. 2139, pp. 19–40. Springer, Heidelberg (2001). https://doi.org/10.1007/3-540-44647-8_2

14. Canetti, R., Lindell, Y., Ostrovsky, R., Sahai, A.: Universally composable two-party and multi-party secure computation. In: 34th ACM STOC, pp. 494–503. ACM Press, May (2002)

15. Cascudo, I.: On squares of cyclic codes. IEEE Trans. Inf. Theor. **65**(2), 1034–1047 (2019)

16. Cascudo, I., Damgård, I., David, B., Döttling, N., Nielsen, J.B.: Rate-1, linear time and additively homomorphic UC commitments. In: Robshaw, M., Katz, J. (eds.) CRYPTO 2016. LNCS, vol. 9816, pp. 179–207. Springer, Heidelberg (2016). https://doi.org/10.1007/978-3-662-53015-3_7

17. Cascudo, I., Damgård, I., David, B., Giacomelli, I., Nielsen, J.B., Trifiletti, R.: Additively homomorphic uc commitments with optimal amortized overhead. In: Katz, J. (ed.) PKC 2015. LNCS, vol. 9020, pp. 495–515. Springer, Heidelberg (2015). https://doi.org/10.1007/978-3-662-46447-2_22

18. Cascudo, I., Damgård, I., David, B., Döttling, N., Dowsley, R., Giacomelli, I.: Efficient UC commitment extension with homomorphism for free (and applications) [full version]. Cryptology ePrint Archive, Report 2018/983 (2018). https://eprint.iacr.org/2018/983

19. Damgård, I., David, B., Giacomelli, I., Nielsen, J.B.: Compact VSS and efficient homomorphic UC commitments. In: Sarkar, P., Iwata, T. (eds.) ASIACRYPT 2014. LNCS, vol. 8874, pp. 213–232. Springer, Heidelberg (2014). https://doi.org/10.1007/978-3-662-45608-8_12

20. Fiat, A., Shamir, A.: How to prove yourself: practical solutions to identification and signature problems. In: Odlyzko, A.M. (ed.) CRYPTO 1986. LNCS, vol. 263, pp. 186–194. Springer, Heidelberg (1987). https://doi.org/10.1007/3-540-47721-7_12

21. Frederiksen, T.K., Pinkas, B., Yanai, A.: Committed MPC. In: Abdalla, M., Dahab, R. (eds.) PKC 2018. LNCS, vol. 10769, pp. 587–619. Springer, Cham (2018). https://doi.org/10.1007/978-3-319-76578-5_20

22. Frederiksen, T.K., Jakobsen, T.P., Nielsen, J.B., Trifiletti, R.: On the complexity of additively homomorphic UC commitments. In: Kushilevitz, E., Malkin, T. (eds.) TCC 2016. LNCS, vol. 9562, pp. 542–565. Springer, Heidelberg (2016). https://doi.org/10.1007/978-3-662-49096-9_23

23. Frederiksen, T.K., Jakobsen, T.P., Nielsen, J.B., Nordholt, P.S., Orlandi, C.: MiniLEGO: efficient secure two-party computation from general assumptions. In: Johansson, T., Nguyen, P.Q. (eds.) EUROCRYPT 2013. LNCS, vol. 7881, pp. 537–556. Springer, Heidelberg (2013). https://doi.org/10.1007/978-3-642-38348-9_32

24. Garay, J.A., Ishai, Y., Kumaresan, R., Wee, H.: On the complexity of UC commitments. In: Nguyen, P.Q., Oswald, E. (eds.) EUROCRYPT 2014. LNCS, vol. 8441, pp. 677–694. Springer, Heidelberg (2014). https://doi.org/10.1007/978-3-642-55220-5_37

25. Goldwasser, S., Kalai, Y.T., Rothblum, G.N.: Delegating computation: interactive proofs for muggles. In: Ladner, R.E., Dwork, C. (eds.) 40th ACM STOC, pp. 113–122. ACM Press, May (2008)

26. Kiayias, A., Zhou, H.-S., Zikas, V.: Fair and robust multi-party computation using a global transaction ledger. In: Fischlin, M., Coron, J.-S. (eds.) EUROCRYPT 2016. LNCS, vol. 9666, pp. 705–734. Springer, Heidelberg (2016). https://doi.org/10.1007/978-3-662-49896-5_25

27. Lindell, Y.: Highly-efficient universally-composable commitments based on the DDH assumption. In: Paterson, K.G. (ed.) EUROCRYPT 2011. LNCS, vol. 6632, pp. 446–466. Springer, Heidelberg (2011). https://doi.org/10.1007/978-3-642-20465-4_25

28. Randriambololona, H.: Asymptotically good binary linear codes with asymptotically good self-intersection spans. IEEE Trans. Inf. Theor. $59(5)$, 3038–3045 (2013)

29. Reingold, O., Rothblum, G.N., Rothblum, R.D.: Constant-round interactive proofs for delegating computation. In: Wichs, D., Mansour, Y. (eds.) 48th ACM STOC, pp. 49–62. ACM Press, June (2016)

30. Vadhan, S.P., Zheng, C.J.: Characterizing pseudoentropy and simplifying pseudorandom generator constructions. In: Karloff, H.J., Pitassi, T. (eds) 44th ACM STOC, pp. 817–836. ACM Press, May (2012)

31. Wahby, R.S., Tzialla, I., Shelat, A., Thaler, J., Walfish, M.: Doubly-efficient zkSNARKs without trusted setup. In: 2018 IEEE Symposium on Security and Privacy, pp. 926–943. IEEE Computer Society Press, May (2018)

Scalable Private Set Union
from Symmetric-Key Techniques

Vladimir Kolesnikov[1]([✉]), Mike Rosulek[2], Ni Trieu[2], and Xiao Wang[3]

[1] Georgia Institute of Technology, Atlanta, USA
kolesnikov@gatech.edu
[2] Oregon State University, Corvallis, USA
{rosulekm,trieun}@eecs.oregonstate.edu
[3] Northwestern University, Evanston, USA
wangxiao@cs.northwestern.edu

Abstract. We present a new efficient protocol for computing private set union (PSU). Here two semi-honest parties, each holding a dataset of known size (or of a known upper bound), wish to compute the union of their sets without revealing anything else to either party. Our protocol is in the OT hybrid model. Beyond OT extension, it is fully based on symmetric-key primitives. We motivate the PSU primitive by its direct application to network security and other areas.

At the technical core of our PSU construction is the reverse private membership test (RPMT) protocol. In RPMT, the sender with input x^* interacts with a receiver holding a set X. As a result, the receiver learns (only) the bit indicating whether $x^* \in X$, while the sender learns nothing about the set X. (Previous similar protocols provide output to the opposite party, hence the term "reverse" private membership.) We believe our RPMT abstraction and constructions may be a building block in other applications as well.

We demonstrate the practicality of our proposed protocol with an implementation. For input sets of size 2^{20} and using a single thread, our protocol requires 238 s to securely compute the set union, regardless of the bit length of the items. Our protocol is amenable to parallelization. Increasing the number of threads from 1 to 32, our protocol requires only 13.1 s, a factor of 18.25× improvement.

To the best of our knowledge, ours is the first protocol that reports on large-size experiments, makes code available, and avoids extensive use of computationally expensive public-key operations. (No PSU code is publicly available for prior work, and the only prior symmetric-key-based work reports on small experiments and focuses on the simpler 3-party, 1-corruption setting.) Our work improves reported PSU state of the art by factor up to 7,600× for large instances.

1 Introduction

Private set union (PSU) is a special case of secure two-party computation. PSU allows two parties holding sets X and Y respectively, to compute the union $X \cup Y$, without revealing anything else, namely what are the items in the intersection of X and Y.

© International Association for Cryptologic Research 2019
S. D. Galbraith and S. Moriai (Eds.): ASIACRYPT 2019, LNCS 11922, pp. 636–666, 2019.
https://doi.org/10.1007/978-3-030-34621-8_23

1.1   Motivation

PSU (like the well-researched private set intersection, PSI) has numerous applications in practice, and tailored efficient solutions are highly desirable. Consider the following use cases. (We note that these use cases cover a wide range of PSU settings, such as multi-party or shared-output PSU. Our work does not address all of the settings, of course, but provides a building block and a baseline for the entire research direction.)

Cyber Risk Assessment and Management via Joint IP Blacklists and Joint Vulnerability Data. As noted in [28,38], organizations aim to optimize their security updates to minimize vulnerabilities in their infrastructure. Crucial role in the above is played by *joint* lists of blacklisted IP addresses, characteristic network traces and other associated data, as well as *joint* lists of data points reported by vulnerability scanners. At the same time, organizations are understandably reluctant to reveal details pertaining to their current or past attacks or sensitive network data. As convincingly argued in [28], the use of MPC in computing set unions of the above data sets will mitigate the organizations' concerns. [28] implements the computation of such set union and related data aggregation as generic MPC in the VIFF framework. As noted by the authors, the major performance bottleneck in their work is private computation of set union. Our tailored PSU algorithms will be applicable to this computation as the main building block.

More generally, privacy-preserving data aggregation is a well-appreciated goal in the network security and other communities. For example, SEPIA [8] is a library aimed to optimize generic MPC to securely and in real-time compute event correlation and aggregation of network traffic statistics. Our PSU protocol can potentially be helpful in that setting too.

Other Applications and Use Cases. Imagine two Internet providers considering a merger, and they would like to calculate how efficient the resulting joint network would be without revealing the information of their existing networks [7]. Another application of combining set-intersection and set-union is the following scenario discussed in [34]. A social services organization wants to determine the list of cancer patients who are on welfare. Some patients may have cancer treatment at multiple hospitals. By using a private set union protocol, the union of each hospital's lists of cancer patients can be computed (while removing duplicate patients without leaking the details of the patients), then a secure set intersection operation between the resulting union and the welfare rolls can be performed.

More generally, PSU is an essential building block for private DB supporting full join. Suppose there are two tables owned by two principals, say DMV (Department of Motor Vehicles) and SSA (Social Security Administration). With a PSU-based implementation, a query such as

SELECT ssn, dob

FROM dmv_db FULL JOIN ssa_db

ON $dmv\_db.ssn = ssa\_db.ssn$ WHERE $dob \geq Jan\ 1, 1980$

will allow the players to learn the two columns of the union, but not learn whether the other player has the matching record.

Malicious model is of course the ultimate goal in this line of research. At the same time, we believe semi-honest guarantee is sufficient in many scenarios. Further, our work may serve as a stepping stone to the malicious-secure solution where it is required. We believe that our performance improvement of *four* orders of magnitude is surprising for a reasonably researched problem, and sets the baseline for the PSU performance.

1.2 Contribution

Over the last decade, there has been a significant amount of work on private set intersection [3,10–13,15–18,20,27,29,32,36,37,43,45–47,49,51]. However, there has been little work on PSU, with current PSU state-of-the-art not scalable for big data. Despite similarities between the two functionalities, many effective PSI techniques do not directly apply to PSU. We give a brief discussion about the unsuitability for PSU of several popular PSI techniques in Sect. 5.4 as well as throughout the paper.

We design a truly scalable PSU protocol, building on newly developed building blocks. In detail, our contribution can be summarized as follows:

1. We identify that existing fast private membership tests, used in leading PSI protocols are not immediately applicable for computing PSU (cf. Sect. 2.1), and a richer PMT of [13] carries 125× performance penalty (cf. Sect. 1.3). We propose a new building block *reverse private membership test* (PMT) in Sect. 4. We present an efficient instantiation of this building block, which serves as the basis of our symmetric-key based PSU protocol.
2. We apply the bucketing technique to further reduce the computation and communication overhead. We identify and overcome several new challenges unique to bucketing in the context of PSU (but not PSI). Details can be found in Sect. 5.
3. Integrating the above two components, we build a truly scalable system for PSU computation that is *three orders of magnitude* faster than the current reported performance for large two-party PSU instances. Specifically, we are ≈7,600× faster than [14], which is the current best reported numbers for larger sets of 1 million elements. [5] consider an easier setting with three parties and one corruption. Although our protocol works in a stronger model than [5], we are still 30× faster in terms of running time on sets of 2^{12} elements and have 100–125× smaller communication (cf. Table 3). Our protocol evaluates PSU of two million-element datasets in about a minute on WAN and 13 s on a LAN.
4. Our implementation is released on Github: https://github.com/osu-crypto/PSU. To our knowledge, this is the first publicly available PSU implementation.

1.3 Related Work

We start by reviewing previous PSU protocols, with particular emphasis on the semi-honest model.

Table 1. Asymptotic communication (bits) and computation costs of two-party PSU protocols in the semi-honest setting. Pub-key: public-key operations; symm: symmetric cryptographic operations. n is the size of the parties' input sets. ℓ is the bit-length item. λ is statistical security parameters. In [5] and our protocol, $\kappa = 128$ is computational security parameter, while $\kappa = 2048$ is the public key length in other protocols. We ignore the pub-key cost of κ base OTs.

Protocol	Comm. (bits)	Comp. [#Ops symm/pub-key]
[34]	$O(\kappa^3 n^2)$	$O(n^2)$ pub-key
[23]	$O(\kappa n)$	$O(n \log \log(n))$ pub-key
[5]	$O(\kappa \ell n \log(n))$	$O(n \ell \log(n))$ symm
[14]	$O(\kappa \lambda n)$	$O(\lambda n)$ pub-key
Ours	$O(\kappa n \log(n))$	$O(n \log(n))$ symm

Kissner and Song [34]. To our knowledge, the first PSU protocol was proposed by Kissner and Song [34]. The PSU of [34] is based on polynomial representations and additively homomorphic encryption (AHE). The core idea of their protocol is that if the sets X (respectively, Y) is represented as a polynomial f (respectively, g) whose roots are the set's elements, then the polynomial representation of the union $X \cup Y$ is $f \times g$. An important property is that an item x is in the set X if and only if $f(x) = 0$. Consequently, for each item e that appears in either set X or Y, it holds that $(f \times g)(e) = f(e) \times g(e) = 0$. The players compute the polynomial $f \times g$ under AHE, and figure out the set of elements based on a procedure called "element reduction", which can reduce the degree of the roots.

Frikken [23]. Relying on the polynomial representation, Frikken [23] proposed a faster PSU protocol with linear communication complexity in the size of the dataset. At the high level, it proceeds as follows. Suppose that $E(f)$ is an encrypted polynomial representation for the set X, a tuple of the form $(xE(f(x)), E(f(x)))$ achieves the specific property that this tuple will be $(0; 0)$ if $x \in X$. In other words, $x \notin X$ can be recovered from the decrypted tuple values. Therefore, instead of computing the encrypted $f \times g$ in [34], Bob just computes the above tuples after receiving the encrypted polynomial representation $E(f)$ from Alice, and sends them back to Alice in random order. Alice now decrypts the tuples and learns the value that is not in the intersection. The work of Frikken [23] requires $O(n\kappa)$ communication, where n is the size of the parties' input sets and κ is the length of public-key/ciphertext. Computational cost of generating each tuple is $O(n)$, thus this protocol requires $O(n^2)$ computation. Moreover, their protocol [23] is expensive due to the multi-point evaluation on the encrypted polynomial, which requires the depth of the arithmetic circuit (leveled fully homomorphic encryption) to be logarithmic of the input set size. The authors claimed that the computation of their protocol can be reduced to $O(n \log \log(n))$ by using the bucketing technique with minor modifications to their protocol, but it is not clear how to modify it. Indeed, using bucketing is quite tricky for PSU until our work. Based on the polynomial representation,

Hazay and Nissim [26] extended the Frikken's protocol in the presence of malicious adversaries.

Blanton and Aguiar [5]. In 2012, Blanton and Aguiar [5] proposed a faster PSU protocol based on oblivious sorting and generic MPC protocols. The core idea of their protocol consists of combining the input sets into a new set, then sorting the resulting set, and comparing adjacent items of the sorted set in order to eliminate duplicates. They focus on constructing the circuit for PSU (and several other set operations) and relegate its evaluation to generic protocols. Their paper provides experimental results on small input set in a three-party and honest majority setting for 32−bit sized elements. Their largest experiment, $n \le 2^{11}$, runs in 25 s; our $n \le 2^{12}$ experiment on larger element sizes runs in 1.42 s. Importantly, they run the experiments in the three-party setting, where evaluation is much faster as wire secrets can be 1-bit long.

We sketch approximate communication cost of their two-party garbled-circuit-based protocol based on state-of-the-art OT extension and half gates [54]. Oblivious sorting of 2^{22} elements per player involves sorting a 2^{23} array. Considering 32-bit elements, such 2PC will require approximately $23 \cdot (2^{23}) \cdot (32 + 32) \cdot 256 = 3,161,095,929,856$ bits $= 395$ GB. Here 256 is the half-gates garbled table size and 32 is the element size. Subsequent duplicate elimination will cost approximately the same as oblivious sort, so total communication cost is ≈790 GB. Considering larger element size, say, 128 bits, results in the corresponding 4× cost increase, bringing total to ≈3.1 TB. Transferring 3 TB over a 400 Mbps WAN will take $\frac{3 \cdot 8 \cdot 10^6}{400} = 60000 \text{ s} = 16.67 \text{ h}$. For comparison, our protocol for this size runs in 250 s, a 240× improvement.

[5] should perhaps be seen as an improvement over current public key-based protocols. As discussed above, our tailored solution outperforms [5] by a large factor even in the setting that is the most unfavorable for us. Because there is no reported data on the performance of [5] on larger set sizes and no existing generic MPC/2PC system supports large circuits generated by [5], we use calculated numbers in our comparison to [5] in Table 3.

Davidson and Cid [14]. Recently, Davidson and Cid [14] proposed an efficient protocol based on an encrypted Bloom filter and additively homomorphic encryption (AHE). In the [14] protocol, the receiver represents its input set Y using Bloom Filter (BF) with k hash functions, and inverts this filter by flipping the bit value of each entry. It then encrypts the inverted Bloom filter by using an IND-CPA secure AHE scheme, and sends it to the sender. For each item x of its input set X, the sender uses the k hash functions to retrieve k encrypted BF entries corresponding to x. He then uses AHE homomorphism to sum up under encryption the k retrieved ciphertexts. Let c be the obtained (AHE-encrypted) sum. The sender sends (AHE-encrypted) pairs $\{cx, c\}$ to receiver. Receiver decrypts them and is able to obtain x iff $c \neq 0$. Indeed, if $x \in Y$, all k entries of x are not set in the inverted BF, resulting in $c = 0$. Therefore, the receiver only obtains $X \setminus Y$, from which it computes $X \cup Y$. [14] requires $O(\kappa \lambda n)$ communication and $O(\lambda n)$ modular exponentiations, where λ is the statistical security parameters, and κ is the length of public-key/ciphertext, which is in the range 1024–2048 due to their use of public-key primitives. In concrete terms, encrypted BF for

PARAMETERS: Set sizes m and n; two parties: sender $\mathcal{S}$ and receiver $\mathcal{R}$

FUNCTIONALITY:

- Wait for an input $X = \{x_1, x_2, \ldots, x_n\} \subseteq \{0,1\}^*$ from sender $\mathcal{S}$, and an input $Y = \{y_1, y_2, \ldots, y_m\} \subseteq \{0,1\}^*$ from receiver $\mathcal{R}$
- Give output $X \cup Y$ to the receiver $\mathcal{R}$.

Fig. 1. Private set union functionality $\mathcal{F}_{\mathsf{psu}}^{m,n}$.

PARAMETERS: A set size n; two parties: sender $\mathcal{S}$ and receiver $\mathcal{R}$

FUNCTIONALITY:

- Wait for an input $x^* \in \{0,1\}^*$ from sender $\mathcal{S}$, and an input $X = \{x_1, x_2, \ldots, x_n\} \subseteq \{0,1\}^*$ from receiver $\mathcal{R}$
- Give the receiver $\mathcal{R}$ output 1 if $x^* \in X$ and 0 otherwise.

Fig. 2. Reverse private membership test functionality $\mathcal{F}_{\mathsf{rpmt}}^n$.

the set size $n = 2^{20}$ requires 8.05 GB and 16.1 GB when using a $\kappa = 1024$ bit and $\kappa = 2048$ bit key length, respectively.

Other Related Work. We note that recent work [13] proposed private membership test with shared output, which can be used to instantiate our reverse private membership test. Our RPMT is much faster. For specific parameters used in our work (bucket size 61, bit length 128), [13] requires 80 KB communication per test while our RPMT construction only needs 0.64 KB, a 125× improvement in terms of communication. In addition, our construction requires 140× fewer symmetric-key operations than [13]. Because we work with small bucket sizes, our polynomial-based RPMT is fast computationally as well.

Outsourcing PSU was considered in the work of Canetti et al. [9]. In this problem, users outsource their encrypted data and computation to an untrusted cloud server, while keeping their data private. The main purpose is to minimize the computational overhead of the users by utilizing the powerful resources of the cloud server.

Table 1 provides a brief comparison to the prior highest-performing PSU protocols in the semi-honest setting. We emphasize that public-key operations are the workhorse of all prior work, while we do only $\kappa = 128$ such operations to initiate OT extension. This is the main reason for 7,600× performance improvement over prior work we observe. We report in detail the performance results and comparisons in Sect. 6.

2   Overview of Our Results and Techniques

We start with a special case. Suppose that the sender has only one item y in its set Y and the receiver holding the set X will receive the resulting union $\{y\} \cup X$.

The protocol must satisfy the following:

(1) if $y \notin X$, the receiver is allowed to learn y as it is implied by the output. The sender learns nothing.
(2) if $y \in X$, the receiver knows that $y \in X$ (implied by the output), but not allowed to learn which is the sender's item y. Sender learns nothing.

Receiver learns which of the cases (1) or (2) occurs. Based on the case, the sender's item y can be conditionally sent to the receiver using a "one-sided" OT, a version of OT that requires transfer of a single encrypted secret, rather than the usual transfer of two encrypted secrets, exactly one of which the receiver can decrypt.

2.1 Reverse Private Membership Test (RPMT)

We formalize the above basic functionality as the RPMT functionality (cf. Fig. 2) and design a corresponding tailored efficient protocol, which we believe to be of independent interest. RPMT is related to the traditional Private Membership Test (PMT) [46], which is a two-party protocol in which the party with input y learns whether or not its item is in the input set X of other party (who learns nothing). In a RPMT, the output is given to the opposite party, i.e. the party holding the set X will learn whether $y \in X$ (and nothing else). We formally describe the ideal RPMT functionality in Fig. 2.

We emphasize that, unlike PSI, use of PMT is not very natural for PSU. This is because the PMT output receiver holds an element, and gets the answer in plaintext whether the element belongs to a set held by the sender. This is implied by the PSI output, and hence can be used there. However, this is extra information in the PSU functionality. We don't know of a natural way to efficiently use PMT with PSU.

This seemingly simple functionality adjustment (PMT→RPMT) doesn't seem to be fixable by a small tweak of PMT. This is because the underlying primitive used to implement fast PMT [36] is a variant of OT extension, and the role of OT receiver naturally belongs to the player with a single-element input y; it is not clear how to amend the protocol to allow (only) the other player to receive the output.

The basic idea for our RPMT is to have the receiver represent a dataset X as a polynomial $\widetilde{P}(x)$ whose roots are its elements, and send the (plaintext) coefficients of the polynomial $P(x) = \widetilde{P}(x) + s$ to the other party, where s is a secret value chosen at random by the receiver. The sender evaluates the received polynomial on y and obtains $P(y) = s'$. It is easy to see that $s' = s$ if $y \in X$, i.e. y is a root of $\widetilde{P}(x)$. At this point, the receiver could compare s' and s in the clear and learn the output of RPMT. However, if $y \notin X$, the value $P(y)$ may leak partial information about y. To prevent this, instead of the receiver sending s to the sender, the parties perform a *private equality test* (PEQT) to determine whether two strings s and s' are equal. The PEQT guarantees that the sender learns nothing about whether $y \in X$ while the polynomial presentation allows receiver to determine whether $y \in X$ but not the value of y (beyond what is implied by $y \in^? X$).

We note that full PEQT is actually not required, and a weaker and slightly efficient subprotocol is sufficient. For uninterrupted flow, we return to this observation in Sect. 2.2.

This brief overview of RPMT ignores an important security issue. In particular, suppose $y \in X$, so the sender can evaluate $P(y) = s$. Then he/she can compute $P(\cdot) - s$: a polynomial whose roots are all of the elements of X! To address this issue, the parties invoke oblivious PRF (OPRF) on their inputs, and use the OPRF's outputs for the polynomial interpolation/evaluation. Recall that OPRF is a 2-party protocol in which the OPRF sender learns a PRF key k and the OPRF receiver learns $F_k(z)$, where F is a pseudorandom function (PRF) and z is the receiver's input. In RPMT, the RPMT sender acts as the OPRF receiver to receive $F_k(y)$ and the RPMT receiver acts as the OPRF sender to obtain the PRF key k. Now, the receiver interpolates a polynomial P over points[1] $\{(x, s \oplus F_k(x))\} \ \forall x \in X$, and sends the coefficients of this polynomial to the other party, who evaluates it on y, and outputs $P(y) \oplus F_k(y)$. Thanks to OPRF, the important properties needed for RPMT still hold: (i) $F_k(y) = F_k(x)$ if $x = y$. Therefore, the sender obtains the secret value s chosen by the receiver; (ii) even if $y \in X$, other elements of X can no longer be inferred from $P(\cdot)$ and $P(y)$. This is intended to make finding roots of $P(\cdot) - P(y)$ useless to the sender. Moreover, to learn X, the sender has to know its OPRF value $F_k(x)$, which is not possible because of the OPRF guarantees. A detailed overview of the RPMT protocol is presented in Sect. 4.

We note that RPMT and OPRF are fast cryptographic tools. Recently, Kolesnikov et al. [36] proposed an efficient protocol which performs many OPRF or PEQT with amortized cost of 5 μs. Therefore, the main computation cost of our RPMT is the multiplication/evaluation of the polynomial, which requires time $O(n \log^2(n))$ using FFT or $O(n^2)$ using a more straight-forward algorithm. This is expensive for large set size $n = |X|$. We avoid the need to work with high-degree polynomials by hashing/bucketing (see below). The communication overhead is small and is equal to $O(n)$.

We can summarize the above gadget for the simple case of PSU (union of a set X and a single element y) as follows: using RPMT on X and y, the receiver learns a bit $b \in \{0, 1\}$ indicating whether $y \in X$. Next, the parties perform one-sided OT to allow receiver obliviously obtain y if $b = 0$ (i.e. $y \notin X$), nothing otherwise.

2.2 An Efficiency Optimization

Going back to the discussion of our RPMT protocol in the previous section, while it uses a PEQT protocol to compare the output of the polynomial, this is in fact overkill for our application to PSU.

Indeed, suppose the sender instead just sends the output of the polynomial s' in the clear to the receiver. Consider the two cases. First, if $y \in X$, we have

---

[1] Of course, $x \in \{0, 1\}^*$ needs to be "hashed down" to an element of the field we are working with. This can be done, e.g., by applying a collision resistant hash function. For simplicity, here we mention, but don't formalize this step.

$\{A, B\} \cap \{C, D\} = \{\}$ $\{A, B, C\} \cap \{C, D\} = \{C\}$

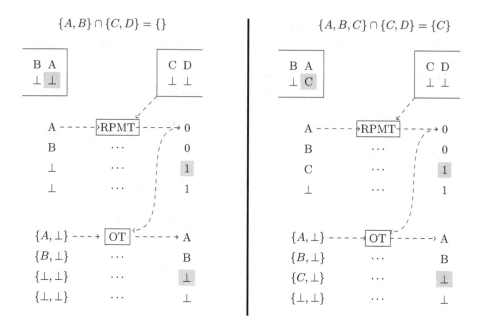

Fig. 3. Illustration of the main idea behind our protocol: using RPMT and oblivious transfer to perform PSU on a sample bin. The left-hand side illustrates that the sender's bin contains 2 real items $\{A, B\}$ and the receiver's bin contains 2 real items $\{C, D\}$, these sets are disjointed. The right-hand side shows that the sender's bin contains 3 real items $\{A, B, C\}$ and the receiver's bin contains 2 real items $\{C, D\}$, these sets have a common item C. An item $\bot$ denotes the global item known by both parties.

$s' = s$, so no information about y would be leaked, as desired. In the other case that $y \notin X$, we want (in the overall PSU protocol) the receiver to learn y anyway! So even if s' leaks information about y, this is fine. Hence, for the purpose of PSU, our protocol can conclude by a plaintext comparison, where the sender sends s' to the receiver.

As it turns out, this optimization, while elegant, is not substantial in terms of overall performance, providing 3–5% improvement in running time and ~10% improvement in communication. This can be seen by sketching relative costs of our subprotocols, and is also supported by our experiments (See Table 5 in the Appendix for more details). Because of this, we chose to present the paper in terms of the more general and conceptually simpler RPMT primitive.

However, we did formalize and prove secure the improved protocol. It is presented, together with a proof of security and experimental results in Appendix A. We feel that this presentation structure allows to focus our main presentation on the simpler primitives, while at the same time devote sufficient attention to an interesting optimization.

2.3 General Case from RPMT

We now discuss how to extend the above approach to the general case of PSU with $|Y| > 1$. The idea is natural: for each item $y \in Y$ simply execute the above gadget on y and X. As a result, the receiver obliviously obtains all items in $Z \leftarrow Y \setminus X$ which directly allows him to learn the union $X \cup Y = X \cup Z$. However, this approach requires n instances of RPMT and n instances of OT (here, we assume that $|X| = |Y| = n$). This results in communication and computation complexity of $O(n^2)$ and $O(n^2 \log(n))$, respectively. Therefore, this PSU construction is only efficient when n is small. Our next trick is to use a hashing technique to overcome this limitation.

At the high level, the idea is that the parties use a hashing scheme to assign their items into bins, and then perform the quadratic-cost PSU on each bin efficiently. By applying a balls-into-bins analysis and minimizing the overall cost, our hashing scheme has $O(n/\log n)$ bins, where each bin contains $O(\log n)$ items. We review the hashing scheme in detail in Sect. 5.2. This optimization reduces costs to $O(n \log n)$ in communication and $O(n \log n \log \log n)$ in computation. However, bucketing introduces a challenge specific to the PSU – the receiver learns additional information on the intersection items, namely, the bucket where the match occurred/did not occur. Consider an example where the receiver's first bin X_1 contains three items and the sender's first bin contains y_1. In our protocol, parties perform RPMT on X_1 and y_1. Suppose $y_1 \in X$, which means, because of bucketing, that $y_1 \in X_1$. From RPMT output, the receiver learns that $y_1 \in X_1$, which cannot be inferred from just the PSU output.

To address this issue, both parties add dummy items $\perp$ into each of their bins to fill them to their maximal size prior to executing RPMT on the bins. Then even if the output of RPMT on $(X_1 \cup \perp)$ and y_1 gives the receiver a bit $b = 1$ (i.e. indicating that $y_1 \in X_1$), the receiver will not learn any information on y_1 since y_1 may be the dummy item $\perp$. We note that this high-level description of the use of dummy items hides some technical nuance, which is explained in detail in Sect. 5.

Figure 3 illustrates the main idea behind our protocol. It is easy to see from the Fig. 3 that the receiver's view in both important cases (two bins are disjoint or two bins have a common item) are exactly same. As noted above, each bin must be padded with $\perp$ to the maximum number of items expected in a bin. In Fig. 3, the maximum bin size is 4. Section 5 formally describes the full construction of our PSU.

2.4 Efficiency

Our PSU protocol requires only $O(\kappa)$ public-key operations to perform base OT (which can run in the offline phase). In the online phase, our protocol consists of $O(n)$ OPRF instances, $O(n)$ PEQT instances, and $O(n)$ OT instances. These building blocks are based on symmetric-key operations, and can use the same base OTs. In terms of communication, our protocol requires $O(\kappa n \log(n))$, where κ is the computational security parameter.

PARAMETERS: A PRF F, and two parties: sender and receiver

FUNCTIONALITY:

- Wait for input x from the receiver.
- Sample a random PRF seed k and give it to the sender. Give $F_k(x)$ to the receiver.

Fig. 4. OPRF ideal functionality.

PARAMETERS: Two parties: sender and receiver

FUNCTIONALITY:

- Wait for input $x_0 \in \{0,1\}^*$ from the sender, and input $x_1 \in \{0,1\}^*$ from the receiver.
- Give the receiver $\mathcal{R}$ output 1 if $x_0 = x_1$ and 0 otherwise.

Fig. 5. The private equality test ideal functionality $\mathcal{F}_{\mathsf{peqt}}$

Our protocol is 3–4 orders of magnitude faster than previous state-of-the art. We present detailed performance analysis and comparisons in Sect. 6.

2.5 Using Padding to Hide Input Set Sizes

If desired, it is easy to add padding to our protocol so as to hide the actual sizes of players input sets. This is done simply by setting the protocol parameters (number of bins, maximal bin size) based on the known upper bound of set size. It is easy to verify that this (higher parameter values) do not cause correctness or security violations. Intuitively, players will process more bins with higher maximal bin sizes, but fewer actual items. However, the number of actual items per bin is hidden by our protocol.

3 Preliminaries

3.1 Oblivious Transfer

Oblivious Transfer (OT) is a ubiquitous cryptographic primitive and is a foundation for almost all efficient secure computation protocols. In OT, a sender with two input strings (x_0, x_1) interacts with a receiver who has an input choice bit b. The result is that the receiver learns x_b without learning anything about x_{1-b}, while the sender learns nothing about b.

The very first OT protocol was proposed by Rabin [48]. While several OT protocols were proposed, they all essentially relied on public key operations (necessarily so, due to lower bound [31]). The OT extension protocols [4,33] went around the lower bound by considering a *batched* OT evaluation. OT extension protocol works by evaluating a small number (namely, computational security parameter κ) of expensive OTs that are used as a base for performing many OTs using only cheap symmetric-key operations. In 2013, Kolesnikov and Kumaresan [35] proposed an optimization and generalization of the IKNP OT extension, which achieved $O(\log \kappa)$ factor performance improvement in communication and computation. In the same year, Asharov et al. [2] proposed several IKNP optimizations (some overlapping with [35]) and provided optimized implementation of OT extension. In this work, we are interested in an specific variant of OT (one-sided OT), in which the sender has only one message to send, and which is received by the receiver based on its choice bit. The sender remains oblivious as to whether or not the receiver received the message. With this OT variant, one can reduce the bandwidth requirement by sending only one secret instead of two. As a result, we can perform many OT instances with amortized cost of 50 ns and 129 bits transmitted.

3.2   Oblivious PRF and Private Equality

Oblivious PRF: An oblivious PRF (OPRF) [21] is a 2-party protocol in which the sender learns (or chooses) a random PRF key k and the receiver learns $F_k(x_1), \ldots, F_k(x_t)$, where F is a PRF and $(x_1, \ldots, x_t) \subseteq \{0, 1\}^*$ are inputs chosen by the receiver. Here, we consider a slightly weaker variant of OPRF due to [36] where the PRF keys are related. We describe the ideal functionality for an OPRF in Fig. 4.

OPRF and BaRK-OPRF Instantiation. While many OPRF protocols exist, we focus on the protocol (BaRK-OPRF) of Kolesnikov et al. [36]. This protocol has the advantage of being based on oblivious-transfer (OT) extension. As a result, it uses only inexpensive symmetric-key cryptographic operations (apart from a constant number of initial public-key operations for base OTs). The protocol efficiently generates a large number of OPRF instances, which makes it a particularly good fit for our eventual PSI application that uses many OPRF instances. Concretely, the amortized cost of each OPRF instance costs roughly 500 bits in communication and a few symmetric-key operations.

Technically speaking, the protocol of [36] achieves a slightly weaker variant of OPRF than what we have defined in Fig. 4. In particular, (1) PRF instances are generated with *related keys*, and (2) the protocol reveals slightly more than just the PRF output $F_k(q)$. We stress that the resulting PRF of [36] remains a secure PRF even under these restrictions. More formally, let $leak(k, q)$ denote the extra information that the protocol leaks to the receiver. [36] gives a security definition for PRF that captures the fact that outputs of F, under related keys $k_1, \ldots, k_n$, are pseudorandom even given $leak(k_i, q_i)$. This guarantee is sufficient for our purpose.

For the ease of presentation and reasoning, we work with the cleaner security definitions that capture the main spirit of BaRK-OPRF. We emphasize that, although cumbersome, it is possible to incorporate all of the [36] relaxations into the definitions. We stress that our eventual application of PSU is secure *in the standard sense* when built from BaRK-OPRF, and we make corresponding remarks in the proof of security outlining how security holds for BaRK-OPRF.

PARAMETERS:

- Two parties: sender $\mathcal{S}$ and receiver $\mathcal{R}$
- Set X is of size n of elements.
- The bit-length of field elements $\sigma = \lambda + \log(n)$.
- Ideal OPRF, $\mathcal{F}_{\mathsf{peqt}}$ primitives specified in Figure 4, and Figure 5, respectively. Let $F_k(x) : \{0,1\}^* \mapsto \{0,1\}^\sigma$ be the underlying OPRF function.
- A collision-resistant hash function $h(x) : \{0,1\}^* \mapsto \{0,1\}^\sigma$.

INPUT OF $\mathcal{S}$: $x^* \in \{0,1\}^*$

INPUT OF $\mathcal{R}$: $X = \{x_1, x_2, \ldots, x_n\} \subseteq \{0,1\}^*$

PROTOCOL:

1. $\mathcal{S}$ acts as OPRF *receiver*, sends x^* to OPRF. $\mathcal{S}$ receives $q^* = F_k(x^*)$ and receiver $\mathcal{R}$ receives k.
2. $\mathcal{R}$ chooses $s \leftarrow \mathbb{F}(2^\sigma)$ at random. $\mathcal{R}$ interpolates a $\mathbb{F}(2^\sigma)$-polynomial $P(x)$ over points $\{(h(x_i), s \oplus q_i)\}$, where $q_i = F_k(x_i), \forall i \in [n]$. Here $s \oplus q_i$ is computed as operation on σ-bit strings.
3. $\mathcal{R}$ sends the coefficients of P to $\mathcal{S}$.
4. $\mathcal{S}$ computes $s^* = P(h(x^*)) \oplus q^*$.
5. $\mathcal{S}$ and $\mathcal{R}$ invoke the $\mathcal{F}_{\mathsf{peqt}}$-functionality:
   - $\mathcal{R}$ acts as *receiver* with input s.
   - $\mathcal{S}$ acts as *sender* with input s^*.
   - $\mathcal{R}$ receives output from $\mathcal{F}_{\mathsf{peqt}}$.

Fig. 6. Reverse private membership test protocol $\mathcal{F}_{\mathsf{rpmt}}^n$.

Private Equality Test (PEQT). Fagin, Naor, and Winkler [19] introduced one of the first PEQT protocols. PEQT is a 2-party protocol in which a receiver who has an input string x_0 interacts with a sender holding an input string x_1. The result is that the receiver learns a bit indicating whether $x_0 = x_1$ and nothing else, whereas the sender learns nothing. We formally define the PEQT functionality in Fig. 5. [19] protocol was based on public-key cryptography. A long list of follow-up works [6,36,39,40,44,46,47] improved the efficiency of PEQT. Some of them were introduced in the context of PSI. PEQT can be immediately obtained from BaRK-OPRF by computing and comparing BaRK-OPRF output in the clear, cf. [36] (i.e., one party learns $F_k(x_1)$; the other party learns k and sends $F_k(x_0)$). We will use the latter most efficient instantiation.

4   Reverse Private Membership Test (RPMT)

We describe our efficient construction of Reverse Private Membership Test (RPMT), which is a semi-honest secure protocol for the functionality specified in Fig. 2. Throughout the paper we use the notations κ, λ for the computational and statistical security parameters, respectively. Our RPMT protocol is described in

Fig. 6. The formal protocol follows the intuition presented in the first part of Sect. 2. Polynomial arithmetic is done in field $\mathbb{F}(2^\sigma)$ for some appropriate σ. We discuss using smaller field size in Sect. 5.3.

RPMT protocol is presented in Fig. 2. We next argue it computes $\mathcal{F}_{\mathsf{rpmt}}^n$ correctly. Afterwards, we state and prove the security properties of the protocol.

Correctness. The main observation of OPRF is that the RPMT sender (acting as OPRF's receiver) obtains the output q^* which is equal to q_i, if $x^* = x_i$. In this case, it is not hard to see that $s^* = P(h(x^*)) \oplus q^* = P(h(x_i)) \oplus q^* = s$. From the $\mathcal{F}_{\mathsf{peqt}}$-functionality, the receiver outputs 1. In case $x^* \notin X$, the OPRF functionality gives the sender q^* which is not in $\{q_i \mid i \in [n]\}$, thus $s^* \neq s$ and the receiver gets 0 from the $\mathcal{F}_{\mathsf{peqt}}$-functionality.

We remark that our RPMT protocol is correct except in case of a collision $P(h(x^*)) = P(h(x_i))$ for $x^* \neq x_i$, which occurs with probability is $2^{-\sigma}$. By setting $\sigma = \lambda + \log(n)$, a union bound shows probability of collision is negligible $2^{-\lambda}$.

Security. We now state and prove security properties of RPMT.

Theorem 1. *The construction of Fig. 6 securely implements functionality $\mathcal{F}_{\mathsf{rpmt}}^n$ in the semi-honest model, given the OPRF and Private Equality Test primitives defined in Fig. 4, and Fig. 5, respectively.*

Proof. We exhibit simulators $\mathsf{Sim}_{\mathcal{R}}$ and $\mathsf{Sim}_{\mathcal{S}}$ for simulating corrupt $\mathcal{R}$ and $\mathcal{S}$ respectively, and argue the indistinguishability of the produced transcript from the real execution.

Corrupt Sender. $\mathsf{Sim}_{\mathcal{S}}(x^*)$ simulates the view of corrupt $\mathcal{S}$, which consists of $\mathcal{S}$'s randomness, input, output and received messages. $\mathsf{Sim}_{\mathcal{S}}$ proceeds as follows. It first chooses $q' \in_R \{0,1\}^\sigma$, calls OPRF simulator $\mathsf{Sim}_{\mathcal{S}_{\mathsf{OPRF}}}(x^*, q')$, and appends its output to the view. We note that BaRK-OPRF is behaving the same as OPRF with respect to the security guarantee needed for simulating this step, namely that q^* obtained in Step 1 is pseudorandom. This is the only direct use of BaRK-OPRF in this protocol, and hence the rest of the argument made w.r.t. OPRF applies to our instantiation as well.

$\mathsf{Sim}_{\mathcal{S}}$ simulates Step 3 as follows. It generates random $s' \in \{0,1\}^\sigma$, and n random points $(x_i', q_i') \in_R (\{0,1\}^*, \{0,1\}^\sigma)$. $\mathsf{Sim}_{\mathcal{S}}$ then interpolates polynomial P over points $\{h(x_i'), s' \oplus q_i'\}$ and appends its coefficients to the generated view.

Finally, to simulate Step 5, $\mathsf{Sim}_{\mathcal{S}}$ runs simulator $\mathsf{Sim}_{\mathsf{PEQT}}$ on input $(s' = P(h(x^*)) \oplus q')$ and appends the output of $\mathsf{Sim}_{\mathsf{PEQT}}$ to its output of the view.

We now argue that the output of $\mathsf{Sim}_{\mathcal{S}}$ is indistinguishable from the real execution. For this, we formally show the simulation by proceeding the sequence of hybrid transcripts T_0, T_1, T_2, T_3, where T_0 is real view of $\mathcal{S}$, and T_3 is the output of $\mathsf{Sim}_{\mathcal{S}}$.

Hybrid 1. Let T_1 be the same as T_0, except that the OPRF execution is replaced as follows. By the OPRF/BaRK-OPRF pseudorandomness guarantee and the indistinguishability of the output of $\mathsf{Sim}_{S_{\mathsf{OPRF}}}$, we replace $F(k, x^*)$ and $F(k, x_i), \forall i \in [n]$, with q' and $q'_i, \forall i \in [n]$, respectively. We note that if $x^* = x_i$, then $q' = q'_i$. It is easy to see that T_0 and T_1 are indistinguishable.

Hybrid 2. Let T_2 be the same as T_1, except that the polynomial is a uniform polynomial of degree $n - 1$ (sampled by interpolating over random points). Consider two following cases:

- $x^* \notin X$: Since all values q'_i are uniformly random from the $\mathcal{S}$'s point of view, so are the $s \oplus q'_i$.
- $x^* = x_i$ (consequently, $q' = q'_i$): Since other values $q'_{j \in [n]}, \forall j \neq i$, are uniformly random from $\mathcal{S}$'s point of view, we replace these $s \oplus q'_j$ with random. Then s is used only in the expression $s \oplus q'_i$. Since s is uniform, $s \oplus q'_i$ is also uniformly random from the $\mathcal{S}$'s view even though the adversary knows $q' = q'_i$.

In summary, the polynomial from the real execution can be replaced with polynomial P sampled over random points. T_1 and T_2 are indistinguishable.

Hybrid 3. Let T_3 be the same as T_2, except the PEQT execution is replaced with running the simulator $\mathsf{Sim}_{R_{\mathsf{PEQT}}}(s')$. Because $\mathsf{Sim}_{R_{\mathsf{PEQT}}}$ is guaranteed to produce output indistinguishable from real, T_3 and T_2 are indistinguishable.

Corrupt Receiver. $\mathsf{Sim}_{\mathcal{R}}(x_1, ..., x_n, out)$ simulates the view of corrupt $\mathcal{R}$, which consists of $\mathcal{R}$'s randomness, input, output and received messages. $\mathsf{Sim}_{\mathcal{R}}$ proceeds as follows. It chooses a random $k' \in_r \{0, 1\}^\kappa$, calls OPRF simulator $\mathsf{Sim}_{S_{\mathsf{OPRF}}}(\perp, k')$, and appends its output to the view. Finally, to simulate Step 5 (Fig. 6), $\mathsf{Sim}_{\mathcal{S}}$ runs simulator $\mathsf{Sim}_{\mathsf{PEQT}}$ on input (k', out) and appends the output of $\mathsf{Sim}_{\mathsf{PEQT}}$ to its output of the view.

The view generated by $\mathsf{Sim}_{\mathcal{R}}$ in indistinguishable from a real view because of the indistinguishability of the transcripts of the underlying simulators.

Communication Cost. Ignoring the fixed cost of base OTs for OT extension, the PMT communication cost (prior to further optimizations discussed in Sect. 5.3) includes:

- OPRF in Step 1: ρ bits, where ρ is the width of the pseudorandom code defined in Table 2 by referencing parameters from [36].
- Sending the coefficients of P in Step 3: $(n + 1)\sigma$ bits
- $\mathcal{F}_{\mathsf{peqt}}$ in Step 5: $\rho + \lambda$ bits

Therefore, the overall communication cost of our PMT protocol is

$$\Phi(n) = 2\rho + \lambda + (n + 1)\sigma \tag{1}$$

5   Private Set Union

We now present our main result, an application of our RPMT to PSU. The construction closely follows the high-level overview presented in the second part of Sect. 2. Recall, the RPMT functionality allows the receiver to learn one-bit output indicating whether the sender's item is in its (receiver's) set, while keeping this item secret (i.e. the receiver will not know which sender's item is among its set). The performance of our RPMT protocol is linear in the size of the receiver's set, resulting in a quadratic costs for PSU.

PARAMETERS:

- Set sizes n_1 and n_2, and two parties: sender $\mathcal{S}$ and receiver $\mathcal{R}$
- A bit-length ℓ. Let $n = \max(n_1, n_2)$.
- Number of bins $\beta = \beta(n)$, and max bin size m, suitable for our hashing scheme (Table 2)
- Ideal $\mathcal{F}_{\mathsf{rpmt}}$ primitive defined in Figure 2, and ideal OT primitive.
- A special dummy item $\perp \in \{0,1\}^*$

INPUT OF $\mathcal{S}$: $X = \{x_1, x_2, \ldots, x_{n_1}\} \subseteq \{0,1\}^\ell$

INPUT OF $\mathcal{R}$: $Y = \{y_1, y_2, \ldots, y_{n_2}\} \subseteq \{0,1\}^\ell$

PROTOCOL:

1. Randomly pick a hash function H from all hash functions with domain $\{0,1\}^\ell$ and range $[\beta]$.
2. $\mathcal{S}$ and $\mathcal{R}$ hash elements of their sets X and Y into β bins under hash function H. Let $B_{\mathcal{S}}[i]$ and $B_{\mathcal{R}}[i]$ denote the set of items in the sender's and receiver's i-th bin, respectively.
3. $\mathcal{S}$ pads each bin $B_{\mathcal{S}}[i]$ with (several copies, as needed) the special item $\perp$ up to the maximum bin size $m + 1$, and randomly permutes all items in this bin.
4. $\mathcal{R}$ pads each bin $B_{\mathcal{R}}[i]$ with one special item $\perp$ and (several, as needed) different dummy items to the maximum bin size $m + 1$.
5. $\mathcal{R}$ initializes set $Z = \{\}$.
6. For each bin $i \in [\beta]$, for each item $x_j \in B_{\mathcal{S}}[i]$:
   (a) $\mathcal{S}$ and $\mathcal{R}$ invoke the $\mathcal{F}_{\mathsf{rpmt}}$-functionality:
      - $\mathcal{S}$ acts as *sender* with input x_j
      - $\mathcal{R}$ acts as *receiver* with input set $B_{\mathcal{R}}[i]$
      - $\mathcal{R}$ obtains bit b_j.
   (b) $\mathcal{S}$ and $\mathcal{R}$ invoke the OT-functionality:
      - $\mathcal{S}$ acts as *sender* with pair-input $\{x_j, \perp\}$
      - $\mathcal{R}$ acts as *receiver* with bit input b_j
      - $\mathcal{R}$ obtains the OT output z_j and sets $Z = Z \cup z_i$.
7. $\mathcal{R}$ outputs $Y \cup Z$.

Fig. 7. Private set union protocol $\mathcal{F}_{\mathsf{psu}}^{n_1, n_2}$.

Next, in Sect. 5.1, we show how to use a hashing/bucketing technique to overcome this limitation. At the high level, the idea is that each party maps their items into bins using a public hash function. Each bin contains a small number of items which allows the two parties to evaluate RPMT on the elements of each bin separately.

Let m denote the maximum sender's bin size when mapping n items to β bins with no (expected) overflow. Within each bin, the protocol requires $(m+1)$ invocations of RPMT. Section 5.2 analyses hashing parameters to minimize the overall cost of our PSU.

5.1   PSU Construction

As described above, in our PSU protocol we place players' elements into β buckets of maximum size $m+1$ each.

We describe the main construction of PSU in Fig. 7. Correctness of our PSU protocol follows from the fact that the RPMT functionality gives the receiver the zero-bit output if its set does not contain the sender's item. In Step 6b, the receiver obliviously receives that item from OT functionality.

We now state and prove security of our PSU construction.

Theorem 2. *The construction of Fig. 7 securely implements the Private Set Union functionality $\mathcal{F}_{psu}^{n_1, n_2}$ of Fig. 1 in the semi-honest model, given the OT and Reverse Private Equality Test primitives defined in Fig. 2.*

Proof. We exhibit simulators $\mathsf{Sim}_{\mathcal{S}}$ and $\mathsf{Sim}_{\mathcal{R}}$ for simulating corrupt $\mathcal{S}$ and $\mathcal{R}$ respectively, and argue the indistinguishability of the produced transcript from the real execution.

Corrupt Sender. When employing the abstraction of the RPMT and OT functionalities, simulating corrupt $\mathcal{S}$ is elementary. $\mathsf{Sim}_{\mathcal{S}}(X)$ simulates the view of corrupt $\mathcal{S}$, which consists of $\mathcal{S}$'s randomness, input, output and received messages. The simulator simulates an execution of the protocol in which $\mathcal{S}$ receives nothing from the PTM and OT ideal functionality in Step 4. Thus, it is straightforward to see that the simulation is perfect.

Corrupt Receiver. $\mathsf{Sim}_{\mathcal{R}}(Y, Z)$ simulates the view of corrupt $\mathcal{R}$, which consists of $\mathcal{R}$'s randomness, input, output and received messages. We will view $\mathsf{Sim}_{\mathcal{R}}$'s input Z as the set $Z = Y \setminus X$, i.e. the set of elements that X "brings to the union." $\mathsf{Sim}_{\mathcal{R}}$ proceeds as follows.

$\mathsf{Sim}_{\mathcal{R}}$ simulates protocol of Fig. 7 bucket-by-bucket. Consider the i-th bucket. Let X_i (respectively Y_i, Z_i) be the set of elements of X (respectively, Y, Z) that are mapped to the i-th bucket. $\mathsf{Sim}_{\mathcal{R}}$ pads Y_i to $m+1$ elements as is done in Step 4. Now, $\mathsf{Sim}_{\mathcal{R}}$ has all the information to simulate Step 6. $\mathsf{Sim}_{\mathcal{R}}$ constructs the sequence simulating when $\mathcal{R}$ discovers new elements in the union. This is an m-element sequence S, where $\mathsf{Sim}_{\mathcal{R}}$ puts $|Z_i|$ elements z_i at randomly chosen slots, and fills the remaining $m - |Z_i|$ elements of the sequence with $\bot$.

$\mathsf{Sim}_{\mathcal{R}}$ then goes through the elements of S. Consider the j-the such element S_j. $\mathsf{Sim}_{\mathcal{R}}$ sets $out_j = 0$ if $S_j = \bot$, and otherwise sets $out_j = 1$. $\mathsf{Sim}_{\mathcal{R}}$ invokes the simulator of $\mathcal{F}_{\mathsf{rpmt}}$ with input (Y_i, out_j), and appends the output of the simulator to its own output. This simulates Step 6a.

$\mathsf{Sim}_{\mathcal{R}}$ proceeds by simulating Step 6b, as follows. $\mathsf{Sim}_{\mathcal{R}}$ invokes the simulator of OT with input (out_j, S_j). This corresponds to $\mathcal{R}$ providing input out_j and receiving output S_j from OT. $\mathsf{Sim}_{\mathcal{R}}$ appends the output of the simulator to its own output.

$\mathsf{Sim}_{\mathcal{R}}$ proceeds simulating each of β bins and terminates. This completes the description of the simulator.

We now argue that the output of $\mathsf{Sim}_{\mathcal{R}}$ is indistinguishable from the real execution. This is easy to see. $\mathsf{Sim}_{\mathcal{R}}$'s reconstruction of how/when the elements of $Z = Y \setminus X$ are discovered by $\mathcal{R}$ is distributed identically to the real execution. The remainder of the simulation refers to simulators of implementations of ideal functionalities.

5.2   Hashing Parameters

A natural first attempt is to hash n items into n bins, where each bin will contain $O(1)$ items on average. If we could have $O(1)$ items per bin in PSU, this would result in $O(n)$ total RPMT instances, a low cost. However, we must hide the actual number of items in each bin, and hence all bins must be padded to an upper bound m. Gonnet [25] showed $m = \frac{\ln(n)}{\ln \ln(n)}(1 + o(1))$. The coefficient of little-o is not specified in [25]; Pinkas $et\ al.$ [47] empirically determined the concrete m given the number of bins β. In our case, n bins is not an optimal strategy. For example, hashing $n = 2^{20}$ elements into n bins, bin size $m = 20$ is required to ensure that overflow occurs with probability $\leq 2^{-40}$. As a result, for $n = 2^{20}$ our PSU protocol performs $21n$ RPMT instances in total, which requires 2^{28} OPRF ciphertexts sent and received. We can do better.

In the following, we analyze the effect of the number of bins β and maximum bin size $m + 1$ on the communication overhead of our protocol, and choose the best parameters to minimize our cost. We recall that the overall communication cost of our PSU protocol is equal to $\beta m \Phi(m+1) + \beta m(\kappa + \sigma)$, where $\Phi(m+1)$ is the RPMT communication cost specified in Eq. (1). To guarantee that mapping n items to β bins with no overflow, we compute the probability that there exists a bin with more than m items:

$$\Pr(\exists \text{bin with } \geq m \text{ items}) \leq \beta \sum_{m+1}^{n} \binom{n}{i} \left(\frac{1}{\beta}\right)^i \left(1 - \frac{1}{\beta}\right)^{n-i} \tag{2}$$

Bounding (2) to be negligible in the statistical security parameter $\lambda = 2^{40}$, we obtain the required bin size m without overflow for a given n and β. To minimize the overall communication cost, we choose $\beta = O(n/\log n)$. According to standard balls-and-bins argument, the maximum bin size is $O(\log(n))$. To determine the coefficients in the big "O", we first fix the number of bins with an initialization value $\beta = \epsilon n = 0.01n$, evaluate Eq. (2) to obtain the necessary m, and calculate the required communication cost given β and m. In order to find "sweet spot" for our communication cost, we increase the scale ϵ by 0.001

after each time. We observe that our protocol yields the lowest communication when ϵ is in a range $[0.4, 0.6]$. Figure 8 shows the result for $n = 2^{16}$: we choose $\beta = \epsilon n = 0.058n$ and require $m = 60$ to achieve 2^{-40} hashing failure probability. We also report the set of our hashing parameters in Table 2.

5.3   Discussion and Optimization

In our RPMT protocol described in Fig. 6, the receiver computes a polynomial of degree $(n-1)$ with the field of $\mathbb{F}(2^\sigma)$, where $\sigma = \lambda + \log(n)$. With hash-to-bin technique used in PSU, we are able to reduce the degree from $(n-1)$ to $O(\log(n))$, which avoids an expensive computation at the cost of manipulating polynomials with high degree. However, we increase the field size by $10\%-12\%$.

Recall that our PSU protocol requires $\beta(m+1)$ RPMT instances in total. For each RPMT protocol, its correctness is violated when a collision event occurs: $P(h(x_i)) = P(h(y_j))$ for $x_i \neq y_j$. To yield collision probability 2^λ over all bins, which is suited for most applications, the size of q_i values is $\sigma = \lambda + \log(\beta(m+1)^2)$. For example, for $n = 2^{20}$, we use the polynomial field size $\mathbb{F}(2^{68})$.

Polynomials with Dummy Points. In Step 4, Fig. 7, receiver pads each bin with one special item $\perp$ and additional different dummies to the maximum bin size $m+1$. This padding serves the purpose of hiding the number of items that were mapped to a specific bin, which would leak some information about the input set. In RPMT protocol (Step 2, Fig. 6), the receiver generates the polynomial over points $\{h(y_i), s \oplus q_i\}$ where some of q_i are the OPRF of the dummy items d_i. Therefore, we simply replace these $q_i = F_k(d_i)$ by random values.

Another optimization, inspired by [37], is that the receiver computes $P(x)$ by first interpolating the polynomial over the non-dummy items only. That is, receiver interpolates P_0 over $m' \leq (m+1)$ points $\{h(y_i), s \oplus F_k(y_i)\}$, and also computes $P_1(x) = \prod_{i=1}^{m'}(x - h(y_i))$ over m' roots $h(y_i)$, where y_i are real items.

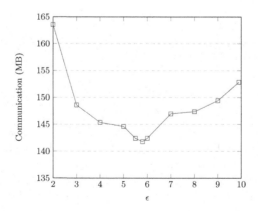

Fig. 8. Communication cost (MB) of our PSU protocol for $n = 2^{16}$ given the number of bins $\beta = 10^{-2}\epsilon n$

Table 2. Hashing parameters for different set sizes n, and our PSU's communication cost (MB). ρ is OT extension matrix width in OPRF ($\approx$ number of bits required per OPRF call) as reported in Table 1 [36], β is the number of bins, $m + 1$ is max bin size PSU with n elements per party. Total PSU communication reported in MB and excludes the fixed cost of base OTs for OT extension.

parameters & comm.	set size n							
	2^8	2^{10}	2^{12}	2^{14}	2^{16}	2^{18}	2^{20}	2^{22}
ρ	424	432	432	440	440	448	448	448
β/n	0.043	0.055	0.05	0.053	0.058	0.052	0.06	0.051
m	63	58	63	62	60	65	61	68
Comm. cost (MB)	0.39	1.81	7.84	33.43	141.78	602.20	2544.7	10748

Then receiver chooses a random polynomial $P_r(x)$ of degree $(m - m' + 1)$; and computes $P(x) = P_0(x) + P_1(x)P_r(x)$. It is easy to see that $P(h(x_i)) = s \oplus F_k(x_i)$, $\forall x_i \in X$. Using hashing parameters from Table 2, the expected value of m' is only 18 for $n = 2^{18}$, while the worst-case $m = 65$. This optimization reduces the cost of expensive polynomial generation (by approximately 200% in our implementation).

Relaxing RPMT. Finally, as discussed in Sect. 2, the use of full-fledged RPMT for PSU is slightly overkill. It would suffice to use an RPMT protocol which leaked some information about the sender's item (in the case that $x^* \notin X$), since the PSU protocol will release that value anyway. In Appendix A we describe a simple change to the RPMT protocol that remains secure in the context of our PSU protocol. Basically, instead of using PEQT to compare polynomial outputs, the sender just sends it polynomial output in the clear. This is safe in the context of PSU since the PSU simulator will have access to the sender's RPMT input x^* whenever the polynomial output leaks information about x^*.

5.4 Discussion: Difficulties in Applying Other PSI Techniques

In addition to the optimizations mentioned above, we also explored other commonly used techniques developed in the context of PSI [11,22,27,36,37,44,47]. Interestingly, we found that many standard techniques for PSI do not directly work for our PSU paradigm, despite the apparent similarity of the two problems. In the following, we will discuss PSU-specific obstacles in applying these techniques. The reader may safely skip this section on the first reading as we discuss here only techniques that we *did not use* in our protocol.

Cuckoo Hashing. This hashing scheme was introduced by Pagh and Rodler [42]. It is the standard hashing scheme in current PSI protocols. At the high level, the receiver uses two (optionally, more) public hash functions h_1, h_2 to store its item in one of the bins $\{h_1(x), h_2(x)\}$. The hashing process uses eviction and the choice of which of the bins is used depends on the entire set. Using the same hash functions and simple hashing, the sender maps its item y into *both* bins $\{h_1(y), h_2(y)\}$ (i.e., item y appears twice in the hash table). Then the parties evaluate PSI bin-by-bin. This is efficient since the receiver has only one item per bin. This hashing scheme avoids a quadratic-cost PSI within a bin.

Unfortunately, this hashing scheme (and the corresponding performance improvement) does not immediately fit in the PSU case. The reason is that the receiver may learn the Cuckoo hash positions of the sender's items, which may reveal information about sender's entire input. Concretely, suppose that in our protocol the sender uses Cuckoo hashing to map its item x into bin $h_1(x)$. If $x \notin Y$, the receiver will learn which bin x is mapped to. As noted above, the bin storing x depends on the whole input set of the sender and this leaks some information about the party input set that cannot be simulated.

Phasing. Permutation-based hashing (phasing) was introduced by Arbitman *et al.* [1] to reduce the bit length of the items that are mapped to bins (in our PSU, this would help reduce the polynomial field size). Phasing was used in [11, 27, 44, 52] to improve PSI performance when input items has short bit length. The idea is to view each item x as two parts: first $\log(\beta)$ bits used to define the bin to which the item is mapped, and the last bits used as a representation to store the item in the bin.

Concretely, the item x can be presented as $x = x_L | x_R$, where x_L has $\log(\beta)$ bit-length. The item x is mapped into bin $x_L \oplus f(x_R)$, where f is a random function that maps arbitrary strings to a range of $[0, \beta]$. That bin will store x_R as a representative of x. Clearly, x_R has $\log(\beta)$ bits shorter than the original item x. This permutation-based hashing technique achieves significant savings,

Table 3. Comparison of total runtime (in seconds) and communication (in MB) between our protocol, [14] and [5]. Both parties have n 128-bit elements as input, except [5] running time is based on 32-bit elements. [14] implementation is in Go, using 8 threads. Our implementation is in C++, 8 threads. [14] and us use fast emulated LAN (10 Gbps, 0.02 ms RTT). Cryptographic strength refers to the computational security of the protocol, according to NIST recommendations. [5] runtime is taken from their 3-party experiments, and [5] communication is calculated by us for 2PC and 128-bit elements. Best results are marked in bold.

	Protocol	Bit key length	Cryptographic strength	Set size n					
				2^8	2^{10}	2^{12}	2^{14}	2^{16}	2^{18}
Time	[14]	1024	Legacy	11.78	44.73	175.7	702.4	2836.5	11341.2
		2048	112	78.02	312.44	1233.59	4952.94	19881.51	79272.48
	[5]	128	128	2.41	11.88	24.88	–	–	–
	Ours	128	128	**0.57**	**0.66**	**0.83**	**1.15**	**2.65**	**10.42**
	Speedup			4×	18×	30×	4306×	7502×	7607×
Comm.	[14]	1024	Legacy	2.83	11.32	45.28	181.12	724.49	2897.97
		2048	112	4.06	16.25	65.01	260.04	1040.18	4160.74
	[5]	128	128	75.5	369.1	1744.83	8053.06	36507.22	163208.76
	Ours	128	128	**0.45**	**2.05**	**8.48**	**34.98**	**144.65**	**652.09**
	Speedup			9.02×	7.92×	7.66×	7.43×	7.19×	6.38×

especially when the original item x has small length (e.g. 32 bits or 64 bits). For instance, assume that the item x has 32-bit length, the set size is $n = 2^{20}$. Then bin elements are only 17 bits long, instead of 32 bits. As a result, we might hope to use the polynomial field size of only $\mathbb{F}(2^{17})$ in RPMT, yielding a significant improvement.

Unfortunately, this general phasing technique does not yield any performance benefit in our PSU paradigm. The underlying reason is that the items in each bin are first given as input to an OPRF for that bin, however the state-of-the-art OPRF protocol that we use [36] is insensitive to the item length. It is only the OPRF output length that determines the field size for polynomial interpolation. Since the OPRF outputs are random, their length must be chosen to avoid collisions with probability $1 - 2^{-\lambda}$.

6   Implementation

Our protocol requires the receiver to generate a polynomial of degree m, and the sender to evaluate it on one point, where $m + 1$ is the maximum bin size. Since the degree $m = O(\log(n))$ of the polynomial is relatively small, we use the straightforward Lagrange interpolation and evaluation algorithm which requires $O(m^2)$ field operations. As parties use the bit-string output of the OPRF as input to the polynomial operations, it is natural to interpolate and evaluate the polynomial over $GF(2^\sigma)$. Our polynomial implementation uses the NTL library [53] with GMP library and GF2X [24] library installed for speeding up the running time. Inspired by Huang et al. [30], we applied pipelining optimization when the receiver sending all polynomials to the sender. In more detail, we find that by sending polynomial coefficients for 2^8 bins in a batch to the sender, we can minimize the overall wall-clock time of the execution.

As detailed in Sect. 2, our PSU protocol builds on a specific OPRF variant [36] and OT extension. We do $\kappa = 128$ Naor-Pinkas OTs [41]. We use the source code (OPRF and OT) from [36,50]. Our complete implementation will freely available on GitHub.

Table 4. Scaling of our protocol with set size and number of threads. Total running time is in seconds. n elements per party, 128-bit length element, and threads $T \in \{1, 4, 16, 32\}$ threads. LAN setting with 10 Gbps network bandwidth, 0.02 ms RTT. WAN setting with 400 Mbps network bandwidth, 40 ms RTT.

Setting	T	Set size n							
		2^8	2^{10}	2^{12}	2^{14}	2^{16}	2^{18}	2^{20}	2^{22}
LAN	1	0.66	0.86	1.42	3.54	12.41	61.34	238.88	1039.64
	4	0.59	0.69	0.98	1.46	4.03	17.94	69.07	301.76
	16	0.55	0.66	0.78	0.97	1.82	6.29	21.9	90.99
	32	0.53	0.63	0.69	0.84	1.56	4.1	13.09	54.63
WAN	1	1.38	1.73	2.61	6.96	23.29	102.5	406.15	1679.85
	4	1.33	1.56	1.99	3.29	8.58	31.05	118.79	463.51
	16	1.25	1.39	1.76	2.55	5.61	18.67	70.55	280.15
	32	1.22	1.33	1.57	2.4	5.02	17.08	62.96	250.97
Speedup		1.13–1.24×	1.3–1.36×	1.66–2.06×	2.9–4.22×	4.64–7.98×	6–14.9×	6.5–18.2×	6.7–19.1×

We implement our protocol in C++, and run our protocol on a single Intel Xeon with 2.30 GHz and 256 GB RAM. We emulate the network by using Linux tc command. In the following, we compare our protocol to the state-of-the-art PSU protocol [14] which provides empirical experiments for a larger set, and the work of [5] which reports experimental numbers for PSU of small sets $n \leq 2^{12}$). Additionally, we demonstrate the scalability and parallelizability of our protocol by evaluating it on sets of up to 2^{22} 128-bit items each.

All comparisons are total running time. We note that our protocols are very amenable to pre-computation (by precomputing and pre-sending OT extension and OPRF matrices).

6.1 Comparison with Prior Work

Since implementation of [14] and [5] are not publicly available, we use their reported experimental numbers. We perform a comparison on the range of set sizes $n = \{2^8, 2^{10}, 2^{12}, 2^{14}, 2^{16}, 2^{18}\}$ to match the parameters used in [14, Tables 3 & 4] and [5, Table 3]. [14] ran experiments on Intel Xeon 3.30 GHz 256 GB RAM and 10 Gbps LAN; we use a similar (1.32× slower) machine as reported above and same LAN. [5] reports running on 2.4 GHz AMD Opteron.

Runtime Comparison. In the [14] protocol, a Bloom filter (BF) of $44n$ elements is used to yield the false-positive probability 2^{-30}. Each element requires expensive encryption, decryption and further manipulation under an additively-homomorphic encryption (AHE).

We report detailed comparisons in Table 3, and here we highlight some numbers. Our protocol runs in 0.94 s for $n = 2^{10}$, while [5] requires 11.88 s, a factor of 18× improvement; and [14] requires 312.44 s with 2048-bit key length (which corresponds to the security level considered in our protocol), a factor of 332× improvement. As the set size n increases, [14] runs correspondingly slower. When increasing the set size to $n - 2^{18}$, [14]'s overall running time is 79, 272.48 s while ours is only 10.42 s.

This is a 7607× improvement in running time compared to [14] (2048-bit key length). A higher improvement factor as we move to higher set size likely indicates that non-protocol-essential system overheads take a higher fraction of resources in smaller set size executions in our protocol. In Sect. 6.2, we demonstrate the scalability and parallelizability of our protocol.

Bandwidth Comparison. The receiver in [14] sends a large encrypted BF. For $n = 2^{20}$, BF size is 8.05 GB and 16.1 GB when encrypted with 1024-bit and 2048-bit key, respectively. [5] relies on generic 2PC/MPC to run their protocol. We sketch approximate communication cost of their protocol in the two-party setting based on state-of-the-art OT extension and half gates (cf. discussion in Sect. 1.3). Oblivious sorting of n elements per party involves sorting an array of size $2n$. Considering ℓ-bit elements, this will require approximately $2n \cdot \log(2n) \cdot 2\ell \cdot 256$ bit. Here 256 is the half-gates garbled table size. The communication complexity of the duplicate elimination [5] costs approximately the same as oblivious sort.

For the bandwidth comparison, we only report the [5]'s communication cost of oblivious sorting and duplicate elimination, which is in favor of their protocol.

We compare bandwidth for the set sizes explored in [14], and summarize their and our results in Table 3. The communication cost of our protocol is significantly less than that of the prior work. Concretely, for $n = 2^{18}$, our protocol requires 652.09 MB of communication, a 6.38× improvement. For very small set size $n = 2^8$, our protocol requires only 0.45 MB while [14] needs 4.06 MB and [5] requires at least 75.5 MB.

Correctness Error Probability. In [14] protocol, Bloom filter introduces a false positive error in the output. Recall, the false positive rate (FPR) is the probability that a *single* element is mistakenly marked as being in the set. The [14]'s implementation chooses FPR of 2^{-30}. Thus, computing the set union of 2^{-18} items each, the probability that the entire output includes a false positive is 2^{-12}. We use simple hashing with probability of existence of an overflowed bin of 2^{-40}. Thus, in our protocol, the correctness error probability 2^{-40} is per *whole* set, not per single item.

6.2   Scalability and Parallelizability

We demonstrate the scalability and parallelizability of our protocol by evaluating it on set sizes $n = \{2^8, 2^{10}, 2^{12}, 2^{14}, 2^{16}, 2^{18}, 2^{20}, 2^{22}\}$. We run each party in parallel with $T \in \{1, 4, 16, 32\}$ threads. We report the performance of our protocol in Table 4, showing running time in both LAN/WAN settings: a LAN setting with 10 Gbps network bandwidth and 0.02 ms round-trip latency; a WAN setting with 400 Mbps network bandwidth and a simulated 40 ms round-trip latency.

Our protocol indeed scales well. Small-size problems are sub-second; medium-size problems ($n = 2^{14}$) are 3.54 s and larger sizes ($n = 2^{20}$) is under 4 min, all single-threaded. Increasing the number of threads runs the $n = 2^{20}$ instance in 13.09 s, a four orders of magnitude improvement over prior work. Benchmarking our implementation in the WAN setting, our protocol also scales well due to the fact that the communication cost is reasonable (for $n = 2^{18}$, our protocol needs 652.09 MB of communication).

Our protocol is very amenable to parallelization. Specifically, our algorithm can be parallelized at the level of bins. For example, when increasing T from 1 to 32, our protocol shows a factor of 19× improvement as the running time reduces from 1039.64 s to 54.63 s for an input of $n = 2^{22}$ elements.

Of particular interest is the last row, which presents the ratio between the runtime of the single thread and 32 threads. Our protocol yields a better speedup when the set size is larger. For smallest set size of $n = 2^8$, the protocol achieves a moderate speed up of about 1.13. When considering the larger database size $n = 2^{22}$, the speed up of 3.4–3.6 is obtained at 4 threads and 6.7–19.1 at 32 threads.

Acknowledgments. We thank all anonymous reviewers and Brice Minaud for insightful feedback.

Vladimir Kolesnikov was supported in part by Sandia National Laboratories, a multimission laboratory managed and operated by National Technology and Engineering Solutions of Sandia, LLC., a wholly owned subsidiary of Honeywell International, Inc., for the U.S. Department of Energy's National Nuclear Security Administration under contract DE-NA-0003525. He was also supported in part by the Office of the Director of National Intelligence (ODNI), Intelligence Advanced Research Projects Activity (IARPA), via 2019-1902070008. The views and conclusions contained herein are those of the authors and should not be interpreted as necessarily representing the official policies, either expressed or implied, of ODNI, IARPA, or the U.S. Government. The U.S. Government is authorized to reproduce and distribute reprints for governmental purposes notwithstanding any copyright annotation therein.

Mike Rosulek and Ni Trieu were partially supported by NSF awards #1617197, a Google faculty award, and a Visa faculty award.

A   RPMT Optimization

In the RPMT protocol, the receiver computes a polynomial P with special output s. The sender computes $s^* = P(h(x^*)) \oplus q^*$, where q^* is its OPRF output. Then the parties use PEQT to securely compare s to s^*.

In the context of PSU, it is not necessary to use PEQT for this step. Instead, the sender can simply send s^* to the receiver. The logic is as follows: If $x^* \in X$, the sender should learn only this fact (and nothing about x^*). This is still the case after the optimization because the sender will compute the same polynomial output s^* for any such $x^* \in X$. If $x^* \in X$, it means that the receiver will eventually learn x^* as part of the PSU output (and the sender can infer that x^* was contributed by the receiver). The PSU simulator will therefore have the value x^*, and it can perfectly simulate the polynomial output $s^* = P(h(x^*)) \oplus q^*$.

We now formalize the details of this modification. Rather than define a weaker/leaky version of RPMT, we instead introduce a protocol for 1-vs-n PSU. Such a functionality is quite similar to RPMT, which can be thought of as revealing only the cardinality of $|\{x^*\} \cup X|$, which is equivalent to revealing the cardinality of $|\{x^*\} \setminus X|$ (either 0 or 1).

The details of the 1-vs-n PSU protocol are given in Fig. 9. Now, using 1-vs-n PSU as a building block instead of RPMT, our full-fledged PSU protocol can be written as in Fig. 10.

The security proof of the full-fledged PSU protocol is essentially the same as in the pre-optimization protocol. The security of the 1-vs-n protocol is given below:

Theorem 3. *The construction of Fig. 9 securely implements functionality $\mathcal{F}_{psu}^{1,n}$ in the semi-honest model, given the OPRF primitive defined in Fig. 4.*

Proof. We exhibit simulators $\mathsf{Sim}_\mathcal{R}$ and $\mathsf{Sim}_\mathcal{S}$ for simulating corrupt $\mathcal{R}$ and $\mathcal{S}$ respectively, and argue the indistinguishability of the produced transcript from the real execution.

PARAMETERS:

- Two parties: sender $\mathcal{S}$ and receiver $\mathcal{R}$
- Set X is of size n of elements.
- The field size $\sigma = \lambda + \log(n)$.
- Ideal OPRF primitive specified in Figure 4. Let $F_k(x) : \{0,1\}^* \mapsto \{0,1\}^\sigma$ be the underlying OPRF function.
- A collision-resistant hash function $h(x) : \{0,1\}^* \mapsto \{0,1\}^\sigma$

INPUT OF $\mathcal{S}$: $x^* \in \{0,1\}^*$
INPUT OF $\mathcal{R}$: $X = \{x_1, x_2, \ldots, x_n\} \subseteq \{0,1\}^*$
PROTOCOL:

1. $\mathcal{S}$ acts as OPRF *receiver*, sends x^* to OPRF. $\mathcal{S}$ receives $q^* = F_k(x^*)$ and receiver $\mathcal{R}$ receives k.
2. $\mathcal{R}$ randomly picks $s \leftarrow \mathbb{F}(2^\sigma)$. $\mathcal{R}$ interpolates a $\mathbb{F}(2^\sigma)$-polynomial $P(x)$ over points $\{(h(x_i), s \oplus q_i)\}$, where $q_i = F_k(x_i), \forall i \in [n]$. Here $s \oplus q_i$ is computed as operation on σ-bit strings.
3. $\mathcal{R}$ sends the coefficients of P to $\mathcal{S}$.
4. $\mathcal{S}$ computes $s^* = P(h(x^*)) \oplus q^*$ and sends it to $\mathcal{R}$.
5. $\mathcal{S}$ and $\mathcal{R}$ invoke the oblivious transfer functionality:
   - $\mathcal{R}$ acts as *receiver* with input 1 if $s^* = s$ and input 0 otherwise.
   - $\mathcal{S}$ acts as *sender* with input $(x^*, \perp)$.
6. If $s^* = s$, then $\mathcal{R}$ gives output X. Otherwise, it learned x^* as output from the oblivious transfer, and outputs $X \cup \{x^*\}$.

Fig. 9. 1-vs-n PSU protocol.

Corrupt Sender. $\mathsf{Sim}_\mathcal{S}(x^*)$ simulates the view of corrupt $\mathcal{S}$, which consists of $\mathcal{S}$'s randomness, input, output and received messages. $\mathsf{Sim}_\mathcal{S}$ proceeds as follows. It first chooses $q' \in_R \{0,1\}^\sigma$, calls OPRF simulator $\mathsf{Sim}_{\mathcal{S}_{\mathsf{OPRF}}}(x^*, q')$, and appends its output to the view.

$\mathsf{Sim}_\mathcal{S}$ simulates Step 3 as follows. It generates random $s' \in \{0,1\}^\sigma$, and n random points $(x_i', q_i') \in_R (\{0,1\}^*, \{0,1\}^\sigma)$. $\mathsf{Sim}_\mathcal{S}$ then interpolates the polynomial P over these points $\{h(x_i'), s' \oplus q_i'\}$ and appends its coefficients to the generated view.

We argue that the output of $\mathsf{Sim}_\mathcal{S}$ is indistinguishable from the real execution. For this, we formally show the simulation by proceeding the sequence of hybrid transcripts T_0, T_1, T_2, where T_0 is real view of $\mathcal{S}$, and T_2 is the output of $\mathsf{Sim}_\mathcal{S}$.

Hybrid 1. Let T_1 be the same as T_0, except that the OPRF execution is replaced as follows. By the OPRF/BaRK-OPRF pseudorandomness guarantee and the indistinguishability of the output of $\mathsf{Sim}_{\mathcal{S}_{\mathsf{OPRF}}}$, we replace $F(k, x^*)$ and $F(k, x_i), \forall i \in [n]$, with q' and $q_i', \forall i \in [n]$, respectively. We note that if $x^* = x_i$, then $q' = q_i'$. It is easy to see that T_0 and T_1 are indistinguishable.

Hybrid 2. Let T_2 be the same as T_1, except that the polynomial is an uniform polynomial of degree $n - 1$. Consider two following cases:

PARAMETERS:

- Set sizes n_1 and n_2, and two parties: sender S and receiver R
- A bit-length ℓ. Let $n = \max(n_1, n_2)$.
- Number of bins $\beta = \beta(n)$, hash function $H : \{0, 1\}^\ell \to [\beta]$, and max bin size m, suitable for our hashing scheme (Table 2)
- A special dummy item $\perp \in \{0, 1\}^*$

INPUT OF S: $X = \{x_1, x_2, \ldots, x_{n_1}\} \subseteq \{0, 1\}^\ell$
INPUT OF R: $Y = \{y_1, y_2, \ldots, y_{n_2}\} \subseteq \{0, 1\}^\ell$
PROTOCOL:

1. S and R hash elements of their sets X and Y into β bins under hash function H. Let $B_S[i]$ and $B_R[i]$ denote the set of items in the sender's and receiver's i-th bin, respectively.
2. S pads each bin $B_S[i]$ with (several copies, as needed) the special item $\perp$ up to the maximum bin size $m + 1$, and randomly permutes all items in this bin.
3. R pads each bin $B_R[i]$ with one special item $\perp$ and (several, as needed) different dummy items to the maximum bin size $m + 1$.
4. R initializes set $Z = \{\}$.
5. For each bin $i \in [\beta]$, for each item $x_j \in B_S[i]$:
   (a) S and R invoke the $\mathcal{F}_{\text{psu}}^{1,m+1}$-functionality:
      - S acts as *sender* with input x_j
      - R acts as *receiver* with input set $B_R[i]$
      - R obtains output $Z_{i,j}$ and sets $Z = Z \cup Z_{i,j}$.
6. R outputs Z.

Fig. 10. Private set union protocol $\mathcal{F}_{\text{psu}}^{n_1, n_2}$.

- $x^* \notin X$: Since all values q_i' are uniformly random from the S's point of view, so are the $s \oplus q_i'$.
- $x^* = x_i$ (consequently, $q' = q_i'$): Since other values $q_{j \in [n]}'$, $\forall j \neq i$, are uniformly random from S's point of view, we replace these $s \oplus q_j'$ with random. Then s is used only in the expression $s \oplus q_i'$. Since s is uniform, $s \oplus q_i'$ is also uniformly random from the S's view even though the adversary knows $q' = q_i'$.

In summary, the polynomial from the real execution can be replaced with a polynomial P over random points. T_1 and T_2 are indistinguishable.

Corrupt Receiver. $\text{Sim}_R(x_1, \ldots, x_n, out)$ simulates R's view, which includes R's randomness, input, output and received messages. Sim_R proceeds as follows.

First, if $out = \{x_1, \ldots, x_n, x^*\}$ for some x^*, then the simulator knows S's input x^* and can trivially simulate all of S's actions honestly. This case of simulation is clearly perfect.

Otherwise, Sim_R chooses a random $k' \in_r \{0, 1\}^\kappa$, calls OPRF simulator $\text{Sim}_{S_{\text{OPRF}}}(\perp, k')$, and appends its output to the view. It simulates a message $s^* = s$ from S in Step 4. Finally, to simulate Step 5, Sim_S runs simulator Sim_{OT} on input $(1, \perp)$ and appends the output of Sim_{OT} to its output of the view.

Table 5. Comparison of total runtime (in seconds) and communication (in MB) between the RPMT version and the optimized version of our protocol. n elements per party, 128-bit length element, and single thread in LAN setting with 10 Gbps network bandwidth, 0.02 ms RTT.

	Our Protocol	Set size n							
		2^8	2^{10}	2^{12}	2^{14}	2^{16}	2^{18}	2^{20}	2^{22}
Time	with PEQT	0.66	0.86	1.42	3.54	12.41	61.34	238.88	1039.64
	without PEQT	0.65	0.86	1.41	3.51	11.02	49.12	229.22	1015.23
Comm.	with PEQT	0.45	2.05	8.48	34.98	144.65	652.09	2693.30	11,077.83
	without PEQT	0.41	1.86	7.72	31.80	131.16	600.62	2470.10	10,233.27

The view generated by $\mathsf{Sim}_{\mathcal{R}}$ in indistinguishable from a real view because of the indistinguishability of the transcripts of the underlying simulators.

References

1. Arbitman, Y., Naor, M., Segev, G.: Backyard cuckoo hashing: constant worst-case operations with a succinct representation. In: 51st FOCS, pp. 787–796. IEEE Computer Society Press, October 2010

2. Asharov, G., Lindell, Y., Schneider, T., Zohner, M.: More efficient oblivious transfer and extensions for faster secure computation. In: Sadeghi, A.R., Gligor, V.D., Yung, M. (eds.) ACM CCS 2013, pp. 535–548. ACM Press, New York (2013)

3. Ateniese, G., De Cristofaro, E., Tsudik, G.: (If) Size matters: size-hiding private set intersection. In: Catalano, D., Fazio, N., Gennaro, R., Nicolosi, A. (eds.) PKC 2011. LNCS, vol. 6571, pp. 156–173. Springer, Heidelberg (2011). https://doi.org/10.1007/978-3-642-19379-8_10

4. Beaver, D.: Correlated pseudorandomness and the complexity of private computations. In: 28th ACM STOC, pp. 479–488. ACM Press, May 1996

5. Blanton, M., Aguiar, E.: Private and oblivious set and multiset operations. In: Youm, H.Y., Won, Y. (eds.) ASIACCS 2012, pp. 40–41. ACM Press, New York (2012)

6. Boudot, F., Schoenmakers, B., Traoré, J.: A fair and efficient solution to the socialist millionaires' problem. Discrete Appl. Math. **111**, 2001 (2001)

7. Brickell, J., Shmatikov, V.: Privacy-preserving graph algorithms in the semi-honest model. In: Roy, B. (ed.) ASIACRYPT 2005. LNCS, vol. 3788, pp. 236–252. Springer, Heidelberg (2005). https://doi.org/10.1007/11593447_13

8. Burkhart, M., Strasser, M., Many, D., Dimitropoulos, X.: SEPIA: privacy-preserving aggregation of multi-domain network events and statistics. In: Proceedings of the 19th USENIX Conference on Security, USENIX Security 2010, p. 15. USENIX Association, Berkeley (2010)

9. Canetti, R., Paneth, O., Papadopoulos, D., Triandopoulos, N.: Verifiable set operations over outsourced databases. In: Krawczyk, H. (ed.) PKC 2014. LNCS, vol. 8383, pp. 113–130. Springer, Heidelberg (2014). https://doi.org/10.1007/978-3-642-54631-0_7

10. Cerulli, A., De Cristofaro, E., Soriente, C.: Nothing refreshes like a RePSI: reactive private set intersection. In: Preneel, B., Vercauteren, F. (eds.) ACNS 2018. LNCS, vol. 10892, pp. 280–300. Springer, Cham (2018). https://doi.org/10.1007/978-3-319-93387-0_15

11. Chen, H., Laine, K., Rindal, P.: Fast private set intersection from homomorphic encryption. In: Thuraisingham, B.M., Evans, D., Malkin, T., Xu, D. (eds.) ACM CCS 2017, pp. 1243–1255. ACM Press, New York (2017)

12. Cho, C., Dachman-Soled, D., Jarecki, S.: Efficient concurrent covert computation of string equality and set intersection. In: Sako, K. (ed.) CT-RSA 2016. LNCS, vol. 9610, pp. 164–179. Springer, Cham (2016). https://doi.org/10.1007/978-3-319-29485-8_10

13. Ciampi, M., Orlandi, C.: Combining private set-intersection with secure two-party computation. In: Catalano, D., De Prisco, R. (eds.) SCN 2018. LNCS, vol. 11035, pp. 464–482. Springer, Cham (2018). https://doi.org/10.1007/978-3-319-98113-0_25

14. Davidson, A., Cid, C.: An efficient toolkit for computing private set operations. In: Pieprzyk, J., Suriadi, S. (eds.) ACISP 2017, Part II. LNCS, vol. 10343, pp. 261–278. Springer, Cham (2017). https://doi.org/10.1007/978-3-319-59870-3_15

15. De Cristofaro, E., Kim, J., Tsudik, G.: Linear-complexity private set intersection protocols secure in malicious model. In: Abe, M. (ed.) ASIACRYPT 2010. LNCS, vol. 6477, pp. 213–231. Springer, Heidelberg (2010). https://doi.org/10.1007/978-3-642-17373-8_13

16. De Cristofaro, E., Tsudik, G.: Practical private set intersection protocols with linear complexity. In: Sion, R. (ed.) FC 2010. LNCS, vol. 6052, pp. 143–159. Springer, Heidelberg (2010). https://doi.org/10.1007/978-3-642-14577-3_13

17. Demmler, D., Rindal, P., Rosulek, M., Trieu, N.: PIR-PSI: scaling private contact discovery. In: Proceedings on Privacy Enhancing Technologies (2018)

18. Dong, C., Chen, L., Wen, Z.: When private set intersection meets big data: an efficient and scalable protocol. In: Sadeghi, A.R., Gligor, V.D., Yung, M. (eds.) ACM CCS 2013, pp. 789–800. ACM Press, New York (2013)

19. Fagin, R., Naor, M., Winkler, P.: Comparing information without leaking it. Commun. ACM **39**, 77–85 (1996)

20. Falk, B.H., Noble, D., Ostrovsky, R.: Private set intersection with linear communication from general assumptions. Cryptology ePrint Archive, Report 2018/238 (2018). https://eprint.iacr.org/2018/238

21. Freedman, M.J., Ishai, Y., Pinkas, B., Reingold, O.: Keyword search and oblivious pseudorandom functions. In: Kilian, J. (ed.) TCC 2005. LNCS, vol. 3378, pp. 303–324. Springer, Heidelberg (2005). https://doi.org/10.1007/978-3-540-30576-7_17

22. Freedman, M.J., Nissim, K., Pinkas, B.: Efficient private matching and set intersection. In: Cachin, C., Camenisch, J.L. (eds.) EUROCRYPT 2004. LNCS, vol. 3027, pp. 1–19. Springer, Heidelberg (2004). https://doi.org/10.1007/978-3-540-24676-3_1

23. Frikken, K.: Privacy-preserving set union. In: Katz, J., Yung, M. (eds.) ACNS 2007. LNCS, vol. 4521, pp. 237–252. Springer, Heidelberg (2007). https://doi.org/10.1007/978-3-540-72738-5_16

24. Gaudry, P., Brent, R., Zimmermann, P., Thomé, E.: https://gforge.inria.fr/projects/gf2x/

25. Gonnet, G.H.: Expected length of the longest probe sequence in hash code searching. J. ACM **28**(2), 289–304 (1981)

26. Hazay, C., Nissim, K.: Efficient set operations in the presence of malicious adversaries. J. Cryptol. **25**(3), 383–433 (2012)

27. Hazay, C., Venkitasubramaniam, M.: Scalable multi-party private set-intersection. In: Fehr, S. (ed.) PKC 2017. LNCS, vol. 10174, pp. 175–203. Springer, Heidelberg (2017). https://doi.org/10.1007/978-3-662-54365-8_8

28. Hogan, K., et al.: Secure multiparty computation for cooperative cyber risk assessment. In: 2016 IEEE Cybersecurity Development (SecDev), pp. 75–76, November 2016

29. Huang, Y., Evans, D., Katz, J.: Private set intersection: are garbled circuits better than custom protocols? In: NDSS 2012. The Internet Society, February 2012

30. Huang, Y., Evans, D., Katz, J., Malka, L.: Faster secure two-party computation using garbled circuits. In: USENIX Security 2011 (2011)

31. Impagliazzo, R., Rudich, S.: Limits on the provable consequences of one-way permutations. In: 21st ACM STOC, pp. 44–61. ACM Press, May 1989

32. Ion, M., et al.: On deploying secure computing commercially: private intersection-sum protocols and their business applications. Cryptology ePrint Archive, Report 2019/723 (2019)

33. Ishai, Y., Kilian, J., Nissim, K., Petrank, E.: Extending oblivious transfers efficiently. In: Boneh, D. (ed.) CRYPTO 2003. LNCS, vol. 2729, pp. 145–161. Springer, Heidelberg (2003). https://doi.org/10.1007/978-3-540-45146-4_9

34. Kissner, L., Song, D.: Privacy-preserving set operations. In: Shoup, V. (ed.) CRYPTO 2005. LNCS, vol. 3621, pp. 241–257. Springer, Heidelberg (2005). https://doi.org/10.1007/11535218_15

35. Kolesnikov, V., Kumaresan, R.: Improved OT extension for transferring short secrets. In: Canetti, R., Garay, J.A. (eds.) CRYPTO 2013. LNCS, vol. 8043, pp. 54–70. Springer, Heidelberg (2013). https://doi.org/10.1007/978-3-642-40084-1_4

36. Kolesnikov, V., Kumaresan, R., Rosulek, M., Trieu, N.: Efficient batched oblivious PRF with applications to private set intersection. In: Weippl, E.R., Katzenbeisser, S., Kruegel, C., Myers, A.C., Halevi, S. (eds.) ACM CCS 2016, pp. 818–829. ACM Press, New York (2016)

37. Kolesnikov, V., Matania, N., Pinkas, B., Rosulek, M., Trieu, N.: Practical multi-party private set intersection from symmetric-key techniques. In: Thuraisingham, B.M., Evans, D., Malkin, T., Xu, D. (eds.) ACM CCS 2017, pp. 1257–1272. ACM Press, New York (2017)

38. Lenstra, A., Voss, T.: Information security risk assessment, aggregation, and mitigation. In: Wang, H., Pieprzyk, J., Varadharajan, V. (eds.) ACISP 2004. LNCS, vol. 3108, pp. 391–401. Springer, Heidelberg (2004). https://doi.org/10.1007/978-3-540-27800-9_34

39. Lipmaa, H.: Verifiable homomorphic oblivious transfer and private equality test. In: Laih, C.-S. (ed.) ASIACRYPT 2003. LNCS, vol. 2894, pp. 416–433. Springer, Heidelberg (2003). https://doi.org/10.1007/978-3-540-40061-5_27

40. Naor, M., Pinkas, B.: Oblivious transfer and polynomial evaluation. In: Proceedings of the Thirty-First Annual ACM Symposium on Theory of Computing, STOC 1999 (1999)

41. Naor, M., Pinkas, B.: Efficient oblivious transfer protocols. In: Kosaraju, S.R. (ed.) 12th SODA, pp. 448–457. ACM-SIAM, January 2001

42. Pagh, R., Rodler, F.F.: Cuckoo hashing. J. Algorithms **51**(2), 122–144 (2004)

43. Pinkas, B., Rosulek, M., Trieu, N., Yanai, A.: SpOT-light: lightweight private set intersection from sparse OT extension. In: Boldyreva, A., Micciancio, D. (eds.) CRYPTO 2019. LNCS, vol. 11694, pp. 401–431. Springer, Cham (2019). https://doi.org/10.1007/978-3-030-26954-8_13

44. Pinkas, B., Schneider, T., Segev, G., Zohner, M.: Phasing: private set intersection using permutation-based hashing. In: Proceedings of the 24th USENIX Conference on Security Symposium, pp. 515–530. USENIX Association (2015)

45. Pinkas, B., Schneider, T., Tkachenko, O., Yanai, A.: Efficient circuit-based PSI with linear communication. In: Ishai, Y., Rijmen, V. (eds.) EUROCRYPT 2019. LNCS, vol. 11478, pp. 122–153. Springer, Cham (2019). https://doi.org/10.1007/978-3-030-17659-4_5

46. Pinkas, B., Schneider, T., Zohner, M.: Faster private set intersection based on OT extension. In: Proceedings of the 23rd USENIX Conference on Security Symposium, pp. 797–812. USENIX Association (2014)

47. Pinkas, B., Schneider, T., Zohner, M.: Scalable private set intersection based on OT extension. ACM Trans. Priv. Secur. **21**(2) (2018)

48. Rabin, M.O.: How to exchange secrets by oblivious transfer. Aiken Computation Laboratory, Harvard U. (1981)

49. Resende, A.C.D., Aranha, D.F.: Faster unbalanced private set intersection. In: Meiklejohn, S., Sako, K. (eds.) FC 2018. LNCS, vol. 10957, pp. 203–221. Springer, Heidelberg (2018). https://doi.org/10.1007/978-3-662-58387-6_11

50. Rindal, P.: libOTe: an efficient, portable, and easy to use Oblivious Transfer Library. https://github.com/osu-crypto/libOTe

51. Rindal, P., Rosulek, M.: Improved private set intersection against malicious adversaries. In: Coron, J.-S., Nielsen, J.B. (eds.) EUROCRYPT 2017, Part I. LNCS, vol. 10210, pp. 235–259. Springer, Cham (2017). https://doi.org/10.1007/978-3-319-56620-7_9

52. Rindal, P., Rosulek, M.: Malicious-secure private set intersection via dual execution. In: Thuraisingham, B.M., Evans, D., Malkin, T., Xu, D. (eds.) ACM CCS 2017, pp. 1229–1242. ACM Press, New York (2017)

53. Shoup, V.: http://www.shoup.net/ntl/

54. Zahur, S., Rosulek, M., Evans, D.: Two halves make a whole: reducing data transfer in garbled circuits using half gates. In: Oswald, E., Fischlin, M. (eds.) EUROCRYPT 2015. LNCS, vol. 9057, pp. 220–250. Springer, Heidelberg (2015). https://doi.org/10.1007/978-3-662-46803-6_8

Author Index

Printed in the United States
By Bookmasters